D0907162

Land-Ocean Systems in the Siberian Arctic

Springer
Berlin
Heidelberg
New York
Barcelona
Hong Kong
London
Milan
Paris
Singapore
Tokyo

H. Kassens, H.A. Bauch, I.A. Dmitrenko, H. Eicken,
H.-W. Hubberten, M. Melles, J. Thiede, L.A. Timokhov (Eds.)

Land-Ocean Systems in the Siberian Arctic

Dynamics and History

With 301 Figures, 6 in color and 85 Tables

Springer

Editors

Dr. Heidemarie Kassens
GEOMAR Forschungszentrum
für marine Geowissenschaften
Wischhofstraße 1-3
D-24148 Kiel / Germany

Dr. Henning A. Bauch
GEOMAR Forschungszentrum
für marine Geowissenschaften
Wischhofstraße 1-3
D-24148 Kiel / Germany

Dr. Igor A. Dmitrenko
State Research Center
of Russian Federation, the Arctic
and Antarctic Research Institute
38, Bering Str.
199397 St. Petersburg / Russia

Dr. Hajo Eicken
University of Alaska Fairbanks
Geophysical Institute
903 Koyukuk Dr., P. O. Box
757320 Fairbanks, AK 99775-7320 /
USA

Prof. Dr. Hans-Wolfgang Hubberten
Alfred-Wegener-Institut
für Polar- und Meeresforschung
Forschungsstelle Potsdam
Telegraphenberg A 43
D-14473 Potsdam / Germany

Dr. Martin Melles
Alfred-Wegener-Institut
für Polar- und Meeresforschung
Forschungsstelle Potsdam
Telegraphenberg A 43
D-14473 Potsdam / Germany

Prof. Dr. Jörn Thiede
Alfred-Wegener-Institut
für Polar- und Meeresforschung
Postfach 120161
D-27515 Bremerhaven / Germany

Prof. Dr. Leonid A. Timokhov
State Research Center
of Russian Federation, the Arctic
and Antarctic Research Institute
38, Bering Str.
199397 St. Petersburg / Russia

ISBN 3-540-65676-6 Springer-Verlag Berlin Heidelberg New York

CIP data applied for

Die Deutsche Bibliothek - CIP-Einheitsaufnahme
Land-ocean systems in the Siberian Arctic : dynamics and history / H. Kassens ... - Berlin ; Heidelberg ; New York , Barcelona ; Hong Kong ; London ; Milan ; Paris ; Singapore ; Tokyo : Springer, 1999
ISBN 3-540-65676-6

Typesetting: Camera-ready by editors
Cover layout: Struve & Partner, Heidelberg
SPIN: 10717803 32 / 3020 - 5 4 3 2 1 0 - Printed on acid -free paper

Preface

The Arctic comprises some of the most sensitive elements of the global environment, which are considered to respond rapidly to climate change. However, our knowledge of the processes driving the Arctic system today and in the past is still very limited, thus making it difficult to predict future climate scenarios. In this context the Laptev Sea and its Siberian hinterland are of particular interest. River discharge into the Laptev Sea contitutes a key source for the Arctic halocline's freshwater budget, and the shallow Laptev Sea Shelf is a major ice production area, linking the Siberian shelves to the Arctic Ocean and the Nordic seas. Based on a multidisciplinary approach, Russian and German scientists are describing in this book the natural processes behind short- and long-term changes in the Laptev Sea and its Siberian hinterland, using paleo-records and modern data which were collected during the past six years. These marine and terrestrial datasets provide important new insights into the causes, impacts, and feedback mechanisms which determine the Arctic climate system today.

The papers included in the volume provide a comprehensive overview of the environmental system of the Laptev Sea and its Siberian hinterland. Although the content of many of the papers cut across disciplines, each was assignedto a specific theme:

A: Modern Ocean and Sea-Ice Processes
B: The Marine Ecosystem
C: Land-Ocean Interactions and Pathways
D: Terrestrial Environment - Past and Present
E: Marine Depositional Environment - Past and Present

The book contains scientific results of the Russian and German projects "LAPEX", "Laptev Sea System", "Late Quaternary Environmental Evolution of Central Taymyr" and "Ecology of the Marginal Seas of the Eurasian Arctic", which were coordinated on the German side by the GEOMAR Research Center for Marine Geosciences (Kiel), the Alfred-Wegener-Institute for Polar and Marine Research (Potsdam and Bremerhaven) and the Institute for Polar Biology (Kiel), as well as on the Russian side by the State Research Center - Arctic and Antarctic Research Institute (St. Petersburg). Many other research institutions in Russia and Germany were involved, and it would not have been possible to carry out the projects without their benevolent assistance. Financial and logistic support has been mainly acquired from the German and Russian Ministries of Science and Technology. Both ministries provided invaluable assistance particularly during the expeditions and workshops.

The remote location and harsh weather conditions made logistics and work in northern Central Siberia and in the Laptev Sea extremely difficult. Hence, the success of the projects would not have been possible without the support of many institutions and authorities in Russia and Germany. In particular, we would like to express our appreciations to the crews of the research vessels "Ivan Kireev", "Kapitan Dranitsyn", "Lotsman", "Polarstern", "Professor Makkaveev", "Professor Multanovsky", and "Sarya 9" as well as to the team of the Lena-Nordenskiöld Station and the helicopter pilots for their tireless efforts above and beyond their duties. Furthermore we would like to thank various authorities in St. Petersburg, Moscow, Murmansk, Yakutsk, Archangel'sk, Tiksi, Khatanga, Norilsk, and Dickson as well as in Berlin, Bonn, Bremerhaven, Hamburg, Kiel, and Potsdam for their good collaboration.

As editors we wish to thank all contributing authors. In this respect we particularly acknowledge the support of the critical referees (who are too numerous to mention by names) who have gone through the papers of this book trying to help to maintain a scientific standard of an international volume. We are further grateful to B. Rohr for her tenacity as editor's assistant, P. P. Overduin for proofreading most of the papers, and H. Cremer for helping to improve the layout of the book.

The political changes which have occurred since the late 80's have made this project possible and we are extraordinarily grateful that those who carry the responsibility for political and administrative decisions have used this time of political change to make such bilateral Russian-German projects possible, and that they had confidence in the participants to establish the required close and intense cooperation, both in the field as well as at their home institutes.

We are devoting this book to the memory of the explorers who lost their lives in this part of the Arctic in their courageous attempts to fulfil their devotion to science. They range from members of the American "Jeanette" expedition under their leader George W. De Long (1879-1881) to several participants of the Russian "Sarja" expedition (1900 - 1902) under Baron Eduard von Toll, and include the tragic disappearance (1997) of the head of the international biological Lena-Nordenskiöld Station to the North of Tiksi Semen Semenovich Isakov from Yakutsk, Rep. Sakha.

The Editors

Kiel, Bremerhaven, Potsdam, St. Petersburg, Fairbanks

Table of Contents

Section A: Modern Ocean and Sea-Ice Processes

Features of Seasonal and Interannual Variability of the Sea Level and Water Circulation in the Laptev Sea
V.K. Pavlov and P.V. Pavlov 3

Numerical Modelling of Storm Surges in the Laptev Sea Based on the Finite Element Method
I. Ashik and A. Novakov 17

Large-Scale Variations of Sea Level in the Laptev Sea
G.N. Voinov and E.A. Zakharchuk 25

Extreme Oscillations of the Sea Level in the Laptev Sea
I. Ashik, Y. Dvorkin and Y. Vanda 37

Internal Waves in the Laptev Sea
E.A. Zakharchuk 43

The Composition of the Coarse Fraction of Aerosols in the Marine Boundary Layer over the Laptev, Kara and Barents Seas
V.P. Shevchenko, A.P. Lisitzin, R. Stein, V.V. Serova, A.B. Isaeva and N.V. Politova 53

New Data on Sea-Ice Albedo in the Laptev and Barents Seas
B.V. Ivanov 59

Possible Causes of Radioactive Contamination in the Laptev Sea
V.K. Pavlov, V.V. Stanovoy and A.I. Nikitin 65

Oceanographic Causes for Transarctic Ice Transport of River Discharge
I. Dmitrenko, P. Golovin, V. Gribanov and H. Kassens 73

Step-Like Vertical Structure Formation Due to Turbulent Mixing of Initially Continuous Density Gradients
A. Zatsepin, S. Dikarev, S. Poyarkov, N. Sheremet, I. Dmitrenko, P. Golovin and H. Kassens 93

Dissolved and Particulate Major and Trace Elements in Newly Formed Ice from the Laptev Sea (Transdrift III, October 1995)
J.A. Hölemann, M. Schirmacher and A. Prange 101

Particle Entrainment into Newly Forming Sea Ice – Freeze-Up Studies in October 1995
F. Lindemann, J.A. Hölemann, A. Korablev and A. Zachek 113

Frazil Ice Formation during the Spring Flood and its Role in Transport of Sediments to the Ice Cover
P. Golovin, I. Dmitrenko, H. Kassens and J.A. Hölemann 125

Section B: The Marine Ecosystem

Pelagic-Benthic Coupling in the Laptev Sea Affected by Ice Cover
C. Grahl, A. Boetius and E.-M. Nöthig 143

Chlorophyll *a* Distribution in Water Column and Sea Ice during the Laptev Sea Freeze-Up Study in Autumn 1995
K. v. Juterzenka and K. Knickmeier 153

Composition, Abundance and Population Structure of Spring-Time Zooplankton in the Shelf-Zone of Laptev Sea
E.N. Abramova 161

Macrobenthos Distribution in the Laptev Sea in Relation to Hydrology
V.V. Petryashov, B.I. Sirenko, A.A. Golikov, A.V. Novozhilov, E. Rachor, D. Piepenburg and M.K. Schmid 169

Carepoctus solidus sp.n., a New Species of Liparid Fish (Scorpaeniformes, Liparidae) from the Lower Bathyal of the Polar Basin
N.V. Chernova 181

Spring Stopover of Birds on the Laptev Sea Polynya
D.V. Solovieva 189

Section C: Land-Ocean Interactions and Pathways

Major, Trace and Rare Earth Element Geochemistry of Suspended Particulate Material of East Siberian Rivers Draining to the Arctic Ocean
V. Rachold 199

Carbon Isotope Composition of Particulate Organic Material in East Siberian Rivers
V. Rachold and H.-W. Hubberten 223

Distribution of River Water and Suspended Sediment Loads in the Deltas of Rivers in the Basins of The Laptev and East-Siberian Seas
V.V. Ivanov and A.A. Piskun 239

Dissolved Oxygen, Silicon, Phosphorous and Suspended Matter Concentrations During the Spring Breakup of The Lena River
S.V. Pivovarov, J.A. Hölemann, H. Kassens, M. Antonow and I. Dmitrenko 251

Distribution Patterns of Heavy Minerals in Siberian Rivers, the Laptev Sea and the eastern Arctic Ocean: An Approach to Identify Sources, Transport and Pathways of Terrigenous Matter
M. Behrends, E. Hoops and B. Peregovich 265

The Role of Coastal Retreat for Sedimentation in the Laptev Sea
F.E. Are 287

Section D: Terrestrial Environment - Past and Present

Seasonal Changes in Hydrology, Energy Balance and Chemistry in the Active Layers of Arctic Tundra Soils in Taymyr Peninsula, Russia
J. Boike and P.P. Overduin 299

The Landscape and Geobotanical Characteristics of the Levinson-Lessing Lake Basin, Byrranga Mountains, Central Taimyr
M.A. Anisimov and I.N. Pospelov 307

Studies of Methane Production and Emission in Relation to the Microrelief of a Polygonal Tundra in Northern Siberia
V.A. Samarkin, A. Gundelwein and E.-M. Pfeiffer 329

Carbon Dioxide and Methane Emmissions at Arctic Tundra Sites in North Siberia
M. Sommerkorn, A. Gundelwein, E.-M. Pfeiffer and M. Bölter 343

The Features of the Hydrological Regime of the Lake-River Systems of the Byrranga Mountains (by the Example of the Levinson-Lessing Lake)
V.P. Zimichev, D.Yu. Bolschyanov, V.G. Mesheryakov and D. Gintz 353

Lead-210 Dating and Heavy Metal Concentration in Recent Sediments of Lama Lake (Norilsk Area, Siberia)
B. Hagedorn, S. Harwart, M.M.R. van der Loeff and M. Melles 361

Late Weichselian to Holocene Diatom Succession in a Sediment Core from Lama Lake, Siberia and Presumed Ecological Implications
U. Kienel 377

Climate and Vegetation History of the Taymyr Peninsula since Middle Weichselian Time - Palynological Evidence from Lake Sediments
J. Hahne and M. Melles 407

Laminated Sediments from Levinson-Lessing Lake, Northern Central Siberia - A 30,000 Year Record of Environmental History?
T. Ebel, M. Melles and F. Niessen 425

High-Resolution Seismic Stratigraphy of Lake Sediments on the Taymyr Peninsula, Central Siberia
F. Niessen, T. Ebel, C. Kopsch and G.B. Fedorov 437

Archaeological Survey in Central Taymyr
V.V. Pitul'ko 457

Marine Pleistocene Deposits of the Taymyr Peninsula and their Age from ESR Dating
D. Bolshiyanov and A. Molodkov 469

Paleoclimatic Indicators from Permafrost Sequences in the Eastern Taymyr Lowland
C. Siegert, A.Yu. Derevyagin, G.N. Shilova, W.-D. Hermichen and A. Hiller 477

Section E: Marine Depositional Environment - Past and Present

Stable Oxygen Isotope Ratios in Benthic Carbonate Shells of Ostracoda, Foraminifera, and Bivalvia from Surface Sediments of the Laptev Sea, Summer 1993 and 1994
H. Erlenkeuser and U. von Grafenstein 503

Determination of Depositional Beryllium-10 Fluxes in the Area of the Laptev Sea and Beryllium-10 Concentrations in Water Samples of High Northern Latitudes
C. Strobl, V. Schulz, S. Vogler, S. Baumann, H. Kassens, P.W. Kubik, M. Suter and A. Mangini 515

Spatial Distribution of Diatom Surface Sediment Assemblages on the Laptev Sea Shelf (Russian Arctic)
H. Cremer 533

R.N. Djinoridze, G. I. Ivanov, E. N. Djinoridze, and R. F. Spielhagen
Diatoms from Surface Sediments of the Saint Anna Trough (Kara Sea) 553

Distribution of Aquatic Palynomorphs in Surface Sediments from the Laptev Sea, Eastern Arctic Ocean
M. Kunz-Pirrung 561

Distribution of Pollen and Spores in Surface Sediments of the Laptev Sea
O.D. Naidina and H.A. Bauch 577

Clay Mineral Distribution in Surface Sediments of the Laptev Sea: Indicator for Sediment Provinces, Dynamics and Sources
B.T. Rossak, H. Kassens, H. Lange and J. Thiede 587

Planktic Foraminifera in Holocene Sediments from the Laptev Sea and the Central Arctic Ocean: Species Distribution and Paleobiogeographical Implication
H.A. Bauch 601

Holocene Diatom Stratigraphy and Paleoceanography of the Eurasian Arctic Seas
Y. Polyakova 615

Late Quaternary Organic Carbon and Biomarker Records from the Laptev Sea Continental Margin (Arctic Ocean): Implications for Organic Carbon Flux and Composition
R. Stein, K. Fahl, F. Niessen and M. Siebold 635

Late Pleistocene Paleoriver Channels on the Laptev Sea Shelf - Implications from Sub-Bottom Profiling
H.P. Kleiber and F. Niessen 657

Main Structural Elements of Eastern Russian Arctic Continental Margin Derived from Satellite Gravity and Multichannel Seismic Reflection Data
S.S. Drachev, G.L. Johnson, S.W. Laxon, D.C. McAdoo and H. Kassens 667

High Resolution Seismic Studies in the Laptev Sea Shelf: First Results and Future Needs
B. Kim, G. Grikurov and V. Soloviev 683

Section F: Summary

Dynamics and History of the Laptev Sea and its Continental Hinterland: A Summary
J. Thiede, L. Timokhov, H.A. Bauch, D. Bolshiyanov, I. Dmitrenko, H. Eicken, K. Fahl, A. Gukov, J. Hölemann, H.W. Hubberten, K. v. Juterzenka, H. Kassens, M. Melles, V. Petryashov, S. Pivovarov, S. Priamikov, V. Rachold, M. Schmid, C. Siegert, M. Spindler, R. Stein and Scientific Party 695

Section A

Modern Ocean and Sea-Ice Processes

Features of Seasonal and Interannual Variability of the Sea Level and Water Circulation in the Laptev Sea

V.K. Pavlov[1] and P.V. Pavlov[2]

(1) State Research Center - Arctic and Antarctic Research Institute, 38 Bering St., 199226 St. Petersburg, Russia

(2) INTAARI Ltd., 38 Bering St., 199226 St. Petersburg, Russia

Received 3 March 1997 and accepted in revised form 3 March 1998

Abstract - Water circulation and spatial structure of the surface level are up to the present time the least investigated processes in the physical oceanography of the Laptev Sea.
Based on the analysis of the available sea level observations and the results of numerical modeling, estimations of the spatial-temporal variability of the surface level and water circulation of the Laptev Sea at seasonal and climatic time scales were obtained.
Probable causes for the interannual variability of sea level and water circulation in the Laptev Sea are considered.

Introduction

The Laptev Sea has the strongest water dynamics of all the marginal Arctic Seas. As a result of extensive open boundaries with the Arctic Basin, water circulation of the Laptev Sea is part of the general water circulation of the Arctic Ocean. However, a complicated bottom topography (more than 71.7% of the Laptev Sea area are shallower than 200 m and more than 6.5% are deeper than 1000 m) (Terms. Notions. Reference Table, 1980), an intense water exchange, baroclinicity, the presence of year-round drifting ice govern in many respects, the water circulation features and the related level regime of the Laptev Sea at seasonal and climatic scales.

Rivers discharge (about 726 km^3 a year) and water exchange through the straits and the open boundaries (about 32000 km^3 a year) play a special role in the formation of water circulation in the Laptev Sea.

The water exchange of the Laptev Sea is subject to a strong seasonal variability. Thus for example, the runoff of the Lena river in winter is 10 times less, as compared to the summer season. The runoff of such rivers as Khatanga, Olenek, Yana and Anabar in the wintertime is actually completely absent. The seasonal variability of oceanographic conditions and general water circulation of the Arctic Ocean, is obviously reflected in the seasonal variability of the water exchange of the Laptev Sea through the straits.

Scarce data of direct observations of currents in the Laptev Sea and practically a complete absence of direct observations of the open sea level do not allow even a rough reconstruction of the mean spatial structure of the level surface and water circulation, not to mention their seasonal and climatic variability.

Hence mathematical modeling is one of the real possible tools for investigating the level regime and water circulation.

This work combines observational data and numerical calculations in order to describe the spatial-temporal structure of the surface level, as well as of the wind-driven and thermohaline water circulation of the Laptev Sea.

The factors causing the variability of sea level and currents at seasonal and climatic time scales are analyzed.

In: Kassens, H., H.A. Bauch, I. Dmitrenko, H. Eicken, H.-W. Hubberten, M. Melles, J. Thiede and L. Timokhov (eds.) Land-Ocean Systems in the Siberian Arctic: Dynamics and History. Springer-Verlag, Berlin, 1999, 3-16.

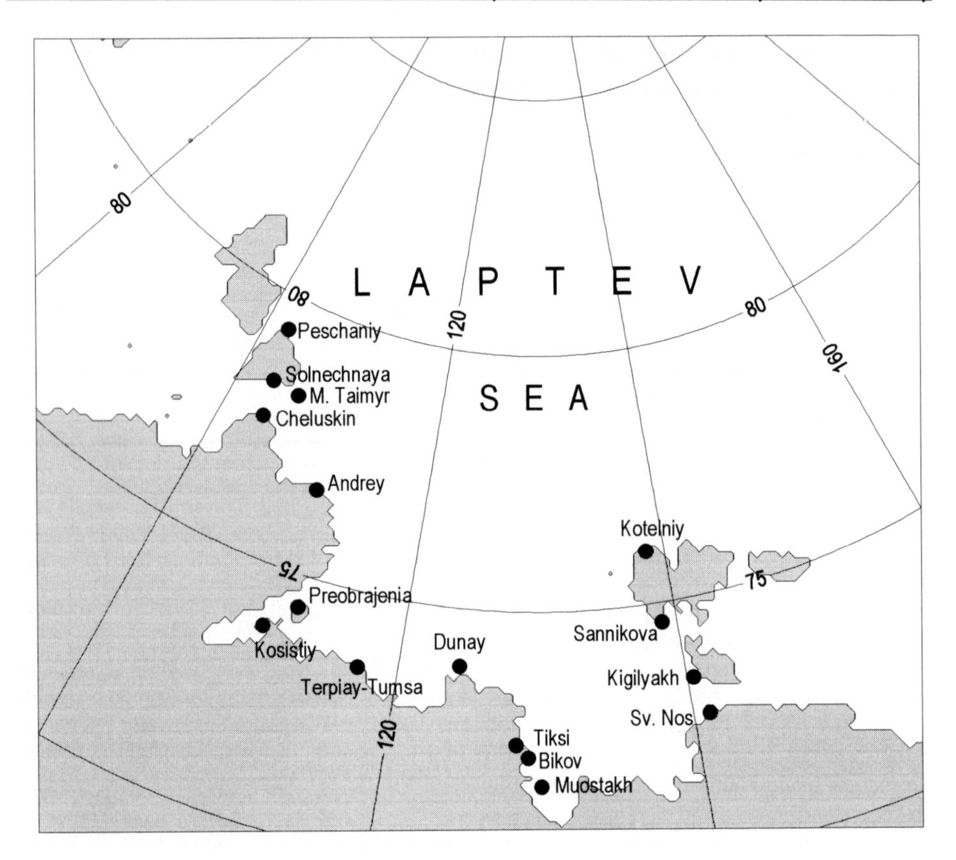

Figure 1: Location of coastal and island observation stations in the Laptev Sea.

Data

The Arctic and Antarctic Research Institute (AARI) carried out regular hydrological and meteorological studies of the Laptev Sea and its rivers for more than 50 years. The hydrographic surveys of the ice-free area of the Laptev Sea from drifting ice were made practically on an annual basis in summer and in some years in the wintertime. At 16 stations along the coast and the islands of the Laptev Sea were carried out observations of sea level and some other oceanographic characteristics (Figure 1).

We employed the following data in this work:

- atmospheric pressure fields over the Laptev Sea area from 1960 to 1980 with a 4 h time interval.
- ice conditions - monthly means from data in the Arctic Ocean Atlas (1980), Romanov (1992), Arctic and Antarctic Sea Ice 1978 - 1987 (1992).
- daily discharges of the main rivers (Khatanga, Anabar, Olenek and Yana) emptying into the Laptev Sea.
- daily observations of sea level at 16 coastal and island observation stations.
- daily observations of water temperature and salinity in the surface layer at island and coastal stations.

- gridded three-dimensional temperature and salinity fields averaged for the summer and winter seasons(Polyakov and Timokhov, 1994).
- gridded three-dimensional temperature and salinity fields for the winter seasons from 1950 to 1990, averaged for the each 10 years within a this period (kindly provided by Prof. L.A. Timokhov).
- gridded three-dimensional temperature and salinity fields for individual years, characterized by the largest coverage of the Laptev Sea area.
- gridded bottom topography field of the Laptev Sea with a spatial interval of 14 km.

Monthly averaging of the pressure fields, sea level and sea water density at the coastal and island stations was performed and mean climatic values of these parameters were obtained. Data of direct current measurements were not used in the work due to their small amount and short time scan(as a rule less than 1 month).

Method

For modeling the spatial-temporal variability of the surface level and water circulation of the Laptev Sea a three-dimensional baroclinic model was used. The model is briefly described below.

The model basic equations are the following equations written in the Bussinesque and quasistatic approximations. Equations of motion:

$$\frac{\partial u}{\partial t}+u\frac{\partial u}{\partial x}+v\frac{\partial u}{\partial y}+w\frac{\partial u}{\partial z}-fv=-\rho_0^{-1}\frac{\partial P}{\partial x}+\frac{\partial}{\partial z}A_z\frac{\partial u}{\partial z}+\frac{\partial}{\partial x}A_L\frac{\partial u}{\partial x}+\frac{\partial}{\partial y}A_L\frac{\partial u}{\partial y} \tag{1}$$

$$\frac{\partial v}{\partial t}+u\frac{\partial v}{\partial x}+v\frac{\partial v}{\partial y}+w\frac{\partial v}{\partial z}+fu=-\rho_0^{-1}\frac{\partial P}{\partial y}+\frac{\partial}{\partial z}A_z\frac{\partial v}{\partial z}+\frac{\partial}{\partial x}A_L\frac{\partial v}{\partial x}+\frac{\partial}{\partial y}A_L\frac{\partial v}{\partial y}$$

Equations of hydrostatics and continuity for non-compressible fluid:

$$\frac{\partial P}{\partial z}=\rho g \tag{2}$$

$$\frac{\partial u}{\partial x}+\frac{\partial v}{\partial y}+\frac{\partial w}{\partial z}=0 \tag{3}$$

The above equations are written in the right-hand system of Cartesian coordinates (axis z - directed downward from undisturbed surface) relative to the Northern Hemisphere.

u, v, w - represent the components of the current speed vector on the axes x, y, z - respectively.

$f=1.4\ 10^{-4}$ - Coriolis parameter, P- pressure, ρ - water density,

A_Z, A_L - coefficients of turbulent exchange by momentum.

Let us assume the following boundary conditions by the vertical coordinate. At sea surface:

$$z=\xi(x,y,t)$$

$$\rho_0 A_z\frac{\partial u}{\partial z}=-\tau_x;\qquad \rho_0 A_z\frac{\partial v}{\partial z}=-\tau_y \tag{4}$$

$$w = -(\frac{\partial \xi}{\partial t} + u\frac{\partial \xi}{\partial x} + v\frac{\partial \xi}{\partial y}) \quad (5)$$

Here: $\tau_{x,y}$ - are the vector components of tangential friction stress at sea surface,

ξ - is the deviation of sea surface from undisturbed state.

On the bottom : z=H(x,y) the speed vector is assumed to be equal to zero:

$$u = v = w = 0 \quad (6)$$

To reduce computation time we have used the method of "splitting" the equations into baroclinic and barotropic components, as suggested in (Killworth et al., 1987; Marchuk and Sarkisyan, 1988):

Coefficient of the horizontal turbulent exchange was determined according to the Smagorinsky formula that was successfully used in the practice of such calculations (Oye et al., 1985):

$$A_L = C(\Delta L)^2[(\frac{\partial u}{\partial x})^2 + \frac{1}{2}(\frac{\partial v}{\partial x} + \frac{\partial u}{\partial y})^2 + (\frac{\partial v}{\partial y})^2]^{1/2}$$

Where ΔL represents the grid area interval and the empirical constant C = 0.1.

Let us determine the coefficient of vertical turbulent exchange by momentum using the following parameterization that is a specific case of solving the turbulent energy balance equation (Marchuk and Sarkisyan, 1980):

$$A_Z = (bh_Z)^2[(\frac{\partial u'}{\partial z})^2 + (\frac{\partial v'}{\partial z})^2 - \frac{g}{\rho_o}\frac{\partial \rho'}{\partial z}]^{1/2}$$

Where: h_Z - is the thickness of the upper quasistationary layer that was chosen in such a way that within it $A_Z \leq$ 1cm/s.

The empirical constant b was prescribed to be equal to 0.05.

The problem of prescribing tangential stresses at the ocean surface includes two parameterizations: at ice-free surface and under ice. Let us use in the first case well known formulas :

$$\tau_x = \frac{1}{2a'}(\frac{\partial P_a}{\partial x} + \frac{\partial P_a}{\partial y}); \qquad \tau_y = \frac{1}{2a'}(\frac{\partial P_a}{\partial x} - \frac{\partial P_a}{\partial y})$$

where : $a' = \sqrt{\frac{f}{2\upsilon'}}$; Pa - atmospheric pressure at sea level.

υ' - coefficient of vertical turbulent air viscosity, the most used value of which is equal to $\upsilon' = 10^4$ cm^2/s.

For calculating tangential friction at the ice-water boundary several ways can be used. The most simple is given in (Pavlov and Kulakov, 1994).

We shall write down the momentum balance equations for the sea ice in the following form:

$$\frac{\partial u_i}{\partial t}+u_i\frac{\partial u_i}{\partial x}+v_i\frac{\partial u_i}{\partial y}-fv_i=-\frac{g}{\rho_i}\frac{\partial \xi}{\partial x}+\frac{\tau_{xa}}{\rho_i h_i}-\frac{\tau_{xw}}{\rho_i h_i}+\frac{\partial}{\partial x}\mu_L\frac{\partial u_i}{\partial x}+\frac{\partial}{\partial y}\mu_L\frac{\partial u_i}{\partial y}$$

$$\frac{\partial v_i}{\partial t}+u_i\frac{\partial v_i}{\partial x}+v_i\frac{\partial v_i}{\partial y}+fu_i=-\frac{g}{\rho_i}\frac{\partial \xi}{\partial y}+\frac{\tau_{ya}}{\rho_i h_i}-\frac{\tau_{yw}}{\rho_i h_i}+\frac{\partial}{\partial x}\mu_L\frac{\partial v_i}{\partial x}+\frac{\partial}{\partial y}\mu_L\frac{\partial v_i}{\partial y} \quad (7)$$

here u_i, v_i - components of the ice velocity vector by axes x, y - respectively; ρ_i, h_i - ice density and thickness, μ_L - coefficients of turbulent exchange by momentum along horizontal axes.

While accepting condition of equal velocities of ice drift and of water at the ice bottom surface, and $w_{z=0}=0$, $A_L=\mu_L$, we shall subtract the equations (7) from the equations (1). Then we shall obtain while representing the term describing the vertical turbulent exchange as a finite difference

$$\tau_{xw}^{t+\delta t}=\tau_{xa}^{t+\delta t}R_w-\rho A_z\frac{\partial u^t}{\partial z}R_i,$$

$$\tau_{yw}^{t+\delta t}=\tau_{ya}^{t+\delta t}R_w-\rho A_z\frac{\partial v^t}{\partial z}R_i,$$

where:

$$R_w=\frac{\rho\delta z}{\rho\delta z+\rho_i h_i}, \qquad R_i=\frac{\rho_i h_i}{\rho\delta z+\rho_i h_i}$$

δt - time step, δz - spatial step for vertical coordinate.

As the drift ice occupies only some part of the water area represented by ice concentration ($0\le a\le 1$), we shall finally obtain

$$\tau_{x0}^{t+\delta t}=\tau_{xa}^{t+\delta t}[aR_w+(1-a)]-a\rho A_z\frac{\partial u^t}{\partial z}R_i \quad (8)$$

$$\tau_{y0}^{t+\delta t}=\tau_{ya}^{t+\delta t}[aR_w+(1-a)]-a\rho A_z\frac{\partial v^t}{\partial z}R_i$$

The formulas (8) allow us to calculate the tangential stress under the drift ice cover with accuracy of spatial-temporal resolution of the model.

The following boundary condition is accepted under the fast ice:

$$u_{z=0}=v_{z=0}=0.$$

for the barotropic component of current

$$\tau_{x0}=C_0\overline{u}(\overline{u}^2+\overline{v}^2)^{1/2}, \qquad \tau_{y0}=C_0\overline{v}(\overline{u}^2+\overline{v}^2)^{1/2}$$

We used this model for diagnostic calculation of three-dimensional fields of wind-driven, thermohaline water circulation and sea level in the Laptev Sea.

Horizontal spatial resolution for our calculation was 55.6 km. We have simulated currents at following 15 levels:

0, 5, 10, 15, 20, 25, 30, 50, 100, 150, 200, 300, 400, 500 and 1000 m.

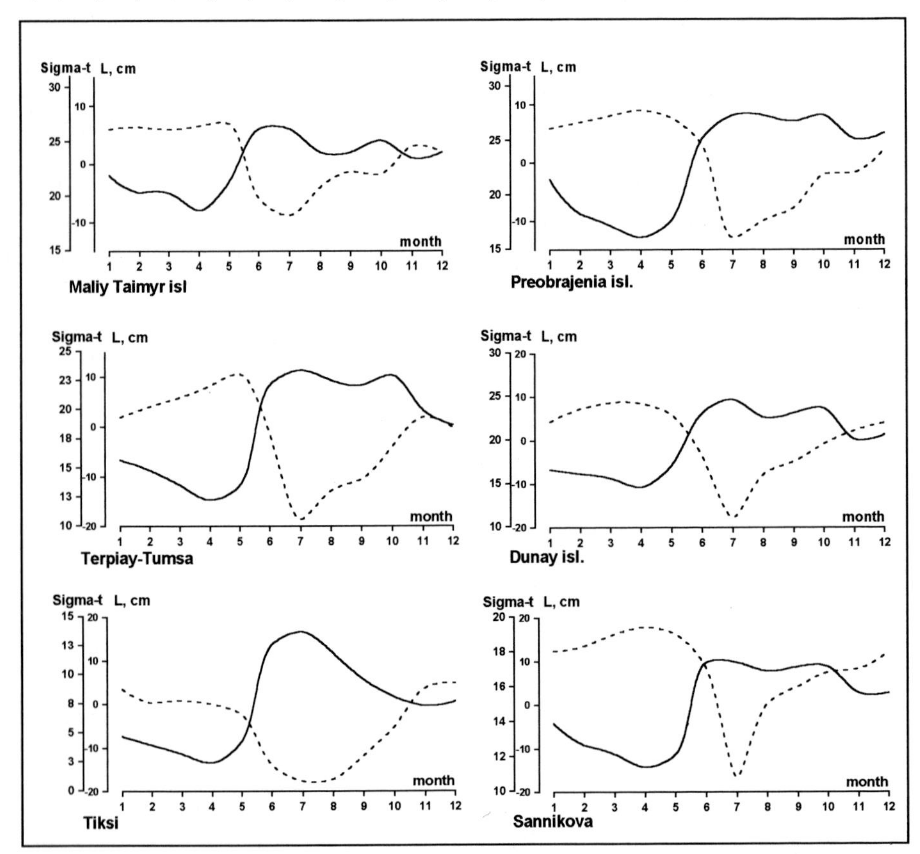

Figure 2: ______ Seasonal variability of the Laptev Sea level; - - - - - Seasonal variability of the Laptev Sea water density (sigma t).

Seasonal variability of the level surface and water circulation

The variability of the Laptev Sea level within a year from observational data at some coastal and island stations is given in Figure 2, has the features, that are also typical for other Siberian shelf seas. In particular, the minimum sea level values are recorded everywhere in April and the maximum in July. The difference in the level between these extremes ranges from 15 to 30 cm, depending on the location of the stations. The general tendency is that amplitudes of seasonal level variations are rising from north to south (Figure 3). Maximum amplitudes of seasonal level variations are observed at the island stations, in inlets and river mouths and the minimum amplitudes are recorded at the island stations in the open sea area. The most intense level increase occurs in May.

In August-September at most coastal and island stations there is a relatively small sea level drop and in October almost everywhere the second local extreme value is observed. At this time the level can reach July values. From December to April the sea level slowly decreases everywhere.

The seasonal variability of sea level correlates well with the seasonal density variability (Figure 2). This suggests the probable causes for the seasonal level differences in the shelf zone of the Laptev Sea to be summer heating and intense freshening of sea water by a strong continental outflow and melting of drifting and fast ice. The hydrostatic component of the level, connected with the local density change, is not large, being not more than 5 cm at small depths (Dvorkin et al., 1978). However changes in the integral circulation of the Laptev Sea due to the seasonal modification in the thermohaline water circulation pattern, induced by a sharp increase in sea water density in summer, can result in significant seasonal level oscillations up to several tens of centimeters.

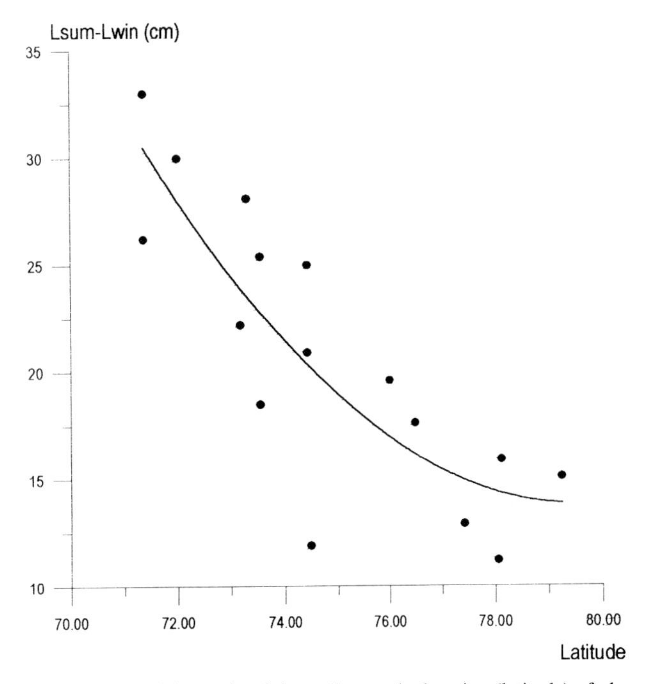

Figure 3: The seasonal difference of the sea level depending on the location (latitude) of observation stations in the Laptev Sea.

This is clearly shown by diagnostic calculations of the surface level of the Laptev Sea, based on averaged density fields for the summer and winter season (Figure 4 A).

The results from modeling seasonal variability of the surface level of the Laptev Sea by use of mean monthly wind stress fields (Figure 4 B) show that the values of the barotropic level components are much less than baroclinic. The contribution of each of these components to the total calculated sea level and its comparison with the real seasonal variability at 6 points of the Laptev Sea is presented in Figure 5. Calculation results revealed predominant influence of thermohaline processes on the seasonal variability of the sea level. General features of the seasonal variability obtained in the result of calculations nearly everywhere correspond to the observation data (see Figure 5) with the exception of the Severnaya Zemlya Archipelago (the Maliy Taimyr station), where the seasonal variability of the sea level associated with thermohaline ocean circulation is reverse to the observation data.

This result was, probably, caused by the insufficient full-scale temperature and salinity observations in this region.

The thermohaline component of the sea level shows the best correlation with observation data rather than the calculated total sea level (Figure 5). Conceivably it is determined by exclusion of the wind component by seasonal averaging of the sea level observation data.

The maximum horizontal gradients of the baroclinic component of the surface level are observed, both in winter and in summer, in the area of the continental slope (Figure 4 A). In summer, the maximum gradients of the barotropic component of the surface level are near the shores. In winter, they are observed around the New-Siberian Islands, especially near the western and northern coast. The maximum levels near the coasts of the islands contribute to the ice outflow. This is probably the cause for recurring polynyas and fractures in this region in winter.

As follows from few data of direct observations and the analysis of the drift of ships, buoys and ice, the total (wind-driven and thermohaline) water circulation of the Laptev Sea has a

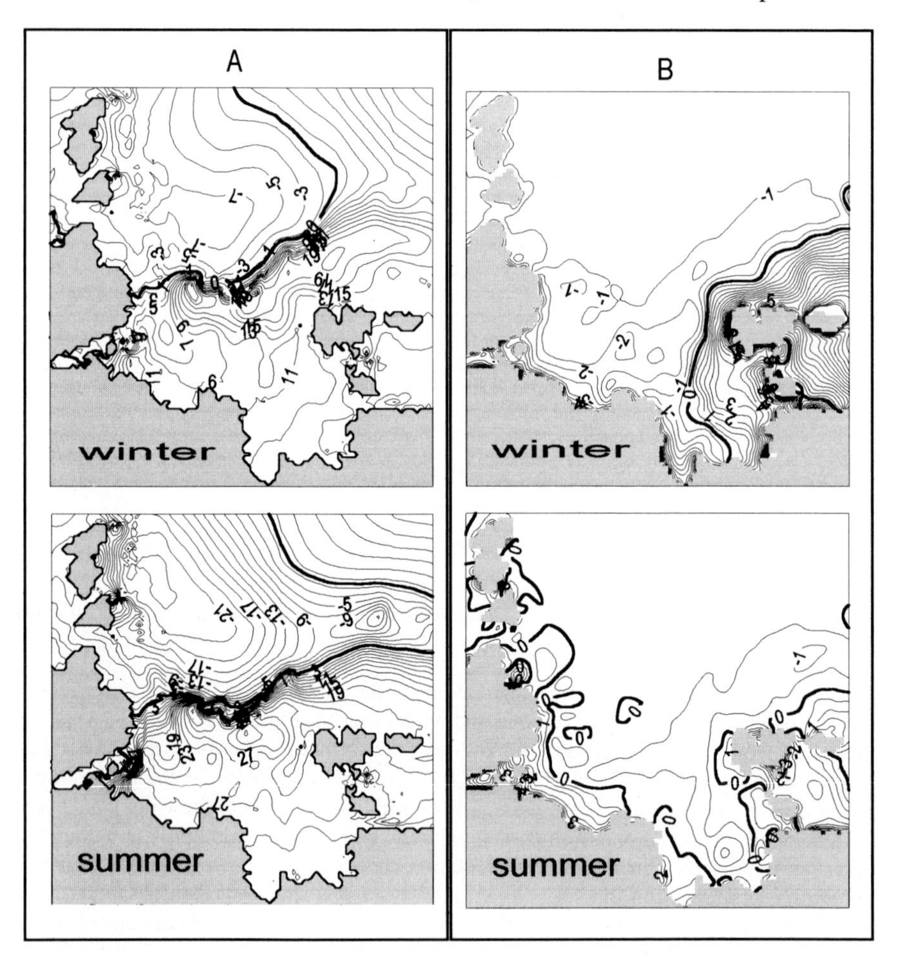

Figure 4: A - The results of diagnostic calculations (based on averaged water density fields) of the surface level in the Laptev Sea for summer and winter; B - The results of modeling the wind-driven surface level in the Laptev Sea for summer and winter.

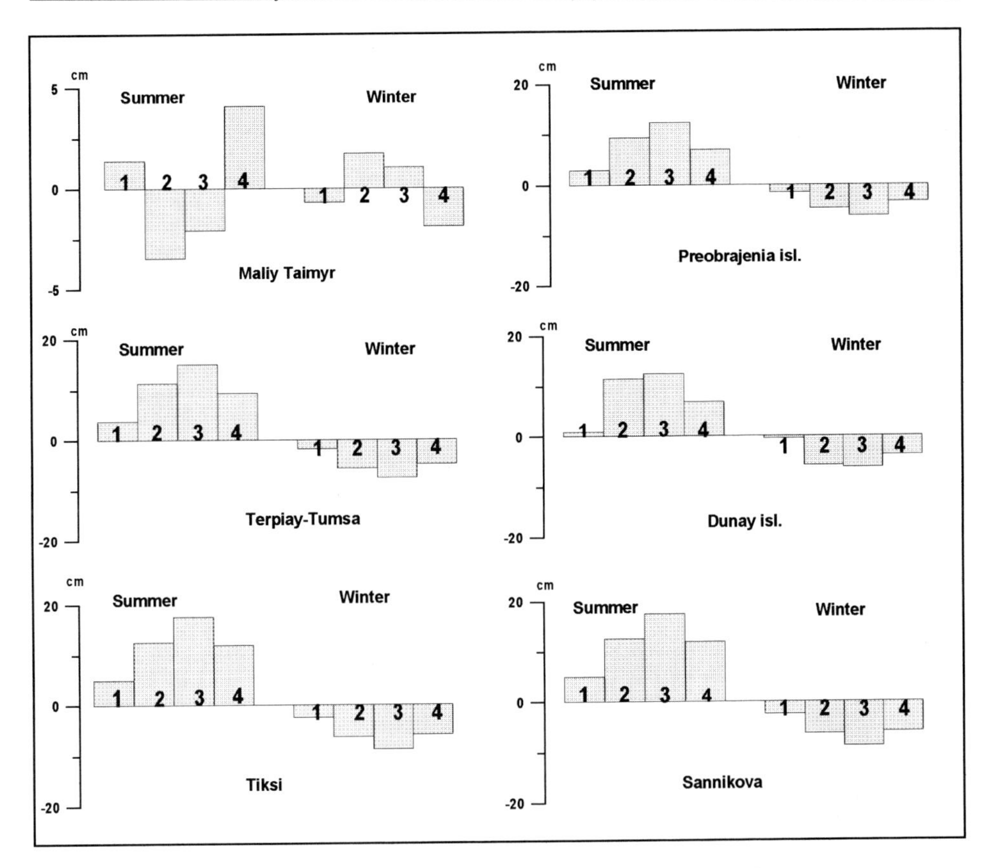

Figure 5: Seasonal difference of the Laptev Sea level: 1 - barotropic component of the sea level (modeling result); 2 - baroclinic component of the sea level (modeling result); 3 - summary of 1 and 2 (modeling result); 4 - observations.

predominantly cyclonic character in the surface layer (Dobrovolsky and Zalogin, 1982; Pavlov et al., 1996). This is confirmed by the results of modeling presented in Figure 6.

In winter, currents of western and north-western directions are observed at the surface in the northern Laptev Sea. This zone is the south-western periphery of the Canadian anticyclonic gyre. A cyclonic eddy is observed in the central part of the sea. Its intensity increases with depth.

In the summertime, the cyclonic water circulation of the Laptev Sea increases and an intensive water inflow from the Kara Sea along the islands of Severnaya Zemlya is observed. Maximum current speeds of 5-10 cm/s are observed at the southern periphery of the cyclonic gyre at the shelf slope and in the northern part of the sea in the zone of the Transarctic current. In the shallow south-eastern Laptev Sea the anticyclonic gyre is formed.

Interannual variability of surface level and water circulation in the Laptev Sea

As follows from the analysis of multiyear observations of sea level in the Laptev Sea at coastal and island stations, there is a steady sea level rise (Figure 7). This is also typical for other

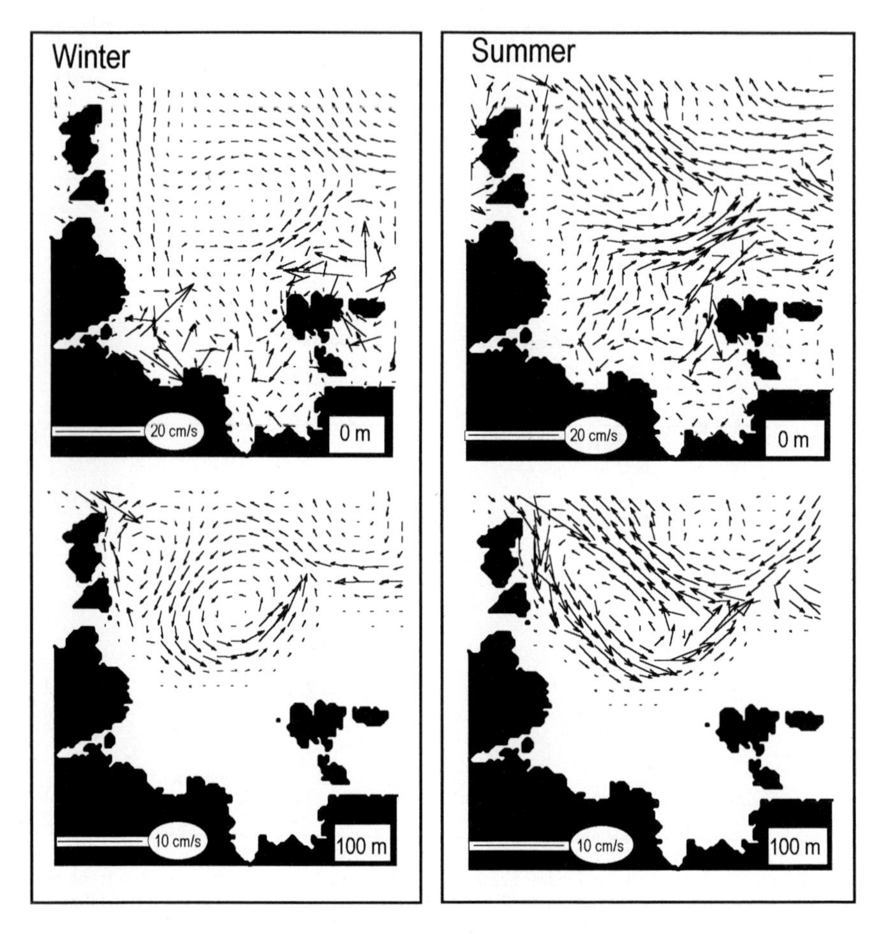

Figure 6: Modeling results of the combined wind-driven and thermohaline water circulation of the Laptev Sea.

Siberian shelf seas (Pavlov et al., 1996). Dvorkin et al. (1978) suggest that the level rise is caused by the sinking of the coast and the shelf of the Arctic Seas. In our opinion, there can be also another cause for the level rise, which is related to a steady water salinity decrease and water temperature increase due to warming in the Arctic (Carmack et al., 1995). Unfortunately, short observation series for water temperature and salinity at coastal and island observation stations do not allow to obtain the climatic variability of these parameters. However the hypothesis may be partly confirmed by the presence of positive trends in the interannual variability of Lena river discharge (Figure 8).

According to spectral analysis of sea level observation series, periods in the range of 13 - 15 years from the second significant peak after the annual one (Pavlov et al, 1996). This is probably related to long-term fluctuations of the thermal state of the Arctic Ocean (Gordienko and Karelin, 1945; Gudkovich, 1961; Nikiforov and Shpaykher, 1980). According to Proshutinsky and Johnson (1996) arctic fluctuations have similar periods.

For analyzing the climatic variability of the level fields and thermohaline water circulation of the Laptev Sea, diagnostic calculations were performed. They were based on the three-dimensional gridded temperature and salinity fields for winter time, smoothed by decades, that

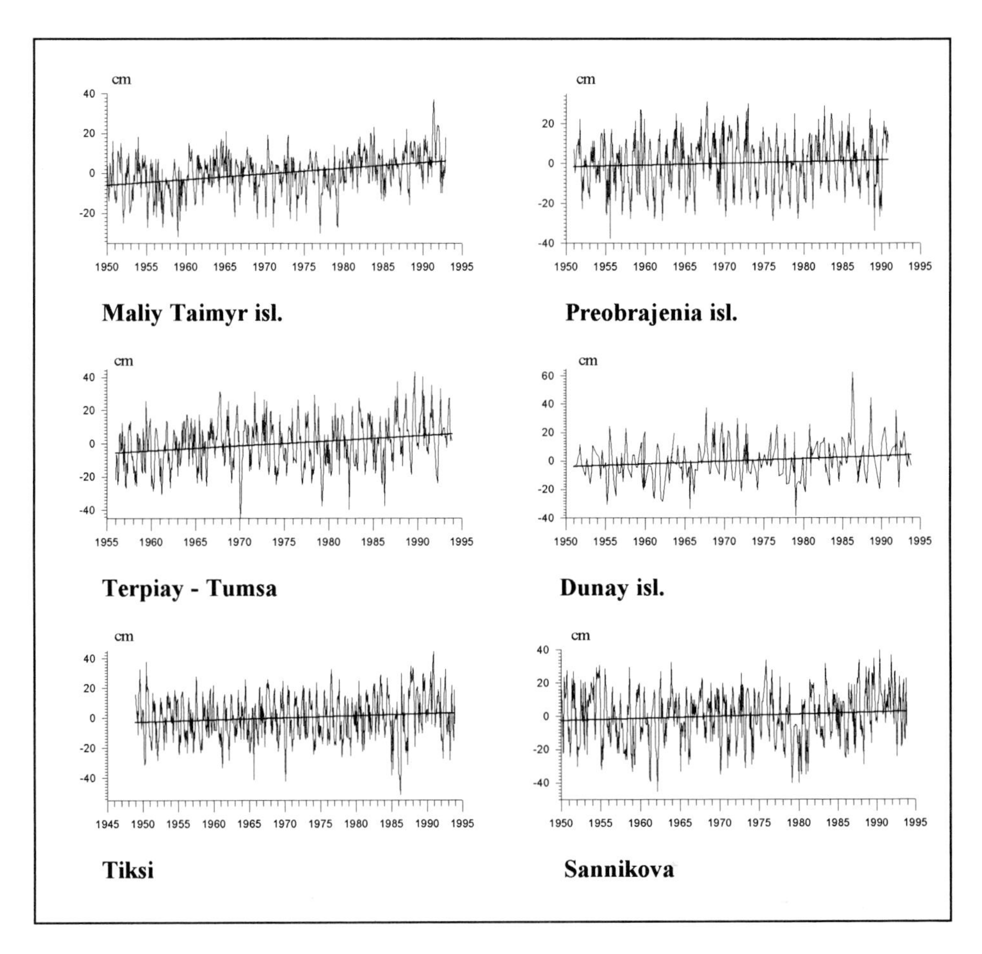

Figure 7: Interannual variability of the Laptev Sea level (observations at the coastal and island stations).

were prepared within the framework of the Russian-American Project (Joint-U.S.-Russian Atlas of the Arctic Ocean, 1997). An analysis of the results (Figure 9) shows that the general cyclonic character of the circulation in the Laptev Sea is preserved. However, the positions of the vorticity centers, their intensity and the vertical structure of currents vary significantly. In the 1950s and 1970s, the cyclonic water circulation in the Laptev Sea is strongly developed (Figure 9), whereas in the 1960s and 1980s it is more weak. In the 1980s, in the eastern Laptev Sea, the anticyclonic water circulation begins to prevail.

For illustrating the interannual variability of the sea level surface, the diagnostic calculations were performed for the 1970s. They were based on the gridded three-dimensional temperature and salinity fields prepared by Polyakov and Timokhov (1994) from data of the most complete winter CTD observations during the "Sever" expeditions. The results show (Figure 10) that the spatial structure of the sea level fields in the Laptev Sea change insignificantly from 1973 to 1978. In 1979 however a sharp modification in the sea level field begins. The center of the area of low sea-level is displaced to the north and a sea level rise north and west of the New-Siberian Islands occurs. Thus, in 1979 the process forming the anticyclonic circulation in the eastern Laptev Sea begins.

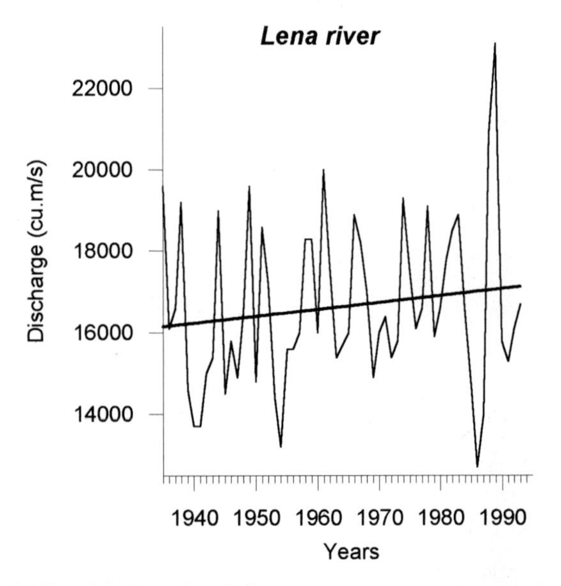

Figure 8: Interannual variability of the Lena river discharge.

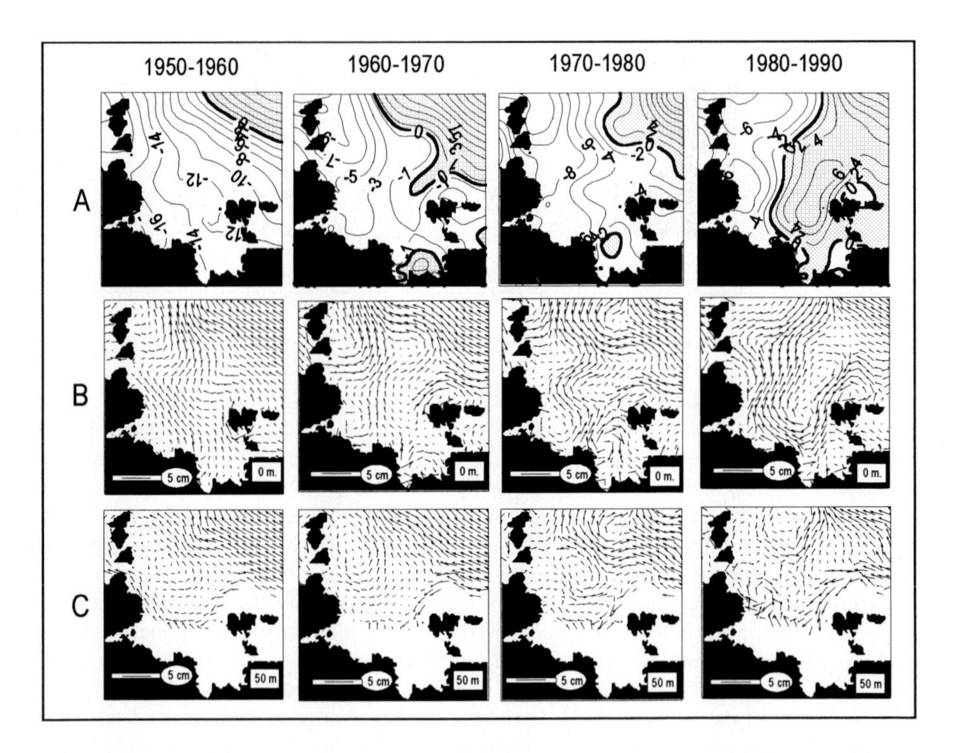

Figure 9: Results of diagnostic calculations of climatic variability of the sea surface level and thermohaline water circulation. A - sea surface level; B - water circulation at the surface; C - water circulation at the depth 50 m.

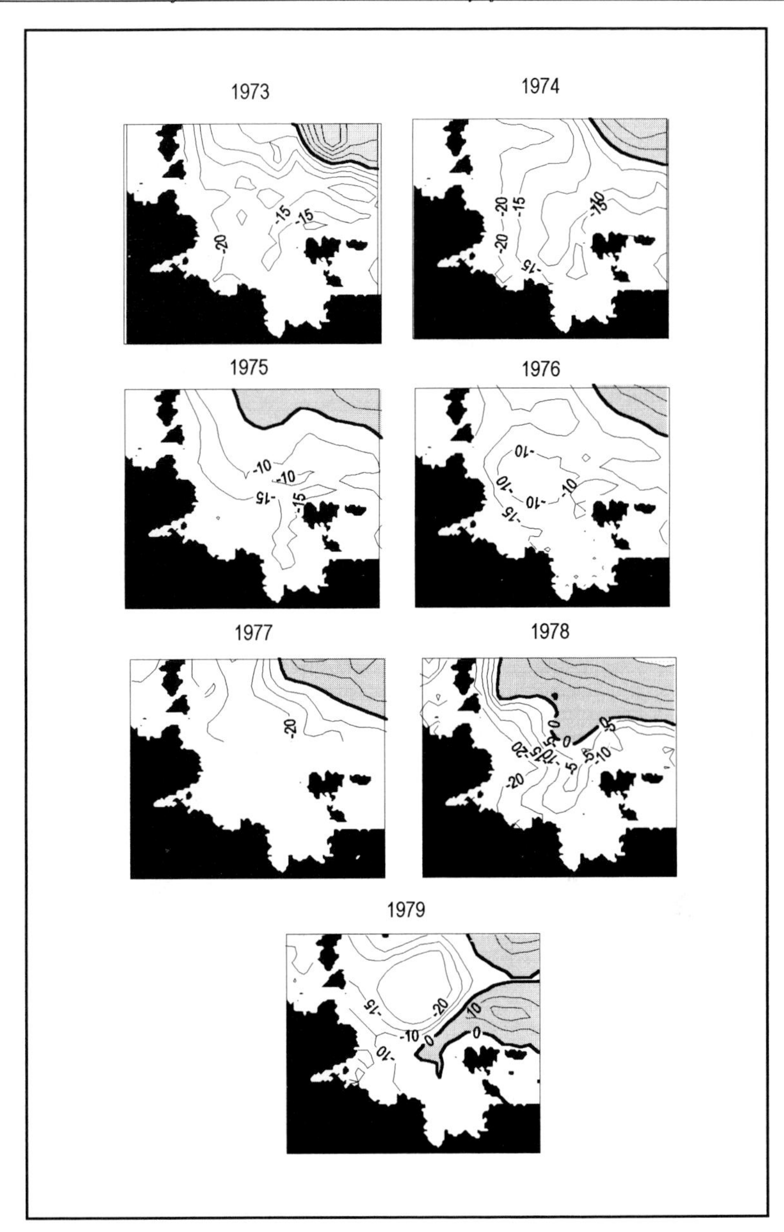

Figure 10: Results of diagnostic calculation of interannual variability of the sea surface level from 1973 to 1979.

Conclusion

The sea level of the Laptev sea has strong seasonal variability. The amplitudes of the seasonal changes of the sea level are 20-30cm.

The main reasons of seasonal changes of the sea level and water circulation in the Laptev sea are significant variances of the atmospheric circulation and the three-dimensional density fields.

The results of separate modeling of wind-driven and thermohaline water circulation showed that the main contribution (60 - 80 %) to formation of the sea level surface and general water circulation in the seasonal and climatic time scales are thermohaline currents.

The minimum sea level values (and currents) for the Laptev Sea are in April - May and the maximum in July during maximum ice melting, increasing of water temperature and intensive river run-off. The second maximum of seasonal sea-level variability is recorded in October due to intensification of western winds over the Laptev sea.

The main features of the spatial structure of the sea level are relatively high sea level in coastal zone and relatively low sea level in center part of the Laptev sea.

The main feature of the water circulation is a cyclonic gyre in the central part of the Laptev Sea. Its position and intensity showed an essential seasonal and interannual variability.

There is a positive trend in the interannual variability of the sea level in the Laptev sea area. The authors believe that a possible reason of such climatic sea level increasing is the shift of the general water circulation of the Laptev sea from a cyclonic to an anticyclonic type due to climatic changes in the Arctic Ocean.

References

Arctic Ocean Atlas (1980) Atlas of the Oceans. The Arctic Ocean, Gorshkov, S.G (ed.) (in Russian). Izd-vo MO SSSR, VMF, Moscow, 190pp.

Arctic and Antarctic Sea Ice, 1978-1987 (1992) Satellite Passive-Microwave Observations and Analysis. NASA, SP-511, 290pp.

Carmach, E.C., R.W. Macdonald, R.G. Perkin, F.A. McLaughlin and R.J. Pearson (1995) Evidens for warming of Atlantic Water in the Southern Canadian Basin of the Arctic Ocean : Results from the Larsen-93 Expedition. Geophys. Res. Letters, 22, (9), 1061-1064.

Gordienko, P.A. and D.B. Karelin (1945) Problems of the ice shift and distribution in the Arctic Basin (in Russian). Probl. Arctik., 3, 5-35.

Dobrovolsky, A.D. and B.S. Zalogin (1982) Seas of the USSR (in Russian). Moskovskiy Universitet, Moscow, 192pp.

Dvorkin, E.N., Yu.V Zakharov and N.V. Mustafin (1978) On the causes of seasonal and multiyear variability of the level of the Laptev and East-Siberian Seas (in Russian). Trudi AARI, 349, 60-68.

Joint-U.S.-Russian Atlas of the Arctic Ocean (1997) Joint-U.S.-Russian Atlas of the Arctic Ocean (on CD-ROM). National Snow and Ice Data Center, Environmental Working Group, Boulder, Colorado.

Nikiforov, E.G. and A.O. Shpaykher (1980) Formation of large scale oscillation of hydrological regime of the Arctic Ocean (in Russian). Gidrometeoizdat, Leningrad, 270pp.

Romanov, I.P.(1992) The ice cover of the Arctic Ocean (in Russian). Gidrometeoizdat, AARI, St.Petersburg, 211pp.

Killworth, P.D., B. Stamforth, B.J. Webb and S.M. Peterson (1987) A free surface Bryan-Cox-Semtner model. J.Phys.Ocean, 17:7

Marchuk, G.I. and A.S Sarkisyan (1980) Mathematical models of the circulation in the ocean (in Russian). Izd-vo Nauka, Novosibirsk, 167pp.

Marchuk, G.I. and A.S Sarkisyan (1988) Mathematical modeling of the oceanic circulation (in Russian). Izd-vo Nauka, Moscow, 302pp.

Oey, L.-Y., G.L. Mellor and R.I. Nires (1985) A three-dimensional simulation of the Hudson-Raritan Estuary. Part I: Description of the model and model simulation. J. Phys. Oceanogr., 15:4.

Pavlov, V.K., L.A. Timokhov, G.A. Baskakov, M.Yu. Kulakov, V.K. Kurazhov, P.V. Pavlov, S.V. Pivovarov and V.V. Stanovoy (1996) Hydrometeorological Regime of the Kara, Laptev, and East-Siberian Seas (with an Introduction by Jamie Morison). Technical Memorandum APL-UW TM1-96, January 1996, APL, Un. of Washington, Seattle, 179pp.

Pavlov, V.K. and M.Yu. Kulakov (1994) Development and implementation of a computer code for modeling the dispersion pollutants in the Arctic Ocean. IAEA, Technical Report 7459/RB/TC, Vienna, 97 pp.

Polyakov, I.V. and L.A. Timokhov (1994) Mean fields of temperature and salinity of the Arctic Ocean (in Russian). Meteorol.& Gidrolog., 7, 68-74.

Proshutinsky, A. and M. Johnson (1996) Two regimes of Arctic Ocean circulation from ocean models and observations. Abstract of poster presented at the 1996 Ocean Sciences Meeting, San Diego, CA. EOS (supplement), 76, (3), OS29.

Terms. Notions. Reference Tables, (1980) Annex for Atlas of the Oceans, Gorshkov, S.G (ed.) (in Russian). Izd-vo MO SSSR, VMF, Moscow, 156pp.

Numerical Modelling of Storm Surges in the Laptev Sea based on the Finite Element Method

I. Ashik and A. Novakov

State Research Center - Arctic and Antarctic Research Institute, 38 Bering St., 199226 St. Petersburg, Russia

Received 3 March 1997 and accepted in revised form 3 March 1998

Abstract - A two-dimensional numerical model is used to simulate storm surges in the Laptev Sea. The model is based on the finite-element method applied to the shallow water equations. The applied method allows for the building of boundary-adapted grids. For time integration the simple explicit method is used. To illustrate the quality of the calculations, surges that occurred in September 1989 and August 1992 are compared to numerical simulations.

Introduction

At the St. Petersburg AARI, numerical hydrodynamic models based on the finite difference method have been used to provide navigation and cargo operations with forecasts and calculations of sea level elevations for several years. The use of finite difference methods leads to regularization of the boundary of a domain. It can have a sufficient influence on the whole solution, particularly in coastal zones, which are often the most important zones in applications. The high spatial resolution is the obvious way to improve description of a basin's morphometry, however, it will require additional computer resources.

In this paper we consider the two-dimensional model based on the finite element method. We use the same basic dynamical formulation as Ashik at al. (1989). To approximate the domain, triangle elements with quadratic basis functions were used.

The sea model

We neglect the sphericity of the earth's surface and use a system of rectangular Cartesian coordinates. A linearized two-dimensional model was used to simulate storm surges in the Laptev Sea:

$$\begin{aligned}
\frac{\partial U}{\partial t} &= fV - g(H+\xi)\frac{\partial \xi}{\partial x} - \frac{(H+\xi)}{\rho}\frac{\partial P_a}{\partial x} + \frac{1}{\rho}\left(\tau_x^s - \tau_x^b\right) \\
\frac{\partial V}{\partial t} &= -fU - g(H+\xi)\frac{\partial \xi}{\partial y} - \frac{(H+\xi)}{\rho}\frac{\partial P_a}{\partial y} + \frac{1}{\rho}\left(\tau_y^s - \tau_y^b\right) \qquad (1) \\
\frac{\partial \xi}{\partial t} &= -\frac{\partial U}{\partial x} - \frac{\partial V}{\partial y}
\end{aligned}$$

where $U = \int_0^{H+\xi} u dz$, $V = \int_0^{H+\xi} v dz$ - the components of the whole-current; f - the Coriolis parameter; x - elevation of the sea surface above the undisturbed level; g - acceleration due to gravity; H - the undisturbed depth of water; P_a - atmospheric pressure at the sea surface;

In: Kassens, H., H.A. Bauch, I. Dmitrenko, H. Eicken, H.-W. Hubberten, M. Melles, J. Thiede and L. Timokhov (eds.) Land-Ocean Systems in the Siberian Arctic: Dynamics and History. Springer-Verlag, Berlin, 1999, 17-23.

τ_x^s , τ_y^s - components of the wind stress; τ_x^b , τ_y^b - components of the bottom stress; r - the density of sea water.

Boundary conditions on the solid boundary are:

$$U = V = 0 \tag{2}$$

And on the open sea boundary:

$$\xi = 0 \tag{3}$$

The bottom stress is parametrized in terms of a quadratic low:

$$\tau_x{}^b = K\rho U\left(U^2 + V^2\right)^{\frac{1}{2}}, \qquad \tau_y{}^b = K\rho V\left(U^2 + V^2\right)^{\frac{1}{2}} \tag{4}$$

where $K = \frac{K_W}{(H+\xi)^2}$; K_W - the coefficient of bottom stress, takes the value 0.0026.

Wind stress on the sea surface was computed from:

$$\tau^s = C^s \rho_a W^2 \tag{5}$$

where r_a - density of air; W - the wind velocity; $C^s = (1.76+0.81W)10^{-3}$.

Solution

Using algorithm of finite element's, we divide the horizontal space domain W into N arbitrary triangle elements. We can write for each element:

$$\hat{U} = \sum_{i=1}^{6} U_i \varphi_i; \; \hat{V} = \sum_{i=1}^{6} V_i \varphi_i; \; \hat{\xi} = \sum_{i=1}^{6} \xi_i \varphi_i \tag{6}$$

where U_i, V_i and x_i - values of U, V and x respectively in i-points; j_i - quadratic basis function (Connor and Brebbia, 1979)

For the whole domain:

$$\hat{U} = \sum_{i=1}^{M} U_i F_i \tag{7}$$

$$\hat{V} = \sum_{i=1}^{M} V_i F_i \tag{8}$$

$$\hat{\xi} = \sum_{i=1}^{M} \xi_i F_i \tag{9}$$

where M - the number of the points all over the domain; F_i - functions, defined as:

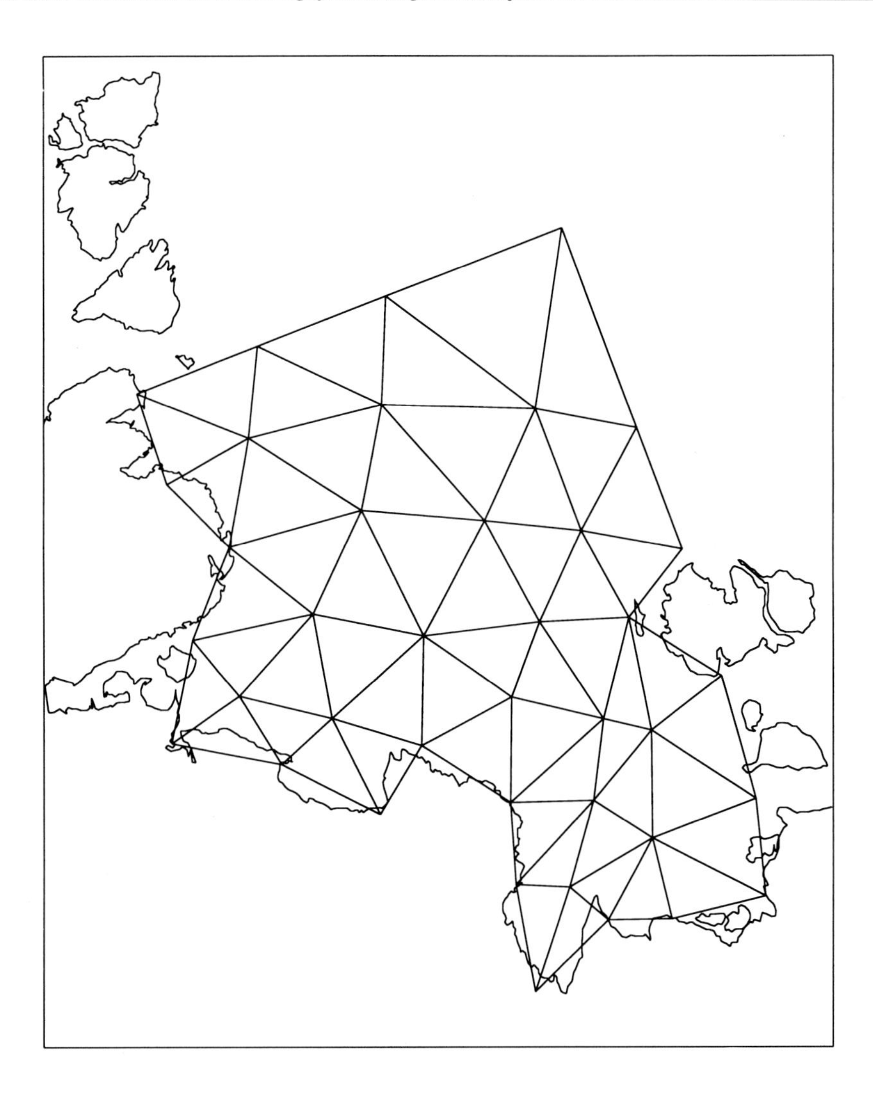

Figure 1: The grid of the finite elements.

$$F_i = \sum_{j=1}^{n} \varphi_j \qquad (10)$$

n is the number of the elements joined in i-th point.

Let us define $F_i = 0$ for the points of the solid boundary in equations (7) and (8) and $F_i= 0$ for the points of the open sea boundary in (9) to satisfy the boundary conditions (2), (3).

Now we can obtain gradients of the variables in the domain, as:

$$A_{i,j}\frac{\partial \varsigma_i}{\partial x} = B_{i,j}\varsigma_i$$
$$A_{i,j}\frac{\partial \varsigma_i}{\partial y} = C_{i,j}\varsigma_i \qquad (11)$$

here A, B and C - following matrixes:

$$A_{i,j} = \iint_{\Omega} F_i F_j dxdy$$
$$B_{i,j} = \iint_{\Omega} F_i \frac{\partial F_j}{\partial x} dxdy \qquad (12)$$
$$C_{i,j} = \iint_{\Omega} F_i \frac{\partial F_j}{\partial y} dxdy$$

Thus, after descretization of the domain, equations (1) can be rewritten:

$$\frac{\partial U_i}{\partial t} = fV_i - g(H_i + \xi_i)\frac{\partial \xi_i}{\partial x} - \frac{(H_i + \xi_i)}{\rho}\frac{\partial P_{ai}}{\partial x} + \frac{1}{\rho}\left(\tau^s_{xi} - \tau^b_{xi}\right)$$
$$\frac{\partial V_i}{\partial t} = -fU_i - g(H_i + \xi_i)\frac{\partial \xi_i}{\partial y} - \frac{(H_i + \xi_i)}{\rho}\frac{\partial P_{ai}}{\partial y} + \frac{1}{\rho}\left(\tau^s_{yi} - \tau^b_{yi}\right) \qquad (13)$$
$$A_{i,j}\frac{\partial \xi_i}{\partial t} = -B_{i,j}U_i - C_{i,j}V_i$$

Gradients $\frac{\partial \xi}{\partial x}$, $\frac{\partial \xi}{\partial y}$, $\frac{\partial P_a}{\partial x}$ and $\frac{\partial P_a}{\partial y}$ in equations (13) were obtained from (11).

This finite element formulation strongly differs from those suggested in (Connor and Brebbia, 1979; Peyret and Taylor, 1986). It appeared to be more stable, and allowed the usage of longer time steps (from Courant - Friedrichs - Lewy condition Δt=600sec).
Integration through time started from a state of zero displacement and motion, expressed by:

$$\xi(x, y; t) = 0,\ U(x, y; t) = V(x, y; t) = 0 \qquad (14)$$

For time integration a simple explicit scheme was applied.
Fields of atmospheric pressure on the sea surface (defined in points of regular geographic grid with step $\Delta\varphi = \Delta\lambda = 5^0$) are used as source information for the calculations. The atmospheric pressure input is interpolated linearly in time between the daily specified values.
The wind velocity and direction are computed from:

$$W = k_1 W_g$$
$$A = A_g - \alpha \qquad (15)$$

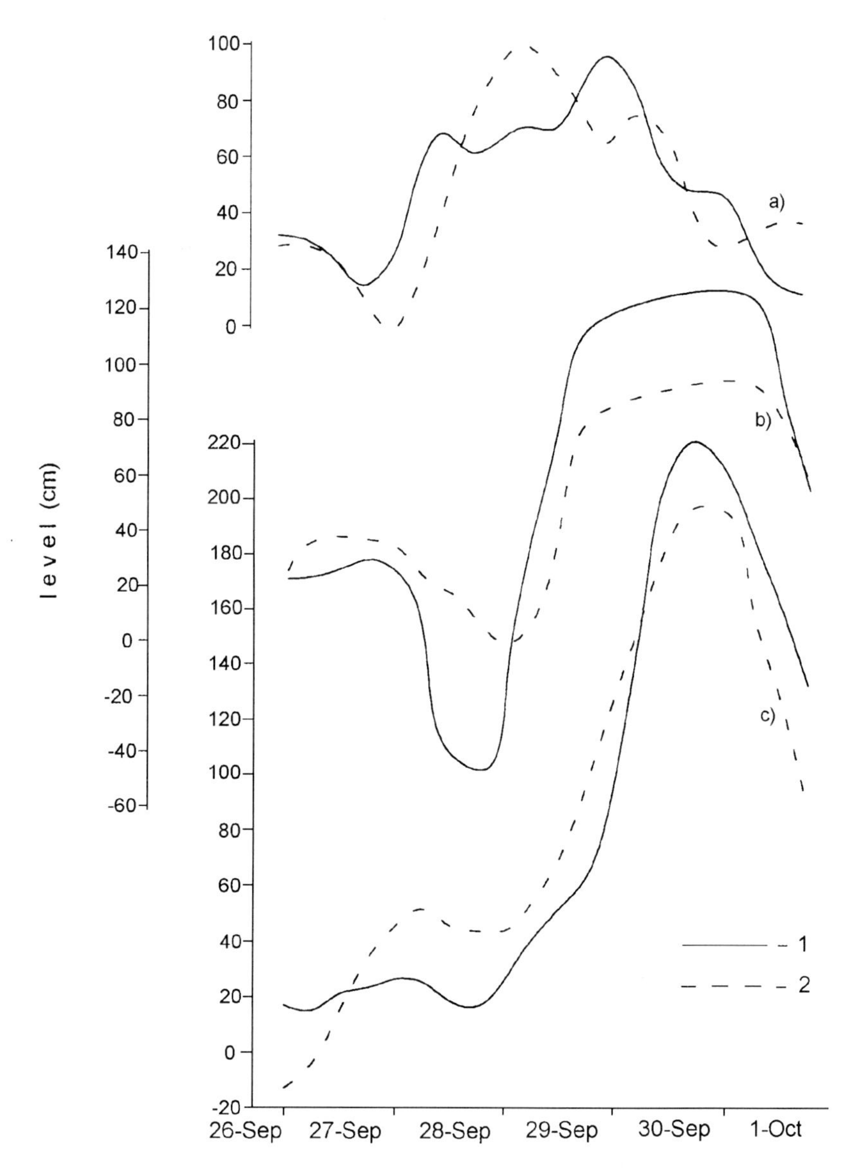

Figure 2: The oscillations of the sea level (1 - observed, 2 - computed) in September 1989, at Terpiay-Tumsa Cape (a), Tiksy Bay (b) and Nayba (c).

where $k_1 = \frac{h}{h+1}$, $h = 0.25 * 1.212^{W_g}$,

$$\alpha = 41.26\exp(-0.07W_g) - 11.27\exp(-0.48W_g).$$

The matrix equations (11) were solved by the Gauss elimination method and the successive

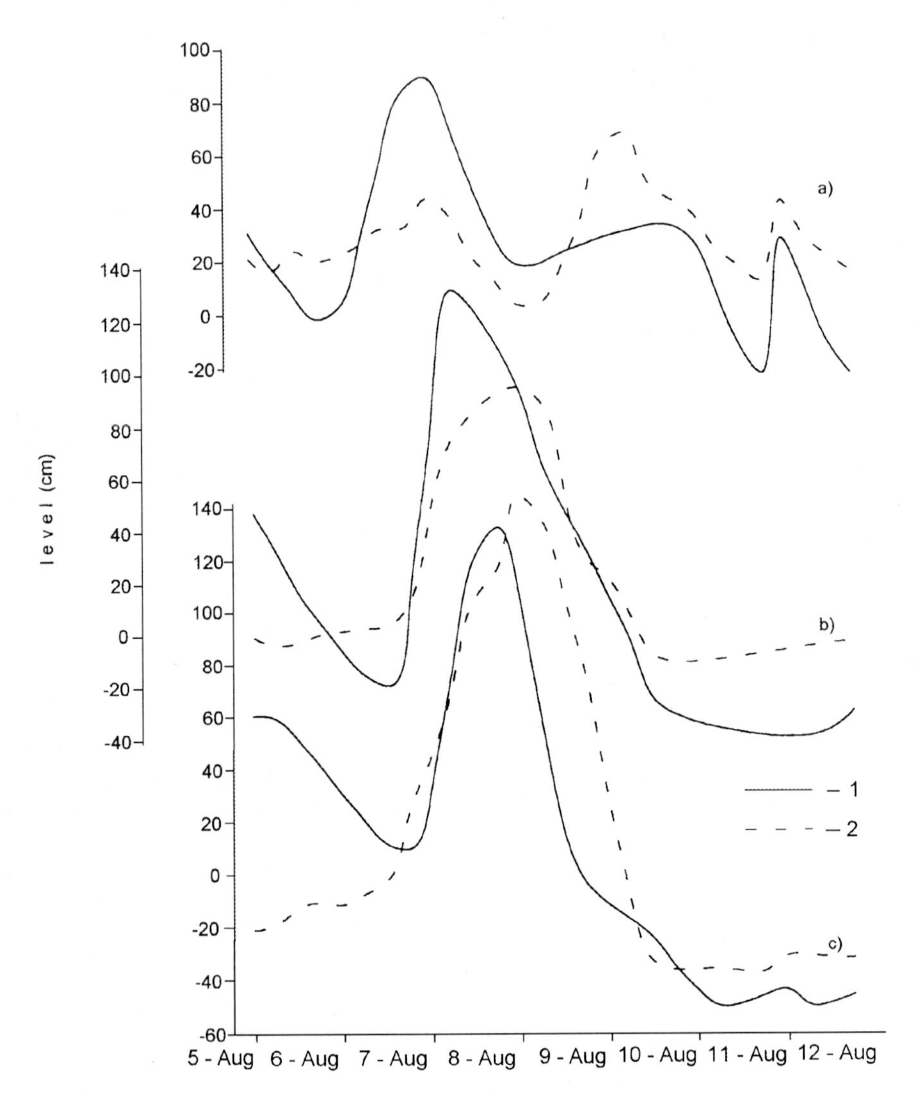

Figure 3: The oscillations of the sea level (1 - observed, 2 - computed) in August 1992, at Terpiay-Tumsa Cape (a), Tiksy Bay (b) and Nayba (c).

overrelexation method. In second case to reduce number of iterations the results of the previous time step were used as a first approximation for the system.

To test this scheme, calculations for a rectangular basin were made. The results were compared with (Gudkovich and Proshutinskiy, 1988) and appeared encouraging.

Results

To simulate storm surges in the Laptev Sea the grid, which consists of 54 elements and 131 points (Figure 1) was used. The open sea boundary was located near 200-meter depth. The influence of the ice cover on forming level evaluations was not taken into account.

Let us consider the results for level oscillations in the southern part of the Laptev Sea during surges in September 1989 and August 1992.

A surge that occurred in September 1989 was caused by a powerful cyclone with core of gales. The genesis area of the cyclone was in the northern Atlantic from where it moved eastwards through the northern areas of the Barents, Kara and Laptev seas. The atmospheric pressure in the centre of the cyclone was estimated to be below 970 Pa. The edge of the drifting ice was situated near 75^0N, so, the south-eastern part of the Laptev Sea was free of ice. An elevation of more than 2m above mean sea-level was observed at Nayba.

In Figure 2 we give the time variation of the calculated and observed sea-surface elevations at the Terpiay-Tumsa Cape, Tiksy Bay and Nayba. The average absolute error of the calculation is 15-20cm, RMS error is 20-25cm, the correlation factor is 0,83.

After a maximum level (1 -1,5 meter) a significant minimum (approximately 2 meters below mean sea-level) was observed in August 1992 along the southern coast of the Laptev Sea. This situation was caused by an intense baric evolution in that area. On August 8th a cyclone was located in the Laptev Sea region, on August 10th it was substituted by an anticyclone. By that time fast ice in the Laptev Sea was broken, the edge of the drifting ice was situated between the Peschany Cape and Shirokostan peninsula, so, the southern part of the sea was mainly free of ice.

The results of the calculation are shown in Figure 3. The average absolute error of this calculation is 20-30cm, RMS error is 25-45cm, the correlation factor - 0,85.

Discussion

Finite element hydrodynamic models using elements with quadratic basis functions are considered to be of less numerical efficiency than finite-difference ones. However the use of modern iteration methods for solution of sparse matrix equations reduce this difference of efficiency to a minimum.

In spite of the poor spatial resolution of the model in this paper the results are encouraging and the coastal line approximation is reliable. We believe that such small flexible models describing coastal zones can become a useful part of compound hydrodynamic models.

Conclusion

A numerical model for solving the shallow water equations, using the finite element method in the horizontal space domain was presented in this paper. The basis functions of the second order were used. The model has been applied to the calculation of the surges that occurred in September 1989 and August 1992 in the Laptev Sea. Computed level oscillations were in agreement with observations.

References

Ashik, I.M., A.Yu. Proshutinskiy and V.A. Stepanov (1989) Some results and outlooks of the numeric forecasts of sea level oscillations in the Arctic Seas (in Russian). Meteorology and Hydrology, 8, 74-82

Gudkovich, Z.M. and A.Yu. Proshutinskiy (1988) Modelling of oscillations of the level in the ice covered seas (in Russian). Proc AARI, v. 413, 85-95

Connor, J.J. and C.A. Brebbia (1979) Finite element techniques for fluid flow (in Russian). Sudostroenie, Leningrad, 264 p.

Peyret, R. and T.D. Taylor (1986) Computational methods for fluid flow (in Russian). Gidrometizdat, Leningrad, 352 p.

Large-Scale Variations of Sea Level in the Laptev Sea

G.N. Voinov and E.A. Zakharchuk

State Research Center - Arctic and Antarctic Research Institute, 38 Bering St., 199226 St. Petersburg, Russia

Received 28 October 1997 and accepted in revised form 3 March 1998

Abstract - Large-scale sea-level oscillations were investigated at 10 coastal stations in the Laptev Sea. Daily mean sea-level heights for the years 1962-1984 were used. Spectral analysis of time series reveals significant energetic peaks at 1 year, 28-30, 7-8, 5.2-5.3, 4.3, 3.0-3.3, 2.4-2.5 and 2.1-2.2 days. Long-period components were extracted using harmonic analysis on the basis of the least square method. The observed monthly tide exceeds the equilibrium tide by a factor of 1.45. Variance analysis of non-tidal (residual) time series shows that the maximum intensity of the sea-level oscillations is observed for periods of less than 1 year (from 66 to 81 % of the total variance). From 13 to 24% of the total variance is accounted for by the long-period tides and 5 - 18% of the variance is due to interannual variability. An analysis of the two-dimensional distribution of temporal variations of amplitudes at the synoptic scale reveals their non-steady-state character for this range. Cross-spectral analysis of the non-tidal level oscillations indicates high coherence (0.68 - 0.98) between sea-level variations at the synoptic scale at different stations. The spatial distribution of phases shows that these sea level variations are mainly progressive waves in the range of less than 60 days. Calculated velocities of these waves are close to phase velocities of shelf waves and have a magnitude of 1.3 - 5.2 m/s for periods from 4 to 46 days.

Introduction

Analysis of the daily mean sea-level oscillations at various coastal stations in the Laptev Sea shows that maximum variations here can be as high as two meters and above. It seems doubtless that such significant perturbations of sea-level can exert influence on dynamics of ice cover, variability of currents, regime of polynyas, transference of sediments. However, the large-scale sea-level variations have not been adequately explored. Study of these processes has been mainly devoted either to the global sea-level variations (for the periods of a year and above) using monthly mean data (Dvorkin, et al, 1989; Bannov-Baykov, 1974), or to storm surges with periods from 1 to 6 days, which have been investigated in most cases using observational data for the navigation period July - September (Alekseyev and Mustafin, 1972). However, as it was shown by Cartwright (1983) results of spectral analysis of the monthly mean data are blurred because of irregular lengths of calendar months. Analysis of the daily mean data has been performed only for the navigation period (Krutskikh, 1974; Vanda, 1989). Low-frequency regions of the spectrum at the synoptical scale for the periods from 10 to 60 days are poorly known. It is also interesting to estimate power contribution of processes of various temporary scales into total variability of the large-scale level variations.

The harmonic analysis of the long-period tides for long time series by the least square method was not made previously. So far studies of the long-period tides in Arctic seas by Vorobyev (1966,1969,1976) were mainly based on data of yearly sea-level series and not on the least square method. Vorobyev's results did often not exceed noise level and were of little reliability. Shelf waves and their contribution to total variability of the level oscillations and currents are completely unexplored in the Arctic shelf zone.

This paper describes the long-period tides and sea-level oscillations for the synoptic scale on the basis of analysis of long series of daily mean sea-level values in the Laptev Sea.

In: Kassens, H., H.A. Bauch, I. Dmitrenko, H. Eicken, H.-W. Hubberten, M. Melles, J. Thiede and L. Timokhov (eds.) Land-Ocean Systems in the Siberian Arctic: Dynamics and History. Springer-Verlag, Berlin, 1999, 25-36.

Data and methods

In order to investigate the above-mentioned aspects of the problem, time series of the daily mean values of the sea level records for the years 1962-1984 at 10 coastal stations in the Laptev Sea (Peschaniy Cape, Maliy Taimyr Island, Andrey Island, Preobrazhenia Island,Terpiay-Tumsa Cape, Dunay Island, Tiksi port, Kotelniy Island, Sannikova station, Kigilyakh Cape) were analyzed (Figure 1). Time series of the daily mean sea level at coastal stations of Mariya Pronchishchevaya Bay and the Stolbovoy Island (3-6- years long) were used additionally with the aim to study various parameters of low-frequency waves.

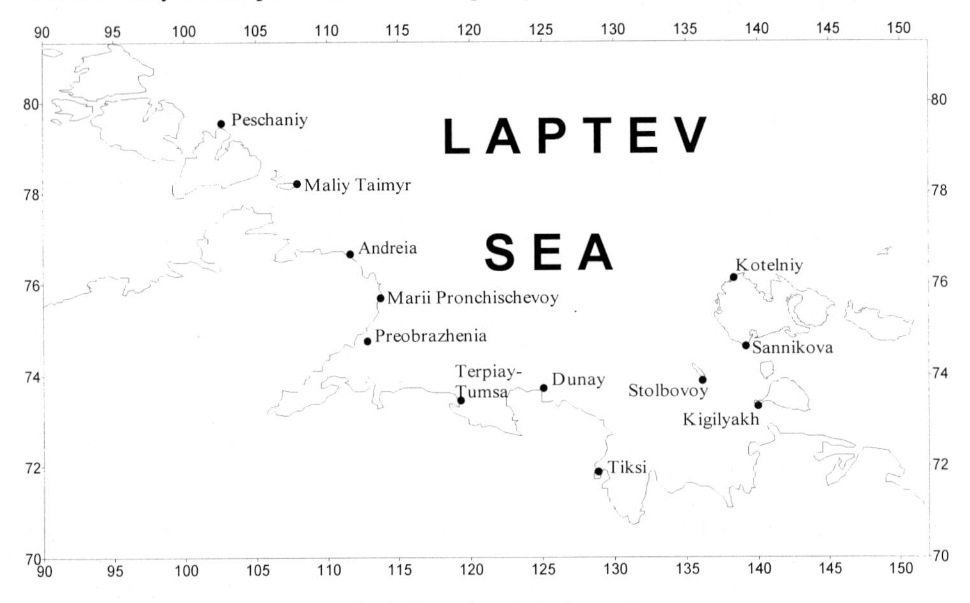

Figure 1: Location of stations of the sea level observations in the Laptev Sea.

Sea level observations were carried out using an automatic tide gauge only at Mariya Pronchishchevaya Bay, Preobrazhenia Island, Tiksi port and Sannikova. At all other stations a tide staff conducted the measurements. The observations were accurate to within ± 1 cm. The visual measurements were conducted at 6 hour intervals.

These data were reduced to daily averages as the arithmetical mean of 4 consecutive observations. The daily mean is actually a low-pass filter. This filter separates out characteristic *c(f)*(or amplitude response factor)

$$c(f) = \sin(m\,\pi f\,\Delta t)/m\,\sin(\pi f\,\Delta t) \qquad (1)$$

where $m = 4$, Δt is the time interval (6 hours), f - the frequency (cycles per hour). The results of the spectral and harmonic analyses were corrected in accordance with (1).

The harmonic analysis of the long-period tides by the least square method was performed using the long-term daily mean sea level at these stations over the entire period of observations at a specific station with a defined datum. Harmonic constants of a long-period tide for 24 constituents were extracted within a range from the nineteen-year nodal to week lunar-solar tide. Technique of the harmonic analysis of these data is given in Voinov (1996). It is significant that contributions in the daily mean of the constituents M_2, MS_4 were removed in advance. Contributions of the components O_1 and Q_1 does not exceed 0.4 cm, and they are free from aliasing for frequencies of the long-period constituents. Contributions of N_2 and M_4 peaks in the daily mean are respectively 0.8 cm and about 1 cm at Preobrazhenia, but are of no practical importance for the lack of aliasing in the long-tidal constituents. Contribution of all the semidiurnal or diurnal constituents is less than 0.1 cm.

Gaps in data were revealed at some points (Peschaniy Cape, Andrey Island, Terpiay-Tumsa Cape, Dunay Island), and in two cases this gap was up to 1 year long. To a first approximation, these gaps were omitted in the harmonic analyses. In the subsequent analysis the gaps were restored using the predicted long-period tides, and the harmonic analyses was recirculated.

The non-tidal oscillations and the long-period tides were separated with the aim to study the large-scale level variations. The predicted long-period tide was obtained on the basis of 10 constituents with the most significant magnitude (Mn, Sa, Ssa, Sta, Msm, Mm, Msf, Mf, Mtm and Msw). Non-tidal (residual) values of the daily mean sea level were derived after subtraction the predicted long-period tide values from the initial mean daily data. As it is known, the constituents Sa (365.26 days) and to a smaller degree Ssa (182.62 days) also are determined mainly by hydrometeorological means such as change of temperature and salinity of water and by various processes in the atmosphere. However, the tidal component of the tide Sa is almost inseparable from annual variations of hydrometeorological factors. However, physical mechanisms causing the Sa and Ssa constituents remain in the aggregate fixed in time, and interannual variations of amplitudes and phases of these constituents are, therefore, not great. Hence, the Sa and Ssa constituents are predictable for most practical purposes.

Spectra for initial series of the level oscillations were obtained using the orthodox method of spectral analysis by computing the cosine transforms of the autocorrelations (Jenkins and Watts, 1969). Amplitude and phase of more than 4000 harmonics were obtained in the result of the Fourier analysis for each of the residual series. Variance of the global oscillations of the sea level (interannual variability) was calculated by taking the sum of each of the variance from 1 to 22 Fourier harmonics. The variance of the oscillations for the periods less than 1 year was derived after subtracting the variance of global oscillations from the variance of the residual series.

For complete investigation of the oscillations of the synoptical scale (periods from days to months) the residual series were divided into semi-annual successive sections (46 sections of realization for each of the station). The Fourier analysis was carried out for each section. Two-dimensional distributions of temporal variations of the amplitudes of the synoptical scale oscillations were constructed on the basis of this analysis. We excluded the oscillations with periods more than half a year from the residual series by polynomial smoothing for investigation of temporal variability of the variance of the synoptic scale level oscillations. The thus obtained series were divided into monthly successive realizations. The variance was calculated for each realization.

The cross- spectral analysis was carried out between sea level oscillations of synoptical scale at different stations in the Laptev Sea for estimating parameters of low-frequency waves. Spatial distribution of the phases indicates that these oscillations represent mainly progressive waves in the range less 60 days. Estimating the wave parameters was performed for typical and uniform cases in the Laptev Sea using the results of the Fourier analysis. Calculation of wave propagation velocities was done using the formula

$$V = l / \tau \quad (2)$$

where l is the distance between stations, τ is the delay period in days equal $\Delta\varphi * T/360°$, $\Delta\varphi$ is the phase difference in degrees, T is the oscillation period in days.

Results

The long-period tides

The results of the analysis of the long-period tides are shown in Table 1. The observed nodal tide Mn (18.6 years) has considerable variations in the amplitude (H) and the phase lag between the stations. Vectorial average values of H and g for the nodal tide Mn in the Laptev sea are $H = 2.6$ cm, $g = 160°$ (without Andrey). The equilibrium nodal tide at the latitude 75° with the factor 0.69 (effect of the yielding Earth) gives 1.09 cm (Cartwright et al., 1971), and values of g (the phase lag) is close to zero. These data do not confirm the consistency of the observed nodal tide with the equilibrium theory. Time of energy dissipation of tides according to Proudman (1960) is equal to 5 years, signifying that the tides with larger periods should correspond to the equilibrium theory. Unfortunately, the shallow-water tidal constituents can produce perturbations of the nodal tide. For example, tri-linear interaction exists between the tide M_2, the tide K_1 and the tide O_1 (Rossiter, 1967).

Table 1: Amplitudes and phases of the long period constituents in the Laptev sea. The phase lag in zone GMT

Station	Period of analyses	Mn H (cm), g (deg.)	Sa H (cm), g (deg.)	Ssa H (cm), g (deg.)	Mm H (cm), g (deg.)	Mf H (cm), g (deg.)
Peschaniy	1962-1984	2.46, 100.5	6.39, 246.5	4.28, 216.2	1.76, 226.4	1.79, 223.9
Maliy Taimyr	1949-1984	3.81, 109.4	5.04, 242.7	3.30, 197.4	1.60, 224.5	1.93, 221.3
Andreia	1954-1984	4.52, 145.2	9.32, 258.4	3.11, 206.2	2.14, 222.6	1.88, 234.6
Preobra-zhenia	1954-1984	4.22, 173.1	12.29 262.1	4.38, 211.9	1.81, 226.1	0.98, 255.6
Terpiay-Tumsa	1960-1984	4.33, 194.0	12.04, 245.3	3.78, 231.1	1.87, 225.6	1.21, 280.3
Dunay	1961-1984	4.19, 171.3	11.22, 249.4	2.56, 195.2	2.23, 228.8	0.78, 266.0
Tiksi	1949-1984	2.18, 130.5	11.87, 223.5	6.19, 215.8	2.30, 240.2	1.14, 282.2
Kotelniy	1956-1984	3.45, 135.2	12.30, 265.3	4.50, 206.9	2.42, 240.5	1.05, 260.5
Sannikova	1952-1984	3.49, 177.1	11.94, 248.5	4.34, 207.1	1.85, 252.8	1.14, 281.0
Kigilyakh	1954-1984	2.95, 209.9	11.23, 244.0	4.51, 188.1	1.97, 263.8	1.17, 304.5

Vorobyev (1969) has plotted the mean annual sea level for 18 stations in Arctic seas for the years 1946-1964. Results for the nodal tide are as follows: the mean amplitude is 6.5 cm, and phase lag is 155°. The method used by Vorobyev gives large value of the amplitude because of an undefined datum with regard to the mean sea level was measured.

The fact that the phase lags are not close to zero does not support existence of the equilibrium nodal tide. Apparently, the observed nodal tide associated with the large contribution from the shallow-water long-period constituents.

The annual harmonic dominates at all the stations. The amplitude of the Sa harmonic equals 5 - 6 cm in the region off the islands of Severnaya Zemlya increasing towards the coast up to 11-12 cm. The annual harmonic is mainly related with the hydrometeorological factors such as the seasonal steric change in the local ocean (estimated as 20 % by Bannov-Baykov, 1974), the

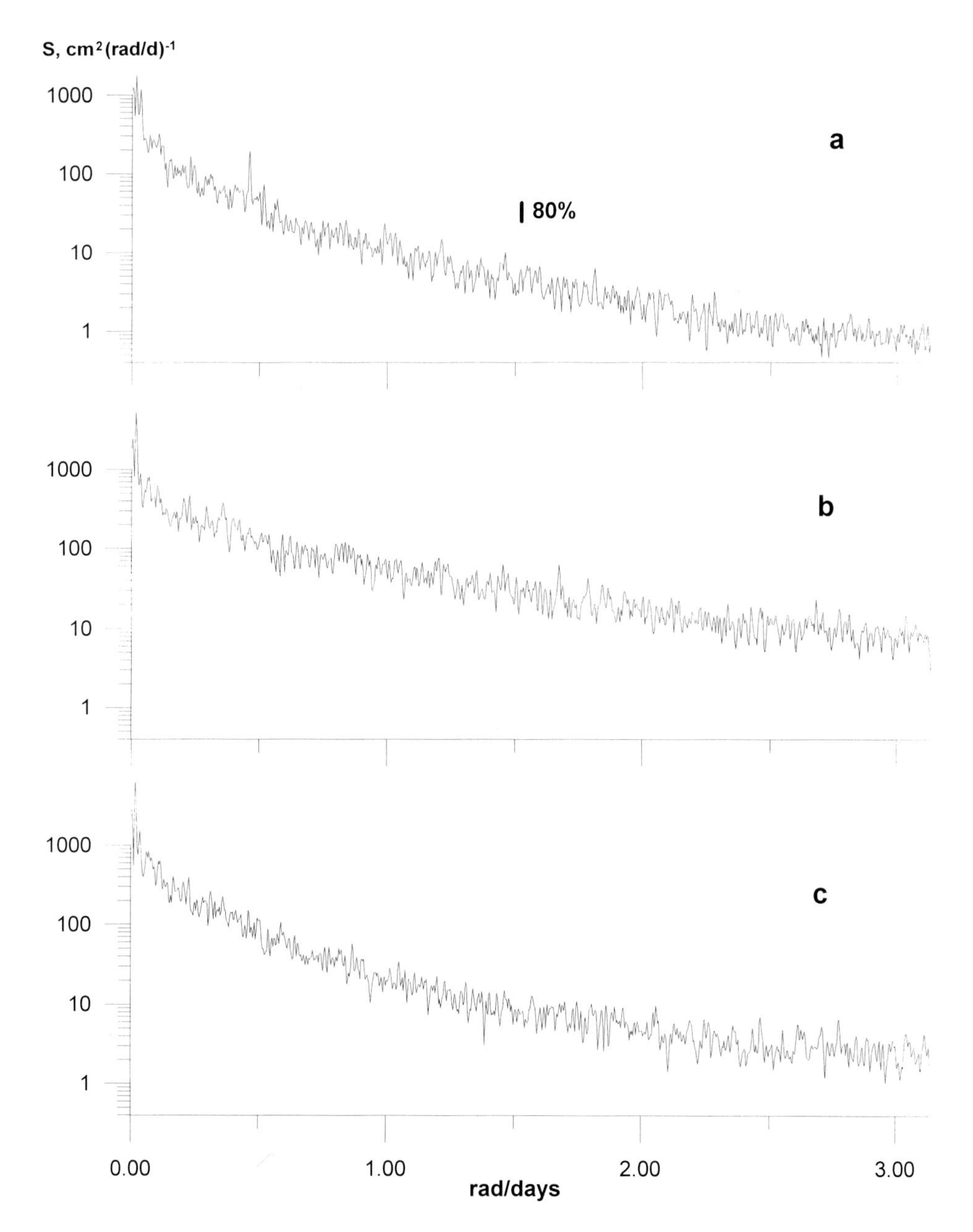

Figure 2: Spectra of variations of sea level at points Peschaniy (a), Dunay (b) and Kotelnyy (c).

annual change in atmospheric pressure at a given point (estimated as 20-30% by Bannov-Baykov, 1974) and the effects of atmospheric processes (estimated as about 50% by Bannov-Baykov, 1974).

The solar semi-annual tide Ssa has a uniform character of propagation of the phase lags, but the amplitides vary within 2.6-6.2 cm. The maximum amplitide (6.2 cm) is observed in Tiksi which is probably connected with the influence of runoff variations of the Lena river. The mean values of H and g for all the stations are H = 4.0 cm, g = 208°. The average tide Ssa for all the stations gives the amplitude exceeding the predicted one based on the equilibrium theory by a factor of 3.3. The average phase lag differs from the theoretical value (180°) by 28°. These estimations of the tide Ssa suggest its existence being close to the dynamical form. The perturbations of the observed tide Ssa are likely associated with the shallow-water constituents such as $S_2 - K_2$, $T_2 - R_2$, $K_1 - P_1$. Maksimov (1970) also represented the semi-annual tide as a gravitational rather than a meteorological tide.

Vorobyev (1976) extracted the semi-annual tide using Fourier analysis from the yearly series of sea level at 26 stations in Arctic seas for the years 1950-1970. Vorobyev's averages in the Laptev sea are 4.5 cm for the observed amplitude and 188° for the phase lag. These results are close to our mean values of the semi-annual tide and suggest that the harmonic constants of the tide Ssa are stable.

Results were obtained for the solar third-annual tide Sta (121.75 days). The amplitude of the tide Sta varies between 0.6 and 4.2 cm. The third-annual tide represents a meteorological rather than gravitational tide.

The lunar elliptic tide Mm (27.55 days) has amplitude of 1.6 to 2.4 cm with the phase changing eastward. The observed mean amplitude tide Mm (2.00 cm) in the Laptev sea is 1.45 times more than the mean equilibrium amplitude (with the factor 0.69). The observed phase lags deviated from equilibrium (phase 180°) roughly by 50-80 degrees.

Vorobyev (1966) calculated the tides Mm and Mf using Darvin's method from yearly series of the sea level at 4 stations in the Laptev sea. These results are, in general, less reliable than ours given in Table 1. It has been found experimentally that a reliable extraction of these tides is possible by analyzing uninterrupted daily mean series with duration of no less than 10 years. Vorobyev' values for Tiksi (averaged for 14 years) of the tide Mm are 3.6 cm for the amplitude and 239° for the phase. Our results are 2.30 cm for the amplitude and 240.2° for the phase. Vorobyev's large value of the amplitude was caused mainly by noise.

Values of the admittance amplitudes and phases indicate that the Mm tide deviates much from equilibrium tide. Apparently, the resonant conditions are observed in the Laptev Sea for the period of about a month. In the Kara Sea according to our results for 20 stations the mean admittance amplitude of the Mm tide constitutes only 0.98. It is possible that the monthly tide may be influenced by local Rossby-like oscillations or the response driven by local or basin-scale forcing by wind or atmospheric pressure (Miller et al., 1993).

The distribution of the observed amplitude and phase of the Mf tide (13.66 days) in the Laptev Sea exhibits deviations from equilibrium that are as large as those for the tide Mm. The amplitude of the fortnightly tide in the central part of the sea constitute 0.8-1.2 cm and increases up to 1.9 cm off the islands of Severnaya Zemlya. The phase has a progressive character of eastward motion. The mean admittance amplitude of the tide Mf in the Laptev Sea equals 0.50.

Schwiderski (1982) computed the fortnightly tide in the Arctic Ocean. Schwiderski's model results disagreed with the observed values because of his calculations had been based on meagre data.

Analyses of spektra

Analysis of the spectral density maximum revealed that significant peaks were observed in spectra of the daily mean sea-level for the periods 1 year, 28-30, 7-8, 5.2-5.3, 4.3, 3.0-3.3, 2.4-2.5 and 2.1-2.2 days (Figure 2). However, half-year peak and third- year peak indicated in Dvorkin et al. (1989), Bannov-Baikov (1974) are poorly identified in our spectra. All the stations a peak on the monthly period at is revealed in our results.

In the field of low frequencies (less than 0,00274 c/d or period at 1 year) resolution the spectrum does not allow us to investigate interannual variability of the oscillations. Values of the peaks obtained using the Fourier analysis are not so stable as the spectral ones, however, they have a higher frequency resolution.

Results of the Fourier analysis of the residual time series revealed the following. In the range of interannual variability the peaks for the periods 4.6, 2.9, 1.9-2.1, 1.2-1.3 and 1.05 years were reliably expressed. On a half of the stations the fluctuations for the periods of about 11 years and 7.7 years are not present. As one can see, our results partially coincide with previous studies (Bannov-Baykov, 1974), however, the oscillations for the periods of 5-7 and 3-4 years are absent at most stations. It is significant that the peaks at 1.2-1.3 and 1.05 years were first discovered in the spectra of the residual series with removed annual harmonic. We have shown common features, but there is also a considerable variability of the spectra between the stations because of different environmental conditions.

As it can be seen in Table 2, the greatest variance of the sea level variations is related with the variability for the periods less than 1 year (from 66 up to 81 % of the total variance).

13 to 24% of the total variance acccount for the long-period tides and 5 - 18% of the variance is due to interannual variability. For the periods less than 1 year the basic energy is related to processes of synoptic scale (periods from days to months).

Table 2: Estimates of variances in differents ranges of the sea level variations.

Station	Total variance in sea level (cm^2, %)	Long period tide (cm^2, %)	Variations with periods more 1 year (cm^2, %)	Variations with periods less 1 year (cm^2, %)
Peschaniy	213, 100	37, 17	24, 11	152, 72
Maliy Taimyr	198, 100	26, 13	18, 9	154, 78
Andreia	354, 100	58, 16	64, 18	232, 66
Preobrazhenia	395, 100	95, 24	21, 5	279. 71
Terpiay-Tumsa	501, 100	91, 18	33, 7	377, 75
Dunay	562, 100	73, 13	41, 7	448, 80
Tiksi	617, 100	95, 15	27, 5	495, 80
Kotelniy	476, 100	94, 20	51, 11	331, 69
Sannikova	443, 100	91, 20	39, 9	313, 71
Kigilyakh	568, 100	75, 13	34, 6	459, 81

Daily to monthly variations of sea-level.

We will consider now in greater detail the oscillations of the synoptic scale. Analysis of two-dimensional densities of temporal distribution of the sea-level amplitudes showed that the level variations for the whole range of the synoptical scale were non-stationary (Figure 3). There are differences from one station to the other. Temporal variations of the variance obtained for monthly intervals at 10 stations reveal strong interannual and innerannual variability. This variability increases while moving to the coast, reaching its maximum values at the Dunay and Tiksi stations (Figure 4). In most of the cases maximum development of the synoptical scale oscillations occur in autumn and winter months. This is supported by conclusions made by Vanda (1989) derived for the mean daily sea level during summer period.

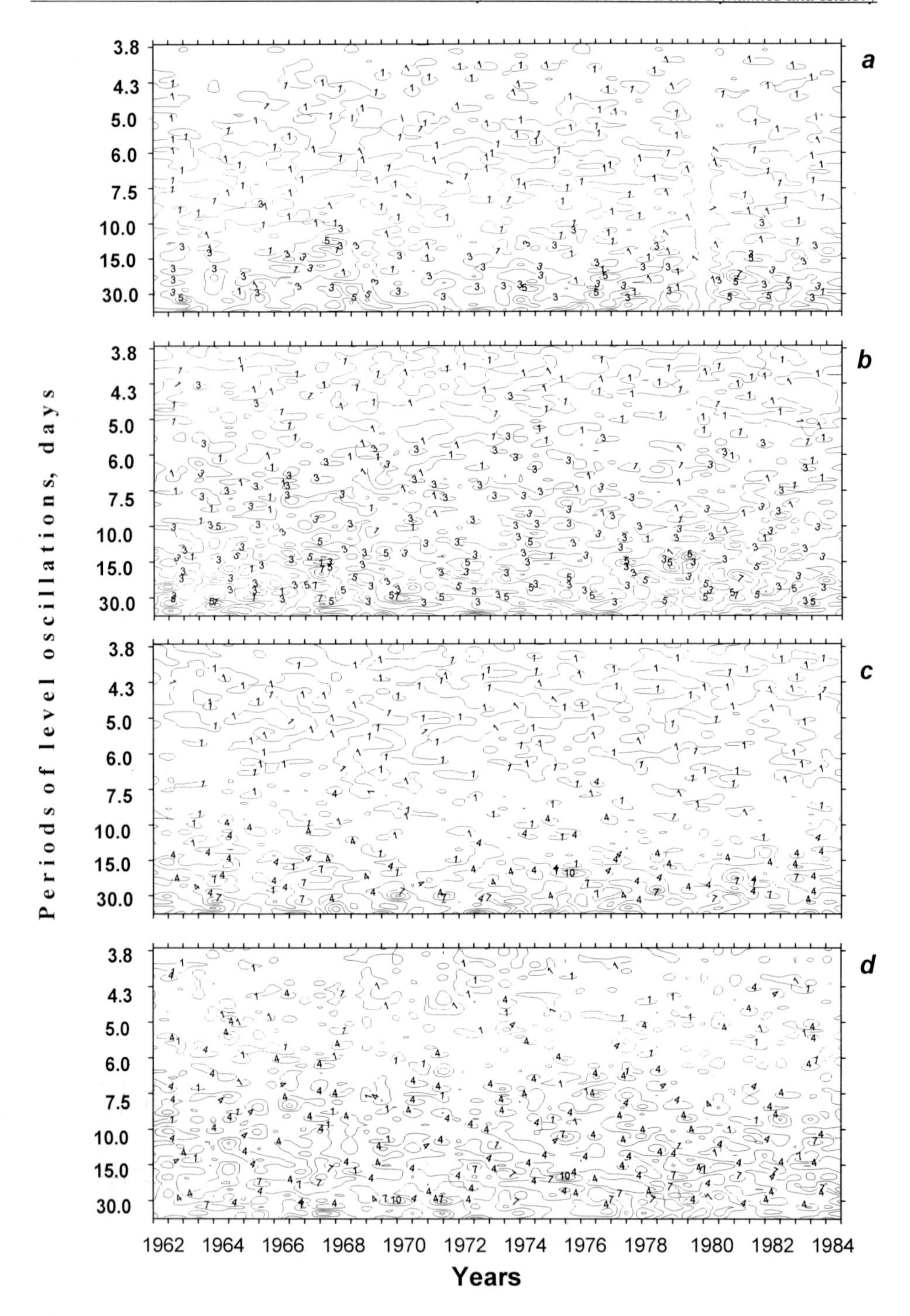

Figure 3: Two-dimensional probability density of amplitudes (cm) of sea level variations for the synoptical scale variability in points Peschaniy (a), Preobrazhenia (b), Tiksi (c), Kotelnyy (d).

Shelf waves

The cross- spectral analysis of the daily mean residual levels has revealed high coherence (0,68 - 0,98) between the level variations of the synoptical scale at various stations. Character of spatial changes of the phases indicates that these sea level oscillations represent predominantly progressive waves in the range of less than 60 days.

Phase speeds, propagation directions and lengths of progressive waves for different periods were estimated using the Fourier analysis. Results of the received estimations are presented in Table 3. The question arises: what is the type of these waves? According to the theory (LeBlond and Mysak, 1978; McWilliams, 1978), only 3 types of waves can exist at frequencies lower than the inertial frequency. These are Kelvin waves, topographic Rossby waves (shelf waves, double Kelvin waves) and shear waves whose special case is jet waves.

The general circulation models of the Arctic Ocean indicate that there are no intense jet currents in the shelf zones of Arctic seas. Mean current speeds here are very weak (1-2 cm/s). It is thus unlikely that the shear wave mode is very developed here.

The Kelvin waves are known to be the gravitation waves. This is the only type of waves which can exist at the frequencies higher and lower than the inertial frequency. Their phase velocity (c) is determined by a simple ratio $c = (gh)^{1/2}$ where g is the acceleration of gravity and h is the sea depth. Sence, the mean depth of the Laptev Sea is 533 m, the phase velocity of the Kelvin waves will be here about 72 m/s. This value significantly exceeds the calculated phase velocities of low-frequency waves. Also, unlike low-frequency waves of the two other types, the Kelvin waves do not have a horizontal mode structure. Synchronous distribution of mean diurnal vectors of current speeds from data of moorings at different points of the Laptev Sea indicates the presence of such a horizontal mode structure.

The double Kelvin waves are formed when the wave energy is trapped by zones with very sharp changes in the large-scale bottom topography. There are no such zones on the Laptev Sea shelf. Theoretically, a significant development of the mode of double Kelvin waves can be expected only near the shelf edge of the Laptev Sea.

Based on these considerations, we have identified the non-periodic levels oscillation of synoptic scale as shelf waves. The empirically derived characteristics of low-frequency waves also confirm this suggestion and agree with the theoretical understanding of the shelf waves.

As one can see in Table 3, values of the wave speeds are too large for Rossby waves and too are small as speeds of the long gravity waves.

Table 3: The parameters of progressive waves in the Laptev sea

Periods (days)	Direction (degrees)	Phase velocity (m/s)	Length (km)
46	165	1.28	5050
36	60	1.48	4659
30	84	1.51	3969
17	66	2.94	4318
15	83	3.14	4111
9	116	4.47	3513
6	56	4.95	2598
5	58	5.18	2267
4	103	4.95	1853

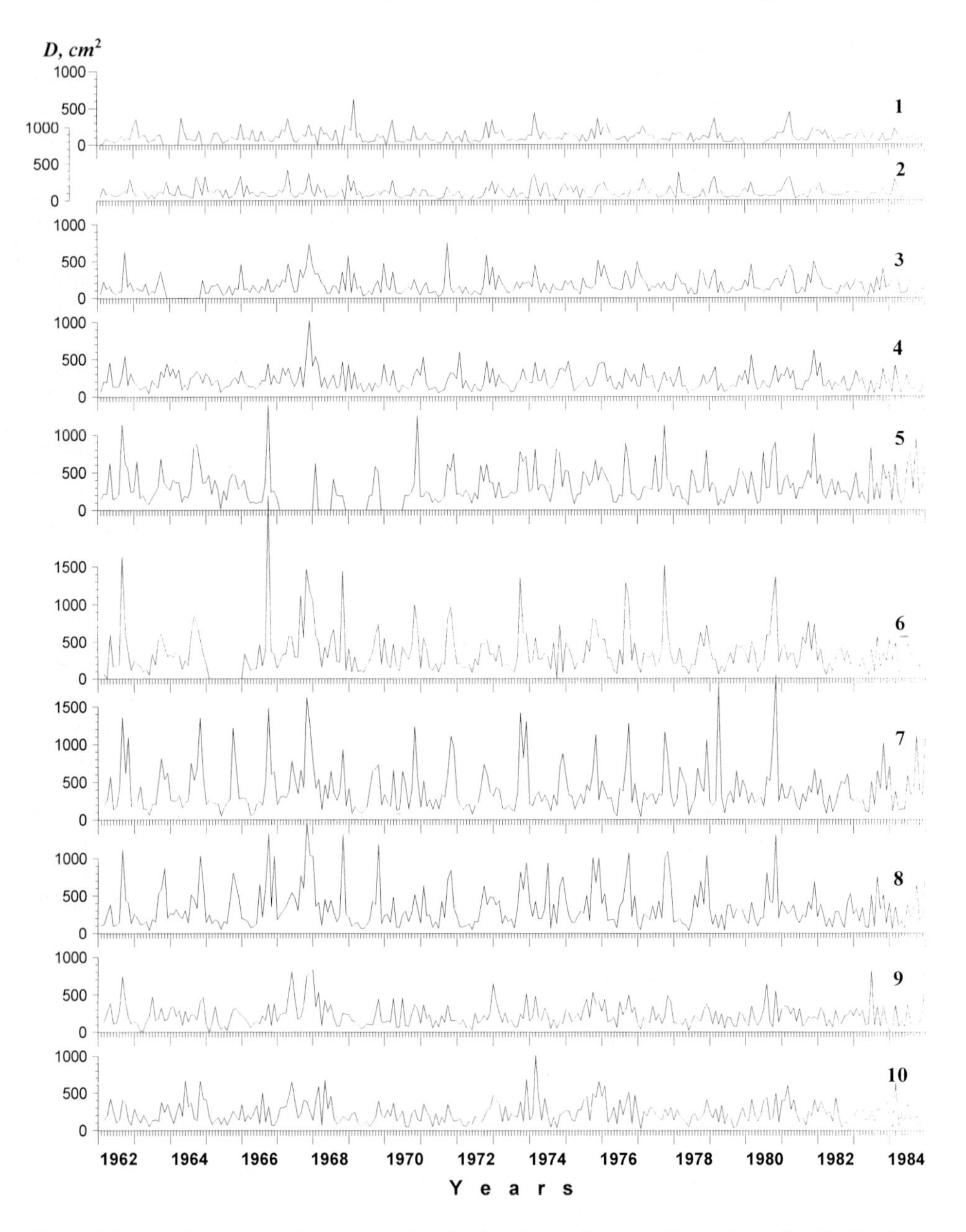

Figure 4: Temporal changes in the variance of sea level variations from monthly sections of residual series at stations in the Laptev sea. 1- Peschaniy; 2- Maliy Taimyr; 3- Andreia; 4- Preobrazhenia; 5- Terpiay-Tumsa; 6- Dunay; 7- Tiksi; 8- Kigilyakh; 9- Sannikova; 10- Kotelniy.

These speeds are more close to the phase speeds of the shelf waves (Efimov et al, 1985). Tendency of increase of the wave speed with decrease of their periods is appreciable (Table 3). The wave lengths are commensurable with the sizes of atmospheric cyclones and anticyclones. The waves propagate eastward. Alekseev and Mustafin (1972) studied the sea level oscillations in Arctic seas on the basis of data for the summer period, and they also indicated a similar fact. It will be coordinated with theoretical representations according to which the shelf waves are always distributed in cyclonical direction with reference to the open ocean (Efimov et al, 1985).

Provided that limit of the phase speed for the shelf waves is equal in magnitude to $c \leq f \cdot L$ and width of the shelf (L) in the Laptev Sea varies from 400 up to 800 kms and f (the inertial frequency) from $1.380 \cdot 10^{-4}$ up to $1.432 \cdot 10^{-4}$ rad/s, we shall obtain that $c \leq 9 \div 18$ m/s. The calculated phase speeds of the shelf waves (Table 3) do not exceed this limit, so it does not contradict to our hypothesis.

In the zonal low-frequency waves, two directions of currents, 0° and 180°, must prevail. Two-dimensional probability density of vectors of current velocities will have two modes in this case. Since the shelf waves in the Laptev Sea propagate eastward, the two-dimensional probability density of the vectors of current velocities should be similar to those of the zonal low- frequency waves.

Estimations of the two-dimensional probability densities of the daily mean vectors of current velocity at moorings obtained in different years (1959-1980) in the open part of the Laptev Sea showed in most cases a two-mode structure (Ipatov, 1997, unpublished data). This result to some extent confirms wave nature of current perturbations for the synoptical scale.

Conclusions

The maximum variance of large-scale level variations in the Laptev Sea is associated with the sea level variations for the periods less than 1 year (from 66 up to 81 % of the total variance).

13 to 24% of the total variance acccount for the long-period tides and 5 - 18% of the variance is due to interannual variability. The sea level oscillations in the synoptical range are a non-stationary modulated process. Intensity of the oscillations of the synoptical scale reveals strong interannual and innerannual variability. This variability is increases landwards, reaching the maximum quantity at Dunay and Tiksi stations. In most of the cases the maximum of the intensity of the oscillations for the synoptic scale occurs in autumn- winter months.

The observed nodal tide is not in agreement with the equilibrium theory. Investigations for other seas are necessary for further evidence. The values of the admittance amplitudes and phases of the monthly tide deviate much from the equilibrium tide. The observed amplitude and phase of the fortnightly tide deviates from the equilibrium one as in the case of the tide Mm. The mean admittance amplitude of the tide Mf in the Laptev Sea is equal to 0.50. The observed mean amplitude tide Mm in the Laptev Sea is 1.45 times greater than the mean equilibrium amplitude. Apparently the resonant conditions are observed in the Laptev Sea at a monthly period.

The level variations in the synoptical range represent predominantly progressive waves in the range of less than 60 days. Propagation speeds of these waves for the periods 5 - 46 days has shown that their magnitudes (1.3 - 5.2 m/s) correspond to the phase speeds of the shelf waves. The wave lengths are comparable with spatial scales of atmospheric cyclones.

Two-dimentsional distribution density of current vectors for the years 1959 - 1980 confirms a wave nature for synoptical scale currents.

References

Alekseyev, G.V. and N.V. Mustafin (1972) On statistical structure of non-periodic fluctuations of a level of the Arctic seas (in Russian). Probl. Arktik., 40, 13- 22.

Bannov-Baykov, Yu.L. (1974) About statistical structure of the large-scale sea level variations in high latitudes

of the Northern Hemisphere (in Russian). Probl. Arctik., 45, 21-26.
Cartwright, D.E. and R.J. Tayler (1971) New computations of the tide-genereting potential. Geophys. J. R. Astron. Soc., 23, 45-74.
Cartwright, D.E. (1983) On the smoothing of climatological time series, with application to sea level at Newlyn. Geophys. J. R. astr. Soc., 75, 639-658.
Dvorkin, E.N., J.V. Zakharov, N.V. Mustafin and E.N. Uranov (1989) Statistical structure of seasonal and interannual fluctuations of sea level of the Arctic seas and adjacent regions of the Atlantic and the Pasific oceans (in Russian). Trudy Arktik. and Antarct. Inst., 417, 6-18.
Efimov, V.V., E.A. Kulikov, A.B. Rabinovich and I.V. Fine (1985) Waves in boundary areas of the ocean (in Russian). Gidrometeoizdat , St. Peterburg, 280 pp.
Jenkins, G.M. and D.G. Watts (1969) Spectral analysis and its applications. (translat. from english). Mir, 1971, v.1, 316 pp, 1972, v.2, 287 pp.
Krutskikh, B.A. (1978) The fundamentals laws of inconstancy of the regime of the Arktic seas in the natural hydrological periods (in Russian). Gidrometeoizdat, Leningrad, 91 pp.
LeBlond, P. H. and L. A. Mysak (1978) Waves in the ocean. Elsevier Oceanography Series 20.
Maksimov, I.V. (1970) The geophysical forses and the waters of ocean (in Russian). Gidrometeoizdat, Leningrad, 447 pp.
Mc Williams J. C (1979). Stable jet modes: a special case of eddy and mean flow interaction. J. Phis. Oceanogr., 3, 344-362.
Miller, A.J., D.S. Luther and M.C. Hendershott (1993) The fortnightly and monthly tides: resonant Rossby waves or nearly equilibriam gravity waves? J. Phys. Oceanogr., 23, 879-897.
Proudman, J. (1960) The condition that a long period tide shall follow the equilibrium law. Geophys. J. R. Astron. Soc., 3, 244-249.
Rossiter, J.R. (1967) An analysis of annual sea level variations in European waters. Geophys. J. R. Astron. Soc., 12, 259-299.
Schwiderski, E.W. (1982) Global ocean tides, 10, the fortnightly lunar tide (Mf). Atlas of Tidal Charts and Maps, Rep. TR 82-151, Naval Surface Weapons Center, Dahlgren, Virginia.
Vanda, Ju.A. (1989) Statistical temporal irregularity of the variations of sea level at the sinoptical scale (in Russian). Trudy Arct. and Antarct. Inst., 414, 68-72.
Voinov, G.N. (1996) Harmonic analyses of tides of long time series of sea level (in Russian). Itogovay sessiy Uchenogo sov. AANII, eks.-press inform, 4, 22-23.
Vorobyev, V.N. (1966) Lunar-solar half-monthly and monthly tides in the Soviet Arctic seas (in Russian). Doklady Akad. Nauk SSSR, 167, 1039-1041.
Vorobyev, V.N. (1969) On the study of nineteen-year tidal variations of the mean sea level in the Earth's high latitudes (in Russian). Oceanology, J. Akad. Nauk SSSR, 9, 959-965.
Vorobyev, V.N. (1976) A semi-annual tide and the drift of the ices of the Arctic basin (in Russian). Trudy Arktic. and Antarkt. Inst., 319, 101-105.

Extreme Oscillations of the Sea Level in the Laptev Sea

I. Ashik, Y. Dvorkin and Y. Vanda

State Research Center - Arctic and Antarctic Research Institute, 38 Bering St., 199226 St. Petersburg, Russia

Received 3 March 1997 and accepted in revised form 3 March 1998

Abstract - Based on an analysis of monitoring data from coastal polar stations, the character of sea-level oscillations in the Laptev Sea is analysed. In addition, sea-level variability and regions displaying a similar character of sea-level oscillations are identified. The synoptic situations typically related to significant negative and positive storm surges along southern coast of the Laptev Sea are described.

Introduction

Shelf zone of the Laptev Sea is rich in mineral resources. In recent years the activity for search of sea oil-gas-bearing fields in this region has increased. Planning and provision of safety of shipping, designing and construction of marine complexes, as well as hydro-meteorological support of their exploitation require comprehensive evidence on different elements of the hydrometeorological regime, in particular, information on sea level oscillations (Kort, 1941; Mustafin, 1961; Ashik et al., 1995). This is a factor which can actively influence human activities in the coastal Arctic zone. Let us note that multiyear observations of the sea level oscillations at island and coastal stations are especially valuable, as among other indicators of dynamic processes in the hydrosphere of the shelf zone of the Laptev Sea information on sea level is most fully and objectively presented.

Methods

The main initial data used for investigating feature of extreme level oscillations of the Laptev Sea were multiyear series (more than 35 years) of level observations at 4 synoptic times at 12 coastal and island stations (Figure 3). Considering that marine activities in the Laptev Sea can be carried out all-year-round, extreme level values were analyzed for all the months of the annual cycle.

The level oscillations in the Laptev Sea are of an extremely complicated character and are governed by a combination of the dynamic and conservative factors (Ashik, 1995). Dimensions and position of the Laptev Sea allow us to consider action of the dynamic factor sufficiently uniform over its area. Thus, differences in character of the level oscillations, the magnitude of extreme oscillations in some regions of the Laptev Sea are primarily governed by the morphometry of these regions: bottom topography, coastline position, etc. (Ashik and Vanda, 1995; Mustafin, 1961). River runoff also plays a considerable role in the regime of the extreme level oscillations in the Laptev Sea.

The extreme sea level oscillations of the Laptev Sea are formed, basically, under influence of deep extensive cyclones, fascinating under his influence whole areal of the sea. During the significant negative surges, over the sea prevalence a forward part of a cyclone, the centre of which at this time places over peninsula Taymir (Figure 1a). The significant positive surges are usually connected with back part of a cyclone, the centre of which at this time places over Novosibirski islands (Figure 2a).

In: Kassens, H., H.A. Bauch, I. Dmitrenko, H. Eicken, H.-W. Hubberten, M. Melles, J. Thiede and L. Timokhov (eds.) Land-Ocean Systems in the Siberian Arctic: Dynamics and History. Springer-Verlag, Berlin, 1999, 37-41.

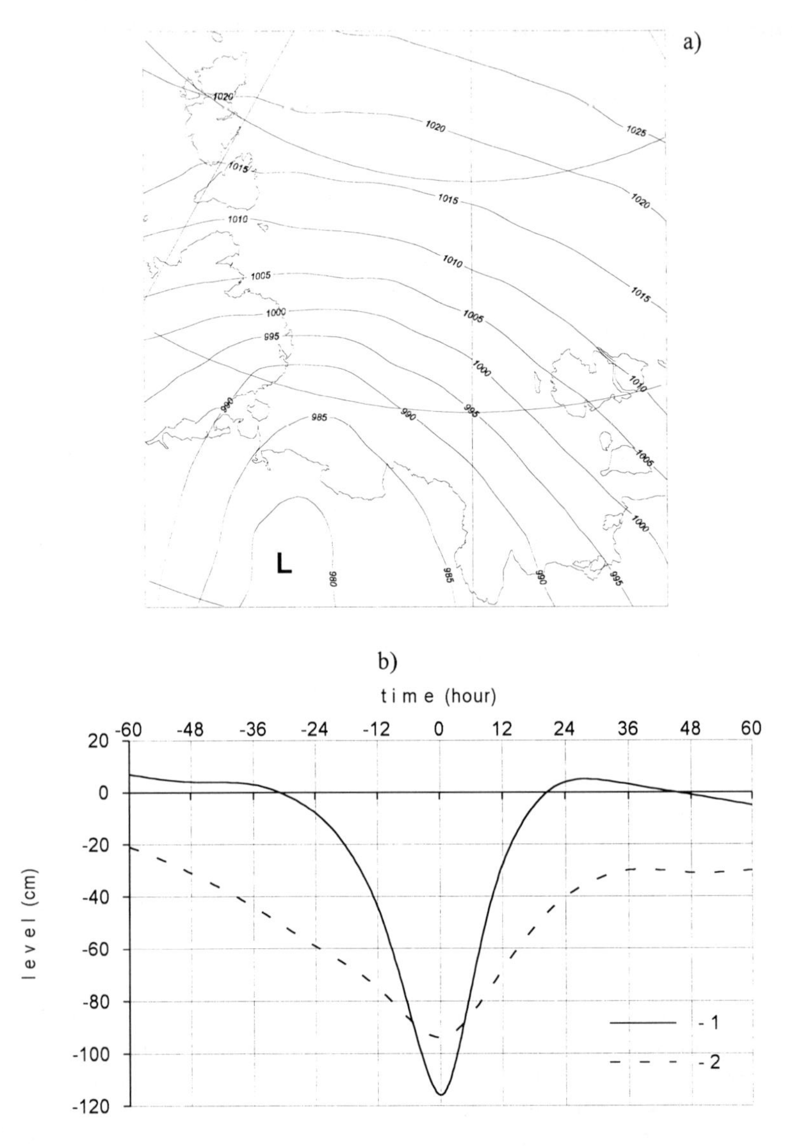

Figure 1: The significant synoptic condition (a) and mean level oscillations (b) for the negative storm surges by the Tiksi (1) and Janski bay (2).

Results

By morphometric indications the Laptev Sea can be divided into three regions: the northern, central and southern part. The depth in the northern part is rather large and amplitude of extreme level oscillations at stations located in this region (Maliy Taymir) is rather small and not exceeding 200 cm. Absolute maximums and minimums at most of the stations of this region do not exceed 100 cm.

With decrease in depths of the regions, the magnitude of the extreme level oscillations begins

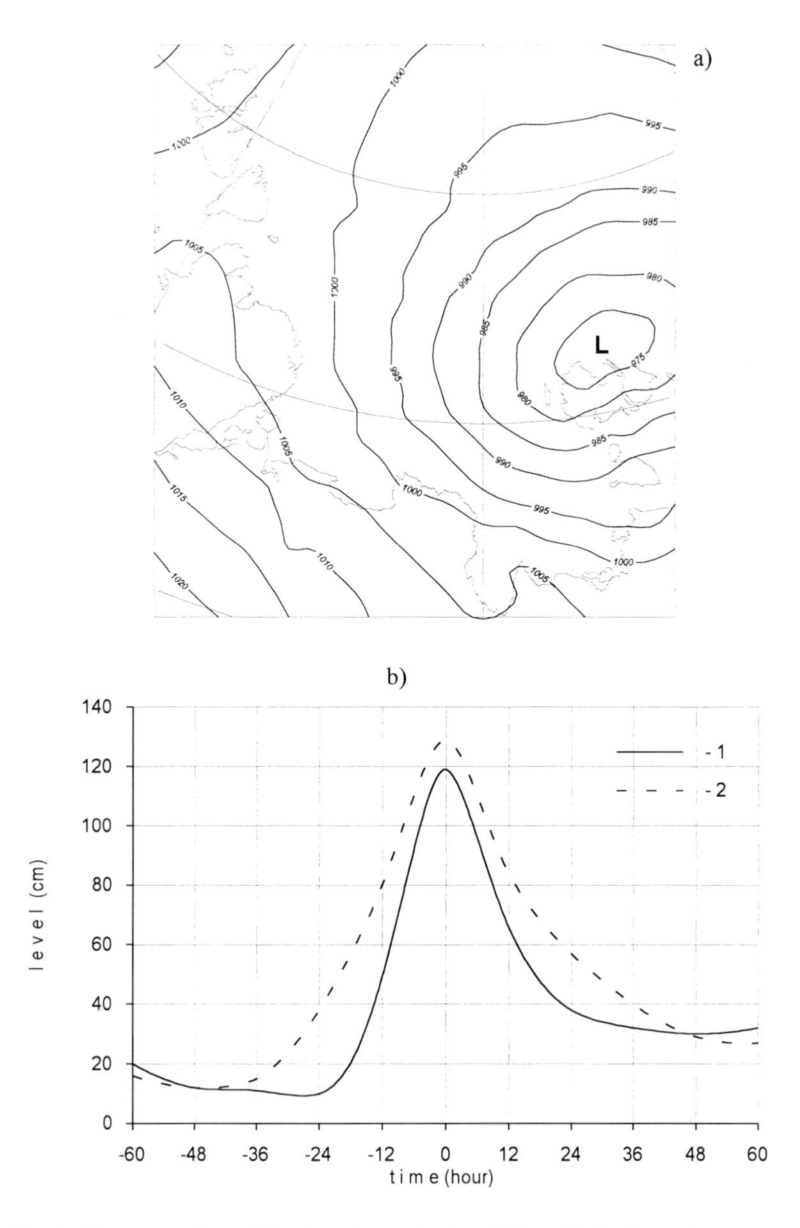

Figure 2: The significant synoptic condition (a) and mean level oscillations (b) for the positive storm surges by the Tiksi (1) and Janski bay (2).

to increase comprising 200—220 cm for the stations Andrey Island, Island Kotelniy, Sannikov Strait, and 250—300 cm for the stations Preobragenia Island and Kosistiy Cape. Whereas the absolute maximum value for the stations of the first group varies from +100 cm to +120 cm and of the minimum from —90 cm to —100 cm, these values comprise from +130 cm to +160 cm and from —130 to —140 cm for the second group, respectively.

A characteristic feature of the regime of the extreme level oscillations in the Laptev Sea is

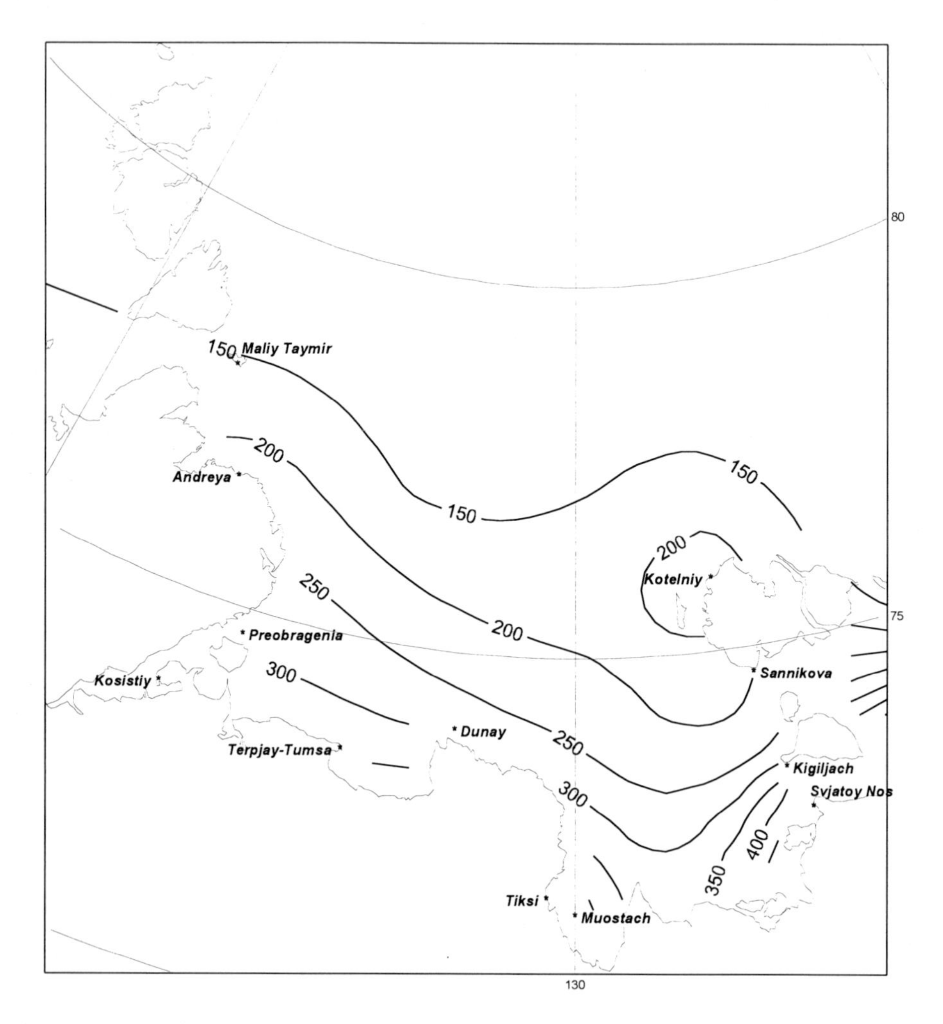

Figure 3: The size of the level oscillations at the Laptev Sea.

excess of the values of absolute maximums over the values of the absolute minimums. The only exception is one station: Cape Svjatoy Nos where the value of the absolute minimum exceeds that of the absolute maximum by 46 cm. At some stations (Maliy Taymir Island, Kosistiy Cape, Sannikova Strait) the difference in the values of the absolute maximum and the absolute minimum is comparatively small being 5—10 cm. At the stations Andreja Island, Preobragenija Island, Kotelniy Island, Tiksi Bay, Dunay Island, Muostach Island the excess of the absolute maximum value over that of the absolute minimum is15—40 cm. The most significant difference between the values of the absolute maximum and minimum governed by sea factors is observed for the stations Terpjay-Tumsa Cape and Kigiljach Cape where it is equal to 50—55 cm.

During significant storm surges, the rise of a level (Figure 2b), as well as its fall (Figure 1b), in Guba Buor-Haja occurs much more intensively, than in Janskiy zaliv. That is connected to distinctions in morphometric of these regions.

Conclusion

It is obvious that existence of this feature in the regime of the extreme level oscillations indicates a uniform character of effective surge directions for most stations of the Laptev Sea and the excess of the northern winds over the southern ones which is confirmed by the analysis of the wind regime of separate regions of the Laptev Sea.

Performed study can be considered as a basis for preparing a specialized regime-reference handbook on the extreme level oscillations in the Laptev Sea.

References

Ashik, I.M. (1995) Numerical prediction of sea surges and ice conditions in the Laptev and the East Siberian Seas. Berichte zur Polarforschung, 176, 47-54.

Ashik, I.M. and Yu.A. Vanda (1995) Catastrophic storm surges in the southern part of the Laptev sea. Berichte zur Polarforschung, 176, 43-46.

Ashik, I.M., E.I. Makarov and Yu.A. Vanda (1995) The operative hydrometeorological providing of petroleum and gas production on the Arctic seas shelfs of the Russia (in Russian). In: Simakov, G.V. (ed.), The second international conference on development of the russian arctic offshore: St.Petersburg State Technic University, 142-143.

Kort, V.G. (1941) Unperiodic sea level oscillations on the Arctic seas and methods for their forecast (in Russian). Proc./Arct. and Antarct. Research Institute, 1941, 175, 103 pp.

Mustafin, N.V. (1961) On catastrophic storm surges in the south-eastern part of the Laptev sea (in Russian). Probl. of Arctic and Antarctic, 7, 3-36.

Internal Waves in the Laptev Sea

E.A. Zakharchuk

State Research Center - Arctic and Antarctic Research Institute, 38 Bering St., 199226 St. Petersburg, Russia

Received 3 March 1997 and accepted in revised form 9 February 1998

Abstract - The results of research on internal waves in the northern part of the Laptev Sea are presented. Significant fluctuations of temperature, salinity and currents with periods from 2 to 20 minutes within the seasonal pycnocline. These fluctuations are identified as high-frequency internal gravity waves, the frequency of which is close to the Brunt-Väisälä frequency. Current meter data were used to calculate the linear invariant of the spectral tensor-function of current velocity. Significant spectral density peaks for cycles of 2.0, 2.2, 3.3, 3.6 and 14.8 min were established. It is suggested that the high-frequency internal gravity waves are generated by destruction of an internal tidal wave in the region of the continental slope. This supposition is confirmed by temporal variability of the linear invariant of the variance tensor of current velocity at 10 m depth. It is suggested that the internal waves play an important role in mixing processes and the formation of mesoscale (tidal) frontal zones close to the continental slope.

Introduction

Internal waves play an important role in a regime of the Arctic Ocean. An increase of the solar heating, of river runoff and of drift ice melting in the Arctic Ocean in the summer period, leads to formation of a very sharp seasonal pycnocline which hinders exchange between surface and deep waters. The shielding influence of sea ice also hinders the wind mixing of the surface layers. However, observations show that thickness of the top quasi-uniform layer in the Arctic Ocean can reach 20-30 m (Nikiforov and Shpaikher, 1980) in the summer period. It is unlikely that such a mixed layer can be formed only by the drift and decay of ice fields. It is possible that the internal waves could make a significant contribution to the mixing of the upper layer of the Arctic Ocean in the summer period by generating turbulence through their instability (Turner, 1973). The problem of heat exchange between the surface and Atlantic waters underlaying them in the Arctic Ocean is poorly investigated. In particular, it is not clear at which space-time scales there is largest heat exchange between the surface and Atlantic waters, and also what oceanographic processes are responsible for this exchange. Some researchers consider that their is a significant contribution to the vertical heat exchange through the destruction of internal waves (Alekseev et al., 1974; D'Asaro and Morison, 1992; Muench et al, 1996; Perkin and Lewis, 1978; Padman, 1995). Questions of effect of the internal waves on ice deformation and their contribution to formation of mesoscale frontal zones in the Arctic Ocean remain poorly investigated. Doubtless, all these problems require deeper and comprehensive research of the internal waves in the Arctic Ocean both experimental and theoretical level.

Internal wave field in the Arctic Ocean is poorly studied. Most of internal wave observations were made in the Greenland-Norwegian Sea (e.g. Plueddemann, 1992; Marmorino and Trump, 1991), Fram Strait region (Sandven and Johannessen, 1987;), Arctic Basin (D'Asaro and Morison, 1992) and continental slope and deep basin of the Laptev Sea (Muench et al., 1996). These works show that internal wave energy was greatest at the shelf break and near rougher topography (e.g. Nansen-Gakkel Ridge, Yermak Plateau), then decreased rapidly with increasing water depths reaching of lowest values over the abyssal plains.

However, these studies described mainly the tidal and near-inertial internal waves and werebased mostly on data from the permanent pycnocline.

In: Kassens, H., H.A. Bauch, I. Dmitrenko, H. Eicken, H.-W. Hubberten, M. Melles, J. Thiede and L. Timokhov (eds.) Land-Ocean Systems in the Siberian Arctic: Dynamics and History. Springer-Verlag, Berlin, 1999, 43-51.

At present, very few information is received on the high-frequency internal gravity waves in the Siberian Shelf Seas. Special experiments here were not carry out but there are only simultaneous observations. For example, temperature and salinity fluctuations with periods several minutes in the seasonal pycnocline layer were recorded in the coastal points of the Laptev and Kara Seas during the Russian-Swedish Expedition "Tundra Ecology-94". These fluctuations were most intensive in the north of the Baidaratskaya Gulf where in the seasonal pycnocline there were recorded very intensive variations reaching more 5°C and 2 in salinity. Temperature and salinity fluctuations in the pycnocline have well-pronounced non-linear wave structure. Period of most intencive fluctuations was equal to 10-12 minutes. These fluctuations were identified as non-linear gravity waves (Zakharchuk and Presniakova, 1997).

This paper describes an observations of high-frequency internal gravity waves in the seasonal pycnocline on the north of the Laptev Sea.

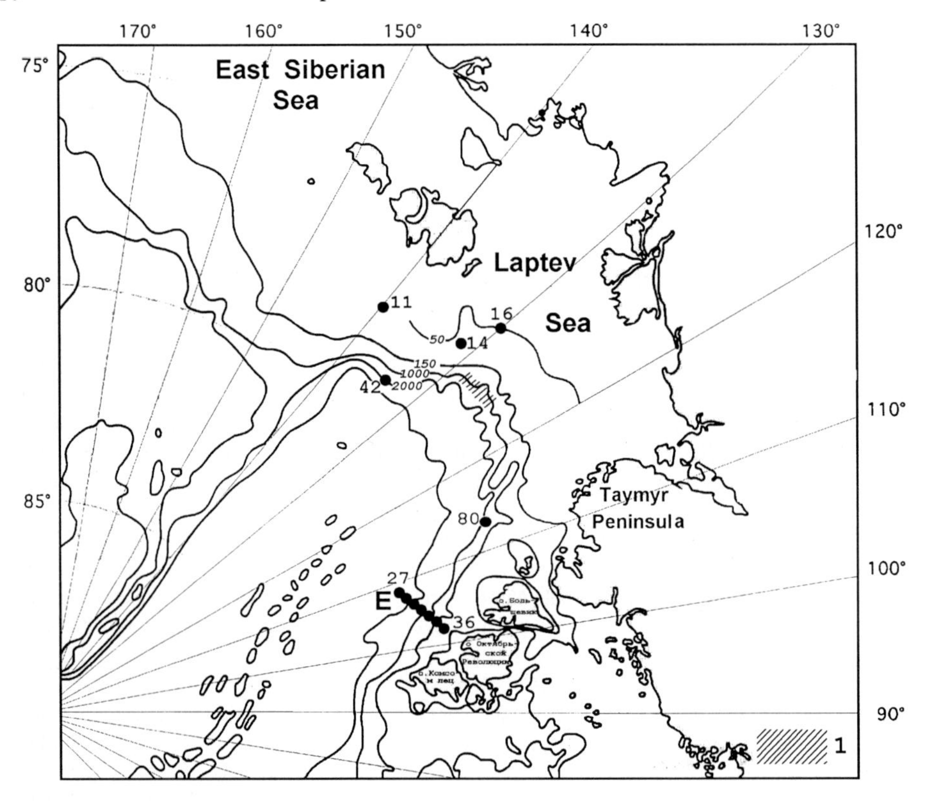

Figure 1: Location of oceanographic stations (the solid circles) in which observations of the internal waves were carried out. E - oceanographic section. 1 - shaded area is a location of the bands of the ice brash near of the ice edge revealed by radio-location photo.

Data analysis and results

During the summer 1995 cruise ARK XI/1 of research vessel *Polarstern*, small-scale fluctuations of temperature, salinity and currents were recorded in the north parth of the Laptev Sea (Figure 1). Temperature and salinity at fixed depth were measured using Seabird SBE-19

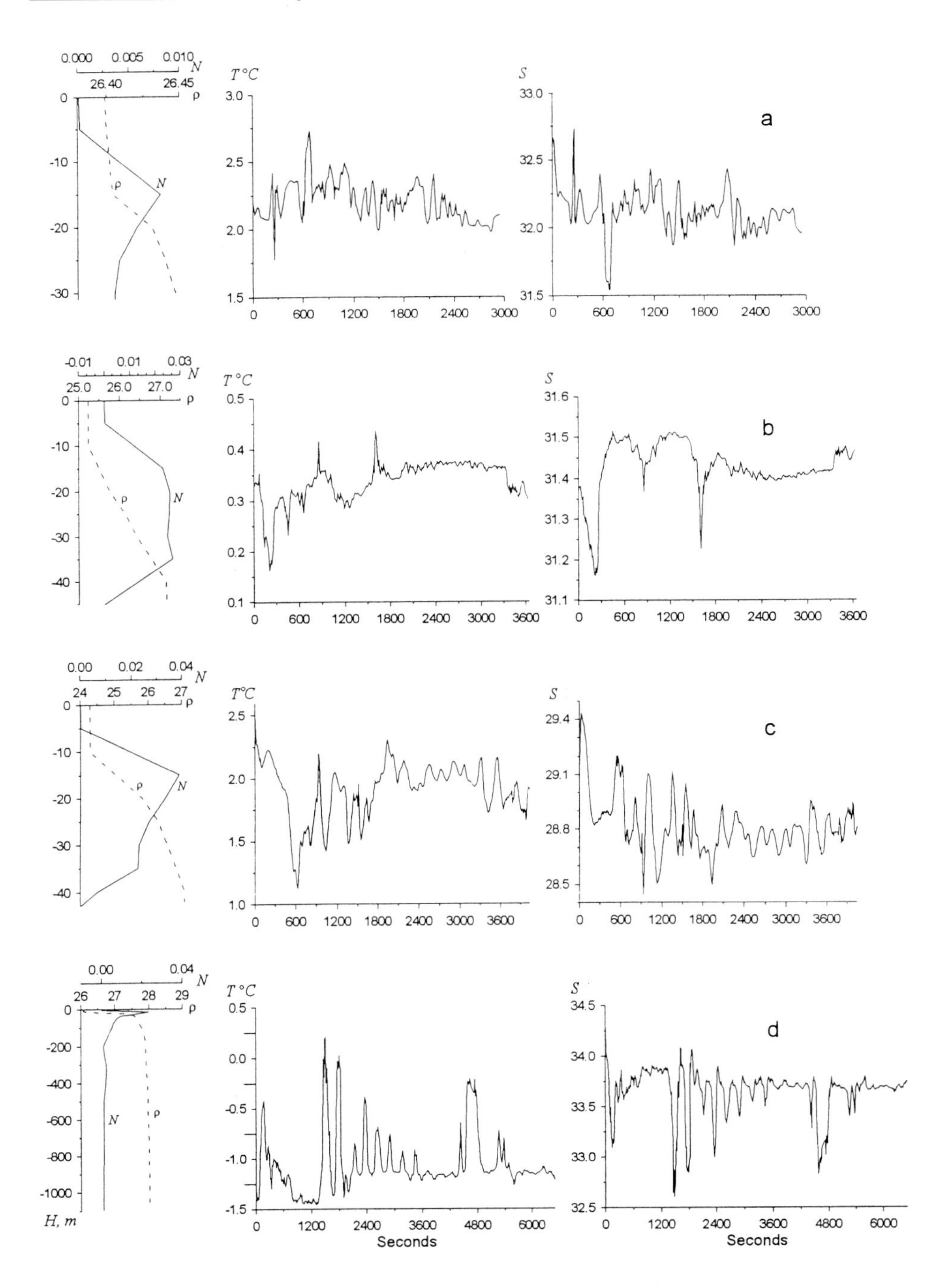

Figure 2: Vertical profiles of density (ρ, c.u.), Brunt-Väisälä frequency (N, sec^{-1}) and smal-scale fluctuations of temperature (T°C) and salinity (S) in the seasonal pycnocline at depths a) 25 m (stations 11), b) 15 m (station 14), c) 25 m (station 16 and d) 23 m (station 80) in different points of the Laptev Sea.

profiler from the drifting ship (Figure 2). Time serie of the horizontal currents at station 42 was measured using Aanderaa current meter.

The results show that fluctuations of temperature and salinity with periods from 2 to 15 minutes are presented in the seasonal pycnocline (Figure 2). Maximum fluctuation were 1.6°C and 1.5. These fluctuations are identified as high-frequency internal gravity waves with frequencies close to the Brunt-Väisälä frequency. The most interesting observation was made at station 80 (Figure 2d), where a packet consisting of 9 waves with 5 minute period and one solitary wave with 9 minute period were recorded.

Current measurements were made at station 42 from a drifting ice-floe at a fixed depth of 10 m, with recording time step of 30 seconds during 10.5 hours in the central part continental slope of the Laptev Sea. These measurements also show high-frequency fluctuations of currents with periods several minutes. The maximum current fluctuations were 13 cm s^{-1}.

The current meter data were used to calculate a spectrum. Since a current is the vector process, its statistical estimates most be invariant respect to a system of coordinates. Therefore the linear invariant $I_1(\omega)$ of the spectral tensor-function (1) of current velocity were colculated (Belyshev et al., 1983).

$$\mathbf{S_v}(\omega)=\begin{pmatrix} S_{\upsilon_1\upsilon_1}, S_{\upsilon_1\upsilon_2} \\ S_{\upsilon_2\upsilon_1}, S_{\upsilon_2\upsilon_2} \end{pmatrix}, \qquad (1)$$

where $\mathbf{V}=\upsilon_1+\upsilon_2$ - vector horizontal current; ω - frequency;

$$S_{\upsilon_1\upsilon_2}(\omega)=C_{\upsilon_1\upsilon_2}(\omega)+iQ_{\upsilon_1\upsilon_2}(\omega);$$

$$S_{\upsilon_2\upsilon_1}(\omega)=C_{\upsilon_2\upsilon_1}(\omega)+iQ_{\upsilon_2\upsilon_1}(\omega);$$

$$C_{\upsilon_1\upsilon_2}(\omega)=C_{\upsilon_2\upsilon_1}(\omega);Q_{\upsilon_1\upsilon_2}(\omega)=-Q_{\upsilon_2\upsilon_1}(\omega).$$

Linear invariant $I_1(\omega)$ equal to the track of matrix $\mathbf{S_v}(\omega)$:

$$I_1(\omega)=S_{\upsilon_1\upsilon_1}(\omega)+S_{\upsilon_2\upsilon_2}(\omega)$$

This invariant denote modulus distribution of changings currents velocity intensity in the frequency band.

Estimate of $I_1(\omega)$ show the significant spectral density peaks for cycles 2.0, 2.2, 3.3, 3.6 and 14.8 min were established (Figure 3).

Phase velocities of high-frequency internal waves may calculate using dispersion relations received in linear approximation for two-layer ocean (Lamb, 1932). Acording to this model, dispersion relation for internal waves having lengths much more than thickness each of ocean layers assumed the following form:

$$c=\sqrt{g(\Delta\rho/\rho)Hh/(H+h)} \qquad (2)$$

where c - phase velocity of internal waves; h - thickness of upper layer;

H - thickness of lower layer; $\Delta\rho = \rho_2 - \rho_1$, ρ_1 - density of upper layer; ρ_2 - density of lower layer; ρ- mean value of density; g - acceleration of gravity.

Dispersion relation for internal waves having lengths much more than a thickness of upper ocean layer but essentially less then a thickness of lower layer assumed the following form:

$$c = \sqrt{g(\Delta\rho / \rho)h} \qquad (3)$$

The phase velocities of internal waves on the stations 11, 14 and 16 were calculated by of dispersion relation (2). The phase velocities of internal waves on the stations 42 and 80 were calculated using of dispersion relation (3). Because we know periods of internal waves (T) and their the phase velocities (c), lengths of internal waves (λ) may be calculate as $\lambda = c \cdot T$. The calculated that parameters of internal waves are given in the Table 1.

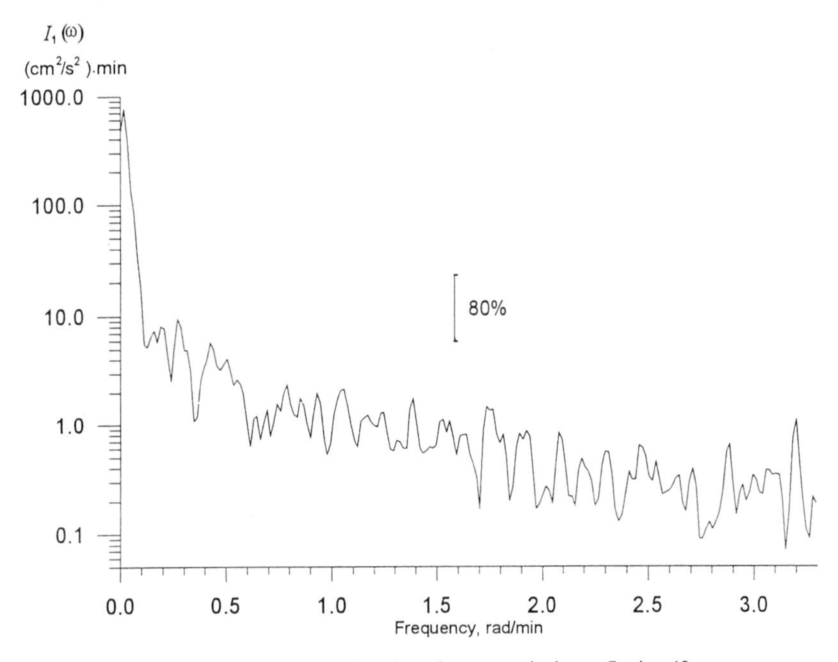

Figure 3: Linear invariant of the spectral tensor-function of current velocity on Station 42.

Since tidal motions are present in the Laptev Sea (Kowalik and Proshutinsky, 1994) one of possible mechanisms for the generation of the small-scale internal waves can be proposed. When the tidal wave arrives at the continental slope, internal waves are generated under certain conditions. Vertical shear of the horizontal current velocity in the internal waves and their instability lead to generation of small-scale turbulence and will intensify mixing in the continental slope-shelf zone. This mechanism must lead to the formation of a mesoscale (tidal) frontal zone between the more mixed waters at the shelf edge and deeper stratified waters situated at the continental slope. This generation mechanism could explain the wide spreading of fine stucture in temperature and salinity and the formation of a frontal zone observed on the section E northeast of Severnaya Zemlya (Figure 4). This figure shows that fine structure is weakly in the deep-water in section E. Wide spreading fine structure is registered approaching to the continental slope and shelf. Temperature and salinity in the warm water core of the Atlantic layer change from 2.7°C, 34.97 at station 27 to 0.7°C, 34.77 at station 33. The largest horizontal gradients of the temperature and salinity are between stations 29 and 31. Weather was fine at this time and long narrow bands of the brash ice were observed on the sea surface. These bands appear to be connected with the surface manifestation of the high-frequency internal waves. Distance between these bands was 150 - 200 m approximately. This distance are close to lengths of high-frequency internal waves which were calculated using of dispersion relations (2) and (3) (see Table 1).

Table 1: Parameters of high-frequency internal gravity waves in the Laptev Sea

Number of stations	Wave periods (min.)	Phase speed (cm s^{-1})	Wave length (m)
11	4.2	6.2	15
14	1.8; 12.0	38.1	40; 274
16	4.2; 7.8	46.1	117; 215
42	2.0; 2.2; 3.3; 3.6; 14.8	77.0	92; 102; 152; 166; 684
80	4.8; 9.0	55.0	152; 297

If our hypothesis on the tidal nature of the high-frequency internal waves is correct, 12-hour periodicity should be observed in the change of intensity of the small-scale fluctuations of the oceanographic parameters. To verify this supposition, temporal variability of the linear invariant of variance tensor of the high-frequency fluctuations of currents on Station 42 was investigated. The low-frequency component of the current velocity was initialy excluded from the data. Then current time serie was divided in ten of 1-hour parts. The linear invariant of variance tensor $I_1(0)$ is calculated for each of these parts.

$$I_1(0) = D_{\upsilon_1}(0) + D_{\upsilon_2}(0) \tag{4}$$

where $D_{\upsilon_1}(0)$ - variance of eastward velocity, $D_{\upsilon_2}(0)$ - variance of northward velocity.

One peak in the temporal variability of the linear invariant of the variance tensor can be seen in Figure 5. This fact confirms the tidal nature of these small-scale internal waves. The time of maximum value of variance differs more than 4 hours from the M_2 period high tide at this point of the Laptev Sea. This temporal difference could suggest that the high-frequency internal waves are generated not by the barotropic tide, but by an internal tide wave.

There is further evidance for the existence of internal tide waves near the continental slope of the Laptev Sea. Narrow bands of brash ice stretching perpendicular to the ice edge and 10-15 km from this region are seen on a radio-location photo made on 20 July 1995 from an aeroplane at height 7 km in the central area of the continental slope of the Laptev Sea (see Figures 1 and 6). The local and quasi-periodic character of these ice bands suggests to idea on wave and perhaps solitary nature of this phenomenon. Distance between the bands was 5-15 km. The spatial scales between the bands of 5 - 15 km are close to lengths of internal tide waves which have also been observed in different regions of the World Ocean.

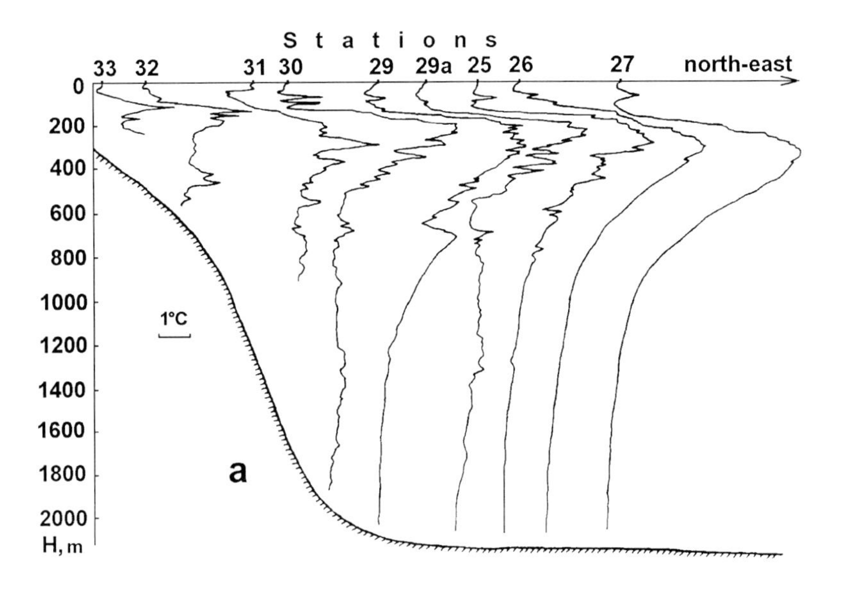

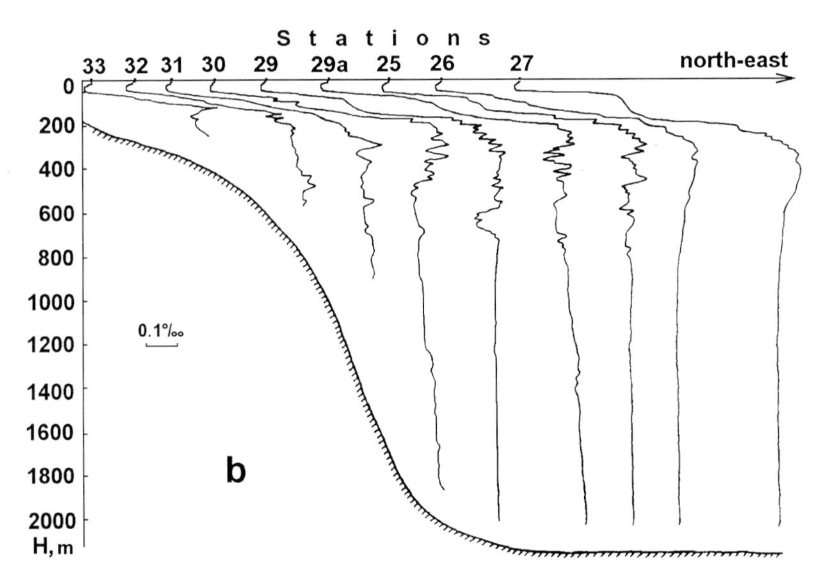

Figure 4: Vertical profiles of temperature (a) and salinity (b) on stations of section E.

Summary

This study describes an observations of high-frequency internal gravity waves in the seasonal pycnocline in the Laptev Sea. The principal conclusions are as follows:

1 High-frequency internal waves having frequensies close to the Brunt-Väisälä frequency are observed in the different regions of the Laptev Sea.

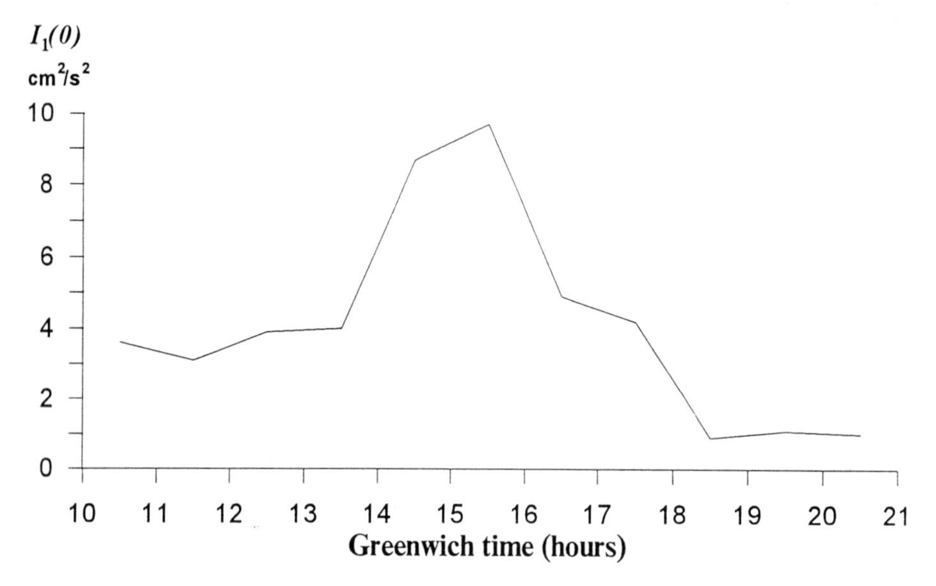

Figure 5: Temporal variability of the linear invariant of variance tensor of high-frequency fluctuations of currents on Station 42.

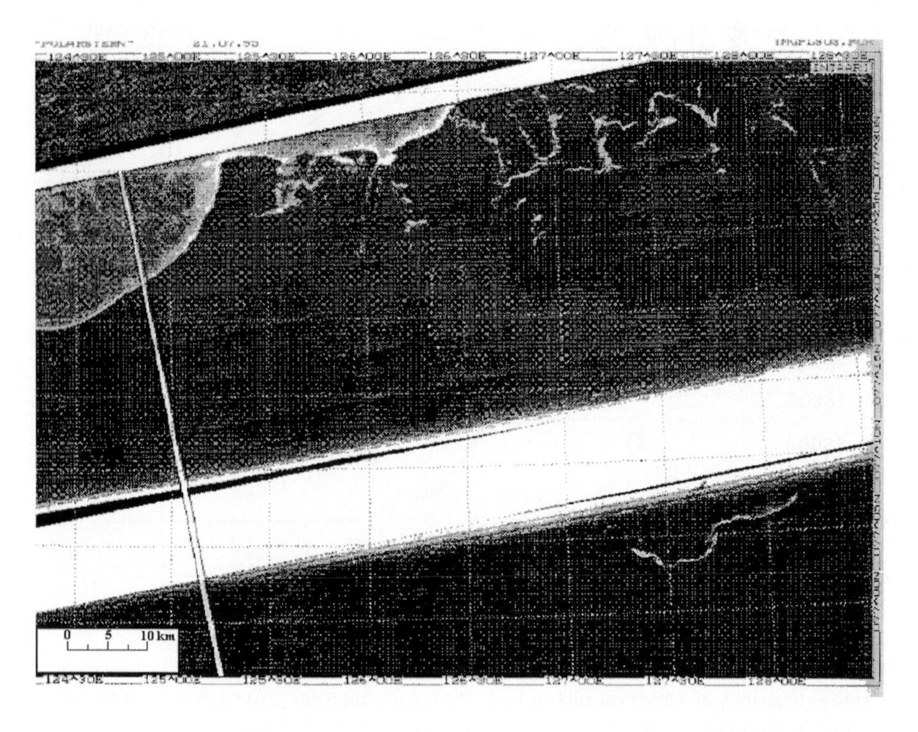

Figure 6: Radio-location photo made on 20 July 1995 from an aeroplane at height 7 km in the central area of the continental slope of the Laptev Sea.

2 High-frequency internal waves probably are generated by an internal tide waves.
3 Vertical shear of the horizontal current velocity in the high-frequency internal waves and their instability lead to generation of small-scale turbulence and increases mixing in the continental slope-shelf zone. This lead to the formation of a mesoscale (tidal) frontal zone between the more mixed waters at the shelf edge and deeper stratified waters situated at the continental slope near Severnaya Zemlya Islands.

References

Alekseev, G.V., A.P. Nagurniy and V.G. Savchenko (1974) Instability of internal waves as the mechanism of transfer of a heat from Atlantic waters in Arctic Basin (in Russian). Probl. Arktik., 45, 94-99.

Belyshev, A.P., Iu.P. Klevantsov and V.A. Rozhkov (1983) Probability analysis of sea currents (in Russian). Gidrometeoizdat. Leningrad. Russia. 264 p.

D'Asaro, E.A. and J.H. Morison (1992) Internal waves and mixing in the Arctic Ocean. Deep-Sea Research, Vol. 39, Suppl. 2, pp. S459-S484.

Kowalik, Z. and A.Y. Proshutinsky (1994) The Arctic Ocean Tides. The Polar Oceans and Their Role in Shaping the Global Environment. Geophysical Monograph 85, pp. 137-158.

Marmorino,G.O. and C.L. Trump (1991) "Turbulent mixing" induced by upgoing near-inertial waves in the seasonal thermocline of the Norwegian Sea, J. Geophys. Res., 96 (C4), 7137-7143.

Muench, R.D., R.K. Dewey and U. Schauer (1996) Internal waves and vertical mixing over the Laptev Sea slope. Proceedings of the acsys conference on the dynamics of the Arctic Climate System (Göteborg, Sweden, 7-10 November 1994). WMO/TD-No. 760, pp. 441-445.

Nikiforov, E.G. and A.O. Shpaikher (1980) Lows of a formation of the large-scale oscillations of the hydrological regime of the Arctic Ocean (in Russian). Gidrometeoizdat. Leningrad. Russia. 270 p.

Padman, L. (1995) Small-Scale Physical Processes in the Arctic Ocean. Arctic Oceanography: Marginal Ise Zones and Continental Shelves Coastal and Estuarine Studies, Volume 49, pp. 97-129.

Perkin, R.G. and E.L. Levis. Mixing in an arctic fiord (1978) J. Phys. Oceanogr., v. 8, N5, 873 - 880.

Plueddemann, A.J. (1992) Internal wave observations from the Arctic Environmental Drifting Buoy, J. Geophys. Res., 97, 12,619-12,638.

Sandven, S. and O.M. Johannessen (1987) High frequency internal wave observations in the marginal ice zone. J. Geophys. Res., 92(C7), 6911-6920.

Turner, J.S. (1973). Buoyancy Effects in Fluids. Cambridge University Press, New York.

Zakharchuk, E.A. and G.E. Presnyakova (1995) High-Frequency Internal Waves in the Kara Sea. Scientific Seminar "Nature Conditions of the Kara and Barents Seas", St. Petersburg, Russia, p. 14.

The Composition of the Coarse Fraction of Aerosols in the Marine Boundary Layer over the Laptev, Kara and Barents Seas

V.P. Shevchenko[1], A.P. Lisitzin[1], R. Stein[2], V.V. Serova[1], A.B. Isaeva[1] and N.V. Politova[1]

(1) P.P. Shirshov Institute of Oceanology, 23 Krasikova, 117851 Moscow, Russia

(2) Alfred-Wegener-Institut für Polar- und Meeresforschung, Postfach 120161, D 27515 Bremerhaven, Germany

Received 10 April 1997 and accepted in revised form 3 March 1998

Abstract - During the 49-th cruise of R/V *Dmitry Mendeleev* in August-October 1993 and PFS *Polarstern* cruise ARK XI/1 in July-September 1995, 22 samples of aerosols were collected in the Laptev, Kara and Barents Seas. Aerosols were studied for particle size, morphology, and composition. New results of this work are presented in this paper. The average size of particles in samples collected by meshes in different areas differ insignificantly (4.18 to 5.54 μm). In most samples mineral particles and organic matter (fibers of vegetation, pollens, diatoms) were the main components, organic carbon content was 17.6 % in average. The individual anthropogenic combustion spheres with diameter from 1 to 10 μm mostly consist of: (1) Si, Al, K, and Fe in the South-Western Laptev Sea; (2) Si, Al, Fe, and K in the Central Kara Sea; (3) Fe, Ni, and Si in the Southern Kara Sea, and (4) Si, Al, Fe, and Ni in the Southern Barents Sea.

Introduction

Traditionally riverine input was assumed to be the main geochemical pathway of terrestrially and anthropogenically derived compounds from their sources to the aquatic environment, but there is much evidence that atmospheric inputs contribute significantly to marine areas (Duce et al., 1991; Lisitzin, 1996). Numerous studies have shown that aerosols in the Arctic are of importance for atmospheric chemistry and climate (Rahn, 1981; Barrie, 1986; Leck et al., 1996). But up to now aerosols of the Russian Arctic were studied little. There are only few articles devoted to aerosol composition studies on the land and islands (Rovinsky et al., 1995; Vinogradova, 1996). We began aerosol research in the marine boundary layer over the seas of the Russian Arctic in 1991; first results have been published elsewhere (Shevchenko et al., 1995; Smirnov et al., 1996; Serova and Gorbunova, 1997). In this paper we present and discuss new data on particle size and composition of coarse (>1 μm) insoluble fraction of Arctic aerosols.

Material and methods

During the 49-th cruise of R/V *Dmitry Mendeleev* in August-October 1993 and PFS *Polarstern* cruise ARK XI/1 in July-September 1995, 22 samples of aerosols were collected by nylon meshes in the Laptev, Kara and Barents Seas (Figure 1). In order to collect the aerosols, 10 nylon meshes (1 m^2 each; pores 0.8 mm) were raised on the mast above the bow of the ship. To exclude contamination from the ship, sampling was interrupted when the relative wind direction was not opposite to the ship movement. After 5 to 24 hours the meshes were fetched down, and we removed the particles by washing the meshes in bidistilled water. Then the water with the particles was passed through Nucleopore filters (0.45 μm), and the filters with thesamples were dried at 40-45 °C. This sampling method is described in more detail by Chester and Johnson (1971).

In: Kassens, H., H.A. Bauch, I. Dmitrenko, H. Eicken, H.-W. Hubberten, M. Melles, J. Thiede and L. Timokhov (eds.) Land-Ocean Systems in the Siberian Arctic: Dynamics and History. Springer-Verlag, Berlin, 1999, 53-59.

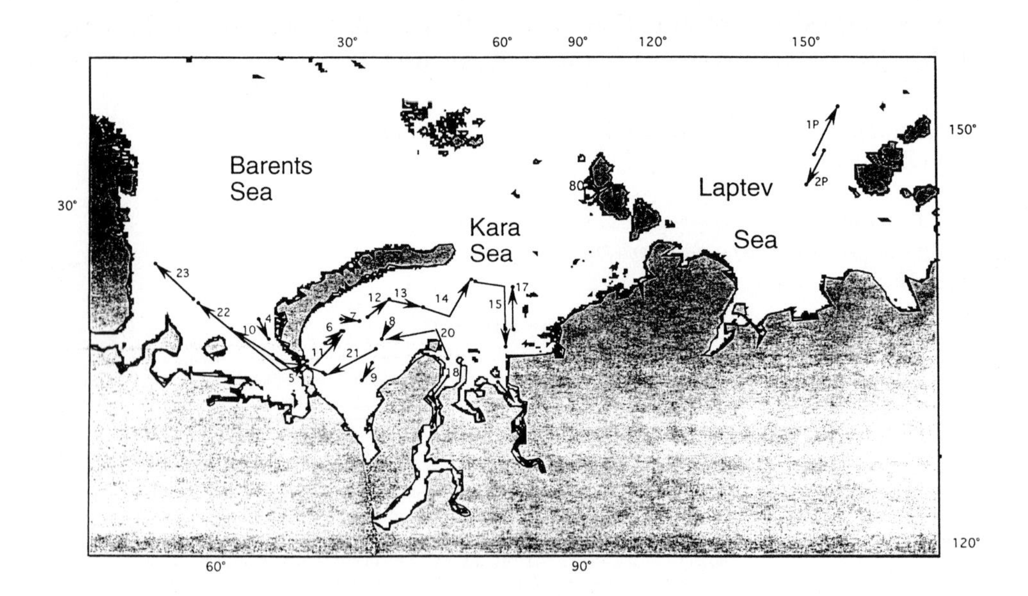

Figure 1: Location of aerosol samples collected in the 49-th ctruise of R/V *Dmitry Mendeleev* (sample numbers from 4 to 23) and in PFS *Polarstern* cruise ARK XI/1 (sample numbers 1P and 2P).

After resuspension of each subsample in distilled water using an ultrasonic probe, particle size analysis was carried out by a computerized electro-optical counter CIS-1 (Galai, Israel). Particles with size from 2 to 100 μm have been counted. Morphology of the particles was studied by scanning electron microscope JSM-U3 of Jeol (Tokyo, Japan) with maximum magnification of 10000 times. Organic carbon (C_{org}) was determined by the gas method using automatic titration on a AN 7529 carbon analyzer, which registered CO_2 resulting the high-temperature (900°C) combustion in a pure oxygen flow after carbonate CO_2 expelling by a 3-M HCl treatment (Gurvich et al., 1995). The analysis accuracy was ±2% relative. The content of SiO_2 and Al_2O_3 were determined by a respective color reaction with a colorimetric endpoint using a KFK 3 photocolorimeter. The sample treatment was as follows: the samples were alloyed at 900°C with a mixture of borax and soda (Na_2CO_3) depleted of silicic acid in a SNOP muffle furnace (Gurvich et al., 1995). The analysis accuracy was ±2% relative. Mineral composition of samples was studied by X-ray diffractometry described in more detail by Serova and Gorbunova (1997). Qualitative elemental composition of selected particles was studied by scanning electron microscope SEM-515 with X-ray microprobe EDAX PV9900 (Philips, USA).

Results and discussion

In August-October 1993, the mass concentration of the coarse fraction of Kara and Barents Seas aerosols which are not soluble in distilled water, varied from 0.02 to 0.48 $\mu g/m^3$ (0.15 $\mu g/m^3$ in average); in the Laptev Sea concentration of insoluble aerosol particles was 0.04 - 0.09 $\mu g/m^3$ at the end of July 1995 (Table 1). These values are similar to those measured in the North Atlantic (Duce et al., 1991). In most of the samples, organic matter (fibres of vegetation, pollens, diatoms) is the main component by mass, as has already been shown earlier (Shevchenko et al., 1995). Content of organic carbon varies from 7.54 to 26.9 % (Table 1) (17.6 % in average, n=7 samples). A number of investigations carried out in the recent years confirm that organic matter is one of the main components of the atmospheric aerosols (Aston et al., 1973; Isidorov, 1990; Matthias-Maser and Jaenicke, 1995).

Mineral particles also play an important role. The contents of different minerals vary significantly from sample to sample (Table 1). In the mineral non-amorphous part of the coarse fraction of aerosols in August-October 1993 the content of quartz ranged from 15 to 56 % (33 % in average), feldspars - from 4 to 20 % (in average 12 %), illite - from 12 to 38 % (in average 23 %), chlorite plus caolinite - from 12 to 48 % (in average 31 %). High variability of mineral composition can probably be explained by the different origin of the air masses.

In most samples small spherical particles (diameter from 1 to 10 μm) presumably of anthropogenic origin (fly ash) were detected. As it has been shown that these particles may be

Table1: Concentrations and composition of insoluble aerosol particles larger than 1 μm in the marine boundary layer over the Laptev, Kara and Barents Seas in summer. *DM49 - samples collected in 49-th cruise of R/V *Dmitry Mendeleev;* ARK XI/1- samples collected in expedition ARK XI/1 of PFS *Polarstern.*

Samples*	Conc., mg/m^3	Average size, μm	$C_{org.}$	SiO_2	Al_2O_3	Quartz	Feldsp.	Illite	Chl.+ Kaol.
			% from total mass			% from mineral phase			
DM49-4	0.26	4.59				38	8	27	25
DM49-5	0.06	4.58				17	8	38	36
DM49-6	0.66	5.39	7.5	56.6	14.0	36	13	15	35
DM49-7	0.20	5.23				27	8	22	42
DM49-8	0.14	5.54				35	17	18	26
DM49-9	0.48	4.42	19.0	34.3	9.5	56	16	16	12
DM49-10	0.06	5.00	26.9	16.2	3.5				
DM49-11	0.03	4.35				23	8	34	34
DM49-12	0.18	4.21				34	12	22	31
DM49-13	0.15	4.85				23	4	25	48
DM49-14	0.05	4.18				36	11	24	28
DM49-15	0.03	4.46				22	7	28	43
DM49-16	0.21	4.76	15.4	29.5	6.5	46	16	16	22
DM49-17	0.21	4.68	9.4	51.4	13.0	43	14	16	26
DM49-18	0.06		20.0	31.0	6.2	36	20	22	22
DM49-19	0.18	6.15				48	18	12	22
DM49-20	0.05					30	7	20	41
DM49-21	0.03	4.90				15	20	29	35
DM49-22	0.02					37	7	28	28
DM49-23	0.05	4.76				31	12	24	32
ARK XI/1-1	0.09	4.86							
ARK XI/1-2	0.09	4.35	25.1	13.9	3.8				

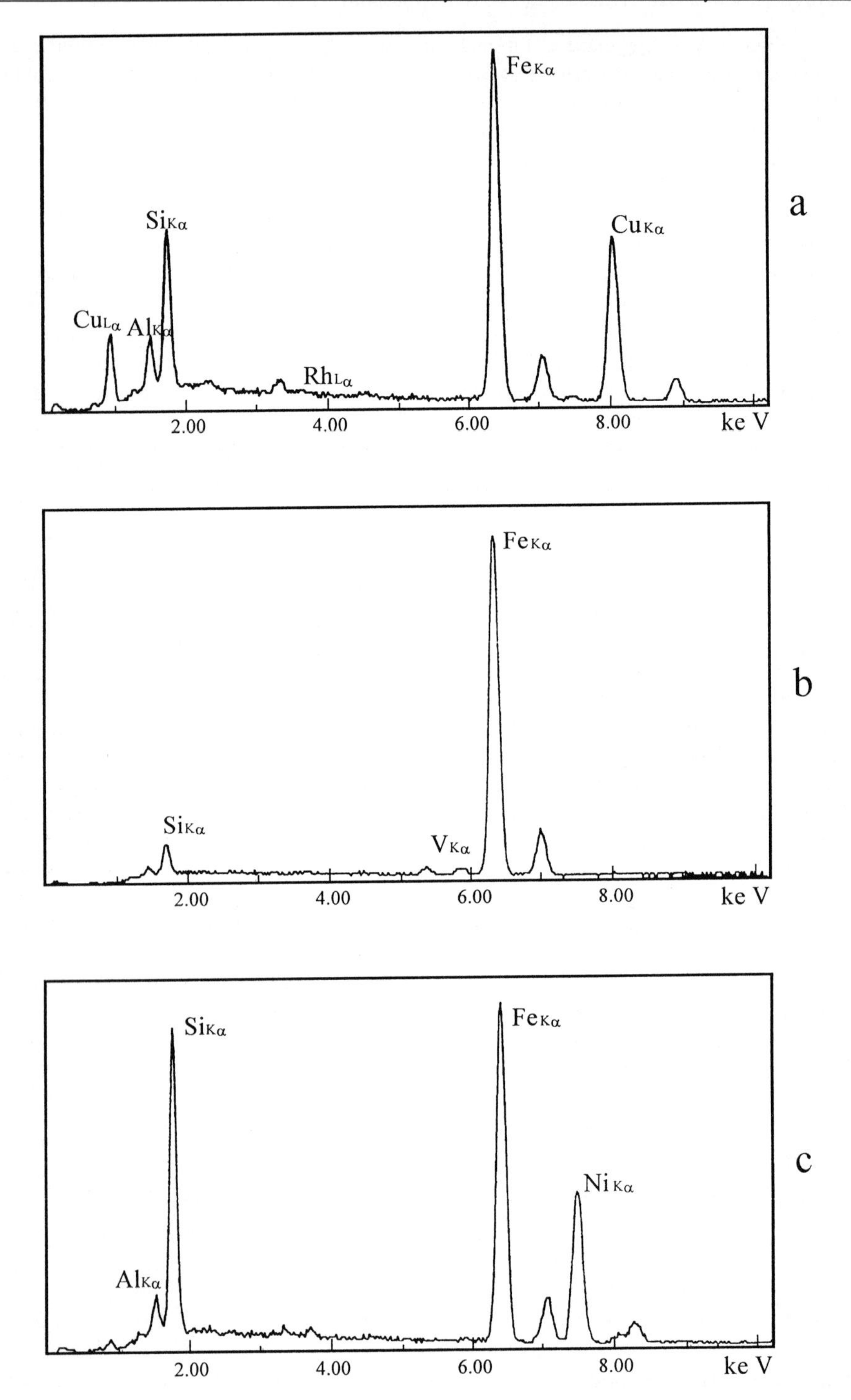

Figure 2: Typical spectra of combustion spheres (data of EDAX PV9900 X-ray microprobe analysis) from samples, collected in 49-th cruise of R/V *Dmitry Mendeleev*: a - DM49-20; b, c - DM49-23.

Table 2: Qualitative composition of fly ash particles (Ø 1-10 μm) in the marine boundary arctic aerosols accordingly to the results of EDAX analysis. Numbers of groups correspond to relative abundance of the groups. In brackets there are elements relatively less abundant in particulates from the given group.

The South-Western Laptev Sea (n=2)	The Southern Kara Sea (n=3)
1) Si, Al, K, Fe	1) Fe, Ni, Si (Cu, Al)
2) Si, Al, Fe (K, Mg)	2) Fe, Si, Ni (Cu, Al, Mg, Ca)
3)P, Si, Al, S (Mg, Fe, Ca)	3) Si, Fe, Cu, Al
4)Fe, Al, P (Cu)	4) Si, Al, K (Fe)
	5) Fe, Cu, Ni, Si
The Central Kara Sea (n=1)	**The Southern Barents Sea (n=1)**
1) Si, Al (K, Fe)	1) Si, Al (Ca, Fe, K, Mg)
2) Fe, Si, Al	2) Fe, Si, Ni (Al, Cu, Mg, Ca)
3) Fe, Ni, Si (Cu)	3) Fe, Ni, Si (Cu, Ca, Al)

formed in high-temperature processes (Van Malderen et al., 1992), they are called combustion spheres. In our work qualitative elemental composition of 110 individual spherical particles were studied. The elemental composition of individual combustion spheres differs from region to region (Table 2). In the South-Western Laptev Sea combustion spheres mostly consist of Si, Al, K, and Fe (samples N 1 and 2 from *Polarstern* expedition ARK XI/1). In the open part of the Kara Sea (sample N 17 from the 49-th cruise of R/V *Dmitry Mendeleev*) Si and Al are the main elements in the combustion spheres. In the southern part of the Kara Sea (samples N 16, 18, 20 of the 49-th cruise of R/V *Dmitry Mendeleev*) Fe, Ni, Si, Cu are dominant presumably due to influence of the Norilsk metallurgic industry. Norilsk occupies the first place in the Russian Arctic with respect to the discharge of pollutants into the atmosphere (Vilchek et al., 1996). Norilsk nickel-copper smelter complex emits into the atmospher much iron, nickel, copper, lead, zink, arsenic, fluorine, mercury, antimony. In the Southern Barents Sea (sample N 23 of 49-th cruise of R/V *Dmitry Mendeleev*) the combustion spheres mainly consist of Si, Al, Fe, and Ni (due to the metallurgic industry of the Kola Peninsula and Scandinavia). The Kola Peninsula is region of intense atmospheric pollution by iron, nickel, copper, zink, lead mostly by Severonikel and Pechenganikel smelters (Vilchek et al., 1996). Typical spectra of combustion spheres are presented in Figure 2.

Conclusions

The results from this study allow the following conclusions:

1 The mass concentration of insoluble aerosol particles in the marine boundary layer over the Russian Arctic in summer is similar to that recorded over the North Atlantic (Ø 0.15 $\mu g/m^3$).

2 The coarse fraction (> 1μm) of the aerosols mainly consists of minerals, but the content of biogenic particles of continental origin (rests of plants, polens etc.) is also very high (the organic carbon content is 17.6 % in average).

3 Anthropogenic combustion spheres of 1 to 10 μm in diameter are present in all samples. Their quantity and contents of Fe, Ni, and Cu increase near large industrial centers (Norilsk, Murmansk etc.).

Acknowledgements

We are grateful to the crews of reseach vessels *Dmitry Mendeleev* and *Polarstern* for all assistance. We thank A.A. Burovkin, A.F. Kuleshov, O.V. Severina for help in sampling, U. Bock, L.V. Demina, R. Frohlking, V.A. Karlov, and G. Kuhn for assistance in analyses, and reviewers M. Kriews and B. Schnetger for their valuable criticism of this paper. This study was financially supported by the Russian Foundation of Basic Research (grants RFBR 96-05-00043) and DFG (grants STE-412/10 and 436 RUS 113/170).

References

Aston, S.R., R. Chester, L.R. Johnson and R.C. Padgham (1973) Eolian dust from the lower atmosphere of the Eastern Atlantic and Indian Oceans, China Sea and Sea of Japan. Marine Geology, 14, 15-28.

Barrie, L.A. (1986) Arctic air pollution: an overview of current knowledge. Atmos. Environ., 20, 643-663.

Chester, R. and L.R. Johnson (1971) Atmospheric dust collected off the West African coast. Nature, 229, 105-107.

Duce, R.A., P.S. Liss, J.T. Merrill et al. (1991) The atmospheric input of trace species to the world ocean. Global Biogeoch. Cycles, 5, 193-259.

Gurvich, E.G., A.B. Isaeva, L.V. Dyomina, M.A. Levitan and K.G. Muravyov (1995) Chemical composition of bottom sediments from the Kara Sea and estuaries of the Ob and Yenisey Rivers. Oceanology, 34, 701-709.

Isidorov, V.A. (1990) Organic Chemistry of the Earth's Atmosphere. Springer-Verlag, Berlin- Heidelberg, 210 pp.

Leck, C., E.K. Bigg, D.S. Covert, J. Heintzenberg, W. Maenhaut, E.D. Nilsson and A.Wiedensohler (1996) Overview of the atmospheric research program during the International Arctic Ocean Expedition of 1991 (IAQE-91) and its scientific results. Tellus, 48B, 136-155.

Lisitzin, A.P. (1996) Oceanic Sedimentation. Lithology and Geochemistry. American Geophysical Union, Washington, D.C., 400 pp.

Matthias-Maser, S. and R. Jaenicke (1995) The size distribution of primary biological aerosol particles with radii >0.2 μm in urban/rural influenced region. Atmospheric Research, 39, 279-286.

Rahn, K.A. (1981) Atmospheric, riverine and oceanic sources of seven trace constituents to the Arctic ocean. Atmos. Environ., 15, 1507-1516.

Rovinsky, F., B. Pastukhov, Y. Bouwolov and L. Bourtseva (1995) Present day state of background pollution in the natural environment in the Russian Arctic in the region of the Ust-Lena reserve, The Science of the Total Environment, 160/161, 193-199.

Serova, V.V. and Z.N. Gorbunova (1997) Mineral composition of soils, aerosols, suspended matter, and bottom sediments of the Lena River estuary and Laptev Sea. Oceanology, 37, 131-135.

Shevchenko, V.P., A.P. Lisitzin, V.M. Kuptzov, G.I. Ivanov, V.N. Lukashin, J.M. Martin, V.Yu. Rusakov, S.A. Safarova, V.V. Serova, R. Van Grieken and H. Van Malderen (1995) The composition of aerosols over the Laptev, the Kara, the Barents, the Greenland and the Norwegian Seas. In: Russian-German Cooperation: Laptev Sea System, Kassens, H., D. Piepenburg, J. Thiede, L. Timokhov, H.W. Hubberten and S.M. Priamikov (eds.), Berichte zur Polarforschung, 176, 7-16.

Smirnov, V.V., V.P. Shevchenko, R. Stein, A.P. Lisitzin, A.V. Savchenko and ARK XI/1 Polarstern Shipboard Scientific Party (1996) Aerosol size distribution over the Laptev Sea in July-September 1995: First results. In: Surface-sediment composition and sedimentary processes in the central Arctic Ocean and along the Eurasian Continental Margin, Stein R., G.I. Ivanov, M.A. Levitan and K. Fahl (eds.), Berichte zur Polarforschung, 212, 139-143.

Van Malderen, H., C. Rojas and R. Van Grieken (1992) Individual giant aerosol particles above the North Sea. Environ. Sci. Technol., 26, 750-756.

Vilchek, G.E., T.M. Krasovskaya, A.V. Tsyban and V.V. Chelyukanov (1996) The environment in the Russian Arctic: status report. Polar Geography, 20, 20-43.

Vinogradova, A.A. (1996) Elemental composition of atmospheric aerosol in Eastern Arctic region. - Izvestiya, Atmospheric and Oceanic Physics, 32, 440-447.

New Data on Sea-Ice Albedo in the Laptev and Barents Seas

B.V. Ivanov

State Research Center - Arctic and Antarctic Research Institute, 38 Bering St., 199226 St. Petersburg, Russia

Received 3 March 1997 and accepted in revised form 9 February 1998

Abstract - Results of recent sea-ice albedo measurements carried out during the summer of 1993 in the Laptev and Barents Seas are reported. Ground-truth data were obtained with hand-held pyranometers during a ship expedition. A relationship between the albedo of melt puddles and their depths was revealed. The temporal variation of albedo for the main surface types of sea ice (snow, puddles, bare and dirty ice) and a relationship between the albedo and thickness of young ice were determined.

Introduction

The sea ice cover is an important component in the climate system. The distribution of drifting ice and the state of its surface during summer and autumn have a pronounced influence on the energy exchange processes in the surface layer of the atmosphere. The most important sea ice variables, relevant for energy exchange and melting are as follows: concentration, thickness, surface temperature and albedo. Albedo is one of the main characteristics (Chernigovskiy, 1963; Buzuev et. al., 1965; Grenfell and Maykut, 1977). Within summer period the parameters of ice surface and air temperature are very close, and compact cloudiness (10/10 Stcu or St) limits the minimum values of the long-wave radiation balance during late summer to not more than 10-20 W/m^2. So, daily means of the turbulent fluxes and the long-wave radiation balance are approximately equal zero. Therefore, the total heat balance of the ice surface as well as melting processes is determined by incoming short-wave solar radiation and reflected radiation, i.e. albedo.

At present, the several simple summer albedo parametrizations are used in thermodynamic ice models (Semtner, 1976; Ivanov and Makshtas, 1990; Curry and Ebert, 1993). These parametrizations are based on the relationship between albedo and ice/snow thickness or relative area of puddles. But this information is not enough. Our ground true observations revealed a large variation of ice surface types (for example, wet snow, dirty and bare ice, puddles, etc.) during late summer. The determination of integral albedo, taking into consideration the contribution of all types of ice surface, allows to improve estimations of energy exchange, ice thickness variability and concentration.

Methods and observations

The weather and ice conditions were rather constant during our field activities in the Barents Sea to the north-west from Franz Iosef Land as it is presented in Table 1.

It was a typical situation for late summer in this region. On average, the ice surface temperature was near or slightly below zero. Melting dominated on the ice cover. Turbulent (sensible and latent) heat fluxes lay from -5 W/m^2 to +10 W/m^2. After a detailed analysis of the ice surface one can distinguish four main types of ice surface as follows: puddles, wet snow, bare ice, dirty ice. Mean values of albedo are presented in Table 2.

Basing on the results, presented in Table 2, we can obtain a new simple parametrization for the integral albedo over a larger ice covered area. This parametrization can be used for all arctic

----

In: Kassens, H., H.A. Bauch, I. Dmitrenko, H. Eicken, H.-W. Hubberten, M. Melles, J. Thiede and L. Timokhov (eds.) Land-Ocean Systems in the Siberian Arctic: Dynamics and History. Springer-Verlag, Berlin, 1999, 59-63.

seas during intensive melting of drifting sea ice.

$$A = A_1*S_1 + A_2*S_2 + A_3*S_3 + A_4*S_4,$$

where: $A_{1,2,3,4}$ - albedo of wet snow, bare and dirty ice, puddles; $S_{1,2,3,4}$ - relative square of different types of ice cover.

Table 1: Weather and ice conditions. T_a - air temperature, h - relative humidity, H-low boundary of clouds, N - type and quantity of clouds, Q - solar radiation.

Ta (C)	h (%)	H (ft)	N (type/th)	Q (W/m^2)	Puddles (%)	Dirty ice (%)	Bare ice (%)	Ice thickness (m)
0.4	98	300	Stcu/10	80-150	30	15-20	5-10	2-3

Table 2: Sea ice albedo, calculated on the base of data (August 12-21, 1993). z - depth of puddles, h_i - thickness of ice rind, h_s - snow depth.

Type I. Albedo of puddles:	
a) without sediments on the bottom and ice rind	
5 cm < z < 20 cm	A = 30 +/- 5 %
20 cm < z < 30 cm	A = 30 +/- 5 %
30 cm < z < 50 cm	A = 26 +/- 3 %
b) with ice rind on surface	
h_i=5 mm	A = 34 +/- 5 %
h_i=10 mm	A = 38 +/- 6 %
c) with sediments on the bottom (patches, cryoconites, thaw holes)	A = 14 +/- 6 %
Type II. Albedo of wet snow:	
blue (h_s<3 cm)	A = 64 +/- 8 %
white (3 cm< h_s < 5 cm)	A = 68 +/- 2 %
white (10 cm< h_s < 15 cm)	A = 70 +/- 4 %
fresh dry snow	A = 78 +/- 8 %
Type III. Albedo of bare ice:	
blue	A = 56 +/- 6 %
white	A = 64 +/- 3 %
Type IV. Albedo of dirty ice:	
slightly	A = 55 +/- 6 %
medium	A = 34 +/- 9 %
dark	A = 14 +/- 2 %

As it is shown in Table 2, the albedos for the shallow (z < 30 cm) ponds are nearly independent of pond depth. These estimations reasonably agree with observations and calculations, made by other investigators (Figure1). In particular, Makshtas and Podgorny (1996) performed measurements in the north-east Svalbard during an expedition onboard RV "*Lance*" in August 1993 and calculated averaged albedos of clean ponds: A=0.56 for 0.02-0.35 cm water depth and A=0.35 for 0.50 cm water depth. In spite of the fact that albedo values are higher than those presented in Table 2 for shallow ponds, the difference may be partly explained by using the LI-190 SB quantum sensor model with a spectral window of 400 - 700 nm, by which Makshtas and Podgorny (1996) performed the measurements. Very close results were obtained by Morassutti and LeDrew (1995) too. In our opinion this phenomenon is related to multiple re-reflection of penetrating solar radiation between lower and upper puddles boundary.

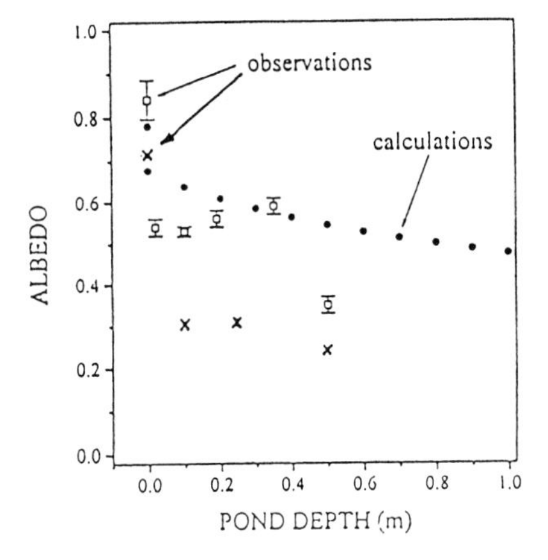

Figure 1: Measured and calculated albedos for melt ponds and bare ice in the wave length range of: 400-700 nm (■) and 300-780 (x) nm. Vertical bars represent standard deviations. ● - bare ice (Makshtas and Podgorny, 1996), x - melt ponds (Ivanov), ■ - melt ponds (Makshtas and Podgorny, 1996)

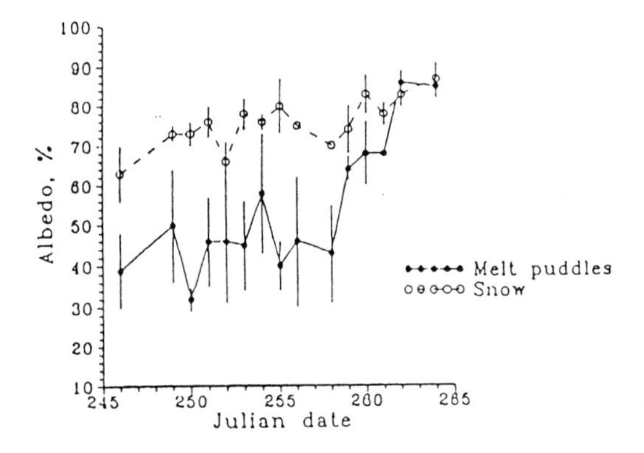

Figure 2: Time series of snow and puddle albedos, obtained during observations in the Laptev Sea (September 03-21, 1993). Vertical bars represent standard deviations.

The measured albedo of ponds of 50 cm depth and ponds with sediments on the bottom are as low as 0.26-0.14 compared to 0.30 observed at 5 cm < z < 30 cm depths. That is clearly related to cryoconite holes and patches observed on the deepest ponds bottom. According to Eicken et al. (1994), the occurrence of cryoconite holes is a consequence of sediments deposited on the pond bottom and is linked, therefore, to a decreased value of the pond albedo.

It was difficult to analyze the data, obtained in the Laptev Sea as autumn conditions started. Air temperature varied from 0°C to -10°C. We observed fresh snow on ice surface after snowfall events and short-time periods with thaw. For many days the sky was cloudless and incoming short-wave radiation exceeded 300 W/m^2, but during the nights the long-wave balance amounted to more than 70 W/m^2. Therefore, during the day we observed the radiation melting on the surface, and during the night-radiation freezing. Ice-cover albedo is a complicated function of this process. As an example of that we show a time series of snow and puddle albedos during our observations in the Laptev Sea region (within the period from September 03 to September 21, 1993) in Figure 2. The difference between puddle and snow albedos was practically constant (20-30 %) for the period from 3 to 15 of September, 1993 before ice thickness becomes approximately equal 20 cm over freezing puddles (Table 3). In this case, absolute values of surface puddles albedos were very close to the albedo of nilas (see Figure 3), which forms in fractures and polynyas. Then, during increasing of ice thickness in the puddles up to 25 cm (Table 3), the albedo of their surfaces became very close to albedo of gray and gray-white ice (see Figure 3) and that of old ice with fresh snow on the surface (Figure 2).

Therefore, the albedos of all types of ice surface (puddles, snow, young ice) were the same for the final period of our observations, ranging between 85-87 %.

Table 3: Puddle ice thickness variation within the period September 13 - 21, 1993.

Date	13.09	15.09	16.09	17.09	18.09	21.09
Puddle ice, thickness (cm)	13.6	17.2	20.4	22.0	23.0	26.0
Standard deviation (cm)	2.6	1.4	3.0	3.0	2.0	2.0

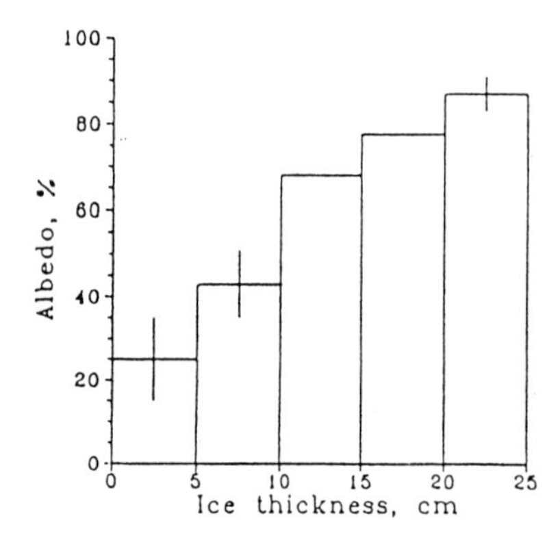

Figure 3: Albedo of young ice of different thickness in the Laptev Sea (September 03-21, 1993). Vertical bars represent standard deviations.

Conclusion

The basic conclusion, which can be made after the preliminary analysis of observations in the Laptev Sea, is that albedo of the basic types of ice cover surface (snow, puddles, young ice) in the second half of September come nearer to 0.85-0.87 size, what is very important for ice cover modeling as, in the first approximation, the period is determined, during which it is necessary to use the albedo parametrization taking into consideration the variety of ice surface condition and period, when it is possible to use the same albedo meanings for various types of sea ice cover surface.

Thus the investigations, performed in summer 1993 in the Barents and Laptev Seas, as a whole, have confirmed present data on reflective ability of the basic surface types of sea ice cover in summer-autumn period. So, the constancy of puddles albedo of depth from 5 cm up to 35-50 cm was found, which had not been described till now, and which was caused by multiple re-reflection of incoming short-wave solar radiation between the upper and lower borders of puddles. The performed observations and processes modeling of carrying of solar radiation in puddles and snow-ice cover, based on their results, will allow to describe processes of formation and destruction of sea ice cover within spring-summer-autumn period more clear when there are its most radical changes.

Acknowledgments

The author is sincerely grateful to organizers of Transdrift-IV expedition for the possibility to study the ice cover of the Laptev and Barents Seas. The author is also thankful to Dr. V. Aleksandrov (State Research Center of the Russian Federation -Arctic and Antarctic Research Institute) for his contribution to the field measurements performance.

References

Buzuev, A.Ya. (1965) Albedo of ice in the Arctic Seas to airborn observations data. (in Russian). Probl. Arktik., 20, 49-54.

Chernigovskiy, N.T. (1963) Radiative properties of ice cover in the Central Arctic (in Russian). Trudy AANII, 253, 86-94.

Curry, J.A. and E.E. Ebert (1993) An intermediate one-dimensional thermodynamic sea ice model for investigating ice-atmosphere interaction. J. Geophys. Res., 98, 85-109.

Eicken, H., V. Alexandrov, R. Gradinger, G. Ilyin, B. Ivanov, A. Luchetta, T. Martin, K. Olsson, E. Reimnitz, R. Pac, P. Poniz and J. Weissenberg (1994) Distribution, structure and hydrography of surface melt puddles. Ber. Polarforsch., 149, 73-76.

Grenfell, T.C. and G.A. Maykut (1977) The optical properties of ice and snow in the Arctic Basin. J. Glacial., 18, 445-463.

Ivanov, B.V. and A.P. Makshtas (1990) Quasi-stationary zero-dimensional model of arctic ice cover. Trudy AANII, 420, 71-80.

Makshtas, A.P. and I.A. Podgorny (1996) Calculation of melt pond albedos on arctic sea ice. Polar Research, 15(1), 43-52.

Morassutti, M.P. and E.F. LeDrew (1995) Melt pond data set for use in sea-ice and climate-related studies. ISTS-EOLTR95-001, Earth-Observations Lab., ISTS and Dept. Geog., University of Waterloo, Ontario, Canada., 55 pp.

Semtner A.J. (1976) A model for the thermodynamic growth of sea ice in numerical investigations of climate. J. Phys. Oceanogr., 6, 379-389.

Possible Causes of Radioactive Contamination in the Laptev Sea

V.K. Pavlov[1], V.V. Stanovoy[1] and A.I. Nikitin[2]

(1) State Research Center - Arctic and Antarctic Research Institute, 38 Bering St., 199226 St. Petersburg, Russia

(2) Scientific Production Association "Typhoon", 82 Lenin St., Obninsk, Russia

Received 3 March 1997 and accepted in revised form 3 March 1998

Abstract - Within the framework of the joint Russian-German summer expedition to the Laptev Sea in 1993 relatively high concentrations of ^{137}Cs and ^{90}Sr were found in the water and sediment environments of the sea. In the present paper, possible mechanisms of the radioactive contamination from adjacent marine and land areas are considered. It is shown that the main cause for enhanced levels of concentration of anthropogenic radionuclides in the Laptev Sea is transport by water and ice from the Kara Sea and adjacent parts of the Arctic Ocean.

Introduction

During the joint Russian-German expedition "Lapex-93" in summer water samples and samples from the upper layer of the bottom sediments were collected to analyse the contents of the anthropogenic radionuclides ^{137}Cs and ^{90}Sr for the first time. These samples were collected by L. Bovkun from the Scientific Production Association "Typhoon" and the observations methods and materials were descripted in expeditional reports (Preliminary report, 1993; Bovkun, 1994). The analyses of these samples, performed at SPA "Typhoon" by A. Nikitin, L. Bovkun et al., have shown, that the ^{137}Cs and ^{90}Sr concentrations in the Laptev Sea are very similar to the concentrations in the Kara Sea. So far the Laptev Sea has been considered less polluted by radioactivity than the Kara Sea, where liquid and solid radioactive wastes have been dumped. Therefore, it is necessary to determine the possible transport pathways and sources of radioactive contamination of the Laptev Sea.

Results of the observations

In Figure 1 the distribution of concentrations ^{137}Cs and ^{90}Sr in the surface water of the Laptev Sea is presented. Figure 1 illustrates the enhanced concentration of ^{137}Cs in the central and northern parts of the sea up to 6-9 Bq/m^3 and lower concentration in the south-eastern coastal region of the sea to 1 Bq/m^3 in the Gulf Buorkhaya. In the south-western coastal part of the Laptev Sea a wide range of concentrations - from 2.1 up to 7.3 Bq/m^3 is observed. Distribution of the ^{90}Sr concentration in the surface water of the Laptev Sea is more monotonous and the values change in the range from 3.7 to 6.6 Bq/m^3. The highest values were measured in the south-western part of the sea.

The distribution of the deposition of ^{137}Cs (Bq/m^2) in the upper 10cm layer of the bottom sediments is presented in Figure 2. The values varies considerably from 50 up to 1874 Bq/m^2, thus the lowest values are observed in a central part of the Laptev Sea.

Possible causes of the radioactive contaminants of the Laptev Sea

In Figure 3 the scheme of potential sources and transport pathways of the radioactive contamination in the Arctic seas is presented.

In: Kassens, H., H.A. Bauch, I. Dmitrenko, H. Eicken, H.-W. Hubberten, M. Melles, J. Thiede and L. Timokhov (eds.) Land-Ocean Systems in the Siberian Arctic: Dynamics and History. Springer-Verlag, Berlin, 1999, 65-72.

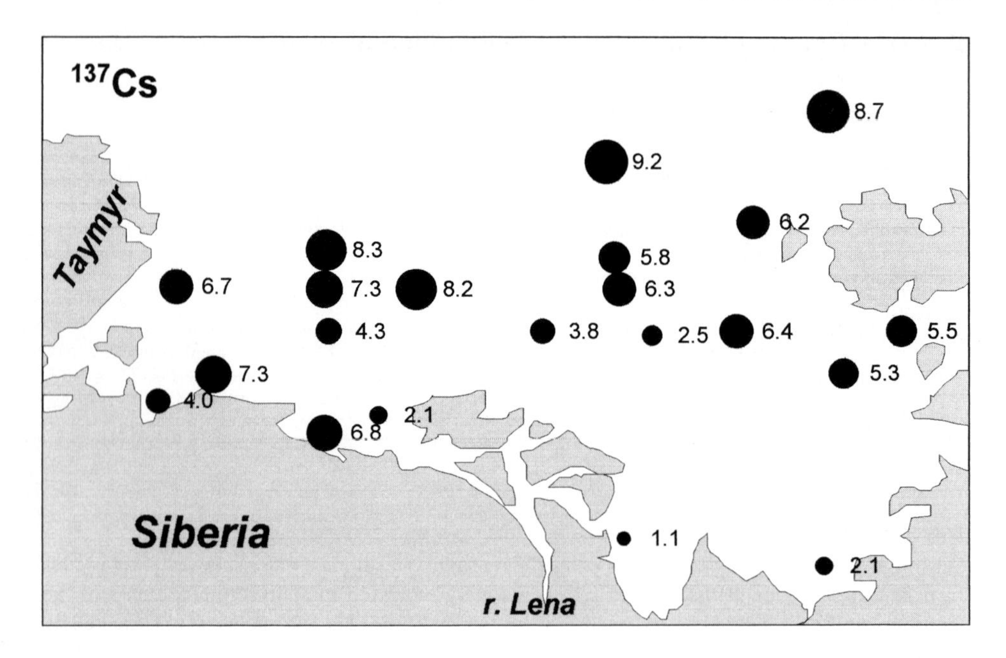

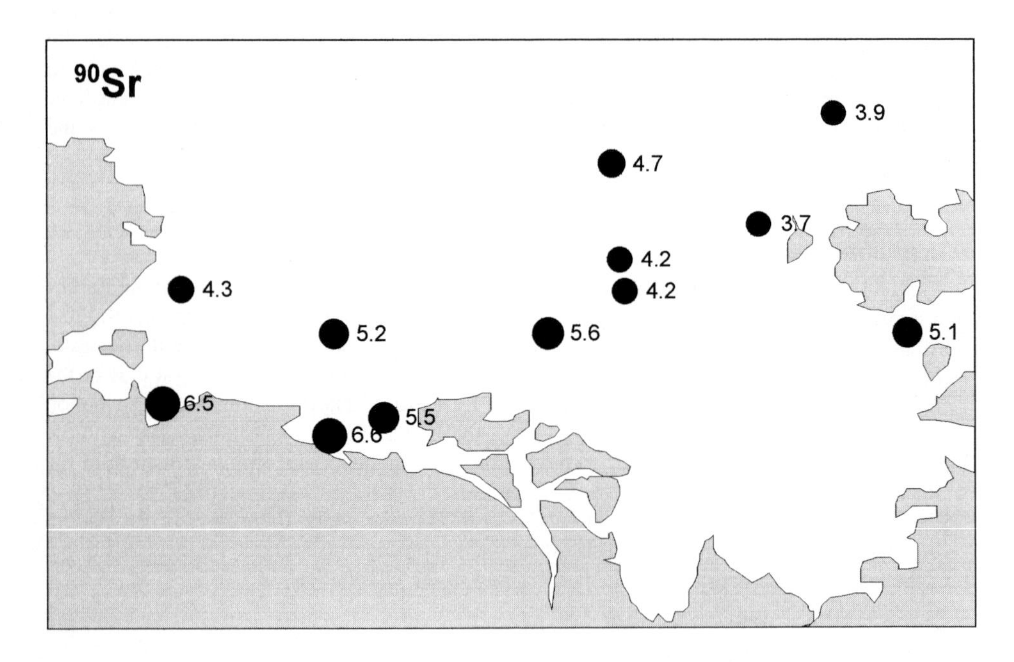

Figure 1: Distribution of the ^{137}Cs and ^{90}Sr concentration (Bq/m^3) at the suface in the Laptev Sea in summer 1993.

Transport by currents from the Kara Sea and the Arctic Basin

Currents transport radionuclides and at the same time there were the processes of mixing, diffusion and sedimentation of the radionuclides. Certainly, the advective transport of pollutants is the most important.

Results from the joint Russian-Norwegian expedition during the summer of 1992 to the Kara Sea show that ^{137}Cs and ^{90}Sr concentrations in the surface water range from 3-8 Bq/m^3 and 3-12 Bq/m^3 respectively. Thus, the observed values of ^{137}Cs and ^{90}Sr concentration in the surface water of the Laptev Sea in summer 1993 are at the same order, as in the Kara Sea in summer 1992.

Similar levels of radionuclide concentrations in the Kara and the Laptev Seas are in agreement with the possibility of advective transport of the radioactive contaminants from the Kara Sea. The resulting water exchange between the Kara and the Laptev Seas through the Vil'kitsky Strait is directed to the east and is estimated from 4,900 km^3/year (0.16 Sv) up to 11,000 km^3/year (0.3 Sv) (Pavlov et al., 1996). In addition a significant flow of water from the Kara Sea exists in the north-western part of the Laptev Sea, with is confirmed by model results (Pavlov, 1996; Figure 4).

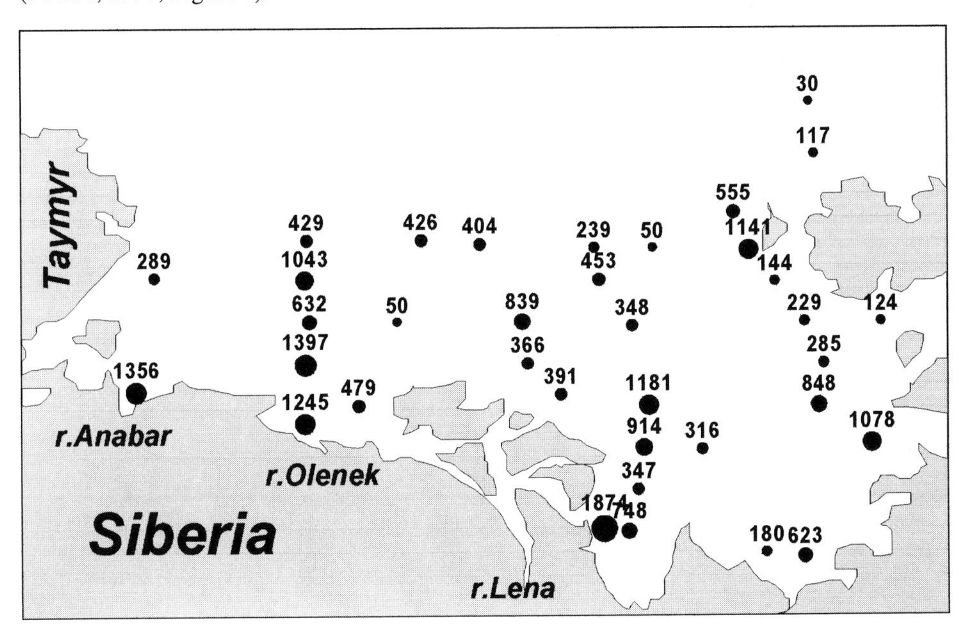

Figure 2: Distribution of the deposition of ^{137}Cs (Bq/m^2) in the upper 10cm layer of the bottom sediments in the Laptev Sea in summer 1993.

The observations in the Kara Sea have shown that the dominant source of the radioactive contamination of the sea area at the present time is the radioactive discharge at the Sellafield plant in former years. This pollution is transported by currents to the Kara Sea from the Barents Sea and from the Arctic Basin (A survey of artificial radionuclides, 1993). The numerical simulation of the consequences of radioactive discharges at Sellafield in the Arctic Ocean has confirmed this conclusion and also has shown the transfer of radioactive contaminants from the Kara Sea and Arctic Basin farther into the Laptev Sea (Kulakov and Pavlov, 1996).

Spatial distribution of the water temperature and salinity obtained during the summer expedition 1993 ("LAPEX-93"; Preliminary Report, 1993) has shown good agreement with the distribution of artificial radionuclide concentration in the Laptev Sea. Thus, more cold and salt waters correspond to the higher concentration of ^{137}Cs (Figure 5). The coefficient of correlation between the water salinty and ^{137}Cs concentration is equal to 0.95.

Very cold bottom water with high salinity was observed in the south-western part of the Laptev Sea and main reason of its formation is the inflow of the waters from the north into the shallow coastal zone during wintertime (Timokhov and Churun, 1994). The same water with temperature -1.766 °C and salinity of 34.338 was observed in the bottom layer on the CTD-station in the south-western part of the sea in summer 1993. Measurements on this station have shown the increase of ^{137}Cs concentration from 6.7 at the surface up to 16.4 Bq/m^3 at the bottom.

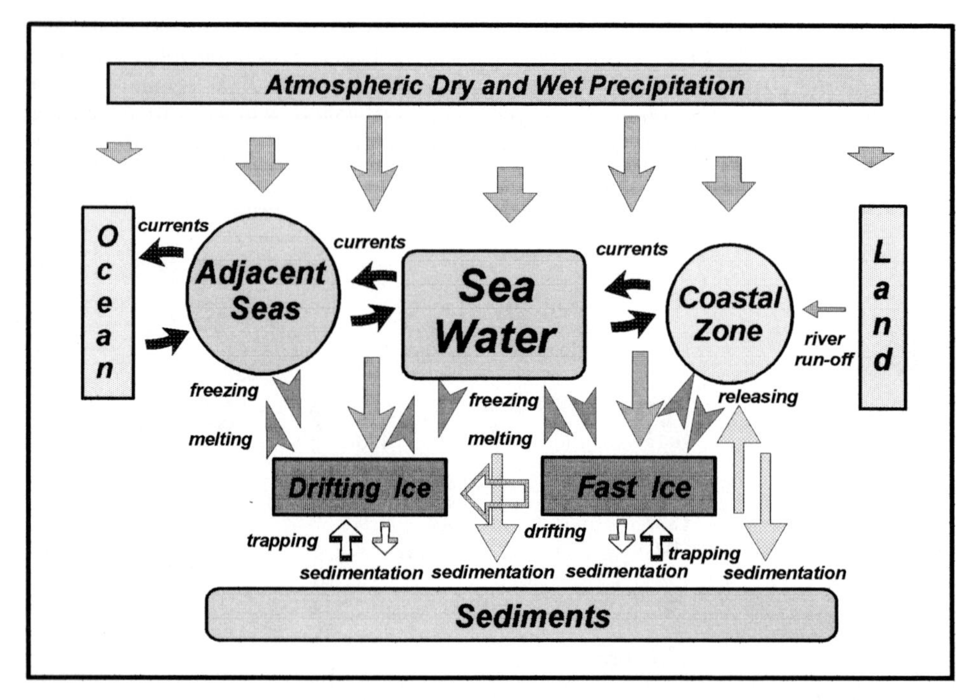

Figure 3: The scheme of potential sources and transport pathways of the radioactive contamination in the Arctic seas.

Transport of radionuclides by drifting ice

By approximate estimations, an ice floe with an area of 1 km^2 and 1.5 m thick in the spring 1993 could carry from the Kara Sea up to 7.4 MBq of ^{137}Cs and 8.4 MBq of ^{90}Sr (Pavlov and Stanovoy, 1996). In accordance with multiyear observations, the ice export through the Vil'kitsky Strait into the Laptev Sea is estimated about 50 km^3 (Pavlov et al., 1996). In addition to this, in July and August 1993 in the western part of the Laptev Sea the southward ice drift was observed (Vanda and Yulin, 1994), that is the part of the contaminated ice from the Kara Sea could drift into the Laptev Sea from the north-west in the summer 1993.

At the formation of the fast ice in the shallow coastal zone, ice is polluted by incorporation of contaminated surface layer of sediments. During the drift of this ice the particles of soil remain at the bottom surface of the ice and are transported with the ice floe and are released to the water at melting of the ice. Also, contamination of sediments may occur when drifting ice moves to

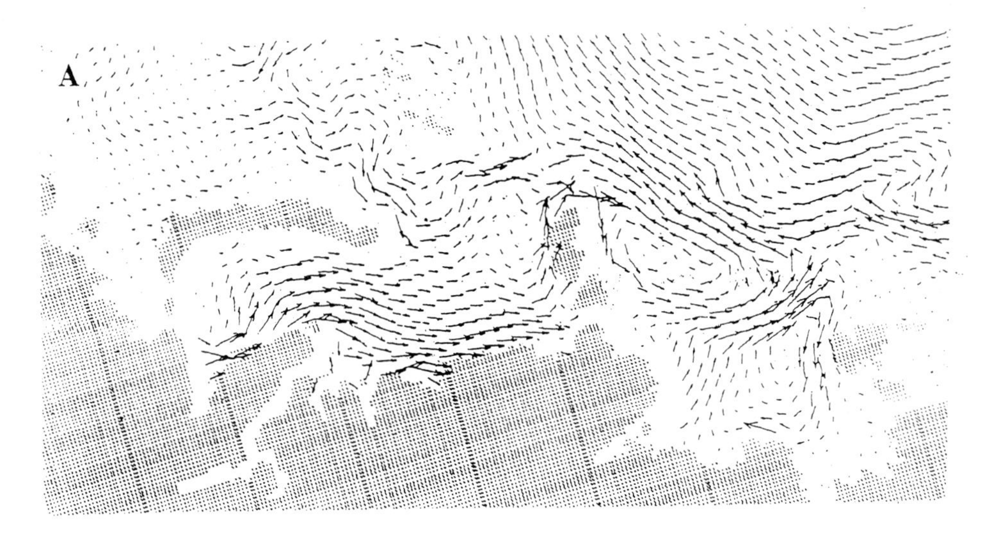

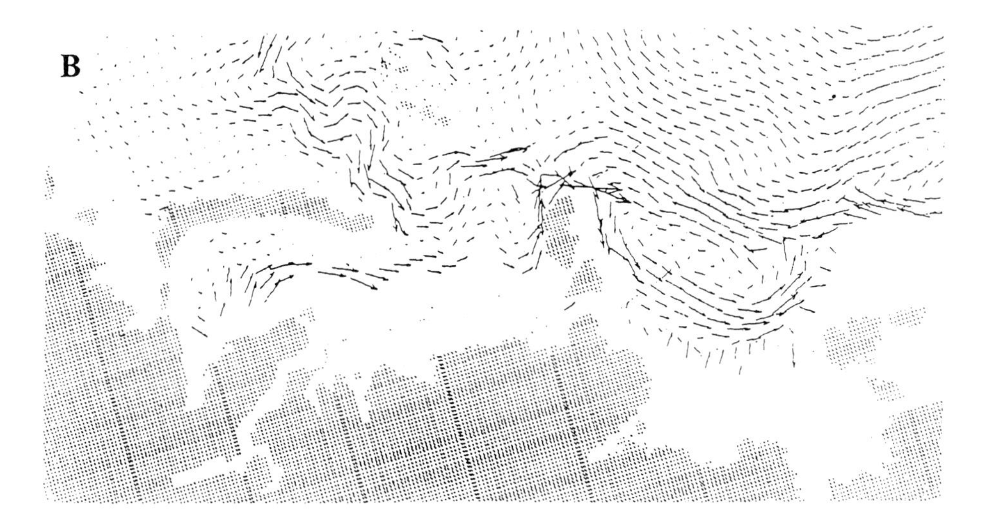

Figure 4: Calculated water circulation in the Laptev Sea in the summertime at the surface (a) and at the depth 100 m (b) (Pavlov, 1996).

the shallow regions.

At melting of the fast ice in the region of its formation the concentration of the radionuclides in the water and in the bottom sediments can increase. This increasing of the concentrations is the result of the accumulating during the long wintertime of radioactive atmospheric precipitations on the upper fast ice surface.

Bottom sediments

The concentration of the ^{137}Cs in the sediments (Figure 2) has quite good agreement with the bottom topography of the Laptev Sea and water circulation (Figure 4) which determine the location of errosion and deposition zones. Some increased values in the coastal zone are the result of the transport of the radionuclides by rivers and of the sedimentation of the radionuclides, released during fast ice melting.

Some increase of the ^{90}Sr and ^{137}Cs concentrations in the surface water of the coastal zone of the Laptev Sea may be attributed to resuspension of radionuclides from the upper layer of the sediments by waves and bottom currents which velocities can reach critical values in this region.

Atmospheric precipitations

Radionuclides are deposited from the atmosphere onto land and water surfaces, upper surface of the drifting and fast ice. On the ice surface radionuclides remain and accumulate, increasing radioactive contamination of the ice cover.

In accordance with the data of Rosgidromet (for the Kara Sea), about 2.1 Bq/m^2 year of ^{137}Cs and 1.4 Bq/m^2 year of ^{90}Sr fell out with precipitations in 1993.

Rivers

There are no sources of anthropogenic radionuclides in the water catchment area of the rivers of the Laptev Sea basin. The only source is the global fallout onto the territory of the water catchment area and subsequent transport by the rivers into the sea. Thus, the contribution of the rivers to the radioactive contaminants in the Laptev Sea is quite small at the present time. The rivers influence the level of radioactive contaminants only near the river mouths.

Discussion and conclusions

During the joint Russian-German expedition in summer 1993 water samples and samples from the upper layer of the bottom sediments were collected to analyse the contents of the anthropogenic radionuclides Discussion and conclusions ^{137}Cs and ^{90}Sr for the first time. So far the Laptev Sea has been considered less polluted by radioactivity than the Kara Sea, therefore the results of the analysis were rather unexpected.

Results from the expedition "Tundra Ecology-94" in summer 1994 confirmed the enhanced levels of the radioactive contamintion of the Laptev Sea. Obtained in summer 1994 in the Laptev Sea fewness data of the ^{137}Cs concentration are at the same level and a similar spatial distribution as formed in 1993. Moreover, the ^{137}Cs concentration in the Vil'kitsky Strait was relatively high (10.6 Bq/m^3) (Josefsson et al., 1995).

The spatial distribution of the ^{137}Cs concentration has a good agreement with the spatial distributions of the water temperature and salinity and with the general water circulation in the Laptev Sea. The coefficient of correlation between the water salinity and ^{137}Cs concentration is equal to 0.95. Therefore the main cause of the enhanced levels of the ^{137}Cs concentration in the water of the Laptev Sea in 1993 and 1994 is the transport of the radionuclides by the currents and the drifting ice from the Kara Sea and the Arctic Basin.

The relation of the ^{90}Sr concentration to the salinity is not so close and the coefficient of correlation is equal to -0.32. Therefore the possible causes of the enhanced levels of the ^{90}Sr concentration in the water of the Laptev Sea in 1993 is not only the transport of the radionuclides by the currents and the drifting ice from the Kara Sea and the Arctic Basin, but also the runoff by rivers.

Some higher level of the ^{90}Sr and ^{137}Cs concentrations in the surface water of the coastal zone of the Laptev Sea may be attributed to resuspension of radionuclides from the upper layer

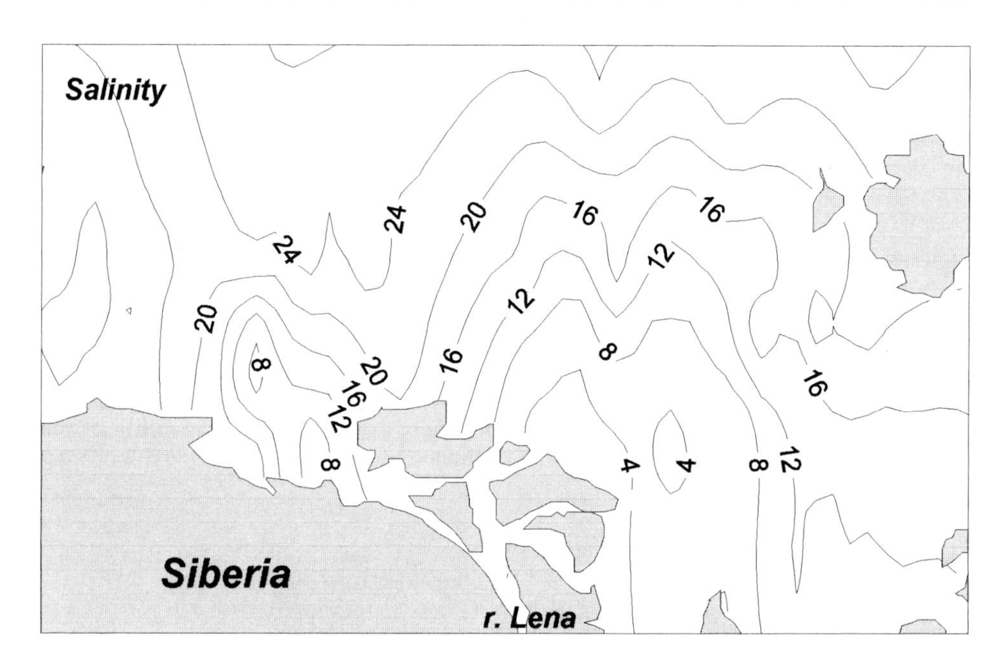

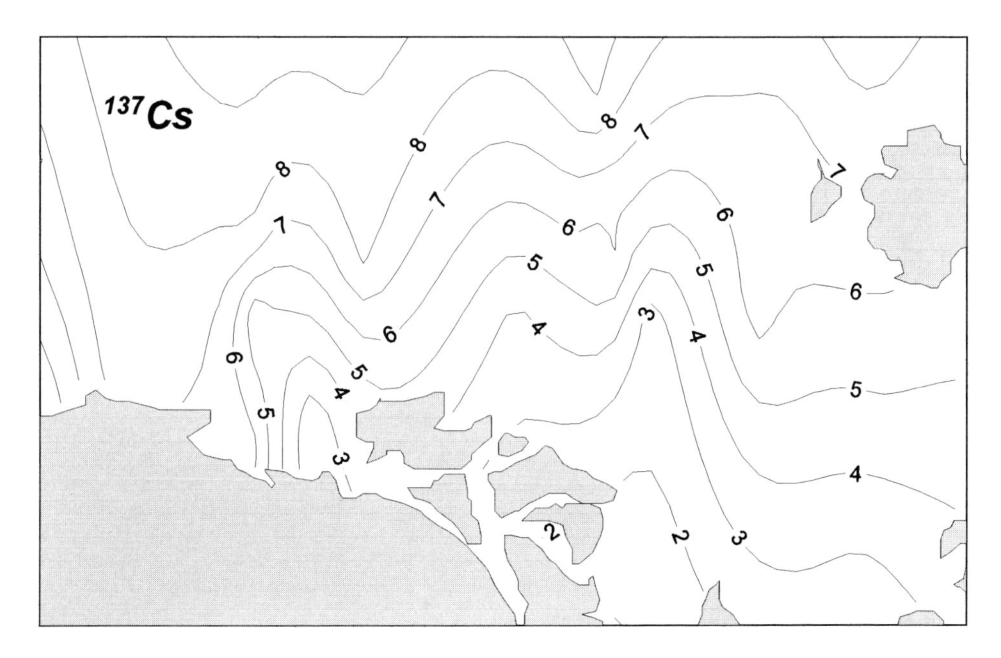

Figure 5: Distribution of the water salinity and ^{137}Cs concentration (Bq/m^3) at the suface in the Laptev Sea in summer 1993.

of the sediments by waves and bottom currents which velocities can reach critical values in this region.

The concentration of the ^{137}Cs in the sediments has quite good agreement with the bottom topography of the Laptev Sea and water circulation which determine the location of errosion and deposition zones.

The ice radioactivity is a topic for further investigations and discussions. According to measurements, the concentration of ^{137}Cs in the ice of the Chukchi Sea was equal to about 1 Bq/m^3, i.e. of the same level as the concentration of ^{137}Cs in water. Also, according to analysis of soil particles at the bottom surface of the ice, values of 4.9-5.6 Bq/kg dry weight were obtained (at maximum values up to 73 Bq/kg dry weight) (Meese et al., 1995). Pavlov and Stanovoy (1996) roughly estimated the possible export of the radionuclides by drifting ice from the Kara Sea into the Arctic Basin.

There are not enough data of the concentration of anthropogenic radionuclides in the Laptev Sea to obtain some conclusions about possible sources and temporal variability of the radionuclide distribution. It is necessary to continue the field observations and modeling investigations in this problem.

References

Bovkun, L. (1994) Radiochemical observations. In: Russian-German Cooperation: The Transdrift I Expedition to the Laptev Sea, Kassens, H. and V.Y. Karpiy (eds.), Ber. Polarforsch. 151, 52-53.

Josefsson, D., E. Holm, B. Persson and P. Roos (1995) Anthropogenic radioactivity along the Russian coast. Preliminary results from the Swedish-Russian Tundra Ecology Expedition, 1994. In: Proc. Arctic Nuclear Waste Assessment Program Workshop, Woods Hole, USA, May 1995.

Kulakov, M.Yu. and V.K. Pavlov (1996) Modelling of the consequences of radioactive waste discharges from the Sellafield plant (in Russian). Trudy AANII (in press).

Meese, D., L. Cooper, I.L. Larsen, W. Tucker, E. Reimnitz and J. Grebmeier (1995) Radionuclides in the arctic sea ice. In: Proc. Arctic Nuclear Waste Assessment Program Workshop, Woods Hole, USA, May 1995.

Pavlov, V.K. (1996) Seasonal variability of the water circulation in the Arctic Ocean. (in Russian). Trudy AANII (in press).

Pavlov, V.K. and V.V. Stanovoy (1996) To the problem of transfer of radionuclide pollution by sea ice. (in Russian). Trudy AANII (in press).

Pavlov, V.K., L.A. Timokhov, G.A. Baskakov, M.Yu. Kulakov, V.K. Kurazhov, P.V. Pavlov, S.V. Pivovarov and V.V. Stanovoy (1996) Hydrometeorological regime of the Kara, Laptev and East-Siberian Seas. - Tecchnical Memorandum APL-UW TM 1-96, Seattle, USA. 179 pp.

Preliminary Report (1993) Preliminary Report of Expeditions to Taymir Peninsula and Laptev Sea in summer 1993 on a board of R/V "Ivan Kireev" and R/I "Polarstern" in the framework of Program Russian-German Cooperation. (in Russian) AARI, St.-Petersburg, Russia. 384 pp.

Timokhov, L.A. and V.N. Churun (1994) Cold bottom water in the southern Laptev Sea (in Russian). In: L. Timokhov (ed.), Nauchnye Rezultaty ekspedicii LAPEKS-93. Gidrometeoizdat, St.-Petersburg, 83-90

Vanda, Yu.A. and A.V. Yulin (1994): Seasonal variability features of ice conditions of the Laptev Sea in 1993(in Russian). In: L. Timokhov (ed.), Nauchnye Rezultaty ekspedicii LAPEKS-93. Gidrometeoizdat, St.-Petersburg, 148-157.

Oceanographic Causes for Transarctic Ice Transport of River Discharge

I. Dmitrenko[1], P. Golovin[1], V. Gribanov[1] and H. Kassens[2]

(1) State Research Center - Arctic and Antarctic Research Institute, 38 Bering St., 199226 St. Petersburg, Russia

(2) GEOMAR Forschungszentrum für marine Geowissenschaften, Wischhofstrasse 1-3, D 24148 Kiel, Germany

Received 3 March 1997 and accepted in revised form 8 February 1998

Abstract - The influence of river discharge on ice-hydrological conditions was investigated during expeditions in the Laptev Sea in 1994, 1995 and 1996 during different seasons of the year within the framework of the Russian-German project "Laptev Sea System". A combined analysis of both ice satellite and CTD observations has shown that the formation and distribution of the fast ice edge is dependent on vertical heat exchange processes with the warm subsurface water layer underlying river water over a depth range of 10 to 25 m. It is formed during the summertime in areas affected by river discharge, which spreads as a result of warm surface water converging at the discharge fronts. Calculations show that advection of heat and double-diffusive convection are the most efficient modes of heat transport to the growing ice at the periphery of the freshened zone. Their values are sufficient to reduce the ice thickness at the periphery of the discharge zone by more than half. This leads to a fast ice edge much further south than the northern limit of the freshened zone. As a result, a considerable amount of riverine dissolved and suspended matter is incorporated into drifting ice and hence into transarctic ice transport.

Introduction

One of the main objectives of the Russian-German multidisciplinary research project the "Laptev Sea System" is to investigate the pathways and mechanisms of suspended matter and sediment transport from the Laptev Sea to the Arctic Ocean. The transarctic ice transport of sediments and suspended matter incorporated into sea ice at freezing was suggested (e.g. Eicken et al., 1997). Since the Lena River is one of the main sources of dissolved and suspended matter supply to the Laptev Sea, it is especially important to investigate the fate of river water.

Previous expedition studies have shown that the majority of suspended matter is distributed in accordance with the spreading of river water and that it is centered in the upper 7-9 m. In this case, it is expected that the most intense incorporation of suspended matter coincides with regions where river discharge was located at the onset of freezing (Dmitrenko et al., in press). Under the influence of river discharge, the upper sea layer is significantly diluted („freshened„). The resulting strong density stratification in most regions (except in shallow water) prevents penetration of sea water mixing processes down to the sea floor. This in turn inhibits the transport of bottom sediments to the sea surface. However, suspended matter concentrated in the surface freshened layer can be incorporated into the forming ice after the onset of freezing. This ice forms a massif of fast ice or drifting first-year ice located beyond the fast ice limits (including flaw polynyas). The latter is extremely important in terms of potential ice transport of sediments. Drift trajectories of GPS-buoys set up on fast ice north-east of the Lena delta in the spring of 1996 confirmed that fast ice can melt in place. However, first-year drifting ice can be entrained to the transarctic drift. Thus, one factor determining the transarctic sea ice transport of river discharge is the type of ice cover into which the river discharge is incorporated at freezing. Hence, this study focuses on the influence of river discharge on the extent of the fast ice cover.

In: Kassens, H., H.A. Bauch, I. Dmitrenko, H. Eicken, H.-W. Hubberten, M. Melles, J. Thiede and L. Timokhov (eds.) Land-Ocean Systems in the Siberian Arctic: Dynamics and History. Springer-Verlag, Berlin, 1999, 73-92.

Materials and methods

The analysis and calculations were based on CTD-soundings during the expeditions Transdrift II (September 1994), Transdrift III (October 1995) and during the winter portion of the Transdrift IV expedition (May 1996; see Figure 1). On the Transdrift II expedition, observations were made from aboard the R/V "Professor Multanovsky" in the ice-free region of the Laptev Sea. During the Transdrift III, expedition work was carried out aboard the icebreaker "Kapitan Dranitsyn" or from the ice cover at a distance of not more than 100 m from the icebreaker. The stations west and north-west of Kotelny Island were on open water and all other stations were under conditions of intense ice formation and ice thickness of 5 to 30 cm. On the Transdrift IV expedition, observations were made from fast ice up to 2.10 m thick. CTD-soundings were conducted using a sounding set OTS-PROBE Serie 3 (Meerestechnic Electronic GmBH, Germany).

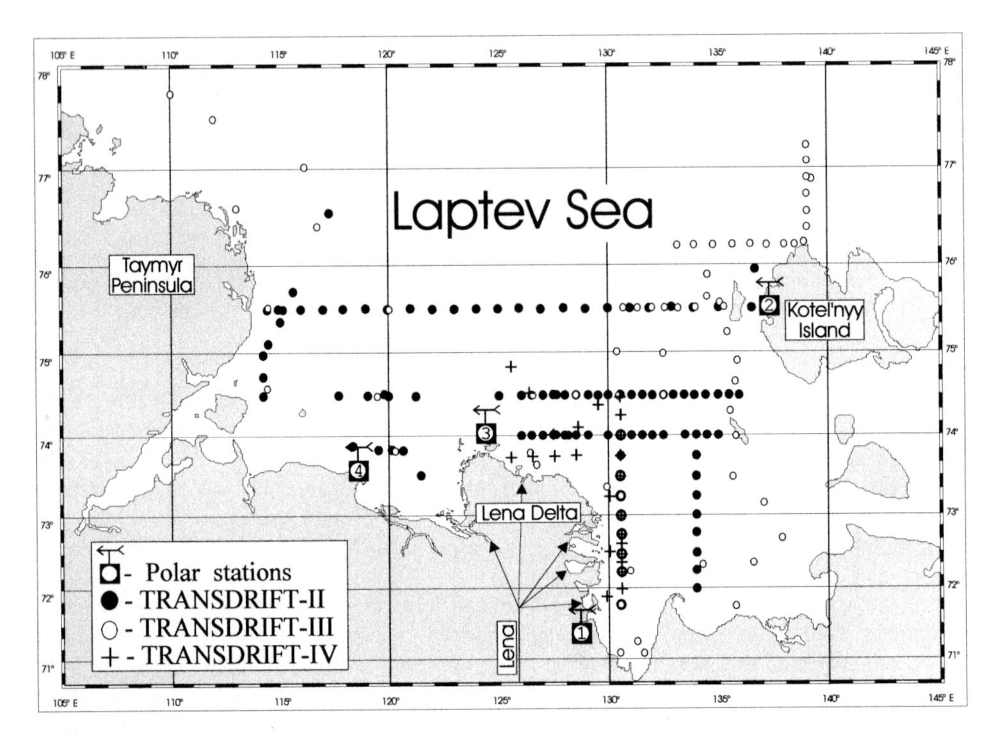

Figure 1: A scheme of the comprehensive oceanographic survey in the Transdrift II, Transdrift III and in the winter phase of the Transdrift IV expedition. Polar stations whose data were used: 1 - Tiksi, 2 - Kotel'ny, 3 - Dunai, 4 -Terpyai-Tumus.

The averaged daily data for standard meteorological air temperature observations at the polar stations "Kotelny", "Dunay", "Terpyai-Tumus", and "Tiksi" (Figure 1) were used for estimating the rate of ice formation in the winter season 1995-1996. The calculated values of ice thickness and salination were verified using CTD-data obtained at the same points in Transdrift IV expedition. The fast ice edge position was taken from the composite 10-day ice charts prepared in AARI on the basis of NOAA and METEOR satellite imagery.

The heat content of the water layers was calculated using actual CTD-data using the formula:

$$Q = \int_{h_1}^{h_2} C_p \rho (T_w - T_{fr}) dz \qquad (1)$$

where: C_p is heat capacity of sea water [J kg^{-1} °C^{-1}], ρ is sea water density [kg m^{-3}], h_1, h_2 are the lower and the upper boundaries of the water layer [m], T_w is water temperature [°C], and T_{fr} is the freezing point of the water [°C]. The formulae for calculating C_p, ρ, T_{fr} were taken from Fofonoff and Millard (1983). The integral in (1) was solved using the trapezoid formula. To obtain the specific heat content, the calculated values were normalized to the thickness of the corresponding layer.

Results

Comprehensive investigations in the summer of 1994 and autumn of 1995 resulted in a large body of information on the spatial and temporal (between summer to autumn) variations of hydrophysical properties within the zones of river water spreading.

The outflow situations in 1994 and 1995 differed considerably. In summer 1994, the outflow of river water was restricted to the south-eastern part of the sea (Figure 2A). In October 1995, the river outflow was spreading predominately towards the north-east (Figure 2B). Nevertheless, the outflow streams followed the sea bottom relief. The vertical salinity distribution in the river water outflow zones, in both 1994 and 1995, was characterized by the presence of two water layers that were vertically quasi-uniform. The upper layer was composed of freshwater, predominantly river discharge. The lower layer was Arctic bottom sea water. The upper quasi-layer was underlain by an intermediate slightly stratified water layer extending up to the main pycnocline. It was formed through the interaction between surface and bottom water. The intermediate water layer directly under the outflow channels was characterized by high concentrations of dissolved oxygen, chlorophyll *a* fluorescence, maximum values of the light transmission coefficient and minimum values of dissolved silicon (Dmitrenko et al., 1995; Golovin et al., 1995; Kassens and Dmitrenko, 1995; Kassens et al., 1997).

Calculations of the water heat content based on CTD-sounding data yielded the following results. The heat content of the intermediate water layer situated between the seasonal and the main pycnocline does not differ significantly from that of the upper well-heated quasi-uniform layer (Figure 3A) in the river water outflow zones during summer. In autumn, especially after the onset of ice formation, the heat content of the upper layer is actually zero. The heat content of the intermediate layer is much larger and remains relatively constant when compared to the summer season (Figure 3B). Hence, the thermal evolution of the water layer from summer to autumn in river discharge outflow zones can be represented as shown in Figure 3. The high heat content of the upper quasi-uniform layer is governed by radiation heating and wind and wave driven mixing in summer. However, the large heat content of the intermediate layer, given the strong density stratification at its upper boundary, requires a different explanation. The spatial variability of the water heat content in the intermediate layer was analyzed in the summer of 1994 and autumn of 1995 (Figure 4). If the heat content of the intermediate layer (Figure 4) is compared to the surface salinity distribution (Figure 2), it is evident that the distribution of higher heat content zones is determined by river water spreading. Regions with high intermediate layer heat content were located along the northern periphery of the spreading

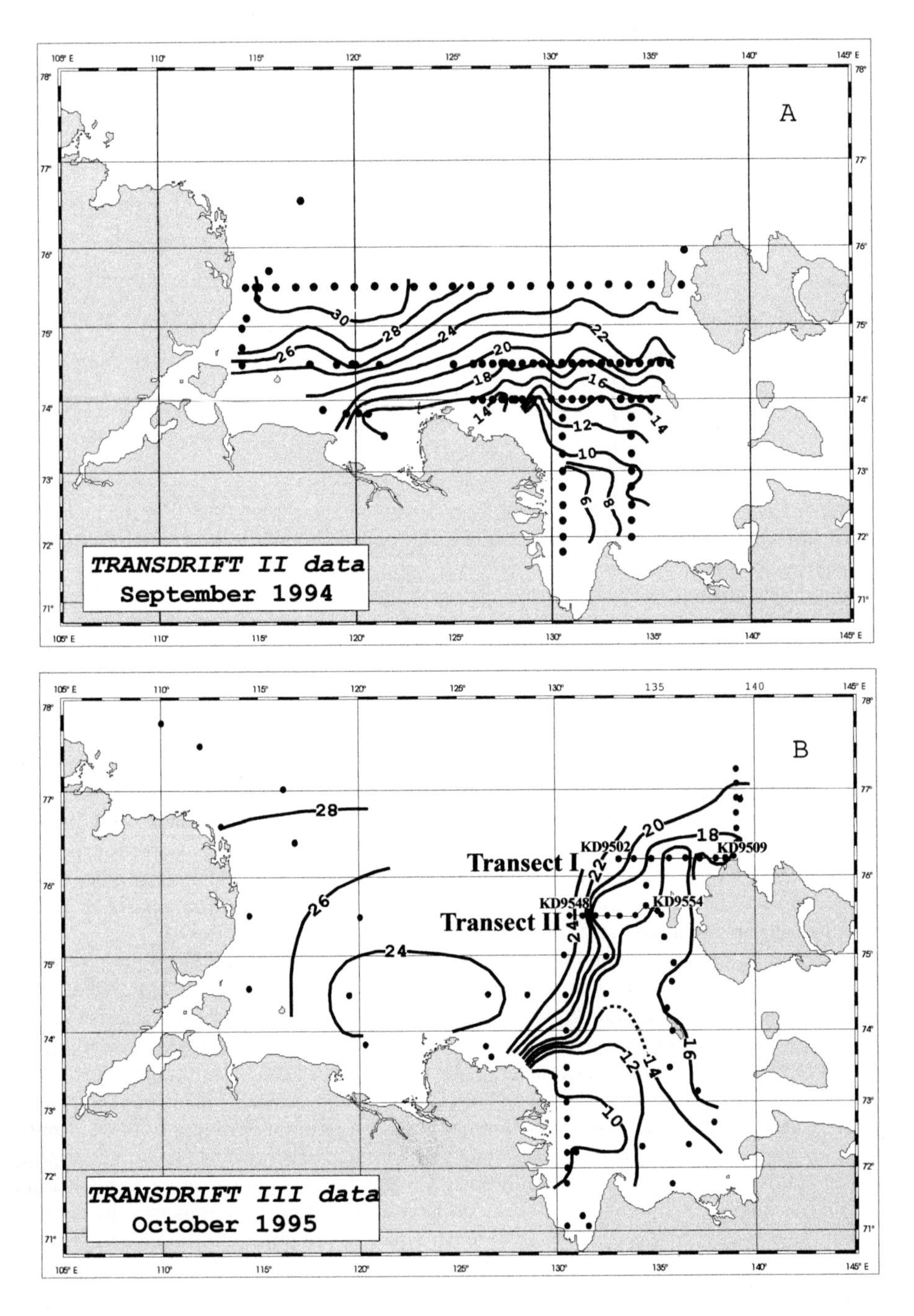

Figure 2: The surface salinity distribution in the Laptev Sea; September 1994, Transdrift II data (A) and October 1995, Transdrift III data (B).

river water. These results allow us to formulate the following questions:

- how is the warm intermediate water layer under the seasonal pycnocline formed?
- how does the intermediate water layer under the growing young ice (with a heat content comparable to that of the surface layer in summertime) influence the development of the winter ice and hydrological processes?

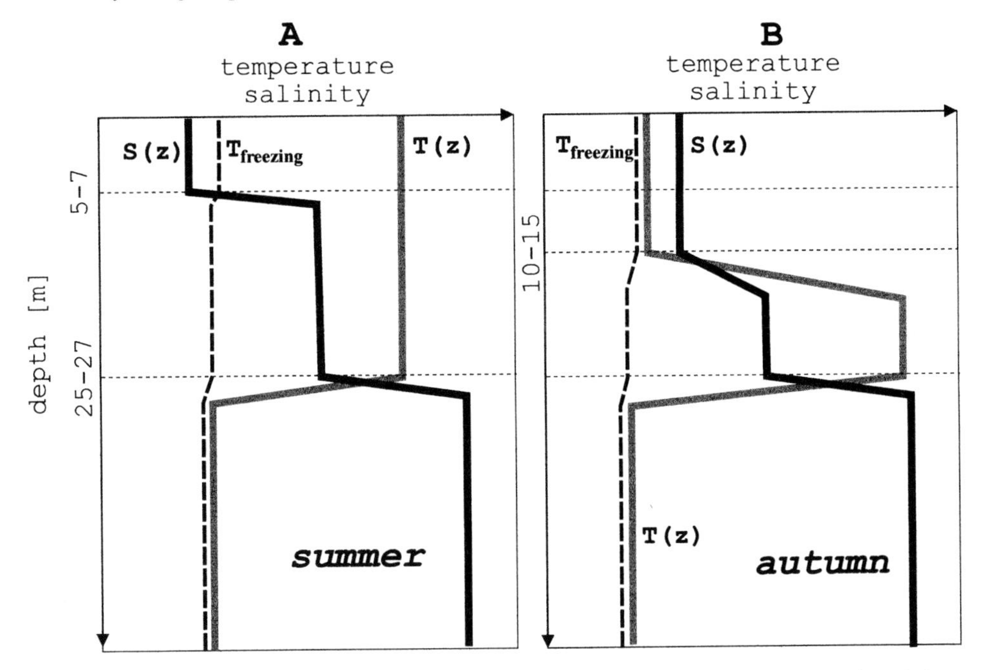

Figure 3: Typical evolution of thermohaline characteristics in the eastern Laptev Sea in the region of river discharge from summer (A) to autumn (B).

Frontal processes and the formation of the thermal subsurface layer

During spring and summer, the intensive river discharge to the Laptev Sea results in the formation of discharge hydrological fronts bounding the freshened water lens. Down to the depth of the discharge lens, the vertical isopycnals, isotherms and isohalines remain parallel to each other, intersected by equal pressure surfaces (baroclinic front; Figure 5A). Another type of front is formed in the lower quasi-isothermal water layer, especially at the periphery of the outflow zone between the seasonal and main pycnocline directly under the baroclinic front. The isopycnals and isohalines, which are approximately parallel to equal pressure surfaces, intersect the isotherms at an angle of up to 90° (Golovin et al., 1995). This is called the thermoclinic front (Figure 5A). The terms "baroclinicity" and "thermoclinicity" are used to indicate that isopycnic surfaces are intersected by equal pressure and temperature surfaces (Fedorov, 1991; Woods, 1980). Thus the discharge fronts in the Laptev Sea have a two-layer structure. From the surface to the depth of penetrating river water (seasonal pycnocline), they are baroclinic. In the lower layer, extending down to the main pycnocline, they are thermoclinic (Golovin et al., 1995). Two questions arise concerning the mechanisms of warm subsurface intermediate layer

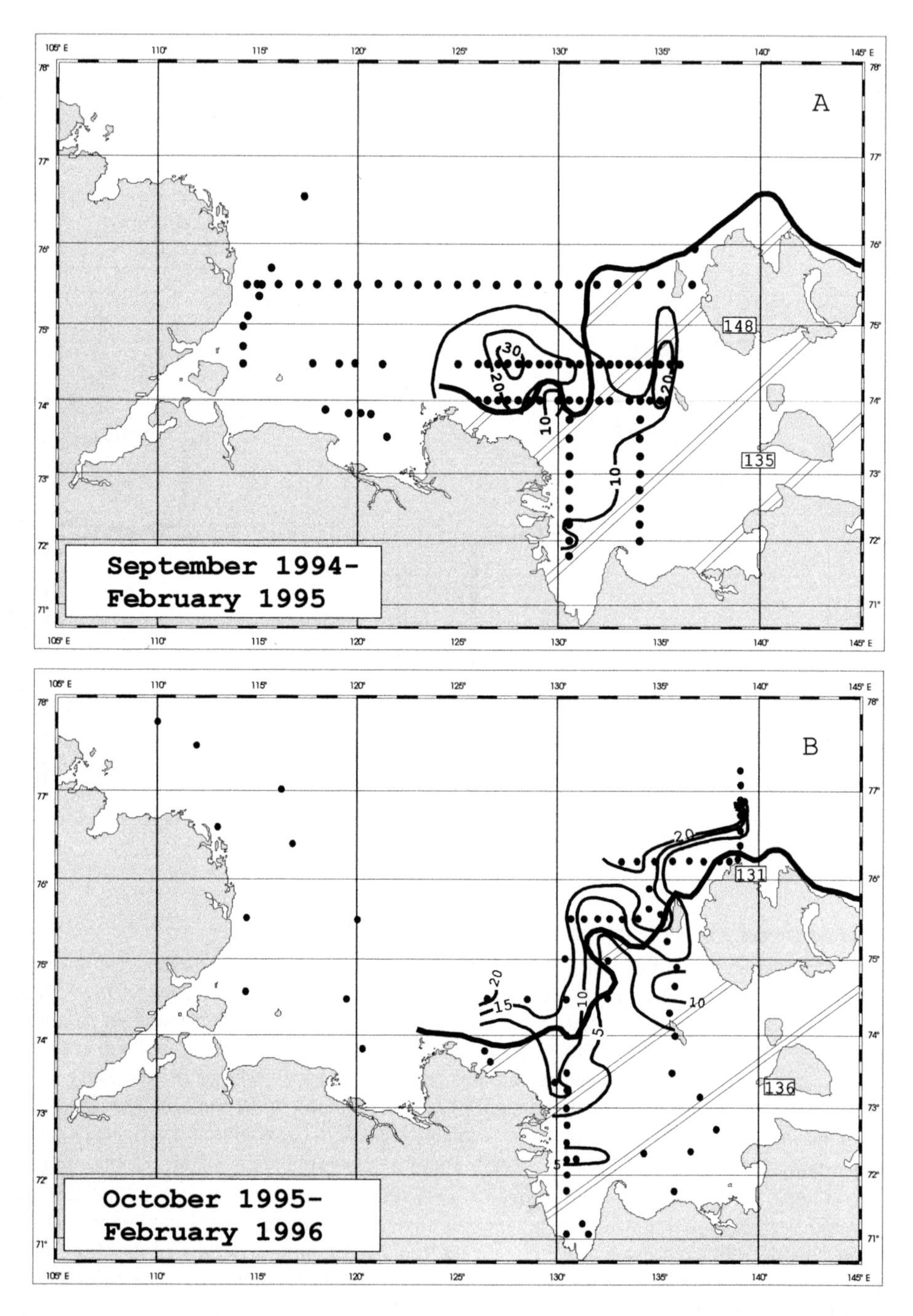

Figure 4: Spatial distribution of the heat content (in 10^4 KJ m^{-2}) of the intermediate water layer during September 1994 (Transdrift II data) (A) and October 1995 (Transdrift III data) (B) and a fragment of the composite ice chart for mid-February 1995 (A) and 1996 (B) (AARI data). ▬ - fast ice edge, ▭ - fast ice field, 123 - fast ice thickness, cm.

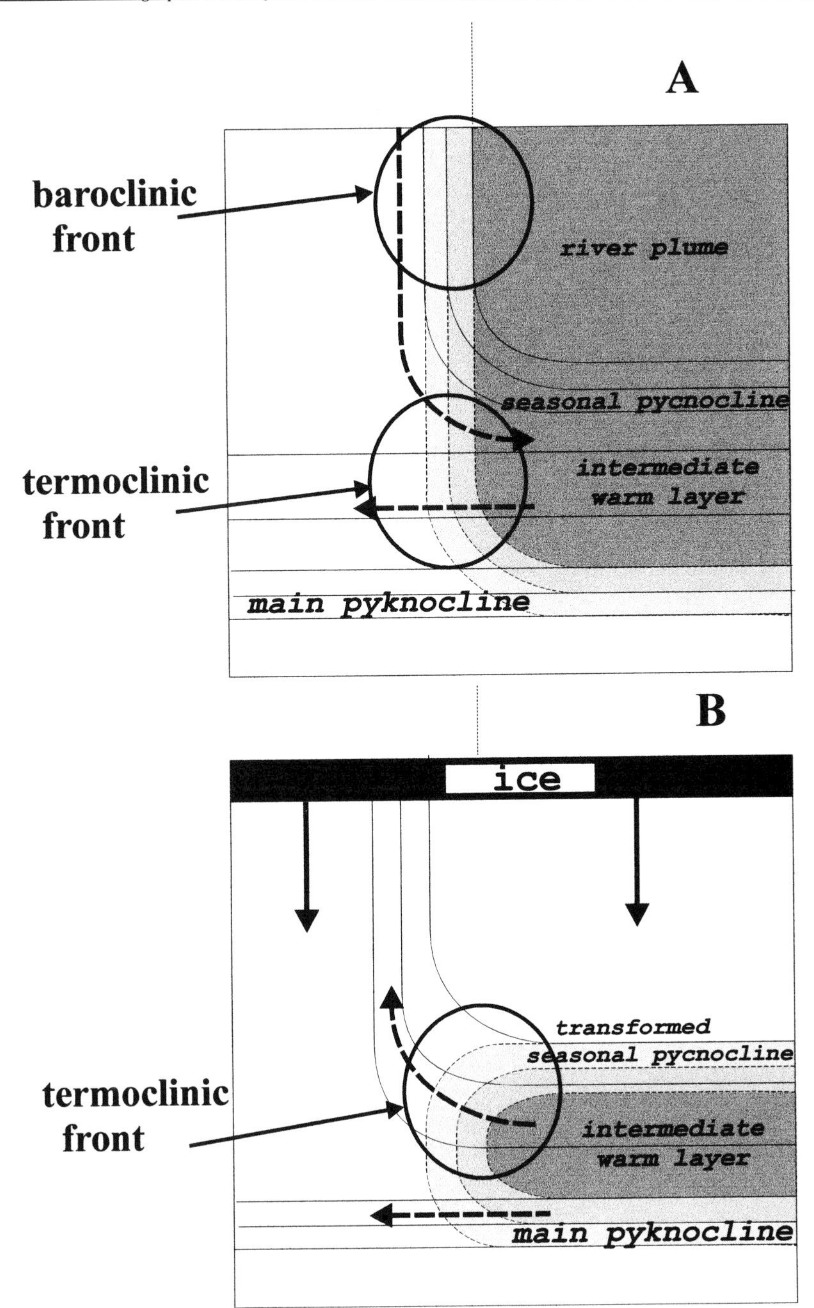

Figure 5: The structure of river discharge front and its evolution from summer (A) to winter (B). —— - isopycnals and isohalines, - - - - - - isotherms, ——▶ - salt flux, ▪ ▪ ▪▶ - heat flux.

formation: a) how do the discharge fronts spread to depths which exceed the depth of freshened water penetration by 4 to 5 times? and b) what mechanisms form the two-layer discharge fronts? The most simple mechanism for the formation of the secondary thermoclinic discharge front relates to the isopycnic convergence of the solar-heated surface water beneath the river water lens. The lighter fresh water flows onto saline sea water. This mechanism is probably most important when the river discharge extends far northward from the river mouths. This must be accompanied by earlier ice melting in the same area in order for the water to have enough time to be heated (as occurred in the summer of 1995, for example). Obviously, the presence of a flaw polynya in these regions during winter and spring will increase heat accumulation (Zakharov, 1966), and thus warmer water will be "sucked" under the river water lens.

Another mechanism is connected with the possible deflection of the thermocline under the influence of frontal convergence in those cases when temperature does not influence the density field formation. This mechanism may be regarded as a passive admixture (Kuz'mina, 1980; MacVean and Woods, 1980). This process can be the most efficient when isotherms are initially inclined to isopycnals. This mechanism probably occurs within a short distance from the river mouth regions. It is most efficient during years when the main front-forming processes occur at a relatively small distance from the mouth regions (for example, the summer of 1994). Both mechanisms form the secondary thermoclinic front beneath the upper baroclinic one in the zone of frontal convergence.

An anomalous lowering of the upper well-heated quasi-thermal layer can result from the development of this type of convergence circulation at the front. The seasonal thermocline under the river water lens is located between 25 and 27 m, reaching the main pycnocline (Figure 3A), though its usual depth corresponds to the depth of the seasonal pycnocline (5-7 m). As the autumn progresses, radiative cooling, and wave and wind driven mixing occur, and the upper water layer is quickly cooled. However, a thicker heated water layer remains "buried" at depths of 10-12 to 25-27 m (Figures 3B, 5B). This leads to the generation of a thick anomalous warm water interlayer at the northern periphery of the zone of river water spreading during summer. Its stability in autumn results from the coincidence of the upper warm layer boundary with the position of the seasonal pycnocline.

A significant redistribution of oceanographic characteristics occurs under the influence of convergent processes in the region of the discharge fronts and under the lens of river water (Figure 6). The lowest concentration of dissolved silicon (less than 320 mg/l) was recorded in the intermediate layer directly under the outflow axes (Kassens and Dmitrenko, 1995; Kassens et al., 1997). In the surface and bottom layers, silicon concentration was 650 and 980 mg/l respectively. These layers coincided spatially with oxygen maxima, with enhanced chlorophyll *a* fluorescence, which reached values typical of the sea surface, and with the absolute maxima of the light transmission coefficients in water. These vertical distribution patterns typical of the river water outflow zone were observed both in summer 1994 and autumn 1995. It is obvious that water yielding such characteristics must originate at the surface. It is formed outside the river outflow zone (low values of dissolved silicon and high transparency). Then, as a result of frontal convergence in the spring and summer, it sinks, thus forming a sufficiently thick intermediate water layer with anomalous hydrochemical and hydro-optical characteristics (Figure 6).

Influence of the surface intermediate layer on the formation of ice-hydrological conditions in the autumn-winter season

The existence of a warm (up to 4°C) and thick (up to 20 m) water layer is expected to significantly alter ice formation conditions and subsequent ice growth processes in regions affected by river discharge. The heat flux from this layer to the surface can probably limit ice formation via convection and may to some extent govern the position of the fast ice edge and the flaw polynya, presumably through some efficient mode of heat transfer to the surface.

Figure 4 provides a basis for testing for a correlation between the fast ice edge position and the heat content of the subsurface layer at a time when the edge position was quasi-steady (mid-February 1995 and 1996). Clearly, fast ice in the eastern Laptev Sea was stable during the autumn of 1994 and was congruent with the spread of river water and with heat content isolines for the subsurface layer (Figure 4A). Fast ice edge formation during the 1995-1996 autumn and winter seasons displayed similar congruencies: the configuration of the fast ice edge correlates well with the heat content isolines in the intermediate layer (Figure 4B). Thus, it seems that the quasi-steady position of the fast ice edge depends on the heat supply from the intermediate water layer. In the river water discharge zones, the fast ice edge was located much further south (up to 150 km) than the northern boundary of river water spreading. Analysis of weekly ice charts has shown that the northward spreading of fast ice was in agreement with the heat content of the intermediate water layer. Investigations of the annual variability of the mid-winter fast ice edge position between 1990 and 1996 shows that it is governed to some extent by the direction of river water spreading in the preceding summer season.

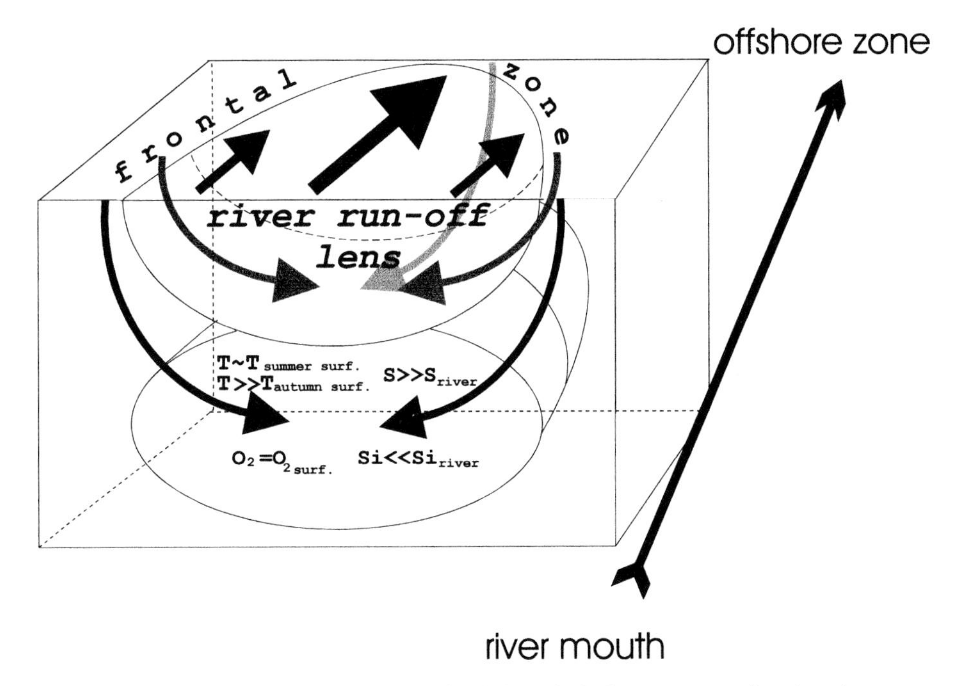

Figure 6: A scheme of the frontal convergence at the discharge hydrofronts; ⟶, direction of convergent motions, ⟶, direction of river water spreading.

The features mentioned above suggest that there exist mechanisms for heat exchange between the intermediate layer and the water surface under the ice cover. The incoming heat restricts ice formation and thus directly influences ice formation conditions. Let us assume that the heat content of the intermediate water layer is $20 \cdot 10^4$ KJ m^{-2} (data from the Transdrift III expedition), the mean salinity of the ice core is 4 (Transdrift IV expedition), and the ice temperature is -2°C at the ice-water boundary and -25°C at the ice-air boundary. Ice core temperature distribution is considered to be linear (Doronin and Kheisin, 1975). According to

Oceanographic Tables (1975), if all heat from the intermediate layer is transferred to the surface, the sea ice core growth is reduced by 68 cm. Is this sufficient to control the ice conditions to the extent as shown in Figure 4? Observations carried out on the ESARI-92 expedition in the eastern Laptev Sea showed that the thickness of fast ice in the vicinity of the fast ice edge was about 1.9 m (Dethleff et al., 1993). This value differs from the thickness of the first-year drifting ice behind the flaw polynya by a value of the same magnitude.

As follows from Figure 4, by the time the fast ice edge attained a quasi-steady character, its position was to a great extent governed by the heat content of the intermediate water layer. If this is the case, then there must have been some mechanisms for an extremely effective heat exchange with the surface. Let us discuss the entire complex of mechanisms responsible for heat transfer from the intermediate warm layer to the lower ice boundary.

Theoretical considerations suggest that the heat exchange between the lower ice surface and the warm intermediate layer, limited by the seasonal pycnocline from above and by the main pycnocline from below, can via the following:

- molecular heat exchange;
- turbulent heat exchange;
- convective heat exchange caused by salination of the surface layer at ice formation;
- double-diffusion convection through the seasonal pycnocline with the surface sub-ice layer;
- intrusion stratification through the lateral boundaries of the thermocline front.

Molecular heat exchange

The molecular heat fluxes calculated during the onset of ice formation in autumn of 1996 using TRANSDRIFT III data vary within 0.1 - 0.5 J $m^{-2} \cdot s^{-1}$. For this calculation, a simple formula (2) has been used (Turner, 1973):

$$Ft_m = Kt_m \cdot DT/DZ, \qquad (2)$$

where DT is the temperature difference through the pycnocline with a thickness DZ, and Kt_m is the coefficient of molecular temperature conductivity ($1.3 \cdot 10^{-7}$ m^2 s^{-1} for sea water). In the mass form the heat flux can be presented as:

$$Q_m = Ft_m \cdot r \cdot Cp \qquad (3)$$

where r is the sea water density and Cp is the heat capacity of sea water (approximately 4.19 J·g $°C^{-1}$. In the three months during which the fast ice edge becomes quasi-steady, about $2.26 \cdot 10^{-3}$ to $21.60 \cdot 10^{-3}$ KJ m^{-2} are transferred to the upper layer. This leads to a decrease in ice thickness of 2 to 7 cm (process 1, Figure 7).

Turbulent heat exchange

The criterion for different regimes of turbulent exchange is the Richardson gradient number. For polar regions, it can be written in the form

$$Ri = g \cdot bDS \cdot H_P \cdot DU^{-2} \qquad (4)$$

where g is gravitational acceleration, $DU \cdot H_P^{-1}$ is the gradient of the horizontal component of the current speed in the pycnocline, DS is the salinity difference in the seasonal pycnocline and b is the salinity compression coefficient. Calculations of the Richardson gradient number for the seasonal pycnocline presented in Table 1 were made for the stations on transect II (Transdrift

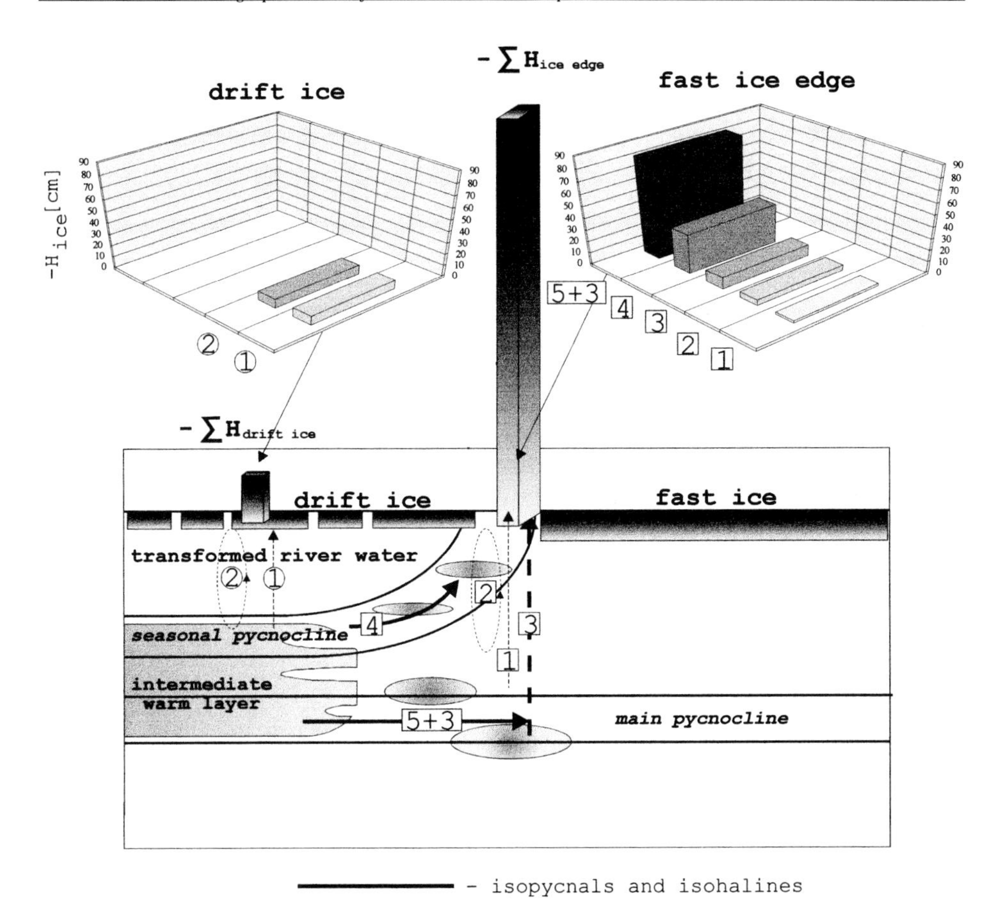

Figure 7: A scheme of the influence of the heat fluxes from the intermediate layer to the surface on the formation of ice conditions in the zone of river water spreading. 1 - molecular heat exchange, 2 - convective heat exchange at ice formation, 3 - double-diffusion convective heat exchange, 4 - heat exchange governed by isopycnic advection of intrusions to the surface, 5+3 - heat exchange governed by the double-diffusion convection at the upper periphery of isopycnic intrusions. The upper diagrams present the efficiency of each of the enumerated processes calculated to centimeters of ungrown ice.

Table 1: Estimates of the Richardson gradient number for different shear values for the horizontal component of current speed (Transdrift III, transect II data).

Station (KD95..)	48	49	50	51	52	53	55	54
DS	7.2	7.7	13.3	11.8	12.9	13.7	14.2	13.2
DU=2 cm/s	619	409	1121	1022	1194	2230	1858	5804
DU=15 cm/s	11	7	20	18	20	40	103	33

III) across the zone of river water spreading (Figure 2B). These estimates were based on the data collected by oceanographic stations during the onset of intensive ice formation and were calculated for two speed shear values. As a rule, exchange through the density boundary occurs at the molecular scale even at DU = 15 cm/s: Ri>15, (Narimousa et al., 1986; st. KD9550-KD9554, Table 1). Heat exchange through the pycnocline via this molecular-turbulent mode becomes possible only at significant shear speeds: 1.5<Ri<15 (st. KD9548-KD9549, Table I). The heat exchange coefficient increases from around 10^{-7} m^2 s^{-2} at the molecular level up to around 10^{-4} m^2 s^{-1} at the molecular-turbulent level of exchange (Foster, 1974; Krylov and Zatsepin, 1992). With salination during ice formation, turbulent heat transfer from the warm intermediate layer to the lower ice surface becomes possible (Ri<1.5). In this case, the exchange coefficient can increase up to about 10 m^2 s^{-1} (Foster, 1974). The absence of reliable observations of currents during autumn and winter prevents estimation of the turbulent heat exchange efficiency. However, it should be noted that in spite of the significant increase in the efficiency of the heat exchange at the turbulent-molecular and turbulent scales, the process is only of a local and occasional character. It is hardly effective at time scales of about 100 days.

Convective heat exchange

To estimate the convective heat exchange between the intermediate layer and the surface, we followed Zubov,s method (Zubov, 1963), which is based on the supposition that, at the density of the upper expanding layer, the stratification tends to become stable. This results in mixing that penetrates to the depth at which the upper layer density is equal to that of the lower one. As this takes place, the heat accumulated by the lower layer is equally redistributed between layers involved in mixing. Calculations of discrete approximations of the functions $S_k = S(I_k)$ and $I_k = I(H,S)$ for the stage of salinity convection were made using the following formulae:

$$I_k = \sum_{i=1}^{k} \frac{1.1(H_{i-1} - H_i) \cdot (S_{i-1} - S_i)}{S_{i-1}} \quad (5)$$

$$S_k = f(\rho_k, T_w) \quad (6)$$

$$\rho_k = \sum_{i=2}^{k} \left(\frac{\sum_{m=1}^{i-1} \rho_m}{2^{i-1}} + \frac{\rho_i}{2} \right) \quad (7)$$

$$H_k = H(I_k) \quad (8)$$

where i is the level number from the sea surface in CTD-sounding; k is the level up to which the convective mixing spreads; H_i, S_i, r_i are the depth, salinity and density at level i; H_k is the depth of convection penetration during spreading up to the level k; S_k and r_k are salinity and density of the convectively mixed layer; I_k is the ice thickness increase sufficient for convection penetration up to level k; T_w is the water's freezing temperature at a given salinity; the function f is the equation of sea water state solved numerically relative to salinity by the method of successive approximations.

Salinity, temperature and the depth of the convective mixing layer were selected by means of the discrete analogue of functions S_k, T_k, $H_k = F(I_k)$ calculated from formulas (5) - (8) and from the "actual" ice thicknesses obtained from Zubov's empirical formula (9) (Doronin and Kheisin, 1975; Zubov, 1963):

$$I_k = -25 + [(25 + I_0)^2 + 8S(-T)]^{1/2} \quad (9)$$

The sum of freezing degree-days S(-T) was calculated using mean daily air temperature from October to June 1994-1995 and 1995-1996 at the polar stations Terpyai-Tumus, Dunai, Tiksi, and Kotel'nyy (Figure 1). The convective heat exchange was estimated using the ratio given in (1).

The amounts of heat passing through the surface due to convective processes are rather small. Calculations show that density stratification in the river water outflow zone prevents the development of convective heat exchange (st. KD9502 and KD9503 in Figure 8A). However, at the periphery of the outflow zone, where density stratification is weaker, the convective heat exchange becomes much more efficient. In the outflow zone, about 10% of the heat accumulated in the intermediate layer can be transferred to the surface due to convective heat exchange. At the periphery, this value increases up to 40-60% (st. KD9504 in Figure 8A). As a rule, the convective heat exchange is more efficient at the beginning of the winter season up to the time the fast ice edge becomes quasi-steady. On average, during the first three months of ice formation, the amount of transferred heat is four times as large as during the next four months up to the time the ice growth ends (Figure 8A, position I, II). When the fast ice edge becomes stable, the ice growth is reduced by 27 and 7 cm at the periphery of the river water outflow zone and beneath it, respectively (process 2, Figure 7). At the time the ice growth ends, these values increase by 15 and 3 cm, respectively. On average, these values are even less at the periphery of the outflow zone and do not exceed 10 cm at the time the ice edge becomes stable.

Heat exchange governed by double-diffusion

The criterion for the possible occurrence of double-diffusion convection is a certain value of the density ratio, $Rr = bDS \cdot aDT^{-1}$, for the density boundary, where temperature makes an unsteady contribution to the density gradient. Here $b = 8 \cdot 10^{-4}$ is the coefficient of salinity compression, with dimensions inverse to salinity, $a = 7 \cdot 10^{-5}$ [°C^{-1}] is the temperature expansion coefficient, and DT and DS are the salinity and temperature differences at the boundary. In our case, these are the temperature and salinity differences across the upper boundary of the warm intermediate layer (the upper part of the seasonal pycnocline). It was found that Rr = 15 is the upper limit for the development of double-diffusion convection and accompanying step-like thermohaline structures (Fedorov, 1976; Huppert, 1971; Neshyba et al., 1971). At Rr > 15, exchange through the boundary surface becomes purely molecular. Estimates of the actual density ratio at the upper boundary of the seasonal halocline at the time preceding ice formation, Rr (only for transect I); at the beginning of ice formation, Rr_f; and at the time the fast ice edge becomes quasi-steady, Rr_i, are presented in Table 2. To calculate the density ratio, Rr_i, the thermohaline characteristics of the upper sub-ice layer calculated using formulae (4) - (8) were used.

Our estimates show that double-diffusion convection before ice formation is impossible (st. KD9502-KD9509, Table 2). During the initial period of ice formation, the double-diffusion convection is possible only at the periphery of the river outflow (st. KD9505, KD9548 in Table 2). When the fast ice edge becomes quasi-steady, the area expands (st. KD9502, KD9503, KD9505, KD9507, KD9509, KD9548, KD9549). However, in the central part of the river outflow zone, the double-diffusion is either extreme or absent during the entire winter season due to the high stability of the seasonal pycnocline.

The heat exchange values governed by double-diffusion have been calculated using a ratio obtained experimentally by Turner (1965, 1973). Based on the Turner´s experiments, Huppert (1971) suggested a formula for calculating the convective heat flux which expresses the dependence of this flux on the Rr value:

$$Ft = b \cdot Ft^{*} \cdot Rr^{-2}, \qquad (10)$$

where b = 3.8, and Ft^* is the heat flux through the boundary at a given temperature difference, DT, which is determined in Fedorov (1976) as:

$$Ft^* = 0.085\ (g \cdot a \cdot Kt_m{}^2/n)^{1/3} \cdot (DT)^{4/3} = g_m \cdot (DT)^{4/3}, \quad (11)$$

where Kt_m is the coefficient of molecular temperature conductivity, n is the kinematic coefficient of molecular viscosity, and g is gravitational acceleration. It follows from (10) and (11) that:

$$Ft = b \cdot g_m \cdot (DT)^{4/3} \cdot Rr^{-2} \quad (12)$$

Formula (12) has been used to estimate the heat fluxes when Rr<15.

Table 2: Estimates of the density ratio in the inversion area of the warm intermediate layer along transects I and II of the Transdrift III expedition.

Station (KD95..)	02	03	04	05	06	07	08	09
Rr	32.0	56.1	45.0	-	88.5	43.3	92.3	34.2
Rr_f	13.8	23.3	22.6	5.0	23.7	5.7	40.1	12.0
Rr_i	15.8	16.9	19.4	3.9	18.6	1.3	37.0	6.5
Station (KD95..)	48	49	50	51	52	53	55	54
Rr_f	13.4	22.7	63.4	92.8	42.9	29.4	46.7	27.6
Rr_i	2.2	11.3	43.6	64.1	27.9	20.1	35.3	21.1

Figure 8B, position I, presents the heat exchange values calculated from ratio (12) during the transition to a quasi-steady fast ice edge (transect I, Transdrift III expedition). Calculations were performed for the density ratios, Rr_f, which corresponded to the initial stage of ice formation. Actually, they represent the lower estimate of the possible heat exchange values for this period. The upper estimate corresponds to the value of the density ratio, Rr_i (position III, Figure 8B). On average, the latter is 1.5-2 times greater. But at the periphery of the river water outflow zone, double diffusion convection is capable of not only "transferring" practically all heat - the potential heat transfer is 5-20 times greater than the actual transfer (st. KD9505, KD9507 in Figure 8B, position III). Taking the periphery stations KD9507 and KD9548 as an example, by February 10th the upper estimates of the heat which can be transferred from the intermediate layer to the surface are 20 times larger than the lower estimates ($520 \cdot 10^4$ and $246 \cdot 10^4$ KJ/m^2, respectively). For the assumed ice cover characteristics, the actual ice growth at the periphery of the outflow zone is reduced by 20 cm when the fast ice edge becomes stable (process 3, Figure 7). The potential ice growth in the presence of a continuous heat flux is reduced by 84 cm (lower estimate). The upper estimates of the same values are 44 and 130 cm, respectively.

Let us draw some conclusions. For the strong density stratification observed in the river water outflow zone, the convective heat transfer is comparable to the molecular processes. Since double-diffusion is impossible in these regions, only these two processes, with an

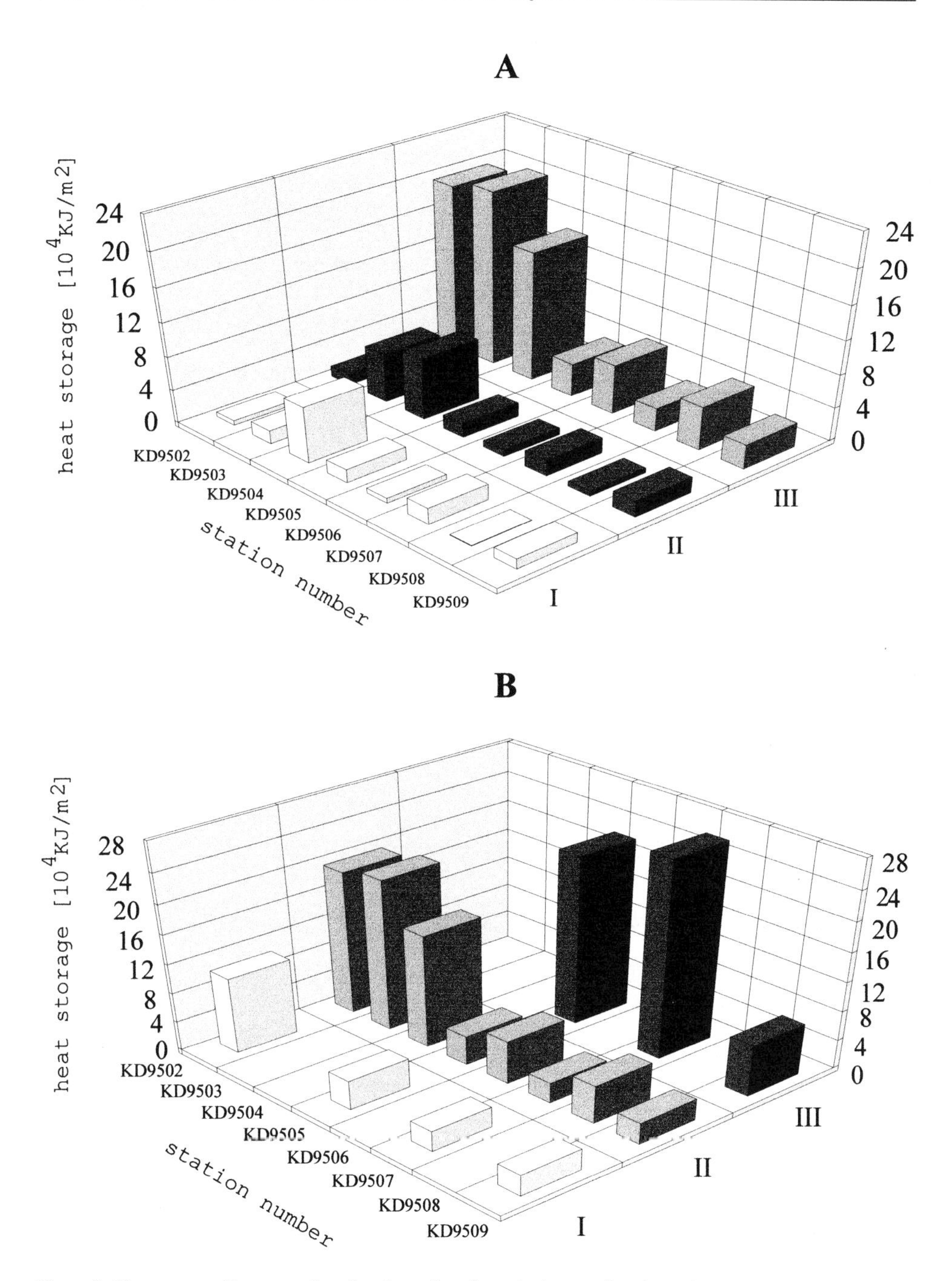

Figure 8: The amount of heat transferred to the surface from the intermediate layer due to both convection at ice formation (A) and double-diffusion convection (B), transect I of the Transdrift III expedition. A: I - by the time the fast ice edge attains a quasi-steady character, II - by the time the ice growth ends, III - the total heat supply of the intermediate water layer; B: I - by the time the fast ice edge attains a quasi-steady character (lower estimate), II- the total heat supply of the intermediate water layer, III - the upper estimate of a potential transfer.

insignificant contribution from the turbulent heat transfer, govern heat flux through the seasonal pycnocline to the surface (Figure 7, left diagram). The amount of heat transferred to the surface by these processes cannot influence the ice cover formation to the extent shown in Figure 4.

At the periphery of the outflow zone, convection is usually 5 times (and, as a maximum, an order of magnitude) more efficient than molecular heat exchange. However, the efficiency of the heat exchange governed by double-diffusion convection is twice as large as the salinity convection (right diagram, Figure 7). So, even when the fast ice edge becomes quasi-steady, all heat from the intermediate layer at the periphery of the outflow zone is transferred to the surface. In reality, however, the heat flux from the intermediate layer is small, on average $5 \cdot 10^4$ KJ m^{-2}, *i.e.* equivalent to 16 cm of unformed ice. However, one should remember that the potential heat exchange with the surface is extremely large. Its realization requires heat advection from the adjacent parts of the river discharge zone.

The existence of such advection was tested by numerous CTD-measurements performed during the expeditions Transdrift I (Kassens et al., 1994), Transdrift II (Kassens and Dmitrenko, 1995) and Transdrift III (Kassens et al., 1997). They demonstrated that the temperature distribution at the periphery of the river water outflow zone under the seasonal pycnocline was characterized by numerous inversions. The thickness of such warm and cold interlayers reached 10 m, and the temperature gradients at the boundaries of the interlayers reached 2-2.5 °C m^{-1} (Golovin et al., 1995). The most pronounced isopycnic temperature inversions were observed at the periphery of the freshened zone in the region west of Kotel,nyy Island during the autumn of 1995 (Figures 1, 9). Their formation was shown to be caused by the existence of the secondary thermoclinic front under the baroclinic surface discharge front (Golovin et al., 1995; Kassens and Dmitrenko, 1995; Figure 5). The mechanisms of thermoclinic front formation were considered in Discussion I. In the absence of horizontal density stratification, but in the presence of strong temperature stratification, any horizontal disturbance leads to the formation of intrusions. It is obvious that intrusion stratification of the thermoclinic secondary hydrofront and further advection of heat transferred by isopycnically spreading warm intrusions are the processes "pumping" heat from beneath the discharge lens to its periphery. This process is equally efficient during both the summer-autumn season (Figures 5A, 9A) and the period of young ice growth (Figures 5B, 9B).

The formation of intrusions at the front is a typically ageostrophic process. The upper estimate of its spatial scale is the Rossby baroclinic deformation radius: $\lambda = \frac{ND_0}{\pi f}$, where N is the Vaisala frequency, D_0 is the fluid layer thickness, and f is the Coriolis parameter. For the Laptev Sea, we have estimated λ=11.8 km. Actual observations of the spatial scale of warm intrusions near the thermoclinic front during the Transdrift II expedition (Kassens and Dmitrenko, 1995) showed that their horizontal dimensions could reach 4 kilometers and more. This estimate is comparable to the width of the river water outflow zone (about 150 km), since the process of intrusive destruction goes on continuously until complete degradation of the warm water layer. The high efficiency of this process permits the assumption that intrusive destruction of the warm layer is complete by the time the fast ice edge attains a quasi-steady character. All heat except for the heat already transferred to the surface via convective, turbulent, and molecular heat exchanges will be redistributed by advection to the periphery of the outflow zone.

Figure 5 displays the evolution of the outflow front from summer (A) to winter (B). Salination of the surface layer occurs at ice formation. As a result, some heat from the upper part of the intermediate layer can be isopycnically exchanged with the surface (Figure 5B, Figure 7, process 4). In order to estimate the intensity of this type of heat exchange, we used the ratios (1), (5) - (9) for the time during which the fast ice edge stabilizes. When calculating,

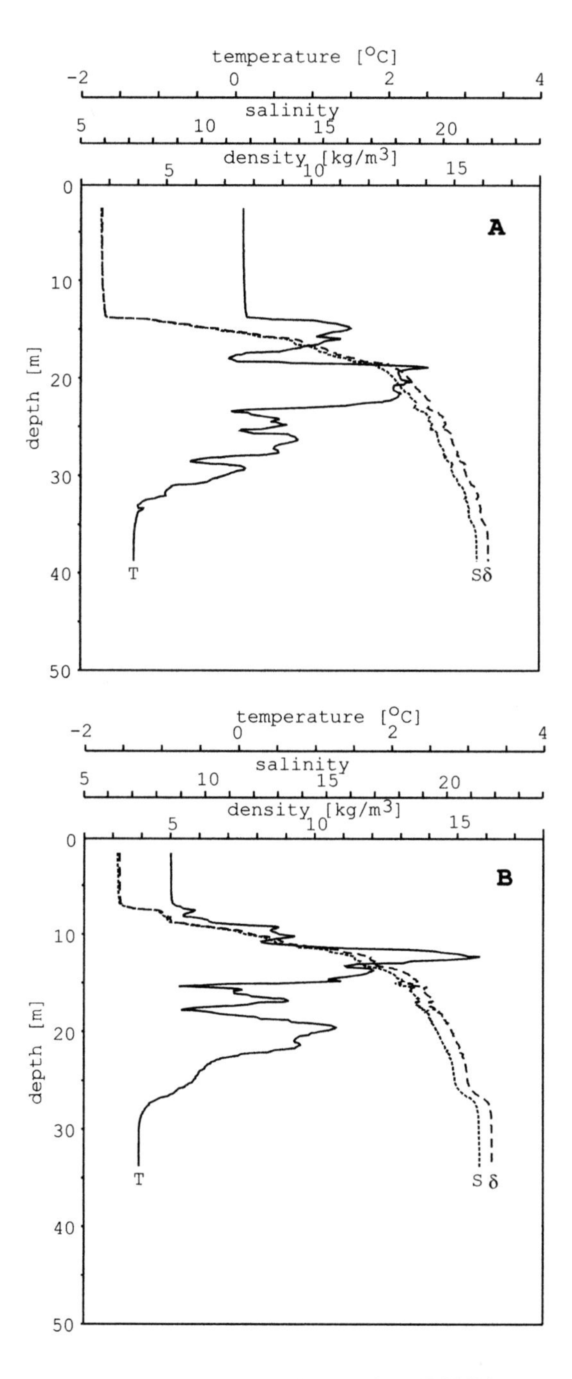

Figure 9: Vertical distribution of thermohaline characteristics at stations KD9519 in open water (A) and KD9554 under the young ice up to 25 cm thick (B).

we assumed the average advection rate to be 4 cm s^{-1}. It was also assumed that advection occurs in all directions with equal probability. It has been established that up to 25% of heat accumulated in the intermediate layer under the lens of freshened water can be transferred directly to the surface by the time the fast ice edge becomes quasi-steady at the periphery of the outflow zone. This results in a concentration of heat in the peripheral zone. The cumulative effect of all vertical heat exchanges with the intermediate layer (in the regions of the peripheral zone itself) amounts to, on average, $15 \cdot 10^4$ KJ m^{-2}. This is equivalent to 47 cm of unformed ice. The amount of heat transferred to the surface by isopycnic advection is twice as large as the amount transferred to the surface via vertical heat exchange.

The heat which remained in the intermediate layer after vertical and quasi-horizontal heat exchange was isopycnically transferred by horizontal advection beneath the periphery of the freshened water zone. Here the double-diffusion heat exchange processes are extremely efficient. Let us assume that, as shown above, the complete destruction of the intermediate warm layer via intrusive stratification occurs even prior to the fast ice edge becoming quasi-steady. Then the heat concentration beneath the periphery of the freshened zone can reach $26 \cdot 10^4$ KJ m^{-2} and more. This corresponds approximately to the lower estimate of the possibility of a double-diffusion heat transfer to the surface (process 5+3, Figure 7). The upper estimate was, as we recall, an order of magnitude greater.

The total heat transfer to the surface can reach $41 \cdot 10^4$ KJ m^{-2} and more by the time the fast ice edge becomes quasi-steady due to all acting processes: molecular and turbulent heat exchange, salinity convection, double-diffusion, "quasi-horizontal" advection of warm intrusions to the surface, horizontal advection of warm intrusions with "secondary" double-diffusion. This would be equal to 129 cm of unformed ice (Figure 7). It is thus not surprising that, in the three months after the onset of ice formation, the fast ice edge already becomes quasi-steady precisely at the periphery of the zone freshened by river discharge.

Conclusions

Figure 7 presents a scheme that summarizes the possible heat exchange mechanisms between the intermediate warm layer and the lower ice boundary. The diagrams above the scheme present the estimates of heat exchange efficiency in cm of unformed ice for all processes under study. A comparison with the actual observed ice cover types confirms our suggestions and provides justification for the following conclusions.

1 Formation of the fast ice edge in the eastern Laptev Sea depends to a great extent on processes determined by the spreading of river discharge. The quasi-steady position of the fast ice edge is confined to the periphery of the transformed river water spreading, and is governed by the large intensity of the heat exchange in this zone.
2 The fast ice edge is formed much further southward than the northern limit of the river discharge spreading. River water rich in suspended matter is located within the zone of drifting ice and the flaw polynya. Its incorporation into drifting ice intensively produced in the polynya triggers a complex of processes usually referred to as the "Transarctic ice transport of river discharge".

Acknowledgements

This article is one result of 4 years of research and expedition studies under the Russian-German Project "Laptev Sea System". The work of many Russian and German scientists in various disciplines of natural science served as a basis for obtaining the results; the authors express their most sincere gratitude to all of them. The results presented here could not have been obtained without the selfless work, under heavy weather and difficult ice conditions, of the crews of the R/V "Professor Multanovsky", the icebreaker "Kapitan Dranitsyn" and their captains, V. Danilenko and O. Agofonov, and the MI-8 helicopter crews of the Tiksi Joint Air Team headed by I. Litovchenko. The authors are grateful to all these people for their fruitful work. Much of this article was initiated and written by the Russian and German authors in 1995 and 1996 in GEOMAR, Kiel. The authors gratefully acknowledge assistance of the German Ministry of Education, Science, Research and Technology (BMBF Grant No. 525 4003 0G0517A). The Russian contribution was also supported by the Russian Ministry for Science and Technical Policy (LAPEX). The authors greatly appreciate the help of the Russian and German scientific coordinators of the project "Laptev Sea System" - Prof. J. Thiede and Prof. L. Timokhov, whose attentive and friendly attitude has enabled the authors to successfully fulfill these studies. Special thanks are due to Dr. V. Zelensky, Head of the Tiksi Territorial Administration for Hydrometeorology for providing meteorological information used in the article and Dr. V. Grishenko, Deputy Director of the AARI for the composite ice charts required for this study. The English manuscript was kindly improved by Pier Paul Overduin.

References

Dethleff, D., D. Nürnberg, E. Reimnitz, M. Saarso and Y.P. Savchenko (1993) East Siberian Arctic Region Expedition ,92: The Laptev Sea - Its Significance for Arctic Sea ice Formation and Transpolar Sediment Flux. Ber. Polarforsch., 120, 1-44.

Dmitrenko, I. and TRANSDRIFT Shipboard Scientific Party (1995) The Distribution of River Run-Off in the Laptev Sea: The Environmental effect. In: Russian-German Cooperation: Laptev Sea System, Kassens, H., D. Piepenburg, J. Thiede, L. Timokhov, H.-W. Hubberten and S. Priamikov (eds.), Ber. Polarforsch. 176, 114-120.

Dmitrenko, I, J. Dehn, P. Golovin, H. Kassens and A. Zatsepin (in press) Influence of sea ice on under-ice mixing under stratified conditions: potential impacts on particle distribution. Estuarine, Coastal and Shelf Science, 46.

Doronin, Y.P. and D.E. Kheisin (1975) Sea Ice (in Russian). Gidrometeoizdat, Leningrad, 317 pp.

Eicken H., E. Reimnitz, V. Alexandrov, T. Martin, H. Kassens and T. Viehoff (1997) Sea-ice processes in the Laptev Sea and their importance for sediment export, Continental Shelf Research, 17, 205-233.

Golovin, P.N., V.A. Gribanov and I.A. Dmitrenko (1995) Macro- and Mesoscale Hydrophisical Structure of the Outflow Zone of the Lena River Water to the Laptev Sea. In: Russian-German Cooperation: Laptev Sea System, Kassens, H., D. Piepenburg, J. Thiede, L. Timokhov, H.-W. Hubberten and S. Priamikov (eds.), Ber. Polarforsch. 176, 99-106.

Fedorov, K.N. (1976) Fine Thermohaline Structure of the Oceanic Waters (in Russian). Leningrad, Gidrometeoizdat, 184 pp.

Fedorov, K.N. (1991) About Thermohaline Characterisrics of Oceanic Fronts (in Russian). In: K.N. Fedorov. Izbrannye Trudy po Fizicheskoi Okeanologii, Gidrometeoizdat, Leningrad, 106-111.

Fofonoff, N.P. and R.C.Jr. Millard (1983) Algorithms for computation of fundamental properties of seawater. Unesco technical paper in marine science, 44, 53 pp.

Foster, T. D. (1974) Thi hierarchy of convection. In: Processus de formation des eaux oceaniques profondes en particulier en Mediterranee Occidentale, Colloques Intern., du CNRS, Paris, 215, 237.

Huppert, H.E. (1971) On the stability of double-diffusive layers. Deep-Sea Res., 18, 10, 1005-1022.

Kassens, H. & I. Dmitrenko (1995) The TRANSDRIFT II Expedition to the Laptev Sea. In: Laptev Sea System: Expeditions in 1994, Kassens H. (ed.), Ber. Polarforsch. 182, 1-180.

Kassens, H., V. Karpiy (eds.) and the Shipboard Scientific Party (1994) Russian-German Cooperation: The TRANSDRIFT I Expedition to the Laptev Sea, Ber. Polarforsch. 151, 168 pp.

Kassens, H., I. Dmitrenko, L. Timokhov and J. Thiede (1997) The TRANSDRIFT III Expedition: Freeze-up Studies in the Laptev Sea. In: Laptev Sea System: Expeditions in 1995, Kassens H. (ed.), Ber. Polarforsch. 248, 1-192.

Krylov, A.D. and A.G. Zatsepin (1992) Frazil ice formation due to difference in heat and salt exchange across a density interfase. J. Marine Systems, 3, 497-506.

Kuz,mina, N.P. (1980) About Oceanic Frontogenesis (in Russian). Izv. Akad.Nauk SSSR, Fizika atmosfery i okeana, 16, 10, 1082-1090.

Mac Vean, M.K. and J.D. Woods (1980) Redestribution of scalars during upper ocean frontogenesis. Quart. J. Roy. Met. Soc., 106, 448, 293-311.

Narimousa, S., R.R. Long and S.A. Kitaigorodskii (1986) Entrainment due to turbulent shear flow at the interface of a stably stratified fluid. Tellus, 38A, 1, 76-87.
Neshyba, S., V.T. Neal and W.W. Denner (1971) Temperature and conductivity measurements under Ice Island T-3. J. Geophys. Res., 76, 33, 8107-8119.
Oceanographic Tables (in Russian) (1975) Gidrometeoizdat, Leningrad, 476 pp.
Turner, J.S. (1965) The coupled turbulent transport of salt and heat across a sharp density interfase. Int. J. Heat and Mass Transfer, 8, 5, 759-767.
Turner, J.S. (1973) Bouyancy effects in fluids. Cambridge Univ. Press, 367 pp.
Woods, J.D. (1980) The generation of thermohaline finestructure at fronts in the ocean. Ocean modelling, 32, 1-4.
Zakharov, V.F. (1966) The Role of Trans-Fast Ice Polynyas in Hydrological Regime of the Laptev Sea (in Russian). Okeanologiya, Moscow, 6, 6, 1014-1022.
Zubov, N.N. (1963) Arctic Ice. U.S. Naval Oceanographic Office and Amer. Meteor. Soc., 491 pp.

Step-Like Vertical Structure Formation Due to Turbulent Mixing of Initially Continuous Density Gradients

A. Zatsepin[1], S. Dikarev[1], S. Poyarkov[1], N. Sheremet[1], I. Dmitrenko[2], P. Golovin[2] and H. Kassens[3]

(1) P.P. Shirshov Instute of Oceanology, 23 Krasikova, 117851 Moscow, Russia

(2) State Research Center - Arctic and Antarctic Research Institute, 38 Bering St., 199226 St. Petersburg, Russia

(3) GEOMAR Forschungszentrum für marine Geowissenschaften, Wischhofstrasse 1-3, D 24248 Kiel, Germany

Received 3 March 1997 and accepted in revised form 9 February 1998

Abstract - The results of a simple laboratory experiments on the stirring of continuously stratified fluid by oscillating vertical rods are described and analyzed. It is discovered that, if turbulent stirring is rather weak, the strong linear stratification is transformed into a step-like structure during each experimental run. This structure consists of nearly homogeneous layers separated by thin density interfaces. The initial average thickness of the layers depends quasi-inversely on the buoyancy frequency and is growing with time during the experiment. Thus, the number of layers decreases with time mainly because of merging of layers. The analysis of our laboratory results combined with the analysis of previous studies give strong support for the suggestion that the "staircase" structure formation may be not an exotic phenomenon in the shallow summer pycnocline of Arctic seas under the influence of drifting ice floes (Golovin et. al, 1996). The disintegration of such a pycnocline into a series of turbulent layers separated by thin density interfaces may enhance the vertical transport of sediments and increase the rate of frazil ice formation (Krylov and Zatsepin, 1992).

Introdution

In the polar Arctic seas a shallow (3-10 m) and strong (10-25 sigma-t units) halo-pycnocline is formed during the summer due to ice melting processes and/or river run-off. The drifting ice floes often produce a considerable velocity shear and turbulent mixing across such a pycnocline. On the base of ship observations in Kara sea, it was shown (Golovin et al., 1996) that during these events the vertical density structure of the shallow pycnocline may be changed from "monolith" to the "step-like" form. Such changes in density structure should influence the exchange of properties between the upper and the lower layers, particularly, the heat, salt and sediment fluxes across the pycnocline zone. As a result, the rates of surface and underwater (frazil) ice formation (or melting) may be also changed.

It was suggested that the transformation of the initially continuous density stratification of the halocline into a series of homogeneous sublayers divided by extremely sharp density interfaces is due to the turbulence instability mechanism in strongly stratified fluid (Phillips, 1972; Posmentier, 1977). The qualitative explanation of this instability mechanism is based on the nonlinear dependence of vertical buoyancy flux on the density gradient in turbulent stratified flow. If the stratification is strong (Richardson number is enough high), the slight local enhancement of the density gradient considerably reduces the turbulent exchange coefficient, so that the buoyancy flux in this region is decreased. The local decrease of the buoyancy flux will tend to further increase of the density gradient. So the perturbation will amplify. As the result of the amplification of small disturbances (both positive and negative) initially continuous stratification may disintegrate into a series of quasi-homogeneous layers separated by sharp density interfaces. If the stratification is rather weak (Richardson number is below the critical

----

In: Kassens, H., H.A. Bauch, I. Dmitrenko, H. Eicken, H.-W. Hubberten, M. Melles, J. Thiede and L. Timokhov (eds.) Land-Ocean Systems in the Siberian Arctic: Dynamics and History. Springer-Verlag, Berlin, 1999, 93-99.

one) the dependence of buoyancy flux on the density gradient is quasi-linear and local perturbations of density profile tends to be smoothen by turbulent diffusion. In this case layers does not form.

This effect was demonstrated in different laboratory and numerical experiments. It was shown that in both cases of shear (Barenblatt et al., 1993; Krylov, 1993; Kan and Tamai, 1994) and shear-free (Ruddick et al., 1989; Park et al., 1994) turbulent flows the formation of step-like density profile structure may occur. In order to explain the mentioned above observations made in the Kara sea the results of laboratory experiments with shear turbulent flows were used by Golovin et al. (1996). It was obtained that in order to satisfy the laboratory criteria of the pycnocline splitting $0.02 < Ri_u < 2$ (Krylov, 1993) (here $Ri_u=(g' \cdot h)/U^2$ -the Richardson number, g' - the reduced gravity, based on the density difference across the pycnocline, h - the thickness of the pycnocline, U - the horizontal velocity difference across it), U must be approximately twice of the ice drift velocity. In other words there must be enough strong current in the water layer below the pycnocline of opposite direction to the direction of ice drift. Although it is quite possible (there was no direct velocity measurements), the suggestion was made that the critical Richardson number Ri_{crit} for step-like structure formation may be a monotonously growing function of the Reynolds number for turbulence (the higher is the Reynolds number, the higher is the value of critical Richardson number), at least for low values of the Reynolds number (Nishida and Yoshida, 1994). In real ocean conditions the Reynolds number for turbulence is larger then in the most of laboratory experiments, so if the mentioned above suggestion is true, the critical Richardson number may also be larger. In these circumstances the shear across the pycnocline, required for the step-like structure formation may be smaller.

In order to check the assumption of the Ri_{crit} dependence on Re and to study more about conformities of the step-like structure formation in turbulent stratified fluid, we provided a series of shear-free experiments described below.

Experemental set-up

The scheme of preliminary experimental set-up is shown on the Figure 1. Here (1) is the tank ($25*16*30$ cm^3), made from 2.0 cm thick organic glass sheet, and filled by linearly stratified salt (NaCl) water solution. The turbulent mixing is produced by the system of horizontally oscillating grids (2). Approximately similar method of mixing was used first by Ruddick et al. (1989). In most of experimental runs there were six grids situated on the same oscillating rod (3) at the distance of about 3.5 cm from each other and from opposite small side walls of the tank. Each grid consists of six cylindrical vertical glass rods 0.7 cm in diameter with the distance of about 2.8 cm from each other. The oscillations are produced by electric motor with eccentric drive (4). The period of oscillations is fixed: T = 2 s, the amplitude is changed from one run to another. The larger is the amplitude of oscillations the higher is the intensity of turbulence. In order to make visible the density inhomogeneities in the turbulent stratified fluid the simple shadowgraph device (5) is used. The shadowgraph picture is monitored by video camera and photo camera.

Observation and results

Twenty five experimental runs were provided with six grids on the rod in order to observe different regimes of turbulent mixing in initially linearly stratified fluid. The initial salinity gradient was changed from 1.0 to 12.5 unit/cm and the amplitude of the oscillations - from 0.5

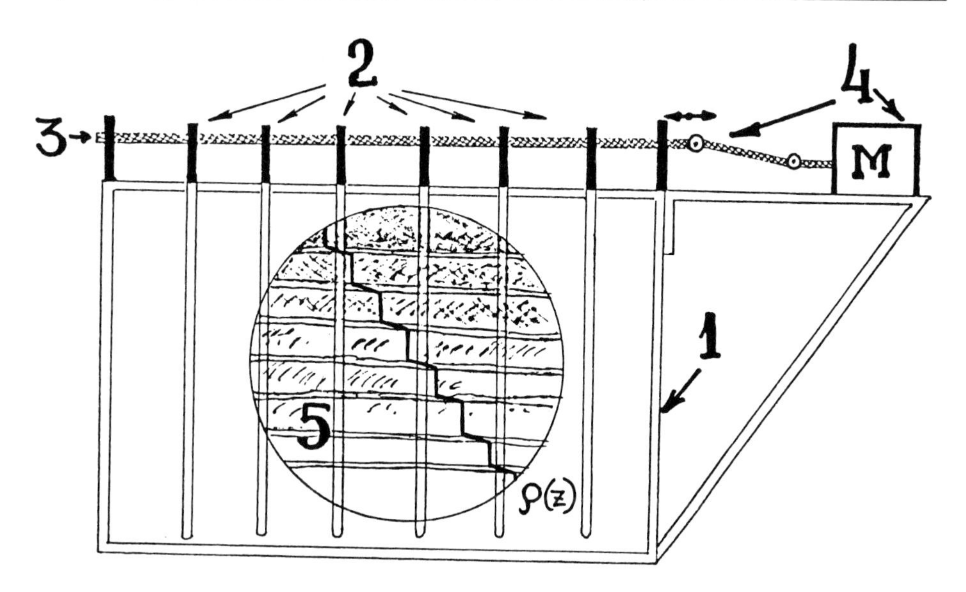

Figure 1: The scheme of experimental set-up (see the text for its detailed description).

to 1.7 cm. It has been discovered that, if turbulent mixing is rather weak, the linear stratification is transformed into the step-like structure (Figures 2 a,b) during each experimental run. This structure consists of nearly homogeneous layers separated by thin density interfaces. The initial thickness of the layers quasi-inversely depends on the density gradient and is growing with time during the experiment. Thus, the number of the layers decrease with time (Figure 2c) mainly because of merging of adjacent layers (vanishing of interfaces) or vanishing of thinner layers (merging of interfaces). Due to the no flux boundary conditions the upper and bottom layers are the most rapidly growing layers. So, the sub-final stage of mixing is the two-layered stratification (Figure 2d).

The main dimensional and non-dimensional parameters are presented in Table1. Here $N = (g\rho^{-1}\delta\rho/\delta z)^{0.5}$ and $\delta\rho/\delta z$ - the buoyancy frequency and the density gradient of the initial stratification, g - the gravity acceleration, $\rho \approx 1$ - the water density, H - the initial thickness of homogeneous layers for each experimental run. The absence in the Table 1 of H value for some of experimental runs means that no distinct layers were observed during these runs.

In order to make quantitative analysis of the data obtained to compare it with the results of Park et al. (1994) who also measured the initial thickness of homogeneous layers (unfortunately the results of Ruddick et al. (1989), are mostly qualitative so it is impossible to compare them with ours), we have analysed our data in terms of non-dimensional parameters. In accordance with the mentioned above authors we choose the Reynolds number of the rod, Re, and the overall Richardson number, Ri, defined as

$$Re = UD/\nu, \quad Ri = (ND/U)^2$$

where $U = 4A/T$ - the velocity scale, A - the amplitude of oscillations, D - the diameter of the rod, $\nu \approx 0.01$ cm^2/s - the kinematic viscosity of the fluid, as the most important non-dimensional parameters. We divided our experimental data and the data of Park et al. (1994) into two parts: the runs with the formation of layers and without them. In our experiment most runs was with layer formation, because our aim was only to determine the boundary in parametric space between to different regimes of mixing, but not to go deeply into the regime with no layer formation. All the points for both experiments are presented on the Figure 3 in the (Ri,Re) plane. For both experiments (that basically are related to different ranges of the Reynolds number) it is possible to separate roughly the points with layering from the points without it by the strait solid line on the Figure 3. This line results in the following power dependence:

$$Ri_{crit} = 1.97 \cdot 10^{-3} \cdot Re^{0.88} \qquad (1)$$

Another important result of our analysis is the parameterization of the initial layer thickness (see Figure 4). It follows from this figure, that

$$H = 2.0 \cdot (U/N) \qquad (2)$$

This relation corresponds well with similar parameterization of Park et al. (1994). It should be mentioned that the parameter U/N is physically analogous to the well-known Ozmidov lengthscale (Ozmidov, 1965). This lengthscale characterize the largest vertical size of three dimensional vortices in the stably stratified fluid, that is the largest vertical size of overturning. However, we did not measured the turbulent energy dissipation and are unable to confirm that the layer thickness is proportional to the Ozmidov length scale, although it seems to be so.

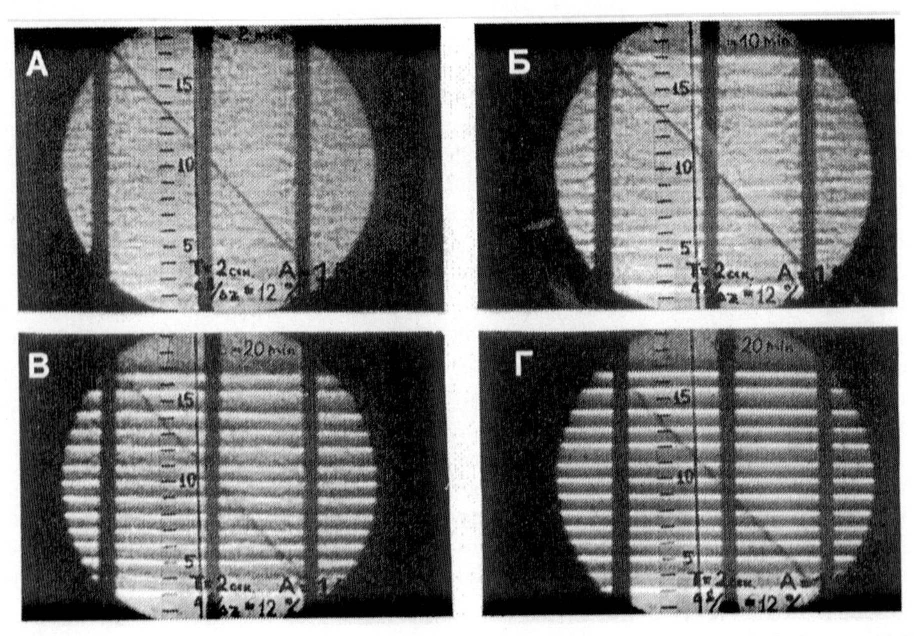

Figure 2: The successive shadowgraph pictures of turbulent stratified fluid in the tank during one of the experimental runs with the step-like structure formation (N = 3.0 rad/s; A = 0.75 cm). A) t = 10 min; Б) 30 min; B) 240 min; Г) 633 min.

In order to find out the role of spatial homogeneity of stirring in the process of layering, we provided the supplementary series of experimental runs in which the number of grids was changed from 6 to 1 for the same other conditions. It was obtained, that the layering event and the initial thickness of the layers does not depend on the number of grids and thus, on the spatial homogeneity of stirring. The decreasing of a number of grids was impressed only in the increasing (quasi-linear) of time period during witch the step-like structure was formed. Next step will be to provide experimental runs with intermittent stirring in order to investigate is there or not any critical time interval of stirring which controls the layer formation process.

Table 1: The dimensional and non-dimensional basic experimental parameters.

Rep	U/N,cm	N,rad/c	U,cm/c	H,cm	Re	Ri
1	0.80	1.8	1.45	1.2	101	0.75
2	0.41	3.2	1.33	0.7	93	2.83
3	0.45	3.2	1.45	0.9	101	2.38
4	0.51	3.1	1.6	1.1	112	1.83
5	0.58	3	1.75	1.8	122	1.44
6	0.68	2.9	2	3	140	1.03
7	0.79	2.8	2.23	3	156	0.77
8	1.03	2.6	2.7	4	189	0.45
9	0.67	2	1.35	1	94	1.07
10	0.84	1.9	1.6	1.5	112	0.69
11	1.05	1.8	1.9	2.2	133	0.43
12	1.46	1.6	2.35	4	164	0.22
13	0.85	1.4	1.2	-	84	0.66
14	1.03	1.4	1.45	1.3	101	0.45
15	1.23	1.3	1.6	2.4	112	0.32
16	1.72	1.1	1.9	3	133	0.16
17	1.3	1	1.3	1.4	91	0.28
18	1.61	0.9	1.45	1.7	101	0.18
19	1.85	0.7	1.3	2.4	91	0.14
20	3	0.5	1.5	-	105	0.05
21	2	0.8	1.6	-	112	0.12
22	0.38	2.6	1	-	70	3.31
23	0.6	2	1.2	-	84	1.36
24	1.8	1.5	2.7	-	189	0.15
25	1.32	2.5	3.3	-	231	0.28

Discssion and conclusions

One of the necessary conditions of the density step-like structure formation is that the energy supply is enough low and only sufficient to overturn a part of the stratified column. But even in this condition the formation f layered density structure from continuous one occurs only when the turbulent vertical mass flux negatively depends on the density gradient. Only in this case the small disturbances of density profile produced by turbulent stirring may grow up (Phillips, 1972; Posmentier, 1977). Another very important factor of layered structure formation and keeping out, is the existence of molecular diffusivity. Due to it the layers may be mixed to the quasi-homogeneous state. Moreover, the quasi-equilibrium state of the whole layered structure is apparently achieved due to molecular diffusivity, that controls the minimal thickness of the density interface and have influence on the mass and admixture fluxes between the mixed layers (Krylov and Zatsepin, 1992).

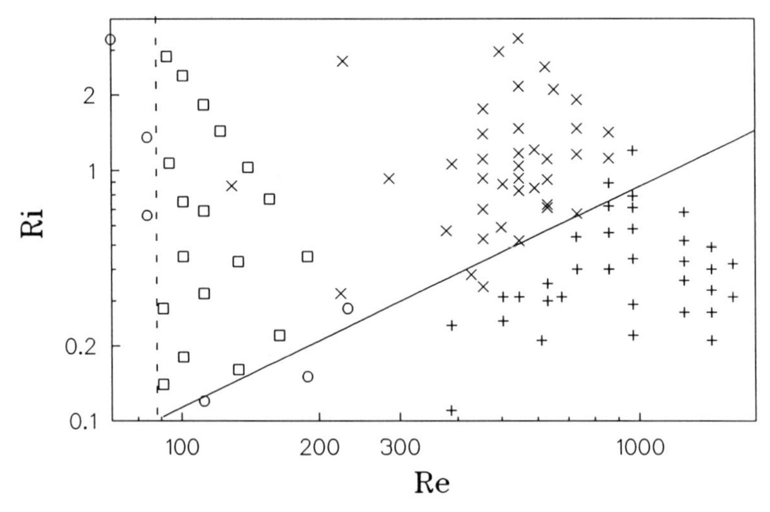

Figure 3: The diagram of the experimental runs in Re - Ri plane: - with layers, o - without layers (our experiment), X - with layers, + - without layers (Park et al., 1994). The dashed vertical line separates the experimental runs without turbulence and layers (Re < 90) from those with turbulence and layers. The inclined solid line separates experimental runs with turbulence and without layers from those with turbulence and layers.

The results of simple experiments described above basically confirm the arguments by Phillips and Posmentier for the turbulence instability and step-like structure formation in the turbulent stratified fluid. It was observed visually that the formation of the initial quasi-periodical disturbances on the density profile is due to the instability of turbulent vertical exchange process. The local overturning first of all occurs near the oscillating rods, where the formation of mixed layers begins. The initial vertically organized structure spreads laterally in the form of quasi-homogeneous intrusions. Sooner or later after the beginning of stirring the layered structure reaches the quasi-stationary stage. During this stage the continuous flux of mass and admixture through the whole water column is maintained due to turbulent mechanism in the mixed layers and predominantly due to molecular one across the density interface.

When the Richardson number is low and the Reynolds high no layering is observed because Phillips and Posmentier mechanism does not work in weakly stratified fluids. The critical Richardson number of layered structure formation is a growing function of the Reynolds number at least for Re = 10^2 - 10^3. Further experimental studies are required in order to obtain $Ri_{crit}(Re)$ for larger values of turbulent Reynolds number (10^3 - 10^4, based on the r.m.s. velocity and the integral lengthscale of turbulence) which seems to be more typical for the real ocean conditions. If the similar dependence will be obtained the application of turbulent instability mechanism for the interpretation of the observed pycnocline splitting (Golovin, et al., 1996) may become more convincing. The vertical scale of homogeneous layers expressed by semi-empirical formula (2) for U = 0.1-1 cm/s and N = $10^{-1}s^{-1}$, gives the realistic estimate of typical layer thickness in the step-like pycnocline: H = 0.2-2 m.

The results of our laboratory experiment combined with the results of previous studies give

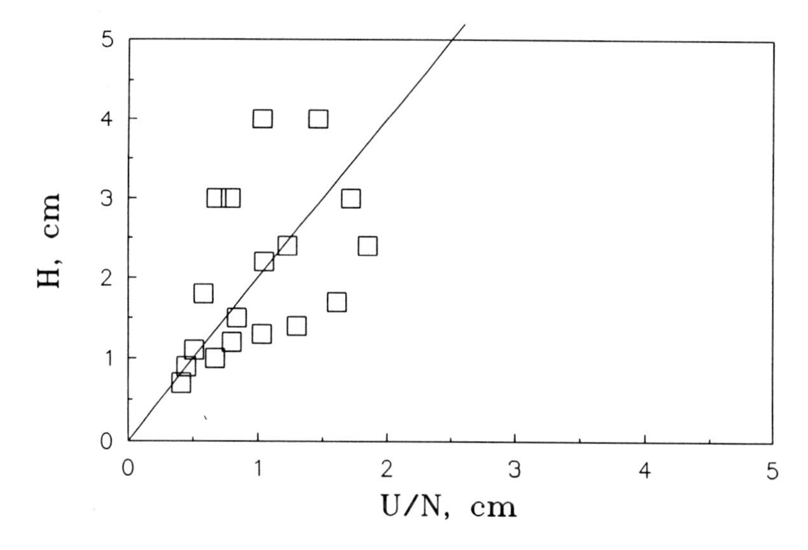

Figure 4: The dependence of the initial layer thickness H on the scaling parameter U/N.

strong support for the suggestion that the step-like structure formation does not depend on the nature of turbulence (shear or shear-free) and it may be not an exotic phenomenon in the shallow summer pycnocline of Arctic seas under the influence of drifting ice floes. The disintegration of such a pycnocline into a series of turbulent layers separated by thin density interfaces may enhance the vertical transport of heat, salt and sediments and increase the rate of underwater frazil ice formation (Krylov and Zatsepin, 1992).

References

Barenblatt, G.I., M. Bertsch, R. Dal Passo, V.M. Prostokishin and M. Ughi (1993) A mathematical model of turbulent heat and mass transfer in stably stratified shear flow. J.Fluid Mech., 253, 341-358.

Golovin, P., I. Dmitrenko,.A. Zatsepin and A. Krylov (1996) The "splitting" of seasonal pycnocline due to the ice edge motion (in Russian). Meteorologiya i Gidrologiya, 6, 82-91.

Kan, K. and N. Tamai (1994) Direct measurements of the mutual- -entrainment velocity at a density interface. In: International symposium on the stratified flows, Grenoble, France, pre-print 4, v.2.

Krylov, A.D. (1993) Laboratory study of the density interface structure in a shear flow. In: The annual international session of the WG "Laboratory modelling of dynamic processes in the ocean", Abstracts, Moscow, Institute for problems in mechanics, Russian Academy of Sciences, 47.

Krylov, A.D. and A.G. Zatsepin (1992) Frazil ice formation due to differential heat and salt exchange across a density interface. J. of Marine Systems, 3, 497-506.

Nishida, S. and S. Yoshida (1994) Stability criterion of a stratified two layer shear flow with hyperbolic-tangent velocity profile. In: Preprints of fourth symp. on stratified flows, V.2, Grenoble, France.

Ozmidov R.V. (1965) On the turbulent exchange in the stably stratified ocean (in Russian). Izvestiya Akademii Nauk SSSR, Phisika Atmosphery i Okeana, 1, 853-860.

Park, Y-G., J.A. Whitehead.and A. Gnanadeskian (1994): Turbulent mixing in stratified fluids: layer formation and energetics. J. of Fluid Mech., 279, 279-311.

Phillips,O.M (1972) Turbulence in a strongly stratified fluid is unstable? Deep-Sea Research 19, 79-81.

Posmentier, E.S. (1977) The generation of salinity fine structure by vertical diffusion. J. Phys. Ocean. 7, 298-300.

Ruddick,B.R., T. McDougall and J.S. Turner (1989) The formation of layers in a uniformly stirring density gradient. Deep-Sea Research, 36, 597-609.

Dissolved and Particulate Major and Trace Elements in Newly Formed Ice from the Laptev Sea (Transdrift III, October 1995)

J.A. Hölemann[1], M. Schirmacher[2] and A. Prange[2]

(1) GEOMAR Forschungszentrum für marine Geowissenschaften, Wischhofstrasse 1-3, D 24148 Kiel, Germany

(2) GKSS Forschungszentrum, Max-Planck-Strasse, D 21502 Geesthacht, Germany

Received 14 February 1997 and accepted in revised form 13 February 97

Abstract - Dissolved and particulate elements were determined in new ice, nilas and young ice of the Laptev Sea. Sampling was carried out during the autumn freeze-up period in 1995.
The median contents of metals in ice-bound particles are comparable to those found in unpolluted riverine and marine sediments and, thus, give no indication of anthropogenic heavy metal pollution.
In contrast, dissolved trace metal concentrations observed in the new ice north of the Lena delta (Tumatskaya branch) were highly enriched in Mn, Fe, Zn, Cd and Pb in comparison to average concentrations in freshwater of the Lena river. We suggest that remobilization of metals from the particulate phase is the cause of elevated dissolved metal concentrations in the young ice. The release of dissolved metals within the ice should also have an effect on the ice ecology and the river-sea transport of heavy metals within the Arctic.
The observed spatial variations within the major-element vs. aluminum ratios (e.g Mg/Al, Ca/Al, K/Al and REE/Al) of ice-rafted sediments could be directly related to the geochemical signatures of different fluvial sediment sources. These results suggest that riverine particulate trace elements are effectively incorporated into the ice and so can be used as tracer for the identification of shelf source areas of ice rafted sediments within the Arctic.

Introduction

Sediments incorporated into newly formed ice can strongly influence the transport and distribution of trace elements. In particular, particle reactive substances like heavy metals may adsorb onto organic and inorganic ice-bound particles and, thus, can be effectively transported from the inner shelf to the Arctic Ocean. As a consequence, the Siberian Arctic shelves, compared to ice free shelves, are less effective barriers for the land-ocean transport of man-made and natural substances. This fact has consequences also for the pollution assessment of Siberian river systems and coastal waters and their possible influence on the Arctic ecosystem. Most recently this aspect has been discussed in several publications mainly focusing on the transport of radioactive substances by ice (Barrie et al., 1992; Weeks, 1994; Pfirman et al., 1995). However, published data on actual contaminant levels are sparse. As an example, data on heavy metal concentrations in sea ice are given by Melnikov (1991), but the published data are only mean values for non-specified regions in the Russian Arctic.

The complex structure and history of sea ice within the Transpolar Drift is another difficulty for the study of contaminant transport by ice. Atmospheric deposition of metals, depletion and enrichment processes caused by surface melting and freezing and biological cycling are some of the processes that make it nearly impossible to differentiate between contaminant sources that are responsible for the pollution of multi-year pack ice. As a result, the assessment of contaminant inputs during ice formation can only be made in source areas of the Arctic pack ice cover.

Recent investigations have shown that the Laptev Sea is one of the key regions for the understanding of processes connected to the incorporation of sedimentary particles into the ice

In: Kassens, H., H.A. Bauch, I. Dmitrenko, H. Eicken, H.-W. Hubberten, M. Melles, J. Thiede and L. Timokhov (eds.) Land-Ocean Systems in the Siberian Arctic: Dynamics and History. Springer-Verlag, Berlin, 1999, 101-111.

(Pfirman et al. 1990; Nürnberg et al., 1994; Eicken et al. 1996; Lindemann et al., this volume, Golovin et al., this volume). First estimations give export sediment rates of 4 x 10^6 t yr^{-1} (Eicken et al., 1996). The off-shelf transport of ice-rafted sediments (IRS) mainly depends on the initial ice formation and sediment entrainment in shallow water during fall freeze-up, the ice growth regime in the flaw lead during the course of the winter and the volume of ice exported beyond the limit of the minimum summer ice extent (Eicken et al., 1996).

The Russian-German Transdrift III expedition into the Laptev Sea that was carried out during the fall freeze-up in 1995 provided the opportunity to study the initial phase of ice formation in the Laptev Sea.

Sea ice formation off the Lena delta started in early-October 1995. The median content of particles in ice measured in this region was 148.9 mg/l (Lindemann et al., this volume). Compared to the suspended matter concentration in the underlying water column this is a more than 20 fold enrichment of particles in the newly formed ice (Lindemann et al., this volume).

Our primary purpose was (1) to assess heavy metal contamination levels in new ice (2) to advance our understanding of trace element cycling connected to the formation of new ice (3) to provide information about the value of trace elements in Arctic Sea ice as a source-area tracer for the reconstruction of ice drift patterns.

Material and methods

Sampling was carried out in different ways, depending on the type of ice. Thin surface ice like pancake ice, dark nilas, slush and grease ice were sampled with a net from the bow of the ships. Surface ice of more than 1 cm but less than 20 cm thickness was sampled using a diving platform. Samples of thin ice were taken with the help of Teflon-chisel, plastic scoop and cash. All samples were melted on board in a clean air cabinet (class 100) in polystyren boxes. Prior to sampling the boxes were rinsed with ultrapure water and leached with diluted HCl for two days.

Surface ice with more than 20 cm thickness was sampled tens off the windward side of the ship meters off the ship with ice coring devices made of plastic and stainless steel. To avoid contamination, the outer layer of the ice core was removed with a ceramic knife. The cleaned cores were stored and allowed to melt in cleaned polystyrene boxes before filtration.

Nirogen pressure filtration was carried out in a clean air cabinet on board Kapitan Dranitzyn. For the determination of particulate trace elements we used 0.4 μm pre-cleaned polycarbonate filters (leached 3 times for 6 hours in 3% subboiled HCl at 60° C temperature). The filtrate was acidified to pH 2 and stored in acid cleaned polyethylene bottles.

Final analysis was performed in the Research Center Geesthacht including salt-matrix separation and pre-concentration of the water samples followed by analysis with total reflection X-ray fluorescence (TXRF), a modified procedure of the conventional energy dispersive XRF (Prange, 1983). Coupled plasma mass spectrometry (ICP-MS) completed the investigations. All analytical procedures in Geesthacht were carried out in clean rooms (class 1000).

Elements in sea water measured by TXRF include V, Mn, Fe, Ni, Cu, Zn, Se, Mo, Cd, U and Pb.

About 30 particle bound elements including heavy metals and the REE were measured in sea-ice by means of TXRF and ICP-MS. For the digestion procedure filters were transferred to Teflon vessels and 4 ml HNO_3, 0,5 ml H_2O_2,1 ml HF and an internal standard (Ga) were added. After running a temperature program (stepwise escalation from 200 W to 800 W) for 20 minutes samples were evaporated to dryness. After 2 ml HCL were added the sample was heated again.

The accuracy and precision of our analytical methods for the determination of dissolved metals were tested using CASS and NASS water standards (near shore and open ocean seawater references, supplied by the National Research Council of Canada) and BCSS and MESS-1 sediment reference material. The ratio of the results of our replicate analyses to the certified values are in the range 0.95-1.13 for the dissolved elements. For Ni some excess was obtained (1.23). However, the values for all other elements are similar to the statistical uncertainties of the concentrations in the standards themselves. Quality assurance for the analysis of particulate elements relied on standard reference materials from the National Research Council of Canada (BCSS, MESS-1). Non-certified elements of the reference material were measured with neutron activation analysis (NAA). For most elements the recovery was between 90 and 110 %. Analytical precision was (i) better than 5% for: K, Ca, Sc and Fe, (ii) 5% - 10% for: Be, Na, Mg, Al, Ti, V, Mn, Co, Ni, Zn, Rb, Sr, Nb, Mo, and Cs, (iii) 10% - 15% for: Li, S, P, Cu, As, Cd, Sb, W, Th and U (iv) 15% - 18% for: Cr, Sn and the REE.

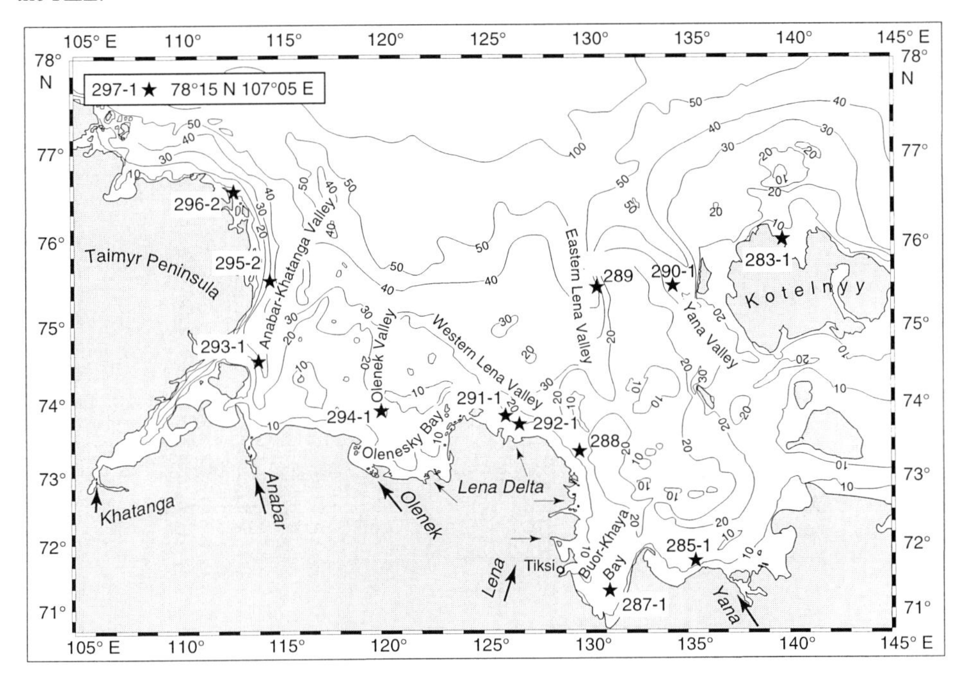

Figure 1: Sampling stations in the Laptev Sea in October 1995.

Results

Particulate major and trace elements in newly formed sea ice (Laptev Sea, October 1995)

Sampling was carried out at 14 stations in the Laptev Sea (Figure 1). Pancake ice, nilas and young ice (nomenclature according to WMO, 1970) were the most common ice types sampled during the autumn freeze-up period. Ice with low contents of ice-rafted particles (< 5 mg kg^{-1}) was sampled in the northern and eastern Laptev Sea (Stations 297, 289, 290, 285, 287 and 288). Higher contents between 5 and 20 mg kg^{-1} were found in the Khatanga-Anabar- region (Stations 295, 293). Young ice sampled at Stations 291 and 292 near the Tumatskaya branch, one of the major branches of the Lena river, and in the Olenesky Bay (Station 294) had high particle contents in excess of 100 mg kg^{-1}, reaching maximum contents above 500 mg kg^{-1}.

Box plots of the particulate element content from all samples are given in Figure 2. To provide information about spatial differences within the trace element composition two regions, the Khatanga-Anabar-Olenek (SW Laptev Sea) and the Lena-Yana region (SE Laptev Sea), were separately documented in Table 1.

Maximum concentrations of Ni, Cu and Pb were usually measured in ice with low particle content (< 2 mg kg^{-1}). The highest arsenic content (> 0.53 μmol g^{-1}) was found in ice covering the northern Yana Valley (Station 290), a region that is also characterized by arsenic enrichment within the surface seafloor sediments (Hölemann et al., 1995, Hölemann et al., subm.). In contrast, cadmium contents above 3 μg g^{-1} were only found in the particle loaden ice near the Tumatskaya branch.

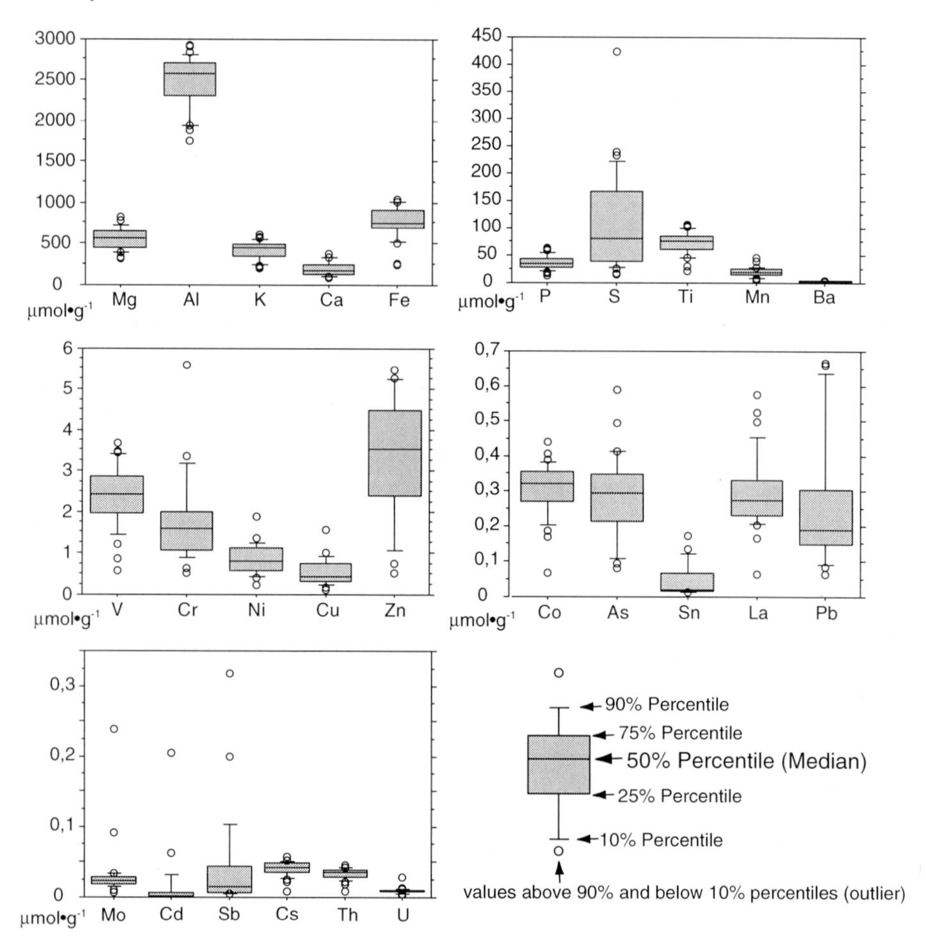

Figure 2: Box plots of particulate element concentrations in ice-rafted sediments.

Dissolved trace elements in newly formed ice (Laptev Sea, October 1995)

Molten ice samples were filtered through polycarbonate filters with a pore size of 0.4 μm. For comparative studies additional sampling of water and suspended matter from the underlying water column was carried out. Median concentrations for selected trace metals are given in Table 2. It is evident that particle loaden young ice sampled at stations 291 and 292 was strongly enriched in dissolved metals compared to concentrations observed in the water column at the station and in the river water. Extremely strong enrichment of dissolved Fe, Mn, Cd and Zn were only found in ice with high particle contents north off the Tumatskaya branch. New and young ice from other regions show only moderate enrichments of Cd, Zn and Pb and low concentrations of Fe, Mn, Cu and Ni.

Table 1: Median concentrations and regional distribution of particulate matter in ice (IRS) and particulate major and trace elements in new and young ice. Italic numbers give the median absolute derivation (MAD) observed for each variable and the number of analyses. Regional differences in the mineralogical composition of ice-rafted sediments are best documented by the element aluminium ratios of Mg, Ca, K and the Rare Earth Elements (REE).

	IRS mg kg^{-1}	**Mg** µmol g^{-1}	**Al** µmol• g^{-1}	**P** µmol• g^{-1}	**S** µmol• g^{-1}	**K** µmol• g^{-1}	**Ca** µmol• g^{-1}	**Cr** µmol• g^{-1}	**Mn** µmol• g^{-1}	**Fe** µmol• g^{-1}
SW Laptev Sea	13.0	659	2649	30	73	407	255	1.65	20.5	889
(n=6)	*7.9*	*11*	*66*	*4*	*8*	*38*	*30*	*0.21*	2.2	*57*
SE Laptev Sea	87.0	547	2572	38	105	489	142	1.65	23.2	713
(n=16)	*85.5*	*57*	*161*	*9*	*68*	*38*	22	*0.33*	*4.3*	*55*
Total	12.8	563	2584	34	80	443	174	1.61	20.5	747
(n=30)	*12.2*	*89*	*133*	*8*	*50*	*69*	*49*	*0.47*	*5.4*	*125*

	Co µmol• g^{-1}	**Ni** µmol• g^{-1}	**Cu** µmol• g^{-1}	**Zn** µmol• g^{-1}	**As** µmol• g^{-1}	**Rb** µmol• g^{-1}	**Sr** µmol• g^{-1}	**Mo** µmol• g^{-1}	**Cd** µmol• g^{-1}	**Sn** µmol• g^{-1}
SW Laptev Sea	0.37	0.89	0.43	3.76	0.30	1.07	1.91	0.020	0.001	0.022
(n=6)	*0.04*	*0.07*	*0.03*	*0.78*	*0.01*	*0.08*	0.04	*0.002*	*0.0002*	*0.008*
SE Laptev Sea	0.32	0.77	0.54	4.05	0.31	1.21	1.91	0.025	0.004	0.021
(n=16)	*0.03*	*0.20*	*0.21*	*0.97*	*0.06*	*0.23*	0.20	*0.005*	*0.002*	*0.003*
Total	0.32	0.80	0.43	3.53	0.29	1.15	1.93	0.023	0.002	0.022
(n=30)	*0.03*	*0.31*	*0.21*	*1.13*	*0.07*	*0.18*	0.16	*0.005*	*0.001*	*0.008*

	REE µmol• g^{-1}	**Pb** µmol• g^{-1}	**Th** µmol• g^{-1}	**U** µmol• g^{-1}	**Mg/Al** mol/mol	**Ca/Al** mol/mol	**REE/Al•10^3** mol/mol	**K/Al** mol/mol
SW Laptev Sea	1.27	0.18	0.030	0.010	0.25	0.096	0.495	0.149
(n=6)	*0.11*	*0.03*	*0.001*	*0.002*	*0.009*	*0.010*	*0.031*	*0.014*
SE Laptev Sea	1.42	0.24	0.039	0.009	0.21	0.058	0.556	0.194
(n=16)	*0.18*	*0.09*	*0.002*	*0.001*	*0.014*	*0.012*	*0.055*	*0.012*
Total	1.38	0.19	0.035	0.009	0.22	0.073	0.539	0.180
(n=30)	*0.21*	*0.06*	*0.006*	*0.001*	*0.023*	*0.022*	*0.077*	*0.022*

The salinity of the sea ice was measured on split samples of the ice cores. Because the salinity within ice cores can be highly variable, the Mo concentration was used as a more accurate method to determine the salinity of the sample in which trace elements were analyzed (Prange and Kremling, 1985). This method is based on the conservative mixing behaviour of

Mo that is reflected by a close relationship between the salinity of water and ice and the concentration of Mo (in the Laptev Sea; r^2=0,839, n=122) (Figure 3.a).

Plotting the concentration of iron, manganese and cadmium against molybdenum (Figures 3 b-d) it is evident that concentration maxima of Fe, Mn and Cd are connected to low salinities. The measured salinities in the ice cores were about 4 (station 291) and 6 (station 292). This points to the river Lena (Tumatskaya branch) as the dominant source for low salinity, particle loaden ice. This is further supported by the $\partial^{18}O$ isotopic signature of the ice at these stations (Eicken, person. commun.).

Near the river mouths new ice is preferably formed at the contact between river water near the point of freezing and more saline water with a temperature below 0°C (Golovin et al., this volume). The Laptev Sea was ice free at the end of September 1995 (Lindemann et al., this volume). New ice formation started in early October. Thus, the fast floating ice field north off the delta was younger than 3 weeks. The meteorology during this period was characterized by calm conditions without precipitation. These circumstances let us suggest that the ice floe moved within a short period of time from its area of origination near the river mouth offshore regions with a higher salinity of the underlying water column (29.1 at Station 291).

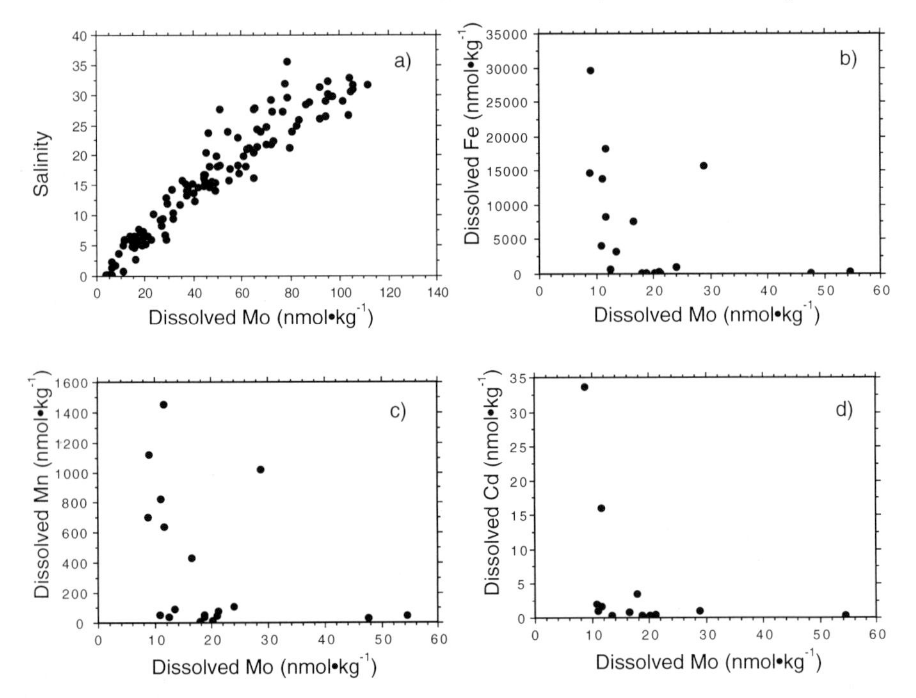

Figure 3, a-d: The good correlation between the salinity of a water sample and the dissolved molybdenum concentration (3 a) allow to use the Mo content as a proxy for the salinity of an ice sample. Figures 3 b-d show that high concentrations of dissolved Fe, Mn and Cd are usually observed in ice with Mo concentrations below 15 nmol kg^{-1}, representing a salinity below 10. New and Young ice with low salinities was characteristic for the region north off the Tumatskaya branch (Lena river).

Discussion

Trace elements in new ice, nilas and young ice.

With the exception of Zn and Cd the median concentration levels of particulate trace metals were close to values typical for surface sediments and suspended matter from the Laptev Sea

(Nolting et al., 1996; Hölemann et al., subm.). Compared to natural background values for marine coastal sediments (Chester, 1990) the ice-rafted sediments (IRS) show no indication for anthropogenic metal enrichment. Contents for Zn and Cd are higher than the median suspended matter and sediment contents that were measured during the expeditions Transdrift I-III (Hölemann, unpubl. data), but close to the published world average contents for river particulate material (Chester, 1990). Elevated cadmium contents of more than 0.027 μmol g^{-1} in suspended matter were also recorded near the Lena river delta in summer 1994 (Transdrift II). Especially the Zn and Cd content of newly formed ice north off the Lena delta suggest that riverine particulate matter from the Tumatskaya branch is the main source for IRS at Stations 291 and 292.

High particle contents in ice from Stations 291 and 292 are associated with high concentrations of dissolved metals, e.g. Fe, Mn, Cd, Zn and Pb. Comparing the median to average concentrations measured in the Lena river water (Table 2) the approximate enrichment factors within the dissolved phase are: 7-25 (Fe), 7 (Mn), 15-30 (Cd), 30-90 (Zn) and 6-40 (Pb). Cu and Ni are only enriched by a factor of 2. Ice from other regions in the Laptev Sea shows a depletion in Fe, Mn, Pb, Cu and Ni. Dissolved metal concentrations as high as the concentrations found in the newly formed ice are only known from polluted rivers like the Rhine (Salomons and Förstner, 1984). The observed high metal concentrations in new ice near the Tumatskaya branch - with no industrial discharge nearby - can only be explained by remobilization processes from the particulate phase within the ice.

Table 2: Dissolved (< 0.4 μm) concentrations of Fe, Mn, Cd, Zn, Pb, Cu and Ni in newly formed ice, water of the Lena river, and the water column at Station 291, north off the Tumatskaya branch (Lena river). Dissolved concentrations given for ice samples are median values.

	Fe nmol•kg^{-1}	Mn nmol•kg^{-1}	Cd nmol•kg^{-1}	Zn nmol•kg^{-1}	Pb nmol•kg^{-1}	Cu nmol•kg^{-1}	Ni nmol•kg^{-1}
Lena river water (Martin et al., 1993)	410	_	0.037-0.07	5.4	0.08	9.4	5.1
Lena river water (Transdrift III, Station 41H)	1494	99.3	_	15.4	0.51	11.9	5.9
Water column at Station 291 (2m, salinity = 29.1)	428	52.0	_	9.7	0.23	6.5	8.2
Ice at Stations 291 and 292	10585 *5497* *(n=9)*	706 *317* *(n=9)*	1.09 *0.67* *(n=7)*	479.1 *353* *(n=9)*	3.08 *1.91* *(n=9)*	17.03 *11.3* *(n=9)*	10.94 *7.3* *(n=9)*
Ice all other stations	281 *143* *(n=10)*	51.9 *23* *(n=10)*	0.44 *0.08* *(n=6)*	47.7 *28* *(n=10)*	0.20 *0.06* *(n=10)*	4.93 *1.22* *(n=10)*	4.56 *2.38* *(n=10)*

New ice formation and trace metal cycling

Highest particle contents in newly formed ice (gray ice and gray-white ice) were found north off the Lena delta near the freshwater outflow of the Tumatskaya branch. The median value for the relative particle content in ice from this region was 148.9 mg kg^{-1} (Lindemann et al., this volume). The maximum particle content in ice from this region analyzed in this study was 250 mg kg^{-1}. Higher relative contents of particulate Zn, S and Cd and lower contents of As together with the isotopic signature of the ice suggest that the ice bound particles north off the Tumatskaya branch are of riverine origin. This is supported by the observation that formation of frazil ice and subsequent incorporation of suspended matter is most intensive at the contact zone between cold freshwater and saline water masses near the freezing point (Dmitrenko et al., 1996). Rising frazil can effectively scavenge suspended matter from a turbid water column and subsequently transport it to the newly forming sea ice cover (Weeks and Ackley, 1982; Reimnitz et al., 1992).

During the initial phase of the freeze up period sea ice can float rather fast from coastal areas to offshore regions with higher salinities of the underlying water column. This would increase the salinity within the porous young ice and as a consequence influence the adsorption processes of trace elements like Cd. (Salomons, 1980).

However, we suggest that mobilization of iron and manganese from the dissolution of metal-hydroxide grain coatings is the cause of high dissolved concentrations of Fe, Mn, Zn, Pb, Cd, etc. (Olsen et al., 1982). One possible explanation for the initiation of this mechanism is the decomposition of organic matter - mainly limnic diatoms - and a subsequent change from an oxic to a reducing microenvironment (low pH or Eh) within the brine channel system. This would result in a dissolution of the metal-hydroxide grain coatings. Unfortunately, we were not able to carry out pH-Eh measurements within the brine. An example for the close connection between the concentration of dissolved iron and other metals in the newly formed ice is given in Figure 4.

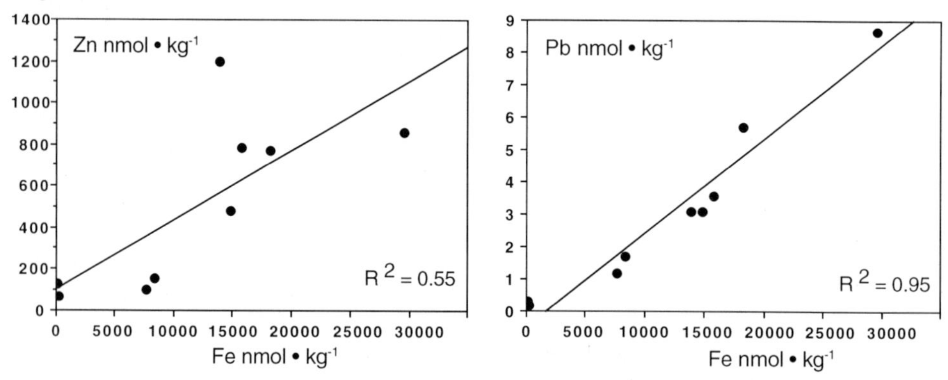

Figure 4: Concentration of dissolved Fe, Zn and Pb in newly formed ice from stations 291 and 292 (north off the Lena delta). The good correlation between Fe and Pb and supports the assumption that high concentrations of dissolved heavy metals in the particle loaden ice are caused by a dissolution of iron oxyhydrate mineral coatings. The correlation between Fe and Zn is not so pronounced but still statistically significant (Rank Spearman Correlation $R^2 = 0.58$; $P = 0.03$).

Although, a control of trace metal sorption by the organic matter was not observed in the Laptev Sea (Garnier et al., 1996), high amounts of dissolved organic matter (DOC) within the brine could be another possible cause for the desorption of metals. The average DOC concentrations (650 µmol) measured in the Lena river are among the highest values reported in the world's rivers (Cauwet and Sidorov, 1996). Riverine humic substances (which comprise about 60 to 80% of the dissolved organics) have a strong capacity for binding metals, in particular Cu and Ni (reviewed in Olsen et al., 1982). Both metals only show a slight enrichment in the ice samples. This stands in contrast to the assumption that DOC plays a major role in the desortption of particulate metals. On the contrary, the data indicate that perhaps organic complexation of Cu and Ni might result in the binding of these metals to particulate organic matter and, thus, to the removal from the dissolved phase.

Based on our data we have so far not been able to distinguish one of these processes as a principle cause for the observed high dissolved metal concentrations within the young ice. However, it can be concluded that the pulse like release of heavy metals to the brine, ice and the subsequent drainage of the brine to the underlying water column has a strong impact on the biological community living in and under the ice. In addition, ice related trace element cycling can significantly modify the transport of metals from the rivers to the sea and the dispersion of hazardous metals and metalloids in the Arctic environment.

Geochemistry of ice-rafted sediments as a tracer for ice drift patterns

The modern sedimentary regime in the Laptev Sea is dominated by the fluvial input of the rivers draining the Siberian platform. Especially the sediment inputs by the Lena river in the east and the Khatanga river in the west control the mineralogical and geochemical compositions of suspended matter and surficial sediments in the southern Laptev Sea. The Khatanga, like the Yenisey, drains a plateau with Triassic flood basalts. Basalts have a unique geochemical signature compared to granite, shales or average continental crust, because they are enriched in e.g. Mg, Ca and depleted in elements like K, and the REE. This should make the normalized element contents (e.g. Mg/Al, Ca/Al, K/Al and REE/Al ratio) a good tracer for the fluvial input from the Khantanga river. This assumption is supported by geochemical and mineralogical field data from the Laptev Sea. As an example, highest Mg/Al and Ca/Al ratios in the surficial sediments were observed in the SW Laptev Sea near the mouth of the Khatanga (Hölemann et al., unpubl. data). Suspensions from this river system are also characterized by high amounts of Mg-bearing pyroxene (Lisitzyn, 1996) that are also found in the sediments of the SW Laptev Sea (Behrens et al., 1996).

The geochemical data from the SW and SE Laptev Sea (Table 1) indicate that the Mg/Al and Ca/Al ratios observed in ice-rafted sediments reflect this general distribution pattern with higher ratios in the western and lower ratios in the eastern Laptev Sea. In contrast K/Al and REE/Al are higher in ice-rafted sediments from the SE Laptev Sea (Figure 5). This supports the hypothesis that the geochemical signature of ice-rafted sediments from the Laptev Sea mainly reflect the chemical composition of river-borne sediments and, thus, make the geochemical fingerprint of ice-rafted sediments a useful tracer for the reconstruction of ice drift patterns in the Laptev Sea and the Transpolar Drift.

The major difference between ice from the southern Laptev Sea and the northern Laptev Sea is that IRS contents in newly formed ice in offshore regions are generally lower. At low particle contents (< 2 mg kg^{-1}) the particle assemblage is characterized by a higher proportion of autochthonous organic matter, like ice algae. In contrast to mineral particles, organic particles have different chemical compositions and a higher capacity for complexation and chelation. This can explain the higher content of particulate S, Cu, Ni and Pb in new ice with low particle contents.

As a consequence the geochemical fingerprint of "clean", newly formed ice from the northern, deeper parts of the Laptev Sea seems to be determined mainly by biological processes. However, a maximum content of As (> 0.47 µmol g^{-1}) was measured in low-particle new ice in the northern Yana valley, a region that is characterized by As enrichment in surficial seafloor sediments (Hölemann et al. 1995; Hölemann et al. subm.). Elevated contents of arsenic (> 0.80 µmol g^{-1}) were also measured within the pack ice of the Transpolar Drift (Wollenburg, unpubl. data) and in sea ice north off the Laptev Sea (~ 0.40 µmol g^{-1}; Hölemann, unpubl. data) during the Polarstern cruise ARK XI/1. In contrast, pack ice from stations near the East Siberian Sea stations showed contents lower than 0.13 µmol g^{-1}

This gives further evidence that resuspension of seafloor sediments in the northern Laptev Sea is an active entrainment process resulting in the incorporation of sediments into the growing ice cover. The mechanism should be more pronounced during winter storm periods with strong resuspension of sediments in the ice-free polynya region and frazil ice formation due to supercooling of the water column.

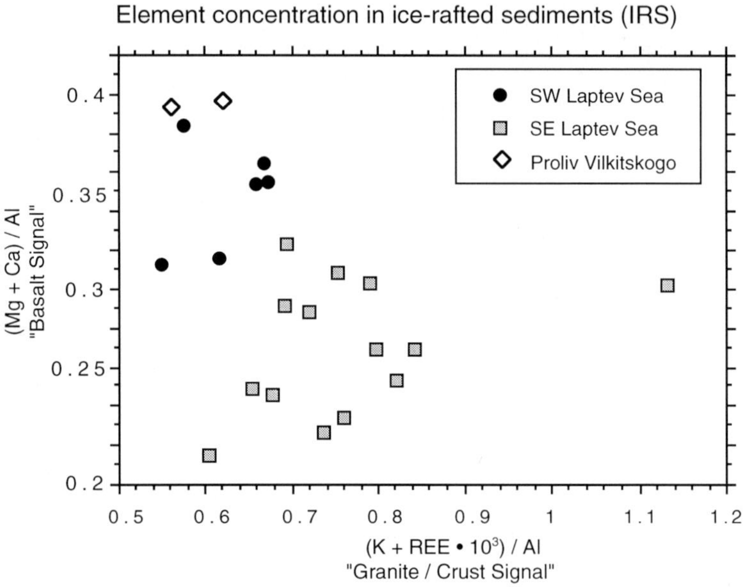

Figure 5: Regional differences in the Major and minor element composition of ice rafted seiments from the NW and SE Laptev Sea. The distinct differences are caused by different IRS source areas. The NW Laptev Sea is influenced by the discharge of rivers draining the Putorana basalt plateau. The geochemical signature of sediments, IRS and suspended matter from the SE Laptev Sea is a mixture of a large number of different lithologies (Lena river runoff plus coastal erosion) that resembles the signature of average weathered earth crust.

Acknowledgements

This work was carried out within the framework of the German-Russian cooperative program "Laptev Sea System" funded by the Russian and German ministries for science and technology. Dr. H. Kassens and Dr. I. Dmitrenko are greatly acknowledged as chief scientists during the expedition into the Laptev Sea. We wish to thank the captain and the crew of the Russian icebreaker Kapitan Dranitzyn for their support. We also thank our Russian and German colleagues, especially Konstantin Tyshko who helped us to walk safely on thin ice. Dr. K. Kremling and Dr. M. Kriews are thanked for valuable comments on the manuscript.

References

Barrie, L. A., D. Gregor, B. Hargrave, R. Lake, D. Muir, R. Shearer, B. Tracey and T. Bidleman (1992) Arctic contaminants: sources, occurrence and pathways. Sci. Tot. Environ., 122, 1-74.

Behrends, M., B. Peregovich and R. Stein (1996) Terrigenous sediment supply into the Arctic Ocean: Heavy mineral distribution in the Laptev Sea. Ber. Polarforsch., 212, 37-42.

Cauwet, G. and I. Sidorov (1996) The biogeochemistry of Lena river: organic carbon and nutrients distribution. Mar. Chem., 53, 211-227.

Chester, R. (1990) Marine geochemistry. Unwin Hyman Ltd, London.

Dmitrenko, I., V. Gribanov, P. Golovin and H. Kassens (1996) Hydrological causes for sea-ice transport of river discharge. Terra Nostra, 9, 1-2.

Eicken, H., E. Reimnitz, V. Alexandrov, T. Martin, H. Kassens and T. Viehoff (1996) Sea-ice processes in the Laptev Sea and their importance for sediment export. - Cont. Shelf Res., 12, 2 , 205-233.

Garnier, J.-M., J.-M. Martin, J.-M. Mouchel and K. Sioud (1996) Partitioning of trace metals between the dissolved and particulate phases and particulate surface reactivity in the Lena River estuary and the Laptev Sea (Russia). Mar. Chem., 53, 269-283.

Hölemann, J. A., M. Schirmacher and A. Prange (1995) Transport and distribution of trace elements in the Laptev Sea: First results of the Transdrift expeditions. Ber. Polarforsch., 176, 297-302.

Hölemann, J. A., M. Schirmacher, A. Prange and H. Kassens (subm.) Major and trace element distribution in surficial and ice rafted sediments of the Laptev Sea (Siberia). Est. Coast. Shelf Sci.

Lindemann, F. and J. A. Hölemann, (this volume): - :

Lisitzyn, A. P., (1996) Oceanic sedimentation: lithology and geochemistry. American Geophysical Union, Washington, D.C., 1-400.

Martin, J. M., D. M. Guan, F. Elbaz-Poulichet, A. J. Thomas and V. V. Gordeev (1993) Preliminary assessment of the distributions of some trace elements (As, Cd, Cu, Fe, Ni, Pb and Zn) in a pristine aquatic environment: the Lena River estuary (Russia). Mar. Chem., 43, 185-199.

Melnikov, S. A. (1991) The state of the Arctic Environment .- Report on heavy metals. Arctic Centre Publications No. 2., Arctic Centre, Rovaniemi, 82-153

Nolting, R. F., M. van Dalen and W. Helder (1996) Distribution of trace and major elements in sediment and pore waters of the Lena Delta and Laptev Sea. Mar. Chem., 53, 285 - 299.

Nürnberg, D., I. Wollenburg, D. Dethleff, H. Eicken, H. Kassens, T. Letzig, E. Reimnitz and J. Thiede (1994) Sediments in Arctic sea ice: Implications for entrainment transport and release. - Mar. Geol., 119, 185-214.

Olsen, C. R., N. H. Cutshall and L. L. Larsen (1982) Pollutant-particle associations and dynamics in coastal marine environments: a review. Mar. Chem., 11, 501-533.

Pfirman, S., M. A. Lange, I. Wollenburg and P. Schlosser (1990) Sea ice characteristics and the role of sediment inclusions in deep-sea deposition: Arctic - Antarctic comparison. In: Geological History of the Polar Oceans: Arctic Versus Antarctic, U. Bleil and J. Thiede (eds.), Netherlands:. Kluwer Academic Publishers, 187-211.

Pfirman, S. L., H. Eicken, D. Bauch and W. F. Weeks (1995) The potential transport of pollutants by Arctic sea ice. Sci. Total Environ. 159, 129-146.

Reimnitz, E., J. L. Marincovich, M. McCormick and W. M. Briggs (1992) Suspension freezing of bottom sediment and biota in the Northwest Passage and implications for Arctic Ocean sedimentation. Can. J. Earth Sci., 29, 693-703.

Salomons, W. (1980) Adsorption processes and hydrodynamic conditions in estuaries. Environ. Technol. Lett. 1, 356-365.

Salomons, W. and U. Förstner (1984) Metals in the Hydrocycle. Springer Verlag, Berlin, 1-349.

Weeks, W., and Ackley, S. F. (1982) The growth, structure, and properties of sea ice. U.S. Army, Cold Regions Research, Monograph No. 82-1, 1-136

Weeks, W. (1994) Possible roles of sea ice in the transport of hazardous material. In: Workshop on Arctic Contamination 1993, 8, I. A. R. p. Committee (ed.), Anchorage, Alaska: Arctic Research of the United States, 34-52.

WMO, (1970) WMO Sea-ice nomenclature. WMO Secretariat of the Meteorological Organization, Monograph No.259. TP.145, Geneva-Switzerland.

Particle Entrainment into Newly Forming Sea Ice – Freeze-Up Studies in October 1995

F. Lindemann[1], J.A. Hölemann[1], A. Korablev[2] and A. Zachek[2]

(1) GEOMAR Forschungszentrum für marine Geowissenschaften, Wischhofstrasse 1-3, D 24148 Kiel, Germany

(2) State Research Center - Arctic and Antarctic Research Institute, 38 Bering St., 199226 St. Petersburg, Russia

Received 14 February 1997 and accepted in revised form 17 October 1997

Abstract - Particulate matter of lithogenic and biogenic origin is often highly enriched in Arctic sea ice. Entrainment of particles is most effective during fall freeze-up over shallow Arctic shelves.
Particle entrainment into newly forming sea ice was investigated during the Transdrift III expedition in October 1995, conducted in the Laptev Sea, Siberian Arctic. Considering the prevailing calm meteorological conditions, processes related to propagating wave fields are identified as possible mechanisms for particle entrainment into the forming sea ice, where particles were significantly enriched as compared to the underlying water column.
Due to high drift velocities and a loose ice cover during the observed freeze-up, sediment export by sea ice from near-shore areas towards the deep basins of the Arctic Ocean was likely.

Introduction

Coastal waters of the Arctic marginal seas are covered by sea ice for more than six months a year. Therefore, classical erosion processes like resuspension of bottom sediments due to storm-induced waves during winter times are mostly prevented. It is assumed that sea ice related processes act as an important factor influencing the sediment regime. For example, regression of shallow seas due to tectonic uplifts or eustatic sea level changes. The withdrawal causes a steepening of the shelf gradient. In order to reestablish the shelf-gradient equilibriuim, sediment erosion in near-shore areas, and sediment deposition further off shore is required. Sediment-laden sea ice can transport sediments from coastal areas across the shelves towards the deep ocean basins, and thus, influencing the overall sediment budget in northern polar regions (Kempema et al., 1989). Sea-ice sediments decrease the albedo when accumulating on the ice surface and therefore accelerate ice melting (Reimnitz and Bruder, 1972; Barnes et al., 1982). Hazardous substances, e.g. heavy metals and persistent organic substances, might be associated with particles. If these particles are incorporated into sea ice, the hazardous substances could be transported over wide areas. Due to ice melting in marginal ice zones, e.g. off eastern Greenland, these substances are released into the water column, where they might enter the food web (Pfirman et al., 1995).

During the last decades, considerable work has been done on particle entrainment into sea ice. Biologists and geologists discuss how lithogenic and bogenic particles are entrained into sea ice (Campbell and Collin, 1958; Reimnitz and Bruder, 1972; Barnes et al., 1982; Weeks and Ackley, 1982; Osterkamp and Gosink, 1984; Kempema et al., 1986; Ackley et al., 1987; Ackermann et al., 1990; Shen and Ackermann, 1990; Reimnitz et al., 1992; Dethleff, 1995). It is assumed that most of the particles are entrained when sea ice is forming in autumn or beginning winter (Medcov and Thomas, 1974; Barnes et al., 1982; Osterkamp and Gosink, 1984). Previous field work concerning particle entrainment into sea ice during freeze-up in autumn was conducted in the Beaufort Sea and the Antarctic (Reimnitz and Dunton, 1979; Ackley et al., 1987; Reimnitz and Kempema, 1987; Reimnitz et al., 1987; Kempema et al.,

In: Kassens, H., H.A. Bauch, I. Dmitrenko, H. Eicken, H.-W. Hubberten, M. Melles, J. Thiede and L. Timokhov (eds.) Land-Ocean Systems in the Siberian Arctic: Dynamics and History. Springer-Verlag, Berlin, 1999, 113-123.

1989). The present study reports on particle entrainment into newly forming sea ice, documented during freeze-up in the Laptev Sea (Siberian Arctic) in October 1995.

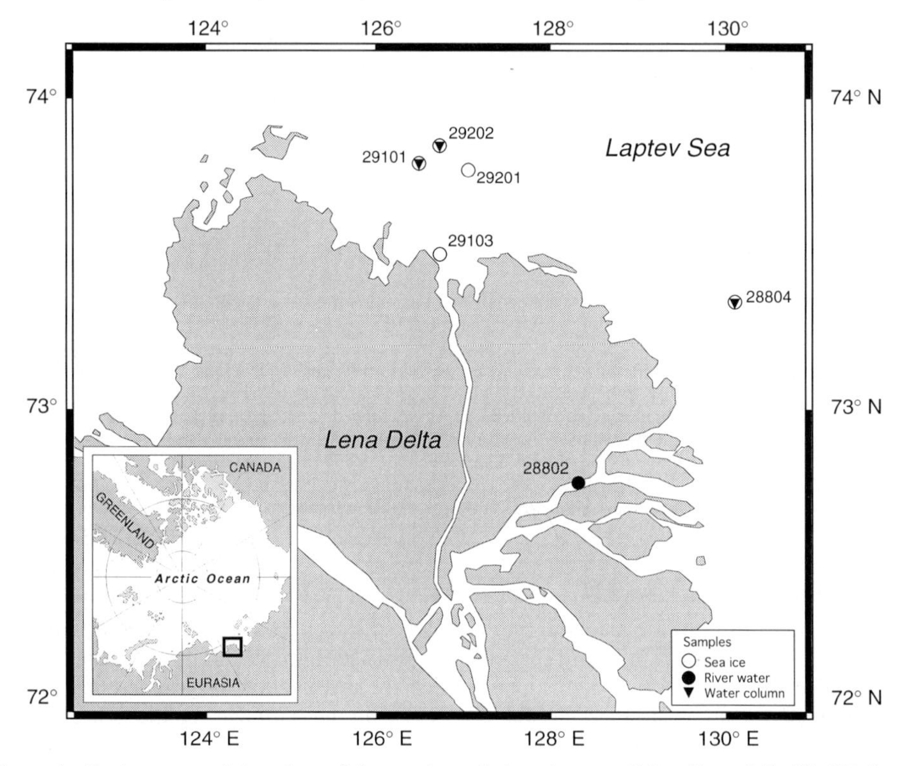

Figure 1: Study area and location of ice stations during the expedition Transdrift III (27. Sept. - 31. Oct. 1995). The first three digits of the station numbers are days of the year.

Methods

One aim of the Transdrift III expedition (27. Sept. - 31. Oct. 1995) conducted in the Laptev Sea aboard the Russian I/B "Kapitan Dranitsyn", was to study sediment entrainment into sea ice. In this paper, we focus on samples taken in and adjacent to the Lena Delta (Figure 1). A total of seven cores of young ice (13 to 20 cm thick) were sampled by means of an ice auger (9 cm diameter) on one helicopter and two ship stations. At two additional ship stations, three samples of nilas were taken by hand from a diving plattform, lowered to the ice surface. The ice cores were split into smaller sections of about 10 cm thickness, and all ice samples were melted at room temperature and vacuum filtered over 0.45 μm Durapore and Whatman GF/F filters. Afterwards, the filters were desalinated with distilled water and freeze-dryed.

The water column was sampled on three ship stations, using a 5 L water sampler, from which 1 L was subsampled. A total of seven water samples were taken in 0, 5, 10 and 18 m water depth. Additionally, four water samples were taken from directly beneath the ice underside. At one helicopter station, three samples of the Lena River water were taken with a 1 L plastic bottle. The suspended particulate matter (SPM) of all water samples was treated the same way as the melted ice samples.

Meteorological readings were made at 0, 3, 6, 9, 12, 15, 18 and 21 UTC time. Wind speed and direction as well as air temperature were recorded by a meteorological device M-49 (Factory of Meteorological Devices, Russia). While the vessel operated in open water, the wave height was estimated visually from the ship's bridge. While the vessel was anchored, the flow metre of the ship was used to measure surface-current speeds. Ice drift velocities were estimated visually.

The equality of the data sets was tested, computing a Mann-Whitney U-test (Sachs, 1984).

Results

Ice conditions

The shelf area of the Laptev Sea was ice free at the end of September 1995 (Figure 2). New ice formation started in early October, and continued throughout the entire expedition. According to the international sea-ice nomenclature (WMO, 1970), different types of floating ice, from grease ice to white-grey ice, were observed during the field work. Adjacent to the Lena Delta, different types of congealed ice were sampled (Table 1). At the end of October 1995, the entire Laptev Sea was covered by sea ice (Figure 3), mainly composed of new ice, nilas (< 10 cm thick) and young ice (10 - 30 cm thick).

Ice drift velocities > 0.5 m s^{-1} were estimated from the anchored ship.

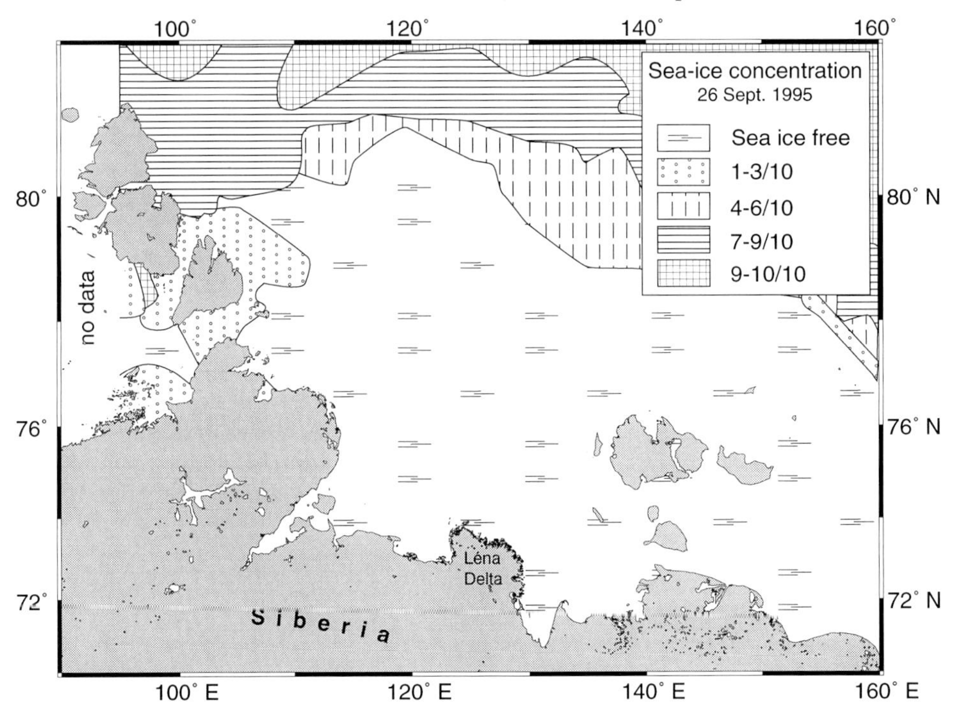

Figure 2: Ice chart of the Laptev Sea at the end of September 1995 (modified from Naval/NOAA Joint Ice Center, unpubl.).

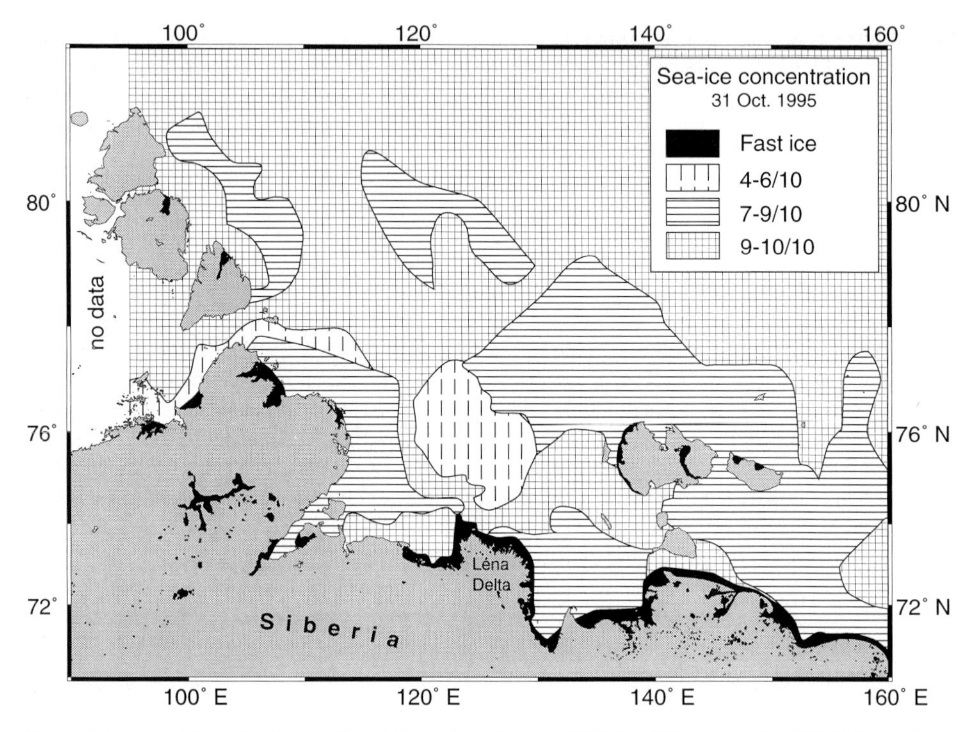

Figure 3: Ice chart from the Laptev Sea at the end of October 1995 (modified from Naval/NOAA Joint Ice Center, unpubl.).

Meteorological conditions

Figure 4 provides a general overview of air temperature, wind speed and wind direction during the course of the expedition (note that the day of the year corresponds to the first three digits of the station number). The meteorology was characterized by calm weather conditions. During the first two October weeks, south-easterly winds predominated. In the second half of October 1995, the wind field changed several times from south-west to north-west and vice versa (Figure 4). As shown in Figure 5, the prevailing wind speed was 5 Beaufort, with a mean of 8.2 ± 3.3 m/s (Table 2). Over a time period of about one week before the first sampling adjacent to the Lena Delta (station 288xx), the wind speed was at the mean value or less (Figure 4). The air temperature in October 1995 decreased more or less continuously (Figure 4), with a minimum value of -14.1 °C (Table 2).

More than half of the observed wave heights ranged between 0.4 and 1.2 m (Fig 5). The overall mean wave height was 1.0 ± 0.7 m (Table 2).

Particle concentration in sea ice and water

Particulate matter, including biota, was found in all ice and water samples. The particle concentrations are shown in Figure 6. The average particle concentration in sea ice was 23 times higher, as compared to the SPM of the underlying water column. No SPM gradients were measured over the water column, with maximum water depth of 22 m. For river water and sea water, the SPM concentrations were within the same range (Figure 6, Table 3). The median value of the SPM for river- and sea water, was 7.8 mg/L, and 6.4 mg/L, respectively (Tabel 3). Comparing these two data sets, no significant differences were found.

Table 1: Samples taken in and adjacent to the Lena Delta during the expedition Transdrift III (27. Sept. - 31. Oct. 1995) into the Laptev Sea. Note that the first three digits of the station number correspond to the day of the year.

Latitude ° N	Longitude ° E	Station	Sample	depth [m]	Type
72.751	128.129	28802	61	0	River water
72.751	128.129	28802	62	0	River water
72.751	128.129	28802		0	River water
73.345	129.922	28804	60	ice underside	water column
73.345	129.922	28804	61	ice underside	water column
73.345	129.922	28804		0	water column
73.345	129.922	28804		5	water column
73.345	129.922	28804		10	water column
73.345	129.922	28804		18	water column
73.345	129.922	28804	62		dark nilas
73.345	129.922	28804	63		dark nilas
73.799	126.284	29101		0	water column
73.799	126.284	29101		5	water column
73.799	126.284	29101		10	water column
73.799	126.284	29101	60		grey-white ice
73.799	126.284	29101	61		grey-white ice
73.507	126.445	29103	60		grey-white ice
73.778	126.864	29201	60		light nilas
73.854	126.527	29202	64	ice underside	water column
73.854	126.527	29202	65	ice underside	water column
73.854	126.527	29202	60		grey-white ice
73.854	126.527	29202	61		grey-white ice
73.854	126.527	29202	62		grey ice
73.854	126.527	29202	63		grey ice

The particle concentrations in sea ice varied by more than one order of magnitude (Table 3, Figure 6). The median value of particle concentration in sea ice was 148.9 mg/L. this value is significantly higher ($U = 3.5$; $m=11$, $n=10$; $p < 0.00$) than that for the underlying water column.

Discussion

Particle entrainment into sea ice

Suspension freezing is a widely accepted particle incorporation process into sea ice. This process requires strong winds, intense turbulence in open water of a shallow sea, and extreme

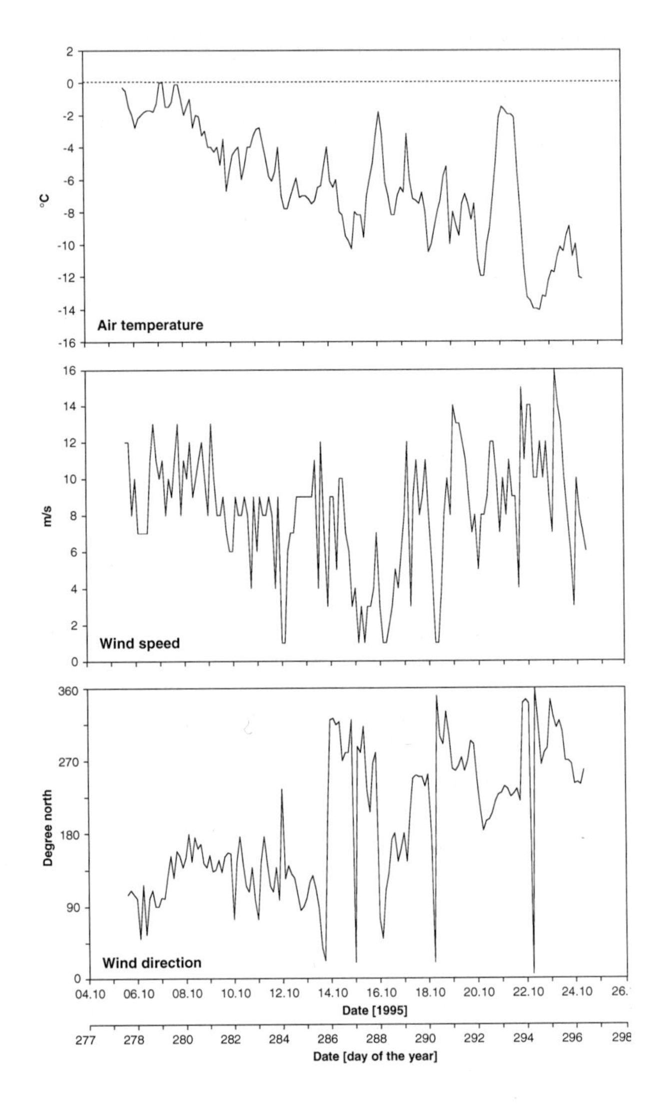

Figure 4: Meteorological readings made aboard I/B "Kapitan Dranitsyn" during the expedition Transdrift III.

Table 2: Meteorological conditions during the expedition Transdrift III (27. Sept. - 31. Oct. 1995) into the Laptev Sea.

Parameter	n	Minimum	Maximum	Mean	Standard deviation
Air temperature [°C]	151	-14.1	0.0	-6.2	3.6
Wind speed [m/s]	151	1.0	16.0	8.2	3.3
Wave high [m]	63	0.1	3.0	1.0	0.7

subfreezing air temperatures (Reimnitz et al., 1992). After ceasing of the storm, frazil and anchor ice with entrapped sediment rise to the water surface, forming a layer of grease ice with turbid interstitial water. After congelation, this ice cover develops into turbid ice. Concerning the calm weather conditions recorded during the present expedition, suspension freezing probably was not the sediment entrainment process for the significant particle concentrations measured in the sea ice.

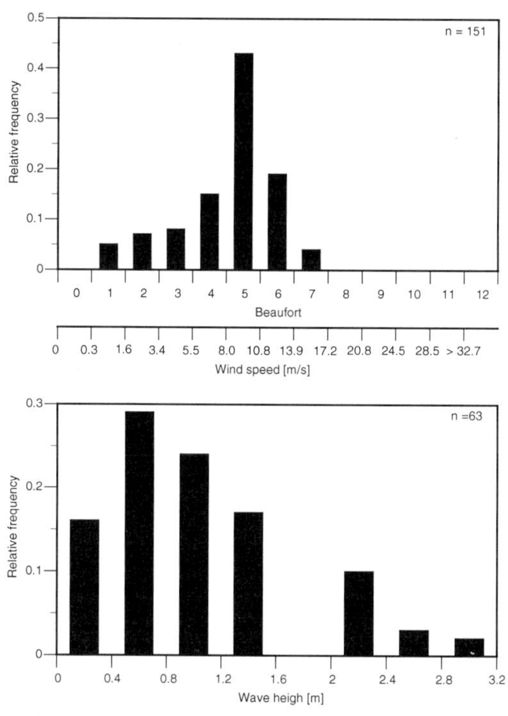

Figure 5: Relative frequency distribution of wind speed and wave heights during the freeze-up of the Laptev Sea in October 1995.

Other possible entrainment processes are linked to wave fields propagating into sea ice at different stages of development, that is congealed and uncongealed sea ice (Martin and Kauffman, 1981; Ackley et al., 1987; Shen and Ackermann, 1990; Ackermann et al., 1990; Weissenberger, 1992; Ackermann et al., 1994). Entrainment into uncongealed sea ice is discussed for biota (Ackley et al., 1987; Weissenberger, 1992). After entrainment and congelation, biota are often enriched in sea ice by some orders of magnitude as compared to the underlying water column (Spindler, 1994). For example, mean numbers of the foraminifer *Neogloboquadrina pachyderma* in Antarctic sea ice are up to 70 times higher than in the upper 60 m of the underlying water column (Dieckmann et al., 1991). On the other hand, Martin and Kauffman (1981) observed an internal downward movement of 1-2 mm sized plastic chips within grease ice during tank experiments. Derived from field observations, Reimnitz and Kempema (1987) supposed a sediment release, and hence a cleaning of grease ice, due to pressure oscillations caused by propagating wave fields.

Little work has been done yet on particle entrainment into congealed sea ice, by propagating wave fields (Ackermann et al., 1990; Ackermann et al., 1994). In tank experiments it was shown that high rates of sediment entrainment can occur during 15 minutes of wave action. It was found that some degree of flexural rigidity of the ice to vertical motion is required for sediment enrichment by waves into the ice cover. Such conditions would exist within a continuous ice sheet or single floes with dimensions equal to or greater than the length of passing waves (Ackermann et al., 1994). Hence, wave fields propagating into a rigid, congealed ice cover provide a possible and effective mechanism for particle entrainment into sea ice.

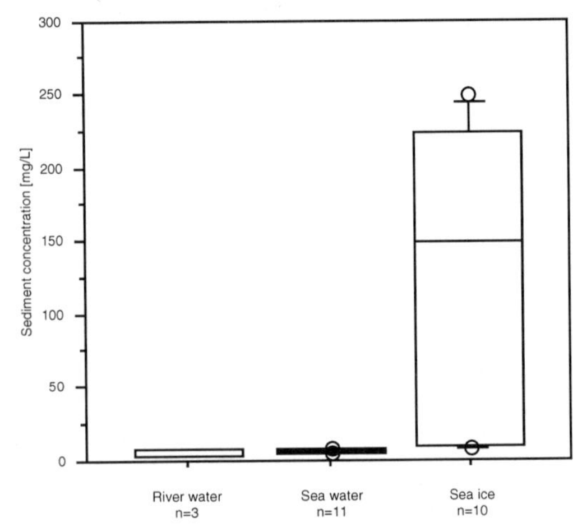

Figure 6: Box-plot of particle concentrations measured in the Lena River, sea water and ice samples adjacent to the Lena Delta.

Table 3: Calculated sedimentological parametres during the expedition Transdrift III (27. Sept. - 31. Oct. 1995) into the Laptev Sea.

Parameter	n	Minimum	Maximum	Median
SPM river water [mg/L]	3	2.5	7.8	7.8
SPM water column [mg/L]	11	4.3	8.4	6.4
Sea ice [mg/L]	10	7.2	249.5	148.9

Considering the calm conditions during the freeze-up in October 1995 in the Laptev Sea, resuspension processes were unlikely. Resuspension of sediments due to wave action or thermohaline circulation would have resulted in higher SPM content in the sea water as compared to river water. But this was not the case. Hence, it is concluded that the SPM concentration in the sea water adjacent to the Lena Delta was determined by the SPM delivered by the River Lena. Pressure oscillations, linked to the crests and troughs of propagating waves, plus related filtration processes within the ice texture, are likely processes for the entrainment of SPM from the underlying water column into the newly forming sea ice during the freeze-up in October 1995. Because of the river controlled SPM content in the sea water, it is possible that particles may have been incorporated into the ice without having been deposited before on the sea floor. Unfortunately, we can not determine, whether the ice canopy was already congealed during the particle entrainment or not, because no time series were sampled.

Sea-surface salinity, measured at sampling locations, shows brackish to marin values (Dmitrenko et al., 1997), but these measurements provide no information of the oceanographic conditions from the place and/or moment of ice formation and sediment entrainment. Because

(i) the sampled ice was already formed, (ii) the ice was drifting and therefore not in place of its formation and (iii) low $\partial^{18}O$ values of the ice samples suggest an origin near the river mouth (H. Eicken, pers. comm. 1997). The latter one also supports the assumption that the sediments were incorporated into the ice without beeing deposited before on the sea floor. Although resuspension of bottom sediments due to wave action and/or bottom currents, for instance tidal currents, cannot be completely excluded, resuspension processes do not appear neccessary for sediment entrainment into sea ice. Subsequently, rafting of sediment-laden young ice will lead to ice thickening and to stacked sediment layers, each of which some centimetres thick. Such sediment layers have been observed from vessels during ice breaking (Figure 7), or in ice cores of thick first-year ice.

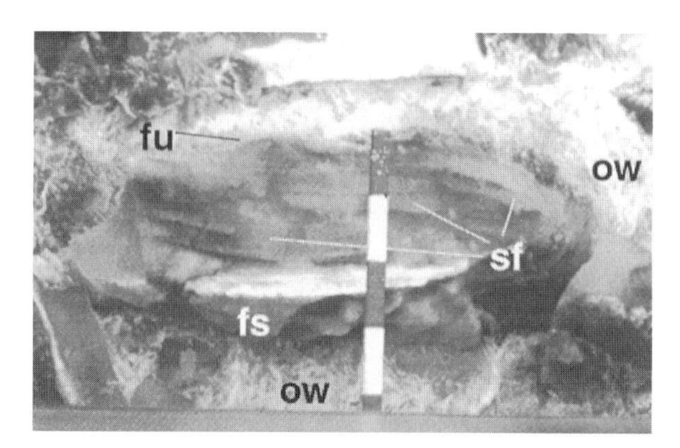

Figure 7: Cross section of a vertically tilted first-year ice floe in the Laptev Sea. Rafting of former young ice thickened, and built up the entire floe. Each bar segment is 0.5 m in length. The photograph was taken vertically to the ice surface from aboard R/V "Polarstern" during the cruise ARK XI/1. Fu = floe underside, fs = floe surface, sf = rafted sediment-laden floes of former young ice (dark layers represent sediment inclusions), ow = open water.

Sediment transport by sea ice

Although the observed high drift velocities of the young sea ice are only estimates, they suggest that incorporated sediments, may be transported over long distances in relatively short time periods.

In order to export sea ice and incorporated sediments out of the Laptev Sea shelf area, the ice must drift in a northerly direction. Off-shore winds from south-east to south-west will force ice motion into such a direction. As shown in Figure 4, such wind directions were observed over several days during the freeze-up in October 1995. Because the ice cover was mainly composed of thin, shallow-draught ice, including patches of open water, conditions for high drift velocities were favourable. Hence, export of sea ice and incorporated sediments from near-shore areas of the Laptev Sea towards the deep Arctic Ocean is most likely during the freeze-up in autumn, even under calm conditions.

Conlusion and Perspectives

1 In October 1995, the freeze-up period in the Laptev Sea was characterized by calm weather. Nevertheless, significant particle concentration in the newly formed ice as compared to the underlying water column were found .
2 Pressure oscillations, linked to wave fields propagating into sea ice, and additionally

filtration processes within the ice texture are possible particle entrainment processes during the observed freeze-up.

3 Even under calm weather conditions, the freeze-up seems to be an important time period for sediment transport by sea ice from shallow shelf areas towards the deep basins of the Laptev Sea, and probably for sediment export to the Arctic Ocean.

4 In order to fully understand sediment entraining processes into sea ice, more field work has to be conducted during freeze-up, including high-resolution time series of oceanographic and meterological parameters, and all disciplines of ice work, including biology, geology, ice physics, meteorology, oceanology and remote sensing. Additionally, tank experiments have to be done, studying particle incorporation in more detail and under controlled condition.

Acknowledgements

We thank captain and crew aboard I/B "Kapitan Dranitsyn" for their support and cooperation. We are especially grateful to E. Reimnitz and K.P. Tyshko for organizing this crowd, and J. Freitag for assistance during field work. The manuscript benefittet substantially from the constructive comments made by H. Eicken and I. Werner, as well as by the referees E. Reimnitz and L.H. Smedsrud. This study was carried out as part of the project "German-Russian Cooperation: System-Laptev Sea", funded by the German Ministry of Education, Science, Research and Technology (grant no. 52540030G0517A).

References

Ackermann, N.L., H.T. Shen and B.E. Sanders (1990) Sediment enrichment of coastal ice covers. Proceedings of the international Association of Hydraulic Research Ice Symposium, 86-96.

Ackermann, N.L., H.T. Shen and B.E. Sanders (1994) Experimental studies of sediment enrichment of arctic ice covers due to wave action an frazil entrainment. J. Geophys. Res., 99(C4), 7761-7770.

Ackley, S.F., G. Dieckmann and H. Shen (1987) Algal and foram incorporation into new sea ice. EOS 68(50), 1736.

Barnes, P.W., E. Reimnitz and D. Fox (1982): Ice rafting of fine-grained sediment, a sorting and transport mechanism. J. Sediment. Petrol., 52(2), 493-502.

Chambell, N.J. and A.E. Collin (1958): The discoloration of Foxe Basin Ice. J. Fish. Res. Board Can., 15(6), 1175-1188.

Colony, R. and A.S. Thorndike (1984) An estimate of the mean field of Arctic sea ice motion. J. Geophys. Res., 89(C6), 10623-10629.

Dethleff, D. (1995) The Laptev Sea – A key region for particle entrainment into Arctic sea ice (in German). PhD thesis, University Kiel, 111 pp.

Dieckmann, G.S., M. Spindler, M.A. Lange, S.F. Ackley and H. Eicken (1991) Antarctic sea ice: a habitat for the foraminifer Neogloboquadrina pachyderma. J. Foram. Res., 21(2), 182-189.

Dmitrenko, I., L. Timokhov, P. Golovin and N. Dmitriev (1997) Oceanographic processes in the Laptev Sea during autumn. In: Laptev Sea System: expeditions in 1995, Kassens, H. (ed.), Ber. Polarforsch. 248, 44-61.

Kempema, E.W., E. Reimnitz and P. Barnes (1989) Sea ice sediment entrainment and rafting in the Arctic. J. Sediment. Petrol., 59(2), 308-317.

Kempema, E.W., E. Reimnitz and R.E. Hunter (1986) Flume studies and field observations of the interaction of frazil ice and anchor ice with sediments. Open-File Report 86-515, 1-48.

Martin, S. and P. Kauffman (1981) A field and laboratory study of wave damping by grease ice. J. Glaciol., 27(96), 283-313.

Medcov, J.C. and M.L.H. Thomas (1974) Surfacing on ice of frozen-in marine bottom materials. Journal Fisheries Research Board of Canada, 31, 1195-1200.

Osterkamp, T.E. and J.P. Gosink (1984) Observations and analyses of sediment-laden sea ice. In: The Alskan Beaufort Sea: ecosystems and environments, Barnes, P.W., D.M Schell and E. Reimnitz (eds.), Academic Press, Orlando, 73-93.

Pfirman, S., H. Eicken, D. Bauch and W.F. Weeks (1995) The potential transport of pollutants by Arctic sea ice. Sci. Total Environ., 159, 129-146.

Reimnitz, E. and K.F. Bruder (1972) River discharge into an ice-covered ocean and releated sediment dispersal, Beaufort Sea, coast of Alaska. Geol. Soc. Am. Bull., 83, 861-866.

Reimnitz, E. and K. Dunton (1979) Diving observations on the soft ice layer under the fast ice at DS-11 in the Stefansson Sound Boulder Patch. Nat. Oceanic and Atmos. Admin., Principal investors' annual report Vol. IX, Boulder, CO, 210-230 .

Reimnitz, E. and E.W. Kempema (1987) Field observations on slush ice generated during freeze-up in Arctic coastal waters. Mar. Geol., 77, 219-231.

Reimnitz, E., E.W. Kempema and P.W. Barnes (1987) Anchor ice, seabed freezing, and sediment dynamics in shallow Arctic Seas. J. Geophys. Res., 92(C13), 14671-14678.

Reimnitz, E., L. Marincovich Jr., M. McCormick and W.M. Briggs (1992) Suspension freezing of bottom sediment and biota in the Northwest Passage and implications for Arctic Ocean sedimentation. Can. J. Earth Sci., 29, 693-703.

Sachs, L. (1984) Angewandte Statistik. Springer Verlag, Berlin, 552 pp.

Shen, H.T. and N.L. Ackermann (1990) Wave-induced sediment enrichment in coastal ice covers. In: Sea ice properties and processes, Ackley, S.F. and W.F. Weeks (eds.), US-Army, Hannover NH, (CRREL-Monograph 90-1), 100-102.

Spindler, M. (1994): Notes on the biology of sea ice in the Arctic and Antarctic. Polar Biol., 14, 319-324.

Weeks, W. and S.F. Ackley, S.F. (1982) The growth, structure, and properties of sea ice. US Army, Hanover NH (CRREL-Monograph 82-1), 1-130 .

Weissenberger, J. (1992) The environmental conditions in the brine channels of Antarctic sea-ice (in German). Ber. Polarforsch. 111, 1-159 .

WMO (1970): WMO sea-ice nomenclature - terminology, codes, illustrated glossary and symbols. WMO Part I-Suppl. No. 5, Geneva, Switzerland.

Frazil Ice Formation during the Spring Flood and its Role in Transport of Sediments to the Ice Cover

P. Golovin[1], I. Dmitrenko[1], H. Kassens[2] and J.A. Hölemann[2]

(1) State Research Center - Arctic and Antarctic Research Institute, 38 Bering St., 199226St. Petersburg, Russia

(2) GEOMAR Forschungszentrum für marine Geowissenschaften, Wischhofstrasse 1- 3, D 24148, Kiel, Germany

Received 3 March 1997 and accepted in revised form 18 November 1997

Abstract - The article describes full-scale experimental studies performed in the Transdrift-IV expedition from fast ice in the near-delta of the Lena river in the Laptev Sea during the period immediately preceding the flood (late May) and during the peak of the flood (early and mid-June). Processing of data has revealed the presence of supercooled water layers 5 to 150 cm thick in the zone of river-sea water contact (in the upper part of the seasonal pycnocline). The supercooling value was observed to be -0.8 ^{0}C. Together with the thickness of supercooled fluid it depended upon both the time (before or during the flood) and the site of measurements (in the zone of main branches or beyond). At one of the stations a conglomerate of frazil ice was found attached to the cable of a bottom temperature meter at the depth of the pycnocline. Using the known conditions, the probability for supercooling and further frazil ice formation at all stations was determined. The results of observations have allowed the local Richardson numbers to be calculated for the river-sea water contact zone - the layer of supercooled fluid. Based on the theory of entrainment at the flat turbulent jet margin and a semi-empirical turbulence theory, it was possible to correctly relate the mean current velocity U in the upper freshened layer to the dynamic velocity U_* (root-mean-square velocity of turbulent variations) and present the entrainment at the river-sea water boundary as a kind of entrainment at the flat turbulent jet margin. Using ratios from laboratory studies of frazil ice formation, the actual rates of frazil ice formation in the river/sea water contact zone were estimated. They were calculated for the different mean motion velocities in the freshened layer during the different periods of the flood development in the near-mouth region of the Lena River. Based on the known concentrations of suspended sediments in the layer freshened by river water, the fluxes of suspended matter to the bottom ice surface, governed by the process of frazil ice formation, were estimated.

Introduction

Many large rivers that carry a vast amount of freshwater and sediments, especially during the spring floods, empty onto the Siberian Shelf Seas (and, in particular into the Laptev Sea). The temperature of fresh river water during the winter-spring season is close to its freezing point. Entering the ice-covered sea, freshwater spreads over the surface of colder, saline sea water. Nansen (1956) found that at this contact some portion of fresh river water becames supercooled and forms frazil ice, especially in the near-mouth sea regions. It should be noted that formation of frazil ice is a phenomenon typical of the entire Arctic Basin, including the summer season (Golovin et al., 1993; Golovin et al., 1996; Timokhov, 1989). Formation of frazil ice at the interface of fresh and saline water leads to the growth of essentially freshwater ice crystals (Golovin et al., 1996). As a result, ice with a chaotic crystals orientation ice formed. This allows us to identify the interlayers of this ice 2 cm to 2 m thick within the general structure and texture of multiyear drifting and fast ice both in the Arctic and Antarctic (Petrov, 1971; Cherepanov, 1972; Cherepanov and Kozlovsky, 1972; Weeks and Ackley, 1982; Martin, 1971).

In: Kassens, H., H.A. Bauch, I. Dmitrenko, H. Eicken, H.-W. Hubberten, M. Melles, J. Thiede and L. Timokhov (eds.) Land-Ocean Systems in the Siberian Arctic: Dynamics and History. Springer-Verlag, Berlin, 1999, 125-140.

The main cause for supercooling of water freshened by river runoff and the subsequent formation of frazil ice is a more effective loss of heat to the underlying cold and saline sea water, than gain of salt via the countering salt flux. The different efficiencies of heat and salt exchanges are governed by the fact that the molecular temperature conductivity coefficient Kt_m exceeds the molecular salt diffusion coefficient Ks_m by two orders of magnitude ($Kt_m/Ks_m=10^2$). This allows us to call the mechanism of supercooling and frazil ice formation under consideration „double-diffusion" (Martin and Kauffman, 1974; Krylov and Zatsepin, 1992) or contact mechanism (Timokhov, 1989).

The article is devoted to the study of possibility of frazil ice formation at the river sea water interface during flood in the near-mouth region of the Lena River, the Laptev Sea. The rate of frazil ice formation in natural environments has been estimated on the basis of laboratory experimental evidence. This has allowed for estimating the intensity of riverine suspended matter supply towards the ice cover due to the double-diffusion mechanism of frazil ice formation.

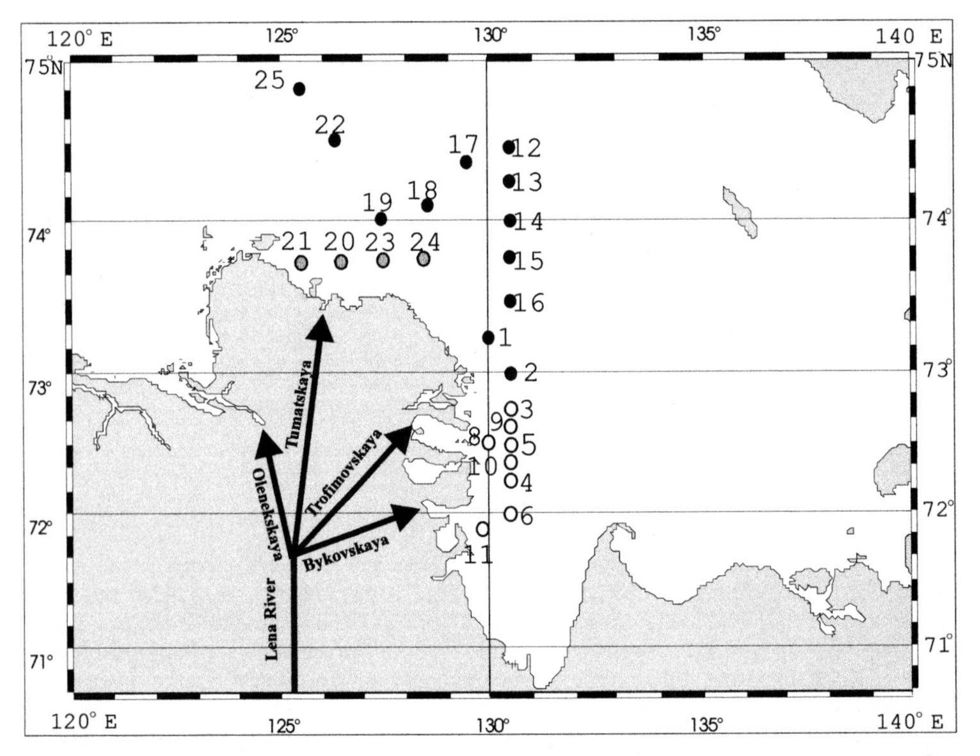

Figure 1: An array of comprehensive oceanographic stations in the Transdrift IV expedition (LN96..); ● - stations taken only during the period before the flood (for example LN9610);. ● - stations taken twice: before the flood and after its onset (for example LN9610, LN9610a); O - stations taken three times: before the flood and twice after its onset (for example LN9610, LN9610a, LN9610b).

Materials and methods

For our studies the materials of hydrological observations collected during the Russian-German expedition Transdrift IV were used. The work was carried out on fast and drifting ice in the

Laptev Sea in the near-delta region of the Lena river from mid-May to mid-June during the period that directly precedes the spring flood, and during the period of its development. The array of comprehensive oceanographic stations is presented in Figure 1. Observations of water temperature and salinity were conducted by means of a fine-structure CTD sonde. We used an OTS-PROBE Serie 3 sonde (Meerestechnik Electronic CmBH, Germany). The main technical characteristics of an OTS-PROBE are the following:

- water temperature: accuracy 0.01°C ; time costant 160 ms;
- electric conductivity: accuracy 0.02 mS/cm; time constant 100 ms;
- hydrostatic pressure: accuracy 0.1%, time constant 40 ms;
- information exchange rate 1200 bit/s.

Temperature and salinity sensors of the sonde are situated close to each other. They are of the same size. Multi-channel sonde is able to perform a parallel cross-examination of sensors. The rate of vertical sounding is about 0.5-1.0 m/s, i.e. about 12-24 parameter values per 1 meter of sea water body (average 15-17). All values of temperature and salinity parameters have been used for analyzing the principal possibility and the value of supercooling in the extremely sharp halo-pycnocline during flood period. Data filtration with depth has not been performed.

Observation results

Processing of temperature and salinity observational data has shown the existence of zones of supercooled water 5 cm to 1.5 m thick in the pycnocline (the interface between river and sea water). Supercooling in some thin layers of these zones $\Delta T_f = T - \tau_s$, where T is the measured (actual) water temperature and τ_s is the water freezing temperature at the given salinity, was calculated using standard algorithms (Fofonoff and Millard, 1983). At a first glance, it was reaching the improbable values of -0.6 to -0.8°C (Figure 2c). Mean super-cooling values varied from -0.01°C up to -0.1°C. The persistence of such supercooling under natural conditions has not been earlier observed. At least the authors are not aware of cases where similar values were directly determined in such an extensive region.

The depth of the supercooled layer, its thickness characterizing the extent of penetration of cooling to the pycnocline and the value of supercooling itself depended on the time and place of measurements. Prior to the spring flood (end of May), supercooling was only observed at stations located at the traverse of the main discharge branches Trofimovskaya and Bykovskaya. The upper boundary of supercooled water is confined to the lower boundary of freshened surface water (Figure 2). Similar position of the location of the upper boundary of the supercooled layer was observed at all stations where this layer was recorded. The supercooled layer itself at the weak inflow of river water occupies a small upper part of the pycnocline (Figure 2a). At the time of maximum discharge from early to mid-June it occupied either much of the pycnocline or the entire pycnocline (Figures 2b, c). Prior to the flood, the value of supercooling in the upper part of the pycnocline was small and rarely exceeded 0.05°C (Figure 2a). With the rising flood the intensity of freshwater inflow sharply increased, causing the lense of fresh water to greatly expand its area below the fast ice. The thickness of the fresh river water layer gradually increased with entrainment and mixing of the underlying layer of saline and cold water (Figure 3). This is especially well seen at the stations located along the traverse crossing the outflow from the main branches (Figure 3). Finally, river water at some comparatively shallow stations occupied the entire water column between the fast ice and the seafloor. With increasing thickness of the river water layer (Figure 3), the upper boundary of the supercooled layer is lowered (Figure 2). The increased rate of river water inflow, associated

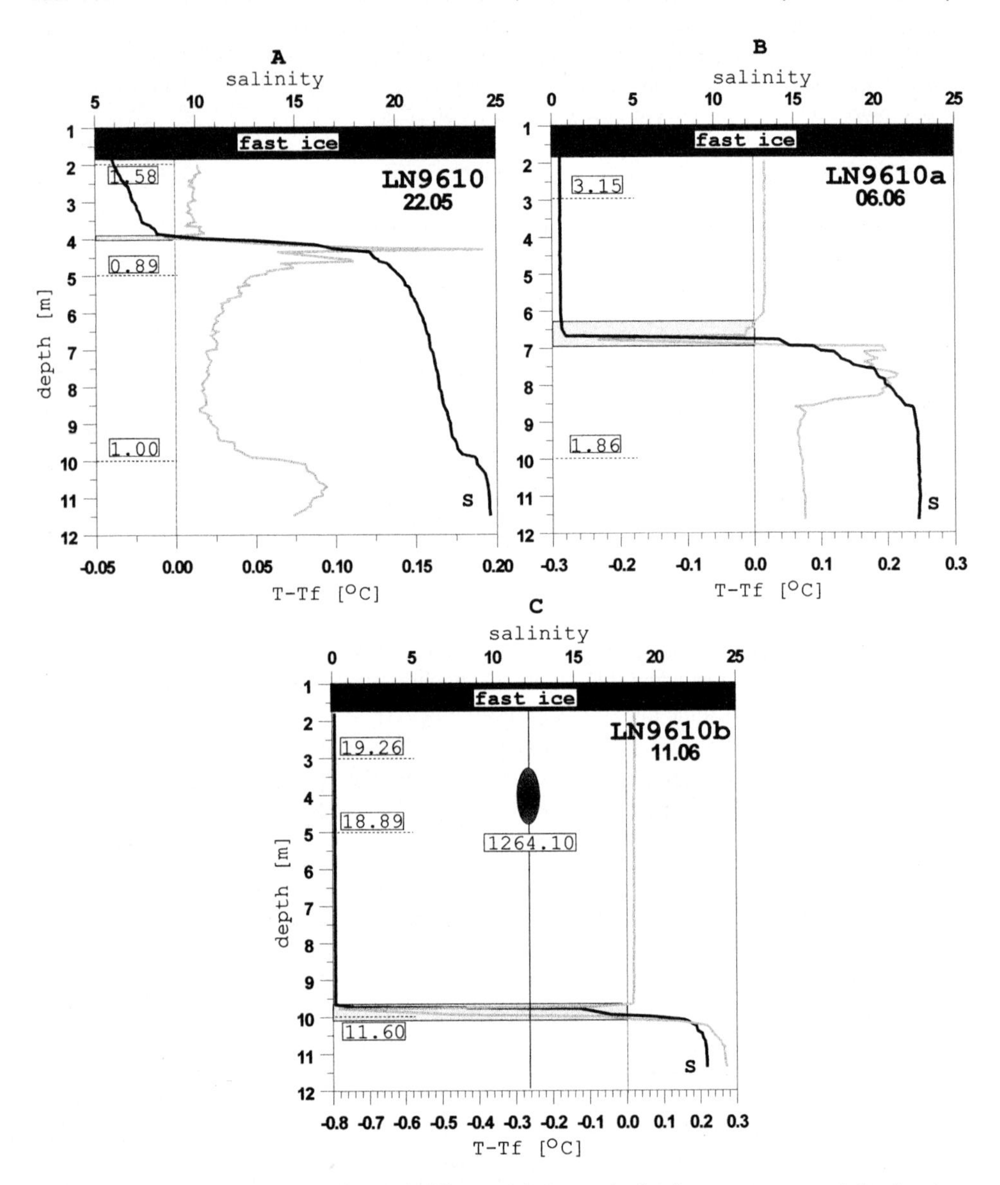

Figure 2: Vertical salinity distribution and differences between the in situ temperature and the freezing temperature at a given salinity for stations LN9610(A), LN9610a(B), LN9610b(C). 2.25 - contamination of suspended particulate matter into the water and frazil ice samples (mg/l); [shaded box] - zone of supercooling; [oval] - an agglomerate of frazil ice.

with increased horizontal velocity and turbulence results in a thickening supercooled layer and a greater absolute value of supercooling. This is due to a more intense turbulent entrainment at the river-sea water boundary (Figure 2).

It should be especially stressed that maximum supercooling rather occurs within pycnocline (Figures 2b, c) than at the freshwater/pycnocline boundary, as has been, for instance,

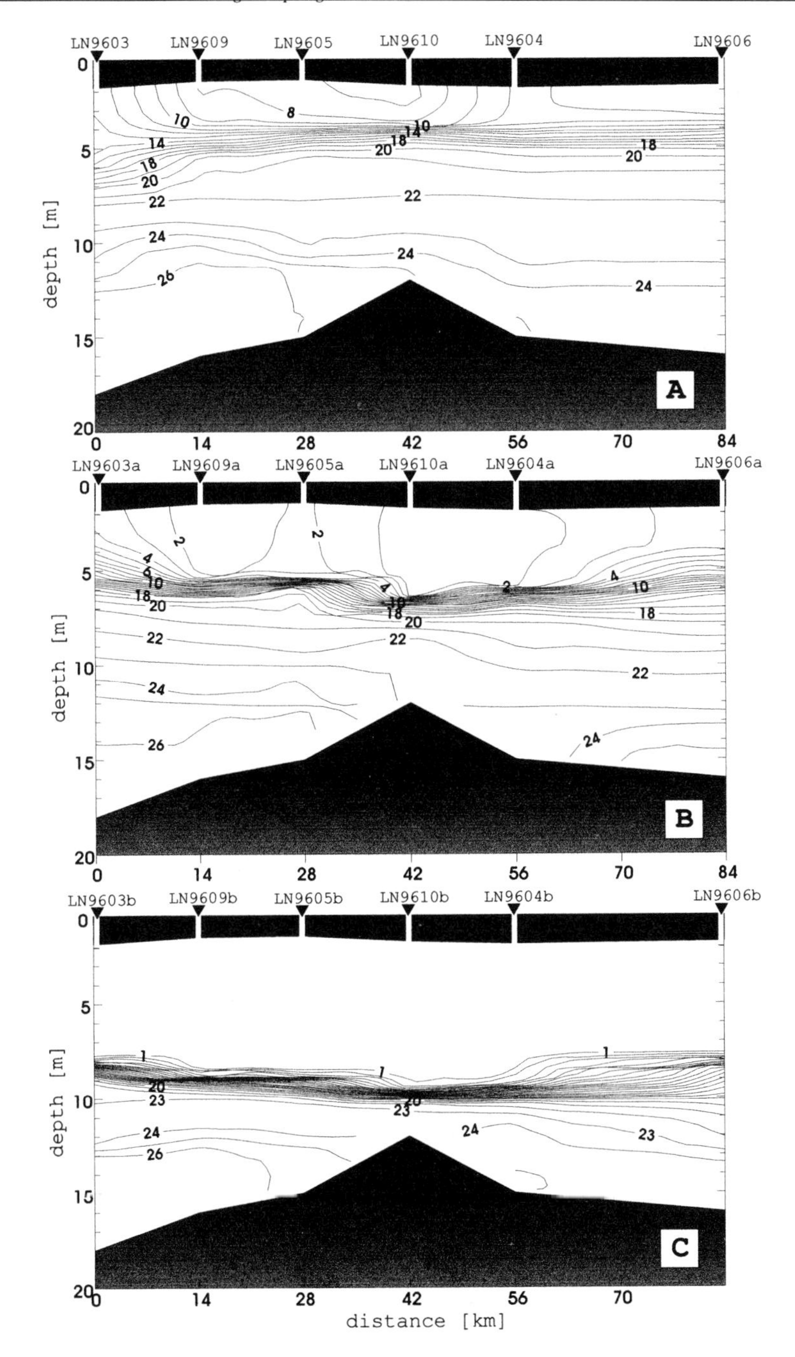

Figure 3: Evolution of vertical water salinity distribution at the onset of the spring flood at the transect along meridian 130 30'E from 72 to 72 45'N; (A) - May 18-22, (B) - June 3-6, (C) - June 10-11.

suggested by Martin and Kauffman (1974). This is determined by a specific process of turbulent entrainment at the pycnocline/freshened layer boundary. In course of laboratory experiments Kan and Tamai (1994) described the mechanism of inner waves initiation at the pycnocline boundary due to velocity shift together with their subsequent collapse due to Kelvin-Helmgolz instability. This results in formation of vortical structure at the pycnocline/fresh water boundary (Kan and Tamai, 1994). The latter favours more effective local penetration of fresh and warm water from the upper layer to pycnocline represented by more saline and cold water. In this case maximum supercooling occurs at that level within pycnocline where non-turbulent freshened water has already penetrated. This happens if loss of heat is more effective than heat exchange at molecular and molecular-turbulent levels (Krylov and Zatsepin, 1992; Voropayev et al., 1995). Buoyancy force hinders fast mixing within pycnocline, whereas part of more saline and cold water from pycnocline rapidly gets mixed when penetrates into the strongly turbulent upper layer (through this mechanism of turbulent entrainment). Hence the supercooling at the pycnocline/upper water layer boundary is considerbly weaker than that observed within pycnocline. If during the period preceding flood intensive turbulent entrainment at this boundary is absent, weak supercooling is observed only at the very boundary between the freshened water layer and pycnocline, and not within pycnocline (Figure 2a). This is in a good accordance with observational data (Martin and Kauffman, 1974).

The presence of supercooled water layers in the pycnocline caused the active formation of frazil ice in these layers. A bottom - temperature and pressure meter was suspended from cable at st. LN9610 at the traverse of the Trofimovskaya branch (Figure 1). It remained there from May 22 to June 11. Thus the period from before through the flood was covered. When the instrument was recovered, an agglomeration of frazil ice was found on the cable at the depth of the pycnocline. Its weight was about 1.5-2 kg and its height was about 50-60 cm. It consisted of transparent chaotically oriented crystals of frazil ice wither which inclusions of sediments transported by river water were clearly seen (Figure 4). The cable in this case served both as a nucleus and a place where crystals of frazil ice were trapped.

Thus as a result of the experimental studies in the Lena delta during spring flooding breakup, the existence of supercooling in the river-sea water contact zone, as well as that of frazil ice formation in this zone were established.

Discussion

Let us estimate the probability of supercooling in the river/sea water contact zone and the rate of possible frazil ice formation at the different regimes of the heat-mass exchange through the pycnocline using the results of laboratory studies.

Features of supercooling in the zone of river-sea water contact

As is known, supercooling of fresh river water from the underlying saline and cold sea water is the necessary physical condition for frazil ice formation. Stable formation of frazil ice is possible only when the freshened layer is close to its freezing point (Golovin et al., 1996; Krylov and Zatsepin, 1992; Voropayev et al., 1995,), i.e. when the following condition is fulfilled:

$$T_1 \sim -a \cdot S_1 , \qquad \text{or } T_1 / S_1 \sim -a \qquad (1)$$

Here T_1 and S_1 are temperature and salinity of the layer, freshened by river runoff; a=0.055 ^oC/ppt is the linear relation coefficient between the freezing temperature and salinity. The

Figure 4: A agglomerate of frazil ice observed at st. LN9610b (Photo by Dr.J.Hölemann).

most favorable conditions for frazil ice formation at the lower boundary of the freshened layer exist when the temperature is also close to its freezing point in both the lower and the upper layers (Krylov and Zatsepin, 1992). Thus supercooling at the lower boundary of freshened water (the upper part of the pycnocline) (Figure 2) is possible at the condition:

$$\Delta T < a \cdot \Delta S, \qquad \text{or } \Delta T/\Delta S < a \qquad (2),$$

where ΔT and ΔS are the temperature and salinity differences through the density interface between the upper layer freshened by the river runoff and the underlying layer of cold and saline water (Figures 2, 3) ($\Delta T > 0$, $\Delta S > 0$). Since supercooling and formation of frazil ice are possible only at a more intense heat exchange through the density interface, as compared to salt (double-diffusion) (Krylov and Zatsepin, 1992; Voropayev et al. 1995; Golovin et al., 1996), the following condition should be fulfilled:

$$Ft > a \cdot Fs \qquad (3)$$

or in the density expression:

$$\alpha\, Ft / \beta Fs = (Kt/Ks) \cdot R\rho^{-1} > a \cdot (\alpha/\beta) \qquad (4)$$

where Ft and Fs are the heat and salt fluxes through the pycnocline, respectively. They have different signs, since Ft is directed downward from the warm freshened layer and Fs is directed upward from the saline underlying layer. Here $\alpha = -1/\rho \cdot (\partial\rho/\partial T) = 7 \cdot 10^{-5}\,(^{\circ}C)^{-1}$ is the thermal expansion coefficient, $\beta = 1/\rho \cdot (\partial\rho/\partial S) = 8 \cdot 10^{-4}$ is the salinity compression coefficient whose dimension is inverse to salinity and Kt and Ks are the effective heat and salt exchange coefficients through the density interface. Two important conditions for supercooling and frazil ice formation were obtained from (4) (Golovin et al., 1996):

$$R\rho > \beta/(\alpha \cdot a) \sim 200 \qquad (5)$$

$$Kt/Ks > 1 \qquad (6).$$

Here $R\rho = \beta\Delta S/\alpha\Delta T$ is the density ratio for the pycnocline between freshened and saline waters. The condition (6) reflects the fact that the most efficient frazil ice formation is possible with molecular-turbulent heat and salt exchange through the pycnocline (Krylov and Zatsepin, 1992). With purely turbulent exchange when $Kt/Ks \approx 1$, the effect of buoyancy is negligible. This occurs either when the pycnocline is weak or external turbulenc is very strong. This lowers the rate of frazil ice formation (Krylov and Zatsepin, 1992; Voropayev et al., 1995).

Under conditions of a purely molecular exchange through the pycnocline when the "molecular core" through which turbulence does not penetrate is preserved (Krylov and Zatsepin, 1992), the following condition should be fulfilled (Golovin et al., 1996) for supercooling and frazil ice formation to be possible:

$$\alpha\, Ft_m / \beta Fs_m < (\alpha Kt_m/\beta Ks_m) \cdot a = 0.48 \qquad (7).$$

Here $\alpha\, Ft_m$ and βFs_m are buoyancy flux associated with the molecular heat and salt fluxes through the pycnocline, and Kt_m and Ks_m are the molecular coefficients of temperature conductivity and salt diffusion respectively. The condition (7) can be presented in the form:

$$R\rho^{-1} \cdot (Kt_m/Ks_m) < 0.48 \qquad (8).$$

Let us check the conditions for supercooling and frazil ice formation (1), (2), (5) and (8) at the upper boundary of the pycnocline. For this purpose we will use the results of measurements at station LN9610 where measurements were conducted prior to the flood on May 22, on June 6, and on 11 - the flood period. Table 1 presents the corresponding estimates.

Table 1: Results of checking the conditions for supercooling and frazil ice formation in the zone of river-sea water contact at st. LN9610, LN9610a and LN9610b.

Station (date)	LN9610 (May 22)	LN9610a (June 6)	LN9610b (June 11).
$T_1 / S_1 \sim -a$	0.051	0.058	0.045
$DT / DS < a$	0.042	0.035	0.045
$Rr > 200$	272	303	255
$Rr^{-1}(Kt_m/Ks_m) < 0.48$	0.37	0.30	0.39

As is seen from the table, all conditions for the occurrence of supercooling and frazil ice formation at the lower boundary of the freshened layer and in the upper part of the pycnocline in the region of the Trofimovskaya branch are fulfilled. These conditions are also fulfilled in the other near-delta regions where an intense inflow of river water was observed. Thus, even if there were no fine temperature and salinity measurements allowing us to directly determine the layers with supercooled water (Figure 2), then based on ratios (1), (2), (5) and (8), we would still be able to estimate the probability of supercooling in the river-sea water contact zone.

The dynamic aspect of frazil ice formation

Studies of double-diffusion frazil ice formation under laboratory conditions have shown that if the external turbulent effect in both layers of a two-layer system is absent, and the heat and salt exchange through the density interface between fresh and saline water is at the molecular level, then the value of supercooling is not large and the density interface is "diffused" by molecular diffusion (Krylov and Zatsepin, 1992; Martin and Kauffman, 1974; McClimans et al., 1978; Stigebrandt, 1981). Here less than several millimeters of frazil ice can form in a day. This important result is in good agreement with frazil ice growth observed in a summer lead filled with melt freshwater (Golovin et al., 1996). Obviously, such small ice formation rates cannot produce meter-layers of frazil ice observed near Arctic river mouths and in the coastal regions of Antarctica (Bulatov, 1963; Kozlovsky, 1971; Cherepanov and Kozlovsky, 1972).

Laboratory experiments showed (Krylov and Zatsepin, 1992) that under artificially created turbulence of a two-layer system and the increased heat-salt exchange through the pycnocline, the rate of frazil ice formation strongly increases. This was confirmed by later laboratory experiments (Voropayev et al., 1995). Turbulence in the layers and turbulent entrainment in the contact zone between freshened and saline waters provide for intense frazil ice formation, with ice crystals subsequently surfacing and sticking together (Krylov and Zatsepin, 1992, Voropayev et al., 1995).

Both-scale studies and laboratory experiments show that supercooling and frazil ice occur in a comparatively thin contact zone at the upper boundary of the pycnocline (Golovin et al., 1996, Krylov and Zatsepin, 1992, Voropayev et al., 1995). Hence, for investigating the rate of frazil ice formation depending on the level of the heat-mass exchange, the parameters and structure of turbulence near the density interface should be known.

In some laboratory experiments turbulence was generated by velocity shear (Turner, 1973; Kantha and Phillips, 1977; Kato and Phillips, 1969; Kan and Tamai, 1994). Such experiments were carried out in circulate flumes. Shear flow was either generated by a rotating plastic screen (Turner, 1973) or by belts installed (Kan and Tamai, 1994) on the bottom and surface of on of the flume straight parts. Roughness elements were attached on the belts. In other laboratory experiments turbulence was generated by oscillating grids (Turner, 1973; Krylov and Zatsepin, 1992; Voropayev et al., 1995). The grids attached to a single rod were mechanically activated. Turbulence induced by these means is called "grid" turbulence. Its structure near pycnocline differs from that of the turbulence generated by velocity shear (the latter predominates in natural environments). Hence, as noted by Turner (1973), this makes comparison of the results of these experiments difficult. However, since "grid" turbulence is more easily parametrizated, this way of turbulence generation is more frequently used in laboratory experiments studying the processes of properties transmission through pycnocline.

Studies of the rate of frazil ice formation (Krylov and Zatsepin, 1992; Voropayev et al., 1995) were performed by means of "grid" turbulence. In these experiments, parameterization of the non-dimensional entrainment velocity U_e/U_* through the pycnocline at turbulent mixing between the layers and interpretation of the results, including the rate of frazil ice formation, were performed by means of the local Richardson number $Ri_* = g \cdot (\Delta\rho/\rho) \cdot L/U_*^2$. The local Richardson number is an analogue of the global number, but the external scale of mean horizontal velocity U is replaced by the root-mean-square velocity of turbulent variations U_* near the density interface. Also, the thickness of the mixed layer Z is substituted to the integral scale of turbulence L. The latter characterizes the mean scale of the most energy-carrying eddies near the density interface which then penetrate the pycnocline and participate in turbulent entrainment (Prandtl, 1949; Turner, 1973; Shlikhting, 1974).

In order to use the ratios obtained by Krylov and Zatsepin (1992) for estimating the rate of frazil ice formation in the river/sea water contact zone based on full-scale studies, we should correctly relate the mean horizontal velocity in the freshened layer U to the root-mean-square velocity of turbulent variations U_*, as well as determine the integral turbulence scale L near the pycnocline. Turner (1973) suggested considering entrainments at the upper pycnocline boundary due to turbulent motion of the upper layer to be entrainment at flat turbulent jet margin. In this case the momentum and mass balance between layers would be probably reached, and the rate of turbulent entrainment would be quasisteady.Current laboratory studies of the transfer processes through the density interface confirm this interpretation (Kan and Tamai, 1994). This allows us to use the experimentally well-tested theory of turbulent entrainment at the expanding margin of a flat turbulent jet (Prandtl, 1949; Shlikhting, 1974) for relating the external mean velocity of the layer U to the mean-root-square velocity of turbulent variations U_* near the pycnocline.

Where a turbulent flat jet mixes with the ambient unmoving fluid the length of the mixing distance corresponding to the integral scale of turbulence L, according to the theoretical and laboratory studies (Prandtl, 1949; Shlikhting, 1974), is proportional in each cross-section to the width of the jet b in this section:

$$L = \alpha_v \cdot b \qquad (9)$$

where α_v is the entrainment constant, varying from 0.08 to 0.12 in different experiments (Turner, 1973). Prandtl (1949) and Shlikhting (1974) determined its value to be 0.125. In our case, where entrainment at the boundary of river and sea water is quasi-steady, the width of the jet b will represent the thickness of the freshened layer (Figure 3). The mean velocity shear and eddies at the upper pycnocline boundary produce the internal waves that become unstable thus generating vorticity in the pycnocline (Turner, 1973; Kan and Tamai, 1994). As a result, part of the fresh water penetrates the more saline and cold water and part of the more saline water is entrained into the turbulent, swiftly moving river water, where it is rapidly mixed (Kan and Tamai, 1994). Thus, in the vorticity area of the pycnocline, there is active contact of saline and freshwater leading to supercooling. The thickness of this layer and the rate of supercooling depend on the characteristic scale of the energy carrying eddies which move in the mean flow generating vorticity in the pycnocline. In turn, the scale of these eddies depends on the external scale of the horizontal velocity in the freshened layer (Phillips, 1977). It follows from the above that the thickness of the supercooled layer in the pycnocline is actually the integral scale of turbulence L penetrating the pycnocline. We can determine it directly from the observational materials by analyzing the profiles $\Delta T_f = T - \tau_s$ (Figure 2).

According to the theory of Prandtl (1949), an additional tangential stress is created for free turbulence at the entrainment boundary of the parallel turbulent jet with unmoving fluid:

$$\tau = \rho \cdot \alpha_v{}^2 \cdot U_{max}{}^2 \qquad (10),$$

where U_{max} is the maximum speed in a flat turbulent jet. In the case under consideration it is a typical scale of the horizontal velocity ($U_{max} = U$) in the upper freshened layer (Figures 2, 3). Based on the semi-empirical turbulence theory (Prandtl, 1949), let us determine the friction velocity (or the dynamic velocity) in the contact zone of freshened and saline waters from (10) as:

$$U_* = (\tau / \rho)^{1/2} = \alpha_v \cdot U \qquad (11).$$

Prandtl (1949) showed that Reynolds stresses $U_* \cong (\overline{u \cdot v})^{1/2}$ characterize the intensity of turbulent (vertical - u' and horizontal - v') velocity variations and depend on the external velocity scale. Hence U_*, represented by the expression (11) determines the required velocity scale for the local Richardson number expressed through the external velocity scale U. As is seen from (11), U_* corresponding to $(\overline{u \cdot v})^{1/2}$, is approximately 10 times as small as the characteristic external velocity scale in the turbulent layer U, which is fully consistent with direct measurements of $(\overline{u})^{1/2}$, $(\overline{v})^{1/2}$ and Reynolds stresses $(\overline{u \cdot v})^{1/2}$ obtained from them (Kan and Tamai, 1994). The maximum value of turbulent velocity variations is observed in the thin upper and lower parts of the pycnocline where vorticity is generated and turbulent entrainment takes place (Kan and Tamai, 1994). Hence the expression for the local Richardson number for the pycnocline between river and sea water under conditions when the salinity field is governed by the density field ($\Delta\rho/\rho \cong \beta\Delta S$), will have the following form:

$$Ri_* = g \cdot \beta\Delta S \cdot L / \alpha_v{}^2 \cdot U^2 \qquad (12).$$

The above considerations concerning the interpretation of entrainment at the river-sea water boundary as entrainment at the boundary of a flat turbulent jet can be tested by a completely different method. Observations of the vertical hydrological structure at LN9610a and LN9610b stations performed during the time span between June 6 and 11 have been strictly bound to the same point on the fast ice. During the period of observation the upper pycnocline boundary displayed considerable downward shift (Figure 2). If considering the turbulent entrainment at the pycnocline/upper layer boundary to be responsible for such changes in the vertical hydrological structure, then its rate U_e can be directly determined. Intensive supply of flood water into the near-deltaic sea area during this time period (especially via the main branches) resulted in gradual entrainment of the underlying saline water into the riverine one together with their subsequent mixing. The U_e value was about $6 \cdot 10^{-4}$ cm/s. It should be noted that entrainment process is usually thought to be related to temperature and salinity variations in either the upper layer (if it is more turbulent than the lower one) or in both layers if their turbulence characteristics are similar (Turner, 1973). However, it should be remembered that indirect identification of the turbulent entrainment process at the density interface has been applied in an enclosed volume of water in course of the laboratory experiments (Turner, 1973). It is apparent that it is impossible to apply it for natural conditions since the boundaries are not closed. Moreover, in our case, when the upper layer remains fresh due to continuous flood water inflow to the sea, identification of such kind is inapplicable.

According to basic experiments by Turner (1973), at large values of the local Reynolds number $Re_* \cong \alpha_v \cdot U \cdot L/\nu$ (in our case it changes from 8400 to 84000 if U changes from 2 cm/s to 20 cm/s), at the mean thickness of the supercooled fluid in the pycnocline $\overline{L} \cong 35$ cm and at large Peclet numbers for salt $Pe_{*s} = Re_* \cdot (\nu/Ks_m)$ and heat $Pe_{*t} = Re_* \cdot (\nu/Kt_m)$, the non-dimensional ratio for the entrainment velocity U_e/U_* is function only of Ri_*. It approaches the constant value at $Ri_* \to 0$ (purely turbulent exchange regime). At the intermediate values of Ri_* (molecular-turbulent exchange regime), the change occurs according to the fundamental law determined by Turner (1973): $U_e/U_* \sim Ri_*^{-3/2}$. In experiments investigating frazil ice formation (Krylov and Zatsepin, 1992) the non-dimensional entrainment velocity for salt is well approximated by the same dependence:

$$U_e / U_* = A \cdot Ri_*^{-3/2} \qquad (13),$$

where A=0.56 is an empirical constant. From (13), knowing the mean thickness of the supercooled water layer $\overline{L}\cong35$ cm during the June 6 to 11 period, mean salinity difference in the pycnocline $\overline{\Delta S}\cong18.44$ and U_e (given above), and also considering (11), we can determine the scale of horizontal velocity in the upper freshened layer U. It equals 10-15 cm/s. Hence, according to (11), U* is about 1.3-1.9 cm/s. The estimated scale of horizontal velocity U in the upper layer seems to be high, but it has been obtained at the station situated at the traverse of the Bykovskaya branch (see Figure 1) which is the main channel for the Lena river fresh water runoff. Hence U value is quite realistic. Such independent evaluation of U and U* confirms our interpretation of the process of turbulent entrainment at the river/sea water boundary as an entrainment process at the flat turbulent jet margin.

The above given description of the approach used for determination of the local turbulence characteristics at the freshened layer-pycnocline boundary does not include evaluation of the Coriolis force influence, since it is insignificant and can be avoided. Similar approach to determination of the rate of entrainment at the upper pycnocline boundary (geophysical application) was used by Turner (1973). Besides this, according to Phillips (1977), at $Z< U_* / \Omega$, the influence of the Coriolis force can be neglected. Here Z is depth, Ω is angular speed of the Earth´s rotation. The U_* / Ω value for the above U_* estimates is about ≈260 m. Sea water depths in the region investigation (Figure 1) are only 6-20 m. So, it is obvious that during flood the Coriolis force does not influence the local turbulence patterns at the freshened layer/pycnocline boundary in the shallow sea area around the Lena river delta.

Estimate of the rate of frazil ice formation

As a result of laboratory experiments of frazil ice formation (Krylov and Zatsepin, 1992), the expression for estimating the rate of its formation was obtained as:

$$V_i = B \cdot U_* \cdot \Delta S \cdot Ri_*^{-1/2} \cdot \{(\Delta T/\Delta S - (7 \cdot a/Ri_*)\} \quad (14),$$

where $B = 3.3 \cdot 10^{-3}$ (°C)$^{-1}$ at the density of frazil ice $\rho_i = 0.1$-0.3 g/cm^3 (Weeks and Ackley, 1982). An approximately similar, rough estimate of frazil ice density (≈0.2 g/cm^3) was obtained by measuring the water volume after melting a sample of frazil ice of known volume (at st. LN9610b). Here $\Delta T=T_1-T_2>0$ and $\Delta S=S_2-S_1>0$ are the temperature and salinity differences through the pycnocline. Based on the above interpretation of entrainment at the interface, let us determine from (12) the value of the local Richardson number Ri_* for the pycnocline between river and sea water at st. LN9610, LN9610a and LN9610b for various values of the external velocity scale in the freshened layer U at actually determined L and ΔS. The estimates are presented in Table 2. According to the classification by Krylov and Zatsepin (1992), at the local Richardson number $Ri_* >>10^2$ the exchange through the pycnocline occurs at the molecular level. At $Ri_*\approx10^2$ with the decrease in the density difference in the pycnocline or the increase in the external velocity scale in the turbulent layers, the "molecular core" converges. This leads to the molecular-turbulent regime of mass- and heat exchange. At this regime the exchange of properties through the pycnocline mainly occurs due to sporadic outbursts of turbulence (eddy formation) in the area of a strongly sharpened density interface. However, turbulence in the interface area is significantly influenced by buoyancy and is not pronounced. This is probably the reason for the different effective exchange coefficients through the pycnocline at this regime. At $Ri_* <10$ there is a purely turbulent exchange regime. At $Ri_* < 2$ when stratification does not influence turbulence, the regime becomes unsteady resulting in a rapid mixing of layers.

Based on this classification, one see can from Table 2 that during the period prior to the flood, at mean velocity in the freshened layer of only about 5 cm/s, the regime of the molecular-turbulent mass-heat exchange already appears in the pycnocline, and at 15-20 cm/s there is a purely turbulent exchange regime. During the flood outflow of river water, when the upper layer becomes strongly freshened (Figures 2, 3), the molecular-turbulent exchange regime through the pycnocline contributing to intensification of frazil ice formation, occurs only at $U \approx 15$-20 cm/s and the turbulent regime at $U > 40$-50 cm/s.

Table 2: Estimates of the local Richardson number (Ri_*) for the pycnocline between river and sea water for stations LN9610, LN9610a and LN9610b at different velocities of river water transport in the upper layer

Station (date)	LN9610 (22.05)	LN9610a (06.06)	LN9610b (11.06)
U = 5 cm/s	106	1360	1167
U = 10 cm/s	26	340	292
U = 15 cm/s	12	151	130
U = 20 cm/s	6.6	85	73
U = 40 cm/s	-	21	18
U = 50 cm/s	-	13.6	11.7
U = 60 cm/s	-	9.4	8.1

Using Ri_* values from Table 2 and determining U_* from (11), let us estimate by means of (14) the real rate of frazil ice formation V_i for different values of the typical motion velocity of river water U in the upper freshened layer. Figure 5 presents the estimates of the rates of frazil ice formation in the contact zone between river and sea water at stations st. LN9610, LN 9610a and LN9610b.

Analysis of the calculated rates of frazil ice formation V_i shows that before the flood the rates are insignificant though sometimes reach 20 cm/day.This rate is typical for the outflow regions of the Trofimovskaya and Bykovskaya branches. Approaching peak river discharge, the rate of frazil ice production increases considerably and can reach 1.7 m a day (Figure 5). This equivals to ≈34 cm of regular sea ice formation if porosity of frazil ice layer is 80 % (Weeks and Ackley, 1982). The maximum rate of frazil ice formation is observed at U ~ 40-50 cm/s. With the increase of U the value V_i is sharply reduced (Figure 5). This is governed by the fact that at certain actual density stability of the pycnocline, a large increase in U leads to the turbulent ($Ri_* < 10$, see Table 2) and then to the unsteady regime ($Ri_* < 2$). Buoyancy is no longer influence the character of the heat -mass exchange through the pycnocline. As a result, the rate of frazil ice formation decreases as the values of the effective salt and heat exchange coefficients become equal ($Kt \approx Ks$). This reduces the efficiency of double-diffusion supercooling or makes it impossible. The conditions (3) and (6) are not fulfilled. A similar conclusion also follows from laboratory experiments (Krylov and Zatsepin, 1992; Voropayev et al., 1995).

A frazil-ice production rate of 1.5 m/day seems to be high, but the observed supercooling in the pycnocline at this time is also very large (Figures 2b, c). Such high ice-production rates will lead to speculations that fast ice thickness could increase accordingly. However, much of this new ice probably also melts locally, since toward the conclusion of our field work, river water with above zero temperature was already spreading below fast ice (Figure 2). A portion of the frazil ice probably is exported with river water farther to the sea, where it may add to the bottom surface of fast or drifting ice.

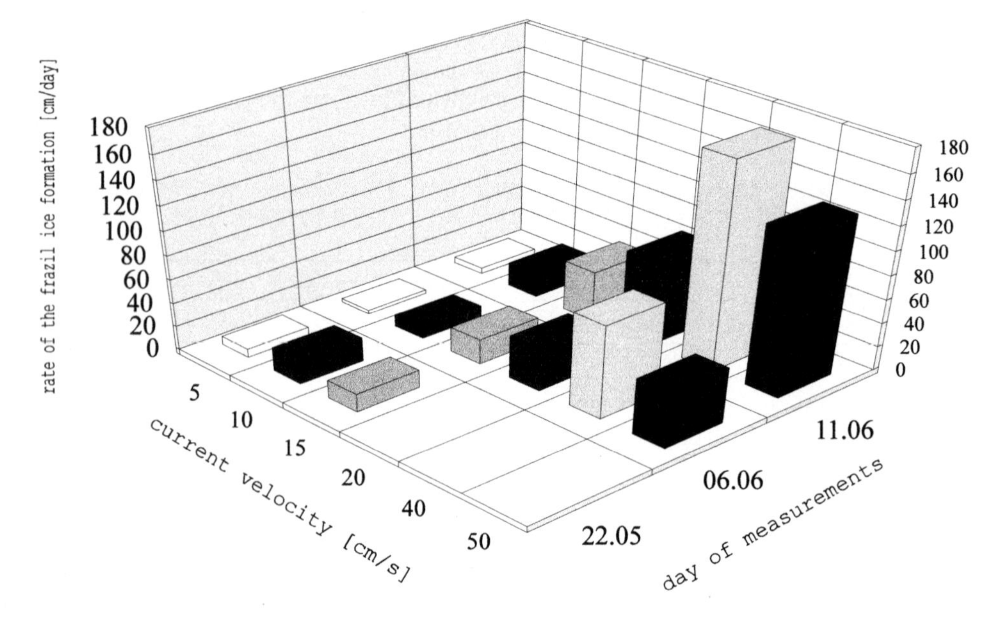

Figure 5: Estimates of the rate of frazil ice formation at st. LN9610 (22.05), LN9610a (06.06), LN9610b (11.06). at different current velocities in the upper freshened layer.

Along with the formation of frazil ice, sediments transported by river water are trapped. The mechanism of frazil ice formation plays quite a significant role for sediment transport to the surface of both fast and drifting ice. Thus at the rate of frazil ice formation of 170 cm a day at st. LN9610b (see Figure 5), the sediment flux to the fast ice can reach 7 g/m^2 a day at the actually measured sediment level in river water of 19.2 mg/l. In this case sediments were assumed to be entirely incorporated into frazil ice at its formation. Laboratory experiments have shown (Reimnitz et al., 1993) that for the actually observed granulometric composition of suspended sediments their significant increase in crystals of surfacing frazil ice can take place, as compared to the surrounding river water. Thus the rate of the sediment flux to the surface might be much higher. This fact was experimentally recorded during the Transdrift IV expedition. The level of suspended and dissolved matter in an agglomeration of frazil ice was 1264.1 mg/l, whereas in the upper freshened layer it was 18.89 mg/l and 19.26 mg/l at 5 m and 3 m depth, respectively.

Conclusions

Application of formula (14), obtained for a specific laboratory model for estimating the rates of frazil ice formation under natural conditions, can cause justified criticism. However, the results of studies in the Transdrift IV expedition made their interpretation possible on the basis of the laboratory experiments. This has permitted us to obtain quite important quantitative estimates of the rates of frazil ice formation and gain understanding of their spatial-temporal variability during the developing spring flood in the pro-delta region of the Lena river. These approaches can also be applied to other Arctic Seas subjected to high river discharge.

During the peak of its production frazil ice also entraps sediments transported by river water. When rising to the surface, frazil ice crystals can be incorporated into surface ice. The formation of frazil ice in the region where Arctic rivers discharge into the seas can be more or less intense during the whole year. This makes the mechanism of frazil ice formation and associated incorporation of sediments into the ice cover very important sediment transport and deposition in these regions. Future studies of the physical and sediment entrainment in pro-deltas also contribute to long-range transport in the Transpolar Drift.

Acknowledgments

This article was initiated and written during the joint work of Russian and German authors in 1996 at GEOMAR, Kiel. We are grateful to the German Ministry of Education, Science, Research and Technology, who provided this possibility. From the Russian side this work was supported by the Ministry for Science and Technology under the Russian-German project „Laptev Sea System".

References

Bulatov, R.P. (1963) Some results of ice studies in the Yenisey Gulf (in Russian). Voprosy Geografii, 62, 192-197.

Cherepanov, N.V. (1972) Systematizing of crystal ice structures in the Arctic (in Russian). Problemy Arktiki i Antarktiki, 40, 78-83.

Cherepanov, N.V. and A.M. Kozlovsky (1972) Frazil ice in coastal water of the Antarctic (in Russian). Inform. Bul. SAE, 84, 61-65.

Fofonoff, N.P. and R.C.Jr. Millard (1983) Algorithms for computation of fundamental properties of sea water. Unesco technical paper in marine science, 44, 53.

Golovin, P.N., S.V. Kochetov and L.A. Timokhov (1993) Features of thermohaline structure of fractures in summer in the Arctic ice (in Russian). Okeanologiya, 33, 6, 833-838.

Golovin, P.N., V.V. Lukin and A.G. Zatsepin (1996) Frazil ice formation in the summer arctic fracture (in Russian). Okeanologiya, in press.

Kan, K. and N. Tamai (1994) Direct measurements of the mutual-entrainment velocity at a density interface. In: Preprints of Fourth Symp. on Stratified Flows, Vol. 4, Grenoble, France.

Kantha, L.and O.M. Phillips (1977) On turbulent entrainment at a stable density interface. J.Fluid Mech., 79, 753-768.

Kato, H. and O.M. Phillips (1969) On the penetration of a turbulent layer into stratified fluid. J. Fluid. Mech., 37, 643-655.

Kozlovsky, A.M. (1971) Frazil ice in the Alasheyev Gulf (in Russian). Proc./ SAE, 47, 222-224.

Krylov, A.D. and A.G. Zatsepin (1992) Frazil ice formation due to difference in heat and salt exchange across a density interface. J. Marine Systems, 3, 497-506.

Martin, S. (1971) Frazil ice in rivers and oceans. Ann. Rev. Fluid Mech., 13, 379-397.

Martin, S. and P. Kauffman (1974) The evolution of under ice melt ponds, or double-diffusion at the freezing point. J. Fluid Mech., 64, 507-527.

McClimans, T.A., C.E. Steen and G. Kjeldgaard (1978) Ice formation in fresh water cooled by a more saline underflow. In: Proc. IAMR Symp. ice problems, Pt.2, 331-336.

Nansen, F. (1956) "Fram" in the polar sea.(in Russian). Gosizd. geograf. literatury, Moscow, 384 pp.

Petrov, I.G. (1971) Experience of regioning of the ice cover of the Arctic Seas by structure (in Russian). Proc./AARI, 300, 39-55.

Phillips, O.M. (1977) The dynamics of the upper ocean. Cambridge university press, 380 pp.

Prandtl, L. (1949) Hydroaeromechanics (in Russian). Foreign literature Publishers, Moscow, 520 pp.

Reimnitz, E., J.R.Clayton, E.W. Kempema, J.R. Payne and W.S. Weber (1993) Interaction of rising frazil with suspended particles: tank experiments with applications to nature. Cold Regions Science and Technology, 21, 117-135.

Shlikhting, G. (1974) The theory of the boundary layer (in Russian). Nauka, Moscow, 712 pp.

Stigebrandt, A. (1981) On the rate of ice formation in water cooled by a more saline sub-layer. Tellus, 33, 6, 604-609.

Timokhov, L.A.(ed.) (1989) Vertical structure and dynamics of the sub-ice layer of the ocean (in Russian). Gidrometeoizdat, Leningrad, 141 pp.

Turner, J.S. (1973) Buoyancy effects in fluids. Cambridge Univ. Press, 367 pp.
Voropayev, S.I., H.J.S. Fernando and L.A. Mitchell (1995) On the rate of frazil ice formation in polar regions in the presence of turbulence. J. of Physical Oceanography, 25, 6, Part II, 1441-1450.
Weeks, W.F. and S.F. Ackley (1982) The growth, structure and properties of sea ice. In: CREL Monogr., 82-1, U.S. Cold Region Research and Engineering Lab., Hanover; N.H., 130 pp.

Section B

The Marine Ecosystem

Pelagic-Benthic Coupling in the Laptev Sea Affected by Ice Cover

C. Grahl[1], A. Boetius[2] and E.-M. Nöthig[1]

(1) Alfred-Wegener-Institut für Polar- und Meeresforschung, Postfach 120161, D 27515 Bremerhaven, Germany

(2) Institut für Ostseeforschung, Postfach 301161, D 18112 Rostock/Warnemünde, Germany

Received 1 August 1997 and accepted in revised form 14 November 1997

Abstract - During two expeditions with RV *Polarstern* to the Laptev Sea (ARK IX/4 in September/October 1993 and ARK XI/1 in July/August 1995), sediment samples were collected from the continental shelf edge to the deep Arctic basins from water depths between 37 m and 3831 m. Pigment concentrations in the surface sediments were determined as a biomarker for the input of phytoplankton material by sedimentation (Pfannkuche 1992). Furthermore, hydrolytic activities of the extracellular enzyme β-glucosidase in the sediments were measured as an indicator of microbial heterotrophic activity linked to the availability of plankton detritus (Boetius and Lochte 1994).
In summer 1993, the eastern Laptev Sea was ice-free and the western section was ice-covered. In contrast, in summer 1995, the shelf edge of the Laptev Sea was ice-free during the whole summer and only the eastern part of the slope was ice-covered. However, in both years, chlorophyll *a* equivalents (chlorophyll *a* + phaeopigments) and enzyme activities in the surface sediments (0-1 cm) decreased from the shelf edge to the bottom at the continental slope more than 10fold with increasing water depth. Highest pigment concentrations and microbial enzymatic activitiy was found at the shelf edge of the Eastern Laptev Sea in summer 1993, in an area that had been crossed several times by the ice edge. However, in general, variations in the ice cover during the Arctic summer seem to have a lesser effect on phytodetritus deposition at the continental slope than water depth and/or distance from the shelf area.

Introduction

Life at the bottom of the ocean mainly depends on the input of organic material from the euphotic zone (Gage and Tyler, 1991). Depending on the climatic and hydrographic regimes in an oceanic region, this food supply can be considered as a rain of organic material with little variation or may be subject to strong seasonal events (Wefer, 1989). In polar regions, primary productivity and thus the export of plant detritus to the seafloor is largely restricted to the summer season when light is available and when the ice cover is receding. The relation between pelagic and benthic processes in Arctic polar waters was summarized by Grebmeier and Barry (1991). They found a direct impact of the input of organic material (influenced by variability in hydrography, sea ice cover, light supply, and pelagic food web structure) on the abundance and biomass of benthic communities. However, their Arctic studies were restricted to the shallow shelf areas of the highly-productive Bering and Chukchi Seas. To date, little is known on pelagic-benthic coupling in the Eastern Arctic Seas (East Siberian, Laptev, Kara Seas) which are characterized by low nutrient concentrations and high ice coverage during summer.

The influence of water-depth, ice cover, and distance from the coast on the input of phytoplankton detritus to the sediment was studied during two cruises to the Laptev Sea, ARK IX/4 in 1993 (Fütterer, 1994) and ARK XI/1 in 1995 (Rachor, 1997). The data of the cruise in 1993 introduced new perspectives to the discussion of the factors controlling benthic microbial activities and biomass in deep-sea sediments (Boetius et al., 1996; Boetius and Damm, 1997; Boetius, in press). Here data from the two expeditions are compared and implications for pelagic-benthic coupling in the outer Laptev Sea are discussed. In our study, processes related to pelagic-benthic coupling via transport of organic matter from the euphotic zone were

In: Kassens, H., H.A. Bauch, I. Dmitrenko, H. Eicken, H.-W. Hubberten, M. Melles, J. Thiede and L. Timokhov (eds.) Land-Ocean Systems in the Siberian Arctic: Dynamics and History. Springer-Verlag, Berlin, 1999, 143-152.

investigated from a "benthic view", focussing on pigment concentrations and microbial activities in the surface sediments of the Laptev Sea slope.

Material and methods

During two Arctic expeditions (ARK IX/4 in 1993 and ARK XI/1 in 1995) with RV *Polarstern* to the Laptev Sea, sediment samples were collected in summer 1993 and 1995 from 23 stations along three transects and from 40 stations along six transects respectively. Water depths ranged from 37 to 3831 m. Samples were taken with a multiple corer (Barnett et al., 1984) and were transferred immediately to a refrigerated laboratory (0 °C) after recovery. Subsamples from the surface sediment (0-1 cm) were shock-frozen at -80 °C for later analyses of the pigment concentrations. After extraction with acetone (90 %) from the sediments, chlorophyll *a* and phaeopigments were determined fluorometrically with a Turner fluorometer according to Shuman and Lorenzen (1975). The sum of chlorophyll *a* and phaeopigments is expressed as chlorophyll *a* equivalents as described by Thiel (1982). Activity of the enzyme β-glucosidase, cleaving β-glycosides like cellulose, was measured immediately after subsampling at *in situ* temperature (0 °C). The maximum velocity (V_{max}) of the enzyme was determined at saturation level of the methylumbelliferone (MUF) labelled substrate MUF-β-glucoside according to Boetius and Lochte (1994).

Results and discussion

During summer in polar waters, the light availability is controlled by snow and ice cover. Primary production is drastically reduced under ice in comparison to the open water (Sakshaug and Slagstad, 1991; Wassmann et al., 1991; Smith 1995). In the Laptev Sea, the extent of the ice cover varies strongly between seasons and also interannually (Eicken et al., 1997). On an annual average, the eastern section has less ice cover during summer because of the generally prevailing wind direction and by the inflow of warmer water from the river Lena. In 1993, the eastern Laptev Sea was ice-free and the western section was ice-covered during summer (Boetius and Damm, 1997). In 1995, the shelf edge of the Laptev Sea was ice-free during the whole summer and only the eastern part of the slope was ice-covered (Figure 1). Data from the investigations in 1995 in the Laptev Sea indicate that the 1-2 m thick ice cover prevented the build-up of high phytoplankton standing stocks in the water column: Primary production in the ice-free waters reached 340 mg C m^{-2} d^{-1}; total production under one square meter of sea ice was less than 20 mg C d^{-1} (Grossmann and Gleitz, 1997). Accordingly, nutrient concentrations were mostly higher under sea ice than in open water (Nalbandov, 1997). Nutrient exhaustion was measured only at the shelf and slope stations along transect H in 1993 (Luchetta et al., 1994), where a phytoplankton bloom occurred along the receding ice edge. In 1995, no such bloom was observed during the cruise (Bartel, 1997), but at some of the shallower stations nutrient concentrations were low (Nalbandov, 1997).

Average concentrations of chlorophyll *a* equivalents in the sediments of the ice-free Arctic slope were comparable to those from continental slopes in temperate oceanic regions (Pfannkuche, 1985; Bovée et al., 1990; Scheibe 1990). However, they were considerably lower under the permanent ice cover in the central Arctic basin than those reported from the central NE-Atlantic abyssal plains at similar depths (Pfannkuche, 1992). During both Laptev Sea expeditions, chlorophyll *a* equivalents (Figure 2) and the potential activity of the enzyme β-glucosidase (Figure 3) in the surface sediments (0-1 cm) decreased with increasing water depth (100-3000 m) ten-fold and six-fold along the ice-free transects in 1993 and 1995 respectively.

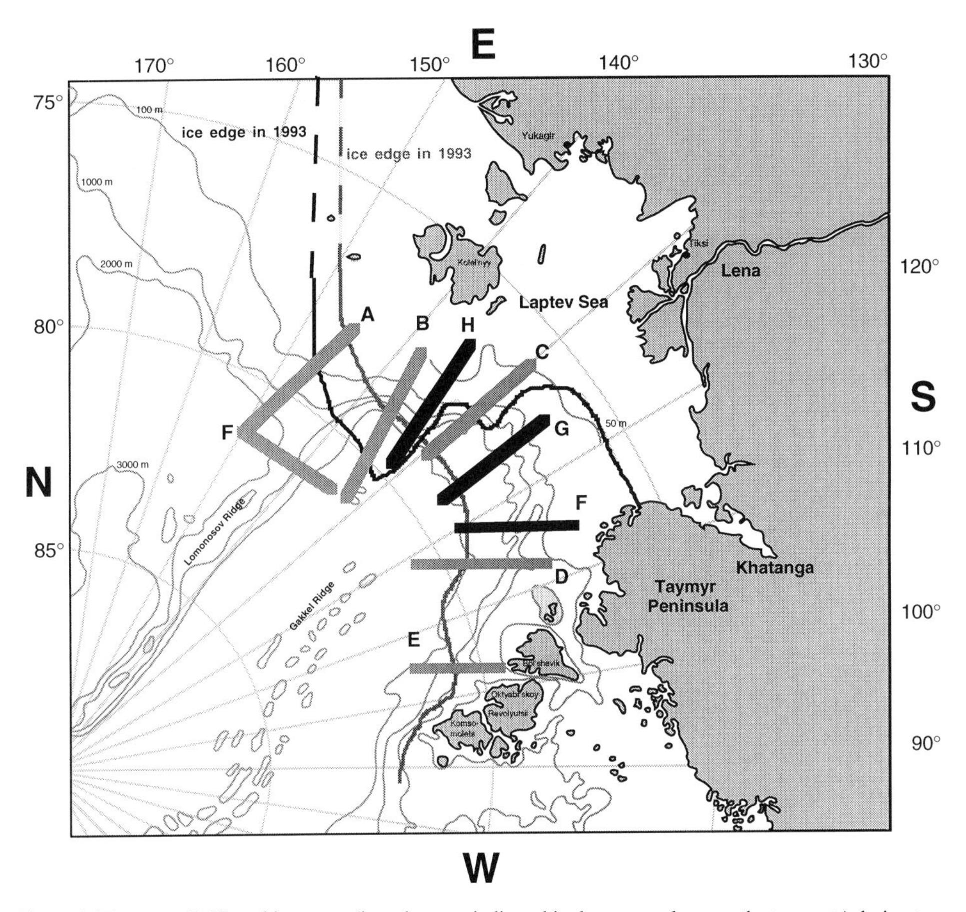

Figure 1: Transects (A-H) and ice cover (ice edges are indicated in the same colours as the transects) during two arctic expeditions in the Laptev Sea. ARK IX/4 (6 August-5 October 1993); F, G, H (1.9.-18.9.). ARK XI/1 (7 July-20 September 1995); A, B, C, D, E, F (24.7.-8.9.).

Along the ice-covered transects, a four-fold decrease in both parameters was observed in 1993 as well as in 1995. Chlorophyll *a* equivalents were 13.15-0.92 µg cm^{-3} in 1993 and 5.54-1.41 µg cm^{-3} in 1995 at the ice-free stations. V_{max} of the enzyme β-glucosidase ranged from 3.02-0.12 µM h^{-1} in 1993 and 2.9-0.36 µM h^{-1} in 1995. At the ice-covered stations chlorophyll *a* equivalents were 3.58-0.56 µg cm^{-3} in 1993 and 3.74-0.19 µg cm^{-3} in 1995 and the β-glucosidase activities ranged from 0.9-0.02 µM h^{-1} in 1993 and 1.87-0.08 µM h^{-1} in 1995 (Table 1). Although the chlorophyll *a* equivalents were twice as high in 1993 compared to 1995 at the ice-free shelf edge stations, V_{max} of the enzyme β-glucosidase was at the same level in both years. The half-life of chlorophyll pigments is approximately three weeks in polar sediments (Graf et al., 1995), however, nothing is known on the persistence of extracellular enzymes in deep-sea sediments. Experiments on the regulation of β-glucosidase of natural microbial assemblages in Arctic sediments have shown that after a single input of substrate the enzyme activity increases substantially within 10 days and might stay at a high level for at least 50 days (Boetius and Lochte, 1996). Thus, changes in enzyme activity and pigment concentrations could occur at different time scales. Nevertheless, the correlation between both

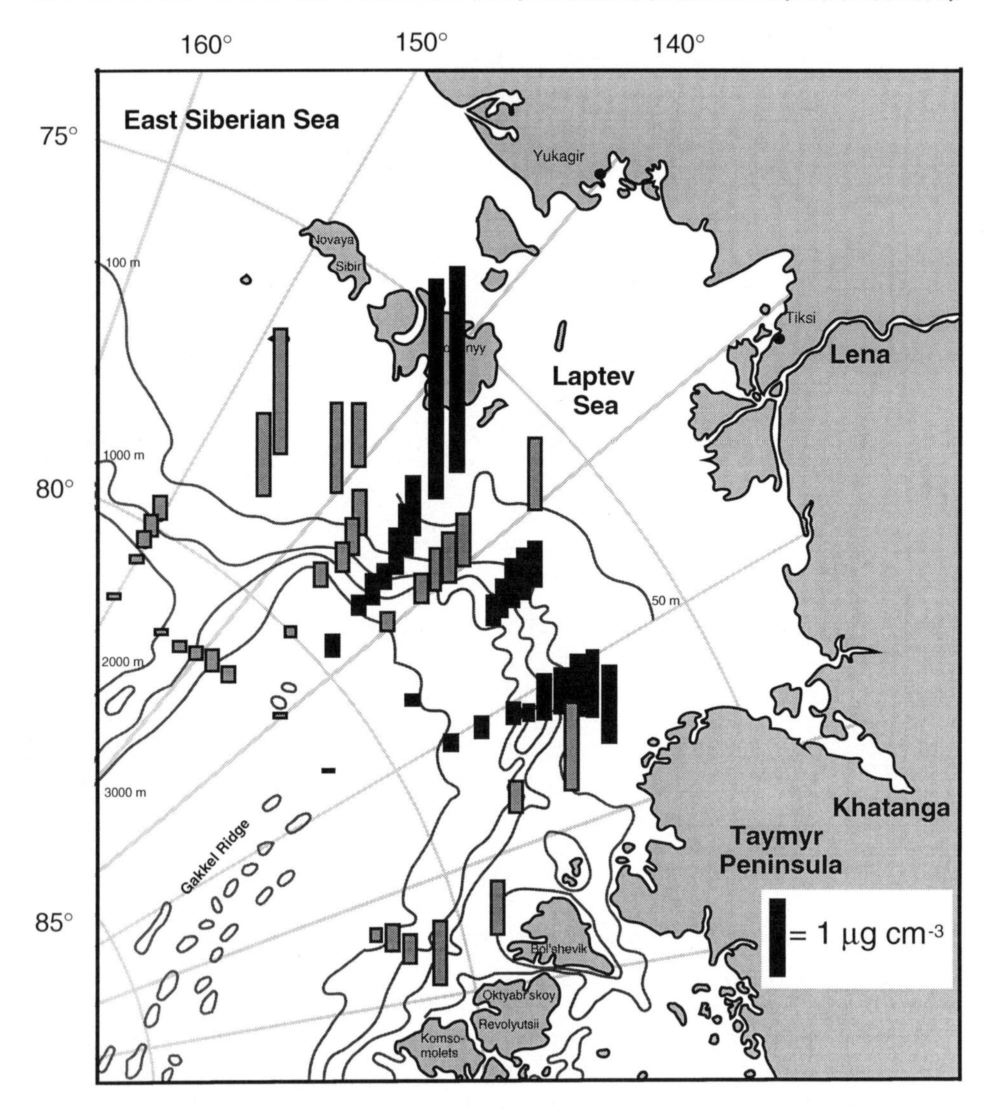

Figure 2: Distribution of chlorophyll *a* equivalents (chlorophyll *a* and phaeopigments) in the Laptev Sea. Columns represent values in the surface sediments (0-1 cm). ARK IX/4. ARK XI/1.

parameters was as high as r=0.70 for all samples (n=58).

Only 10-20 % of the organic matter produced in the euphotic zone is exported since most of it is remineralized (Martin et al., 1987). During sinking the particles are further degraded due to zooplankton grazing, bacterial lysis and chemical processes (Karl et al., 1988). Therefore, particulate organic matter flux to the sediment decreases exponentially in quantity and additionally in quality with increasing water depth (Honjo et al., 1988; Suess, 1980). These processes could be responsible for the large reduction in pigment concentrations and microbial activity in the Arctic slope sediments, and Wheeler et al. (1996) found that heterotrophic degradation processes in the Arctic water column function as efficiently as in other oceanic

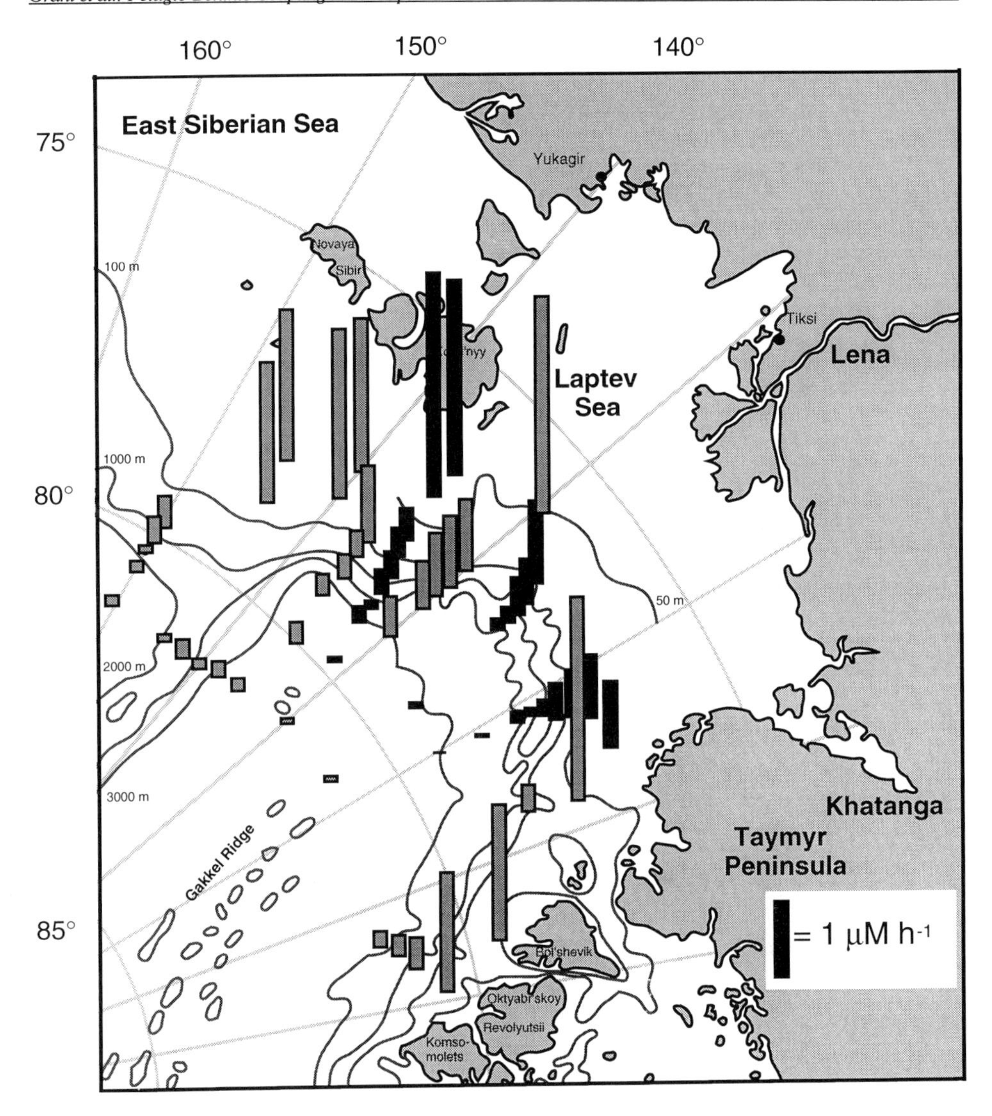

Figure 3: Distribution of enzymatic activity of β-glucosidase in the Laptev Sea. Columns represent values in the surface sediments (0-1 cm). ARK IX/4. ARK XI/1.

regions. However, neither particulate organic matter (POM) flux rates in the deeper water column nor the probably substantial contribution of laterally advected organic matter from the more productive, extremely large shelf areas of the Arctic Ocean have been investigated yet (see Walsh et al., 1988). A regression analysis of the data presented in Table 1 indicated that depth, i.e. the loss in POM flux with increasing water depth, was the most important factor limiting food supply to the benthos in bathyal and abyssal regions of the slope, explaining 88% of the variability, whereas the ice cover at the time of sampling would only explain 46% of the variability. Furthermore, the effect of the ice cover on pigment concentrations in the sediments was not significant when the values at ice-covered and ice-free stations at the three slope sections upper (<250 m, p=0.37), middle (250-2500 m, p=0.10) or bottom of the slope

(>2500 m, p=0.15) were compared. However, since ice conditions might quickly change during spring and summer in the Laptev Sea, a better relationship might only be detectable if the variation in ice coverage over periods of weeks to months could be taken into account (see Figure 4).

If open water areas persist long enough, nutrients become the limiting factor in the oligotrophic outer Laptev Sea (Luchetta et al., 1994). In the Laptev Sea nitrate concentrations are rather low and can be depleted by the phytoplankton within 1-3 weeks due to the stable stratification of the surface water layers. Thus, in late summer, plankton biomass and flux of plankton detritus to sediments below persistent open water areas might be as low as in completely ice-covered areas of the Laptev Sea due to nutrient exhaustion. In contrast, the

Table 1: Chlorophyll *a* equivalents (CPE) and β-glucosidase activity (β-Glu) at the different transects sampled in 1993 and 1995. Data are sorted according to transects and increasing water depth.

Transect	**CPE**	**β-Glu**	**Depth**	**Ice cover**	**Date**
1993	(μg cm^{-3})	(μM h^{-1})	(m)	(%)	(D/M)
H	9.35	2.61	37	0	01.09.
	13.15	3.02	54	0	06.09.
	2.73	0.45	213	0	05.09.
	2.36	0.41	532	10	05.09.
	2.12	0.36	796	0	05.09.
	1.19	0.33	2019	0	04.09.
	1.36	0.12	2534	0	03.09.
	0.92	0.25	2976	0	02.09.
	1.05	0.09	3427	80	03.09.
G	2.07	1.14	93	0	07.09.
	2.39	0.61	240	90	09.09.
	2.25	0.53	561	90	09.09.
	1.74	0.23	1009	90	08.09.
	1.34	0.15	1985	90	10.09.
	0.56	0.08	3237	100	12.09.
F	3.58	0.90	65	80	18.09.
	3.13	0.89	107	100	18.09.
	2.83	0.65	230	90	17.09.
	2.08	0.52	534	100	17.09.
	2.06	0.22	1104	90	17.09.
	0.82	0.14	1517	100	15.09.
	1.07	0.16	1935	90	15.09.
	1.07	0.06	2620	90	14.09.
	0.84	0.02	3051	80	13.09.

Table 1 (continued):

Transect 1995	CPE ($\mu g\ cm^{-3}$)	β-Glu ($\mu M\ h^{-1}$)	Depth (m)	Ice cover (%)	Date (D/M)
A	5.54	2.04	54	0	24.07.
	3.74	1.87	100	90	28.07.
	1.18	0.46	577	50	30.08.
	1.06	0.39	997	70	29.08.
	0.76	0.13	1619	70	29.08.
	0.46	0.15	2009	90	28.08.
B	2.87	2.08	39	0	30.07.
	4.06	2.29	49	0	30.07.
	2.11	1.05	550	0	02.09.
	1.66	0.36	978	0	01.09.
	1.33	0.33	1900	70	15.08.
	1.11	0.31	2758	80	18.08.
	0.52	0.30	3358	80	19.08.
	0.26	0.08	3831	90	21.08.
C	3.25	2.90	52	0	31.07.
	2.41	0.98	94	0	01.08.
	2.33	0.98	512	0	02.08.
	1.97	0.88	1787	0	03.08.
	1.41	0.65	2324	0	04.08.
	0.93	0.54	3147	80	05.08.
D	3.97	2.74	140	0	07.09.
	1.48	0.39	1183	0	06.09.
	0.19	0.08	3596	80	04.09.
E	2.36	1.81	264	0	12.08.
	2.94	1.60	254	30	08.09.
	1.37	0.43	1142	40	11.08.
	1.23	0.31	2770	70	08.08.
	0.60	0.22	3260	70	09.08.
F	0.74	0.19	2861	70	22.08.
	0.99	0.24	2026	80	23.08.
	0.65	0.16	1810	80	23.08.
	0.54	0.27	1276	80	24.08.
	0.36	0.12	1600	80	25.08.
	0.36	0.15	2589	90	27.08.

movement of the ice edge can lead to upwelling processes near the ice edge (Fennel, 1997) and thus to nutrient replenishment in the surface layer. Wind stress at the ice edge is mediated to the water below and can cause Ekman transport showing similarities to the processes at coastal boundaries. During summer 1993, several shifts of the ice edge across the upper slope section of transect H were detected by comparing satellite images provided by J. Kolatschek and T. Martin (AWI Physik I). The differences in the pigment concentrations and enzyme activity at the stations on the shelf edge of the Laptev Sea are shown in Figure 4. In contrast, pigment concentrations and enzyme activities in the sediments below the permanently open-water area at the transects A and C were as low as, or only slightly higher than, below the completely ice-covered and thus light-limited area of transects G and F. We suggest that the high pigment concentrations and microbial activity at the shallow stations of transect H could reflect such a positive effect of moving ice edges on primary productivity and thus POM export. Although higher enzyme activities indicate an earlier input of pigments (see below; Boetius and Lochte, 1996).

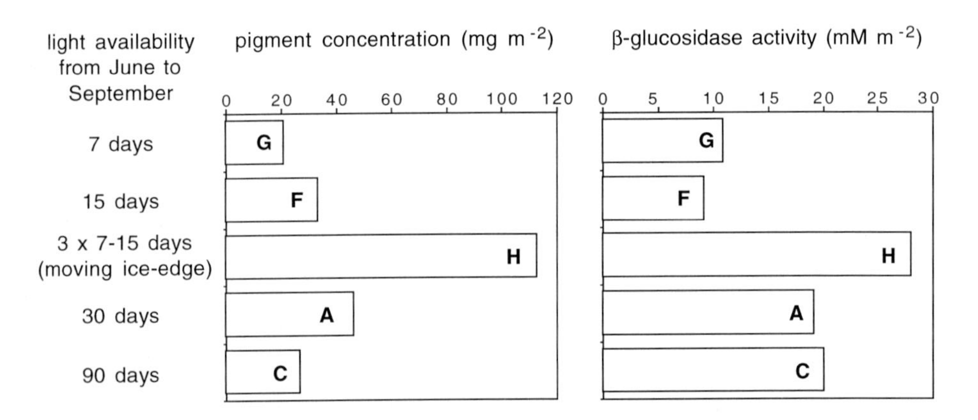

Figure 4: Pigment concentrations and microbial activities in surface sediments (0-1cm) of the shelf edge (37-107 m). The values are averages of 2 stations at each transect (1993: G, F, H; 1995: A, C). Duration of open water areas were estimated from DMSP (Defense Meteorological Satellite Program, USA) satellite pictures of ice concentrations kindly provided by J. Kolatschek and T. Martin (AWI Physik I).

The data from our expeditions to the outer Laptev Sea during two years of different ice coverage indicate that pelagic-benthic coupling is closely geared at the upper continental slope, the extent and persistence of the ice cover having a substantial effect on the input of POM to the Arctic sea floor. A deposition of chlorophyll *a* equivalents of 100 mg m^{-2} during summer might represent the upper limit for the input of phytodetritus to the outer shelf area of the Laptev Sea. Our data show that POM flux declines with increasing water depth down the slope, yet it remains an open question whether this is caused by the degradation of particles during sinking through the water column or by a decreasing advection of POM with increasing distance from the more productive shelf area of the Laptev Sea.

Acknowledgements

We thank the captains and the crew of the RV POLARSTERN for their valuable assistance at sea during both expeditions. We appreciate helpful discussion and comments on the manuscript from Karin Lochte and an unknown reviewer. Thanks to Anja Bartel, Christiane Lorenzen and

Karin Springer for technical assistance on the ship and in the lab. This is publication no. 1506 of the Alfred-Wegener-Institute for Polar and Marine Research.

References

Barnett, P.R.O., J. Watson and D. Connelly (1984) A multiple corer for taking virtually undisturbed samples from shelf, bathyal and abyssal sediments. Oceanologica Acta, 7, 399-408.

Bartel, A. (1997) Phytoplankton- and protozooplankton ecology and vertical particle flux. In: Scientific Cruise Report of the Arctic Expedition ARK-XI/1 of RV "Polarstern" in 1995, Rachor, E. (ed.), Reports on Polar Research, 226, 89-92.

Boetius, A. and K. Lochte (1994) Effect of organic enrichments on hydrolytic potentials and growth of bacteria in deep-sea sediments. Marine Ecology Progress Series, 140, 239-250.

Boetius, A. and K. Lochte (1996) Regulation of microbial enzymatic degradation of organic matter in deep-sea sediments. Marine Ecology Progress Series, 104, 299-307.

Boetius, A., C. Grahl, I. Kröncke, G. Liebezeit and E.-M. Nöthig (1996) Distribution of plant pigments in surface sediments of the eastern Arctic. In: Surface-sediment composition and sedimentary processes in the central Arctic Ocean and along the Eurasian Continental Margin, Stein, R., G.I. Ivanov, M.A. Levitan and K. Fahl (eds.), Reports on Polar Research, 212, 213-218.

Boetius, A. and E. Damm (1997) Benthic oxygen uptake, hydrolytic potentials and microbial biomass at the Arctic slope. Deep-Sea Research (in press).

Boetius, A. (in press) Availability of organic matter and microbial degradative activities in sediments of the Arctic continental slope. Deep-Sea Research.

Bovée, F. de, L.D. Guidi and J. Soyer (1990) Quantitative distribution of deep-sea meiobenthos in the northwestern Mediterranean (Gulf of Lions). Continental Shelf Research, 10, 1123-1145.

Eicken, H., E. Reimnitz, V. Alexandrov, T. Martin, H. Kassens and T. Viehoff (1997) Sea-ice processes in the Laptev Sea and their importance for sediment export. Continental Shelf Research, 17 (2), 205-233.

Fennel, W. (1997) Wind forced oceanic responses near ice edges. Journal of Marine Systems (in press).

Fütterer, D.K. (1994) The Expedition ARCTIC '93 Leg ARK-IX/4 of RV "Polarstern" 1993. Reports on Polar Research, 149, 244pp.

Gage, J.D. and P.A. Tyler (1991) Deep-sea biology. Cambridge University Press, Cambridge, 504pp.

Graf, G., S.A. Gerlach, P. Linke, W. Queisser, W. Ritzrau, A. Scheltz, L. Thomsen and U. Witte (1995) Benthic-pelagic coupling in the Greenland-Norwegian Sea and its effect on the geological record. Geologische Rundschau, 84, 49-58.

Grebmeier, J.M. and J.P. Barry (1991) The influence of oceanographic processes on pelagic-benthic coupling in polar regions: A benthic perspective. Journal of Marine Systems, 2, 495-518.

Grossmann, S. and M. Gleitz (1997) Primary and micro-heterotrophic productivity within ice-associated habitats. In: Scientific Cruise Report of the Arctic Expedition ARK-XI/1 of RV "Polarstern" in 1995, Rachor, E. (ed.), Reports on Polar Research, 226, 73-83.

Honjo, S., S.J. Manganini and G. Wefer (1988) Annual particle flux and a winter outburst of sedimentation in the northern Norwegian Sea. Deep-Sea Research, 35, 1223-1234.

Karl, D.M., G.A. Knauer and J.H. Martin (1988) Downward flux of particulate organic matter in the ocean: a particle decomposition paradox. Nature, 332, 438-441.

Luchetta, A., P. Poniz and G. Ilyin (1994) Nutrients and oxygen. In: The expedition Arctic '93 LEG-IX/4 of RV "Polarstern" 1993, Fütterer, K.D. (ed.), Reports on Polar Research, 149, 37-39.

Martin, J.H., G.A. Knauer, D.M. Karl and W.W. Broenkow (1987) VERTEX: carbon cycling in the northeast Pacific. Deep-Sea Research, 34, 267-285.

Nalbandov, Y. (1997) Hydrochemistry data. In: Scientific Cruise Report of the Arctic Expedition ARK-XI/1 of RV "Polarstern" in 1995, Rachor, E. (ed.), Reports on Polar Research, 226, A28-A38.

Pfannkuche, O. (1985) The deep-sea meiofauna of the Porcupine Seabight and abyssal plain (NE Atlantic): population structure, distribution, standing stocks. Oceanologica Acta, 8, 343-353.

Pfannkuche, O. (1992) Organic carbon flux through the benthic community in the temperate abyssal Northeast Atlantic. In: Deep-sea food chains and the global carbon cycle, Rowe, G.T. and V. Pariente (eds.), Kluwer Academic Publishers, Dordrecht, 183-198.

Rachor, E. (1997) Scientific Cruise Report of the Arctic Expedition ARK-XI/1 of RV "Polarstern" in 1995. Reports on Polar Research, 226, 157pp.

Sakshaug, E. and D. Slagstad (1991) Light and productivity of phytoplankton in polar marine ecosystems: a physiological view. Polar Research, 10, 69-85.

Scheibe, S. (1990) Untersuchungen zur Verteilung der Meiofauna und sedimentchemischer Parameter im Nordost-Atlantik. Diplomarbeit, Institut für Hydrobiologie und Fischereiwissenschaft, Universität Hamburg, 126pp.

Shuman, F.R. and C.J. Lorenzen (1975) Quantitative degradation of chlorophyll by a marine herbivore. Limnology and Oceanography, 20, 580-586.

Smith, W.O. (1995) Primary productivity and new production in the Northeast Water (Greenland) Polynya during summer 1992. Journal of Geophysical Research, 100, 4357-4370.

Suess, E. (1980) Particulate organic carbon flux in the oceans-surface productivity and oxygen utilisation. Nature, 288, 260-263.

Thiel, H. (1982) Zoobenthos of the CINECA area and other upwelling regions. Rapports et Procés-verbaux des Réuninos du Conseil international pour l'exploration de la Mer, 180, 323-334.

Walsh, J.J., C.P. McRoy, L.K. Coachman, J.J. Goering, J.J. Nihoul, T.E. Whitledge, T.H. Blackburn, P.L. Parker, C.D. Wirick, P.G. Shuert, J.M. Grebmeier, A.M. Springer, R.D. Tripp, D.A. Hansell, S. Djendi, E. Deleersnijder, K. Henriksen, B.A. Lund, P. Andersen, F.E. Müller-Karger and K. Dean (1988). Carbon and nitrogen cycling within the Bering/Chukchi Seas: Source regions for organic matter effecting AOU demands of the Arctic Ocean. Progress in Oceanography, 22, 277-359.

Wassmann, P., R. Peinert and V. Smetacek (1991) Patterns of production and sedimentation in the boreal and polar Northeast Atlantic. Polar Research, 10, 209-228.

Wefer, G. (1989) Particle flux in the ocean: Effects of episodic production. In: Productivity of the Ocean, Berger, W.H., V. Smetacek and G. Wefer (eds.), John Wiley and Sons, New York, 139-154.

Wheeler, P.A., M. Gosslin, E. Sherr, D. Thibault, D.L. Kirchman, R. Benner and T.E. Whitledge (1996). Active cycling of organic carbon in the Central Arctic Ocean. Nature, 380, 697-699.

Chlorophyll *a* Distribution in Water Column and Sea Ice during the Laptev Sea Freeze-Up Study in Autumn 1995

K. v. Juterzenka and K. Knickmeier

Institut für Polarökologie, Universität Kiel, Wischhofstrasse 1-3, D 24148 Kiel, Germany

Received 20 February 1997 and accepted in revised form 19 January 1998

Abstract - During the Transdrift III expedition to the Laptev Sea in October 1995, samples were taken from the water column as well as various types of newly formed sea ice and analysed for Chlorophyll *a* (Chl *a*) as indicator of algal biomass. Ice samples included grease ice, nilas and cores from young ice (≤ 25 cm). The phytoplankton biomass in the water column (water depth 14 m to 40 m) as expressed by integrated Chl *a* values ranged between 2.3 mg m^{-2} north of the Lena Delta and 15.9 mg m^{-2} in a polynya at Buor Khaya Bay southeast of Tiksi. Pigment concentrations in the ice samples varied considerably (0.3 to 2.36 mg m^{-3} in grease ice and 0.2 to 14.6 mg m^{-3} in nilas, respectively). However, the Chl *a* content increased as soon as ice floes were formed from grease ice or pancake ice. Greenish slush ice and overlying water flooding the ice near cracks showed low pigment concentrations, indicating that the colouring was not caused by algae.
The Chl *a* distribution in water column and sea ice during autumnal freeze-up was compared with data from August/September 1993 (Transdrift I). The observed patterns are discussed with respect to the highly variable sediment load of sea ice ("clean" and "dirty" ice).

Introduction

Freezing in the Laptev Sea starts in October and results in the development of a coastal belt of fast ice and a zone of pack ice, separated from each other by a polynya (an overview of Laptev Sea characteristics was given by Timokhov (1994)). As soon as ice crystals form, suspended particles and organisms such as microalgae from the water column are incorporated into the ice sheet by various mechanisms, e.g. scavenging by frazil ice crystals or wave pumping (Ackley et al., 1987; Reimnitz et al., 1993; Ackley and Sullivan, 1994; Gradinger and Ikävalko, 1998). Several studies reported on temporal and spatial variability with regard to the distribution and primary production of sea-ice microalgae (e.g. Gosselin et al., 1986; Hsiao, 1988; Legendre et al., 1992a, b). Ackley and Sullivan (1994) presented the variation of Chl *a* concentrations in the Antarctic Weddell Sea ice as a function of pack ice texture. In the Canadian Arctic, the Chl *a* - content of first year ice parallels an inshore-offshore salinity gradient on a large scale whereas the variation of snow-ice cover and therefore irradiance at the bottom of the ice seemed to be responsible for patchiness on a small-scale (Gosselin et al., 1986; Smith et al., 1988). However, information on algal biomass in newly formed sea ice in Arctic shelf seas is still limited and mechanisms causing distribution patterns on different scales are not well understood as yet.

The Laptev Sea is one of the most important sea-ice producing areas in the Arctic, and it is known that sediment can be entrained into the ice in large amounts and transported via the Transpolar Drift (e.g., Eicken et al., 1997). Consequently, the shallow Laptev Sea may be considered as "starting point" for the development of the sea ice communities found in other regions of the Arctic Ocean. The ice-algal biomass provides not only food for the microbial community within the ice, but can be consumed at the ice underside as well. The interface between sea ice and underlying seawater provides a habitat for a distinctive flora and fauna as it was reported especially from multi-year ice (Melnikov and Bondarchuk, 1987; Gulliksen and Lønne, 1989; Lønne and Gulliksen, 1991; Werner, 1997).

In: Kassens, H., H.A. Bauch, I. Dmitrenko, H. Eicken, H.-W. Hubberten, M. Melles, J. Thiede and L. Timokhov (eds.) Land-Ocean Systems in the Siberian Arctic: Dynamics and History. Springer-Verlag, Berlin, 1999, 153-160.

Areas showing different stages of new ice formation were studied in the Laptev Sea during the late freeze-up period in October 1995. In the present study, special emphasis was layed on the following questions: Which pigment concentrations (Chl *a* and phaeopigments) as indicators of algal biomass occur in the water column, new ice, nilas and young ice during the freeze-up period? Which biomass can be found in the newly forming ice sheet compared to the standing stock in the water column?

Material and methods

During the Transdrift III expedition on I/B "Kapitan Dranitsyn" (1.10.- 30.10.1995), water samples as well as samples of new and young ice were collected. Informations about ice cover during the expedition are given in Aleksandrov and Kolatschek (1997) and Lindemann et al. (this volume). Different ice types were classified according to the WMO Sea-Ice Nomenclature (1970). Discrimination between "clean" and "dirty" sea ice was carried out by optical examination, whereas particle-stained "dirty" ice showed sediment concentrations of 10 mg l^{-1} at least during this expedition (F. Lindemann, pers. comm.). Further informations on ship and ice stations are listed in the cruise report (Kassens, 1997).

Water samples were taken at selected depth (0 m, 5 m, 10 m, 15 m, 20 m, 30 m; down to the sea floor) using a Niskin sampler. Sub-ice water was sampled by means of 1l plastic containers opened under the ice from drilled or hacked holes. Ice samples were taken by a ladle or 1l plastic containers (grease ice) or collected in plastic buckets (pancake ice, nilas) and allowed to melt in the dark at 4°C. Cores from ice floes thicker than 10 cm were taken with an ice auger and cut into sections for subsequent processing (sections were measured individually). All samples were filtered on GF/F filters, stored frozen and analysed for Chl *a* and phaeopigments according to Arar and Collins (1992) using a Turner design fluorometer. Water samples taken during Transdrift I (2.8.-5.10.1993; R/V "Ivan Kireyev") were obtained with identical procedures but analysed photometrically according to Jeffrey and Humphrey (1975). Further informations about this cruise are given in Kassens and Karpiy (1994).

Results and discussion

Standing stock

Phytoplankton biomass expressed as Chl *a* integrated over the entire water column was lower than 7.3 mg m^{-2} at most stations but reached a maximum in a polynya in Buor Khaya Bay (15.9 and 9.7 mg m^{-2} at St. 32 and 33, respectively; Figure 1). In comparison, values of the standing stock in August/September 1993 ranged from 1.2 mg m^{-2} northwest off the island of Belkovsky to 25.7 mg m^{-2} northeast of the Lena outflow. The observed pattern cannot be explained by different integration depth (overall 14 - 40 m in autumn 1995), since the areas showing high biomass were shallow (14 and 15 m st St. 32 and 33).

A maximum of Chl *a* in the Buor Khaya area was also reported by Heiskanen and Keck (1996) in September 1991, although they suggested because of low Chl *a* and high phaeopigment concentrations that the phytoplankton community was already in a senescent phase or affected by grazing in other areas of the Laptev Sea. Our findings showed that Chl *a* concentrations as high as measured in the summer period can also be found in October. Similar Chl *a* biomass had been reported from the end of the growth season in ice-covered areas of the Barents Sea (10.1 to 14.7 mg m^{-2}; Hegseth, 1997). Summer and autumn values from the Laptev Sea lay within the lower range of Chlorophyll concentrations from productive regions of the Canadian Arctic, integrated over the photic zone (summerized by Demers et al., 1986).

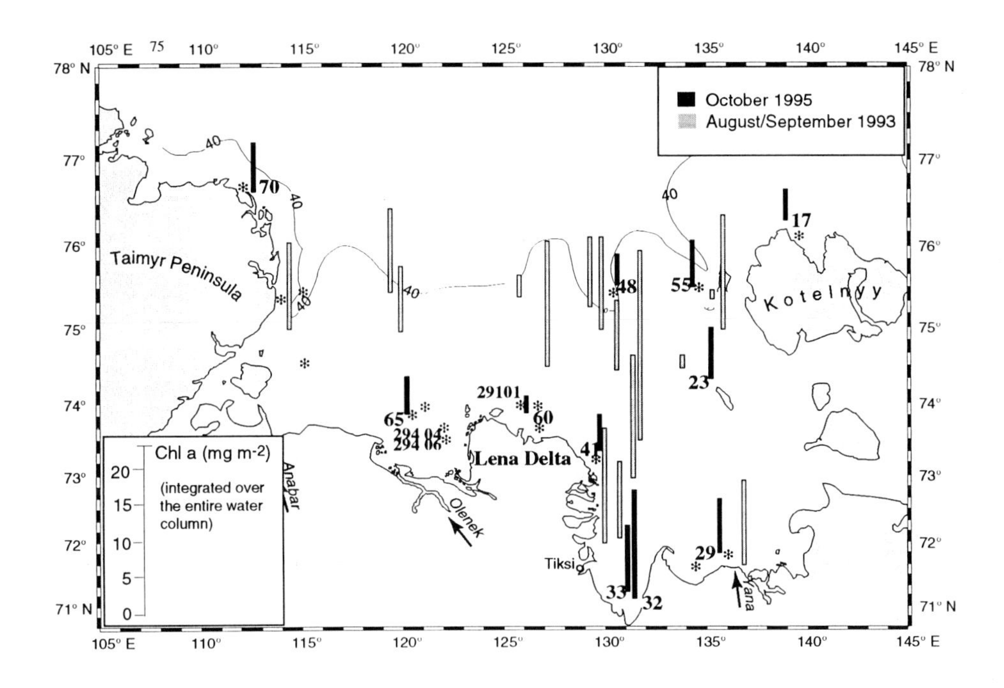

Figure 1: Horizontal distribution of Chl *a* on the Laptev Sea shelf in August/September 1993 and October 1995. Values are integrated over the entire water column. Numbers represent Transdrift III stations (October 1995), asterics mark ice stations. at three stations where ice cores were analysed in detail, numbers of the ice station were given (first three digits indicate day of the year). the line indicates the 40m isobath.

Distribution in water column and sea ice

At four locations, the vertical distribution of Chl *a* and phaeopigments was compared with concentrations measured in overlaying sea ice (Figure 2). Ice samples were taken from the dominating ice types, which had been pancake ice at St. 48, dark nilas at St. 55, young ice/light nilas at St. 60 (ice cover 10/10) and light/dark nilas at St. 65 (ice cover 9/10). Chl *a* concentrations in the water column ranged from 0.05 mg m^{-3} to 0.30 mg m^{-3}, concentration of phaeopigments from 0.04 mg m^{-3} to 0.26 mg m^{-2}. At all stations, the concentration of algal pigments in ice samples were considerably higher than those measured in the water column, indicating an enrichent of microalgae in the new ice.

At St. 65, a difference between two floes of light nilas was observed: In contrast to the other ice floe sampled, the "dirty" nilas floe showed a phaeopigment concentration about four times higher than the Chl *a* concentration (30.6 mg m^{-3}).

Chl *a* concentrations in samples from various ice types varied considerably and ranged from 0.05 mg m^{-3} in "green slush" up to 30.0 mg m^{-3} in the bottom layer of young ice (Figure 3). The Chl *a* content as well as the observed range of Chl *a* concentration increased as soon as ice floes are formed from grease ice layers or pancake ice (Figure 2, Figure 3). Despite the data base presented in this study is very limited, these findings corroborate the view that enrichment processes possibly begin at a very early stage of sea ice formation (Ackley et al., 1987; Garrison et al., 1983, 1990), but an enrichment may not lead to a significant increase in biomass until the ice is consolidated in pancakes or nilas sheets. Enrichment of algal biomass

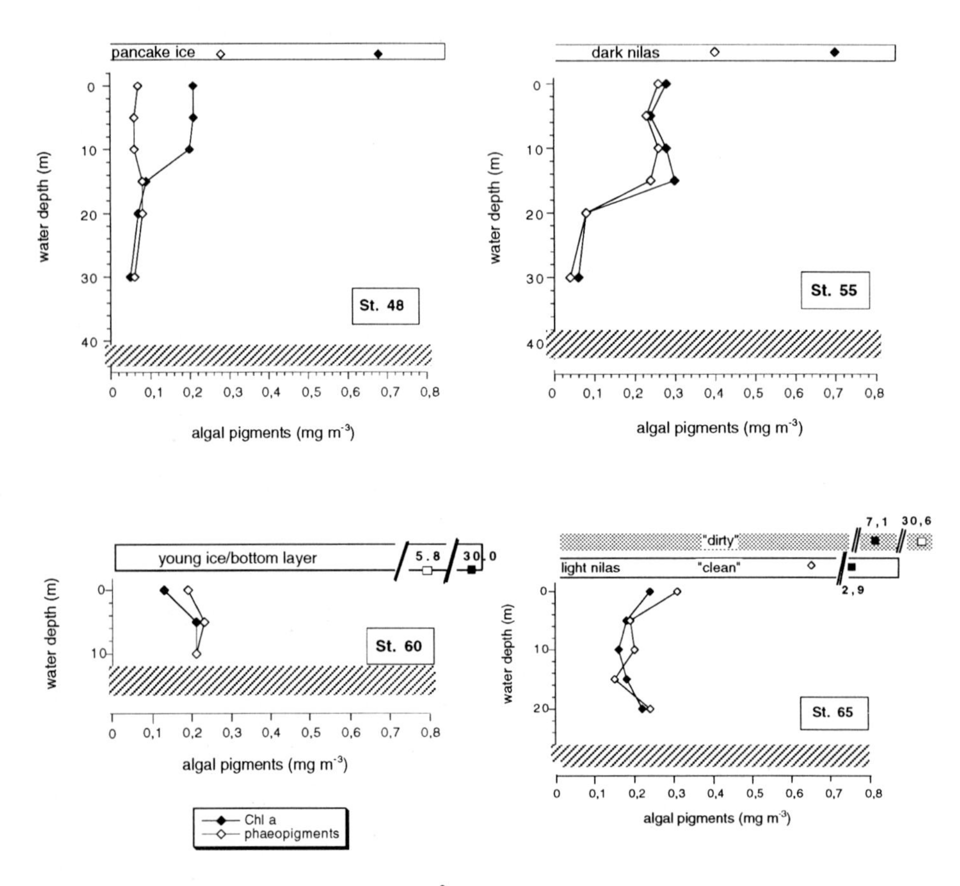

Figure 2: Vertical distribution of Chl *a* (mg m^{-3}) in water column and sea ice at stations 48 (ice station:* 28901), 55 (* 29001), 60 (* 29101) and 65 (* 29401). At St. 65, two ice floes are represented. Particular water depth is marked by striped areas.

and organisms in grease ice, pancake ice and nilas has been described in other studies from the Greenland Sea and the Antarctic (Gradinger and Ikävalko, 1998; Garrison et al., 1990), but especially Chl *a* concentrations found in grease ice appear to be highly variable and can reach values >20 times higher than in the underlying water column (Garrison et al., 1990).

Greenish appearing slush ice and water flooding the ice near cracks showed low pigment concentrations, indicating that the colouring was not caused by algae (Figure 3). Greenish coloured water flooding over pack ice floes was also observed by Gradinger (1996) along the cruise track of RV "Polarstern" and interpreted as evidence for the presence of highly productive under-ice ponds. In this study, it may be possible that cracking and deforming young ice is flooded from the floe edges and the greenish colour is an effect of light refraction. The occurrence of flooding is supported by a relatively high salinity of the melted slush and water samples (S = 12.6 - 18.4).

Distribution within growing sea ice

The vertical Chl *a* distribution in ice cores was analysed in detail at 3 stations. Although section length varied considerably, close relations to at least the sediment load can be seen (Figure 4). The vertical distribution of algal pigments within ice cores 29101 and 294 04 showed high concentrations of algal pigments (Chl *a* and phaeopigments) in sediment-loaden layers of "dirty ice" (2.3 mg m^{-2}, 3.85 mg m^{-2}) whereas the amount of phaeopigments varied (Figure 4a, b). At location 294 06 the pigment maximum was measured in the bottom layer (0 -2 cm) of a "clean" core (Figure 4b).

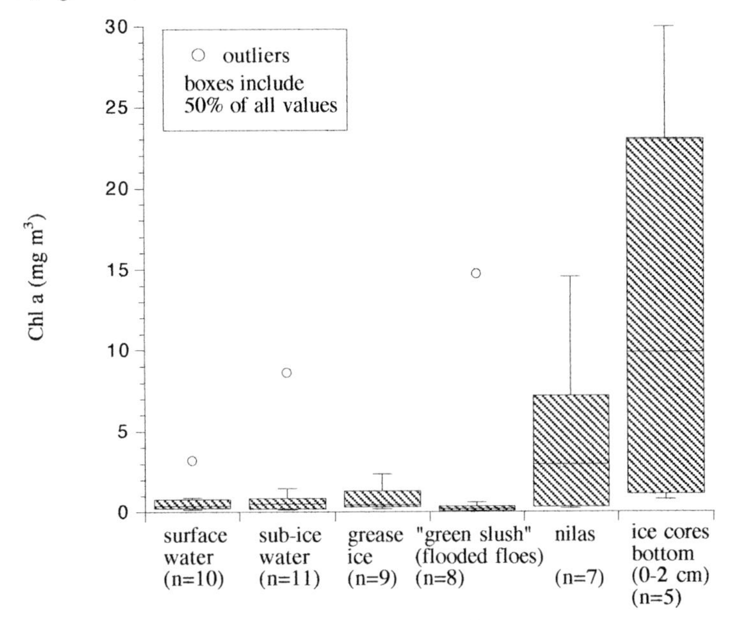

Figure 3: Range of Chl *a* and phaeopigment concentrations (mg m^{-3}) in water and ice samples (Box-Whisker-plot; water samples: n=21; ice samples: n=29). Top and bottom of the boxes mark +/- 25% variation (upper and lower quartiles), lines extending from the boxes mark minimum and maximum values. Horizontal lines within boxes are median values. Where the distamce between data point and nearest quartile exceeds the sum of quartile and 1.5xdistance between quartiles, values are defined as outliers.

The last-mentioned pigment distribution corresponds with patterns often observed in Arctic multi-year ice floes, which are housing a well developed bottom community with high Chl *a* concentrations in the lowermost centimetres (Gradinger, 1995). The high amount of algal pigments in "dirty ice" layers at the other locations may be caused by algal cell incorporation into the ice together with resuspended sediment particles, although other explanations are conceivable. This patterns give no hint which mechanisms are responsible for particle enrichment in the Laptev Sea ice, but point to a similar mechanism for suspended sediment and algae. The high amount of phaeopigments in the "dirty ice" layer at location 294 04 was possibly caused by the incorporation of sedimentated and resuspended phytoplankton cells. The observed pigment composition can also be caused by algal degradation within the ice or may be explained by an overestimation of phaeopigments due to the presence of ice-transported chlorophyll *b*, which is a common pigment in freshwater chlorophytes (Heiskanen and Keck, 1996)

The total amount of Chl *a* within the forming ice sheet (mg pigments per m^{-2} sea ice) can exceed 4 mg m^{-2} (e.g. at location 291 01), which lies within the range of the Chl *a* standing stock in the water column in October 1995 (Figure 1). Chl *a* biomass of ice-associated communities in other Arctic and Subarctic shelf regions had been reported from < 0.5 to 300 mg m^{-2}, whereas the ice thickness in the areas studied ranged from 30 - 300 cm (discussed by Booth, 1984; Smith, 1988 and Cota et al., 1991).

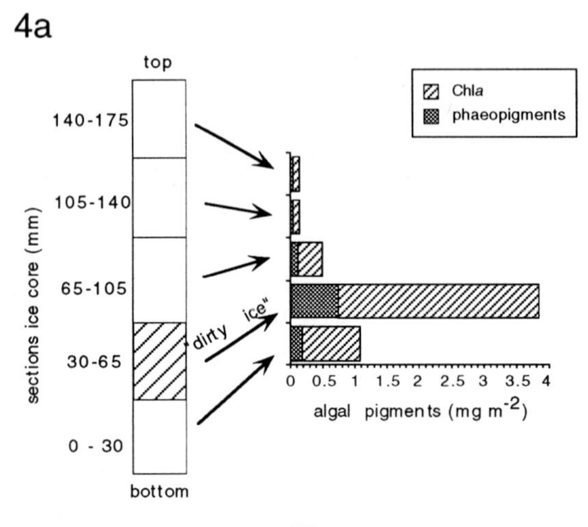

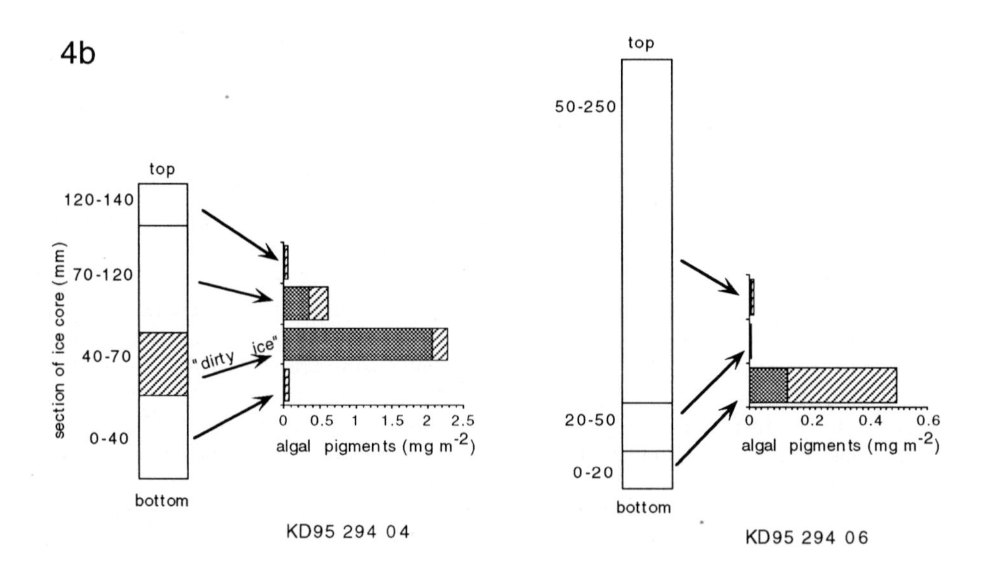

Figure 4: Chl *a* and phaeopigment content (mg m^{-2}) in three ice cores; Chl *a* concentrations were measured for each core section separately and the amount of Chl *a* for each layer of the ice sheet were calculated. Shaded areas indicate sections of sediment-loaden "dirty ice". Note different scales for pigment concentration. a) Ice station *291 01; b) Ice stations *294 04, *294 06 (KD95: I/B "Kapitan Dranitsyn", 1995).

Conclusions

The results of this study indicate that the variability of Chl *a* concentrations within newly forming sea ice in the Laptev Sea is high. It is likely that the observed patterns are caused by different physical and biological processes, which vary due to the hydrographical characteristics of the Laptev Sea water masses as well as the individual history of different ice floes during their growth. However, total algal biomass within the ice is relatively high and within the range of the water column standing stock. Therefore it may serve as food for the developing micro- and meiofaunal ice community. The partially high autumnal biomass in the water column could be used by remaining zooplankton as last food source before the winter period. In order to fully understand the highly complex and dynamic processes of enrichment and entrainment of living cells as well as sediment particles into sea ice, biologists, sedimentologists and ice physicists are challenged to continue their joint studies in field work and experiments.

Acknowledgements

We would like to thank all collegues and crew members on board I/B Kapitan Dranitsyn for their support and cooperation. We are especially indepted to K.P. Tyshko and the russian ice group, K. Tuschling and F. Lindemann, which assisted in taking ice samples at any time and even in exceptional circumstances. R. Gradinger gave helpful comments during the data analysis. The manuscript was improved by the comments of two anonymous referees. This study was carried out as part of the project "German-Russian Investigations of Marginal Seas of the Eurasian Arctic" and funded by the German Ministry of Education, Science, Research and Technology (BMBF).

References

Ackley, S.F., G. Dieckmann and H. Shen (1987) Algal and foram incorporation into new sea ice. Eos, 68, 1736.

Ackley, S.F. and C.W. Sullivan (1994) Physical controls on the development and characteristics of Antarctic sea ice biological communities - a review and synthesis. Deep-Sea Res.I, 41, 1583-1604.

Aleksandrov, V. and J. Kolatschek (1997) Sea ice conditions during the Transdrift III expedition. In: Laptev Sea System: Expeditions in 1995, H. Kassens (ed.), Ber. Polarforsch., 19-20.

Arar, E.J. and G.B. Collins (1992) In vitro determination of Chl *a* and phaeophytin *a* in marine and freshwater phytoplankton by fluorescence. U.S. EPA, Cincinnati, 14pp.

Booth, J.A. (1984) The epontic algal community of the ice edge zone and its significance to the Davis Strait Ecosystem. Arctic, 37, 234-243.

Cota, G.F., L. Legendre, M. Gosselin and R.G. Ingram (1991) Ecology of bottom ice algae: I. Environmental controls and variabilty. J. Mar. Syst., 2, 257-278.

Demers S., L. Legendre, J.C. Therriault and R.G. Ingram (1986) Biological production at the ice-water ergocline. In: Marine interfaces ecohydrodynamics, J.C.J. Nihoul. (ed.), Elsevier, Amsterdam, 31-54

Eicken, H., E. Reimnitz, V. Alexandrov, T. Martin, H. Kassens and T. Viehoff (1997) Sea-ice processes in the Laptev Sea and their importance for sediment export. Cont. Shelf Res., 17, 205-233.

Garrison, D.L., S.F. Ackley and K.R. Buck (1983) A physical mechanism for establishing algal populations in frazil ice. Nature, 306, 363-365.

Garrison, D.L., A.R. Close and E. Reimnitz (1990) Microorganisms concentrated by frazil ice. In: Sea ice properties and processes, Ackley, S.F. and W.F. Weeks (eds), U.S. Army Corps of Engineers - CRRL Monograph 90-1 (Proc. of W.F. Weeks Sea Ice Symposium), 92-96.

Gosselin, M., L. Legendre, J.-C. Therriault, S. Demers and M. Rochet (1986) Physical control of the horizontal patchiness of sea-ice microalgae. Mar. Ecol. Prog. Ser., 29, 289-298.

Gulliksen, B. and O.J. Lønne (1989) Distribution, abundance, and ecological importance of marine sympagic fauna in the Arctic. Rapp. P.-v. Reun. Cons. int. Explor. Mer., 188, 133-138.

Gradinger, R. (1995) Climate change and biological oceanography of the Arctic Ocean. Phil. Trans. R. Soc. Lond. A, 352, 277-286.

Gradinger, R. (1996) Algal bloom under Arctic pack ice. Mar. Ecol. Prog. Ser., 131, 301-305.

Gradinger, R. and J. Ikävalko (1998) Organism incorporation into newly forming Arctic sea ice in the Greenland Sea. J. Plankton Res., in press.
Heiskanen, A.-S.and A. Keck (1996) Distribution and sinking rates of phytoplankton, detritus, and particulate biogenic silica in the Laptev Sea and Lena River (Arctic Siberia). Mar. Chem., 53, 229-245.
Hegseth, E.N. (1997) Phytoplankton in the Barents Sea - the end of a growth season. Polar Biol., 17, 235-241.
Hsiao, S.I.C. (1988) Spatial and seasonal variations in primary production of sea ice microalgae and phytoplankton in Frobisher Bay, Arctic Canada. Mar. Ecol. Prog. Ser., 44, 275-285.
Jeffrey, S.W and G.F. Humphrey (1975) New spektrophotometric equations for determining chlorophylls a, b and c1 and c2 in higher plants, algae and natural phytoplankton. Biochem. Physiol. Pflanz., 167, 191-194.
Kassens, H. (ed.) (1997) Laptev Sea System: Expeditions in 1995. Ber. Polarforsch, 248, 210pp.
Kassens, H. and V. Karpiy (eds.) (1994) Russian-German Cooperation: The Transdrift I Expedition to the Laptev Sea. Ber. Polarforsch., 151, 168pp.
Legendre, L., S.F. Ackley, D. Dieckmann, B. Gulliksen, R. Horner, T. Hoshiai, I.A. Melnikov, W.S. Reeburgh, M. Spindler and C.W. Sullivan (1992a) Ecology of sea ice biota. 2. Global significance. Polar Biol., 12, 429-444.
Legendre, L., M.-J. Martineau, J.-C. Therriault and S. Demers (1992b) Chlorophyll *a* biomass and growth of sea-ice microalgae along a salinity gradient (southeastern Hudson Bay, Canadian Arctic). Polar Biol, 12, 445-453.
Lindemann, F., J. Hölemann, J.A. Korablev and A. Zachek (1998) Particle entrainment into forming sea ice - freeze-up studies in October 1995 (this volume).
Lønne, O.J. and B. Gulliksen (1991) On the distribution of sympagic macro-fauna in the seasonally ice covered Barents Sea. Polar Biol., 11, 457-469.
Melnikov, I.A. and L.L. Bondarchuk (1987) Ecology of mass accumulations of colonial diatom algae under drifting Arctic ice. Oceanology, 27, 233-236.
Reimnitz, E., J.R. Clayton, E.W. Kempema, J.R. Payne and W.S. Weber (1993) Interaction of rising frazil with suspended particles: tank experiments with applications to nature. Cold Reg. Sci. Technol., 21, 117-135.
Smith, R.E.H., J. Anning, P. Clement and G. Cota (1988) Abundance and production of ice algae in Resolute Passage, Canadian Arctic. Mar. Ecol. Prog. Ser., 48, 251-263.
Timokhov, L.A. (1994) Regional characteristics of the Laptev Sea and the East Siberian Sea: Climate Topography, ice phases, thermohaline regime, circulation. In: Kassens, H., H.-W. Hubberten, S.M. Pryamikov and R. Stein (eds.), Russian-German Cooperation in the Siberian Shelf Seas: Geo-System Laptev Sea, Ber. Polarforsch., 144, 15-31.
Werner, I. (1997) Ecological studies on the Arctic under-ice habitat - colonization and processes at the ice-water inetrface. Ber. Sonderforschungsbereich 313, Universität Kiel, 70, 1-167.
World Meteorological Organization (1970) WMO sea-Ice Nomenclature.- WMO/OMM/BMO - Secretariat of the World Meteorological Organization, Geneva, No. 259, TP. 145.

Composition, Abundance and Population Structure of Spring-Time Zooplankton in the Shelf-Zone of Laptev Sea

E.N. Abramova

Lena-Delta State Reserve, 28 Fiodorova St., 678400 Tiksi, Yakutia, Russia

Received 1 May 1997 and accepted in revised form 9 February 1998

Abstract - New evidence was received concerning the composition, distribution and abundance of zooplankton in May-June 1996 in the shelf area of Laptev Sea. The minimal depth of area is 6 m, the maximum 33 m. As measured at various stations, ice thickness ranges between 1.20 and 2.07 m. Species diversity of spring-time zooplankton is represented by 23 taxa of marine, brackishwater and freshwater species. Average abundance of zooplankton across the examined area in May amounted to 155 individuals per m^3. From records of most stations, brackishwater plankton (mass species *Drepanopus bungei*) was predominantly abundant, being at the time in the stage of reproduction. Induced by spring flood, variations in hydrologic conditions in June have resulted in quantitative and qualitative changes in zooplankton community. Brief analysis of the population structure and development biology of some mass species in the shelf zone of Laptev Sea was made on the basis of previous and newly obtained data.

Introduction

Previous investigations were effective to accumulate in the literature a sufficiently comprehensive evidence on summer-time zooplankton of the shelf area of Laptev Sea. Species composition, distribution, quantitative parameters characterizing plankton organisms are comparatively well understood ecological aspects of plankton existence are touched upon. However, knowledge of the structure and functioning of zooplankton community in the shelf area of Laptev Sea in extended winter periods is insufficient numbering only several papers (Pirozhnikov, 1958; Lutsik et al., 1981; Abramova, 1996) with complete lack of evidence on interSeasonal and annual dynamics of zooplankton in Laptev Sea. Thus, mostly intriguing is the evidence concerning behavior of pelagic communities under ice cover. The results received by joint Russian-German expedition Transdrift-IV between 16 May and 16 June 1996 proved useful to permit to somewhat close the information gap.

Material and methods

In the total, the spring expedition of 1996 collected 44 zooplankton samples from 24 stations located in 3 transects: one vertical and two horizontal (Figure 1). One-time sampling at all the stations was made prior to spring flood (between 18.05 and 29.05); after the beginning of the flood (3.06 - 12.06) the second sampling was made twice at stations 3-6, 8-11 and once at st. 20, 21, 23 and 24. Complete sampling was made using small Jeddy net (hole input diameter 0.17 m, seive mesh 86μm). At each stations nets were dipped three times permitting to receive a representative sample. Arrangement of depths, temperatures and salinity at each station in May is summarized in Table 1. After June 10, positive temperatures bellow 1°C were recorded on the southern stations of surface horizon of vertical transect of and along the whole water column at shallow water stations. With the increase of freshwater runoff in June, evident salinity reduction was reported across the whole of the observation area with the exception of station 21 where near surface salinity was reported to increase to 37‰ and near bottom salinity to 38. Mass species development biology was studied with attraction of the previous evidence

----

In: Kassens, H., H.A. Bauch, I. Dmitrenko, H. Eicken, H.-W. Hubberten, M. Melles, J. Thiede and L. Timokhov (eds.) Land-Ocean Systems in the Siberian Arctic: Dynamics and History. Springer-Verlag, Berlin, 1999, 161-168.

on zooplankton of Buorkhaya Bay (Figure 1) received by the author in various seasons between 1993-1996.

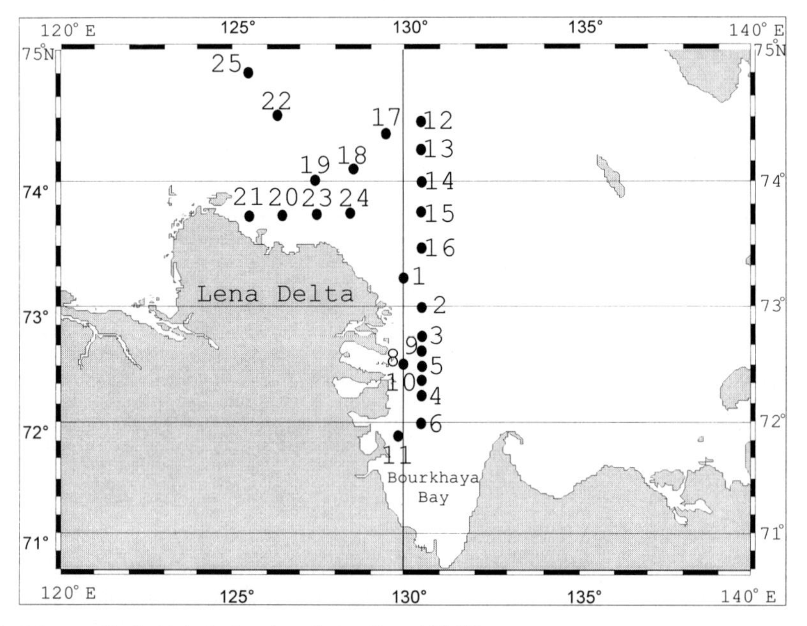

Figure 1: Scheme of hydrobiological stations during Transdrift IV.

Results and discussion

Species composition, population structure, development biology

Quantitative sample analysis has permitted to identify the presence of 2 Rotatoria species, 1 Cladocera, 13 Copepoda and representatives of another 5 types of invertebrates (Table 2). Calanoida of estuarine-arctic complex was prevailing in the composition of pelagic fauna. Among them mostly distributed in May-June proved to be *Drepanopus bungei*, *Acartia longiremis* and representatives of *Pseudocalanus* genus. Two species of Harpacticoida (*Danielssenia sibirica* and *Ectinosoma arcticum*) have not been previously listed among zooplankton of Laptev Sea. By our evidence, *E.arcticum* is a species common for summer-time plankton of major streams mouth in the Lena-Delta with 0-4 salinity range though it may occur at salinity up to 16. Recorded at station 20 in June *D. sibirica* proved to be our first finding. In literature this species is reported from the coast line and mouth of the river on the island B. Lyakhovsky (Borutsky, 1952). Sexual and age structure of *D. bungei* population in May (one of mostly abundant species of the shelf area of Laptev Sea) is shown in Table 3.

At study period the species was in the stage of reproduction. *D. bungei* seems to have more than one generation in the Laptev Sea annually (Abramova, 1996). The onset of spawning of *D. bungei* in Buorkhaya Bay was observed to fall on February. Mature males and females, copepodite and nauplii stages of the species in various interrelations occur in plankton till October. Late in October-November the mass of *D. bungei* population is represented by copepodites of 5 - 6th stages containing considerable fat inclusions. In January-February over 70% of recorded males and females again proved to be mature, some females were reported to have spermatophores and 4-7 large-sized eggs attached to genital segment. Nauplii of new

Table 1: Temperature, salinity, depth and ice cover distribution in the Laptev Sea during the Transdrift IV expedition in May 1996

Station	Date	Latitude °N	Longitude °E	Depth m	Temperature (°C) surface / bottom	Salinity surface / bottom	Ice sm
1	18.05	73°15.60	130°00.40	21.5	-0.45 / -1.46	13.53 / 29.23	143
2	18.05	73°00.50	130°30.50	22.0	-0.73 / -1.43	13.84 / 27.62	153
3	18.05	74°44.80	130°29.50	17.5	-0.69 / -1.32	12.97 / 26.81	177
4	20.05	72°14.15	130°29.77	13.0	-0.55 / -1.20	10.73 / 24.65	182
5	20.05	72°29.78	130°34.08	14.0	-0.38 / -1.21	7.20 / 26.29	120
6	20.05	72°00.30	130°29.10	15.0	-0.46 / -1.31	9.04 / 24.87	164
8	20.05	72°30.93	130°05.49	6.0	-0.09 / -0.80	2.12 / 20.48	170
9	22.05	72°37.40	130°29.47	15.0	-0.42 / -1.35	8.15 / 26.63	145
10	22.05	72°22.61	130°29.62	12.0	-0.37 / -1.24	5.88 / 24.63	175
11	22.05	71°53.04	129°51.60	7.0	-0.25 / -0.75	5.28 / 21.31	196
12	23.05	74°30.52	130°29.42	25.0	-1.62 / -1.53	29.13 / 30.11	159
13	23.05	74°14.82	130°30.70	25.0	-1.56 / -1.59	29.33 / 29.56	170
14	23.05	74°00.08	130°30.11	25.0	-1.43 / -1.48	26.47 / 30.79	162
15	23.05	73°45.02	130°30.36	24.0	-1.34 / -1.46	19.48 / 29.79	167
16	25.05	73°30.35	130°30.41	19.0	-1.13 / -1.46	21.11 / 27.74	140
17	25.05	72°24.44	129°29.08	20.0	-1.59 / -1.51	29.26 / 30.73	148
18	25.05	74°07.02	128°28.57	26.0	-1.67 / -1.32	30.89 / 32.15	207
19	25.05	74°01.74	127°30.10	28.0	-1.58 / -1.48	29.28 / 32.42	123
20	27.05	73°44.95	126°30.64	13.0	-1.52 / -1.60	27.16 / 28.79	144
21	27.05	73°45.00	125°29.66	6.0	-1.95 / -1.89	35.52 / 36.32	204
22	29.05	74°32.72	126°26.58	33.0	-1.68 / -1.63	31.07 / 32.63	134
23	29.05	73°45.03	127°30.68	9.0	-1.20 / -1.34	24.48 / 25.16	145
24	29.05	73°44.69	128°32.30	22.0	-1.24 / -1.61	23.13 / 31.95	159
25	29.05	74°51.54	125°29.73	33.0	-1.69 / -1.68	31.90 / 32.63	143

generation appear in February. Conventionally two well-distinguished reproduction peaks are observed in Buorkhaya Bay, accompanied by increased abundance of mature males in the population and mass outcome of nauplii in March-April and September. Depending on particular conditions of different years, spawning peaks may be shifted to earlier or later periods. In May-June 1996, mainly mature females, individual males and copepodits of 5th stage were recorded in *A. longiremis* population, some of females having spermatophores. The content of this species in the samples was varying between 0.5 and 18.5% of the total density of organisms. The available material has permitted to outline the following dynamic pattern of the species development: copepodits of 2-5 stages were recorded in the samples from Buorkhaya Bay in February and March, copepodits of 4-5 stages predominated in April, scarce mature male and female were recorded in May-June. Evidently, part of population starts

Table 2: Zooplankton species collected in the Laptev Sea in spring 1996

Taxa /Station	**1**	**2**	**3**	**4**	**5**	**6**	**8**	**9**	**10**	**11**	**12**	**13**	**14**	**15**	**16**	**17**	**18**	**19**	**20**	**21**	**22**	**23**	**24**	**25**
Hydrozoa					+									+					+		+			+
Rotatoria:																								
Keratella quadrata	+		+																					
Kellicottia longispina																						+		
Polychaeta																								
Polychaeta larvae	+	+	+	+		+	+	+			+	+	+		+	+	+	+	+	+		+	+	
Maxillopoda:																								
Calanus sp.juv.															+							+		
Pseudocalanus major			+											+	+						+	+	+	+
P. minutus											+				+	+	+	+		+	+			+
Pseudocalanus sp. juv.	+	+	+	+	+	+	+	+	+	+	+	+	+		+	+	+	+	+	+		+	+	+
Microcalanus pygmaeus				+		+					+	+			+			+	+		+	+		+
Drepanopus bungei	+	+	+	+	+	+	+	+	+	+	+	+	+	+	+	+	+	+	+		+	+	+	+
Acartia longiremis	+	+	+	+	+	+	+	+	+	+	+	+		+	+	+	+	+	+	+	+	+	+	+
Oithona similis	+	+	+		+	+					+		+	+	+		+	+	+		+	+	+	+
Oncaea borealis		+			+			+				+	+				+	+	+	+	+	+	+	+
Cyclops sp. juv.			+																					
Microsetella norvegica											+							+						+
Danielssenia sibirica																			+					
Ectinosoma arcticum			+					+																
Harpacticoida sp.			+	+	+	+		+		+				+	+				+					+
Bosmina sp. juv.	+																							
Chaetognatha:																								
Sagitta sp. juv.	+			+	+	+		+						+	+	+	+	+		+	+	+	+	+
Echinodermata:																								
Ophiuroidea larvae														+										
Appendicularia																		+						
Mollusca														+							+			+

reproducing at this time; however, in our opinion, reproduction peak falls assumingly on July. Unfortunately, no data for this month is so far available. In August only occasional occurrence of nauplii has been reported for this population while copepodits of 1-3 stages were dominating. In the samples taken from Buorkhaya Bay in October-November this species was not observed. In May-June 1996 copepodits of 1-3 stages of genus *Pseudocalanus* are showing frequent occurrence in mesozooplankton compositions. So far we failed to make their species identification. Abundance of these copepods in the samples was varying between 1.5 and 16% of the overall organismic abundance in the samples. Among mature individuals of this genus females *P. minutus* and *P. major* and mature males *P. minutus* were recorded. Assuming that copepodits belonged to one species, particularly *P. minutus*, the structure of his population in May could be represented as following (Table 4).

Judging from the Table 4, reproduction of this species had started prior to our observations and proceeded in May-June. McLaren et al. (1987) had described the life cycle of *P. minutus* from Bedford basin in Nova Scotia, Canada, pointing to the fact that male specimens of this species had been recorded in samples only in late winter-early summer. The authors concluded that *P. minutus* is monocyclic, spending most part of the year, with the exception of the above period, in a diapause at the 5th copepoditic stage. Our data also suggests that in the shelf of Laptev Sea *P. minutus* is reproducing in late winter- early summer. However, as evidenced by the literature, this species is reported to reproduce in August-September in New-Siberian Islands shallows (Pavshtiks, 1990; Abramova, 1996). The above evidences allow us to suggest the existence of two generations of *P. minutus* in Laptev Sea in a year. Accounting for possible errors in identification of representatives of this genus, one can definitely assume in any case the fact of reproduction of a certain species of *Pseudocalanus* genus in May-June in the shelf zone of Laptev Sea.

Table 3: Sex-age structure of *D. bungei* population at Laptev Sea shelf, May 1996

Development stage	Females	Males	C5	C4	C3	C2-C1	Nauplii
% of population	18.1	6.5	5.5	0.8	0.3	1.3	67.5

Table 4: Sex-age structure of *P. minutus* population at Laptev Sea shelf, May 1996

Development stage	Females	Males	C5	C4	C3	C2-C1	Nauplii
% of population	7.9	1.5	2.3	0.4	1.1	68.1	18.2

Small copepod *O. similis* whose population consisted of copepodits of 5th-6th stages, larvae of Polychaeta belonging to at least two species and youngsters of Sagitta may be also attributed to widespread forms of spring-time zooplankton.

Elements of freshwater fauna, especially the Rotifera are sufficiently common for plankton in surface water of low salinity of the shelf of Laptev Sea and are directly associated with arrival and distribution of Lena-River water. Species diversity of spring-time zooplankton is comparatively modest. Part of the species of estuarine-arctic complex characteristic for the shelf zone of Laptev Sea, such as *Jaschnovia tolli*, *Limnocalanus macrurus* were lacking at this season. By our observations, mature specimens *J. tolli* are occurring in samples from Buorkhaya Bay between September and January, those of *L. macrurus* between August and February; occurrence of single mature specimens may be reported sometimes even in March. In the beginning of this period, *L.macrurus* population is represented by nauplii and copepodits of various stages - among which C5 is predominating in abundance. In August mature individuals account for less than 10% of the population abundance. Gradual maturing of the remaining part of population proceeds until November. Later only mature males and females are encountered, evidently producing resting eggs. Early nauplii of new generation appear in June. Life cycle of *J. tolli* may be probably similar.

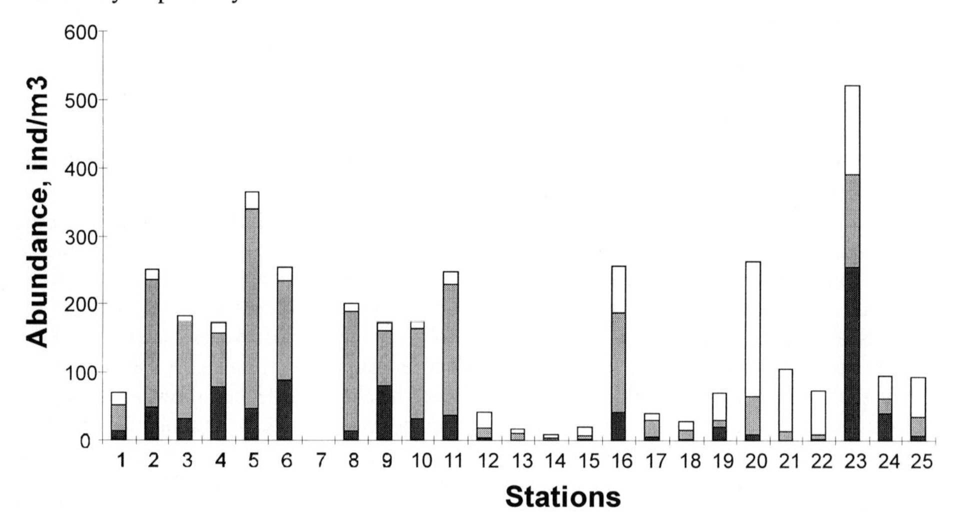

Figure 2: Horizontal distribution of zooplancton abundance in May 1996 and percentage of *Drepanopus bungei* against total abundance. Dark grey: copepoditen stages of *D. bundei*, light grey: nauplii of *D. bundei*, white: other species.

Abundance of zooplankton

In May average abundance of zooplankton in study area amounted to 155 individuals/m^3. Figure 2 shows distribution of zooplankton across the stations in May. Stations located eastward of the Lena-Delta basically represented by *D. bungei* in reproductive stage showed richer abundance and more uniform distribution pattern. In this area salinity over the whole water column did not exceed 27. In May average density reading at stations 3-11 (221 ind./m^3) proved to exceed average value for the whole area, decreasing in early June to 164 ind./m^3 and to 37 ind./m^3 by the end of the first decade of June. Abundance chandes recorded at stations 3-11, being fulfill three times during the observation period, as well as variations in water salinity are shown in Figure 3 (samples taken at st. 5 in June were lost). In June flood-induced changes in hydrological conditions, primarily surface salinity variation, impacted the abundance of zooplankton as well as the structure of pelagic community. Samples taken from stations located eastward of the Lena-Delta started to demonstrate the presence of freshwater fauna such as *Cyclops sp.* juv. and *Bosmina sp.* juv. However, *D. bungei* remained a predominant species in the area even in June despite the fact that water salinity at some stations of 5-6 m depth became practically zero. By all evidence, *D. bungei* is well adopted to wide range of salinity variations which is also confirmed by its occurrence in summer in the mouth of large streams where more than 50% of plankton is represented by freshwater species.

Northward from the Lena-Delta (st. 12-25) in higher salinity zones (Table 1), average zooplankton abundance in May accounted for 116.5 ind./m^3. At stations 12-19, 23, 24 the same predominance of nauplial and copepoditic stages of *D. bungei* was recorded, though in diminished relative number.

At stations 20-22, 25 *O. similis* and Polychaeta larvae were dominating in abundance with increased abundance of *Pseudocalanus* sp. In June samples from these stations proved practically empty, average abundance of zooplankton dropped down to 8 ind./m^3 as a result of

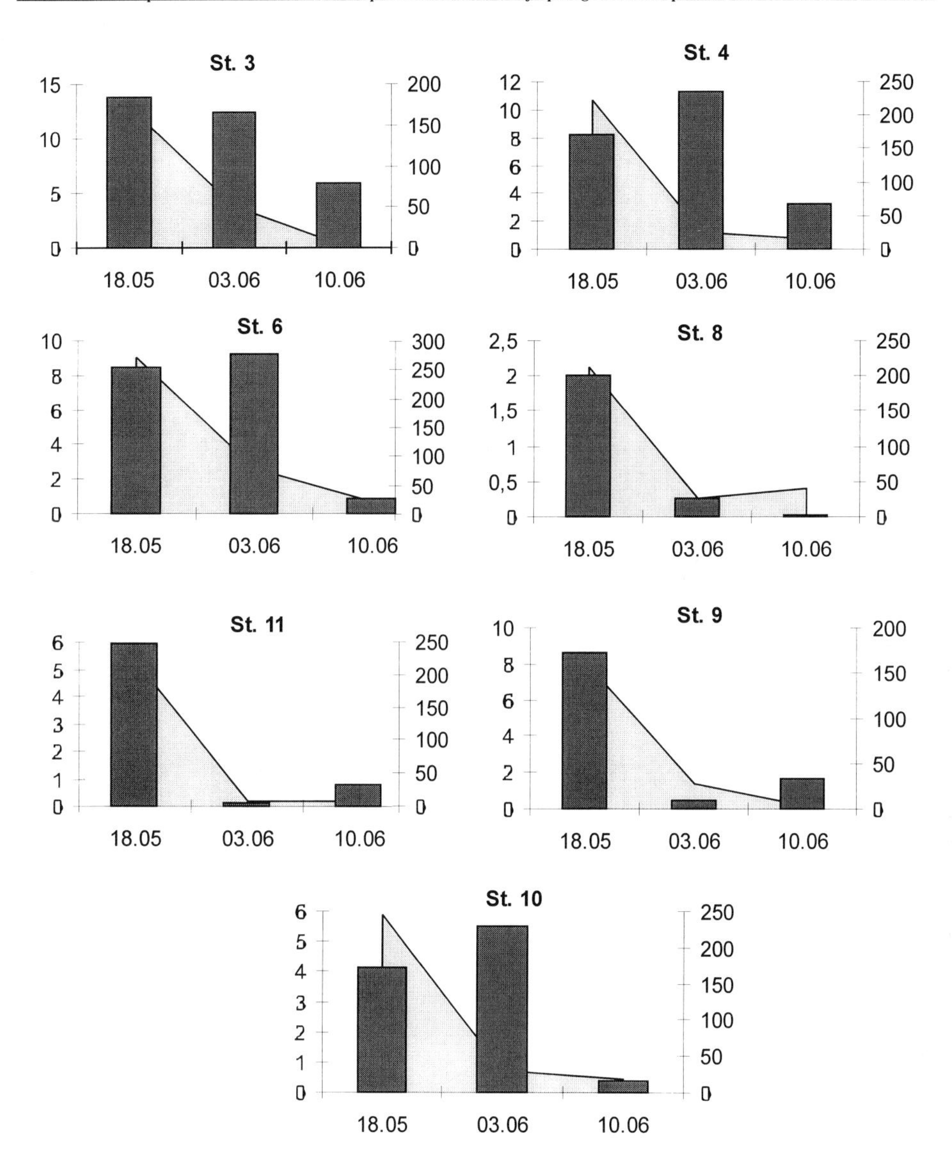

Figure 3: Change in surface salinity and zooplancton abundance at selected stations. Light gray: surface salinity (left scale), dark grey: abundance (ind./m^3, right scale).

negative impact of freshwater runoff, primarily on marine community prevailing in this area. In a sample taken on 11.06 from st.21 no presence of plankton was reported.

Conclusion

Species diversity of zooplankton in May-June in the shelf area of Laptev Sea may be characterized as comparatively moderate with estuarine-arctic complex dominating in number of species. Such mass species as *L. macrurus* and *J. tolly*, characteristic of the summer zooplankton of the area, were not found in spring evidently due to specific features of their life cycle. The observation period was time of reproduction of *D. bungei* dominating in abundance at most stations. In May-June spawning of one of the species of *Pseudocalanus* genus (assumingly *P. minutus*) was observed. The distribution pattern of zooplankton in the observation area is highly affected by water salinity. Mass brackishwater species (*D. bungei, A. longiremis, Pseudocalanus* sp) were occurring in spring-time zooplankton along the whole range of salinity. In June the enrich of plankton by freshwater elements to the east of the Lena-Delta is one more improve of the fact that the area receives maximum of river runoff.

Acknowledgments

The author is sincerely grateful to the organizers of the Transdrift-IV expedition for the possibility to study the Laptev Sea in a most intriguing season. The author is also thankful to Dr. A.Yu. Gukov (Tiksi Hydrometeorological Department) for his direct participation in zooplankton sampling during the expedition.

References

Abramova, E.N. (1996) Study of zooplankton in New-Siberian Islands shallow water zone of Laptev Sea. Marine Biology, 22 (2), 89-93.

Borutsky, E.V. (1952) Freshwater Harpacticoida Fauna of USSR (in Russian) V.3 (4), 434 pp.

Gukov, A.Yu. (1995) Hydrobiological research in Lena Polynya. Reports on Polar Research, 176, 228-229.

Lutsik, A.I., N.I. Silina and N.K. Lutsik (1981) Zooplankton and fish fauna of South-East part of Laptev Sea (in Russian). Oceanologia, 21(2), 370-374.

McLaren, I.A., E. Laberge, C.J. Corkett and J.-M. Sevigny (1989) Life cycles of four species of *Pseudocalanus* in Nova Scotia, Canad.Journ.Zool., 67, 552-558.

Pavshtiks, E.A. (1990) Composition and quantitative distribution of zooplankton in the vicinity of Novosibirsk Islands (in Russian). In: Ecosistems of New-Siberian Islands shallow water zone and fauna of Laptev Sea and adjacent water basins (Investigation of fauna of the seas), V. 37(45), 89-104.

Pirozhnikov, P.L. (1958) Range of inhabitance and ecology of copepod *Senestrella calanoides* Juday (in Russian), Zool.Journal, 37(4), 625-629.

Macrobenthos Distribution in the Laptev Sea in Relation to Hydrology

V.V. Petryashov[1], B.I. Sirenko[1], A.A. Golikov[1], A.V. Novozhilov[2], E. Rachor[3], D. Piepenburg[4] and M.K. Schmid[4]

(1) Zoological Institute, Russian Academy of Sciences, 1 Universitetskaya, 199034 St. Petersburg, Russia

(2) State Research Centre - Arctic and Antarctic Research Institute, 38 Bering St., 199226 St. Petersburg, Russia

(3) Alfred-Wegener-Institut für Polar- und Meeresforschung, Postfach 120161, D 27515 Bremerhaven, Germany

(4) Institut für Polarökologie, Universität Kiel, Wischhofstrasse 1-3, D 24148 Kiel, Germany

Received 3 March 1997 and accepted in revised form 5 December 1997

Abstract - As a results of six joint expeditions to the Laptev Sea in 1993, 1994 and 1995, 150 invertebrate species were added to the list of species known from this high-Arctic region, comprising now 1235 species in total. Most of these species (987) are macrozoobenthic. The geographic distribution of macrobenthic species numbers and biomass exhibited a pronounced increase to the north up to 77°N. In terms of biogeographic composition, Boreal-Arctic species accounted for most of the species, primarily widespread Boreal-Arctic species in the southern and central Laptev Sea at depths from 10 to 75 m and primarily Atlantic Boreal-Arctic species in the northern Laptev Sea at depths of 90 to 1100 m. Only in depths of more than 2000 m, as well as in estuarine and brackish coastal waters, Arctic forms had significant species shares. On sandy and silty bottoms, nine major communities and two trophic zones were distinguished. The distribution of species numbers, biomass and faunal composition is discussed in relation to water depth, seabed sedimentology and near-bottom water salinities. The latter are supposed to be the major environmental factor regulating the distribution and structure of macrobenthic communities. An ecological-biogeographical zonation scheme is proposed, combining the current information about the distribution of macrobenthic communities and brackish water masses.

Introduction

The Laptev Sea, which occupies a central position among the Eurasian seas, is of particular scientific interest owing to its unique features. For instance, it is the only Eurasian sea which includes large deep-sea basins with depths of more than 3000 m. Moreover, the Gakkel Ridge, which is the continuation of the mid-Atlantic ridge, meets the continental slope of the Laptev Sea. The hydrography is characterized by pronounced thermohaline gradients, caused by the massive inflow of riverine waters in summer, and by a diverse ice regime with quasi-stationary polynyas in winter. Large water masses of Atlantic and Pacific origin do apparently not penetrate directly into the southern and central epicontinental parts of the Laptev Sea.

The fact that the Laptev Sea is difficult to reach from both east and west, as well as its rigorous ice conditions, account for the insufficient knowledge about it. It was not until the last few years that it became possible to conduct several expeditions of Russian and German research vessels to this remote Arctic region (RV "Ivan Kireev" in 1993; ice-breaking RV "Polarstern" in 1993 and 1995; RV "Professor Multanovsky" in 1994; icebreaker "Kapitan Dranitsyn" and RV "Jakov Smirnitsky" in 1995).

Rich benthological material collected from grab samples and dredge catches during these expeditions (Table 1) permitted us not only to improve the previous knowledge on species diversity in the Laptev Sea, but also to obtain new information on benthos distribution (Rachor et al., 1994; Anisimova et al., 1997; Piepenburg and Schmid, 1997). The present paper aims to synoptically summarize the results of these investigations.

In: Kassens, H., H.A. Bauch, I. Dmitrenko, H. Eicken, H.-W. Hubberten, M. Melles, J. Thiede and L. Timokhov (eds.) Land-Ocean Systems in the Siberian Arctic: Dynamics and History. Springer-Verlag, Berlin, 1999, 169-180.

Table 1: List of expeditions. Gears used: tr - trawl, gr - grab (van Veen or Okean type), dq - quant. sample, obtained by divers, bs - benthopelagic net; also given is number of samples taken with each gear.

Expedition	Years	Stations	Samples	depth range	gears used
Exp. KIN/ZISP	1973	52	210	0-35 m	200dq + 10tr
Ivan Kireev	1993	38	144	9-45	108gr + 36tr
Polarstern	1993	19	29	38-3081	19tr + 10bs
Prof. Multanovsky	1994	14	41	13-38	38gr + 3tr
Polarstern	1995	39	80	40-3833	59gr + 14tr + 7bs
Yakov Smirnitzky	1995	19	60	7-192	57gr + 3tr
Kapitan Dranitsyn	1995	15	38	15-47	38gr
TOTALS		196	602	0-3833	519 quant. samples + 83 qual. samples

Species numbers

During the taxonomic studies conducted mainly by scientists of the Zoological Institute RAS in St. Petersburg, 150 species previously not known for the Laptev Sea were found in the material collected since 1993. The numbers of invertebrate and algal species reported from the Laptev Sea now total 1234 and 146, respectively. Most of these are macrobenthic (987 animals and 27 macrophytes). For comparison: only 192 meiobenthic and 56 zooplankton species are currently known from this sea. In 1994, a total of 1084 animal species had been reported to occur in the Laptev Sea (Sirenko and Piepenburg, 1994), while only 522 species were mentioned by Zenkevich (1963). Local species richness (i.e. species numbers per station) of the macrobenthos generally increased along a depth gradient from the coast to the shelf break and declined farther northward to minimum values in the deep Nansen and Amundsen Basins.

Biomass

The distribution of macrobenthic biomass (total wet mass, including shells etc.) in the central and southern Laptev Sea (south of about 77°N, water depths < 50 m) exhibited a similar geographic and bathymetric pattern as species numbers: figures generally increased from south to north and with water depths. Three latitudinal zones of different macrobenthic biomass were distinguished in this region (Figure 1): < 50 g m^{-2}, 50 to 100 g m^{-2} and > 100 g m^{-2}. The maximum width of the first zone was observed in the eastern Laptev Sea where the riverine influence is most intensive. In contrast, the other two zones were broadest in the western Laptev Sea. North of these three zones biomass generally declined with water depths to less than 25 g m^{-2}.

At some sites near the river mouths, macrobenthos biomass was as high as 200 to 300 gm^{-2} in fine-grained sediments which were characterized by the highest content of total organic carbon in the Laptev Sea (s. Fahl & Stein, 1997). Such high figures were not yet noted for areas influenced by the rivers Olenek and Khatanga. However, this seems primarily to be due to the lack of data. Relatively high biomass was also recorded under the cyclonic meso-scale gyre north of the Yana Bay, as well as in the straits off the New Siberian Islands, the area between the southwestern and northeastern Khatanga Bay and a small depression south of Vasilyevskaya Bank. Moreover, biomass generally increased in relatively fine-grained sediments of the northern parts of the Yana, Eastern Lena and Anabar-Khatanga paleovalleys. In the latter, a maximum biomass of 255 g m^{-2} was recorded. Near the continental slope on the northern Laptev Sea shelf, biomass values are between 10 and 30 g m^{-2}. Only in the north of the New Siberian Islands, in the area bordering the East Siberian Sea, biomass amounts up to about 70 g m^{-2}. Down the continental slope, total wet mass of macrofauna decreases to less than 1 g m^{-2}.

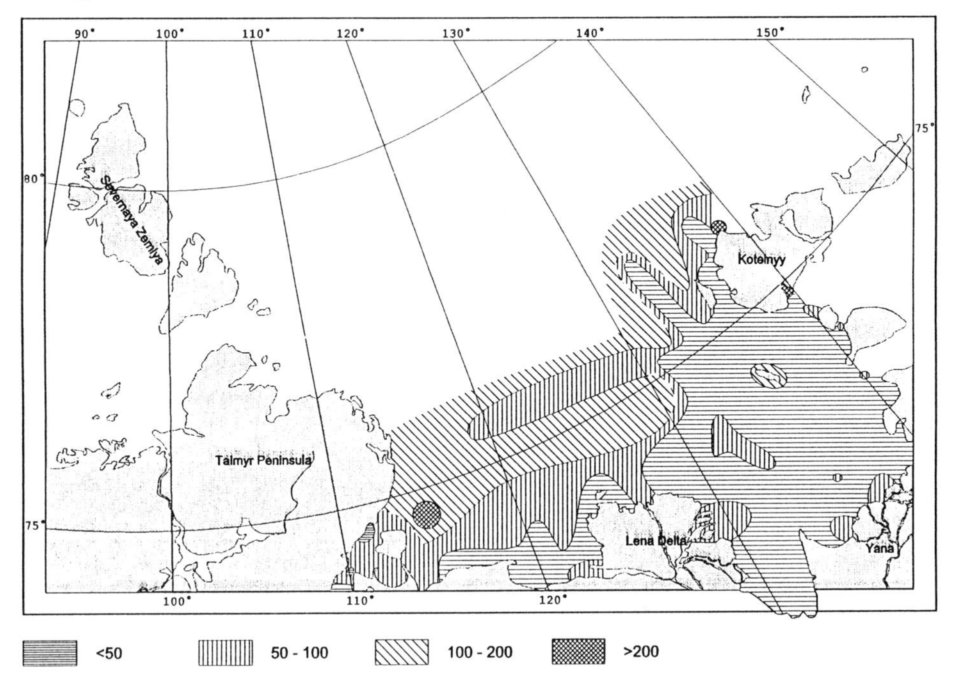

Figure 1: Distribution of macrozoobenthic biomass (g wet weight m^{-2}, incl. shells) in the southern Laptev Sea.

Biogeography

Biogeographic analyses, based exemplarily on the knowledge of the global distribution of 192 species of bivalves, gastropods and crustaceans (isopods, mysids), suggest that the benthic fauna in nearly all parts of the Laptev Sea consists primarily of Boreal-Arctic species (Gorbunov, 1946; Golikov et al., 1990; original data). In terms of species numbers, Arctic species constitute, on the average, slightly more than one quarter of the fauna. However, they play a more significant role (33 to 50%) in the brackish waters of estuaries and adjacent areas in the south-eastern Laptev Sea (Figure 2). Towards the central parts of the sea, their species proportions declined to 25 to 27% while those of widespread Boreal-Arctic species increased to 44 to 64%. In the northern Laptev Sea, Atlantic Boreal-Arctic species were dominant (36 to 51%). Pacific Boreal-Arctic species were recorded in depths of less than 250 m, but they did not account for more than 14% of the fauna in any region studied. Wide-spread subtropical-Arctic and bipolar species occurred in depths of more than 300 m, but they are even less important in terms of species proportions.

In the northern Laptev Sea, widespread Boreal-Arctic species dominated the benthic fauna in

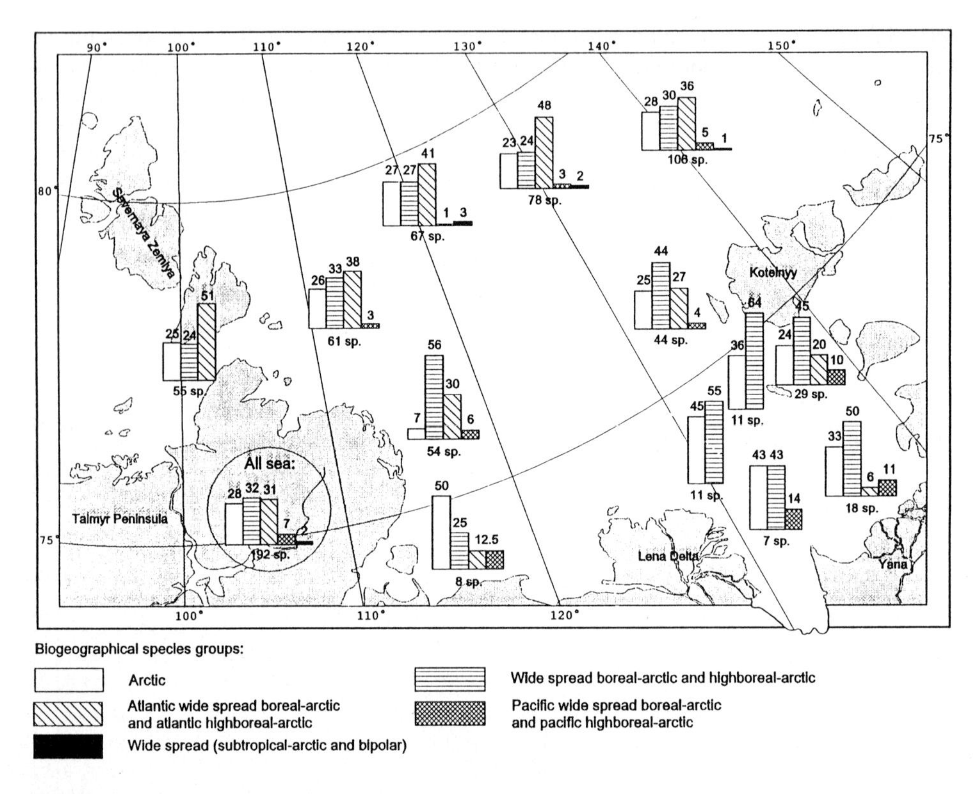

Figure 2: Species percentages of different biogeographic groups in various parts of the Laptev Sea.

depths of 50 to 75 m (Figure 3). In greater depths (up to 1100 m), Atlantic Boreal-Arctic species became more important, generally constituting more than 50% of the macrobenthic species in the samples taken in the depth influenced by Atlantic intermediate waters. In the near-bottom Arctic waters of depths of more than 1100 m, so-called "pseudo-abyssal" Arctic species and subspecies (terminology, see in Golikov, 1985) predominate, attaining species proportions of up to 70% in depths of more than 3000 m.

The following conclusions can be drawn from the biogeographic analyses: Firstly, the composition of the benthic fauna is affected much more by North Atlantic than by North Pacific elements, particularly in the northern Laptev Sea. Secondly, the low level of species endemism, resulting from the geologically young age of the modern fauna, indicates that the Arctic shelf fauna is still in the process of being formed after the climatic changes in the Pleistocene and Holocene. Only the abyssal fauna in near-bottom Arctic waters is dominated by endemic-Arctic species and may be regarded as true Arctic.

Communities

Based on differences in biomass dominants (s. Vorob'ev, 1949), nine major macrobenthic communities were distinguished in the Laptev Sea (Figure 4). In its southern parts, communities are characterized by bivalve molluscs: (1) *Portlandia aestuariorum*, frequently associated with the subdominant species *Cyrtodaria kurriana*; (2) *Portlandia arctica* (synonymous with *P. siliqua*); (3) *Tridonta borealis* and *Nicania montagui* (in polyhaline waters with salinities of 15 to 30); (4) *Leionucula bellotii*; and (5) *Tridonta borealis* (further north-westward). In the northern Laptev Sea, macrobenthic assemblages characterized by the following species were found in different depth zones: by the brittle stars (6)*Ophiocten sericeum*; (7) *O. sericeum*, associated with *Ophiacantha bidentata* and *Ophiopleura borealis*; (8) by polychaetes of the family Maldanidae, associated with the polychaete *Spiochaetopterus*; and (9) by the holothurian *Kolga hyalina*, associated with the sea urchin *Pourtalesia jeffreysi* and the sea cucumber *Elpidia* sp.

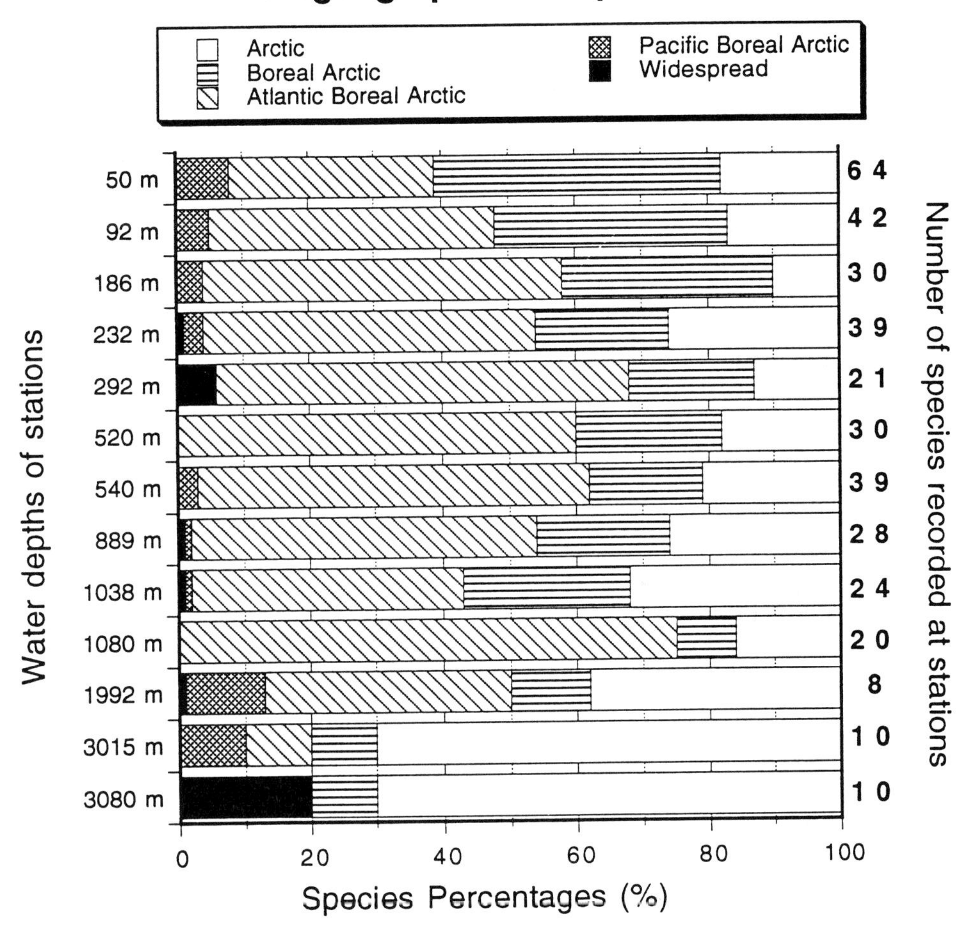

Figure 3: Species percentages of different biogeographic groups at stations in depths ranging from 50 to 3080 m.

Trophic structure

Two major trophic zones were distinguished (Figure 5): one dominated by deposit feeders, and the other dominated by suspension feeders. The first covers the larger part of the Laptev Sea, encompassing nearly all regions in which a transformation of riverine waters occurs: the bottom depression of the Sadko Trough, the continental slope, as well as probably most of the outer shelf. The zone of suspension feeders is located mainly in the western and central Laptev Sea where dense waters from the Arctic Basin enter the near-bottom layers. Moreover, isolated smaller areas with a predominance of sestonophages are located west of Stolbovoi Island, in the straits between the New Siberian Islands, and under the cyclonic gyre north of Yana Bay.

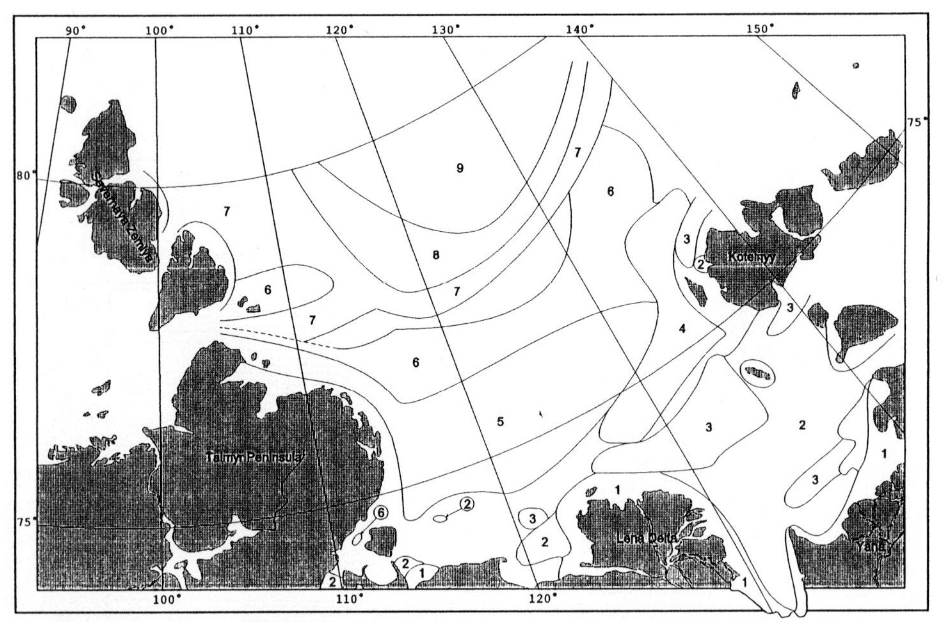

1 - Portlandia aestuariorum; 2 - P. siliqua; 3 - Tridonta borealis + Nicania montagui; 4 - Leionucula bellotii; 5 - Tridonta borealis; 6 - Ophiocten sericeum; 7 - Ophiocten sericeum + Ophiacantha bidentata + Ophiopleura borealis; 8 - Maldanidae gen.sp. + Spiochaetopterus sp.; 9 - Kolga hyalina + Pourtalesia jeffreysi + Elpidia sp.

Figure 4: Distribution of nine major macrozoobenthic communities distinguished in the Laptev Sea. For each assemblage, the names of dominant (and, partly, sub-dominant) indicator species are given.

Principal environmental factors

The distribution and structure of macrobenthic communities are known to be affected by a variety of environmental factors, such as water depth, seabed sedimentology, sea ice cover and near-bottom water salinity.

Water depth

The lowest number of macrobenthic species per station was observed in summer at a depth of 8 m (Figure 6). At depths of 45 to 120 m, species numbers increased to more than 140. The absolute maximum (214 species per station) was noted in the adjoining part of the East Siberian Sea at 54 m depth. Species numbers declined in depths of more than 120 m. Below 2000 m, figures ranged from 24 to 38.

Seabed sedimentology

Significant differences were also observed between benthic communities inhabiting stony and rocky grounds on the one hand and those occurring on sandy and silty sediments on the other. For the Laptev Sea, there are unfortunately few data on the benthos of stony and rocky

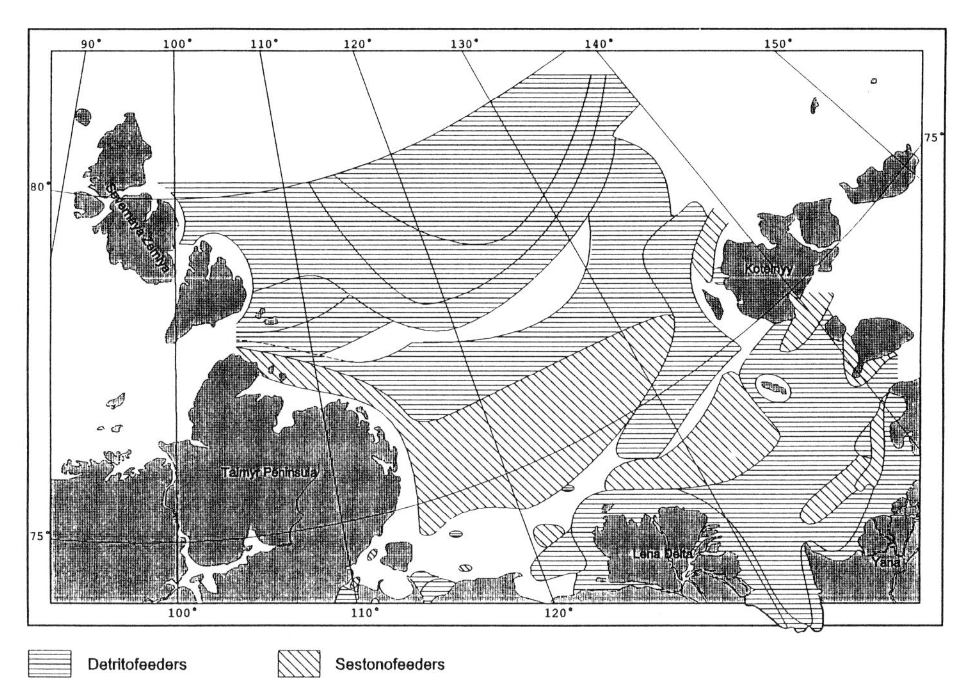

Figure 5: Distribution of macrozoobenthic trophic zones in the Laptev Sea.

grounds, i.e. only from sites off the islands Stolbovoi, Kotelnyi and Belkovskii (Golikov et al., 1990). There, sedentary organisms, such as the algae *Phyllophora truncata*, the sponge *Suberites domuncula*, the hydroid *Laphoeina maxima*, the barnacle *Balanus crenatus* and the bivalve *Musculus corrugatus*, predominated the communities at depths of 3 to 22 m. On sediment grounds (sand and silt, without stones) in adjoining areas at the same depths, where burrowing organisms, such as the bivalve molluscs *Portlandia aestuariorum*, *P. arctica*, *Tridonta borealis* and *Nicania montagui* predominated, species numbers were 1.2 to 6 times higher. In the southern and central Laptev Sea, the maximum numbers of species (50 to 100) were noted mainly in sandy sediments (weight proportion of sand 45 to 70% and that of silt 9 to 31%). Besides faunal composition, macrobenthic biomass is also strongly affected by seabed granulometric characteristics, albeit in the opposite direction: biomass values were found to be 2 to 6.5 times higher on rocky and stony grounds than those (100 g m^{-2} at maximum) reported from adjacent sandy and silty sediment in the same depths.

Near-bottom water salinity

The strong salinity gradients caused by the strong riverine inflow into the Laptev Sea apparently have the most important effect on macrobenthic distribution patterns. It is evident that the indicator species of the major biocenoses, as outlined above (Figure 5) and also presented by Sirenko et al. (1995), are confined to certain salinity ranges. This implies that the distribution of the communities primarily reflects the extent of the spreading of brackish near-bottom waters during summer.

The bivalve *Portlandia aestuariorum*, for instance, as well as the isopod *Saduria entomon glacialis* and the mysid *Mysis relicta*, occur in the estuaries of the large rivers and are closely bound to the very brackish surface waters (estuarine-Arctic waters with salinities of 5 to 18) that may reach down to depths of 10, rarely 15 m. Echinoderms are absent in this zone. Further north and/or at greater depths, salinities in the near-bottom water range between 18 and 30 (polyhaline-Arctic water) due to the admixture of riverine water during summer. In this zone, the bivalve *Portlandia arctica* and the isopod *Saduria sibirica* replace their above-mentioned sibling species, locally forming mass occurrences. The cumacean *Diastylis sulcata* is another typical inhabitant of polyhaline waters. Only five echinoderm species, however, seem to penetrate this brackish-water zone (*Myriotrochus rinkii, Leptasterias groenlandica, Urasterias linki, Stegophiura nodosa, Ophiocten sericeum*). The effect of riverine inflow gradually diminishes towards the north and with increasing depth. In regions with near-bottom water salinities between 30 and 32 (poly-euhaline water), more echinoderm species thrive. The indicator species of this region (where also the extensive winter polynyas of the Laptev Sea are situated) is the bivalve *Leionucula bellotii.* In regions and/or depths where riverine influences are negligible, near-bottom water salinities are always above 32 (euhaline water). Typical species of this zone are the crustaceans *Munnopsis typica* and *Munnopsurus giganteus*, as well as the echinoderms *Ophiura sarsi* and *Ctenodiscus crispatus.*

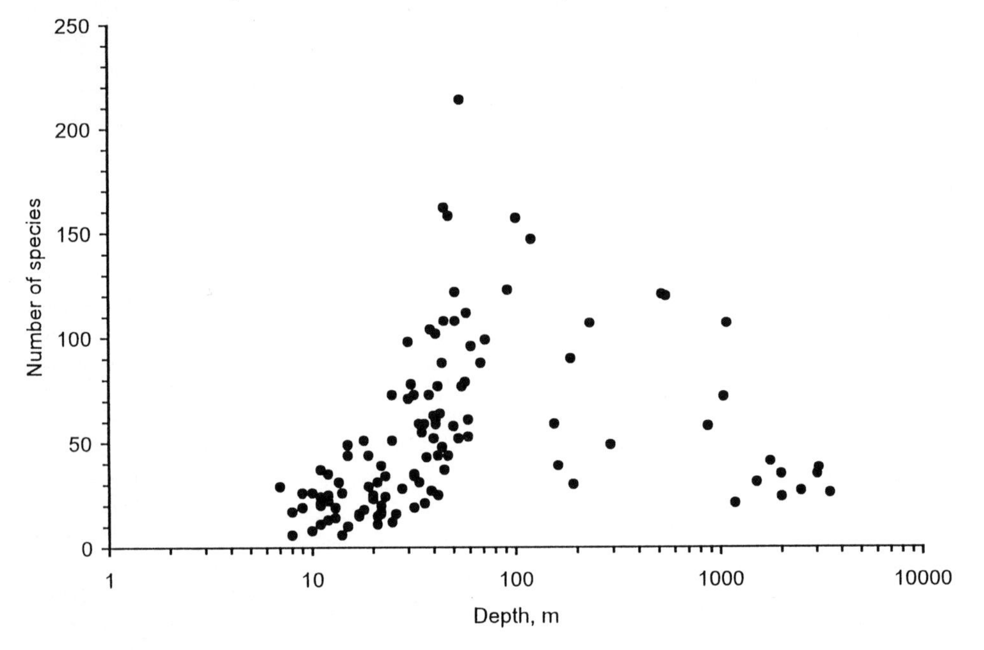

Figure 6: Relation between macrozoobenthic species numbers, recorded at stations sampled during expeditions in 1993-1995, and water depth in the Laptev Sea.

Synoptically interpreting the distribution of both macrobenthic communities and brackish water masses, two ecological-biogeographical districts can be distinguished: estuarine-Arctic and marine-Arctic. The latter district can be further divided into three ecological-biogeographical regions: polyhaline-Arctic, poly-euhaline-Arctic (or mixo-euhaline-Arctic), and euhaline-Arctic (Figure 7).

Besides salinity gradients, the pattern of sea ice cover may affect the ecological-biogeographical zonation. During winter, both the estuarine-Arctic district and the polyhaline-Arctic region are covered by fast ice, large parts of the poly-euhaline region coincide with the positions of the extensive quasi-stationary winter polynyas, and most of the euhaline region is

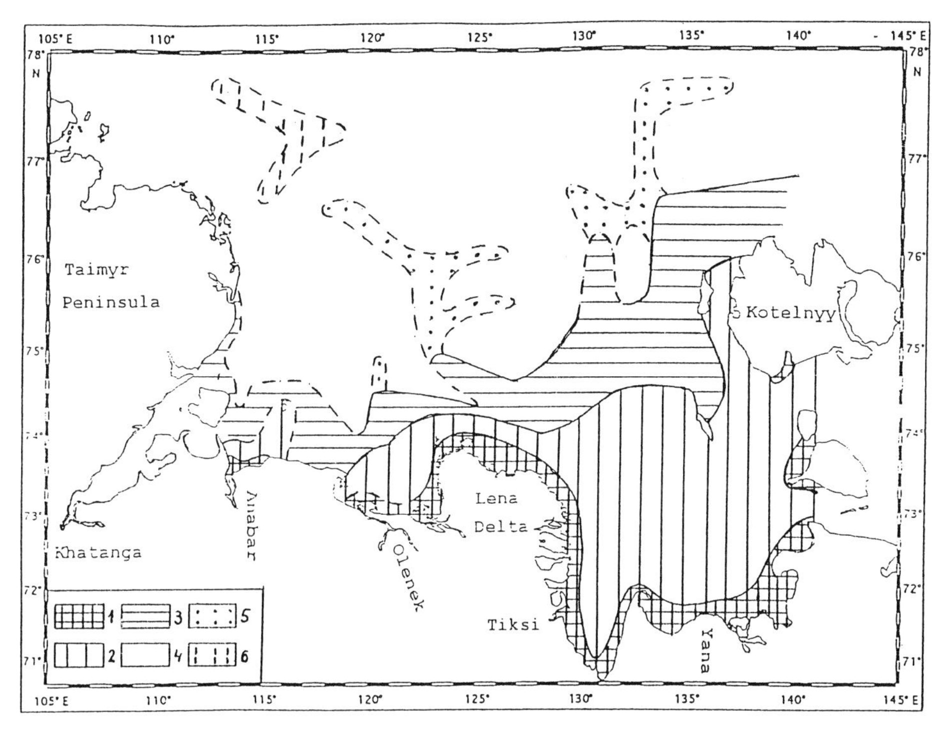

Figure 7: Ecological-biogeographical zonation in the Laptev Sea. 1) Estuarine-Arctic district; 2-6) Marine-Arctic district: 2) Polyhaline-Arctic region; 3) Poly-Euhaline-Arctic region; 4) Euhaline-Arctic region; 5+6) Euhaline-Arctic subregions of periodic inflow of brackish water.

characterized by drift ice cover. Moreover, the primarily latitudinal pattern of the ecological-biogeographical zonation is disturbed by a northward transport of brackish water masses and their associated fauna (e.g. *Portlandia arctica*, *Saduria sibirica*) along the paleovalleys, especially along their eastern slopes (Golovin et al., 1995; Dmitrenko et al., 1995). Hoevev, the extent of these northward movements varies both between valleys in relation to the intensity of river run-off and between years in relation to interannual oscillations in the hydrological regime. Therefore, these sub-regions of periodic inflow of brackish water can be regarded as distinct parts of the euhaline-Arctic region. Specimens of typically brackish-water species that are regularly recorded in these areas are probably expatriated individuals of southern populations and may serve as markers of the spreading of riverine water masses in the northern Laptev Sea.

Species numbers and biomass were also strongly related to near-bottom water salinities (Figures 8 and 9). Generally, there is a trend that values of both parametes increase with salinity, being especially obvious for biomass. However, local minima of both species numbers and biomass are discernible near the salinity boundaries between estuarine and polyhaline (16 to 18), as well as between polyhaline and euhaline waters (approximately 30). This pattern provides further evidence for the validity of the discrimination of ecological-biogeographical zonation outlined above.

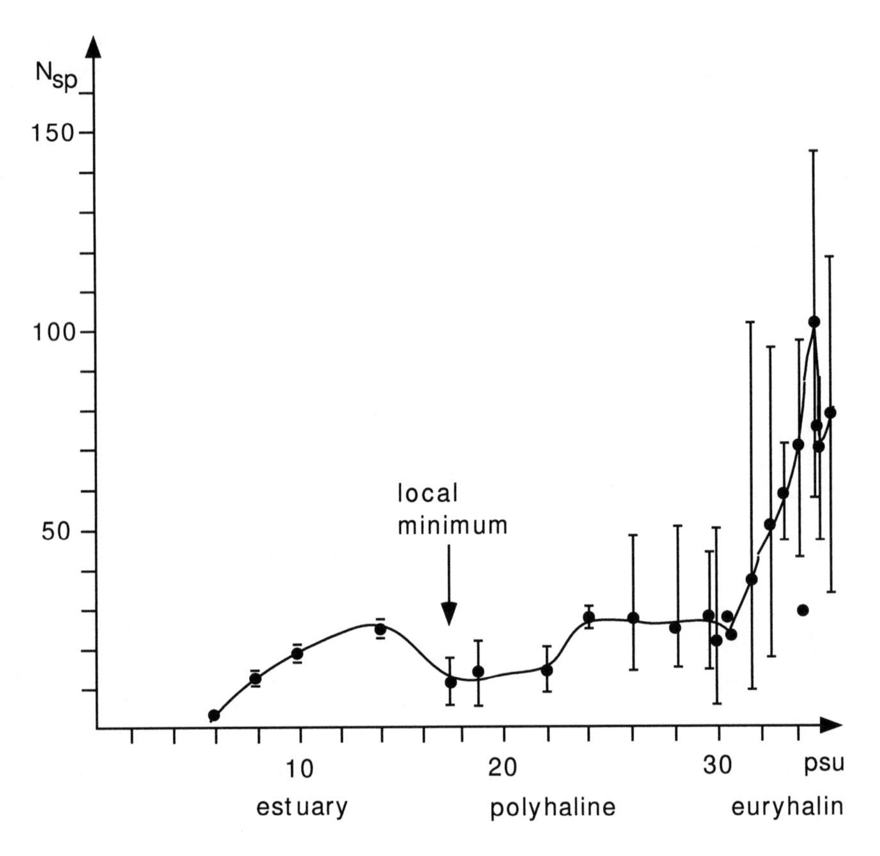

Figure 8: Relation between salinities of near-bottom waters and macrozoobenthic species numbers (mean ± s.d.), recorded at stations sampled during expeditions in 1993-1995 in the Laptev Sea.

Acknowledgements

We wish to thank our colleagues in the Zoological Institute RAS for the identification of species without which this communication would not be possible. Financial support for the work was obtained from the RFFI foundation (N 96-04-48111), as well as from the German Ministry of Education, Science, Research and Technology in the framework of projects 03F11GUS, 03F14GUS, 03PL009A and "Laptev Sea System". This contribution is publication no. 1343 of the Alfred-Wegener-Institut Bremerhaven.

References

Anisimova, N., H. Deubel, S. Potin and E. Rachor (1997) Zoobenthos, in: Rachor, E. (ed.): Scientific cruise report of the Arctic expedition ARK-XI/1 of RV "Polarstern" in 1995. Ber. Polarforsch., 226, 103-110.

Derjugin, K.M. (1993) Benthos of the Lena River estuary (in Russian). Issled. morei SSSR, 15: 63-66.

Dmitrenko, I.A. and Transdrift II Shipboard Scientific Party (1995) The distribution of river run-off in the Laptev Sea: the environmental effect. in: Kassens, H., D. Piepenburg, J. Thiede, L. Timokhov, H.-W. Hubberten, S.M. Priamikov (eds.), Russian-German cooperation: Laptev Sea System. Ber. Polarforsch., 176, 114-120.

Fahl, K. and R. Stein (1997) Modern organic carbon deposition in the Laptev Sea and the adjacent continental slope: surface water productivity vs. terrigenous input. Org. Geochem., 26, 379-390.

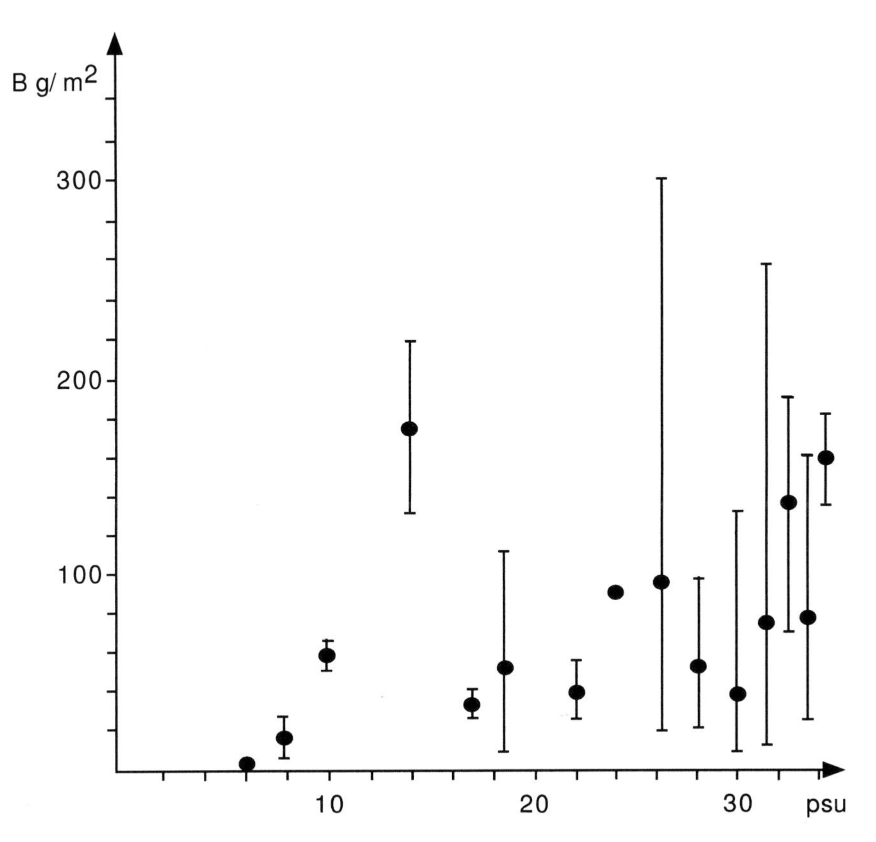

Figure 9: Relation between salinities of near-bottom waters and macrozoobenthic biomass (g wet weight m^{-2}, mean ± s.d.), recorded at stations sampled during expeditions in 1993-1995 in the Laptev Sea.

Gukov, A.Yu. (1989) Bottom biocoenoses of the Buor-Haya Bay of the Laptev Sea (in Russian). Okeanologiya, 29(2), 316-317.

Gukov, A.Yu. (1991) On the bottom fauna of the Yansky Bay of the Laptev Sea (in Russian). Okeanologiya, 31(3), 454-456.

Gukov, A.Yu. (1992) On the study of the Anabar Bay of the Laptev Sea (in Russian). Okeanologiya, 32(3), 506-509.

Gogorev, R.M. (1994) Some features of the horizontal distribution of phytoplankton in the Laptev Sea (August-September 1993) (in Russian). in: L.A. Timokhov (ed.), Scientific results of the expedition LAPEX-93, S.Pb., 337-352.

Golikov, A.N. (1985) Zonations and organismic assemblages: comments on the comprehensive review by Pérès (1982). Mar. Ecol. Prog. Ser., 23, 203-206.

Golikov, A.N., O.A. Scarlato, V.G. Averincev, T.V. Menshutkina, O.K. Novikov and A.M. Sheremetevsky (1990) Ecosystems of the New Siberian shoals, their distribution and functioning (in Russian). in: Golikov, A.N. (ed.), Ecosystem of the New Siberian shoals and the fauna of the Laptev Sea and adjacent waters: Explorations of the Fauna of the Seas, 37(45), 4-79.

Golovin, P.N., V.A. Gribanov and I.A. Dmitrenko (1995) Macro- and meso-scale hydrophysical structure of the Laptev Sea. in: Kassens, H., D. Piepenburg, J. Thiede, L. Timokhov, H.-W. Hubberten, S.M. Priamikov (eds.), Russian-German cooperation: Laptev Sea System. Ber. Polarforsch., 176, 99-106.

Gorbunov, G.P. (1946) Bottom population of New Siberian shoals and central part of the Arctic Ocean (in Russian). In: Tr. dreif. exp. Glavsevmorputi na "G.Sedaov" 1937-1940: M., L. 3, 30-138.

Hinz, K. and M.K. Schmid (1995) Deutsch-Russische Untersuchungen des Benthos des Laptewmeeres. Ber. Polarforsch., 155, 66-67.

Petryashov, V.V., B.I. Sirenko, E. Rachor and K. Hinz (1994) Distribution of the macrobenthos in the Laptev Sea from materials of the expeditions of RV *Ivan Kireev* and ice-breaker RV *Polarstern* in 1993 (in

Russian). Scientific results of the expedition LAPEX-93. S.-Pb.: 277-288.
Piepenburg, D. and M.K. Schmid (1997) A photographic survey of the epibenthic megafauna of the Arctic Laptev Sea shelf: distribution, abundance, and estimates of biomass and organic carbon demand. Mar. Ecol. Prog. Ser., 147, 63-75.
Popov, A.M. (1932): A hydrobiological review of the Laptev Sea (in Russian). Issled. fauny morei SSSR, 15, 189-229.
Rachor, E., K. Hinz and B. Sirenko (1994) Macrofauna. in: Fütterer, D.K. (ed.), The expedition Arctic '93, leg ARK-IX/4, of RV "Polarstern" 1993. Ber. Polarforsch., 149, 97-106.
Sirenko, B.I., V.V. Petryashov, E. Rachor and K. Hinz (1995) Bottom biocenoses of the Laptev Sea and adjacent seas. in: Kassens, H., D. Piepenburg, J. Thiede, L. Timokhov, H.-W. Hubberten, S.M. Priamikov (eds.), Russian-German cooperation: Laptev Sea System. Ber. Polarforsch., 176, 211-221.
Sirenko, B.I. and D. Piepenburg (1994) Current knowledge on biodiversity and benthic zonation patterns of Eurasian Arctic shelf seas, with special reference to the Laptev Sea. In: Kassens, H. (ed.), Russian-German Cooperation in the Siberian Shelf Seas: Geo-System Laptev Sea. Ber. Polarforsch., 144, 69-77.
Stuxberg, A. (1880): Evertebratefaunen i Sibiriens Ishav. Bihang K. Svenska Vet. Akad. Handl., 5(22), 1-76.
Vorob'ev, V.P. (1949): The benthos of the Sea of Azov. Trudy AzTscherNIRO, 13, 1-193.
Zenkevich, L.A. (1963): Biology of the Seas of USSR. Allen and Unwin, London [translation of: Zenkevich, L.A. (1947): The seas of the USSR II: Fauna and biological productivity of the sea (in Russian). AN SSSR Publ., Moscow, 739 pp.]

Carepoctus solidus sp.n., a New Species of Liparid Fish (Scorpaeniformes, Liparidae) from the Lower Bathyal of the Polar Basin

N.V. Chernova

Zoological Institute, Russian Academy of Sciences, 1 Universitetskaya, 199034 St. Petersburg, Russia

Received 3 March 1997 and accepted in revised form 3 March 1998

Abstract - A new species of snailfish, *Careproctus solidus* (Scorpaeniformes, Liparidae), is described from lower bathyal depth of the Polar Basin. The holotype and only known specimen is deposited in the Zoological Institute, Russian Academy of Sciences, St. Petersburg (ZISP 50626, immature female SL 117.5 mm). The fish was collected by the German-Russian expedition by RV "Polarstern" in 1993 on the continental slope of the northern part of the Laptev Sea (78°23.4 N, 133°09,5 E) at a depth of 2151-1934 m. *C. solidus* resembles the Arctic shelf species *C. reinhardti* (Krøyer) in counts, prickled skin, 0+2 pores in temporal canal, but differs in the oblique mouth, the pectoral fin in a low position, 3 (2+0+1) radials in the pectoral girdle, a small triangular opercular flap, round (not longitudinally oval) pupil, and a solid (ungelatinous) body. The relantionships of this unusially deep-living *Careproctus* species are unclear, but it is the second example of a unique liparid fish from the deep Polar Basins.

Introduction

Fish investigations in the Laptev Sea are still at the stage of faunal exploration. The German-Russian expeditions of 1993-1995 there provided new and valuable ichthyological results, and the list of fish species confirmed as occuring in the area has been increased by about 25 %; the number of marine species recorded has doubled (Neyelov and Chernova, 1994; Chernova and Neyelov, 1995). This report contains the description of a new species of liparid fish which has been collected during those expeditions by the R/V "Polarstern" in the northern part of the Laptev Sea.

Only three species of deep-sea liparids have so far been found in the Central Arctic Basin at a depth in excess of 1000 metres: *Paraliparis bathybius* (Collett, 1879), *P. violaceus* Chernova, 1991 and *Rhodichthys regina* Collett, 1879 (Collett, 1879; Andriashev, 1948, 1954; Chernova, 1991; Andriashev and Chernova, 1993). Liparids of the genus *Careproctus* Krøyer, 1862 are known in the marginal Arctic seas only from shelf waters in neighbouring parts of the North Atlantic; in the Barents and Kara Seas their vertical distribution is restricted to depths of 50-628 m (Able and Irion, 1985; Chernova, 1991). Therefore, the finding of a specimen of the genus *Careproctus* at depth approaching 2000 m in the lower bathyal of the Laptev Sea is quite unexpected and very noteworthy. Comparison of the specimen with other known *Careproctus* reveals that it is a new species, which is described below.

Careproctus solidus new species - Thickset tadpole (Figures 1-4)

Material examined

Holotype. ZISP 50626. Immature female, standart length (SL) 117.5 mm. R/V "Polarstern", cruise ARK 27, St. 35: 78°23.4 N, 133°09.5 E; Sept.4, 1993.; depth 1151-1934 m. Collector Katja Hinz.

In: Kassens, H., H.A. Bauch, I. Dmitrenko, H. Eicken, H.-W. Hubberten, M. Melles, J. Thiede and L. Timokhov (eds.) Land-Ocean Systems in the Siberian Arctic: Dynamics and History. Springer-Verlag, Berlin, 1999, 181-188.

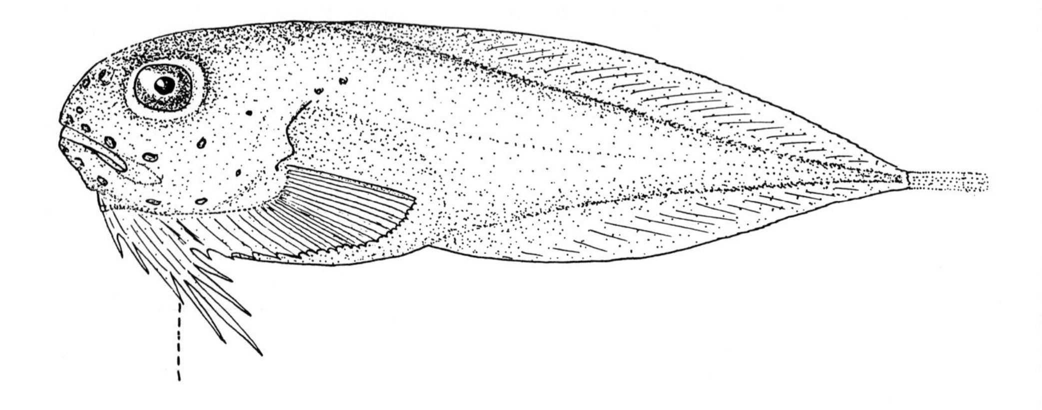

Figure 1: *Careproctus solidus* sp.n. Holotype ZISP 50626. Immature female SL 117.5 mm. Laptev Sea, depth 2151-1934 m.

Diagnosis

Vertebrae 60 (10+50), dorsal fin rays (D) 55. Pectoral fin rays (P) 31. Radials in pectoral girdle (Rad. pect.) 3 (2+0+1), the upper two notched. Two pairs of pleural ribs present. Hypural plate single, with a posterior slit. Mouth terminal, oblique. Teeth simple. Gill slit situated above pectoral fin base, its length about 1/3 of head length (HL). Posterior end of operculum below level of lower margin of eye. Two suprabranchial pores present; post-ocular pore absent. Pectoral fin located low. Body quite firm, robust and almost ungelatinous. Skin with cactus-like prickles. Head 27.4 % of SL, preanal length 44 %, disk length 5.8 %. Colour light, with greyish tint; peritoneum pale.

Description

Vert. 60, D 55, A 50, P 31. Rad. pect. 3. Gill rakers 7. Burke's pore formula: 2-5-7-2. Pyloric caeca 21.

Head large (its length 3.6 times in SL), high and compressed, with its width much less than its depth (53 and 87 % of HL); tapering downward on the sides so that in ventral view the contours of the eyes and bony rim of the interorbital space are visible (Figure 2). Upper head contour widely rounded in front of eye. Snout deep, not projected. Eye large, 4 times in HL, placed inside of large (1/3 of HL) orbit. Distance between orbital rim and contour of eye about 1/3 of eye diameter. Pupil round, about 1/3 of eye diameter. Suborbital space (from lower margin of orbit to end of mouth cleft) equal to eye. Nostril ending in a short tube situated slightly anterior to eye center. Mouth oblique, terminal. Mouth cleft slants downward posteriorly, its end reaching to vertical of anterior margin of eye; posterior end of upper jaw reaching to vertical of eye center. Teeth simple, 21 (24) oblique rows on each half of upper (lower) jaw; about 7 teeth in a full row near symphysis; symphysial gap absent. Lips wide, not fleshy. Chin not massive, rounded in ventral view. Upper and lower pharyngeal teeth present. Gill slit situated above pectoral fin base, extending downward to the level of the 2nd pectoral ray; gill slit quite large (its length nearly 1/3 of HL), and located rather low, its upper end level with pupil and lower end reaching down almost to level of posterior end of mouth cleft. Opercular flap small and triangular, placed at lower part of the gill slit; its posterior end below the level of the lower margin of the eye. Gill rakers knob-like, with a few needle-like appendages on their tops. Cephalic pores not contoured. Circum-oral pores large, size similar to nostril, oval, not submerged. Symphysial mandibular pores half as large as others, distance between them (pm1-pm1) half of distance pm1-pm2 (Figure 2). Postorbital pore absent. Suprabranchial pores two, small.

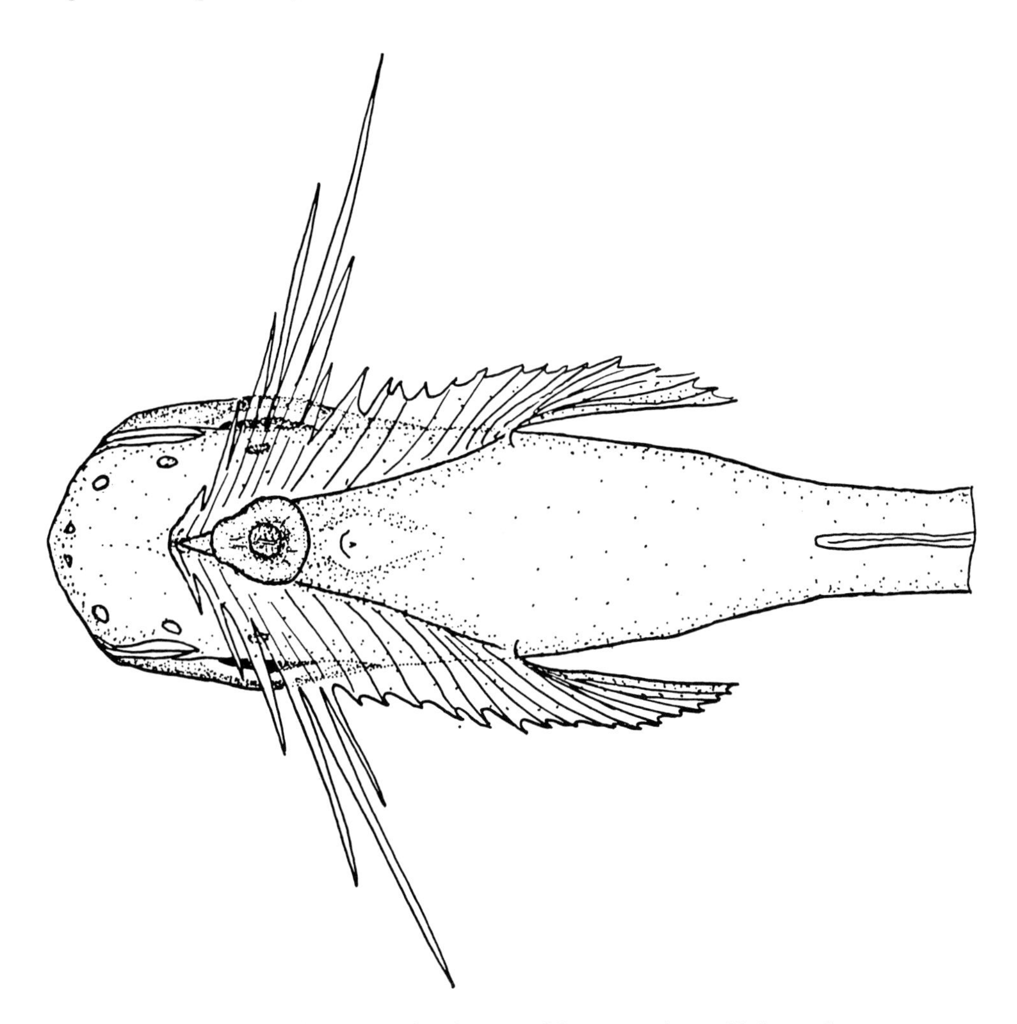

Figure 2: Holotype in ventral view. Pm 1 and pm 2 - pores of the preoperculo-mandibular canal.

Body quite deep, compressed, not elongate. Maximum body depth (above dorsal fin origin) 4 in SL. Trunk long; preanal length 2.3 in SL. Dorsal and anal fins low, at half of fin length equal to about 2/3 of body depth at this point. Skeleton from radiograph. Vertebrae 60: abdominal 10, caudal 50, distinctly recognizable. Vertebral column with a slight curve only along its anterior part. Interneural of the first dorsal ray between 3rd and 4th vertebrae; one free interneural (without ray) anteriorly. Pleural ribs in two pairs, thin and long but not curved. Hypural plate single, with a posterior slit. Caudal fin rays missing.

Pectoral fin situated low on the body, uppermost ray almost on level of end of mouth cleft. Fin deeply notched; upper lobe consisting of 23 rays (including the shortest ray at the notch; its length about half of upper lobe length), reaching almost to anal fin origin; lower lobe with 8 rays, extending to 2/3 of distance from disk to anal fin origin (the longest ray about 1.7 times longer than upper lobe rays). Base of lowermost pectoral ray on a vertical just behind nostril. Notch rays somewhat more widely spaced than the others. Pectoral girdle with 3 (2+0+1) radials, medium-sized; the two upper radials narrowly notched from below (Figure 3), with rudimentary fenestrae. Scapula notched, with fenestra and a strong helve (=shaft). Coracoid with short helve framed by wide ribs. Coracoidal part of pectoral girdle clearly protruding beyond ventral profile of body (Figure 1).

Disk small, about 5 times in HL and 0.8 of eye diameter, pearl-shaped, with bowl-like margins. Disk center small, round, skeletal rays not visible through skin. Anterior lobe of disk gradually passing into marginal ring, its length equal to disk center. Marginal ring of disk smooth, not sculptured by segmental pads on its inner rim, with width about equal to disk center. Distance of mandible from disk equalling two disk diameters. Anus situated on a vertical through posterior margin of eye. Distance from disk to anus about equal to disk center. Small ural papilla present. Pyloric caeca 21, banana-shaped, length of all nearly similar, about 8-9 % SL.

Skin with cactus-like prickles, also present on snout, circumorbital space and unpaired fins, obviously also scattered on other parts of head and body but lost in the specimen. Prickles look like rosettes of 2-10 sharp needles (without a knob-like base) protruding beyond the integuments. The skin is not speckled where prickles have been lost. Subcutaneous gelatinous tissue weakly developed, and therefore the body is rather firm and solid.

Measurements given as percentage of SL (in brackets as percentage of HL): head 27.4, its width 14.5 (52.8) and depth 23.8 (86.9); snout length 10.4 (37.9); eye 6.9 (25.1), orbit 9.4 (29.2); post-orbital head length 12.9 (47.2), gill slit 8.7 (31.7), interorbital width 12.8 (46.6); body depth 24.7, depth above anal fin origin 20.4, predorsal 27.4, preanal 44.3, mandible to disk 11.9, mandible to anus 18.9, disk 5.8 (21.1), disk to anus 1.8, anus to A 28.9; pectoral length: upper lobe 17.0 (62), shortest notch ray 8.1, lower lobe 28.1.

Colour in life not known but obviously pink with dusky tint. Specimen in alcohol light, some greyish colour, slightly darker on top of head, along dorsal fin base and sides of body above pectoral fins. Body below integument dusky, with scattered melanophores, slightly denser than on skin. Orobranchial cavity, stomach, peritoneum pale. No silvery guanine layer on belly below integument, or on lower half of eye. The holotype is an inmature female contaning unripened ovarian eggs (diameter less than 1 mm).

Distribution

The only known specimen was collected between 1934 and 2151 m water depth (temperature near sea floor -0.8 C, salinity 34.86) at the continental slope of the Laptev Sea.

Etymology

The name is derived from the Latin "solidus" (solid, not gelatinous, firm); it refers to the comparatively hard consistence of the body, caused by poor development of subcutaneous gelatinous tissue, a situation which is quite unusual for the genus *Careproctus*.

Comparative notes.

Only three species of *Careproctus*, *C. reinhardti* (Krøyer, 1862), *C. ranula* (Goode et Bean, 1880) and *C. micropus* (Günther, 1887) are presently recognized from the Arctic area (Able and Irion, 1985; Chernova, 1991). Comparative material of these species examined in this study was collected in the region extending from the Barents to Laptev Sea (see Chernova,

1991). *C. solidus* resembles *C. reinhardti* in some characters (counts, cactus-like prickles on the skin, 0+2 pores in the temporal sensory canal), but it differs greatly in having an oblique (versus horizontal) mouth, the pectoral fin situated low on the body (vs. high), with the uppermost ray level with the end of the mouth cleft (vs. lower half of eye); the opercular flap small and triangular (vs. widely rounded) and ending below the level of the eye (vs. level with upper half of eye), the pupil round (vs. longitudinally-oval), the eye lacking a silvery guanine layer on its lower half (Figure 4); pectoral girdle with 3 (2+0+1) versus 4 (3+1) notched radials, fenestrae rudimentary (vs. large as in *Liparis*), and the absence of thick subcutaneous gelatinous tissue (which entirely covers the anterior half of the dorsal fin in *C. reinhardti*). These differences are sufficient to separate *C. solidus* from *C. reinhardti* and all other *Careproctus* as a distinct new species. Two other *Careproctus* from the Arctic area, *C. ranula* and *C. micropus*, are small-sized dwarf species (specimens mature at a length of 55-70 mm SL), with smooth skins, 1+1 pores in the temporal canal, and 4 (3+1) unnotched radials in pectoral girdle. *C. solidus* differs from these species in having a prickled skin, 0+2 pores in temporal canal, 3 (2+0+1) radials in the pectoral girdle, and in the larger size of the adults (with maturate at length of more than 118 mm SL).

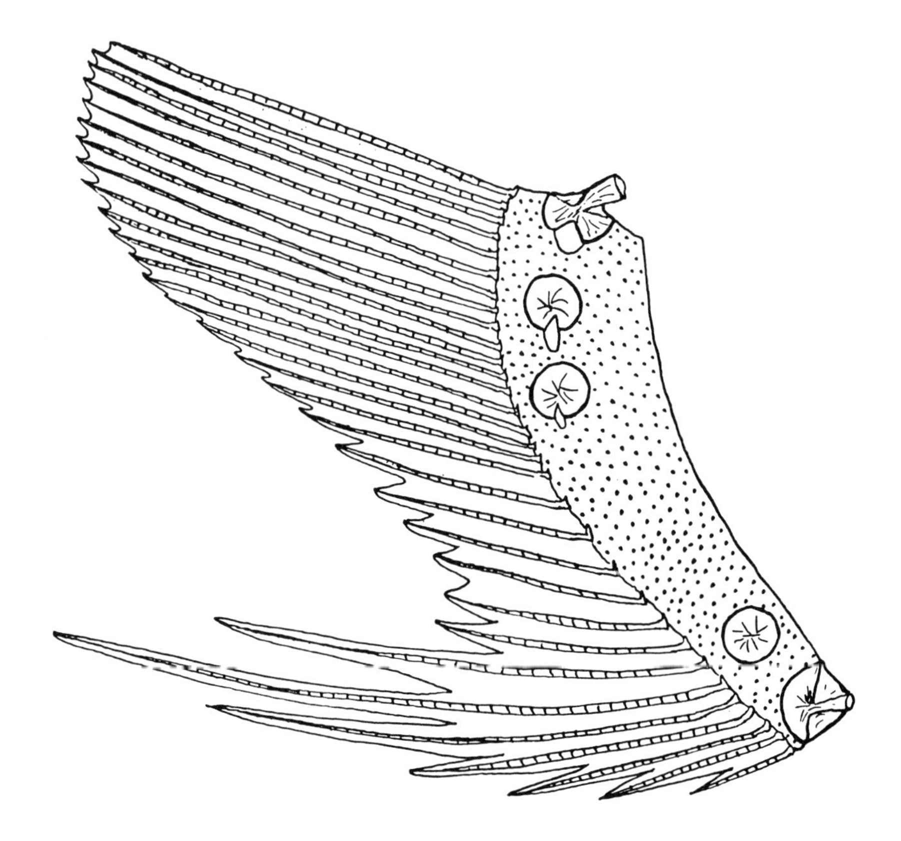

Figure 3: Pectoral girdle of *C. solidus*. Holotype.

The deep vertical distribution of *C. solidus* in comparison with the three species mentioned above also supports its distinctiveness. *C. reinhardti* does not occur below 700 m, (being usually found at depth of 150-350 m). *C. ranula* is distributed between 95-628 m (mostly 250-430 m) (Chernova, 1991). *C. micropus* has been recorded in the Faroe Trench from 988-1099 m (Günther, 1887), and in the Davis and Dutch Straits, near Jan Mayen Island, Faroe Islands and Iceland from 100-1785 m (Lütken, 1898), but in the Barents and Kara Seas it is known only from 150-325 m (Chernova, 1991).

The deepest occurring species of *Careproctus* appear to be *C. merretti* and *C. aciculipunctatus* described from the Porcupine Seabight (eastern North Atlantic) at a depth of 3990-3920 m (Andriashev and Chernova, 1988, 1997). *C. merretti* is characterized by a small and depressed head, naked skin, and a black peritoneum; *C. aciculipunctatus* has a black body and its skin prickles appear as a single needles. Both species have two radials in the pectoral girdle situated opposite each other (1+0+0+1), one postorbital and one suprabranchial pore in the temporal sensory canal, and a single hypural plate with no slit; thus they belong to another morphotype than does *C. solidus*.

Figure 4: Eye of *C. solidus* (left) and *C. reinhardti* (right), the last one with longitudinally oval pupil and silvery guanine layer on lower half.

In the rich and diverse *Careproctus* fauna of the north Pacific (Kido, 1988; Pitruk, 1991), *C. mederi* seems to be the most similar to *C. solidus* in such characters as the prickled skin, simple teeth, oblique mouth, and pectoral fin located low on the body, but it differs significantly in that the disk is much more reduced, the gill slit is very large (reaching down to the level of 15-20 pectoral rays), the lower lobe of the pectoral fin is short, and the subcutaneous gelatinous tissue is greatly developed. The species of the "*C. rastrinus*"-group (Pitruk, 1991) which have a prickled skin and comparatively large disk as does *C. solidus*, differ from the latter in having a *Liparis*-like pectoral girdle with 4 notched radials, 3 (vs. 2) pairs of pleural ribs, the mouth not oblique, and the pectoral fin not placed low on the body.

Discussion

Plesiomorphic morphological characters of fishes in the genus *Careproctus* (with *Liparis* as an outgroup) are the presence of four unequally spaced notched radials in the pectoral girdle and 3 large fenestrae in the cartilaginous basal lamina. More derived species possess rounded radials lacking notches and fenestrae; the number of radials may be reduced to 2 (1+0+0+1) (Andriashev and Stein, in press). In the hypural complex a high degree of fusion of elements is a derived character; lesser fusion is plesiomorphic. Thus, *C. solidus* is a moderately derived species having 3 (2+0+1) radials, the fenestrae reduced in size, and the hypural plate partially divided by a slit. However, *C. solidus* appears to be more derived than are *C. reinhardti*, *C. mederi* and the "*C. rastrinus*"-group, but is less derived than *C. merretti* and *C. aciculipunctatus.*

The occurrence of *C. solidus* at lower bathyal depth in the Polar Basin invites a discussion of some zoogeographical problems. Possible pathways for *Paraliparis*, *Careproctus* and some other North Pacific secondary deep-sea fishes into the North Atlantic and Arctic depths have been suggested (Andriashev, 1990, 1991). The view is that because the Bering Strait has been shallow for a long time it probably did not allow deep-living genera such as *Paraliparis* and *Careproctus* (which have no planctonic larvae) to penetrate from the North Pacific directly into the Arctic. For these secondary deep-sea liparids another dispersal route, much more complicated, was proposed: (1) dispersion from their place of origin in the North Pacific south to the Patagonian-West Antarctic area and there evolving to form a rich secondary centre of speciation; (2) after the tectonic opening of the Drake Passage these southern hemisphere liparid fishes were able to penetrate into the western South Atlantic and from there disperse again northward into the depths of the North Atlantic and Polar Basins where some 11 endemic species have evolved. This scenario is supported by the discovery of a new species of *Paraliparis* (*P. abyssorum*) in the eastern North Atlantic (Andriashev and Chernova 1997) that is morphologically closely related to the Arctic *P. bathybius*.

The morphological type represented by *C. solidus* does not allow relating it to any known *Careproctus* from either the Arctic shelf or north Atlantic waters. It resembles *C. reinhardti* in some characters, but differs in many significant features; it is also quite far away from the north Pacific group of prickled species (*C. mederi*, *C. rastrinus* etc.); so these too should not be regarded as closely related. Thus the question of the origin of *C. solidus* at depth in the Polar Basin is still open. It is, however, the second such example among the deep-sea liparids of the Polar Basins; the other is the morphologically unique *Rh. regina* for which closely related species are also unknown.

References

Able, K.W. and W. Irion (1985) Distribution and reproductive seasonality of snailfishes in the St. Lawrence River estuary and the Gulf of St. Lawrence. Can. J. Zool., 63 (7), 1622-1628.

Andriashev A.P. (1948) To the knowledge of the fishes of the Laptev Sea (in Russian). Trudy Zool. Inst. AN SSSR, 7, 76-100.

Andriashev, A.P. (1954) Fishes of the northern seas of the USSR (in Russian). Izd-vo AN SSSR, Moscow-Leningrad. 556 pp. English translation: (1964) Israel Progr. Sci. Transl. (836), 617 pp.

Andriashev, A.P. (1990) On a probability of the transocean (non-Arctic) pathways of some North Pacific secondary deep-sea fishes into the North Atlantic and Arctic depths (Family Liparidae as an example) (in Russian). Zool. J. (Moscow), 69 (1), 61-67.

Andriashev, A.P. (1991) Possible pathways of *Paraliparis* (Pisces: Liparidae) and some other North Pacific secondary deep-sea fishes into North Atlantic and Arctic depth. Polar Biol., 11, 213-218.

Andriashev, A.P. and N.V. Chernova (1988) New species of the genus *Careproctus* (Liparididae) from the Porcupine Seabight at a depth of 4 km (in Russian). Voprosy ikhtiologii (Moscow), 28 (6),1023-1026.

Andriashev, A.P. and N.V. Chernova (1993) Annotated list of fish-like vertebrates and fishes of the Arctic seas and adjacent waters (in Russian). Voprosy ikhtiologii (Moscow), 34 (4), 435-456. English translation: J.Ichthyol. (1995), 35 (1), 81-123.

Andriashev, A.P. and N.V. Chernova (1997) Two new species of snailfishes (Liparidae, Scorpaeniformes) from the abyssal depth of the eastern North Atlantic (in Russian). Voprosy ikhtiologii (Moscow), 37 (4), 437-443.

Andriashev, A.P. and D.L. Stein (in press) Review of the snailfish genus *Careproctus* (Liparidae, Scorpaeniformes) in the Antarctic. Contrib. Sci. Los Angeles County Mus.

Chernova, N.V. (1991) Snailfishes (Liparididae) from the Eurasian Arctic (in Russian). Kol'sky nauchn. Tzentr, Apatity, 111 pp.

Chernova, N.V. and A.V. Neyelov (1995) Fish caught in the Laptev Sea during the cruise of RV "Polarstern" in 1993. Ber. Polarforsch., 176, 222-227.

Collett, R. (1879) Fiske fra Nordhavs-Expeditionens sidste Togt, sommeren 1878. Forh. Vidensk. Selsk. Krist., (1878), 14, 1-106.

Goode, G.B. and T.N. Bean (1880) Description of the new species of *Liparis* (*L. ranula*) obtained by the United States Fish Commission off Halifax, Nova Scotia. Proc.U.S.Nat. Mus., 2, 46-48.

Günther, A. (1887) Report on the deep-sea fishes collected by H.M.S. Challenger during the year 1873-76. Challenger Reports, Zool., 22, 66-70.

Kido K. (1988) Phylogeny of the family Liparididae, with the taxonomy of the species found around Japan. Mem. Fac. Fish. Hokkaido Univ., 35 (2), 125-256.

Krøyer, H. (1862) Nogle bidrag til nordisk Ichthyologie. Naturhistorisk. Tidskrift, 3 (1), 252-310.

Lütken, Ch.F. (1898) The ichthyological results. The Danish Ingolf Expedition, 2, 14-18.

Neyelov, A.V. and N.V. Chernova (1994) Preliminary check-list of fishes collected in the Laptev Sea during the expedition aboard the icebreaker "Polarstern" in 1993 (in Russian). In: Nauchn. resul'tati expeditsii LAPEX-93, Timokhov L.A. (ed.), Gidrometeoizdat, St.Peterburg, 272-276.
Pitruk D.L. (1991) Morphological bases of the system of liparid fishes (Scorpaeniformes, Liparididae) of the "*Careproctus*"-type from the Okhotsk Sea (in Russian). Avtoref. cand. diss. Inst. Biol. Morya, Vladivostok, 22 pp.

Spring Stopover of Birds on the Laptev Sea Polynya

D.V. Solovieva

Lena-Delta State Reserve, 28 Fiodorova St., 678400 Tiksi, Yakutia, Russia

Received 3 March 1997 and accepted in revised form 9 February 1998

Abstract - Four species of marine birds were found to occur in spring-time on the Laptev Sea Polynya. All birds were observed migrating westward above ice-free waters and drift ice fields. Inspection of King Eider *(Somateria spectabilis)* bioenergetics on the Polynya shows that this species daily pays 3.17 BMR in males and 3.44 BMR in females for the occurring here. Eider's impact on bottom fauna is 0.10-0.17% of biomass.

Introduction

The first suggestions that Great Siberian Polynya (Figure 1) is inhabited by birds came from the very beginning of the century. The early spring migration of „dark“ ducks over Chukchi Sea fast ice was noted by the Zarya expedition (Birulya, 1907). The information from weather stations' staff, as well as the fact that some species do migrate from sea to land in springtime when arriving at the breeding grounds, generally supported this idea. The Transdrift IV expedition at the Laptev Sea Polynya (Figure 1) enabled to obtain the data on bird fauna and abundance here. Surveys, being conducted in one week at the moment of intensive spring migration, don't allow to understand either birds stay on the Polynya for a long time, or they use it only as a fly-way, providing stopover for the rest and feeding. Periods and purposes for why birds visit Polynya are still under discussion: from our point of view, the food sources of the bottom and the ice-free water of polynya, not used by birds during the entire winter, provide good conditions for overspringing of many marine birds. Ducks were noted to appear offshore the New-Siberian Islands already in April (Rakhimov, pers. com.).

Is it difficult for birds to survive in low ambient temperatures aggravating by unstable ice conditions? We try to answer by the estimation of daily energy budget (DEB) in birds and their impact on the fauna of invertebrates in the Polynya. We use one species - the King Eider (*Somateria spectabilis*) - for these proposes because additional data to calculate it's DEB are available. Complete picture of bird community existence on the Polynya is a subject for the future investigations.

Methods

Aerial surveys and observation of migration

We surveyed the Laptev Sea Polynya adjacent to the Lena-Delta on 25, 27, and 29 of May 1996. Helicopter surveys were conducted from an altitude of 100m at 180 km/hr. Transect width was 250m on one side of MI-8 helicopter. A tape recorder was used to note all birds observations. Total area of the Polynya was estimated from the ice-map. As all birds were found to use the ice-free parts of Polynya only, the ice-free area was taking into account to estimate bird abundance (Figure 2). Spring bird migration was registered from the edge of both fast and drifting pack ice. Totally we spent six hours observing migration on 25 and 29 of May.

In: Kassens, H., H.A. Bauch, I. Dmitrenko, H. Eicken, H.-W. Hubberten, M. Melles, J. Thiede and L. Timokhov (eds.) Land-Ocean Systems in the Siberian Arctic: Dynamics and History. Springer-Verlag, Berlin, 1999, 189-195.

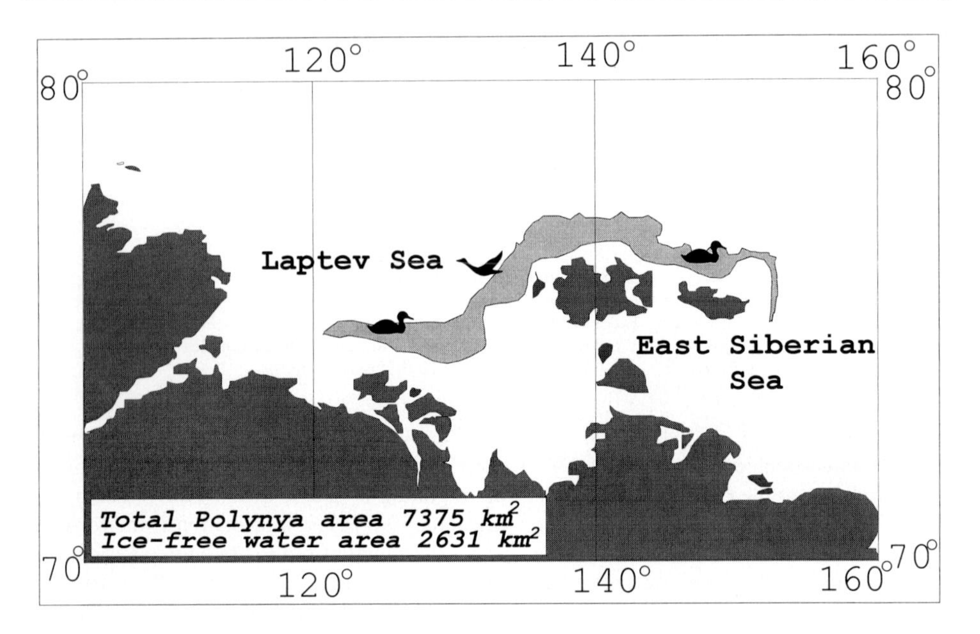

Figure 1: Location of the Laptev Sea Polynya.

Bioenergetics

Body mass and basal metabolic rate (BMR)

Body mass used in this study was obtained from 16 (9 males and 7 females) King Eiders, been shot in early June 1996 when arriving on breeding grounds in the north of Lena-Delta from the Polynya. The mass lost during the 85km-flight was neglected because of its insignificant value.

We used the experimental value of BMR, obtained by Jenssen et al. (1988) in winter for the closely related Common Eider (*S. mollissima*), to estimate King Eider's BMR under the same conditions. BMR in King Eider was calculated with exponent 0.723, the slope BMR equation for non-passerine birds in winter (Kendeigh et al., 1977).

Thermoregulation

Given that King Eiders spend the entire May on the Polynya, mean air temperature during May equalling -13.9°C (after data of Dunay weather station), is used as an ambient air temperature (T_A). The water temperature (T_W) on the sea surface in the period of observations was -1.7°C. The termoregulatory heat production (TR) was calculated from the equations for Common Eider (Jenssen et al., 1988) without regard to differences between species:

$$TR\,(J / g \cdot day) = 7.78\,(T_{LC} - T);\; T_{LC} = 16^{\circ}C \qquad (1)$$

We supposed, that birds were spending all the time on the water surface except for the time devoted to diving and flying. Hence half of the bird body was exposed to air while the other one was exposed to water. Therefore TR might be found as a median value between TR_A , calculated under T_A, and TR_W , calculated under T_W. We neglect termoregulation under the water because Eiders use wings to dive, leading to extra heat production, which in turn compensates cooling.

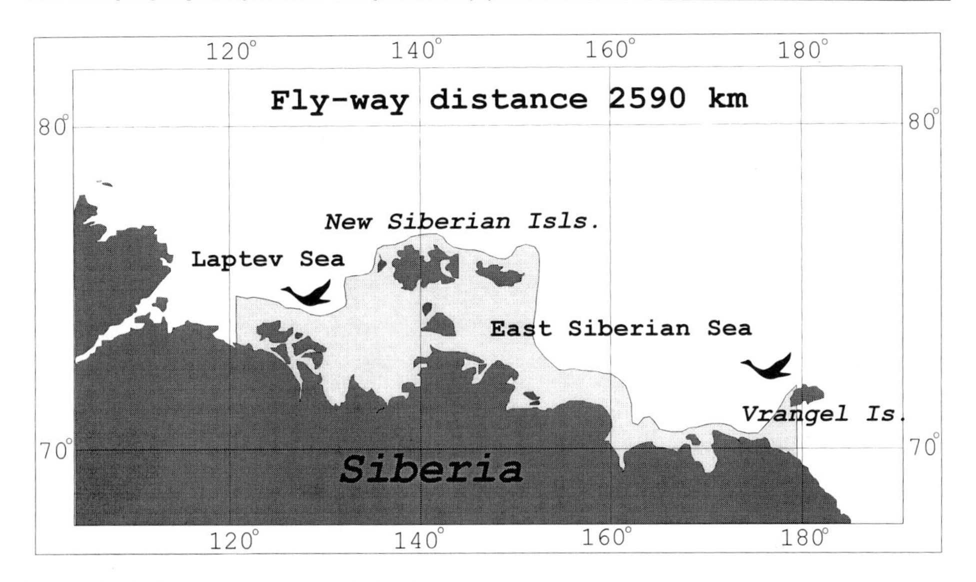

Figure 2: Birds fly-way along fast ice edge in Siberian seas.

Daily time budget

The method of the DEB calculation using the DTB is widespread in bioenergetic studies. Each activity distinguished inside the DTB was given by energy cost in units of BMR: diving - 5 (Woakes and Butler, 1983), pausing - 1.21, swimming - 1.5, comfort - 1.3, courtship - 1.6, rest - 1.15, and sleeping - 1 (Dolnik, 1995). We calculated the portion, which each activity brings into the DEB, by multiplying the time spent in this activity to its energy cost. DTB of wintering King Eiders was obtained by Systad (1996) for the light period of day. His equations (2, 3) allow us to find the time devoted to feeding (diving and pausing) for maintenance under Polar day (24 hours):

$$\text{Diving (min)} = -8.184 + 0.215 \cdot \text{day length} \quad (2)$$

$$\text{Pausing (min)} = -1.684 + 0.165 \cdot \text{day length} \quad (3)$$

We used the DTB (Systad, 1996) with enlarging of feeding time according to (2, 3) and supposing the rest of daytime as sleeping. Time devoted to flying was found from distance of migration and average velocity of duck flight in 60km/hr.

Productive energy (PE)

We assumed three productive processes taking place in King Eider when on the Polynya: (1) fat accumulation as a compensation of the mass lost during migration to the Polynya; (2) migration westward above the Polynya; and (3) gametogenesis and accompanying fat deposition in female.

Cost of migration was found from the equation (Dolnik, 1995) for Non-Passerines:

$$PE_{mg}\ \text{(KJ/km)} = 0.075m^{0.53} \quad (4)$$

So the total value of migration PE_{MG} is:

$$PE_{MG}\ (KJ) = 1.54 \cdot (PE_{mg} \cdot L_1 + PE_{mg} \cdot L_2) \quad (5)$$

where L_1 is distance of migration from the Chukchi Sea Polynya to the Laptev Sea Polynya, L_2 is distance of migration above Laptev Sea Polynya, coefficient 1.54 shows the efficiency of lipogenesis. Females deposit into clutch was calculated according to the equation of Dolnik (1995):

$$PE_{el}\ (KJ) = 32.49 m_f^{0.626} \quad (6)$$

Impact on the bottom fauna

Five stomachs of King Eiders from the Polynya provide the data on diet. Caloric density of main food (Gastropoda and Bivalvia) is 17.8 KJ/g, metabolic efficiency coefficient is 0.81, ratio between dry and wet weight is 0.25 in molluscs (Kendeigh et al., 1977), while the shell accounts for 30% of the wet weight (Alimov, 1981). Dive duration in King Eider of 91 seconds was used (Systad, 1996).

Results

Aerial surveys and observation of migration

We found four marine birds occurring on the Polynya in significant numbers (Table 1), as follows: King Eider, Long-tail Duck (*Clangula hyemalis*), Kittiwake (*Rissa tridactyla*), and Pomarine Skua (*Stercorarius pomarinus*). One more species - the Snow Bunting (*Plectrophenax nivalis*) - was registered once.

Intensive westward migration passed on 25 and 29 of May (Table 1). Ducks migrated in large flocks sometimes up to 300 individuals in King Eider, some flocks flew across drift ice fields from one lead to another. Kittiwakes and Pomarine Skuas migrated individually or in loose flocks up to 6, their flight passed above leads, following its configurations. The most intensive migration was observed from fast ice edge (74 24'N 129 29'E) on 25 of May when we counted 2627 King Eiders in 68 minutes.

Bioenergetics

Body mass and basal metabolic rate (BMR)

Mean body mass in King Eider males was 1790 (SD, 184)g, in females 1700 (SD, 209)g. Basal metabolic rate in winter was 607.7KJ/day and 584.5KJ/day accordingly.

Thermoregulation

Thermoregulation, calculated for non-diving and non-flying time of day (Table 3), was estimated at 0.23 BMR for males and 0.24 BMR for females.

Daily time budget

Winter DTB of King Eider's maintenance (Systad, 1996), transformed according to equations (2, 3)for polar day and added by extra feeding to cover productive needs, is shown in Table 3. We also add sleeping and flying activities. Summarising values of each EA we get the total energy of activity as 2.24 BMR for males and 2.32 BMR for females.

Table 1: Abundance estimates (individuals) without visibility correction factors.

Date	May 25	May 27	May 28	Average
Route length (km)	64	82	205	
Area surveyed (km^2)	16	20.5	50.3	
per cent of Polynya area	0.001	0.001	0.003	
KING EIDER	8240	22470	28810	19840
LONG-TAILED DUCK	30840	-	25050	27940
DUCK sp.	106900	28890	12200	49300
KITTIWAKE	-	6100	-	6100

Table 2: Intensity of migration (individuals/hour).

Date	King Eider	Long-tailed Duck	Kittiwake	Pomarine Skua
May 25	2325	-	-	3
May 29	29	17	15	24

Table 3: Daily time budget and energy of activity (EA) in King Eider on the Polynya.

Activities	Diving	Pausing	Flying*	Swiming	Rest	Comfort	Sleeping
Time (min) in female	399	312	66	142	98	28	498
EA (BMR) in female	1.39	0.26	-	0.11	0.10	0.03	0.35
Time (min) in male	424	332	66	142.2	98.4	27.6	473
EA (BMR) in male	1.47	0.28	-	0.11	0.10	0.03	0.33

* energy of flying activity is included into bellow PE_{mg} .

Productive energy (PE)

Males spend 3.98KJ per km of fly-way (calculated according to (4)), in females this value is 3.87KJ/km. The L_l distance is 1400 km. Therefore productive energy for this migration is 5572KJ in males and 5418KJ in females, in terms of body fat it is 142.9g and 138.9g correspondingly. It takes 8563KJ in males and 8326KJ in females to restore such amount of fat loss, in units of BMR it is 14.09 and 14.24. In consideration that birds restore the fat lost during overspringing on the Polynya (31 days), the daily energy devoted to fat regeneration is calculated at 0.45 and 0.46 BMR. Migration over the Polynya with distance L_2=1190km takes 4748KJ of energy from males and 4617KJ from females, equal to 0.25 BMR daily in both sexes. So the total value of daily PE_{MG} (see (5)) is 0.71 BMR and doesn't differ between sexes. Female deposited 3420KJ into clutch, equal to 0.18 her daily BMR. Total PE in females is 0.89 BMR.

Daily metabolised energy (EM), as a sum of TR, EA, and PE, consists 3.17 BMR in King Eider males and 3.45 BMR in females.

Impact on the bottom fauna

Stomachs in King Eiders coming from the Polynya contain remains of Gastropoda and Bivalvia shells. Two mollusc species were identified from the remains as *Leonucula bellotii* and *Cryptonatica clausa.* Biocoenoses, been dominated by *Leonucula bellotii* with average biomass in 111-188g/m^2 (Sirenko et al., 1995), was found directly beneath the place where Polynya was situated in spring 1996. We used the average between sexes ME, calculated from average BMR, for the purpose of impact estimation. Average ME in bird is 1973 KJ/day. So intake by one bird in 782g of shell food was calculated from ME using metabolic efficiency coefficient and value of caloric density of food. Number of diving acts during daily feeding is 271, therefore 2.9g of shell food must be taken by each dive. Intake of entire Eider population (Table 1) during 31 days springing is 480961kg of shells, or 0.183g per m^2 of bottom fauna, situated beneath the ice-free area of the Polynya. So the impact on benthic fauna is 0.10-0.17% of its biomass.

Discussion

All bird species we found on the Polynya, except the Snow Bunting, are marine birds. Absence of the birds such as Brunnich's Guillemot (*Uria lomvia*) and Black Guillemot (*Cepphus grylle*) was in a good accord with early spring migration of these species to breeding colonies situated westward of our study area (Uspensky, 1969). Kittiwakes, breeding in the same colonies, usually migrate later than auks. The Kittiwakes observed migrating westward in this study, might breed along the Eastern Taimyr coast. Ducks and Pomarine Skua widespread breeding in the maritime tundra, migrate to the breeding grounds close to Polynya.

Our study of King Eider bioenergetics shows that the value of metabolised energy in 3.17 BMR (males) and 3.44 BMR (females) doesn't enlarge maximal permissible level for this value in 4 BMR. Living on the Polynya Eiders take 2.8g of shell food by one dive, but the average body mass in *Leonucula bellotii* here is 0.12g (Petriashov, pers. com.). This allows us to conclude, that Eiders feed by selecting the large prey, which doesn't dominate in biocoenoses. The selective feeding may be the main reason for King Eiders to leave relatively warm waters of the Bering Sea, where many species of sea-ducks winter, reducing the source of large prey, occur on the Great Siberian Polynya for overspringing.

Acknowledgements

I am cordially thankful to Dr. H. Kassens (GEOMAR, Kiel) for the possibility of field work this study is based on. The stomach contents were identified by Prof. Ya. I. Starobogatov and Dr. B. I. Sirenko, Dr. V. V. Petriashov consulted on bottom fauna. Dr. I. A. Dmitrenko and V. Yu. Karpiy provided technical support. Special thanks are due to Prof. V. R. Dolnik for his invaluable expertise on bioenergetics approaches to birds ecology. He also reviewed the preliminary draft.

References

Alimov, A.F. (1981) Functional ecology of freshwater Bivalves (in Russian). Nauka, Leningrad, 248 pp.

Birulya, A.A. (1907) Notes on the life of birds along the polar coast of Siberia (in Russian). Zapiski Imperatorskoy Academii Nauk po fiz.-mat. sekcii 18(2), 1-157.

Dolnik, V.R. (1995) Energy and time resources in free-living birds (in Russian). Nauka, St.Petersburg, 360pp.

Jenssen, B.M., M. Ekker and C. Bech (1989) Thermoregulation in winter-acclimatized common eiders (*Somateria mollissima*) in air and water. Can. J. Zool., 67, 669-673.

Kendeigh, S.G., V.R. Dolnik and V.M. Gavrilov (1977) Avian energetics. Granivorous birds in ecosystems, 12, 127-204.

Sirenko, B.I., V.V. Petriashov, E. Rachor and K. Hinz (1995) Bottom biocoenoses of the Laptev Sea and adjacent areas. In: Berichte zur Polarforshchung, 176, Kassens, H., D. Piepenburg, J. Thiede, L. Timokhov, H.W. Hubberten and S. Priamikov (eds.), Alfred Wegener Institute for Polar and Marine Research, Bremenhaven, 211-221.

Systad, G.H. (1996) Effects of reduced day length on the activity pattern of wintering sea ducks. Cand. Scient. thesis, University of Tromso, Tromso, Norway.

Uspenskiy, S.M. (1969) Life in High Latitudes. Demonstrated mainly on birds (in Russian). Myisl', Moskva, 463pp.

Woakes, A.J. and P.J. Butler (1983) Swimming and diving in Tufted Ducks, *Aythya fuligula*, with particular reference to heat rate and gas exchange. J. exp. Biol., 107, 311-329.

Section C

Land-Ocean Interactions and Pathways

Major, Trace and Rare Earth Element Geochemistry of Suspended Particulate Material of East Siberian Rivers Draining to the Arctic Ocean

V. Rachold

Alfred-Wegener-Institut für Polar- und Meeresforschung, Forschungsstelle Potsdam, Telegrafenberg A43, D 14473 Potsdam, Germany

Received 23 May 1997 and accepted in revised form 5 March 1998

Abstract - The study presents data on the major, trace, and rare earth element (REE) composition of suspended particulate material (SPM) of the majors rivers draining to the Laptev Sea (the Lena, Yana, and Khatanga), as well as their major tributaries. The chemical data can be used to distinguish between material supplied by individual rivers and to fingerprint the sediment transport to the Laptev Sea and further to the Arctic Ocean.
Due to the size of the drainage basin, the average chemical composition of Lena SPM is very similar to average shale, representing the weathering product of the upper continental crust. Small variations in the chemical composition along the course of the river can be related to material input through major tributaries. The SPM of these tributaries draining more restricted basins of specific lithologies exhibits significant differences from the Lena which is seen in the trace element composition and the REE pattern. The Yana SPM is characterized by the strongest variations in chemical composition, which are related to the heterogeneity of the Yana basin. In particular, the Yana SPM shows higher Zr concentrations and an enrichment in As and Cs. The latter is attributed to granitic intrusions forming the source of Au and Sn ore deposits in the Yana catchment. As a result of the dominance of basaltic rocks of the Siberian Trap in the Khatanga basin, the chemical composition of Khatanga SPM differs significantly from both the Lena and the Yana. The Khatanga SPM is strongly enriched with elements that are associated with basalts, i.e. Ca, Co, Cu, Fe, Mg, Ni, Ti, and V. The shale-normalized REE pattern shows a depletion in light REE and a positive Eu anomaly, which corresponds with the Siberian Trap. However, due to weathering, a heavy REE depletion relative to light REE is superimposed on this pattern. Factor analysis of the chemical data set identifies the major processes controlling the SPM composition and clearly differentiates the SPM of the three studied rivers.

Introduction

The detrital component of Arctic marine sediments is mainly derived from material supplied by the major Siberian rivers Lena, Yenisey, and Ob. Today, the Lena is thought to be of importance because large amounts of sediment loaded sea ice feeding the Transpolar Drift are formed in the Laptev Sea shelf area, into which the Lena drains (Nürnberg et al., 1994; Reimnitz et al., 1994; Reimnitz et al., 1995). Recent studies demonstrated that sea ice is an important agent for sediment transport in the Arctic Ocean and Northern Seas (Bischof et al., 1990; Dethleff et al., 1993; Wollenburg, 1993; Eisenhauer et al., 1994; Eicken et al., 1996). For this reason, the interdisciplinary Russian-German project, "The Laptev Sea System", concentrates on geological, biological, chemical, and oceanographic processes in the Laptev Sea shelf area (Kassens and Karpiy, 1994; Kassens et al., 1994; Kassens et al., 1995). Within this project, the present study focuses on fluvial sediment transport to the Laptev Sea. The objectives are to characterize the material supplied by each river and to find individual chemical signals allowing the identification of this material in Arctic marine sediments.

Numerous recent studies focused on different chemical aspects of the Lena, such as the dissolved major ion geochemistry (Gordeev and Sidorov, 1993), the dissolved and particulate

---

In: Kassens, H., H.A. Bauch, I. Dmitrenko, H. Eicken, H.-W. Hubberten, M. Melles, J. Thiede and L. Timokhov (eds.) Land-Ocean Systems in the Siberian Arctic: Dynamics and History. Springer-Verlag, Berlin, 1999, 199-222.

trace metal concentrations (Martin et al., 1993; Guieu et al., 1996), the partitioning of trace metals between dissolved and particulate phases (Garnier et al., 1996), the biogeochemistry (Cauwet and Sidorov, 1996; Lara et al., in press), the Sr isotope geochemistry (Rachold et al., in press [a]), dissolved Si concentrations and stable oxygen isotopes (Letólle et al., 1993), inorganic geochemistry of the suspended particulate material (Rachold et al., 1996), and the origin of particulate organic material (Rachold and Hubberten, this volume). However, the other rivers draining to the Laptev Sea are scarcely studied. This article presents a complete data set of the inorganic geochemistry of suspended particulate material (SPM) of all major rivers draining to the Laptev Sea, i. e. the Lena, Yana, and Khatanga, as well as their major tributaries. These data can be used to differentiate between material supplied by individual rivers and to fingerprint the sediment transport to the Laptev Sea and further to the Arctic Ocean.

Geology and hydrology of the catchments

This chapter presents a short overview of the hydrological characteristics and the most important lithologies of the Lena, Yana, and Khatanga basins (Figure 1). Detailed information is given by Alabyan et al. (1995), Gordeev et al. (1996), and Rachold et al. (1996).

In terms of water discharge, the Lena belongs to the eight largest rivers in the world and is the second largest among the arctic rivers after the Yenisey (Milliman and Meade, 1983). Average annual sediment and water discharge data are presented in Figure 1 (Rachold et al., 1996). Due to extreme seasonal differences in temperature and precipitation (up to 100° C temperature difference between summer and winter), monthly water discharge and sediment transport exhibit strong variations. The central part of the Yana basin represents the coldest spot in the northern hemisphere with winter temperatures as low as -70 ° C in Verkhoyansk. On average, the Yana freezes down to its riverbed once every four years. For this reason, 99% of the clastic material is transported during summer (June to September). The water discharge of the Yana is approximately 20 times smaller than the Lena. Its sediment supply to the Laptev Sea, however, can be important (Alabyan et al., 1995) since the concentration of total SPM may be 25 times higher than that of the Lena (up to 1000 mg/l) during high water stage. Although the Khatanga is the second largest river draining to the Laptev Sea in terms of water discharge, its sediment transport does not exceed 50 % of the Yana's (Figure 1).

The drainage areas of the Lena, Yana, and Khatanga basins consist of three major tectonic units: the Siberian platform, the Baikal folded region, and the Verkhoyano-Kolymean folded region (Figure 1). The Siberian Platform is composed mainly of Cambrian and Precambrian limestones, terrigenous sediments of Jurassic to Cretaceous age, and Quaternary alluvial deposits. The north-western part is formed from Permian/Triassic volcanic rocks (Siberian Trap) while the south-eastern region (Aldan highland) is composed of Archean and Proterozoic crystalline and metamorphic rocks. The Baikal folded region consists of gneiss, shists, quartzites, and marbleized limestones of Proterozoic age. In the Verkhoyano-Kolymean folded region, Permian and Carboniferous terrigenous sediments as well as volcanic rocks and granitoids of Triassic and Jurassic age are more common (Gordeev and Sidorov, 1993; Alabyan et al., 1995; Rachold et al., 1996).

Samples and methods

Sampling was performed during the expeditions "Lena 94" (Rachold et al., 1995) "Lena/Yana 95" (Rachold et al., 1997), and "Khatanga 96" (Rachold et al., in press [b]) from July 9 to July 28, 1994, June 26 to September 7, 1995 and July 14 to August 20, 1996, respectively. Water, sediment, and SPM samples were collected at 80 stations. The Lena samples were collected

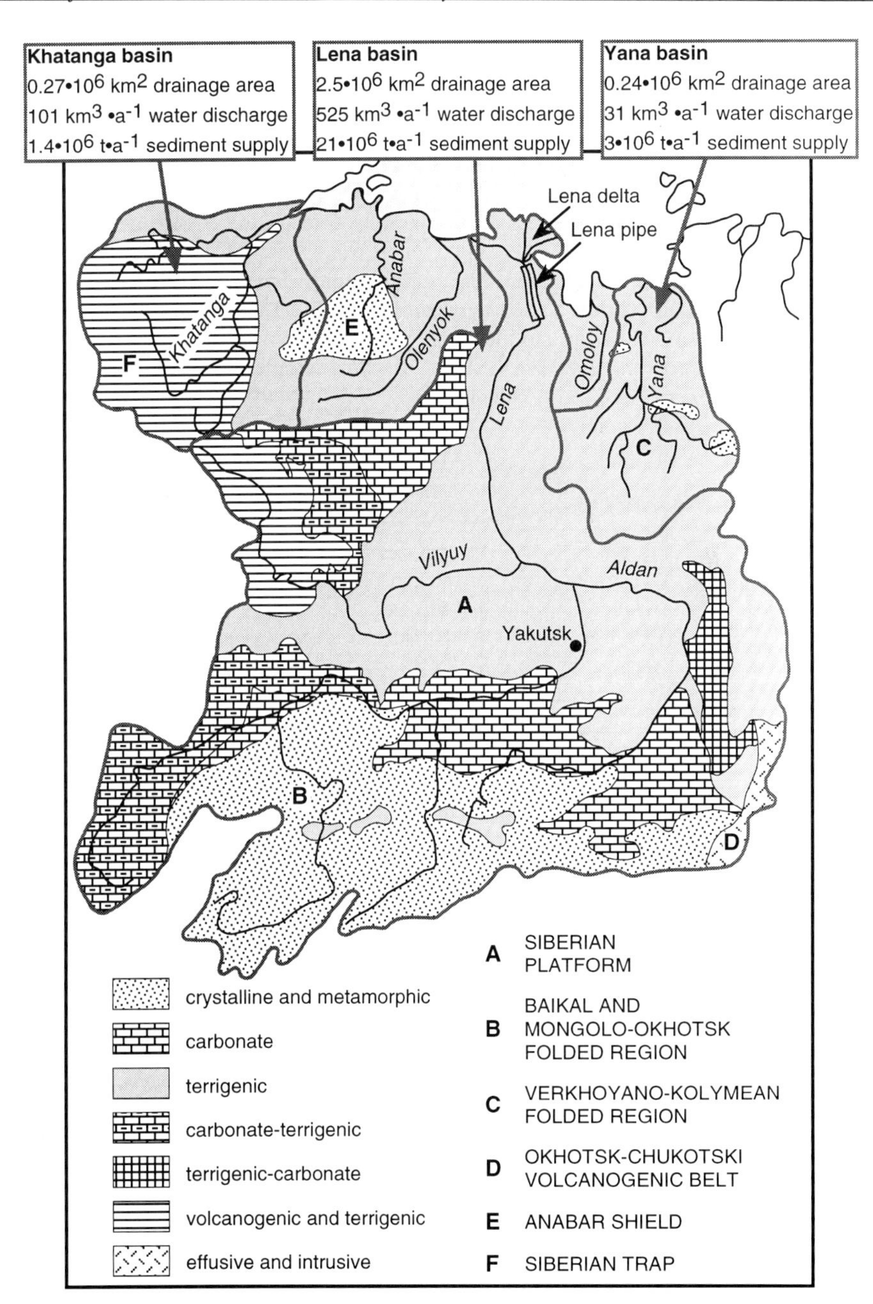

Figure 1: Simplified geological map of the Lena, Yana, and Khatanga catchments (modified from Gordeev and Sidorov, 1993). Data on water and sediment discharge are taken from Rachold et al. (1996).

during July when the water level and sediment load of the Lena were still high (maximum values were reached immediately after ice breakup in the beginning of June). In the Yana, water level and sediment discharge in 1995 reached maximum values in August just at the time of sampling. Samples of the Khatanga were collected at the end of July, three weeks after ice breakup.

Figure 2 shows the locations of stations where SPM samples analyzed within the present study were collected. The Lena, Yana, and Khatanga tributaries are marked by numbers L1 to L5, Y1 to Y9, and K1 to K3 throughout all figures and tables. Since the catchment of the Omoloy (Y9) and its tributary the Kyugyulyur (Y8) are very similar to the Yana's, they are treated as Yana tributaries although the Omoloy drains directly to the Laptev Sea (Figure 2). Samples of river water were retrieved using a PTFE water sampler. The SPM was separated immediately after sampling by vacuum filtration through Nucleopore filters of 0.4 μm pore size. The filtered water volume varied between 0.25 and 2 l depending on sediment load. Except for the smaller tributaries and the Yana, three samples were collected at each station, equally distributed across the river. On the Lena, the three samples were combined to one SPM sample that was later analyzed for its chemical composition. At the Khatanga stations, the three samples were analyzed separately.

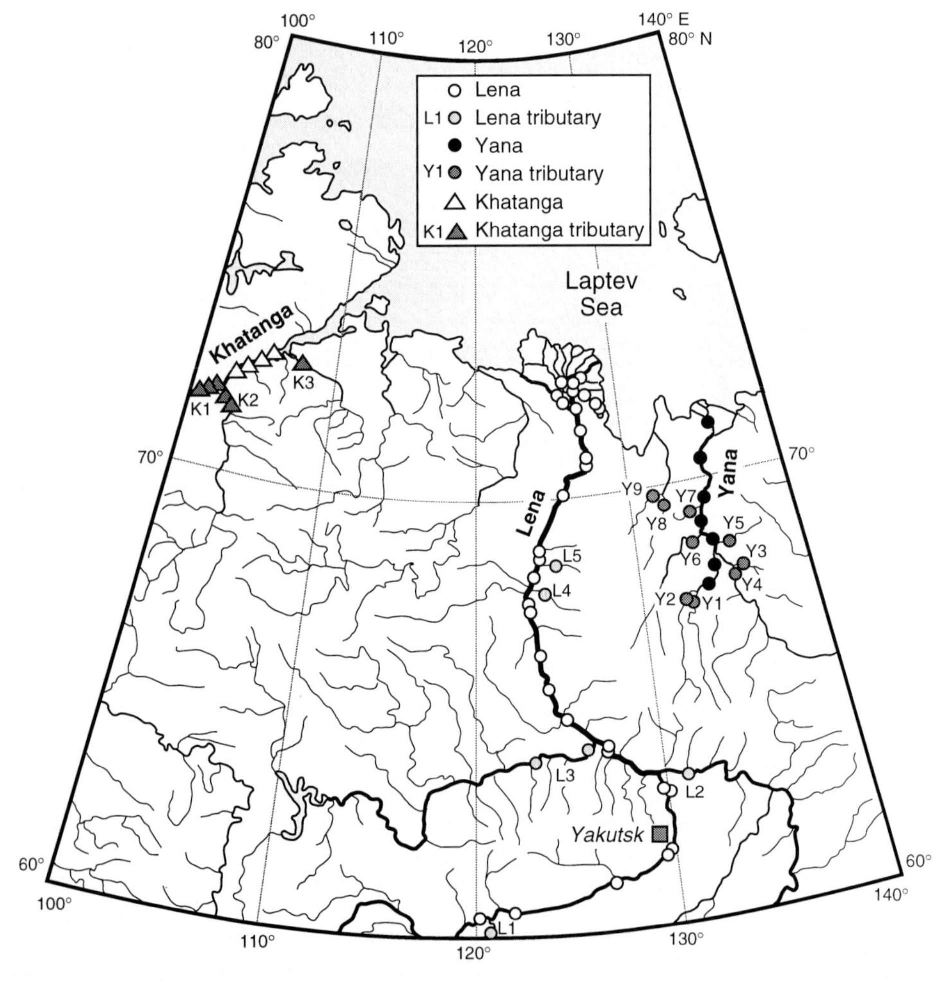

Figure 2: Sampling locations during the expeditions "Lena 94" (Rachold et al., 1995), "Lena/Yana 95" (Rachold et al., 1997), and "Khatanga 96" (Rachold et al., in press [b]).

Total SPM concentrations were estimated from the weight difference between the original and the sediment-loaded filters after freeze-drying. Acid digestions of the sediment loaded filters were performed in PTFE vessels using ultrapure HNO_3, $HClO_4$, and HF (Heinrichs and Herrmann, 1990). The residues were redissolved in HNO_3 and diluted with H_2O to a final volume of 20 to 50 ml depending on the sediment load. Al, Ti, Fe, Mn, Mg, Ca, Na, K, Ba, and Sr were analyzed by ICP-OES (Perkin Elmer Optima 3000 XL), Bi, Cd, Co, Cs, Cu, Ga, Hf, Li, Mo, Nb, Ni, Pb, Sb, Sn, Ta, Th, Tl, U, V, W, Y, Zn, Zr, and rare earth elements (REE) by ICP-MS (Fisons Plasma Quad), and As by GF-AAS (Perkin Elmer SIMAA 6000). Accuracy of the analytical methods was checked by parallel analysis of an international standard reference material (MESS-1, BCSS-1, PACS-1, GSD-7, GSD-9, GSD-10) for every method. The analytical precision was better than 5 % for major elements and better than 10 % for trace and rare earth elements.

Earlier studies demonstrated that heavy rare earth elements (HREE) that are commonly carried by zircon and other heavy minerals can remain in resistant phases during HF-$HClO_4$ dissolution (c.f. Condie, 1991). Within the present study, international reference material was analyzed for Zr concentrations in order to quantify the amount of resistant phase present. The observed Zr concentrations amounted to 70 % of the certified values at minimum, indicating that up to 30 % of the zircon was not dissolved during HF-$HClO_4$-digestion. However, further investigations indicated that the amount of resistant phase strongly depended on both digestion time and grain-size of the sample. Zr concentrations in standard reference material which was pulverized to < 35 μm or treated with HF-$HClO_4$ for up to 48 hours yielded 80-90 % of the certified values. The SPM analyzed within the present study consists mainly of fine-grained material < 6 μm (see below). Therefore, a significant deficit in Zr and REE concentrations resulting from incomplete dissolution can be ruled out.

Some samples showed extraordinarily high As, Cd, Cu, Sn, Pb, and Zn concentrations which could not be explained by natural processes or anthropogenic pollution and these high values were attributed to contamination from the ship. All data where the concentrations of the aforementioned elements were more than 2-times higher than the normal values were eliminated.

Six selected SPM samples were analyzed for grain-size distribution by laser-granulometry. The mineralogical composition of these samples was semi-quantitatively analyzed from smear-slides of the total SPM by x-ray diffraction (XRD). In addition, the clay mineral distribution was studied in the fraction < 6.3 μm by x-ray diffraction after Atterberg grain-size separation (Gibbs, 1965).

Results and discussion

Table 1 presents the average chemical composition of SPM of the Lena, Yana, and Khatanga and their major tributaries. The labels of the tributaries are indicated in the location map (Figure 2) and in the other figures. The minerals identified in the SPM from XRD are listed in Table 2. In the following section the composition of the SPM from each river will be discussed separately.

Lena SPM

The Lena exhibited total SPM concentrations of 6 to 77 mg/l, with an average of 28 mg/l. In general, Lena SPM was characterized by high amounts of fine-grained material although some variations within the river could be observed. The grain-size distribution of two stations is illustrated in the upper left corner of Figure 3. The major fraction of the SPM consisted of quartz, feldspars (K-feldspars and plagioclase), and muscovite-type illite. Amphibole and illite-

Table 1: Average chemical composition of SPM from Lena, Yana, and Khatanga rivers and their major tributaries. Major elements in weight % and trace elements in ppm. The column n corresponds to the number of samples. The labels of the tributaries are indicated in the location map (Figure 2).

	n	Al_2O_3	TiO_2	Fe_2O_3	MnO	MgO	CaO	Na_2O	K_2O	As	Ba	Bi	Cd	Co	Cs	Cu	Ga	Hf
		%	%	%	%	%	%	%	%	ppm	ppm	ppm	ppm	ppm	ppm	ppm	ppm	ppm
Lena basin																		
Lena river	**31**	**13.43**	**0.618**	**5.44**	**0.209**	**2.30**	**2.06**	**1.81**	**2.56**	**9.1**	**718**	**0.24**	**0.65**	**18**	**4.4**	**35**	**23**	**3.8**
std. dev.		1.53	0.073	0.54	0.057	0.50	0.74	0.30	0.34	1.6	75	0.03	0.22	2	0.9	9	4	0.4
min		9.87	0.400	4.27	0.115	1.36	0.94	1.19	1.82	6.5	567	0.18	0.30	15	3.4	26	17	3.0
max		15.85	0.752	6.72	0.309	3.37	3.70	2.25	3.12	12.8	892	0.33	0.98	24	6.5	57	32	4.5
Chara (L1)	1	10.79	0.471	5.24	0.163	3.25	2.69	1.23	1.96	12.2	658	0.32	0.51	21	3.5	45	21	3.2
Aldan (L2)	1	8.79	0.365	4.44	0.356	1.36	1.31	1.08	1.68	16.6	523	0.27	-	16	4.1	28	22	2.9
Vilyuy (L3)	2	13.45	0.632	5.09	0.202	2.91	3.04	1.18	2.56	7.8	601	0.15	0.41	22	3.2	34	24	3.2
Menkere (L4)	1	14.84	0.687	6.21	0.100	1.99	0.70	1.75	2.82	15.0	712	0.18	-	26	5.2	34	26	5.6
Dzhardzhan (L5)	1	13.76	0.673	5.00	0.084	1.35	0.50	2.00	2.64	12.0	623	0.22	-	13	5.4	28	24	5.6
Yana basin																		
Yana river	**7**	**12.98**	**0.734**	**5.43**	**0.097**	**1.37**	**0.75**	**2.10**	**2.62**	**26.7**	**617**	**0.23**	**0.32**	**17**	**5.7**	**30**	**22**	**5.4**
std. dev.		1.49	0.057	0.57	0.022	0.30	0.28	0.27	0.42	15.9	52	0.06	0.12	3	0.8	3	6	0.6
min		9.74	0.648	4.61	0.068	0.81	0.37	1.83	2.20	13.1	529	0.15	0.14	14	4.7	26	18	4.6
max		14.78	0.848	6.62	0.136	1.75	1.17	2.65	3.38	59.6	698	0.36	0.46	24	7.3	34	34	6.4
Sartang (Y1)	1	11.19	0.717	5.17	0.076	0.97	0.36	2.32	4.95	34.7	623	0.22	0.12	14	6.0	27	23	5.9
Dulgalakh (Y2)	1	12.07	0.682	5.51	0.091	1.06	0.43	2.12	1.87	34.7	561	0.28	0.19	15	6.7	26	23	5.2
Tuostakh (Y3)	1	12.05	0.618	5.29	0.095	2.99	4.24	1.52	2.10	27.2	820	0.37	0.50	21	6.7	37	18	3.9
Adicha (Y4)	1	16.24	0.857	7.42	0.111	1.55	0.61	1.78	2.56	24.0	578	0.30	0.21	19	7.0	59	22	4.7
Olbye (Y5)	1	14.15	0.703	5.46	0.078	1.33	0.67	2.02	2.31	17.8	702	0.22	0.26	15	5.6	32	22	4.6
Bytaktay (Y6)	1	8.68	0.847	7.52	0.121	0.99	0.69	2.26	2.52	62.3	608	0.40	0.30	21	8.2	40	44	6.5
Baky (Y7)	1	12.66	0.541	8.54	0.102	1.31	0.75	0.74	1.88	25.0	550	0.31	0.46	17	7.8	51	27	5.4
Kuygyulyur (Y8)	1	16.38	0.709	7.89	0.066	1.26	0.51	0.68	2.21	32.0	626	0.34	0.28	19	8.3	45	29	4.8
Omoloy (Y9)	1	12.93	0.663	6.70	0.090	1.26	0.57	1.59	2.14	17.1	625	0.21	0.28	16	5.0	34	24	4.6
Khatanga basin																		
Khatanga river	**12**	**13.07**	**0.929**	**8.75**	**0.156**	**4.21**	**4.33**	**1.35**	**1.26**	**9.3**	**334**	**0.13**	**0.22**	**35**	**2.5**	**82**	**16**	**3.2**
std. dev.		0.50	0.051	0.33	0.012	0.31	0.57	0.12	0.07	2.1	22	0.04	0.03	4	0.4	6	1	0.1
min		12.17	0.855	8.15	0.130	3.54	3.29	1.15	1.16	6.4	274	0.09	0.15	31	2.0	73	15	3.0
max		13.78	1.037	9.27	0.175	4.68	5.21	1.59	1.37	13.2	370	0.22	0.29	48	3.2	89	18	3.3
Kheta (K1)	7	13.38	1.013	8.87	0.145	4.44	5.31	1.56	1.16	7.8	325	0.08	0.20	33	1.8	80	15	3.1
Kotuy (K2)	4	11.97	0.852	9.18	0.137	5.27	4.16	1.28	1.22	11.0	367	0.13	0.27	39	2.5	103	17	3.1
Popigay (K3)	3	11.21	0.711	6.40	0.137	2.77	2.94	1.06	1.60	9.1	427	0.19	0.37	22	3.2	50	15	3.5

Table 1 (continued):

	n	Li	Mo	Nb	Ni	Pb	Rb	Sb	Sn	Sr	Ta	Th	Tl	U	V	W	Y	Zn	Zr
		ppm	ppm	ppm	ppm	ppm	ppm	ppm	ppm	ppm	ppm	ppm	ppm	ppm	ppm	ppm	ppm	ppm	ppm
Lena basin																			
Lena river	**31**	**43**	**1.23**	**13**	**53**	**24**	**103**	**0.57**	**1.9**	**194**	**0.80**	**12.2**	**0.52**	**3.2**	**97**	**1.41**	**24**	**141**	**132**
std. dev.		5	0.36	1	13	3	15	0.23	0.2	36	0.14	1.1	0.08	0.5	12	0.17	2	19	17
min		32	0.84	10	35	20	82	0.31	1.7	120	0.54	9.6	0.40	2.3	78	1.10	20	120	96
max		55	1.98	15	88	33	134	1.33	2.3	274	1.05	15.0	0.68	4.8	129	1.71	29	195	163
Chara (L1)	1	65	1.80	10	65	27	90	0.45	2.1	162	0.49	14.0	0.52	3.5	89	-	23	-	103
Aldan (L2)	1	38	1.20	9	37	30	73	0.71	-	122	0.51	9.3	0.43	2.1	68	1.20	22	131	101
Vilyuy (L3)	2	46	0.87	12	51	17	81	0.40	1.7	181	0.56	8.3	0.42	2.2	105	1.10	21	136	118
Menkere (L4)	1	54	1.02	16	44	30	112	-	2.2	144	0.52	12.4	0.62	2.8	116	1.48	28	-	180
Dzhardzhan (L5)	1	46	1.08	16	38	22	110	0.62	1.9	134	0.98	11.2	0.54	2.8	98	1.54	22	130	186
Yana basin																			
Yana river	**7**	**47**	**0.96**	**14**	**39**	**23**	**82**	**2.10**	**1.6**	**138**	**0.80**	**8.5**	**0.50**	**2.9**	**110**	**1.88**	**23**	**130**	**187**
std. dev.		6	0.12	2	4	3	5	0.58	0.7	16	0.14	0.9	0.07	0.3	11	0.53	2	14	34
min		37	0.80	13	33	19	77	1.05	0.3	117	0.59	7.0	0.43	2.4	92	1.20	19	117	158
max		53	1.16	20	45	28	89	2.80	2.3	171	1.04	9.9	0.63	3.3	129	2.70	25	156	241
Sartang (Y1)	1	39	1.00	16	37	20	115	1.09	-	127	0.71	8.3	0.52	2.7	110	1.45	19	109	220
Dulgalakh (Y2)	1	41	0.97	15	35	26	74	1.07	0.2	125	0.74	9.2	0.55	3.0	109	1.46	22	113	193
Tuostakh (Y3)	1	49	1.40	12	50	20	81	2.90	2.9	141	0.74	12.0	0.51	3.1	107	1.70	26	-	130
Adicha (Y4)	1	52	0.97	14	56	23	87	2.30	2.4	147	0.86	9.4	0.49	2.8	155	1.60	24	163	158
Olbye (Y5)	1	45	1.00	14	40	18	87	1.40	2.4	158	0.84	10.0	0.45	3.1	113	1.70	26	132	162
Bytaktay (Y6)	1	49	1.34	17	54	32	87	1.90	1.7	139	0.98	8.8	0.70	3.2	146	1.90	25	164	235
Baky (Y7)	1	42	1.40	11	58	25	108	1.20	2.4	108	0.75	10.0	0.50	3.0	134	1.70	31	187	207
Kuygyulyur (Y8)	1	64	1.50	14	59	30	113	0.87	2.7	96	0.86	10.0	0.56	3.7	154	1.70	23	200	189
Omoloy (Y9)	1	51	1.00	13	44	21	91	0.77	2.3	118	0.79	8.5	0.46	2.6	109	1.40	24	198	179
Khatanga basin																			
Khatanga river	**12**	**27**	**0.88**	**10**	**84**	**12**	**55**	**0.43**	**1.5**	**192**	**0.49**	**5.3**	**0.26**	**1.5**	**349**	**0.86**	**22**	**104**	**123**
std. dev.		6	0.15	1	9	3	8	0.18	0.2	9	0.18	0.6	0.03	0.1	17	0.16	1	8	4
min		20	0.63	8	75	8	41	0.31	1.2	178	0.23	4.3	0.22	1.3	327	0.63	20	91	115
max		36	1.16	11	112	19	68	1.00	1.8	208	0.89	6.6	0.31	1.6	376	1.09	24	121	128
Kheta (K1)	7	17	0.66	8	79	8	41	0.32	1.1	201	0.42	4.2	0.21	1.3	377	0.60	20	92	115
Kotuy (K2)	4	38	1.06	9	101	12	60	0.37	1.7	178	0.57	4.6	0.25	1.4	324	0.78	23	112	122
Popigay (K3)	3	33	1.39	11	54	29	75	0.48	1.8	155	0.65	7.8	0.35	1.8	233	1.13	21	132	139

Table 1 (continued):

	n	La	Ce	Pr	Nd	Sm	Eu	Gd	Tb	Dy	Ho	Er	Tm	Yb	Lu
		ppm	ppm	ppm	ppm	ppm	ppm	ppm	ppm	ppm	ppm	ppm	ppm	ppm	ppm
Lena basin															
Lena river	**31**	**44**	**95**	**10.2**	**37**	**6.6**	**1.6**	**6.4**	**0.80**	**4.6**	**0.85**	**2.6**	**0.34**	**2.4**	**0.36**
std. dev.		5	10	1.1	4	0.6	0.1	0.5	0.08	0.4	0.08	0.2	0.04	0.3	0.04
min		31	68	7.1	27	4.5	1.4	5.2	0.61	3.4	0.65	2.0	0.24	2.0	0.23
max		54	124	12.8	45	7.7	1.9	7.7	0.97	5.2	1.00	3.0	0.42	2.9	0.44
Chara (L1)	1	51	110	12.0	45	8.0	1.8	7.0	0.79	4.6	0.81	2.1	0.33	2.1	0.34
Aldan (L2)	1	33	74	8.1	29	5.6	1.5	5.5	0.68	3.9	0.72	2.3	0.32	2.0	0.35
Vilyuy (L3)	2	30	65	6.9	26	4.8	1.5	5.0	0.61	3.6	0.70	2.2	0.30	2.0	0.33
Menkere (L4)	1	38	80	9.0	34	6.2	1.9	6.6	0.82	5.0	0.92	2.8	0.40	2.8	0.42
Dzhardzhan (L5)	1	32	70	7.6	28	5.4	1.6	5.8	0.66	4.0	0.74	2.4	0.36	2.4	0.38
Yana basin															
Yana river	**7**	**31**	**68**	**7.8**	**29**	**5.8**	**2.1**	**5.9**	**0.80**	**4.4**	**0.87**	**2.5**	**0.36**	**2.4**	**0.42**
std. dev.		2	3	0.4	2	0.4	0.8	0.3	0.06	0.3	0.06	0.2	0.02	0.2	0.02
min		27	63	6.7	25	4.8	1.2	5.3	0.71	3.7	0.74	2.2	0.32	2.2	0.39
max		34	73	8.1	31	6.1	3.5	6.2	0.91	4.7	0.91	2.8	0.39	2.8	0.45
Sartang (Y1)	1	28	60	6.9	26	5.0	1.2	5.6	0.77	3.7	0.78	2.2	0.33	2.3	0.38
Dulgalakh (Y2)	1	32	71	7.9	29	5.8	1.5	6.9	0.89	4.4	0.92	2.6	0.37	2.6	0.42
Tuostakh (Y3)	1	31	67	8.0	31	6.5	1.7	6.4	0.87	4.9	0.91	2.8	0.38	2.6	0.46
Adicha (Y4)	1	33	74	8.2	31	6.0	1.7	5.8	0.82	4.8	0.91	2.7	0.38	2.5	0.42
Olbye (Y5)	1	33	72	8.0	31	5.9	1.7	6.0	0.82	4.7	0.92	2.7	0.41	2.5	0.42
Bytaktay (Y6)	1	32	71	7.9	28	5.7	1.5	7.3	0.96	4.7	0.96	2.8	0.43	3.0	0.47
Baky (Y7)	1	29	58	7.2	28	5.7	1.6	6.5	0.86	5.3	1.00	3.1	0.46	3.1	0.54
Kuygyulyur (Y8)	1	26	61	6.8	26	5.1	1.6	5.4	0.71	4.1	0.78	2.4	0.36	2.6	0.45
Omoloy (Y9)	1	27	61	6.8	26	5.3	1.6	5.5	0.72	4.2	0.81	2.4	0.33	2.3	0.39
Khatanga basin															
Khatanga river	**12**	**22**	**48**	**5.6**	**22**	**4.5**	**1.3**	**4.7**	**0.74**	**4.3**	**0.86**	**2.4**	**0.35**	**2.2**	**0.37**
std. dev.		2	5	0.5	2	0.3	0.1	0.4	0.04	0.2	0.04	0.1	0.01	0.1	0.02
min		18	40	4.8	18	3.9	1.2	4.0	0.67	4.0	0.77	2.2	0.32	2.1	0.34
max		26	58	6.5	25	4.9	1.4	5.5	0.82	4.6	0.90	2.6	0.37	2.4	0.42
Kheta (K1)	7	18	39	4.6	18	3.9	1.2	4.3	0.70	4.1	0.82	2.4	0.34	2.2	0.37
Kotuy (K2)	4	19	41	5.0	19	4.2	1.3	4.8	0.75	4.4	0.89	2.5	0.37	2.3	0.38
Popigay (K3)	3	30	66	7.5	27	5.2	1.3	5.4	0.75	4.0	0.79	2.2	0.32	2.0	0.35

smectite mixed-layer minerals could be identified and were exclusively observed in the Lena (Table 2).

Figure 3 presents the average Al-normalized major and trace element concentrations of Lena SPM in comparison to average shale (Wedepohl, 1971 and 1991). Al normalization was applied to compensate for dilution by organic material and quartz. The average chemical composition of Lena SPM compared well with average shale, except for Ba, Bi, Cd, Mn, Sb, and Zn which will be discussed below. In general, average shale can be regarded as the weathering product of the upper continental crust after subtracting the "soluble" elements. The agreement of Lena SPM and average shale therefore suggests that the average composition of all lithologies in the Lena basin is comparable to the upper continental crust. Since the Lena drains a large basin, the agreement of the average composition of the rocks in this area and the upper crust is not surprising. The same holds true for other major rivers (Gaillardet et al.,

1995). The average trace metal concentrations presented here are in excellent agreement with data from the Lena delta reported by Martin et al. (1993), except for Ni (Table 3).

Table 2: Mineral phases observed in the SPM from Lena, Yana, and Khatanga rivers.

	Lena SPM	Yana SPM	Khatanga SPM
major component	quartz K-feldspar plagioclase illite	quartz K-feldspar plagioclase	quartz
minor component	illite-smectite mixed-layer amphibole	chlorite illite	plagioclase pyroxene quartz illite

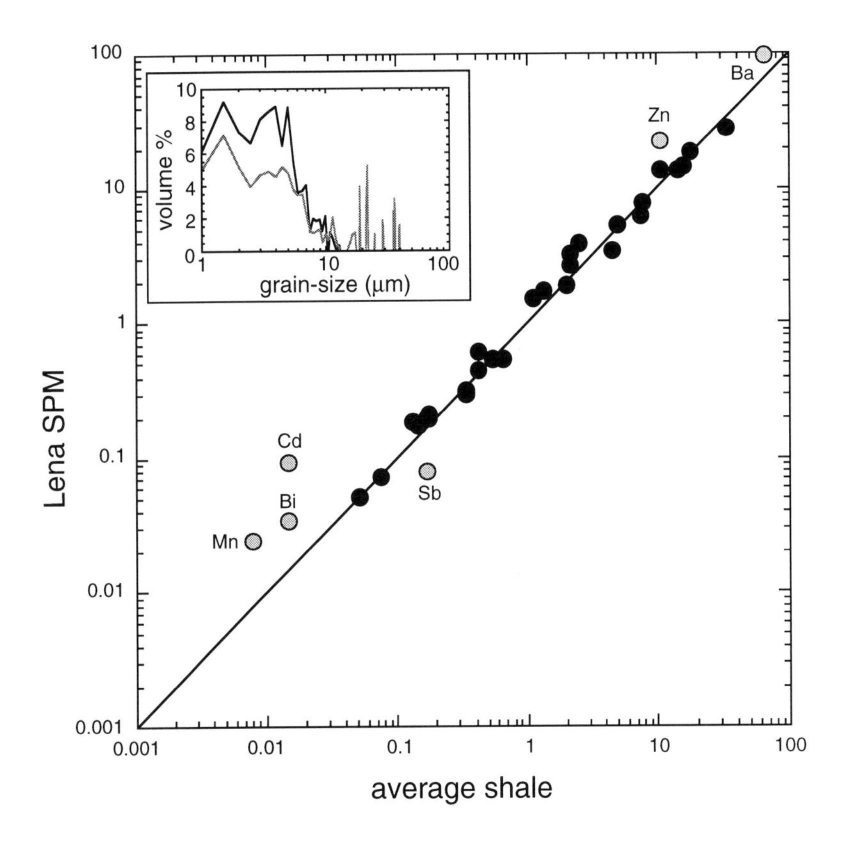

Figure 3: Average chemical composition of Lena SPM. The x-axis corresponds to the composition of average shale (Wedepohl, 1971 and 1991). The diagonal represents the straight line of identical values. Al normalized concentrations based on weight ratios (major elements) and weight ratio • 10^4 (trace elements) are displayed. The grain-size distribution (volume %) is shown in the upper left corner.

Table 3: Average trace metal concentrations of Lena SPM.

	Cu ppm	Ni ppm	Pb ppm	Zn ppm
this study	35	53	24	141
Martin et al., 1993	24-33	20-35	22-26	106-181

Generally, polar regions are characterized by reduced denudation rates (Ugolini, 1986). In the Lena basin denudation rates are additionally reduced as a result of low precipitation levels (Gordeev and Sidorov, 1993). However, recent Sr isotope studies of the dissolved load of Siberian rivers did not corroborate the reduced denudation rate (Huh and Edmond, 1997). The authors argue that frost action in periglacial environments prevents the accumulation of a weathered mantle and thus exposure is maintained. This appears to counteract the climatic effects of low temperatures and precipitation (Huh and Edmond, 1997). Although the Lena SPM data of the present study do not clearly reflect the reduced denudation rate, significant amounts of feldspar in the SPM - which are also seen in high Ba/Al ratios (Figure 3) - indicate reduced chemical weathering. The purpose of this study, however, is not to estimate denudation rates but to fingerprint sediment transport. To study weathering processes in detail, data on the dissolved load are essential.

In the Lena SPM, the Al-normalized Bi, Cd, Mn, and Zn concentrations were higher average shale, whereas the Al-normalized Sb concentration was slightly lower than (Figure 3). Low Sb/Al and high Bi/Al ratios cannot be explained but seem to be related to the geology of the Lena catchment. Slight enrichments in Cd and Zn relative to average shale were observed in all three rivers. Previous investigations described the Lena as a pristine aquatic environment (Martin et al., 1993). Furthermore, the central part of Yakutia is far away from any industrial areas which might emit atmospheric pollutants. For this reason, the enrichments in Cd and Zn cannot be explained by anthropogenic pollution. The enrichment in these elements may be related to the adsorption onto Mn oxihydroxides which occur as coatings around silicate grains or as discrete grains. Mn oxihydroxides have extremely high adsorption affinities for trace metals (Morel and Hering, 1993). Furthermore, recent studies showed that Cd may be enriched due to adsorption onto organic material (Elbaz-Poulichet et al., 1996). This also holds true for Zn in the Lena, although for both elements the relationship was not very pronounced (Garnier et al., 1996). Organic carbon concentrations of Lena river SPM were in the range of 3 to 11 % (Rachold and Hubberten, this volume). However, only a weak correlation between Cd and Zn and both Mn and organic carbon was observed in the present data set. The enrichment in Cd and Zn was not unambiguously related to one of these phases but it is assumed that organic particulates played a major role.

Although the chemical composition of Lena SPM was very homogeneous throughout the river, a trend along the course of the river was identified. In order to document the processes controlling this evolution, the development of U/Al and Th/Al ratios along the Lena was studied in detail (Figure 4a and 4b). Both elements are very sensitive to geological provenance (Condie et al., 1995). While U is highly mobile during weathering and characterized by high dissolved concentrations in river water (Palmer and Edmond, 1993; Pande et al., 1994), Th is one of the least mobile elements during weathering (Langmuir and Hermann, 1980).

Due to the rather strong flow of the Lena and the resultant short residence time for the SPM, alteration processes within the river seem to be of minor importance. Based on an average current of 4 km/h, the 2500 km transport along the Lena requires a time period of 3 to 4 weeks. This is extremely short compared to the dissolution period of silicates of the dominant grain-

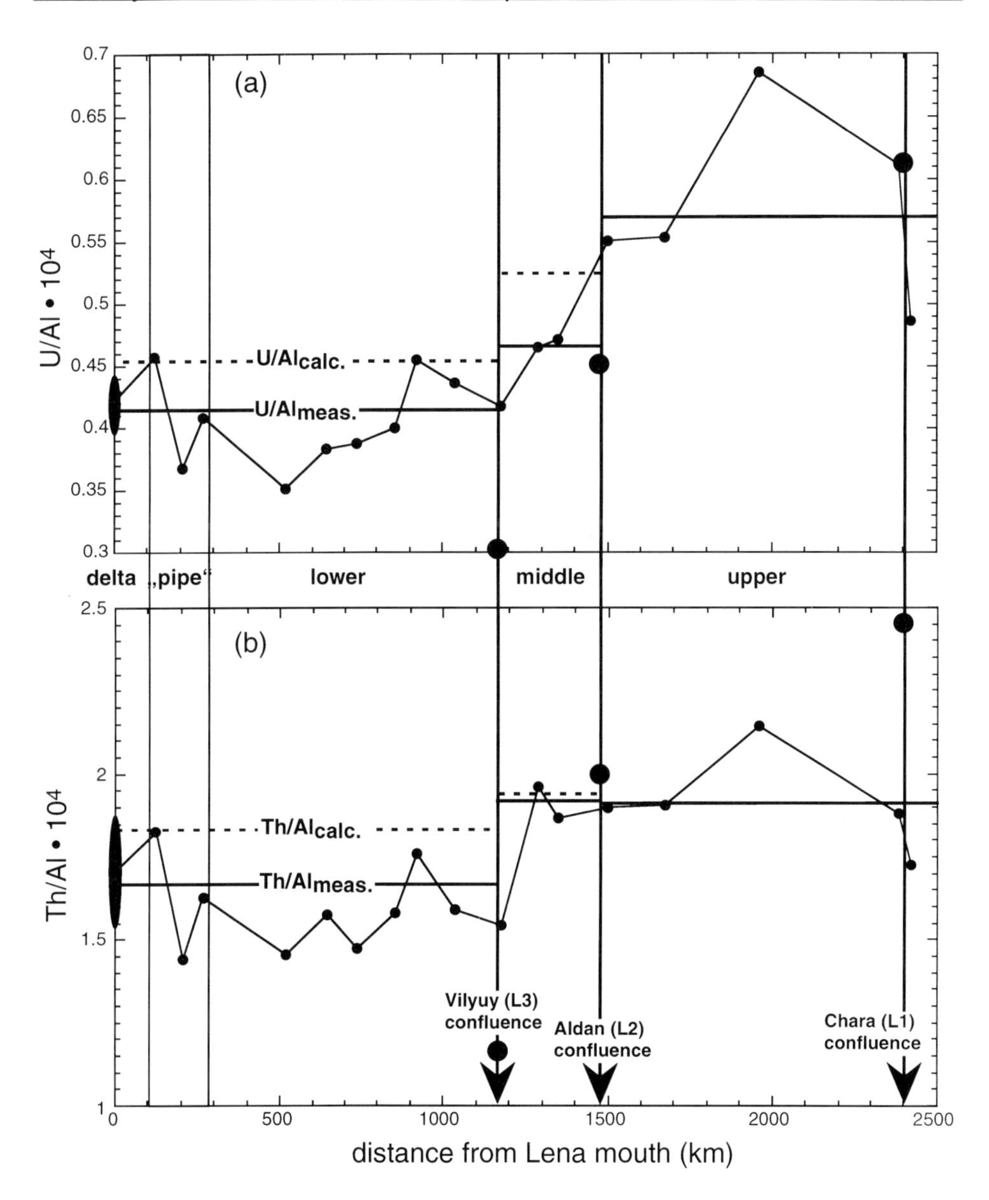

Figure 4: Evolution of Al normalized U (a) and Th (b) concentrations along the course of the Lena. Element ratios are based on weight ratios • 10^4. The scale on the x-axis corresponds to the distance of the station from the Lena mouth. The values of the major tributaries Chara (L1), Aldan (L2), and Vilyuy (L3) are indicated by larger circles, according to the position of their confluences (black arrows). The Lena has been subdivided into the upper Lena (south of the Aldan confluence), the middle Lena (between Aldan and Vilyuy confluence) and the lower Lena (north of the Vilyuy confluence). The bold, black lines represent the average U/Al ratios (a) and Th/Al ratios (b) analyzed in these sections. The dashed lines correspond to the U/Al ratios (a) and Th/Al ratios (b) estimated from mixing calculations (see text).

size, which ranges from 100 years to 3 million years (Lasaga, 1984 and Lasaga et al., 1994). For this reason, the chemical composition of SPM, produced by weathering and directly related to both the geology and the weathering regime of the catchment, is assumed to remain constant during the short period of riverine transport. Furthermore, previous studies clearly documented that the Lena is not affected by pollution (Martin et al., 1993; Guieu et al., 1996).

Additional variations in the chemical composition of the SPM could be related to changes in transport energy. An increase in flow velocity may result in the resuspension of coarse-grained bottom sediments whereas a reduction in transport energy may correspond to the deposition of fine-grained SPM. Two prominent changes in transport energy occur in the Lena. North of the settlement Kyusyur, the Lena cuts through the Verkhoyansk range and narrows from a width of approximately 20 km to less than 2 km in the so-called Lena pipe. The narrow channel of the Lena pipe opens directly into the wide delta plain. The position of the Lena pipe and the Lena delta are indicated in Figure 1 and Figure 4. Although the transitions from the broad Lena to the Lena pipe and from the Lena pipe to the Lena delta represent drastic changes in transport energy, the chemical composition of the SPM was almost unaffected (Figure 4). Compared to variations in U/Al and Th/Al ratios in the upper and middle Lena, the slight increase in the Lena delta appears negligible.

The evolution observed in the course of the Lena is therefore largely attributed to the input of SPM by larger tributaries, i.e. the Chara (L1), the Aldan (L2), and the Vilyuy (L3). The values of the tributaries L1-L3 are indicated in Figure 4 according to the position of their confluences. Based on the U and Th data, the Lena can be subdivided into three sections: the upper Lena (south of the Aldan confluence), the middle Lena (between Aldan and Vilyuy confluence), and the lower Lena (north of the Vilyuy). The upper Lena and the Chara were characterized by the highest U/Al and Th/Al ratios, under the influence of the Baikal and Mongolo-Okhotsk folded region (Figure 1). North of the confluence of the Aldan, which drains Archean and Proterozoic crystalline and metamorphic rocks of the Aldan shield, a decrease in U/Al ratios was seen, while Th/Al ratios remained constant. North of the confluence of the Vilyuy, which drains the sedimentary basin of the Siberian platform and the Siberian Trap basalts, a pronounced decrease in both U/Al and Th/Al ratios was observed. Unfortunately, accurate data on the SPM discharge of the Aldan and the Vilyuy are not available. It is estimated that the SPM discharge of the Aldan accounts for about 60 % and the Vilyuy for about 10 % of the Lena's discharge (Zaitsev, pers. comm.). On the basis of these data and using average U/Al and Th/Al ratios of the upper Lena and of the Aldan, U/Al and Th/Al ratios resulting from the mixing of these two sources could be calculated and compared to the observed values in the middle Lena. Accordingly, based on the observed ratios of the middle Lena and of the Vilyuy, the expected U/Al and Th/Al ratios of the lower Lena can be estimated. Observed and estimated U/Al and Th/Al ratios of the middle and upper Lena are displayed in Figure 4a and 4b. The observed values are in general agreement with the calculated data but the decreases in U/Al and Th/Al ratios in both the middle and the upper Lena were more pronounced than calculated. However, since the sediment discharge data are not very accurate, it seems reasonable that the chemical evolution along the Lena is controlled mainly by material input through tributaries. Alteration processes within the river and variations in transport energy play a minor role in comparison. Since U and Th are affected in the same way, local differences in the weathering regime can also be neglected.

REE patterns of river SPM are sensitive to the drainage basin geology. Rivers draining basaltic island arcs are characterized by flat or heavy REE-enriched, shale-normalized patterns, while rivers draining Archean crustal rocks show a strong enrichment in light REE (Goldstein and Jacobsen, 1988). Recent studies, however, have shown that the REE pattern of source rock and its weathering product are not identical (Braun et al., 1993; Condie et al., 1995). The SPM of major rivers draining large basins of heterogeneous geology display similar REE patterns characterized by a slight enrichment in light REE and a deficiency of heavy REE (Martin et al., 1976; Goldstein and Jacobsen, 1988). The enrichment in light REE relative to heavy REE in the particulate fraction is balanced by a strong depletion in light REE in relation to heavy REE in the dissolved fraction (Goldstein and Jacobsen, 1988; Elderfield et al., 1990; Sholkovitz, 1995) where Fe-organic colloids are major REE carriers (Sholkovitz, 1995).

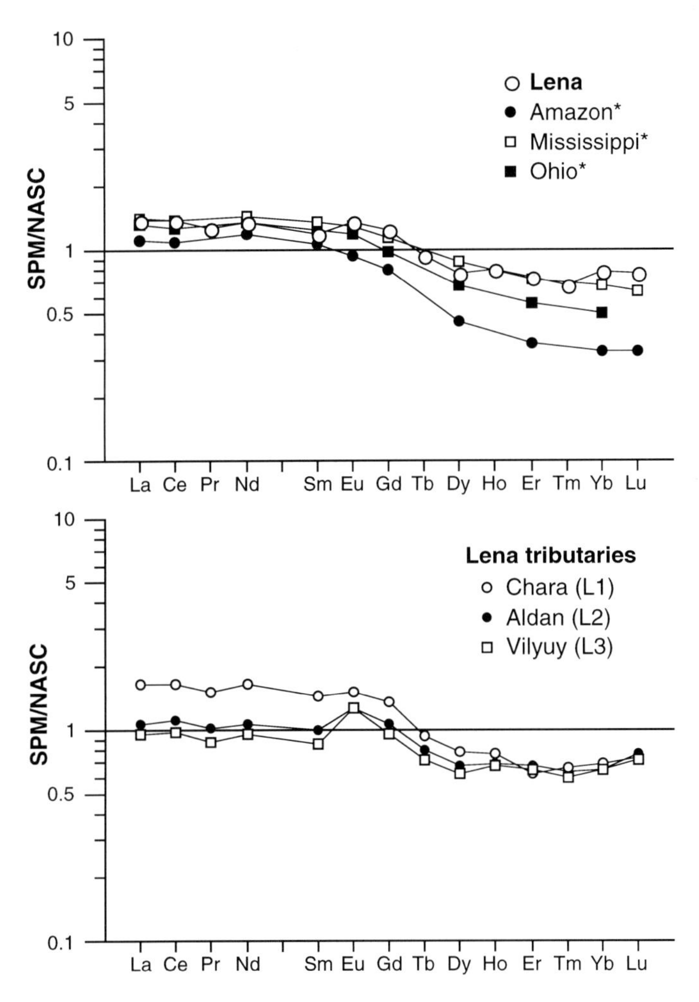

Figure 5: Shale normalized REE patterns of SPM from the Lena and its major tributaries Chara (L1), Aldan (L2), and Vilyuy (L3). Amazon, Mississippi, and Ohio SPM (Goldstein and Jacobsen, 1988) are shown for comparison. Shale data (North American Shale Composite, NASC) are taken from Condie (1993).

Figure 5 presents the REE patterns of SPM of the Lena and other major rivers, i.e. the Amazon, the Mississippi, and the Ohio (data are taken from Goldstein and Jacobsen, 1988). The concentrations are normalized to the North American Shale Composite (NASC, Condie, 1993). As a result of the size of the Lena basin, the REE pattern of Lena SPM proves to be very similar to other major rivers, for example the Mississippi. In contrast, the SPM of the Lena tributaries Chara, Aldan, and Vilyuy, which drain smaller basins of less heterogeneous geology, is characterized by different REE patterns (Figure 5). Chara SPM showed a more pronounced enrichment in light REE and a stronger depletion in heavy REE in accordance with REE patterns reported for rivers draining Archean crustal rocks (Goldstein and Jacobsen,

1988). Heavy REE concentrations of Vilyuy and Aldan SPM were similar to the Chara while light REE concentrations were significantly lower. In addition, Vilyuy SPM was characterized by a positive Eu anomaly. The catchment of the upper Vilyuy consists of basaltic rocks of the Siberian Trap. Due to the magmatic fractionation of Ca-rich feldspars, the upper continental crust and average shale are characterized by a negative Eu anomaly relative to the primitive earth mantle or chondrite, respectively. Continental flood basalts, on the other hand, do not exhibit this pronounced Eu anomaly since they are less fractionated. Therefore, the average shale-normalized REE patterns of material derived from basaltic rocks display a positive Eu anomaly. The SPM of the Vilyuy reflected the REE composition of the Trap basalts. Since the Khatanga almost exclusively drains the Siberian Trap, a detailed discussion of the major, trace, and REE composition of the basalt is given in the Khatanga chapter. Summarizing, it can be concluded that REE patterns of the Lena tributaries clearly reflect the geology of their corresponding catchments. The Lena integrates the SPM derived from different lithologies and imported via tributaries. For this reason, the average REE pattern of the Lena cannot be distinguished from other major rivers.

Yana SPM

In general, the Yana was characterized by the highest total SPM concentrations observed in the studied rivers, ranging from 57 to 750 mg/l and averaging 285 mg/l. Consequently, the SPM of the Yana contained higher amounts of comparatively coarse-grained material (upper left corner of Figure 6). Due to the frequent occurrence of silt-sized particles, the SPM consisted mainly of quartz and feldspars. Plagioclase and K-feldspars could be identified. The clay minerals chlorite and illite were observed; smectite and mixed-layer minerals did not occur (Table 2). However, in comparison to the other rivers studied, the clay minerals were less abundant.

The chemical composition of SPM of the Yana and its tributaries exhibited the most pronounced variations observed in this study. On the one hand, these variations must be attributed to the heterogeneity of the catchment, and on the other hand, to the abundance of coarse-grained material. The average Al-normalized major and trace element concentrations of Yana SPM in comparison to average shale (Wedepohl, 1971 and 1991) are presented in Figure 6. In contrast to average Lena SPM, the Al-normalized concentrations of several elements differ from average shale. In particular, As, Ba, Bi, Cd, Cs, Na, Sb, W, Zn, and Zr are characterized by higher, and Ca and Sr by lower, Al-normalized concentrations. In order to explain the deviations, several processes must be considered.

(1) As with the Lena, Cd and Zn may be enriched due to adsorption onto organic material. However, since the grain-size distribution of Yana SPM was displaced towards coarser material, the enrichment is less pronounced.

(2) The enrichment in Zr in Yana SPM may be related to the dominance of silt-sized particles. The mineral zircon is strongly enriched in this fraction as can be seen from the high Zr concentrations in loess (Schnetger, 1992). Zr and Hf concentrations in the bedload are consequently significantly higher than in the suspended load (Dupré et al., 1996), as they are, for example, in the Congo river.

(3) Low Ca/Al, Sr/Al, and high Na/Al ratios must be attributed to the lithology of the drainage basin, which consists mainly of Mesozoic terrigenous sediments and volcanic rocks as well as granitic intrusions of Triassic and Jurassic age. As for the Lena, high Ba/Al ratios can be explained by the occurrence of feldspar.

(4) Pronounced enrichments in As, Bi, Cs, Sb, and W indicate the influence of ore deposits in the Yana basin. Au and Sn ore deposits originating from granitic intrusions are a typical feature of the Yana basin (Efremenko, 1976). The source rocks consist of granite and granodiorite, tourmaline granite, and aplite. They exhibit very high concentrations of As (up to

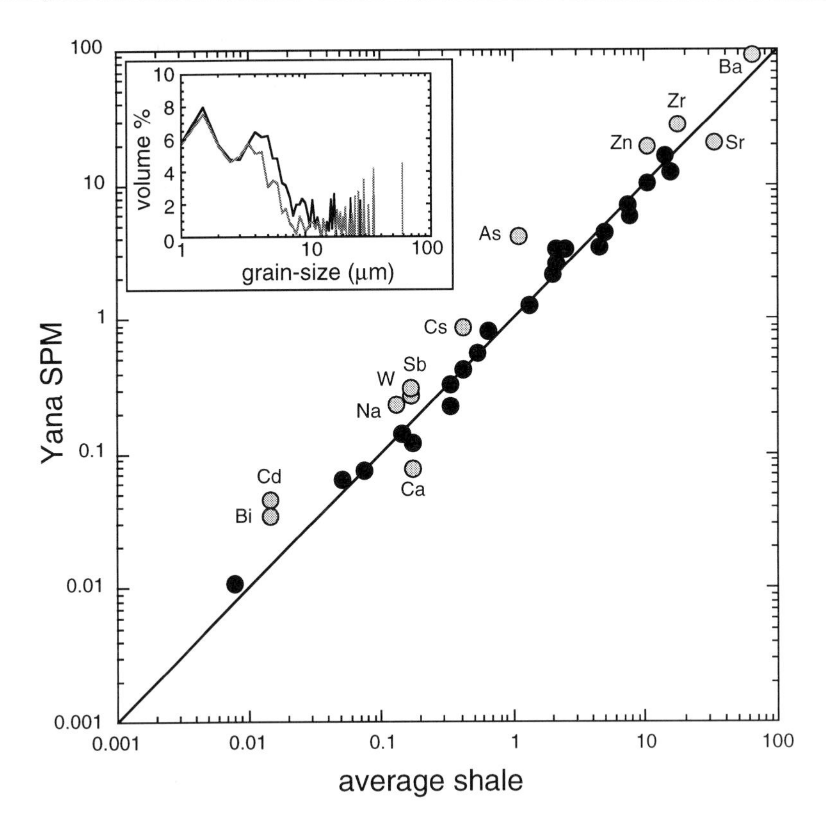

Figure 6: Average chemical composition of SPM from the Yana. The x-axis corresponds to the composition of average shale (Wedepohl, 1971 and 1991). The diagonal represents the straight line of identical values. Al normalized concentrations based on weight ratios (major elements) and weight ratio • 10^4 (trace elements) are displayed. The grain-size distribution (volume %) is shown in the upper left corner.

350 ppm) and Cs (up to 90 ppm) (Efremenko, 1976). Consequently, even a small contribution of material derived from these intrusions results in significant trace element enrichments in the Yana SPM. The maximum As concentrations (62 ppm) were observed in the Bytaktay (Y6), the maximum Cs concentrations (more than 8 ppm) in the Bytaktay and the Kyugyurlyur (Y8), both of which are left bank tributaries of the Yana.

REE patterns of SPM of the Yana and its tributaries are shown in Figure 7. The average Yana SPM was characterized by a flat REE distribution and a pronounced positive Eu anomaly. The only tributary that displayed a similar pattern was the Aditcha (Y4) draining the southeastern part of the Yana catchment. In agreement with hydrological data (Korotaev, pers. comm.), it can be concluded that a major fraction of Yana SPM is derived from this tributary. The other tributaries were characterized by flat REE patterns and a slight enrichment in middle REE which may be attributed to the occurrence of silt-sized particles containing significant amounts of heavy minerals. Heavy minerals, such as zircon, garnet, apatite, epidote, and tourmaline, may have high abundances of REE and exhibit strong variabilities in REE patterns (Taylor and McLennan, 1985).

In contrast to the trace element enrichment, the REE pattern of Yana SPM, in particular the positive Eu anomaly, cannot be explained by the influence of granitic intrusions. Volcanic rocks of Triassic and Jurassic age, which are also present in the Yana catchment, seem to control the

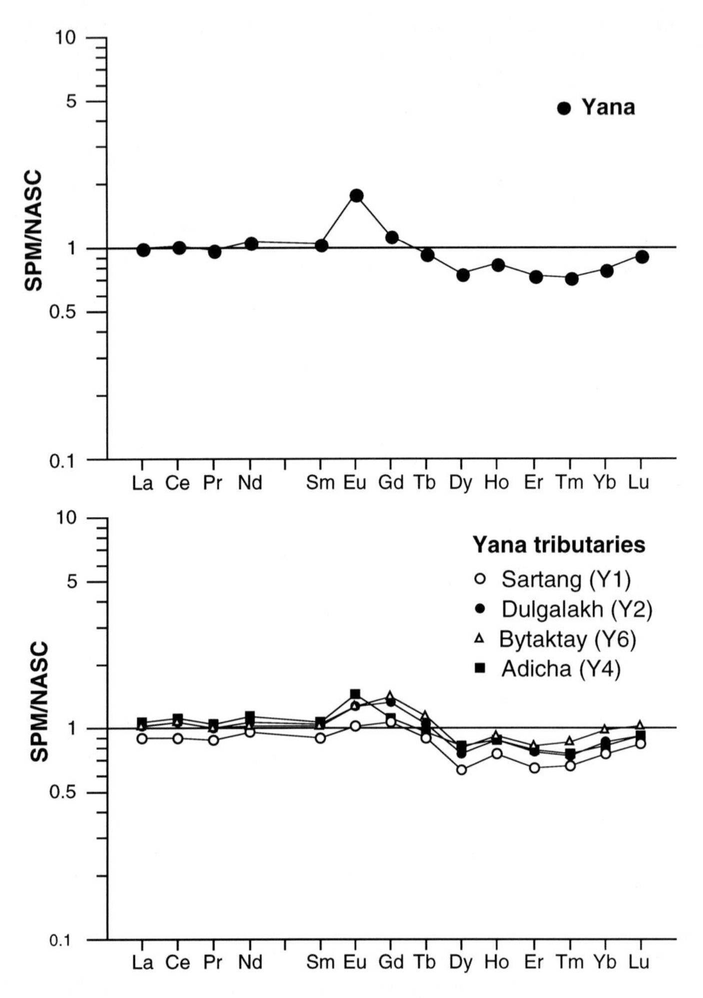

Figure 7: Shale normalized REE patterns of SPM from the Yana and its major tributaries Sartang (Y1), Dulgalakh (Y2), Bytaktay (Y6), and Adicha (Y4). Shale data (North American Shale Composite, NASC) are taken from Condie (1993).

REE pattern of the SPM. As a consequence of the heterogeneity of the Yana basin and the lack of chemical data for the rocks in the catchment, it is not possible to relate the SPM of individual tributaries to specific lithologies. However, the average Yana SPM can be clearly separated from Lena SPM.

Khatanga SPM

Total SPM concentrations of the Khatanga ranged from 19 to 59 mg/l with an average of 32 mg/l. The grain-size distribution of two samples of Khatanga SPM is presented in the upper left corner of Figure 8. Compared to the Yana, the clay mineral fraction was more abundant. The SPM consisted mainly of the clay mineral smectite. The quartz content was significantly lower than in Yana and Lena SPM, and plagioclase and pyroxene were observed. Furthermore, illite and K-feldspars were less abundant (Table 2). The catchment of the Khatanga consists mainly of basaltic rocks of the Siberian Trap (Figure 1). Plagioclase and pyroxene observed in the Khatanga SPM represent the primary mineral content of the Trap basalts. Smectite, on the other hand, typically forms from weathering of basaltic rocks. For that reason, it can be stated that the mineralogical composition of Khatanga SPM is closely related to the Siberian Trap basalts which dominate the drainage area.

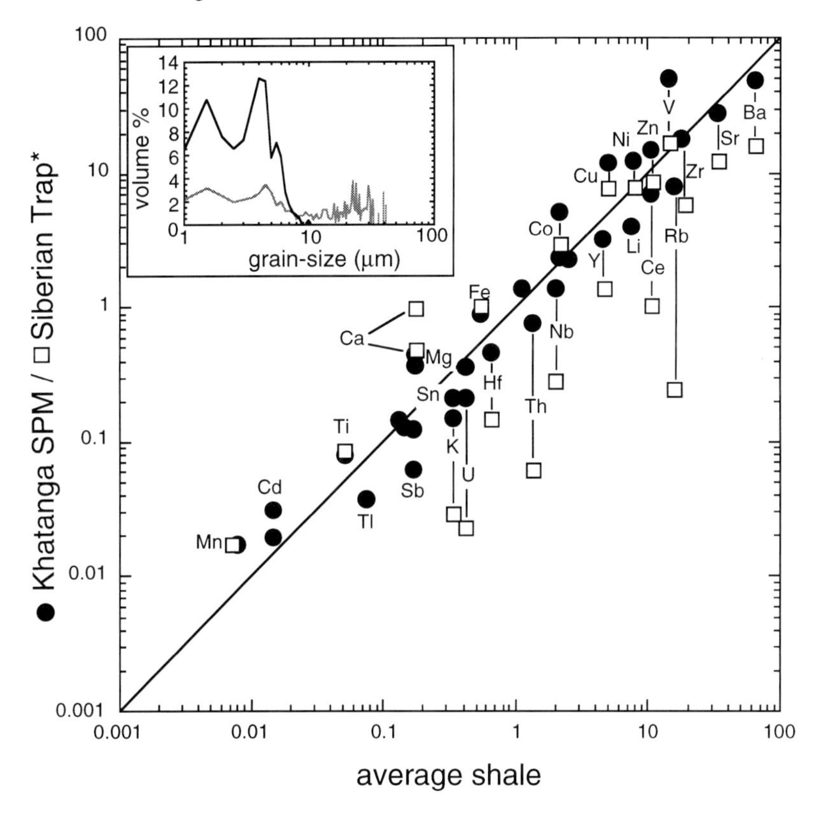

Figure 8: Average chemical composition of SPM of the Khatanga and the Siberian Trap basalt (Lightfoot et al., 1993). The x-axis corresponds to the composition of average shale (Wedepohl, 1971 and 1991). The diagonal represents the straight line of identical values. Al normalized concentrations based on weight ratios (major elements) and weight ratio • 10^4 (trace elements) are displayed. The grain-size distribution (volume %) is shown in the upper left corner.

The average Al-normalized major and trace element concentrations of Khatanga SPM and the Siberian Trap basalts (Lightfoot et al., 1993) in comparison to average shale are illustrated in Figure 8. In contrast to Lena and Yana SPM, most of the elements in the Khatanga SPM showed significant deviations from average shale. Significantly higher Al-normalized concentrations compared to average shale were observed for Ca, Co, Cu, Fe, Mg, Mn, Ni, Ti, and V. The enrichment in these elements can directly be related to the influence of the Siberian Trap, except for Ni and V (Figure 8). Al-normalized concentrations of Ce, Hf, K, Li, Nb, Rb, Sb, Sn, Th, Tl, U, and Y were significantly lower than in average shale which is also attributed to the Trap basalts (Figure 8). Note that for the Trap basalts there are no Cd, Li, Sb, Sn, and Tl data available. In contrast to the Lena and Yana SPM, Ba/Al ratios in the Khatanga SPM were not higher than in average shale due to smaller amounts of feldspars.

The dominance of the Siberian Trap basalts is also seen in the REE pattern of Khatanga SPM.

In Figure 9, REE patterns of Khatanga SPM and Siberian Trap basalts (Lightfoot et al., 1993) are displayed. Except for the enrichment in light REE in the Khatanga SPM, basalts and SPM showed a very similar distribution. The same held true for the southern tributaries Kheta (K1) and Kotuy (K2). In the SPM of the Popigay (K3), which is a tundra river draining the northeastern part of the Khatanga basin, the influence of the Siberian Trap was less pronounced. The enrichment in light REE relative to heavy REE is a general feature observed in river SPM and is balanced by the depletion in light REE relative to heavy REE in the dissolved fraction.

In summary, it can be concluded that the chemical composition of Khatanga SPM is very closely related to the Siberian Trap basalts, except for Cd and Zn. As for the Lena and Yana SPM, enrichment in these elements may be related to adsorption onto organic particulates.

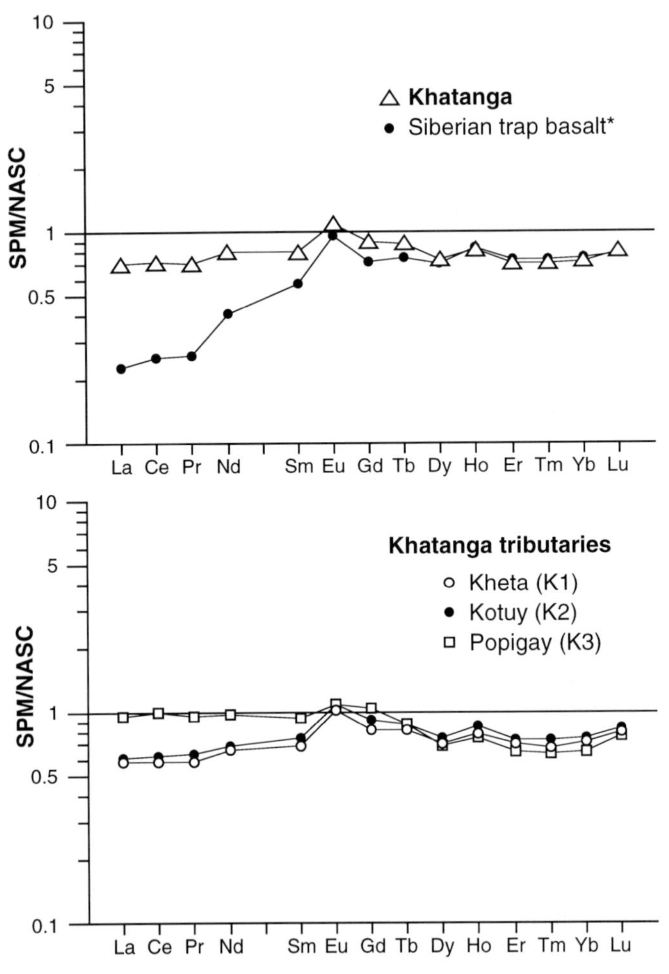

Figure 9: Shale normalized REE patterns of SPM from the Khatanga and its major tributaries, the Kheta (K1), Kotuy (K2), and Popigay (K3). REE data of the Siberian Trap basalts (Lightfoot et al., 1993) are shown for comparison. Shale data (North American Shale Composite, NASC) are taken from Condie (1993).

Factor analysis of the SPM data

Factor analyses of river water chemical data, such as principle component analysis, has been applied to identify the mixing of different sources in numerous studies (c.f. Christophersen and Hooper, 1992). In the present study, principle component analysis (orthogonal transformation) of the chemical data set was used in order to: (1) reduce the large number of variables to a small number of factors, (2) identify the general interrelationship between major and trace elements, and (3) distinguish between SPM of individual rivers. The concentrations of 31 elements (Al, Ti, Fe, Mn, Mg, Ca, Na, K, Ba, Bi, Ce, Co, Cs, Ga, Hf, La, Li, Mo, Nb, Ni, Rb, Sb, Sr, Ta, Th, Tl, U, V, W, Y, and Zr) in 64 samples were included in the calculation. As, Cd, Cu, Pb, Sn, and Zn were omitted from the computation due to contamination effects.

The general variations in the chemical composition of the SPM can be explained by two factors accounting for 71.4 % of the total variance. Figure 10 presents the factor loadings of these two factors. The factor scores are illustrated in Figure 11. Factor 1 (59.6 % of the variance) clearly separates elements predominant in mafic rocks (high negative factor loadings for Ca, Co, Fe, Mg, Ni, Ti, and V) from elements that are more abundant in acidic rocks (high positive factor loadings for Ba, Bi, Ce, Cs, Hf, K, La, Li, Mo, Na, Nb, Rb, Sb, Ta, Th, Tl, U, W, Y, and Zr). This factor discriminates the Khatanga SPM from the Lena and Yana SPM. While the geochemistry of Khatanga SPM is closely related to the basaltic rocks of the Siberian Trap, the composition of Lena and Yana SPM is similar to average shale, except for some elements. As a consequence, high negative factor scores can be observed in Khatanga SPM and positive factor scores in Lena and Yana SPM (Figure 11). As has been discussed above, the REE pattern of Vilyuy (L3) SPM indicates the influence of Siberian Trap basalts. This is also seen in the factor score diagram where the Vilyuy plots between the Lena and Khatanga (Figure 11). In the Popigay (K3) SPM, the dominance of the Siberian Trap is less pronounced than in the other Khatanga tributaries.The Popigay consequently plots very close to the Vilyuy in the factor score diagram (Figure 11).

Factor 2 (11.8 % of the variance) is characterized by high positive factor loadings for Cs, Hf, Sb, Ti, and Zr and a negative Mn factor loading (Figure 10). This factor may be related to two different processes: (1) The negative factor loading for Mn and the high positive factor loadings for Hf, Ti, and Zr suggest that differences in grain-size distribution are responsible for this factor. Due to the dominance of fine-grained material, which is characterized by high amounts of Mn oxihydroxides, Lena SPM exhibits high Mn concentrations. Yana SPM, on the other hand, displays higher amounts of coarse-grained material resulting in high Hf, Ti, and Zr contents. (2) High positive loadings for Cs and Sb are related to granitic intrusions in the Yana catchment (see above). As a result of both processes, factor 2 clearly differentiates between Lena and Yana SPM. It can be concluded that factor analysis serves as a helpful tool to identify the interrelationship of chemical elements in river SPM and to separate between SPM transported by individual rivers.

Summary and conclusions

In order to unravel the sediment transport of East Siberian rivers draining to the Laptev Sea, grain-size distribution, mineralogy, and inorganic chemical composition of SPM from the Lena, Yana, and Khatanga rivers and their major tributaries were analyzed. Based on these data the following conclusions can be drawn:

1 The average chemical composition of Lena SPM compares well with average shale, except for Mn-enrichment in the fine-grained SPM and Cd and Zn, which seem to be associated with organic material. This finding indicates that the average composition of all lithologies in the Lena catchment is similar to the upper continental crust. High feldspar contents

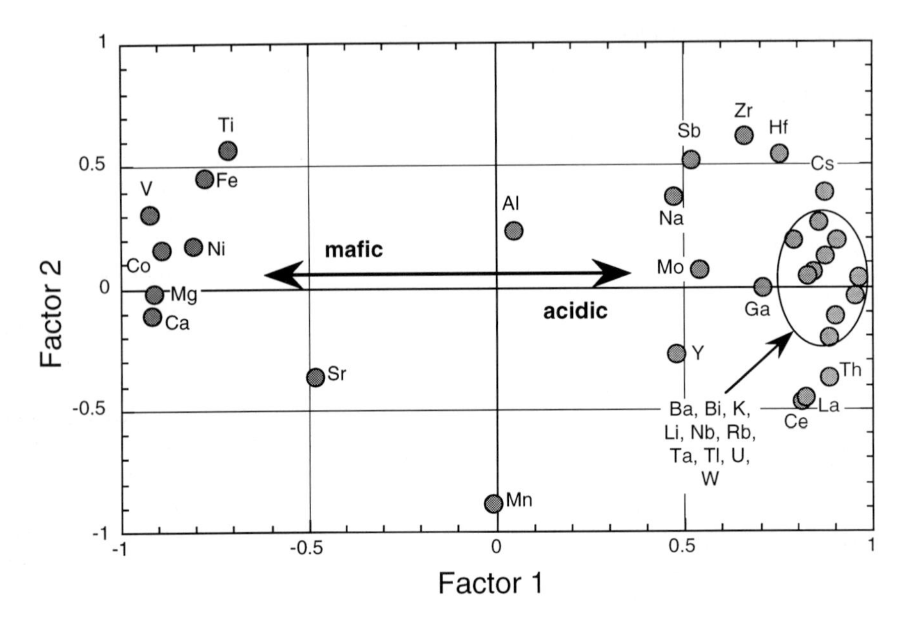

Figure 10: Factor loadings of principle component analysis (64 samples, 31 chemical elements). Factor 1 (x-axis, 59.6 % of the variance) includes high positive factor loadings for Ba, Bi, Ce, Cs, Hf, K, La, Li, Mo, Na, Nb, Rb, Sb, Ta, Th, Tl, U, W, Y, and Zr and high negative factor loadings for Ca, Co, Fe, Mg, Ni, Ti, and V. Factor 2 (y-axis, 11.8 % of the variance) is characterized by high positive factor loadings for Cs, Hf, Sb, Ti, and Zr and a negative Mn factor loading.

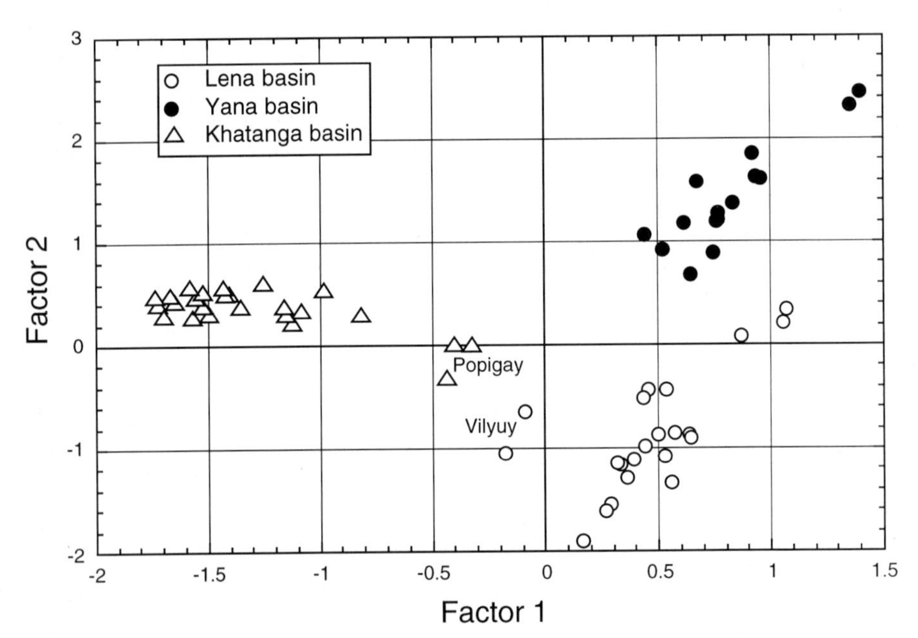

Figure 11: Factor scores of principle component analysis (64 elements, 31 chemical elements). Factor 1 (x-axis) clearly separates the Khatanga basin, whereas factor 2 differentiates between Lena and Yana basin.

corresponding to elevated Ba/Al ratios indicate that denudation rates are noticeably reduced in the subarctic climate of the Lena basin.

2 Although the chemical composition of Lena SPM is very homogeneous throughout the river, a trend along the course of the river could be identified. Since alteration processes within the river, regional variations in the weathering regime, and changes in transport energy play a minor role, this evolution is attributed mainly to material input through tributaries. The SPM of Lena tributaries draining more restricted catchments clearly reflect the regional geology.

3 Although the Yana is characterized by the strongest variations in SPM composition as a result of the heterogeneity of its catchment and the frequent occurrence of coarse-grained material, Yana SPM can be clearly distinguished from Lena SPM. The enrichments in As, Bi, Cs, Sb, and W are typical features and are related to granitic intrusions forming the source of Au and Sn ore-deposits in the Yana basin. Furthermore, Yana SPM shows significantly lower Ca/Al and higher Na/Al ratios.

4 The most prominent differences were observed in the Khatanga SPM, which is strongly influenced by basaltic rocks of the Siberian Trap. Consequently, the SPM is enriched in elements associated with basalts, i.e. Ca, Co, Cu, Fe, Mg, Ni, Ti, and V, but depleted in incompatible elements, i.e. K, Rb, U, and Th.

5 REE patterns of the SPM are closely related to the geology of the catchment. However, a general enrichment in light REE relative to heavy REE is observed and is related to weathering. While the REE distribution of Lena SPM is very similar to other major rivers due to the size of the drainage basin, REE patterns of the Yana and Khatanga SPM and of their tributaries represent characteristic signals of the lithology of the individual catchment.

6 Factor analysis of the chemical data serves as a helpful tool to distinguish between the SPM of individual rivers. The variations within the data set can mainly be explained by two factors accounting for more than 70 % of the variance. Factor 1 discriminates elements associated with basalts and consequently separates the Khatanga SPM from the Lena and Yana SPM. Factor 2 describes the Yana basin and includes elements that are enriched in coarse-grained material and elements originating from granitic intrusions which formed ore-deposits in the Yana catchment.

As a result of differences in the lithology of the catchments, rivers supplying material to the Laptev Sea can clearly be distinguished by means of their inorganic geochemical SPM composition. The chemical data can thus be used to clarify sediment transport. Analyses of Laptev Sea sediments, ice-rafted detritus, and Arctic Ocean sediments which are currently carried out by other working groups of the "Laptev Sea" project will help to identify the pathways of sediment transport from the Eurasian continent to the Laptev Sea and further to the Arctic Ocean.

Acknowledgments

The author thanks M. Zimmer and H.-G. Plessen from the Geoforschungszentrum in Potsdam, M. Tintelnot from the Forschungsinstitut Senckenberg in Wilhelmshaven, G. Kuhn from the Alfred-Wegener-Institut in Bremerhaven, and A. Eulenburg from the Alfred-Wegener-Institut in Potsdam for supporting the analytical work. Special thanks to A. Alabyan, V. N. Korotaev, and A. Zaitsev from Moscow State University, A. Ufimzev from the Arctic and Antarctic Research Institute in St. Petersburg, and to the captains and crew members of "Prof. Makkaveev", "Sarya 9", and "Lotsman". Critical comments of B. Hagedorn and the reviewers H.-J. Brumsack and K. Wallmann are greatly appreciated. This study is part of the German-Russian project "The Laptev Sea System" which is funded by the German Ministry of Education, Science, Research and Technology (BMBF) and the Russian Ministry of Research and Technology. This is publication No. 1358 of the Alfred-Wegener-Institute for Polar and Marine Research.

References

Alabyan, A.M., R.S. Chalov, V.N. Korotaev, A.Y. Sidorchuk and A.A. Zaitsev (1995) Natural and technogenic water and sediment supply to the Laptev Sea. Reports on Polar Research, 176, 265-271.

Bischof, J., J. Koch, M. Kubisch, R.F. Spielhagen and J. Thiede (1990) Nordic Seas surface ice drift reconstructions: evidence from ice rafted coal fragments during oxygen isotope stage 6. In: Glacimarine Environments: Processes and Sediments, Dowdeswell, J.A. and J.D. Scourse (eds.), Geological Society, London (Geological Society Special Publ. No 53), 235-251.

Braun, J.-J., M. Pagel, A. Herbillon and C. Rosin (1993) Mobilization and redistribution of REEs and thorium in a syenitic lateritic profile: a mass balance study. Geochim. Cosmochim. Acta, 57, 4419-4434.

Cauwet, G. and I.S. Sidorov (1996) The biogeochemistry of Lena River: organic carbon and nutrients distribution. Mar. Chem., 53, 211-227.

Christophersen, N. and R.P. Hooper (1992) Multivariate analysis of stream water chemical data: the use of principle components analysis for the end-member mixing problem. Water Resources Res., 28, 99-107.

Condie, K.C. (1991) Another look at rare earth elements in shales. Geochim. Cosmochim. Acta, 55, 2527-2531.

Condie, K.C. (1993) Chemical composition and evolution of the upper continental crust: contrasting results from surface samples and shales. Chem. Geol., 104, 1-37.

Condie, K.C., J. Dengate and R.L. Cullers (1995) Behavior of rare earth elements in a paleoweathering profile on granodiorite in the Front Range, Colorado, USA. Geochim. Cosmochim. Acta, 59, 279-294.

Dethleff, D., D. Nürnberg, E. Reimnitz, M. Saarso and Y.P. Savchenko (1993) East Siberian Arctic Region Expedition '92: The Laptev Sea - Its significance for Arctic sea-ice formation and transpolar sediment flux. Reports on Polar Research, 120, 1-44.

Dupré, B., J. Gaillardet, D. Rouseau and C.J. Allegre (1996) Major and trace elements of river-borne material: The Congo Basin. Geochim. Cosmochim. Acta, 60, 1301-1321.

Efremenko, E.A. (1976) Granitoid rock series of high-aluminum type in Yana-Borulak ore-bearing area in Yakutia (in Russian). Geology and Geophysics, 3, 33-44.

Eicken, H., E. Reimnitz, V. Alexandrov, T. Martin, H. Kassens and T. Viehoff (1996) Sea-ice processes in the Laptev Sea and their importance for sediment transport. Continental Shelf Res., 17, 205-233.

Eisenhauer, A., R.F. Spielhagen, M. Frank, G. Hentzschel, A. Mangini, P.W. Kubik, B. Dittrich-Hannen and T. Billen (1994) ^{10}Be record of sediment cores from high northern latitudes: implications for environmental and climatic changes. Earth Planet. Sci. Lett., 124, 171-184.

Elbaz-Poulichet, F., J.-M. Garnier, D.M. Guan, J.-M. Martin and A.J. Thomas (1996) The conservative behaviour of trace metals (Cd, Cu, Ni and Pb) and As in the surface plume of stratified estuaries: example of the Rhone River (France). Estuarine, Coastal and Shelf Science, 42, 289-310.

Elderfield, H., R. Upstill-Goddard and E.R. Sholkovitz (1990) The rare earth elements in rivers, estuaries, and coastal seas and their significance to the composition of ocean waters. Geochim. Cosmochim. Acta, 54, 971-991.

Gaillardet, J., B. Dupre and C.J. Allegre (1995) A global geochemical mass budget applied to the Congo Basin rivers: erosion rates and continental crust composition. Geochim. Cosmochim. Acta, 59, 3469-3485.

Garnier, J.-M., J.-M. Martin, J.-M. Mouchel and K. Sioud (1996) Partitioning of trace metals between the dissolved and particulate phases and particulate surface reactivity in the Lena River estuary and the Laptev Sea (Russia). Mar. Chem., 53, 269-283.

Gibbs, R.J. (1965) Error due to segregation in quantitative clay mineral X-ray diffraction mounting techniques. Amer. Mineralogist, 50, 741-751.

Goldstein, S.J. and S.B. Jacobsen (1988) Rare earth elements in river water. Earth Planet. Sci. Lett., 89, 35-47.

Gordeev, V.V. and I.S. Sidorov (1993) Concentrations of major elements and their outflow into the Laptev Sea by the Lena River. Mar. Chem., 43, 33-45.

Gordeev, V.V., J.M. Martin, I.S. Sidorov and M.V. Sidorova (1996) A reassessment of the Eurasian river input of water, sediment, major elements, and nutrients to the Arctic Ocean. Amer. J. Sci., 296, 664-691.

Guieu, C., W.W. Huang, J.-M. Martin and Y.Y. Yong (1996) Outflow of trace metals into the Laptev Sea by the Lena River. Mar. Chem., 53, 255-267.

Heinrichs, H. and A.G. Herrman (1990) Principles of analytic geochemistry (in German). Springer, Heidelberg, 669pp.

Huh, Y. and J.M. Edmond (1997) Chemical weathering yields and Sr isotope systematics from major Siberian rivers. 7th annual Goldschmidt Conference, Tucson, AZ, 2-6 June 1997.

Kassens, H. and V. Karpiy (1994) Russian-German cooperation: the Transdrift 1 Expedition to the Laptev Sea. Alfred Wegener Institute, Bremerhaven (Reports on Polar Research 151), 168pp.

Kassens, H., H.-W. Hubberten, S.M. Pryamikov and R. Stein (1994) Russian-German cooperation in the Siberian shelf seas: Geo-System Laptev Sea. Alfred Wegener Institute, Bremerhaven (Reports on Polar Research 144), 133pp.

Kassens, H., D. Piepenburg, J. Thiede, L. Timokhov, H.-W. Hubberten and S.M. Pryamikov (1995) Russian-German Cooperation: Laptev Sea System. Alfred Wegener Institute, Bremerhaven (Reports on Polar Research 176), 387pp.

Korotaev, V.N. (1995) pers. comm.

Langmuir, D. and J.S. Hermann (1980) The mobility of Th in natural waters at low temperature. Geochim. Cosmochim. Acta, 44, 1753-1766.

Lara, R., V. Rachold, G. Kattner, H.-W. Hubberten, G. Guggenberger and A. Skoog (in press) Dissolved organic matter and nutrients in the Lena river, Siberian Arctic: characteristics and distribution. Mar. Chem.

Lasaga, A.C. (1984) Chemical kinetics of water-rock interactions. Geophys. Res. B., 89, 4009-4025.

Lasaga, A.C., J.M. Soler, J. Ganor, T.E. Burch and K.L. Nagy (1994) Chemical weathering rate laws and global geochemical cycles. Geochim. Cosmochim. Acta, 58, 2361-2386.

Létolle, R., J.M. Martin, A.J. Thomas, V.V. Gordeev, S. Gusarova and I.S. Sidorov (1993) ^{18}O abundance and dissolved silicate in the Lena delta and Laptev Sea (Russia). Mar. Chem., 43, 47-64.

Lightfoot, P.C., C.J. Hawkesworth, J. Hergt, A.J. Naldrett, N.S. Gorbachev, V.A. Fedorenko and W. Doherty (1993) Remobilisation of the continental lithosphere by a mantle plume: major-, trace-element, and Sr-, Nd-, and Pb-isotope evidence from picritic, and tholeiitic lavas of the Noril´sk District, Siberian Trap, Russia. Contr. Mineral. Petrol., 114, 171-188.

Martin, J.M., O. Hogdahl and J.C. Philippot (1976) Rare earth element supply to the oceans. J. Geophys. Res., 81, 3119-3124.

Martin, J.M., D.M. Guan, F. Elbaz-Poulichet, A.J. Thomas and V.V. Gordeev (1993) Preliminary assessment of the distributions of some trace elements (As, Cd, Cu, Fe, Ni, Pb, and Zn) in a pristine aquatic environment: the Lena River estuary (Russia). Mar. Chem., 43, 185-199.

Milliman, J.D. and R.H. Meade (1983) World-wide delivery of river sediment to the oceans. J. Geol., 91, 1-21.

Morel, F.M.M. and J.G. Hering (1993) Principles and applications of aquatic chemistry. Wiley, New York, 588pp.

Nürnberg, D., I. Wollenburg, D. Dethleff, H. Eicken, H. Kassens, T. Letzig, E. Reimnitz and J. Thiede (1994) Sediments in Arctic sea ice: Implications for entrainment, transport and release. Mar. Geol., 119, 185-214.

Palmer, M.R. and J.M. Edmond (1993) Uranium in river water. Geochim. Cosmochim. Acta, 57, 4947-4955.

Pande, K., M.M. Sarin, J.R. Trivedi, S. Krishnaswami and K.K. Sharma (1994) The Indus river system (India-Pakistan): major-ion chemistry, uranium and strontium isotopes. Chem. Geol., 116, 245-259.

Rachold, V., J. Hermel and V.N. Korotaev (1995) Expedition to the Lena River in July/August 1994. Reports on Polar Research, 182, 181-195.

Rachold, V., A. Alabyan, H.-W. Hubberten, V.N. Korotaev and A.A. Zaitsev (1996) Sediment transport to the Laptev Sea - hydrology and geochemistry of the Lena River. Polar Research, 15, 183-196.

Rachold, V., E. Hoops, A. Alabyan, V.N. Korotaev and A.A. Zaitsev (1997) Expedition to the Lena and Yana Rivers June-September 1995. Reports on Polar Research, 248, 193-210.

Rachold, V., A. Eisenhauer, H.-W. Hubberten and and H. Meyer (in press [a]) Sr isotopic composition of suspended particulate material (SPM) of East Siberian rivers - sediment transport to the Arctic Ocean. Arctic and Alpine Research.

Rachold, V., E. Hoops and A.V. Ufimzev (in press [b]) Expedition to the Khatanga River July-August 1996. Reports on Polar Research.

Rachold, V. and H.-W. Hubberten (this volume). Carbon isotope composition of particulate organic material of East Siberian rivers.

Reimnitz, E., D. Dethleff and D. Nürnberg (1994) Contrasts in Arctic shelf sea-ice regimes and some implications: Beaufort Sea versus Laptev Sea. Mar. Geol., 119, 215-225.

Reimnitz, E., H. Kassens and H. Eicken (1995) Sediment transport by Laptev Sea ice. Reports on Polar Research, 176, 71-77.

Schnetger, B. (1992) Chemical composition of loess from a local and worldwide view. N. Jb. Miner. Mh., 1992-1, 29-47.

Sholkovitz, E.R. (1995) The aquatic chemistry of Rare Earth Elements in rivers and estuaries. Aquatic Chemistry, 1, 1-34.

Taylor, S.R. and S.M. McLennan (1985) The continental crust: its composition and evolution. Blackwell, Oxford, 312pp.

Ugolini, J.C. (1986) Processes and rates of weathering in cold and polar desert environments. In: Rates of chemical weathering of rocks and minerals, Colman, S.M. and D.P. Dethier (eds.), Acad. Press, London, 193-235.

Wedepohl, K.H. (1971) Environmental influences on the chemical composition of shales and clays. In: Physics and Chemistry of the Earth, Vol. 8, Ahrens, L.H., F. Press, S.K. Runcorn and H.C. Urey (eds.), Pergamon, New York, 305-333.

Wedepohl, K.H. (1991) The composition of the upper earth´s crust and the natural cycles of selected metals. Metals in natural raw materials. In: Metals and their compounds in the environment, Merian, E. (ed.), VCH-Verlagsgesellschaft, Weinheim, 3-17.
Wollenburg, I. (1993) Sediment transport by Arctic Sea Ice: the recent load of lithogenic and biogenic material (in German). Alfred Wegener Institute, Bremerhaven (Reports on Polar Research 127), 159pp.
Zaitsev, A.A.. (1995) pers. comm.

Carbon Isotope Composition of Particulate Organic Material in East Siberian Rivers

V. Rachold and H.-W. Hubberten

Alfred-Wegener-Institut für Polar- und Meeresforschung, Forschungsstelle Potsdam, Telegrafenberg A43, D 14473 Potsdam, Germany

Received 14. February 1997 and accepted in revised form 5 March 1998

Abstract - Stable carbon isotope methods have been applied to study the origin of particulate organic material (POM) in the Siberian rivers Lena, Yana, and Khatanga, all of which drain to the Laptev Sea. The carbon isotope composition ($\delta^{13}C$ values) of POM in the studied rivers exhibits a range of -25 to -31 ‰ vs. V-PDB. The negative correlation between sediment load and total organic carbon (TOC) concentration of the suspended particulate material (SPM) can be explained by a two-endmember mixing model for the POM, which is also confirmed by the carbon isotope data. A linear correlation between $\delta^{13}C$ values and reciprocal particulate organic carbon (POC) concentrations is observed. Mixing calculations indicate that the major fraction originates from an isotopically heavier, detrital endmember, while a smaller amount is formed from an isotopically lighter, autochthonous phytoplankton endmember. Annual carbon fluxes of the studied rivers are presented in order to demonstrate the significance of riverine carbon export to the Laptev Sea.

Introduction

Within the framework of the German-Russian project "The Laptev Sea System" (Kassens and Karpiy, 1994; Kassens et al., 1994; Kassens et al., 1995), our study concentrates on the material transport of East Siberian rivers discharging into the Laptev Sea. Today, large amounts of sea ice feeding the Transpolar drift are formed above the wide Laptev Sea shelf (Dethleff et al., 1993; Nürnberg et al., 1994; Eicken et al., 1996). Material transported by the Siberian rivers, mainly by the Lena, is to some extent incorporated into sea ice entering the Transpolar drift and contributes to sedimentation in the central Arctic Ocean and the North Atlantic (Bischof et al., 1990; Reimnitz et al., 1994).

The oceanic carbon budget is strongly influenced by fluvial carbon input (Degens et al., 1991). To expand our knowledge of the origin of riverine carbon and its transfer to the oceans, stable carbon isotope methods have been applied in numerous studies (Fontugne and Jouanneau, 1987; Cai et al., 1988; Mariotti et al., 1991; Quay et al., 1992; Tan and Edmond, 1993; Thornton and McManus, 1994). However, previous investigations concentrated mainly on tropical rivers. Existing data on Arctic rivers have been summarized by Telang et al. (1991). Recent studies reported on the biogeochemistry of the Lena (Cauwet and Sidorov, 1996; Lara et al. in press). In the present article, we focus on the particulate organic material (POM) transport of the Lena, Yana, and Khatanga (Figure 1). The objectives are to identify the sources of the POM in the Siberian rivers using carbon isotope ratios and to quantify the POM transport to the Laptev Sea.

Hydrology and Geology of the Lena, Yana, and Khatanga basins

The following section presents a short overview of the hydrological characteristics of the Lena, Yana, and Khatanga catchment areas. Detailed information has been presented elsewhere (Alabyan et al., 1995; Rachold et al., 1996). In terms of water discharge, the Lena is the eighth

In: Kassens, H., H.A. Bauch, I. Dmitrenko, H. Eicken, H.-W. Hubberten, M. Melles, J. Thiede and L. Timokhov (eds.) Land-Ocean Systems in the Siberian Arctic: Dynamics and History. Springer-Verlag, Berlin, 1999, 223-238.

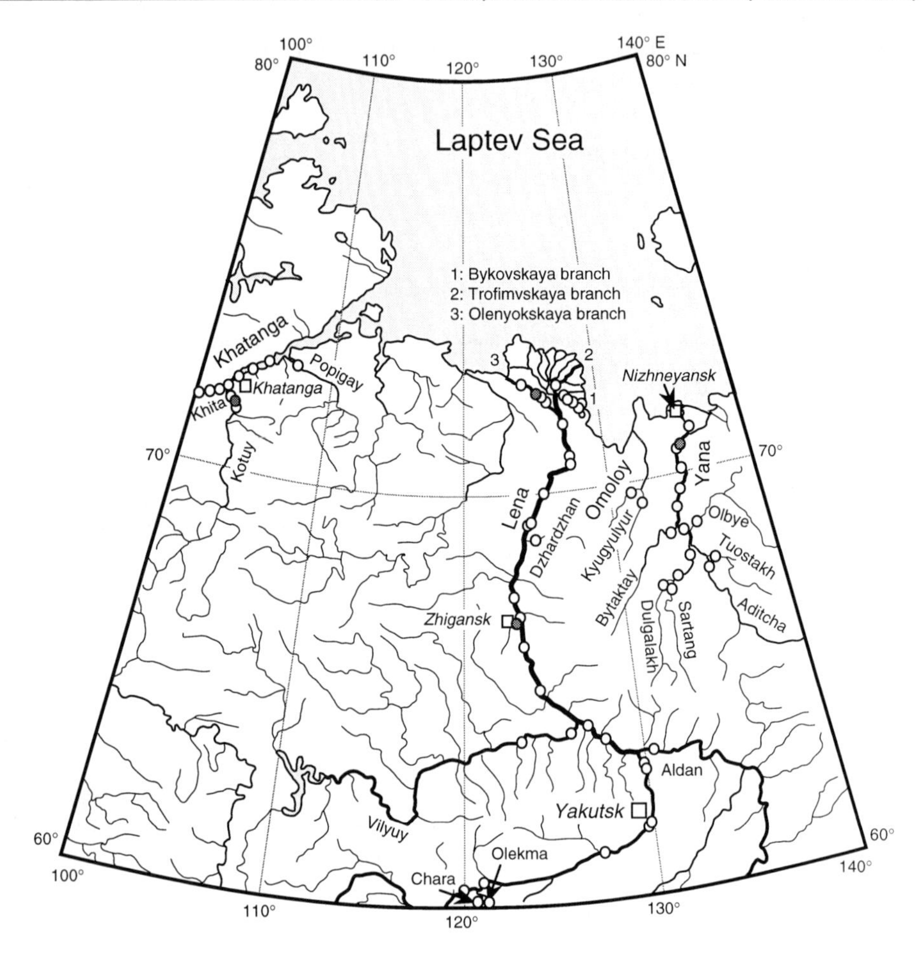

Figure 1: Locations of POM samples collected during the expeditions "Lena 94" (Rachold et al., 1995), "Lena/Yana 95" (Rachold et al., 1997), and "Khatanga 96" (Rachold et al., in press). White circles refer to POM samples, gray circles to peat and coal samples.

largest river in the world and the second largest among the Arctic rivers after the Yenisey (Milliman and Meade, 1983). A total of 520 km^3 of water and 21 • 10^6 tons of suspended particulate material (SPM) are transported per year (Alabyan et al., 1995). The Lena has a length of 4400 km and drains an area of 2.49 • 10^6 km^2. The length of the Yana is 1170 km, the catchment area has a size of 0.24 • 10^6 km^2. Water discharge of the Yana (31 km^3 • a^{-1}) is of minor importance for the water mass formation in the Laptev Sea. However, since SPM concentrations can be much higher than that of the Lena, the sediment transport of the Yana (3 • 10^6 tons • a^{-1}) must be taken into account (Alabyan et al., 1995). Although the Khatanga is the second largest river draining to the Laptev Sea in terms of catchment area (of 0.27 •10^6 km^2) and water discharge (101 km^3 • a^{-1}), its sediment transport (1.4 • 10^6 tons • a^{-1}) does not exceed 50 % of the Yana's. Compared to the Lena, Yana, and Khatanga, the Omoloy contribution seems to be negligible. Alabyan et al. (1995), however, described a 10-fold enrichment of SPM concentrations due to gold mining activities in the catchment area.

East Siberia is characterized by long and cold winters (minimum temperature -45 to -50°C for three months duration) and short, hot summers (maximum temperature +30 to +35°C for two months duration). The average annual precipitation in the Lena basin is 330 mm of which 70 to 80 % occurs during the summer (Gordeev and Sidorov, 1993). Due to this extreme climate, most of the water and sediment is transported during summer, between June to September. The seasonal variations in water discharge of the Lena, Yana, and Khatanga are presented in Figure 2. Average daily water discharges measured at hydrometeorological stations located in the lower reaches of the rivers are displayed (Russian Federal Service for Hydrometeorology and Environmental Monitoring, unpublished data).

The majority of the study area is covered by Siberian taiga. Only the northernmost regions, mainly the Lena and Yana deltas and the upper reaches of the Khatanga, are located within the tundra zone.

Samples and methods

Sampling was carried out by the Potsdam branch of the Alfred Wegener Institute, in co-operation with the Geographical Faculty of the Moscow State University (1994 and 1995) and the Arctic and Antarctic Research Institute, St. Petersburg (1996) on three expeditions to the Siberian rivers. For detailed information on the expeditions, the reader is referred to the cruise reports (Rachold et al., 1995, 1997, and in press). Samples of the Lena and its tributaries were collected from July 9 to July 28, 1994 and from July 1 to August 2, 1995. At this time of the year, water level and SPM concentrations of the Lena were still high but lower than the maximum water and sediment discharge immediately after ice breakup in the beginning of June. The Yana and its tributaries, as well as the Omoloy, were sampled from August 8 to August 23, 1995. Yana water level and sediment discharge in 1995 reached maximum values in August, during the sampling period. Samples of the Khatanga were collected at the end of July 1996, three weeks after ice breakup. The sampling periods are indicated in Figure 2. The results reported in the present article reflect the summer situation of the Siberian rivers. Due to the severe climate in the drainage area, the winter season can almost be neglected in terms of mass fluxes. During the spring flood, on the other hand, water and sediment discharges may increase dramatically. This is especially true for the Lena whereas the spring flood is less pronounced for the Yana and Khatanga. However, the increase in total organic carbon (TOC) concentrations in the Lena is of minor significance. Average TOC concentrations during spring flood amounted to 1050 μmol/l and during summer and autumn to 620 μmol/l (Cauwet and Sidorov, 1996). Since our samples were collected in early summer, the data can be regarded as average for the warm season.

Water samples were obtained by a glass water sampler (1994) and a Teflon water sampler (1995 and 1996). The sampling locations are indicated in Figure 1. In general, water samples were retrieved from the middle of the river channel at a water depth of 1.5 m. Stations along the main rivers in 1995 and 1996 were sampled as river cross sections. At these stations, two or three samples distributed equally across the river were collected. In the following, we refer to average values calculated for each station. River cross sections are treated as field replicates. Samples taken at river cross sections exhibited small variations in carbon isotope ratios. The standard deviation amounted to less than ± 0.4 ‰ (Table 1).

The SPM was separated by vacuum filtration through precombusted 0.7 μm Whatman glass-fiber filters (GF/F) immediately after sampling. The filtered water volume varied between 1 and 4 l, depending on sediment load. After air-drying under dust-free conditions, the filters were stored dry and cool. A total of 90 samples corresponding to 61 stations were analyzed for total SPM content, organic C and N concentrations of the SPM, and for the stable carbon isotope composition of the POM at the Alfred Wegener Institute, Research Department Potsdam. In

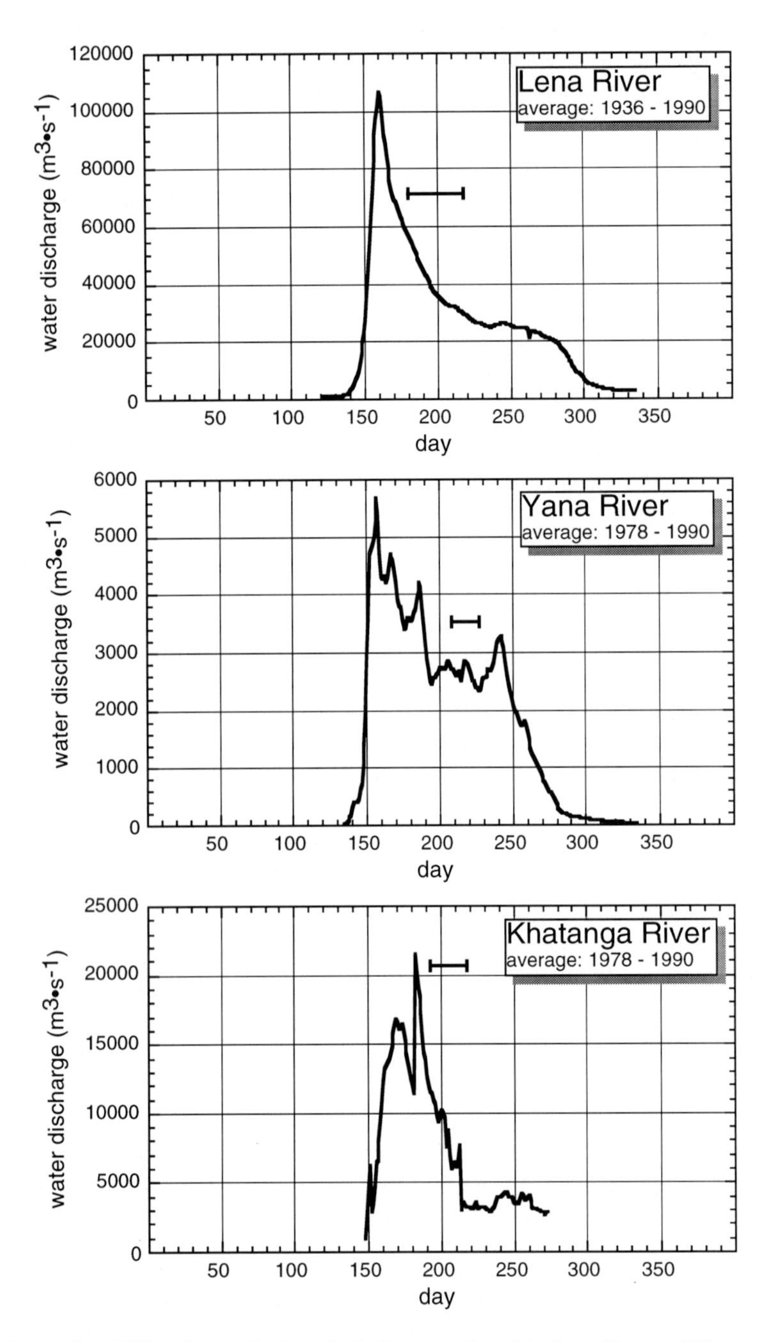

Figure 2: Seasonal variability of water discharge in the lower reaches of the Lena, Yana, and Khatanga. Average daily water discharge ($m^3 \cdot s^{-1}$) measured at hydrometeorological stations in the lower reaches of the rivers is presented (Russian Federal Service for Hydrometeorology and Environmental Monitoring, unpublished data). The periods of sampling for POM analyses are indicated by bars.

Table 1: Total SPM content, C, N concentrations of SPM, and $\delta^{13}C$ values of POM of rivers in the Lena, Yana, Omoloy, and Khatanga basins. Since some locations were sampled during two expeditions (1994 and 1995), station number and sampling year are indicated. The "location" column describes the distance (river-km) of the station from the river mouth. Sampling date, coordinates, number of field replicates (n), and standard deviation of $\delta^{13}C$ values are presented.

station, year	date	river	location	Lat	Long	Sed. Load	C	N	C/N	δ13C of POM	std. dev.	n
			river km	N	E	mg/l	%	%		‰ vs. V-PDB		
8, 1995	07/06/1995	Lena	km 2420	60°19.5´	120°15.8´	15.6	6.18	0.67	9.2	-26.9	0.02	2
5, 1995	07/05/1995	Lena	km 2384	60°27.4´	120°37.8´	19.0	7.69	0.78	10.2	-27.1	0.38	3
3, 1995	07/02/1995	Lena	km 1957	61°08.1´	126°58.3´	14.6	6.93	0.75	9.2	-26.3	0.15	3
1, 1994	07/09/1994	Lena	km 1670	62°01.3´	129°49.5´	14.4	5.15	0.55	9.3	-27.5	-	1
10, 1995	07/07/1995	Lena	km 1670	61°47.8´	129°39.4´	20.8	6.62	0.62	10.6	-27.9	0.15	3
2, 1994	07/13/1994	Lena	km 1497	63°13.3´	129°38,4´	9.7	8.83	0.96	9.2	-27.4	-	1
11, 1995	07/11/1995	Lena	km 1497	63°16.2´	129°34.9´	25.5	4.24	0.43	9.9	-27.4	-	1
4, 1994	07/14/1994	Lena	km 1347	63°55.6´	127°23.5´	27.3	4.08	0.39	10.4	-26.6	-	1
14, 1995	07/17/1995	Lena	km 1289	64°22.6´	126°30.0´	15.9	5.63	0.75	7.5	-28.6	0.27	2
7, 1994	07/16/1994	Lena	km 1174	64°47.2´	125°24.7´	65.9	3.09	0.29	10.5	-25.7	-	1
8, 1994	07/17/1994	Lena	km 1036	65°42.2´	124°18.1´	65.6	2.84	0.29	9.9	-25.7	-	1
16, 1995	07/21/1995	Lena	km 921	66°41.1´	123°29.5´	22.9	4.65	0.59	7.9	-28.8	0.19	3
10, 1994	07/20/1994	Lena	km 852	67°14.6´	123°12.8´	73.2	3.14	0.31	10.2	-26.0	-	1
12, 1994	07/20/1994	Lena	km 644	68°51.1´	124°00.3´	62.3	3.05	0.30	10.2	-26.2	-	1
19, 1995	07/24/1995	Lena	km 644	68°51.1´	123°57.8´	27.0	4.65	0.55	8.4	-27.6	0.01	3
13, 1994	07/21/1994	Lena	km 518	69°45.9´	125°03.2´	35.1	3.27	0.34	9.5	-26.9	-	1
15, 1994	07/22/1994	Lena	km 264	70°52.4´	127°28.5´	42.5	3.88	0.37	10.6	-27.0	-	1
20, 1995	07/26/1995	Lena	km 264	70°81.5´	127°33.1´	27.5	4.68	0.58	8.1	-28.7	-	1
16, 1994	07/22/1994	Lena	km 207	71°45.1´	127°13.6´	18.9	4.26	0.45	9.5	-27.0	-	1
17, 1994	07/23/1994	Bykovskaya branch	Lena delta	72°23.9´	126°53.4´	14.5	4.40	0.56	7.9	-27.3	-	1
18, 1994	07/23/1994	Bykovskaya branch	Lena delta	72°16.6´	127°52.7´	12.5	5.14	0.62	8.3	-27.2	-	1
20, 1994	07/24/1994	Bykovskaya branch	Lena delta	72°01.4´	129°08.3´	9.9	5.73	0.74	7.7	-28.3	-	1
25, 1995	07/30/1995	Bykovskaya branch	Lena delta	72°01.6´	129°07.8´	5.8	9.41	1.41	6.7	-31.3	-	1
26, 1995	07/31/1995	Bykovskaya branch	Lena delta	72°20.9´	127°28.7´	14.7	6.79	0.94	7.2	-29.0	-	1
21, 1994	07/25/1994	Trofimoskaya branch	Lena delta	72°26.2´	126°43.4´	39.5	3.73	0.40	9.4	-26.5	-	1
22, 1994	07/25/1994	Olenyokskaya branch	Lena delta	72°21.2´	126°32.9´	35.1	4.49	0.48	9.3	-26.7	-	1
22, 1995	07/27/1995	Olenyokskaya branch	Lena delta	72°19.3´	126°00.0´	11.6	6.54	1.01	6.6	-29.0	-	1
31, 1995	08/01/1995	Olenyokskaya branch	Lena delta	72°36.2´	124°27.3´	5.4	11.63	1.75	6.6	-29.0	-	1
6, 1995	07/05/1995	Chara	km 2398	60°16.1´	120°59.2´	5.7	9.65	0.91	10.6	-27.4	-	1
7, 1995	07/05/1995	Olekma	km 2395	60°16.1´	121°09.0´	1.2	36.82	3.66	10.1	-30.1	-	1
12, 1995	07/15/1995	Aldan	km 1474	63°25.1´	129°36.4´	7.8	9.15	1.17	8.0	-30.1	0.35	3

Table 1 (continued):

station, year	date	river	location	Lat	Long	Sed. Load	C	N	C/N	δ13C of POM	std. dev.	n
			river km	N	E	mg/l	%	%		‰ vs. V-PDB		
6, 1994	07/16/1994	Vilyuy	km 1266	64°15.4´	126°24.5´	27.6	4.12	0.52	7.9	-28.4	-	1
15, 1995	07/18/1995	Vilyuy	km 1266	63°57.8´	123°17.5´	14.5	5.79	0.72	8.1	-28.6	-	1
18, 1995	07/24/1995	Dzhardzhan	km 656	68°47.0´	124°06.7´	46.0	2.02	0.24	8.2	-25.0	-	1
44, 1995	08/13/1995	Yana	km 791	67°32.9´	134°01.9´	735.9	1.62	0.15	11.1	-25.5	-	1
43, 1995	08/12/1995	Yana	km 633	68°12.3´	134°54.7´	290.1	2.00	0.18	11.3	-25.4	-	1
42, 1995	08/12/1995	Yana	km 560	68°36.4´	134°39.6´	368.1	2.55	0.20	13.1	-26.3	-	1
41, 1995	08/11/1995	Yana	km 506	68°57.2´	134°27.8´	315.9	1.86	0.19	10.0	-25.2	-	1
40, 1995	08/10/1995	Yana	km 395	69°43.1´	135°04.4´	44.9	2.33	0.25	9.2	-26.1	-	1
39, 1995	08/09/1995	Yana	km 297	70°13.4´	135°06.9´	61.2	2.20	0.21	10.5	-25.9	0.00	2
38, 1995	08/09/1995	Yana	km 154	70°45.5´	136°04.0´	102.0	2.12	0.21	10.2	-26.0	0.03	2
36, 1995	08/09/1995	Yana	km 123	70°58.4´	136°29.4´	70.6	2.23	0.22	10.1	-26.0	-	1
45, 1995	08/13/1995	Sartang	km 810	67°27.3´	133°13.1´	898.4	1.83	0.18	10.4	-25.7	-	1
46, 1995	08/13/1995	Dulgalakh	km 810	67°29.6´	133°14.5´	1142.8	1.90	0.17	11.3	-25.5	-	1
47, 1995	08/17/1995	Tuostakh	km 625	67°51.9´	135°40.4´	31.6	3.66	0.25	14.8	-26.2	-	1
48, 1995	08/17/1995	Adicha	km 625	67°48.7´	135°27.3´	167.8	2.69	0.24	11.3	-26.2	-	1
49, 1995	08/18/1995	Olbye	km 580	68°24.3´	134°58.9´	114.9	2.73	0.23	12.0	-26.5	-	1
50, 1995	08/18/1995	Bytaktay	km 545	68°44.6´	134°20.4´	472.0	2.20	0.18	12.1	-25.7	-	1
59, 1995	08/21/1995	Omoloy	km 85	70°42.3´	133°28.6´	5.8	5.72	0.51	11.3	-27.2	-	1
58, 1995	08/21/1995	Kyugyulyur	km 80	70°42.6´	133°24.1´	8.6	7.09	0.55	12.9	-26.6	-	1
6, 1996	07/24/1996	Khatanga	km 205	71°58.4´	102°23.8´	52.2	2.58	0.27	9.5	-26.8	0.22	2
10, 1996	07/25/1996	Khatanga	km 160	72°16.4´	103°05.2´	32.6	2.98	0.35	8.5	-27.6	0.19	3
9, 1996	07/25/1996	Khatanga	km 98	72°36.6´	104°27.7´	22.5	3.33	0.41	8.2	-28.3	0.10	3
7, 1996	07/24/1996	Khatanga	km 62	72°46.2´	105°14.3´	22.7	3.68	0.42	8.9	-28.0	0.25	3
1, 1996	07/22/1996	Khita	km 320	71°34.1´	99°37.7´	66.7	2.38	0.26	9.0	-26.9	0.08	3
2, 1996	07/23/1996	Khita	km 270	71°45.1´	100°57.5´	68.2	2.73	0.28	9.6	-26.9	0.23	2
3, 1996	07/23/1996	Khita	km 225	71°54.1´	102°00.6`	112.7	2.42	0.22	11.1	-27.2	0.17	2
4, 1996	07/23/1996	Kotuy	km 295	71°30.6´	103°11.7´	22.4	3.00	0.33	9.2	-27.1	0.04	2
5, 1996	07/24/1996	Kotuy	km 225	71°51.9´	102°07.7´	13.8	3.10	0.32	9.8	-25.0	-	1
8, 1996	07/24/1996	Popigay	km 50	72°58.7´	106°08.8´	9.8	6.09	0.76	8.0	-28.9	0.25	2

addition, two peat samples collected in the Lena and Yana deltas and two coal samples from a river terrace near Zhigansk at the Lena and from a Khatanga tributary - the Kotuy - were included in the isotope analysis. Total SPM concentrations were calculated from the weight difference between the clean and the sediment loaded filters after freeze-drying. C, N concentrations were measured using a LECO elemental analyzer. $^{13}C/^{12}C$ isotope ratios were measured using a FINNIGAN DELTA S mass spectrometer after removal of carbonate with 1 N HCl in Ag-cups and combustion to CO_2 in a Heraeus elemental analyzer (Fry et al., 1992). Sample volumes were chosen to contain about 0.2 mg C. Accuracy of the analytical methods was checked by parallel analysis of international standard reference material (IAEA-CH-7). Results are expressed vs. V-PDB in the form:

$$\delta^{13}C\ (‰) = [(^{13}C/^{12}C_{sample} - {}^{13}C/^{12}C_{standard})/(^{13}C/^{12}C_{standard})] \bullet 1000.$$

The precision of the SPM determination is better than ± 1%, and that of the C and N analyses better than ± 4 %. The analytical precision of the carbon isotope analyses is ± 0.2 ‰.

Results

The results are presented in Table 1. Since some locations were sampled on two expeditions, both the station number and the sampling year are indicated. The "location" column indicates the distance (river-km) of the station from the river mouth in the Laptev Sea. For the tributaries, the distance of the confluences to the river mouth is given. If sampling was performed on cross sections, average values, standard deviation of $\delta^{13}C$ values, and number of samples are listed.

Total SPM content (turbidity) and organic C concentrations of the SPM of the three studied rivers varied within the wide range of 1.2 mg/l (one Lena tributary) to 1140 mg/l (Yana) and 1.6 % C (Yana) to 36.8 % C (one Lena tributary), respectively. Figure 3 displays the correlation between turbidity and organic C concentrations of the SPM. In general high SPM content corresponded to low organic C concentrations in all of the studied rivers. This general relationship between turbidity and particulate organic carbon (POC) has been reported for several other rivers (Meybeck, 1982; Cauwet and Mackenzie, 1993) including the Lena (Cauwet and Sidorov, 1996).

Figure 4 presents the $\delta^{13}C$ values of the POM along the Lena (a), Yana (b), and Khatanga (c) as well as along their tributaries. The scale on the x-axis corresponds to the distance (river-km) of the station from the river mouth. The stable carbon isotope ratio of Lena POM ranged from -25.7 ‰ to -28.8 ‰ with an average value of -27.1 ‰ and a standard deviation of ± 0.9 ‰ (n = 19) (Figure 4a). In the Lena delta, the POM was isotopically lighter (average: -28.2 ‰, standard deviation: ± 1.4 ‰, n = 9) ranging from -26.5 ‰ to -31.3 ‰. The data of most of the tributaries fell within the range of the Lena itself. The Dzhardzhan was characterized by a ^{13}C enrichment ($\delta^{13}C$ = -25.0 ‰) and the Olekma and Aldan by a depletion in ^{13}C ($\delta^{13}C$ = -30.1 ‰ in both cases). Except for the lowermost Lena, the variations between samples collected at the same locations in 1994 and 1995 were small. In the upper Lena and the Vilyuy the differences did not exceed 0.4 ‰. The stations of the lowermost Lena (12-1994, 19-1995 and 15-1994, 20-1995) exhibited more pronounced displacements of 1.4 and 1.7 ‰. Our Lena average of -27.1 ‰ is in general agreement with carbon isotope data of marine sediments located within the main Lena outflow into the Laptev Sea. While surface sediments of the southern Laptev Sea close to the Lena delta are characterized by $\delta^{13}C$ values of -26.4 ‰ that may be attributed to POM exported by the Lena, in the northern Laptev Sea $\delta^{13}C$ values increase to -23.7 ‰, reflecting the marine POM contribution (Erlenkeuser, 1996). C/N ratios in the Lena basin are in the range of 6.5 to 19.5 with an average of 9.2.

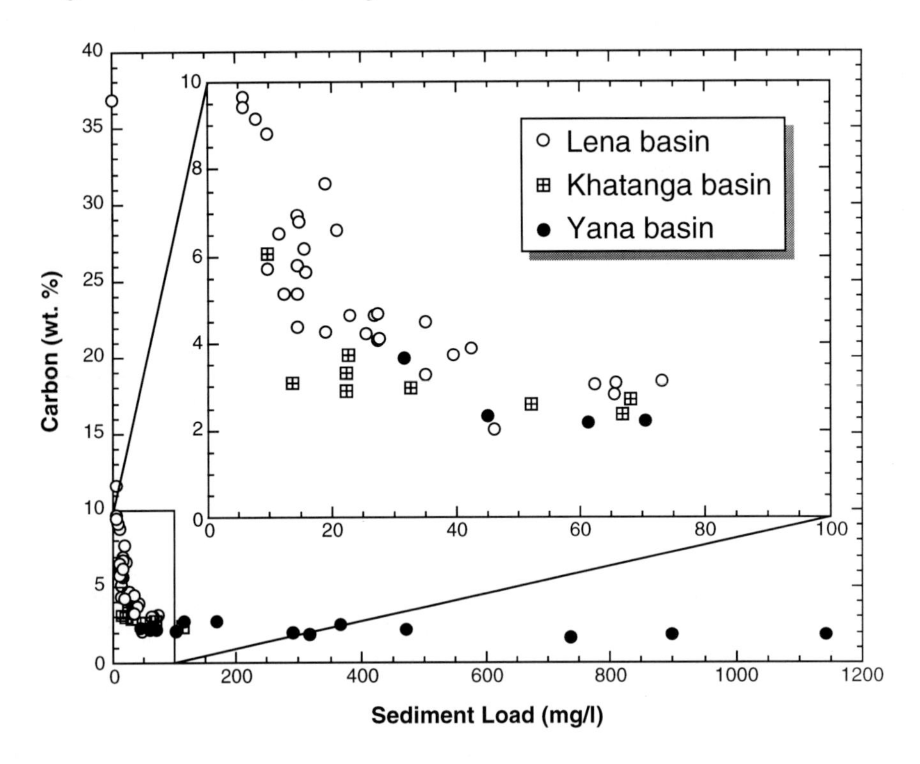

Figure 3: Relationship between total SPM content (x-axis, mg/l) and organic carbon concentrations of the SPM (y-axis, wt. %) in the Lena, Yana, and Khatanga and their major tributaries.

The Yana POM exhibited heavier $\delta^{13}C$ values within the narrow range of -25.2 ‰ to -26.3 ‰ and an average of -25.7 ‰ (standard deviation: ± 0.4 ‰, n = 8) (Figure 4b). The tributaries of the Yana did not differ significantly from the Yana. C/N ratios of the Yana POM averaged 11.2.

The carbon isotope ratios of POM of the Omoloy and its tributary, the Kyugyulyur, were -27.2 ‰ and -26.6 ‰, and the C/N ratios were 11.3 and 12.9, respectively.

The Khatanga was characterized by C/N ratios of 8.0 to 11.1 with an average of 9.2. The average $\delta^{13}C$ value of the Khatanga was -27.7 ‰, ranging from -26.8 to -28.3 ‰ (standard deviation ± 0.6 ‰, n = 4). While one tributary - the Popigay - showed a ^{13}C depletion ($\delta^{13}C$ = -28.9 ‰), one station on another tributary - the Kotuy - was characterized by a comparably heavy $\delta^{13}C$ value of -25.0 ‰ (Figure 4c).

In general, $\delta^{13}C$ values of coal and peat did not significantly differ from that of the SPM: coal from the Lena basin, -25.7 ± 0.1 ‰ (n = 3); coal from the Khatanga basin, -24.8 ± 0.4 ‰ (n = 8); peat from the Lena delta, -26.7 ± 0.4 ‰ (n = 2); and peat from the Yana delta, -26.9 ± 0.1 ‰ (n = 2).

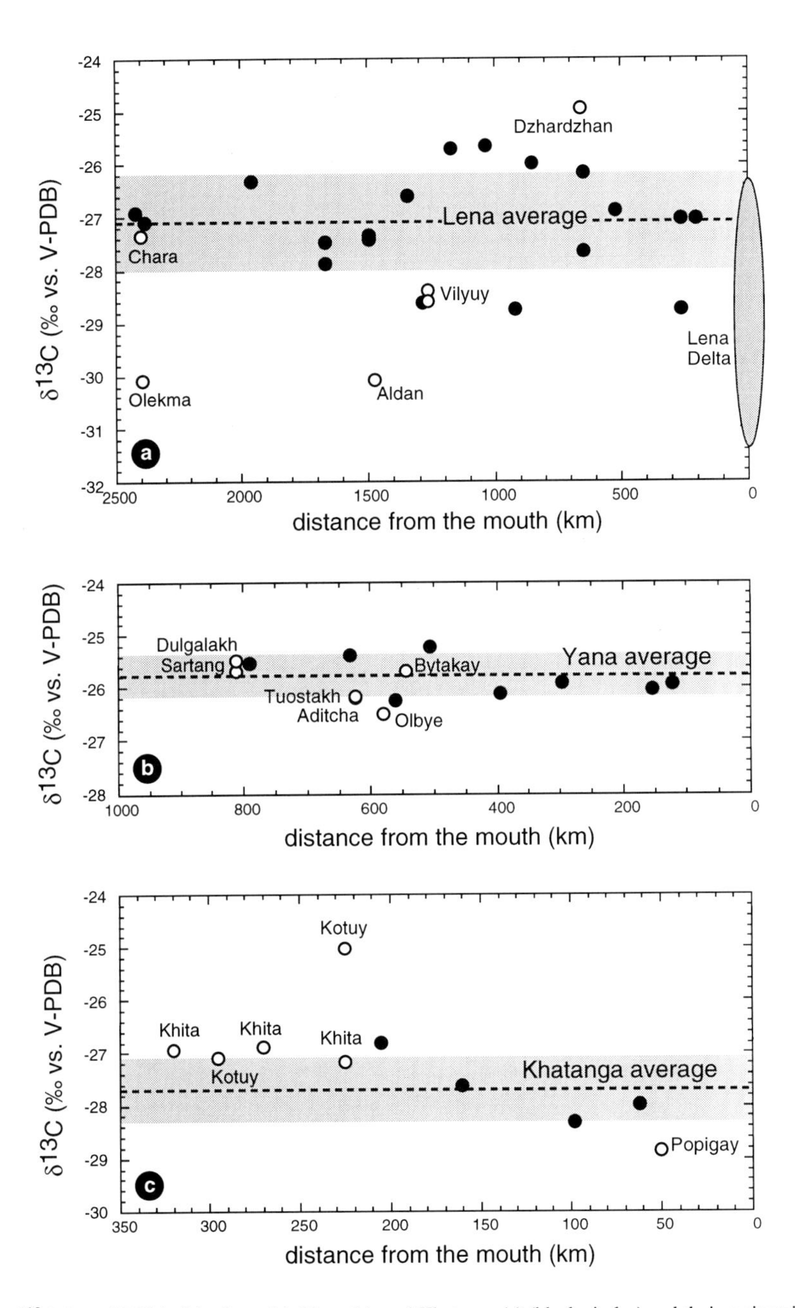

Figure 4: $\delta^{13}C$ data of POM of the Lena (a), Yana (b), and Khatanga (c) (black circles) and their major tributaries (white circles). The scale on the x-axis corresponds to the distance (river-km) of the station from the river mouths. For the tributaries the distance between confluence and mouth is given. Average $\delta^{13}C$ values of the Lena, Yana, and Khatanga are indicated by dotted lines. The shaded areas represent the standard deviation of $\delta^{13}C$ values within the rivers except for the tributaries.

Discussion

The relationship between TOC concentrations of the SPM and total sediment load (Figure 3) indicates that the observed variations in TOC can be explained by a two-component mixing model. In the following the stable carbon isotope composition of the POM will be discussed in order to identify endmembers and to quantify their contribution to the POM.

Two-component mixing model

Figure 5a presents the relationship between $\delta^{13}C$ values of the POM and reciprocal organic C concentrations in the SPM. In general, low C concentrations correspond to heavy $\delta^{13}C$ values and the data for each river studied plot linearly. The Yana is characterized by intermediate $\delta^{13}C$ values while the Lena exhibits slightly heavier and the Khatanga slightly lighter $\delta^{13}C$ values. However, the differences between the three rivers are small. C/N ratios and reciprocal organic C concentrations of the SPM exhibit a similar relationship. An increase in C concentrations is paralleled by a decrease in C/N ratios.

The linear relation between $\delta^{13}C$ values and reciprocal C concentrations of the SPM suggests that the variations in carbon isotope composition of POM can be attributed to mixing of two approximately constant endmembers (Faure, 1986). The hypothetical endmembers are indicated in Figure 5a. Endmember [d], which was extrapolated from the lowest observed C concentration, is characterized by a $\delta^{13}C$ value of -25 ‰. Most probably this endmember, which is almost constant within the three studied rivers, represents a continuous background organic C component. Endmember [a], which was extrapolated from 1/C -> 0, has a value of ca. -31 ‰. However, the isotope composition of endmember [a] varies slightly among the three rivers. As shown in Figure 5a, each river has a noticeably individual character. In order to fit all rivers into one parameter configuration of the model, the relationship between $\delta^{13}C$ values of POM and reciprocal POC concentrations in the water volume (as calculated from the sediment load and its TOC concentration: POC [mg/l] = SPM [mg/l] • TOC [%] • 10^{-2}) must be studied (Figure 5b). Figure 5b clearly shows that the variations in carbon isotope composition can be explained by two-component mixing of the hypothetical endmembers [a] and [d], except for one sample (station 5, 1996). Station 5, 1996 is located close to a coal mine at the Kotuy, a tributary of the Khatanga. The $\delta^{13}C$ value of -25.0 ‰ observed at station 5, 1996 is very similar to the carbon isotope ratio of the Kotuy coal, which averages -24.8 ‰. It thus seems apparent that the POM at this station is formed mainly from coal.

Quantification of endmember contributions

Before discussing possible sources for the hypothetical endmembers of our mixing model, we present a mixing calculation that enables us to quantify the contribution of the two endmembers in each sample. The relative proportions can be evaluated from

$$\delta^{13}C\ (POC) = f \cdot \delta^{13}C\ (\text{endmember [a]}) + (1-f) \cdot \delta^{13}C\ (\text{endmember [d]})$$

where f is the relative amount of endmember [a].

The results show that the studied POM is generally dominated by endmember [d], which comprises more than 50 % of most of the samples. The highest proportions of endmember [d], ranging from 79 to 96 % (average 87 %), are observed in the Yana POM, where total SPM concentrations are > 100 mg/l. In the Lena POM, endmember [d] ranges from 37 to 89 % (average 65 %), while in the POM of the Lena delta, at total SPM concentrations of < 10 mg/l, the contributions of endmember [d] only amount to 0 to 76 % (average 46 %). In the Khatanga POM, endmember [d] averages 55 %, ranging from 44 to 70 %.

Possible endmember sources

The riverine POC reservoir originates from two different sources: (1) detrital organic carbon which is formed from soil-derived terrigeneous material, eroded coal deposits, and reworked peat and (2) *in situ* produced riverine plankton.

In general, autochthonous phytoplankton is characterized by C/N ratios averaging ca. 6 to 7, while the C/N ratio of terrestrial plants is generally higher than 10. C/N ratios are controlled by several other parameters and therefore cannot be applied alone to identify the origin of organic material (Wetzel, 1983; Lerman et al., 1995). However, the general trend observed from our data suggests that samples of low organic C concentrations and high turbidity are dominated by terrestrial organic material whereas samples of high organic C concentrations and low turbidity are additionally influenced by autochthonous phytoplankton.

The carbon isotope composition of terrestrial plants making use of the C_3 photosynthetic pathway (trees are C_3 plants) represents an average of -25 to -26 ‰. Aerial plants utilizing the C_4 cycle, which are mainly grasses, are less depleted in ^{13}C and exhibit an average carbon isotope composition of -12 ‰ (O´Leary, 1981). Almost the entire catchment areas of the studied rivers are covered by Siberian taiga. Only the northernmost regions, i.e. the coastal zone and the Lena delta, are located within the tundra zone. For this reason, it is suggested that the detrital organic material within the rivers is strongly dominated by C_3 plants with an average $\delta^{13}C$ of -25 to -26 ‰. The $\delta^{13}C$ values of Lena coal exhibit a mean value of -25.7 ‰ and the Khatanga coal of -24.8 ‰. Peat samples from the Lena and Yana deltas are characterized by a carbon isotope composition of ca. -27 ‰. Peat deposits, however, are only observed in the river deltas and do not contribute to the detrital organic material in the upper reaches of the rivers. Terrestrial organic material and coal have very similar $\delta^{13}C$ values. Pooling these two components, the average carbon isotope composition of the detrital organic material in the catchment areas amounts to ca. -25 ‰. It therefore seems reasonable to regard endmember [d] of the mixing model indicated in Figure 5a and 5b as the detrital organic carbon fraction.

The carbon isotope composition of autochthonous riverine POM depends on the isotope fractionation between phytoplankton and the various fractions of dissolved inorganic carbon (DIC), i.e. dissolved CO_2, HCO_3^-, and CO_3^{2-}. The relative amounts of CO_2, HCO_3^-, and CO_3^{2-} dissolved in water are related to temperature and pH. The major sources of carbon contributing to the DIC pool are (a) CO_2 derived from the decay of organic material in soils and (b) CO_2 released during the dissolution of carbonates. In general, the uptake of atmospheric CO_2 in surface waters is negligibly small; in contrast, CO_2 loss by degassing is more common (Mook and Tan, 1991).

The carbon isotope composition of soil derived CO_2 yields values of around -26 ‰. For this reason, $\delta^{13}C$ values of DIC as low as -26 ‰ have been observed in drainage areas that are characterized by very low limestone content, large amounts of vegetation and a low pH (Longinelli and Edmond, 1983). However, most rivers belong to the Ca bicarbonate type and their isotope compositions are determined by the reaction of limestone and soil derived CO_2 that produces DIC $\delta^{13}C$ values of about -12 ‰ (Tan and Edmond, 1993). In the absence of vegetation, on the other hand, exchange with atmospheric CO_2 may affect the DIC and its $\delta^{13}C$ values can be as heavy as -7 ‰. The degassing of CO_2 may cause a further increase in $\delta^{13}C$ values (Mook and Tan, 1991).

During summer, the Lena belongs to the Ca bicarbonate type of river and its geochemistry is controlled mainly by weathering of limestones and by groundwater (Gordeev and Sidorov,

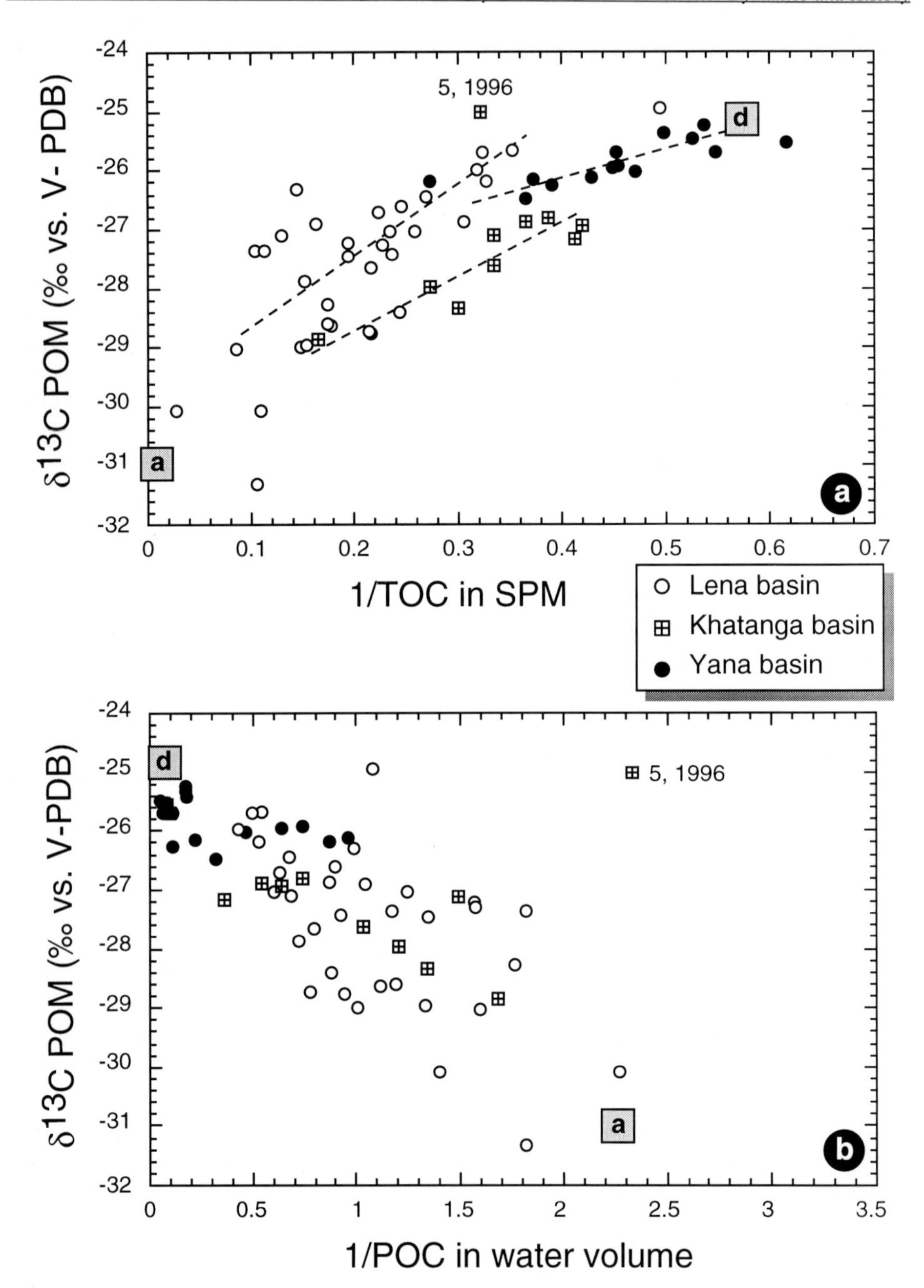

Figure 5: $\delta^{13}C$ values of POM vs. (a) 1/TOC concentrations in SPM and (b) 1/POC concentrations in the water volume (POC [mg/l] = SPM [mg/l] • TOC [%] • 10^{-2}) of the Lena, Yana, and Khatanga basins. The linear correlations suggest two-endmember mixing. Hypothetical detrital [d] and autochthonous [a] endmembers are presented. See text for explanation.

1993). The pH of the main river is about 8 (Zaitsev, pers. comm.) and DIC is mainly formed from HCO_3^-. Summer water temperatures are cold to moderate, averaging 10 to 13°C in the Lena delta during September (Létolle et al., 1993), with maximum temperatures of 18°C in the upper reaches during July. In September 1994, the DIC concentrations reached 700 µmol/l, corresponding to $\delta^{13}C$ values of -6 ‰ (Erlenkeuser, 1995). Similar DIC concentrations (about 500 to 600 µmol/l) were reported for September 1989 and 1991 by Cauwet and Sidorov (1996). The DIC data refer to the lowermost Lena and the Lena delta. There are no data available for the upper Lena. The heavy isotope ratio of Lena DIC may be attributed to the dominance of carbonate weathering. Moreover, CO_2 exchange with the atmosphere may further increase the $\delta^{13}C$ values of DIC downriver. No DIC data are available for the Yana and Khatanga. However, since $\delta^{13}C$ values of POM do not significantly differ from the Lena, we assume similar isotopic ratios for the DIC (although carbonates are less common in the Khatanga and Yana basins).

The isotope fractionation between plankton and the various fractions of DIC depends on their concentrations and on temperature, species and growth rates. The fractionation between dissolved CO_2 and plankton amounts to -10 ‰ to -15 ‰, depending on temperature, which is equivalent to a fractionation between HCO_3^- and plankton of -20 to -25 ‰ (Tan and Strain, 1983). For this reason, the $\delta^{13}C$ values of riverine plankton are generally lighter than -30 ‰. At very light $\delta^{13}C$ values of the DIC, the autochthonous plankton can be isotopically very light, up to -40 ‰ (Cai et al., 1988; Mariotti et al., 1991; Tan and Edmond, 1993).

Based on $\delta^{13}C$ values of -6 ‰ for Lena DIC, the dominance of HCO_3^-, and on a fractionation of ca. -15 ‰ between dissolved CO_2 and plankton in the cold to moderate temperatures of the Lena, $\delta^{13}C$ values of the autochthonous POM are expected to be around -31.0 ‰. This isotope ratio is in agreement with the postulated endmember [a] indicated in Figure 5a and 5b.

Unfortunately, no data are available on the seasonal variation of plankton development in the Lena. However, chemical data indicate that seasonal variations may be significant. For this reason, our data represent the summer situation only.

Organic carbon flux to the Laptev Sea

Based on the average organic C concentrations of SPM in the Lena, Yana, and Khatanga, which were 5.4 %, 2.1 %, and 3.1 %, respectively and on the annual sediment discharge as given by Rachold et al. (1996), we are able to evaluate the POC flux to the Laptev Sea. Since seasonal variations in C concentrations are not included in this calculation, the results must be regarded as rough extrapolations. By adding the dissolved organic carbon (DOC) flux of the Lena reported by Lara et al. (in press), we can also give an estimate of the TOC export of the Lena. Table 2 presents the results and several other estimates. While TOC, POC, and DOC data are available for the Lena, only very incomplete studies exist for the Yana and Khatanga . The available data, however, underline the importance of the riverine carbon fluxes of the Siberian rivers for the Laptev Sea, and especially the contribution of the Lena.

Conclusions

Carbon isotope data indicate that POM of Siberian rivers is generally formed from two endmembers. The first endmember [d] represents detrital organic material and is characterized by a $\delta^{13}C$ value of ca. -25.0 ‰. The second endmember [a] is isotopically lighter, averaging ca. -31.0 ‰, and may be attributed to autochthonous riverine phytoplankton. Mixing calculations indicate that riverine POM is formed mainly from detrital organic material. This is in general agreement with the small amount of total phytoplankton biomass in the Lena (Cauwet and Sidorov, 1996; Sorokin and Sorokin, 1996). However, in the Lena delta, at total SPM concentrations < 10 mg/l, the autochthonous endmember becomes more important. Accordingly, higher amounts of phytoplankton biomass have been observed in the delta (Sorokin and Sorokin, 1996).

Although the data base of the TOC, POC, and DOC export of the Siberian rivers is still very limited, existing data indicate the significance of Siberian river runoff for the Laptev Sea.

Table 2: Organic carbon fluxes of Siberian rivers draining to the Laptev Sea: (1) this study; (2) Lara et al., in press; (3) Romankevich and Artemyev, 1985; (4) Cauwet and Sidorov, 1996; (5) Telang et al., 1991.

	POC[1] $t \cdot yr^{-1}$	DOC[2] $t \cdot yr^{-1}$	TOC[1] $t \cdot yr^{-1}$	TOC[3] $t \cdot yr^{-1}$	TOC[4] $t \cdot yr^{-1}$	TOC[5] $t \cdot yr^{-1}$
Lena River	$1.2 \cdot 10^6$	$3.6 \cdot 10^6$	$4.8 \cdot 10^6$	$5.06 \cdot 10^6$	$5.3 \cdot 10^6$	$4.8 \cdot 10^6$
Yana River	$0.06 \cdot 10^6$	-	-	-	-	$0.2 \cdot 10^6$
Khatanga River	$0.04 \cdot 10^6$	-	-	-	-	-

Acknowledgments

The authors wish to thank A. M. Alabyan, V. N. Korotaev, and A. A. Zaitsev for organizing the "Lena 94" and "Lena/Yana 95" expeditions and A. W. Ufimzev, who arranged most of the "Khatanga 96" expedition. Special thanks to the captains and crew members of RV "Prof. Makkaveev", "Sarya 9", and RV "Lotsman". L. Schönicke and M. Stapke are gratefully acknowledged for supporting the analytical work. The comments of U. Wand, H. Oberhänsli, of the reviewer M. E. Böttcher, and of one anonymous reviewer helped to improve the manuscript. This study is funded by the German Ministry of Education, Science, Research and Technology (BMBF) within the "Verbundvorhaben System Laptev See". This is publication No 1357 of the Alfred Wegener Institute for Polar and Marine Research.

References

Alabyan, A.M., R.S. Chalov, V.N. Korotaev, A.Y. Sidorchuk, and A.A. Zaitsev (1995) Natural and technogenic water and sediment supply to the Laptev Sea. Reports on Polar Research, 176, 265-271.

Bischof, J., J. Koch, M. Kubisch, R.F. Spielhagen, and J. Thiede (1990) Nordic Seas surface ice drift reconstructions: evidence from ice rafted coal fragments during oxygen isotope stage 6. In: Glacimarine environments: processes and sediments, Dowdeswell, J.A. and J.D. Scourse (eds.), Geological Society, London (Geological Society Special Publ. No 53), 235-251.

Cai, D., F.C. Tan and J.M. Edmond (1988) Sources and transport of particulate organic carbon in the Amazon river and estuary. Estuarine, Coastal and Shelf Science, 26, 1-14.

Cauwet, G. and F.T. Mackenzie (1993) Carbon inputs and distribution in estuaries of turbid rivers: the Yang Tze and Yellow rivers (China). Mar. Chem., 43, 235-246.

Cauwet, G. and I.S. Sidorov (1996) The biogeochemistry of Lena River: organic carbon and nutrient distribution. Mar. Chem., 53, 211-227.

Degens, E.T., S. Kempe and J.E. Ritchey (1991) Biogeochemistry of major world rivers. Wiley, New York (SCOPE 42), 356pp.

Dethleff, D., D. Nürnberg, E. Reimnitz, M. Saarso and Y.P. Savchenko (1993) East Siberian Arctic Region Expedition ´92: The Laptev Sea - Its significance for Arctic sea-ice formation and transpolar sediment flux. Reports on Polar Research, 120, 1-44.

Eicken, H., E. Reimnitz, V. Alexandrov, T. Martin, H. Kassens and T. Viehoff (1996) Sea-ice processes in the Laptev Sea and their importance for sediment transport. Continental Shelf Res., 17, 205-233.

Erlenkeuser, H. (1995) Stable carbon isotope ratios in the waters of the Laptev Sea/Sep. 94. Reports on Polar Research, 176, 170-177.
Erlenkeuser, H. (1996) Stable carbon isotope composition of the Laptev Sea sediments. Terra Nostra, 96/9, 87.
Faure, G. (1986) Principles of isotope geology, 2nd Edition. Wiley, New York, 589pp.
Fontugne, M.R. and J.M. Jouanneau (1987) Modulation of the particulate organic flux to the ocean by a macrotidal estuary: evidence from measurements of carbon isotopes in organic matter from the Gironde system. Estuarine, Coastal and Shelf Science, 24, 377-388.
Fry, B., W. Brand, F.J. Mersch, K. Tholke and R. Garritt (1992) Automated analysis system for coupled $\delta^{13}C$ and $\delta^{15}N$ measurements. Anal. Chem., 64, 288-291.
Gordeev, V.V. and I.S. Sidorov (1993) Concentrations of major elements and their outflow into the Laptev Sea by the Lena River. Mar. Chem., 43, 33-45.
Kassens, H. and V. Karpiy (1994) Russian-German cooperation: the Transdrift 1 Expedition to the Laptev Sea. Alfred Wegener Institute, Bremerhaven (Reports on Polar Research 151), 168pp.
Kassens, H., H.-W. Hubberten, S.M. Pryamikov and R. Stein (1994) Russian-German cooperation in the Siberian shelf seas: Geo-System Laptev Sea. Alfred Wegener Institute, Bremerhaven (Reports on Polar Research 144), 133pp.
Kassens, H., D. Piepenburg, J. Thiede, L. Timokhov, H.-W. Hubberten and S.M. Pryamikov (1995) Russian-German Cooperation: Laptev Sea System. Alfred Wegener Institute, Bremerhaven (Reports on Polar Research 176), 387pp.
Lara, R., V. Rachold, G. Kattner, H.-W. Hubberten, G. Guggenberger and A. Skoog (in press) Dissolved organic matter and nutrients in the Lena river, Siberian Arctic: characteristics and distribution. Mar. Chem.
Lerman, A., D. Imboden and J. Gat (1995) Physics and chemistry of lakes. Springer, Berlin, 334pp.
Létolle, R., J.M. Martin, A.J. Thomas, V.V. Gordeev, S. Gusarova and I.S. Sidorov (1993) ^{18}O abundance and dissolved silicate in the Lena delta and Laptev Sea (Russia). Mar. Chem., 43, 47-64.
Longinelli, A. and J. M. Edmond (1983) Isotope geochemistry of the Amazon basin: a reconnaissance. J. Geophys. Res., 88, 3703-3717.
Mariotti, A., F. Gabel, P. Giresse and Kinga-Mouzeo (1991) Carbon isotope composition and geochemistry of particulate organic matter in the Congo River (Central Africa): Application to the study of Quaternary sediments off the mouth of the river. Chem. Geol., 86, 345-357.
Meybeck, M. (1982) Carbon, nitrogen, and phosphorous transport by world rivers. Amer. J. Sci., 282, 401-450.
Milliman, J.D. and R.H. Meade (1983) World-wide delivery of river sediment to the oceans. J. Geol., 91, 1-21.
Mook, W.G. and F.C. Tan (1991) Stable carbon isotopes in rivers and estuaries. In: Biogeochemistry of major world rivers, Degens, E.T., S. Kempe and J.E. Ritchey (eds.), Wiley, New York (SCOPE 42), 245-264.
Nürnberg, D., I. Wollenburg, D. Dethleff, H. Eicken, H. Kassens, T. Letzig, E. Reimnitz and J. Thiede (1994) Sediments in Arctic sea ice: Implications for entrainment, transport and release. Mar. Geol., 119, 185-214.
O'Leary, M. (1981) Carbon isotope fractionation in plants. Phytochemistry, 20, 553-567.
Quay, P.D., D.O. Wilbur, J.E. Richey, J.I. Hedges and A.H. Devol (1992) Carbon cycling in the Amazon River: Implications from the ^{13}C compositions of particles and solutes. Limnol. Oceanogr., 37, 857-871.
Rachold, V., J. Hermel and V.N. Korotaev (1995) Expedition to the Lena River in July/August 1994. Reports on Polar Research, 182, 181-195.
Rachold, V., A.M. Alabyan, H.-W. Hubberten, V.N. Korotaev and A.A. Zaitsev (1996) Sediment transport to the Laptev Sea - hydrology and geochemistry of the Lena river. Polar Research, 15, 183-196.
Rachold, V., E. Hoops, A.M. Alabyan, V.N. Korotaev and A.A. Zaitsev (1997) Expedition to the Lena and Yana Rivers June-September 1995. Reports on Polar Research, 248, 193-210.
Rachold, V., E. Hoops and A.V. Ufimzev (in press) Expedition to the Khatanga River July-August 1996. Reports on Polar Research.
Reimnitz, E., D. Dethleff and D. Nürnberg (1994) Contrasts in Arctic shelf sea-ice regimes and some implications: Beaufort Sea versus Laptev Sea. Mar. Geol., 119, 215-225.
Romankevich, E.A. and V.E. Artemyev (1985) Input of organic carbon into seas and oceans bordering the territory of the Soviet Union. In: Transport of carbon and minerals in major world rivers, Part 2, Degens, E.T., S. Kempe and R. Herrera (eds.), Mitt. Geol. Palaeontol. Inst. Univ. Hamburg (SCOPE 58), 459-469.
Russian Federal Service for Hydrometeorology and Environmental Monitoring (Roshydromet), State Hydrological Institute, St. Petersburg (unpublished data).
Sorokin, Y.I. and P.Y. Sorokin (1996) Plankton and primary production in the Lena River Estuary and in the South-eastern Laptev Sea. Estuarine, Coastal and Shelf Science, 43, 399-418.
Tan, F.C. and P.M. Strain (1983) Sources, sinks and distribution of organic carbon in the St Lawrence Estuary. Geochim. Cosmochim. Acta, 47, 125-132.
Tan, F.C. and J.M. Edmond (1993):Carbon isotope geochemistry of the Orinoco Basin. Estuarine, Coastal and Shelf Science, 36, 541-547.

Telang, S.A., R. Pocklington, A.S. Naidu, E.A. Romankevich, I.I. Gitelson and M.I. Gladyshev (1991) Carbon and mineral transport in major North American, Russian Arctic, and Siberian Rivers: The St. Lawrence, the Mackenzie, the Yukon, the Arctic Alaskan Rivers, the Arctic Basin Rivers in the Soviet Union, and the Yenisei. In: Biogeochemistry of major world rivers, Degens, E.T., S. Kempe and J.E. Ritchey (eds.), Wiley, New York (SCOPE 42), 75-104.

Thornton, S.F. and J. McManus (1994) Application of organic carbon and nitrogen stable isotopes and C/N ratios as source indicators of organic matter provenance in estuarine systems: Evidence from the Tay Estuary, Scotland. Estuarine, Coastal and Shelf Science, 38, 219-233.

Wetzel, R.G. (1983) Limnology. Saunders, Philadelphia. 753pp.

Distribution of River Water and Suspended Sediment Loads in the Deltas of Rivers in the Basins of The Laptev and East-Siberian Seas

V.V. Ivanov and A.A. Piskun

State Research Center - Arctic and Antarctic Research Institute, 38 Bering St., 199226 St. Petersburg, Russia

Received 3 March 1997 and accepted in revised form 3 March 1998

Abstract - The article presents estimates of the annual and seasonal distribution of river water and suspended sediment load and their variability in complicated fork river deltas of the Laptev and East-Siberian Seas. The estimates are based on the methodology, which includes a complex of developed or improved methods for hydraulic-morphometric parameterization of channels, hydraulic calculations and physical and numerical modeling. A particular emphasis is given to rivers Lena and Kolyma deltas that belong to different types according hydrologic regime. In the Lena delta only its lower part is influenced by the sea. In the Kolyma delta the whole of the water area including the head of the delta experiences the influence of the sea. The main factor affecting water runoff and sediments distribution in the Lena delta branches is river runoff. For the Kolyma delta under the conditions of steady water motion the water content ratios of the main arms remain quite stable at different combinations of water discharges in the river and background sea level. At large surge level rises there is a considerable redistribution of water discharges over the delta area, reverse currents are observed (negative discharge values). A comparative analysis of estimates of the water runoff distribution in the Kolyma delta obtained by different methods shows that hydraulic calculations are sufficient for most applied tasks. Further studies should be focused on developing methods to calculate sediment transport for assessing the distribution of pollutants in deltas and their export to the sea.

Introduction

The river runoff is the main factor that governs the hydrological, hydrochemical and hydrobiological processes occurring in the deltas. It also significantly influences the water balance, water exchange, ice thermal conditions, formation of structure and chemistry of water and bottom sediments of sea areas, primarily in the regions of inflow of rivers and large delta arms.

Distribution of water discharges and levels in the deltas of rivers of the Laptev and East-Siberian Seas depend on the water and channel regime of these deltas (especially on the character of the runoff distribution within a year) and on the sea regime (primarily on the background sea level fluctuations). From this viewpoint the deltas of large rivers of the seas mentioned, can be divided into two groups. The deltas where the sea influence extends over the entire delta, including its head, i.e. the water runoff and levels distribution depends on the sea regime, belong to the first group. These are the Kolyma and Olenek rivers. The second group includes the deltas whose water and channel regime depend on the sea influence only in the lower reaches. These are the Lena and Yana and Indigirka rivers which are subjected to backwater from the sea only during the short catastrophic surges. For all indicated deltas the influence of tides on the redistribution of water and sediment load discharges is insignificant (Ivanov, 1970). The mouths of the Khatanga and Anabar Rivers do not have developed deltas and are not considered here.

The regime of deltas of the first group and of the lower reaches of the arms of deltas of the second type, situated in the sea influence zone, depends on the seasonal variations of the runoff

----

In: Kassens, H., H.A. Bauch, I. Dmitrenko, H. Eicken, H.-W. Hubberten, M. Melles, J. Thiede and L. Timokhov (eds.) Land-Ocean Systems in the Siberian Arctic: Dynamics and History. Springer-Verlag, Berlin, 1999, 239-250.

of water and sediment load. It also depends on sea level oscillations within seasonal and synoptic time scales. Depending on the combination of the water content and background sea level, the distribution of water and sediment load discharges does not remain constant. In the general case, the water and sediment load discharges in some branches can change not only the value, but also the direction of the water flow. In channel forked of the second type beyond the sea influence zone, the distribution of water and sediment load discharges completely depends on the water content of the river. Due to difficulties in measuring water and sediment load discharges in the intensive sea influence zone, systematic observations of these regime elements were organized and are only carried out at the downstream sections of rivers and at the fork nodes of the delta arms beyond the sea influence. Table 1 presents water and sediment load discharges, based on downstream measuring sections (Figure 1) and Figure 2 their distribution within a year (Hydrological Yearbooks, 1936-1993). As follows from Table 1, these observations provide fairly complete information on the multiyear and within a year variability of the water and sediment load discharge to the deltas. Observations of the water and sediment load discharge at permanent sites in the main delta arms of Lena are carried out systematically and are published in the editions of the State Water Cadastre on river mouths (Yearly data, 1977-1993). As to the other deltas only data of the expedition observations of the AARI (Ivanov, 1963; Nalimov, 1965; Ivanov, Piskun, 1995), Arktikproject (Turanov, 1960), MGU and other institutions (Bogomolov et al., 1979; Chalov et al., 1995) are available. These observations were as a rule, of reconnaissance character, frequently without a reliable plan-altitude reference, which did not allow reliable comparisons of measurements made in the delta arms by different institutions in various years. Any data on rivers Alaseya and Olenek deltas run-off is absent.

Table 1: Knowledge of water runoff and sediment discharge to the mouths of rivers in the basins of the Laptev and East-Siberian Seas.

River	Downstream section	Distance, km		Observation period of	
		from the mouth	from the delta head	water runoff	sediment discharge
Olenek	7.5km below the Buur river mouth	236.5	209.5	1964-present	1968-present
Lena	Kusyur	317	211	1934-present	1960-present
Lena	4.7km up the Stolb island	111	4.7	1950-present	1970-present
Yana	Dzhangky	391	247	1938-present	1962-present
Yana	Yubileinaya	158	14	1972-present	1973-present
Indigirka	Vorontsovo	350	223	1936-present	1956-86
Alazeya	Andryushkino	529	521	1968-71, 1978-present	1978-present
Kolyma	Srednekolymsk	641	536	1927-present	1966-present
Kolyma	Kolymskoye	273	168	1977-present	1977-present

This article presents the results of applying a comprehensive methodology developed by the authors for estimating the distribution and redistribution of the water runoff and sediment load discharge for sparsely investigated complicated fork deltas of the rivers in the arctic zone by the example of the Kolyma delta. This methodology allows not only averaged characteristics to be obtained, but also the typical features of their change depending on the river runoff and sea level regime. Also, some estimates of the other river deltas in the Laptev Sea Basin that were not included into a previous publication (Ivanov and Piskun, 1995) are presented. The distribution of water and sediment load discharge in the arms of the Indigirka delta based on full-scale studies are given. Yana delta is not examined in this paper as estimations of water run-off and suspended sediment load distribution in its branches made on the base of full-scale observations data and hydraulic calculations were stated earlier by the authors (Ivanov and Piskun, 1995).

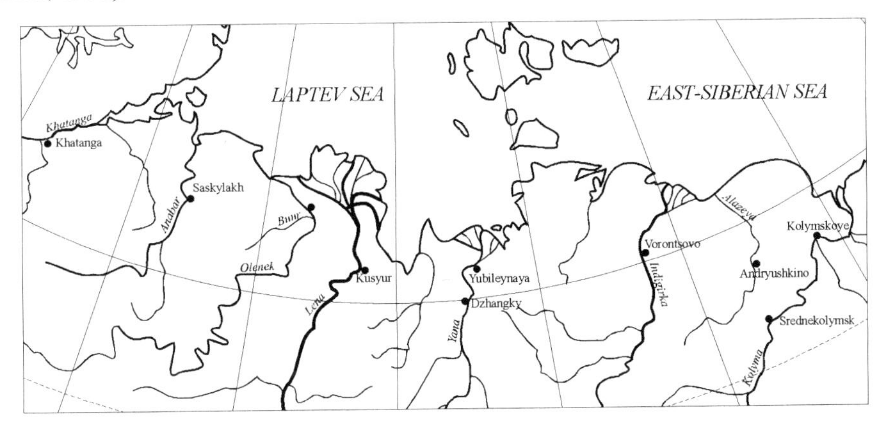

Figure 1: A general layout of river deltas in the basins of the Laptev and east Siberian Seas.

Methods and materials

In the general case, data of measurements in sparsely investigated river deltas are always insufficient for estimating the distribution and redistribution of water and sediment load discharges. This is attributed to a complicated character of the delta forks in the zone of river and sea influence, severe climatic conditions making expedition studies difficult. Hence, full-scale studies should be minimized to an extent sufficient to obtain initial information for hydraulic calculations, physical and numerical modeling of water and sediment load dynamics.

Based on multiyear experience of AARI studies of the water and channel regime of river deltas in the arctic zone, the following is undertaken in succession for estimating the distribution of water and sediment load discharges: collection, generalization and analysis of historical information on water discharges and levels, sediments and hydrographic surveys of channels in the deltas; calculation of hydraulic-morphometric characteristics and hydraulic resistance by delta arms (as initial information for hydraulic calculations and hydrodynamic modeling) (Grishanin, 1967; Ivanov, 1968; Piskun, 1978). If initial data on the resistance of arms at different channel filling are insufficient, full-scale measurements of water discharges and levels and hydrographic surveys are organized and small-scale aerodynamic modeling is performed (Gilyarov and Ivanov, 1960, 1967, 1975). Also, aerodynamic models are used for verifying different design delta regulation variants for improving shipping conditions, and lately for addressing ecological problems. Depending on the aims and goals for which estimates of the distribution of water discharges and levels are made, hydraulic calculations are performed (Ivanov, 1968; Ivanov et al., 1983, 1990; Piskun, 1992) and, if necessary, hydrodynamic modeling (Ivanov and Kotrekhov, 1976). The methods of hydraulic calculations are addressed to engineering problems or to the collection of initial information for hydrodynamic models, including models for transfer of pollutants.

According to the methodological approach presented above, a number of estimates and studies of the distribution and redistribution of the water and suspended sediment load discharge in the river deltas in the arctic zone were made. This enables us to have now quite a

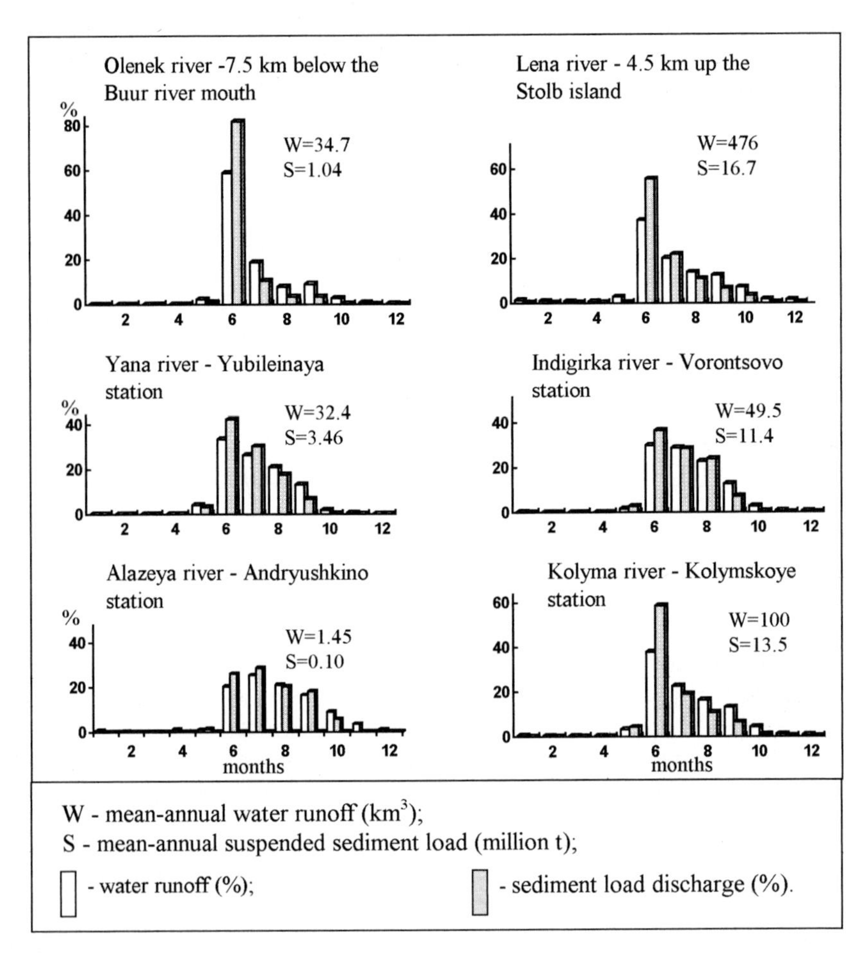

Figure 2: Distribution within a year of water and suspended sediment load discharges at the downstream sections of rivers in the basins of the Laptev and East Siberian Seas.

clear understanding of all the main aspects of the problem under consideration.

Let us consider the estimates of the distribution of the water and suspended sediment load discharges in the river deltas of the basins of the Laptev and East-Siberian Seas taking into account the level of our knowledge.

Results

The history of the studies of river deltas of the Laptev Sea Basin and the water and sediment load discharge distribution by their arms has been presented in (Ivanov and Piskun, 1995). In the present article we consider new results, in particular, more detailed estimates of the water and suspended sediment load discharge by the main arms of the Lena delta. Figure 3 presents general understanding of the distribution of water and suspended sediment load discharges over the Lena delta and Table 2 gives their statistical characteristics.

As follows from Table 2 at the variation coefficients of mean monthly water discharges (in

June-September) for the main channel of Lena (section I in Figure 3) of 0.14-0.26, their value for the Bykovskaya and Trofimovskaya branches (section II and III, respectively) is 0.16-

Table 2: Characteristics of mean monthly and annual water discharges and suspended sediment load in the Lena delta. Note: Q,R - are discharges of water and suspended sediment loads; C_{vQ}, C_{vR} - are variation coefficients of water discharges.

Carac-teristic	I	II	III	IV	V	VI	VII	VIII	IX	X	XI	XII	year
						Main Channel (section I)							
Q,m^3/s	2740	2220	1710	1440	4610	62600	38700	26000	22400	12900	3340	2860	15100
C_vQ	0.21	0.29	0.34	0.33	1.12	0.14	0.20	0.24	0.26	0.24	0.25	0.22	0.12
R,kg/s						3450	1320	623	437	208			517
C_vR						0.43	0.52	0.77	0.70	0.95			0.32
						Bykovskaya branch (section II)							
Q,m^3/s	509	376	267	209	1000	19200	10500	6500	5670	3160	676	550	4070
C_vQ	0.34	0.42	0.50	0.51	1.32	0.16	0.23	0.28	0.28	0.28	0.27	0.30	0.14
R,kg/s						834	279	140	93.5	44			120
C_vR						0.36	0.59	0.74	0.66	0.86			0.30
						Trofimovskaya branch (section III)							
Q,m^3/s	2360	2060	1700	1470	3830	37300	24300	17300	14600	8650	2380	2260	9860
C_vQ	0.12	0.16	0.20	0.19	0.78	0.16	0.18	0.23	0.28	0.25	0.23	0.18	0.14
R,kg/s						2160	992	574	326	148			363
C_vR						0.41	0.39	0.58	0.64	-			0.36
						Tumatskaya branch (section IV)							
Q,m^3/s	22.6	9.31	3.31	0.96	318	6010	2740	1470	1100	580	74.8	32.8	1030
C_vQ	1.72	2.09	2.22	2.96	1.53	0.24	0.32	0.41	0.50	0.53	0.82	1.32	0.26
R,kg/s						442	189	72.2	30.9	17.1			62.2
C_vR						0.49	1.17	1.07	0.92	-			0.44
						Olenekskaya branch (section V)							
Q,m^3/s	98.3	60.1	25.9	10.4	292	5910	2800	1630	1280	710	163	134	1080
C_vQ	0.85	1.12	1.27	1.90	1.61	0.28	0.29	0.34	0.40	0.42	0.62	0.76	0.19
R,kg/s						304	80.6	44.3	25.1	15.1			39.9
C_vR						1.08	0.42	0.59	0.80	-			0.68

0.28, for Tumatskaya branch (section IV) - 0.24-0.50 and for Olenekskaya branch (section V) - 0.28-0.40.

Of interest are the results of comparison of initial data on water turbidity. It turned out that variations of daily water turbidity and its values for the main channel - 4.7 km upstream Stolb Island (section I in Figure 3) and in the Bykovskaya branch (section II) were practically equal. Thus in 1969, close to a small water content year, and in 1974, close to a large water content year, the correlation coefficients of relations of daily water turbidity in June-September for sections I and II turned out to be practically equal to 1. This indicates that it is appropriate to use the turbidity values in the main channel for estimating the sediment load discharge in the Bykovskaya branch. Accordingly, a sufficiently close consistency of the variations of mean monthly turbidity values at permanent hydrometric sections in the other delta arms with that of the main channel (section I) can be observed. However, for the Tumatskaya (section IV) and Olenekskaya (section V) branches in separate cases (July, August 1983 and June 1984) a multiple excess of monthly turbidity compared to its values at the hydrometric section in the main channel was observed. The reason for this must be in intensive local erosion of plots located higher than the above mentioned sections, but lower than section I. In any case, it is necessary to take into account an approximate character of data on the runoff of suspended sediment load at the delta sections. This is constantly stressed in the editions of the "Annual Data on the Regime and Water Quality of the Seas and Marine River Mouths".

In view of the observed consistency between the variations of mean monthly water turbidity at the hydrometric sections in the Lena delta arms and in its main channel, this feature can be used for calculating the runoff of suspended sediment load at these sections, based on the actual turbidity data at the hydrometric section in the main channel - 4.7 km upstream Stolb Island and data on water discharges at the sections in the arms. The comparison of thus calculated mean monthly discharges of sediment loads at the sections in the delta arms with the measured sediment load values over the 1977 to 1993 period has shown the correlation coefficients to be 0.89 to 0.96.

The relations of mean monthly suspended sediment load and water discharges at permanent hydrometric sections in the Lena delta for June-September are subordinated to the power type dependencies with correlation coefficients within 0.88-0.92. This allows these dependencies to be used for approximate estimates of the sediment load discharge at the indicated sections (in case the observations of both water turbidity and discharges are interrupted), based on the level data at the gauge of Stolb Island and multiyear curves of water discharges for these sections.

It should be stressed that the presented data characterize the discharge of suspended sediment load at the head of the Lena delta. They cannot serve as a direct indicator of the amount of suspended material exported to the sea, especially on a seasonal scale. This problem needs special research as delta processes are rather complicated.

The level of knowledge of the distribution of water and suspended sediment load discharge in the river deltas of the East-Siberian Sea Basin is very non-unoform. In terms of full-scale observations all deltas are poorly studied. Only the Kolyma delta in view of the largest commercial value was investigated using the entire complex of means: full-scale observations, hydraulic calculations, physical and numerical modeling. This allowed the estimation of the typical spatial-temporal features of the distribution of water and sediment load discharge at different characteristics of the water content and sea regime.

The Kolyma delta begins 3 km downstream the Chersky settlement where the river bifurcates into two branches - Kamennaya and Chernousovskaya. In 10 km downstream from the Chersky settlement to the left of Kamennaya there is a cross-branch Markhoyanovskaya which merging with Chernousovskaya, creates the main left arm - the Pokhodskaya branch (Figure 4). The Kamennaya branch is the main delta arm with sea shipping over its entire length. The Kamennaya and Pokhodskaya branches are connected by the Poperechnaya I and

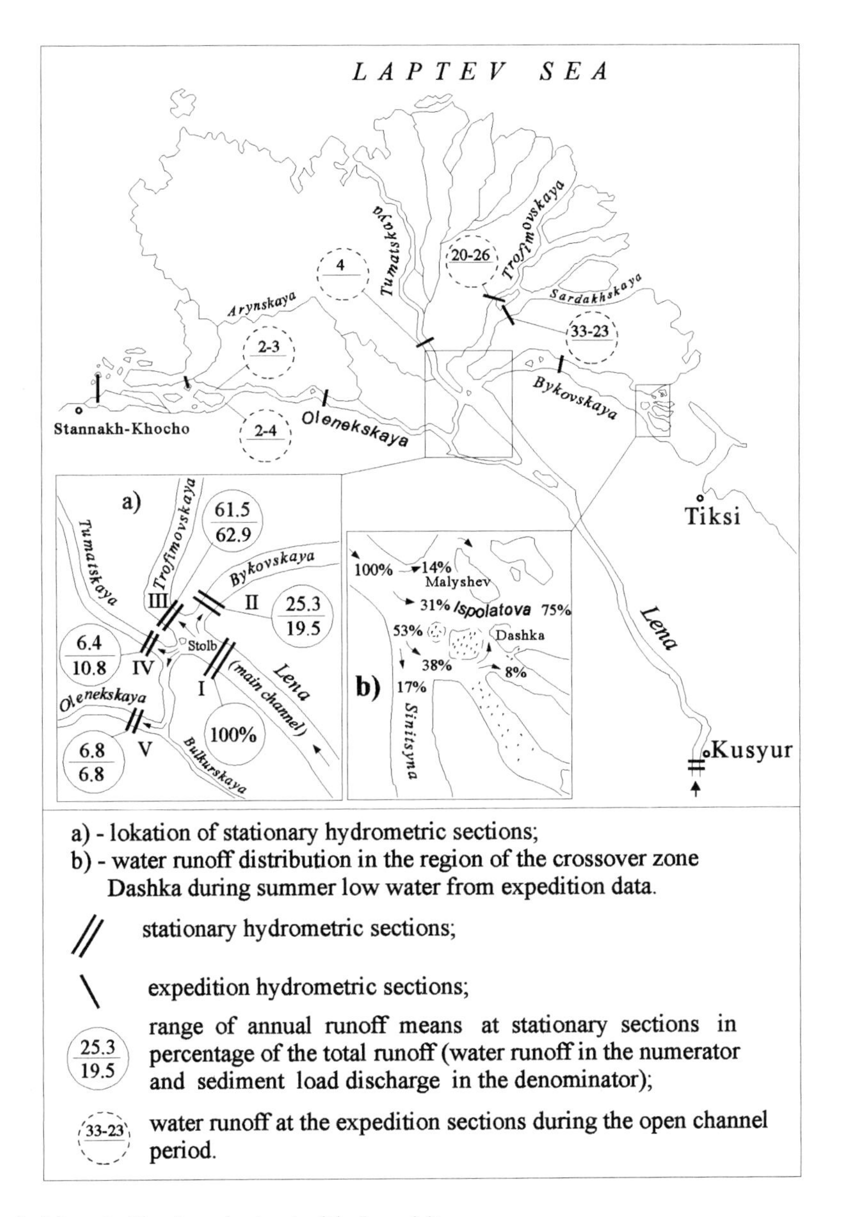

Figure 3: A layout of the channel network of the Lena delta.

Selivanovskaya branches. In the lower reaches of the Kamennaya branch, the Poperechnaya II branch separates from it independently exiting to the delta front. The water regime of the delta over its entire length is influenced both by the river runoff and sea level oscillations.

The runoff to the Kolyma delta is usually characterized on the basis of data of the Kolymskoye section which has a long observations series but does not take into account the water content of the Stadukhinskaya flood-plain branch. Stadukhinskaya branch by-passes the main hydrometric section situated in the main channel. Hence it is more acceptable to estimate the distribution of water and sediment load discharges relative to the expedition section located 3 km above the delta head near the Chersky settlement, rather than to the Kolymskoye section. Unfortunately, limited runoff observations at this section do not permit obtaining seasonal and multiyear parameters. Most expedition measurements of the runoff in the Kolyma delta were made by the All-Union Arctic Institute in 1934, 1935 and 1937, the Arktikproject in 1953 and 1954 and the AARI in 1991. However, they were of occasional character covering only small parts of some delta branches. The measurements were most numerous in the Kamennaya and Lesnaya branches near the Krai Lesa settlement. At the other sections 1-2 discharges per section were measured. A comparison of 1991 measurements in the delta branches with previous measurements at close conditions of summer low water has shown the differences of up to 10%. This can be attributed to different observational methods or changes in the channel net of the delta. Since the necessary measurements of water discharges of the Kolyma river at the entrance to the delta were absent, the obtained data have not allowed a reliable determination of the water content of the branches relative to the total river runoff. However, these data (except for 1991) were used as initial data for the design and regulation of the aerodynamic model of the Kolyma delta. Using this model, the distribution of water discharges in the individual branches was obtained for two regimes: at the river runoff of 4500 m^3/s and sea level 0.75 m lower than the mean level and at the runoff of 7700 m^3/s and mean sea level (Gilyarov and Ivanov, 1967). The results of aerodynamic modeling have served as a basis for parameterizing the delta branches for hydraulic calculations and mathematical modeling (Ivanov, 1968; Ivanov and Kotrekhov, 1974; Piskun, 1992). Thus correlated data on the water and channel regime of the Kolyma delta and the dynamics of the seasonal and rain floods and surge waves at a wide range of water discharges in the river and sea level were obtained.

Table 3: Comparative data on the distribution of water discharges (m^3/s) in the Kolyma delta, based on modeling and calculations.

Branch	Regime I			Regime II		
	Aerodynamic model	Hydraulic calculations	Numerical modeling	Aerodynamic model	Hydraulic calculations	Numerical modeling
Kolyma at the delta head	4500	4500	4500	7700	7700	7700
Kamennaya at the delta head	3800	3780	3790	6450	6460	6350
Chernousova	700	710	706	1250	1230	1360
Kamennaya upstream the head of Pope- rechnaya I	2670	2630	2720	4470	4500	4530
Pokhodskaya at the head	1830	1860	1780	3230	3190	3170
Kamennaya downstream the head of Poperechnaya I	2610	2600	2520	4390	4410	4010
Kamennaya upstream the head of Pope- rechnaya II	2610	2560	2520	4150	4160	4070
Kamennaya at the exit to the sea	1310	1280	1260	2150	2100	2080
Poperechnaya II	1300	1270	1260	2100	2050	1990

Table 3 presents the generalized results of studies for two typical regimes. Calculations have shown that the ratios of water discharge of individual delta arms are quite stable at different combinations of the water content and background sea level (Table 4). At unstable water motion there is a significant redistribution of water discharges over the delta area. Reverse currents are observed (negative discharge values) being especially significant at large surges (Ivanov and Kotrekhov, 1976). A comparison of the calculated water discharges in the Kolyma delta with

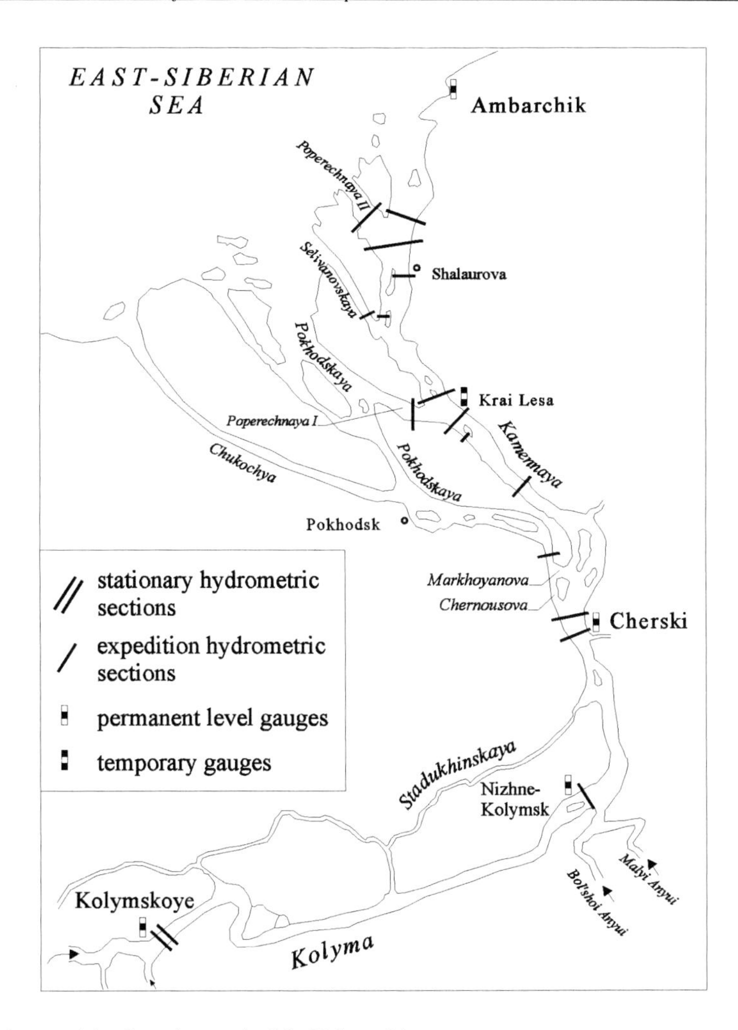

Figure 4: A layout of the channel network of the Kolyma delta.

full-scale and model values has shown their satisfactory agreement.

The ratio of the transported suspended matter discharges in the delta arms differs from water discharges. More than 85% of the sediment load is transported along the Kamennaya branch at the delta head. About 2% is transported to the Selivanovskaya branch and about 17% to the Poperechnaya branch. A characteristic feature of the sediment load discharge regime in the Kolyma delta, found during the expeditions in the 1930s and the 1950s is an approximate equality of the water turbidity values measured simultaneously along the length of the Kamennaya branch. This feature was used for data parameterizing in hydraulic calculations of the water and channel regime of the Kolyma delta (Piskun, 1992). Table 5 presents the calculated values of sediment load discharges in the main delta arms at the channel-forming water discharge in the Kolyma river equal to 23600 m^3/s and mean navigation sea level. The suspended sediment load discharges were determined as a product of water turbidity prescribed from actual data at the entry to the calculation area, and the calculated water discharges in the arms. The bottom sediment load discharges were determined as a fraction of the suspended sediment load discharge assumed to be equal to 0.05 taking into account the experience of such calculations for other deltas (Ivanov et al., 1983). The transportation ability of the flow was determined from formulas (Grishanin, 1979): for bottom sediments from Shamov's formula and for suspended sediments from Karaushev's formula.

Table 4: Water discharge in the Kamennaya branch of the Kolyma delta (in % of the water discharge in the river) at its exit to the bar from the results of hydraulic calculation for different boundary conditions from the side of the river and the sea at stable regime.

Water discharge	**Sea level mark near Ambarchik (m)**			
in the section Kolymskaya (m^3/s)	-0.75	-0.50	-0.25	0.00
2000	30.0	29.0	28.8	28.3
6000	28.4	28.2	28.1	28.1
10000	28.1	28.1	28.0	27.9
14000	28.0	27.9	27.8	27.4
18000	27.9	27.7	27.3	27.3
24000	27.8	27.4	27.3	27.3
32000	27.6	27.6	27.5	27.1

Table 5: The calculated discharges of suspended and bottom sediment loads and the flow transporting capacity (kg/s) in the main Kolyma delta arms at the channel-forming water discharge in the river and mean navigation level at the delta front.

Branch, zone	**Bottom sediment load**		**Suspended sediment load**	
	discharge	flow transporting capacity	discharge	flow transporting capacity
Kamennaya				
-at the delta head	133	121	6660	24100
-downstream Krai Lesa	92.4	65.8	4620	8040
-near Shalaurov	86.2	28.0	4310	1570
Pokhodskaya				
-upstream the Pokhodsk settlment	66.0	12.9	3300	1080
-upstream the Poperechnaya I mouth	65.9	40.7	3290	3640
Chernousova	27.9	127	1400	27400
Poperechnaya II	42.8	62.6	2140	2240

The Indigirka delta belongs to the second type of deltas where the sea influence on the runoff regime does not extend to the head. The place of furcation of the Russko-Ust'inskaya branch is considered the delta head (Figure 5). A little downstream the most strong Srednyaya branch separates. The Kolyma branch is the extreme right branch from which the Ularov branch separates. There are also many small branches in the delta creating around one hundred islands.

As found during the navigation period of 1958 and 1959 when the AARI expedition carried out first measurements of water discharges in the delta (Turanov, 1960), the Srednyaya branch has the largest water content transporting in its head about 54% of the Indigirka runoff. The second place belongs to the Russko-Ust'inskaya branch (33%) and the fraction of the Kolymskaya branch is 8%. Approximately a similar ratio of the water content of the branches was obtained by the MGU expedition in 1974 (Bogomolov et al., 1979). Data on the runoff were partly supplemented for the main branches depending on the water content and made detailed for most of the branches, including the exit of the Srednyaya branch to the bar. Simultaneously, data on the discharges of suspended sediment load were obtained for some sections. As is seen from Table 6 which presents generalized evidence on the distribution of water and sediment load discharges in the main arms of the Indigirka delta, the fraction of suspended matter transported by the main branches approximately corresponds to their water content.

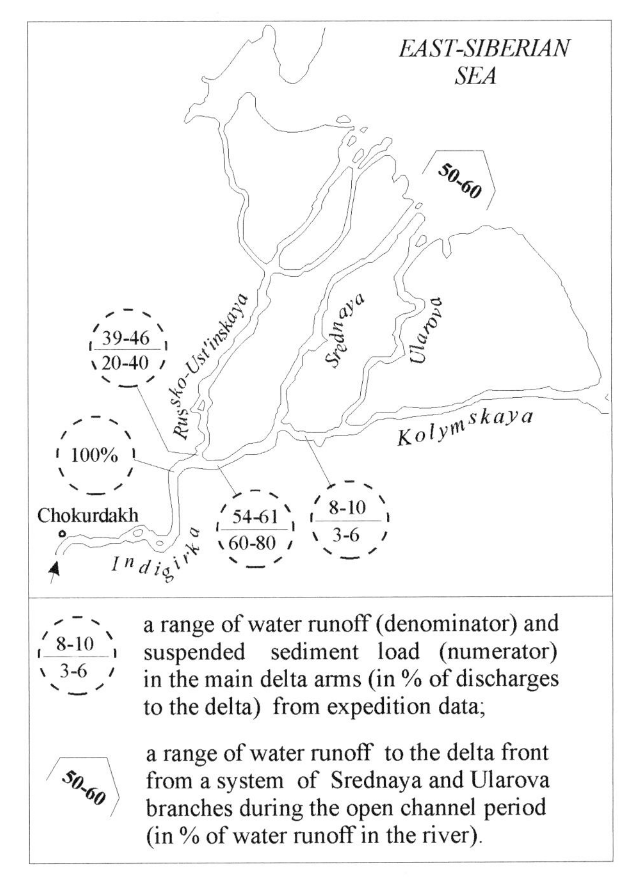

Figure 5: A layout of the channel network of the Indigirka delta.

Conclusion

In conclusion, it should be noted that the studies performed, have shown the use of a complex of full-scale, model and calculated data to be fruitful for estimates of the water and channel regime in complicated river delta forks of the Arctic under conditions of complicated river/sea interaction. Further development of methods is to be directed toward improving calculations of the sediment load transport as a basis for estimates of transfer of pollutants, their spreading in the deltas and export to the sea.

Table 6: Ranges of ratios of water and suspended sediment load discharges at the heads of the main arms of the Indigirka delta during summer low water from expedition data.

Element	Water flow			
	Main channel of the Indigirka river	Russko-Ust'inskaya	Kolymskaya	Srednaya
Water discharge (%)	100	39-46	54-61	46-51
Sediment load discharge (%)	100	20-40	60-80	57-74

References

Annual data on the regime and water quality of the seas and marine river mouths. Part 2, V. 5 and V.6 for 1977-1993.

Bogomolov, A.L., G.M. Zayets, V.N. Korotayev, R.V. Lodina, V.A. Miloshevich, A.Ye. Mikhinov, A.Yu. Sidorchuk and R.S. Chalov (1979) Main processes forming the Indigirka delta. In: Erosion of soils and channel processes, MGU, M., 7, 146-159.

Chalov, R.S., V.M. Panchenko, and S.Ya. Zernov (1995) Water courses of the Lena basin. MIKIS, 600pp.

Gilyarov, N.P. and V.V. Ivanov (1960) Modeling of mouths of the arctic rivers. Probl. Arktiki I Antarktiki, 2, 35-42.

Gilyarov, N.P. and V.V. Ivanov (1975) Results of laboratory studies of the water regime of the mouth areas of large rivers of Siberia. Proc. of the 4th All-Union Hydrological Congress, 5, 381-387.

Gilyarov, N.P. and V.V. Ivanov (1967) Water regime of the Kolyma river delta from laboratory studies. Trudy AANII, 278, 22-38.

Grishanin, K.V. (1967) Hydraulic calculation of the water regime elements in the deltas of rivers in the arctic zone. Trudy AANII ,278, 5-21.

Grishanin, K.V. (1979) Dynamics of channel flows. L., Gidrometeoizdat, 312pp.

Hydrological Yearbooks. V.8, No. 0-7 and V.8 No. 8 for 1936-1993.

Ivanov, V.V. (1968) A method for hydraulic calculation of the water regime elements in river deltas. Trudy AANII, 283, 30-63.

Ivanov, V.V. (1970) On the temporal variability of the runoff and levels in river deltas. Trudy AANII, 290, 6-17.

Ivanov, V.V. (1963) Runoff and current of the main delta branches of Lena. Trudy AANII, 234, 76-85.

Ivanov, V.V. and Ye.P. Kotrekhov (1976) Experience of numerical modeling of unstable water motion in a multiarm river delta. Trudy AANII, 314, 16-35.

Ivanov, V.V. and A.A. Piskun (1995) Distribution of river water and suspended sediments in the river deltas of the Laptev Sea. Reports on Polar Research. Russian-German Cooperation: Laptev Sea System. Ber. Polarforsch, 176, 142-153.

Ivanov, V.V., M.A. Mikhalev, A.S. Marchenko, A.A. Piskun and K.Ye. Chernin (1983) A hydraulic method for calculating the water and channel regime in multiarm river channels. Trudy AANII, 378, 5-22.

Ivanov, V.V., A.A. Piskun, M.A. Mikhalev and A.S. Marchenko (1990) A method for hydraulic calculation of the water regime in multiarm river deltas taking into account the channel process. Proc. of the 5th All-Union Hydrological Congress, 9, 74-78.

Nalimov, Yu.V. (1965) Hydrological characteristics of the Main Channel Branch in the Yana delta. Trudy AANII, 268, 57-77.

Piskun, A.A. (1978) Assessment of the calculation accuracy of resistance module of the river channel. Probl. Arktiki I Antarktiki, 53, 55-60.

Piskun, A.A. (1992) A method for hydraulic calculation of the water and channel regime in complicated fork river deltas (by the example of the Kolyma delta). AARI, St. Petersburg., Dep. v VNIIGMI-MCD, No.1127-gm92 of 8.11.92, 16pp.

Turanov, I.M. (1960) Examination of the Indigirka river mouth in 1958-1959. Probl. Arktiki I Antarktiki, 5, 77-78.

Dissolved Oxygen, Silicon, Phosphorous and Suspended Matter Concentrations During the Spring Breakup of the Lena River

S.V. Pivovarov[1], J.A. Hölemann[2], H. Kassens[2], M. Antonow[3] and I. Dmitrenko[1]

(1) State Research Center - Arctic and Antarctic Research Institute, 38 Bering St., 199226 St.Petersburg, Russia

(2) GEOMAR Forschungszentrum für marine Geowissenschaften, Wischhofstrasse 1-3, D 24148 Kiel, Germany

(3) TU Bergakademie Freiberg, Institut für Geologie, Bernhard-von-Cotta-Str. 2, D 09596 Freiberg, Germany

Received 11 March 1997 and accepted in revised form 3 March 1998

Abstract - In spring of 1996 hydrochemical and sedimentological measurements were carried out in the Lena river and along transects in the SE Laptev Sea off the delta. The investigations were conducted during the season of the spring thaw in which the freshwater discharge of the Lena river increases dramatically and the ice on the river breaks up.
River water during the low discharge winter period showed high concentrations of silicon but low concentrations of oxygen and suspended matter. During the first phase of the spring breakup this high-silicon winter water was flushed from the river into the Laptev Sea. This is reflected by elevated silicon concentrations in the water column off the Lena delta. In the course of the breakup, the silicon concentration in the river water and the low salinity freshwater layer off the delta decreased, while the turbidity increased significantly.
The strong inflow of river water was also reflected by a steadily thickening freshwater layer that formed in the SE Laptev Sea off the Lena delta. This turbid water mass was separated by a strong pycnocline from the underlying, less turbid brackish water. This phenomenon suggests that during the breakup period riverine suspended matter is laterally advected above the pycnoline far into the SE Laptev Sea.

Introduction

Five large Siberian rivers yearly discharge 714 km^3 of fresh water to the Laptev Sea. Almost 505 km^3 come from the Lena river, one of the largest river systems on earth with a drainage area of 2.46 x 10^6 km^2. At the river mouth, the Lena forms an extensive delta covering an area of 32000 km^2 crossed by many distributaries (Figure 1). The largest distributaries are the Trofimovskaya (61.5 % of the total Lena river discharge), the Bykovskaya (25.3 %), the Olenekskaya (6.8 %) and the Tumatskaya distributary (6.4 %) (Ivanov and Piskun, 1995).

Freshwater discharge from the Lena is highly variable throughout the year. During winter the discharge can be as low as 366 m^3/s (April 27, 1940 observed near the Kyusyur station). The highest water discharge ever observed occurred in the spring of 1944 (194000 m^3/s on June 11). Thus, the peak draining can be 500 times higher than the discharge during winter. This is accompanied by a ten to twentyfold increase in suspended particulate matter (SPM) concentration (Cauwet and Sidorov, 1996). Russian data indicate that, in June alone, about 200 km^3 of fresh water and more than 10 million tons of SPM flow into the Laptev Sea (Ivanov and Piskun, 1995).

Despite of the fact that the river breakup in spring is obviously the most important period of the hydrologic cycle of Siberian rivers, it is also the least investigated period. Only few investigations in Siberia were carried out during the spring thaw (Tasakov, 1955; Antonov, 1967; Ivanov and Nalimov, 1981; Nalimov, 1995). Usually these investigations comprise visual observations of ice conditions. A more detailed description of breakup processes in Arctic rivers was carried out in the delta of the Colville river in Alaska (Hamilton et al., 1974; Walker, 1973).

In: Kassens, H., H.A. Bauch, I. Dmitrenko, H. Eicken, H.-W. Hubberten, M. Melles, J. Thiede and L. Timokhov (eds.) Land-Ocean Systems in the Siberian Arctic: Dynamics and History. Springer-Verlag, Berlin, 1999, 251-264.

To advance our understanding of the processes connected to the river breakup and peak-stream flow in the Siberian Arctic the multidisciplinary Transdrift IV expedition started in spring of 1996 (Kassens, 1996). Primary investigation areas were the Bykovskaya distributary and the Laptev Sea off the major river outlets (i.e. Bykovskaya, Trofimovskaya mouths).

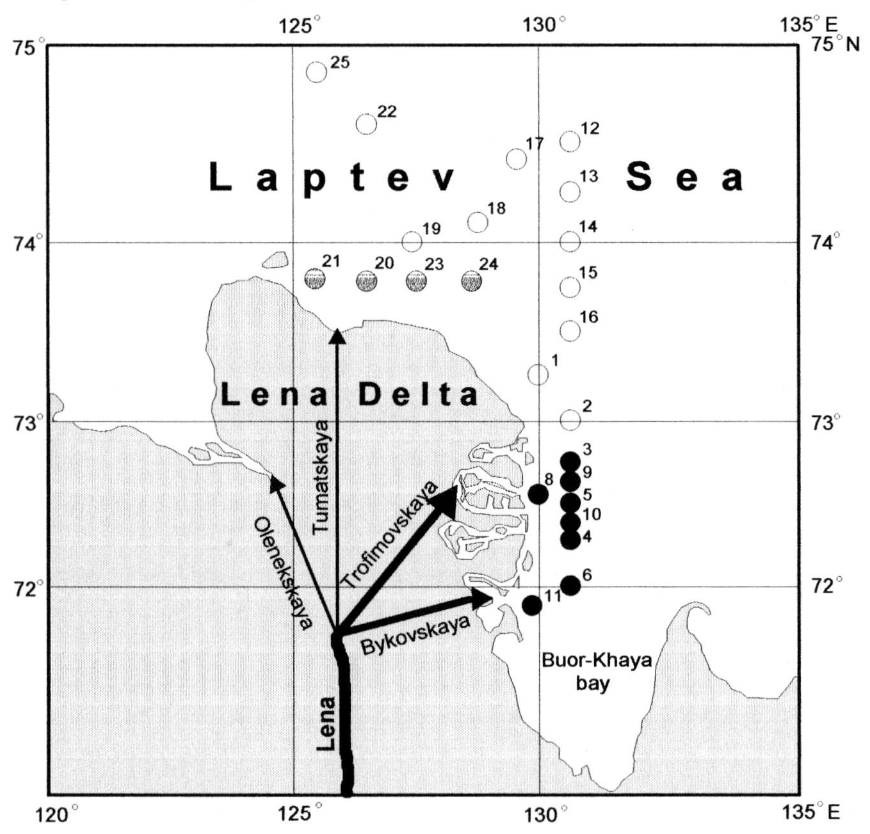

Figure 1: Water sampling stations in the Laptev Sea and major distributaries of the Lena river during Transdrift IV.

Results

Sampling and methods

In order to obtain data about the hydrologic and sedimentologic regime in the Laptev Sea sampling and observations were carried out from the fast ice along transect in the ice covered SE Laptev Sea. This study focuses on investigations in the Bykovskaya channel and along a N-S transect (130°30' E) approximately ten nautical miles east off the mouths of the Trofimovskaya and Bykovskaya distributaries (stations 3, 4, 5, 6, 9 and 10). Two additional stations (8 and 11) were situated near (≈ 5 nautical miles) the two major river outlets (Figure 1).

Surveys at these eight stations (Figure 1, black circles) were repeated three times to cover the (i) winter situation before the breakup on May 18-22 and (ii) the beginning of the spring breakup and the peak-draining period (June 3-6 and June 10-11).

The program was completed by measurements in the ice free polynya region north off the delta (stations 22 and 25).

Samples from river water were taken near the "Lena-Nordenskjøld" station (72°11'N,

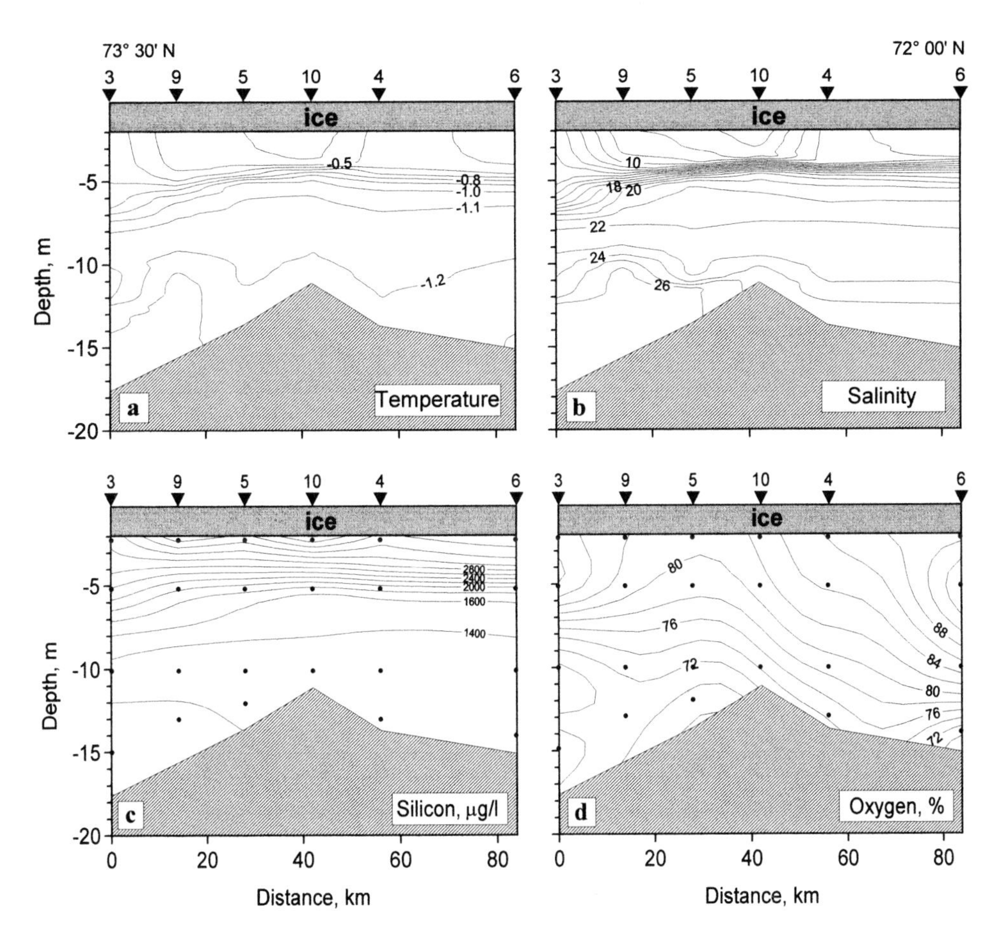

Figure 2: Temperature (a), salinity (b), silicon (c, μg/l) and oxygen (d, % saturation) distribution along the 130° 30' E transect before the flood 18-22.05.1996.

128°03'E) on the right bank of the Bykovskaya channel during the period from May 17 to June 13. When the river was completely ice covered sampling was carried out from the ice at a distance of about 50 m from the bank. The depth of the river at this point was approximately 10 m. During the initial phase of the breakup the existence of water on the river ice and water strips between the bank and the ice (zakraina) made it impossible to take samples from the central part of the river channel. Water samples were collected near the river bank in water depth between 1 and 2 m. The river bed near the bank was frozen and ice covered, and thus, local resuspension of sediments near the banks does not significantly influence the SPM measurements. Nevertheless, SPM measurements in a river system as big as the Lena river can be modified strongly by the local hydrodynamic regime.

After profiling the water column with a CTD probe (salinity in this paper is given as practical salinity units, PSU) sea water samples were collected by titanium half-litre samplers. Chemical analyses were performed generally within 3-6 hours after sampling. The dissolved oxygen was determined by means of the Winkler's method and dissolved inorganic phosphorus and inorganic silicon was measured colorimetrically (Oradovsky, 1993). The results are given as silicon (Si) or phosphorus (P) concentrations.

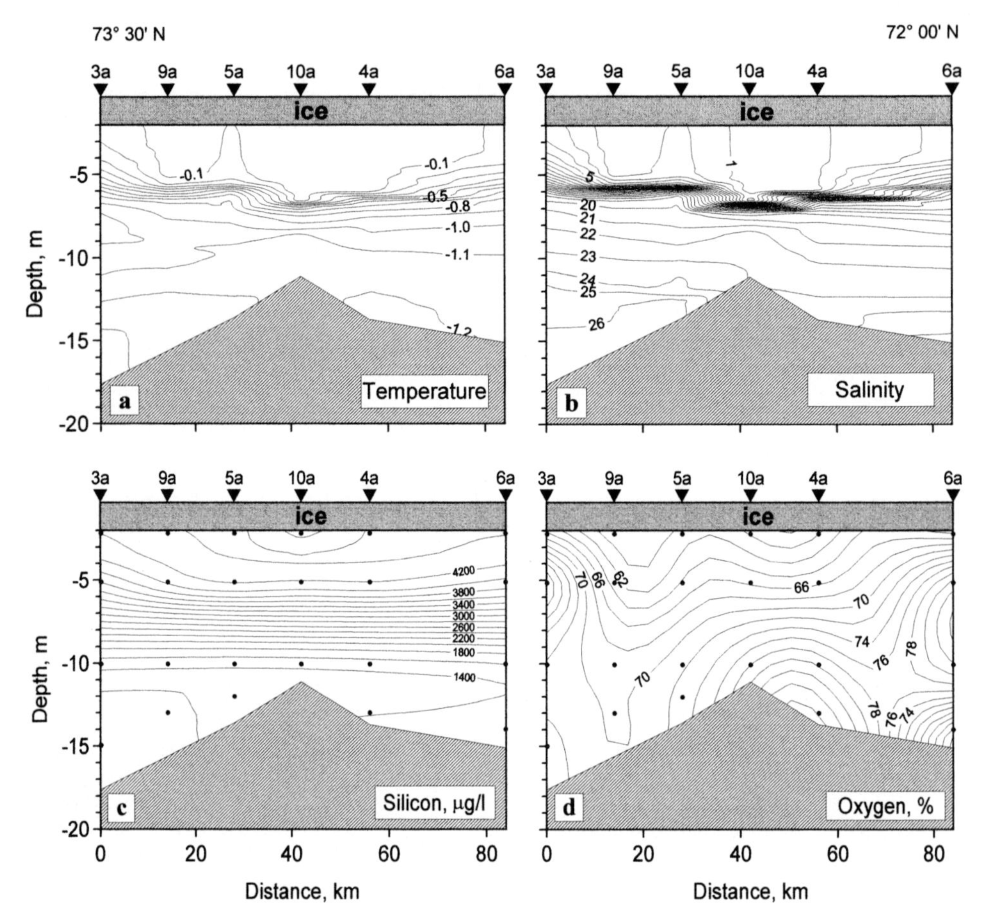

Figure 3: Temperature (a), salinity (b), silicon (c, µg/l) and oxygen (d, % saturation) distribution along the 130° 30' E transect during the first stage of breakup 03-06.06.1996.

Samples for the determination of SPM concentrations were taken with 5 litre water samplers. Samples of river water were taken with polyethylene bottles at the water surface in ice free parts of the river. Water was filtered through pre-weighted polyvinylidenfluorid membrane filters (Millipore HVLP, 0.45 µm). If necessary filters were washed with demineralized water to remove sea salt.

The turbidity of the water column was determined by means of optical backscatter sensors (OBS). This nephelometer unit belongs to a computerised modular probe system which was designed by ADM Elektronik GmbH (Warnau, Germany). Function and deployment routines of the system has been described by Hass et al. (1995). Measurements were carried out in depth intervals of at least one meter.

Distribution of salinity and temperature in the course of the breakup

All surveys along the transect revealed a distinct two-layer stratification of the water column (Figures 2, 3, 4). During the first survey (May 18-22, Figures 2 a, b) the upper layer - characterised by a salinity between 5 and 10 - had a thickness of 2-3 m. Between 4 m and 6 m water depth the salinity increased to values above 20. Also the water temperature changed from

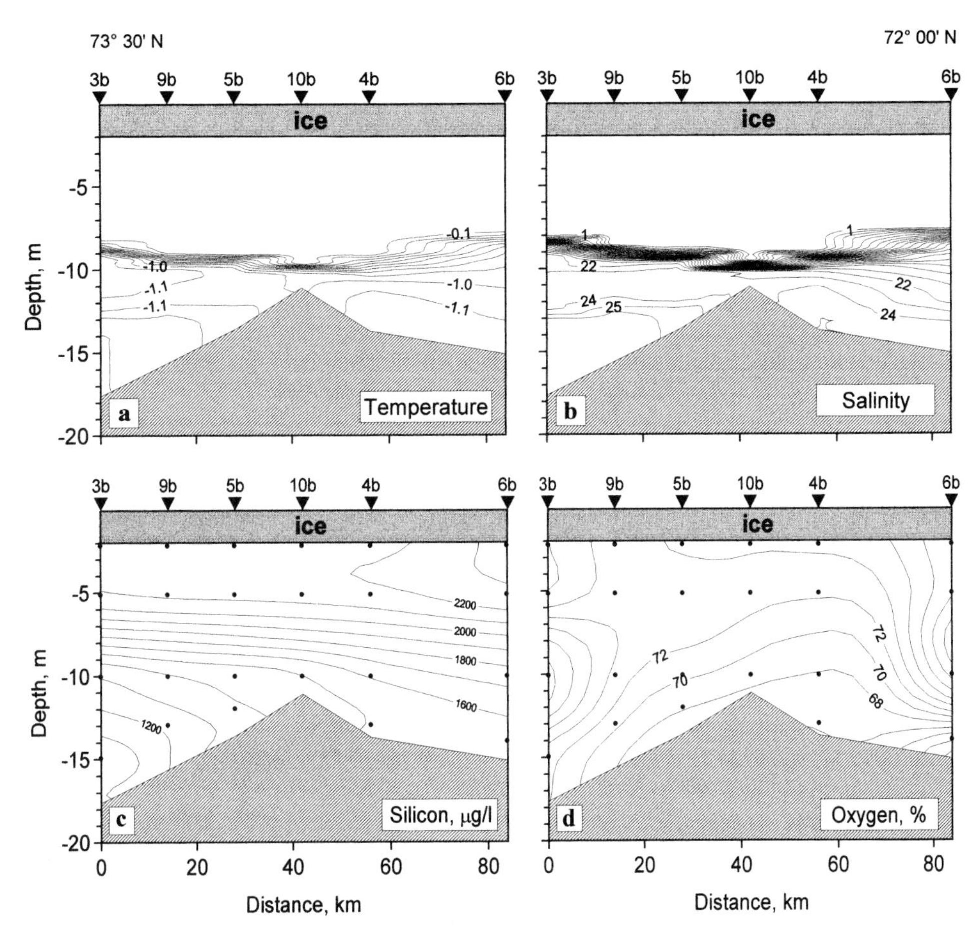

Figure 4: Temperature (a), salinity (b), silicon (c, µg/l) and oxygen (d, % saturation) distribution along the 130° 30' E transect during the second stage of breakup 10-11.06.1996.

-0.3 °C at the surface to -1.0 °C at 6 m. Only a weak gradient could be observed in the water column below this pycnocline.

The observations during the second survey (June 3-6) have shown significant changes in the vertical structure of the water column (Figures 3 a, b). A layer of river water with a salinity of 0.5 to 0.8 could be found under the ice between stations 5 and 4. At the other stations the salinity had decreased approximately 3fold, compared to the winter values. In addition, the pycnocline shifted downwards to depth ranging from 5 to 7 m.

The thickness of the upper freshwater layer further increased until June 10-11 (the third survey) and formed a 6-7 m thick layer under the fast ice (Figure 4 a, b). At station 10 the fresh water flow reached a depth of 9.5 m. The temperature in this layer was about 0 °C with a salinity between 0.15-0.22.

Distribution of dissolved silicon and phosphorus

Between May 17 and May 31 high silicon concentrations above 4000 µg/l with maxima of 4500 µg/l were measured in the Bykovskaya Channel (Figure 5). In contrast, concentrations of phosphorus in winter river water were low, 6-9 µg/l. A slightly reduced concentration of silicon on May 26, may reflect an admixture of melted snow because the sample was not taken under the ice, but from the water layer spreading over the surface of the ice-bound river.

In the period from May 31 to June 5, the concentration of silicon in the river water decreased to 2100 µg/l and remained at this level until the end of the observations at June 12. At this time the water level in the river still increased, but the Bykovskaya channel remained ice-covered.

During the first survey along the N-S transect east off the delta the maximum silicon concentration of 3840 µg/l was measured in the layer above the pycnocline at a depth of 2 m at Station 10. At this station, the silicon concentration was only about 700 µg/l lower than the silicon concentration recorded in the Lena river during the same time period. Like salinity and temperature, the vertical concentration of silicon also showed a strong gradient from 3800-2500 µg/l in the surface layer to 1400-970 µg/l below the pycnocline (Figures 2 c).

During the second survey (3-6 June, Figure 3 c) the concentration of silicon in the surface layer was 4100-4470 µg/l . This concentration range resembles the silicon signature typical for winter river water (Figure 5).

The third survey revealed a decrease in silicon concentration above the pycnocline (Figure 4 c). Concentration were in the range between 2140 and 2340 µg/l. This corresponds to concentration measured in river water during the period of high freshwater discharge. The water column beneath the pycnocline only showed a small increase in silicon concentration suggesting that input of remineralized silicon plays no significant role during this period.

Oxygen concentration

The concentration of dissolved oxygen in winter water of the Lena River was 5.7-5.8 ml/l, or 56 % of the theoretical saturation. Water above the pycnocline showed an oxygen saturation ranging from 77 to 94 %. The highest oxygen saturation (85 %) was observed directly beneath the ice cover (Figure 2 d). This is almost 30 % higher than the observed saturation percentage in the river. During the second survey along the transect (3-6 June) minimum oxygen concentration of 5.7-5.8 ml/l (56-57 % of saturation) in 2 m water depth at stations 10 and 4 reflect an inflow of winter river water (Figure 3 d). Low oxygen saturation in bottom water were usually observed at the northern flanks of shoals.

The water column below the pycnocline was characterised by oxygen concentrations between 4.2-7.8 ml/l. The oxygen distribution shown in Figure 6 gives evidence that oxygenated water from the polynya region (Station 12) sinks and flows towards the Lena delta where it ventilates the bottom water layer.

Concentration of particulate suspended matter

River water in winter was clear with SPM values below 1 mg/l (Figure 5). During the initial phase of the breakup the SPM concentration increased to > 20 mg/l, reaching maximum concentrations observed around June 8. A nearly twofold increase in SPM concentrations (> 35 mg/l) was observed on June 13 in the course of the strong ice drift on the river .

During the first survey on the N-S transect the observed SPM concentrations in the water column ranged between 0.5 and 3 mg/l (Figure 7). A two- to threefold increase in SPM was observed during the second survey. Both stations near the river channels (8 and 11) showed SPM concentrations between 7-10 mg/l. A strong increase in SPM concentrations above the pycnocline was observed during the third survey. Concentrations between 16 and 29 mg/l were recorded at the northern stations of the transect (3, 5, 10). The highest amount of suspended matter (73 mg/l) was measured near the Trofimovskaya branch (Station 8).

Below the pycnocline SPM concentration changed only slightly (below 5 mg/l), sea water remained transparent although the turbidity of the surface layer was about 20 mg/l with

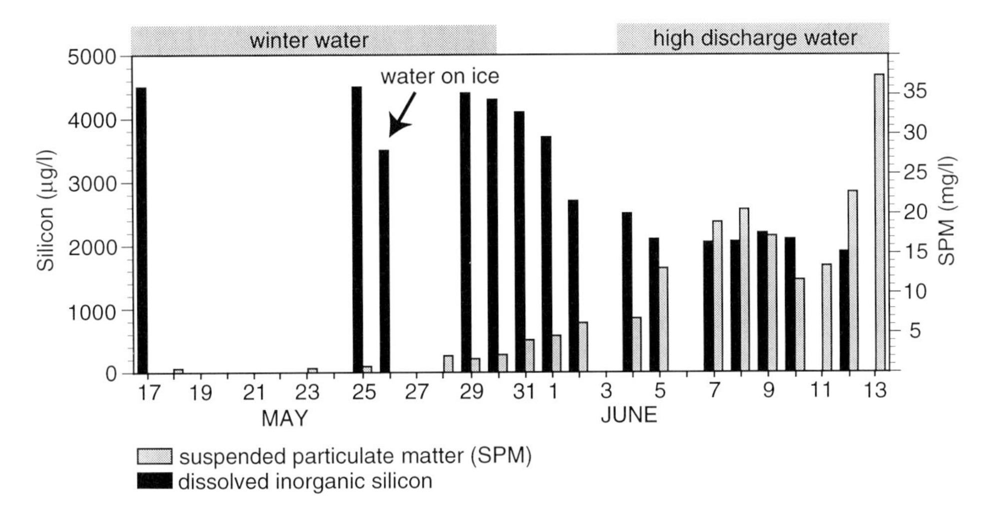

Figure 5: Silicon and suspended matter concentration in the Lena river near research station "Lena-Nordenskjøld" (Bykovskaya distributary) in May-June 1996.

maximum concentrations of more than 70 mg/l (Station 8). This suggest that erosion of seafloor sediment does not contribute to the high SPM concentrations above the pycnocline.

Another process affecting the dispersion of riverine SPM was observed during the third survey. Frazil ice formed at the contact between the freshwater and the saline water masses (Golovin et al., this volume). The rising frazil ice formed large conglomerates of ice platelets under the fast ice. Particulate matter concentrations above 1000 mg/kg in the ice suggest that frazil ice formation is an effective SPM scavenging mechanism during the river breakup (Golovin et al., this volume).

Discussion

Spreading of river water deduced from silicon distribution in the Laptev Sea

The dissolved silicon concentration in Siberian rivers is high, compared to the surface layer of the Arctic Basin. Thus, the distribution of silicon can be used as a tracer for the identification of river water in Arctic shelf seas (Rusanov, 1974; Rusanov and Ivanov, 1978; Ivanov et al., 1984). The 250 µg/l isoline is assumed to be the boundary of water masses influenced strongly by freshwater input and marine water masses (Rusanov et al., 1979; Buynevich et al., 1980).

Figure 8 describes the general silicon distribution pattern (summer and winter) in surface water of the Laptev Sea. The mean silicon distribution based on data of 11 oceanographic surveys during summer (August-September) is illustrated in Figure 8b. The influence of river water at this time of the year is predominant in the eastern Laptev Sea. In the western region, the influence of water masses with low silicon content from the Arctic Ocean is predominant and only the coastal regions are affected by river runoff. The winter (March-May) distribution (5 m) is shown in Figure 8B. This spatial distribution is based on data from 5 oceanographic surveys carried out during airborne expeditions from 1980 to 1986. Comparing the two distribution patterns, it becomes evident that the spatial distribution of freshwater in the upper water layer within the Laptev Sea remains approximately the same during summer and winter. However, silicon concentration during winter is twice as high as during summer.

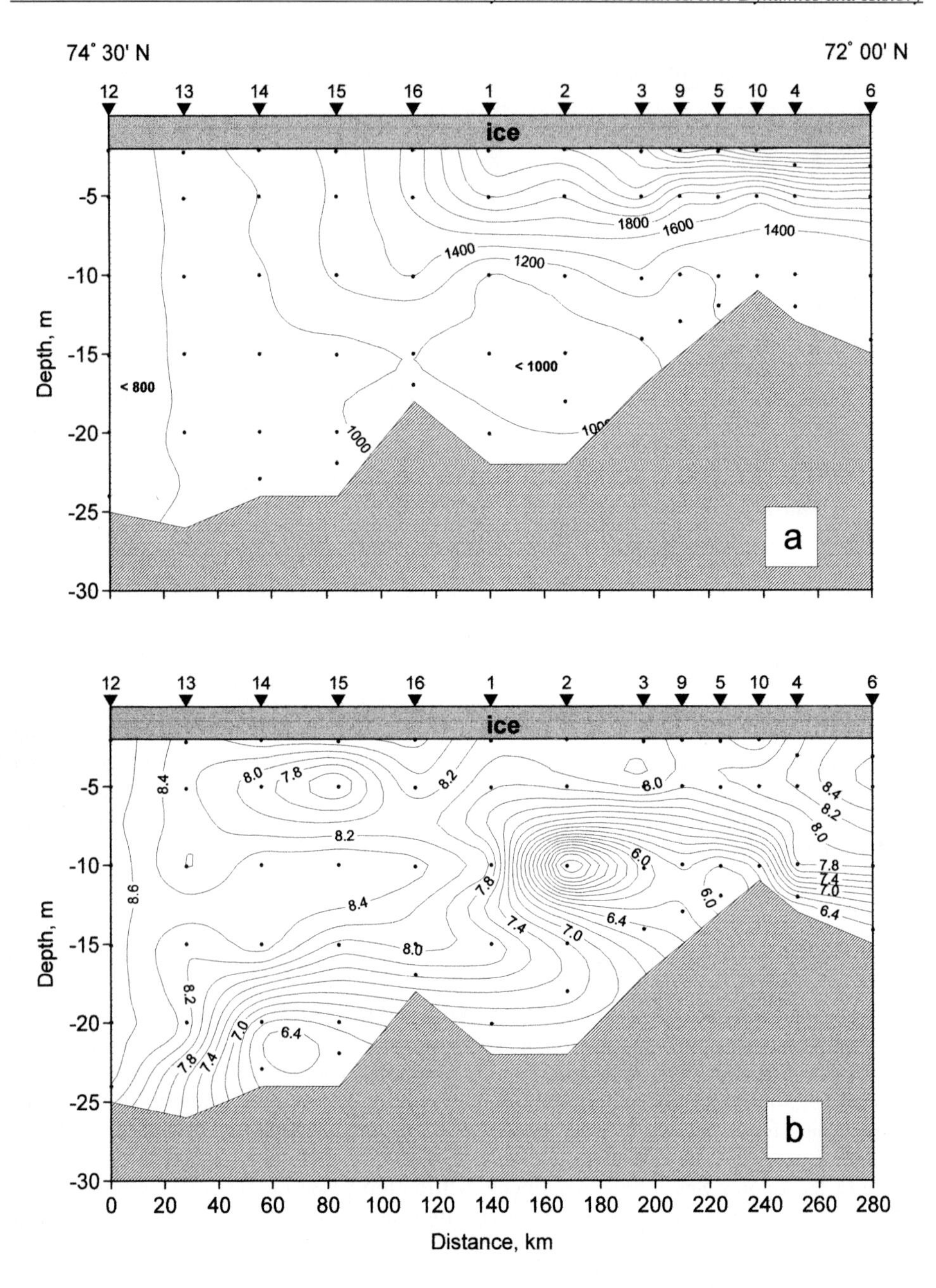

Figure 6: Silicon (a, μg/l) and oxygen (b, ml/l) distribution along the 130° 30' E transect (May 18-25.1996).

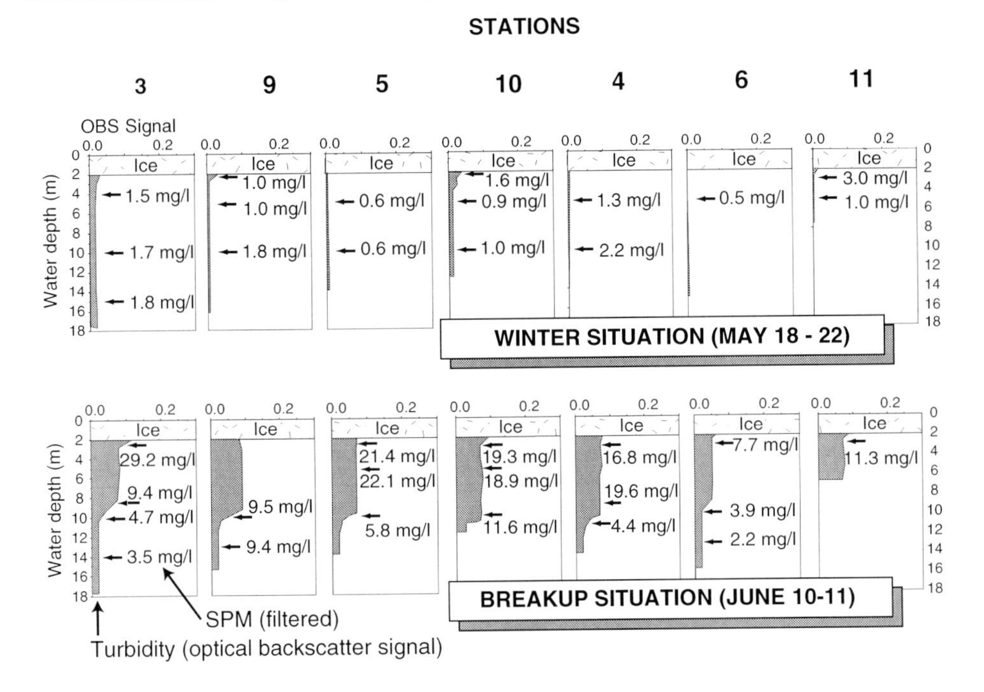

Figure 7: In situ optical backscatter record and measured SPM content along the N-S transect (130°30' E). Stations 3, 9, 5, 10, 4, 6 and 11 show the typical low content in the water column before the spring breakup. The stations were sampled again between June 10-11. At this time the upper water layer (between approximately 2-10 m water depth) was characterised by high a concentration of SPM and strong optical backscatter signals.

The discrepancy between the summer and winter concentration of silicon is most obvious in the shallow coastal regions. This is caused by an inflow of silicon-rich freshwater during winter - when the major source of river water is silicon enriched ground water -, and the absence of algae which consume dissolved inorganic silicon (Buynevich et al., 1980). Because winter convective mixing is usually not able to destroy the halocline, the distribution of silicon in the surface water off the delta reflects the input by the Lena river.

In other regions where stratification of the water column breaks down due to high turbulence (e.g. the polynya), elevated concentrations of silicon may also be caused by the input of remineralized silicon from bottom sediments.

Oxygen, nutrient and suspended matter distribution during winter (lean-flow)

Before the breakup the oxygen saturation in the Bykovskaya distributary was 56 %. This is in agreement with earlier winter observations in the Lena (Buynevich et al., 1980; Cauwet and Sidorov, 1996). The comparatively high oxygen saturation distinguishes the Lena river from other large Siberian rivers. As an example the water in the estuary of the Ob river shows a complete consumption of oxygen almost every winter (Yudanov, 1929; Rusanov et al., 1979).

The oxygen distribution under the fast ice cover in the SE Laptev Sea is influenced by the processes occurring in the polynya region. Intensive mixing in the ice-free polynya results in the formation of ventilated, cold, dense water. The distribution of oxygen suggests that oxygen enriched sea water sinks in the polynya region and flows towards the northern Lena delta.

Although, no complete consumption of oxygen was observed in 1996, the possibility of sporadic occurrences of hydrogen sulphide near the river mouths in regions near the delta was described by Sidorov and Gukov (1992). This is an indication of a complete consumption of oxygen caused by a stable stratification of the water column and decomposition of organic matter.

The measured low concentration of dissolved inorganic phosphorus is a feature typical of all large Siberian river systems. In contrast, silicon concentrations are high. This reflects the inflow of phosphorous-depleted and silicon-enriched ground water which is the major water source during the winter season (Simanchuk, 1938; Zubakina, 1974, 1979; Rusanov et al., 1979). The very low concentration of SPM in the river (< 1 mg/l) can be explained by low current velocities and the complete absence of sediment input from runoff.

Although, the freshwater discharge during the winter season is low, inflow of freshwater influences the thermohaline structure of the Laptev Sea near the delta. This is indicated by the presence of a water layer with reduced salinity below 10 that is separated by a pycnocline from the underlying more brackish water. Like salinity, the silicon concentration above the pycnocline is also a result of the mixing of river water with the brackish water of the SE Laptev Sea. Low suspended matter concentrations in river water and the missing turbulence in the ice-covered SE Laptev Sea can explain the low SPM concentrations in the water column off the delta.

The beginning of the spring thaw

The first indications of the beginning spring thaw and the resulting increase in water discharge of the Lena river were observed during May 31 to June 5 in the Bykovskaya Channel. During this period, the water level within the river increased and the concentration of silicon decreased to its typical summer concentration of approximately 2100 μg/l. The decrease in silicon concentration is caused by a change in the sources of the river water. During winter the major inflow comes from ground water while, on the contrary, terrestrial runoff is the dominate source in summer (Gordeev and Sidorov, 1993). In the course of the summer, silicon is also consumed to some extend by freshwater diatoms. Increasing turbulence at the riverbed and the input of particles with the runoff are the causes for an increase in SPM concentrations recorded during this time period.

While the silicon concentration in the river decreased the silicon concentration in the surface water layer near the river mouth increased significantly. This is accompanied by a salinity below 1 and the formation of a distinct halocline in the central part of the N-S transect near the river mouth.

Low salinity in combination with high concentrations of silicon, low SPM concentrations andmedium oxygen saturation levels suggest that with the onset of the breakup winter water is flushed from the river channels to the sea, forming a wedge of freshwater between the ice cover and the brackish water of the SE Laptev Sea.

The hydrochemical and thermohaline structure of the water column beneath the pycnocline showed no significant changes. Only the oxygen saturation in the near bottom layer increased. This might be caused by an increased turbulence in the saline water which is also supported by a slight increase of suspended matter concentrations beneath the pycnocline. However, the main oxygen distribution pattern did not change.

Beginning of the high discharge period (spring-summer situation)

The third oceanographic survey along the 130°30'E transect revealed a completely new situation. The thickness of the freshwater wedge increased by 4-5 m. At station 10 near the Trofimovskaya river mouth the freshwater layer - with a salinity between 0.15-0.22 and a temperature about 0 °C - reached a thickness of 9.5 m. The suspended matter concentrations measured in the freshwater layer off the delta near the Bykovskaya branch and within the

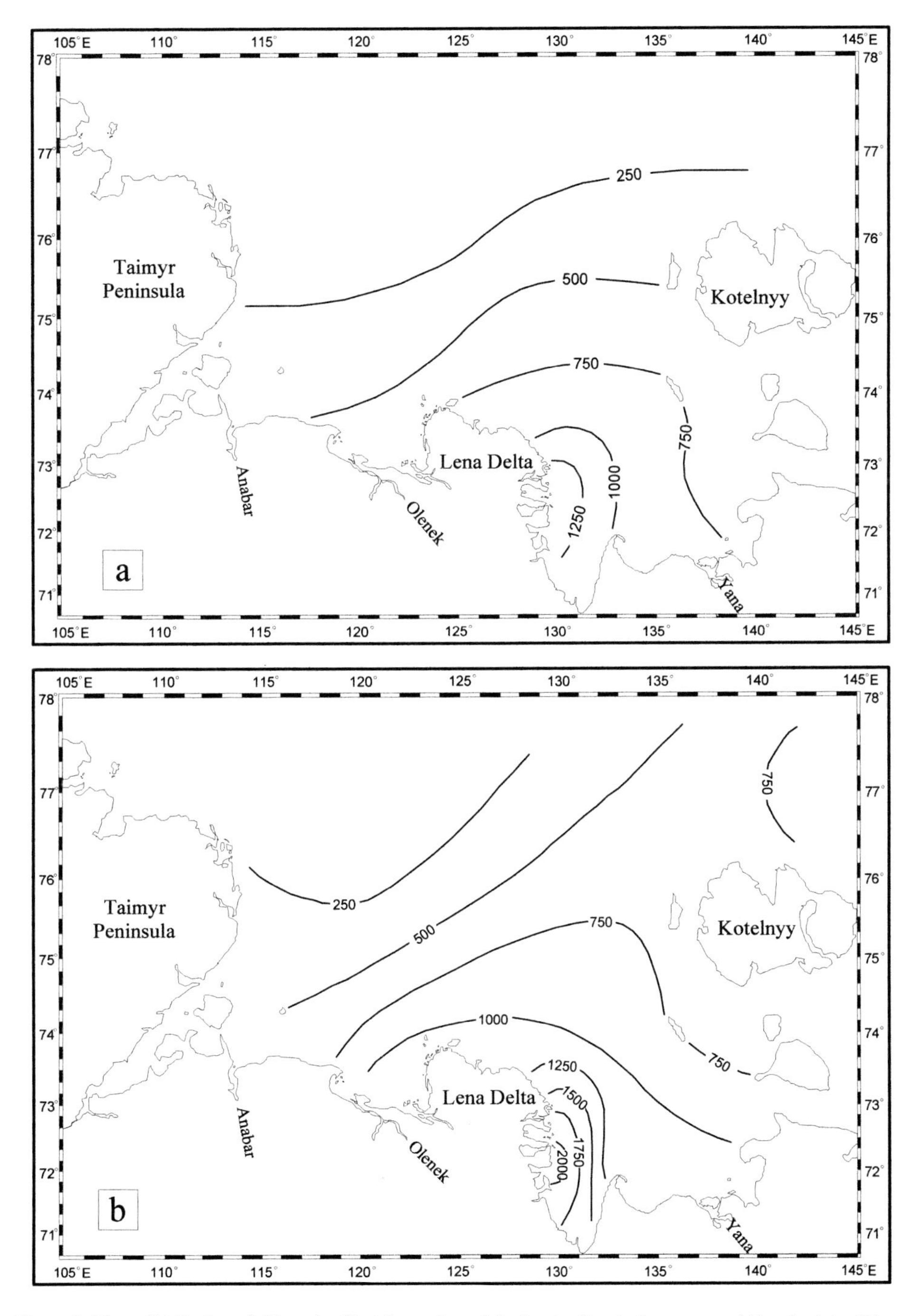

Figure 8: Mean distribution of silicon (µg/l) at the surface of the Laptev Sea during summer (A) and winter (B).

channel itself were nearly the same. This suggests that the stratification effectively prevents the deposition of riverine SPM within and near the delta. This is further supported by SPM concentrations above 70 mg/l measured at station 8 seaward off the Trofimovskaya distributary. These high concentrations correspond well with earlier observations in the Trofimovskaya channel during peak-draining (50-70 mg/l; Cauwet and Sidorov, 1996). The effective river-sea transport during the breakup also has implications for the sediment budget of the Laptev Sea because the major proportion of the annual SPM export to the Laptev Sea happens during this short time period (Cauwet and Sidorov, 1996).

Below the pycnocline the situation only changed little. The temperature and salinity structure of the water layer below 12 m depth was almost the same as during the first surveys. Also the spatial distribution of oxygen saturation appeared unchanged. The concentration of suspended matter in the near-bottom layer changed only moderately, showing SPM concentration below 5 mg/l. Sea water below the pycnocline remained transparent, although the turbidity of the surface layer was high (20 mg/l with maximum concentrations of more than 70 mg/l at station 8).

Another important mechanism for the transport and dispersal of suspended matter in the Laptev Sea could be observed during the last survey (Golovin et al., this volume). Particle loaden frazil ice aggregates formed at the contact zone of the 0 °C-freshwater and brackish water with a temperature below 0 °C. The frazil ice rises and formed a layer of ice platelets under the fast ice. Thus, riverine SPM is coupled to the decay, movement and melting of the fast ice cover within the Laptev Sea.

Conclusion

The unique hydrographic regime that accompanies breakup of the Lena river is important in the hydrologic and sedimentologic development of the SE Laptev Sea. In spring 1996, the concentrations of dissolved oxygen, silicon, phosphorous and suspended matter during the could be described for the first time in detail.

During the onset of the breakup, floodwater moved down the river and flushed the winter river water to the sea. In the course of the spring thaw the floodwater formed a wedge of freshwater within the ice covered Laptev Sea. A sharp pycnocline was established between the spring floodwater and the underlying brackish water. During June 10-11 the freshwater layer below the fast ice reached a thickness of 6-7 m. Silicon concentration and high suspended matter load indicate that river water from the Lena passed nearly unchanged through the delta and the near shore area. This spring floodwater completely filled the layer above the pycnocline pushing out the brackish water and winter river water from the southern part of the SE Laptev Sea. Because sedimentation of riverine SPM was impeded by a strong pycnocline, large amounts of particulate matter can be advected far into the SE Laptev Sea.

The results presented in this study only describe the environmental conditions during spring of 1996. The mode of freshwater discharge during the breakup depends strongly on the hydrologic conditions, i.e. the magnitude and timing of water discharge, the duration of the flood, the presence of ice dams in the Lena river etc.. As an example, in the pro-delta area river water can also spread over the fast ice. Thus, different combinations of hydrologic and ice conditions are possible. This illustrates that the conclusions drawn from the study of the 1996 breakup are only a first step towards a better understanding of the complicated processes that shape the environment in the Siberian Arctic.

Acknowledgments

This work was supported by the Ministry for Science and Technical Policy of the Russian Federation and the German Ministry of Education, Science, Research and Technology of the Federal Republic Germany. The authors gratefully acknowledge the assistance of the Transdrift-IV participants Bettina Rohr, Martina Schirmacher, Nikolay Fomitchev, Aleksey Visnevsky, etc., who under heavy Arctic conditions helped to collect the samples and prepare them for analysis. Erk Reimnitz and Venuoplan Ittekkot are thanked for valuable comments on the manuscript.

References

Alekin, O.A. and Yu.I. Lyakhin (1984) Chemistry of the ocean (in Russian).Gidrometeoizdat, Leningrad, 343 pp.

Antonov, V.S. (1967) The mouth region of Lena River (hydrological assay) (in Russian). Gidrometeoizdat, Leningrad, 107 pp.

Buynevich, A.G., V.P. Rusanov and V.M. Smagin (1980) Spreading of river water in the Laptev Sea by the hydrochemical elements (in Russian). Tr. AANII, 358, 116-125.

Cauwet, G. and I.S. Sidorov (1996) The biogeochemistry of Lena River: organic carbon and nutrient distribution. Mar. Chem., 53, 211-227.

Golovin, P., I. Dmitrenko, H. Kassens and J. Hölemann (1996) Frazil ice formation during the spring flood and its role in transport of sediments to the ice cover. In this volume.

Gordeev, V.V. and I.S. Sidorov (1993) Concentrations of major elements and their outflow into the Laptev Sea by the Lena river. Mar. Chem., 43 (1-4), 33-45.

Hamilton, R.A., C.L. Ho and H.J. Walker (1974) Breakup flooding and nutrient source of Colville river delta during 1973. In: The Coast and Shelf of the Beaufort Sea, Reed J.C. and J.E. Sater (eds.), The Arctic Institute of North America 3426 N., Washington Blvd., Arlington, VA 22201, 637-648.

Hass, C., M. Antonow and Shipbord Scientific Party (1995) Movement of Laptev Sea Shelf Waters during the Transdrift II Expedition. In: Russian-German Cooperation: Laptev Sea System, H. Kassens et al. (eds.), Reports on Polar Research, 176, 121-134.

Ivanov, V.V., and Yu.V. Nalimov (1981) Results of air expeditions in the downstream and mouth areas of Arctic rivers (in Russian). Prob. Arktiki i Antarktiki, 51, 79-91.

Ivanov, V.V., A.A. Piskun and R.A. Korabel (1983) Run-off distribution by the main arms of the Lena delta (in Russian). Tr. AANII, 378, 59-71.

Ivanov, V.V. and A.A. Piskun (1995) Distribution of river water and suspended sediments in the deltas of the Laptev Sea. In: Russian-German Cooperation: Laptev Sea System, H. Kassens et al. (eds.), Reports on Polar Research, 176, 142-153.

Ivanov, V.V., V.P. Rusanov, O.I. Gordin and I.V. Osipova (1984) Interannual variability of river water distribution in the Kara Sea (in Russian). Tr. AANII, 368, 74-81.

Kassens, H., et al. (eds.) (1995): Russian-German Cooperation: Laptev Sea System.- Reports on Polar Research 176, 387 pp.

Nalimov, Yu.V. (1995) The ice-thermal regime at front deltas of rivers of the Laptev Sea. In: Russian-German Cooperation: Laptev Sea System, H. Kassens et al. (eds.), Reports on Polar Research, 176, 55-61.

Oradovsky, A.G. (ed.) (1993) A Handbook on Chemical Analysis of Sea Waters (in Russian). Gidrometeoizdat, St. Petersburg, 264 pp.

Rusanov, V.P. (1974) Silicon distribution in surface waters of the Arctic Basin in winter (in Russian). Okeanologiya, 14 (5), 823-829.

Rusanov, V.P. and V.V. Ivanov (1978) Peculiarities of determining the sea boundaries of the mouth areas of Arctic rivers (in Russian). Tr. GOIN, 142, 122-125.

Rusanov, V.P., N.I. Yakovlev and A.G. Buynevich (1979) Hydrochemical regime of the Arctic Ocean (in Russian). Tr. AANII, 355, 144 pp.

Sidorov, I.S. and A.Yu. Gukov (1992) Influence of the oxygen regime on the zoobenthos existence conditions in the coastal regions of the Laptev Sea (in Russian). Okeanologiya, 32 (5), 902-904.

Simanchuk, A.O. (1938) The hydrochemical characteristics of water of the Lena and Ebetem rivers (in Russian). Tr. ANII, 105, 75-99.

Tasakov, P.D. (1955) On the winter regime and the ice breakup process in the Lena river mouth (in Russian). Tr. AANII, 72, 105-129.

Walker, H.J. (1973) Spring discharge of an Arctic river determined from salinity measurements beneath sea ice. Water Resources Res., 9 (2), 474-480.

Vasilyev, A.N. (1976) The interaction of riverine and marine waters in the Ob' mouth area (in Russian).Tr. AANII, 314, 183-196.

Yudanov, I.G. (1929) On the cognition of mortality in the river Ob (in Russian). Trudy Sibirskoi nauchnoi rybopromyslovoi stantsii, 3, 8-15.
Zubakina, A.N. (1974) Some hydrochemical features of mouth areas of the Anabar, Lena, Yana and Indigirka rivers (in Russian). Tr. GOIN, 118, 44-55.
Zubakina, A.N. (1979) The specific features of the hydrochemical regime of the Lena mouth area (in Russian). Tr. GOIN, 143, 69-76.

Distribution Patterns of Heavy Minerals in Siberian Rivers, the Laptev Sea and the Eastern Arctic Ocean: An Approach to Identify Sources, Transport and Pathways of Terrigenous Matter

M. Behrends[1], E. Hoops[2] and B. Peregovich[3]

(1) Alfred-Wegener-Institut für Polar- und Meeresforschung, Postfach 120161, D 27568 Bremerhaven, Germany

(2) Alfred-Wegener-Institut für Polar- und Meeresforschung, Forschungsstelle Potsdam, Telegrafenberg A43, D 14473 Potsdam, Germany

(3) GEOMAR Forschungszentrum für marine Geowissenschaften, Wischhofstrasse 1-3, D 24148 Kiel, Germany

Received 11 March 1997 and accepted in revised form 21 January 1998

Abstract - Heavy minerals of sediments from the rivers draining into the Laptev Sea, the Laptev Sea shelf, the central Arctic Ocean and additional potential source areas were investigated to identify transport pathways of terrigenous matter. The western part of the Laptev Sea is dominated by clinopyroxene, delivered by the river Khatanga, which derived from the sheet basalt complexes of the Taimyr Peninsula, and from Severnaya Zemlya. The eastern part of the Laptev Sea is characterized by very high amounts of amphibole, supplied by the rivers Lena and Yana from the Siberian hinterland and/or the erosion products from Kotelny Island. Orthopyroxene, present on the whole Laptev Sea shelf but not found elsewhere in comparable amounts in the study area, seems to be characteristic for this region. This heavy mineral composition is indicative for a Laptev Sea source and can be found in the sediments of the central Arctic Ocean. Sediment input from the western Arctic via the Beaufort Gyre is indicated by the occurrence of detrital carbonate and significant amounts of opaque minerals.

The clay assemblages of sea-ice sediments and clay mineral assemblages of underlying surface sediments of the central Arctic Ocean reveal different patterns due to winnowing processes of clay particles after release. Heavy minerals may fill the gap as indicator for sediment supply by sea ice in surface sediments from the sources up to the Fram Strait.

Introduction

Large amounts of sea ice are formed in the broad and shallow Eurasian shelf regions, of which the Laptev Sea is believed to be the most important source of sea ice traversing the central Arctic Ocean with the Transpolar Drift into the Fram Strait area (Dethleff et al., 1993; Wollenburg, 1993; Kassens & Karpij, 1994; Nürnberg et al., 1994; Eicken et al., 1995; Reimnitz et al., 1995). Estimates of ice export from the Laptev Sea range from 400 km^3yr^{-1} (Eicken et al., 1996) to 540 km^3yr^{-1} (Timokhov, 1994). In comparison the estimated ice export from the Barents Sea is 40 km^3yr^{-1}, Kara Sea 240 km^3yr^{-1}, East Siberian Sea 150 km^3yr^{-1} and the Chukchi Sea 10 km^3yr^{-1} (Timokhov, 1994).

Investigations of particulate matter in sea ice from the Transpolar Drift suggest its source originating from the Laptev Sea (Pfirman et al., 1990, 1997; Abelmann, 1992; Wollenburg, 1993; J. Wollenburg, 1995; Nürnberg et al., 1994). Clay mineral distribution of surface sediments in the central Arctic show different pattern due to winnowing processes (Stein et al., 1994b) and transport in the water column after release of particles. Heavy minerals reach the open ocean after entrainment into sea ice. Because of their high density they settle down where they are released from sea ice and are less affected by currents. Thus, heavy minerals have proven to be a useful tool to identify source areas and sea-ice pathways in the Arctic Ocean (e.g. Lapina, 1965; Lisitzin, 1996; Silverberg, 1972; Naugler, 1967; Luepke and Escowitz, 1989).

In: Kassens, H., H.A. Bauch, I. Dmitrenko, H. Eicken, H.-W. Hubberten, M. Melles, J. Thiede and L. Timokhov (eds.) Land-Ocean Systems in the Siberian Arctic: Dynamics and History. Springer-Verlag, Berlin, 1999, 265-286.

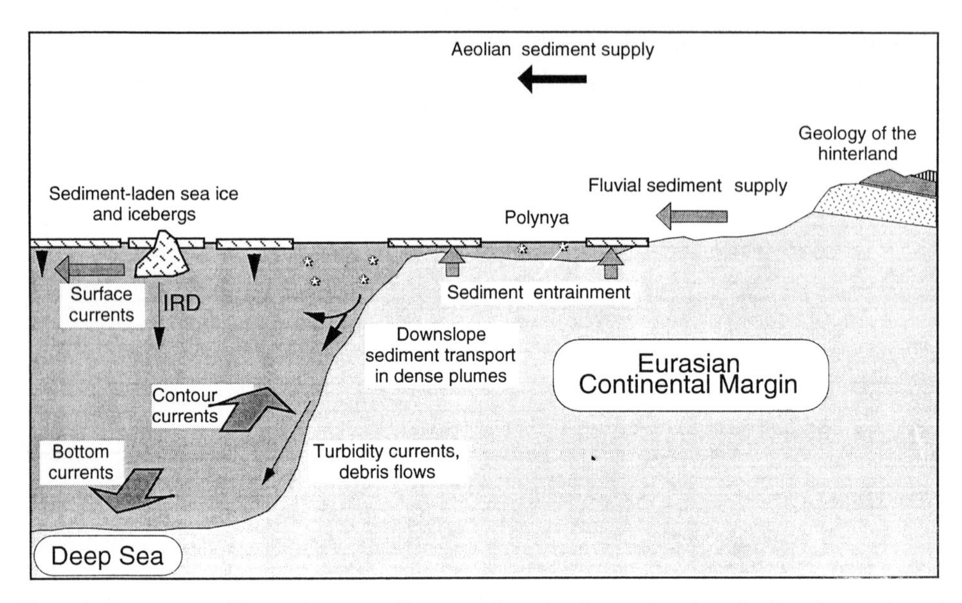

Figure 1: Factors controlling terrigenous sediment supply and sedimentation along the Eurasian continental margin and in the deep-sea environment (Stein and Korolev, 1994).

The most important factors controlling sediment supply along the Eurasian continental margin and deep-sea areas are fluvial input, drift of sea ice (Dethlef, 1995; Eicken et al., 1997), ocean currents, and gravitational mass flows (Fütterer, 1994; Stein and Korolev, 1994) (Figure 1). Input of sediment and water delivery onto the shelf are controlled by the Siberian river systems (Gordeev and Sidorov, 1993). In the western part of the Laptev Sea, the rivers Khatanga, Olenek, and Anabar are responsible for the sediment supply. In the Eastern Laptev Sea, the rivers Lena and Yana control the sedimentation. The total annual sediment discharge of the Siberian rivers accounts for ca. 27 Mio tons (Rachold et al., 1996). The major portion (approximately 21 Mio tons per year) is transported by the Lena River. Due to the extreme climate in the drainage area with winter temperatures down to -45 to -50°C and summer temperatures reaching +35°C, 90% of the sediment are transported during the warm season (Rachold et al., 1996). Different processes are controlling the entrainment of particles into sea ice (Dethlef et al., 1993; Wollenburg, 1993; Reimnitz et al., 1994; Eicken et al., 1997). First the content of coarse-grained sediment in sea ice is triggered by the availability of coarse-grained material in shelf sediments. Fine-grained particles in sea ice seem to be enriched due to formation processes, such as suspension freezing (Reimnitz et al., 1992). Open shallow seas connected with subfreezing temperatures, strong winds and turbulence cause a supercooled watercolumn, where frazil ice forms. The underwater ice crystals interact with sedimentary particles in the water column of the supercooled sea (frazil ice) and on the sea floor (anchor ice) and lift particulate matter to the sea surface, forming a layer of slush ice. Nevertheless, the fine sand fraction is found in sea-ice sediments on the Laptev Sea shelf (Dethlef et al., 1993; Eicken et al., 1997) as well as in the multi-year ice in the central Arctic Ocean (Nürnberg et al., 1994).

The ice export from the Laptev Sea into the Arctic Ocean is dominated by the drift pattern during summer, between 120 - 140°E (Eicken et al., 1997). Ice export over the entire width of the shelf occurs during winter time (Timokhov, 1994). After incorporation into sea ice sediment is leaving the Laptev Sea transported by the Transpolar Drift through the central Arctic Ocean to the Fram Strait.

Methods

To study the present sedimentary environment, surface sediments were taken with a giant boxcorer during RV Polarstern cruises ARK-IV/3 (Thiede, 1988) ARK-VIII/3 (Fütterer, 1992), ARK-IX/4 (Fütterer, 1994), and ARK-XI/1 (Rachor, 1997), during RV Ivan Kireyev cruise Transdrift I (Kassens, 1994), ESARE (Dethleff et al., 1993) and RV Prof. Multanowsky cruise Transdrift II (Kassens et al., 1995), RV "Prof. Logachev" and RV "Akademik Golitzin". In addition, river sediments were collected during the expeditions Lena 94 (Rachold et al., 1995), Lena/Yana 95 (Rachold et al., 1997) and Khatanga 96 (Rachold et al., in prep.) with a heavy pale (Rachold et al., 1997) (Figure 2).

The heavy minerals of the subfraction 63 - 125μm were investigated. The grains were separated using sodium metatungstate ($Na_6[H_2W_{12}O_{40}]$) with a density of 2.83g/cm^3. After 20min centrifugation, the heavy fraction was frozen in liquid nitrogen (Fessenden, 1959; Scull, 1960). The light minerals were decanted, and the heavy minerals were rinsed on filters and dried. The heavy fraction was mounted with Meltmount (n=1,68) (Mange and Maurer, 1991), optically identified, and counted using a polarisation microscope. On average about 220 grains per sample were counted on the strewslides (Table 1). The average content of heavy minerals in the analysed sand fraction is 6%.

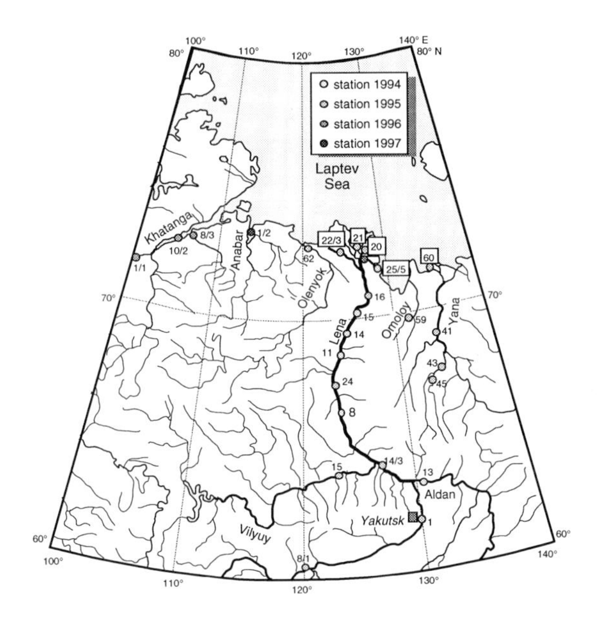

Figure 2: Location map of the river samples. Station numbers are indicated.

All heavy minerals including dolomite (detrital carbonate) and opaques are counted and calculated as 100 grain-%. All informations given in % are grain-%. The group of opaque minerals is undifferentiated.

Concerning calcite the specific weight is between 2.7 and 2.94 g/cm^3, which can be lower than the density solution of sodium metatungstate. Dolomite, magnesite and siderite (chalybite), however, have densities of 2.86-2.93, 2.96 and 3.78-3.96 g/cm^3, respectively (Boenigk, 1983). Thus, they are interpreted as heavy minerals (Boenigk, 1983; Mange and Maurer,

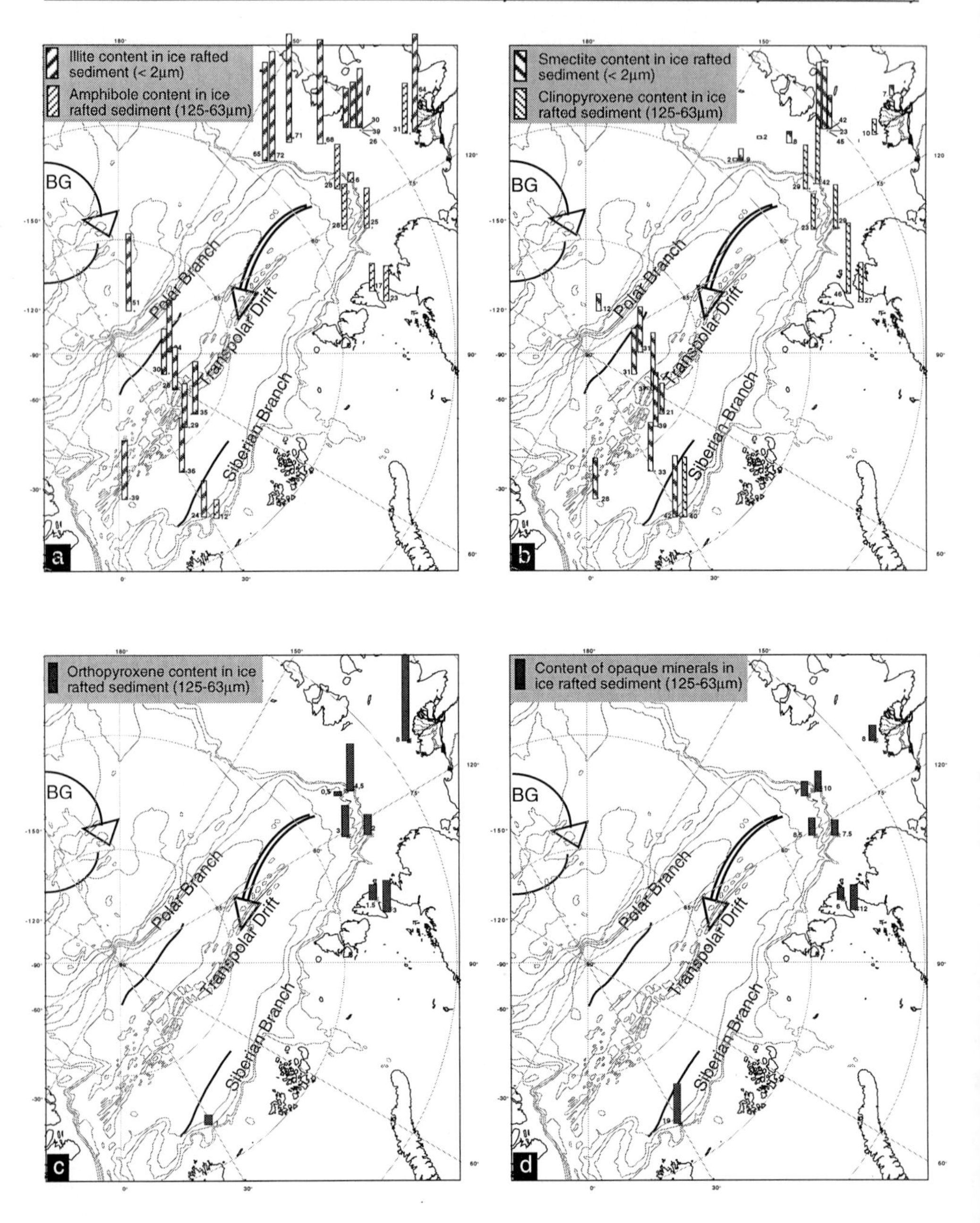

Figure 3: Heavy- and clay mineral data in sea-ice sediments in the Laptev Sea and the central Arctic Ocean (clay data from Wollenburg, 1993; Nürnberg et al., 1994; BG = Beaufort Gyre).

1991). Because an exact identification of these minerals is not possible using a polarisation microscope, they are grouped into "detrital carbonate".

To show the average error depending on the amount of counted grains, a calculation of confidence intervals (after Boenigk, 1983) is listed in Table 1.

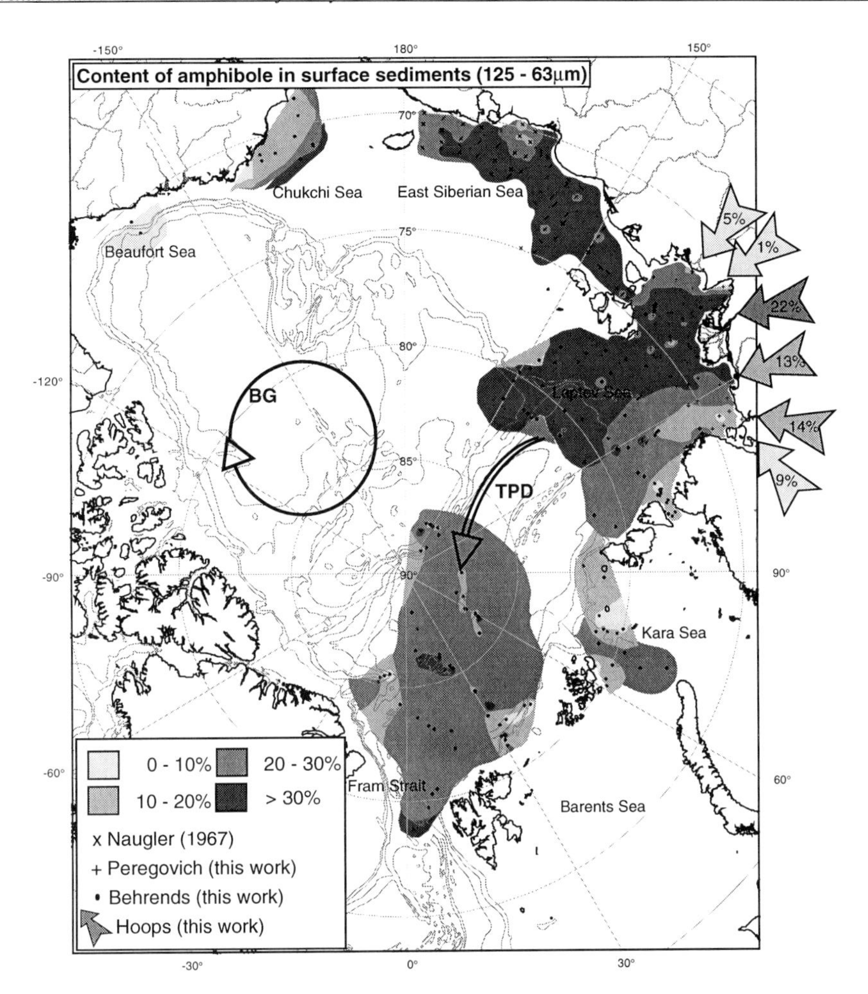

Figure 4: Distribution of amphibole (TPD = Transpolar Drift System; BG = Beaufort Gyre).

Results

Sea ice sediments

For this study, only a few sea ice sediment samples with heavy mineral data were available. The amphibole distribution on the Laptev Sea shelf shows highest values off the river mouth and in the central Laptev Sea. The content of amphibole in the sample north of Svalbard is 12 % (Figure 3a). Near the outflow of the Lena river, the content of clinopyroxene is 10 %, while toward the north, the values are increasing. The multi-year ice of the Transpolar Drift shows a high content of 40 % clinopyroxene (Figure 3b). The amount of orthopyroxene is low (0.5 and 8 %); the highest values are found in front of the Lena outflow (Figure 3c). The values of opaque mineral values are 0.5 and 8 % on the Laptev Sea shelf, but 19 % in the sample north of Svalbard (Figure 3d). Detrital carbonates are not found in the investigated sea-ice sediments.

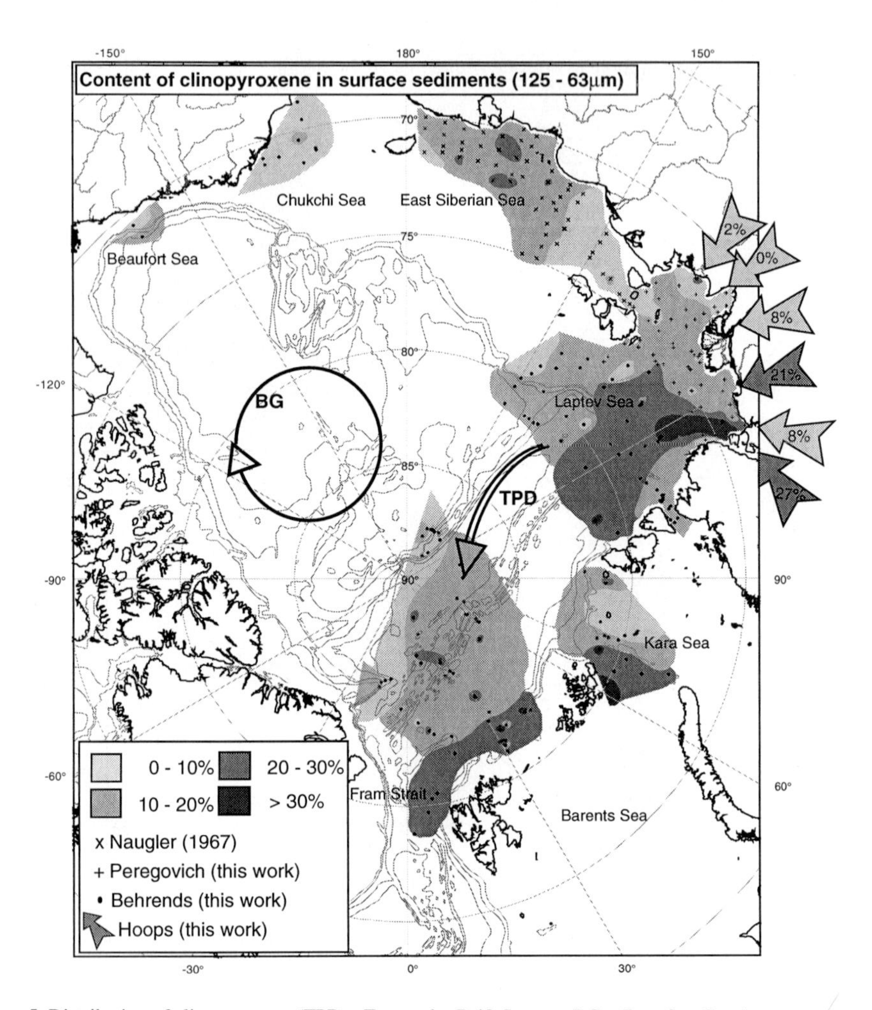

Figure 5: Distribution of clinopyroxene (TPD = Transpolar Drift System; BG = Beaufort Gyre).

Surface sediments from east Siberian rivers, the Laptev Sea shelf, the central Arctic Ocean and adjacent shelf areas

Distribution of amphibole

High amounts of amphibole were found in the East Siberian Sea as well as in the eastern Laptev Sea, with values >30 %, decreasing to the west and to the north. In the rivers draining into the Laptev Sea, the highest contents occur in the Lena delta sediments (22 %). The amphibole distribution in the central Arctic Ocean sediments generally show values between 10-20 % (Figure 4).

Distribution of clinopyroxene

The distribution of clinopyroxene displays its highest values in the western Laptev Sea. The maximum values seem to be related to the Khatanga outflow, where river sediments contain 27 % clinopyroxene. Other potential source areas (Beaufort Sea, Chukchi Sea, East Siberian Sea)

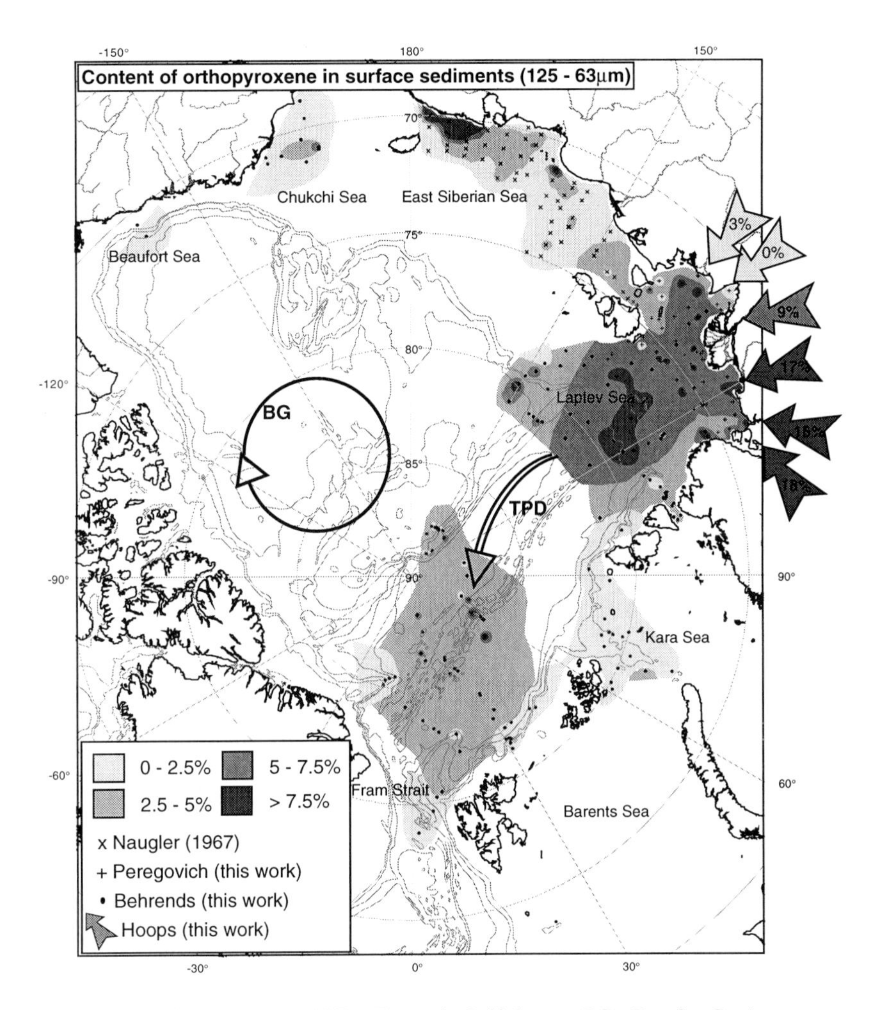

Figure 6: Distribution of orthopyroxene (TPD = Transpolar Drift System; BG = Beaufort Gyre).

are characterized by lower clinopyroxene values (<30 %), except for the Kara Sea where sediments show slightly higher amounts (Figure 5).

Distribution of orthopyroxene

The values of orthopyroxene are generally low in the investigation area. Local maximum values only appear in the eastern East Siberian Sea, in contrast to the Laptev Sea where generally higher contents (≥7.5 %) were found. From the Laptev Sea the distribution of orthopyroxene shows decreasing trends to the east and west and northward to the central Arctic Ocean (Figure 6).

Distribution of detrital carbonate

The Beaufort and the Chukchi Sea sediments contain up to 16 % of detrital carbonate. The rivers draining into the Laptev Sea, on the other hand, carry only very small amounts of detrital

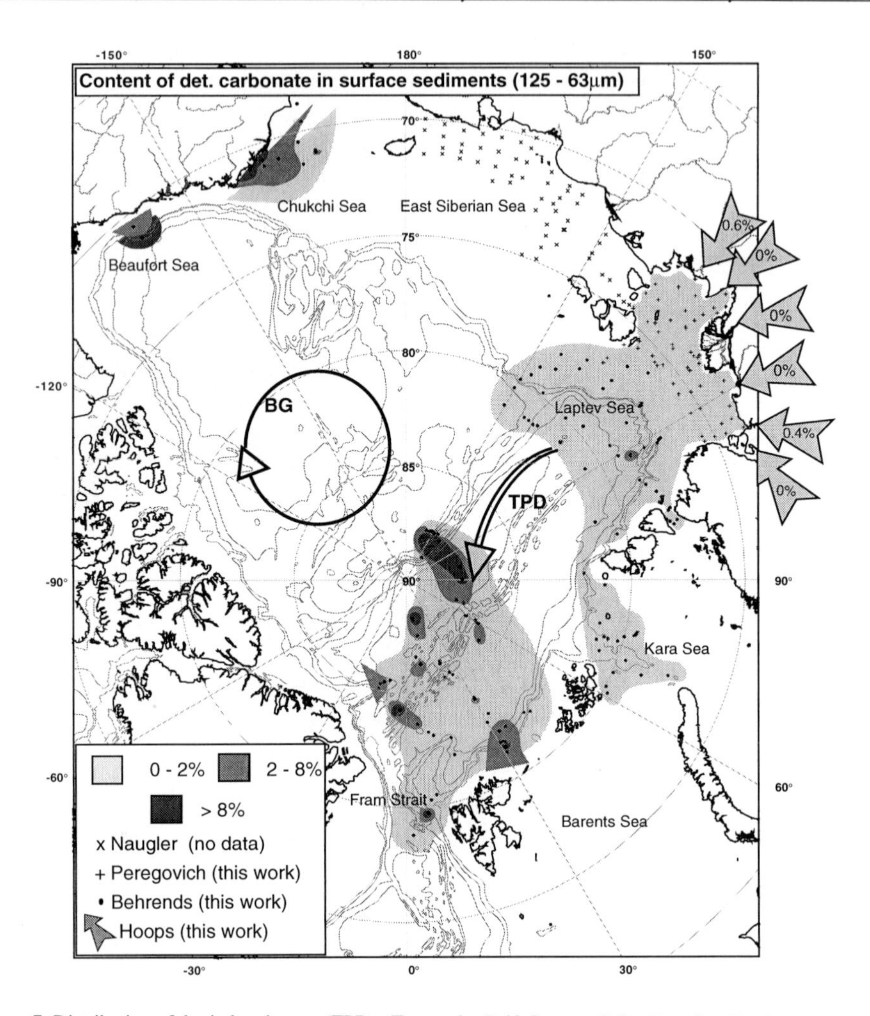

Figure 7: Distribution of detrital carbonate (TPD = Transpolar Drift System; BG = Beaufort Gyre).

carbonate in their freight. Thus, detrital carbonate contents <2 % were determined in the shelf sediments of the Laptev Sea and the Kara Sea. In the Central Arctic values of >2 % are common; at some locations up to 8 % and more are reached (Figure 7).

Distribution of opaque minerals

Generally, the amount of opaque minerals is high in the investigation area, varying between 20 and 30 %, except for the Laptev Sea shelf where minimum values of <10% occur. In contrast, the rivers Yana and Anabar contain 42 % and 29 %, respectively. From the Laptev Sea toward the central Arctic Ocean a trend of decreasing amounts of opaque minerals was observed (Figure 8).

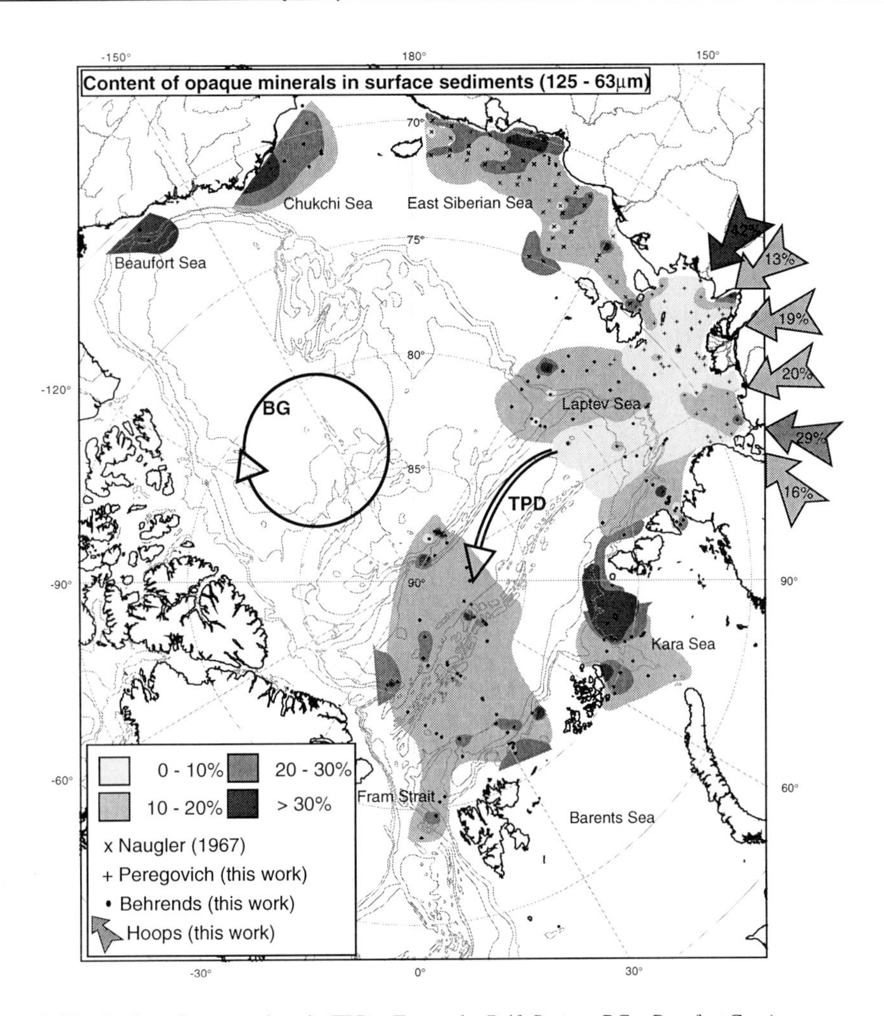

Figure 8: Distribution of opaque minerals (TPD = Transpolar Drift System; BG = Beaufort Gyre).

Discussion

The general circulation pattern of the Transpolar Drift system was reflected in the smectite and illite distribution in multi-year sea ice from the Arctic Ocean (Nürnberg et al., 1994; Figures 3a, 3b). Smectite concentrations found in the Laptev Sea correlate to those in central Arctic Ocean sea ice, which provides evidence for a ice drift branch north of Svalbard, linking source areas in the Laptev Sea to sediment-laden ice in the central Arctic Ocean. Distribution of freshwater diatoms and shallow-water benthic foraminifera in sea ice show a comparable pattern (Abelmann, 1992; Wollenburg, 1995). In this study heavy minerals in surface sediments should be tested as tracer for sea-ice transport and surface water circulation pattern. Until now, the heavy mineral analysis of sea-ice sediments only give an idea about the distribution, but shows clear trends that can be combined and compared with the clay mineralogy of sea-ice sediments. Small scattering of mineralogical parameters in sea ice on the shelf are probably due to the small data base and/or the annual variability of sediment transported by sea ice, because suspension freezing is expected to occur only once a year or every few years for a short time

(Reimnitz et al., 1992). This is the reason why sediment samples from first-year ice of the Laptev Sea (Wollenburg, 1993) are probably not showing expected clear gradients, which were found in mineralogical parameters in sediments from the shelf floor (clay: Wahsner 1995, Wahsner et al., 1996, 1998; heavy minerals: Lapina, 1965; Silverberg, 1972; Behrends et al. 1996). Nevertheless, mineralogical data display an eastern-Laptev-Sea-influenced Transpolar Drift and Polar branch in the central Arctic Ocean, while the Siberian branch can be related to the western shelf area of the Laptev Sea.

Concerning the distribution of heavy minerals in the Arctic Ocean, a clear relation between the mineralogical characteristics of the central Arctic Ocean and the heavy mineral inventory of the Laptev Sea shelf and slope is obvious. Underneath the Siberian branch, a clinopyroxene maximum can be related to the western Laptev Sea. The sea-ice-sediment input from the Kara Sea is probably rather small due to deeper water depth and therefore worse conditions for sediment entrainment into sea ice (Kempema et al., 1989; Nürnberg et al., 1995; Pfirmann et al., 1989, 1997). Based on the low orthopyroxene concentrations the Kara Sea can be excluded as a main source. The generally low orthopyroxene concentration and, compared with the Laptev Sea, lower contents of clinopyroxene also suggest the East Siberian Sea not to be a potential source area.

Underneath the Transpolar Drift as well as the Polar Branch the mineralogical characteristics show a heavy mineral pattern related to the eastern Laptev Sea shelf and slope. The values of heavy mineral components are highly variable on the Lomonosov Ridge, probably due to mixing and export of sediment-laden sea ice of the Beaufort Gyre into the Transpolar Drift System. This is indicated by a higher content of detrital carbonate, which does not occur in such high amounts on the Siberian shelves. Until now, detrital carbonate is not found in the sea-ice sediment samples. On the East Siberian Sea shelf detrital carbonate is not listed as a separate component (Figure 7), but included in "traces", showing an irregular distribution with only two samples with high values of 4 % (Naugler, 1967).

Detrital carbonate is not only indicating the influence of the Beaufort Gyre and therefore the sediment input from the Amerasian shelves into the eastern Arctic Ocean (Darby et al., 1989; Nørgaard-Pedersen, 1996), but also shows the input from Svalbard along the Barents Sea continental slope (Solheim and Elverhøi, 1996; Vogt 1997). In this area, however, the interpretation is more difficult due to mixing of signals, caused by different transport mechanisms of sediment, such as ice rafting and melting, and downslope transport of sediment.

Another mineral group typical for the Amerasian shelves are opaque minerals. Other mineral combinations already excluded the Kara Sea (see above), where also higher amounts of opaques were found, as a main source. Thus, the ice-rafted sediment from the Beaufort Sea and the Chukchi Sea transported via the Beaufort Gyre to the Lomonosov Ridge, are probably causing the slightly higher concentration of opaque minerals in the Arctic Ocean sediment in comparison to the Laptev Sea shelf sediments. The discrepancy of concentrations of investigated heavy minerals found on the shelf in comparison with those in the river sediments are probably due to different maximum output of suspended particular matter of the single rivers and/or the different hydromorphology.

Conclusion

Heavy minerals in the Arctic Ocean were studied with respect to their relevance for tracing transport pathways. This study shows that heavy minerals can be used to identify source areas and to reconstruct sea-ice sediment input and surface water circulation patterns. The investigation of heavy minerals on core material will give new important information about past sea-ice drift pattern.

Acknowledgements

This study was performed within the bilateral Russian-German multidisciplinary research project "System Laptev Sea". We would like to thank R. Stein for improving the manuscript and for generous support. We thank the reviewers J. Bischoff and R.F. Spielhagen for numerous constructive suggestions and comments. The financial supports by the Alfred Wegener Institute and the German Ministry of Education, Science, Research and Technology (BMBF) are gratefully acknowledged. This is contribution No. 1371 of the Alfred-Wegener-Institute for Polar and Marine Research.

References

Abelmann, A. (1992) Diatom assemblages in Arctic sea ice - indicator for ice drift pathways. Deep Sea Res., 39 (2), 525-538.

Behrends, M., B. Peregovich and R. Stein (1996) Terrigenous sediment supply into the Arctic Ocean: Heavy mineral distribution in the Laptev Sea. In: Surface-sediment composition and sedimentary processes in the Central Arctic Ocean and along the Eurasian Continental Margin, R. Stein, G. Ivanov, M. Levitan and K. Fahl (eds.), Reports on Polar Res., 212, 37-42.

Boenigk, W. (1983) Schwermineralanalyse. Ferdinand Enke, Stuttgart, 158 pp.

Darby, D.A., A.S. Naidu, T.C. Mowatt and G. Jones (1989) Sediment composition and sedimentary processes in the Arctic Ocean. In: The Arctic Seas - Climatology, Oceanography, Geology, and Biology, Herman, Y. (ed.), van Nostrand Reinhold, New York, 657-720.

Dethlef, D., D. Nürnberg, E. Reimnitz, M. Saarso and Y. P. Savchenko (1993) East Siberian Arctic Region Expedition ´92: The Laptev Sea - Its Significance for Arctic Sea ice Formation and Transpolar Sediment Flux. Reports on Polar Res., 120, 3-37.

Dethlef, D. (1995) Die Laptevsee - eine Schlüsselregion für den Fremdstoffeintrag in das arktische Meereis. Ph.D. thesis, Christian-Albrechts-University, Kiel.

Eicken, H., T. Viehoff, T. Martin, J. Kolatschek, V. Alexandrov and E. Reimnitz (1995) Studies of clean andsediment-laden ice in the Laptev Sea. In: Russian-German Cooperation: The Laptev Sea System, Kassens, H., D. Piepenburg, J. Thiede, L. Timokhov, H.-W. Hubberten and S. M. Priamikov (eds.), Reports on Polar Res.,176, 62-70.

Eicken, H., E. Reimnitz, T. Alexandrov, T. Martin, H. Kassens and T. Viehoff (1997) Sea-ice processes in the Laptev Sea and their importance for sediment export. Cont. Shelf. Res., 17 (2), 205-233.

Fessenden, F.W. (1959) Removal of heavy liquid separates from glass centrifuge tubes. J. Sed. Petrol., 29, 269-280.

Fütterer, D.K. (1992) ARCTIC '91: The Expedition ARK-VIII/3 of RV "Polarstern" in 1991. Reports on Polar Res., 107, 267 pp.

Fütterer, D.K. (1994) The expedition ARCTIC´93, Leg ARK-IX/4 of RV "Polarstern" 1993. Reports on Polar Res., 149, 244 pp.

Gordeev, V.V. and I.S. Sidorov (1993) Concentrations of major elements and their outflow into the Laptev Sea by the Lena River. Mar. Chem., 43, 33-45.

Hermel, J. (1995) Sedimentpetrographische Untersuchungen an Flußsedimenten der Lena (Yakutsk - Laptev-See), Ostsibirien. Unpubl. diploma thesis, University Braunschweig, 120 pp.

Kassens, H. and V. Karpiy (1994) Russian German Cooperation: The Transdrift I Expedition to the Laptev Sea. Reports on Polar Res., 151, 168 pp.

Kassens, H., D. Piepenburg, J. Thiede, L. Timokhov, H.-W. Hubberten and S. M. Priamikov (1995) Russian-German Cooperation: The Laptev Sea System. Reports on Polar Res., 176, 387 pp.

Kassens, H. (1995) Laptev Sea System: Expeditions in 1994. Reports on Polar Res., 182, 195 pp.

Kempema, E.W., E. Reimnitz and P.W. Barnes (1989) Sea-ice sediment entrainment and rafting in the Arctic. J. Sed. Petrology, 59, 308-317.

Lapina, N. N. (1965) The determination of distribution paths of sediments, based on mineralogical investigations of marine deposits (example Laptev Sea). Uchennye Zapiski NIIGA, Ser. Region. Geol., 7, 139-157 (in Russian).

Lisitzin, A. P. (1996) Oceanic Sedimentation. American Geophysical Union, Washington, 400 pp.

Luepke, G. and E. Escowitz (1989) Grain-Size, Heavy-Mineral, and Geochemical Analyses of Sediments from the Chuckchi Sea, Alaska. U.S. Geological Survey Bulletin, 1896, 8 pp.

Mange, M. A. and H.F.W. Maurer (1991) Schwerminerale in Farbe. Ferdinand Enke, Stuttgart, 148 pp.

Naugler, F.P. (1967) Recent sediments of the East Siberian Sea. Ph.D. thesis, University of Washington, 70 pp.

Nürnberg, D., I. Wollenburg, D. Dethlef, H. Eicken, H. Kassens, T. Letzig, E. Reimnitz and J. Thiede (1994) Sediments in Arctic sea ice: Implications for entrainment, transport and release. In: Arctic Ocean Marine Geology, Thiede, J., T.O. Vorren and R.F. Spielhagen, Mar. Geol., Elsevier Science B.V., Amsterdam,185-214.

Nürnberg, D., M.A. Levitan, J.A. Pavlidis and E.S. Shelekhova (1995) Distribution of clay minerals in surface sediments from the eastern Barents and southwestern Kara Seas. Geol. Rundsch., 84, 665-682.

Nørgaard-Pedersen, N. (1996) Late Quaternary Arctic Ocean Sediment Records: Surface Ocean Conditions and Provenance of Ice-rafted Debris. Ph.D. thesis, Christian-Albrechts-University, 115 pp.

Pfirman, S. L. and A. Solheim (1989) Subglacial meltwater discharge in the open marine tidewater glacier environment: Observations from Nordaustlandet. Mar. Geol., 86, 283-319.

Pfirman, S., M.A. Lange, I. Wollenburg and P. Schlosser (1990) Sea ice characteristics and the role of sediment inclusions in deep-sea deposition: Arctic and Antarctic comparisons. In: Geological History of the Polar Oceans: Arctic versus Antarctic, Bleil U. and J. Thiede (eds.), Kluwer Academic Publishers, Dordrecht, 187-211.

Pfirman, S.L., R. Colony, D. Nürnberg, H. Eicken and I. Rigor, (1997) Reconstructing the origin and trajectory of drifting Arctic sea ice. J. of Geophys. Res., 102 (C6), 12575-12568.

Rachold, V., J. Hermel and V.N. Korotaev (1995) Expedition to the Lena River in July-August 1994. In: Laptev Sea System: Expeditons in 1994, Kassens, H. (ed.), Reports on Polar Res., 182, 181-195.

Rachold, V., E. Hoops, A.M. Alabyan, V.N. Korotaev and A.A. Zaitsev (1997) Expedition to the Lena and Yana Rivers. In: Expeditions in 1995, Kassens, H. (ed.), Reports on Polar Res., 248, 193-210.

Rachold, V., E. Hoops, and A.W. Ufimtsev Expedition to the Khatanga River July-August 1996. In: Expeditons in 1996, Kassens, H. and V. Churun (eds.), Reports on Polar Res. (in prep.).

Rachold, V., A. Alabyan, H.-W. Hubberten, V.N. Korotaev and A.A. Zaitsev (1996) Sediment transport to the Laptev Sea - hydrology and geochemistry of the Lena River. Polar Research, 15(2), 183-196.

Rachor, E. (1997) Scientific Cruise Report of the Arctic Expedition ARK-XI/1 of RV "Polarstern" 1995. Reports on Polar Res., 266, 157 pp.

Reimnitz, E., L. Marincovich, Jr., M. McCormick and W.M. Briggs (1992) Suspension freezing of bottom sediment and biota in the Northwest Passage and implications for Arctic Ocean sedimentation. Can. J. Earth Sci., 29, 693-703.

Reimnitz, E., D. Dethleff and D. Nürnberg (1994) Contrasts in Arctic shelf sea-ice regimes and some implications: Beaufort Sea versus Laptev Sea. In: Arctic Ocean Marine Geology, Thiede, J., T.O. Vorren and R.F. Spielhagen (eds.), Mar. Geol., Elsevier Science B.V., Amsterdam, 215-225.

Reimnitz, E., H. Eicken and T. Martin (1995) Multiyear fast ice along the Taymyr Peninsula, Siberia. Arctic, 48, 359-367.

Scull, B.J. (1960) Removal of heavy liquid separates from glass centrifuge tubes. Alternativ method. J. Sed. Petrol., 30, 626.

Silverberg, N. (1972) Sedimentology of the surface sediments of the East Siberian and Laptev Seas. University of Washington, 184 pp.

Stein, R., C. Schubert, H. Grobe and D.K. Fütterer (1994a) Late Quaternary changes in organic carbon and carbonate deposition in the central Arctic Ocean: First results of the ARCTIC ´91 Expedition. In: Proceedings of the 1992 International Conference on Arctic Margins, Anchorage, Alaska. D.K. Thurston and K. Fujita, U.S. Depart. Interior, Miner. Mang. Surv., Alaska Outer Continental Shelf Region, 363-368.

Stein, R. and S. Korolev (1994) Shelf-to-basin sediment transport in the eastern Arctic Ocean. In: Russian-German Cooperation in the Siberian Shelf Seas: Geo-System Laptev-Sea, Kassens, H., H.-W. Hubberten, S. M. Pryamikov and R. Stein (eds.), Reports on Polar Res., 144, 87-100.

Solheim, A. and A. Elverhøi (1996) Structure and lithological composition of Quaternary sediments of the Kara Sea. In: Surface-sediment composition and sedimentary processes in the Central Arctic Ocean and along the Eurasian Continental Margin, R. Stein, G. Ivanov, M. Levitan and K. Fahl (eds.), Reports on Polar Res., 212, 144-158.

Thiede, J. (1988) Scientific Cruise Report of Arctic Expedition ARK-IV/3. Reports on Polar Res., 43, 237 pp.

Timokhov, L.A. (1994) Regional characteristics of the Laptev and the East Siberian Seas: Climate, topography, ice phases, thermohaline regime, circulation. In: Russian-German Cooperation in the Siberian Shelf Seas: Geo-System Laptev-Sea, Kassens, H., H.-W. Hubberten, S.M. Pryamikov and R. Stein (eds.), Reports on Polar Res., 144, 15-31.

Vogt, C. (1997) Zeitliche und räumliche Verteilung von Mineralvergesellschaftungen in spätquartären Sedimenten des Arktischen Ozeans und ihre Nützlichkeit als Klimaindikatoren während der Glazial/Interglazial-Wechsel. Reports on Polar Res., 251, 309 pp.

Wahsner, M. (1995) Mineralogical and sedimentological characterization of surface sediments from the Laptev Sea. In: Russian-German Cooperation: The Laptev Sea System, Kassens, H., D. Piepenburg, J. Thiede, L. Timokhov, H.-W. Hubberten and S. M. Priamikov (eds.), Reports on Polar Res., 176, 303-313.

Wahsner, M., G. Tarasov and G. Ivanov (1996) Marine geological investigations of surface sediments in the Franz-Josef-Land area and the St. Anna Trough. In: Surface-sediment composition and sedimentary processes in the Central Arctic Ocean and along the Eurasian Continental Margin, Stein R., G. Ivanov, M. Levitan and K. Fahl (eds.), Reports on Polar Res., 212, 172-184.

Wahsner, M., C. Müller, R. Stein, M. Levitan, G. Ivanov, E. Shelekova and G. Tarasov (1998) Clay mineral distributions in surface sediments from the central Arctic Ocean and the Eurasian continental margin as indicator for source areas and transport pathways - A synthesis. Boreas, subm.

Wollenburg, I. (1993) Sedimenttransport durch das Arktische Meereis: Die rezente lithogene und biogene Materialfracht. Reports on Polar Res., 127, 159 pp.

Wollenburg, J. (1995) Benthische Foraminiferenfaunen als Wassermassen-, Produktions- und Eisdriftanzeiger im Arktischen Ozean. Reports on Polar Res., 179, 227 pp.

Table 1: Sediment samples (surface samples 0-1 cm, fraction 63-125 µm in grain %) from the Beaufort Sea, Chukchi Sea, Laptev Sea, Kara Sea, central Arctic Ocean, Siberian rivers and sea ice.

Latitude °N	Longitude °E	Water depth (m)	Sample No.	Amphibole	Estimation of error +/-*	Orthopyroxen	Estimation of error +/-*	Clinopyroxen	Estimation of error +/-*	Epidote	Estimation of error +/-*	Apatite	Estimation of error +/-*	ZTR (Zircon/Tourmaline/Rutile)	Sphen	Estimation of error +/-*	Garnet	Estimation of error +/-*	Opaque Minerals	Estimation of error +/-*	Detrital Carbonate	Estimation of error +/-*	Biotite and Chlorite	Kyanite and Staurolite	Others	Sum of counted grains
Barents Sea continental slope																										
81.591	31.550	1384	PS1516	16.1	6.3	1.5	2.1	32.1	8.0	10.9	5.3	0.7	1.4	0.7	0.7	1.4	2.9	2.9	19.7	6.8	9.5	5.0	5.1	0.0	0.0	137
81.645	31.510	1822	PS1517	23.0	6.8	2.6	2.6	25.7	7.1	7.2	4.2	2.6	2.6	1.3	0.7	1.4	2.6	2.6	21.1	6.6	5.3	3.6	6.6	1.3	0.0	152
81.678	31.200	2572	PS1518	16.8	5.5	0.5	1.0	24.3	6.3	14.1	5.1	1.6	1.8	1.5	2.7	2.4	6.5	3.6	13.5	5.0	5.4	3.3	12.4	0.5	0.0	185
82.155	31.900	3045	PS1520	11.3	5.0	1.3	1.8	36.9	7.6	7.5	4.2	3.1	2.7	2.5	3.1	2.7	4.4	3.2	18.8	6.2	7.5	4.2	3.8	0.0	0.0	160
83.083	31.900	3789	PS1521	33.6	6.2	6.5	3.2	15.9	4.8	11.2	4.1	3.9	2.5	0.0	3.0	2.2	4.7	2.8	15.1	4.7	0.4	0.8	5.6	0.0	0.0	232
84.023	30.380	4045	PS1522	12.1	4.6	4.5	2.9	24.7	6.1	10.6	4.4	5.1	3.1	1.0	6.1	3.4	8.1	3.9	20.2	5.7	0.5	1.0	7.1	0.0	0.0	198
85.358	26.360	3634	PS1524	30.2	6.9	3.4	2.7	20.1	6.0	12.3	4.9	3.9	2.9	0.0	2.8	2.5	6.7	3.7	11.7	4.8	0.6	1.2	7.3	1.2	0.0	179
85.510	25.270	3137	PS1525	29.3	6.2	4.7	2.9	11.2	4.3	14.4	4.8	3.3	2.4	2.3	2.8	2.3	7.0	3.5	19.1	5.4	0.5	1.0	5.1	0.5	0.0	215
86.035	22.130	3727	PS1527	32.8	6.2	3.5	2.4	20.5	5.3	8.7	3.7	3.5	2.4	2.1	3.9	2.6	7.0	3.4	13.5	4.5	0.0	0.0	3.4	0.9	0.0	229
86.138	23.020	4029	PS1528	26.8	5.9	3.1	2.3	20.5	5.4	13.4	4.6	2.7	2.2	2.2	3.6	2.5	7.6	3.5	12.5	4.4	0.4	0.8	7.1	0.0	0.0	224
85.381	21.920	2875	PS1529	31.9	5.8	3.1	2.2	17.7	4.8	12.6	4.2	1.6	1.6	1.6	4.3	2.5	8.7	3.5	11.4	4.0	0.8	1.1	6.3	0.0	0.0	254
73.646	22.920	459	PS2439	30.5	6.1	0.0	0.0	2.2	2.0	9.7	3.9	7.1	3.4	5.7	5.3	3.0	10.2	4.0	17.3	5.0	1.3	1.5	9.7	0.4	0.4	226
81.210	30.600	187	PS2440	29.8	5.9	2.5	2.0	19.7	5.2	9.2	3.7	5.5	3.0	0.8	1.7	1.7	1.3	1.5	21.4	5.3	5.5	3.0	1.6	0.4	0.4	238
81.465	30.890	560	PS2441	17.1	4.5	0.3	0.6	14.0	4.1	5.6	2.7	1.0	1.2	0.9	1.7	1.5	0.7	1.0	16.1	4.3	23.8	5.0	18.1	0.3	0.0	286
81.685	30.330	2806	PS2442	14.0	6.5	1.8	2.5	20.2	7.5	8.8	5.3	1.8	2.5	1.8	1.8	2.5	1.8	2.5	8.8	5.3	5.3	4.2	34.2	0.0	0.0	114
82.201	34.590	2463	PS2443	6.8	7.6	0.0	0.0	18.2	11.6	4.5	6.3	2.3	4.5	0.0	0.0	0.0	9.1	8.7	27.3	13.4	4.5	6.3	27.2	0.0	0.0	44
82.163	42.040	1013	PS2447	12.9	4.5	1.8	1.8	31.3	6.2	11.6	4.3	1.8	1.8	0.8	3.1	2.3	6.7	3.3	17.0	5.0	0.4	0.8	11.6	0.9	0.0	224
82.023	43.570	287	PS2449	7.7	3.4	0.0	0.0	23.2	5.4	6.1	3.1	0.8	1.1	7.7	3.7	2.4	5.7	3.0	42.7	6.3	0.0	0.0	0.8	1.2	0.4	246
Central Arctic Ocean																										
82.776	29.930	3800	PS2158	18.8	5.7	3.2	2.6	22.6	6.1	10.2	4.4	2.2	2.2	1.6	2.2	2.2	6.5	3.6	25.3	6.4	0.5	1.0	5.4	0.5	1.1	186
83.960	30.370	4055	PS2159	22.3	5.9	4.1	2.8	23.4	6.0	8.6	4.0	3.0	2.4	1.5	2.5	2.2	9.6	4.2	16.8	5.3	3.6	2.7	2.5	1.0	1.0	197
85.794	50.830	3981	PS2162	16.9	7.9	7.9	5.7	23.6	9.0	12.4	7.0	2.2	3.1	0.0	1.1	2.2	5.6	4.9	16.9	7.9	6.7	5.3	5.6	1.1	0.0	89
86.241	59.220	3040	PS2163	32.8	6.6	3.4	2.5	19.1	5.5	9.8	4.2	3.4	2.5	1.0	1.0	1.4	4.9	3.0	15.7	5.1	4.4	2.9	4.5	0.0	0.0	204
86.337	59.180	2004	PS2164	14.9	4.2	2.5	1.9	14.9	4.2	15.2	4.3	2.1	1.7	0.4	1.8	1.6	7.1	3.1	24.8	5.1	1.4	1.4	13.8	0.4	0.7	282
86.446	59.960	2011	PS2165	15.5	5.0	4.2	2.7	13.1	4.6	10.3	4.2	3.8	2.6	5.6	2.3	2.1	9.9	4.1	24.4	5.9	2.3	2.1	7.5	1.0	0.0	213
86.860	59.700	3618	PS2166	27.0	7.3	8.8	4.7	14.2	5.7	10.8	5.1	0.7	1.4	0.7	4.1	3.3	7.4	4.3	12.2	5.4	2.0	2.3	11.5	0.7	0.0	148
86.945	59.020	4434	PS2167	19.4	9.3	4.2	4.7	6.9	6.0	5.6	5.4	0.0	0.0	0.0	0.0	0.0	4.2	4.7	38.9	11.5	0.0	0.0	20.9	0.0	0.0	72
87.585	68.980	4384	PS2171	27.5	7.1	1.3	1.8	15.6	5.7	0.6	1.2	1.3	1.8	1.3	5.0	3.4	8.8	4.5	18.1	6.1	3.1	2.7	15.0	1.9	0.6	160
87.257	68.380	4384	PS2172	18.6	6.2	6.4	3.9	16.0	5.9	16.0	5.9	2.6	2.5	1.9	1.9	2.2	7.1	4.1	19.2	6.3	3.8	3.1	5.1	1.2	0.0	156

Table 1 (continued):

Latitude °N	Longitude °E	Water depth (m)	Sample No.	Amphibole	Estimation of error +/-*	Orthopyroxen	Estimation of error +/-*	Clinopyroxen	Estimation of error +/-*	Epidote	Estimation of error +/-*	Apatite	Estimation of error +/-*	ZTR (Zircon/Tourmaline/Rutile)	Sphen	Estimation of error +/-*	Garnet	Estimation of error +/-*	Opaque Minerals	Estimation of error +/-*	Detrital Carbonate	Estimation of error +/-*	Biotite and Chlorite	Kyanite and Staurolite	Others	Sum of counted grains
87.494	90.500	4427	PS2174	14.0	6.3	2.5	2.8	19.0	7.1	18.2	7.0	4.1	3.6	1.6	0.8	1.6	2.5	2.8	20.7	7.4	9.9	5.4	6.6	0.0	0.0	121
87.567	103.600	4378	PS2175	29.9	7.1	2.4	2.4	16.5	5.8	9.8	4.6	0.6	1.2	1.2	1.8	2.1	3.0	2.7	11.0	4.9	12.2	5.1	10.3	1.2	0.0	164
88.005	159.200	4009	PS2178	31.4	6.2	2.7	2.2	12.1	4.4	8.5	3.7	4.0	2.6	3.1	2.7	2.2	4.9	2.9	9.0	3.8	9.0	3.8	11.7	0.9	0.0	223
87.746	138.000	1230	PS2179	22.2	6.0	5.7	3.3	9.3	4.2	15.5	5.2	4.1	2.8	0.5	3.6	2.7	2.6	2.3	16.5	5.3	11.3	4.5	8.3	0.0	0.5	194
87.626	156.700	4005	PS2180	25.8	5.9	5.0	2.9	11.3	4.3	13.1	4.5	1.4	1.6	1.8	1.4	1.6	8.6	3.8	14.0	4.7	5.0	2.9	11.7	0.0	0.9	221
87.595	153.100	3112	PS2181	18.8	5.1	1.7	1.7	7.7	3.5	12.8	4.4	3.0	2.2	4.3	5.1	2.9	9.0	3.7	27.8	5.9	1.3	1.5	6.4	1.3	0.9	234
87.572	151.100	2489	PS2182	23.4	6.0	3.6	2.7	14.7	5.0	12.7	4.7	2.0	2.0	4.0	5.6	3.3	12.7	4.7	12.7	4.7	1.5	1.7	7.1	0.0	0.0	197
87.602	148.800	2016	PS2183	13.7	4.2	0.4	0.8	9.9	3.7	10.6	3.8	3.0	2.1	4.2	5.7	2.9	13.7	4.2	30.0	5.7	2.3	1.8	5.3	1.1	0.0	263
87.611	148.100	1640	PS2184	23.4	5.4	2.0	1.8	9.7	3.8	9.3	3.7	4.4	2.6	1.2	5.2	2.8	4.0	2.5	23.8	5.4	5.2	2.8	10.9	0.8	0.0	248
87.529	144.200	1073	PS2185	31.7	5.7	4.2	2.5	13.6	4.2	10.2	3.7	1.9	1.7	1.9	3.8	2.3	8.3	3.4	15.5	4.4	1.9	1.7	6.4	0.4	0.4	265
88.512	139.900	1867	PS2186	25.3	7.0	4.5	3.3	10.4	4.9	7.1	4.1	7.8	4.3	1.2	1.9	2.2	5.8	3.8	18.8	6.3	9.7	4.8	7.1	0.0	0.0	154
88.780	144.600	1001	PS2189	14.7	4.4	1.9	1.7	6.0	2.9	9.4	3.6	1.9	1.7	3.4	4.2	2.5	7.9	3.3	43.4	6.1	0.8	1.1	5.3	0.4	0.8	265
88.260	9.856	4375	PS2192	28.2	10.2	5.1	5.0	21.8	9.4	3.8	4.3	5.1	5.0	0.0	2.6	3.6	6.4	5.5	15.4	8.2	9.0	6.5	2.6	0.0	0.0	78
87.511	11.480	4399	PS2193	27.8	7.5	2.8	2.7	13.9	5.8	10.4	5.1	4.9	3.6	2.1	4.2	3.3	4.9	3.6	20.8	6.8	4.9	3.6	2.8	0.0	0.7	144
86.593	7.487	4326	PS2194	20.4	5.7	7.0	3.6	26.9	6.3	6.0	3.4	1.5	1.7	0.0	3.0	2.4	4.5	2.9	24.9	6.1	2.0	2.0	3.0	0.5	0.5	201
86.253	9.616	3793	PS2195	31.4	6.2	2.7	2.2	14.6	4.7	10.2	4.0	4.9	2.9	0.0	3.1	2.3	5.3	3.0	17.7	5.1	4.0	2.6	5.3	0.9	0.0	226
85.564	-9.060	3820	PS2198	8.8	3.8	0.4	0.8	8.8	3.8	12.8	4.4	7.0	3.4	2.6	5.7	3.1	21.1	5.4	29.1	6.0	0.4	0.8	2.6	0.4	0.0	227
85.434	-11.900	1789	PS2199	7.1	3.1	1.1	1.2	5.7	2.8	8.5	3.3	4.3	2.4	2.5	5.0	2.6	26.6	5.3	34.8	5.7	4.3	2.4	0.0	0.4	0.0	282
Fram Strait																										
85.327	-14.000	1073	PS2200	16.2	4.3	1.4	1.4	16.2	4.3	7.2	3.0	2.4	1.8	2.7	1.7	1.5	10.7	3.6	27.9	5.3	7.2	3.0	3.4	1.0	1.7	290
85.108	-14.400	1081	PS2202	20.1	4.7	1.0	1.2	10.0	3.5	8.7	3.3	2.4	1.8	3.1	2.1	1.7	8.7	3.3	26.3	5.2	7.6	3.1	3.1	0.3	6.6	289
84.262	-2.560	3020	PS2206	18.8	6.5	2.8	2.7	15.3	6.0	10.4	5.1	2.8	2.7	2.8	3.5	3.1	4.9	3.6	18.1	6.4	9.0	4.8	11.2	0.7	0.0	144
83.640	4.603	3681	PS2208	24.2	7.0	3.4	3.0	9.4	4.8	16.8	6.1	4.0	3.2	1.4	6.0	3.9	2.7	2.7	20.1	6.6	5.4	3.7	6.0	0.7	0.0	149
83.225	8.573	4046	PS2209	29.9	6.6	4.6	3.0	21.6	5.9	10.3	4.4	2.1	2.1	2.0	4.6	3.0	7.7	3.8	9.8	4.3	2.1	2.1	4.7	0.5	0.0	194
83.045	10.120	3949	PS2210	25.0	5.6	3.8	2.5	17.1	4.9	16.7	4.8	2.1	1.9	1.2	4.2	2.6	5.0	2.8	15.4	4.7	1.3	1.5	6.2	2.1	0.0	240
82.023	15.670	2531	PS2212	20.9	5.1	3.5	2.3	23.6	5.3	12.4	4.1	3.1	2.2	2.4	2.3	1.9	5.0	2.7	17.8	4.8	1.9	1.7	4.7	2.4	0.0	258
80.473	8.205	897	PS2213	24.9	5.5	2.9	2.1	26.9	5.7	13.5	4.4	2.0	1.8	2.8	2.0	1.8	6.9	3.2	13.5	4.4	1.6	1.6	1.6	1.2	0.0	245
80.268	6.626	552	PS2214	28.1	6.5	1.6	1.8	24.0	6.2	15.1	5.2	2.1	2.1	1.0	1.6	1.8	4.2	2.9	15.6	5.2	1.0	1.4	5.7	0.0	0.0	192
79.694	5.340	2045	PS2215	20.2	5.6	1.9	1.9	17.8	5.3	13.0	4.7	4.3	2.8	1.9	1.9	1.9	4.8	3.0	20.7	5.6	9.1	4.0	4.3	0.0	0.0	208
82.796	16.070	1376	PS1535	35.7	6.3	2.1	1.9	23.4	5.5	5.5	3.0	3.8	2.5	0.8	0.9	1.2	5.5	3.0	16.6	4.9	0.9	1.2	4.7	0.0	0.0	235

Table 1 (continued):

Latitude °N	Longitude °E	Water depth (m)	Sample No.	Amphibole	Estimation of error +/-*	Orthopyroxen	Estimation of error +/-*	Clinopyroxen	Estimation of error +/-*	Epidote	Estimation of error +/-*	Apatite	Estimation of error +/-*	ZTR (Zircon/Tourmaline/Rutile)	Sphen	Estimation of error +/-*	Garnet	Estimation of error +/-*	Opaque Minerals	Estimation of error +/-*	Detrital Carbonate	Estimation of error +/-*	Biotite and Chlorite	Kyanite and Staurolite	Others	Sum of counted grains
78.796	1.666	2544	PS1532	23.5	5.0	1.7	1.5	15.2	4.2	13.5	4.0	5.9	2.8	2.4	3.5	2.2	4.5	2.4	22.8	4.9	0.3	0.6	6.6	0.0	0.0	289
Laptev Sea																										
78.030	102.300	148	PS2450	13.3	4.4	2.1	1.9	10.7	4.1	7.7	3.5	2.6	2.1	1.3	2.6	2.1	3.4	2.4	8.2	3.6	0.0	0.0	48.1	0.0	0.0	233
77.706	102.300	143	PS2451	12.4	4.5	3.7	2.6	16.5	5.0	9.6	4.0	4.6	2.8	2.3	3.2	2.4	10.6	4.2	21.6	5.6	0.0	0.0	13.3	2.3	0.0	218
77.891	101.600	132	PS2452	16.0	4.0	3.0	1.9	26.4	4.8	8.9	3.1	1.5	1.3	1.5	3.0	1.9	11.3	3.4	25.5	4.7	0.0	0.0	2.1	0.9	0.0	337
76.508	133.400	38	PS2453	31.8	6.7	6.3	3.5	7.8	3.9	10.4	4.4	5.2	3.2	0.5	1.0	1.4	4.7	3.1	4.7	3.1	0.0	0.0	27.6	0.0	0.0	192
79.651	130.500	3429	PS2455	34.8	6.8	6.1	3.4	8.6	4.0	16.2	5.2	4.5	2.9	1.5	3.0	2.4	11.1	4.5	7.1	3.7	0.0	0.0	7.1	0.0	0.0	198
77.910	133.600	73	PS2461	27.2	5.2	6.1	2.8	20.7	4.7	11.6	3.7	3.4	2.1	2.0	3.1	2.0	3.4	2.1	15.0	4.2	0.0	0.0	7.2	0.3	0.0	294
77.405	133.600	54	PS2462	33.3	6.9	10.2	4.4	4.3	3.0	12.4	4.8	4.8	3.1	4.3	2.7	2.4	9.7	4.3	11.8	4.7	0.0	0.0	4.8	1.1	0.5	186
77.030	126.400	92	PS2463	30.3	6.7	8.5	4.1	5.3	3.3	8.5	4.1	5.9	3.4	2.1	8.0	4.0	14.9	5.2	15.4	5.3	0.0	0.0	1.1	0.0	0.0	188
77.183	126.200	1026	PS2465	29.1	6.5	11.2	4.5	16.8	5.3	14.3	5.0	4.6	3.0	1.0	5.1	3.1	7.7	3.8	4.6	3.0	0.0	0.0	5.1	0.0	0.5	196
77.083	126.200	284	PS2467	35.3	5.5	7.9	3.1	12.5	3.8	10.9	3.6	4.3	2.3	3.4	4.6	2.4	8.6	3.2	8.9	3.3	0.0	0.0	2.3	1.3	0.0	303
78.060	125.000	2332	PS2469	29.0	6.5	8.3	4.0	21.8	5.9	11.4	4.6	1.6	1.8	0.5	3.1	2.5	8.3	4.0	7.8	3.9	0.0	0.0	7.8	0.0	0.5	193
79.216	122.900	3233	PS2470	27.9	5.7	10.8	3.9	19.5	5.0	9.2	3.6	5.2	2.8	2.0	3.6	2.4	7.2	3.3	10.8	3.9	0.0	0.0	3.6	0.4	0.0	251
79.155	119.800	3048	PS2471	43.2	7.8	8.0	4.3	20.4	6.3	9.3	4.6	0.0	0.0	2.5	3.1	2.7	4.9	3.4	5.6	3.6	0.0	0.0	3.1	0.0	0.0	162
78.666	118.700	2620	PS2472	21.3	9.2	13.8	7.7	28.8	10.1	11.3	7.1	1.3	2.5	2.5	2.5	3.5	0.0	0.0	7.5	5.9	6.3	5.4	5.0	0.0	0.0	80
77.981	118.600	1927	PS2473	30.5	10.2	6.1	5.3	26.8	9.8	12.2	7.2	0.0	0.0	4.9	1.2	2.4	3.7	4.2	8.5	6.2	1.2	2.4	4.9	0.0	0.0	82
77.391	118.200	524	PS2476	25.6	6.2	6.5	3.5	24.6	6.1	8.5	4.0	4.5	2.9	5.0	4.5	2.9	5.5	3.2	7.5	3.7	0.0	0.0	7.5	0.0	0.0	199
77.246	118.600	193	PS2477	18.1	4.5	3.0	2.0	55.0	5.8	7.4	3.0	1.7	1.5	0.6	2.0	1.6	2.7	1.9	8.4	3.2	0.0	0.0	1.0	0.0	0.0	298
77.171	118.700	101	PS2478	15.4	4.1	6.2	2.8	49.8	5.7	9.8	3.4	2.0	1.6	0.6	3.9	2.2	3.6	2.1	5.9	2.7	0.3	0.6	1.7	0.6	0.0	305
78.261	109.200	51	PS2480	4.5	3.1	0.0	0.0	15.3	5.4	8.5	4.2	0.6	1.2	13.1	2.3	2.3	11.9	4.9	42.0	7.4	0.0	0.0	0.0	1.1	0.6	176
78.761	112.700	1216	PS2483	20.2	6.2	4.2	3.1	29.8	7.1	12.5	5.1	1.2	1.7	2.4	1.2	1.7	7.7	4.1	20.2	6.2	0.0	0.0	0.6	0.0	0.0	168
78.581	111.400	235	PS2484	24.1	5.2	2.2	1.8	40.7	6.0	8.5	3.4	2.2	1.8	0.7	1.1	1.3	2.2	1.8	11.1	3.8	0.4	0.8	4.5	0.0	2.2	270
77.900	105.100	229	PS2485	14.3	4.4	2.8	2.1	13.9	4.4	11.2	4.0	2.0	1.8	1.2	1.6	1.6	3.2	2.2	8.0	3.4	0.8	1.1	39.1	1.6	0.4	251
78.008	144.900	54	PS2721	39.3	5.6	4.2	2.3	13.6	3.9	12.3	3.7	5.2	2.5	0.9	2.6	1.8	7.1	2.9	11.7	3.7	0.0	0.0	2.6	0.0	0.3	308
78.976	147.300	85	PS2722	17.2	4.6	2.2	1.8	8.2	3.4	7.1	3.1	2.2	1.8	0.0	0.7	1.0	2.2	1.8	52.4	6.1	1.5	1.5	5.2	0.4	0.4	267
79.458	148.100	224	PS2723	38.6	6.6	5.5	3.1	8.6	3.8	10.0	4.0	2.7	2.2	1.4	3.2	2.4	5.5	3.1	18.2	5.2	0.5	1.0	5.5	0.5	0.0	220
78.656	144.100	77	PS2725	33.2	5.5	3.4	2.1	17.1	4.4	13.4	4.0	2.7	1.9	1.4	2.1	1.7	6.2	2.8	14.0	4.1	0.3	0.6	5.9	0.3	0.0	292
77.998	140.000	44	PS2726	35.3	6.2	5.5	3.0	11.1	4.1	10.6	4.0	0.9	1.2	4.7	2.1	1.9	14.5	4.6	11.9	4.2	0.0	0.0	2.6	0.9	0.0	235
77.496	140.000	40	PS2727	33.8	5.6	5.3	2.7	11.7	3.8	14.6	4.2	2.1	1.7	1.4	3.6	2.2	12.1	3.9	14.2	4.2	0.4	0.8	0.4	0.0	0.4	281

Table 1 (continued):

Latitude °N	Longitude °E	Water depth (m)	Sample No.	Amphibole	Estimation of error +/-*	Orthopyroxen	Estimation of error +/-*	Clinopyroxen	Estimation of error +/-*	Epidote	Estimation of error +/-*	Apatite	Estimation of error +/-*	ZTR (Zircon/Tourmaline/Rutile)	Sphen	Estimation of error +/-*	Garnet	Estimation of error +/-*	Opaque Minerals	Estimation of error +/-*	Detrital Carbonate	Estimation of error +/-*	Biotite and Chlorite	Kyanite and Staurolite	Others	Sum of counted grains
75.998	130.000	54	PS2729	40.6	6.4	5.0	2.8	21.8	5.3	7.5	3.4	3.8	2.5	1.2	4.2	2.6	5.9	3.0	5.0	2.8	0.4	0.8	4.6	0.0	0.0	239
81.221	106.600	3129	PS2740	23.7	6.2	4.2	2.9	32.6	6.8	10.5	4.4	1.6	1.8	1.1	0.5	1.0	6.3	3.5	11.6	4.6	1.6	1.8	6.3	0.0	0.0	190
80.415	102.100	255	PS2745	23.3	5.3	1.6	1.6	26.1	5.5	13.8	4.3	0.4	0.8	2.0	1.6	1.6	4.3	2.6	21.3	5.1	1.6	1.6	4.0	0.0	0.0	253
79.988	134.900	3353	PS2750	37.2	5.5	5.8	2.7	14.2	4.0	9.4	3.3	2.9	1.9	1.9	3.6	2.1	7.1	2.9	13.9	3.9	0.0	0.0	3.5	0.3	0.0	309
80.883	131.100	3846	PS2752	34.9	5.4	6.0	2.7	16.2	4.2	10.8	3.5	1.3	1.3	1.5	2.9	1.9	5.1	2.5	9.8	3.4	0.3	0.6	10.8	0.0	0.3	315
81.071	138.900	1755	PS2756	27.9	6.7	2.8	2.5	13.4	5.1	15.1	5.4	3.9	2.9	2.9	3.9	2.9	3.9	2.9	19.0	5.9	0.0	0.0	6.1	0.6	0.6	179
81.163	140.200	1241	PS2757	35.7	5.9	6.1	3.0	14.8	4.4	11.8	4.0	3.8	2.4	1.2	1.9	1.7	3.0	2.1	12.5	4.1	0.8	1.1	8.4	0.0	0.0	263
81.158	141.900	920	PS2758	36.4	6.1	4.9	2.7	15.4	4.6	11.7	4.1	2.4	1.9	2.0	2.8	2.1	5.7	3.0	10.1	3.8	0.4	0.8	6.9	1.2	0.0	247
81.151	143.400	1590	PS2759	33.8	6.2	3.4	2.4	12.4	4.3	10.7	4.0	4.7	2.8	1.7	3.4	2.4	4.3	2.7	11.1	4.1	1.3	1.5	12.4	0.0	0.9	234
81.201	150.100	2569	PS2761	30.5	5.8	5.6	2.9	14.1	4.4	10.8	3.9	2.8	2.1	1.6	5.6	2.9	5.6	2.9	14.1	4.4	1.2	1.4	7.6	0.0	0.4	249
80.311	150.100	1643	PS2763	32.3	6.9	8.1	4.0	11.3	4.6	8.1	4.0	3.8	2.8	0.0	1.6	1.8	5.4	3.3	11.8	4.7	1.6	1.8	15.6	0.5	0.0	186
79.883	149.800	509	PS2765	31.6	5.7	3.7	2.3	11.5	3.9	11.5	3.9	2.6	1.9	0.8	2.6	1.9	6.7	3.0	14.9	4.3	1.5	1.5	11.9	0.8	0.0	269
79.816	143.100	990	PS2768	32.6	6.7	4.1	2.9	13.0	4.8	12.4	4.7	3.1	2.5	3.1	4.7	3.0	8.8	4.1	10.4	4.4	0.0	0.0	6.8	0.5	0.5	193
80.681	121.300	3489	PS2774	30.2	5.3	6.0	2.7	26.2	5.1	11.0	3.6	2.7	1.9	1.0	5.3	2.6	4.7	2.4	6.3	2.8	0.0	0.0	6.6	0.0	0.0	301
82.076	91.970	998	PS2787	16.9	5.1	4.7	2.9	35.7	6.6	8.0	3.7	2.8	2.3	1.0	0.9	1.3	5.6	3.2	11.7	4.4	0.0	0.0	12.6	0.0	0.0	213
81.165	88.470	304	PS2791	16.7	4.7	0.4	0.8	22.2	5.2	9.1	3.6	1.2	1.4	3.2	1.2	1.4	9.5	3.7	25.4	5.5	0.0	0.0	11.1	0.0	0.0	252
	Kara Sea																									
80.665	63.990	33	PL94-3	2.5	2.0	1.7	1.7	28.8	5.9	3.4	2.4	1.7	1.7	8.9	2.1	1.9	9.7	3.9	33.5	6.1	2.1	1.9	1.6	3.8	0.0	236
81.159	72.800	567	PL94-13	26.5	5.9	0.9	1.3	18.1	5.1	14.2	4.6	3.5	2.4	3.5	3.5	2.4	4.9	2.9	18.1	5.1	0.0	0.0	6.6	0.0	0.0	226
81.182	77.500	110	PL94-16	9.4	4.2	0.5	1.0	4.7	3.1	4.2	2.9	1.6	1.8	5.8	3.1	2.5	11.5	4.6	40.3	7.1	1.0	1.4	11.0	5.7	1.0	191
80.672	73.830	292	PL94-19	10.8	3.7	0.0	0.0	13.9	4.1	12.2	3.9	0.7	1.0	6.2	2.8	1.9	9.7	3.5	37.8	5.7	0.0	0.0	1.7	4.1	0.0	288
80.873	70.030	590	PL94-22	25.8	6.1	2.4	2.1	30.6	6.4	12.9	4.6	1.0	1.4	1.5	1.4	1.6	4.8	3.0	14.4	4.9	0.0	0.0	3.4	1.9	0.0	209
79.659	70.030	511	PL94-32	23.2	5.6	0.4	0.8	27.6	5.9	12.7	4.4	3.9	2.6	1.2	3.1	2.3	2.6	2.1	19.7	5.3	0.0	0.0	5.2	0.0	0.0	228
78.778	68.080	400	PL94-51	24.3	5.4	2.7	2.0	25.1	5.4	14.9	4.5	3.1	2.2	2.0	1.6	1.6	4.7	2.7	17.3	4.7	1.6	1.6	2.4	0.0	0.4	255
77.670	70.040	312	PL94-63	25.5	5.5	1.2	1.4	25.5	5.5	18.2	4.9	1.6	1.6	1.6	2.0	1.8	4.0	2.5	16.6	4.7	0.8	1.1	1.6	1.2	0.0	247
79.788	60.890	265	AG80/1-1	13.0	4.2	0.4	0.8	46.2	6.2	5.3	2.8	2.3	1.9	3.0	1.5	1.5	4.2	2.5	21.0	5.0	1.5	1.5	1.2	0.4	0.0	262
80.371	55.870	26	AG80/11-1	0.0	0.0	1.4	1.6	80.2	5.4	0.0	0.0	0.0	0.0	0.0	1.8	1.8	0.0	0.0	15.2	4.9	0.9	1.3	0.0	0.5	0.0	217
79.982	62.230	251	AG80/2-1	10.7	4.3	0.0	0.0	36.9	6.7	12.6	4.6	1.0	1.4	1.5	2.4	2.1	4.4	2.9	26.7	6.2	1.0	1.4	2.5	0.5	0.0	206
79.968	66.140	350	AG80/3-1	13.8	5.7	0.0	0.0	26.9	7.4	11.0	5.2	3.4	3.0	1.4	4.1	3.3	6.2	4.0	28.3	7.5	1.4	2.0	2.8	0.7	0.0	145
80.010	75.490	143	AG80/6-1	7.5	3.4	0.8	1.2	9.6	3.8	23.3	5.5	7.1	3.3	4.6	6.3	3.1	15.4	4.7	23.8	5.5	0.0	0.0	1.7	0.0	0.0	240

Table 1 (continued):

Latitude °N	Longitude °E	Water depth (m)	Sample No.	Amphibole	Estimation of error +/-*	Orthopyroxen	Estimation of error +/-*	Clinopyroxen	Estimation of error +/-*	Epidote	Estimation of error +/-*	Apatite	Estimation of error +/-*	ZTR (Zircon/Tourmaline/Rutile)	Sphen	Estimation of error +/-*	Garnet	Estimation of error +/-*	Opaque Minerals	Estimation of error +/-*	Detrital Carbonate	Estimation of error +/-*	Biotite and Chlorite	Kyanite and Staurolite	Others	Sum of counted grains
80.238	74.000	256	AG80/7-1	15.4	4.5	0.8	1.1	17.8	4.8	19.7	4.9	5.0	2.7	3.5	4.6	2.6	13.9	4.3	18.1	4.8	0.4	0.8	0.8	0.0	0.0	259
80.871	74.140	179	AG80/8-1	6.6	3.3	0.0	0.0	11.8	4.3	7.5	3.5	3.5	2.4	6.7	3.1	2.3	11.0	4.1	44.7	6.6	0.0	0.0	2.2	3.0	0.0	228
Laptev Sea shelf																										
71.753	135.660	16	IK93 01-5	26.7	6.7	4.0	3.0	8.5	4.2	13.1	5.1	4.0	3.0	3.4	4.0	3.0	11.4	4.8	15.3	5.4	0.0	0.0	1.7	1.7	6.3	176
72.010	130.987	18	IK93 06-5	35.4	7.1	6.1	3.5	9.9	4.4	7.2	3.8	3.3	2.6	2.2	4.4	3.0	11.0	4.6	10.5	4.5	0.0	0.0	6.6	0.0	3.3	182
72.550	131.297	21	IK93 07-3	34.6	7.0	7.0	3.8	13.5	5.0	12.4	4.8	6.5	3.6	1.6	3.2	2.6	4.9	3.2	8.6	4.1	1.6	1.8	2.7	0.0	3.2	185
72.492	136.590	24	IK93 09-2	30.4	6.7	9.4	4.2	9.9	4.3	14.1	5.0	3.7	2.7	1.6	4.2	2.9	11.5	4.6	9.4	4.2	0.0	0.0	1.0	0.5	4.2	191
73.072	139.430	16	IK93 13-7	30.8	6.8	2.7	2.4	10.8	4.6	10.8	4.6	3.2	2.6	3.8	4.9	3.2	4.3	3.0	20.5	5.9	0.0	0.0	3.2	0.0	4.9	186
73.003	133.467	18	IK93 15-2	43.9	7.2	6.3	3.5	18.5	5.6	7.9	3.9	5.8	3.4	0.5	3.2	2.6	3.7	2.7	6.3	3.5	0.0	0.0	0.0	0.5	3.2	189
73.002	131.502	28	IK93 16-8	35.6	7.0	8.0	4.0	14.9	5.2	14.9	5.2	4.3	3.0	0.5	2.1	2.1	4.3	3.0	7.4	3.8	0.0	0.0	3.7	0.0	4.3	188
73.498	137.523	24	IK93 18-3	47.6	7.3	2.1	2.1	17.5	5.5	6.3	3.5	5.3	3.3	2.1	4.2	2.9	1.6	1.8	7.4	3.8	0.0	0.0	2.1	0.0	3.7	189
73.495	131.520	18	IK93 20-1	26.2	6.3	5.6	3.3	10.8	4.4	10.3	4.3	3.1	2.5	5.1	3.6	2.7	15.4	5.1	14.9	5.1	0.0	0.0	1.0	0.5	3.6	197
73.482	131.650	25	IK93 21-4	38.9	7.2	5.9	3.5	15.7	5.3	13.0	4.9	5.9	3.5	2.7	2.7	2.4	4.3	3.0	4.9	3.2	0.0	0.0	2.2	0.5	3.2	185
73.630	139.653	17	IK93 23-5	29.5	6.7	6.6	3.7	9.3	4.3	13.7	5.1	4.4	3.0	2.2	2.7	2.4	6.0	3.5	15.8	5.4	1.1	1.5	0.0	2.7	6.0	183
73.502	121.668	13	IK93 24-3	43.3	7.4	5.6	3.4	13.5	5.1	12.9	5.0	3.4	2.7	1.7	3.9	2.9	6.7	3.7	4.5	3.1	0.0	0.0	0.6	0.0	3.9	178
73.832	117.872	10	IK93 25-bg	26.6	6.4	6.3	3.5	16.1	5.3	11.5	4.6	1.0	1.4	5.2	1.6	1.8	16.7	5.4	9.9	4.3	0.5	1.0	0.0	0.5	4.2	192
74.000	115.963	13	IK93 26-bg	16.2	5.3	5.1	3.1	18.3	5.5	6.6	3.5	1.0	1.4	6.6	2.0	2.0	18.8	5.6	22.3	5.9	0.0	0.0	0.5	0.5	2.0	197
74.000	127.503	27	IK93 30-4	32.0	6.9	7.2	3.8	16.0	5.4	17.1	5.6	1.7	1.9	3.9	3.3	2.7	5.0	3.2	5.5	3.4	0.6	1.1	0.0	0.6	7.2	181
74.490	115.983	16	IK93 36-3	9.2	4.3	3.3	2.6	44.0	7.3	7.1	3.8	0.0	0.0	2.7	2.2	2.2	10.9	4.6	17.4	5.6	0.0	0.0	0.0	0.0	3.3	184
73.502	117.848	17	IK93 37-bg	21.8	5.9	10.4	4.4	33.7	6.8	5.7	3.3	3.1	2.5	4.7	2.6	2.3	5.2	3.2	9.3	4.2	0.0	0.0	0.0	0.0	3.6	193
74.500	122.993	16	IK93 40-5	30.9	6.7	6.3	3.5	12.0	4.7	8.4	4.0	4.7	3.1	3.1	4.7	3.1	7.3	3.8	16.8	5.4	0.0	0.0	0.0	0.5	5.2	191
74.402	131.005	30	IK93 44-10	33.2	6.8	7.3	3.7	11.4	4.6	7.3	3.7	4.7	3.0	2.1	1.6	1.8	14.5	5.1	15.0	5.1	0.0	0.0	0.0	0.0	3.1	193
74.495	134.038	14	IK93 46-5	44.3	7.3	4.9	3.2	15.3	5.3	15.3	5.3	4.4	3.0	2.2	3.8	2.8	3.3	2.6	3.3	2.6	0.0	0.0	1.1	0.0	2.2	183
74.500	139.670	24	IK93 49-7	28.3	6.5	3.7	2.7	3.1	2.5	7.3	3.8	1.6	1.8	6.8	2.1	2.1	13.1	4.9	29.8	6.6	0.0	0.0	0.5	1.0	2.6	191
74.997	123.010	32	IK93 56-1	42.4	7.3	6.0	3.5	7.6	3.9	14.1	5.1	2.2	2.2	2.7	4.3	3.0	10.9	4.6	7.1	3.8	0.5	1.0	0.5	0.0	1.6	184
75.017	119.888	34	IK93 58-5	27.5	6.7	10.1	4.5	20.2	6.0	10.1	4.5	3.4	2.7	2.8	1.1	1.6	6.2	3.6	14.0	5.2	0.0	0.0	0.0	0.0	4.5	178
74.993	114.548	45	IK93 61-9	15.1	5.3	3.8	2.8	40.9	7.2	11.8	4.7	1.6	1.8	2.2	2.7	2.4	5.4	3.3	11.8	4.7	0.0	0.0	0.5	1.1	3.2	186
75.465	119.967	43	IK93 65-8	20.3	6.0	5.1	3.3	31.6	7.0	7.3	3.9	2.3	2.3	1.7	5.1	3.3	5.1	3.3	11.9	4.9	0.0	0.0	0.0	1.1	8.5	177
75.482	123.842	44	IK93 67-1	23.4	6.2	8.2	4.0	14.1	5.1	11.4	4.7	4.9	3.2	3.3	4.3	3.0	6.0	3.5	15.8	5.4	0.0	0.0	0.0	1.6	7.1	184
75.428	125.830	41	IK93 68-8	33.2	6.8	6.3	3.5	14.2	5.1	13.7	5.0	2.6	2.3	4.2	5.3	3.3	9.5	4.3	7.9	3.9	0.0	0.0	0.5	0.0	2.6	190

Table 1 (continued):

Latitude °N	Longitude °E	Water depth (m)	Sample No.	Amphibole	Estimation of error +/-*	Orthopyroxen	Estimation of error +/-*	Clinopyroxen	Estimation of error +/-*	Epidote	Estimation of error +/-*	Apatite	Estimation of error +/-*	ZTR (Zircon/Tourmaline/Rutile)	Sphen	Estimation of error +/-*	Garnet	Estimation of error +/-*	Opaque Minerals	Estimation of error +/-*	Detrital Carbonate	Estimation of error +/-*	Biotite and Chlorite	Kyanite and Staurolite	Others	Sum of counted grains
75.283	129.477	44	IK93 70-6	38.8	7.2	8.2	4.1	14.2	5.2	7.1	3.8	2.7	2.4	5.5	3.3	2.6	3.3	2.6	8.7	4.2	0.0	0.0	0.0	1.1	7.1	183
75.383	131.828	20	IK93 71-3	31.9	6.9	6.6	3.7	11.0	4.6	8.8	4.2	1.6	1.9	2.7	4.9	3.2	9.9	4.4	17.0	5.6	0.0	0.0	0.0	0.5	4.9	182
75.343	135.155	47	IK93 73-7	29.2	7.0	7.0	3.9	13.5	5.2	6.4	3.7	5.3	3.4	3.5	1.8	2.0	7.0	3.9	8.2	4.2	9.4	4.5	0.0	1.8	7.0	171
77.112	137.225	33	IK93 84-1	31.1	6.9	6.1	3.6	12.2	4.9	12.8	5.0	1.1	1.6	5.0	5.0	3.2	7.8	4.0	12.8	5.0	0.0	0.0	0.0	0.0	6.1	180
75.940	136.708	20	IK93 K1-1	44.7	7.2	7.4	3.8	13.7	5.0	15.8	5.3	3.2	2.5	1.1	1.6	1.8	2.6	2.3	3.7	2.7	0.0	0.0	1.6	0.0	4.7	191
76.835	137.295	30	IK93 K2-1	37.0	7.0	7.3	3.8	10.9	4.5	15.6	5.2	1.0	1.4	7.3	3.1	2.5	6.8	3.6	6.8	3.6	0.0	0.0	0.0	0.5	3.6	192
73.665	113.997	9	IK93 Z2-4	24.7	6.5	12.1	4.9	24.7	6.5	9.2	4.4	2.9	2.5	2.3	6.3	3.7	6.3	3.7	5.2	3.4	0.0	0.0	2.3	0.0	4.0	174
72.033	130.127	14	IK93 Z4-3	34.3	7.0	3.3	2.6	8.8	4.2	11.6	4.7	4.4	3.0	3.3	1.1	1.5	10.5	4.5	8.8	4.2	0.0	0.0	5.5	0.6	7.7	183
75.503	130.014	51	PM94 17-4	39.1	7.0	3.6	2.7	15.6	5.2	16.7	5.4	4.2	2.9	2.6	1.0	1.4	7.3	3.7	7.3	3.7	0.5	1.0	1.0	0.0	1.0	193
74.000	125.988	14	PM94 41-4	37.0	6.8	4.0	2.8	11.0	4.4	7.0	3.6	3.5	2.6	7.0	6.5	3.5	9.5	4.1	12.5	4.7	0.0	0.0	0.5	0.0	1.5	200
74.499	126.003	40	PM94 42-3	33.3	6.9	7.9	3.9	20.1	5.8	12.7	4.8	4.8	3.1	3.2	3.7	2.7	4.2	2.9	2.6	2.3	0.0	0.0	2.6	0.0	4.8	189
74.503	130.495	25	PM94 51-7	19.1	5.6	7.2	3.7	12.4	4.7	9.8	4.3	2.1	2.1	6.2	1.0	1.4	13.9	5.0	27.8	6.4	0.0	0.0	0.5	0.0	0.0	194
74.502	136.383	27	PM94 62-1	39.9	7.1	4.8	3.1	12.2	4.8	7.4	3.8	5.3	3.3	1.1	2.7	2.4	2.1	2.1	14.4	5.1	0.0	0.0	5.3	0.0	4.8	188
74.504	126.582	36	PM94 63-8	36.6	6.9	7.2	3.7	19.1	5.6	11.3	4.5	2.1	2.1	3.1	3.6	2.7	8.8	4.1	6.2	3.5	0.0	0.0	1.0	0.5	0.5	194
72.250	133.995	21	PM94 75-3	24.3	6.3	9.2	4.2	9.7	4.4	11.4	4.7	4.3	3.0	5.9	3.2	2.6	8.6	4.1	18.4	5.7	0.0	0.0	0.5	0.0	4.3	185
73.750	134.004	17	PM94 81-2	44.2	7.1	6.6	3.5	14.2	5.0	16.8	5.3	5.1	3.1	3.6	1.0	1.4	3.6	2.7	3.0	2.4	0.0	0.0	1.0	0.5	0.5	197
73.999	128.175	27	PM94 82-1	37.3	6.9	8.8	4.1	10.9	4.5	13.0	4.8	2.6	2.3	5.7	2.6	2.3	7.8	3.9	9.3	4.2	0.5	1.0	1.6	0.0	0.0	194
74.501	119.835	34	PM94 92-3	26.2	6.4	8.4	4.0	26.7	6.4	11.0	4.5	4.2	2.9	3.7	2.1	2.1	8.4	4.0	3.7	2.7	0.5	1.0	1.0	0.0	4.2	191
74.501	114.284	37	PM94 94-5	11.3	4.5	5.6	3.3	44.1	7.1	10.8	4.4	3.1	2.5	1.5	1.5	1.7	5.6	3.3	12.3	4.7	0.0	0.0	0.5	1.0	2.6	195
75.501	115.545	48	PM94 99-1	13.1	4.9	5.2	3.2	47.1	7.2	11.5	4.6	1.0	1.4	1.6	2.6	2.3	6.3	3.5	9.4	4.2	0.5	1.0	1.0	0.0	0.5	191
74.308	135.446	32	KD95 23-7	30.6	6.7	4.3	3.0	10.2	4.4	15.6	5.3	2.7	2.4	4.8	0.5	1.0	2.2	2.1	6.5	3.6	0.0	0.0	22.6	0.0	0.0	187
71.753	135.737	15	KD95 29-x	15.7	5.3	4.9	3.2	48.1	7.3	11.4	4.7	2.2	2.2	3.2	1.1	1.5	5.4	3.3	6.5	3.6	0.0	0.0	1.6	0.0	0.0	186
71.234	131.344	15	KD95 33-10	6.0	3.9	0.7	1.4	3.4	3.0	1.3	1.9	0.7	1.4	22.1	2.7	2.7	3.4	3.0	54.4	8.2	0.0	0.0	5.4	0.0	0.0	149
75.479	130.696	39	KD95 48-11	50.3	7.2	9.4	4.2	12.0	4.7	9.4	4.2	2.1	2.1	2.1	1.0	1.4	6.3	3.5	5.2	3.2	0.0	0.0	1.6	0.5	0.0	191
75.601	134.522	36	KD95 55-10	41.1	7.1	2.7	2.3	15.7	5.3	13.0	4.9	1.6	1.8	3.2	1.1	1.5	7.6	3.8	5.9	3.4	0.5	1.0	7.6	0.0	0.0	191
73.903	126.914	23	KD95 61-3	38.3	7.1	3.3	2.6	10.6	4.5	8.9	4.2	2.2	2.2	2.2	2.2	2.2	5.0	3.2	8.3	4.1	0.0	0.0	18.9	0.0	0.0	185
73.846	120.317	21	KD95 65-11	26.8	6.4	10.0	4.3	16.8	5.4	9.5	4.2	1.6	1.8	1.1	3.2	2.5	16.3	5.3	12.6	4.8	0.0	0.0	0.5	0.5	1.1	191
75.485	114.482	34	KD95 68-7	47.8	7.3	8.2	4.0	15.2	5.3	10.9	4.6	1.6	1.8	2.7	3.3	2.6	2.7	2.4	4.3	3.0	0.0	0.0	2.7	0.5	0.0	186
Beaufort Sea/Chukchi Sea																										
70.733	-161.000	45	BC 2	13.1	3.617	2.7	1.738	13	3.61	10	3.22	2.3	1.61	4.1	0	0	9.5	3.14	30	4.913	14	3.72	1.9	0	4.1	348

Table 1 (continued):

Latitude °N	Longitude °E	Water depth (m)	Sample No.	Amphibole	Estimation of error +/-*	Orthopyroxen	Estimation of error +/-*	Clinopyroxen	Estimation of error +/-*	Epidote	Estimation of error +/-*	Apatite	Estimation of error +/-*	ZTR (Zircon/Tourmaline/Rutile)	Sphen	Estimation of error +/-*	Garnet	Estimation of error +/-*	Opaque Minerals	Estimation of error +/-*	Detrital Carbonate	Estimation of error +/-*	Biotite and Chlorite	Kyanite and Staurolite	Others	Sum of counted grains
70.883	-163.000	42	BC 11	19.8	4.138	2.9	1.742	16	3.81	14	3.6	2.9	1.74	1.8	1.7	1.34	9.5	3.04	23	4.37	4	2.035	4.3	0.3	3.4	371
70.980	-168.000	44	BC 15	20.2	4.02	5.9	2.359	15	3.58	13	3.37	2.7	1.62	3.5	1.6	1.26	5.4	2.26	20	4.005	2	1.402	10.2	0	4.3	399
70.410	-166.000	43	BC 17	16.3	4.969	2.8	2.219	22	5.57	13	4.52	2.3	2.02	3	0.8	1.2	5	2.93	23	5.662	2	1.883	10.3	0	3.8	221
70.960	-159.000	49	BC 20	5.8	3.005	1.2	1.4	13	4.32	11	4.02	3.7	2.43	5.7	1.2	1.4	8.7	3.62	35	6.132	13	4.324	2.5	0.4	3.7	242
71.020	-161.000	48	BC 21	11.5	3.128	1.4	1.152	15	3.5	8	2.66	1.7	1.27	3.3	0.7	0.82	11	3.07	39	4.783	4	1.922	3.4	1.5	1.2	416
71.408	-166.000	43	BC 34	31.9	5.116	1.2	1.195	11	3.43	13	3.69	2.1	1.57	2.1	0	0	1.2	1.2	11	3.434	2	1.537	24.3	0	1.8	332
70.915	-168.000	43	BC 35	26.8	5.73	1.3	1.465	15	4.62	19	5.08	2.9	2.17	0.8	0.8	1.15	3.3	2.31	12	4.204	4	2.535	14	1	2.9	239
69.578	-167.000	46	BC 38	17.7	4.254	1.2	1.214	14	3.87	13	3.75	1.9	1.52	2.2	0.3	0.61	6.5	2.75	30	5.108	1	1.109	12.3	0	0.6	322
68.820	-167.000	41	BC 40	19.3	4.002	0.5	0.715	11	3.17	7	2.59	2.1	1.45	1.6	0.8	0.9	1.8	1.35	14	3.519	4	1.987	39	0.3	1.5	389
71.178	-142.000	2728	P 19	10.8	3.85	0.4	0.783	13	4.17	5	2.7	1.2	1.35	0.4	0.8	1.1	3.1	2.15	40	6.076	16	4.547	10	0	0.4	260
70.551	-142.000	405	P 45	13.1	4.539	0.5	0.949	10	4.04	9	3.85	3.2	2.37	0.5	0.9	1.27	0.9	1.27	34	6.373	6	3.195	21.4	1.5	0	221
River surface samples, fraktion 63 - 125µm in grain-% (see Fig. 2)																										
			20 Lena Delta	31.48		10.82		11.81		7.87		2.94		0.29	5.896		10.81		13.72		0		1.098	1.098	2.152	n. a.**
			21 Lena Delta	15.67		9.404		10.97		5.49		0.78		8.07	3.918		11.76		30.56		0		0	0.784	2.586	n. a.**
			22/3 Lena Delta	29.97	5.587	12.65	4.054	6.744	3.06	2.53	1.91	1.27	1.37	4.22	0.416	0.78	11.81	3.94	29.54	5.563	0	0	0	0.416	0.416	269
			25/5 Lena Delta	32.59	5.965	11.44	4.051	10.57	3.91	2.21	1.87	1.33	1.46	0.88	2.653	2.05	11.44	4.05	23.34	5.383	0	0	3.096	0.442	0	247
			Ø Lena Delta	27.4		11.1		10.0		4.5		1.6		3.4	3.2		11.5		24.3		0.0		1.0	0.7	1.3	
			1 Lena	21.43		0.714		10.71		8.57		2.14		5	2.143		12.86		32.86		0		0	1.428	2.143	n. a.**
			8 Lena	21.66		7.736		10.06		6.19		0.77		4.07	2.321		17.02		26.3		0		0	1.547	2.321	n. a.**
			24 Lena	24.37		3.361		8.403		8.4		0.84		3.36	2.521		13.45		30.25		0		0	2.521	2.521	n. a.**
			11 Lena	18.11		5.763		12.35		6.59		0.77		3.25	2.432		16.46		29.64		0		0.332	1.646	2.653	n. a.**
			14 Lena	17.5		7.874		10.5		7		0.87		2.62	2.625		17.5		29.75		0		0	1.75	2.012	n. a.**
			15 Lena	13.12		11.37		9.62		5.25		2.62		2.59	5.283		20.99		22.75		0		0	0.891	5.51	n. a.**
			16 Lena	15.74		10.23		7.87		3.94		2.36		3.98	3.148		24.4		25.97		0		0	0.787	1.574	n. a.**
			8I1 Lena	38.79	5.998	4.664	2.596	7.008	3.14	4.2	2.47	0	0	3.73	1.402	1.45	3.734	2.33	34.11	5.836	0	0	0.942	0	0.942	264
			13 Aldan	18.09	5.109	17.02	4.989	12.77	4.43	5.32	2.98	2.13	1.92	5.31	0	0	22.34	5.53	17.02	4.989	0	0	0	0	0	227
			15 Vilyuy	4.124	2.751	2.062	1.966	28.25	6.23	7.22	3.58	0.31	0.77	16.49	0	0	13.4	4.71	27.84	6.2	0	0	0	0	0.309	209

Table 1 (continued):

Latitude °N	Longitude °E	Water depth (m)	Sample No.	Amphibole	Estimation of error +/-*	Orthopyroxen	Estimation of error +/-*	Clinopyroxen	Estimation of error +/-*	Epidote	Estimation of error +/-*	Apatite	Estimation of error +/-*	ZTR (Zircon/Tourmaline/Rutile)	Sphen	Estimation of error +/-*	Garnet	Estimation of error +/-*	Opaque Minerals	Estimation of error +/-*	Detrital Carbonate	Estimation of error +/-*	Biotite and Chlorite	Kyanite and Staurolite	Others	Sum of counted grains
			14/3 Lena	50.93	6.074	10.93	3.791	11.87	3.93	0.93	1.17	0.93	1.17	1.86	0.267	0.63	7.2	3.14	12.8	4.059	0	0	0	1.333	0.933	271
			Ø Lena (all)	23.6		8.4		11.3		5.4		1.3		4.4	2.3		14.3		25.8		0.0		0.4	1.0	1.7	
			60 Yana outflow	7.595		4.051		3.544		5.32		2.66		1.01	0		5.063		67.09		0		0	0	2.658	231
			41 Yana	17.13	4.875	0	0	4.623	2.72	0	0	1.38	1.51	3.23	0	0	3.696	2.44	65.77	6.138	0	0	3.232	0	0.938	239
			43 Yana	1.626	1.21	0	0	1.931	1.32	0	0	0.71	0.8	20.63	0	0	1.016	0.96	73.07	4.244	0	0	1.016	0	0	437
			45 Dulgalakh	0	0	5.43	2.768	0.652	0.98	1.3	1.38	1.09	1.27	61.98	0	0	1.303	1.39	28.24	5.499	0	0	0	0	0	268
			Ø Yana (all)	6.6		2.4		2.7		1.7		1.5		21.7	0.0		2.8		58.5		0.0		1.1	0.0	0.9	
			59 Omoloy	3.441	2.244	0	0	0	0	0	0	0	0	4.3	0	0	0	0	80.21	4.904	0	0	12.05	0	0	264
			62 Olenjek	15.64	4.187	21.1	4.704	25.09	5	2.18	1.68	0.72	0.97	2.9	0.365	0.7	6.182	2.78	24.73	4.974	0	0	0	0	1.084	301
			1/2 Anabar	14.89	4.757	18.15	5.15	8.368	3.7	3.26	2.37	1.4	1.57	3.72	0.925	1.28	15.35	4.82	31.63	6.214	0	0	0.925	0	0	224
			8/3 Khatanga	10.81	3.765	21.21	4.957	32.01	5.66	1.6	1.52	1.2	1.32	1.59	0.792	1.07	10	3.64	19.6	4.814	0	0	0.396	0	0.396	272
			1/1 Khita	1.66	1.715	18.24	5.184	53.65	6.69	0.51	0.96	0	0	0	0	0	2.21	1.97	23.22	5.668	0	0	0	0	0	222
			10/2 Khatanga	1 401	1.4	14.95	4.247	44.41	5.92	0	0	0	0	1.86	0	0	3.265	2.12	30.38	5.477	0	0	0	0	0.463	282
			Ø Khatanga (all)	4.6		18.1		43.4		0.7		0.4		1.2	0.3		5.2		24.4		0.0		0.1	0.0	0.3	
Sea-ice sediment samples																										
77.405	126.215		ARK-IX/4 25122	6.5	3.5	4.5	2.9	42.3	7.0	6.5	3.5	1.5	1.7	2.0	1.5	1.7	5.0	3.1	10.0	4.2	0.0	0.0	1.5	0.0	20.4	201
77.432	115.958		ARK-IX/4 2621	24.9	6.1	2.0	2.0	28.9	6.4	16.9	5.3	0.5	1.0	2.5	0.0	0.0	2.0	2.0	7.5	3.7	0.0	0.0	0.0	0.0	13.4	201
77.667	100.532		ARK-IX/4 24011	22.5	5.9	3.0	2.4	26.5	6.2	17.5	5.4	2.5	2.2	2.5	0.5	1.0	3.5	2.6	12.0	4.6	0.0	0.0	1.5	0.0	9.5	200
77.690	125.915		ARK-IX/4 25313	28.0	6.3	0.5	1.0	29.0	6.4	19.0	5.5	1.0	1.4	2.0	0.5	1.0	2.0	2.0	7.0	3.6	0.0	0.0	1.0	0.0	9.5	200
78.353	117.900		ARK-IX/4 25812	28.2	6.3	3.0	2.4	23.3	5.9	18.3	5.4	1.0	1.4	4.0	2.5	2.2	2.0	2.0	8.4	3.9	0.0	0.0	0.0	0.0	8.4	202
81.653	30.238		ARK-IX/4 2251	11.8	4.6	1.0	1.4	39.5	7.0	6.7	3.6	1.5	1.7	2.5	2.1	2.1	4.1	2.8	19.0	5.6	0.0	0.0	0.0	0.0	11.8	195
78.167	103.083		ARK-IX/4 23811	17.9	5.5	1.5	1.7	46.4	7.1	11.2	4.5	2.0	2.0	4.6	1.0	1.4	1.0	1.4	6.1	3.4	0.0	0.0	0.0	0.0	8.2	196
73.854	126.527		KD95 63-1	31.3	6.6	8.1	3.9	10.6	4.4	9.6	4.2	0.5	1.0	3.5	1.5	1.7	12.6	4.7	7.6	3.8	0.0	0.0	0.0	0.0	14.6	198

Table 1 (continued):

Remarks:

* Estimation of error after Boenigk, 1983

$$V = Z\sqrt{((p*q)/n)}$$

V = possible range of the real value from the counted value (in %)

Z = Faktor, which depends on how many % of the investigated sample are between max. and min. range of the confidence interval.

p = %-value of mineral x in heavy mineral spectrum

q = (100-p) = %-value of specimen, which is not q

n = sum of counted grains

If Z=2 (95% probability) are 95 of 100 counts of one sample in the estimated range

n. a.** not available; Data from Hermel 1995, unpubl. thesis

The Role of Coastal Retreat for Sedimentation in the Laptev Sea

F.E. Are

St. Petersburg State University of Means of Communications, 9 Moskovskii, 190031 St. Petersburg, Russia

Received 29 January 1997 and accepted in revised form 31 July 1997

Abstract - According to the published data the amount of sediments supplied to the Laptev Sea from shore erosion is an important component of the overall sediment balance of that sea, but at present one can calculate this amount more accurately than before. Data on shore retreat rate, height of coastal cliffs, maximum depth of wave influence on the sea floor, position of the lower boundary of shore face, lithology of sediments building the coast and shore face, volumetric ice content of these sediments and their dry bulk density are needed for such calculations. All these data are available from aerial photographs, geological and various geographical maps and bathimetric monitoring. The methodology of calculations is discussed in this paper. The amount of sediment released to the sea due to erosion of Anabar-Olenyok section of the coast is calculated as an example. This amount is compared with the published data on the sediment discharge of the rivers draining into the Laptev Sea. The results of comparison showed that (1) - the amount of sediments supplied to the Laptev Sea by rivers and retreating shores are of the same order; (2) - the transport of sediments into the sea by rivers is poorly studied and difficult to evaluate by means of direct measurements. To solve this problem the amount of sediments used for the growth of deltas must be determined by means of comparison of aerial photographs and bathymetric profiles taken with sufficient time interval and by subtracting this amount from the river sediment discharge.

Introduction

Thousands of kilometers of Arctic sea coast retreat 2-6 m/y under the action of shore erosion (Are, 1985; Reimnitz et al., 1988). So some tens of square kilometers of Arctic land are consumed by the sea every year. Actually it is a special kind of sea transgression occurring with the rather constant sea level. This process was especially intensive just after the end of the last glacio-eustatic transgression about 5000 years ago. A few islands on the Siberian shelf disappeared due to erosion of their shores during this time (Zigarev and Sovershaev, 1984).

It is evident that shore erosion has to be taken into account by various kinds of modern human activity on the coast. For example Bykovsky settlement on the Laptev Sea coast and Tuktoyaktuk on the Beaufort Sea coast are in danger because of the sea offensive.

The study of shore erosion is needed for proper understanding of the transformation of modern off-shore permafrost which was widespread on the emerged shelf before last sea transgression. The upper tens of meters of shelf permafrost contained 60-80 % ice by volume. The thawing of this ice leads to subsidence of the surface by up to 30-40 m. This transformation would become especially important where oil and gas are produced offshore and are brought across the shifting shoreline in pipes.

Looking into the future we have to consider the influence of global changes on shore erosion. It is known that sea level was rising at a rate between 1 and 2.5 mm/y during last 100 years. It is supposed that this rise will continue in the future and most probably will reach 50 cm in the year 2100 (Houghton et al., 1996). It is evident that such a rise of sea level will accelerate shore erosion considerably and may be dramatically. Special studies are needed to forecast the consequences.

Large quantities of free gases and gas hydrates of local and hypogene origin are preserved in frozen sediments and rocks. The retreat of Arctic shores creates conditions necessary for release

---

In: Kassens, H., H.A. Bauch, I. Dmitrenko, H. Eicken, H.-W. Hubberten, M. Melles, J. Thiede and L. Timokhov (eds.) Land-Ocean Systems in the Siberian Arctic: Dynamics and History. Springer-Verlag, Berlin, 1999, 287-295.

and migration of gases into atmosphere. The gas emission into atmosphere in the areas of rich gas fields may dramatically influence the global changes of climate, favoring development of green house effect.

Shore erosion is a source of sediments coming into the sea from the land. Therefore it plays a part in formation of Arctic sea sediment balance. But the importance of this part is not determined reliably till now. For a long time the sediment discharge of rivers was considered in Russia as the main terrestrial input into the sediment balance of the seas. Suzdalsky (1974) made a very rough comparison of river and shore input into White, Barents and Kara Seas and obtained values of the same order. He pointed out the necessity of further investigations. Shuyski (1983) made a new but also very rough evaluation and came to the conclusion that abrasion of sea shores and shore faces brings to the World Ocean as many sediments as the river transport to their mouths.

At present it is possible to determine the input of shore erosion into the sediment balance of Arctic seas more accurately than in previous attempts. The main goal of this paper is to evaluate the input of coastal erosion into the sediment balance of Laptev Sea. For this purpose we need the data on shore retreat rate (V, m/y), height of the cliffs (H, m), water depth at the lower boundary of the shore face, lithology of sediments composing the coast, volumetric ice content of these sediments and their dry bulk density.

Among the parameters listed the least studied and understood is the lower boundary of the shore face, which corresponds to the average maximum wave base, that is, the maximum depth of wave influence on the bottom (h_{max}, m). This boundary is usually well-defined geomorphologically in deep seas (the depths > h_{max}) along the abrasion coasts composed of sand or coarse-grained sediments or rocks. In such situation the wave base is marked by the brow of a underwater accretion terrace (Figure 1) (Zenkovich, 1962). In shallow seas the lower boundary of shore face may be associated with the izobath where the comparatively steep slope of shore face changes into a rather gentle slope of transition zone. However, in some beach profiles the change of inclination is indistinguishable. In some areas the boundary between shore face and transition zone may be determined by the change of bottom sediments: sandy on the shore face and silty in transition zone (Reineck and Singh, 1980). In some seas, as for example the Laptev Sea, the waves influence the bottom on the major part of its area. In this situation the notion of shore face is useless.

Nevertheless, it is necessary to choose the position of boundary between the area of shore erosion and the area of sea floor erosion for calculating the sediment amount supplied into the

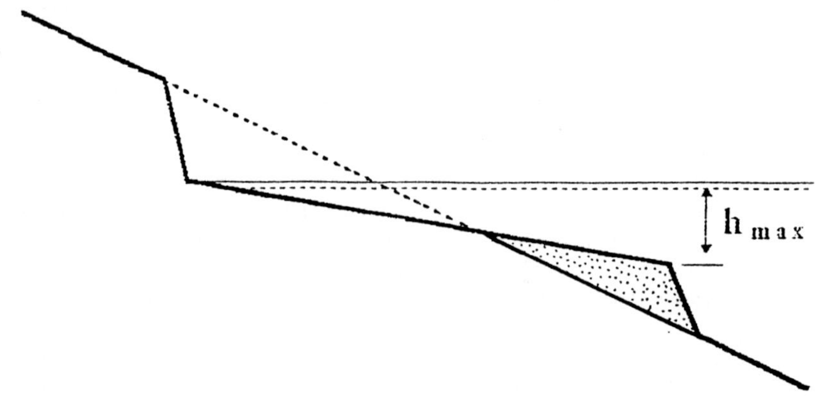

Figure 1: The scheme of bench-like profile for abrasion shore by high bathymetric gradient offshore from accretion terrace. h_{max} - the depth of wave influence on the bottom.

sea by destruction of the shores. At present we can only use an assumed maximum depth for the shore face supplying the sea with sediments. This approach can not be substantiated reliabl, because of a lack of real data. For a better approach repeated bathymetric surveys with sufficiently long time intervals are needed first of all. Such data exist (Kluyev, 1970) and they have to be used in future investigations.

Most data on the shore retreat rate of Arctic seas are published by Are (1985), Reimnitz at al. (1988) and Barnes et al. (1991). The height of coastal bluffs may be obtained from topographic maps at 1:25 000 scale. Satisfactory data on lithology of coastal sediments are presented on geological maps of 1:1 000 000 and larger scales. The Map of Yakutia permafrost landscapes of 1:2 500 000 scale gives but a rough idea of the coastal sediment ice content. The height of the cliffs and near shore bathymetry may be obtained from the topographic maps of 1: 25 000 and other scales. Various additional data may be found in numerous scientific publications. Actual field observations would of course be of great importance.

Methodology of determination of shore erosion input into the sediment balance of the Laptev Sea

The quantity of sediments supplied to the sea by shore erosion may be calculated on the base of two different principal ideas. The classical theory of abrasion supposes that the retreat of abrasion shore has a limit determined by the minimum inclination of the shore face (Zenkovich, 1962). The volume of the coast section eroded according to this idea is shown in Figure 2. Thus the annual volume on 1 m length of the shore may be calculated as follows

$$W = (F_1 + F_2) \cdot 1 = V \cdot H + 0.5\ V \cdot h_{max} = V\ (H + 0.5\ h_{max})$$

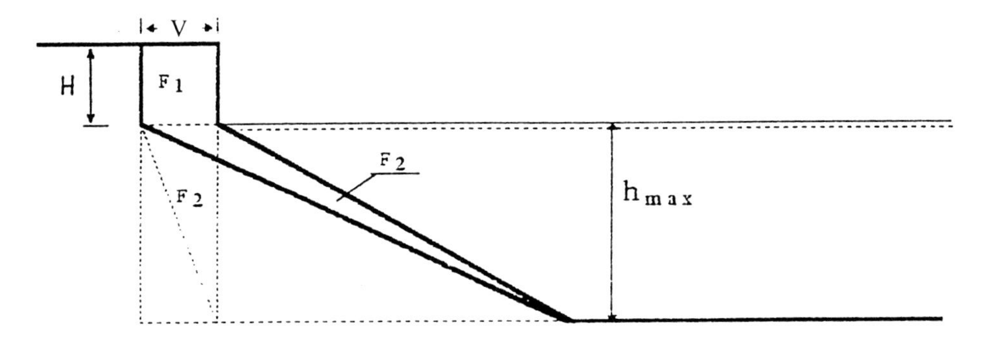

Figure 2: The scheme of flat shore face transformation by limited retreat of shallow sea shore.

The second idea supposes continuous retreat of the shore at comparatively constant rate. Such kind of retreat is observed on many sections of Arctic sea coasts during last decades. The profile of shore face retreats parallel to itself in this case with the same rate as the shore line (Figure 3). The volume of coast section eroded per meter of shoreline now will equal

$$W = (F_1 + F_2) \cdot 1 = V \cdot H + V \cdot h_{max} = V\ (H + h_{max})$$

It is important that by continuous retreat the volume of material coming from the shore face is two times larger than by limited retreat.

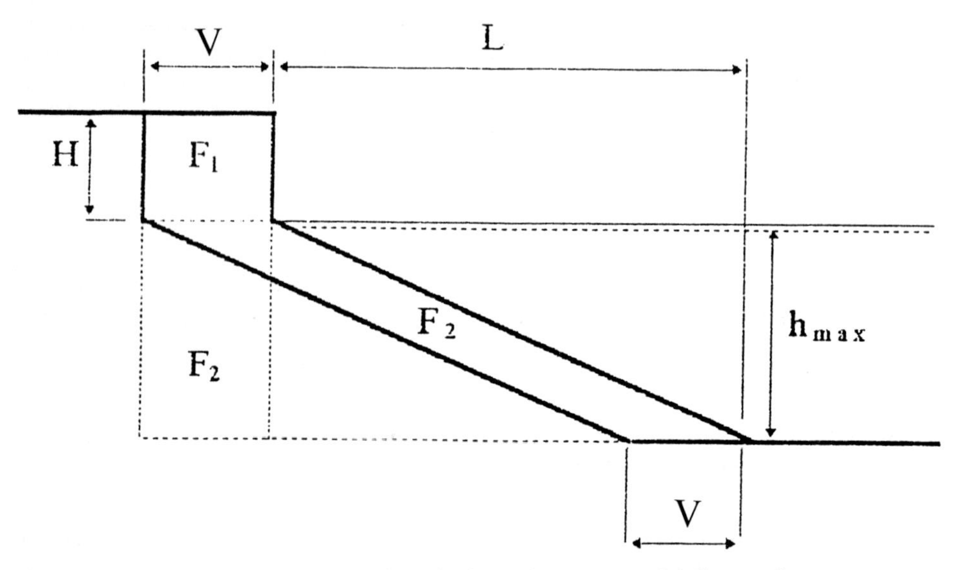

Figure 3: The scheme of flat shore face transformation by continuous retreat of shallow sea shore.

Very simple geometrical considerations show that volume of sediments supplied to the sea by erosion of the shore face do not depend on inclination and form of it. This is true for both continuous and limited retreat. Figure 4 shows the scheme of unlimited retreat of concave shore face. The volume of shore face section eroded according to this scheme is

$$W = V \cdot h_1 + V \cdot h_2 + V \cdot h_3 = V\,(h_1 + h_2 + h_3\,) = V \cdot h_{max}$$

So the result is the same as for the flat shore face retreat (Figure 3).

The scheme of limited retreat of flat and concave shore faces is shown in Figure 5. The volume of coast section eroded for both forms of shore face may be calculated as follows.

Concave shore face:

$$W = 0.5\ (V + a_1)\ h_1 + 0.5\ a_1 + a_2)\ h_2 + 0.5 \cdot a_2 \cdot h_3$$

Flat shore face:

$$W = 0.5 \cdot V \cdot h_{max} = 0.5\ (V + a_1)\ h_1 + 0.5\ (a_1 + a_2)\ h_2 + 0.5\ .\ a_2\ .\ h_3$$

So in both cases the eroded volume is determined by the sum of the areas of two trapeziums and one triangle which are equal in pairs.

In any case the eroded volume is determined by the sum of the areas F_1 and F_2 (Figures 2 and 3) as well as by retreat of a vertical wall having the height equal $H + h_{max}$. So the V, H and hmax are the geometrical parameters needed for calculating of volume eroded. The width of shore face L (Figure 3) is not needed.

The reasonable choice of h_{max} value will depend on local bathymetrical conditions. US scientists calculating the volume of eroded sediments for the western part of Alaska Beaufort Sea coast used $h_{max} = 2$ m at first because a pronounced, and widespread break in slope marks the outer boundary of a shore bench at this depth (Reimnitz et al., 1988). Later on they came to the conclusion that the shore erosion reaches the 6 m isobath and possibly still deeper (Barnes et al., 1991).

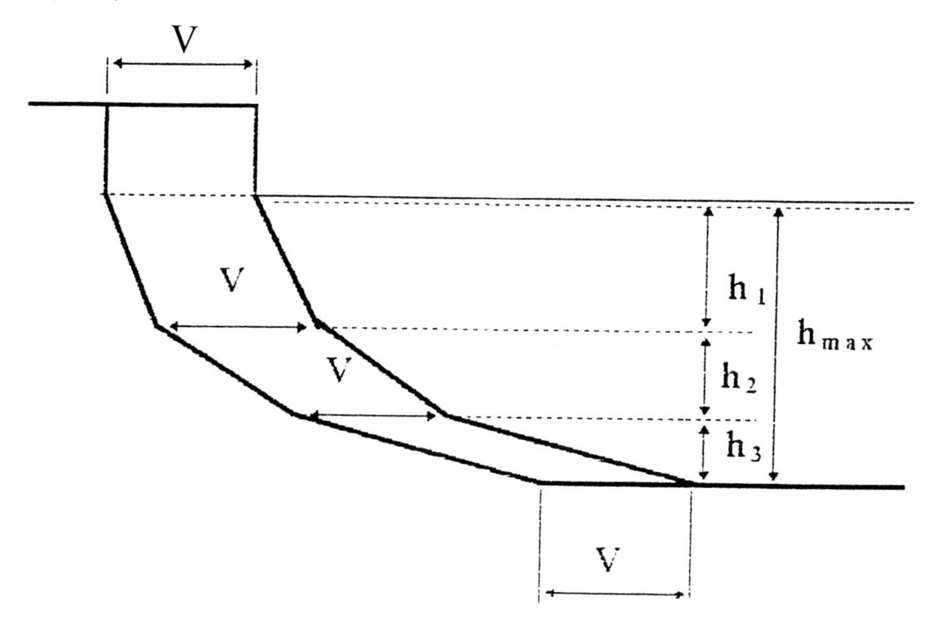

Figure 4: The scheme of concave shore face transformation by continuous retreat of shallow sea shore.

Amount of sediments supplied to the sea by retreat of Anabar-Olenyok section of the coast

Kluyev compared bathymetric profiles till 5 m isobath along the Anabar-Olenyok coast (160 km) obtained in 1942-43 and in 1962-64. This isobath is situated at a distance of 6-8 km from the shore. The results of comparison showed that erosion of the bottom is taking place everywhere in this area at a rate as high as 4 cm/y (Kluyev, 1967, 1970; Kluyev and Kotykh, 1985). About the same situation is observed in the cape Krestovsky area of the East-Siberian Sea out to the 6 m isobath despite of very low bathymetric gradient of 0.0003. Changes of bottom relief in deeper parts of the seas were not investigated.

In Kluyev's opinion, the retreat rate of shores and the form of the underwater slope in above mentioned areas do not change with time. So the slope retreats parallel to itself. This means that the volume of eroded sediments has to be calculated in accordance with the idea of continuous retreat at least for the sections of shores investigated by Kluyev.

Noteworthy is the fact that according to Kluyev the bottom of the seas is composed of sand in the area between 4-6 and 20-25 m isobaths, and is silty outside of this depth range. The coastal plains in the sections investigated by Kluyev are composed by ice complex ("yedoma") - perennially frozen fine-grained Quaternary sediments with high ice content and with inclusions of large ice wedges. The usual total ice content in these sediments is in the range of 60-80 % by volume. The granulometric composition is characterized by 60-80 % of particles less then 0.05 mm in diameter. So the absence of these particles may be caused by wave erosion of the bottom in the above zone.

The Anabar-Olenyok coast is one of the longest and comparatively well studied sections of

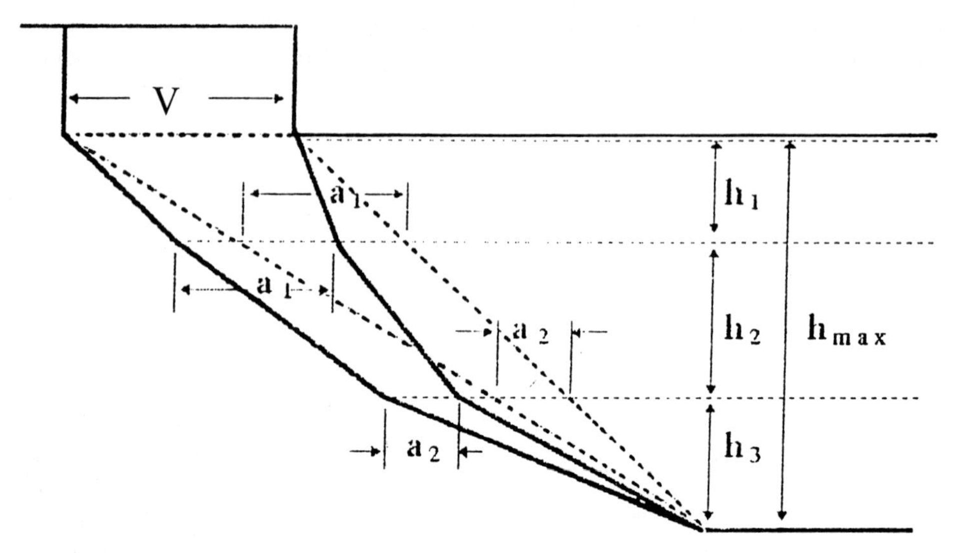

Figure 5: The scheme of concave shore face transformation by limited retreat of shallow sea shore.

retreating shores in the Laptev Sea (Figure 6). The rates of retreat were determined by means of comparison of aerial photographs taken in 1949 and 1971. The scales of photographs were 1:36 000 and 1:84 000. The comparison was performed without any special apparatus. The lowest retreat rate which could be reliably determined using this procedure was about 0.5 m/y. The mean rate of shore retreat for the 85 km of the coast in Figure 6 is about 2 m/y.

A part of the coast 69.5 m long is composed by ice complex or thermokarst depression deposits. The latter are composed mainly of the thawed and subsided ice complex. Therefore the height of the cliffs crossing thermokarst depressions may be considered as the result of thaw settlement of ice complex for the first approach. Accepting this viewpoint means that the quantity of sediments supplied to the sea by erosion of cliffs composed of ice complex and thermokarst depression deposits per meter of shoreline is equal. So there is no need to calculate the quantity of mineral components in ice complex using values of ice content. All shore composed by ice complex or thermokarst depression deposits may be considered as having the height of the cliffs equal to the height of cliffs in thermokarst depressions (10 m).

Little is known about the thickness of ice complex in this area. In some places the lower ends of ice-wedges are below sea level. For the first approach it is assumed that thaw settlement of all deposits which occur below sea level is zero. According to Kluyev the retreat of shores is continuous and the maximum depth of erosion is at least 5 m (actually it is larger). The dry bulk density of silty sands is approximately 1.5 t/m^3.

Using the data listed, the mass of sediments supplied to the sea due to erosion of 69.5 km of the shore will be

$$M = ß \cdot V\,(H + h_{max}) \cdot D = 1.5 \cdot 2\,(10 + 5) \cdot 69\,500 = 3\,127\,500 \text{ t/y},$$

where ß is the dry bulk density of thawed sediments, D - the length of the section of coast.

A 15.5 km long part of the coast presented in Figure 6, with the cliffs up to 1 m high is composed of fine-grained sands with low ice content giving no thaw settlement. The mass of sediments coming from this section will be

$$M = 1.5 \cdot 2 \cdot (0.5 + 5) \cdot 15\,500 = 255\,750 \text{ t/y}$$

Thus the total sediment yield from 85 km long retreating coastal sector is about 3.4 Mt/y.

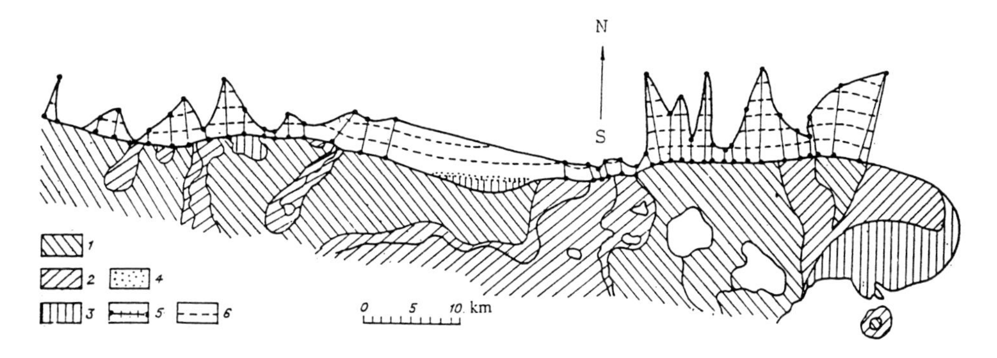

Figure 6: Mean annual shore retreat rate for 22 years (1949-1971) along 85 km of Anabar-Olenyok coast, Laptev Sea (Are, 1985). (1) - ice complex, height of cliffs 30-40 m; (2) - thawed ice complex in the thermocarst lake depressions, height of cliffs 8-12 m; (3) - surfaces up to 1 m high; (4) - inter tidal surfaces; (5) - graph of the shore retreat rate, m/y; (6) - isometric lines of shore retreat rate values at a spacing of 1 m/y.

Sediment discharge of rivers draining into the Laptev Sea

Let us compare the results of calculations of the Anabar-Olenyok coast input into the sediment balance of Laptev Sea with the data on sediment discharge of the rivers. The monitoring of water and sediment discharge of rivers flowing into the Laptev Sea is carried out during last decades. One of the earliest results is presented in Table 1. The points of sediment discharge monitoring listed in the Table 1 are situated upstream from the mouths and deltas of the rivers at the distances as far as 200-300 km. So the numbers in Table 1 do not represent the total riverine sediment supply to the Laptev Sea.

The results of latest generalization are published by Ivanov and Piskun (1995). These investigators evaluate the mean sediment discharge of the Lena river at 18.4 Mt/y in Kyusyur. They believe that the total discharge in delta channels is larger due to erosion of shores and reaches 19.4 Mt/y. The sediment discharge of Olenyok and Yana rivers makes up 1.48 and 4.19 Mt/y respectively. Ivanov and Piskun emphasize that all data available characterize the sediment discharge in the river or its delta but not the amount of sediments coming into the sea. To estimate the input of rivers into sediment balance of the sea special full-scale investigations are needed. The same authors give the value of suspended sediment discharge in the main Lena river channel 4.7 km upstream from Stolb island as large as 517 kg/s which is equal 16.3 Mt/y (this volume).

Recent data of Moscow State University investigators presented in Table 2 give another version of river sediment discharge values. These data belong to the upper margins of deltas. Alabyan et al. state, that only 2.1-3.5 Mt/y from the total Lena river suspended sediment discharge enter the sea. The major part of sediments according to them accretes in the vast delta plain. But unfortunately in the publication cited no reason is given for this extremely important statement.

Noteworthy is that one of the authors of the latter publication, Korotayev, gives the value of Lena sediment discharge at the apex of delta as large as 12 Mt/y in another book (Tshalov et al., 1995).

Table 1: Sediment transport of Yakutian rivers (Doronina, 1962).

River	Point	Years	Mean turbidity		Sediment discharge
			g/m^3	t/km^3	Mt/y
Anabar	Saskylakh	1936	14.3	14 300	0.235
Olenyok	Kuoyka	1938-40	15	15 000	0.535
Lena	Kyusyr	1936-44	23.1	23 100	11.804
Yana	Yuttakh	1938-41	200	200 000	6.360

Table 2: Water and sediment discharge of the rivers flowing into the Laptev Sea (Alabyan et. al., 1995).

River	Water discharge	Sediment discharge
	km^3/y	Mt/y
Anabar	13.2	0.4
Khatanga	101	1.4
Lena	520	21
Yana	30.7	3
All rivers	700	26

Conclusions

The calculation of the sediment supply to Laptev Sea due to Anabar-Olenyok coastal retreat and comparison of this amount with the input of Yakutian rivers shows that these two parts of sediment balance of the Laptev Sea are at least of the same order. The input of shore erosion may be even larger because the total length of erosional shores of the Laptev Sea makes up approximately 1000-1300 km, that is 12-15 times longer than the 85 km Anabar-Olenyok coastal section considered.

Acknowledgments

The author wishes to thank the German Ministry of Education, Science, Research and Technology (BM BF Grant No. 525 4003 OG0517A) who supported this research. It was carried out on the base of German-Russian scientific co-operation and owing to support of GEOMAR, Kiel and AWI, Bremerhaven. The author would like also to thank personally Dr. H. Kassens for numerous discussions.

References

Alabyan, A.M., R.S. Chalov, V.N. Korotayev, A.Yu. Sidorchuk and A.A. Zaitsev (1995) Natural and technogenic water and sediment supply to the Laptev Sea. Berichte zur Polarforschung, 176, AWI, Bremerhaven, BRD, 265-271.

Are, F.E. (1985) Osnovy prognoza termoabrazii beregov (Principles of shore thermoabrasion forecast). -Novosibirsk, Nauka Publ., 172 pp.

Barnes, P.W., E. Reimnitz, and B.P. Rollyson (1991) Map of Beaufort Sea coastal erosion and accretion, north-eastern Alaska. U.S. Geological Survey. Miscellaneous Investigations, Map 1182-4.

Doronina, N. A. (1962) Reki (The rivers). In: Severnaya Yakutia. Transactions of AARI, vol. 236, Leningrad. Morskoy transport Publ, 193-222.

Houghton, J.T. et al. (1996) Climate change 1995, Contribution of WGI to the Second Assessment Report of the Intergovernmental Panel on Climate Change. Cambridge University Press, 572 pp.

Ivanov, V.V. and A.A. Piskun (1995) Distribution of river water and suspended sediments in the river deltas of the Laptev Sea. Berichte zur Polarforschung, 176, AWI, Bremerhaven, BRD, 142-153.

Kluyev, Y.V. (1967) Rol merzlotnyh faktorov v dinamike relyefa dna polyarnyh morey (The role of cryogenic factors in dynamics of bottom relief of polar seas). Author abstract of candidate theses, geographical sciences. Petersburg State University,12 pp.

Kluyev, Y.V. (1970) Termicheskaya abraziya pribreznoy polosy polyarnyh morey (Thermal abrasion of near shore zone of polar seas). Izvestiya vsesoyuznogo geograficheskogo obshchestva, vol. 102, vypusk 2, 129-135.

Kluyev, Y.V. and Kotyukh (1985) Some peculiarities of the dynamics of the relief of the bed of the Laptev Sea. - Polar Geography and Geology, vol. 9/4, 301-307.

Reimnitz , E., S.M. Graves and P.W. Barnes (1988) Beaufort Sea coastal erosion, sediment flux, shoreline evolution, and the erosional shelf profile. U.S. Geological Survey. To accompany Map I-1182-G, 22 pp.

Reineck, H.-E. and I.B. Singh (1980) Depositional sedimentary environments. Springer-Verlag, 549 pp.

Suzdalsky, O.V. (1974) Litodinamika melkovodja Belogo, Barenceva i Karskogo morey (Lythodynamics of White, Barents and Kara Sea shallows). In: Geologiya morya (Sbornik statey), vypusk 4, NII geologii Arktiki, 27-33.

Shuyski, Y.D. (1983) Sovremenny balans nanosov v beregovoy zone morya (Recent sediment balance in the shore zone of the sea). Author abstract of doctor theses, geographical sciences. Moscow State University, 41 pp.

Tshalov, R.S., V.M. Panchenko and S.Y. Zernov (1995) Vodnye puti basseyna Leny (Waterways in the Lena river basin). Moskva, MIKIS, 600 pp.

Zenkovich, V.P. (1962) Osnovy ucheniya o razvitii morskih beregov (The foundations of teaching on sea shore evolution). Moskva, Izdatelstvo Akademii nauk SSSR, 710 pp.

Zigarev, L.A. and V.A. Sovershaev (1984) Termoabrazionnoe razrushenie arkticheskih ostrovov (Thermoabrasional destruction of Arctic islands). In: Beregovye processy v kriolitozone. Novosibirsk: Nauka, 31-38.

Section D

Terrestrial Environment - Past and Present

Seasonal Changes in Hydrology, Energy Balance and Chemistry in the Active Layers of Arctic Tundra Soils in Taymyr Peninsula, Russia

J. Boike[1] and P.P. Overduin[2]

(1) Alfred-Wegener-Institut für Polar- und Meeresforschung, Forschungsstelle Potsdam, Telegrafenberg A43, D 14473 Potsdam, Germany

(2) Department of Geography, York University, 4700 Keele St., North York, Ontario, M3J 1P3, Canada

Revised 9 July 1997 and accepted in revised form 10 October 1997.

Abstract - This study seeks to address gaps in our understanding of the complex coupled energy and water balance of the active layer, the boundary between atmosphere and permafrost. Measurement profiles installed in a variety of landscapes provided microclimatological, hydrological, chemical and physical data, including radiative energy, soil water contents, soil water chemistry and soil temperatures. It was found that the active layer freezes in one of two modes: either from the surface downward or in low water content nodes within the soil profile. The mode of freezing appears to depend on both soil heterogeneity and soil volumetric water content. Even at temperatures of less than -10C a significant volume of water remains unfrozen, an important consideration because the freezing process is the dominant heat source for the active layer in the fall. Solute exclusion has little effect on the soil water concentration profile, and thus little effect on the progress of freezing.

Introduction and objectives

The active layer, the upper ground layer which freezes and thaws annually in permafrost areas, is of hydrological, biological, geomorphological and climatological importance. Conditions for groundwater and surface water flow, gas fluxes, plant growth and soil formation are all limited and to some extent determined by this zone. The moisture and heat transfer characteristics of this layer also determine the boundary layer interactions of the underlying permafrost and the atmosphere and are therefore important input parameters for geothermal or climate modeling. In addition, potential global climate change is expected to have a more severe effect at higher latitudes than in temperate regions (Smith, 1986). Changes in the characteristics of the permafrost (thickness, temperature, moisture content) can indicate climate change, provided the system permafrost-active layer-atmosphere is sufficiently well understood: the polar regions can thus be used as an indicator for global change.

Hydrological processes in the active layer are the synthesis of a complex interplay between hydrological inputs (snow and ground ice melt, rain) and microclimatological factors (e.g. net radiation, evaporation, vegetation and snow cover). Measurements of these variables become especially important during periods of phase change since the thermal and moisture regimes of the active layer are strongly coupled through phase transitions.

During the winter, the active layer is covered with dry snow characterized by a high albedo and low sublimation rate (Ohmura, 1982). With increasing net radiation in spring, the albedo and sublimation rate of the snow increase and snowmelt is initiated. Evaporation rates are highest shortly after the initiation of snowmelt due to water saturation of the soil surface and the occurrence of overland flow (Ohmura, 1982). Soil temperatures increase steadily in the frozen ground and thaw of the active layer is initiated. During phase change, the ground temperatures are stabilized around 0∞C through the consumption of latent heat (‚zero curtain effect‘; Washburn, 1973). With progress of the summer, the thaw depth of the active layer deepens as

----

In: Kassens, H., H.A. Bauch, I. Dmitrenko, H. Eicken, H.-W. Hubberten, M. Melles, J. Thiede and L. Timokhov (eds.) Land-Ocean Systems in the Siberian Arctic: Dynamics and History. Springer-Verlag, Berlin, 1999, 299-306.

energy is transported downward (Hinkel and Outcalt, 1994). The extent of the saturated zone (‚suprapermafrost groundwater') depends on the hydrological inputs (precipitation, lateral inflow, thaw of ice) and outputs (evaporation, lateral discharge) and is limited by the depth of frozen ground. Transport through convection cells, driven by the density anomaly of water, has been postulated for coarse, water saturated soils (Ray et al., 1983; Hallet, 1990). In the fall, net radiation becomes negative, the active layer cools, freeze back is initiated and, as a result, energy is produced and transported towards regions of lower temperatures. Outcalt et al. (1990) suggest that internal distillation - evaporation of water, convection in the vapour phase and subsequent condensation - is the dominant heat transfer process during this time. Although several studies (e.g. Outcalt et al., 1990; Hinkel and Outcalt, 1994) have focused on the

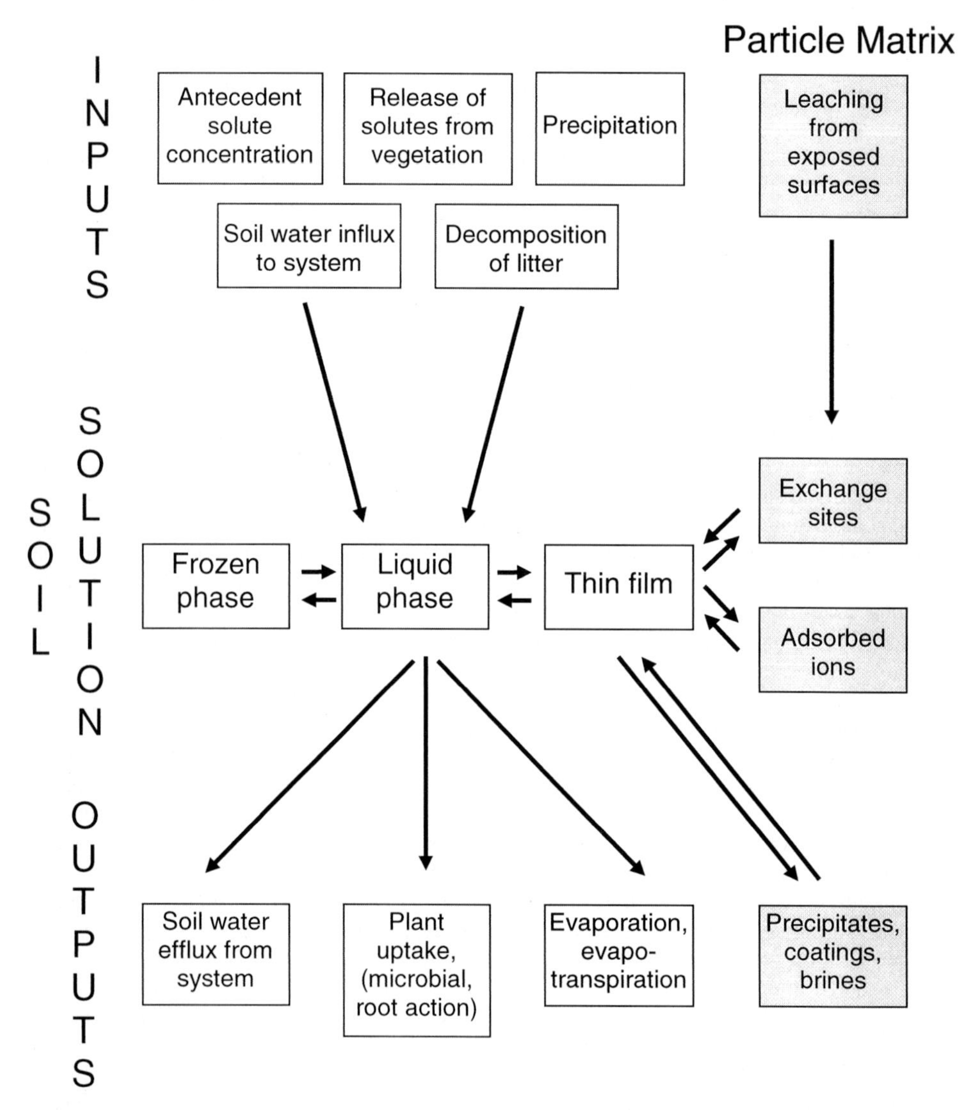

Figure 1. Schematic of solute inputs and outputs to the active layer soil water system. The soil solution contains solutes partitioned amongst the soil particle (right) and the frozen and liquid phases of the soil solution.

detection of these heat transport processes, their relative importance for active layer thermal dynamics is still unknown.

Studies of changes in soil water solute concentrations during freezing provide data for developing our understanding of the freezing processes and their potential effects on the energy balance and water chemistry in the boundary layers of the permafrost region.The inputs and outputs to soil water solutes in a soil profile are outlined in Figure 1. As active layer temperatures drop in late summer, bulk pore water forms an ice matrix, excluding solutes to the unfrozen water layers surrounding soil particles (a movement from the freezing liquid phase to the unfrozen), and potentially into unfrozen soil. In addition to the effects of freezing-induced moisture migration, solute exclusion also leads to solute concentration. The dramatic changes in depth specific solute concentrations can occur concomitant with freezing has been demonstrated in laboratory investigations (Ershov et al., 1992).

The overall objectives of this paper are to present an integrated study of the hydrological and thermal dynamics of the active layer for one complete freeze-thaw cycle. The results of two expeditions are evaluated, the first from August to September 1994 and the second from May to October 1995, carried out in the Levinson-Lessing Lake catchment on the Taymyr Peninsula, the northernmost continental part of the circumpolar arctic (Figure 2).

Methods and instrumentation

In the summer of 1994, transects were instrumented on 3 slopes in the Levinson-Lessing Lake catchment (Figure 2) differing in substrate material and slope aspect and inclination (Table 1). During this and subsequent expeditions in 1995, data were collected to calculate the energy and mass balance of the active layer along the 3 slopes. The data collected included soil volumetric water content, bulk electrical conductivity and temperature, the thaw depth of frozen ground and climatic data at two automatic weather stations (Figure 2). Water samples from the saturated and unsaturated zone, the frozen ground and precipitation were gathered for geochemical analysis. To trace the effects of the seasonal freeze-thaw cycle on the vertical distribution of water and solutes, cores of the active layer were taken during late summer and freeze-back and their soil water was extracted via centrifugation for chemical analysis. Details of the method are given in Overduin and Young (1997).

Changes in active layer water content and solute concentrations are important in determining the mode of heat transfer in the active layer, especially during phase change. A direct method for *in situ* measurements of solute concentrations over time is unavailable; part of this study was devoted to exploring the potential of time domain reflectometry (TDR) as a method for measuring both variables. As a first step, currently applied models were tested for calculation of the unfrozen volumetric water content (q) and bulk electrical conductivity (s_b) of the active layer, based on *in situ* application of TDR. The determination of q was shown to be accurate to within 0.03 m^3 with current models and does not require soil specific calibration. For determination of the soil water electrical conductivity s_w, a model was developed to calculate σ_w from TDR traces for this permafrost setting to within a factor of 2 of the values obtained via direct soil water sampling techniques (Boike and Roth, 1997).

Results and discussion

Seasonal hydrological and thermal dynamics of the active layer

Field experiments carried out for one continuous thaw-freeze cycle gave insight into the thermal and hydrological dynamics and the dominant heat and mass transfer mechanisms in the active

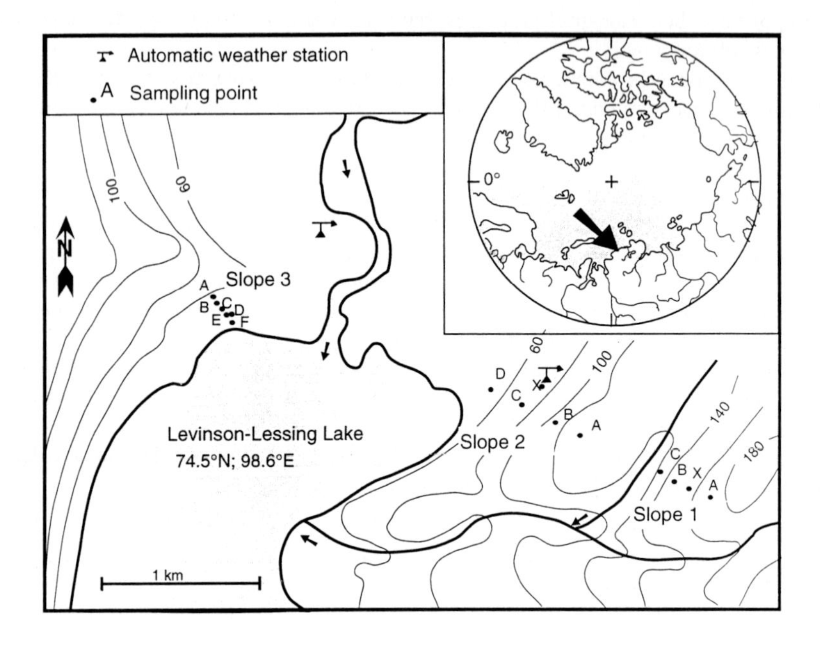

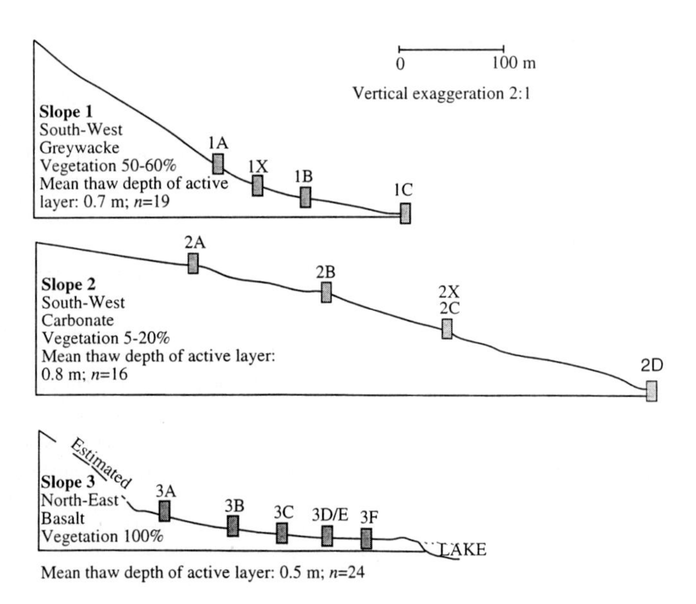

Figure 2. Detailed map of the intensive study area . The box marks the position of the area in the Levinson-Lessing basin.

layer in Siberia (Boike et al., 1997). A complete cycle of the active layer thaw-freeze processes is shown in Figure 3. In May, liquid water comprised between 3 to 9 % of the frozen soil volume even at -12 ∞C. This fraction increased with increasing soil temperatures in the frozen ground during spring. The gradual temperature increase suggests conduction as the major process of heat transport. During autumn freeze-back, a large proportion of latent heat was removed from a zone where small temperature gradients existed with opposite directions. This suggests that convective processes are responsible for efficient energy transfer within the unfrozen region of the active layer (Boike et al., 1998).

A simple water and energy balance model allowed quantification of the individual energy and water balance components through the ground surface. It was found that the dominant heat loss components in this system during spring and summer were sensible and latent heat fluxes into the atmosphere. During fall, the dominant heat source was phase change produced by active layer freeze-back.

The average amount of energy consumed by evaporation during the spring and summer period ranged from approximately 25 to 50 % of the total available energy. The third largest energy sink was the thawing of the active layer which can consume more than 20 % of the total net radiation. This term was largest when the active layer was driest because of the soil's strongly decreased hydraulic conductivity and the loss due to evaporation was therefore reduced. Of the available energy transferred into the ground, between 70 to 100 % was consumed for thawing and the rest for the warming of the ground.

Table 1. Soil characteristics of slopes 1, 2 and 3. Data are averaged from all soil profiles on each slope (number of data points *n*). Marks: + determined using TDR probes at saturation; * determined using bail tests in mini-piezometers (Lee and Cherry, 1978); # after Pfeiffer et al., (1996).

Slope	porosity+	C_{org} [%]	bulk density [kg m^{-3}]	saturated K* [m s^{-1}]	> 2 mm [% weight]	sand	silt	clay	soil tax#
1 average range *n*	0.36 0.3-0.53 16	3.9 3-4.7 6	1470 1340-1560 12	2*10^{-4} 1*10^{-3}-6*10^{-8} 6	59	18	17	6	lower slope: sandy-skeletal pergelic cryaquept; upper slope: sandy-skeletal pergelic cryorthent
2 average range *n*	0.34 0.3-0.44 22	1.6 0.8-2.5 6	1700 1430-1820 18	5*10^{-6} 2*10^{-5}-8*10^{-8} 8	36	41	20	3	loamy skeletal carbonatic calcareous pergelic cryorthent
3 average range *n*	0.5 0.4-0.58 14	4.2 1.9-8.7 5	1160 930-1380 19	2*10^{-5} 3*10^{-5}-3*10^{-7} 4	14	48	29	9	sandy nonacid pergelic cryaquept

Heterogenenity of the active layer

A tracer experiment using Brilliant Blue FCF (Colour Index Food Blue 2) was conducted on slope 2 and permitted visualization of water flow paths in three dimensions (Boike et al., 1998). Preferential flowpaths in the active layer were detected by higher concentrations of the dye. The dye pattern formed troughs, indicating that cryogenic processes such as solifluction and cryoturbation are active on this slope, the geomorphological results of which were also observed in the field. The spatial heterogeneity of the soil (and therefore spatially varying water contents) had a direct effect on the freezing dynamics during the fall. Highly resolved measurements of temperature and moisture profiles gave insights into freeze back mechanisms. In a drier, well-drained, coarse-grained soil profile (2x), it was found that zones with an initially low water content froze earlier relative to regions with higher water contents (Boike et al., 1998). Thus, freezing occurred through the development of ‚cells' within the profile rather than through the descent of a distinct freezing front. In comparison, a wet, poorly-drained,

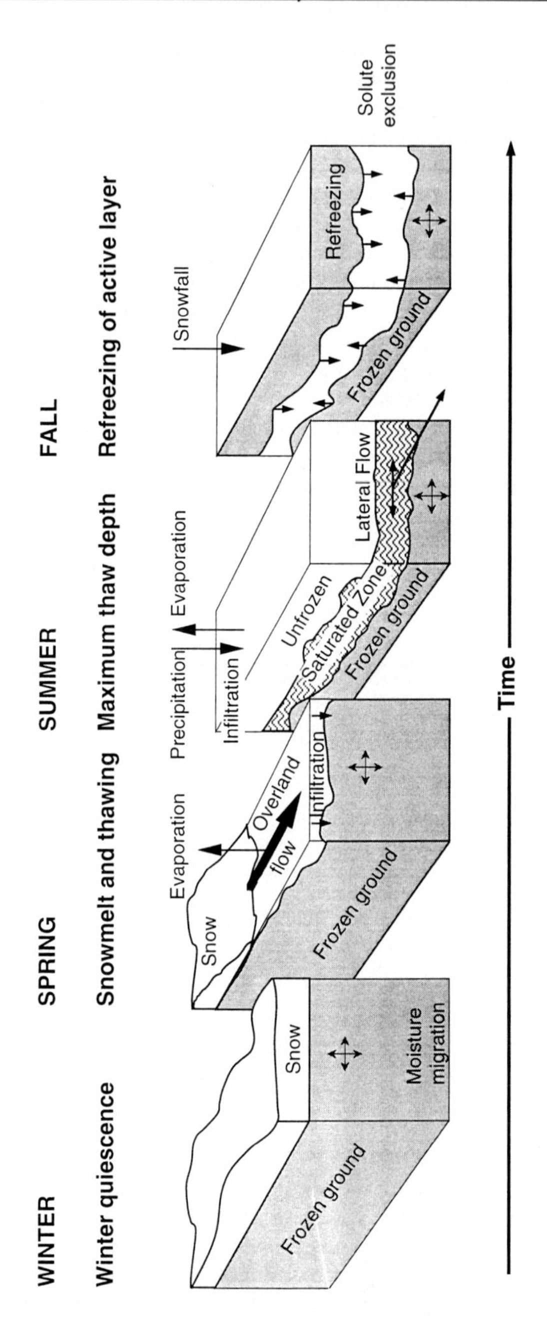

Figure 3. Conventional scheme of hydrological processes in the active layer from spring to fall.

fine-grained soil showed freezing front development; this freezing front moved downward through the active layer as the soil cooled (Overduin and Young, 1997).

Soil water chemistry during freeze-back

At the drier site (2x), solute exclusion resulted in a general increase of soil water electrical conductivity (s_w) in the freezing soil solution. At depths with initially lower water contents where freezing proceeded somewhat faster, solute concentration increases were higher than those at depths with initially higher water contents (Boike et al., 1998). Compared to s_w changes during the summer, variations in s_w as a result of phase change in the fall were an order of magnitude greater (Boike et al., 1997). In our study, however, soil water solute concentration profiles provided no support for observable macroscale solute redistribution as a result of solute exclusion. Dramatic changes in soil solution chemical composition, total solute concentration and vertical distribution did occur in the soil profile as freezing progressed, but were limited to species important in reduction and oxidation processes. Increases in the concentration of Fe^{2+} and Mn^{2+} by a factor of five in the unfrozen portion of the active layer occurred without simultaneous depletion in some other horizon, while species with higher mobilities showed no change in concentration or distribution. The increase is thus clearly not the result of redistributive transport processes, but of release of the ions into the soil solution. The increases were large and rapid enough to require the agency of bacteria, although soil temperatures throughout the soil profiles were less than 1C. We suggest that the formation of a frozen upper soil horizon has the potential to alter the reduction potential of the soil water. The increasingly reducing nature of the soil, as it is separated from atmospheric influence by a frozen layer, can then result in the release of reduced species at depth, particularly in carbonate poor and reduction-oxidation couple rich soils (Overduin and Young, 1997).

Summary

TDR has been successfully applied to measure water content and bulk electrical conductivity in the active layer and thus can be used as an *in situ* technique for studying the temporal dynamics of soil water and solutes in the active layer at these sites. From energy and water balance studies, it was found that most incident radiative energy during spring and summer was consumed by sensible and latent heat fluxes into the atmosphere.

The heterogeneity of the active layer at a drier, well drained site had the effect of (i) routing water along preferential flowpaths and (ii) allowing freezing to occur in cells, rather than through the descent of a distinct freezing front. While freezing generally had the effect of increasing the s_w of the unfrozen soil water, it did not result in redistribution of solutes vertically within the soil profile. Soils which underwent freezing through the formation of a surface frozen layer and a descending freezing front also underwent non-redistributive increases in soil water solute concentration, perhaps as a result of changes in reduction potential.

Acknowledgments

We especially thank the members of the 1994 and 1995 expeditions from the Arctic and Antarctic Research Institute, St. Petersburg who made the field work successful and unique. Funds are provided by the German Ministry of Education, Science, Research and Technology (BMBF Grant # O3PLO14A). This is contribution no. 1354 of the Alfred-Wegener-Institute for Polar and Marine Research, Potsdam and Bremerhaven, Germany.

References

Boike, J. and K. Roth (1997) Time Domain Reflectometry as a field method for measuring water content and soil water electrical conductivity at a continuous permafrost site. Permafrost and Periglacial Processes 8 (in press).

Boike, J., W.K.P. van Loon, P.P. Overduin and H.-W. Hubberten (1997) Solute movement in the active layer, Taymyr, Siberia. In: Proceedings of Int. Symp. on Physics, Chemistry, and Ecology of Seasonally Frozen Soils, Fairbanks, Alaska, June 10-12, 1997, 127-132.

Boike, J., K. Roth and P.P. Overduin (1998) Thermal and hydrologic dynamics of the active layer at a continuous permafrost site (Taymyr Peninsula, Siberia). Water Resources Research (in press).

Ershov, E.D., Yu.P. Lebedenko, E.M. Chuvilin and N.S. Naumova. (1992) Mass transfer in freezing saline soils. In: Proceedings, 1st International Conference on cryopedology, November 10-14, Pushchino, Russia, 115-122.

Hallet, B. (1990) Self-organization in freezing soils: From microscopic ice lenses to patterned ground. Canadian Journal of Physics 68, 842-852.

Hinkel, K.M. and S.I. Outcalt (1994) Identification of heat transfer processes during soil cooling, freezing, and thaw in central Alaska. Permafrost and Periglacial Processes 5, 217-235.

Lee, D.R. and J.A. Cherry (1978) A field exercise on groundwater flow using seepage meters and mini-piezometers, J. of Geological Education, 27, 6-10.

Ohmura, A. (1982) Climate and energy balance of the arctic tundra. J. of Climatology 2, 65-84.

Outcalt, S.I., F.E. Nelson and K.M. Hinkel (1990) The zero-curtain effect: heat and mass transfer across an isothermal region in freezing soil. Water Resources Research 26 (7), 1509-1516.

Overduin, P. and K.L. Young, (1997) The Effect of Freezing on Soil Moisture and Nutrient Distribution at Levinson-Lessing Lake, Taymyr Peninsula, Siberia. In: Proceedings of Int. Symp. on Physics, Chemistry, and Ecology of Seasonally Frozen Soils, Fairbanks, Alaska, June 10-12, 1997, 327-333.

Ray, R.J., W.B. Krantz, T.N. Caine and R.D. Gunn (1983) A model for sorted patterned-ground regularity. J. Glaciol. 29, 317-337.

Pfeiffer, E.-M., A. Gundelwein, T. Nàthen, H. Becker and G. Guggenberger (1996) Characterization of the organic matter in permafrost soils and sediments of the Taymyr Peninsula/Siberia and Severnaya Zemlya/Arctic Region. In: Russian German Cooperation: The expedition Taymyr 1995 and the expedition Kolyma 1995 of the ISSP Pushino group, Bolshiyanov, D.Y. and H.-W. Hubberten (eds.). Reports on Polar Res. 211, Alfred Wegener Institute for Polar and Marine Research, Bremerhaven, 46-63.

Smith, M. (1986) The significance of climate change for the permafrost environment. Proceedings of the Canadian Climate Program Workshop, March 3-5, Geneva Park, Ontario, 67-81.

Washburn, A. (1973) Periglacial processes and environments. St. Martin's Press, New York, 406pp.

The Landscape and Geobotanical Characteristics of the Levinson-Lessing Lake Basin, Byrranga Mountains, Central Taimyr

M.A. Anisimov[1] and I.N. Pospelov[2]

(1) State Research Center - Arctic and Antarctic Research Institute, 38 Bering St., 199226 St. Petersburg, Russia

(2) Taimyrsky State Biosphere Reserve, Krasnoyarski Kray, 18 Sovietskaya Street, Khatanga, Russia

Received 30 June 1997 and accepted in revised form 26 January 1998

Abstract - A landscape and geobotanical map of the Levinson-Lessing Lake basin in the Byrranga mountains of the Taimyr peninsula is presented, accompanied by explanatory text. The lake basin has an area of about 500 km^2 and is situated on the border between the typical tundra and arctic tundra bioregional subzones. The mountainous character of this territory renders this border meaningless and it is most expedient to designate a special province in the geobotanical subdivision system: the Byrranga mountain tundra region.
The map was created at a 1:100.000 scale on the basis of aerial photographs and field investigations by the authors from 1993 to 1996. Thirty-four landscape units were identified, ranging from polygonal bog complexes and shrub communities in the valleys to cold alpine deserts. For each area probable geological genesis, relief characteristics, vegetation and soil types are described.

Introduction

During the years 1993-1996, a Russia-German cooperative project (AARI, AWI, IPO, MGU, ISSP, IFB, MLU) conducted investigations in the Levinson-Lessing Lake basin, Byrranga mountains, Central Taimyr. The lake and a large part of its basin are situated within the Taimyrsky State Biosphere Reserve. Reserve employees also worked in the region, but their research was mainly devoted to the biotic natural environments. Until now the region has received only scant attention from the research community. As a result of this series of studies, Levinson-Lessing Lake has become the most investigated area in the mountains of Central Taimyr. Extensive findings on the geology, hydrology, soils, vegetation and animals of the region have been collected and require integration (Report on Polar Research, No. 175, 1995; No. 211, 1996; No. 237, 1997).

The most convenient form of representation of this material is a complex landscape map of the lake basin territory. In the creation of the map presented here, special attention was given to vegetation, which is not only an indicator for, but also an important factor in determining landscape dynamics. The landscape-geobotanical map, in giving a general overview of the characteristics of the territory covered, can be applied for many purposes.

Materials and methods

The map was made following field survey route observations by M.A.Anisimov from 1993 until 1996 and by I.N.Pospelov during 1993 and 1996, on the basis of topographic maps (1:100,000) and from 1:50,000 scale aerial photographs. About 600 field descriptions in total were made during the project: about 400 are landscape descriptions and 200 are landscape-geobotanical descriptions. During primary analysis, some preliminary maps were created; in particular, the landscape map (Report AARI, 1995) and the permafrost features map (Chronicle of Nature, 1995).

In: Kassens, H., H.A. Bauch, I. Dmitrenko, H. Eicken, H.-W. Hubberten, M. Melles, J. Thiede and L. Timokhov (eds.) Land-Ocean Systems in the Siberian Arctic: Dynamics and History. Springer-Verlag, Berlin, 1999, 307-327.

Field descriptions comprised absolute elevation, ground composition, geological structure,relief, micro- and nanorelief and vegetation type. Soil characteristics were determined using basic soil trenches, which are described in each landscape unit. Dominant species were identified and plant cover (total and by species for dominants) was measured. A complete list of species is given in the landscape-geobotanical descriptions. In arctic landscapes the interrelation of all natural components is displayed in heaviest measure. Of central importance to all components is the permafrost and processes related to it, which create cryogenic relief forms. The complex landscape map therefore reflects the distribution of vegetation, soil and other ecosystem components more accurately and completely than individual thematic maps, constructed on the basis of hierarchical classifications of each component of the environment. These classifications frequently do not correlate with each other. Much experience has been amassed creating similar maps for the Taimyrsky Reserve territory (Pospelov, 1996), although the methods employed in their creation are constantly improving. The bases for the selection of territorial units are assumed to be geological structure and relief. The various forms of micro- and nanorelief, generally cryogenic in origin, are used to differentiate units and determine the spatial distribution of vegetation and soil types.

In the creation of the map we have consciously resisted the application of current Russian landscape taxonomy because it has no practical analogues abroad, where such units as "facia" - a minimal natural territorial complex - and "landscape" - a large taxonomic unit - are accepted. There exists a difference of opinions on the treatment of terms. So, in the territory of the Levinson-Lessing Lake basin, by using the different criteria (Isaczenco, 1965; Armand, 1975, Michailov, 1971), it was possible to allocate from one to four regional units to a landscape (described in the section on physico-geographical landscape characteristics). These cartographic units should not be understood on the basis of landscape-terminology (by Russian classification the majority of them have a rank of "uroczishje", sometimes treated as analogous to "stow"; Unsread, 1935). In the explanatory text, we define the term "unit" as some spatially and originally homogeneous territory. In the majority of cases it is possible to speak of their genetic complexity when, to all appearances, the geosystem was formed by spatially uniform processes. Sometimes, however, the structural elements of a unit are irregular, as with, for example, occasional rock outcrops on the background of a spotty tundral plain. A geologic-geomorphologic principle was applied as the basis for numbering landscape units. Watersheds, slopes, valleys, terraces and bogs are identified. Within these categories, landscape units are classified and numbered in order of: decrease in elevation, increase of the degree of ground patterning, decrease of the angle (for slopes), and stage of development and age of the relief forms (for terraces and bogs). Hence, we refer to all the units representing different landscape types in the following order: denudational (watersheds), transitional (slopes and valleys), accumulative-denudational (terraces) and accumulative (bogs). Attention was given to the geological structure and genesis of micro- and nanorelief for each unit. In the physico-geographical description of the lake basin, the analysis of the cryogenic relief forms and indicators for the developmental dynamics of cryomorphogenetic processes are given; as mentioned, this factor is determining for structures of geosystems. Communities were given names based on dominant species (for example, "herbaceous-dwarf shrub-moss tundra"); in each description, dominants and the most common species were indicated. In some cases indicator species were noted (for example, on limestone there were calciphilous species indicative of such). The names of vascular plants are given as per Arctic Flora USSR (1960-1987); the names of mosses as per Ignatov, Afonina (1992). In practically all units, the heterogeneity of the vegetative cover at the community level was too great to allow classification of vegetation for the entire unit as a whole phytocoenoses. In our opinion, neither of the existing approaches to classification of vegetation for the Arctic, especially Arctic/alpine regions, is acceptable. The vegetation is, in many cases, characterized not by developed communities, but by stable and unstable primitive plant aggregations.

The physico-geographical characteristics of the research area

The Levinson-Lessing Lake basin is situated in the central part of the Taimyr peninsula, within the limits of the Glavnaya Range in the Byrranga mountains. The centre of the territory is located at 74° 30′ N and has an area of about 500 km^2. The elevation ranges from 47 m a.s.l. at the lake surface up to 568 m. a.s.l. The average height of the mountains is about 300 m a.s.l. and mountains occupy not less than 80 % of the territory. To the north, the Krasnaja river enters the lake through a low-lying peatland plain, 5 km wide, influencedby ancient terraces and bedrock outcrops. The lake is drained by the short (less than 1 km) Protoczny stream, which flows into the Ledjanaja river and further on into Ledjanaja Bay of Taimyr Lake.

Geological structure and relief

The Byrranga mountains, running parallel to the northern coast of the Taimyr peninsula for more than 1000 km, arose during the Hercynian (Palaeozoic) orogeny and have, since then, undergone repeated tectonic dislocations, resulting in their current folded block structure. The most common bedrock types are auleurolites. Dolerite, gabbro-dolerite and diabase intrusions are also common, often in the form of weathered outcrops, and comprise the crests of ranges and ledges. There are a few limestone plateaus, hardly weathered as a consequence of the low mechanical durability of the bedrock. Traces of ancient karst are evident. Such deposits are more widely distributed outside the lake basin, especially in the Levly river region, and often adjoin coal slate outcrops. Tectonic dislocations are displayed through faulting, and are easily distinguished on space photographs. The largest of the breaks is the Levinson-Lessing Lake basin. This break is observed further southwards (Nedi Lake), and northwards (the valley of the Zamknutaya river). The second largest break is the valley of the Krasnaja river. It is almost rectilinear and runs parallel to the Glavnaja Range. Some small valleys (for example, the Wresanny stream) may also be related to tectonic structures.

In the Quaternary, the territory was exposed to glaciations. The question of what type of ice cover existed (ice sheet or alpine) and when the last glaciation occurred remains open. The discovery of buried glacial ice with terminal moraine material in the Zamknutaja river valley is a clear indication of glaciation. At modern scales of proluvial accumulation, alluvial fans of some water channels could not have been generated (with respect to their current size and capacity); hence, it is possible to assume that they were formed during deglaciation, when drainage was large enough to explain the depositions. The degree of entrenchment of small valleys, even those not affected by tectonic structures, is indicative of high speed deglaciation. On the other hand, significant moraine and fluvioglacial deposits were not found anywhere (with the exception of the afore-mentioned cases). Ancient terraced platforms are observed along the entire lakeshore. There are up to 13 levels, the most easily distinguished of which lie at an elevation of 100-120 m a.s.l. They usually resemble hills with flat tops, composed of rolled and sometimes calcified pebbles.

Outside the lake basin, a significant portion of the flat surface of these terraces was preserved in the Ledjanaja riverbed. This terrace fragment has a residual block structure (thermokarst with an ancient network of ice lodes), demonstrating that its formation preceded active permafrost processes. Proceeding from the foregoing, it is possible to rather confidently assert that the last glaciation of Central Byrranga was alpine or even exclusively niche glaciers, and that it was preceded by a significant marine transgression. Until now, the last glaciation was thought to have been covering or half-covering (reticulate) (Antropogen of Taimyr, 1982; Strelkov, 1965; Taimyr-Severnaja Zemlja region, 1970). Nevertheless, small niche-glaciers may have existed in the region not long ago; such a structure, similar to a niche glacier with a push-moraine, was found in the upper reaches of one of the Scalisty stream tributaries.

In postglacial time the lake was finally separated from an extensive shallow water reservoir. In the intermountain depression of the northern lakeshore, the large lacustrine-alluvial terrace was generated, on which peat accumulated; the accumulation rate, however, was very low and the current peat depth is not more than 0.5 m. Formation of the ice-lode and of polygonal microrelief occurred simultaneously. The presence of peat-palsa bogs, which cannot be generated under modern conditions in Central Taimyr, indicate a probably gradual drainage. Such bogs are described on the northern coast of Ledjanaja Bay and their azonality is emphasized by a lush dwarf bush (Betula nana) growth, which is currently dominant of more southern tundra. The low pebbly lake terraces drained last. Based on indirect attributes of the terraces, the drainage occurred against a background of climatic cooling: higher elevation polygonal networks are large. Of modern geomorphologic processes, alluvial, permafrost (more detailed description given below), erosional and deluvial (scree processes) are the most advanced. It is interesting that erosional processes in the cleft of the left hand tributary of the Mramornaja river are rather active, where the annual accumulation of rough sediments reaches 1 m in some places. The general relief in the region is middle mountainous (the term "middle mountainous" is used here to indicate altitudinal zonation). The range of relief features decreases with increasing distance from the lake basin. On the north-west lakeshore the relief amplitude attains 400 m, to the north-west of the basin, 100-150 m, and to the north, not more than 50 m.

The mountain structures usually have a table shape, with abrupt slopes and flat tops. At higher elevations the slopes are frequently terraced by nival altiplanation and structural terraces and are complicated by outcrops of dolerite and other intrusions resistant to weathering. The auleurolites are pelitisated to gravel and smaller-sized material. Downslope, profile development of cryogenic micro- and nanorelief is intensified. The valleys of the small mountain rivers usually have V-shaped or U-shaped transverse sections, sometimes canyon-shaped (Mramornaja river, Vodopadny stream). The corresponding rivers have no more than 2 fluvial levels: low and high flood-plain. In the majority of cases alluvial deposits are not differentiated. Deluvial accumulation occurs only on the periphery of the mountain lake kettles. At one location, significant displays of thaw slumping (gelifluction) are observed.

The large valleys and small valley mouths have 3-4 alluvial levels: usually low and high flood-plain, and terrace. The lake terraces were described earlier. In the bogs, thermokarst processes are very intense, and there are several thaw lakes, of which the largest is about 0.7 km^2.

Permafrost, cryomorphogenetic processes and cryogenic relief

The lake basin is in the zone of continuous permafrost and under rather severe conditions. The permafrost reaches to a depth of 500 m and a temperature of -13°C (Geocryologie of USSR, 1989). The depth of seasonal ground thaw ranges from 20-25 cm in peat up to 1 m in gravelly soil without vegetation. Under these conditions, cryogenic processes are the leading factor in determining the structure and functioning of geosystems. Practically all important cryogenic processes, except pingo formation, are displayed within the Levinson-Lessing Lake basin. It is most expedient to consider them together with consequent cryogenic relief forms. We do not adhere to any of the existing classification systems for cryogenic processes and forms; however, as a basis for their characterization, we use the genetic classification of B.I.Vtiyrin (1969), and the morphological classification system of A.L.Washburn (1979, 1988).

Only seldom does any cryogenic process occur alone. They generally occur in conjunction with non-cryogenic morphogenetic processes: aeolian, alluvial and other processes. The majority of cryomorphogenetic processes form dynamic series, in which non-cryogenic phenomena play determining or secondary roles. Many of the issues surrounding the nature of cryogenesis remain unclear; however, it is possible to define the general tendency of their development for a given region.

Almost all cryogenic processes develop only in fine grained soils. It is therefore of primary importance to consider cryogenic weathering for mountain regions. Together with other exogenic processes, cryogenic weathering results in the formation of cryoclastites and cryoclastopelites (Popov et al., 1985); that is, the monolithic bedrock is destroyed. Certainly, the most important factor is the seasonal freezing and thawing of the ground, though meteorological processes (precipitation, wind) play a role as well. As a result of the former process, boulder debris is formed. Further degradation occurs with spatial irregularity, depending on the durability of the bedrock, snow cover depth, intensity of aeolian processes and vegetation. Initial cryogenic forms are sorted soils (or patterned ground). At first, sorted areas form among coarse material (debris islands); stone pavements may also form among coarse materials. In the researched region, these forms were observed at the highest elevation study plots or on stable screes. As secondary weathering processes, cryogenic ground sorting results in stone polygons and in sorted stone stripes on the slopes. When the sorting locations are expanded and leveled, the influence of snow intensifies and gravel circles, circumscribed by coarse material, and gravelly hummocks with cracks between them, faintly pronounced in nanorelief, are formed. These processes improve the conditions for vegetative growth. The role of vegetation in nanorelief formation gradually increases. Plants act as thermal insulation and contribute to the accumulation of water. The vegetative cover has been repeatedly shown to be determining, or at least essential, in the formation of permafrost (Tyrtikov, 1979; Common permafrost science, 1974). Vegetation, and especially moss, promotes the accumulation of moisture in cracks and the formation of ice wedges in autumn and winter. In parallel with these processes, cryogenetic weathering continues, leading to the formation of silt-gravel and silt (fine-earth) soils (cryopelits; Popov et al., 1985). After the soil transformetion to loam, the role of snow and wind erosion is reduced (as these patterned soils are observed only in locations where snow cover depth is significant) and the role of ice-segregation in the thawing layer is increased. Cryoturbation in the centre of sorted circles gradually decreases, but is amplified on their periphery. The following evolutionary stages are (consistently): hummock-spotty, spotty-hummocky and hummocky tundra. Within the limits of the studied region, these features are most widely distributed at low elevations within the lake basin and on degraded fragments of lake terraces. Any subsequent ice-segregation activity results in thermokarst and gelifluction processes.

Examples of these processes are most distinct at low hypsometric levels. First, striped tundra is evident, which we name dellic. The term "dell", however, also has a general geological treatment including non-permafrost territories, and we try, therefore, to avoid it. The shallow (less than 1.2m), wet and flat stripes form drainage channels between their ridges and are one of the most widespread forms of microrelief in the lake basin. They represent one form of thermokarst activity, which is well developed on slopes with an ice-rich horizon. Nevertheless, it is possible to show that the dellic complexes are currently rather stable. Thaw slumping gelifluction was observed at only one location, but its genesis remains subject to examination. Thermokarst processes are distributed rather sporadically throughout the lake basin and occur only at a low hypsometric levels.

In the river valleys, cryogenic forms are rarely found, except in the intermountain depression of the Krasnaja river mouth. Here, on the lacustrine-alluvial terrace, various forms of polygonal bog complexes are found, from homogeneous and typical low-centered polygons to palsa bogs. The latter are relicts of a warmer epoch, as the formation of typical palsa bog complexes is now possible only in the southern part of Taimyr. The formation of such complexes requires considerably high rates of peat accumulation (Popov et al., 1985). Icing mounds are also found among the cryogenic forms in valleys. Three small icing mounds (up to 20 m in diameter and up to 1.5 m thick) were found in the main channel and in one bayou of the Krasnaja river around the mouth of Zamknutaja river. This phenomenon has not previously been observed in the Byrranga mountains. The displays of cryomorphogenetic processes on the lake terraces deserve special attention. The surface structure of the terrace exposures at 100-120 m a.s.l., described above, displays features of ancient permafrost. On the lower lake terrace, replacement of ice-lodes by ground lodes was evident. At the same time, however, these features were not visible on the smaller exposures of higher altitude terraces.

This leads to the conclusion that the cryomorphogenetic processes in the region developed following a basic pattern. In combination with non-cryogenic processes, the general ecosystem development can be regarded as a uniform, dynamic series, based on the general tendency of historical relief development. At different stages of this series, various factors prevail in determining the ecosystem: geomorphologic, cryogenic or biotic (vegetation).

Climatic conditions and hydrology

Peculiarities of the climate and hydrology of the region were preliminary reported severral times (Berichte zur Polarforschung, #175, 1995; #211, 1996). The region's climate is Arctic continental. The average air temperature in January is about -40 °C and in July ranges from 5 to 7 °C. In some years (for example, in 1993) summer temperature do not even exceed + 5 °C. Nevertheless, ground surface temperature under the influence of solar radiation can exceed + 30 °C on clear days (field observation by the authors). The annual precipitation reaches about 250 mm , with the maximum occurring in summer. A feature of the lake basin's microclimate are strong northern winds, passing here as arctic air masses move through the Glavnaja Range of the Byrranga mountains.

In addition to the main water reservoir, there are 5 lakes in intermountain kettles (Krasnoje, Nagornoje and 3 unnamed) and about 10 shallow thaw and bayou lakes in the depression of the Krasnaja river's alluvial plain. The largest rivers in the region are the Krasnaja (about 30 km length) and the Mramornaja. The density of the stream network is rather significant.

Vegetation

In existing schemes of geobotanical subdivision (Alexandrova, 1977; Czernov and Matveeva, 1979), the Byrranga mountains are treated as a border between typical (subarctic) and arctic tundra conditions and are not recognized as a special territorial unit. However, taking into account the mountainous character of the territory, as well as the sharp differences in its vegetation (in particular, in its community species compositions) compared to that of the surrounding tundra of the Taimyr plain, some evidence of altitude zonation and the specific character of the local flora, there is a strong basis for the recognition of this mountain massif as an independent geobotanical province (by analogy with the previously accepted Ural-Paichoi, Charaulach and other provinces of alpine tundra, Alexandrova, 1977) or as a geobotanical district (Andreev, Alexandrova, 1981). Additional support for this proposition is provided by the sheer variety of communities present in the Byrranga range, from cold mountain deserts to extrazonal dwarf birch brushwood on the southern macroslopes of mountains, and by the prevalence of specific alpine tundra flora and the significantly richer floristic composition in comparison with the Taimyr plain. Plant populations present in the range are significantly removed from their main area of distribution (Eremogone formosa, Papaver leucotrichum, Poa jordalii et al.), which serves as interesting material for the historical analysis of the floral development of this region. The investigation of the region's flora included 263 taxa of vascular plants and is rather typical of the Byrranga mountains. The circumpolar species of the cryophyte group (arctic, arctic-alpine, meta-arctic), a significant number, especially in alpine landscapes, of Siberian and East-Siberian cryophytes, form the basis of the investigation. The role of hypoarctic and boreal species was greater in the large valleys (for example, the Krasnaja river valley). The typical Arctic character of the flora is represented by the major role of the families Brassicaceae, Poaceae; genera Draba, Poa, Saxifraga, and, at the same time, typical alpine species are also well represented, through the families Asteraceae, Fabaceae, Rosaceae; generas Pedicularis, Papaver, Potentilla. Thus, on the basis of areal and taxonomic analysis, it is possible to identify the region as typical Siberian-Arctic/alpine flora.

Dwarf shrubs, herbs, mosses and lichens were observed within the area of investigation. The vegetation as a whole is subject to altitude zonation. However, the limits of the belts (zones) so varied with elevation that to give their sharp hypsometric significance is practically unrealistic. On slopes with different aspects and inclinations we found consecutive replacement of the same types of plant cover, but the extent of the belts and their limits with respect to elevation can be distinguished at a scale of 200 m and more. Hence, by distribution over an area of phytocoenotic units, connected with generically consistent surfaces, we can divide them into the background and the local elements of vegetation. The former is represented in communities with mosaic or complex vegetation, whereas the development of the latter is governed by external factors (geomorphologic, hydrothermal). On the map they are included in structural units as separate elements.

At the highest hypsometric levels, the background communities are cool mountain deserts, with sparse lichen-herb and lichen-moss-herb vegetative cover. These communities are formed in areas with low snow accumulation, combined with coarse residual deposits. They are replaced downwards on the profile by gravelly dry moss-forb and forb-moss tundra with solitary plant cushions, frequently without a continuous vegetative cover. Such communities are distinguished from deserts by a richer flora and greater cover (up to 30 %). Depending on drainage, either petrophilous forbs (*Papaver polare, Saxifraga sp. et al.*) or, on more humid, fine-grained sites, Deschampsia borealis, are dominant. A belt of dry herb-dwarf shrub-moss, mesic dwarf shrub-moss and dry herb-dwarf shrub tundra occurs on patterned grounds, from fine-grained-gravel to lower elevation loamy sites. Within this belt, clear sub-belts are identifiable, distinguished by the introduction of separate plant species. At the upper elevation limits of this belt, dwarf willow *(Salix polaris)* plays an essential role, at mid-elevations, the sedge *(Carex arctisibirica)* cover is also significant, and at lower elevations dryas *(Dryas punctata)* begins to dominate. Dryas is practically unique to this sub-belt, with clear altitudinal distribution limits. However, it occurs as the only dominant only on gravelly soils, whereas on sorted soils, it tends to be replaced by other species. Such a wide range of zonation mimics the zonal community distribution of the Taimyr plain: sedge-dwarf willow-moss in arctic tundra subzones and dryas-sedge-moss in subarctic subzones.

This unique complex of slope communities, which we considered to be a background for further distinctions between regions, was advanced on striped (dellic) slopes. As a rule, all of these communities were found below 300 m; on ridges, therefore, the vegetation was mesic herb-dryas-moss, and in drainage stripes it was moist sedge-moss and sedge-dwarf willow-moss.

Superimposed on this background vegetation, communities of polygonal bogs at all stages of development were found, with wet sedge-moss *(Carex concolor*) and moss-sedge depressions and moist herb-shrub-moss on the raised boundaries of polygons or palsas.

Fragments of background communities were sometimes also local elements of complex vegetation units. The most frequent example was moss-forb tundra on decimated auleurolites outcrops and on breaks in altiplanation terraces. The local communities of the territory were rather diverse, and the following were found to be most common. The nival and subnival communities were distributed in all altitude belts. Sparse moss-lichen vegetation found on the periphery of snow-beds with moist nival short herbs *(Phippsia algida, Saxifraga nivalis et al.),* and mesic communities dominated by cassiope *(Cassiope tetragona)* were both consistent with the effects of decreased persistence of spring snow. The latter trailing dwarf shrub is very distinctive in that, although it never appeared as a dominant in the background communities, it abounded where thick snow covers are subject to rapid thawing. The cassiope-moss and moss-cassiope communities thus formed in slope niches, on terrace ledges and among boulder debris.

The communities found on limestones in various parts of the lake basin could also be included as background vegetation. They were common and comprise the dry sedge-forb-dryas *(Carex rupestris)* tundra, with a floristic composition enriched with calciphiles *(Lesquerella arctica, Eritrichium sericeum, Oxytropis putoranica, Carex redowskiana).* The mesic communities of alpine meadows were found only on protected warm slopes in the lowest mountain belt. Only here grew the frequently plentiful alpine meadow plant species: *Delphinium middendorffii, Senecio tundricola, Arnica iljinii, Hedysarum arcticum* et al. Sometimes the vegetation had a dry, steppe character, with grasses dominating (*Poa glauca, Koeleria asiatica, Roegneria villosa et al.).* Flood-plain mesic meadows dominated by *Oxytropis middendorffii, Astragalus tolmatczewii, A. umbellatus* et al. were also distributed locally. The remaining local communities will be described in the explanatory text for particular units.

Soils

The soils of various units were closely connected to the structure and grain size of the parent material, the cryogenic nanorelief, and the degree of development, structure and composition of the vegetation. Three types of soil forming processes were distinguished within the region. All soils have permafrost in parts of the first two meters, and are gelic subunits therefore. It should be noted that soils and soil processes of the area surveyed were reported formerly by the group of the authors participated in the Taimyr project in 1995-96 yrs. But the data obtained by Pfeiffer, Gundelwein, and others (Pfeiffer et al., 1996; Gundelwein et al., 1997) are coming from small key areas of the Levinson-Lessing Lake basin while the conclusions presented below are more general.

The turf forming process was observed mainly in alpine landscapes and sometimes on the flood-plain pebbles and dry terraces. It was associated with drained, well-thawed sites with a prevalence of herbs and dwarf shrubs. The soils formed there were rather rich in organic matte. Depending on the mechanical structure of the parent material, these soils were regosols (on gravely and fine earth-gravely surfaces), or Leptosols (from finer grained material). The names given here use the FAO-UNESCO soil nomenclature, (1990).

Dystri- and Gascari-gelic Rigosols in alpine desert and tundra regions were mostly primitive. The soil profile of Lithic Leptosol under dwarf shrub and herb-dwarf shrub tundra was a little more advanced. Gufri-gelic Regosols were the richest in soil organic matte; they had an advanced Histosols and were found in mountain meadows. At mountain spring fen and moist gravel dell locations, Dystri - or Umbri-gelic Regosol regosols were found. Leptosols were identified beneath similar tundra on loam and loamy clays: primitive Umbri-gelic or Molli-gelic Leptosols, in which humic contents were greater than 1 %. Stagni-gelic Leptosols were formed on barren loamy and loamy-gravelly spots. The turf soils on flood-plain places had converted to fluvisols of which haplic types are the most primitive. A little more humificated were the ochric fluvisols (underlying the grass-forb and dryas-forb communities of the flood plain). Under the influence of constant moistening and seasonal gleization, mollic-gleyic fluvisols developed on cotton grass growing mud banks. The gleying process occurs whereever soils are subject to long soil moisture conditions combined with a thin active layer, often beneath well developed moss beds. These gleysols are rather diverse in character, depending on the degree of moistesre regime, the degree of soil profile differentiation and the presence of humus horizon. At a low degree of gleyization and some humification of a profile (which usually occurs in spotty tundra on the drained borders of sorted circles) Gleysols develop to gelic-mollic. The most widely distributed Gleysols in the region were Umbri-gelic Gleysols, distinguished by the presence of a umbric or histic horizon.

They occurred in the mountain moss spotty tundra and hummock tundra, on ridges of dellic complexes and on lake terraces. Eutri-gelic Gleysols possessed the simplest profile - the gleyed horizon lay directly beneath the moss litter; they were usually found in soil complexes of spotty tundra, beneath moss hollows. In the presence of a thin peat horizon the Gleysols progress to Umbri-gelic. This generally occurred at weakly paludificated sites, in drainage stripes, on moist mossy tundra and on polygonal bogs in the early stages of development. If the peat is covered by a thin layer of humus, the Gleysols progressed to gelic-distric-humic tundra soils. The peat formation process may lead to Histosol formation with thick peat horizons. These were found only in flat palsas of bog complexes on terraces.

Landscape subdivision of the basin territory

The landscape structure of the region is, as a whole, rather monotonous. The greater part of the area tends to the typically middle-alpine landscape of the Glavnaja Range of the Byrranga mountains. The only exception to this is the extreme northern part of the basin (north of the Zamcnutaja river), which can be classified, under all existing schemes of subdivision, as part of the intermountain kettle of Ugolnaja river. Its relief has a flat, faintly hilly character, with substantial development of alluvial and kettle-bog complexes. Watershed vegetation here is similar to arctic herb-dwarf shrub-moss tundra. However, only 2 units in the lake basin conform to this landscape type and they are ecotonal.

At the same time, there are the larger geosystems, for which the application of a landscape type is open to question. There are the intermountain lake denression and mouth part of the Krasnaja river, but also small mountain ridges to the south-east of the basin. The former qualify for the rank of landscape based on such parameters as the specific character of bedrock composition, their genesis (alluvial and lacustrine-marine), mesorelief and vegetation. Their area is small, however, in comparison with the surrounding landscapes. Also, despite territorial integrity, their configuration is rather difficult, and the distribution of background units of the lowest rank is modest. The latter geosystem represents the same mountain massifs, but at the surface it is comprised of more weathered material and outcrops of bedrock are rare. It is isolated territorially from Glavnaja Range and on general appearance resembles the eastern part of a foothill ridge, Nedy, which is more likely to be considered at the rank of landscape type.

Explanatory text accompanying the landscape-geobotanical map of the Levinson-Lessing Lake basin (M 1:100.000)

Mountain watersheds

1 - Platforms of ancient flattened surfaces at elevations of more than 400 meters with striped patterned ground

Alternation of fine grained material with boulder debris from bedrock material more stable to weathering (dolerites, diabases). Patterned ground is widely distributed (debris islands, stone polygons, stone pits, and large spot-hummocks on flat, low-lying soils). These units retain only a slight snow cover during the winter, resulting in sparse vegetation represented by two variants: A) Higher elevations boulder debris with sparse (5-15 % cover) lichen aggregations (*Alectoria ochroleuca, Cetraria sp.),* soils are absent, and B) The tops and flat tops with

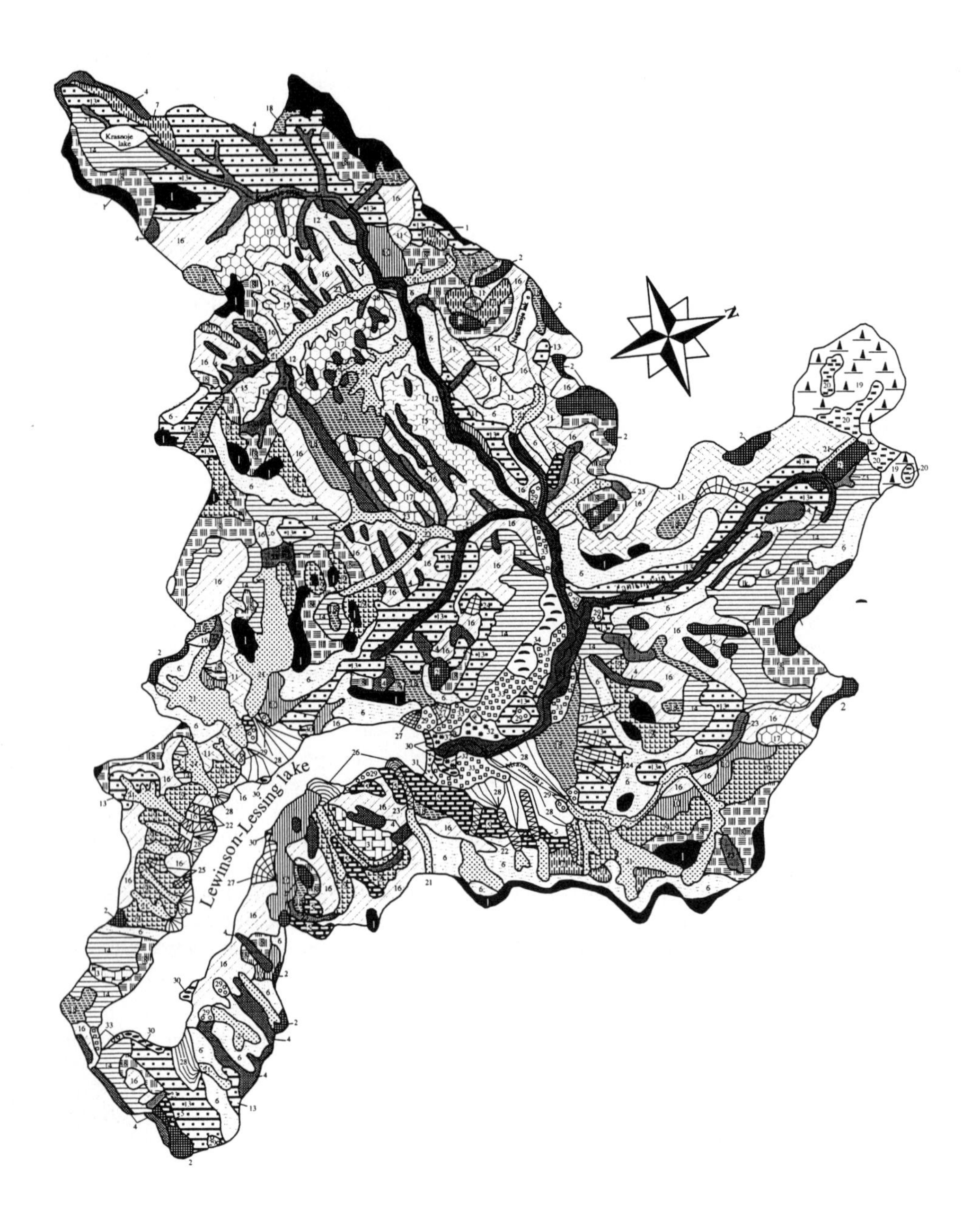

Figure 1: Landscape-geobotanic map of the Levinson-Lessing Lake Basin

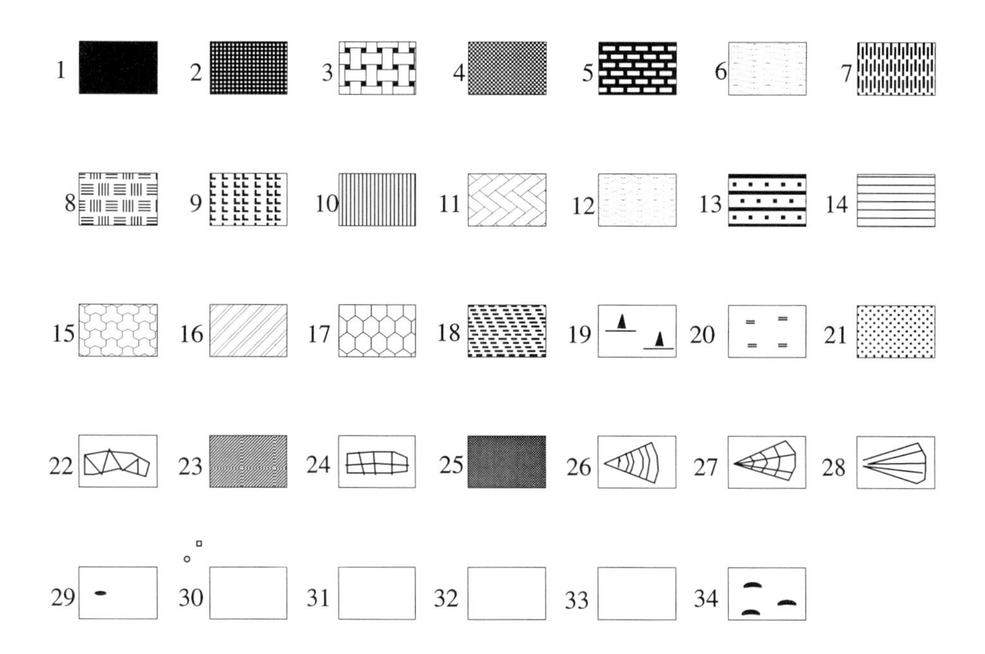

Figure 1: Explanations.

sparsely distributed herbs (Novosieversia glacialis, Poa abbreviata, Potentilla subvahliana, Luzula nivalis et al.) and moss-herb (mosses: Andreaea sp., Dicranoweisia crispula), cool mountain desert with 10-20 % cover on distric regosols.

2 - Terrace tops with boulder debris on the ledges and patterned ground (stone polygons, spots-hummocks) on the surfaces of steppes

These units are influenced by low (up to 5 m) cryoplanation terraces, and are composed of gravelly to boulder sized material. The vegetation is represented by 2 elements, depending on which form of micro- and mesorelief is present: A) Flat terrace surfaces: dry forb-moss (in moss layer: *Rhacomitrium spsp.; vascular: Novosieversia glacialis, Luzula spsp., Eritrichium villosum subsp. pulvinatum , Salix polaris et al.);* structural tundra with 10-30 % cover on the Dystri-gelic Regosols; B) Terraces slopes: boulder debris with fragmented mesic cassiope-moss vegetation on the Distri-gelic Regosols (*Cassiope tetragona, Draba pilosa, Saxifraga spinulosa* et al.; the dominant moss is *Hylocomium splendens var. obtusifolium.*), with about 50 % cover at low elevations. At high elevations, lichen aggregations similar to those of 1A are developed.

3 - Horizontal surfaces with remnants of auleurorolitic ridges and hummock nanorelief

These features are found only in the region north-east of the lakeshore and are probably associated with the proximity of large limestone outcrops. The 2 elements of vegetation (again, depending on which mesorelief elements are present): A) Ridges: dry moss-forb mountain tundra (*Papaver polare, Novosieversia glacialis, Saxifraga spsp.* et al.; the main moss is *Rhacomitrium lanuginosum*) with 10-20 % cover. B) Depressions between ridges: dry forb-dwarf shrub-moss spotty mountain tundra with 20-40 % cover (*Salix arctica, Dryas punctata, Carex rupestris, Papaver polare, Oxytropis nigrescens*; the forb and grasses are various; in the moss layer: *Hylocomium splendens var. Obtusifolium*). The soils are Dystri-gelic Regosols and Eutri-gelic Leptosols; in depressions the turf horizon is better developed.

4 - Structurally denudated mountain ridges with flat and inclined surfaces and patterned ground, which is defined by cryogenic weathering (rings of fine-grained soil) and boulder debris on the ledges of ridges

These features are composed mainly of dolerites and are distributed at all hypsometric levels. Their vegetation, however, varies depending on elevation. The boulder debris of the lower mountain belt (A) differs by having a rather specific petrophilous vegetation. The boulder slopes have a mesic herb-cassiope-moss vegetation with 40-60 % cover *(Cassiope tetragona, Saxifraga hyperborea, Dryopteris fragrans, Rhodiola rosea*; in moss layer *Hylocomium splendens var. obtusifolium dominates).* The soils are Regosols and have developed in the cracks between boulders. In protected locations, which receive more snow in winter, small mountain mesic short grass meadows with *Potentilla nivea subsp. mischkinii, Arnica iljinii, Saussurea tilesii* et al. on Eutri-gelic Regosols are found. For similar features of the upper mountain belt (B) (but also on heavily snowed portions of the lower belt) lichens alone, with 5-20 % cover, are found. On the western half of the central lake basin, the surfaces of the ridges are flat (C), with sparse (5-7 % cover) petrophilus forbs (*Novosieversia glacialis, Luzula nivalis, Saxifraga caespitosa* et al.) on Dystri-gelic Regosols.

5 - Prominent surfaces and plateaus, composed of limestones with rock outcrops and angles of inclination from 0 up to 10°

They have a very complicated relief. We consider them as a complex, consisting of 5 elements of relief mesoform, with specific calciphilous vegetation types. These relatively low (190-270 m a.s.l.) limestone mountains lie in the northern and north-eastern portion of the lake basin. The elements of the complex herein provided are usually grouped in one uniform ecological series: A) Rock outcrops with a very sparse (not more than 1-2 %) cover of crustaceous lichens and petrophilous forbs (*Papaver leucotrichum, Draba cinerea, D. macrocarpa et al.*) on Littic Loptosol. In rock cracks advanced Calcari-gelic Regosols develop. B) The structural mountain deserts with sparse lichens cover *(Thamnolia sp.)* and slightly greater vascular plant cover (*Braya purpurascens, Taraxacum phymatocarpum, Cardamine bellidifolia*). The cover is not more than 1 % and exposed locations are practically lifeless. There is no soil cover. C) Flatter and lower exposures than the previous landscape type, with patterned ground. These regions are occupied by dry willow-moss-forb tundra *(Salix arctica, Braya purpurascens, Poa arctica, Oxygraphis glacialis; in fragmented* moss layers, *Rhacomitrium lanuginosum dominated*). Cover is 1-10 % and the soils are Galcari-gelic Regosols. D) Stone-striped slopes with spotty-hummocks; herb-dryas tundra (*Dryas punctata, Carex rupestris, Astragalus tolmaczewii, Oxytropis putoranica, Eritrichium sericeum subsp.arctisibiricum et al.);* vegetative cover up to 30 %; the soils are Rendzi-gelic Leptosols. E) Feet of stone-striped slopes, moist, with moss-sedge and sedge-moss vegetation (*Carex redowskiana, C. atrofusca, Equisetum variegatum, Juncus castaneus; mosses*: *Tomentypnum nitens, Orthothecium chryseon*; cover up to 60 %) on Rendfi-gelic Leptosols. Moist nival short grass meadows with *Oxygraphis glacialis, Minuartia stricta, Stellaria crassipes* et al develop near snow fields.

Slopes

6 - Abrupt stony slopes with rock outcrops, boulders debris, screes, often complicated by ledges and by structural and cryoplanation terraces

As a rule, these slopes are rather extended and have 2 sharply pronounced altitude belts, though their height and limits vary. The vegetation of the bottom belts is represented by a combination of 3 elements: A) Boulder debris, between which grass-forb-dryas vegetation with 40-80 % cover develops (*Dryas punctata, Hierochloe alpina, Festuca auriculata, Astragalus alpinus, Potentilla uniflora* et al.) on fragmented Eutri-gelic Regosols. B) Alpine forb meadows on scree slopes with 60-80 % cover (grasses: *Poa glauca, P.arctica, Trisetum spicatum* et al.; forbs and legumes: *Hedysarum arcticum, Astragalus alpinus, Papaver pulvinatum, P.angustifolium, Delphinium middendorffii, Potentilla nivea subsp. mischkinii, Pedicularis amoena* et al.) growing on Eutri-gelic Regosols. C) Slope hollows with mesic herb-cassiope-moss vegetation of an almost continuous cover (*Cassiope tetragona, Carex vaginata, Saxifraga nivalis, Ledum decumbens*; mosses: *Hylocomium splendens var. obtusifolium*) growing on Eutri-gelic Leptosols. These 3 elements occur according to the distribution of snow during the winter and to the presence of turf on the slopes, as well as to the intensity of scree processes. The intensity of scree processes increases sharply upslope where the fourth element dominates: D) Screes with sparse grass-forb aggregations (*Arabis petraea s.l., Saxifraga spinulosa, Poa glauca* et al.); cover varies from 10-20 % at lower elevations up to 1-3 % higher up. The soils are found to be Regosols or not presented. The vegetation of similar slopes on the northern side of the Krasnaja river valley has some calciphilous species (*Oxytropis putoranica, Draba pohlei, Eremogone formosa*), despite a complete absence here of limestone bedrock at the surface.

7 - Altiplanation terraces with boulder debris on ledges and surface hummock nanorelief

These landscapes are indicated on the map only when they form large contiguous areas. This occurs only in the northwest portion of the basin. The vegetation types comprise 2 elements: A) Dry hummocky alpine tundra with terraced surfaces and forb-moss vegetation (*Papaver polare, Novosieversia glacialis, Potentilla uniflora* et al.; mosses, concentrated in cracks between hummocks, are *Hylocomium splendens var. obtusifolium, Tomentypnum nitens*); cover up to 30 %; the soils are the ochric leptosols. B) Boulder debris of ledges with mesic herb-lichen-moss vegetation (analogous to 2B), on Dystri-gelic Regosols, which occur as fragments between boulders.

8 - Slope surfaces with inclination 15-20°, complicated by numerous nivation hollows and boulder debris

These landscapes types are marked by a combination of frequent ledges with rock outcrops and of snow-fields. As a result of the scant snow cover on ledges (element A) the dry stony tundra is replaced by sparse (3-5 %) petrophilous forbs. The other elements are: B) Boulder debris with mesic herb- and herb-moss vegetation on the borders of ledges and on slopes (the dominant moss is *Hylocomium splendens var. obtusifolium*, 10-30 % cover) and C) Nivation hollows with moist herb-moss vegetation and with up to 70 % cover (*Phippsia algida, Saxifraga cernua; mosses: Oncophorus wahlenbergii, Bryum criophilum* et al.). The soils in A) are Dystri-gelic Regosols, in the others, Eutri-gelic Regosols.

9 - Stepped slopes with angle of inclination 10-20°

Their structure results from bedrocks with different weathering stabilities forming steps from 1 to 5 meters thick and is similar to that of the previously described region, in which such terraces were formed by nivation processes. The vegetation is classified into 3 elements: A) Bedrock outcrops with crustaceous lichens and sparse (3-5 %) petrophilous herbs (*Saxifraga caespitosa, S. spinulosa, Papaver polare, Hierochloe alpina*) on ochric regosols; B) The lowest portion of microslopes with dry-mesic dwarf shrub-moss and dwarf shrub mountain tundra (30-50 % cover). *Salix polaris* dominates at higher elevations, *Dryas punctata* dominates at lower elevations; the usual forbs are *Lloydia serotina, Arabis petraea s.l., Potentilla uniflora, Draba alpina* et al.; mosses: *Hylocomium splendens var. obtusifolium, Tomentypnum nitens*; the soils are Eutri-gelic Regosols; C) Moist depressions at the rear joints of the steps (spring fens) with practically continuous (cover 90-100%) herb-moss vegetation (*Lagotis minor, Carex arctisibirica, Juncus* biglumis, *Saxifraga cernua, Deschampsia borealis; the moss Bryum criophilum dominates*) on Gelic Regosols.

10 - Surfaces with inclination from 5 to 15°, with separate bedrock outliers and spotty hummocky alpine tundra

These regions share a similar genesis with the previous ones, but differ by an absence of moist places with fens. There are 2 vegetation elements: A) Bedrock outliers with crustaceous lichens and sparse forbs (about 10 % cover) on Dystri-gelic Regosols; and B) Slope surfaces with dry-mesic hummocky forb-moss mountain tundra (*Papaver polare, Myosotis asiatica, Draba alpina, D. macrocarpa, Saxifraga spsp.* et al., mosses: *Hylocomium splendens var. obtusifolium, Aulacomnium turgidum);* cover up to 40 % ; the soils are Dystri-gelic Leptosols.

11- Discontinuous surfaces with bedrock outcrops and irregular nanorelief

These outcrops of auleurolites are distinct from other landscape types only through their weak nanorelief development and vegetation. The latter comprises 2 elements: A) Primitive plant aggregations of petrophilous forbs and grasses (5-15 % cover) develop on the bedrock outcrops ; the soils are ochric regosols; B) -On main slope surfaces - mesic hummocky forb-dwarf willow-moss alpine tundra (*Salix polaris, Novosieversia glacialis, Deschampsia borealis, Luzula confusa; mosses: Hylocomium splendens var. obtusifolium, Aulacomnium turgidum, Tomentypnum nitens*) cover up to 70 %; the soils are Dystri-gelic Leptosols.

12 - Small-block sloped surfaces, slightly disjointed by deep erosion hollows

These structures are generally found on the side slopes of small mountain valleys. The deep erosion hollows are frequently associated with a change in bedrock composition. The vegetation comprises 2 elements: A) On positive mesorelief features: dry moss-forb spotty tundra (*Papaver polare, Saxifraga spinulosa, Myosotis asiatica* et al.) with 30-40 % cover; in the moss layer: Rhacomitrium lanuginosum and Hylocomium splendens var. obtusifolium dominate; the soils are Dystri-gelic Regosols; B) On negative mesorelief features: subnival, mesic herb-dwarf willow-moss, with an almost continuous cover (*Salix polaris, Saxifraga cernua, Cerastium regelii, Deschampsia spsp.*, the moss layer composed of *Sanionia uncinata*) on Eutri-gelic Regosols.

13 - Concave inclined slopes with striped nanorelief, complicated by bedrock outcrops and by outliers of terrace platforms

This area is characterized by a very complicated topography, with moisture levels varying from extremely dry to moderately moist; landscape elements (except for dellic drainage strips) occur at irregular intervals . Taken together, however, the elements form a massif defined by this landscape type. On the bedrock outliers (A), which are frequently used by predatory birds as nesting and feeding sites, one finds well-developed mesic herb meadow communities with 10-

30 % cover (*Bromopsis arctica, Hierochloe pauciflora, Erysimum pallasii* et al.) on Eutri-gelic Regosols. The outliers of the terrace platforms, (B), support dry-mesic forb-dryas alpine tundra (*Dryas punctata, Carex rupestris, Lloydia serotina, Astragalus alpinus* et al.) with a cover of about 30 %, on Dystri-gelic Regosols. Most of the area is occupied by stripe soil (dellic complexes). In the underdeveloped complexes, (C), spotty-striped tundra with mesic sedge-dwarf shrub-moss vegetation forms a 50 % cover; the dominants are *Carex arctisibirica, Salix polaris, Dryas punctata*; in the moss layer: *Aulacomnium turgidum, Tomentypnum nitens*) and the soils are Umbri-gelic and Molli-gelic Gleysols. In the developed complexes, 2 plant communities alternate at regular intervals: D) Mesic hummock-spotty herb-dwarf shrub-moss tundra with 30-50 % cover on the ridges of striped soil *(Carex arctisibirica, Arctagrostis latifolia, Dryas punctata, Salix polaris*; in the moss layer: *Hylocomium splendens var. obtusifolium*) on Umbri-gelic Gleysols; and E) Moist hummock sedge-moss communities in the troughs of the striped soil with continuous (100%) plant cover *(Carex concolor, Eriophorum polystachion*; the dominant moss is *Tomentypnum nitens*) on the Umbri-gelic Gleysols.

14 - Concave slopes (dellic complexes) with drainage channels of 0,3 to 1,2 m depth

These slopes are widely distributed and occupy a significant portion of the region. The drainage channels reach their maximum depth close to the lake. The vegetation comprises 3 regular and repeated elements: A) Mesic spotty sedge-dryas-moss tundra on ridges with 40-70 % cover (*Carex arctisibirica, Luzula confusa, Dryas punctata, Salix polaris*; the main moss is *Hylocomium splendens var. obtusifolium*) on Umbri-gelic Gleysols. The drainage channels of striped patterned ground support one of two type of vegetation, depending on structure and depth: B) Trough-like dells with wet sedge-moss vegetation (*Carex concolor, Eriophorum polystachion, Arctagrostis latifolia*, and sometimes bushes of *Salix reptans*; mosses: *Tomentypnum nitens*), on Umbri-gelic Gleysols ; C) V -shaped dells with wet graminoid vegetation (*Carex concolor, Eriophorum polystachion, Hierochloe pauciflora*) on Fibri-gelic Hystosols. Vegetative cover in the both types of dell is continuous.

15 - Spotty-striped slanting gravel surfaces with occasional rock streams

The vegetation in this region consists of two community types: A) Most surfaces have dry-mesic forb-moss spotty-hummocky alpine tundra with 30-40 % cover *(Papaver polare, Myosotis asiatica, Saxifraga spsp, Hylocomium splendens var. obtusifolium* dominated between hummocks). The soils are Lithic Leptosols; B) Rock streams with sparse (5-7 % cover) herb and primitive plant aggregations; the Distri-gelic Regosols are well-developed only between stones.

16 - Concave slopes with slight striped ground nanorelief in the initial stages of development of dell complexes

In this landscape type the ridges and troughs of patterned ground do not differ in vegetation. Both areas support mesic herb-dwarf shrub-moss and herb-moss communities (*Salix polaris, Luzula confusa, Equisetum variegatum, Draba pauciflora*; at lower altitudes: *Dryas punctata*; the common mosses are *Tomentypnum nitens* and *Orthothecium chryseon*); the soils are Umbri-gelic Gleysols; vegetative cover ranges from about 50 % on the ridges and up to 80 % in the almost flat dells.

17 - Striped horizontal surfaces with spotty, sometimes hummock-spotty nanorelief

This landscape type is common on the low (not more than 250 m a.s.l.) and flat watersheds and on their shallow slopes, where the ground has been subjected to pelitisation (soil material ranges from pelitized to clay loam). The vegetation consists of mesic cotton grass-dwarf willow-moss and sedge-dwarf shrub-moss communities (*Deschampsia borealis, Carex arctisib irica, Salix polaris, Alopecurus alpinus;* a well-developed moss layer of mostly *Tomentypnum nitens and Aulacomnium turgidum*); up to 80 % cover; the soil cover is a composite of Rentzi-gelic Leptosols regions bordered by Molli-gelic Gleysols surrounded by a background Distri-gelic Gleysol.

18 - Saturated surfaces at various hypsometric levels with angles of inclination close to 0°

Here the nanorelief is spotty and weakly striped and the vegetation is sedge-dwarf shrub-moss (*Carex arctisibirica, C.concolor, Salix polaris*, sometimes *S.reptans;* in the moss layer: *Tomentypnum nitens*); up to 90 % cover; the soils are a nanocomplex of Stagni-gelic Leptosol regions surrounded by Molli-gelic Gleysol borders, separated by Distri-gelic Gleysols.

19 - Watersheds on the northern macroslope of Byrranga Glavnaja Ridge, consisting of fine-grained material with occasional rock outcrops

The rudiments of striped patterned ground and thermokarst depressions are present as well. Generally speaking, they belong to another landscape: the intermountain kettles of the Ugolnaja river depression. The vegetation is close to that of the arctic tundra subzones (small role of dryas in community composition and its replacement by *Salix polaris*). The vegetation here comprises 3 elements: A) Weathered bedrock outcrops support sparse (3-5 % cover) forb-lichen vegetation between stones on a fragmented Distri Regosols; B) Mesic hummocky-spotty dwarf willow-sedge-moss tundra with 50-80 % cover (*Salix polaris*, *Carex arctisibirica*; in the moss layer: *Tomentypnum nitens, Ptilidium ciliare*) on Molli-gelic- and Umbri-gelic Gleysols; C) Thermokarst microdepressions and striped patterned ground with continuous wet sedge-moss vegetation (*Carex concolor, Eriophorum scheuchzeri, E.polistachyon*; mosses: *Tomentypnum nitens*) on Dysri-gelic Geysols.

20 - Polygonal surfaces on nearly horizontal slopes of the lower portions of watersheds

These features also occur in the previous landscape type. This palsa bog complex is generated through the natural draining of small lakes. The vegetation comprises 2 elements as a result of differences in microrelief: A) Palsa remnants with moist sedge-moss vegetation (*Carex arctisibirica, Luzula nivalis, Calamagrostis holmii*; moss layer consists of *Polytrichum strictum, Dicranum spsp*.,.); the soils are Dystri-gelic Gleysols; B) Flat depressions with wet moss-sedge-cotton grass vegetation (*Carex concolor, Eriophorum polistachyon, E.medium*; the sparse moss layer consists of *Meesia triquetra, Limprichtia revolvens*); the soils are Fibri-gelic Hystosols; the cover is continuous throughout the complex.

Valleys

21 - Deep V- shaped valleys of large mountain streams with abrupt scree slopes and with rock outcrops on the upper slopes

This is one of the most structurally complicated units. It includes almost all of the unit #6 units (the abrupt slopes). The elements of relief and vegetation vary even more here, because of the variety of exposures; the range of flora supported is much greater. The slope surfaces show evidence of the elements of ecological succession. The main elements are: A) Scree with sparse (cover < 5%) herbs (*Arabis petraea s.l., Cerastium beeringianum, Saxifraga spsp., Poa glauca* et al.) on Dystri-gelic Regosols and Dystri-gelic Regosols; B) Forb-dryas boulder debris (*Dryas punctata, Hierochloe alpina, Festuca auriculata, Astragalus alpinus, Potentilla uniflora* et al.) on fragmented Eutri-gelic Regosols; C) Mountain dry-mesic herbaceous meadows (*Poa glauca, Roegneria villosa., Astragalus alpinus, Senecio tundricola, S.resedifolius, Pedicularis verticillata* et al.) on Eutri-gelic Regosols; D) Niches on slopes with mesic herb-cassiope-moss (*Cassiope tetragona, Saxifraga cernua, S. nelsoniana; Hylocomium splendens var. obtusifolium*) communities, on the same soils. The vegetative cover corresponds exactly to similar communities of unit #6. The alluvial plains of each valley support a further 2 vegetative community types depending on the level of the flood-plain: E) Low altitude pebble banks with sparse (cover 1-3 %) forbs (*Chamaenerion latifolium, Arabis petraea*); soils are not presented; and H) High altitude pebble banks with herbaceous cover up to 40 % (*Oxytropis nigrescens, Papaver polare, Festuca brachyphylla, Luzula confusa* et al.) on Dystri Regosols.

22 - Flat-bottomed valleys of small mountain streams with scree or overgrown slopes

In these regions, wide flood plains are formed and the modern erosion and alluvial processes are still active. Generally, these landscape types occur on the limestone plateau on the north-east lakeshore. Because of this shore's favourable south-east aspect and the consequent fast thaw of even thick snow covers, the fragments of shrub communities observed here are almost unique in the region. The vegetation supported represents a series of ecological communities: A) Scree slopes with sparse (3-7 % cover) dry herbaceous (on limestones: herbaceous and lichen) vegetation: *Saxifraga oppositifolia, Braya purpurascens, Parrya nudicaulis* et al. on Dystric Regosols; B) Pebble banks with sparse forb short grass meadows (*Oxytropis middendorffii, O. nigrescens, Novosieversia glacialis* et al.; cover up to 30 %) growing on soils of fluvitierrestrial origin. C) Rare bushes of *Salix alaxensis* and *S.lanata*, with a layer of grasses, *Cardamine pratensis, Equisetum arvense, E.variegatum* et al.) on Eutric Fluvisols. The cover is either continuous or, in the willows-bushes, close to 30 %. Beneath the limestone slopes and near snow-fields there is usually wet nival turf with *Oxygraphis glacialis, Minuartia stricta, Cochlearia arctica* et al.

23 - U-shaped, nival, upper reaches of streams valleys with large snow accumulation and late thaw

The delay of snow thaw in this region can reach 2 months relative to the background surfaces. The slopes are composed of scree and are occupied (A) by primitive nival herbaceous aggregations (*Saxifraga cernua, S.nivalis*; in moss cushions: *Oncophorus wahlenberghii*) with irregular cover from 1 to 30-40 %; the soils are fragmented distric regosols. B) Lower slope reaches of pebbles at the altitude of alluvial deposits, with sparse vegetation consisting of a few herbs (10-15 % cover): *Cardamine bellidifolia, Arabis petraea, Papaver polare, Eritrichium villosum* et al. The soils are fragmented Eutric Fluvisols.

24 - Erosion valleys with overgrown slopes and with hummock-tussock nanorelief, on the northern macroslope of the lake basin

The vegetation of the valley bottoms is continuous, moist sedge-moss (*Carex arctisibirica, Lagotis minor,* Eriophorum callitrix; in the moss layer: *Tomentypnum nitens*) on moist Umbri-gelic Gleysols.

25 - Flood-plain of the Krasnaja river and its tributaries (large streams in the river delta).

This region is characterized by well-developed alluvial relief, but the limits between alluvial levels are rather variable; they are frequently expressed only in vegetation, which is again present as a series of ecologically distinct communities and includes the following elements: A) Low altitude pebble banks with solitary plants (*Thlaspi cochleariforme, Papaver polare, Leymus interior, Arabis petraea*), cover not more than 1 %, soils are absent.; B) Mid-altitude pebble banks with mesic herbaceous meadows, up to 30-50 % cover (*Oxytropis middendorffii, Astragalus tolmaczewii, A. alpinus, Pedicularis amoena, Festuca rubra subsp. arctica* et al.) on Dystric Fluvisols; C) High altitude sites with dry-mesic moss-herb-dryas short grass meadows (cover up to 90 %). They are present here as fragments only; the species composition is similar to that of the previous level, but here Dryas punctata dominates and there fragments of a moss layer (*Sanionia uncinata*); the soils are Fluvisols; D) High altitude locations with continuous mesic shrub-moss vegetation in fragments (*Salix reptans, S.lanata, Poa alpigena, Saxifraga nelsoniana* et al.; the continuous moss layer is almost only Tomentypnum nitens); the soils are the same; E) Moss-grass homogeneous bogs of the flood plain and terrace recesses (*Carex concolor, Eriophorum medium, E. polistachyon, Dupontia fisheri* et al.; in the fragmented mosslayer: Meesia triquetra, Limprichtia revolvens); the plant cover is continuous; the soils are Fibri-gelic Histosols. Mud banks with thickets of *Carex saxatilis subsp.laxa, C. maritima, Juncus castaneus* et al. occur occasionally; the cover here is not more 30 %; the soils are Mollic Fluvisols.

26- Gravelly alluvial fans with angles of inclination 10-20°, with hummocky nanorelief

The fans belong to ancient watercourses and there are no modern channels. They support mesic herb-dwarf shrub-moss tundra with up to 70 % cover *(Carex arctisibirica, C.misandra, Dryas punctata, Salix polaris, Salix arctica*; in the moss layer: *Hylocomium splendens var.obtusifolium).* The soils are Molli-gelic Gleysols.

27 - Fine grained alluvial fans with angles of inclination 3-8=83 and spotty-hummock nanorelief.

These fans often form part of the lakeshore and are frequently complicated by fragments of low lake terraces. This unit included 2 vegetation elements: A) The main surface with sedge-dwarf shrub-moss tundra, up to 80 % cover (*Carex arctisibirica, C. misandra, Dryas punctata, Salix polaris*; moss layer: *Tomentypnum nitens*); the soils of the hummocky regions are Molli-gelic Gleysols and between hummocks they are Umbri-gelic Gleysols. The presence of such species as *Oxytropis mertensiana, Eriophorum callitrix, Minuartia stricta* indicate a high level of nitrogen and available organic material in the soils; B) Channels with moist forb-dryas or dryas-forb meadows (*Dryas punctata, Poa alpigena, Carex tripartita, Gastrolychnis apetala, Artemisia tilesii* et al.) on Molli-gelic Leptosols with 20-40 % cover.

28 - Alluvial fans with angles of inclination of 0-5°, with pebbly banks and tundra between them

Such fans occur along the larger channels of small rivers. This landscape unit characterizes many of the older channels and indicates intense modern accumulation. Stone-loam spotty-hummocky tundra with small shallow thermocarst depressions and striped ground is found between the older channels. The vegetation forms 3 distinct elements: A) Pebbly sites proximal to channels with mesic forb-dryas vegetation *(Dryas punctata, Oxytropis middendorffii, O. arctica subsp. taimyrensis*, et al.) with about 50 % cover on Fluvisols; B) Mesic spotty-hummocky sedge-dwarf shrub-moss tundra *(Carex misandra, , Dryas punctata, Cassiope tetragona*, the moss layer is polydominant), 60-80 % cover, the soils surrounding spots are Molli-gelic Gleysols and in the depressions between borders they are Skeletic Cryosol; C) Thermokarst depressions with a continuous wet shrub-sedge-moss vegetation (*Salix reptans, Carex concolor, C. arctisibirica*; in the moss layer *Tomentypnum nitens* dominated), the soils are Fibri-gelic Histosols.

Terraces

29 - Fragments of marine accumulation marine terraces at 90-100 m a.s.l., with hummocks, and occasional large polygonal nanorelief

This region is part of the best preserved high altitude terraces, fragments of which can be found all along the shore, but are generally not marked on the map because of their small size. They usually occur as hills with flat or convex tops; in one case, however, in both the watersheds of the lake and of Ledjanaja Bay, the fragment occurs as a great residual block, hilly massif. The vegetation is classified into 3 elements: A) Forb hummocky dry communities of the terrace tops (*Carex rupestris, Potentilla uniflora, Oxytropis nigrescens, Hierochloe alpina* et al.) with 5-30 % cover on Geli-lithic Leptosols; B) Slopes with dry herb-dryas vegetation (composition of herbs is analogous, up to 60 % cover) on similar soils; C) Subnival mesic cassiope-moss tundra of snow covered slopes (*Cassiope tetragona, Carex vaginata, Taraxacum arcticum*; in the moss layer: Hylocomium splendens var.*obtusifolium*) with practically continuous plant cover on similar soils.

30 - Fragments of lacustrine accumulation terraces with large polygonal surface.

These terrace fragments are generally situated 3-5 m above the present lake level. Their surfaces are broken by cracks presumably as a result of ancient ground lodes. The surface of the polygons between the cracks are marked by hummocky nanorelief. The edges of the terraces are frequently marked by ice-shove ridges. The unit comprises 3 vegetation elements: A) Denudation-spotty terrace surface with the foliose lichens and sparse (cover not more 15 %) forb (*Papaver polare, Eritrichium villosum subsp.pulvinatum, Saxifraga oppositofolia* et al.) on Distri-gelic Regosols; B) Cracks with continuous mesic herb-cassiope-moss vegetation (*Cassiope tetragona, Carex misandra, Luzula nivalis, sometimes Vaccinium minus*; the moss *Rhacomitrium lanuginosum* dominated completely) on Dystric Leptosols; C) Ice-shove ridge and beach with herbaceous vegetation (*Papaver lapponicum, P.pulvinatum, Draba subcapitata, Chrysosplenium alternifolium* et al.).

31- Flooded low wet marshes of the lake.

This landscape unit occurs only once in the region, on the north-east lakeshore. The greater portion of its surface is under water for 2-3 weeks after the spring thaw and growth therefore begins late. The soil is composed of lake mud and mud-pebbles and shows a subtle polygonal microrelief, especially in areas somewhat raised. This unit consists of 3 elements, the vegetation of which form an ecological series: A) Low altitude mud and sandy-mud banks with sparse (cover 5-10 %) wet swamp vegetation (*Carex saxatilis sp. laxa, Eriophorum scheuchzeri, E. Polistachyon and Dupontia fisheri* on mollic-gleyic fluvisols; in the water: *Ranunculus gmelinii, Pleuropogon sabinii*); B) Higher level sedge tussocks (*Carex saxatilis ssp. laxa*), also with *Carex concolor, Eriophorum spsp.*, Pedicularis albolabiata; cover about 50%, the soils are the same; C) Highest level moist polygon-formed pebbly plots, covered beyond by mud, with sparse, frequently calciphilous, forbs (*Arabis petraea, Oxygraphis glacialis, Thlaspi cochleariforme, Braya purpurascens, Armeria maritima*) and solitary bushes of *Salix reptans*; cover 10-25 %, the soils are Dystri-gelic Leptosols.

Bogs

32 - Typical polygonal bogs on the northern shoreline region

The polygons are about 10 across and their borders reach a height of 0.5 m above the centers. The vegetation consists of 2 elements from the polygonal bog complex: A) -Wet moss-sedge communities of polygons (*Carex concolor, C. chordorrhiza*; in the fragmented moss layer:

Meesia triquetra, *Aulacomnium palustre*) on Fibri-gelic Histosols; B) Moist herb-moss communities of the polygon borders (*Carex arctisibirica*, sometimes *Astragalus umbellatus, Calamagrostis holmii*; the moss layer consists of *Tomentypnum nitens, Sphagnum spsp.*); the soils are Umbri-gelic Gleysols; the cover is continuous.

33 - Residual-polygonal bogs from the same location, but further from the lakeshore

These bogs are characterized by a significant development of thermokarst on the background flat-polygons and the low center polygon. The outliers of the elevated polygons occupy 5-25 % of the area. The vegetation is classified into 2 elements according to microrelief: A) Polygon outliers with moist shrub-herb-moss communities (*Salix reptans, S.pulchra, Carex arctisibirica, Luzula confusa, Senecio atropurpureus, Calamagrostis holmii*; the dense moss layer consists of *Polytrichum strictum, Dicranum elongatum, Aulacomnium palustre*). The soil is Umbri-gelic Gleysol; B) Thermokarst depressions with vegetation and soils analogous to polygons of unit 32; continuous plant cover.

34 - Palsa bog complexes

There is one large massif in the Krasnaja River mouth valley; it is probably a relict of an ancient warmer epoch. The cracks between palsas are frequently filled with water and the palsas reach 1,5 m In height. Denudation is evident on the palsas. The vegetation comprises 3 elements: A) Mesic shrub-herb-moss communities on palsas with 90-100 % cover (*Salix reptans, Cassiope tetragona, Carex arctisibirica, Luzula confusa, Senecio atropurpureus, Calamagrostis holmii*), the moss layer is thick (*Polytrichum strictum, Dicranum elongatum, Aulacomnium turgidum).* The soils are Umbri-gelic or Dystri-gelic Gleysols; B) Wet continuous sedge-cotton grass-moss communities in cracks (*Eriophorum polistachyon, E. medium, Carex concolor*, the hygrophilous moss Limprichtia revolvens) on Dystri-gelic Gleysols; C) Undergrowth of Arctophila fulva in water-filled cracks; the floating moss Calliergon giganteum is sometimes also found.

Acknowledgements

We thank all of the field group investigators from the AARI, AWI, IPÖ, MGU, ISSP, IfB, MLU expedition and members of the Taimyrsky reserve for their assistance in the field. We especially thank the expedition chief, Dr. D. Yu.Bolshijanov, and the Reserve chief, Yu. M. Karbainov, for the organization of field work. We gratefully acknowledge E.B. Pospelova for her review of the manuscript and for botanical consultation, M.V. Orlov for consultation on soil sciences and Prof. V.D. Vassiljevskaja for assistance in soil nomenclature.

References

Alexandrova, V.D. (1977) Geobotanic subdivision of Arctica and Antarctica (in Russian). Nauka, Leningrad, 188pp.

Andreev, V.N. and V.D. Alexandrova (1981) Geobotanical division of the Soviet Arctic. In: L.C. Bliss (ed.) Tundra ecosystems: a comparative analysis, Cambrige Univ. Press, Cambridge, 1981, 813pp.

Arctic flora of the USSR (1960-1987), Vol. I-X. Nauka, Moscow - Leningrad (in Russian).

Armand, D.L. (1975) Landscape science (in Russian). Mysl, Moscow, 287pp.

Bolshiyanov, D. Yu. and H.-W. Hubberten (1996) Russian-German Cooperation: The Expedition Taymyr 1995 and the Expedition Kolyma 1995 of the ISSP Pushchino group. - Reports on Polar Research. No. 211. Bremerhaven, 215pp.

Chronicle of the Nature State Biosphere reserve "Taimyrsky" (1995) Vol. 10. Chatanga, 11-93 (in Russian).

Czernov, Ju.I. and N.V. Matveeva (1979) The laws of zonal distribution of Taimyr plant communities (in Russian). In: Arctic Tundras and Polar Deserts of Taimyr.Nauka, Leningrad, 166-200.

Ershov, E.D (ed.) (1989) Geocryology of the USSR (in Russian). Nedra, Moscow, 514pp.
Gundelwein, A., H. Becker, T. Muller-Lupp and N. Schmidt (1997) Soils. Melles, M., B. Hagedorn and D.Yu. Bolshiyanov (eds.) Russian-German Cooperation: The Expedition Taymyr/Severnaya Zemlya 1996. - Reports on Polar Research, 237, Bremerhaven, 11- 14.
Ignatov, M.S. and O.M. Afonina (1992) Checklist of mosses of the former USSR (in Russian). Nauka, Moscow, 425pp.
Isachenko, A.G. (1965) Bases of landscape science and physico-geographical subdivision (in Russian). Vysshaya shkola, Moscow, 327pp.
Kind, N.V. and B.N. Leonov, eds. (1982) The Quarternary period of Taimyr (in Russian). Nauka, Moscow, 183pp.
Melles, M., B. Hagedorn and D.Yu. Bolshiyanov (eds.) (1997) Russian-German Cooperation: The Expedition Taymyr/Severnaya Zemlya 1996. - Reports on Polar Research. No.237. Bremerhaven, 178pp.
Melnikov, P.I. and N.I. Tolstikhin (eds.) (1974) General permafrost science (in Russian). Nauka, Novosibirsk, 309pp.
Michailov, I.S. (1971) The analysis of the spatial structure of some landscapes of an arctic region (in Russian). In : Problems of physico-geographical subdivision of polar regions. Leningrad, Gidrometeoizdat (Trudy AANII 304), 147-164.
Pfeiffer, E., A. Gundelwein, T. Nothen, H. Becker and G. Guggenbergen (1996) Characterization of the organic matter in permafrost soils and sediments of the Taimyr Peninsula/Siberia and Severnaya Zemlya/Arctic region. In: Bolshiyanov, D.Yu. and H.-W. Hubberten (eds.), Russian-German Cooperation: The Expedition Taymyr 1995 and the Expedition Kolyma 1995 of the ISSP Pushchino group. Reports on Polar Research, No. 211. Bremerhaven, 46 - 63.
Popov, A.I., G.E. Rosenbaum and N.V. Tumel (eds.) (1985) Cryolithology (in Russian). Izdatel'stvo MGU, Moscow, 239pp.
Pospelov, I.N. (1996) Various scale landscape mapping of SBR Taimyrsky territory as a basis for management ecological monitoring (in Russian). In: Problems of the Natural Reserve Business in Siberia. Shushenskoye, 111-114.
Preliminary report of the A-162 expedition to the Taimyr peninsula in 1995 (in Russian) (1995). AARI, St.-Petersburg, 160pp.
Siegert, C. and D.Yu. Bolshiyanov (1995) Russian-German Cooperation: The Expedition Taymyr 1994. - Reports on Polar Research. No.175. Bremerhaven, 98pp.
Sisko, R.K. (ed.) (1970) The Taimyr-Severnaja Zemlja region (in Russian). Gidrometeoizdat, Leningrad, 375pp.
Soil map of the World (1990). The re-examined legend 1:5 000 000. Food and Agriculture Organization of the United Nations, Unesco, Paris.
Strelkov, S.A. (1965) Siberia. A history of the development of Siberian and Far Eastern relief (in Russian). Nauka, Moscow, 336pp.
Tyrtikov, A.P (1979) Dynamics of the plant cover and the development of cryogenic relief forms (in Russian). Moscow, 147pp.
Unstead, J.F. (1935) The British Isles. University of London Press, London, 292 pp.
Vtyrin, B.I. (1969) Problems of cryogenic relief genesis. In: Geography and geomorphology of Asia (in Russian). Nauka, Moscow, 118-130.
Washburn, A.L. (1979). Geocryology. A survey of periglacial processes and environments. London, 406pp.
Washburn, A.L. (1988). The cold world. The geocryological research (in Russian). Moscow, 294pp.

Studies of Methane Production and Emission in Relation to the Microrelief of a Polygonal Tundra in Northern Siberia

V.A. Samarkin[1], A. Gundelwein[2] and E.-M. Pfeiffer[2]

(1) Institute of Soil Science and Photosynthesis, Russian Academy of Sciences, Pushchino, Russia
(2) Institut für Bodenkunde, Universität Hamburg, Allendeplatz 2, D 20146 Hamburg, Germany

Received 2 March 1997 and accepted in revised form 3 March 1998

Abstract - Methane production and emission in a polygonal tundra on Taimyr Peninsula in North Siberia (75°N, 98°E) were investigated during summer 1996. The in situ methane emission was measured several times per day between July and September at a typical polygonal tundra site. Additionally, measurements were carried out at different plots along a transsect through the investigated valley to determine the variability of methane emissions of the whole polygon area. CH_4 emission rates showed high response on microrelief, which influences water table and soil temperatures. No diurnal variations in methane emissions were found. The emission rates ranged from 0-160 mg $CH_4*d^{-1}*m^{-2}$. Based on detailed mapping of soils and surface structures (patterned grounds) a projection on total methane emission from the investigated polygonal tundra is possible. The mean daily emission rate between July and September is about 50 mg $CH_4*d^{-1}*m^{-2}$, which is about 4% of total gaseous carbon loss from soil to atmosphere. Isotope investigations proof the reduction of CO_2/H_2 to be the major pathway of methane production and fermentation being negligible. At the end of the thawing season methane emissions and methane content of the pore water decrease. Methane oxidation is very effective even when the water table is only a few centimeters below the surface.

Introduction

Tundra wetlands are supposed to be an important source for the greenhouse gas methane. About 120 Tg methane per year are emitted from natural wetlands (IPCC 1994). Northern Siberian wetlands, which cover an area of more than $211*10^6$ ha including $2.9*10^6$ ha peatlands (Botch et al., 1995), are assessed as one major source of methane from wetlands (Harris et al., 1993).

Numerous studies concerning methane emissions from northern high latitude wetlands were carried out in recent years, most of them in Canada, Alaska and Scandinavia (Bartlett et al., 1992; Harriss et al., 1993; Morrissey and Livingstone, 1992; Sebacher et al., 1986; Svensson and Rosswall, 1984; Whalen et al., 1991; Whalen and Reeburgh, 1988, 1990 and 1992). Only a few measurements are done recently in Siberian tundra by Inoue et al. (1995), Nakayama (1995a and b), Rivkina et al. (1993) and Samarkin et al. (1994). This study presents new methane flux measurements and investigations concerning production from a wet polygonal tundra in North Siberia.

Investigation area

The investigation area is located in the southern Byrranga Mountains on Taimyr Peninsula, Northern Siberia (75°N, 98°E). The methane emission measurements were carried out in the valley of Krasnaya River, north of Lake Levinson Lessing.

This river valley is dominated by wet, gleyic soils out of loamy sands, rich in organic matter (Histic Pergelic Cryaquepts). The maximum thickness of active layer is 45-50 cm. The major

----

In: Kassens, H., H.A. Bauch, I. Dmitrenko, H. Eicken, H.-W. Hubberten, M. Melles, J. Thiede and L. Timokhov (eds.) Land-Ocean Systems in the Siberian Arctic: Dynamics and History. Springer-Verlag, Berlin, 1999, 329-342.

patterned ground structure of the valley are low centred ice-wedge polygons (compare Figure 1 and Table 1).

The low centred ice-wedge polygons are characterized by two different microrelief sites: the central depression, with a water table near or above the surface, and the high border, where the water table is distinctly below the surface. Polygon diameters range between 6-25 m, about 33% of the area are polygon walls, 67% are polygon depressions (more informations about the investigation area and the soil properties of the investigated ice-wedge polygons can be found in Pfeiffer et al, 1995; Gundelwein, 1998).

Table 1: Soil description of a typical ice-wedge polygon, Lake Levinson Lessing (according to Soil Survey Stuff 1994)

Polygon wall (site 8):		
Depth (cm)	Horizon	Description
0 - 06	Oi	slightly decomposed peat, black (7.5YR2/1), platy structure.
06 - 45	Bg	loamy sand, yellowish gray (2.5Y5/1), <1 vol% rock fragments, very strong rooted. Max. thawing depth: 45 cm.
> 45	Cf	loamy sand, <1 vol% rock fragments, ice lenses, permafrost boundary.
Polygon depression (site 7):		
Depth (cm)	Horizon	Description
0 - 02	Oi	slightly decomposed peat, black (7.5YR2/1), platy structure.
02 - 06	Oe	intermediate decomposed peat, black (10YR2/1), platy structure.
06 - 35	Bg	loamy sand, yellowish gray (2.5Y5/1), <1 vol% rock fragments, very strong rooted. max. thawing depth: 35 cm.
> 35	Cf	loamy sand, < 1 vol% rock fragments, ice lenses, permafrost boundary.

Methods

Methane emissions

Methane emissions are determined by using closed chamber technique in combination with a small circulation pump and a trace gas analyzer (TGA). The TGA works on the basis of photoaccustic infrared spectroscopy (TGA 1302 from Bruel and Kjaer Company, Denmark). The chambers are sealed against the soil surface by setting on waterfilled, stainless steel frames. The measurements are conducted every 90 seconds within a period of 15 to 30 minutes, so a high resolution is possible. TGA-results are compared with results of gaschromatograph measurements in laboratory (1-4 ppm deviation).

Measurements are repeated 4-6 times a day in the center of the polygon area at a polygon depression (site 7) and an polygon wall site (site 8). Additionally measurements are carried out at 16 different sites (polygon walls and polygon depressions) within a transect through the river valley. The daily and seasonal variations were recorded and the variations between different but similar patterned ground structures were monitored.

To determine the influence of different water tables on methane oxidation and emission a special field experiment was done: the water table of a polygon polygon depression was lowered in two steps from 1 cm above to 18 cm below surface, the methane emission rates being measured simultaneously.

Soil temperatures in 5, 10 and 20 cm depth, thickness of active layer, water table, air pressure and air temperature were monitored during every measurement.

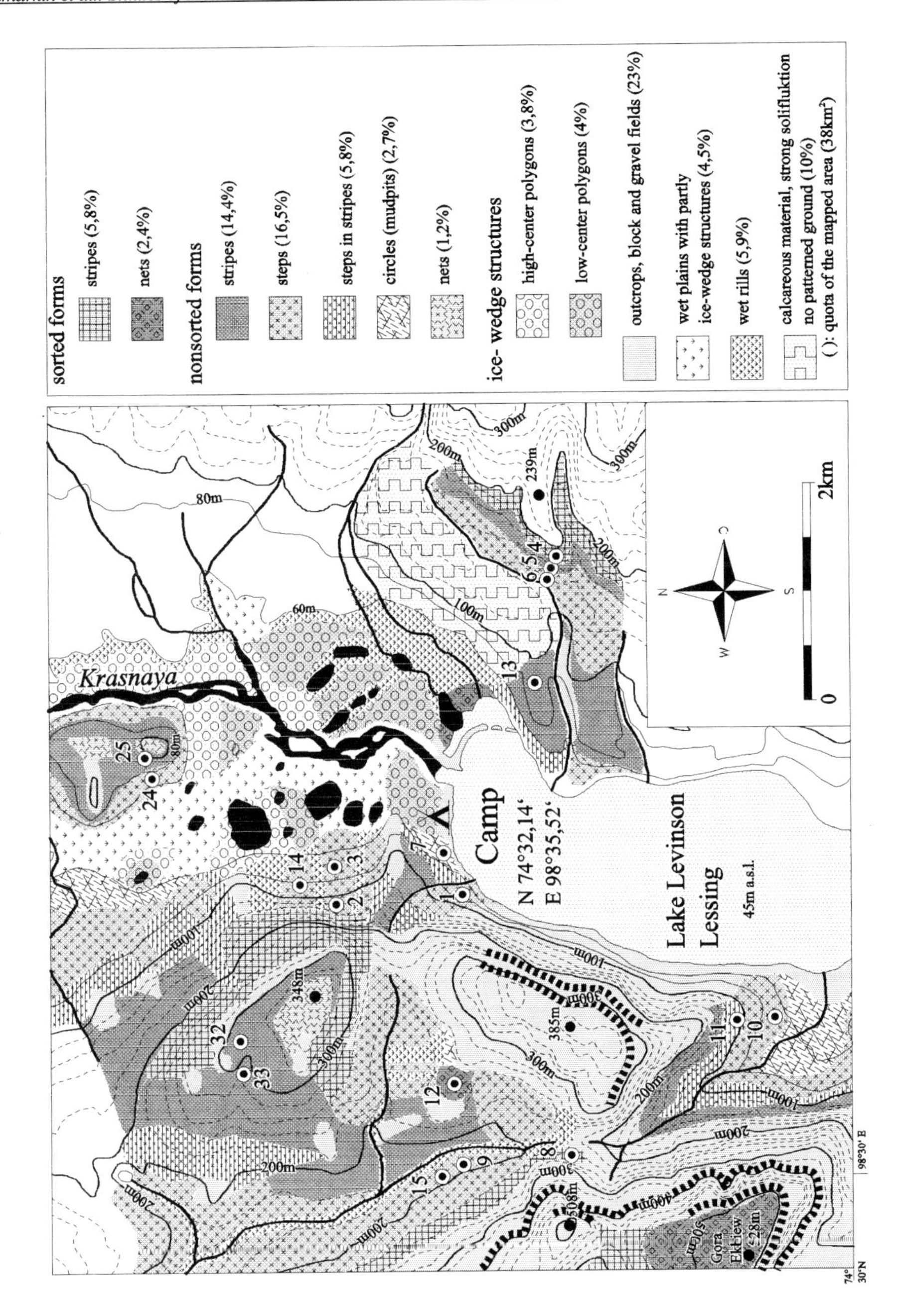

Figure 1: Distribution of patterned grounds, Lake Levinson Lessing (74°N, 98°E).

Methane concentrations in pore water

Pore water samples (10 ml) were taken in 5 cm steps from the active layer, drawn into a pre-vacuumated glass tube (27 ml, Bellco Glass Co.) and immediately analyzed, using headspace technique (McAuliffe, 1971) by a portable XPM-4 Gas Chromatograph (Chromatograph Company, Russia) equipped with a flame ionization detector (FID) and a 2 m long stainless steel column filled with Poropack Q resin.

The methane store [mg CH_4*m^{-2}] in a 30 cm thick active layer (300 l bulk soil) was calculated using the following equation:

methane store in the active layer = Σ *methane store in the 5 cm horizons a' 50 l bulk soil*

$$\Sigma\, mc*(wc*10^{--2})*50$$

where mc is the methane content in the pore water of a 5 cm horizon [mg CH_4* 1000 cm^{-3} bulk soil] and wc is the water content of the bulk soil [in %].

Methane production rates and pathways (tracer experiments)

The microbial methane production rate from ^{14}C-labelled bicarbonate and acetate was measured using tracer technique (Kuivila et al., 1989; Reeburgh, 1983). Soil samples for the experiment were collected from active layer at six depths between 0-30 cm in 5 cm intervals. Subsamples of 5 cm3 were rapidly transferred into test tubes with a cut off tip, stoppered by rubber plunger from cut end, were moved by plunger to the opposite side of tubes and stoppered by buthil rubber stoppers without any gas phase. The soil samples in the test tubes were line injected with a syringe through the rubber stopper with 100 µl of ^{14}C-labelled sodium bicarbonate or 100 µl of 2-^{14}C sodium acetate solution with 0.37 MBq. activity. The samples were incubated at in situ temperatures for 24-72 hours. The experiment was terminated by the injection of 5 ml of 1 M NaOH solution in each tube, control samples were fixed immediately after tracer injection. In the laboratory each tube was inserted into the special rubber stopper through the hole of appropriate diameter. The rubber stopper is equipped with two 3 mm stainless steel tubings with stopcocks for air bubbling. A 100 ml glass jar, containing 50 ml of saturated NaCl solution was closed by this stopper and connected with the apparatus for combustion of CH_4 formed. The soil-NaOH-mixture from these tube was injected into the jar by plunger movement and stripped with air which than passed through (1) a bubbling flask containing 10 ml of 0.5 N NaOH, (2) a quartz tube containing silica granules covered by Co-oxyde catalyst at 800°C to combust CH_4, and (3) a bubble tower with 15 ml of scintillation cocktail containing 10 % phenethylamine to trap the $^{14}CH_4$ derived $^{14}CO_2$. The activity was measured by LS 5000TD, Beckmann Company liquid scintillation counter.

Methane production rates were calculated using the following equation:

Rate = \F(Ap * C * W;As * t) µg CH_4 [C] * dm^{-3} wet soil * d^{-3}

where Ap is the activity of CH_4 formed (dpm), As the activity of the added substrate, C the substrate concentration in pore waters (DIC or methyle carbon in acetate), W the soil pore water content and t the time of incubation.

In the case of methane oxidation the radioactivity of carbon dioxide, microbial biomass and organic exometabolites were measured individually and summarized. For methane oxidation Ap is the activity of $^{14}CO_2$, microbial biomass and organic exometabolites formed from ^{14}C-labelled methane, As the radioactivity (dpm) of $^{14}CH_4$ injected in the sample and C methane carbon concentrations in pore water. A more detailled description of all mentioned methods above can be found in Samarkin et al (1997).

The isotopic composition (^{13}C and D) of the methane gas samples is measured in the Institute of Geology and Geophysics, University of Leipzig.

Other investigations

The cation contents of the pore water samples were analysed at the Alfred-Wegener-Institut Potsdam with an ICP OES Optima 3000 (Perkin Elmer Co.) and NBR Standard SRM 1643d.

Results and discussion

Methane emission

Methane emissions from the polygon wall (site 8) and the polygon depression (site 7) differ strongly from each other: while the mean emission rate of the wet polygon depression was about 75 mg CH_4 $*d^{-1}*m^{-2}$, the mean emission rate of the higher and dryer polygon wall part of the polygon was near zero. The emission rates showed no diurnal variations, they decreased during the measurement period (compare Figures 2a and b).

In the polygon depression the soil temperatures in 5 cm depth decreased between July and September from 8-10°C to 2-3°C. The methane emissions of the investigated polygon depression decreased over time with decreasing temperatures from about 120 mg $CH_4*d^{-1}*m^{-2}$ at the end of July down to 40 mg $CH_4*d^{-1}*m^{-2}$ at the beginning of September. The water table was always near or above the surface. Strong reductive conditions were indicated by high amounts of reduced iron in pore water samples (70-90 mmol/l) in comparison to the polygon wall (0-5 mmol/l). During heavy rainfalls the water table increase and the strong anoxic conditions leaded to increased emission rates.

At the polygon wall the water table position was always below the surface. At the end of July the soil temperatures at the polygon wall were lower than the comparable soil temperatures at the polygon depression. One month later, at the end of August soil temperatures were nearly the same at polygon wall and polygon depression. The contents of dissolved methane in the soil pore water from permafrost boundary up to 10 cm below the surface were the same in polygon wall and polygon depression (5-10 mg CH_4/l), only in the dryer upper part of the polygon wall the methane content of the pore water was lower than in the polygon depression part of the polygon (> 1 mg CH_4/l, compare Figures 3a and b).

The dependence of the methane emission rates on the water table and the temperature was confirmed by a multiple regression analysis. A high response of emission rates on soil temperatures can be shown in 5 cm soil depth (multiple regression coefficient β: 7.5, error probability p: 0.0014, so significance level < 1%) and in 10 cm depth (β: -9.5, p: 0.0251) and less on water table position (β: -2.0, p: 0.0490). The influence of temperature on the methane emission rates is also reported by different authors (Harris et al., 1993; Morrisey and Livingstone, 1992; Svensson and Rosswall, 1984; Whalen et al, 1991; Whalen and Reeburg, 1992). In opposite to the CO_2-Emissions, which show a linear relationship, the CH_4-emissions show a logarithmic relationship concerning temperature changes (Moore and Knowles, 1989). In other studies the water table position (Funk et al, 1994; Moore and Knowles, 1989; Moore and Dalva, 1993; Moore and Roulet, 1993) or the thickness of active layer was more important (Whalen and Reeburgh, 1992).

The influence of the water table position became obviously during the field experiment: after a first phase of high gas fluxes the methane emission rate decreased as a result of artificially lowering of the water table at a polygon depression (compare Figure 4). The high flux might be

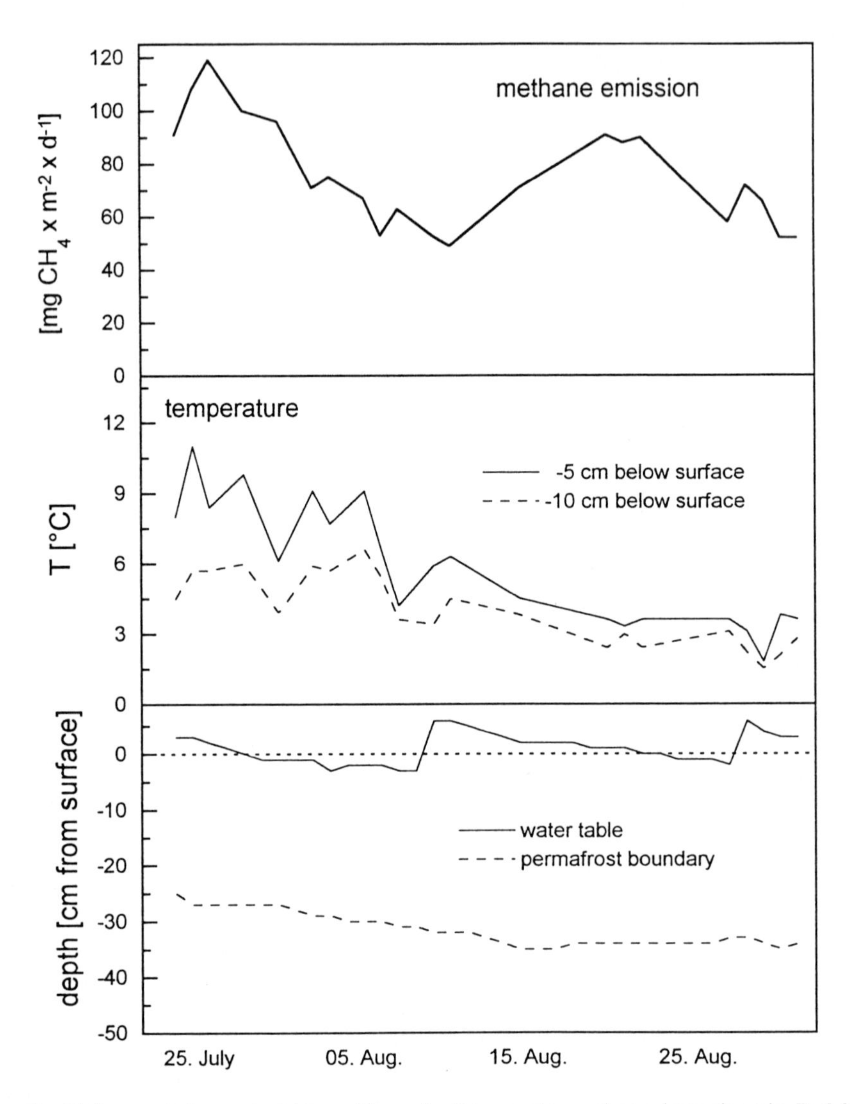

Figure 2a: Methane emission, water table position and soil temperature, polygon depression (site 7), July and August 1996.

the result of better gaseous permeability of the aerated soil horizons. Moore and Roulet (1993) reported similar observations during laboratory experiments with soil columns. After reducing the water table the emission rate decreased. This was interpreted as an increase of oxidation processes by methanotrophic microorganisms or a stop of CH_4-production. The relationship between water table position and methane emission rates was also a logarithmic one (Moore and Roulet, 1993).

Both, methane production and methane consumption, were dependent on water table and soil temperature. When the water table was low enough, the produced methane was almost totally consumed by microorganisms in the upper, well oxigenized soil horizon. This was the situation at the polygon wall.

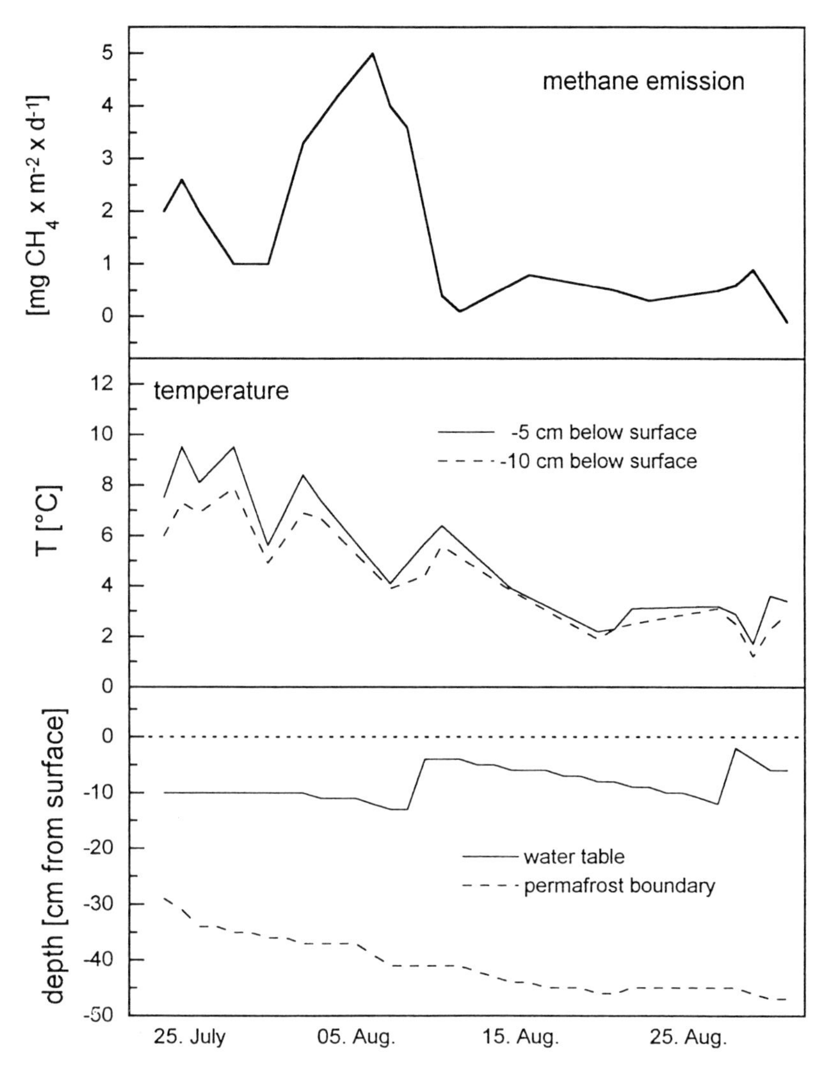

Figure 2b: Methane emission, water table position and soil temperature, polygon wall (site 8), July and August 1996.

The methane emission rates of different sites (Polygon depressions and polygon walls) along a transsect through the river valley ranged between 0-160 mg CH_4*d^{-1}*m^{-2}. They showed the same decreasing trend during the summer season as the emission rates of the polygon sites reported above (sites 7 and 8). The range of methane emissions from the polygon depressions of the transect were 25-160 mg CH_4*d^{-1}*m^{-2} (mean: 74.4 mg CH_4*d^{-1}*m^{-2}) from the polygon polygon walls only 0-18 mg CH_4*d^{-1}*m^{-2} (mean: 2.3 mg CH_4*d^{-1}*m^{-2}). About 33% of the area are polygon walls and 76% polygon depressions. So the mean diurnal methane flux from the polygon area can be calculated as about 50 mg CH_4*d^{-1}*m^{-2} for the time from July until September. That are 4% of the total gaseous carbon loss from these soils to the atmosphere

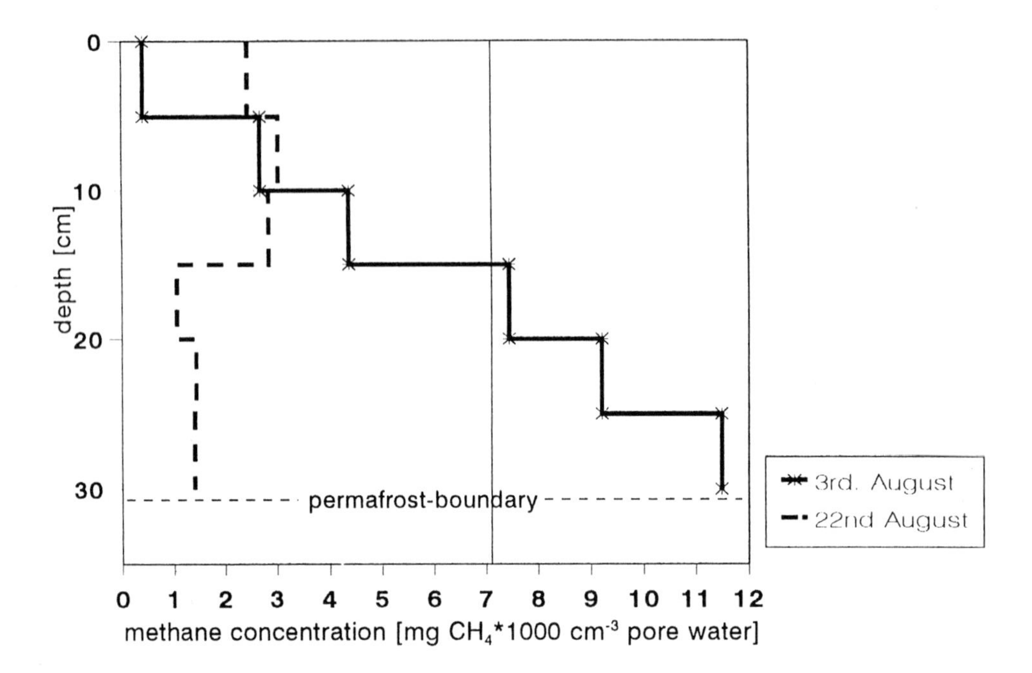

Figure 3a: Methane concentrations in soil pore water, polygon depression, August 1996.

during that period (compare Sommerkorn et al, this volume). Svensson and Rosswall (1984) found 5.5% of total gaseous loss from wet, not always waterlogged areas to the atmosphere were methane.

The presented methane emission rates fall into the range of results from other investigations: wet, subarctic and arctic tundra north of 60°N have mean daily emission rates between 0-600 mg $CH_4*d^{-1}*m^{-2}$. The most of the reported emisssion rates are between 10-150 mg $CH_4*d^{-1}*m^{-2}$ (Bartlett et al, 1992; Harriss et al, 1993; Inoue et al., 1995; Morrissey and Livingstone, 1992; Nakayama, 1995 a,b; Rivkina et al., 1993; Samarkin et al., 1994; Sebacher et al., 1986; Svensson and Rosswall, 1984; Whalen et al., 1991; Whalen and Reeburgh, 1988, 1990, 1992).

Methane production and consumption

The methane concentrations in pore water are presented in Figures 3a and b. The methane store in the 30 cm thick active layer of the polygon depression part of the patterned ground, calculated using data from methane concentrations in pore waters and taking into account the different water contents of the soil horizons (50-90 w/w%, mean 60 w/w%), is about 1070 mg CH_4*m^{-2} at the beginning of August, and 390 mg CH_4*m^{-2} at the end of August. The methane store decreased from 3th to 22th August on 680 mg CH_4*m^{-2} or 36 mg $CH_4*d^{-1}*m^{-2}$. This value - received by calculation - is lower than the mean methane flux measured by chamber in this period.

The amount of stored methane in pore water was high at the end of July and decreased over the time. It was exhausted at the end of August. explanations for this observation may be

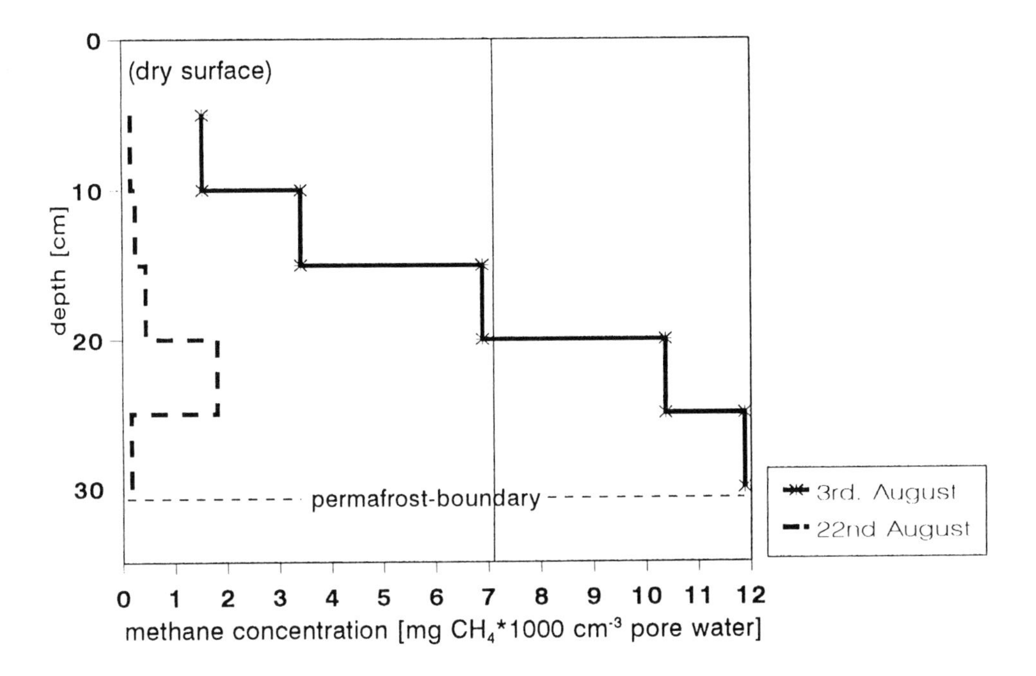

Figure 3b : Methane concentrations in soil pore water, polygon wall, August 1996.

different reactions of methane oxidizing and methane producing bacteria on changing temperatures: the soil temperatures decreased between end of July and end of August from about 7-9°C to 1-3°C. Moore and Dalva (1993) found out, that methane producing bacteria are more sensitive on temperature changes than methane consuming bacteria. That explains the decreasing emission rates at the end of August: less methane was produced, the stored methane in the soil pore water was emitted to the atmosphere and the rate of methane oxidation was still high.

Partly, the high methane contents in soil pore water in July also can be explained by the different behaviour of methanogene and methanotroph bacteria: the soil temperatures increased and also the methane production - faster than the methane consumption. But after snowmelt the thickness of the active layer did only increase slowly. It is unlikely that the whole stored methane was produced during the short time at the beginning of the thawing period in July. Further, methane may have been produced in the year before at the beginning of the winter period. At this time, a thin frost layer forestalled gaseous exchange between the soil and the atmosphere, the soil-gas system was closed from the surface and the bottom. The CH_4-generating bacteria still produced CH_4 in the melted layer. Methane oxidation was low because of low oxygen contents in the soil and strong anaerob conditions. The rhizosphere oxidation of the died plants was low, too. The formed methane would be trapped in the frozen soil until the beginning of thawing in the next summer.

Methane contents in pore water were between 1-11 $\mu g\ CH_4*ml^{-1}$ pore water. Svensson and Rosswall (1984) found 0-20 $\mu g\ CH_4*ml^{-1}$ pore water, Williams and Crawford (1984) found 0-24 $\mu g\ CH_4*ml^{-1}$ pore water.

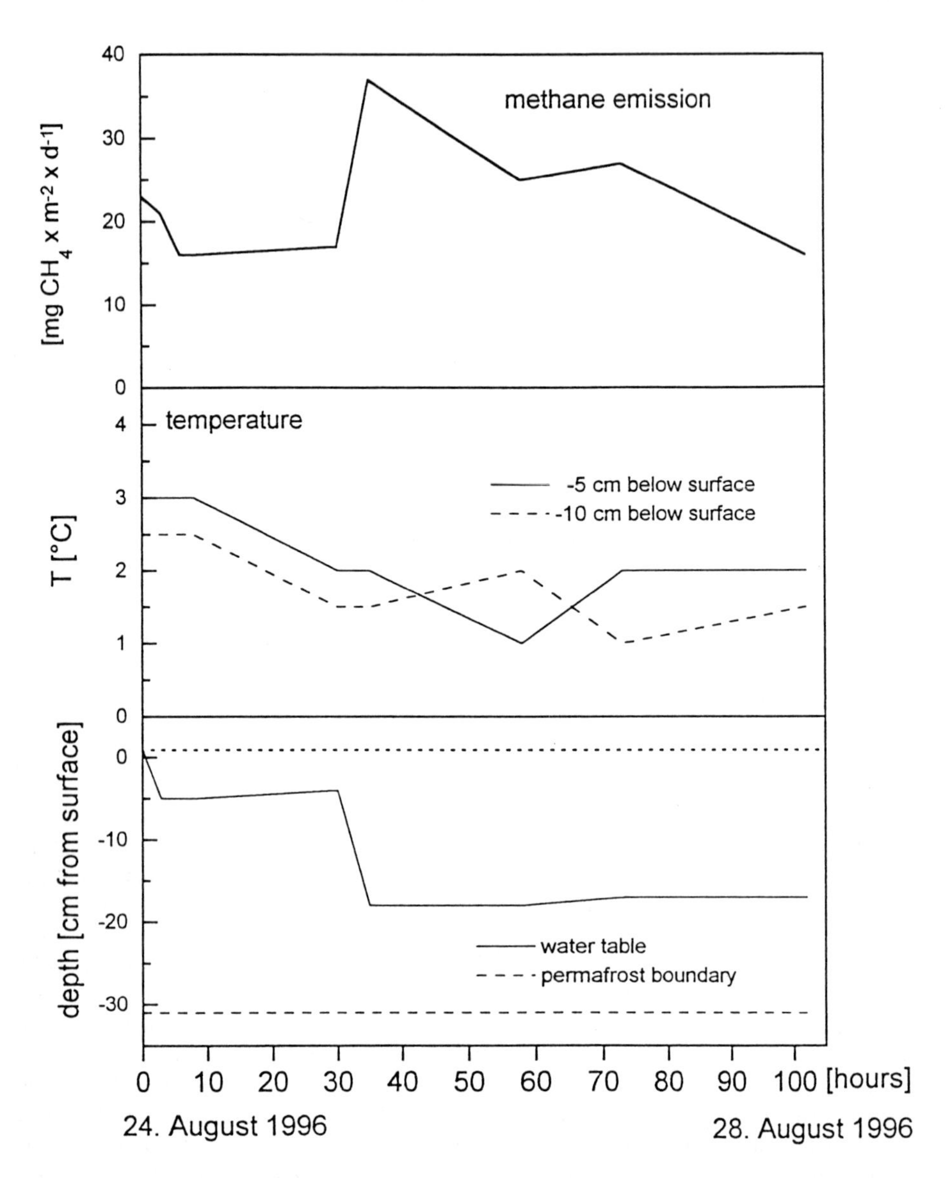

Figure 4: Methane emission from a polygon depression due to artificial water table lowering during a field experiment, August 1996.

To determine the CH_4 balance in soils the methane production was measured in the polygon depression. The production rate in the active layer calculated on the basis of the isotopic experiments (data from 5-8 of August 1996, presented in Figure 5) was 67 mg $CH_4*d^{-1}*m^{-2}$ for the polygon depression and 12.9 mg $CH_4*d^{-1}*m^{-2}$ for the polygon wall. Acetotrophic methane generation at the polygon depression contribute about 3.5 % of total methane production.

The stable isotopic composition of carbon and hydrogen in methane dissolved in pore waters

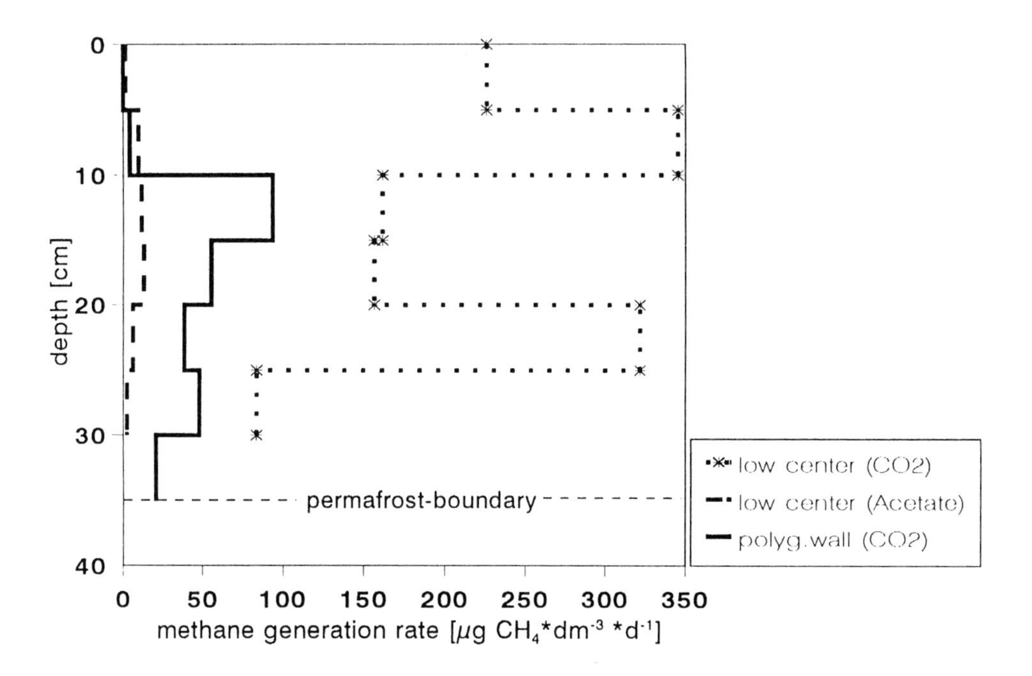

Figure 5: Methane generation rates, calculated by ^{14}C-CO_2 and 2-^{14}C-acetate measurements, polygon depression and polygon wall.

(^{13}C: -60.7‰ PDB, D: -300‰), CO_2 (^{13}C: -15.9 ‰ PDB) and SOM (^{13}C:-24 - -28 ‰ PDB, compare Gundelwein, 1998) confirmed that bacterial CO_2/H_2-reduction is the main methane generation process in these soils (Sugimoto and Wada, 1993, 1995). The minor role of the acetotrophic pathway for methane production was also confirmed by studies of Lansdown et al (1992). In opposite to these results Kotsyurbenko et al (1993) found out that most of the methane produced under cold temperature conditions is generated by the acetotrophic pathway.

A strong enrichment of the heavy carbon isotope ^{13}C in the SOM emphasizes the importance of the horizon between 10-15 cm depth for methane generation (Gundelwein, 1998).

The sum of the diurnal loss of methane store from the polygon depression (36 mg CH_4*d^{-1}*m^{-2}) and the methane production rate (67 mg CH_4*d^{-1}*m^{-2}) show a possible methane flux of about 103 mg CH_4*d^{-1}*m^{-2} if no methane is oxidized. By comparison of this value with the mean chamber derived flux rate from the polygon depression of 73.2 mg CH_4*d^{-1}*m^{-2} it is possible to assume that about 30 % of the produced methane is oxidized. The rate of methane oxidation in soils was measured (see Figure 6).

In the active layer of the water saturated polygon depression 18.4 mg CH_4*d^{-1}*m^{-2} were oxidized by methanotrophic bacteria, that were 27.4 % of the produced methane. At the moss covered polygon wall the oxidation rate was 3 times higher than at the polygon depression. About 59.3 mg CH_4*d^{-1}*m^{-2} were oxidized. The consumption of methane was 4.6 times higher than the production. A so called "active methane biofilter" on the water table level in moss covered tundra soils was described earlier (Vecherskaya et al, 1993).

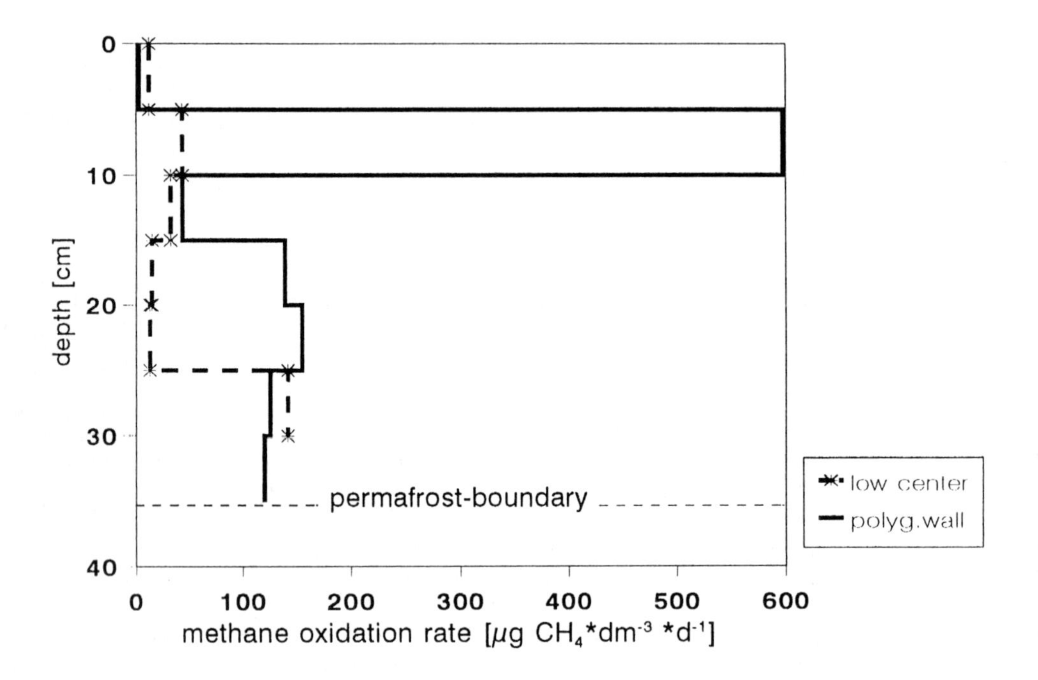

Figure 6: Methane oxidation rates in the active layer, polygon depression and polygon wall.

Account must be taken that methane oxidation rate may be underestimated because activity of the rhizosphere processes was not measured. Oxygen transport to the rhizosphere by plants may stimulate microbial CH_4 consumption around the roots and sufficiently increase the amount of oxidized methane (Gerard and Chanton, 1993).

Conclusions

The fluxes and production rates of the investigated siberian tundra were high and comparable to observed data from other arctic and high arctic tundra ecosystems in Alaska and Canada. The dependence of methane fluxes on soil temperature and water table position and the microrelief position was confirmed. The main pathway of methane production as investigated by isotope analysis was the reduction of CO_2. The process of methane oxidation was very effective. Therefore methane emission rates will strongly decrease with the water table when temperatures and thawing depths of the frozen soils will increase.

More detailed investigations of methane production dynamics in a possibly warmer climate are needed. Tracer experiments and studies of the isotopic composition of all mineral and organic compartiments including methane gas are necessary to understand the future behaviour of SOM.

Acknowledgements

The research work was partly supported by the Russian Foundation for Basic Researches (Grant 95-04-12526a) and by the German Ministry of Education, Science, Research and Technology (Project 03PL014B). We thank Dr. H.-M. Nitzsche, Institute of Geology and Geophysics, University of Leipzig, for isotopic analysis of the gas samples, Dr. B. Hagedorn, Alfred-Wegener-Institute Potsdam for cation analysis of soil pore water and Dr. R. Uerkvitz, University of Hamburg for support in statistics.

References

Bartlett, K.B., P.M. Cril., R.L. Sass, R.C. Harriss and N.B. Dise (1992) Methane emissions from tundra environments in the Yukon-Kuskokwim Delta, Alaska.- Journal Geophys. Research, 97, 16.645-16.660.

Botch, M.S., K.I. Kobak, T.S. Vinson and T.P. Kolchugina (1995) Carbon pools and accumulation in peatlands of the former Soviet Union.- Global Biogeochemical Cycles 9 (1), 37-46.

Funk, D.W., E.R. Pullman, K.M. Peterson, P.M. Crill and W.D. Billings, W.D. (1994) Influence of water table on carbon dioxide, carbon monoxide and methane fluxes from taiga bog microcosms.- Global Biogeochemical Cycles 8, 271-278.

Gerard, G. and J. Chanton. (1993) Quantification of methane oxydation in the rhizosphere of emergent aquatic macrophites: Defining upper limits.- Biogeochemistry, 23, 79-97.

Gundelwein, A. (1998) Composition and decomposition of organic matter in permafrost affected soils of North Siberia (in German). Dissertation Hamburg, Hamburger Bodenkundliche Arbeiten 39, 160p.

Harriss, R., K. Bartlett., S. Frolking. and P. Crill. (1993) Methane Emissions from Northern High-Latitude Wetlands.- In: Biogeochemistry of global change: radiatively active trace gases, Oremland, R.S. (ed.), Chapman and Hall, New York, 449-485.

Inoue, G., S. Maksyutov. and N. Panikov (1995) CO_2 and CH_4-emission from wetlands in west Siberia.- Proceedings of the Second Symposium on the Joint Siberian Permafrost Studies between Japan and Russia in 1993, Tsukuba/Japan, 37-43.

International Panel on Climate Change [IPCC] (1994): Radiative forcing of climate change,the 1994 report of the scientific assessment working group of IPCC, WMO/UNEP, University Press, Cambridge.

Kotsyurbenko, O.R., A.N. Nozhevnokova and G.A.Zavarzin (1993) Methanogenic degradation of organic matter by anaerobic bacteria at low temperature.- Chemosphere 27 (9), 1745-1761.

Kuivila, M., J.W. Murray, A.H. Devol. and P.C. Novelli (1989) Methane production, sulfate reduction, and competition for substrates in the sediments of Lake Washington.- Geochim.Cosmochim.Acta 53, 2597-2599.

Lansdown, J.M., P.D. Quay. and S.L. King (1992) CH_4-production via CO_2-reduction in a temperate bog: A source of ^{13}C-depleted CH_4.- Geochimica et Cosmochimica Acta 56, 3493-3503.

McAulife C. (1971) GC determination of solutes by multiple phase equilibration.- Chem.Tech. 1, 46-51.

Moore, T.R. and M. Dalva, M. (1993): The influence of temperature and water table position on carbon dioxide and methane emissions from laboratory columns of peatland soils.- Journal of Soil Science, 44, 651-664.

Moore, T.R. and R. Knowles. (1989) The influence of water levels on methane and carbon dioxide emissions from peatland soils.- Canadian Journal of Soil Science 69, 33-38.

Moore, T.R. and N.T. Roulet (1993) Methane Flux: water table relations in northern wetlands.-Geophysical Research Letter 20 (7), 587-590.

Morrisey, L.A. and G.P. Livingstone (1992) Methane emission from Arctic tundra: An assessment of local spatial variability.- J. Geophys. Res. 97, 16.661-16.670.

Nakayama, T. (1995a) Estimation of Methane Emission from Natural Wetlands in Siberian
Permafrost Area.- Dissertation Graduate School of Science, Hokkaido University.

Nakayama, T. (1995b): Estimation of methane emission from Siberian tundra wetlands. Proceedings of the Third Symposium on the Joint Siberian Permafrost Studies between Japan and Russia in 1994, i-word, Sapporo, 31-36.

Pfeiffer, E.-M., A. Gundelwein, T. Nöthen, H. Becker and G. Guggenberger (1996) Characterization of the Organic Matter in Permafrost Soils and Sediments of theTaymyr Peninsula/Siberia and Severnaya Zemlya/Arctic Region.- Bereichte zur Polarforschung 211, 46-83.

Reeburgh, W.S. (1983) Rates of biogeochemical processes in anoxic sediments. Am.Rev. Earth.Planet.Sci. 11, 269-298.

Rivkina, Y.E., V.A. Samarkin. and D.A. Gilichinsky (1993) Methane and permafrost soil of the Kolyma-Indigirka lowland. Eurasian Soil Science 25, 50-53.

Samarkin, V.A., D.G. Fedorov-Davydov, M.S. Vecherskaya. and E.M. Rivkina. (1994) CO_2 and CH_4 emissions on cryosols and subsoil permafrost and possible global climate change.- In: Soil processes and greenhouse effect, Lal,, R., J. Kimble and.E. Levine (eds.),. USDA, Soil Conservation Service, National Soil Survey Center, Lincoln.

Samarkin, V.A., A. Gundelwein and E.-M. Pfeiffer (1997) Methane Emissions.- Berichte zur Polarforschung237, 35-44.

Sebacher, D.I., R.C. Harriss, K.B. Bartlett, S.M. Sebacher. and S.S. Grice. (1986) Atmospheric methane sources: Alaskan tundra bogs, an alpine fen and a subarctic boreal marsh.- Tellus, 38, Ser.B, 1-10.

Soil Survey Stuff (1994) Keys to Soil Taxonomy, 6th Ed., Blacksburg/Virginia.

Sugimoto, A. and E. Wada. (1993) Carbon isotopic composition of bacterial methane in a soil incubation experiment: Contributions of acetate and CO_2/H_2.- Geochim Cosmochim. Acta 27, 4015-4027.

Sugimoto, A. and E. Wada (1995) Hydrogen isotopic composition of bacterial methane: CO_2/H_2 reduction and acetate fermentation.- Geochim. Cosmochim Acta 59, 1329-1337.

Svensson, B.H. and T. Rosswall. (1984) In situ methane production from acid peat in plant communities with different moisture regimes in a subarctic mire.- Oikos 43, 341-350.

Vecherskaya, M.S., V.F. Galchenlo., E.N. Sokolova. and V.A. Samarkin. (1993) Activity and Species Composition of Aerobic Methanotrophic Communities in Tundra Soils.- Current Microbiology 27, 181-184.

Whalen, S.C. and W.S. Reeburgh (1988) A methane flux time series for tundra environments.-Global Biogeochemical Cycles, 2, 399-409.

Whalen, S.C. and W.S. Reeburgh (1990) Consumption of atmospheric methane by tundra soils.- Nature 346, 160-162.

Whalen, S.C. and W.S. Reeburgh. (1992) Interannual variations in tundra methane emission: a 4-year time series at fixed sites.- Global Biogeochemical Cycles 6, 139-159.

Whalen, S.C., W.S. Reeburgh. and K.S. Kizer (1991) Methane consumption and emission by Taiga.- Global Biogeochemical Cycles5, 261-273.

Carbon Dioxide and Methane Emmissions at Arctic Tundra Sites in North Siberia

M. Sommerkorn[1], A. Gundelwein[2], E.-M. Pfeiffer[2] and M. Bölter[1]

(1) Institut für Polarökologie, Universität Kiel, Wischhofstrasse 1-3, D 24148 Kiel, Germany
(2) Institut für Bodenkunde, Universität Hamburg, Allendeplatz 2, D 20146 Hamburg, Germany

Received 11 March 1997 and accepted in revised form 13 January 1998

Abstract - Carbon emmissions (CO_2 and CH_4) at an arctic polygonal tundra on Taymyr Peninsula, North Siberia (75°N, 98°E), were measured during summer 1996. The average emissions of carbon dioxide were about 50 times higher (150 mg CO_2*m^{-2}*h^{-1}) than those of methane (3 mg CH_4 *m^{-2}*h^{-1}). Emission rates of carbon dioxide and methane show dependency on water table and soil temperature. Whereas carbon dioxide emmissions appear to be primarily dependent on soil temperature, the water table position plays the major role with respect to methane emissions. Compared to the wet central polygon depression, the methane emissions from the dryer polygon margin practically ceased, while carbon dioxide emissions were slightly higher at the latter site. Thus, gaseous carbon loss from permafrost affected soils to the atmosphere was determined by the position of sites in the microrelief.

Introduction

Carbon dioxide emmissions from soils originate mainly from aerobic microbial metabolism as well as root respiration of vascular plants, whereas methane is the product of anaerobic microbial metabolism. Tundra wetlands can be an important source for both the greenhouse gases carbon dioxide and methane. Northern wetlands - north of 60-70° northern latitude - are especially important as sources for these gases in context with the predicted climate warming (Matthews and Fung, 1987; Aselman and Crutzen, 1989). Between 250-455 petagrams organic carbon (1Pg = 10^{15} g) are present in the permafrost and seasonally thawed soil layers (Miller et al., 1983; Post et al., 1985, Gorham, 1991). About 165 Mio ha peatland area are situated on the territory of the former Soviet Union (FSU), about 30 Mio ha (18 %) in Northern Siberia (Botch et al., 1995). This underlines the importance of carbon flux measurements at wet tundra sites in the Siberian arctic.

Methods

Trace gas measurements (C-emission)

Measurements were carried through in the centre of the polygon area of Krasnaya valley (compare Samarkin et al., 1997; Sommerkorn, 1997). Measuring sites for all investigations (trace gases, microclimate, soil) were closely neighboured.

Carbon dioxide was measured by means of a multichannel Infrared Gas Analyzer (IRGA), operating in open system (Walz Company, Germany). This technique allows continuous measurements of CO_2-emmissions at several sites. Flux data presented here are 15 min. average values of 10 sec. interval readings.

Methane was measured by the closed chamber technique and direct CH_4 determination by a multigasmonitor (Brüel and Kjaer Company, Danmark), working with photoacoustic infrared spectroscopy. The measurements were calibrated against a gaschromatograph (Carlo Erba GC

In: Kassens, H., H.A. Bauch, I. Dmitrenko, H. Eicken, H.-W. Hubberten, M. Melles, J. Thiede and L. Timokhov (eds.) Land-Ocean Systems in the Siberian Arctic: Dynamics and History. Springer-Verlag, Berlin, 1999, 343-352.

6000 with Flame Ionisation Detector [FID] and Temperature Conductivity Detector [WLD], Fisons Company). For detailed description see Samarkin et al. 1997. Soil temperatures at various depths were continuously monitored by means of thermistors connected to data loggers (Grant, Great Britain). Water table position was investigated by boreholes.

Investigation area

Lake Levinson Lessing is located in the southern part of the Byrranga Mountains in northern Taymyr Peninsula, Central Siberia (75°32'N, 98°35'E, see Figure.1). The region is dominated by a cold-continental climate (mean annual air temperature: -15°C, mean July air temperature: 6.6°C, mean annual precipitation: < 281 mm, climate reports from "Taymyr Lake Station" 1963-1992, amount of days > 0°C: about 50, own observations during summer 1996).

Soils

The Levinson Lessing-region is dominated by weakly developed and wet soils, strongly influenced by cryoturbation and their position in the micro- and macrorelief.

The valley of the Krasnaya-River is covered by ice-wedge polygons with an accumulation of weakly decomposed plant material and wet soils of loamy-sandy sediments (Soil Taxonomy: Histic Pergelic Cryaquepts, Pergelic Cryofibrists; compare Gundelwein et al., 1997). The microrelief of the ice-wedge polygons is characterized as depression (wet polygon centre depression) and as apex (dry polygon rise, see also Figure. 2). The polygon diameters range between 5-25 m, the apexes with underlying ice wedges are 10 to 60 cm high. About 34 % of the polygon area are apexes, 66 % are depressions. The main soil parameters of the investigated polygon site are presented in Tables 1 and 2. A more detailed description can be found in Gundelwein et al. (1997) and Pfeiffer et al. (1996).

Vegetation

The main vegetation is characterized by a typical wet tundra in the typical tundra belt of Taymyr Peninsula (Aleksandrova, 1980). The vegetation of the polygon depressions is dominated by vascular plants like *Carex stans* and *Dupontia fisheri* and by mosses like *Plagomnium elatum* and *Drepanocladus uncinatus*. Mosses have a coverage of 100 %, whereas vascular plants cover 80 % of soil surface. On moderately high polygon apexes vegetation shows an increasing rate of dwarf shrubs, especially *Dryas punctata*, accounting for a total of 35 % coverage. Total vascular plant cover here is 55 %, whereas mosses account for 95 %. *Tormenthypnum nitens* dominates, representing as much as 75 % coverage.

Results

Microclimate

During the field-season from July 20th to August 15th a warm period was followed by cool, cloudy and humid weather. Average ambient temperature during the field-stay was 8°C, maximum and minimum were 20°C and 1°C, respectively. Microclimate of the polygonal tundra showed differences between polygon trough and polygon apex. Surface temperature at the apex was up to 10°C higher than at the depression. Temperature in the deeper soil horizons tended to be somewhat more homogenous at the polygon depression than at the apex. The water table position was higher at the polygon depression than at the apex area. The water table at the depression was never found deeper than 7 cm below soil surface, whereas it was observed as deep as 14 cm at the apex. During a heavy rain fall event on August 6th, the water table at the depressions of the polygons increased up to 6 cm above the soil surface.

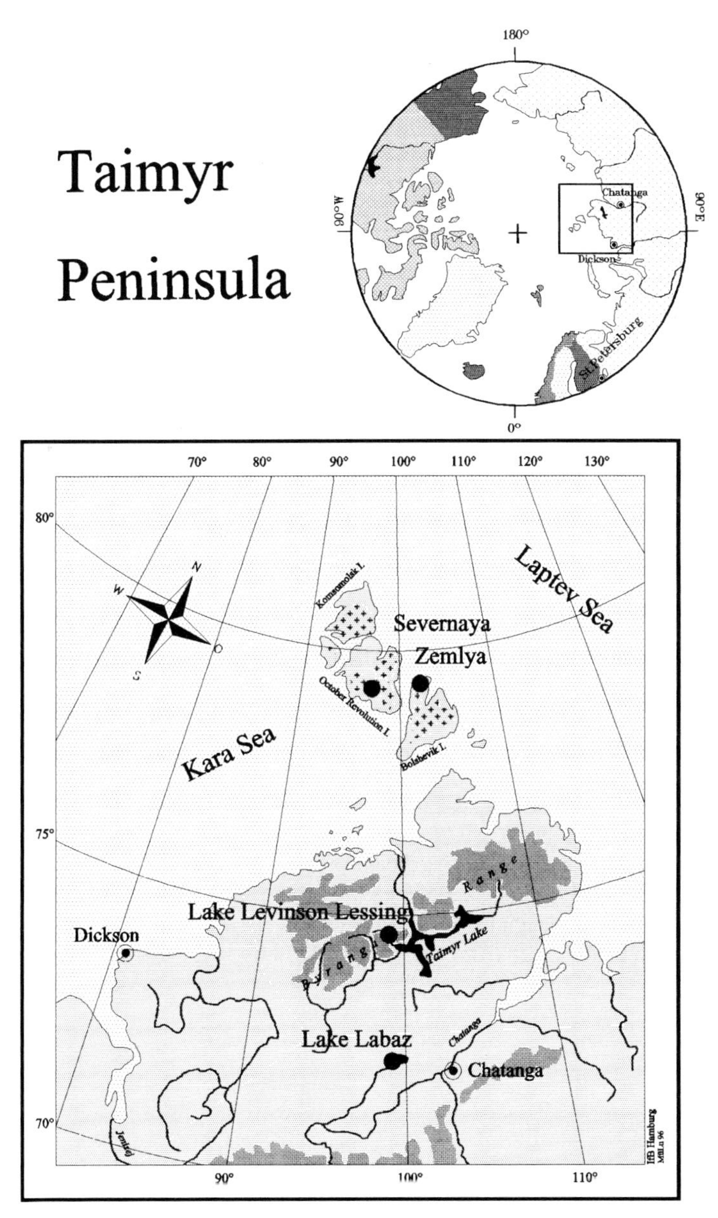

Figure 1: Research area.

Carbon dioxide emissions

At the polygon depressions carbon dioxide emmissions from soils ranged from 20 to 260 mg $CO_2*m^{-2}*h^{-1}$ over the field season, with an average of 120 mg $CO_2*m^{-2}*h^{-1}$ (Figure. 3a). Carbon dioxide emissions from the soils of the polygon apex varied between

30 mg CO_2*m^{-2}*h^{-1} and 300 mg CO_2*m^{-2}*h^{-1} (Figure. 3b). The average CO_2-losses were about 170 mg CO_2*m^{-2}*h^{-1} and thus some 40 % higher than at the polygon depression.

CO_2-emmissions from soil at both sites revealed a strong diurnal course with an amplitude of as much as 160 mg CO_2*m^{-2}*h^{-1} over less than 24 h. The emmissions showed best correlation with soil temperature at 2 cm depth (Table 3). Also, the total magnitude of emmissions were modified by watertable position, especially at the polygon depression. CO_2-emissions increased with falling watertable, especially pronounced if the change of watertable occured close to the soil surface.

In the polygon depressions carbon dioxide emmissions from soils ranged from 20 to 260 mg CO_2*m^{-2}*h^{-1} over the field season, showing an average of 120 mg CO_2*m^{-2}*h^{-1} (Figure.3a). Carbon dioxide emissions of the polygon apex showed a minimum of 30 mg CO_2*m^{-2}*h^{-1} and a maximum of 300 mg CO_2*m^{-2}*h^{-1} (Figure.3b). The average CO_2-losses were about 170 mg CO_2*m^{-2}*h^{-1} and thus some 40 % higher than in the polygon depression.

CO_2-emmissions from soil at both sites revealed a strong diurnal course with a maximum of about 160 mg CO_2*m^{-2}*h^{-1} . The best temperature correlation of the emmissions could be obtained with soil temperatures at 2 cm depth (Table 3). Also, the total magnitude of emmissions were modified by watertable position. CO_2-emissions were increasing with falling watertable.

Methane emissions

The methane emission rates from the polygon depression ranged between 1.8 and 5.2 mg CH_4 *m^{-2}*h^{-1} (mean: 3.1 mg CH_4 *m^{-2}*h^{-1}, compare Figure. 3a). The emmissions varied in dependence of the water table position. Falling of the water table below the soil surface caused a decrease of methane emissions. After heavy rainfalls in the first days of August the water table rose up to 6 cm above the ground. With a delay of 4 days the methane emission rates also rose from about 2.1 up to 3.1 mg CH_4 *m^{-2}*h^{-1} in the polygon depression.

The methane emissions from the dry apex part of the investigated polygon ranged between 0 and 0.2 mg CH_4 *m^{-2}*h^{-1} (mean: 0.1 mg CH_4 *m^{-2}*h^{-1}, Figure.3b), sometimes the emmissions were negative, thus the apex became a sink for methane.

Magnitude of methane emissions decreased during the summer period. The methane emissions showed no diurnal variations. The mean methane emmissions of the whole polygon area is 2.2 mg CH_4 *m^{-2}*h^{-1} between middle of July until beginning of September (see also Samarkin et al., 1997, Samarkin et al., 1997, this volume).

Discussion

The results of this study indicate that CO_2-emmissions from soils of polygon depression and polygon apex differ as a result of microclimatical differences as well as water regime. The higher temperature in 2 cm depth observed at the polygon apex compared to the polygon depression, as well as the permanently about 8 cm lower watertable at the polygon apex lead to CO_2-emissions of an average of 60 % more at the apex than at the depression site. In total, CO_2-emmissions presented here were higher than those measured by Bunnel et al. (1975) in the wet sedge tundra of Barrow (60-140 mg CO_2*m^{-2}*h^{-1}), but within the same range than values from a wet sedge tundra in the southern tundra belt of Taymyr Peninsula measured during summer of 1995 (Sommerkorn, unpublished data). Previous studies have indicated that factors controlling the ecosystem and soil emmissions are complex and site specific, but frequently, as in the present paper, the temperature was identified as the primary environmental factor (e.g.

Peterson and Billings, 1975, Svensson, 1980, Moore, 1986). Several studies indicate that surface temperatures are better correlated with soil CO_2-flux than are temperatures of deeper horizons (Luken and Billings, 1985, Moore, 1986). The data in the present paper shows best correlation with the temperature at 2 cm depth. This suggests that the major source of CO_2 from the soil may be found around this layer, which is only 1 cm or less below the photosynthetic active parts of the mosses and thus still in the layer of living mosstissue.

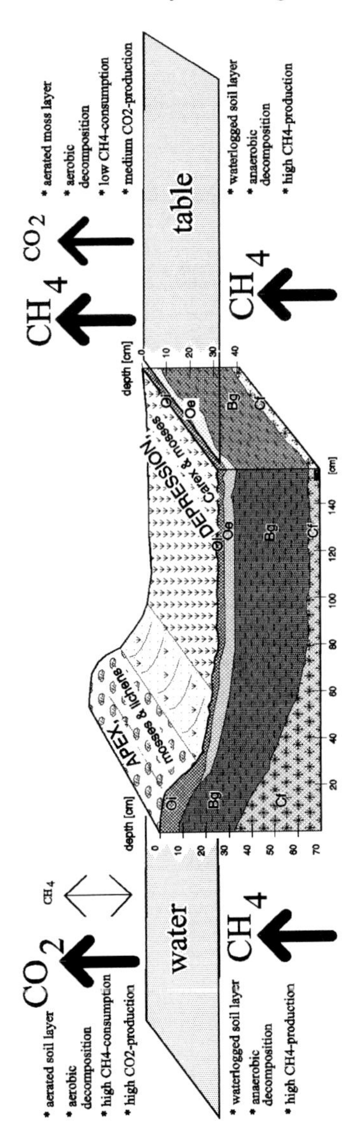

Figure 2: Soils of a typical ice-wedge polygon, Lake Levinson Lessing (with *Oi* = intermediate decomposed plant material, *Oe* = slightly decomposed plant material, *Bg* = gleyic sandy mineral soil material).

In the data presented here a negative trend in CO_2-emmissions can be seen with a rising watertable. Since a rising water table reduces oxygen diffusion rates it seems logical that this process has a negative feedback on soil respiration (Billings et al., 1984; Oberbauer et al., 1991; Silvola et al., 1996). This effect can, however, be modified through density of the soil and lateral flow of oxygen-rich water. Higher watertables could also cause lower CO_2-emmissions due to lowered CO_2-diffusion rates. This fact, however would result in a time-lag in CO_2-emmissions after times of high water levels, which can not be seen from the data presented here. The combined effect of lower temperatures in higher soil horizons due to the higher water table causes some problems to filter out the significance of the factor water from field-data. Future laboratory experiments on microcosms and modeling of the results will help to filter out these different factors. Other factors controlling CO_2-flux may be explained by intrinsic site differences, e.g. litter quality, soil density, vegetation composition and microbial population (Nadelhoffer et al., 1991).

Table 1: Soil characterization of a typical polygon depression (Pergelic Cryaquept)

Horizon	depth [cm]	pH ($CaCl_2$)	C [w/w-%]	N [w/w-%]
Oi (slightly decomposed plant material)	0-2	5.0	18.0	0.6
Oe (intermediate decomposed plant material)	2-6	4.0	10.3	0.6
Bg (gleyic sandy mineral soil material)	6-35	4.3	6.0	0.4

Table 2: Soil characterization of a typical polygon apex (Pergelic Cryaquept)

Horizon	depth [cm]	pH (CaCl2)	C [w/w-%]	N [w/w-%]
Oi (slightly decomposed plant material)	0-6	5.0	13.0	0.5
Bg (gleyic sandy mineral horizon)	6-45	4.6	4.0	0.3

Table 3: Regression coefficients (r^2) of CO_2-emmissions from polygonal tundra microsites with soil temperatures (°C) at various depths using an Arrhenius equation.

	surface	*2 cm*	*5cm*
polygon depression	0.52	0.63	0.59
polygon apex	0.41	0.68	0.64

Although numerous studies have been conducted concerning CO_2-emmissions from tundra (e.g. Poole and Miller, 1982; Luken and Billings, 1985; Oberbauer et al., 1989), it is still not fully understood to which extent both spatial heterogenity and temporal patterns influence the total-system-emmissions of CO_2 in tundra (Waddington and Roulet, 1996). Our data reveal the need for measurements either with the open flow technique or for other techniques with a high time resolution in order to obtain results which can be correlated to the determining factors and thus lead to correct modelling. Another opportunity offered by the open flow technique is the direct correlation of soil respiration data to photosynthetic CO_2-emmissions of mosses and vascular plants. In a next step, these data will be linked in order to receive whole system models for various types of tundra and microsites.

Methane emission rates from wet arctic tundra range - reported in the actual literature - between 0.12-5.9 mg CH_4 $*m^{-2}*h^{-1}$ (summarized in Vourlitis and Oechel, 1997). The presented methane emission rates are similar. The mean carbon loss by methane from the polygon area is about 1.6 mg C $*m^{-2}*h^{-1}$ during the summer period. This represents for only 4% of the total gaseous carbon loss from the polygon area to the atmosphere, which adds to about 38.9 mg $C*m^{-2}*h^{-1}$ (CO_2 and CH_4). Nevertheless, this relative small quota of methane is important with regard to the predicted global warming, for the efficiancy of methane for the retention of longwave radiation is about 27 times higher than that of carbon dioxide (Lelieveld et al., 1993).

Whilst the dependence of methane emissions on the watertable position was shown in numerous studies, the role of soil temperature is less clear (Svensson and Rosswall, 1984; Moore and Knowles, 1989; Whalen et al., 1991; Whalen and Reeburgh, 1992; Harriss et al., 1993). In this study the methane emissions revealed a primary dependency on the water table position, in contrast to the CO_2-emissions (Figure. 3a, 3b). The decrease of methane emissions at the dry apex sites could be explained by the oxidation of methane in the dryer and well aerated upper part of the active layer. The production rates and methane contents in pore water are about the same in the deeper apex horizons as in the deeper horizons of the polygon depressions (compare Samarkin et al., 1997, same volume).

Beside watertable and soil temperature, other parameters like thickness of active layer, structure of vegetation cover, as well as microbial characteristics can influence the methane emissions from arctic soils (Morrissey and Livingstone, 1992; Chanton et al., 1992; Topp and Pattey, 1997; Vourlitis and Oechel, 1997)

Variations in microrelief lead to strong variations in methane emission rates. Therefore, measurements of methane emission rates at only one single site are insufficient, complementary mapping of soils and microrelief, a large number of measurements at different sites as done during our campaign at Levinson Lessing Lake and a mobile, light weight equipment are necessary for reliable flux data. In the next step the methane flux measurements will be combined with stable carbon isotope investigations and studies of soil organic matter quality.

Acknowledgements

The authors wish to thank V. Samarkin for cooperation in the field and valuable discussion. This study was financed by the German Ministry of Education, Science, Research and Technology (BMBF grant 03PLO14B).

References

Aleksandrova, V.D. (1980) The Arctic and Antarctic: Their devision into geobotanical areas. Cambridge University Press, Cambridge, 247 pp.

Aselman, I. and P.J. Crutzen (1989) Global distribution of natural freshwater wetlands and rice paddies, their net primary productivity, seasonality and possible methane emissions. J.Atmos. Chem. 8, 307-358.

Billings, W.D., K.M. Peterson, J.O. Luken, and D.A. Mortensen (1984) Interaction of increasing atmospheric carbon dioxide and soil nitrogen on the carbon balance of tundra microcosms. Oecologia 65, 26-29.

Botch, M.S., K.I. Kobak, T.S. Vinson, T.P. Kolchugina (1995) Carbon pools and accumulation in peatlands of the former Soviet Union. Global Biogeochemical Cycles 9, 37-46.

Bunnel, F.L., S.F. MacLean Jr. and J. Brown (1975) Barrow, Alaska, U.S.A.. In: Rosswall, T. and O.W. Heal (eds.), Structure and function of tundra ecosystems, Ecological Bulletins 20. Stockholm: Swedish Natural Science Research Council, 425-448.

Chanton, J.P., C.S. Martens, C.A. Kelley, P.M. Crill and W.J. Showers (1992) Methane transport mechanisms and isotopic fractionation in emergent macrophytes of an Alaskan tundra lake.J. Geophys.Res. 97, 16681-16688.

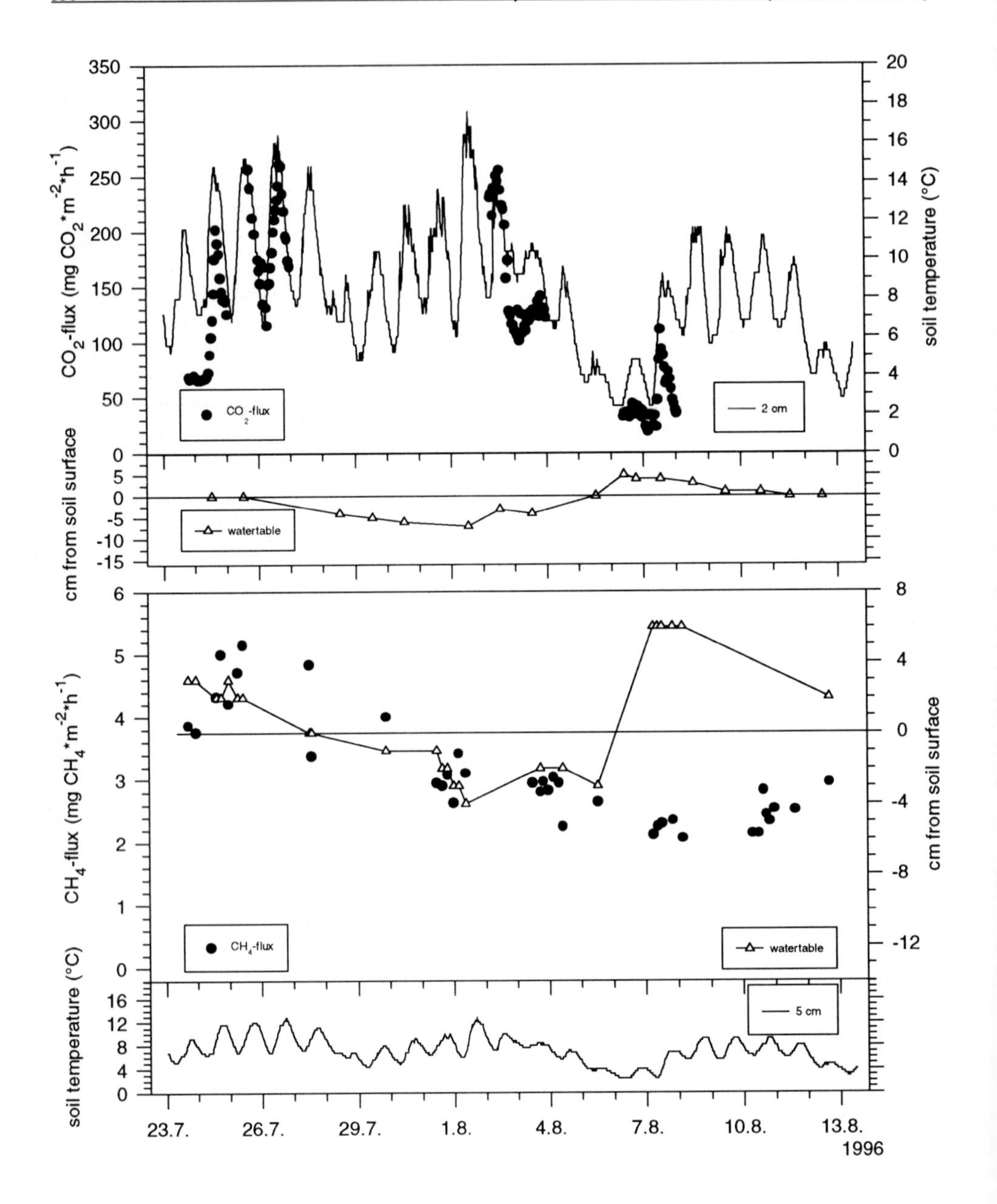

Figure 3a: Polygon depression - water table, soil temperature at 2 cm depth and trace gas emissions (CO_2 and CH_4)

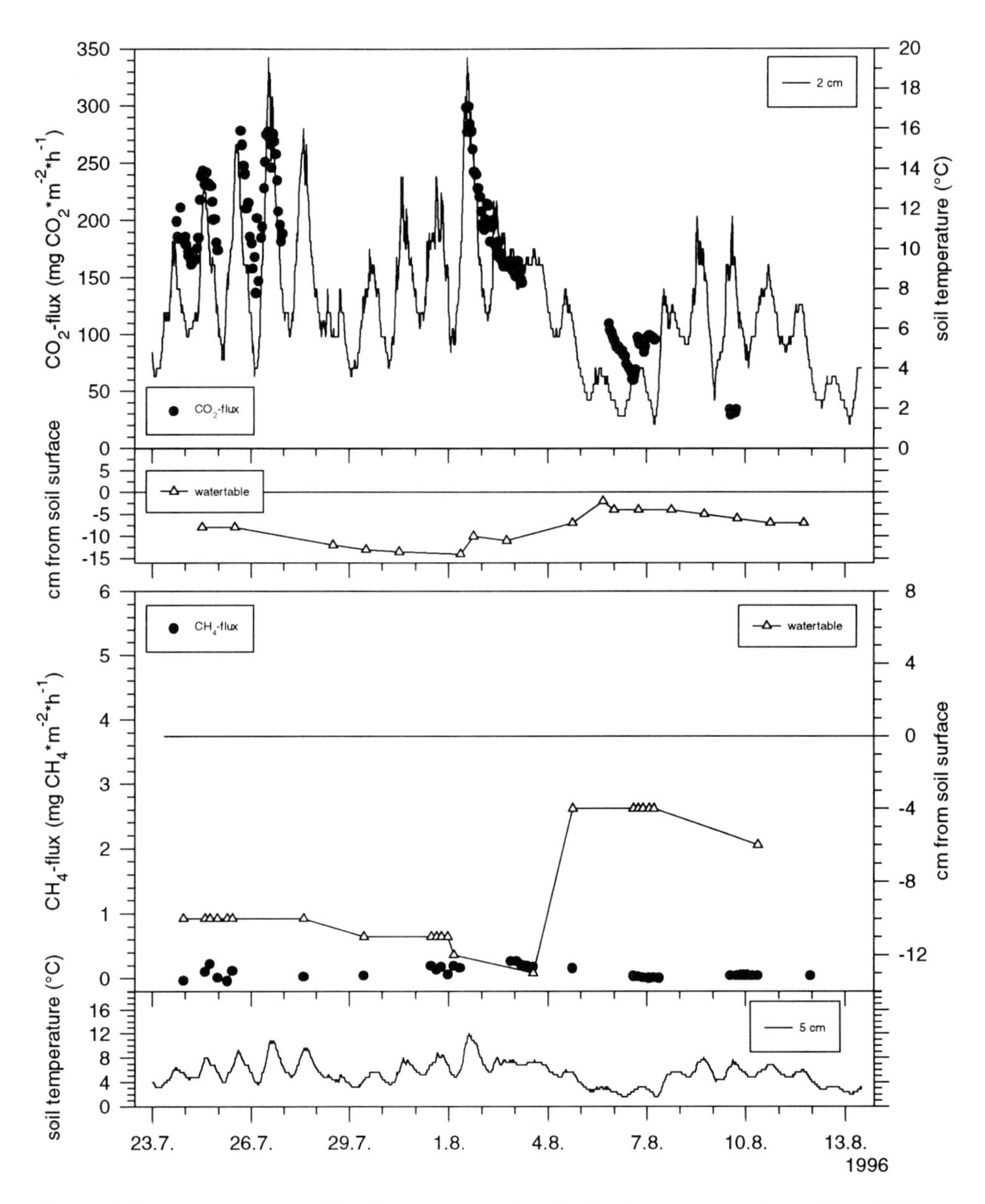

Figure 3b: Polygon apex - water table, soil temperature at 2 cm depth and trace gas emissions (CO_2 and CH_4)

Gorham, E. (1991) Northern peatlands: Role in the carbon cycle and probable responses to climate warming. Ecological Applications 1, 182-195.

Gundelwein, A., H. Becker, T. Müller-Lupp, and N. Schmidt (1997) Lake Levinson-Lessing - the soils. In: The Expedition on Taymyr and Severnaya Zemlya 1996, Melles, M. (ed.), Reports on Polar Research 237, 11-14.

Harriss, R., K. Bartlett, S. Frolking and P. Crill (1993) Methane Emissions from Northern High-Latitude Wetlands. In: Biogeochemistry of global change: radiatively active trace gases, Oremland, R.S. (ed.), Chapman and Hall, New York, 449-485.

Lelieveld, J., P.J. Crutzen, and C. Brühl (1993) Climate effects of atmospheric methane. Chemosphere 26, 739-768.

Luken, J.O. and W.D. Billings (1985) The influence of microtopographic heterogenity on carbon dioxide efflux from a subarctic bog. Holarctic Ecology 8, 306-312.

Matthews, E. and I. Fung (1987) Methane emission from natural wetlands: global distribution, area, and environmental characteristics of sources. Global Biogeochem. Cycles 1, 61-86.

Miller, P.C., R. Kendall and W.C. Oechel (1983) Simulating carbon accumulation in northern ecosystems. Simulation 40, 119-131.

Moore, T.R., (1986) Carbon dioxide evolution from subarctic peatlands in Eastern Canada. Arctic and Alpine Research 19, 189-193.

Moore, T.R. and R. Knowles (1989) The influence of water table levels on methane and carbon dioxide emissions from peatland soils. Can. J. Soil Science 69, 33-38.

Morrisey, L.A. and G.P. Livingstone (1992) Methane emission from Arctic tundra: An assessment of local spatial variability. J. Geophys. Res. 97, 16661-16670.

Nadelhoffer, K.J., A.E. Giblin, G.R. Shaver and J.A. Laundre (1991) Effects of temperature and substrate quality on element mineralization in six arctic soils. Ecology 72, 242-253.

Oberbauer, S.F., S.J. Hastings, J.L. Beyers and W.C. Oechel (1989) Comparative effects of downslope water and nutrient movement on plant nutrition, photosynthesis and growth in Alaskan tundra. Holarctic Ecology 12, 324-334.

Oberbauer, S. F., J. D. Tenhunen and J. F. Reynolds (1991) Environmental effects on CO_2 efflux from water track and tussock tundra in arctic Alaska, U.S.A.. Arctic and Alpine Research 23, 162-169.

Peterson, K.M. and W.D. Billings (1975) Carbon dioxide flux from tundra soils and vegetation as related to temperature at Barrow, Alaska. American Midland Naturalist 94, 88-98.

Pfeiffer, E.-M., A. Gundelwein, T. Nöthen, H. Becker, and G. Guggenberger (1996) Characterization of the Organic Matter in Permafrost Soils and Sediments of theTaymyr Peninsula/Siberia and Severnaya Zemlya/Arctic Region. In: The Expedition on Taymyr 1995, Bolshiyanov, D.Yu. and H.-W. Hubberten (eds.), Reports on Polar Research 211, 46-83.

Poole, D. K. and P.C. Miller (1982) Carbon dioxide flux from three arctic tundra types in North-Central Alaska, U.S.A.. Arctic and Alpine Research 14, 27-32.

Post, W.M., J. Pastor, P.J. Zinke and A.G. Strangenberger (1985) Global patterns of soil nitrogen. Nature 317, 613-616.

Samarkin, V.A., A. Gundelwein and E.-M. Pfeiffer (1997) Methane emissions. In: The Expedition on Taymyr and Severnaya Zemlya 1996, Melles, M.(ed.), Reports on Polar Research 237, 35-44.

Silvola, J., J. Alm, U. Ahlholm, H. Nykänen and P.J. Martikainen (1996) CO_2 fluxes from peat in boeal mires under varying temperature and moisture conditions. Journal of Ecology 84, 219-228.

Sommerkorn, M. (1997) Microbial Activity. In: The Expedition on Taymyr and Severnaya Zemlya 1996, M. Melles (ed.), Reports on Polar Research 273, 30-35.

Svensson, B.H. (1980) Carbon dioxide and methane emmissions from the ombotrophic parts of a subarctic mire. In: Ecology of a Subarctic Mire, Sonesson, M. (ed.), Ecological Bulletin 30. Stockholm: Swedish Natural Science Research Council, 235-250.

Svensson, B.H. and T. Rosswall (1984) In situ methane production from acid peat in plant communities with different moisture regimes in a subarctic mire. Oikos 43, 341-350.

Topp, E. and E. Pattey (1997) Soils as sources and sinks for atmospheric methane. Can. J. Soil Science 77, 167-178.

Vourlitis, G.L. and W.C. Oechel (1997) The role of northern ecosystems in the global methane budget. In: Global change and Arctic Terrestrial Ecosystems, Oechel, W.C., T. Callaghan, T. Gilmanov, J.I. Holten, B. Maxwell, U. Molau and B. Sveinbjörnsson (eds.), Ecological Studies 124, Springer, Berlin, 266-289.

Waddington, J.M. and N.T. Roulet, (1996) Atmosphere-wetland carbon exchanges: Scale dependency of CO2 and CH4 exchange on the developmental topography of a peatland. Global Biogeochemical Cycles 10, 233-245.

Whalen, S.C., W.S. Reeburgh and K.S. Kizer (1991) Methane consumption and emissions by taiga. Global Biogeochemical Cycles 5, 261-273.

Whalen, S.C. and W.S. Reeburgh (1992) Interannual variations in tundra methane emission: a 4-year time series at fixed sites. Global Biogeochemical Cycles 6, 139-159.

The Features of the Hydrological Regime of the Lake-River Systems of the Byrranga Mountains (by the Example of the Levinson-Lessing Lake)

V.P. Zimichev[1], D.Yu. Bolshyanov[1], V.G. Mesheryakov[1] and D. Gintz[2]

(1) State Research Center - Arctic and Antarctic Research Institute, 38 Bering St., 199226 St. Petersburg, Russia

(2) Martin-Luther-Universität Halle-Wittenberg, Domstrasse 5, D 06108 Halle, Germany

Received 3 March 1997 and accepted in revised form 25 March 1998

Abstract - Based on the hydrological and meteorological studies of the joint Russian-German Expedition "Taimyr-Severnaya Zemlya" in the Levinson-Lessing Lake basin, a first attempt is made to estimate the water balance, sediment load balance and the regime of the lake-river systems of the Glavnaya ridge of the Byrranga mountains.

Introduction

The water systems of the mountainous Taimyr peninsula have previously not been investigated with respect to the dynamics of water balance components and the mechanism of solid transport discharged into the lake's basin. In order to collect quantitative evidence on typical natural water bodies, special comprehensive observations of the regime of water balance characteristics in the lake-river system of the Levinson-Lessing-Lake were carried out in the framework of the joint Russian-German comprehensive Expedition "Taimyr-Severnaya Zemlya" (Siegert and Bolshiyanov, 1995).

Study site

The lake-river systems of the Byrranga mountains show a number of common morphogenetic features indicating the hydrological regime, sedimentation conditions and water exchange. The network of streams and rivers in the basin reflects the mountainous topography. The investigated lake-river system is typical for the Central Taimyr mountains. Therefore it can be used to estimate the main hydrological characteristics of a lake-river system in this region.

The drainage basin of the Levinson-Lessing-Lake is located about 50 km west of the Taimyr Lake in the southwestern part of the Byrranga Mountains (Figure 1; 72° 28′ N and 98° 37′ E in the centre of the lake). The basin is a Quaternary tectonic depression with numerous deeply entrenched valleys between more ancient relicts of pediments and slopes of low fault block mountains. The lake trough is a deep fold which is still active.

The drainage basin comprises 496,4 km^2, including the lake itself with 24.55 km^2. There are approximately 50 tributaries of different stream orders flowing into the Levinson-Lessing-Lake. The water leaves the lake via the Protochny river in the south (Figure 1).

The Krasnaya river, coming from the north, is the largest tributary of the Levinson-Lessing-Lake. It drains more than 70% of the total catchment, and supplies the main sediment input. One factor controlling the runoff dynamics of the Krasnaya is the Krasnoye lake, the source of the Krasnaya.

In the upper reaches, the river presents a typical mountainous pattern in a trough-shaped narrow valley. Downstream, the Krasnaya crosses a flat tectonic depression, and between the

---

In: Kassens, H., H.A. Bauch, I. Dmitrenko, H. Eicken, H.-W. Hubberten, M. Melles, J. Thiede and L. Timokhov (eds.) Land-Ocean Systems in the Siberian Arctic: Dynamics and History. Springer-Verlag, Berlin, 1999, 353-360.

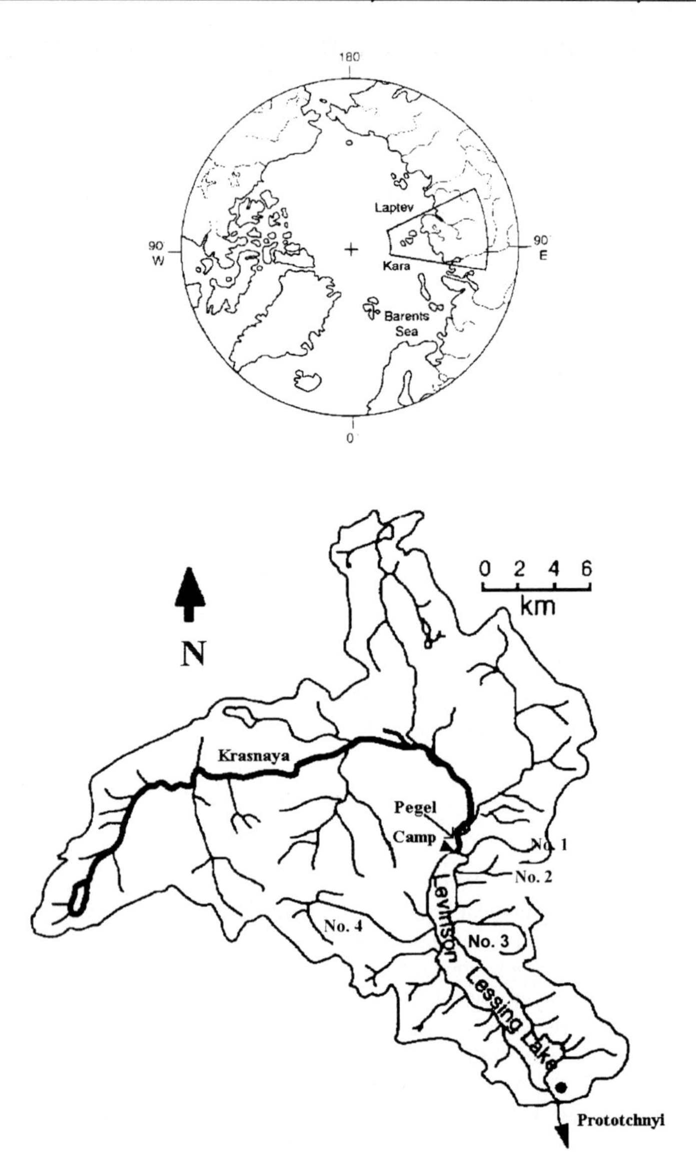

Figure 1: Location map and drainage basin.

low flood-plain and the first terrace exhibits a braided channel system with meandering reaches. Depending on the water level in the channel, one or several branches are active.

Instrumentation

The gauging station in the Krasnaya river was located about 500 meter above the river delta at the Levinson-Lessing Lake. The water discharge calculations were based on automatic water level recordings and correspondent flow velocity measurements with a propeller. This measuring station was equipped with temperature, electric conductivity and pH-probes. These values were continuously recorded with a datalogger. The samples for suspended sediment concentration in the Krasnaya river were collected in the same cross profile. Integrated water samples were collected with a fife-litter bottle, additionally some samples were collected from a boat with a multiple point-sampler (Bley and Schmidt 1994). In the vertical profile three one litre samples in 5, 20 and 35 cm above ground were taken at the same time.

Mobile probes for water temperature, electric conductivity and pH-measurements were also available.

Flow regime and runoff dynamics

The Krasnaya river has an arctic nival flow regime, dominated by the rapid melting of snow and ice in the short winter-spring transition period in late June (Figure 2). The snow accumulation in the watershed is non-uniform. The largest snow supplies are found in snowdrifts usually located in the upper reaches of the mountain ravines, where melting continues during the entire summer season.

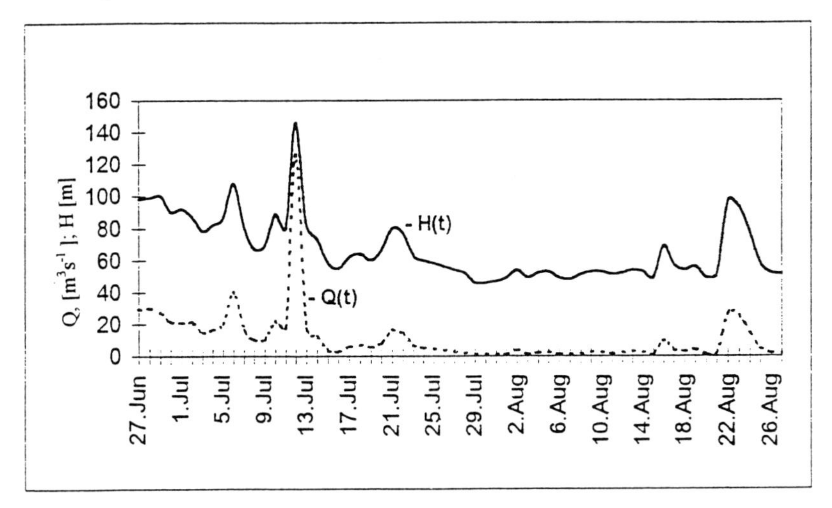

Figure 2: Water level [H; m] and discharge [Q, m^3s^{-1}] in the Krasnaya river 1994.

With the thawing period ice drift begins on the Prototchy stream, and flow velocity can reach values of more than 2 m/s. In contrast to the sharp rapid snow melt hydrograph in the Krasnaya river, the hydrograph of the Prototchy stream is wider , without a sharp peak, due to the regulation effect of the lake (Figure 3). The time lag between the two hydrograph peaks is about one day.

Diurnal water level fluctuations of about 0.10 m in height and corresponding discharge variability occurred in the Krasnaya during the summer season. The discharge variability is controlled by the increase in snowmelt water, soil warming and a deeper thawing of the active layer, during the daily air temperature gradient. In the Protochny stream similar gauge variations are completely absent, due to the lake's dampening storage effect.

At the beginning of the cold season the runoff from the watershed is so depleted that the river has only limited ice supplies in its channel when the water surface starts to freeze. In some zones the ice is completely absent, in others the ice freezes down to the river's pebble bed. In the winter-spring transition period different river ice break-up phenomena occur; such as downstream flow of ice slabs or ice melting at its place of origin. Ice flow under flood conditions can cause river bed and bank erosion, e.g. bottom material is picked up, ice rafted pebbles or floating material which slumped from the river banks onto the ice particles. The transported sediment is deposited in the river delta and in the northern part of the Levinson-Lessing-Lake.

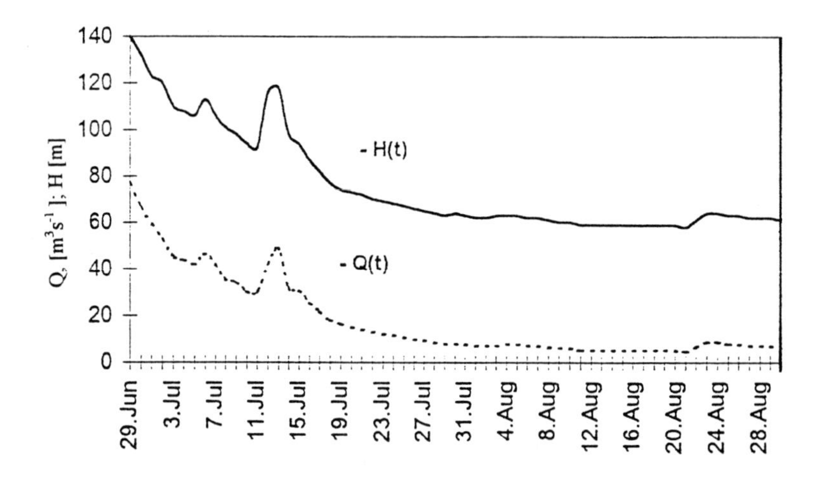

Figure 3: Water level [H; m] and discharge [Q, m^3s^{-1}] in the Prototchny stream 1994.

The braided channel system in the lower part of the Krasnaya valley is about 200 m in width. At the measuring station, about 500 m upstream of the river delta, three channels are developed (Figure 4). The right one with a flow width up to several tens of meters (55 m) is running permanently in the ice free time; the other two are only active during individual flood events. Only in the spring snowmelt flood and in a few summer storm events all three channels are connected in a single channel.

During individual floods the water level rises very rapidly (few hours) and the flow velocities increase significantly, reaching more than 2 m/s. The falling limb of the hydrograph needs about 12 to 48 hours to reach the initial value.

Sediment transport dynamics

In the three year observation period the suspended sediment yield was highest during the flood events induced by snowmelt. In such cases active erosion of the banks occurs and a large amount of detrital material is transported. These floods transported 75 % to 95 % of the total seasonal sediment load. The solid load at this stage can reach 100 kg/s. Generally there is a weak correlation between the suspended sediment concentration and discharge. The turbidity can differ by two to three orders of magnitude. In the observation period 1994 the suspended sediment transport amounted to 10500 tons. About 95 % of this amount was transported during a snowmelt flood on July 12th.

In the 1995 field season, the suspended sediment transport by the Krasnaya reached a quantity of about 5500 t at low discharge, with mean variations in the suspended load of 0.04 to 0.4 kg/s. In comparison, the rain generated flood events transported a total of 5200 t of

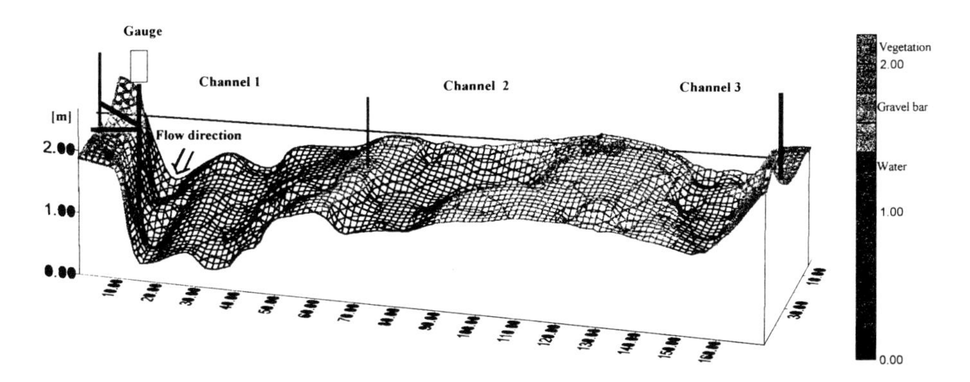

Figure 4: Measuring site at the Krasnaya river.

sediment in a 12 day period. The complete suspended sediment transport during the discharge season accumulated to about 10700 t.

The weak correlation between discharge values and the amount of sediment discharge is a well-known phenomenon (Schmidt et al., 1992) (Figure 5). Here, it can be explained by the influence of the non-alluvial rock slope pebble fans in the lower parts of the Krasnaya river. The material located there cannot be transported by the river at its normal regime. The presence of suspended sediment in the flow is governed mostly by local washout of the composed peat masses near the mouth shore. In these cases, no stable relation between the sediment and water discharges could be observed. The accuracy of reconstructing the gaps in the sediment discharge measurements is only within the given order of measurements. In this case the interpolation of the dependent values in the correlation R(t) is only an approximation.

During the entire measuring seasons the suspended sediment discharge in the Protochny stream was quite small with values between 0.005 to 0.04 kg/s. In the time period when the Levinson-Lessing-Lake is mainly ice covered to the beginning of the freeze up of the Protochny stream, the solid discharge values decreased by an order of magnitude. The solid Lake discharge reached about 170 t in 1993, about 200 t in 1994 and 180 t in 1995.

The sediment load and discharge of the other tributaries to the Levinson-Lessing-Lake vary over a wide range (Table 2). Sediment concentrations from 0.0004 kg/m^3 to 0.25 kg/m^3 and sediment discharge values from 0.00005 kg/s to 0.13 kg/s were observed.

The sediment load of the tributaries, including the Krasnaya river, are strongly controlled by the intensity of the stormflows. In the middle of August the water supplies out of the snow fields fade out to zero for most of the tributaries.

Lake water budget

According to measurements in 1995 (Table 1), about 32 % of the total summer outflow from the lake originates from liquid precipitation of which 0.038 km^3 precipitate over the Krasnaya watershed and 0.01 km^3 over the remaining part of the lake watershed. The precipitation peaks coincide with the Protochny hydrograph without a time lag, in contrast to the Krasnaya hydrograph, where a time lag of five to ten hours occurred (Figure 6).

According to the discharge data from the Krasnaya and the Protochny river it was possible to assign a ratio of inflow and outflow volume for the lake (Table 1). Based on the results of individual measurements (Figure 1, Table 2), some understanding of the direct inflow volume from the remaining part of the lake watershed can be achieved. The water volume in the

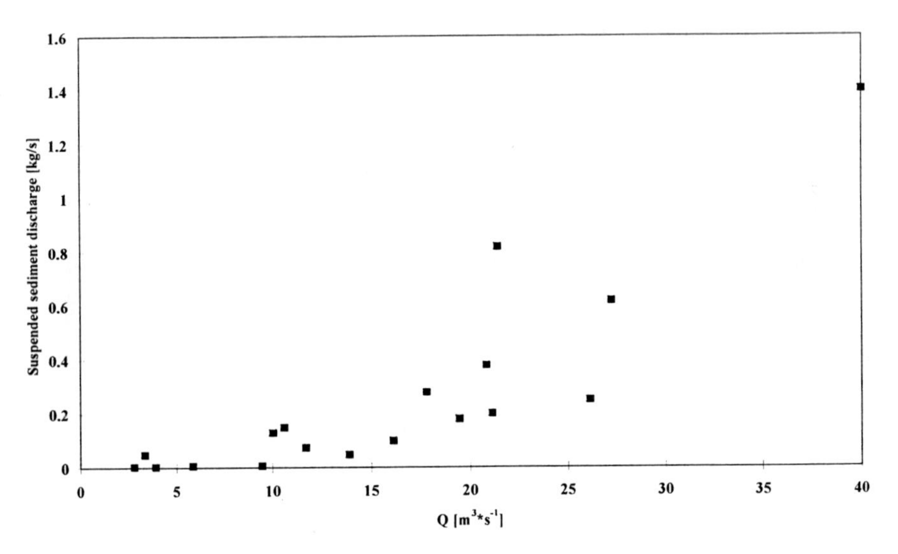

Figure 5: Relation between suspended sediment discharge (kg/s) and water discharge (Q).

Levinson-Lessing-Lake in the pre-spring period is about 1.42 km^3. The runoff of the Protochny stream during the summer season amounts to 0.15 to 0.18 km^3. Thus, a complete water exchange of the lake occurs over a 9 to 10 year period.

To calculate the Levinson-Lessing-Lake water balance for 1995, the following formula can be applied:

$$Y = Y_1 + Y_2 + Y_3 + X + Z - E$$

where
Y = Protochny stream runoff
Y_1 = Krasnaya river runoff
Y_2 = main lake tributaries runoff
Y_3 = remaining stream network runoff
X = precipitation over the lake watershed (without the Krasnaya)
Z = ground component of the outflow
E = evaporation

The calculation approximates the total runoff Y_3 is in the same order of magnitude as Y_2, about 0.012 km^3. The total evaporation from the water surface and the remaining part of the drainage basin, calculated by a certain method, is quite insignificant and less than 0.007 km^3.

Using data of the full-scale measurements of the other components of the lake-river-system, the following results can be obtained:

$$0.18 = 0.12 + 0.018 + 0.012 + 0.01 + Z - 0.007$$

$$100\ \% = 67\ \% + 10\ \% + 6.7\ \% + 5.6\ \% + Z - 3.9\ \%$$

The total incoming component of the Levinson-Lessing-Lake water balance in 1995 was about 0.16 km^3 which was 88.9 % of the Protochny stream runoff. Thus, the value of the ground component of the outflow below the downstream section of the Krasnaya river is about 0.027 km^3 , i.e. about 15 % of the total outflow from the lake basin. In 1994, a year with a low water supply, the values Y (100 %), Y_1 (47 %) and Y_2 (18 %) were measured over the incomplete hydrological cycle (from June 29 to August30).

Table1: Water balance characteristics measured in the streams and rivers of the Levingson-Lessing Lake basin in 1995.

Total water runoff (km^3)	**Total volume of liquid precipitation (km^3)**	**Total runoff of the other streams (km^3)**
0,12 (Krasnaya river)	0,048	0,018
0,18 (Prototchny stream)		

Table 2: Data of measurements of the intermediate inflow of the Levingson-Lessing Lake.

Stream	**Date of measurement**	**Water discharge ($m^3 * s^{-1}$)**	**Sediment load (kg/s)**
No. 1	01.08.93	0,12	0,00006
No. 2	29.07.93	0,05	0,000045
No. 3	29.07.93	0,11	0,00009
No. 4	02.08.93	0,48	0,00106
No. 3	22.08.94	2,5	0,0051
No. 4	22.08.94	0,57	0,133

Conclusion

Based on the observed data, the following conclusions can be made on the typical features of the mountainous lake-river regime of the Taimyr-Peninsular.

1. The hydrological regime can be identified as arctic nival. Only in the three arctic summer months the surface drainage system is active. In some small watersheds the river runoff occurred only for 1,5 month after beginning of the snow melt. A predominant snow melting spring flood event typically takes place around the end of June. The summer-autumn low water period was interrupted by some precipitation indicated flood events with a short but intense character. In the remaining time of the year, the discharge decreases to zero, the stream and rivers freeze completely. The surface drainage dynamics of the Byrranga mountains are determined by the seasonal factor.
2. A diurnal dynamic in river runoff with an amplitude of about 0.1 m, influenced by the daily air temperature oscillation, appeared. The lake regulated discharge of the Protochny stream is mainly influenced by the seasonal factor, the daily oscillation could be ignored. The lake regulated outflow of the Protochny stream is predominantly controlled by the seasonal factor.
3. The solid discharge regime has a random character and is regulated by the sediment supply of the river banks. The sediments entrained from the river bed were almost absent. The relation between water discharge and sediment concentration R(Q) was quite weak and can only give an idea of the magnitude of solid discharge values at certain water levels.
4. During a stormflow, the sediment load can increase by several magnitudes in comparison with low level discharge. The lake functions as a sediment trap for the coarse sediments.

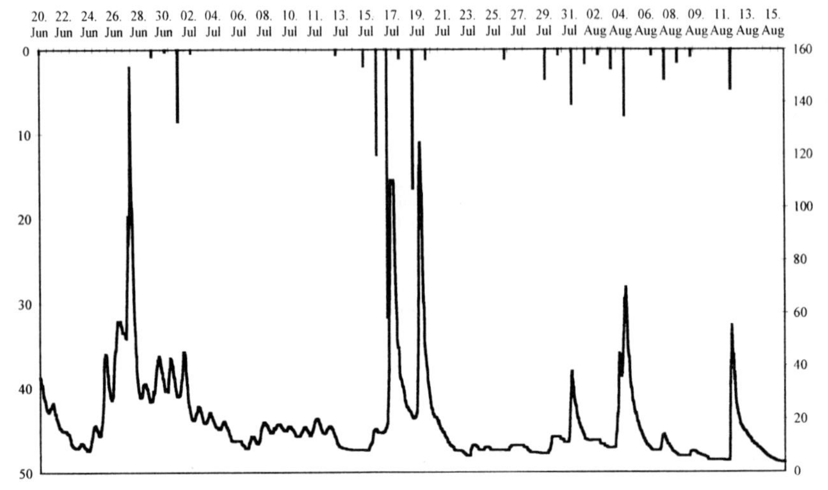

Figure 6: Krasnaya Hydrograph 1995 and precipitation.

They accumulate in several river deltas and in the lake's deep-water zones, and are not exported outside. The lake's sediment output in the summertime has no significant variations and the variability over several years is very small.

5 The thermal regime of the lake's tributaries and the outflow river differs significantly. The water temperature of the lake's tributaries is closely connected to the air temperature variations, both within a day and over the entire season. The water temperature of the Protochny is governed by the thermal regime of the lake, and does not depend on the daily air temperature variations.

6 The complete water exchange of the Levinson-Lessing-Lake occurs within 9 to 10 years.

References

Bley, D. and K.-H. Schmidt (1994) Schwebstofferfassung über die Trübungsmessung in einem Wildbach (Lainbach, Oberbayern). In: Messungen in fluvialen Systemen. Feld- und Labormethoden zur Erfassung des Wasser- und Stoffhaushaltes, Barsch, D., R. Mäusbacher, K.-H. Pörtge and K.-H. Schmidt (Hrsg).

Siegert, C. and D. Bolshiyanov (1995) Russian-German Cooperation: The Expedition TAYMYR 1994. Berichte zur Polarforschung 175.

Schmidt, K.-H., D. Bley, R. Busskamp and D. Gintz (1992) Feststofftransport und Flußbettdynamik in Wildbachsystemen. Das Beispiel des Lainbachs in Oberbayern. Die Erde, 123.

Lead-210 Dating and Heavy Metal Concentration in Recent Sediments of Lama Lake (Norilsk Area, Siberia)

B. Hagedorn[1], S. Harwart[1], M.M.R. van der Loeff[2] and M. Melles[1]

(1) Alfred-Wegener-Institut für Polar- und Meeresforschung, Forschungsstelle Potsdam, Telegrafenberg A43, D 14473 Potsdam, Germany

(2) Alfred-Wegener-Institut für Polar- und Meeresforschung, Postfach 120161, D 27570 Bremerhaven, Germany

Revised 5 May 1997 and accepted in revised form 15 March 1998

Abstract - Lama Lake is situated in the Noril'sk district of the southern Taymyr Peninsula, Siberia (69°N, 88°E) and covers an area of 466 km^2. The geology of the catchment is dominated by Permo-Triassic Continental Flood Basalts (CFB). Cu-Ni sulfides have been mined in the Noril'sk area since 1930. Radiochemical ^{210}Pb and geochemical investigations were carried out on a 0.52 m core from Lama Lake. A mean accumulation rate of 0.04 g cm^{-2} a^{-1} was calculated from the ^{210}Pb data. The chemical composition of the lake sediments identifies the CFB formations as the main source of the sediment. The Cd, Cu, Pb, Ni and Zn concentrations show a progressive increase from 4 cm depth up to the sediment-water interface. With the identification of the geogenic source of the sediments, enrichment factors (EF_M) of the heavy metals can be calculated as: EF_{Cd} (2.5) > EF_{Pb} (2.0) > EF_{Zn} (1.3) = EF_{Cu} (1.3) = EF_{Ni} (1.3). The Cd, Pb and Zn enrichment factors display the same pattern as those determined in other arctic regions (e.g. Greenland), whereas Cu and Ni are more enriched in Lama Lake. According to the ^{210}Pb ages, the onset of heavy metal enrichment took place in 1940, coinciding with the start of mining of Cu-Ni sulfide deposits in the Noril'sk area. This supports the assumption of an anthropogenic source for the enriched heavy metals in the sediment core.

Introduction

Lake sediments generally provide suitable archives to study paleoecological and climatic developments of the past, due to a generally continuous history of accumulation, good preservation and high temporal resolution. ^{210}Pb dating has proven a powerful tool for estimating recent (150 y) sediment accumulation rates, and provides a temporal framework for the environmental history of a lake and its catchment (Koide et al., 1973; Krishnaswami et al., 1971). ^{210}Pb is a product of the natural ^{238}U decay series. Among the other products, ^{222}Rn ($T_{1/2}$ =3.8 d) appears, partly escapes to the atmosphere and rapidly decays to ^{210}Pb ($T_{1/2}$=22.1 a). The ^{210}Pb is scavenged by aerosols, washed out by rain or snow and deposited in the lake and its catchment. The supply of atmospheric ^{210}Pb to lake sediment depends on several factors: 1) atmospheric flux, 2) residence time in catchment soils, 3) lake water residence time, 4) particle settling velocity, 5) fraction of metals bound to particulates, and 6) the extent of sediment focusing in the lake (Appleby and Oldfield, 1992). During the last few years, there have been a number of field-based studies of stable lead and ^{210}Pb cycling in lakes (e.g. White and Driscol, 1985). These studies demonstrate the varied and complex dynamic behavior of lead nuclides. Benoit and Hemond (1991) observed ^{210}Pb mobility under anoxic conditions and indicated the risk of overestimating ^{210}Pb ages from geochemically lowered, apparent accumulation rates. On the other hand, Appleby et al. (1979) and Crusius and Anderson (1995) investigated laminated lake sediments to evaluate the mobility of ^{210}Pb in lake sediments. They found a good agreement between varve counting and ^{210}Pb dating.

In the past 30 years, ^{210}Pb dating, together with heavy metal investigations of lake sediments, has been used to evaluate the temporal and spatial distribution of anthropogenic

In: Kassens, H., H.A. Bauch, I. Dmitrenko, H. Eicken, H.-W. Hubberten, M. Melles, J. Thiede and L. Timokhov (eds.) Land-Ocean Systems in the Siberian Arctic: Dynamics and History. Springer-Verlag, Berlin, 1999, 361-376.

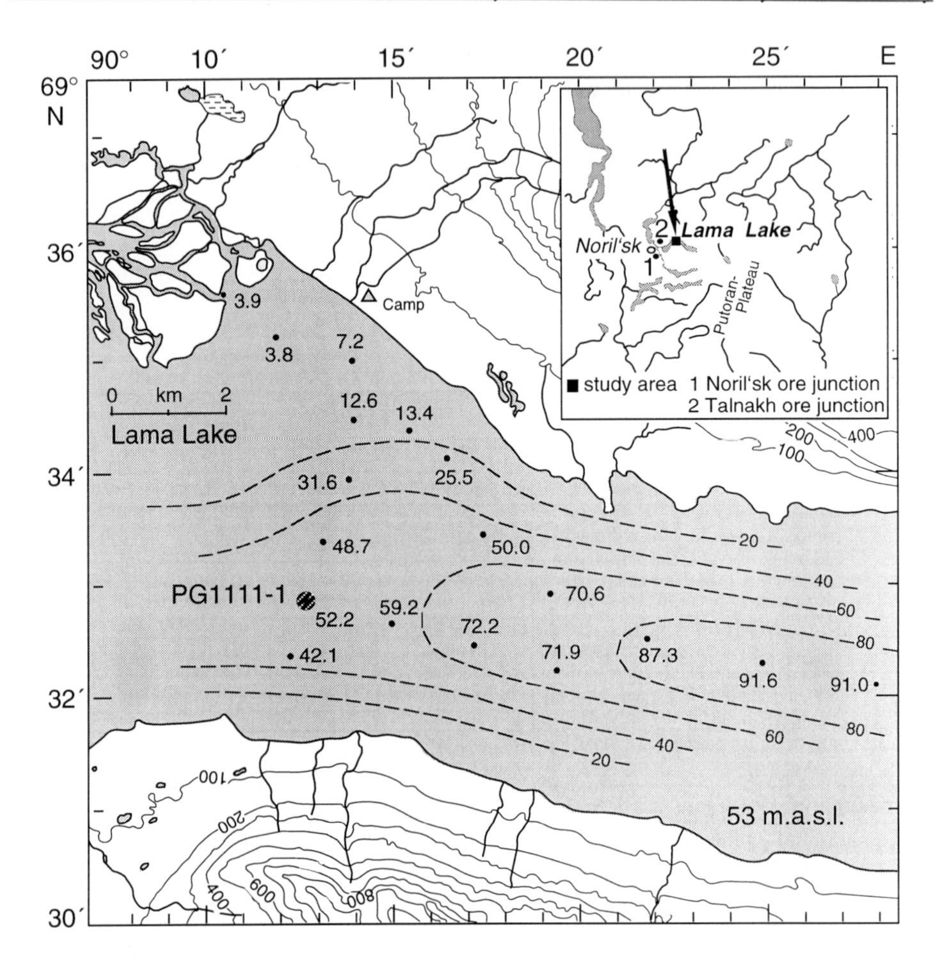

Figure 1: Map of the Lama Lake area and sampling site of the investigated core PG1111-1.

pollution in industrialized zones (e.g. Bollhöfer et al., 1994; Petit et al., 1984). Investigations of heavy metal concentrations on dated ice layers of Greenland snow and ice show pollution of the Northern Hemisphere since 1850 (Boutron et al., 1994; Murozimi et al., 1969). The greatest Pb enrichment occurred during the mid-1960s. Because of a reduction in leaded gasoline use, Pb concentrations have decreased by a factor of 7.5 over the past 20 years. Vinogradova and Polissar (1995) studied the chemical composition, source regions and pathways of atmospheric aerosols of the Central Russian Arctic in 1985, 1986 and 1988 and found an enrichment of Pb, Zn and Cd on the order of 10^2 to 10^3 compared to the mean crustal contents. These authors consider the Noril'sk area to be a possible source of the heavy metals.

Site description and climate conditions

Lama Lake is situated in the Noril´sk district of the southern Taymyr Peninsula, Siberia. Geography and characteristics of the catchment and lake are given in Figure 1 and Table 1, respectively. The lake is located in an east-west trending valley (69°30'N, 88°12'E) of the Putoran Plateau. The catchment, with relief up to 1200 m, is dominated by Permo-Triassic

Continental Flood Basalt (CFB) formations (e.g. Naldrett et al., 1995). Continuous permafrost up to 300 m depth is present below an active layer of 0.5 m to 2.5 m. Mean annual air temperature is sub-zero (-9.8°C) but maximum daily air temperatures reach 15°C in July. Between October and May, the lake is covered by ice and the peak of snowmelt is usually around the end of May and June. Total liquid precipitation amounts to 300-500 mm a^{-1} (Galaziy and Parmuzin, 1981). The main wind direction is east to northeast with maximum speeds of about 40 m/s. About 10 rivers contribute water to the lake and the total runoff is between 500 and 800 mm a^{-1}. Lama Lake itself drains into Pyasino Lake, situated to the northwest (Figure 1). The lake is periodically dimictic with neutral to alkaline conditions (Kienel, 1998). Data on groundwater flow and the water balance of the catchment are not available.

Table 1: Characteristics of Lake Lama and the catchment.

Altitude a.s.l.	53	m
Lake area	466	km^2
Catchment area	6200	km^2
Catchment/lake ratio	13	
Maximum depth	254	m
Month of ice cover	8	
Liquid precipitation	300-500	mm a^{-1}
Trophication	oligotroph	
pH	7.0 - 8.0	

Since 1930, Exploration and mining of Ni-Cu and platinum group element (PGE) deposits have been developed in the Noril'sk area (Kotlyakov and Agranat, 1994). The position of Lama Lake, 50 km east of Noril'sk mining, makes this lake suitable for investigations of anthropogenic pollution of the Taymyr Peninsula. There is no direct surface discharge of wastewater into the lake. Due to the prevailing wind direction, heavy metals from the mining activity and metal smelters in the Noril'sk area can be transported and deposited via atmosphere only.

This study presents the first results of ^{210}Pb-dating and heavy metal analysis in Lama Lake and estimates the temporal evolution of heavy metal fluxes into the lake.

Methods

The investigated sediment core PG1111-1 (52 cm long and 6 cm in diameter) was taken from the central part of Lama Lake in 52 m water depth during an expedition in summer 1993 (Melles, 1994). The sediment was collected with a gravity corer, a technique which ensures the recovery of undisturbed near-surface sediments. After retrieval, the core was stored at 4°C in a plastic liner. In 1995, the core was sectioned at 0.5 cm intervals down to 2.5 cm and at 1 cm intervals below this depth. The ^{210}Pb content was determined down to 17 cm by alpha counting of the decay product ^{210}Po ($t_{1/2}$ = 138 d). Acid digestion of 500 mg of sample for ^{210}Pb, and of 100 mg for trace elements was performed in Teflon autoclaves using ultrapure HNO_3 - HF - H_3PO_4 acids. Before digestion, the freeze-dried samples were spiked with a ^{208}Po-yield tracer. ^{210}Po was plated onto pure (99.99%) Ag-plates in a 2N HCl solution in the presence of ascorbic acid (Fleer and Bacon, 1984). Major and trace element concentrations were determined by ICP-OES (major elements and Ba, Co, Cu, V, Sr, Ni, Zn) and by graphite-furnace AAS

(As, Cd, Pb). The organic carbon content was measured on decarbonated samples by using a columetric method (ELTRA). All instruments have a precision ≤ 1% rel. for the measured concentrations. International standard reference materials (SRM) as well as double analyses were used to check the external precision. These measurements yielded an accuracy of ±5% for major element concentrations (wt.%) and of ±10% for trace element concentrations (ppm) and C_{org} (wt.%). The water content of each core section was determined from the mass difference between wet and dry sediment, assuming a specific density of 1 g cm^{-3} for water. Grain size determination was carried out using the classical method of wet sieving and Atterberg separation.

Results

Coring site and lithology

The coring site was located in the middle part of the lake, about 5 km south of the Mikchangdo river inflow. Depth measurements on various points around the coring site (Figure 1) and new seismic data collected during 1997, indicated pelagic sedimentation at the sampling area. A description of the lithology and change in water content is given in Figure 4. The main mineral components of the sediment were feldspar and pyroxene. The grain size composition (< 0.1% sand, 54% silt and 46% clay) was homogeneous over the investigated sediment column. Three visible boundary layers at 7.5 cm, 12.5 cm and 22 cm depth were observed. Within these layers a strong decrease in water content and change in color from brown to dark-brown occurs.

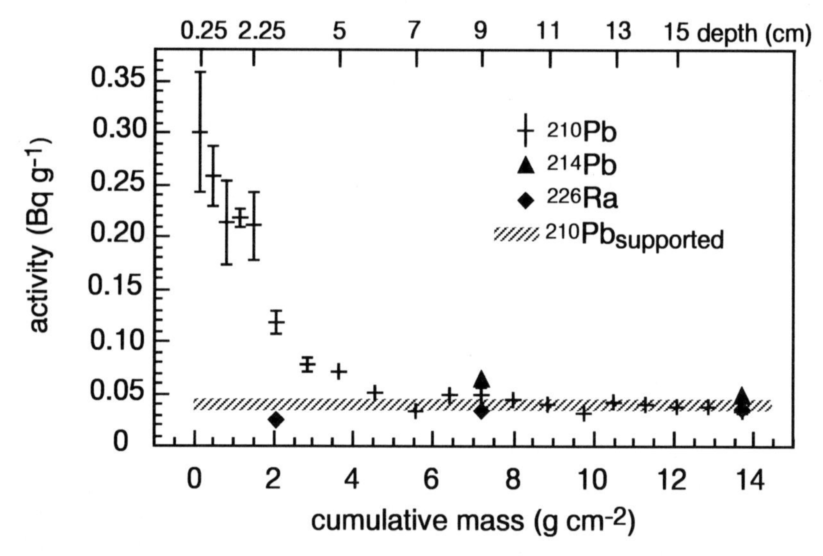

Figure 2: ^{210}Pb, ^{226}Ra and ^{214}Pb activities (Bq g^{-1}) *versus* cumulative mass (core PG1111-1). Also shown is the supported ^{210}Pb activity (Bq g^{-1}) estimated from the lower part of core (below 8 cm).

^{210}Pb dating

The results of alpha and gamma spectroscopy and calculated accumulation rates in the core sections are shown in Figure 2 and Table 2. The measured total ^{210}Pb ($^{210}Pb_{total}$) activity is composed of ^{210}Pb from the atmospheric input ($^{210}Pb_{unsupported}$) and from the *in situ* decay of ^{238}U in the sediment ($^{210}Pb_{supported}$). Since the accumulation rates and ages were calculated from the $^{210}Pb_{unsupported}$, the measured $^{210}Pb_{total}$ have to corrected:

$$^{210}Pb_{unsupported} = {}^{210}Pb_{total} - {}^{210}Pb_{supported} \qquad (1)$$

whereas $^{210}Pb_{supported}$ can be estimated from core depths at which the age of sedimentation deposition is greater than $5t_{1/2}$ of ^{210}Pb. This is the time at which the $^{210}Pb_{unsupported}$ has decayed and the measured ^{210}Pb results from ^{238}U decay in the mineral matter only. This depth was reached below 8 cm, where the ^{210}Pb activity had a constant value of 0.041 ± 0.006 Bq g^{-1} (Figure 2). As a further control, gamma-spectrometrical measurements of ^{226}Ra and ^{214}Pb, which are in secular equilibrium with $^{210}Pb_{supported}$, were carried out (Table 2). Within the error margin, the measured ^{226}Ra and ^{214}Pb activities agreed with the estimated $^{210}Pb_{supported}$ activities from the lower core sections (Figure 2). Therefore a mean value of 0.04 ± 0.01 Bq g^{-1} activities was estimated for the $^{210}Pb_{supported}$ and assumed to be constant for all depths.

Table 2: ^{210}Pb, ^{214}Pb and ^{226}Ra activities and calculated accumulation rate (r) of core PG1111-1.

depth (cm)	cumulative mass (g cm^{-2})	^{210}Pb (Bq g^{-1})	± (Bq g^{-1})	r (g cm^{-2} y^{-1})	^{214}Pb (Bq g^{-1})	^{226}Ra (Bq g^{-1})
					-	-
0.25	0.14	0.300	0.058	0.046	-	-
0.75	0.44	0.258	0.029	0.049	-	-
1.25	0.76	0.214	0.040	0.049	-	-
1.75	1.07	0.217	0.009	0.039	-	-
2.25	1.36	0.210	0.032	0.030	-	-
3	1.89	0.118	0.011	0.044	0.03	-
4	2.64	0.078	0.006	0.053	-	-
5	3.39	0.070	0.005	0.036	-	-
6	4.21	0.051	0.003	0.026	-	-
7	5.18	0.033	0.002	-	-	-
8	5.95	0.050	0.003	-	-	-
9	6.69	0.048	0.003	-	0.03	0.06
10	7,43	0.045	0.003	-	-	-
11	8.19	0.039	0.001	-	-	-
12	9.02	0.032	0.002	-	-	-
13	9.73	0.043	0.003	-	-	-
14	10.43	0.040	0.003	-	-	-
15	11.18	0.039	0.003	-	-	-
16	11.92	0.037	0.002	-	-	-
17	12.67	0.034	0.003	-	0.04	0.05

The sediment accumulation rates and ages were determined using the CRS model which assumes a Constant Rate of ^{210}Pb Supply. The applicability of this model was tested by Appleby et al. (1979), who compared ^{210}Pb profiles to annually laminated sediments, and found a good agreement between varve counting and CRS derived ^{210}Pb data. The sediment accumulation rate and age determinations were calculated according to the CRS model as:

$$A_z = A_0 \, e^{-\lambda t(z)} \qquad (2)$$

$$A_z = {}_{z=0}\int^{z} \rho(1-\phi_x)\, C_x \, dx \qquad (3)$$

$$\phi_x = V_w(x) / V_w(x) + M_s(x)/\rho_s \qquad (4)$$

where:

A_z (Bq cm^{-2})	= integrated activity of unsupported ^{210}Pb at depth (z);
A_0 (Bq cm^{-2})	= integrated activity of unsupported ^{210}Pb at depth z = 0;
λ (a^{-1})	= decay constant of ^{210}Pb (0.031);
t_z (a)	= apparent age at depth (z).
C_x (Bq g^{-1})	= activity of unsupported ^{210}Pb at depth x;
ϕ_x (cm^3 cm^{-3})	= porosity at depth (x)
ρ_s (g cm^{-3})	= density of dry sediment (assumed as 2.6 g cm^{-3}),
V_w (cm^3)	= volume of water at depth (x)
M_s (g)	= mass of dry sediment at depth (x)

The accumulation rate r (g cm^{-2} a^{-1}) was calculated for each depth (z) as:

$$r_z = \lambda A_z / C_x \tag{5}$$

The calculation becomes less precise below 5 cm depth due to the low unsupported ^{210}Pb content. The age and accumulation rate is given in Figure 3. The determined accumulation rates for the dry sediment were between 0.057 and 0.027 g cm^{-2} a^{-1}. The highest accumulation rate was observed at 4 cm depth, which corresponds to 1940. From the ^{210}Pb dates, a ^{210}Pb inventory of 0.426 Bq cm^{-2} and mean ^{210}Pb supply of 0.013 Bq cm^{-2} a^{-1} were derived.

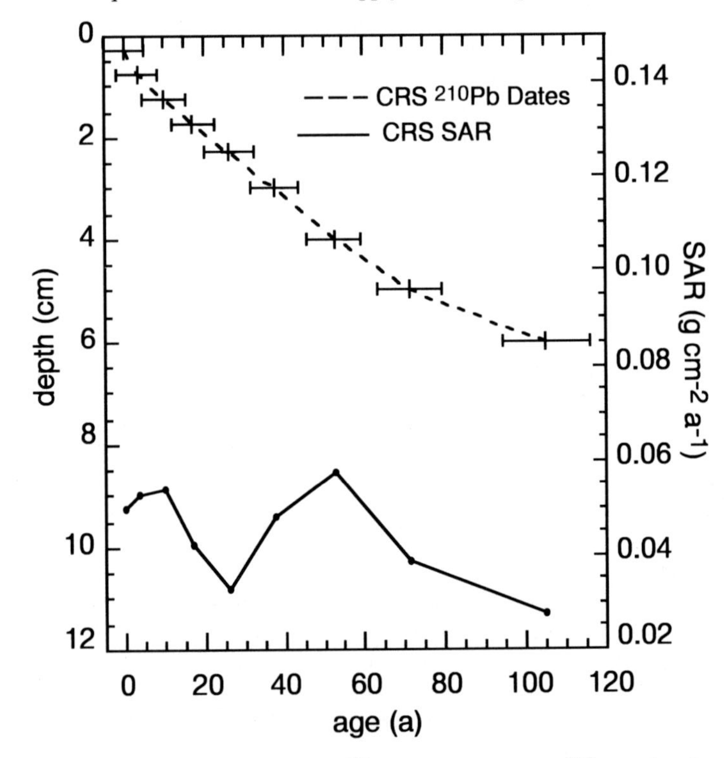

Figure 3: Depth versus age determined from the CRS ^{210}Pb model. Error bars in ^{210}Pb are based on two standard deviations in counting statistics (see Table 2). Also plotted are sediment accumulation rates (SAR) versus age calculated from the CRS ^{210}Pb model.

Geochemistry

The geochemical composition of the core is shown in Table 3. The major elements, with the exception of Mn, had rather constant distribution over the entire core with a mean standard deviation below 10%. To deduce the source of the sediments, a comparison with the chemical composition of the CFB formations was undertaken. This formation has been studied in detail by Lightfoot et al. (1993) and Wooden et al. (1993). There was good agreement with the Mokulaevsky (MK) and Nadezhdinsky (Nd3) formations (Figure 5). Therefore, these formations can be regarded as the geogenic source of the core sediments. Mg and Ca are depleted compared to the CFBs. This may be the result of weathering processes in polar regions as discussed by Ugolini (1986). A strong enrichment of MnO (from 0.2 to 0.5 wt%) was observed at a depth of 6.5 to 7.5 cm (Figure 4). At this depth, a dark brown layer and a decrease in the water content were also observed. Such layers are reported from many lake surface sediments and are regarded as the base of an oxidation zone (Kephkay, 1985; McKee et al., 1989), where Mn as well as Fe oxides precipitate.

The trace metals Cd, Cu, Ni, Pb and Zn showed a progressive increase towards the sediment-water interface within the upper part of the core (0 to 5 cm depth; Figure 4). Due to the good chemical agreement between the CFB formations and the lower part of the core, the geogenic source of the sediment is well established. This is the basis for the calculation of enrichment factors for the trace metals Cd, Cu, Ni, Pb and Zn using the mean composition of the lower part of the core (6-17cm) as a reference for the unpolluted geogenic supply. This calculation is performed for the uppermost 0.5 cm sediment section:

$$EF_M = (M/Al)_{sf} : (M/Al)_{mean} \qquad (6)$$

where EF is the enrichment factor, M represents the metal content (Zn, Pb, Ni Cu), sf refers to the surface sediment, mean refers to the mean for the lower core sediments, and Al is aluminium.

In order to eliminate dilution effects, Al-normalization was applied because Al is regarded as a stable constituent of the mineral matter.

The highest enrichment was found for Cd (EF_{Cd}= 2.5) and Pb (EF_{Pb}= 2.0), whereas the enrichments of Cu (EF_{Cu} =1.3), Ni (EF_{Ni}= 1.3) and Zn ($EF_{Zn,}$= 1.3) were smaller. The heavy metal enrichment (or supplementary heavy metals) presumably reflects pollution. To evaluate the „historical„ enrichment of the supplementary heavy metal concentrations independent of variations in the accumulation rate, we calculated the initial ^{210}Pb activity $C_0{}^z$ (Bq g^{-1})for each depth (z). With this record of $C_0{}^z$, the heavy metal/$C_0{}^z$ (µg Bq^{-1}) ratios were determined: 1) for the total (geogenic + supplementary) heavy metal concentration, and 2) for the geogenic heavy metal concentration. From the difference, the accumulation- corrected temporal distribution of the supplementary heavy metal/$C_0{}^z$ ratios can be calculated. We prefer the ^{210}Pb-normalization over the Al-normalization because ^{210}Pb is transported by the atmosphere and takes the same pathway into the lake sediments that we assume for the heavy metals. The ^{210}Pb-normalization makes the calculation of heavy metal enrichment independent of the accumulation rate. Of course, this was also achieved by Al-normalization but owing to the similar chemical behavior of ^{210}Pb and the heavy metals, the ^{210}Pb-normalization also takes into account further factors, such as: distribution coefficients between suspended matter and lake water, grain size distribution of the lake sediments and catchment and water residence times. The results are shown in Figure 6. The first enrichment of Cd, Pb and Zn occured at 4 cm depth, which corresponds to the deposition year 1940. Above 3 cm, continuous increases in metal accumulation up to the sediment surface were evident. In contrast, Ni and Cu were depleted at 4 cm depth but also increased above 3 cm.

Table 3: Element concentrations, mean concentrations (m) and standard deviation (s.d.) from 6-17 cm depth of core PG1111-1.

depth	cm	0.25	1.25	1.75	2.25	3	4	5	6	7	8	9	10	11	13	14	15	16	17	6-17 m	s.d. abs.	s.d. %
Al_2O	wt %	15.3	15.6	15.3	15.5	15.3	15.7	15.7	15.7	16.5	16.4	16.7	15.5	15.9	15.9	15.6	15.8	15.9	15.9	16.0	0.38	2
Fe_2O		11.5	11.5	11.5	11.5	11.3	11.2	11.3	11.4	11.3	11.5	11.7	11.2	11.2	11.6	11.6	11.6	11.7	12.1	11.5	0.25	2
MnO		0.21	0.18	0.19	0.20	0.19	0.18	0.24	0.35	0.49	0.18	0.16	0.15	0.14	0.18	0.18	0.16	0.20	0.15			
MgO		5.02	4.87	4.86	4.87	4.86	4.79	4.84	4.79	4.69	4.92	4.97	4.83	4.77	4.93	4.94	4.97	5.15	5.10	4.9	0.13	3
CaO		6.73	6.58	6.59	6.72	6.75	6.72	6.79	6.51	6.03	6.61	6.60	6.59	6.50	6.77	6.69	6.79	6.96	6.63	6.6	0.22	3
TiO_2		0.86	0.94	0.85	1.09	1.10	0.80	1.08	1.08	1.06	1.11	1.15	1.05	0.89	1.00	1.12	1.13	1.10	1.05	1.1	0.07	7
Na_2O		1.80	1.80	1.77	1.78	1.84	1.83	1.88	1.83	1.81	1.96	1.89	1.90	1.90	1.93	1.87	1.95	1.98	1.95	1.9	0.05	3
K_2O		0.94	1.00	0.96	0.94	0.93	0.98	0.98	1.04	1.04	1.04	1.01	0.96	1.00	0.98	0.95	0.97	0.94	0.94	1.0	0.04	4
P_2O_5		0.79	0.79	0.64	0.93	0.86	0.68	0.80	0.80	0.81	0.84	0.87	0.81	0.78	0.82	0.87	0.91	0.95	1.03	0.9	0.07	8
Corg		0.94	0.94	0.94	0.97	0.94	0.77	0.74	0.72	0.71	0.90	0.84	0.93	0.88	0.91	1.11	0.81	0.84	0.76	0.9	0.11	13
N		0.11	0.11	0.11	0.11	0.11	0.09	0.09	0.09	0.08	0.10	0.09	0.08	0.09	0.09	0.11	0.10	0.10	0.08	0.1	0.01	10
Ba	ppm	205	212	206	208	207	222	220	222	242	217	241	230	220	229	221	215	220	215	225	9	4
Co		48	47	47	48	47	46	47	46	47	46	49	45	43	46	48	48	48	48	47	2	3
Cu		143	130	123	119	113	108	107	111	116	111	119	106	109	115	118	114	118	125	115	5	4
V		198	196	204	213	210	206	209	201	193	201	205	197	192	191	204	205	212	200	200	6	3
Sr		229	231	230	229	230	230	229	230	241	236	237	229	233	233	232	236	237	239	235	4	2
Ni		100	91	88	88	82	78	80	80	83	81	83	79	79	81	83	82	84	86	82	2	2
Zn		104	98	97	95	92	96	93	92	92	92	93	90	92	92	92	92	93	93	92	1	1
Pb		10.8	10.0	9.8	8.5	6.5	7.1	5.9	6.1	6.3	6.3	6.0	5.9	6.3	5.4	5.6	5.7	4.9	5.2	5.8	0.4	8
Cd		0.30	0.21	0.19	0.17	0.15	0.18	0.13	0.14	0.13	0.12	0.12	0.12	0.13	0.12	0.12	0.13	0.11	0.14	0.13	0.01	8
As		4.1	3.2	5.5	6.3	6.6	3.9	6.1	6.4	6.0	6.0	6.7	5.8	6.1	6.1	6.3	6.2	4.0	6.4	6.0	0.7	11

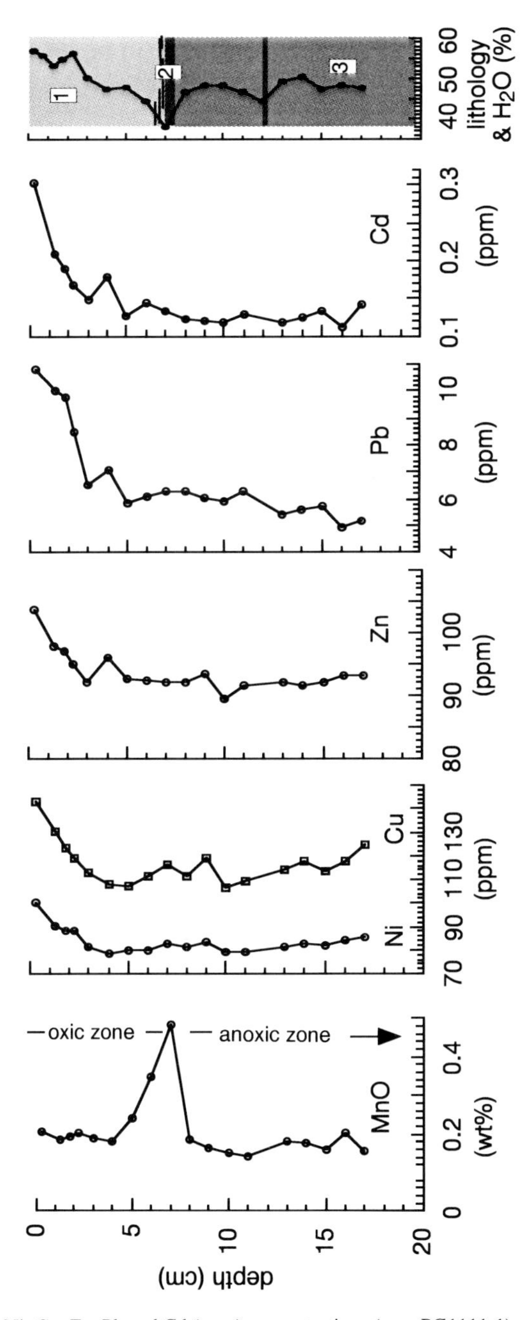

Figure 4: MnO (wt%), Ni, Cu, Zn, Pb and Cd (ppm) concentrations (core PG1111-1) and core lithology with (1) dark yellowish brown mud, (2) dark brown layer, and (3) olive brown mud with a black layer at 12.5 cm. Water content calculated as % of dry sediment.

Discussion

Hydrological investigations in July 1993 yielded a pH of 8.1 (water surface) and 7.8 (water-sediment interface). As shown by Tessier et al. (1985), these conditions cause high adsorption of the heavy metals by natural solids within the water column. The most important carriers of heavy metals are Fe and Mn oxides, clay minerals and organic matter. The oxygen content of 12 mg l^{-1} (water surface) and 2.8 mg l^{-1} (water-sediment interface) indicate oxic conditions over the whole water column during this time (Melles 1994).

According to McKee et al. (1989) and Williams (1992), the Mn peak at 7 cm depth indicates the boundary between oxic (above) and anoxic (below) conditions in the sediment core. Within the oxidized zone, algal degradation and Fe and Mn oxide precipitation occur. Burial in the anoxic layer cause Fe and Mn reduction, with resultant upward migration to the oxic/anoxic boundary where oxide precipitation takes place once again. For these reasons, the heavy metal enrichment in the investigated core occurs in a zone where trace metal distributions are affected by degradation of organic matter and the redox-driven cycling of Mn and Fe. Consequently, the discussion concerning anthropogenic pollution must consider the possible influence of these processes.

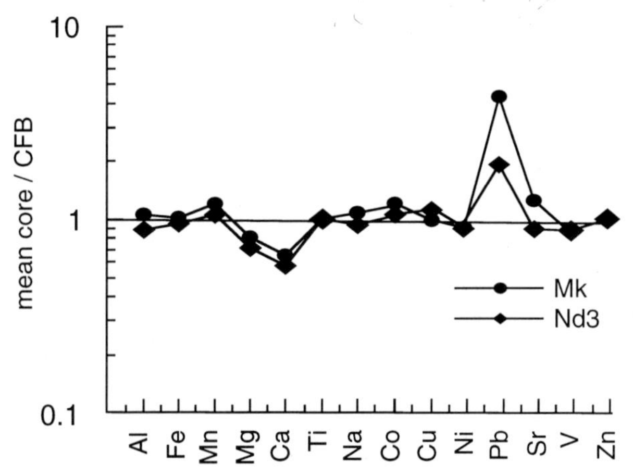

Figure 5: CFB-normalized spidergram of the mean element concentrations of core PG1111-1 from 6-17 cm depth. For the normalization, the chemical composition of CFB Formation Nd3 (Nadezhdinsky) and MK (Mokulaevsky) as reported by Lightfoot et al. (1993) are used.

^{210}Pb accumulation and mobilization

During the last few years, a number of field-based studies of stable lead and ^{210}Pb cycling in lakes (e.g. Benoit and Hemond, 1991; Carpenter et al., 1981; Dominik et al., 1981) demonstrated ^{210}Pb mobility during seasonal anoxia. As shown by Benoit and Hemond (1991), this could cause dating errors especially for sedimentation rates below 0.1 cm a^{-1}. On the other hand, ^{210}Pb dating of laminated sediments from seasonal anoxia show excellent agreement between varve counting and ^{210}Pb dates (Brunskill and Ludlam, 1988; Crusius and Anderson, 1995). Information about seasonal anoxia in Lama Lake are not available. However, using the diffusion model of Benoit and Hemond (1991) suggests that ^{210}Pb mobilization due to seasonal anoxia in Lama Lake is of minor importance. Assuming conditions estimated in Lama Lake - a sedimentation rate of 1 mm a^{-1}, porosity between 0.79 and 0.82 and a partition coefficient between sediment and porewater of more than 10^3 (Tessier et al., 1985) - the sedimentation rate could be overestimated by $\leq 20\%$. This overestimation causes a reduction of the calculated mean sedimentation rate from 1 mm a^{-1} to 0.8 mm a^{-1}.

Besides mobilization processes, ^{210}Pb profiles could also be modified by sediment mixing due to bioturbation or sediment redistribution. This effect was observed by Nittrouer et al. (1983, 1984) on the Washington Continental Shelf, and by Robbins and Edgington (1975) in Lake Michigan. Such mixing processes are particle-selective and could sometimes be identified by constant ^{210}Pb activities over the mixing depth and by deep penetration of short-lived radionuclides like ^{234}Th and ^{137}Cs. In the present sediment core, neither constant ^{210}Pb activities over greater depth intervals nor macroscopic or microscopic indications of bioturbation could be observed. So far, a cross check of the ^{210}Pb data by other physical dating methods (e.g. ^{14}C, ^{137}Cs) is not available.

Trace metal mobilization during organic degradation and redox cycling

Studies of trace metal fluxes across the sediment-water interface in marine waters were performed by Westerlund et al. (1986). Their results show that Cu, Cd, Ni and Zn were released from sediments under oxidizing conditions whereas a release of Pb was not observed. Lapp and Balzer (1993) attributed an observed high pore-water flux of Cd, Cu and Ni in oxic coastal sediments of Kiel Bight to the aerobic degradation of organic matter. Meyers and Ishiwatari (1995) stated that degradation of organic matter in lacustrine bottom sediments is much greater under oxic than anoxic conditions and about 75% of organic carbon can be released to the hypolimnion due to these processes (Meyers and Eadie, 1993). Therefore the enrichment of Cd, Cu and Ni in the oxygenized zone of the sediment could be a steady-state condition (uptake and release by organic matter). The organic carbon contents of the investigated core are between 1.0 and 0.7 wt.% with a typical lacustrine C/N ratio of 8 in the upper part of sediment (Figure 7). A decrease of the carbon content from 1.0 to 0.7 wt.% between 3 and 6 cm depth may reflect degradation processes. On the other hand, since the accumulation rate also increases at this depth (Figure 3), the low C_{org} content could be the result of dilution. Since no change in the C/N ratio occurs at this depth, the lowering in organic carbon could also be the result of low organic production within the water column.

The most important control on Pb and Zn concentration in lake water are pH and redox conditions. The partition coefficients of Pb and Zn between water and Fe and Mn oxides decrease rapidly under acidification, and trace metals are released to the water. In neutral to alkaline waters like Lama Lake, the partition coefficients of Pb and Zn are high ($K_D > 10^3$; Tessier et al., 1985). Furthermore, the reduction of the Mn and Fe oxide carrier phases can result in the mobilization of bound trace metals. Pore water investigations on various Canadian lakes indicate that Zn is depleted under anoxic conditions due to precipitation of Zn sulfide (e.g. Matisoff et al., 1980; Tessier et al., 1989). Investigations of Pb mobility in the lacustrine environments sometimes produce contradictory results. As shown above, some investigations indicate Pb mobility during anoxic conditions. Based on laboratory experiments, Frevert (1987) attributed the efficient and

rapid removal of dissolved Pb under anoxic conditions to sulfide precipitation. Following the results of various field-based studies on Pb and Zn in lacustrine and marine environments, the sediments acted as a sink more than as a source (e.g. Carignan and Nriagu,1985; Hamilton-Taylor and Davison, 1995; Westerlund et al., 1986). This suggests that the observed Pb and Zn enrichment can be attributed to pollution rather than redox phenomena and organic degradation.

Since Cu and Ni are more mobile, the negative values of the ^{210}Pb-normalized profiles below 3 cm (Figure 5) could have been affected by organic degradation. On the other hand, Cd showed the same „enrichment-profile„ as the more immobile trace elements Pb and Zn, which confirms an anthropogenic source of Cd as well. Furthermore, the Cd enrichment seems to be too high for a steady state due to organic degradation.

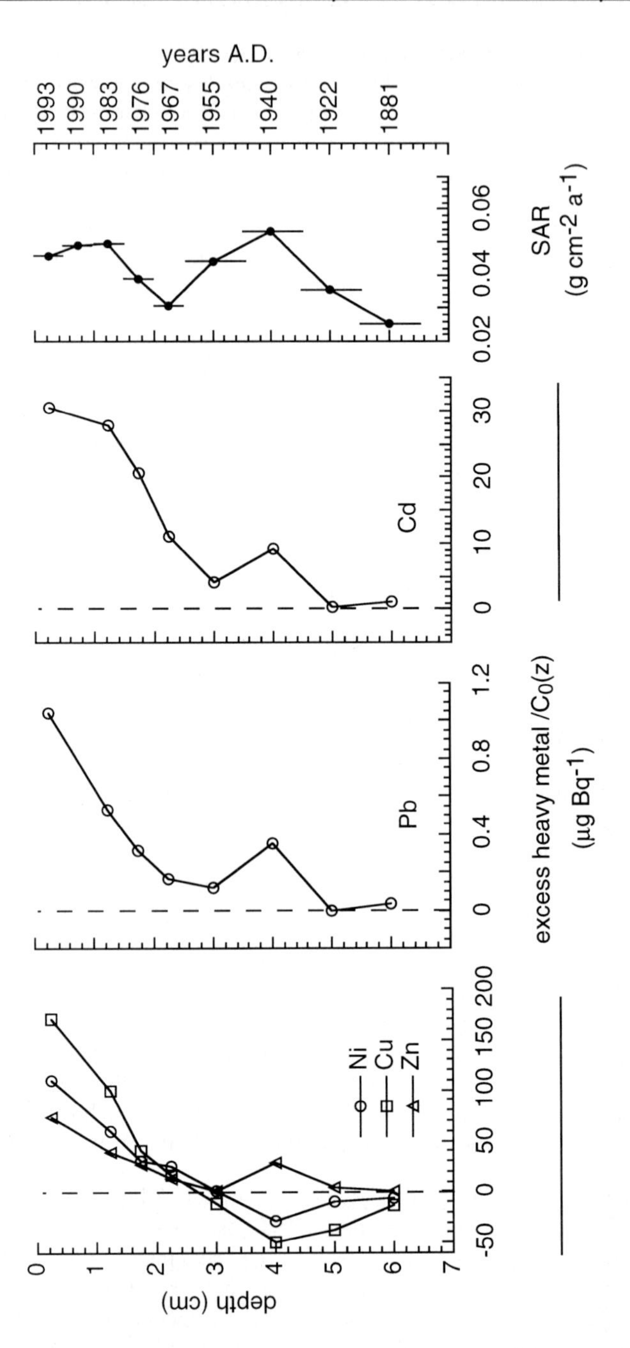

Figure 6: Ratios of excess trace metal concentrations to initial unsupported ^{210}Pb activities. The ratios are corrected for the geogenic component of heavy metals. For explanations see text. Sediment accumulation rates (SAR) and ages (years A.D.) are derived from the ^{210}Pb data using the CRS model.

Based on our data, there are some arguments for an anthropogenic origin of the supplementary heavy metals. Besides Cu-Ni sulfides, Pb and Zn occurred in larger amounts in the PGE deposits of Noril'sk (Genkin and Evstigneeva, 1986) and could therefore have been emitted as a result of mining activities. Additionally, investigations of aerosols in the Noril'sk area from Pacyna et al. (1985) demonstrate that Cd, Zn and Pb were emitted to the atmosphere. According to the ^{210}Pb dating, the onset of enrichment was about 1940. This corresponds to the time when the Mining and Metallurgical Combine of Noril'sk was established (Kotlyakov and Agranat, 1994). Consequently, we conclude that anthropogenic heavy metal pollution has affected the upper 5 cm of the investigated sediments.

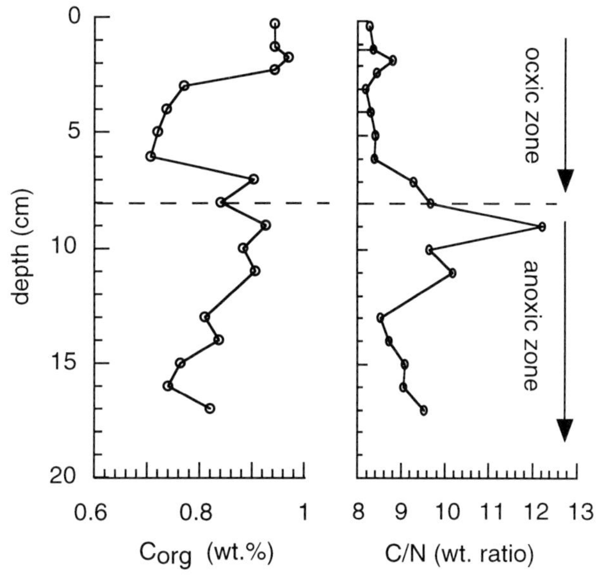

Figure 7: Content of organic carbon (C_{org} wt.%) and C_{org}/N weight ratios of core PG1111-1.

Temporal evolution of trace metal contamination

Anthropogenic heavy metal pollution in northern polar regions is known from investigations of snow and glaciers from Greenland (Boutron et al., 1991) as well as from investigations of aerosols from the Arctic station Vavilon Dome at 79.5°N, 95.4°E (Vinogradova and Polissar, 1995). These studies demonstrate a significant supply of anthropogenic Cd, Pb and Zn in the arctic atmosphere. The Pb contribution to the atmosphere is mainly related to alkyl-leaded petrol. Due to the reduction of lead in gasoline additives, the atmospheric concentration has diminished by more than 90% in Greenland ice and snow as well as in the United States and Europe since the late 70s. The contribution of anthropogenic Cd and Zn originated mainly from industrial processes (Nriagu and Pacyna, 1988). The decrease of these elements in the atmosphere during the last 20 years is less rapid than for Pb (Boutron et al., 1991, 1994). A comparison of the observed enrichment factors for Pb, Cd , Cu, Ni and Zn with those derived from Greenland ice (Boutron et al. 1991, 1994), from arctic aerosols (Vinogradova and Polissar, 1995) and from spider webs in clean-air regions of Germany (Rachold et al., 1992) is shown in Figure 8. The relative abundance of Pb, Cd and Zn in Lama Lake displays the same trend as that of the above-mentioned other regions, whereas Cu and Ni are rather strongly enriched. This can be explained by the position of Lama Lake close to the Cu-Ni mining area. In contrast to Greenland ice and European lakes (e.g. Lake Constance, Bollhöfer et al., 1994), there was no evidence in the Lama Lake core for a decreasing input of Pb over the last 20 years. This could be the result of the ongoing mining activities in the Noril'sk area as well as of the continuous use of Pb additives in Russian fuel until the present.

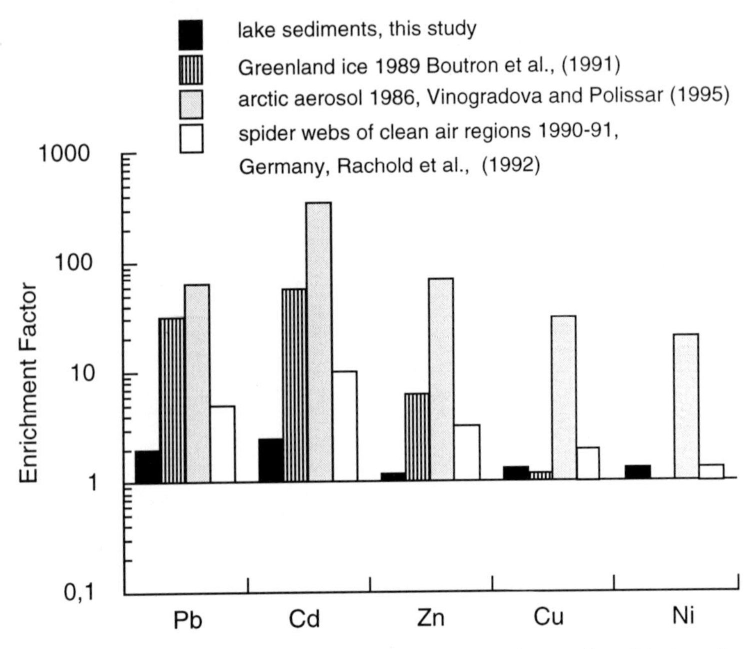

Figure 8: Enrichment factors for different environments of the world. With exception of the investigated core the enrichment factors are calculated for the mean crustal element concentrations from Taylor and McLennan (1985) as unpolluted source.

Conclusions

From our first studies of heavy metal distribution within core PG1111-1 from Lama Lake, southern Taymyr Peninsular, we draw the following conclusions:

1 The ^{210}Pb data revealed sediment accumulation rates between 0.05 to 0.03 mg cm^{-2} a^{-1}, which point to a relatively constant sediment flux during the last 100 years. Anoxic conditions and mobilization phenomena in Lama Lake could result in an overestimation of the sedimentation rate by up to 20%.

2 The geochemical composition of the lower part of the core identifies the CFB formations Mk and Nd3 as the main source of the deposited sediments.

3 With the identification of these geogenic sources, enrichment factors for the heavy metals of Cd, Cu, Ni, Pb and Zn were calculated. The highest enrichment was observed for the elements Cd (EF 2.5) and Pb (EF 2.0). The enrichment of Cd, Pb and Zn can be attributed to anthropogenic contamination rather than to mobilization and migration processes within the sediment.

4 The onset of Cd, Pb and Zn enrichment coincided with the onset of mining activities at Noril'sk, supporting the assumption of an anthropogenic source of the heavy metals. The recent decrease of the anthropogenic Pb contribution to lake sediments as recently documented in the Northern Hemisphere (e.g. Greenland, Europe, United States) was not be observed in Lake Lama.

Acknowledgments

This study was carried out within the framework of the Taymyr Project. We thank the German Ministry of Education, Science, Research and Technology (BMBF) for the financially supporting this project. We are grateful to Damian Gore and one anonymous reviewer for their constructive and critical comments on this paper. We are grateful to Heike Höltzen, Jana Friedrich and Hans-Jürgen Walter for their assistance in ^{210}Pb analyses and Volker Rachold for critical discussion. This is publication No. 1355 of the Alfred Wegener Institute.

References

Appleby, P.G. and F. Oldfield (1992) Application of ^{210}Pb to sedimentation studies, In: Uranium-series disequilibrium: Applications to earth, marine and environmental sciences, M. Ivanovich and R.S. Harmon (eds), Oxford University Press, Oxford, 731-778.

Appleby, P.G., F. Oldfield, R. Thompson, P. Huttunen and K. Tolonen (1979) ^{210}Pb dating of annually laminated lake sediments from Finland. Nature, 280, 53-55.

Benoit, G. and H.F. Hemond (1991) Evidence for diffusive redistribution of ^{210}Pb in lake sediments. Geochim. Cosmochim. Acta, 55, 1963-1975.

Bollhöfer, A., A. Mangini, A. Lenhard, M. Wessels, F. Giovanoli and B. Schwarz (1994) High resolution ^{210}Pb dating of Lake Constance sediments: stable lead in Lake Constance. Environ. Geol., 24, 267-274.

Boutron, C.F., J.-P. Candelone and S. Hong (1994) Past and recent changes in large-scale tropospheric cycles of lead and other heavy metals as documented in the Antarctic andGreenland snow and ice: a review. Geochim. Cosmochim. Acta, 58, 3217-3225.

Boutron, C.F., U. Görlach, J.-P. Candelone, M.A. Bolshov and R.J. Delms (1991) Decrease in anthropogenic lead, cadmium and zinc in Greenland snows since the late 1960's. Nature, 353, 153-156.

Brunskill, G.J. and S.D. Ludlam (1988) The variation of annual ^{210}Pb flux to varved sediments of Fayetteville Green Lake, New York from 1885 to 1965. Verh. Int. Verein Limnol., 23, 848-854.

Carignian, R. and J.O. Nriagu (1985) Trace metal deposition and mobility in the sediments of two lakes near Sudbury, Ontario. Geochim Cosmochim. Acta, 49, 1753-1764.

Carpenter, R., J.T. Bennet and M.L. Peterson (1981) ^{210}Pb activities in and fluxes to sediments of the Washington continental slope and shelf. Geochim. Cosmochim. Acta, 45, 1155-1172.

Crusius, J. and R.F. Anderson (1995) Evaluating the mobility of ^{137}Cs, $^{239+240}$Pu and ^{210}Pb from their distributions in laminated lake sediments. Journ. Paleolimnol., 13, 119-141.

Dominik, P.J., A. Mangini and G. Müller (1981) Determination of recent deposition rates in Lake Constance with radioisotope methods. Sedimentology, 28, 653-677.

Fleer, A.P. and M.P. Bacon (1984) Determination of ^{210}Pb and ^{210}Po in seawater and marine particular matter. Nuclear Instruments and Methods in Physics Research 223, 243-249.

Frevert, T. (1987) Heavy metals in Lake Kinneret (Israel). II. Hydrogen sulfide dependent precipitation of copper, cadmium, lead and zinc. Arch. Hydrobiol., 109, 1-24.

Genkin, A.D. and T.L. Evstigneeva (1986) Associations of Platin-Group minerals of the Noril'sk copper-nickel sulfide ores. Economic Geology, 81, 1203-1212.

Galiziy, G.J., and Parmuzin, Y.P. (eds.) (1981) History of the great lakes in the Central Siberian (in Russian) Novosibirsk, Nauka, 136pp.

Hahne, J. and M. Melles (1997) Late-and post-glacial vegetation and climate history of south-western Taymyr Peninsula, Central Siberia, as revealed by pollen analyses of a core from Lake Lama. Vegetation History and Archaeobotany, 6, 1-8.

Hamilton-Taylor, J. and W. Davison (1995): Redox driven cycling of trace elements in lakes. In: Physics and chemistry of lakes, A. Lerman, D. Imboden and J. Gat (eds.), 2. edition, Springer Verlag Berlin: 218-263

Kephkay, P.E. (1985) Kinetics of microbial manganese oxidation and trace metal binding in sediments. results from *in situ* dialysis technique. Limnol. Oceanogr., 30, 713-726.

Kienel, U. (this volume): Late Weichselian to Holocene Diatom successionin a sediment core from Lama Lake, Siberia and presumed ecological implications.

Koide M., K.W. Bruland and E.D. Goldberg (1973) Th-228/Th-232 and Pb-210 geochronologies in marine and lake sediments. Geochim. Cosmochim. Acta 37, 1171-1187.

Kotlyakov, V.M. and G.A. Agranat (1994): The Russian North: Problems and Prospects. Polar Geography and Geology, 1994, 18, 4, pp. 285-295.

Krishnaswami, S., D. Lal, J.M. Martin and M. Meybeck (1971) Geochronology of lake sediments. Earth and Plan. Sci. Lett. 11, 407-414.

Lapp, B. and W. Balzer (1993) Early diagenesis of trace metals used as an indicator of past productivity changes in coastal sediments. Geochim. Cosmochim. Acta, 57, 4639-4652.

Lightfoot, P.C., C.J. Hawkesworth, J. Hergt, A.J. Naldrett, N.S. Gorbachev, V.A. Federenko and W. Doherty

(1993) Remobilization of continental lithosphere by a mantle plume: major and trace-element, and Sr-, Nd-, and Pb-isotope evidence from picritic and tholeiitic lavas of the Noril'sk District, Siberian Trap, Russia. Contrib. Mineral. Petrol. 114, 171-188.
Matisoff, G., A.H. Lindsay, S. Matis and F.M. Soster (1980) Trace metal mineral equilibria in Lake Erie sediments. J. Great Lakes Res., 6, 353-366.
McKee J.D., T.P. Wilson D.T., Long and R.M. Owen (1989) Geochemical partitioning of Pb, Zn, Cu, Fe and Mn across the sediment-water interface in large lakes. J. Great Lakes Res., 15, 46-58.
Melles, M. (1994) The expeditions Norilsk/Taymyr 1993 and Bunger Oasis 1993/94 of the AWI Research Unit Potsdam. Reports on Polar Research, 148, 3-25.
Meyers, P.A. and R. Ishiwatari (1995) Organic matter accumulation records in lake sediments. In: Physics and chemistry of lakes, A. Lerman, D. Imboden and J. Gat (eds.), 2. edition, Springer Verlag Berlin, 279-323
Meyers, P.A. and B.J. Eadie (1993) Sources, degradation, and recycling of organic matter associated with sinking particles in Lake Michigan. Org. Geochemistry, 20, 47-56.
Murozumi, M., T.J. Chow and C.C. Patterson (1969) Chemical concentrations of pollutant lead aerosols, terrestrial dusts and seasalts in Greenland and Antarctic snow strata. Geochim. Cosmochim. Acta, 33, 1247-1294.
Naldrett, A.J., V.A Fedorenko,.; P.C Lightfoot,. V.I Kunilov,. N.S Gorbachev,. W. Doherty, and Z. Johan (1995) Ni-Cu-PGE deposits of Noril´sk region, Siberia: their formation in conduits for flood basalt volcanism. Applied earth science-Transactions of the Institution of Mining and Metallurgy, 1995, 104, B, pp. B18-B36.
Nittrouer, C.A., D.J. DeMaster, B.A. McKee, N.H. Cutshall and I.L. Larsen (1983/1984) The effect of sediment mixing on Pb-210 accumulation rates for the Washington Continental Shelf. Marine Geology, 54, 201-221.
Nriagu, J.O. and J.M. Pacyna (1988) Quantitative assessment of worldwide contamination of air, water and soils by trace metals. Nature, 333, 138-140.
Pacyna, J.M., B. Ottar, U. Tomza and W. Maenhaut (1985) Long range transport of trace elements to ny-Alesund, Spitzbergen, Atmos. Environ., 19, 6, 857-865.
Petit, D., J.P. Mennessier and L. Lamberts (1984: Stable lead isotopes in pond sediments as tracers of past and present atmospheric lead pollution in Belgium. Atmos. Environm., 18, 6, 1189-1193.
Rachold, V., H. Heinrichs and H.-J. Brumsack (1992) Spiderwebs: natural collectors of atmospheric particulate matter (in German) Naturwissenschaften, 79, 175-178.
Robbins, J.A. and D.N. Edgington (1975) Determination of recent sedimentation rates in Lake Michigan using Pb-210 and Cs-137. Geochim. Cosmochim. Acta, 39, 285-304.
Tessier, A., R. Carignan, B. Dubreuil and F. Rapin (1989) Partitioning of zinc between the water column and the oxic sediments in lakes. Geochim. Cosmochim. Acta, 53, 1511-1522.
Tessier A., F. Rapin, and R. Carignan (1985) Trace metals in oxic lake sediments: possible adsorption onto iron oxyhydroxides. Geochim. Cosmochim. Acta, 49, 183-194.
Ugolini, F.C. (1986) Processes and rates of weathering in cold and polar desert environments. In: Rates of chemical weathering of rocks and minerals, St. M. Colman and D.P. Dethier (eds.), Orlando, Florida Academic Press., 193-235.
Vinogradova, A.A. and A.V. Polissar (1995) Element composition of the aerosol in the atmosphere of the Central Russian Arctic. Atmospheric and Oceanic Physics, english translation 31, 2, 248-257.
Westerlund, S.F.G., L.G. Anderson, P.O.J. Hall, Ä. Iverfeldt, M.M. Rutgers van der Loeff and B. Sundby (1986) Benthic fluxes of cadmium, copper, nickel, zinc and lead in coastal environment. Geochim. Cosmochim. Acta, 50, 1289-1296.
White, J.R. and C.T. Driscoll (1985) Lead cycling in an acidic Adirondack lake. Environ. Sci. Technol., 19, 118-1187.
Williams T.M. (1992) Diagenetic metal profiles in recent sediments of a Scottish freshwater loch. Environ. Geol. Water Sci., 20, 117-123.
Wooden, J.L., G.K. Czamanske, V.A. Fedorenko, N.T. Arndt, C. Chauvel, R.M. Bouse, B-S. W. King, R.J. Knight and D.F. Siems (1993) Isotopic and trace element constraints on mantle and crustal contributions to Siberian continental flood basalts, Noril'sk area Siberia. Geochim. Cosmochim. Acta, 57, 3677-3704.

Late Weichselian to Holocene Diatom Succession in a Sediment Core from Lama Lake, Siberia and Presumed Ecological Implications

U. Kienel

Alfred-Wegener-Institut für Polar- und Meeresforschung, Forschungsstelle Potsdam, Telegrafenberg A43, D 14473 Potsdam, Germany

Revised 5 May 1997 and accepted in revised form 15 March 1998

Abstract - The study presents the first lacustrine diatom record spanning a sequence down to the latest Weichselian (ca. 11.000 y BP) from Taymyr Peninsula, NW-Siberia. The aim of this contribution is to assess the development of Lama Lake (NE Norilsk) from the changes in diatom assemblage composition within the context of local pollen assemblages (Hahne and Melles, this volume) and sedimentological/ geochemical data.
Several features of the diatom assemblages of Lama Lake sequence are related to the establishment of microhabitats in marginal areas: stable ‚background' percentages of alpha-meso to eutrophic taxa throughout the sequence, distinctive fluctuations in diversity indices and the record of acidophilus taxa, all of which rely on occurrence patterns of periphytic taxa. The large water body is considered to act as a buffer against environmental changes which are thus less strongly reflected by the diatom assemblages.
From a general agreement between main features of diatom (ratio of euplanktonic/non-planktonic taxa) and pollen (ratio of arboreal/non arboreal pollen) data trends, changes in lake water surface temperature are inferred.
The distinctive maxima of certain diatom taxa and their specific ecological preferences for thermal and trophic conditions, as well as the pronounced succession of planktonic species maxima, enable us to assess trends in these lake parameters. However, a much higher temporal resolution is needed to resolve timing effects in the responses of terrestrial and aquatic ecosystems.

Introduction

The present study is part of the, Taymyr/Severnaya Zemlya' project, which aims at improving our understanding of the late Quaternary environmental history of northwestern Siberia. Various natural data archives have been sampled along a north-south transect running through the Severnaya Zemlya Archipelago, the Taymyr Peninsula and the northwest Putoran Plateau (Norilsk area), which today represent the ecoclimatic zones of polar desert, tundra and forest tundra (Figure 1).

One basic question regarding the late Quaternary history of the region is the eastern extent of the Late Weichselian ice shield in western Siberia. The maximum glaciation theory (Grosswald, 1977) hypothesizes glaciation of the whole Taymyr Peninsula, and the minimum glaciation theory (Velichko et al., 1984) proceeds from a glaciation restricted to the Putoran Plateau. According to Astakhov (1997), retarded deglaciation was the leading factor in the transition to the Holocene in the Russian Arctic, predetermined by the massive Pleistocene cryolithosphere which prevented the formation of rapid deglaciation features. Only mobile biotic components maintained sufficiently the speed of global change and thus adequately reflected climatic fluctuations in the Holocene climate transition.

Lake sediments contain a suite of such biotic components, and thus have great potential for studies of paleoenvironmental change. The sediment record is often of high quality in terms of its variety, richness and temporal resolution (Battarbee, 1991). High latitude aquatic ecosystems, which have adapted to low natural energy flows (temperature, solar radiation), are

In: Kassens, H., H.A. Bauch, I. Dmitrenko, H. Eicken, H.-W. Hubberten, M. Melles, J. Thiede and L. Timokhov (eds.) Land-Ocean Systems in the Siberian Arctic: Dynamics and History. Springer-Verlag, Berlin, 1999, 377-405.

expected to be particularly sensitive to changes in the magnitude and timing of available energy and to changes in physical and geochemical conditions (e.g. Maxwell, 1992; Vincent and Pienitz, 1996).

Besides cladocera and chironomids, diatoms are the most frequently used fossil group in paleolimnology since they tend to be well preserved, diverse, ubiquitous, and identifiable to a low taxonomic level. Because of their short life cycle, they respond quickly to environmental change (Dixit et al., 1992) and, due to the considerable volume of data on their ecological characteristics, they are a sensitive tool in reconstructions of past environmental conditions. In contrast to palynology, diatom analysis provides an autochtonous signal. The assemblage of diatoms preserved in lake sediments can directly reflect the floristic composition and productivity of lake diatom communities, and can indirectly reflect lake water quality in terms of nutrient status, lake water pH and salinity (Battarbee, 1986).

Multivariate statistical methods (including canonical correspondence analysis and weighted averaging techniques as given in e.g. Birks et al., 1990 and ter Braak and Looman, 1986) have been used to infer these parameters from fossil diatom assemblages (e.g. pH: Birks et al., 1990; Stevenson et al., 1991; nutrients: Hall and Smol, 1992; Anderson et al., 1993; Bennion, 1995; Reavie et al., 1995).

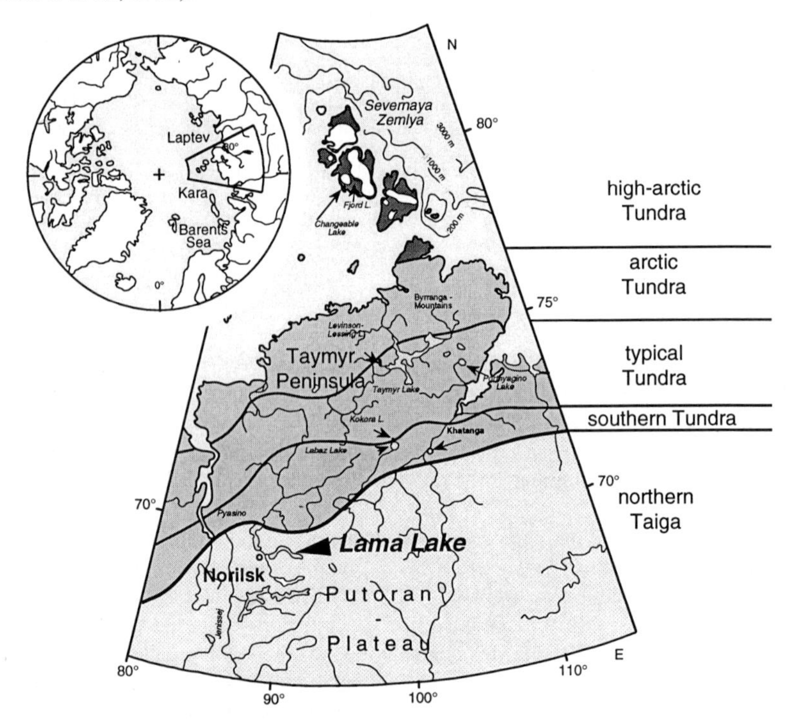

Figure 1: Geographical setting of Lama Lake (NE Norilsk, W Siberia) within the N - S transect spanning Severnaya Zemlya, Taymyr Peninsula and the Norilsk area northwest of Putoran Plateau.

The influence of lake depth, summer surface water temperature and ice cover on the composition of diatom assemblages was recorded from investigations of lakes and ponds arranged along latitudinal and altitudinal transects in arctic and alpine regions (Lemmen et al., 1988; Pienitz, 1993; Wolfe, 1994; Pienitz et al., 1995a, b; Wunsam and Schmidt, 1995; Lotter et al., subm.).

This study presents the first diatom sequence from the northwest Siberian region that spans to the late Weichselian (ca. 11.000 y BP). Little and only very general information about the large lakes of Taymyr Peninsula and the Putoran Plateau has been collected (Aleksyuk and Bekman, 1981; Chernyaeva, 1981; Shur and Sid'ko, 1985). Most of the studies describe modern flora or use lake sediment sequences of less than 1.5 m. In order to infer relative changes in the past development of Lama Lake (northeast Norilsk, northwest Putoran Plateau) from fossil diatom assemblages, the conventional methods using diatom indices and categories of life form, trophic and pH preferences were used. Within the context of palynological (Hahne and Melles, 1997; Hahne and Melles, this volume), sedimentological and geochemical data (Hagedorn et al., this volume; Harwart et al., 1997; Harwart et al., subm.), the diatom-inferred lake history can yield indications of the paleoenvironmental processes influencing Lama Lake and its catchment since the late Weichselian.

Material and methods

Site, Coring

The sediment core PG1111 was recovered during the summer of 1993 from a water depth of 52 m, 2.2 km from the shore of Lama Lake (69°32.9′ N; 90°12.7′ E; Figure 1). To avoid disturbing the soft upper sediments during sampling (PG1111-1, max. depth: 0.54 m), a gravity corer was used. A 10.6 m sediment sequence was recovered by repeated deployment of a piston corer operating from a floating platform. The individual cores from overlapping sediment depths were correlated to the complete sequence on the basis of whole-core physical measurements, core descriptions, and analytical results. For a more detailed description see Melles et al. (1994b).

Chronology, Palynology, Sedimentology and Geochemistry

The chronology for core PG1111 is provided by regional pollen assemblage zones in correlation with radiometrically dated profiles. Together with the combined chronozones of Mangerud et al. (1974) and Khotinsky (1984), they provide the chronological framework (see Hahne and Melles, this volume; Hahne and Melles, 1997).

Generally, grain size remains almost uniform, varying only from silty clay to clayey silt. A major change in sediment colour from olive brown to greyish and black occurs synchronously with a transition in sediment structure from densely laminated to a coarser stratification around 7.2 m depth. According to palynology, this section is placed in the Allerød (Hahne and Melles, 1997). Geochemical analyses of main elements (Harwart et al., 1997) revealed a change from high contents in Fe and Mg to high contents in Si, Al, Ti, Na and K synchronous to the mentioned change in sediment colour and structure. From these results a change in weathering regime is inferred. Prevailing physical weathering is replaced to an increasing degree by chemical weathering processes accompanied by soil formation (Harwart et al., 1997).

Dating

From radioisotopic ^{210}Pb investigations of the uppermost part of PG1111, mean sedimentation rates of 0.62 ± 0.01 mm y^{-1} have been calculated (Hagedorn et al., this volume). ^{14}C dating of the older sediments provided extremely old ages throughout the profile, resulting from the high coal particle content. Even dating of pollen grains was impossible, since not all coal particles could be removed. The chronozones employed, from Oldest Dryas to Subatlantic, are based on comparison of vegetation development with radiometrically dated Siberian pollen profiles (Hahne and Melles, this volume; Hahne and Melles, 1997).

Sampling and laboratory techniques

The sediment core was sampled at least at every 10 cm for diatom analysis. In general, sample processing of 0.5 g of freeze-dried sediment followed the method outlined in Schrader and Gersonde (1978). Briefly, the organic and calcareous compounds were removed by digestion in LÖSOL (washing benzine) followed by boiling with HCl (conc.) and H_2O_2 (30%) at 170°C. Repeated flushing with distilled water (up to 10 times) removed finer particles and residual acid. The remaining suspension was filled to 50 µl and stored in plastic bottles.

Preparation of slides began with 0.4 µl aliquots taken from the stored suspension and followed an evaporation method developed in the laboratory of the Alfred Wegener Institute, Bremerhaven and outlined in Zielinski (1993). For permanent slides, the mounting medium MOUNTEX (n = 1.67) was used.

Counting and Taxonomy

Diatoms were identified using a ZEISS Axioplan microscope (equipped with differential interference contrast) at 1000x magnification. The diatom valves were counted along transects measured by stage micrometer until 40 mm of transect were examined or until at least 400 to 500 valves per sample were counted. In samples extremely poor in diatoms, at least 100 valves were counted. Since aliquots from sample material and suspension were used for all slides, the number of valves encountered per millimetre of microscope transect in each sample gives a semi-quantitative approximation of diatom concentration (Bradbury, 1997). Taxa with less than five occurrences and with a relative maximum abundance of less than 2% were excluded from graphical illustration.

The following literature was used for determination of diatom taxa: Cleve-Euler (1951-1955), Kling and Håkansson (1988), Krammer and Lange-Bertalot (1986, 1988, 1991a, 1991b), Lange-Bertalot (1993), Round et al. (1990).

Table 1: Terminology used for relative abundance of diatom species.

percentage proportion	terminology
1 individual	sparse
< 1%	rare
1 to 5%	few
5 to 10%	frequent
> 10%	abundant

Relative abundance

According to the percentage proportions of species in diatom assemblages, the terminology listed in Table 1 is used.

Sequence splitting

The numerical zonation of the sequence is performed using the CONISS subprogram of the TILIA version 1.10 (stratigraphically CONstrained cluster analysis by the method of Incremental Sum of Squares; Grimm, 1987). Prior to application of the procedure, data are

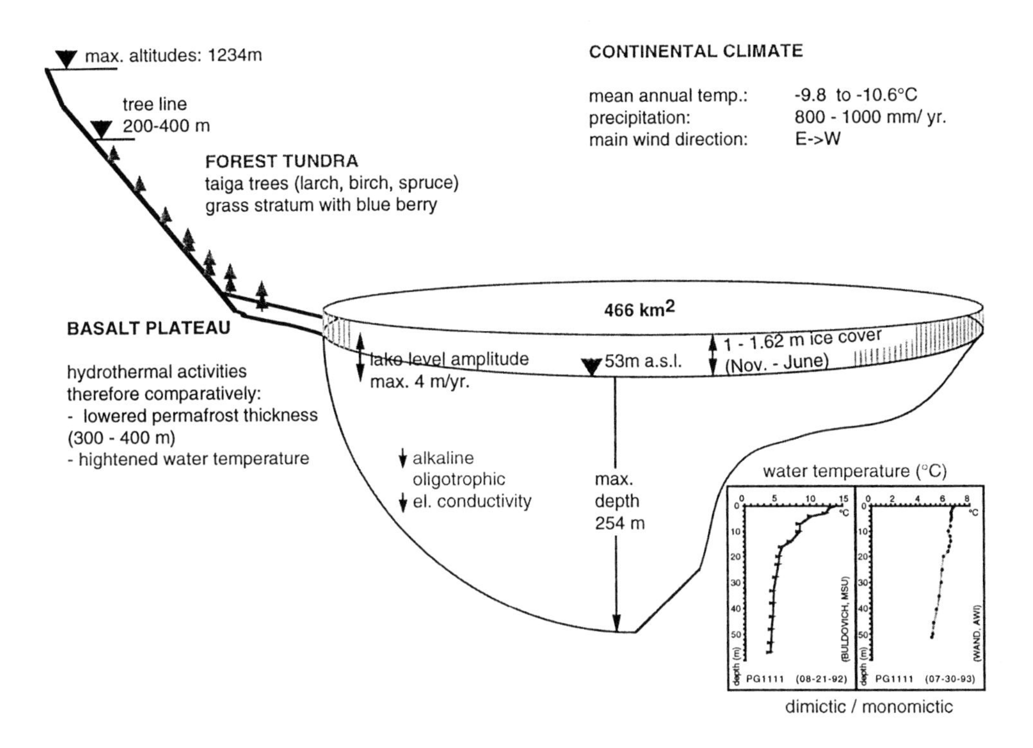

Figure 2: Characteristics of Lama Lake and catchment (NE Norilsk, W Siberia).

culled by eliminating all taxa with maximum percentages below a limiting value of 2.0%. A square root transformation (Edwards and Cavalli-Sforza's square root chord distance) was used as a signal-to-noise ratio Dissimilarity Coefficient - increasing the contribution of minor and less common taxa, but down-weighing abundant types. Finally, a dendrogram (not shown in this publication) was plotted using total dispersion at each stage in order to check the subdivisions (DAZ) of the diatom stratigraphy.

Lama Lake

Lama Lake belongs to the large over-deepened lakes which are of tectonic origin and typical for the northwestern Putoran Plateau, but unique among other subarctic lakes (Galaziy and Paramuzin, 1981). Lake and catchment characteristics are presented in Figure 2. Thermally exceptional conditions, which originate from hydrothermal activities following the uplift of the Basalt Plateau and the formation of stress fractures, have led to the creation of a microclimate. This in turn leads to a lower thickness of permafrost in the catchment area of Lama Lake (300-400 m) as opposed to 700 m generally in the region. Additional results of the microclimatic conditions are the formation of a talik, the development of a thawing zone with higher thickness in the lake's surroundings, and the heightened water temperature.

The lake is bordered by steeply sloped mountains to the north and south, reaching altitudes up to 1234 m, while the shoreline area has gentle morphology. The catchment area is large at 6210 km^2 (Bogdanov, 1985).

The structure and composition of vegetation are not uniform. Vegetation in the immediate Lama Lake surrounding is dense and dominated by shrubs, birch and Taiga trees (larch and less spruce). In the grass-dwarf-shrub layer, grass prevails and blueberry is abundant, and alder and willow dominate in the brushwood. The treeline passes at an altitude between 200 and 400 m a.s.l. The altitudinal succession is not transitional but rather patchy.

Above the tree line, initial soil development is observed while in lower elevations with a closed vegetation cover brown soils were found. In depressions and on beach terraces peat formation (hummocky tundra) occurs (A. Raab, pers. comm.).

Lama Lake has its largest inflow on the northern shore, NW of the sediment sampling site. The lake shore reaches up to 10 m in width and is covered with pebbles, gravel and some sand due to the high annual lake level fluctuation (up to 4 m). During high lake level periods, vegetation near the shoreline is submerged. Sparsely distributed macrophytes (*Chara* sp.) have been observed in well illuminated places.

Temperature profiles (in Figure 2) suggest a thermal regime within the transition from subpolar dimictic to polar monomictic type. The ice free period persists only from early July until late October (Bogdanov, 1985).

The annual maximum of biomass production of the Putoran lakes occurs in fall (August/September) (Aleksyuk and Bekman, 1981; Chernyaeva, 1981; Shur and Sid‘ko, 1985).

According to chemical and hydrological measurements, the lake water is slightly alkaline (Melles et al., 1994), considered to derive from the basaltic bedrock which covers the whole catchment area. Together with the high water volume, this gives the lake a high buffering capacity against external inputs. Conductivity values are around 85 µS/cm. Since the development of the Norilsk smelter complex in the 1930s, concentrations of Cd, Cu, Ni, Pb and Zn in the lake sediments show a progressive increase (Hagedorn et al., this volume).

Results

Diatom record

Diatoms in the sequence are first recorded from a sample taken at 6.68 m sediment depth. Relying on the palynological zonation (Hahne and Melles, this volume; Hahne and Melles, 1997), this first record is placed within the Younger Dryas period.

Over 200 diatom taxa were identified in the Lama Lake core PG1111 (maximum 91 species per sample). Such species-rich assemblages are typical for oligotrophic boreal lakes (e.g. Pienitz and Smol, 1993; Pienitz et al., 1995a; Lange-Bertalot and Metzeltin, 1996). Relative frequencies of the important species are presented in Figures 3 A and B. The Lama Lake sequence is split into eight local diatom assemblage zones (DAZ; see Table 2 and Figure 4).

Diatom assemblage zones (* marks species maxima)

Small centric diatom taxa dominated the diatom assemblages in the Lama Lake sequence. *Cyclotella comensis* was the most abundant diatom species in the sequence associated with *Cyclotella gordonensis* and *Cyclotella* cf. *kuetzingiana* var. *radiosa*. Changes in the main constituents of the planktonic group were remarkable, especially in the lower part of the sequence. In the first assemblages (DAZ-1), *Cyclotella rossii* (*) was the most abundant planktonic diatom co-occurring with frequent *Aulacoseira islandica* (*). Subsequently, in (DAZ-2), a maximum abundance of *C. gordonensis* (*) was observed. In DAZ-3, the proportions of *Cyclotella comensis* and *Cyclotella* cf. *kuetzingiana* var. *radiosa* increased. A

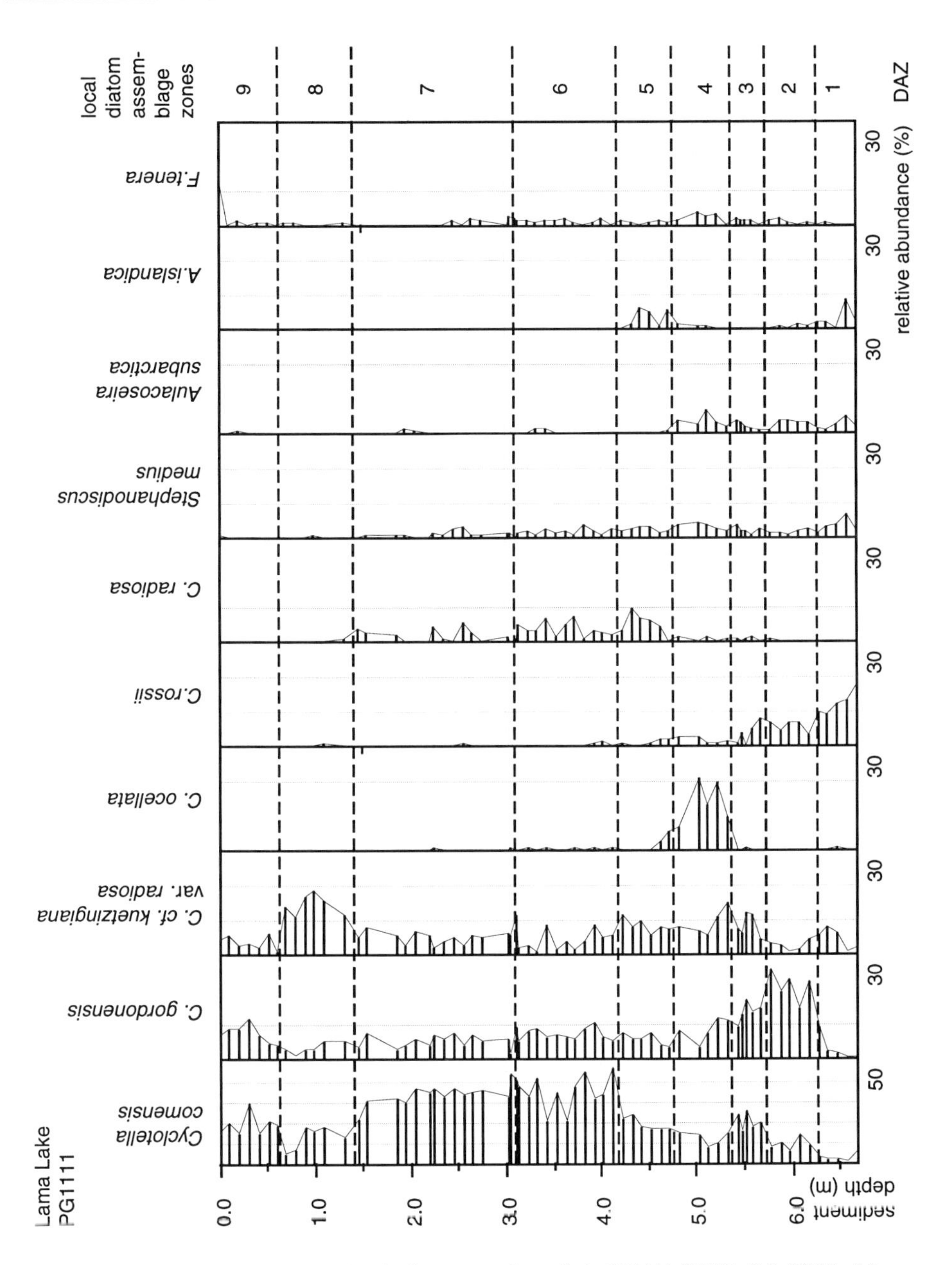

Figure 3A: Relative abundance of euplanktonic diatom taxa - Lama Lake PG1111 (NE Norilsk, W Siberia).

distinctive maximum of *C. ocellata* (*), unique in the sequence and associated with frequent *Aulacoseira subarctica* (*), marked DAZ-4. From DAZ-5 to DAZ-8, *Cyclotella comensis* dominated the assemblages. *C. radiosa* (*) became frequent in co-occurrence with *Aulacoseira islandica* only in DAZ-5. In DAZ-6, *Cyclotella comensis* showed the highest abundance in the sequence, though strongly fluctuating. A uniform ratio of taxa in the planktonic group (mainly

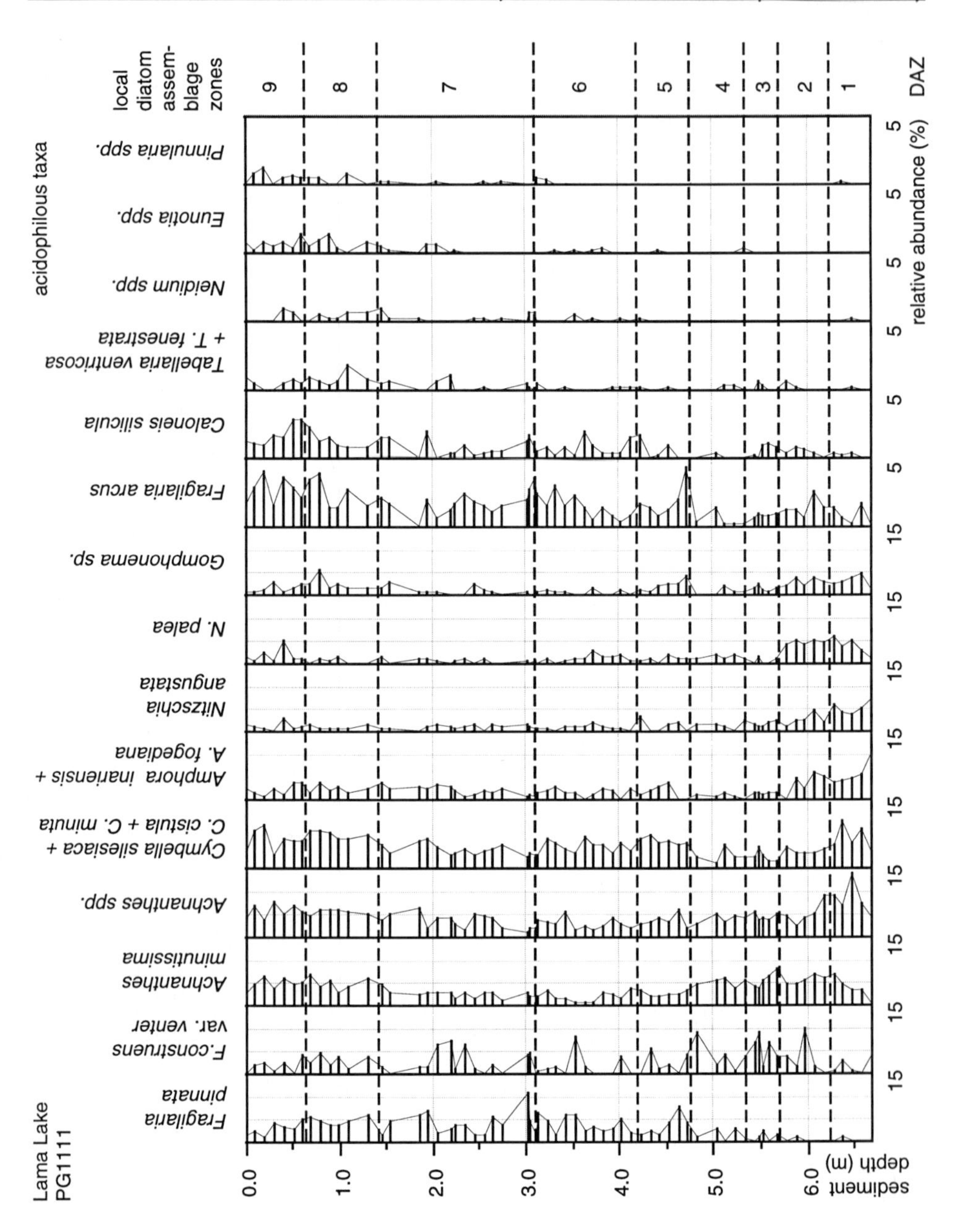

Figure 3B: Relative abundance of tychoplanktonic and non-planktonic (acidophilous) diatom taxa - Lama Lake PG1111 (NE Norilsk, W Siberia).

C. comensis and *C. gordonensis*) characterized DAZ-7. In DAZ-8, *C. kuetzingiana* var. *radiosa* and *C. comensis* contributed 20 % each to the diatom assemblage. Strongly fluctuating proportions of taxa in the planktonic group marked DAZ-9, terminating with a maximum frequency of *Fragilaria tenera* (13 %).

Among the periphytic taxa, *Cymbella minuta* and *C. silesiaca* (abundant), *Achnanthes minutissima* var. *minutissima* (abundant to frequent), *Amphora inariensis, A. fogediana, Fragilaria construens and F. pinnata* (frequent) were present throughout the sequence. Periphytic taxa dominated in the diatom assemblages of the DAZ-1, DAZ-2, DAZ-7 and DAZ-8.

In DAZ-1, the periphytic assemblage consisted of abundant small *Achnanthes* species (*A. calcar, A. clevei, A. oblongella, A. laevis*), *Cymbella* species (*C. minuta**, *C. silesiaca**), *Amphora* species (*A. inariensis**, *A. fogediana*) and *Nitzschia angustata**. Abundant *Achnanthes minutissima* var. *minutissima, Fragilaria construens** and *Nitzschia palea**, together with the frequent small *Achnanthes* species mentioned above, *Amphora inariensis** and *A. fogediana* accounted for the high percentage of periphytic diatoms in DAZ-2.

Again in DAZ-8 and 9, the small *Achnanthes* species and *A. minutissima* var. *minutissima* became abundant and predominated the periphytic diatom assemblages in association with *Cymbella minuta* and *C. silesiaca*. They were associated with few *Gomphonema pumilum, G. olivaceum, Nitzschia palea, Fragilaria arcus* (*) and *Caloneis silesiaca* (*). Taxa belonging to the genera *Eunotia, Neidium* and *Pinnularia*, together with *Tabellaria fenestrata* and *T. ventricosa* comprised up to 3% of the assemblages.

Diatom concentration

Semi-quantitative approximations of diatom valve and chrysophycean cyst concentrations are given in Figure 5. Both parameters show generally similar trends, whilst the ratio of diatom sversus chrysophycean cysts increases upwards in the sequence. Only two significant decreases are visible in DAZ-8.

Table 2: Range of local diatom assemblage zones in PG1111, Lama Lake (NE Norilsk, NW-Siberia).

Diatom assemblage zone	depth range
DAZ-9	(0.70 - 0 m)
DAZ-8	(1.30 - 0.70 m)
DAZ-7	(3.10 - 1.30 m)
DAZ-6	(4.20 - 3.10 m)
DAZ-5	(4.75 - 4.20 m)
DAZ-4	(5.30 - 4.75 m)
DAZ-3	(5.70 - 5.30 m)
DAZ-2	(6.25 - 5.70 m)
DAZ-1	(6.68 - 6.25 m)

Diatom concentrations increased strongly from the first record until they reached a first maximum in DAZ-3. A steep increase from 30 to 140 valves per millimetre of microscope transect (v/mm), which represents the maximum concentration in the sequence, was observed in DAZ-4. An abrupt decrease to 10 v/mm along with deteriorating preservation marks the onset of DAZ-5. Subsequent diatom concentrations tended to increase again until a peak of 80 v/mm marked the transition from DAZ-6 to 7. Comparatively high concentrations with 85 v/mm were reached later in this zone. From this point, a general decrease in diatom concentration was recorded. DAZ-8 and 9 were characterized by low values approximating 20 v/mm.

Diversity

Diversity is a measure of the organizational level of a biocoenosis. The commonly used diversity indices - number of species (S), Shannon-Weaver index (H‘), and the related evenness value (E) - have been calculated for the recorded diatom assemblages (Figure 6). Since the taxonomical diversity of a biostratigraphic sample is a function of the counting sum, this parameter is included as well. Lower counting sums than the recommended 400 to 500 valves (Battarbee, 1980) are derived from samples with low diatom concentrations and probably result in some extreme values in the diversity indices.

The most common diversity index is the number of species per sample S (see Figure 6). The maximum number is reached in DAZ-8 with a total of 91 species. Planktonic taxa contribute a maximum of 23 species (DAZ-4). The minimum of 10 planktonic species was found in the DAZ-1 and 9. An assemblage with 36 species (Smin) was recorded from a sample at 4.72 m (DAZ-5), along with the lowest diatom concentrations in the sequence and strong indications of dissolution. Since it is difficult to determine the real number of species, the commonly used diversity index according to Shannon and Weaver (1949) H‘ has been calculated as follows:

$$H' = -\sum_{i=1}^{s} (p_i \cdot \log p_i) \quad (1)$$

where s is the total number of species, and pi is the relative abundance of species i. H‘ measures the relative proportions of the single species to each other. It tends to increase as more species occur in similar proportions. In contrast to other diversity indices, the Shannon-Weaver index is sensitive to rare species (Kohmann and Schmedtje, 1986) which is considered to be advantageous in the species-rich assemblage recorded from the Lama Lake sequence.

According to the calculated H‘ values the diatom assemblages in the Lama Lake sequence reached their greatest diversity value H‘max=3.72 in DAZ-1 and DAZ-2 and a second maximum H‘=3.69 in DAZ-7. The diversity was lowest in a sample from 4.12 m sediment depth (H‘min=2.54, low DAZ-6).

In addition, the evenness E of the distribution of individuals with respect to species has been calculated as follows:

$$E = H'/H_{max} = H'/\log s \quad (2)$$

Highest evenness values approximating 1 are reached when all individuals are evenly distributed with respect to species. Consequently, E is a measure for the deviation of the given diversity of the assemblage from the optimum (Kohmann and Schmedtje, 1986).

The highest evenness values were calculated for samples at 4.72 m (Emax=0.895) and 4.62 m (E=0.863) sediment depth in the lower part of DAZ-5. High values ranging from 0.76 to 0.87 were calculated for DAZ-1 and 2. They were similarly high in DAZ-8 and 9. Lowest evenness (E=0.616) was reached in the sample at 4.12 m sediment depth, which showed the lowest diversity as well.

Life form

It is generally considered difficult to subdivide periphytic taxa in sediment assemblages into detailed life form groups, since few taxa show exclusive preferences. A general distinction between planktonic and periphytic diatom taxa, however, can usually be made (Battarbee, 1986).

Some taxa often found in plankton can also occur in the periphyton (e.g. *Tabellaria flocculosa, Cyclotella ocellata*). The small chain-forming *Fragilaria* species (such as *F. pinnata* and *F. construens*) are found attached only passively to the substrate and can thus

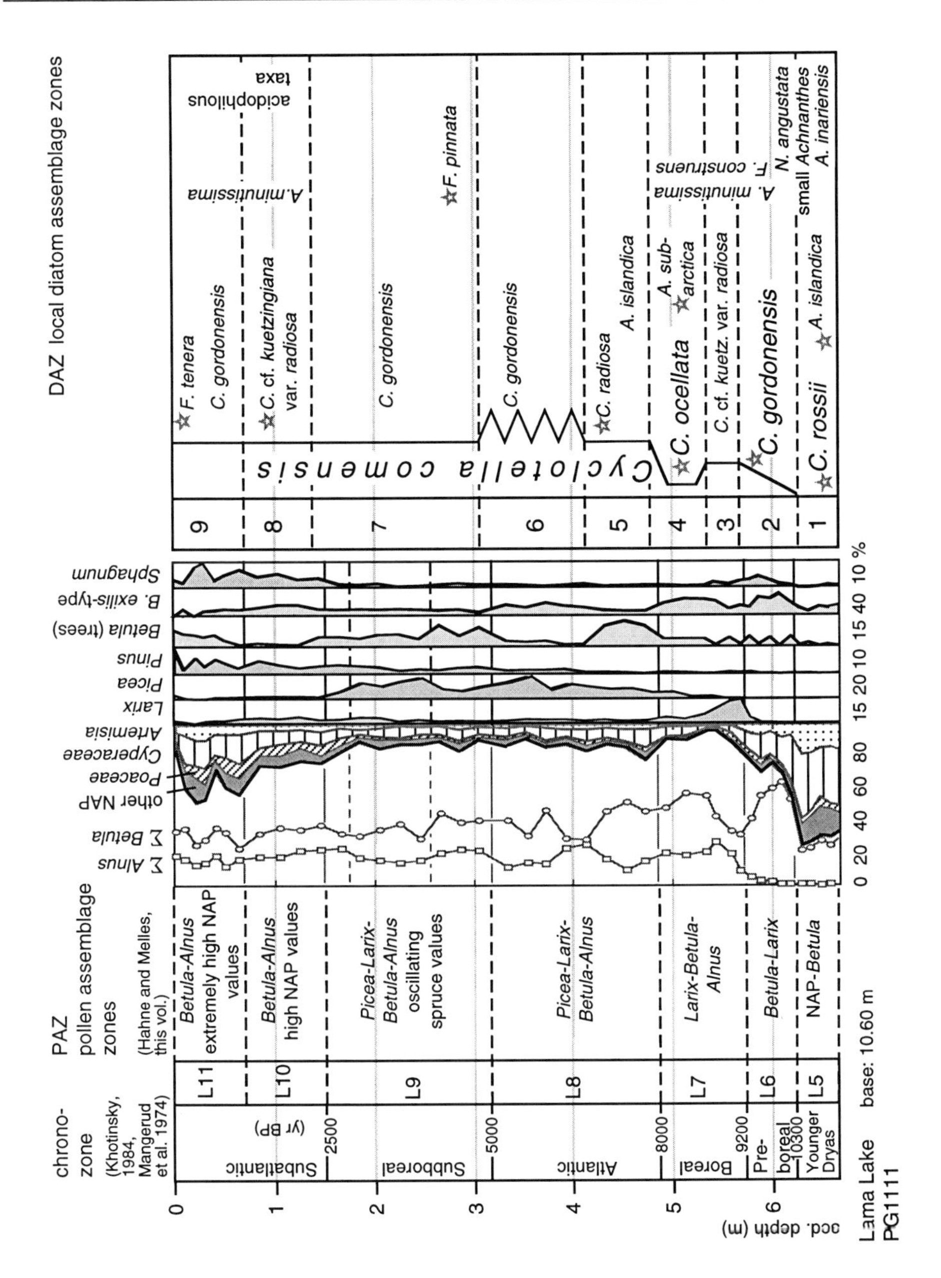

Figure 4: Local diatom assemblage zones in compilation with lithology, chronozones and pollen assemblage zones (* mark location of relative maximum abundances) - Lama Lake PG1111 (NE Norilsk, W Siberia).

easily drift and occur in the plankton. The life form of such species is often described as ‚tychoplanktonic' (see extensive review in Voigt, 1997).

The proportions of the life form groups (based on relative species abundances) in the diatom assemblages, especially the ratio of euplanktonic to non-planktonic diatom taxa (Ep/Np), can be used to clearly subdivide the Lama Lake sequence into three major units (Figure 7).

The first unit with non-planktonic taxa comprising over 50% of the total assemblage coincided in range roughly with DAZ-1. The Ep/Np ratio reached its lowest value (0.46) in the sequence.

The second unit, including the DAZ-2 to 7, was characterized by Ep/Np ratio values above 1. Tychoplanktonic taxa occurred with a relative maximum (33.4%) in DAZ-4 concomitant with the relative maximum of euplanktonic taxa (72.1%). „True„ planktonic taxa, however, reached a maximum in relative abundance in DAZ-3, terminating a strong increase following the proportional maximum of non-planktonic taxa (68.4%) in DAZ-1. After a retreat in DAZ-5 (with values approximating 1), the Ep/Np-ratio increased again in DAZ-6 and had values around 2. A steady decrease to finally equal proportions of euplanktonic and non-planktonic taxa marked DAZ-7. In the third unit, comprising the DAZ-8 and 9, this general trend continued towards assemblages dominated by non-planktonic taxa. This trend was interrupted shortly in DAZ-9 and in the uppermost part the Ep/Np ratio increased, mostly as a result of higher proportions of tychoplanktonic taxa.

Trophic indications

Nitrogen and phosphorus are the most important nutrients governing algal growth (e.g. Hecky and Kilham, 1988). Nutrient enrichment leads to well-defined functional responses of the plankton species present and many of these changes have an autecological basis, as indicated by numerous phytoplankton monitoring and experimental studies (e.g. Reynolds, 1984; Sommer, 1990; van Donk and Kilham, 1990).

The nutrient ratio and limitations (Tilman et al., 1982) and the availability of dissolved silica regulate the composition of the phytoplankton community (species composition). Several studies showed a ‚resource-based competition‘ under changing nutrient ratios (e.g. Tilman et al. 1982; Kilham and Kilham, 1984). For instance, the availability of dissolved silica (e.g. Bradbury, 1975; Bradbury, 1988; Conley and Schelske 1993; Reavie et al., 1995) and phosphorus enrichment (e.g. Carney, 1982; Anderson, 1989; Stoermer, 1993; Sabater and Haworth, 1995), processes coupled with lowering Si:P ratios during lake eutrophication, are responsible for a characteristic replacement of plankton species groups. Pappas and Stoermer (1995) found that inorganic nitrogen enrichment in Lake Huron affected the phytoplankton species composition as well. Most remarkable was an increase in relative abundance of *Cyclotella comensis*.

A number of methods have been developed to infer past trophic conditions in lakes from diatoms. Some examples include the use of indicator species (Battarbee, 1978; Brugam, 1988), ratios such as Araphidineae : Centricaceae (Stockner, 1971) or Centrales : Pennales (Nygaard, 1956), and indices (e.g. diatom inferred trophic index by Agbeti and Dickman, 1989). In recent years, statistical models including ordination methods and weighted-averaging regression and calibration have been developed. Finally, diatom-based transfer functions enabling the reconstruction of lake trophic status have been generated (e.g. Agbeti, 1992; Anderson et al., 1993; Bennion, 1995; Hall and Smol, 1992; Reavie et al., 1995).

In order to assess relative changes in past trophic conditions in Lama Lake, the diatom taxa have been grouped according to the concept of trophic tolerance groups (Hofmann, 1994), based on the Vollenweider model (Vollenweider, 1979), which describes the freshwater trophic status as a function of total phosphorus (TP) concentration. Additional data from the literature (e.g. Stoermer, 1993; Stoermer et al., 1985; Stoermer et al., 1996), an extensive compilation in Voigt (1997) and TP tolerances from Reavie et al. (1995) and Bennion (1995) were included. The relative abundance data of the observed diatom taxa are placed in five categories (Figure 8).

The proportion of tolerant taxa is higher in DAZ-1. Their highest proportion in DAZ-8

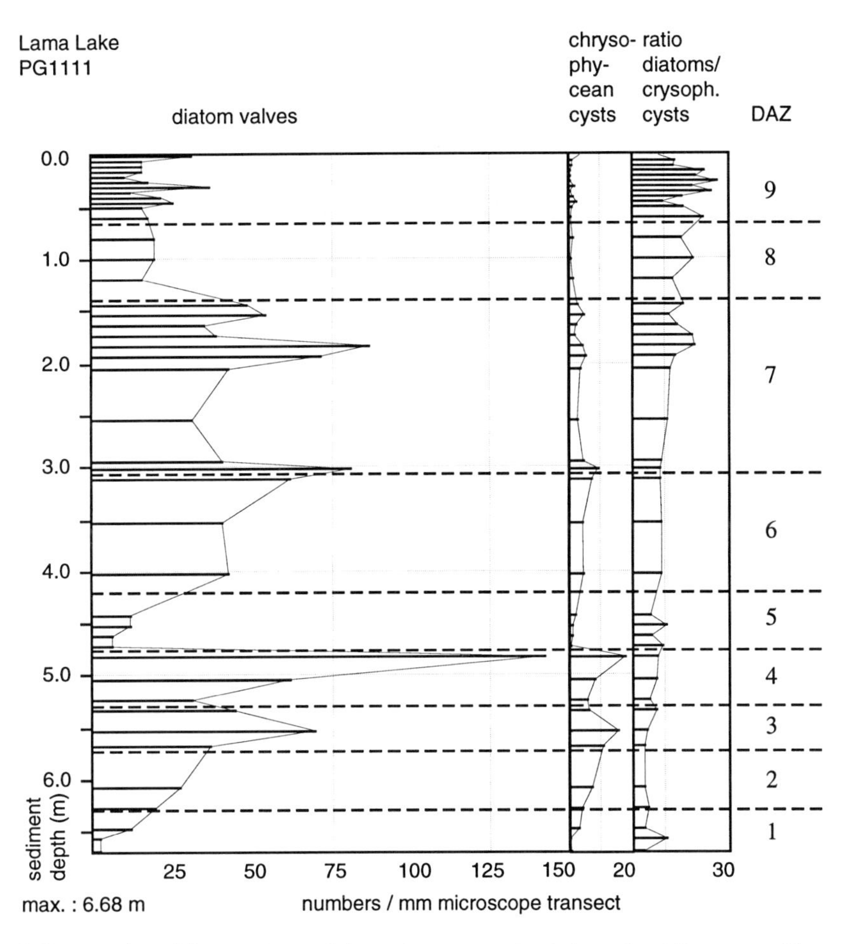

Figure 5: Concentration of diatom valves and chrysophycean cysts in relation to local diatom assemblage zones (DAZ) - Lama Lake PG1111 (NE Norilsk, W Siberia).

coincided with very unstable abundances of the taxa placed in the oligotrophic and oligo/beta mesotrophic tolerance groups. The taxa grouped in the oligotrophic tolerance group comprise 20 to 30 % of the assemblages in DAZ-1. Their proportions increased to a maximum level of 40 % in DAZ-2 and dropped to a relatively stable value of 10 % for the remaining sequence. Only in DAZ-9 proportions around 15 % were reached. Most abundant were taxa placed in the oligo/beta mesotrophic tolerance group. A remarkable increase from percentages approximating 20 in DAZ-1 towards values of 45 % was recorded for DAZ-3. After a drop to 30 % in DAZ-4, percentages increased again to values between 40 and 50 % in DAZ-6 and 7. For DAZ-8 and 9, strongly fluctuating percentages around 30 % were recorded. Taxa placed in the oligo/alpha mesotrophic tolerance group comprised on average 10 % of the assemblages. A distinctive proportion of taxa with higher trophic requirements (alpha-meso/eutrophic group) occurred in DAZ-4 (20 %). Background levels were on the order of 7 %.

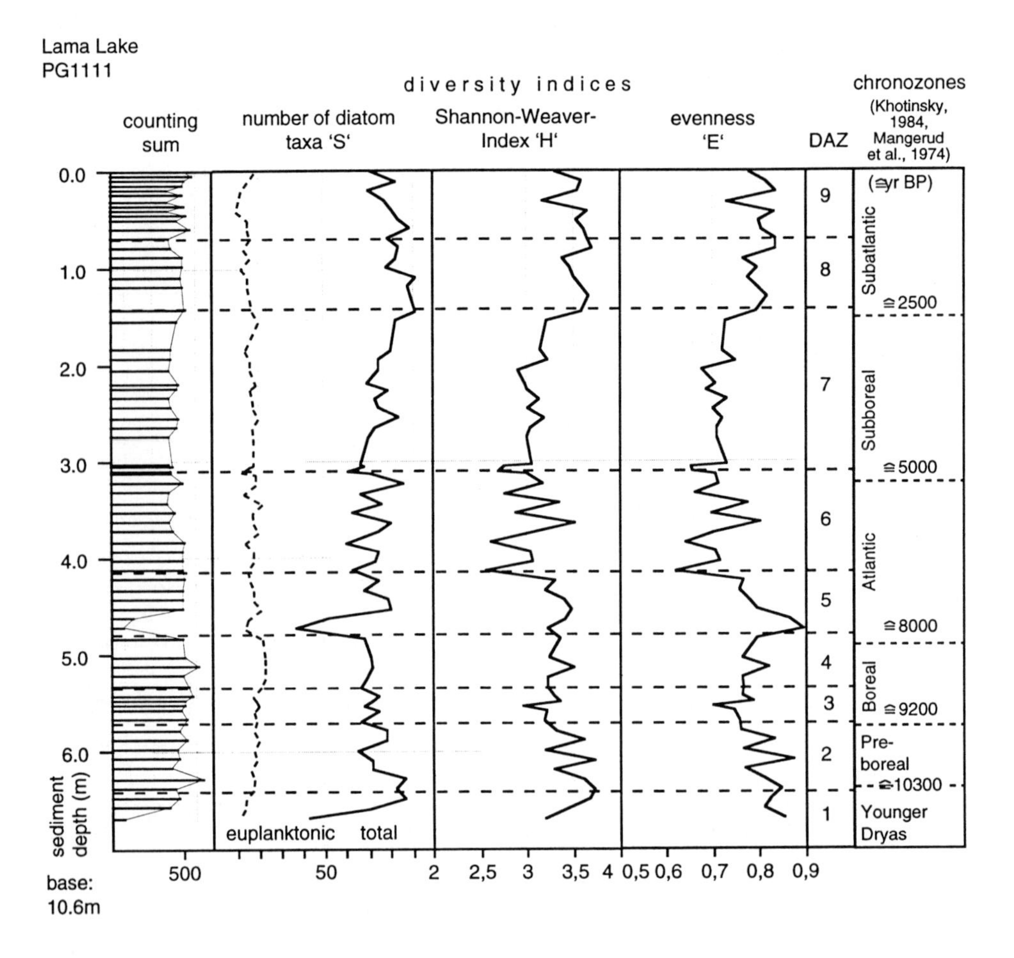

Figure 6: Compilation of diatom counting sum and diversity indices in relation to chronozones and local diatom assemblage zones (DAZ) - Lama Lake PG1111 (NE Norilsk, W Siberia).

pH-indication

Diatoms are sensitive indicators of lake water pH (Battarbee, 1986). A variety of numerical procedures have been developed for quantitative inference of pH from fossil diatom assemblages. Indices by Nygaard (1956) and improved by Renberg and Hellberg (1982) were based on Hustedt's pH spectrum (Hustedt, 1937-39) and have been widely used. The application of several statistical techniques (e.g. Birks et al., 1990) eliminated the subjectivity of preference grouping.

The conventional preference grouping using the Hustedt classification (Hustedt, 1937-39) has been applied to the available data in this study in order to assess trends in Lama Lake water pH (Figure 9). Again, additional data from the literature (e.g. Gasse et al, 1995; Korsman and Birks, 1996; Stevenson et al., 1991, Voigt, 1997) were included.

Alkalibiontic taxa comprise nearly constantly 5 % of the diatom assemblages. The highest proportions of alkaliphilous taxa were recorded for the earliest diatom assemblages (47 %) and in DAZ-4 (45 %), in both cases followed by a strong decrease. From DAZ-6 upwards, they amount on average to 30 % of the assemblages. Taxa preferring circumneutral conditions generally prevailed in the diatom assemblages in Lama Lake. Comparatively low percentages (20 to 30 %) were distinctive in the first recorded assemblages only, coinciding with the maximum in alkaliphilous taxa. Their proportion tended to increase until DAZ-6 where they occasionally formed 60 % of the assemblage. Throughout the zones DAZ-7 to 8 the percentage was relatively stable around 50 %. From the upper DAZ-8 on, the proportion of circumneutral taxa tended to decrease to 35 %. Taxa classified as acidophilous contributed to the assemblages only in DAZ 8 and 9, comprising about 3 %.

Discussion

Diatom occurrence

No diatoms have been found in samples below a sediment depth of 6.68 m. Judging from the basic ecological requirements of diatoms, there are some factors that may inhibit diatom establishment, but there are counter-indications for some.

1) Light limitation due to a high particle content in the water column caused by melt-water inflows and turbulence during thawing periods (Patrick, 1977; Sabater and Haworth, 1995).

2) Light and temperature limitation: due to snow-covered (or very thick) ice cover (Doubleday et al., 1994), contradicted by the presence of pollen in the whole section (Hahne and Melles, this volume; Hahne and Melles, 1997).

3) pH-limitation: A pH above 8.5 is considered to be critical for diatom establishment (Patrick, 1977). Terrestrial input of weathering products from the basaltic bedrock of Lama Lake and its catchment area can increase lake water alkalinity especially during times of active runoff.

4) Nutrient limitation due to a strong restriction of runoff processes and weathering and consequently a restriction of external nutrient supply during cold an dry conditions.

On the other hand, diatom preservation is strongly affected by dissolution processes. Dissolution of the frustules starts immediately after the death of the organism in the water column, since lake water is undersaturated in dissolved SiO_2. The degree of dissolution is found to increase with raising pH and temperatures (Lewin, 1961), their dwelling time in the water column, a lower silicification of the frustules and an increasing surface to volume ratio (Hurd and Birdwhistell, 1983). Dissolution affects diatom assemblages in such a way that small-sized forms with a low SA/V ratio and strongly silicified parts of the larger forms are left (Barker, 1992). Post-depositional dissolution has been related to for example turbulence-induced intense mixing of the water body accompanied by oxygenation of the bottom waters (McMinn, 1995), dissolved Si-diffusion rates (Rippey, 1983) and the silica content of the pore water (Flower, 1993).

Corroboration for dissolution's primary role in causing the lack of diatoms at least in the lower part of the diatom-bearing sequence is provided by the poor preservational stage of the earliest recorded diatoms. The predominant diatom species in the first assemblage (e.g. *Cyclotella rossii, Cymbella silesiaca, Fragilaria construens, Nitzschia angustata,* see Figures. 3 A and B) show alkaliphilous preferences, have small robust forms, and thus may be considered a dissolution-reduced assemblage (Barker, 1992).

After compiling and evaluating the above data, it appears that a combination of two processes which are difficult to disentangle may explain the absence of diatoms: 1) Non-establishment due to light deficiency coupled with low temperature, high pH and a lack of nutrients, and 2) non-preservation as a result of dissolution originating from pore water chemistry (high alkalinity, undersaturation in silica) and/or turbulence-induced oxygenation processes.

Within-basin variations

There are striking peculiarities of the diatom flora of larger lakes that must be regarded, emphasized for instance by Stoermer (1993) from studies on the Great Lakes. The timing and magnitude of limnological and biological responses to changes in temperature, precipitation, runoff processes, external input, and nutrient content, for example, cannot necessarily be assumed by analogy from the smaller lakes. Within-basin variations derive from the establishment of microhabitats and factors influencing the deposition of diatoms such as mixing regime, bottom morphology and transport by bottom currents (Thayer et al., 1983) and top-down effects (Stoermer, 1993; Schelske, 1994).

For instance, the partly inverse trend in the curves of evenness and relative abundance of euplanktonic diatoms could be the result of such variations. The observed differences in diatom assemblage composition and species number result mainly from changes in rare periphytic species. The marginal areas they inhabit are more strongly influenced by variations in ecological parameters; microhabitats are therefore easily formed.

Also, the background level of taxa placed in the alpha-meso/eutrophic tolerance group (on the order of 7 %) is considered to derive from such microhabitats (e.g. shallow regions with higher external nutrient input). This is substantiated by the generally non-planktonic life form of the contributing taxa (e.g. *Achnanthes clevei, Meridion circulare, Diatoms tenuis, Navicula reinhardtii, Nitzschia pseudofonticola, Cymatopleura solea, Rhopalodia gibba, Nitzschia bacillum, Surirella* sp.)

In order to investigate if and how such processes cause within-basin variations and influence the representativity of diatom counts from sediment samples, the study of numerous surface samples, plankton, and benthos samples is underway.

Ecological implications

Diatoms and pollen

The comparison of changes in pollen and diatom assemblages provides useful information concerning the character of environmental changes. The timing of reaction of aquatic and terrestrial ecosystems to changes is determined by different population dynamics, taphonomic processes and the character of the change. The classical concept proceeds from a minor time lag in the reaction of aquatic plants to rapid environmental changes, to terrestrial plants, where such factors as pedogenesis and migration are important (Iversen, 1954; West, 1964; Wright, 1984; Birks, 1986). Long-term climate changes, however, are found to cause a more or less synchronous behaviour of both systems (Lotter et al., 1992; MacDonald et al., 1993; Lotter et al., 1995). A much higher temporal resolution in terms of sample distance is needed to resolve such timing effects.

Nevertheless, from the evident parallels in the development of organisms inhabiting the two ecosystems, some basic conclusions concerning the influencing environmental variables can be drawn. Near the arctic treeline the relative abundance curve of arboreal pollen can indicate the climate, corroborated as well by the correlation with the $\delta^{18}O$ - curve recorded from the Academy of Sciences Glacier on Severnaya Zemlya (Klementev et al., 1991; see Hahne and Melles, this volume). In this context, the good correlation of the curves of relative abundance of arboreal pollen and euplanktonic diatoms is interesting (Figure 10). Pollen counts, however, do not exist for all samples with diatom records, resulting in a deficiency of this compilation. Thus, the calculated correlation coefficient of 0.83 may be too high.

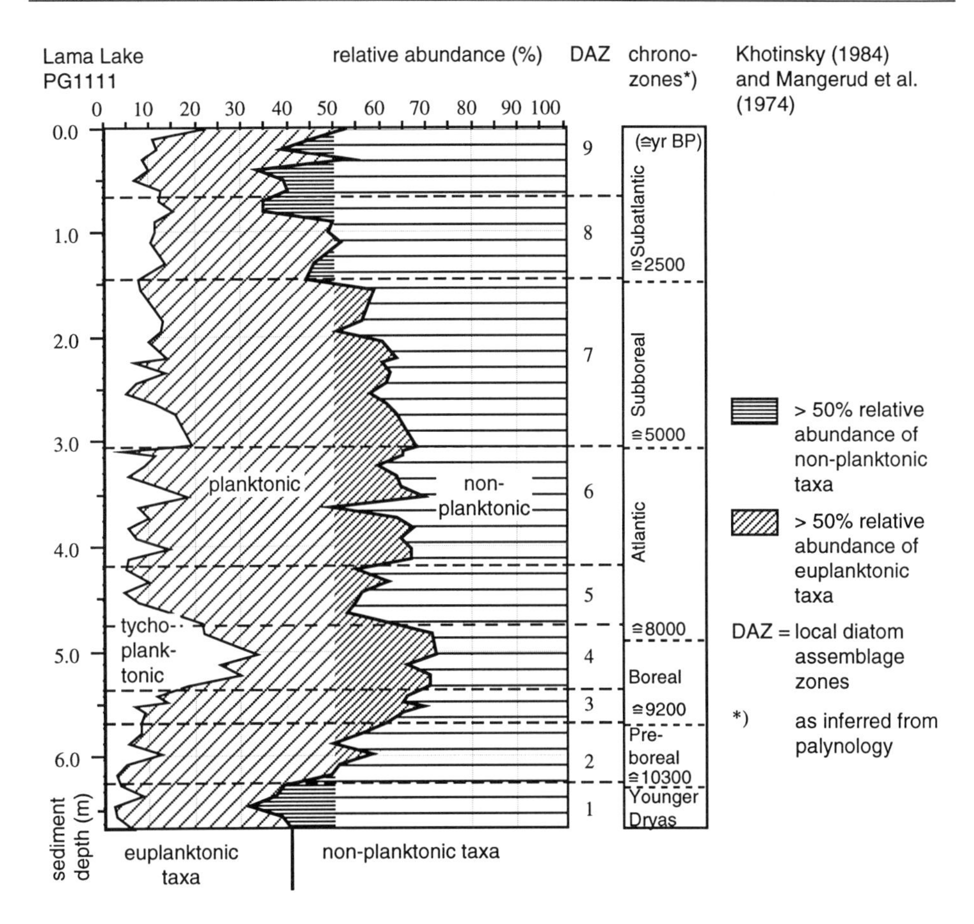

Figure 7: Ratio of euplanktonic / non-planktonic diatom taxa - Lama Lake PG1111 (NE Norilsk, W Siberia).

Diatom life form and trophic implications

In diatom analysis, the changes in the euplanktonic/periphytic ratio were conventionally interpreted in two ways, as the result of a change in the morphology of the lake basin as the lake level changed and/or productivity (Battarbee, 1986).

A lake level lowering is considered to be reflected by low Ep/Np ratios. Similarly, low Ep/Np ratios could result from material transport towards the centre of the lake or an increase in area of shallower marginal regions, since this creates large areas suitable for macrophyte growth, which in turn provides substrates for benthic/periphytic diatom colonization (Bennion, 1995). Sedimentological investigations (Harwart et al., subm.) revealed no indications for rapid changes in lake level or external material input in the diatom-bearing section of the sequence. Thus, lake level-determined changes in the Ep/NP ratio appear subordinated.

Lake summer surface temperature strongly influences the composition of diatom assemblages in arctic and alpine regions (Lotter et al., subm.; Pienitz, 1993; Pienitz et al., 1995; Wunsam et al., 1995, Weckström et al., 1997). In high and mid-arctic sites, the extent of snow and ice cover, and especially the timing of ice break-up and stratification, strongly determine the distribution, growth and composition of diatom assemblages (e.g. Lemmen et al., 1988; Wolfe, 1994). During cold periods with a long ice-cover season, when lakes have only a small „moat„

of open water (Smol, 1988), shallow water or periphytic taxa are favoured. Planktonic taxa characteristic for deeper water appear to be more abundant during warmer periods (see also Pienitz et al., 1995; Wunsam et al., 1995).

Ice cover, summer surface temperature and thermal regime have an overwhelming influence on overall diatom assemblage in polar regions (Smol, 1988), while shifts in individual taxa within these communities may reflect other limnological variables.

Taking these facts into consideration, prominent changes in the ratio of euplanktonic to non-planktonic taxa (Ep/Np) were used to subdivide the Lama Lake sequence (Figure 7). Three sections were clearly distinguished and may indicate subunits in terms of the lake‘s average annual surface temperature and the amplitude of its fluctuation. These parameters are connected with the duration of the ice-cover season and the character of mixing regime.

The first subunit with Ep/Np values below 1 coincided in range roughly with DAZ-1. Hahne and Melles (1997) record high NAP values (non-arboreal pollen), mainly *Artemisia* and *Cyperaceae* and place this unit in the Younger Dryas chronozone (Khotinsky, 1984). The most distinctive feature in the diatom assemblages, besides the high percentage of small periphytic diatoms and the predominance of taxa with alkaliphilous and oligotrophic affinities, is a planktonic complex unique in the whole sequence, consisting of *Cyclotella rossii, Aulacoseira islandica* and *A. subarctica*. The temperature tolerance of *Cyclotella rossii*, accounting for 20% of the diatom assemblage, is high. In a transect spanning subarctic Fennoscandian lakes, it is placed at the lower end of the temperature range (Weckström et al., 1997) while in the North Canadian training set (Pienitz et al., 1995) *C. rossii* increases in abundance with temperature. For the co-occurring *Aulacoseira islandica* and *A. subarctica,* growth under low light and temperature conditions, survival in unfavourable conditions through the formation of resting spores, and a start of annual production under the ice (resulting in a strong vernal peak) are recorded (e.g. Popovskaya, 1977; Stoermer, 1993). Especially in large turbulent lakes (e.g. the Great Lakes, the large Swedish lakes, Lake Baikal), *A. islandica* is observed in considerable abundance (Popovskaya, 1977; Stoermer, 1983; Willén, 1984). Due to the meroplanktonic life form, *Aulacoseira* sp. are favoured by turbulence in the water column (Lund, 1954).

In the second subunit, the Ep/Np values exceeded 1 and reached maximum values of 2.5, meaning a predominance of euplanktonic diatoms. This section comprises the pollen assemblage zones L6 to L9, related to the chronozones Preboreal, Boreal, Atlantic and Subboreal (Hahne and Melles, 1997).

Strong fluctuations of the Ep/Np values along with a distinctive species succession within the euplanktonic complex suggest further subdivisions described by the applied local diatom assemblage zones (DAZ). As emphasized by Smol (1988), such shifts in individual taxa within the diatom communities in polar regions may well reflect changes in environmental variables, such as pH, nutrient content or conductivity, whose influence in such climates are weaker than those of climate and ice cover. The distinctive maxima of individual planktonic diatom taxa could be related to conditions favouring the growth of certain species. When grouping these abundant species according to their ecological preferences, the proportion of the groupings in which they are included is consequently determined to a high degree. Thus, it is considered reasonable to discuss changes in the nutrient content of Lama Lake with respect to the species succession within the planktonic complex (see section on trophy above). Strong similarities to changes in catchment vegetation, as recorded from palynological investigations (Hahne and Melles, 1997), are evident.

Parallel to the Preboreal warming, indicated by a rapid increase of arboreal pollen mainly of the *Betula exilis*-type, diatom concentration increases and planktonic taxa progressively predominate the diatom assemblages in DAZ-2. *Cyclotella gordonensis*, a small-sized cool-water species found in lakes extremely poor in nutrients (Kling and Håkansson, 1988), shows

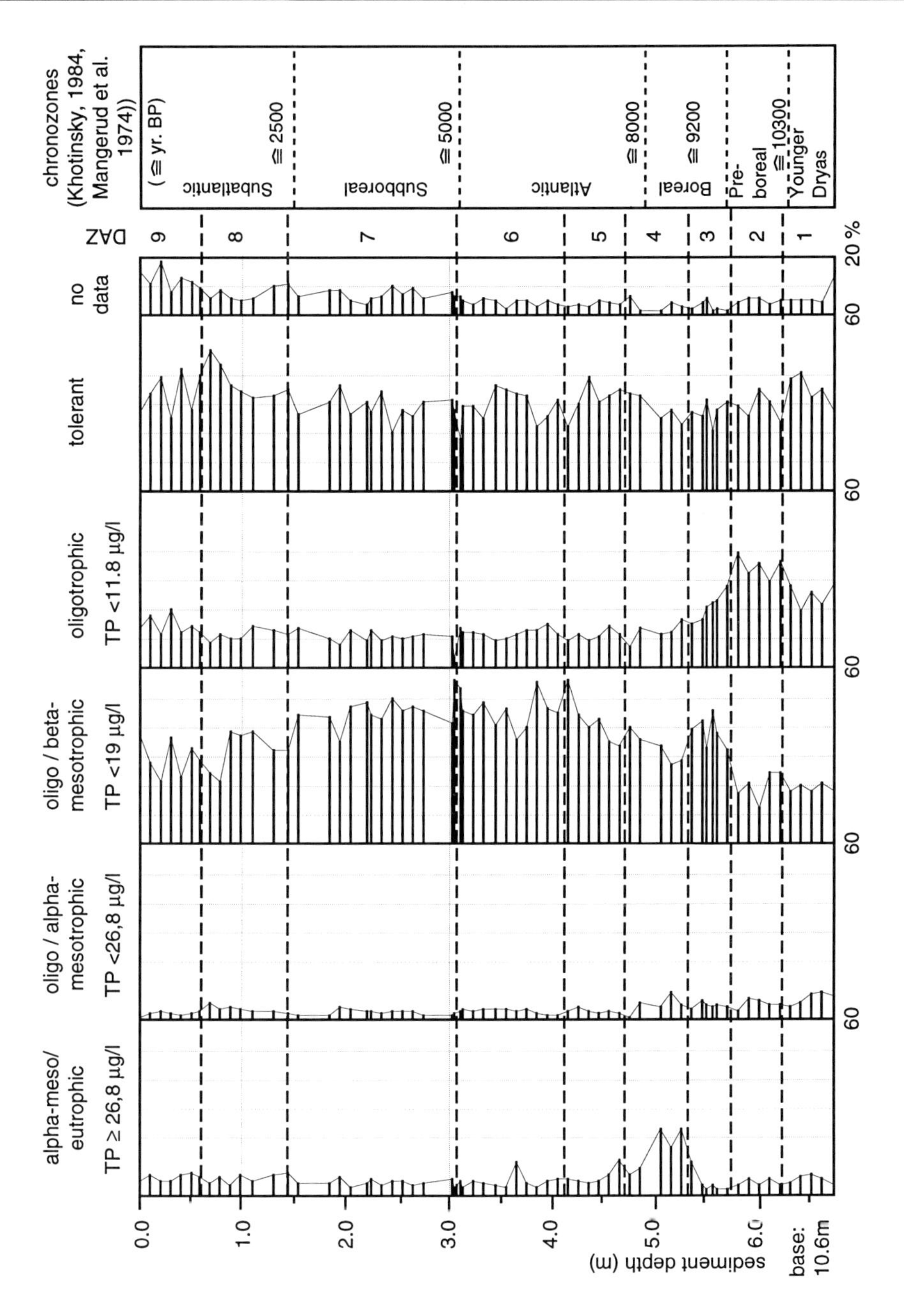

Figure 8: Relative abundance of diatoms in trophic tolerance groups (Hofmann, 1994) basing on the Vollenweider model (1979) (TP = total phosphorus concentration, DAZ = local diatom assemblage zones) - Lama Lake PG1111 (NE Norilsk, W Siberia).

the maximum abundance in the sequence. The co-occurring *Aulacoseira subarctica* and *Fragilaria construens* var. *venter* are both known as opportunistic taxa. The tychoplanktonic

Fragilaria construens var. *venter* is especially flexible and adapted to use high temporary nutrient concentrations occurring also in the pelagic zone of oligotrophic lakes (e.g. Pienitz et al., 1995, Voigt, 1997). It is possible that *F. construens* var. *venter* partly replaces the small planktonic *Cyclotella* sp. during thermally and trophically unfavourable periods, which would corroborate its opportunistic life strategy.

The pollen maxima of *Larix* and *Alnus*, accompanied by a rapid decrease of *Betula exilis*, provide the palynological evidence to infer the onset of a strong warming trend culminating in the optimal climate conditions for the region (Khotinsky, 1984), placed in the Boreal chronozone. Further increasing diatom concentrations, along with higher proportions of oligo/beta-mesotrophic taxa, were recorded for DAZ-3. The planktonic diatom complex is predominated by *Cyclotella comensis*, associated with higher nitrate concentrations (Stoermer et al., 1985; Pappas and Stoermer, 1995) and an annual maximum during summer stratification (Stoermer, 1993).

Contemporary to the highest AP values, diatom concentration and proportions of euplanktonic diatom taxa and diatom taxa placed in the alpha-meso/ eutrophic tolerance group reach their maxima in the sequence. The occurrence of *Cyclotella ocellata* in the sequence is also exceptional, as it is the most abundant planktonic diatom in DAZ-4. In the North Canadian training set, *C. ocellata* plots at the upper end of temperature gradient (Pienitz et al, 1995). Stoermer (1993) records the annual maximum during summer stratification and a requirement for higher phosphorus content of the water. TP optima from phosphorus training sets range from 10.5 (Wunsam and Schmidt, 1995) to 19.5 µg/l (Reavie et al., 1995).

From palynological evidence, the transition to the Atlantic is marked by a slight decrease in arboreal pollen relative to cooler periods (Khotinsky, 1984; Hahne and Melles, 1997). In comparison, the diatom assemblage appears to respond more intensely to early Atlantic ‚Novosagorskiy - cooling' (8000-7900 yr. BP; Khotinsky, 1984), when diatom concentrations decreased rapidly (10-20 v/mm). Abundant *Cymbella* spp. and *Fragilaria pinnata* were responsible for the Ep/Np ratio decreases. The contemporary low species number, high diversity, and maximum evenness of the assemblages can be interpreted in several ways. First, they may simply result from dissolution and the differing susceptibility in different species, given the poor preservation of the observed valves. Second, these features could also be responses to cooler conditions. Diversity decreased while evenness increased, since less species found suitable conditions and co-occurred in more even proportions.

Subsequently, all diatom-inferred parameters stabilized again. On a species level, a second maximum in *Aulacoseira islandica* was followed by the maximum abundance of *Cyclotella radiosa*.

Within DAZ-6, the Ep/Np ratio fell compared to the Boreal diatom assemblages. Species proportions in the planktonic complex fluctuated strongly. However, *Cyclotella comensis* always predominates and attains maximum abundance in the sequence. The proportions of taxa with preferences to oligo/beta-mesotrophic and circumneutral conditions peaked.

The following period is placed in the Subboreal on the basis of pollen records and is characterized by a general decrease in Ep/Np ratio. In the first cool period (4600-4100 y. BP, Khotinsky, 1984), when *Picea* pollen strongly decreased, a strong increase of small *Fragilaria* spp. (mainly *F. pinnata*) was observed. These species have been found to dominate in alkaline, glacial habitats (e.g. Haworth, 1976). Higher abundances may also be related to their tychoplanktonic and thus flexible life form. Beyond this feature, diatom assemblages are stable in terms of diversity and distribution in pH preference and nutrient tolerance groupings.

The early Subatlantic cooling, as indicated by the ‚first Holocene NAP maximum' (Hahne and Melles, 1997), seemed to be slightly predated by changes in the diatom assemblages. The Ep/Np ratio rapidly dropped below '1', marking the third subunit with respect to this

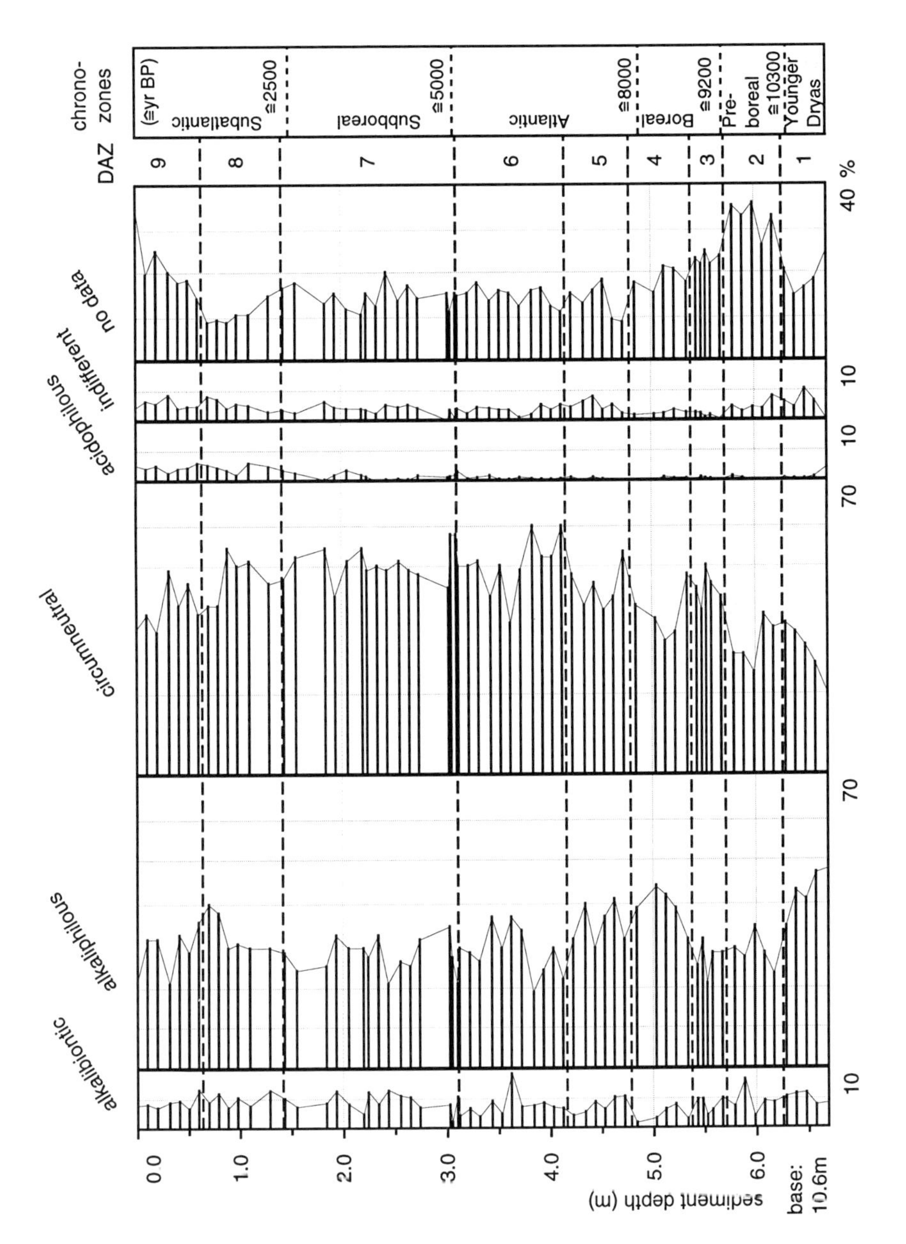

Figure 9: Relative abundances of diatoms in pH-preference groups (Hustedt, 1937-39) in relation to local diatom assemblage zones (DAZ and chronozones according to Khotinsky (1984) and Mangerud et al. (1974) - Lama Lake PG1111 (NE Norilsk, W Siberia).

parameter. Such non-planktonic species as *Cymbella* spp., small *Achnanthes* spp., and *Achnanthes minutissima* var. *minutissima* became abundant. The non-planktonic diatom assemblages were similar to those recorded from Preboreal times (DAZ-2). Even diatom concentration showed similarly low values.

A recovery of euplanktonic diatoms with Ep/Np values approximating 1 was recorded for the early Subatlantic (DAZ-8). In this planktonic diatom complex, *C. comensis* and *C.* cf. *kuetzingiana* var. *radiosa* were abundant. The proportions of taxa with oligo/ beta mesotrophic tolerances decreased.

Contemporary with increasing abundances of *Sphagnum* spores (an indicator for moist conditions commonly occurring in peat and bog communities), acidophilous taxa comprised 3% of the assemblages (Figure 9).

The increasing proportions of *Artemisia* and *Cyperaceae* in the pollen assemblages recorded as ‚second Holocene NAP maximum' mark the higher Subatlantic (Hahne and Melles, this volume; Hahne and Melles, 1997). Towards modern times, however, AP values re-increased, with *Betula*-tree pollen (and *Pinus*, likely a result of long distance transport) as the main contributor.

Similar trends were observed in the Ep/Np ratios. While the non-planktonic assemblage remained almost stable in composition compared to the previous DAZ, the planktonic complex became dominated by *C. comensis* and *C. gordonensis*. The distinctive sub-modern increase in euplanktonic taxa resulted from a sudden abundance in *Fragilaria tenera*, a tychoplanktonic diatom with oligo/beta mesotrophic preferences. Since *Fragilaria tenera* is a very delicate and fragile diatom, this feature could also be a preservational effect. However, based on the contemporary pollen signal, more favourable conditions in terms of nutrients and temperature can be inferred from this feature. Acidophilous species remained present, comprising approximately 3 % together with *Sphagnum* spores (10 %).

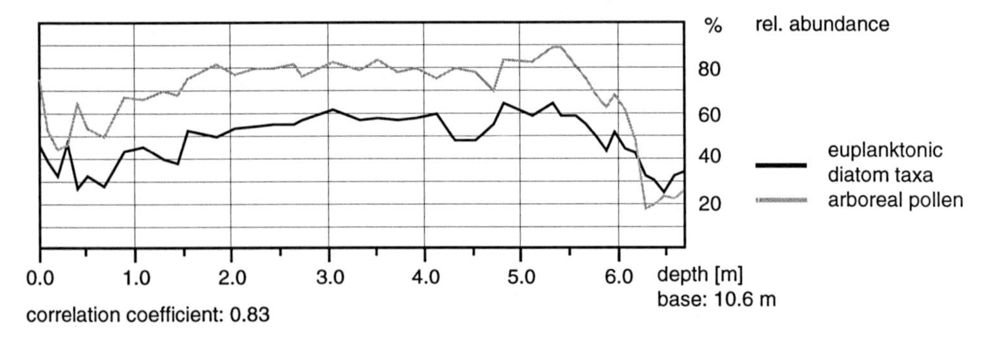

Figure 10: Comparison of relative abundance - curves of euplanktonic diatom taxa and arboreal pollen (Hahne and Melles, this volume) - Lama Lake PG1111 (NE Norilsk, W Siberia).

Conclusions

Several features in the diatom assemblages of the Lama Lake sequence are considered to derive from the establishment of microhabitats in marginal areas: the presence of alpha-meso to eutrophic taxa in stable ‚background' percentages throughout the sequence, the distinctive fluctuations in diversity indices and the record of acidophilous taxa, all of which result from the occurrence patterns of periphytic taxa. Acidophilous diatom taxa were recorded contemporary to *Sphagnum* spores, indicating the formation of swampy depressions in the lake's catchment and moister conditions. The large body of water is considered to act as a buffer against environmental changes which thus are less strongly reflected by the diatom assemblages. Changes in the diatom assemblage composition caused by lake level changes appear subordonated. Sedimentological investigations (Harwart et al., subm.) revealed no indications for changes in the morphology of the lake basin either. The climatic changes for the western Siberian region recorded by, for example, Khotinsky (1984), Velichko et al. (1997) and Astakhov (1997), as compiled in Figure 11, are clearly reflected in the diatom assemblages of Lama Lake.

The correlation of the abundance of euplanktonic diatom taxa with that of arboreal pollen can be supplemented by the sediment‘s total organic matter content. An overwhelming influence of climate (via the lake‘s summer surface temperatures) on the composition of diatom assemblages in subpolar and polar regions is generally recorded (e.g. Smol, 1988). On this basis, the euplanktonic/non-planktonic diatom ratio (Ep/Np) is used to infer trends in the lake's average annual surface temperature (LST) and the amplitude of its variations.

The distinctive maxima of certain diatom taxa and their specific ecological preferences for thermal and trophic conditions, as well as the succession of planktonic species maxima, enable us to assess these parameters in the lake development. These results are substantiated by corresponding results from the palynology, sedimentology and geochemistry of the lake‘s sediments. However, a much higher temporal resolution is needed to resolve timing effects in the responses of terrestrial and aquatic ecosystems.

The lowest LST-average, a high temperature fluctuation amplitude and a low and unstable trophic status in the DAZ-1 (coinciding in range with the Younger Dryas, regionally termed Norilsk Stadial) are inferred from the lowest Ep/Np ratio, an abundance in species with high survival capacity and a maximum percentage in trophy-tolerant species. During DAZ-2 (Preboreal), the LST-average increases while the amplitude remains low because of increased precipitation (Velichko et al., 1997). The trophic status is assessed as oligotrophic according to the proportions in trophic tolerance groupings. A moderate LST-amplitude on the highest average and the highest trophic status of Lama Lake are inferred for DAZ-3 and 4 (Boreal), supported by the high proportions of alpha-meso to eutrophic diatom taxa and maxima in the total organic carbon content of the sediments, arboreal pollen abundance, diatom concentration, and Ep/Np ratio . It is possible that nutrient enrichment occurred in marginal areas of the lake. The high abundance of planktonic taxa with annual maxima in the summer epilimnion could be indicative of a pronounced summer stratification of the lake. A rapid decrease in average LST marks the end of this period and is reflected in a sudden retreat in the Ep/Np ratio and a lower trophic status (DAZ-5). Favourable climatic conditions are inferred for the remaining Atlantic time (DAZ-6), during which average LST and amplitude are both considered high. Diatom concentration and the Ep/Np ratio attain higher values again, while the proportions within the trophic tolerance groupings fluctuate, suggesting unstable trophic conditions. A subsequent decrease in the Ep/Np ratio during the DAZ-7 (Subboreal), indicates a general decreasing trend in the average LST . The species composition of the diatom assemblage remains stable, leading to stable proportions in the ecological preference groupings. These trends are terminated by a rapid decrease in the average LST. The Ep/Np ratio and the diatom concentrations drop to levels comparable to Preboreal times, which is presumably a response to the early Subatlantic cooling. Subsequently, the average LST fluctuates strongly, based on the values of the Ep/Np ratio. The ‚little climate optimum‘ (dated to 1000 yr. BP; Velichko et al, 1997) could coincide with a rapid increase of the Ep/Np ratio in DAZ-9. The high proportions of trophy-tolerant oligotrophic taxa could indicate unstable trophic conditions at a low level.

A recent change in material supply is indicated by a distinct change in the diatom assemblage and may originate from pollution from the Norilsk smelter complex. Heavy metal concentrations have increased in the sediments since the 1930s (Hagedorn et al., this volume).

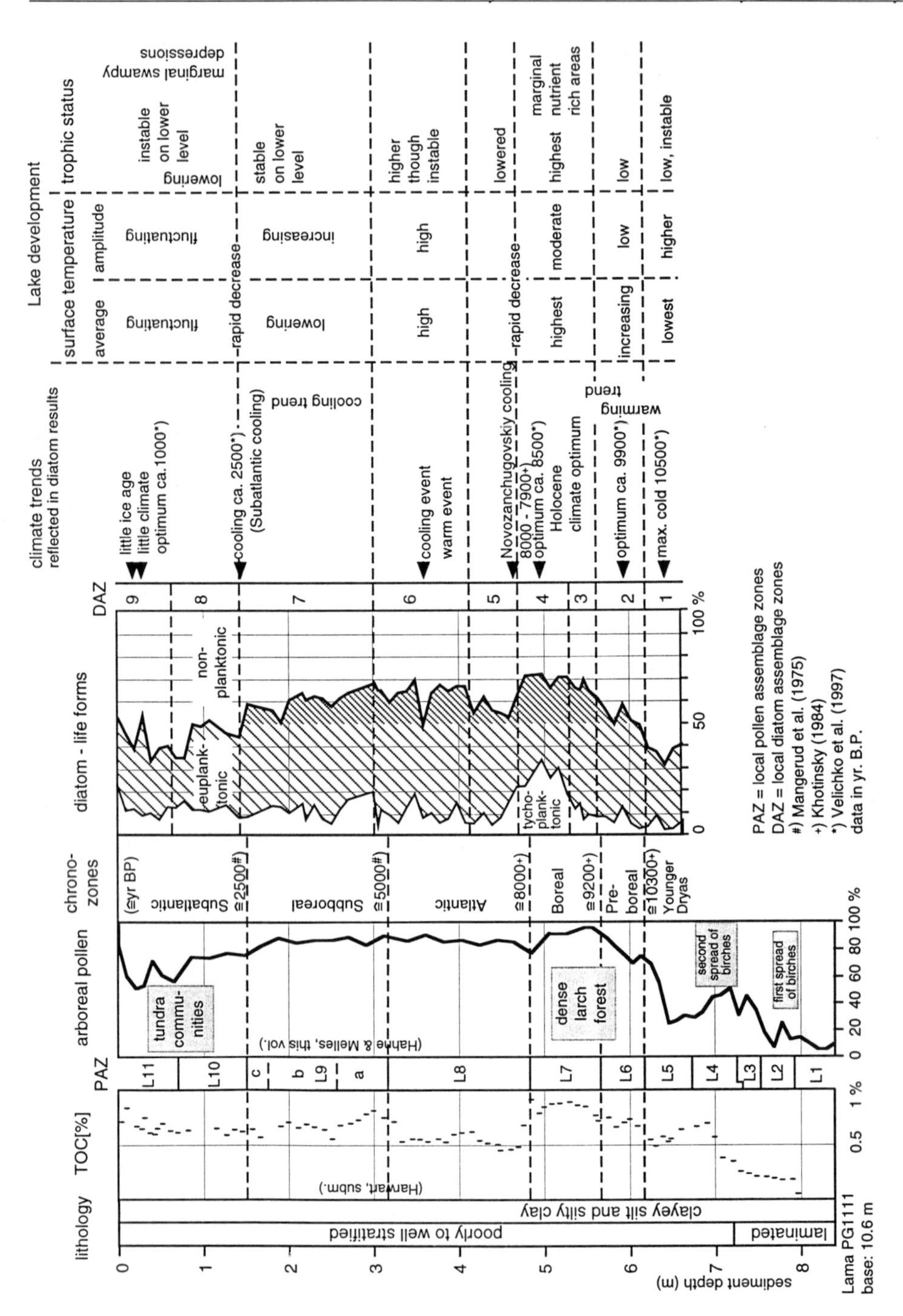

Figure 11: Compilation of the results of lithology, palynology and diatom analysis and inferred climate trends - Lama Lake PG1111 (NE Norilsk, W Siberia).

Acknowledgements

I would like to thank the participants of the expedition Norilsk/Taymyr 1993 for collecting the sample material. C. Siegert, M. Melles, J. Hahne, S. Harwart, A. Raab, and B. Wagner of the Alfred Wegener Institute, Potsdam, Germany and U. Zielinski of the Alfred Wegener Institute, Bremerhaven, Germany are acknowledged for helpful discussions and comments. R. Voigt (University Göttingen) and R. Klee (Bavarian Provincial Authority for Water-supply, Munich) provided help with taxonomical problems, A. Abelmann and R. Gersonde provided start-up help, and U. Bock and R. Cordelair gave me an introduction to laboratory techniques and sample treatment (all are from the Alfred Wegener Institute, Bremerhaven). Special thanks are due to R. Pienitz (University Laval) and an anonymous reviewer for reviewing the manuscript and P.P. Overduin for reviewing the English. This is AWI - contribution No. 1352.

References

Agbeti, M.D. (1992) Relationship between diatom assemblages and trophic variables: a comparison of old and new approaches. Can. J. Fish. Aquat. Sci., 49, 1171-1175.

Agbeti, M.D. and M. Dickman (1989) Use of fossil diatom assemblages to determine historical changes in trophic status. Can. J. Fish. Aquat. Sci., 46, 1013-1021.

Aleksyuk, G.V. and M.Y. Bekman (1981) Phytoplankton of the Putoran Lakes, their conditions and development (in Russian). In: Galaziy, G.J. and Y.P. Paramuzin (eds.), Lakes in the North-West of the Siberian Platform (in Russian), Leningrad, Nauka, 110-123.

Anderson, N. J. (1989) A whole-basin diatom accumulation rate for a small eutrophic lake in Northern Ireland and its palaeoecological implications., J. Ecol., 77, 926-946.

Anderson, N. J., B. Rippey and C.E. Gibson (1993) A comparison of sedimentary and diatom-inferred phosphorus profiles: implications of defining pre-disturbance nutrient conditions. Hydrobiologia, 253, 357-366.

Astakhov, V. (1997) Late Glacial Events in the Central Russian Arctic. Quaternary International, 41/42, 17-26.

Barker, P. (1992) Differential diatom dissolution in Late Quaternary sediments from Lake Manyara, Tanzania: an experimental approach. J. Paleolimnology, 7, 235-251.

Battarbee, R.W. (1980) Diatoms in lake sediments, In: B. E. Berglund (ed.), Paleohydrological changes in the temperate zone in the last 15.000 years, University of Lund, 177-225.

Battarbee, R.W. (1986) Diatom analysis, In: B. E. Berglund (ed.), Handbook of Holocene Palaeoecology and Palaeohydrology. John Wiley and Sons Ldt., 530-570.

Battarbee, R.W. (1978) Relative composition, concentration and calculated influx of diatoms from a sediment core from Lough Neagh, Northern Ireland., Pol. Arch. Hydrobiol., 25, 9-16.

Battarbee, R. (1991) Recent paleolimnology and diatom-based environmental reconstruction. In: Shane, L.C.K. and E.J. Cushing (eds.), Quaternary Landscapes, Minneapolis, University of Minnesota Press, 129-174.

Bennion, H. (1995) Surface-sediment diatom assemblages in shallow, artificial, enriched ponds, and implications for reconstructing trophic status, Diatom Research, 10, 1-19.

Birks, H.J.B. (1986) Late Quaternary biotic changes in terrestrial and lacustrine environments, with particular reference to Northwest Europe. In: Handbook of Holocene Palaeoecology and Palaeohydrology. Berglund, B.E. (ed.), J. Wiley Sons, Chichester, 3-65.

Birks, H.J., J.M. Line, S. Juggins, A.C. Stevenson and C.J.F. ter Braak (1990) Diatoms and pH reconstruction. Phil. Trans. R. Soc. Lond., B 327, 263-278.

Bogdanov, A.L. (1985) Assessment of anthropogenous effects in the lakes of Putoran Plateau (in Russian). In: Adamenko, V.I. and A.N. Egorov (eds.), Geography of the Taymyr Lakes (in Russian), Leningrad, Nauka, 184-194.

Bradbury, J. P. (1975) Diatom stratigraphy and human settlement in Minnesota. Geol. Soc. Am. Special paper, 171, 74 pp.

Bradbury, J.P. (1988) Fossil diatoms and Neogene paleolimnology. Paleogeogrraphy, Paleoecology, Paleoclimatology, 62, 299-316.

Bradbury, J.P. (1997) A diatom record of climate and hydrology for the past 200 ka from Owens Lake , California with Comparison to other Great Basin records. Quaternary Science Review, 16, 203-219.

Brugam, R.B. (1988) Long-term history of eutrophication in Washington lakes. In: Adams, W.J., G.A. Chapman and W.G. Landis (eds.), Aquatic toxicology and hazard assessment, Philadelphia, ASTM STP 971, American Society for Testing and Materials, 63-70.

Carney, H.J. (1982) Algal dynamics and trophic interactions in the recent history of Frains Lake, Michigan. Ecology, 63, 1814-1826.

Chernyaeva, G.P. (1981) Composition, ecology and quantity of diatoms in the sediments of characteristic lakes.

In: Galaziy, G.J. and Y.P. Paramuzin (eds.), Lakes in the North-West of the Siberian Platform (in Russian), Leningrad, Nauka, 103-133.

Cleve-Euler, A. (1951-1955) The diatoms of Schweden and Finnland (in German). K. Svenska Vetenskapsakad. Handl. Fjärde Ser., 2 (1), 3-163, 3 (3), 1-153, 4 (1), 3-158, 4 (5), 3-255, 5 (4), 3-231.

Conley, D.J. and C.L. Schelske (1993) Potential role of sponge spicules in influencing the silicon biochemistry of Florida lakes. Can. J. Fish. Aquat. Sci., 50, 296-302.

Dixit, S.S., J.P. Smol, J.C. Kingston and D.F. Charles (1992) Diatoms: Powerful indicators of environmental change. Environmental Sciences and Technology, 26, 22-33.

Doubleday, N.C., M.S.V. Douglas and J.P. Smol (1994) Paleoenvironmental studies of black carbon deposition in the high Arctic a case study from Northern Ellesmere Island. Science of Total Environment, 160/161, 661-668.

Flower, R.J. (1993) Diatom preservation: experiments and observations on dissolution and breakage in modern and fossil material. Hydrobiologica, 269/270, 473-484.

Galaziy, G.I. and Y.P. Paramuzin (1981a) History of the large lakes of the central Sub-Arctic (in Russian). Nauka, Novosibirsk, 137 pp.

Galaziy, G.I. and Y.P. Paramuzin (1981b) The lakes of the north-western part of the Siberian Platform (in Russian). Nauka, Novosibirsk, 190 pp.

Gasse, F., S. Juggins and L. Ben Khalifa (1995) Diatom-based, transfer functions for inferring past hydrochemical characteristics of African lakes. Paleogeography, Paleoclimatology, Paleoecology, 117, 31-54.

Gavrilova, M.K. (1981) Modern climate and permafrost on continents (in Russian). Nauka, Novosibirsk, 112 pp.

Grimm, E.C. (1987) CONISS: A Fortran 77 program for stratigraphically constrained cluster analysis by the method of incremental sum of squares. Computers and Geosciences, 13, 13-35.

Grosswald, M.G. (1977) The latest Eurasian ice sheet. In: Glacio-glacial Studies Chronicle and Discussions. Avsyuk, G.A. and V.M. Kotlyakov (eds.), 45-60.

Hagedorn, B., S. Harwart, M. Rutgers van der Loeff and M. Melles (this volume) Lead-210 dating and heavy metal concentration in recent sediments of the Lake Lama (Norilsk area, Siberia).

Hahne, J. and M. Melles (1997) Late Weichselian and postglacial vegetation and climate history of the south-western Taymyr Peninsula (Central Siberia) as revealed by pollen analysis of sediments from Lake Lama. Vegetation History and Archaeobotany, 6, 1-8.

Hahne, J. and M. Melles (this volume) Climate and vegetation history on the Taymyr Peninsula since Middle Weichselian times - palynological evidence from lake sediments.

Hall, R.I. and J.P. Smol (1992) A weighted averaging regression and calibration model for inferring total phosphorus concentration from diatoms in British Columbia (Canada) lakes. Freshwater Biology, 27, 417-434.

Harwart, S., B. Hagedorn, M. Melles, M. and H.-W. Hubberten (1997) Geochemical variations as climate indicators in the Late Quaternary, Lama Lake, Taymyr Peninsula, Siberia (in German) (abstract). In: Geochemische Variationen in den Geo- und Umweltwissenschaften, Geochemiker Treffen, TU Bergakademie Freiberg, Freiberg, 57.

Harwart, S., B. Hagedorn, M. Melles, M. and U. Wand (subm.) Physical and geochemical changes in the sediments of Lake Lama related to the Late Pleistocene and Holocene climatic evolution. Boreas.

Haworth, E. Y. (1976) Two late-glacial (late Devensian) diatom assemblage profiles from northern Scotland. New Phycologist, 77, 227-256.

Hecky, R.E. and P. Kilham (1988) Nutrient limitation of phytoplankton in freshwater and marine environments: a review of recent evidence on the effects of enrichment. Limnol. Oceanogr., 33, 796-822.

Hofmann, G. (1994) Epiphytic diatoms in lakes and their suitability as trophy indicators (in German). Bibliotheca Diatomologica, 30, 241.

Hurd, D.C. and S. Birdwhistell (1983) On producing a more general model for biogenic silica dissolution. American Journal of Science, 183, 1-28.

Hustedt, F. (1937-39) Systematical and ecological investigations of diatom floras from Java, Bali, Sumatra (in German). Arch. Hydrobiol., Suppl. 15, 131-506, Suppl. 16, 1-394.

Iversen, J. (1954) The Late-glacial flora of Denmark and its relation to climate and soil., Danm. geol. Unders., II Raekke, 80, 87-119.

Khotinsky, N.A. (1984) Holocene climatic change. In: Late Quaternary Environments of the Soviet Union. Velichko, A.A. (ed.), Minnesota Press, Minneapolis, 305-309.

Kilham, S.S. and P. Kilham (1984) The importance of resource supply rates in determining phytoplankton community structure. In: Meyers, E.G. and J.R. Strickler (eds.), Trophic interactions within Aquatic Ecosystems, AAAS Selected Symp., 7-27.

Klementev, O.L., W.I. Nikolaev, W.J. Potapenko and L.M. Swatyugin (1991) Structure and thermohydrodynamic conditions of the glaciers on Severnaya Zemlya (in Russian). Data of Glaciological Studies, 73, 103-109.

Kling, H. and H. Håkansson (1988) A light and electron microscope study of *Cyclotella* species (Bacillariophyceae) from Central and Northern Canadian lakes. Diatom Research, 3 (1), 55-82.

Kohmann, F. and U. Schmedtje (1986) Diversity and diversity indices (in German). Münchener Beiträge zur Abwasser-, Fischerei- und Flußbiologie, 40, 135-166.
Korsman, T. and H.J.B. Birks (1996) Diatom-based water chemistry reconstructions from northern Sweden: a comparison of reconstruction techniques. J. Paleolimnology, 15, 65-77.
Krammer, K. and H. Lange-Bertalot (1986) The freshwater flora of Central Europe (in German). vol. 2/1, Bacillariophyceae. part 1: Naviculaceae, Fischer, Stuttgart, 876 pp.
Krammer, K. and H. Lange-Bertalot (1988) The freshwater flora of Central Europe (in German). vol. 2/2, Bacillariophyceae. part 2: Bacillariaceae, Epithemiaceae, Surirellaceae, Fischer, Stuttgart, 596 pp.
Krammer, K. and H. Lange-Bertalot (1991a) The freshwater flora of Central Europe (in German). vol. 2/3, Bacillariophyceae. part 3: Centrales, Fragilariaceae, Eunotiaceae, Fischer, Stuttgart, 576 pp.
Krammer, K. and H. Lange-Bertalot (1991b) The freshwater flora of Central Europe (in German). vol. 2/4, Bacillariophyceae. part 4: Achnanthaceae, critical remarks to *Navicula* (Lineolatae) and *Gomphonema*, Fischer, Stuttgart, 437 pp.
Lange-Bertalot, H. (1993) 85 new taxa (in German). Bibliotheca Diatomologica, Bd. 27, Cramer, Berlin, Stuttgart, 454 pp.
Lange-Bertalot, H. and D. Metzeltin (1996) Indicators of Oligotrophy. 800 taxa representative of three ecological distinct lake types. Carbonate buffered- Oligotrophic- Weakly buffered soft water. Iconographica Diatomologica, 2, 1-390.
Lemmen, D. S., R. Gilbert, J.P. Smol and R.I. Hall (1988) Holocene sedimentation in glacial Tasikutaaq Lake, Baffin Island. Can. J. Earth Sci., 25, 810-823.
Lewin, J. (1961) The dissolution of silica from diatom walls. Geochimica Cosmochimica Acta, v. 21, 182-198.
Lotter, A.F., U. Eicher, H.J.B. Birks and U. Siegenthaler (1992) Late glacial climatic oscillations as recorded in Swiss lake sediments., J. Quat. Sci., 7, 187-204.
Lotter, A.F., H.J.B. Birks and B. Zolitschka (1995) Late-glacial pollen and diatom changes in response to two different environmental perturbations: volcanic eruption and Younger Dryas cooling. J. Paleolimnology, 23-47.
Lotter, A.F., R. Pienitz and R. Schmidt (subm.) Diatoms as indicators of environmental change near Arctic and Alpine treeline. In: Stoermer, E.F. and J.P. Smol (eds.), Applied diatom studies, Cambridge, Cambridge University Press.
Lotter, A., H.J.B. Birks, W. Hofmann and A. Marchetto (subm.) Modern Diatom, Cladocera, Chironomid and Chrysophyte cyst assemblages as quantitative indicators for the reconstruction of past environmental conditions in the Alps. II. Nutrients.
Lund, J. W. G. (1954) The seasonal cycle of the planktonic diatom *Melosira italica* (EHR.) ssp. *subarctica* O. Müller. J. Ecol., 42, 151-179.
MacDonald, G.M., W.D. Edwards, K.A. Moser, R. Pienitz and J.P. Smol (1993) Rapid response of treeline vegetation and lakes to past climate warming. Nature, 361, 243-246.
Mangerud, J., S.T. Andersen, B.E. Berglund and J.J. Donner (1974) Quaternary stratigraphy of Norden, a proposal for terminology and classification. Boreas, 3, 109-128.
Maxwell, B. (1992) Arctic climate potential for change under global warming In: Arctic Ecosystems in a Changing Climate. Chapin, F.S., J.F. Reynolds, R.L. Jeffries, G.R. Shaver, J. Svoboda and E.W. Chu (eds.), Academic Press, New York, 11-34.
McMinn, A. (1995) Comparison of diatom preservation between oxic and anoxic basins in Ellis Fjord, Antarctica. Diatom Research, 10, 145-151.
Melles, M., U. Wand, W.-D. Hermichen, B. Bergemann, D.Y. Bolshiyanov and S.F. Khrutsky (1994) The expedition Norilsk/Taymyr 1993 of the AWI research unit Potsdam. Berichte zur Polarforschung, 148, 3-25.
Nygaard, G. (1956) Ancient and recent flora of diatoms and chrysophyceae in Lake Gribsø, Studies on the humic and acid Lake Gribsø. Folia Limnologica Scandinavica, 8, 32-94.
Pappas, J.L. and E.F. Stoermer (1995) Effects of inorganic nitrogen enrichment on Lake Huron phytoplankton: an experimental study. J. Great Lakes Res., 21, 178-191.
Patrick, R. (1977) Ecology of freshwater diatoms and diatom communities. In: The biology of diatoms, Werner, D. (ed.), Botanical Monographs, 13, 284-332.
Pienitz, R. (1993) Paleoclimate proxy data inferred from freshwater diatoms from the Yukon and the Northwest Territories. Ph.D. thesis, Queen's University, 222 pp.
Pienitz, R. and J.P. Smol (1993) Diatom assemblages and their relationship to environmental variables in lakes form the boreal forest-tundra ecotone near Yellowknife, Northwest Territories, Canada. Hydrobiologia, 269/270, 391-404.
Pienitz, R., J.P. Smol. and H.J.B. Birks (1995a) Assessment of freshwater diatoms as quantitative indicators of past climatic change in the Yukon and Northwest Territories, Canada. J. Paleolimnology, 13, 21-49.
Pienitz, R., M.S.V. Douglas, J.P. Smol, P. Huttunen and J. Meriläinen (1995b) Diatom, chrysophyte and protozoan distributions along a latitudinal transect in Fennoscandia. Ecography, 18, 429-439.
Popovskaya, G.I. (1977) Dynamics of the phytoplankton of the pelagial (1964-1974), In: Biological productivity of the pelagial in Lake Baikal and its variability, Beckman, M.Yu. (ed.), Nauka, Novosibirsk, 5-39.

Reavie, E.D., R.I. Hall and J.P. Smol (1995) An expanded weighted averaging model for inferring past total phosphorus concentrations from diatom assemblages in eutrophic British Columbia (Canada) lakes. J. Paleolimnology, 14, 49-67.
Renberg, I. and T. Hellberg (1982) The pH history of lakes in south-western Sweden, as calculated from the subfossil diatom flora of the sediment. Ambio, 11, 30-33.
Reynolds, C.S. (1984) Ecology of freshwater phytoplankton. Cambridge, Cambridge University Press, 384 pp.
Rippey, B. (1983) A laboratory study of silicon release processes from a lake sediment (Lough Neagh, Northern Ireland). Archive of Hydrobiology, 96, 417-433.
Round, F.E., R.M. Crawford and D.G. Mann (1990) The Diatoms, Biology and morphology of the genera. Cambridge University Press, Cambridge, 747 pp.
Sabater, S. and E.Y. Haworth (1995) An assessment of recent trophic changes in Wintermere South Basin (England) based on diatom remains and fossil pigments. J. Paleolimnology, 14, 151-163.
Schelske, C.L. (1994) Did top-down effects amplify anthropogenic nutrient perturbations in Lake Michigan? Comment on Evans (1992). Can. J. Fish. Aquat. Sci., 51, 2147-2149.
Schrader, H.J. and R. Gersonde (1978) Diatoms and Silicoflagellates. In: Micropaleontological counting methods and techniques - an exercise on an eight metres section of the lower Pliocene of Capo Rossello, Sicily, W.J. Zachariasse et al. (eds.), Utrecht Micropal. Bull., 17, 129-176.
Shannon, C.E. and W. Weaver (1949) The mathematical theory of communication. Urbana, Univ. Illinois Press, 117 pp.
Shur, L.A. and F.Y. Sid'ko (1985) The phytoplankton of Taymyr Lake (in Russian). In: Adamenko, V.I. and A.N. Egorov (eds.), Geography of the Taymyr Lakes (in Russian), Leningrad, Nauka, 125-130.
Smol, J.P. (1988) Paleolimnology Recent Advances and Future Challenges. In: Scientific Perspectives in Theoretical and Applied Limnology, De Bernardi, R., G. Giussani and L. Barbanti (eds.), Memorie dell'Istituto Italiano di Idrobiologia, 47, 253-276.
Sommer, U. (1990) Phytoplankton nutrient competition - from laboratory to lake. In: Grace, J. and D. Tilman (eds.), Perspectives on plant competition, New York, Academic Press, 193-213.
Stevenson, A.C., S. Juggins, H.J.B. Birks, D.S. Anderson, N.J. Anderson, R.W. Battarbee, F. Berge, R.B. Davis, R.J. Flower, E.Y. Haworth, V.J. Jones, J.C. Kingston, A.M. Kreiser, J.M. Line, M.A.R. Munro and I. Renberg (1991) The Surface waters acidification project palaeolimnology programme: Modern diatom / lake-water chemistry data set. London, ENSIS Publishing, 86 pp.
Stockner, J. G. (1971) Preliminary characteristics of lakes in the Experimental Area, North-western Ontario, using diatom occurrence in the sediments. J. Fish. Res. Board Can., 28, 265-275.
Stoermer, E. F. (1993) Evaluating diatom succession: some peculiarities of the Great Lakes case. J. Paleolimnology, 8, 71-83.
Stoermer, E.F., J.P. Kociolek, C.L. Schelske and D.J. Conley (1985) Siliceous microfossil succession in the recent history of Lake Superior. Proceedings of the Academy of Natural Sciences of Philadelphia, 137, 106-118.
Stoermer, E. F., G. Emmert, M.L. Julius and C.L. Schelske, (1996) Paleolimnologic evidence of rapid recent change in Lake Eries trophic status. Can. J. Fish. Aquat. Sci., 53, 1451-1458.
ter Braak, C.J.F. and C.W.N. Looman (1986) Weighted averaging of species, logistic regression and the Gaussian response model. Vegetatio, 65, 3-11.
Thayer, V.L., T.C. Johnson and H.J. Schrader (1983) Distribution of diatoms in Lake Superior sediments. J. Great Lakes Res., 9, 497-507.
Tilman, D., S.S. Kilham and P. Kilham (1982) Phytoplankton community ecology: The Role of limiting nutrients. Ann. Rev. Ecol. Syst., 13, 349-372.
van Donk, E. and S.S. Kilham (1990) Temperature effects on silicon- and phosphorus limitad growth and competitive interactions among three diatoms. Journal Phycol., 26, 40-50.
Velichko, A.A., L.L. Isayeva, A. Makeyev, G.G. Matishov and M.A. Faustova (1984) Late Pleistocene glaciation of the arctic shelf and the reconstruction of the Eurasian ice sheets. In: Late Quaternary Environments of the Soviet Union, Velichko, A.A. (ed.), Minnesota Press, Minneapolis, 35-41.
Velichko, A.A., A.A. Andreev and V.A. Klimanov (1997) Climate and vegetation dynamics in the tundra and forest zone during the Late Glacial and Holocene. Quaternary International, 41/42, 71-96.
Velichko, A.A., P.M. Dolukhanov, N.W. Rutter and N.R. Catto (1997) Quaternary of northern Eurasia: Late Pleistocene and Holocene landscapes, Stratigraphy and environments. Quaternary International, 41/42, 191 pp.
Vincent, F.W. and R. Pienitz (1996) Sensitivity of high-latitude freshwater ecosystems to global change: temperature and solar ultraviolet radiation. Geoscience Canada, 23, 231-236.
Voigt, R. (1997) Investigations of palaeolimnology and vegetation-history from sediments of Fuschlsee and Chiemsee (Salzburg and Bavaria) (in German). Diss. Botanicae, 270, Cramer, Berlin, Suttgart, 303 pp.
Vollenweider, R. A. (1979) Nutrient doping as a basis for stressing the eutrophication of stagnant waterbodies (in German). Z. Wasser-Abwasser-Forschung, 12, 46-56.
Weckström, J., A. Korhola and T. Blom (1997) Diatoms as quantitative indicators of pH and water temperature in subarctic Fennoscandian lakes. Arctic and Alpine Research, 29 (1), 75-92.

West, R.G. (1964) Inter-relations of ecology and Quaternary paleobotany. J. Ecology, 52, Suppl., 47-57.
Willén, E. (1991) Planktonic diatoms - an ecological overview. Algological Studies, 62, 69-106.
Wolfe, A. P. (1994) Late Wisconsinian and Holocene diatom stratigraphy from Amarok Lake, Baffin Island, N.W.T., Canada., J. Paleolimnology, 10, 129-139.
Wright, H.E. (1984) Sensitivity and response of natural systems to climatic change in the Late Quaternary. Quaternary Science Review, 3, 91-131.
Wunsam, S. and R. Schmidt (1995) A diatom-phosphorus transfer function for Alpine and pre-alpine lakes. Memorie dell'Instituto Italiano di Idrobiologia, 53, 85-99.
Wunsam, S., R. Schmidt and R. Klee (1995) *Cyclotella*-taxa (Bacillariophyceae) in lakes of the Alpine region and their relationship to environmental variables. Aquatic Sciences, 57, 360-386.
Zielinski, U. (1993) Quantitative estimation of palaeoenvironmental parameters of the Antarctic Surface Water in the Late Quaternary using transfer functions with diatoms (in German). Berichte zur Polarforschung, 126, 148 pp.

Climate and Vegetation History of the Taymyr Peninsula since Middle Weichselian Time - Palynological Evidence from Lake Sediments

J. Hahne and M. Melles

Alfred-Wegener-Institut für Polar- Meeresforschung, Forschungsstelle Potsdam, Telegrafenberg A43, D 14473 Potsdam, Germany

Revised 5 May 1997 and accepted in revised form 15 March 1998

Abstract - Within the scope of the Taymyr Project, detailed palynological investigations were carried out on two long sediment cores (10.6 and 22.4 m) from the lakes Lama and Levinson-Lessing. The results reveal continuous information concerning the vegetation and climate histories of the unglaciated southern and northern Taymyr Peninsula since probably Late and Middle Weichselian times, respectively. The Weichselian was characterized by a cold and dry climate leading to strongly reduced vegetation. The stadial setting was interrupted at least twice by relatively short, warm interstadials recorded in the lower part of the core from Levinson-Lessing Lake. These interstadials may represent parts of the Middle Weichselian. The Pleistocene/Holocene transition in both cores is characterized by a climatic warming trend during the Bølling, Allerød, and Preboreal periods, with interruptions during the Older (post-Bølling) and Younger Dryas events. The Holocene climatic optimum occurred during the late Preboreal and Boreal periods. During this time interval only, larch forests dominated at Lama Lake, and birch and later dense alder boscages at Levinson-Lessing Lake. Starting in the Atlantic period, a climatic deterioration took place, which favoured sedge and sweet grass communities at Levinson-Lessing Lake but did not significantly influence the forest vegetation at Lama Lake. The climatic deterioration continued up to the middle Subboreal period, when present climate and vegetation conditions were probably established. A comparison of these results with preliminary palynological data from the lakes Changeable, Severnaya Zemlya Archipelago, and Kokora, central Taymyr Peninsula, yields new information concerning Weichselian and Holocene zonal vegetation shifts in northern Central Siberia.

Introduction

The multi-disciplinary, German-Russian research project "Taymyr" focuses on the Late Quaternary climatic and environmental history of northern Central Siberia. For this purpose, comprehensive geoscientific investigations are being carried out on natural paleoenvironmental data archives, such as permafrost profiles and lake sediment cores, along a transect of ca. 1400 km length, from the Northern Taiga in the surroundings of the town of Norilsk, via different tundra zones on the Taymyr Peninsula, to the High-arctic Tundra on Severnaya Zemlya (Figure 1).

Palynological investigations play an important role within the scope of this project. Pollen assemblages in sediment sequences reflect the regional vegetation during the time of sediment formation. Since the type of vegetation depends upon the duration and intensity of positive summer temperatures and the amount of precipitation, palynological results represent important proxy data for the climatic evolution of an area. They complement paleoclimatic information derived from other proxies, such as stable isotope data from ground ice bodies or sedimentological and cryolithological data from sediment sequences (Melles et al., 1996).

In this study, we present and discuss detailed palynological results obtained from two long lake sediment cores from the southern and northern Taymyr Peninsula (Figure 1). Along with preliminary data from the central Taymyr Peninsula and Severnaya Zemlya Archipelago, the

In: Kassens, H., H.A. Bauch, I. Dmitrenko, H. Eicken, H.-W. Hubberten, M. Melles, J. Thiede and L. Timokhov (eds.) Land-Ocean Systems in the Siberian Arctic: Dynamics and History. Springer-Verlag, Berlin, 1999, 407-423.

results supply new information concerning the temporal and spatial variations in the position and extent of vegetation zones since Middle Weichselian time. The influence of climate variability on these variations is studied by comparing the palynological results with ice core data from Severnaya Zemlya.

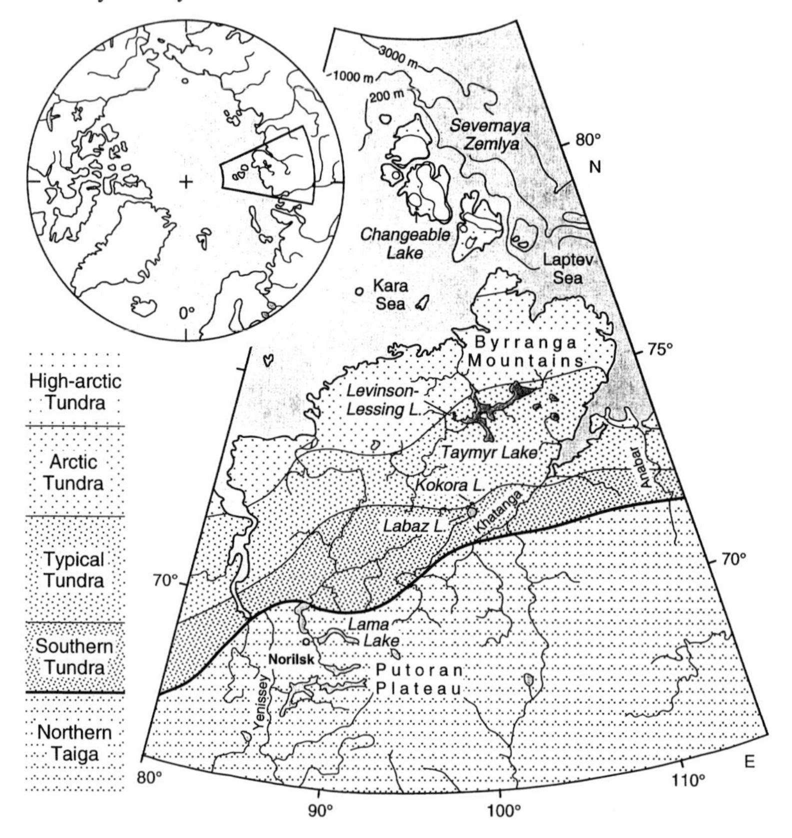

Figure 1: Map of the study area, showing current vegetation zones and geographical terms used in the text.

Study area

The project study area is located in the transition zone between the marine West Siberian and continental East Siberian climates (Gavrilova, 1981; Makeev et al., 1991). It covers the entire spectrum of arctic landscapes and vegetation zones (Figure 1). The four lakes in this study are situated in the High-arctic Tundra (Changeable Lake), in the northern part of the Typical Tundra (Levinson-Lessing Lake), at the boundary between Typical Tundra and Southern Tundra (Kokora Lake), and in the Northern Taiga (Lama Lake). Special emphasis is placed on the palynological results from Lama and Levinson-Lessing Lakes.

Lama Lake lies to the east of the town Norilsk (Figure 1) and fills an oblong, west to east trending depression with a maximum water depth of 254 m. To the north and south, the lake is bordered by steep mountain slopes whose peaks reach altitudes of up to roughly 1000 m a.s.l. The Norilsk area is characterized by a continental climate. Positive temperatures occur 90-100 days per year and snow cover persists for 9.5 months. Annual precipitation at Lama Lake varies between 800 and 1000 mm. The present-day vegetation was described in detail by

Galaziy and Parmuzin (1981). It is characterized by a strong vertical zonation, progressing from larch-spruce or pure spruce forest close to the lake shore, via lichen-herb-dwarf-shrub tundra, to rock-debris tundra and higher-rock tundra in high altitudes.

The 108 m deep Levinson-Lessing Lake is located in the southern Byrranga Mountains to the west of Taymyr Lake (Figure 1). At the latter lake, positive air temperatures were recorded in June, July, and August (30 year means: 0.6, 6.6, and 6.1°C) by the Taymyr Meteorological Station. In December, January, and February mean temperatures are below -30°C. The precipitation ranges between 200 and 300 mm/a. The ground is snow-free in July and August, and in some years also in June. The present-day vegetation in the Levinson-Lessing Lake surroundings is composed of sweet grass, sedge, and herb communities. It shows a vertical zonation in the surrounding hills, which reach altitudes of up to 570 m a.s.l. Specimens of *Betula* and *Alnus* shrubs with dwarfish growths can be found in depressions and on favourably exposed slopes. More detailed descriptions of the present-day vegetation are presented by Zhurbenko (1995) and Anisimov and Pospelov (this volume).

Material and methods

The four lake sediment cores in this study were recovered during different expeditions (Figure 1, Table 1): Lama Lake was sampled in 1993 (Melles et al., 1994a), Levinson-Lessing and Kokora Lakes in 1995 (Bolshiyanov and Hubberten, 1996), and Changeable Lake in 1996 (Melles et al., 1997). Coring was carried out in spring or early summer through holes in the lake ice, with the exception of Lama Lake, which was sampled in summer from a floating platform.

Table 1: Locations, recoveries, water depths, and shortest distances from the shore lines of the four lake sediment cores mentioned in this study.

Core No.	Lake	Geographical position	Recovery [m]	Water depth [m]	Distance from shore [km]
PG1111	Lama	69°32.9´ N; 90°12.7´ E	10.6	52.2	> 2.2
PG1225	Kokora	72°26.2´ N; 99°25.7´ E	5.2	18.2	> 0.4
PG1228	Levinson-Lessing	74°28.4´ N; 98°38.2´ E	22.4	108.0	> 0.6
PG1239	Changeable	79°07.3´ N; 95°06.0´ E	12.7	17.3	> 0.9

The lake sediment sequences were recovered with gravity and piston corers. The gravity corer was used to sample the soft near-surface sediments without disturbance. Lower sediments were recovered by repeated deployment of the piston corer. The individual cores from overlapping sediment depths were correlated to the complete sequence on the basis of whole-core physical property measurements, core descriptions, and analytical results. A more detailed description of the coring technique is given by Melles et al. (1994b).

For preparing the pollen samples, 2 to 8 cm^3 of the bulk dry sediment were used. Silica were removed using HF acid (70 %), followed by ultrasonic sieving (mesh size 6 x 8 µm) and acetolysis treatment. In order to determine the pollen concentrations, Lycopodium spores were added to every second sample of core PG1228 from Levinson-Lessing Lake. In the core from Levinson-Lessing Lake, in contrast, naturally high contents of Lycopodium in most samples excluded this determination. In both cores, arboreal pollen (AP) counts of 500 were attained in most Holocene samples. For the lowest Holocene and the Weichselian interstadials the values are around 250 AP, and for parts of extremely cold sequences they decrease down to 20 - 30 AP.

The palynological results are represented as total pollen diagrams (Figures 2 and 3), i.e. the pollen sums (100 %) include all pollen, with the exception of aquatics, spores, and algae (*Pediastrum*).

In the diagram from Lama Lake (Figure 2), the *Alnus* curve consists of tree alders (*Alnus*) and of shrub alders, which in the Russian literature are generally called "Alnaster". Its pollen grains are very similar to those of *Alnus viridis* (small and oval pores) and are therefore listed in the diagram under „*Alnus viridis*-type“. The *Betula* curve consists of tree birches (*Betula* trees), *Betula nana* (smaller pollen grains) and other shrub birches without clearly visible vestibulum (several species, which are collected under "*Betula exilis*-type"). Not mentioned are very rare finds of *Hippophaë, Ephedra distachya*-type and Ericaceae (*Ledum* and *Vaccinium*-type). An initial, regional interpretation of the palynological results from the Lama Lake core PG1111 was presented by Hahne and Melles (1997).

In the diagram from Levinson-Lessing Lake (Figure 3), the presence of *Larix*, *Picea*, and *Pinus* pollens is probably the result of long-distance transport. The curves of *Alnus* and *Betula* consist predominantly of shrub types as tree pollen contribute less than 5 %. *Hippophaë*, *Ephedra distachya*-type, Primulaceae, Lentibulariaceae, Liliaceae, Dipsacaceae, Polygonaceae, *Epilobium*, Boraginaceae, Scrophulariaceae, *Centaurea*-types, *Pedicularis* and *Lloydia* are not recorded due to their extremely rare occurrence.

Results and stratigraphy

The 10.6 m long sediment core PG1111 from the central part of Lama Lake did not reach the lacustrine sediment base. It consists of clayey silts and silty clays, with sand contents of less than 5 % (Figure 2). Clastic sediment components clearly predominate over biogenic components. Neither terrestrial macrofossils, which are promising for reliable radiocarbon dating, nor time-synchronous marker horizons, such as tephra layers, were found. The sediment color change repeatedly throughout the succession. Within the uppermost 0.3 m and below 7.3 m, brownish and yellowish colors indicate oxic conditions, whereas in the intervening horizon, grayish and dark greenish colors indicate anoxic conditions in the sediment. Down to a depth of 7.2 m the sediment is well to crudely stratified, below 7.2 m it is laminated with individual layers being less than 1 mm thick.

The pollen diagram from core PG1111 was subdivided into eleven regional pollen assemblage zones (PAZs), L1 - L11 (Table 2, Figure 2), whose boundaries reflect distinct vegetation changes in the catchment area of Lama Lake. In order to obtain initial chronological information for this core, the PAZ succession was correlated with the Late Weichselian and Holocene chronozones defined by Mangerud et al. (1974). This correlation can only supply a rough approximation of sediment ages because: 1) the climate and vegetation changes in Siberia could have been time-transgressive relative to those in northern Europe, which have been studied and dated by Mangerud et al. (1974), 2) the ages of the zone boundaries, originally defined by radiocarbon dates measured on bulk sediment, underwent repeated modifications with progress in radiocarbon dating techniques (Wohlfarth, 1996), and 3) some of the zone boundaries, such as the Younger Dryas to Preboreal boundaries, are dated to ^{14}C age plateaus, hampering calibrations and thus accurate age determinations (Björck et al., 1996). Nevertheless, the correlation of the PAZs with the chronozones supplies sufficient information for a discussion of general trends in the climate and vegetation history in the Siberian study area.

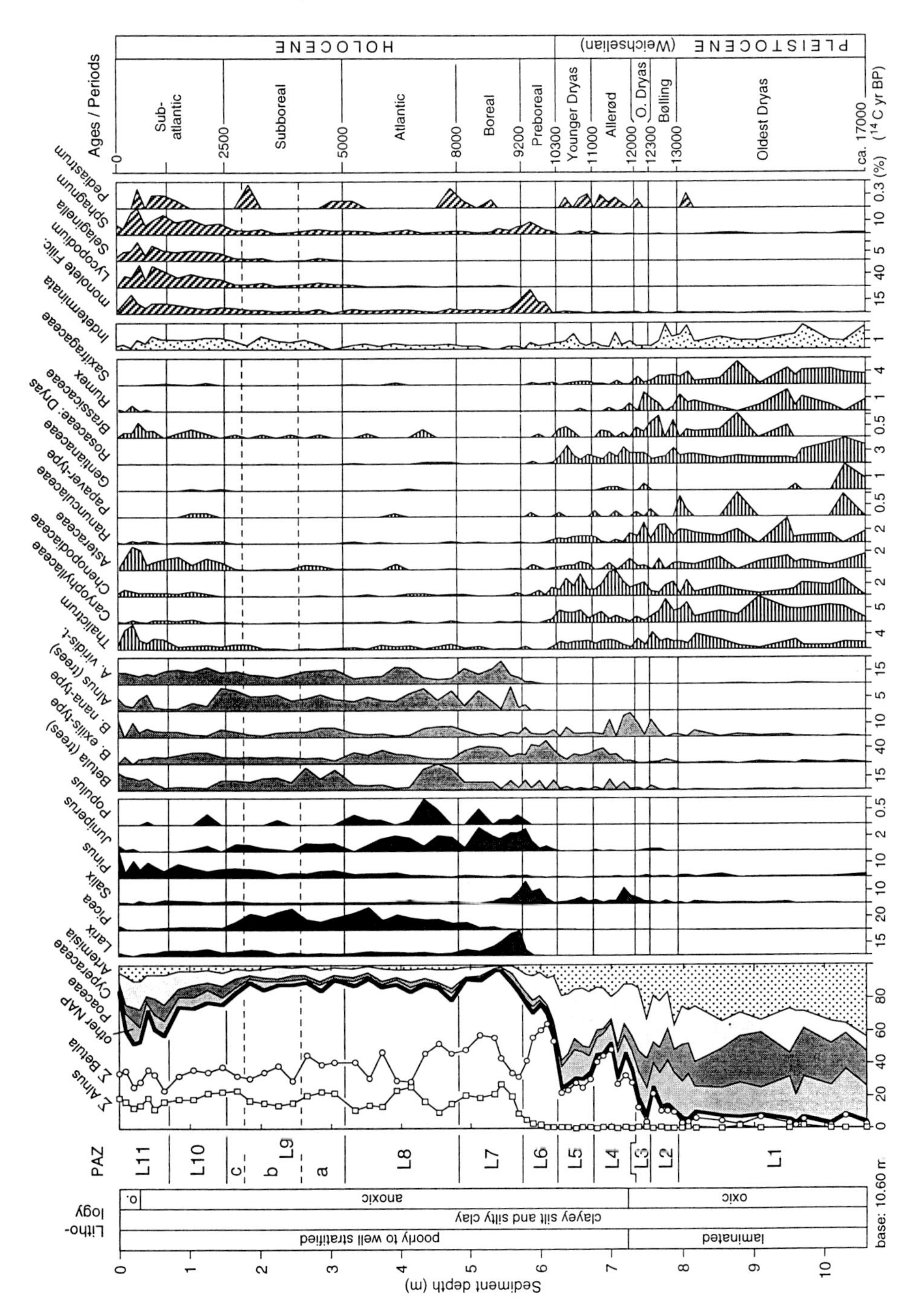

Figure 2: Pollen profile of the sediment core PG1111 from the central part of Lama Lake. The chronology is based on correlations of the regional Pollen Assemblage Zones (PAZ) with radiocarbon -dated pollen records on land (ages of period boundaries according to Mangerud et al., 1974 and Khotinskiy, 1984), supported by ^{210}Pb measurements in near-surface sediments.

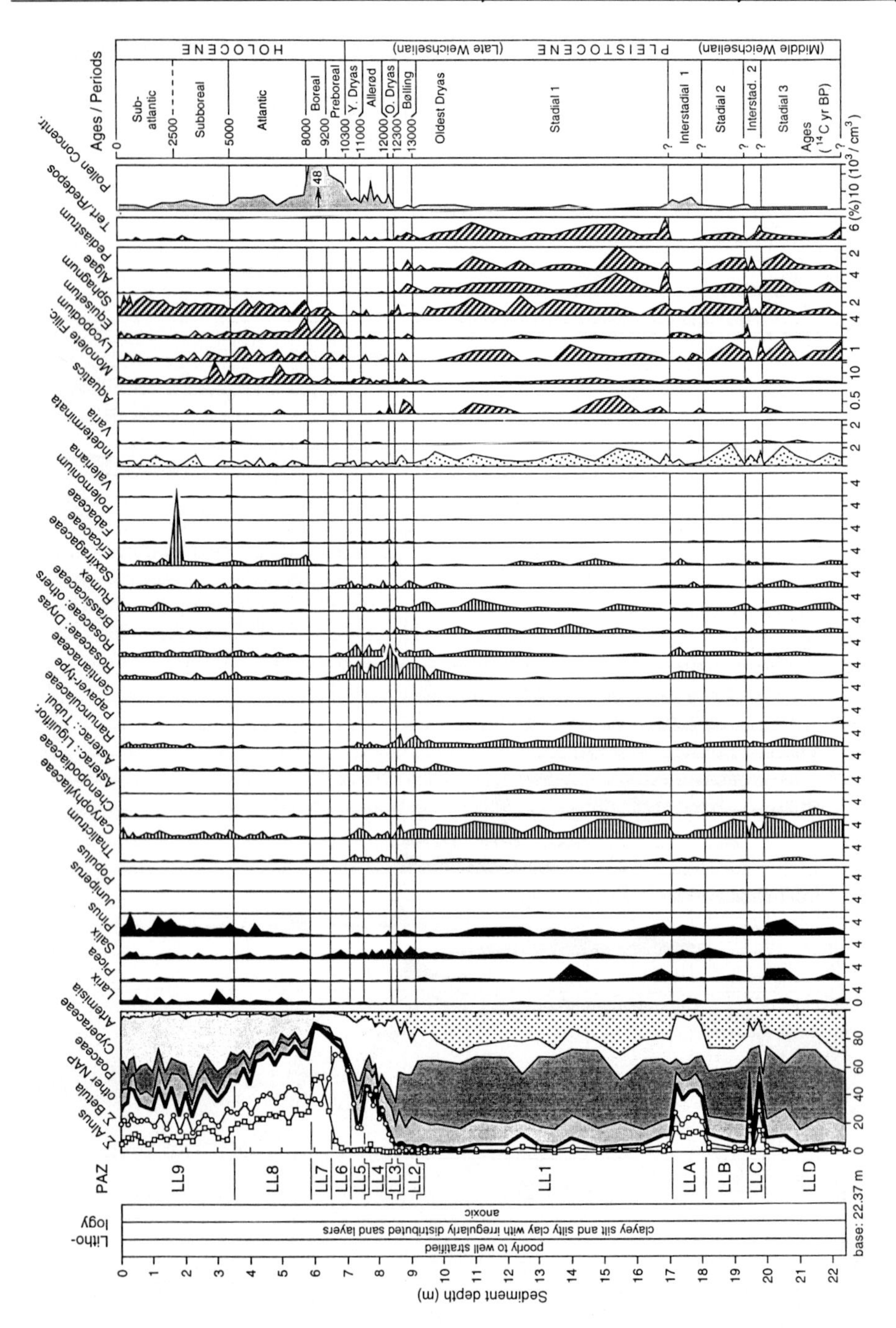

Figure 3: Pollen profile of the sediment core PG1228 from the central part of Levinson-Lessing Lake. The chronology is based on correlations of the regional Pollen Assemblage Zones (PAZ) with radiocarbon-dated pollen records on land (ages of period boundaries according to Mangerud et al., 1974 and Khotinskiy, 1984), supported by ^{210}Pb measurements in near-surface sediments and one radiocarbon dating of terrestrial plant remains.

The PAZs L1 - L5 in the lower part of core PG1111 from Lama Lake clearly indicate the Late Weichselian chronozones Oldest Dryas (DR1, stadial), Bølling (interstadial), Older Dryas (DR2, very short stadial), Allerød (interstadial), and Younger Dryas (DR3, stadial). These climatic events are recorded in many pollen diagrams from Europe (e.g. Hahne et al., 1994), North America (Mott and Stea, 1993), and Siberia (Velichko et al., 1997), as well as in stable isotope data from Greenland ice cores (Daansgard et al., 1993; Stuiver et al., 1995). The occurrence of the prominent Younger Dryas cooling event in northern Central Siberia was first evidenced by the isotopic composition of an ice core from the Severnaya Zemlya Archipelago (Klementyev et al. 1991; Stiévenard et al. 1996; Figures 1 and 4).

The Weichselian/Holocene boundary represents a stratigraphically fixed point. In Russia and in Europe, this boundary was radiocarbon dated to 10300 ^{14}C yr BP (Khotinskiy, 1984), corresponding to a calendar age of ca. 11450 cal. yr BP (Björck et al. 1996). At site PG1111, it occurs at 623 cm sediment depth. This results in a mean sedimentation rate during the Holocene of about 0.54 mm/a, which corresponds well with a sedimentation rate of 0.62 $\pm$ 0.01 mm/a for the upper 50 cm of the core, calculated from ^{210}Pb analyses (Hagedorn et al., this volume). A somewhat higher sedimentation rate can be expected in the near-surface sediments due to the lower consolidation and slightly higher water contents there.

The Holocene sediments in Lama Lake were subdivided into the PAZs L6 - L11. A comparison of the PAZ succession with radiocarbon-dated vegetation changes in Siberia (e.g. Makeev, 1983; Khotinskiy, 1984; Velichko et al., 1996), allows their connection with the Holocene periods Preboreal, Boreal, Atlantic, Subboreal, and Subatlantic. The sediment depths and calibrated ages of the period boundaries indicate that the sedimentation rates throughout the Holocene were relatively constant (0.29 - 0.73 mm/a).

The 22.4 m long sediment core PG1228 from the central part of Levinson-Lessing Lake represents only the upper part of the lacustrine sediment fill. This is evidenced by sub-bottom profiles crossing the coring location (Niessen et al., this volume). The sediments in core PG1228 consist of partly laminated clayey silts and silty clays, with irregularly incised, well sorted sand layers of a few millimetres to some centimetres thick (Figure 3). Grayish and dark greenish colors throughout the sequence indicate anoxic conditions in the sediment. A more detailed description, along with a presentation and discussion of sedimentological results is given by Ebel et al. (this volume).

The pollen diagram from core PG1228 was subdivided into thirteen regional pollen assemblage zones (PAZs), LL1 - LL9 and LLA - LLD (Table 3, Figure 3). The succession of the PAZs LL1 - LL8 in the central part of the core can clearly be correlated with the PAZs L1 - L8 of core PG1111 from Lama Lake (Table 2, Figure 2), based on the individual vegetation developments and considering the different lake locations latitudinally. PAZ LL9 in the upper part of the core from Levinson-Lessing Lake probably corresponds with PAZs L9 - L11 of Lama Lake. A subdivision of PAZ LL9 in Levinson-Lessing Lake is not useful, due to the rather small differences between pollen assemblages in this zone.

The stratigraphy of core PG1228 from Levinson Lessing Lake (Figure 3), as deduced from the PAZs correlation with core PG1111 from Lama Lake, indicates rather constant sedimentation rates of 0.46 to 0.75 mm/a (mean 0.57 mm/a) since the start of the Preboreal period (ca. 11450 cal. yr BP; Björck et al., 1996). This coincides well with ^{210}Pb and ^{137}Cs measurements on near-surface sediments from another coring point 4 km further north. They indicate a recent sedimentation rate 2 to 3 times higher (1.5 mm/a; B. Hagedorn, pers. comm. 1997), which can be expected for this part of the lake based on sediment geometry recorded in a sub-bottom profile along the lake axis (Niessen et al., this volume). In addition, the pollen stratigraphy of core PG1228 is supported by a radiocarbon age of 5650 $\pm$ 90 ^{14}C yr BP, measured on terrestrial plant remains from 467 cm sediment depth (Ebel et al., this volume), which confirms the coincidence of PAZ LL8 with the Atlantic period.

Table 2: Summary of regional pollen assemblage zones (PAZs) in the sediment core PG1111 from Lama Lake.

PAZ	Depth [cm]	Description
L1	1060-792	NAP zone (tree- and shrubless, extremely cold and dry)
L2	792-753	NAP-shrub-birch zone (interstadial character)
L3	753-733	as L1
L4	733-673	birch-NAP zone (interstadial character)
L5	673-623	NAP-birch zone (stadial character)
L6	623-573	birch-larch zone (beginning of a period with interglacial character)
L7	573-483	larch-birch-alder zone
L8	483-319	spruce-larch-birch-alder zone
L9	319-150	spruce-larch-birch-alder zone with alternating spruce values
L10	150-70	birch-alder zone with high NAP values
L11	70-0	birch-alder zone with extremely high NAP values

Table 3: Summary of regional pollen assemblage zones (PAZs) in the sediment core PG1228 from Levinson-Lessing Lake.

PAZ	Depth [cm]	Description
LLD	2237-1955	NAP zone (no shrubs, extremely cold and dry)
LLC	1955-1935	birch-alder-NAP zone (interstadial character)
LLB	1935-1815	as LLD
LLA	1815-1705	birch-alder-NAP zone (interstadial character)
LL1	1705-905	NAP zone (no shrubs, extremely cold and dry)
LL2	905-855	NAP zone with a minor spread of Cyp. and birch shrubs (incr. humidity)
LL3	855-835	as LL1
LL4	835-745	birch-NAP zone (interstadial character)
LL5	745-710	NAP-birch zone (stadial character)
LL6	710-650	birch zone (beginning of a period with interglacial character)
LL7	650-590	alder-birch zone (climatic optimum)
LL8	590-350	birch-alder-NAP zone (AP values above 50%)
LL9	350-0	NAP-birch-alder zone (AP values below 50%)

In the lower part of core PG1228 from Levinson-Lessing Lake two PAZs of interstadial character (LLA and LLC) are incised in PAZs of stadial character (LL1, LLB, and LLD). The core PG1111 from Lama Lake did not penetrate this succession. It probably represents the upper part of the Middle Weichselian. An unstable climate with several rapid temperature rises is recorded in the GRIP ice core from Greenland for the period 23 - 58 ka BP (Daansgard et al., 1993). These interstadials include the most recent prominent Middle Weichselian thermomer recorded in central Europe, the Denekamp Interstadial, which was dated to 32 - 28 ka BP

(Behre and van der Plicht, 1992). In Siberia, an interstadial of similar age (32 - 30 ka BP) was described by Kind (1974); however, interstadial settings were also reported for earlier (Isayeva, 1984) and later (Rybakova, 1989) Middle Weichselian times. Hence, the absolute ages of the interstadials recorded in core PG1228 must be kept open, at least until reliable radiocarbon dates become available.

Vegetation and climate history

In Siberia, both the chronology of the last climatic cycle and the related environmental changes are still poorly known. The last interglacial, called Kasantzev, is believed to occupy a period around 130-100 ka BP (e.g., Archipov, 1989). This period covers the Eemian interglacial, which corresponds to the oxygen isotope stage 5e (Behre and Lade, 1986; Martinson et al., 1987). According to Archipov (1989), the Weichselian glacial is subdivided into an Early Weichselian stadial (Zyrian, 100 - 50 ka BP), a Middle Weichselian interstadial (Kargin, ca. 50 - 22 ka BP) and a Late Weichselian stadial (Sartan, ca. 22 - 10 ka BP).

Pre-Middle Weichselian

During the Eemian (Kasantzev) interglacial, the Siberian vegetation showed a climatically related zonal structure similar to the present one (Gerasimov and Velichko, 1984). The climate is supposed to have been warmer than today, resulting in the location of the forest line further to the north.

The Early Weichselian (Zyrian) is regarded as a glacial period with continental glaciation. According to Ukraintseva (1993), the glaciation in Central Siberia took place in two stages. The first one embraced the North Siberian Lowland between Yenissey and Pogigai Rivers from three centres: Kara Sea, Putoran Plateau, and Anabar River (Figure 1). During the second stage, the ice of the northern and southern accumulation centers did not coalesce in the eastern and central parts of the North Siberian Lowland.

Due to the poor age control on these ice advances, it is not clear whether they correlate with warm or with cold phases recorded for the Early Weichselian in other areas. For example, in central Europe, the Early Weichselian was a period of two long-lasting and warm interstadials with birch and pine forests, the "Brørup" and "Odderade" (Behre and Lade, 1986; Hahne et al., 1994). They correspond with the isotope stages 5c and 5a and have equivalents in western, southern, and northern Europe, recorded in profiles from Grande Pile (Woillard, 1979), Les Echets (DeBeaulieu and Reille, 1984), and Scandinavia (Donner, 1996).

Middle Weichselian

The Middle Weichselian in Siberia (Kargin) is regarded as a cold period with several rather warm interstadials (Grichuk, 1984). During these interstadials, the climate was at least partly warmer than that of today, evidenced by the development of light birch or larch forests in areas where at present typical tundra communities are established. This leads Ukrainseva (1993) to term the Kargin an interglacial. For northern Central Siberia, three interstadials were distinguished based on radiocarbon ages: an early one at 50 - 44 ka BP, a middle one at 42 - 33 ka BP, and a late one at 30 - 24 ka BP (Isayeva, 1984).

The lower part of core PG1228 from Levinson-Lessing Lake probably penetrates into the Middle Weichselian (Figures 1 and 3). The pollen assemblages in this section showed a distinct succession of stadial periods (LLD, LLB, LL1), interrupted by two interstadials (LLC and LLA; Table 3, Figure 3). The latter, termed Levinson-Lessing Interstadials 1 and 2, are still of unknown absolute age (see above). In both interstadials, AP values of more than 40% indicate climatic conditions at least as favourable as those of today.

Interstadial 1 (LLA) is a longer-lasting thermomer. Assuming similar sedimentation rates to those of the Holocene, it may comprise more than 1000 years. High values of *Betula*, *Alnus* and Cyperaceae and low values of *Artemisia* can be regarded as a signal for increased humidity. A shrub tundra composed of *Betula* and *Alnus* species likely spread in the southern parts of the Taymyr Peninsula. However, it probably did not reach the Levinson-Lessing Lake area, because a comparison to the present-day situation indicates a graminoid-herb vegetation for the surroundings of the lake.

Interstadial 2 (LLC) comprises three samples with AP concentrations of 40 to 50 % . An intervening sample with a stadial character (AP <10 %) indicates that Interstadial 2 is built up of two very short thermomers of a few hundred years maximum duration. They are characterized by the presence of birch and alder shrubs. The occurrence of elm and, to a greater degree, hazel (about 3.5 %), are only recorded here and argue for an authochtonous deposition. An advance of *Corylus cornuta* and *Ulmus pumila* can be taken into consideration. They are known to have settled this area in the Middle Weichselian period (Ukraintseva, 1993).

In the intermittent stadial periods (LLD, LLB, LL1) very low pollen concentrations and AP values of less than 5 % indicate extremely cold and dry conditions. *Artemisia*, Poaceae and numerous taxa typical for a high-arctic environment, such as Caryophyllaceae, Asteraceae, *Thalictrum, Saxifragaceae,* and *Rumex arcticus*, covered favourable slopes and depressions in a predominant arctic desert.

The borders between the stadial and interstadial Middle Weichselian periods are characterized by rapid changes rather than gradual transitions in the AP pollen contents and, less pronounced, in the total pollen concentrations (Figure 3). Sample intervals of 10 cm at the corresponding depths, which probably represent about 100 years, indicate abrupt changes in the climatic conditions at the transitions both to and from the interstadials, and a rapid vegetation response.

Late Weichselian

The Late Weichselian in Siberia was charcterized by a very cold climate (Velichko, 1993). This period, as well as all following intervals, are well represented, not only in the core PG1228 from Levinson-Lessing Lake (Figure 3), but also in the lower part of the core PG1111 from Lama Lake (Figure 2). In both cores, the vegetation was strongly reduced during Stadial 1 and Oldest Dryas (LL1 and L1). A cold and dry climate, similar to that of the Middle Weichselian stadial periods, led to arctic desert at Levinson-Lessing Lake and graminoid-herb-*Artemisia* tundra at Lama Lake. At the end of the oldest Dryas, a slight increase in moist habitats is indicated in both cores by a spread of Cyperaceae communities.

The Late Weichselian/Holocene transition (PAZs LL1 - LL6 and L1 - L6) is represented by the thermomers Bølling, Allerød, and Preboreal, interrupted by the stadials Older Dryas and Younger Dryas. The specific PAZ boundaries for these periods have developed more clearly than they do in certain European sequences (e.g., Hahne, 1992). This is probably due to the more continental climate of Central Siberia, leading to more distinct vegetation changes as a result of even small climatic fluctuations during the Late Weichselian.

The Bølling interstadial (LL 2 and L 2) is very well developed in the Lama Lake diagram. In Levinson-Lessing Lake, in contrast, a slight increase of the birch curve is restricted to only one sample. The spread of Cyperaceae and a decrease of the *Artemisia* values in several samples, however, support the correlation. Hence, during the Bølling period a dense graminoid-herb tundra likely developed on the southern Taymyr Peninsula, whilst sparsely growing grass communities existed in a predominant arctic desert on the northern peninsula.

The climatic deterioration of the Older Dryas is recorded in both diagrams (LL3 and L3). This event is often difficult to identify due to its short duration (up to 300 years). In the diagram from Lama Lake, the Older Dryas is characterized by a distinct decrease in AP values, the absence of shrub vegetation, and a spread of *Artemisia*. At Levinson-Lessing Lake, in contrast, *Artemisia* is reduced in favour of high Cyperaceae values, which point to higher precipitation and an increase of moist habitats.

During the Allerød interstadial (LL 4 and L 4), a considerable spread of shrub birches and willows took place in the Lama Lake area on the southern Taymyr Peninsula. Shrub birches also increased at Levinson-Lessing Lake. In the second half of the Allerød period, they may have occupied higher altitudes of the Byrranga Mountains. Pollen concentrations 2 - 5 times higher than those of today in the diagram from Levinson-Lessing Lake evidence favourable climatic conditions. A decrease of AP concentrations in the middle of this period is recorded in both diagrams, thus indicating a short-term climatic deterioration of regional rather than local character.

The Younger Dryas stadial (LL5 and L5) in both profiles is characterized by a distinct decrease in *Betula* values and an increase of Cyperaceae. This may be due to a cold but relatively moist climate which eliminated or, at Lama Lake, at least strongly reduced the birch boscages. The Late Weichselian/Holocene transition in both cores exhibit high values of *Dryas* (*Dryas octopetala* and *D. punctata*), with maxima during the colder periods Oldest, Older, and Younger Dryas. This succession is comparable with Europe, where macrofossils of *Dryas* led to the naming of these events.

Holocene

The beginning of the Preboreal period in the diagrams from both Levinson-Lessing Lake and Lama Lake (PAZs LL6 and L6) is characterized by rapidly increasing birch values. At the end of this period, shrub birches occurred in considerable amounts in the Levinson-Lessing Lake area, and light forests of tree birches were developed at Lama Lake. A small climatic deterioration within the Preboreal is indicated by reduced AP values in the Lama Lake core but was not recorded in Levinson-Lessing Lake. This event may coincide with the "Pereslavl Event" described by Khotinskiy (1984) for northern Central Siberia, or the "Preboreal Oscillation" recorded in the GRIP ice core from Greenland as well as in various terrestrial records in Europe (Björck et al., 1996).

The transition from the Preboreal to the Boreal period (LL7 and L7) in both diagrams is characterized by a decrease of birch pollen in favour of alder and other AP. At Levinson-Lessing Lake, the Boreal shrub-birch tundra gave way to a shrub-alder-birch tundra during the Preboreal period, and the light birch forest at Lama Lake was replaced by a dense larch forest.

The late Preboreal and the Boreal periods clearly represent the Holocene climatic optimum on the Taymyr Peninsula. This is evidenced not only by the pollen assemblages during this time interval, but also by very high pollen concentrations, which are up to an order of magnitude higher than current levels in Levinson-Lessing Lake (Figure 3). This interpretation is in good agreement with a radiocarbon-dated strong warming shortly after 10 ka BP in the Labaz Lake area (Siegert et al., this volume), and a Holocene temperature maximum between 10.2 and 9.0 ka BP reported for the Severnaya Zemlya Archipelago (Makeev, 1983; Figure 1). It also coincides with the occurrence of trees at their range limits in northwestern Canada between 10.0 and 8.5 ka BP (Spear, 1993). In other areas of Russia and in western Europe, in contrast, determinations of the Holocene climatic optimum often place it somewhat later, in the Atlantic period (e.g., Klimanov, 1989).

For northern Scandinavia, summer temperatures of 4°C above current temperatures are suggested for about 9.0 ka BP (Birks, pers. comm. 1997). A similar temperature increase can be expected for the Holocene climatic optimum at Levinson-Lessing Lake from the occurrence of a shrub tundra. It would have resulted in mean temperatures of ca. 10°C for July and August and about 5°C for June on the northern Taymyr Peninsula, insufficient for the establishment of a boreal forest.

The transition from the Boreal to the Atlantic period, as recorded in the cores from Levinson-Lessing and Lama Lakes (LL8 and L8), is characterized by significant decreases in AP values, indicating a slight climatic deterioration. This event presumably coincides with the "Novosanchu-govskyi Cooling" in Central Siberia, which is dated to 8.0 - 7.9 ka BP (Khotinskiy, 1984) and probably also occurred in West Siberia, North America, and Europe (Kind, 1974).

Within the Atlantic period, different vegetation trends are recorded on the northern and southern Taymyr Peninsula. In the Levinson-Lessing Lake profile, slowly but continuously decreasing AP values (shrubs) indicate that the climatic deterioration continued throughout the Atlantic period. As a consequence, the Boreal shrub tundra was gradually eliminated and replaced by a graminoid-herb tundra. Simultaneously rising Cyperaceae values may indicate increasing moisture during this period, whilst rising *Pinus* contents are traced back to increasing long-distance transport from beyond the studied area.

At Lama Lake, in contrast, the Atlantic period was characterized by rather small vegetation changes. The climate was favourable for a continuous advance of spruce (*Picea obovata*) and for the establishment of a closed taiga composed of *Larix*, *Betula*, and *Picea*. Hence, the decrease in summer temperatures of about 1 to 2°C, indicated by the elimination of shrub tundra at Levinson-Lessing Lake, was either lower in the Lama Lake area or too slight to diminish the larch-spruce-birch forests persistently. In the upper part of the Atlantic period, small variations in the *Picea* concentration at Lama Lake may coincide with *Alnus* and *Betula* variations in the Levinson-Lessing Lake profile.

The Atlantic/Subboreal boundary in the diagram from Levinson-Lessing Lake (LL8/LL9) is poorly recorded by the pollen assemblage. It is probably located at 350 cm, where a linear interpolation using mean Holocene sedimentation rates indicates an age of 5.0 ka BP and where NAP values exceed 50 %. The PAZ LL 9 comprises the Subboreal and Subatlantic periods, which cannot be separated from pollen stratigraphic aspects because arboreal indicators are missing. At a depth of about 220 cm (around 3.0 ka BP), a more or less continuous increase of non-arboreal pollen changes to roughly constant NAP values. This is the time when the climatic decline, which had started in the Atlantic period, probably reached its end. The elimination of the birch and alder shrubs was complete, resulting in a nearly pure grass-herb tundra until the present.

In the Lama Lake diagram, in contrast, the Subboreal period (L9; 5.0 - 2.5 ka BP) is clearly expressed. From alternating spruce values it can be subdivided into two cooling events (PAZs L9a and L9c) and one intervening warming event (PAZ L9b), which have been described previously by Khotinskiy (1984). The slight Subboreal temperature variations are best reflected by the spruce values, because the Lama Lake area was probably very close to the northern range of *Picea obovata* during that period. Following the last Subboreal cooling event, spruce values remain below 5 % during Subboreal times (L10; 2.5 - 0 ka BP), whereas NAP values increase distinctly. A spread of heliophytes, such as *Artemisia*, *Thalictrum*, Filicinae, *Lycopodium*, and *Selaginella* suggest a thinning of the previous dense forest communities, and high values of *Sphagnum* may indicate increased precipitation.

Shift of vegetation zones in northern Central Siberia

The detailed palynological investigations on the sediment cores PG1111 and PG1228, presented and discussed above, have illustrated that the Weichselian and Holocene evolution of the summer temperatures on the Taymyr Peninsula is best reflected in the contents of arboreal pollen (AP). Their temporal variations at Lama Lake correlate very well with those at Levinson-Lessing Lake, as well as with the stable oxygen isotope ratios in an ice core from Severnaya Zemlya (Figure 4). These good correlations evidence, firstly, that the general trends in climatic changes since Allerød time were comparable along the entire, more than 1000 km long transect covered by the study area (Figure 1). In addition, the good correlation between the pollen and isotope data, which mirror the summer and annual temperatures, respectively, indicates that the continentality of the climate was widely constant throughout this time interval.

In order to reveal a first impression about the reaction of the vegetation to past climatic changes, the detailed palynological investigations on sediment cores from Lama and Levinson-Lessing Lakes were extended by preliminary pollen analyses on cores from two additional lakes in the project study area: Changeable Lake on the Severnaya Zemlya Archipelago and Kokora Lake in the Taymyr Lowland (Figure 1, Table 1).

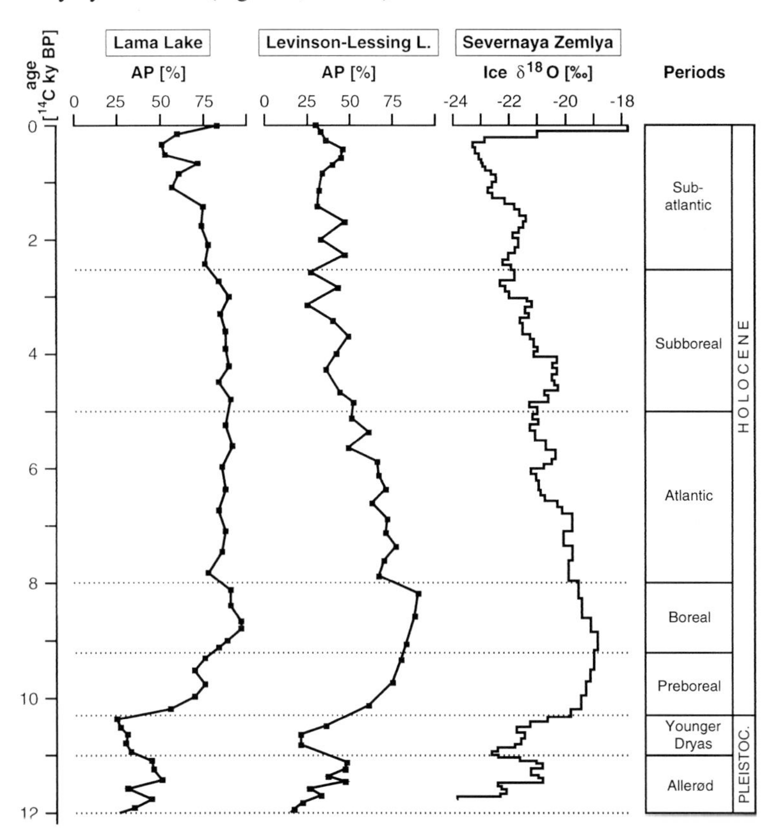

Figure 4: Comparison of the arboreal pollen (AP) contents in the sediment cores PG1111 from Lama Lake and PG1228 from Levinson-Lessing Lake (see Figures 2 and 3) with the stable oxygen isotope ratios in an ice core from the Academy of Sciences ice cap on Severnaya Zemlya (after Klementyev et al., 1991).

Today, the islands of the Severnaya Zemlya Archipelago represent an arctic desert. Plant cover is restricted to lichens and mosses. The palynological results obtained from a 12.7 m long core from Changeable Lake indicate that a similarly poor vegetation existed at the start of lacustrine sedimentation following the ice retreat during latest Weichselian time (Figure 5). Probably until late Preboreal and Boreal times, warming led to the Holocene climatic optimum and the development of a light graminoid-herb tundra. Subsequently, the vegetation was thinned out and the actual vegetation conditions were established, at the latest, by about 5.0 ka BP (Atlantic/Subboreal boundary).

The modern vegetation at Kokora Lake is characterized by predominantly graminoid-herb tundra, with boscages of alder and birch (with *Betula nana*), and with single larches of dwarfish growth restricted to favourably exposed slopes and depressions. A preliminary, 5.2 m long pollen record from Kokora Lake indicates an early Holocene to Atlantic spreading of dense boscages combined with light forests, which were composed of larches and tree birches (forest tundra; Figure 5). Starting in the Atlantic period, this vegetation transformed into a shrub-graminoid-herb tundra. In the lower part of the core, sediments reflecting a graminoid-herb tundra with very high Cyperaceae contents represent the initial lake stage, which was probably formed during the Younger Dryas event. They are underlain by terrestrial, sandy deposits with plant remains and extremely high values of *Larix* (ca. 60 %). For these sediments, a formation during Allerød time is assumed, supported by the overlaying sediment succession and the occurrence of larch trunks in a nearby area which revealed a radiocarbon age of 10.5 ± 0.5 ka BP (Ukraintseva, 1993).

Based on the temporal and spatial distribution of vegetation types in the four sediment cores investigated, the following conclusions can be drawn concerning zonal vegetational shifts in northern Central Siberia since latest Weichselian time (Figure 5).

At the end of the Late Weichselian, the slight warming during the Bølling period led to a significant reduction of *Artemisia* pollen at Lama Lake, and to a shift from arctic desert to

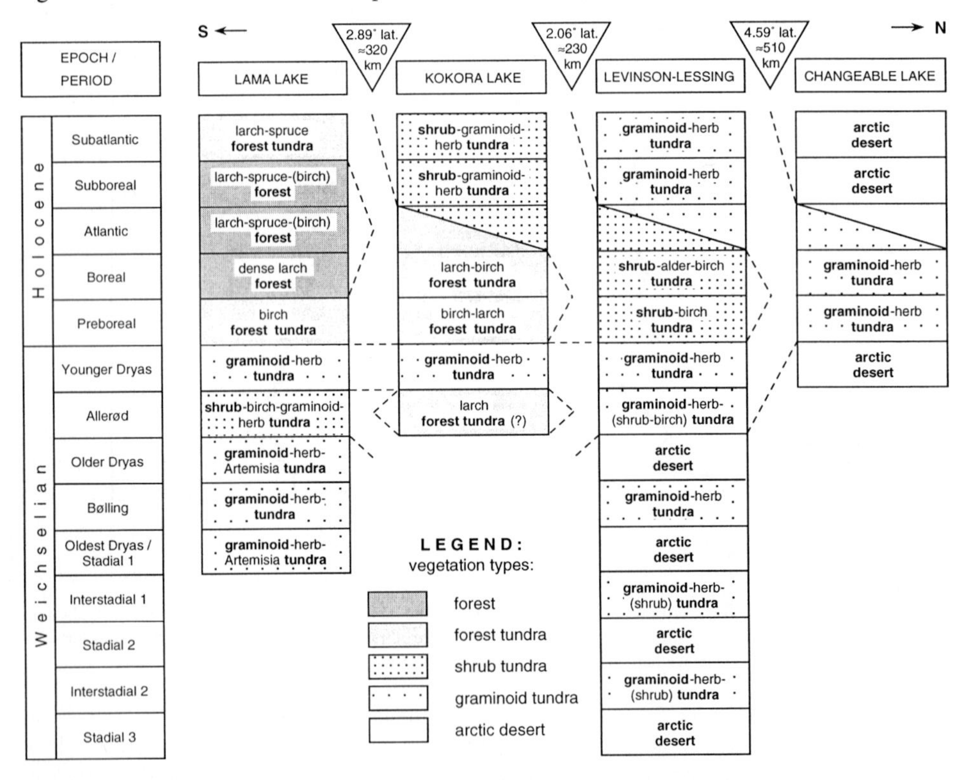

Figure 5: Scheme of Holocene and Weichselian zonal vegetational shifts (duration as far as recorded) in northern Central Siberia, based on detailed palynological results from Lama Lake and Levinson-Lessing Lake, supplemented by preliminary data from the lakes Kokora and Changeable. Latitudinal distances between the lakes are presented in the triangles at the top (for lake locations see Figure 1).

graminoid tundra at Levinson-Lessing Lake (Figure 5). The Allerød warming was even more pronounced. At Lama Lake, a shrub tundra, and at Levinson Lessing Lake, a graminoid tundra with sparsely occurring birch shrubs, were established. A larch forest tundra occupied the Kokora Lake area, presumably at the same time. Possible larch settlement to the north of the still larch-free Lama Lake area may indicate restricted spreading routes (e.g., along the northern rim of the Putoran Plateau) or restricted larch source habitats (e.g., the upper Lena River) at the end of the last glacial, but could also be due to especially favourable local climates in the Taymyr Lowland. This observation agrees with Khotinskiy (1984), who assumed that the Late Weichselian in Russia was characterized by the appearance of a hyperzonality or a "zone mixing". In contrast to today, the major vegetation zones (tundra, taiga, and southern steppes) were not established in global belts but widely distributed over the same areas.

Starting at the Weichselian/Holocene boundary, a clear zonal vegetation pattern developed within a very short time (Figure 5). It led from forest on the southern Taymyr Peninsula, via forest tundra and shrub tundra on the central and northern peninsula, to graminoid tundra on Severnaya Zemlya . This zonality change may be traced back to the establishment of a zonal atmospheric circulation system, as described by Khotinskiy (1984). During the Holocene climatic optimum in late Preboreal and Boreal times, larch and birch trees spread to Kokora Lake forming a forest tundra. Spruce (*Picea obovata*), in contrast, probably did not advance so far northward. The vegetation zones gradually shifted to the south, mainly during the Atlantic period. The current location of vegetation zones was widely established at the beginning of the Subboreal period. Subsequent distinct vegetation changes are recorded in the Lama Lake area only leading to the establishment of the present forest tundra at the Subboreal/Subatlantic boundary.

In summary, both the establishment of the modern zonal vegetation pattern at the Weichselian/Holocene boundary and the Holocene shifting trends are well recorded by the palynological data from the four investigated lake sediment cores. The data give a first impression of the mobility of vegetation zones and their sensitivity to climate changes. For determinations of absolute values and a detailed mapping of vegetation changes through time, however, additional palynological and sedimentological data are required, and especially more accurate dating via reliable radiocarbon dates.

Acknowledgements

We thank the participants of the expeditions Norilsk/Taymyr 1993, Taymyr 1995, and Taymyr/Severnaya Zemlya 1996 for their competent help in collecting the lake sediment cores. S. Harwart, T. Ebel, and A. Raab are acknowledged for the core descriptions and subsampling for pollen analyses. We are also indebted to Profs. S. Björck (Copenhagen) and H.-J. Beug (Göttingen) for their constructive reviews. The latter colleague kindly made his laboratory at the Institute for Palynology and Quaternary Sciences available for the preparation of the pollen samples. Special thanks are due to C. Siegert for translations of Russian articles and very helpful discussions. Financial support was provided by the German Ministry of Education, Science, Research and Technology (BMBF; Grant No. 03PL014A,B). This is contribution No. 1353 of the Alfred Wegener Institute for Polar and Marine Research, Bremerhaven and Potsdam.

References

Anisimov, M. A and I.N. Pospelov (this volume) The landscape and geobotanical characteristics of the Levinson-Lessing Lake basin, Byrranga Mountains, Central Taymyr.

Archipov, S.A. (1989) Chronostratigraphical scale of the glacial Pleistocene in northern Siberia (in Russian). In:

The Pleistocene in Siberia - Stratigraphy and large-scale correlation, Skabichevskaya, N.A. (ed.), Nauka, Novosibirsk, 20-30.

Behre, K.E. and U. Lade (1986) A succession of the Eemian and 4 Weichselian interstadials in Oerel/Niedersachsen and its vegetation development (in German). Eiszeitalter und Gegenwart, 36, 11-36.

Behre, K.E. and J. van der Plicht (1992) Towards an absolute chronology for the last glacial period in Europe: radiocarbon dates from Oerel, northern Germany. Vegetation History and Archaebotany, 1(2), 111-117.

Björck, S., B. Kromer, S. Johnsen, O. Bennike, D. Hammarlund, G. Lemdahl, G. Possnert, T.L. Rasmussen, B. Wohlfarth, C.U. Hammer and M. Spurk (1996) Synchronized terrestrial-atmospheric deglacial records around the North Atlantic. Science, 274, 1155-1160.

Bolshiyanov, D.Y. and H.-W. Hubberten (1996) Russian-German Cooperation: The Expedition Taymyr 1995. In: The Expedition Taymyr 1995 and the Expedition Kolymy 1995 of the ISSP Pushchino Group, Bolshiyanov, D.Y. and H.-W. Hubberten (eds.), Reports on Polar Research, 211, 5-198.

Daansgard, W., S.J. Johnsen, H.B. Clausen, D. Dahl-Jensen, N.S. Gundestrup, C.U. Hammer, C.S. Hvidberg, J.P. Steffensen, S.E. Steinbjörnsdottir, J. Jouzel and G. Bond (1993) Evidence for general instability of past climate from a 250-kyr ice-core record, Nature, 364, 218-220.

DeBeaulieu, J.-L. and M. Reille (1984) Upper Pleistocene pollen record from Les Echets, near Lyon, France. Boreas, 13, 111-113.

Donner, J. (1996) The Early and Middle Weichselian interstadials in the central area of the Scandinavian glaciations. Quaternary Science Reviews, 15, 471-479.

Ebel, T., M. Melles and F. Niessen (this volume) Laminated sediments from Levinson-Lessing Lake, northern Central Siberia - a 30,000 year record of environmental history?

Galaziy, G.I. and Yu. P. Parmuzin (1981) The lakes on the northwestern Siberian Platform (in Russian). Nauka, Novosibirsk, 187pp.

Gavrilova, M.K. (1981) Modern climate and permafrost on continents (in Russian). Nauka, Novosibirsk, 112pp.

Gerasimov, I.P. and A.A. Velichko (1984) Complex paleogeographic atlases - monographs for the Anthropogene and its prediction value (in Russian). In: Reports presented at the 27th International Geological Congress, Moscow, USSR, Aug. 4-14, 1984, Nikiforova, K.V. et al. (eds.), Vol. 3, Nauka, Moscow, 57-66.

Grichuk, V.P. (1984) Late Pleistocene vegetation history. In: Late Quaternary environments of the Soviet Union, Velichko, A.A. (ed.), Longman, London, 155-178.

Hagedorn, B., S. Harwart, M. Rudgers van der Loeff and M. Melles, M. (this volume) Lead-210 dating and heavy metal concentration in recent sediments of the Lake Lama (Noril´sk area, Siberia).

Hahne, J. (1992) Investigations of the late and post-glacial vegetation history in northern Bavaria (Bayerisches Vogtland, Fichtelgebirge, Steinwald) (in German). Flora, 187, 169-200.

Hahne, J., H. Mengeling, J. Merkt and F. Grammann (1994) Eemian, Weichselian and Saalian deposits of the drill core Quakenbrück GE 2 (in German). In: Neuere Untersuchungen an Interglazialen in Niedersachsen. Geologisches Jahrbuch, Reihe A, Heft 134, 9-69.

Hahne, J. and M. Melles (1997) Late- and post-glacial vegetation and climate history of the south-western Taymyr Peninsula, central Siberia, as revealed by pollen analysis of a core from Lake Lama. Vegetation History and Archaeobotany, 6(1), 1-8.

Isayeva, L.L. (1984) Late Pleistocene glaciation of north-central Siberia. In: Late Quaternary environments of the Soviet Union, Velichko, A.A. (ed.), Longman, London, 21-30.

Khotinskiy, N.A. (1984) Holocene vegetation history. In: Late Quaternary environments of the Soviet Union, Velichko, A.A. (ed.), Longman, London, 179-200

Kind, N.V. (1974) Geochronology of the late Anthropogene from isotope data (in Russian). Nauka, Moscow.

Klementyev, O.L., V.I. Nikolaev, V.Yu. Potapenko and L.M. Savatyugin (1991) Structures and thermodynamic conditions of the glaciers on the Severnaya Zemlya Archipelago (in Russian). Data Glaciol. Stud., 73, 103-109.

Klimanov, V.A. (1989) Paleoclimatic reconstructions in the territory of the USSR for the major Holocene temperature maxima (after palynological data) (in Russian). In: The Pleistocene in Siberia, stratigraphy and large-scale correlation, Skabichevskaya, N.A. (ed.), Nauka, Novosibirsk, 131-136.

Makeev, V.M. (1983) The history of ice-proximal lakes on the Severnaya Zemlya Archipelago (in Russian). In: History of Lakes of the USSR, Vol. 1, Tallin, 122-123.

Makeev, V.M., D.Yu. Bolshiyanov and S.R. Verkulich (1991) Holocene air temperatures (in Russian). In: The Arctic climate regime at the boundary between XX and XXI century, Khrutskiy, B.A. (ed.), Gidrometeoizdat, Leningrad, 160-186.

Mangerud, J., S.T. Andersen, B.E. Berglund and J.J. Donner (1974) Quaternary stratigraphy of Norden, a proposal for terminology and classification. Boreas, 3, 109-128.

Martinson, D.G., N.G. Pisias, J.D. Hays, J. Imbrie, T.C.jr. Moore and N.J. Shackleton (1987) Age dating of the orbital theory of the Ice Ages: development of a high-resolution 0 to 300,000-year chronostratigraphy. Quaternary Research, 27, 1-29.

Melles, M., U. Wand, W.D. Hermichen, B. Bergemann, D.Yu. Bolshiyanov and S.F. Khrutsky (1994a) The Expedition Norilsk/Taymyr 1993 of the AWI Research Unit Potsdam. In: The Expeditions Norilsk/Taymyr

1993 and Bunger Oasis 1993/94 of the AWI Research Unit Potsdam, Melles, M. (ed.), Reports on Polar Research, 148: 1-25.

Melles, M., T. Kulbe, P.P. Overduin and S. Verkulich (1994b) The Expedition Bunger Oasis 1993/94 of the AWI Research Unit Potsdam. In: The Expeditions Norilsk/Taymyr 1993 and Bunger Oasis 1993/94 of the AWI Research Unit Potsdam, Melles, M. (ed.), Reports on Polar Research, 148, 27-80.

Melles, M., C. Siegert, J. Hahne and H.-W. Hubberten (1996) Climatic and environmental history of northern Central Siberia - preliminary results (in German). Geowissenschaften, 14(9), 28-32.

Melles, M., B. Hagedorn, D.Yu. Bolshiyanov (1997) Russian-German Cooperation: The Expedition Taymyr/Severnaya Zemlya 1997. Reports on Polar Research, 237, 170pp.

Mott, R.J. and R.R. Stea (1993) Late-Glacial (Allerød/Younger Dryas) buried organic deposits, Nova Scotia, Canada. Quaternary Science Reviews, 12, 645-657.

Niessen, F., T. Ebel, C. Kopsch and G.B. Fedorov (this volume) High-resolution seismic stratigraphy of lake sediments on the Taymyr Peninsula, Central Siberia.

Rybakova, N.O. (1989) Late Quaternary changes in plant cover and climate in the Kolyma Lowland (in Russian). In: The Pleistocene in Siberia, stratigraphy and large-scale correlation, Skabichevskaya, N.A. (ed.), Nauka, Novosibirsk, 137-142.

Siegert, C., A.Yu. Dereviagin, G.N. Shilova, W.-D. Hermichen and A. Hiller (this volume) Paleoclimatic evidence from permafrost sequences in the eastern Taymyr Lowland.

Spear, R.W. (1993) The palynological record of the Late Quaternary arctic tree line in northwest Canada. Review of Paleobotany and Palynology, 79, 99-111.

Stiévenard, M., V. Nikolaev, D.Yu. Bol´shiyanov, C. Fléhoc, J. Jouzel, O.L. Klementyev and R. Souchez (1996) Pleistocene ice at the bottom of the Vavilov ice cap, Severnaya Zemlya, Russian Arctic. Journal of Glaciology, 42(142), 403-406.

Stuiver, M., P.M. Grootes and T.F. Braziunas (1995): The GISP2 $\delta^{18}O$ climate record of the past 16,500 years and the role of the sun, ocean, and volcanoes. Quaternary Research, 44, 341-354.

Ukraintseva, V.V. (1993) Vegetation cover and environment of the "Mammoth Epoch" in Siberia. Hot Springs, South Dakota, 309pp.

Velichko, A.A. (1993) Evolution of landscapes and climates in northern Eurasia, Late Pleistocene-Holocene, elements and prognosis, 1. Regional Paleogeography (in Russian). Nauka, Moscow, 102pp.

Velichko, A.A., O.K. Borisova, O.K., Kremenetski, C.V., Cwymar, L.C. and MacDonald, G.M. (1996) Climate and vegetation change in the area adjacent to the Laptev Sea during the late Pleistocene and Holocene. Abstract for Third Workshop on Russian-German Cooperation: Laptev Sea System. Terra Nostra, 96/9, 75.

Velichko, A.A., A.A. Andreev and V.A. Klimanov (1997) Climate and vegetation dynamics in the tundra and forest zone during the Late Glacial and Holocene. In: Quaternary of northern Eurasia: Late Pleistocene and Holocene landscapes, stratigraphy and environments, Velichko, A.A. et al. (eds.) Quaternary International, 41/42, 71-96.

Wohlfarth, B. (1996) The chronology of the last termination: a review of radiocarbon-dated, high-resolution terrestrial stratigraphies. Quaternary Science Reviews, 15, 267-284.

Woillard, G. (1979) The last interglacial-glacial cycle at Grande Pile in Northeastern France. Bulletin Societe Belgique de Geologie, 88, 51-69.

Zhurbenko, M.P. (1995) Geobotanical studies. In: Russian-German Cooperation: The Expedition Taymyr 1994, Siegert and D.Yu. Bolshiyanov (eds.), Reports on Polar Research, 175, 25-26.

Laminated Sediments from Levinson-Lessing Lake, Northern Central Siberia - A 30,000 Year Record of Environmental History?

T. Ebel[1], M. Melles[1] and F. Niessen[2]

(1) Alfred-Wegener-Institut für Polar- und Meeresforschung, Forschungsstelle Potsdam, Telegrafenberg A 43, D 14473 Potsdam, Germany

(2) Alfred-Wegener-Institut für Polar- und Meeresforschung, Postfach 120161, D 27568 Bremerhaven, Germany

Revised 5 May 1997 and accepted in revised form 15 March 1998

Abstract - Sediment description and initial stratigraphical, geochemical, and physical analyses were carried out on a 22.4 m long sediment core from the arctic Levinson-Lessing Lake, Taymyr Peninsula (northern Central Siberia). The results reveal a continuous sedimentary history of the lake and its dependence on climatic variations since the late Middle Weichselian. The core consists of two major sediment types, clastic varves and sandy layers, which are linked to seasonal sediment supply related to meltwater runoff and episodic events related to turbidity currents, respectively. Higher frequencies of the event deposits during the Weichselian were presumably initiated by a lower lake level. Climatic warming at the Pleistocene-Holocene transition led to increased biogenic accumulation, originating from both aquatic production and terrestrial supply, but to no significant change in sedimentation rates. From the available data, a glaciation in the lake´s catchment area can be excluded during the period since the late Middle Weichselian.

Introduction

Over the last few years there has been an ongoing debate concerning the extent of the Eurasian glaciation during the Last Glacial Maximum (20 - 18 ka BP). Different opinions range from a minimal ice sheet centered in the Barents Sea (Figure 1) with an isolated small glaciation covering the Putorana Plateau (Velichko et al., 1984) to the assumption of a large ice sheet (maximal variation as shown in Figure 1) spreading from the Kara Sea towards the south, southwest and southeast (Grosswald 1980, 1988; Grosswald and Hughes, 1995). This would have affected large areas of northern Central Siberia, including most of the Taymyr Peninsula. Investigations of potential records of environmental development with respect to the influence of glaciation are therefore of special interest for the studies carried out within the scope of the Taymyr Project.

Levinson-Lessing Lake, with 110 m water depth the deepest lake of the northern Taymyr Peninsula, is situated in the southern part of the Paleozoic fold system of the Byrranga Mountains, 50 km northwest of Taymyr Lake (Figure 2). The lake basin is 15 km long and 1 to 2 km wide. It covers an area of approximately 25 km^2, stretching from the main inflow in the north, the Krasnaya River, to the Protochnaya outflow in the south. Numerous small streams drain into the lake from the adjacent eastern and western slopes. The Levinson-Lessing catchment reaches from 47 m up to 570 m a.s.l., covering an area of roughly 515 km^2.

The geomorphological setting of the basin and its catchment reflects a tectonic origin, reshaped by glacial abrasion during Early Weichselian time (Niessen et al., this volume). Today, large solifluction lobes and - along the lake‘s northwestern shore - tors with a relative elevation of up to 400 m characterize a typical permafrost climatic regime within the catchment.

Levinson-Lessing Lake can be described as a cold, monomictic and apparently holomictic lake, with its single annual turnover occurring just after complete melting of the more than 2 m thick winter ice cover. In Figure 3 (taken from Hagedorn et al., 1996), the waterbody‘s basic

---

In: Kassens, H., H.A. Bauch, I. Dmitrenko, H. Eicken, H.-W. Hubberten, M. Melles, J. Thiede and L. Timokhov (eds.) Land-Ocean Systems in the Siberian Arctic: Dynamics and History. Springer-Verlag, Berlin, 1999, 425-435.

characteristics during spring and summer are shown. The lower pH values in the upper 10 meters observed in the beginning of July are caused by acidic meltwater from snow seeping through the ice cover and from melting of the lake ice itself at the ice/water-interface. Oxygen supersaturation occurs in the still stagnant waterbody when phytoplancton production, initiated by increasing light availability due to higher solar radiation and the improving transparency of the exposed ice cover, starts at the end of June. The minor inverse thermal gradient observed in spring time is due to the cooling effect of the ice cover and ceases just after melting of the lake ice is complete in August.

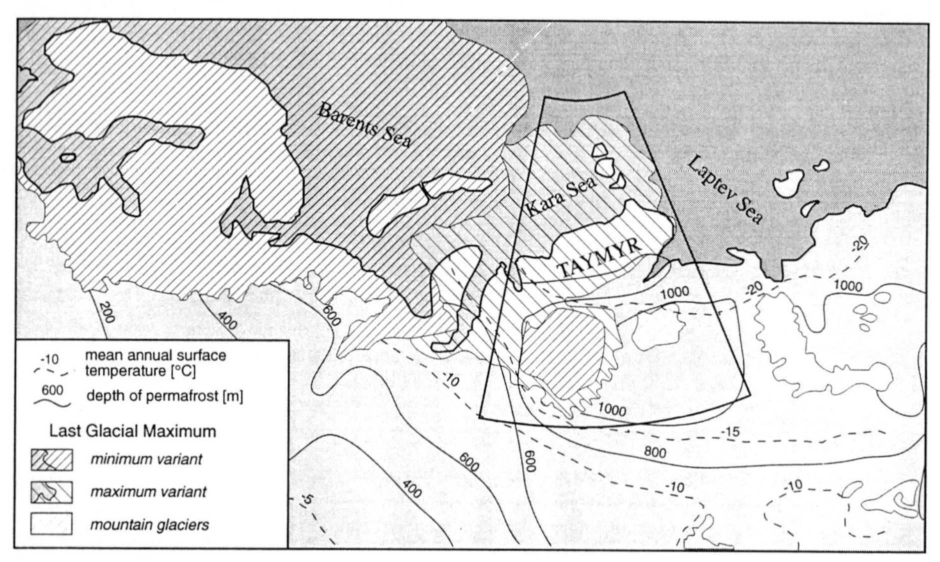

Figure 1: Map of northern Eurasia showing the Taymyr Project study area (encircled) and the extent of the Late Weichselian glaciation (minimum and maximum variants after Velichko et al., 1984)

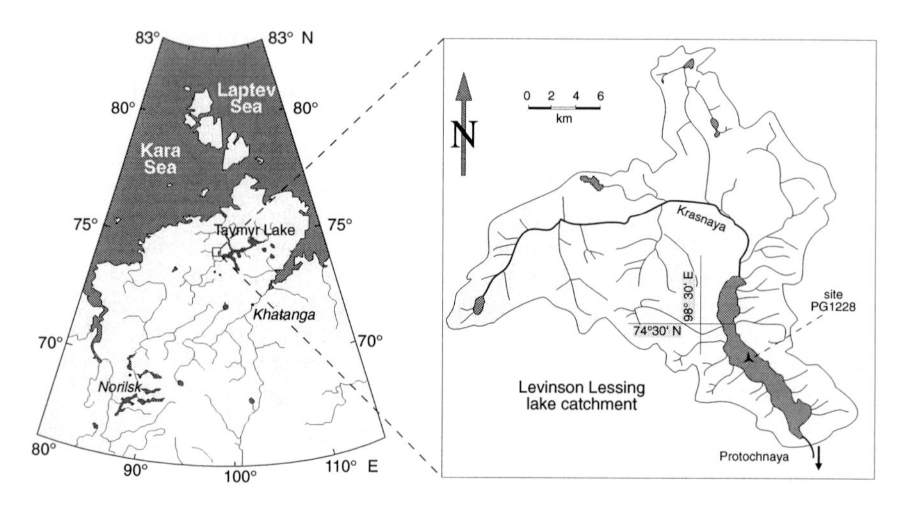

Figure 2: Levinson-Lessing Lake with the location of sediment core PG1228, and the catchment area with the major inflow Krasnaya River in the north, and the outflow, Protochnaya River, in the south. A detailed bathymetric map is published by Niessen et al. (this volume).

Long sediment cores were first recovered from Levinson-Lessing Lake during the 1995 expedition of the multi-disciplinary German-Russian *Taymyr Project* (Overduin et al., 1996). Coring in the central part of Levinson-Lessing lake resulted in a continuous sediment sequence of 22.37 m length. Based on palynological results, the core represents a complete record from the late Middle Weichselian to the present (Hahne and Melles, this volume). According to a seismic survey, it comprises only the upper third of the lacustrine sediments in Levinson-Lessing Lake, and is underlain by at least some additional 35 meters of laminated sediments (Niessen et al., this volume).

In this paper we present a description of the sediment core PG1228, along with sediment stratigraphical, geochemical and physical data. These results provide information concerning the sedimentation processes in Levinson-Lessing Lake and their dependence on climatic variations since the late Middle Weichselian.

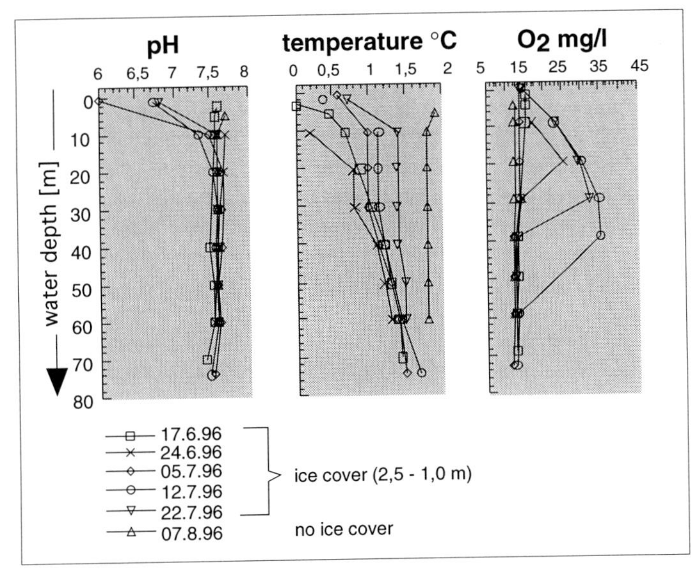

Figure 3: Distribution of pH, temperature, and oxygen content in the water column of Levinson-Lessing Lake at the transition from ice-covered winter and spring conditions to ice-free summer conditions. (Hagedorn et al., 1996).

Material and methods

Lake sediment coring in the central part of Levinson-Lessing Lake (site PG1228; Figure 2) was carried out in spring 1995 from the 2 m thick ice cover at a water depth of 108 m. The 22.37 m long sediment sequence was recovered using two different coring devices (both manufactured by UWITEC, Austria). A light gravity corer, consisting of a PVC core tube (6 cm diameter) and variable weights (4-12 kg) was employed for non-disruptive sampling of the soft near-surface sediments (0 - 27 cm). Longer sediment cores were recovered with a piston corer consisting of a 3 m long steel tube with an exchangeable inner PVC liner (6 cm diameter), equipped with a hydraulic core catcher. The entire gear is operated manually by winches mounted to the legs of a tripod, penetration momentum being supplied by a cylindrical hammer (20 or 40 kg) at the upper end of the coring device. Maximum recovery with every deployment of the piston corer is limited by the tube length to 3 m. Deeper sediments can be sampled by releasing the piston, fixed at the lower end of the corer during its way through the water column and overlying sediment, and thus starting the coring process at defined depths. Nine deployments of the piston corer, with an overlap of approx. 50 cm each, yielded a complete sediment sequence from the sediment/water interface to a depth of 22.37 m. The cores were cut into 1 m sections, sealed with plastic caps and flexible tape and stored in thermostatted boxes at positive temperatures during transport via Khatanga and St. Petersburg to Potsdam. For detailed descriptions concerning the coring technique see Melles et al. (1994).

Prior to opening the cores, physical properties of the sediment were measured using a non-destructive Multi Sensor Core Logging system (Geotek Ltd. UK) at the Alfred Wegener Institute (AWI) in Bremerhaven (described in detail in Weber et al., 1997). Wet bulk densities and porosities were calculated from γ-ray absorption at 1 cm intervals, and susceptibility values were determined correspondingly using a Bartington MS2 loop sensor.

Subsampling of the core segments and additional analyses were carried out at AWI Potsdam. The cores were halved longitudinally; one half was used for sampling and the other for the archive. Following photographic documentation and macroscopic description, subsamples (1 cm thick sediment slices) were taken from the core. They were freeze-dried and ground to < 63 µm particle size. Water content was calculated from the differences between wet and dry weights. Total carbon, sulfur, and nitrogen contents were determined using a CNS-932 Mikro analyzer (LECO Corporation). The total organic carbon content was measured with a CS 100/1000 S (ELTRA) in corresponding samples after treatment with hydrochloric acid (10%) to remove the carbonate-bound carbon. In addition, sediment slabs of 2 x 10 x 1.5 cm were used for the preparation of large-sized thin sections, which were investigated using a petrographic microscope.

Radiocarbon dating was carried out on a sample of terrestrial plant remains found in sufficient amount at a sediment depth of 467 cm in core PG1228. The measurement was conducted by ^{14}C Accelerator Mass Spectrometry (AMS) at the *Research Laboratory for Archaeology and History of Art*, Oxford. The ^{14}C age was calibrated to calendar years BP (before 1950) using the OxCal v2.18 program based on the Stuiver and Reimer (1993) ^{14}C calibration curve.

Results and discussion

Stratigraphy

The stratigraphy of sediment core PG1228, as presented in Figure 5, is based mainly on a correlation of regional pollen assemblage zones (PAZ), determined for this core with radiocarbon dated pollen records from Siberia and other areas of the northern hemisphere (Hahne and Melles, this volume). Holocene sedimentation rates of 0.7 mm/a, calculated from the pollen stratigraphy, are supported by those determined by ^{210}Pb measurements in shallow sediments (Hagedorn, pers. comm.). In addition, the agreement of PAZ LL8 with the Atlantic period is confirmed by the radiocarbon age of 5650 ± 90 ^{14}C yr BP (Lab No.: OxA-6526) of terrestrial plant remains from 467 cm sediment depth, corresponding to a calibrated age of 6670-6280 cal. yr BP.

Lithology and sedimentary structures

The sediment core PG1228 is clearly dominated by minerogenic sediment components. The organogenic components include terrestrial plant detritus, algal material and fossil coal and methane, considerable amounts of which expanded during sediment recovery from about 108 m water depth as a result of a pressure release of far more than 1000 kPa . Vivianite and pyrite accretions occur as mm-scale concretions in the upper 9 m of the sediment column. There is no evidence of postsedimentary destruction of sedimentary structures by bioturbation.

Most terrestrial material is transported into Levinson-Lessing Lake in early summer (June/July) through meltwater supplied from snow fields and the thawing active layer (Bolshiyanov et al., 1995; Gintz et al., 1996). Summer rain events result in irregularly occurring surface runoff which contribute minorly to overall sedimentation. In contrast, direct gravitational or cryogenic sediment supply to the lake (e.g. slumps, solifluction) is negligible, as indicated by the low relief at the lake shores and the absence of ice-rafted gravels in sediment core PG1228.

From the lithology and sedimentary microstructures, two main sediment types are distinguished in core PG1228, reflecting two different sedimentary facies: fine-grained laminae and sandy layers.

The fine-grained laminae (Figure 4), consisting of couplets with silt-sized basal and clay-sized top layers, comprise about 80% of the entire sediment column. Average couplet thickness is 0.7 mm, with small variations throughout the sequence. This corresponds well with the mean annual sedimentation rate of 0.72 mm/a, calculated for the uppermost 467 cm of the sediment sequence from a reliable radiocarbon age from core PG1228 (see above). Similar sedimentation rates are indicated by the pollen stratigraphy, varying between 0.46 and 0.75 mm/a for the Holocene (Hahne and Melles, this volume). ^{210}Pb measurements on a short gravity core about 4 km to the north of site PG1228, towards the Krasnaya inflow, indicate a two- to threefold higher sedimentation rate of ca. 1.5 mm/a (Hagedorn, pers. comm.) This is, however, in agreement with the sediment geometry recorded in sub-bottom profiles (Niessen et al., this volume).

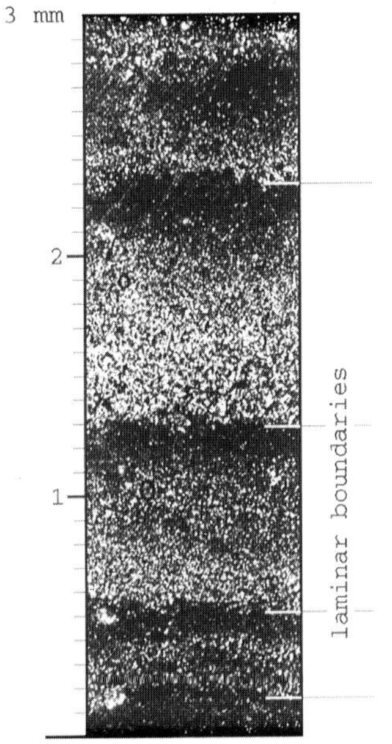

Figure 4: Microscopic photograph of clastic laminations (varves) at 19.85 m sediment depth in core PG1228 under polarized light. Note the sharp lower boundaries and graded bedding of individual laminae.

These regular fine-grained laminae very likely represent annual layers (clastic varves) as described by Sturm and Matter (1978) in alpine lakes. Varve formation in the predominantly monomictic Levinson-Lessing Lake can be explained by the summer sediment supply and winter ice coverage. From the sediment delivered during summer, coarse-grained particles are deposited in the stream channels and their deltas, while most of the fine-grained material

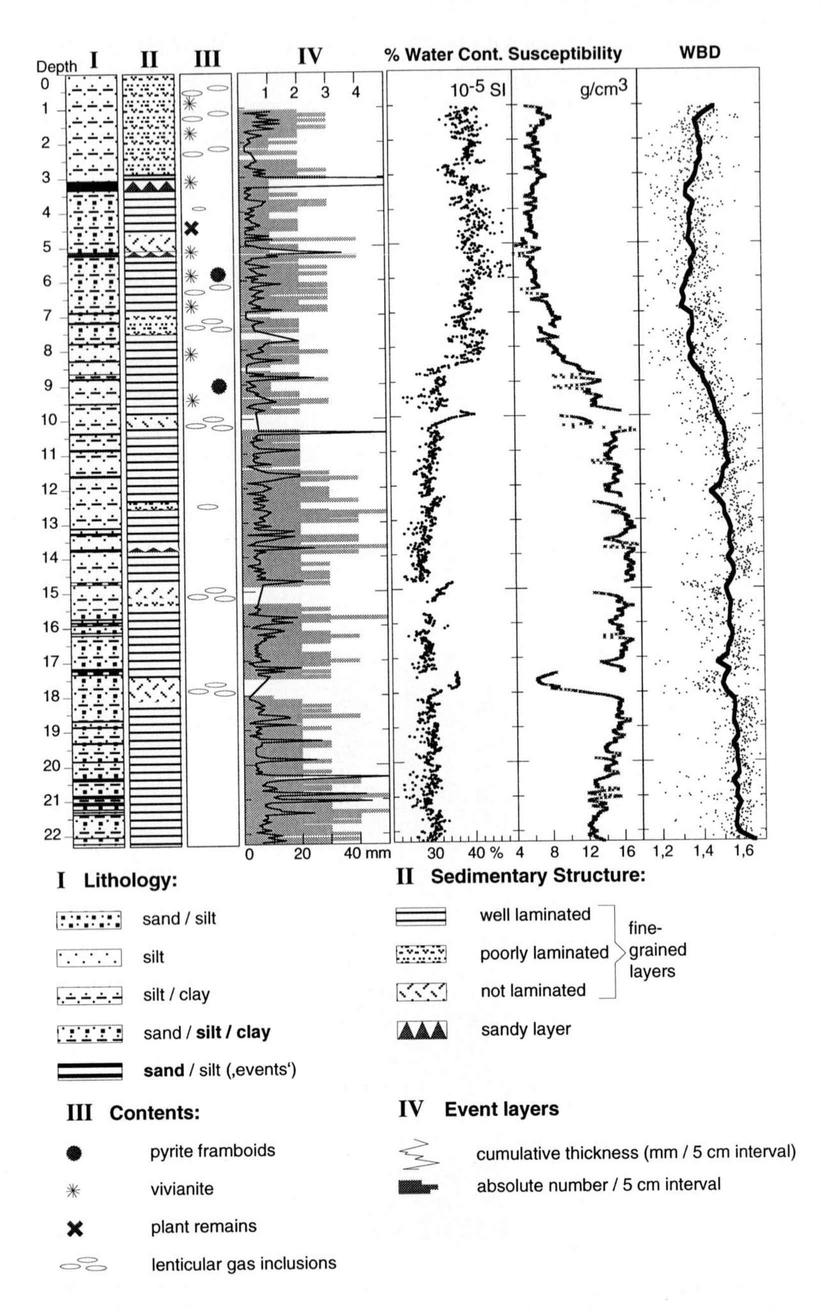

Figure 5: Log of sediment core PG1228 showing **a**) lithology, sedimentary structures, macroscopic description, sand layer frequencies and their cumulative thickness, water contents, susceptibilties, wet bulk densities (WBD) with a running mean (heavy line).

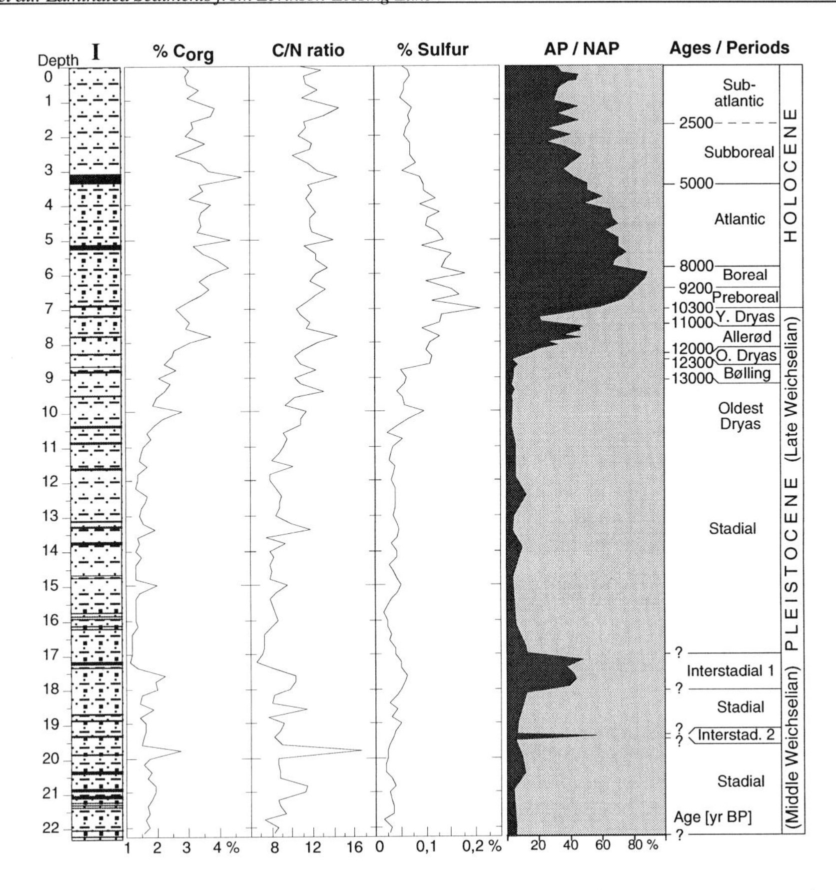

Figure 5: Log of sediment core PG1228 showing **b**) the total organic carbon (C_{org}) contents, total organic carbon to total nitrogen ratios (C/N), sulfur contents, arboreal pollen (AP) contents, and the correlation of regional pollen assemblage zones (PAZs) with corresponding chronozones as defined by Mangerud et al. (1974) and Khotinskiy (1984).

remains in suspension due to continuous circulation forced by wind, wave action and meltwater inflow during that time. In winter, as a result of the lake ice cover, turbulence in the lake water decreases, allowing the settling of finest particles according to Stokes´ Law, and thus completing the annual layer with a clay top.

The sandy layers amount to about 20 % of the sediment core PG1228 and occur irregularly throughout the sequence (Figure 5). Their thickness varies between 2 mm and 20 cm. Homogeneous layers of well sorted sand can be distinguished from layers changing from sand at the base to silty clay at the top. Both types are characterized by sharp lower boundaries. One possible process explaining the formation of these layers could be extreme meltwater freshets during very warm summers, as in 1996, when about 80% of the mean annual Krasnaya River sediment load was discharged within 2 days (Gintz and Meinel, 1997). Events like these could have caused the formation of turbid currents transporting sand-sized particles in suspension to distal areas of the lake. However, the wide range in layer thickness from 2 mm to 20 cm and their rare occurrence, about every 50 years during the Holocene, argue against this mechanism. In addition, cumulative layer thicknesses (Figure 5) show no variation in relation to major climatic changes since the Middle Weichselian.

The most likely process forming the sandy layers are turbidity currents irregularly initiated by collapsing delta front sediments and slumping of sediment material from the steep slopes of the lake bed. As indicated by the absence of erosional channels in sub-bottom profiles crossing the lake (Niessen et al., this volume), such turbidity currents did not lead to a significant resuspension of surface sediments along their flow paths. Differences in the composition of the sandy layers (massive and graded) may be due to different source areas: whilst graded layers may represent distal turbidites from the Krasnaya River, massive layers could represent proximal turbidites or slumping material from littoral areas and delta regions of small rivers entering the lake at its southwestern shore (Figure 2). Additional mineralogical and geochemical data which characterize the source area sediment compositions and allow a comparison with the composition of the sandy layers in the sediment record may support this interpretation.

Similar sediment types (fine-grained laminae and coarse-grained, „surge-type„ layers) and processes were observed in the High Arctic Lake C2, Ellesmere Island, Canada (Zolitschka, 1996; Retelle and Child, 1996). The existence of overflow, interflow, and underflow processes as described in these investigations and by Sturm and Matter (1978) is not likely in Levinson-Lessing Lake due to the very weak stratification of the lake water column.

Whilst the number of sandy layers in core PG1228 decreases slightly in the Holocene sediments, their cumulative thickness is comparable to values in Weichselian sediments (Figure 5). This indicates that turbidites were more frequent during the Weichselian, but had no stronger contribution to the total sediment accumulation, as the individual events were of lower energy than those during the Holocene. Higher turbidite frequencies together with lower current energies during the Weichselian could have been caused by a significantly lower lake level at that time. This interpretation is also supported by sub-bottom profile investigations from Levinson-Lessing Lake (Niessen et al., this volume).

Based on the largely constant contribution of sandy layers to sediment accumulation, and on the with small variations in laminae thicknesses throughout the fine-grained sediments (around 0.7 mm), rather constant long-term sedimentation rates throughout the time interval covered by the core were determined. This implies that no glaciation occurred in the Levinson-Lessing Lake area during that time period. The existence of glaciers within the catchment area would have caused significant changes in both the sedimentation rates and facies.

Moreover, relatively constant sedimentation rates allow an estimation of the ages of Interstadials 1 and 2 occurring in the lower portion of the core (Figure 5). The linear extrapolation of the post-Bølling sedimentation rate of 0.7 mm/a towards the underlying sediments results in ca. 27 and 25 ka BP for Interstadials 1 and 2, respectively. This indicates that those periods with assumed interstadial conditions could correspond with the youngest Middle Weichselian warming in Siberia, which, according to Isayeva (1984), occurred between 24 and 30 ka BP. However, since interstadial conditions in Siberia were also described for several other Middle Weichselian time periods (e.g. Grichuk, 1984; Kind, 1974; Rybakova, 1989) and since a displacement of Holocene material during the coring process cannot be completely excluded, the absolute ages of Interstadials 1 and 2 can only be determined by accurate absolute dating. As bulk sediment ^{14}C dating is regarded to provide unreliable results due to contamination by fossil coal, radiocarbon dating of defined organic fractions (such as humic acids) must be performed and is currently in progress.

Geochemical and physical properties

The geochemical and physical properties of sediments in core PG1228 show a very good correlation with the palynological results, especially with the ratio of arboreal to non-arboreal pollen, which, according to Hahne and Melles (this volume), mirrors variations in Late Pleistocene and Holocene average summer temperatures (Figure 5).

The sedimentary processes influencing both the chemical composition and the physical properties of the sediment have undergone changes dependent on natural climatic variations. The total organic carbon, nitrogen, and sulfur contents show distinct increases at the Pleistocene-Holocene transition (Figure 5). Whilst the values of organic carbon and nitrogen remain at high levels closer to the sediment surface, sulfur is clearly enriched in the Preboreal and Boreal pollen zones, which are believed to represent the Holocene climatic optimum in this area (Hahne and Melles, this volume). Varying contents of all three parameters within the Weichselian sediments are less pronounced, showing somewhat higher values in horizons which exhibit an interstadial character according to pollen data (Interstadial 1 and 2).

The increasing contents of organogenic components in warmer periods are not exclusively the result of enhanced aquatic biogenic accumulation. The total organic carbon contents comprise autochthonous aquatic plant material as well as allochthonous terrestrial plant detritus and fossil coal. Higher C/N ratios during warmer times clearly evidence increasing proportions of vascular plant detritus in the total organic fraction (Meyers and Ishiwatari, 1995). At least for the Holocene, this was probably due to both enhanced availability of plants in the lake´s catchment and enhanced river supply to the lake, as indicated by a denser vegetation cover and the supposedly higher amounts of summer precipitation during that period (Hahne & Melles, this volume).

The higher amounts of organic material, starting at the Pleistocene-Holocene transition, led to chemical changes in the interstitial waters of the sediment. The occurrence of pyrite (FeS) framboids in parts of the post-Oldest Dryas deposits indicates anoxic conditions within the pore water, as well as sufficient sulfate supply for that time, whereas iron oxides are ubiquitous throughout the entire sequence. Autochthonous vivianite ($Fe_3(PO_4)_2 \cdot 8H_20$) points to a complete pyritic fixation of sulfide at these depths, allowing post-sedimentary formation of iron phosphates (Zolitschka, 1990). The anoxic pore waters and increased organic contents are related to increased biological methane formation. Horizons with highest methane contents are rendered macroscopically visible by enrichments of lenticular gas inclusions (Figure 5). Whilst methane remains dissolved and is partly consumed in surface sediments and the water column by methanotrophic bacteria at site PG1228, in the central, deepest part of the lake, in shallow waters it is released to the atmosphere in significant amounts (Samarkin, 1997).

The high gas contents in parts of sediment core PG1228 strongly affected the whole-core measurements of the sediments physical properties - i.e. wet bulk density (WBD) and susceptibility (Figure 5) - resulting in scattered values and under-estimations of both parameters. Nevertheless, their general trends, similarly to that of the water content, show good (positive or negative) correlations with geochemical data. Significant decreases in WBD and susceptibilities towards the upper, Holocene part of the sediment core can be traced back to higher water and methane contents as well as to a higher concentration of organic material with lower density, and related lower concentrations of minerogenic components, including magnetically active mineral grains.

Conclusions

From the description and first analyses of core PG1228 the following conclusions can be drawn on the sedimentation history recorded in the Levinson-Lessing Lake basin since the late Middle Weichselian:

A continuous lacustrine sediment record was recovered from Levinson-Lessing Lake covering at least the last 30,000 years. Sedimentation rates with a mean value of 0.7 mm/a were relatively constant throughout this period.

The sedimentation was dominated (ca. 80% of the whole record) by seasonal fluvial sediment supply, which caused the formation of annual laminations (clastic varves). In addition, episodic events, probably due to turbidity currents originating from littoral and delta areas, led to the accumulation of sandy layers (ca. 20% of the record).

Higher frequencies of lower energy turbidites during the Middle Weichselian can be interpreted as indicators for a lower lake level at that time.

Climate warming during Middle Weichselian interstadials and post-Oldest Dryas led to increased organogenic sedimentation, comprising both autochthonous aquatic plant material and allochthonous terrestrial plant detritus.

No evidence was found for the occurrence of any glaciation in the Levinson-Lessing Lake catchment area since the late Middle Weichselian.

In summary, core PG1228 from Levinson-Lessing Lake provides the longest, laminated sediment sequence recovered from a High Arctic environment. It may be of major importance for the understanding of Late Pleistocene climate-controlled environmental evolution in a part of the world that acts as a key region with regard to recent climatic change debates.

Acknowledgements

Thanks to the members of the 1995 Taymyr Expedition - especially the coring team - for their invaluable help and the warm atmosphere during sometimes difficult conditions. Dr. M. Sturm and Dr. B. Zolitschka are acknowledged for critically reviewing this manuscript and providing helpful commands. Special thanks are due to M. Köhler for thin section preparation at the GeoForschungsZentrum Potsdam. Financial support was provided by the German Ministry of Education, Science, Research and Technology (BMBF; Grant No. 03PL014A,B). This is contribution No. 1351 of the Alfred Wegener Institute for Polar and Marine Research, Bremerhaven and Potsdam.

References

Bolshiyanov, D.Y., D. Ginz and V.P. Zimichev (1995) Surface drainage. In: Russian-German Cooperation: The Expedition Taymyr 1994, Siegert, C. and D.Y. Bolshiyanov (eds.), Ber. Polarforsch. AWI, Bremerhaven, 175, 13-17.

Gintz, D.A. and T. Meinel (1997) Hydrology and suspension transport in the Krasnaya River. In: Russian-German Cooperation: The Expedition Taymyr/Severnaya Zemlya 1996, Melles, M., B. Hagedorn and D.Y. Bolshiyanov (eds.), Ber. Polarforsch. AWI, Bremerhaven, 237, 58-61.

Gintz, D.A., V. Mescherjakov, H. Becker, J. Boike and B. Hagedorn (1996) Hydrological Investigations at the Krasnaja River. In: Russian-German Cooperation: The Expedition Taymyr 1995 and the Expedition Kolyma 1995 of the Pushchino Group, Bolshiyanov, D.Y. and H.W. Hubberten (eds.), Ber. Polarforsch. AWI, Bremerhaven, 211, 92-95.

Grichuk, V.P. (1984) Late Pleistocene vegetation history. In: Late Quaternary environments of the Soviet Union, Velichko, A.A. (ed.). University of Minnesota Press, Minneapolis, 155-178.

Grosswald, M.G. (1980) Late Weichselian ice sheets of northern Eurasia. Quaternary Research, 13, 1-32.

Grosswald, M.G. (1988) An Antarctic-style ice sheet in the Northern Hemisphere: toward a new global glacial theory. Polar Geography and Geology, 12.

Grosswald, M.G. and T.H. Hughes (1995) Paleoglaciology´s grand unsolved problem. Journal of Glaciology, 41(138), 313-332.

Hagedorn, B., J. Boike, T. Ebel, D.A. Gintz and H.-W. Hubberten (1996) Sedimentation processes, hydrology and geochemistry of the Levinson Lessing Lake (Taymyr Peninsula, Siberia). In: Third Workshop on Russian-German Cooperation: Laptev Sea System, Terra Nostra, 96/9, 21-22.

Hahne, J. and M. Melles (this volume) Climate and vegetation development on the Taymyr Peninsula since Middle Weichselian time - palynological investigations on lake sediment cores.

Isayeva, L.L. (1984) Late Pleistocene glaciation of north-central Siberia. In: Late Quaternary environments of the Soviet Union, Velichko, A.A. (ed.). Longman, London, 21-30.

Khotinskiy, N.A. (1984) Holocene Vegetation History. In: Late Quaternary Environments of the Soviet Union, Velichko, A.A. (ed.). University of Minnesota Press, Minneapolis, 179-200.

Kind, N.V. (1974) Geochronology of the late Anthropogene from isotope data [in Russian]. Nauka, Moscow.

Mangerud, J., S.T. Andersen, B.E. Berglund and J.J. Donner (1974) Quaternary Stratigraphy of Norden, a Proposal for Terminology and Classification. Boreas, 3, 109-128.
Melles, M., T. Kulbe, P.P. Overduin and S. Verkulich (1994) The Expedition Bunger Oasis 1993/94 of the AWI Research Unit Potsdam. In: The Expeditions Norilsk/Taymyr 1993 and Bunger Oasis 1993/94 of the AWI Research Unit Potsdam, Melles, M. (ed.), Ber. Polarforsch. AWI, Bremerhaven, 148, 35-38.
Meyers, P.A. and R. Ishiwatari (1995) Organic matter accumulation records in lake sediments. In: Physics and Chemistry of Lakes, Lerman, A., D.M. Imboden and J.R. Gat (eds.). Springer, 279-328.
Niessen, F., T. Ebel, C. Kopsch and G.B. Fedorov (this volume) High-resolution seismic stratigraphy of lake sediments on the Taymyr Peninsula, Central Siberia.
Overduin, P.P., D. Bolshiyanov and T. Ebel (1996) Lake sediment sampling on the Taymyr Peninsula. In: Russian-German Cooperation: The Expedition Taymyr 1995 and the Expedition Kolyma 1995 of the Pushchino Group, Bolshiyanov, D.Y. and H.-W. Hubberten (eds.), Ber. Polarforsch. AWI, Bremerhaven, 211, 111-121.
Retelle, M.J. and J.K. Child (1996) Suspended sediment transport and deposition in a high arctic meromictic lake. J. Paleolimnol., 16, 151-167.
Rybakova, N.O. (1989) Late Quaternary changes in plant cover and climate in the Kolyma Lowland [in Russian]. In: The Pleistocene in Siberia, stratigraphy and large-scale correlation [in Russian], Skabichevskaya, N.A. (ed.), Nauka, Novosibirsk, 137-142.
Samarkin, V.A. (1997) Methane Biogeochemistry in Levinson-Lessing Lake. In: Russian-German Cooperation: The Expedition Taymyr/Severnaya Zemlya 1996, Melles, M., B. Hagedorn and D.Y. Bolshiyanov (eds.). Ber. Polarforsch. AWI, Bremerhaven, 237, 67-70.
Stuiver, M.A. and P.J. Reimer (1993) Extended ^{14}C data base and revised CALIB 3.0 ^{14}C age calibration program. Radiocarbon, 35(1), 215-230.
Sturm, M. and A. Matter (1978) Turbidites and varves in Lake Brienz (Switzerland); deposition of clastic detritus by density currents. In: Modern and ancient lake sediments; proceedings of a symposium, Matter, A. and M. Tucker (eds.). Blackwell, Oxford, 147-168.
Velichko, A.A., L.L. Isayeva, V.M. Makeyev, G.G. Matishov and M.A. Faustova (1984) Late Pleistocene Glaciation of the Arctic Shelf, and the Reconstruction of Eurasian Ice Sheets. In: Late Quaternary Environments of the Soviet Union, Velichko, A.A. (ed.), University of Minnesota Press, Minneapolis, 35-41.
Weber, M.E., F. Niessen, G. Kuhn and M. Wiedicke (1997) Calibration and application of marine sedimentary physical properties using a multi-sensor core logger. Marine Geology, 136, 115-172.
Zolitschka, B. (1990) Late Quaternary annually laminated Lake Sediments from selected Eifel-Maar Lakes [in German]. Documenta Naturae, 60, München, 226 pp.
Zolitschka, B. (1996) Recent sedimentation in a high arctic lake, northern Ellesmere Island, Canada. J. Paleolimnol, 16, 169-186.

High-Resolution Seismic Stratigraphy of Lake Sediments on the Taymyr Peninsula, Central Siberia

F. Niessen[1], T. Ebel[2], C. Kopsch[2] and G.B. Fedorov[3]

(1) Alfred-Wegener-Institut für Polar- und Meeresforschung, Postfach 120161, D 27568 Bremerhaven, Germany

(2) Alfred-Wegener-Institut für Polar- und Meeresforschung, Forschungsstelle Potsdam, Telegrafenberg A43, D 14473 Potsdam, Germany

(3) State Research Center - Arctic and Antarctic Research Institute, 38 Bering St., 199226 St. Petersburg, Russia

Revised 5 May 1997 and accepted in revised form 10 January 1998

Abstract - High-resolution seismic profiles (GeoChirp, 1.5-11.5 kHz) were recorded in two lakes on the Taymyr Peninsula during the Taymyr Expedition 1996 in order to reconstruct changes in the depositional environment and to test different hypotheses of glaciation in central Siberia. Four major seismic units of mostly well-stratified sediments are identified in both Lakes Taymyr (T1-T4) and Levinson-Lessing (L1-L4). The sequential stratigraphic approach is used to explain different unit geometry by lake level changes. The lowermost units L3, L4 and T4 cover hummocky topography and are interpreted as high-stand deposits. Overlying units L2 and T3 occur only in the deeper part of the basin and onlap against the slope. These deposits represent a lower lake level followed by a transgression during the Holocene (L1, T1 and T2). The chronology of the lower units is somewhat speculative. Correlation of our evidence with the results of other studies suggest that, in Lake Taymyr, the high-stand unit T4 was probably deposited during Mid to Late Weichselian, and the low-stand unit T3 during Latest Weichselian times. In Lake Levinson Lessing, high-stand units L4 and L3 are possibly older than Mid-Weichselian overlain by low-stand deposits (L2) of Mid to Late Weichselian age as indicated by a pollen-dated sediment core penetrating into unit L2. In total, the fills of the basins comprise unconsolidated muds and sands of more than 20 m in Taymyr and more than 60 m in Levinson Lessing. The units L4 to L1 are undisturbed, whereas T4 to T1 are partly erosive. None of the units were overconsolidated by glacier ice. This implies that the last major glaciation of the lake basins occurred earlier than the Mid-Weichselian, possibly during the Early Weichselian (Marine Isotope Stage 4).

Introduction

Seismic stratigraphy is commonly used to characterise the variation of depositional environments in space and time (Bally, 1987). Subbottom profiling in lakes is a convenient tool to identify major depositional episodes which usually match lithological units in sediment cores (e.g. Giovanoli et al., 1984; Svendsen et al., 1989; Hubberten et al., 1995). Seismic studies in lakes on the Taymyr Peninsula (Figure 1) are part of an ongoing investigation of environmental and climatic history in northern Mid-Siberia during the late Quaternary (Bolshiyanov and Hubberten, 1996; Melles et al., 1996; Hahne and Melles, this volume; Hahne and Melles, 1997; Melles et al., 1997). During an earlier expedition in 1995, two long sediment cores were retrieved from the deepest parts of Lakes Taymyr and Levinson-Lessing (Overduin et al., 1996). Pollen stratigraphy from one of the cores (PG1228, Levinson-Lessing) suggests a Mid-Weichselian basal age (Hahne and Melles, this volume). None of the cores reached basal moraines or bedrock. At the beginning of this study, therefore, the deeper part of the sediment fills as well as the lateral distributions of sediment units were completely unknown.

Here we present the first seismic results of spatial/temporal stratigraphic pattern from Lakes Taymyr and Levinson-Lessing. The field work is described in Niessen et al. (1997). The aim of this paper is to interpret key profiles in order to define a seismic stratigraphy of the lake fills,

In: Kassens, H., H.A. Bauch, I. Dmitrenko, H. Eicken, H.-W. Hubberten, M. Melles, J. Thiede and L. Timokhov (eds.) Land-Ocean Systems in the Siberian Arctic: Dynamics and History. Springer-Verlag, Berlin, 1999, 437-456.

and to compare the seismic results with sediment core data. The following questions are addressed: (i) How thick is the undisturbed post-glacial fill of the lacustrine basins? (ii) Is there evidence in the seismic stratigraphy for past changes in the sedimentary environments? (iii) What conclusions can be drawn about the timing of the last major glaciations in the area?

The area of the Taymyr Peninsula (Figure 1) is of particular interest for palaeoclimatic research because of the ongoing debate concerning the extent and chronology of glaciations in Central Siberia during Weichselian time. Various reconstructions have been proposed for the last glacial maximum. These include: (i) small isolated ice caps over the archipelagos of the Eurasian shelf and over the Putoran Plateau in Central Siberia but mostly ice-free conditions over the Taymyr Peninsula (Dunayev et al., 1988; Velichko et al., 1984); (ii) a continuous ice sheet over the Barents and Kara Seas including Taymyr (Elverhøi et al., 1993); and (iii) a large ice-sheet of more than 2,500 m thickness which covered the entire Siberian shelf and extended far into continental Eurasia (Grosswald, 1988).

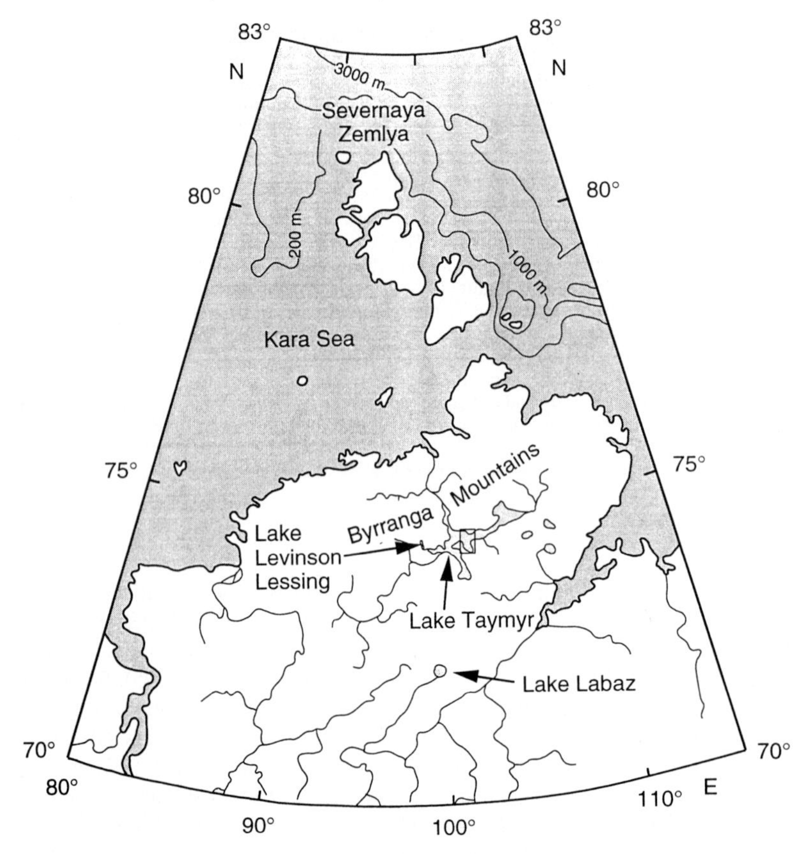

Figure 1: Taymyr Peninsula and location of the investigated lakes Levinson Lessing and Taymyr.

Investigated lakes

Lake Taymyr is the largest water body on the Taymyr Peninsula (Figure 1). The lake is situated south of the Byrranga Mountains and drains northward through the orogeny into the eastern Kara Sea. From west to east the basin is more than 200 km long and mostly between 10 and 20 km wide. It is divided into several sub-basins. Approximately 70% of the lake area has a water depth of less than 3 m (Overduin et al., 1996) although the exact bathymetry of the lake is not known. Seismic studies were carried out only in the central part of the lake (Figure 1) where the water depth reaches more than 20 m.

Levinson Lessing is situated in the southern Byrranga Mountains and drains through the Ledyannaya River into Lake Taymyr. A map of the lake catchment is presented in Ebel et al. (this volume). The lake is about 15 km long and mostly between 1 and 1.5 km wide. The maximum water depth is about 110 m in the central part (Figure 2). Numerous streams drain into the lake which form small fan systems. The largest river, Krasnaya, enters the lake from the north where a relatively large fan system has been built up. The northern part of the lake valley as well as some of the adjacent valleys are characterised by typical U-shaped cross sections suggesting a glacial origin or overprint (Niessen et al., 1997)

Both lakes are dominated by clastic sedimentation. A complete winter freeze-over occurs in September and October and the ice persists until July/August. Normally, the lakes become completely ice-free during the summer months when the main clastic input takes place.

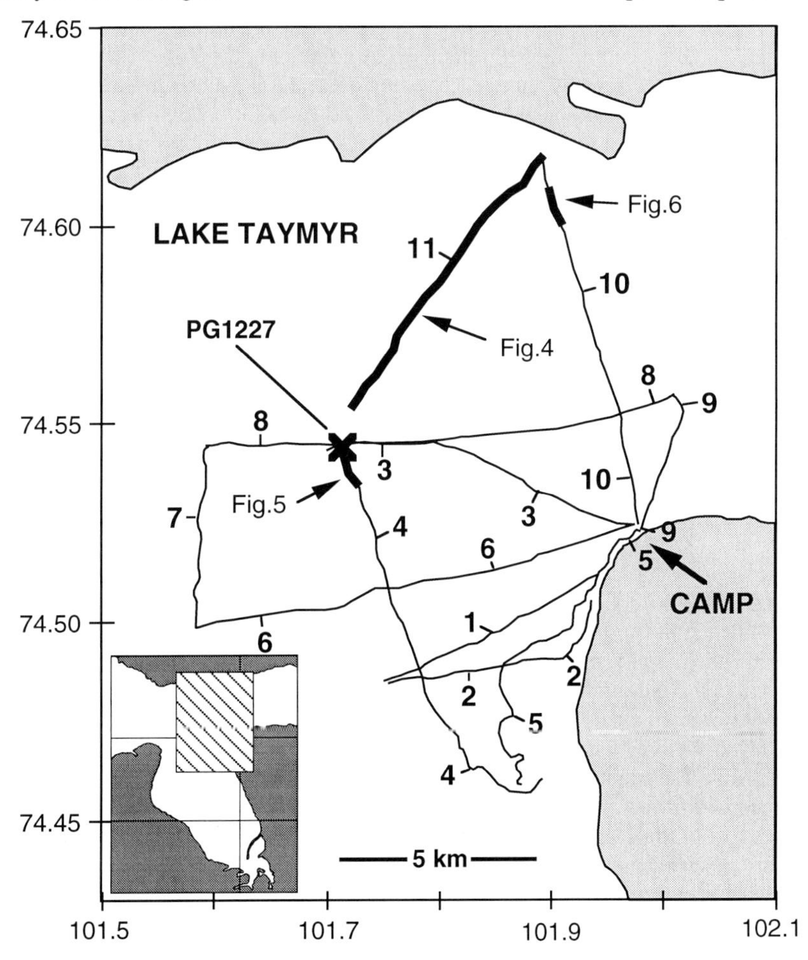

Figure 2: High resolution seismic profiles and coring location (PG1227) in central Lake Taymyr (location of detailed map see Figure 1).

Methods

An inflatable boat was equipped with a portable "Chirp" sediment echo sounding system (GeoChirp 6100A, Geoacoustics, UK) for continuous subbottom profiling during the Taymyr expedition in summer 1996. The GeoChirp was modified at the Alfred Wegener Institute by combining it with a digital delay box (for operation in deeper water), a GPS receiver (Trimble Scoutmaster, for positioning) and a four-channel DAT tape recorder (Sony PC 204A, for data storage of seismic trigger, seismic signal and GPS position). A detailed system description is given by Niessen et al. (1997). The GeoChirp system offers the choice of two different modes: sweeps of 2 to 8 kHz for deeper penetration and of 1.5 to 11.5 kHz for higher resolution. For both modes, the sweep length is 32 ms. The returning sweep of signals is processed in the Chirp Transceiver over a period of 130 ms during which a cross correlation of the received echo pulses is performed. The profiles are plotted on a chart recorder in analog mode. We used a P-wave velocity of 1500 m s-1 to calculate sediment depths from the two-way travel times.

A total of 11 profiles (103 km) and 42 profiles (ca. 59.1 km) were recorded in high resolution mode on Lakes Taymyr and Levinson Lessing, respectively. On Levinson Lessing an additional number of 24 profiles (about 38 km) were recorded in high penetration mode (Figure 2 and Figure 3). GPS positioning was optimised by combining the Trimble data output on the lakes with GPS data received by a reference station on land (GPS-DAN linked to a power book by PCMCIA-port). Data sampling rate and storage was 2 sec for both GPS systems. Since this combination does not provide exact differential GPS data, post processing was carried out at the Institut für Geodäsie der Universität Dresden. The accuracy was improved from +/- 500 m (unprocessed) to +/- 10 m (processed). If not stated otherwise, depths reported in this paper are expressed in metres below present lake level (m b.p.l.l.).

The water content of sediments from core PG1228 was determined from the wet and dry weights of sub-samples. Water contents of PG1227 are based on calculations from the wet bulk densities assuming a constant grain density of 2.65 g cm^{-3}. Wet bulk density is measured by a gamma-ray absorption sensor installed on a Multi-Sensor Core Logger (Geotek Ltd., UK). The of whole-core gamma-ray logging method and the calculation of water contents from logging data are described in detail in Weber et al. (1997).

Description of seismic stratigraphies

Lake Taymyr

In Lake Taymyr, sound penetration is observed to a sediment depth of up to 25 m. Over large areas of the central lake, however, there is strong backscatter from the sediment surface which results in diffraction (no penetration) or a diffuse appearance of deeper reflectors (Figure 4, below 21 m water depth). In many profiles strong reflectors can also be seen subbottom which occur only locally and often terminate abruptly (profile 4 between 38 m and 45 m, Figure 5; profile 10 between 20 m and 28 m, Figure 6). Therefore, lateral correlation of stratification and seismic units over long distances is often not possible. This is particularly true for the deeper part of the fill. Stratigraphic features and distinct variability in unit geometry are best seen in profiles recorded from 15 to 21 m water depth (e.g. profile 11, Figure 4). A link to profile 4 in the deepest part of the lake (Figure 5, location of sediment core PG1127) can only be achieved on the basis of seismic interpretation because there is a lateral gap in the data set over a distance of about 1 km between profiles 10 and 4 due to technical problems in the field (Figure 2).

Unit T1

The topmost unit T1 is present in all areas observed. The unit is stratified although the reflectors

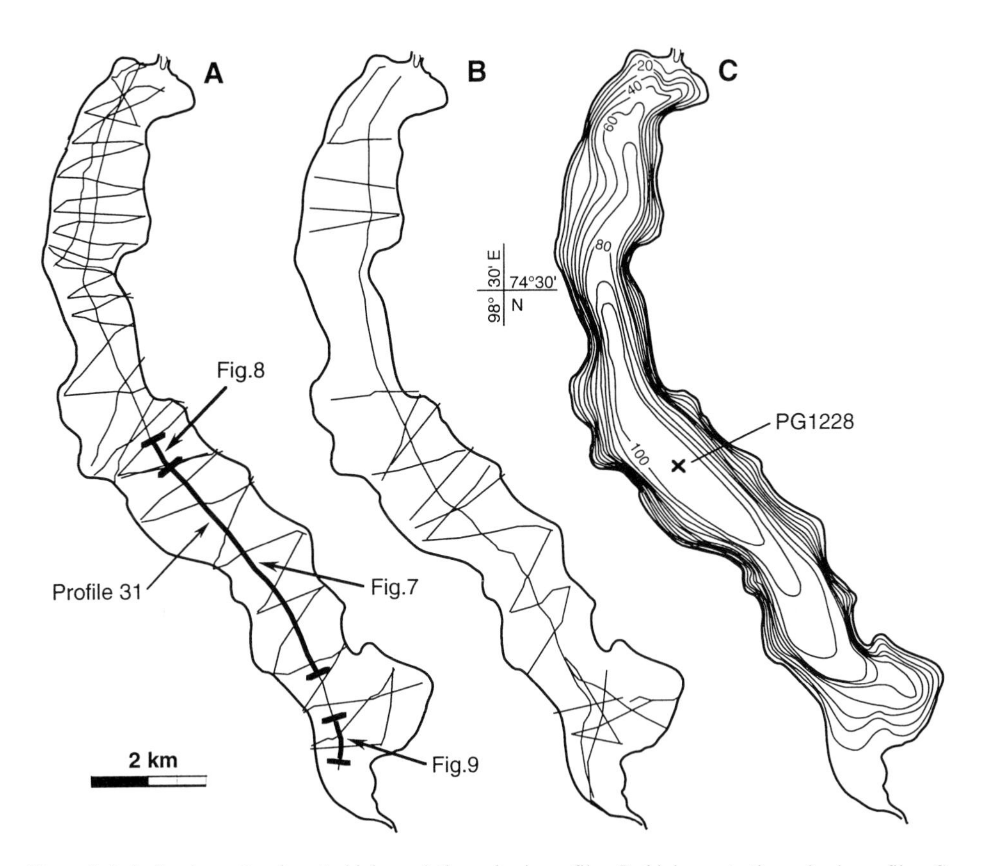

Figure 3: Lake Levinson Lessing. A: high resolution seismic profiles, B: high penetration seismic profiles, C: bathymetry and coring location (PG1227). Geographic coordinates refer to map C.

appear slightly diffuse, in particular in shallower water (Figures 4 and 6). There is a decrease in unit thickness from about 12 m in the deepest part of the lake (Figure 5) to less than 2 m between 10 and 20 m water depth (Figures 4 and 6).

Unit T2

T2 is distinguished from T1 by the fact that T2-sediments are not deposited or strongly condensed in shallow water (Figures 4 and 6). Also, the T1/T2 boundary is often erosive in water depth above about 25 m (Figure 4). In deeper water, thicknesses of T2 rapidly increase to 10 m or more (Figure 4). The backscatter characteristic is similar to that of unit T1. This explains why the T1/T2 boundary is not distinctive in the deepest part of the lake (Figure 5) where no erosion or change in the unit geometry is observed. In Figure 5 the placement of the boundary is based on tracing the reflector to the southerly end of the profile where truncation of T2 sediments becomes visible. There are no significant changes in the sediment core PG1127 at the T1/T2 boundary (Figure 5). Both units comprise massive muds (Ebel pers. comm.) with relatively high water contents (Figure 5).

Unit T3

The top of unit T3 is marked by a strong reflector near the deepest part of the lake. This reflector is oblique to the overlying unit T2 and correlates with a strong decrease in water content and an discontinuous change from mud to poorly bedded sands in core PG1227 (Figure 5). There is no information on the geometry, thickness and backscatter characteristics of unit T3 in the profundal area of the lake because all acoustic energy is reflected at the top of T3 in Profile 4 (Figure 5). As a result of the limited sound penetration, T3 is not seen in the deeper part of profile 11 below 25 m. Above 40 m T3 appears to be condensed to a strong reflector forming the T4/T2 boundary along an unconformity (Figure 4).

Unit T4

Unit T4 is defined only in profiles from relatively shallow water (Figures 4 and 6). T4 is well stratified and the sediments drape subbottom topography (Figures 4 and 6). The reflection geometry indicates an increase in unit thickness towards shallower water. In the shallow part of profile 10 (Figure 6), T4 is characterised by the local occurrence of strong reflectors discontinuous to the stratification and by numerous small-scale diffraction hyperbolae. Both patterns are indicative of upwardly migrating sedimentary gas, locally migrating even into the overlying unit T1, where a pock mark is observed on the lake bottom (Figure 6). The shallow part of profile 10 is located near the delta of a stream draining a catchment to the north of the lake.

Reflectors of T4 are truncated at the top of the unit, which defines a distinct unconformity (Figure 4, Figure 6). Thus, the original thickness of unit T4 cannot be reconstructed. It might have been well in excess of 20 m thick. The lower boundary of T4, however, is clearly seen in shallow water profiles (e.g. profile 11, Figure 4), where it is marked by a strong reflector below 22 m, which defines a hummocky surface. Diffuse reflections from below this strong reflector indicate that a further sedimentary unit of unknown thickness may be underlying T4.

Lake Levinson Lessing

In Lake Levinson-Lessing maximum sound penetration was observed to sediment depths of about 60 m in high-penetration profiles (2-8 kHz). The relatively uniform, well-stratified sediment fill thins toward the southern end of the lake (Figure 7) so that a strong basal reflector becomes visible if the sediment fill is less than 40 m thick (Figure 9). Above the basal reflector there are four seismic units which are best seen in water depths between 40 and 20 m (Figure 9). A characterisation of seismic stratigraphy is not possible in the northern part of the lake because sound penetration is limited to only a few metres, in particular near the Krasnaya delta. Because the entire south-north profile (no. 31, Figure 3) cannot be presented for geometric reasons, unit thicknesses were measured at 22 section intervals between 106 m water depth in the central part (coring location PG1228, Figure 7) and 18 m water depth near the outlet in the most southerly part of the lake (Figure 9). The results are presented in Figure 10.

Unit L1

Unit L1 is well stratified in deeper water and slightly diffuse in the shallower parts of the profile near the outlet (Figure 8 and Figure 9). The reflectors show strong backscatter so that L1 appears as a dark horizon overlying the deeper part of the fill without unconformity (Figure 7). The unit thickness steadily decreases from 8.5 m at the coring location to 5 m at a distance of 4.5 km to the south of the coring location (Figure 7). A strong decrease to less than 2 m thickness is observed in shallow water at the southern end of the profile (Figure 9 and Figure 10). Subbottom topography is draped with sediment below 35 m b.p.l.l.. For unit L1, the

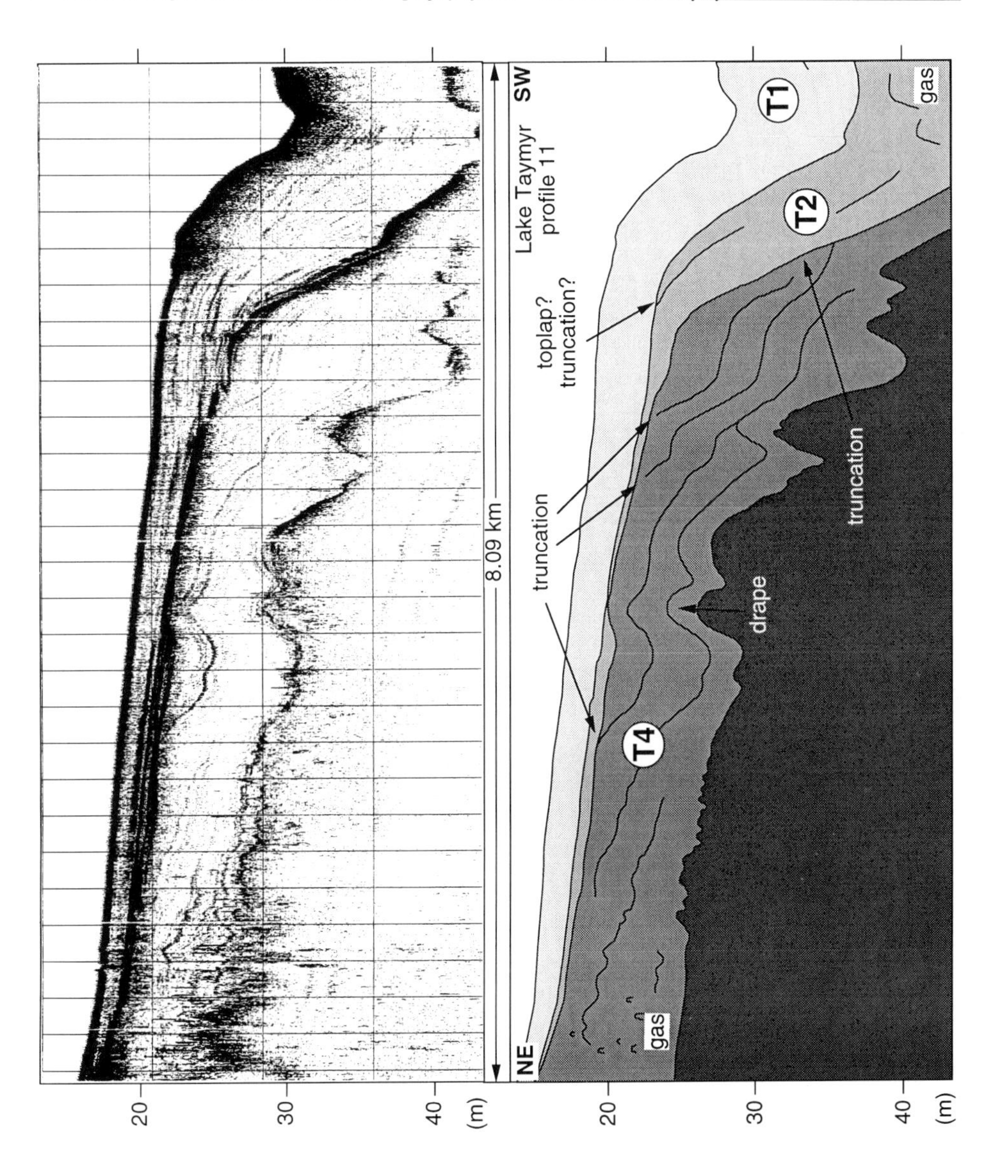

Figure 4: Seismic profile 11 and seismic units from Lake Taymyr.

sediment core PG1228 comprises laminated silt and clay with distinct sand layers. Water content is above 40% and shows relatively strong variability, in particular above 6 m core depth (Figure 8).

Unit L2

In general, unit L2 is characterised by the strongest lateral variability of sediment thickness from about 30 m in the central part to about 2 m at 38 m (Figure 10). Onlaps show that the unit lenses out above 33 m (Figure 9). Similar to the unit L1, L2 sediments drape subbottom topography above small topographic highs in the profundal area of the lake (Figure 7, Figure 8). Like L1, unit L2 is well stratified but reflectors show generally weaker backscatter (Figure 7). In the central part of the lake, at about 18 m subbottom, a set of strong reflectors is present which can also be seen in the "drape" above topographic highs (Figure 8). These reflectors prevent deeper sound penetration in most parts of the profundal area. Abrupt onsets and lateral discontinuity of the reflectors are typical for a high content of sedimentary gas.

Core PG1228 penetrated the top 13 m of the unit L2 (Figure 8). The sediments are laminated and of similar texture to those of unit L1. The L1/L2 boundary is marked by a significant downcore decrease in water content as well as a decrease in the water-content variability (Figure 8). Strong subbottom reflectors (gas) occur over a depth interval where pollen results suggest a rapid alternation of stadial and interstadial periods (Figure 8).

Unit L3

Unit L3 is characterised by draping subbottom topographies even in shallower water above 35 m, and by relatively weak backscatter (Figure 9). The lower boundary can only be traced to a sediment depth of about 40 m where a unit thickness of 17 m is observed (Figure 10). With increasing water depth, the L2/L3 boundary becomes less distinct. In the deepest part of the lake the boundary is not seen in high resolution-mode profiles because sound penetration through unit L2 is limited. Near the coring location, however, the lower boundary of L3 is located deeper than 56 m subbottom because high penetration profiles show stratified deposits down to that sediment depth.

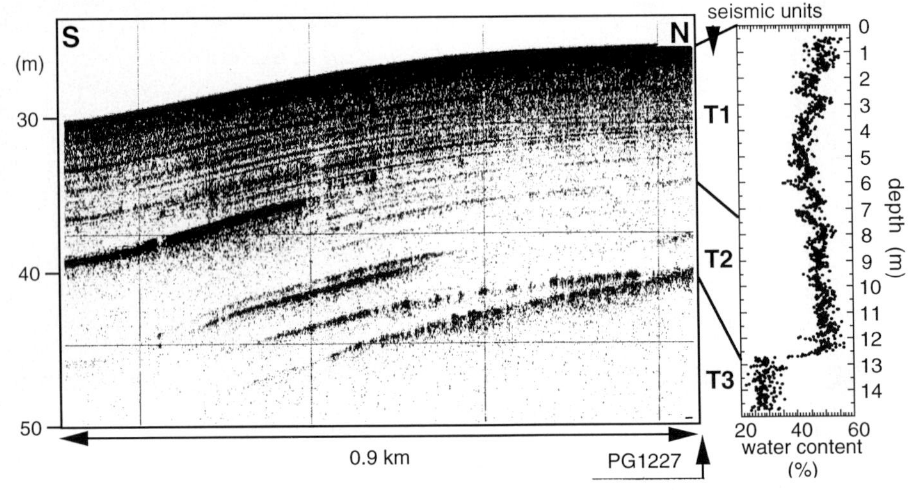

Figure 5: Water content of core PG1227 correlated with seismic units of the northern end of profile 4 from Lake Taymyr.

Unit L4

Unit L4 is largely transparent and characterised by variable thicknesses above an irregular topography marked by a strong basal reflector (Figure 9). The latter can be traced to a water depth of only 40 m. Below that depth the sediment fill is too thick to allow sufficient sound penetration down to the basal reflector. In the shallow area of the lake, the unit thickness is generally a few metres (Figure 9). L4 thickness remains undetermined in the deeper part of the lake and can only be estimated by extrapolation (Figure9). The relative flat top of L3 and its irregular bottom indicate that the sediments tend to fill depressions and level subbottom topography.

Interpretation of sequences, depositional environments and unit chronology

Sequence stratigraphy and lake level changes

The sequential stratigraphic approach to linking unit geometry to sea-level changes is well established for marine records (Bally, 1987). The seismic units of both Lakes Taymyr and Levinson Lessing include geometric patterns very similar to typical High Stand System Tracks (HST), Low Stand System Tracks (LST) and Transgressive System Tracks (TST) as outlined by Bally (1987) for marine shelf and slope environments. Therefore, key profiles from shallow to deeper water from Lake Taymyr (profile 11, Figure 4) and Levinson Lessing (profile 31, Figure 9) are used to interpret past lake level changes (Figure 11).

The drape-geometry and increase of sediment thicknesses of the units L4, L3 and T4 toward the present shore are indicative of a higher lake level during deposition (HST) compared to the present situation. For example, the deposition of large thicknesses of T4 sediments in Lake Taymyr, as observed in profile 11 (Figure 4), is hardly possible unless a higher lake level of at least 20 m above present is assumed. A similar situation is interpreted for Lake Levinson Lessing during the deposition of unit L4 and L3 (Figure 9).

The high-stand phase is followed by a significant drop of lake levels during which parts of the HST deposits (e.g. T4) were eroded. In Lake Taymyr, this unconformity forms a typical system boundary in the sense of Bally (1987) which can be traced down to 40 m b.p.l.l.(Figure 4). Thus, the lake level must have dropped significantly below the present level, possibly to more than 40 m b.p.l.l.. The geometry of a LST unit is not seen in Lake Taymyr due to limited sound penetration. However, sequence stratigraphy suggests that a LST unit T3 must be overlying T4 in the deeper part of the basin (Figure 11) because the truncation of L4 sediments in shallow areas implies deposition in the basin. Unit T3 probably consists of a large amount of reworked T4 sediments. Also, the basal sand of core PG1227, which forms the top of unit T3 (Figure 5), is evidence for a low-stand phase in Lake Taymyr. According to a preliminary interpretation by Overduin et al. (1996) the deposition of the sand is associated with alluvial deposition . This suggests a very low lake level or even non-existence of Lake Taymyr at the time T3 was deposited. In Lake Levinson Lessing, LST deposits are indicated by L2 geometry, in which adjacent sediments indicate that L2 lenses out between 30 and 40 m b.p.l.l. (Figure 9).

The low-stand is followed by a transgressive phase (Figure 11) where sediments successively accumulate on the upper slope (L1, T2). Finally, deposition reappears in present day shallow water areas of the lakes (L1, T1). The lake level remains intermediate (Figure 11) as is indicated by the increase of unit thicknesses toward the basin (and thus sediment focusing) which is not seen in HST deposits (Figures 4 and 9). In Lake Taymyr, TST consists of two units, T1 and T2, of which the boundary is, in places, erosive (Figure 4). This is interpreted to be caused by erosion following a slight drop in lake level during the transgressional phase as indicated in Figure 11. Such an unconforiaty is not seen in TST deposits of Lake Levinson Lessing (L1) so that a lake level fluctuation during the transgression may not have occurred there.

The Holocene - the transgressive phase

The transgressional phase during which the units L1, T1 and T2 accumulated is associated with Holocene deposition. Between 9 and 7 m, core PG1228 from Levinson Lessing has clear

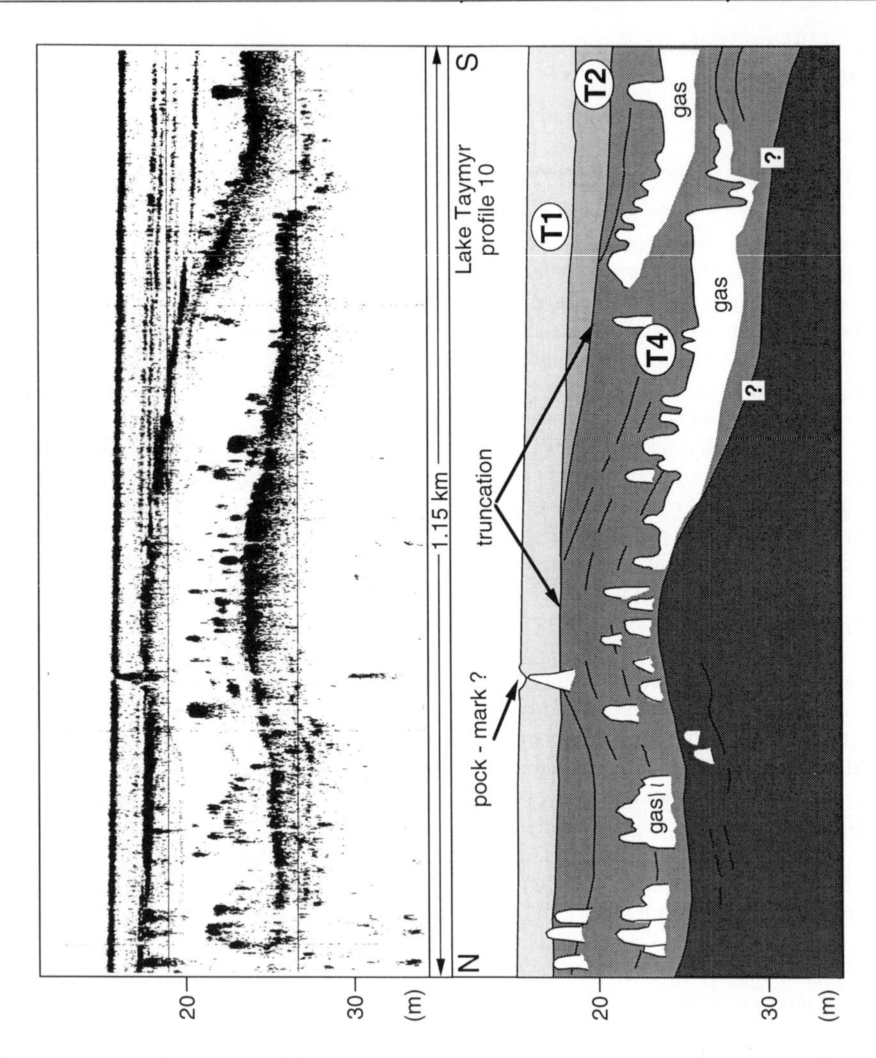

Figure 6: Seismic profile 10 and seismic units from Lake Taymyr.

pollen evidence for the Pleistocene-Holocene boundary, including the Allerød, Bølling and Younger Dryas Chronozones (Hahne & Melles this volume) which correlate with the L1/L2 boundary (Figure 8). In deep water, the draping of L1 sediments over subbottom topography suggests deposition from suspension by which settling from "pelagic rain" is probably more prominent than deposition from turbidity flows. The latter would be associated with slump deposits and distinct onlaps which are not common in the seismic record of the central and southern part of Levinson Lessing. The relatively strong backscatter of the L1 reflectors can be explained by a large variability in the water content and thus wet bulk density, which defines a pattern of strong impedance contrasts in Holocene sediments (Figure 8). The steady decrease of sound penetration associated with increasing Holocene unit thickness from the southern and central part of the lake to the north (Figure 7) suggests that clastic sediment input from the Krasnaya River becomes more important in the northern area of the lake. Sediment analysis will show whether the decrease of penetration is related to an increase in grain size and/or sedimentary gas.

Core PG1227 from Lake Taymyr has not yet been dated. However, it is assumed that T1 and T2 represent the Holocene because both units are built up of massive muds with relatively high water contents between 40 and 60% (core PG1227, Figure 5) similar to water contents measured in Holocene muds in Lake Levinson Lessing (PG1228, Figure 8). In contrast, Levinson Lessing sediments of Late and Middle Weichselian age have water contents of about 30% or less (Figure 8). In Lake Taymyr, a significant decrease in sediment water content is observed at the T2/T3 transition (Figure 5), which we interpret as the Holocene-Pleistocene boundary. This assumption is reasonable because preliminary pollen results from two samples of the basal sand of core PG1227 (top of unit T3) suggest a Pre-Holocene age (Hahne, pers. comm.).

From shallow to deeper water (Figure 4) the increase of unit thicknesses of T1 and, in particular, T2 indicates that sediment focusing towards the central area of the lake is an important process during the Holocene history of Lake Taymyr. This is explained by the morphology of the lake which is characterised by large areas of very shallow water. During storm events the water becomes turbid indicating significant resuspension and lateral transport of fine-grained material into deeper water (Niessen et al., 1997). Lateral transport of organic matter is probably associated with the above process because the entire core PG1227 is rich in gas which led the sediments expand by 1 to 1.5% after recovery (Overduin et al., 1996). The high content of sedimentary gas can explain the diffuse character of the backscatter (Figure 5). In particular, local occurrence of strong reflectors, rapid lateral shifts in penetration and subbottom resolution can often be associated with gas bubbles in the sediments (e.g. Niessen et al., 1993).

We cannot interpret, based on the study of both lakes, whether the transgressive trend during the Holocene is controlled by climatic change. However, a general climatic deterioration after the early Holocene climatic optimum is indicated in the pollen record from Lake Levinson Lessing (Figure 8), which shows decreasing aboreal pollen since the Preboreal to Boreal chronozones (Figure 8; Hahne and Melles, this volume). It has to be tested whether the climatic deterioration also affected the hydrology of the region so that a rise of lake levels was caused.

Middle and Late Weichselian - non-glacial lacustrine deposition during LST and HST conditions

According to the pollen results from core PG1228 from Lake Levinson Lessing (Figure 8), unit L2 was deposited during Middle and Late Weichselian time. This interpretation is based on two interstadials at 17.5 and 19.5 m core depth indicated by a significant increase in arboreal pollen (Figure 8; Hahne and Melles, this volume). Thus, the last glacial maximum is located somewhere between 9 and 17 m core depth which corresponds to the upper part of seismic unit L2. Lack of erosional events and/or deformation or overconsolidation of sediments is clear evidence that no glacial erosion affected the lake during the deposition of the entire unit L2 (Figure 7 and Figure 8). Moreover, except for the water content, the pre-Holocene lithology of PG1228 can hardly be distinguished from Holocene lithology (Ebel et al., this volume). Thus, for Late and Middle Weichselian time there is no indication in the sediment core for glacial activity in the catchment of the lake (such as typical glacial varves; e.g. Leonard, 1985; Leemann and Niessen, 1994). The discovery of a continuous undisturbed record of Middle and Late Weichselian age also implies that the input of terrigenous debris was not interrupted. Therefore, seasonal melting of ice and snow on the lake surface and in the catchment has to be assumed even for the last glacial maximum in order to provide an input of suspended particles

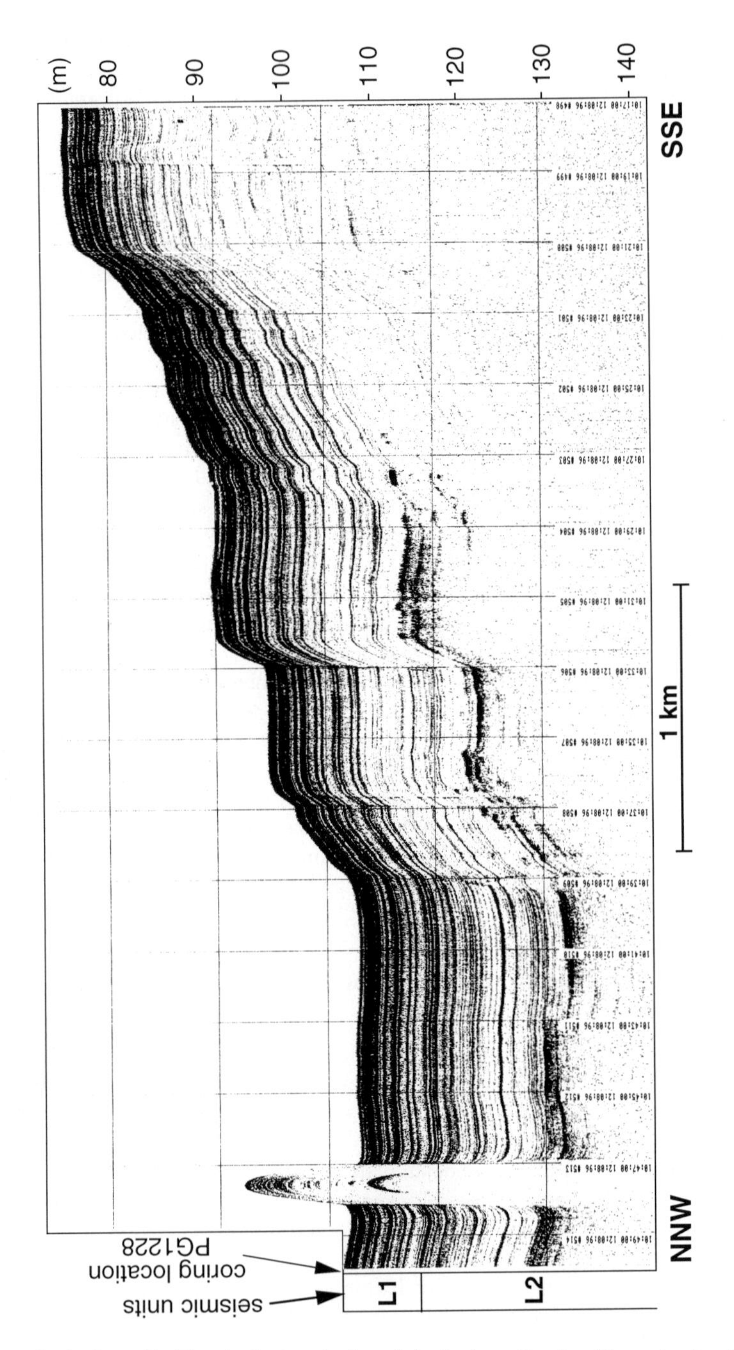

Figure 7: Part of seismic profile 31 and seismic units from Lake Levinson Lessing. The coring location PG1228 is at the left hand side of the profile.

from streams and rivers.

It is interesting to note that the stronger reflectors within unit L2 occur in the same level as the two interstadials in the sediment core (Figure 8; Hahne and Melles, this volume). These reflectors may be associated with an increased gas content in the sediments because the backscatter characteristic (Figure 8) is similar to those from strong reflectors observed in Lake Taymyr (Figures 5, 6). In Levinson Lessing, the presence of strong reflectors in the records on top of topographic highs may also be indicative of upward migration of sedimentary gas (Figure 8). Increased gas content can be interpreted by a higher input of organic matter to the lake during interstadial periods of Mid Weichselian age. Because only 13 m of the 22 m thick unit L2 are cored (Figures 8, 10), the total number of interstadials in unit L2 is not yet known. For the time period between 23 and 58 ka B.P., there are numerous of interstadials recorded from the Greenland Ice Core Project (e.g. Johnsen et al., 1992) and high resolution sediment studies from the North Atlantic (e.g. Bond and Lotti, 1995). However, any correlation between the pre-Holocene record of Levinson Lessing and Mid-Weichselian interstadials from other regions remains speculative until more datings become available.

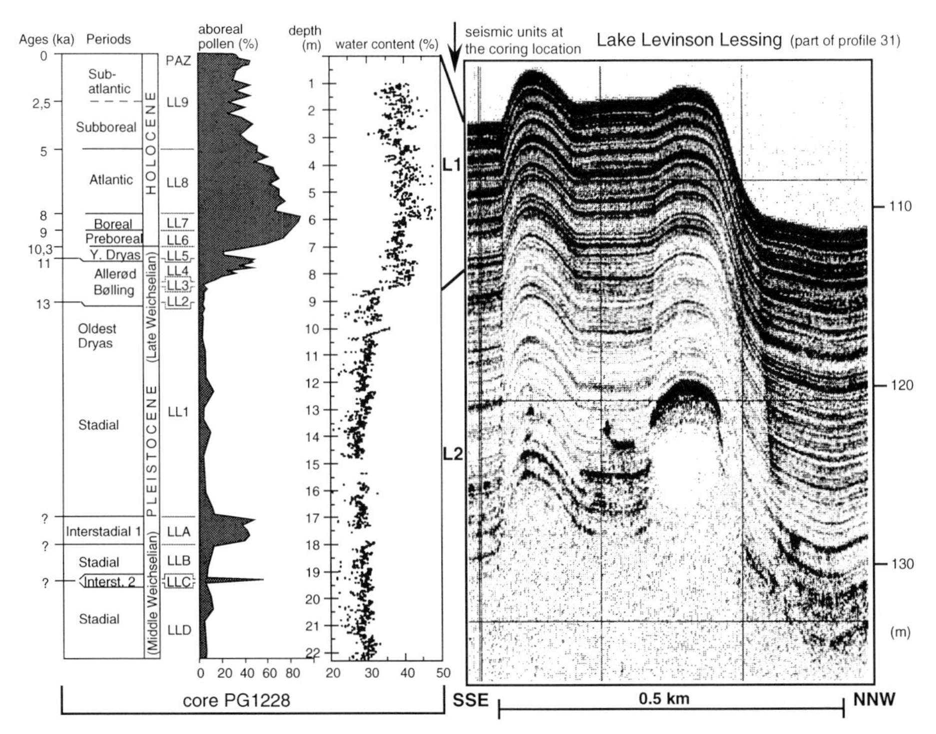

Figure 8: Chronozones, pollen assemblage zones (PAZ), arboreal pollen concentration (all after Hahne & Melles this volume), and water content of core PG1228 correlated with seismic units in profile 31.

The entire period of deposition of unit L2 was characterised by lower than current lake levels in Levinson Lessing (Figure 9 and 11). Melles et al. (1996) and Hahne and Melles (this volume) discuss the possibility of an increased continentality leading to very dry climatic conditions during stadial periods of the Weichselian in Central Siberia as compared to the Holocene. The decrease of this continentality may be associated with an increase in precipitation and a subsequent rise of lake levels on the Taymyr Peninsula at the end of the Pleistocene (transition from L2 to L1). However, the effect of lake level changes on sedimentation in the central part of Levinson Lessing was probably minor because there is no change in the drape geometry visible between units L1 and L2 (Figure 8). This can be explained by the small shallow-water area of the lake compared to the large area of more than 30 m water depth (Figure 3).

In Lake Taymyr, preliminary pollen results from two samples of the basal sand of core PG1227 (top of unit T3) suggest a Pleistocene age. The pollen assemblage is extremely poor in arboreal pollen and thus indicates a cold climate (Hahne, pers. comm.). Therefore, the low-stand phase of Lake Taymyr also occurred in Pre-Holocene time. There is no direct information about the duration and possible causes of very low lake levels of Lake Taymyr. Other evidence (Møller et al., submitted) may indirectly suggest that there is only a relatively short time window of Latest Pleistocene age during which the low stand might have persisted. Møller et al. (submitted) describe a section of massive non-glacial lacustrine silts rich in organic detritus which is exposed along the north-western shore of Lake Taymyr. This section is radiocarbon dated as 33.8 ka BP to 16.8 ka BP and associated with pro-deltaic sedimentation in a palaeo-Lake Taymyr characterised by a much higher lake level of up to 60 m compared to today. This implies that a significant drop of the Taymyr lake level can only have occurred after 16.8 ka BP and may have persisted until the beginning of the Holocene. Post-depositional erosion of unit T4 sediments by glacier ice during the low-stand phase can be excluded because the sound penetration of the Chirp system into T4 and the lack of deformation of T4 sediments suggest that no overconsolidation by moving ice has occurred. Consistently, there is no evidence of glacier-derived deposits in the sedimentary record of PG1227.

Since Møller et al. (submitted) describe sediments and chronological evidence that a palaeo-Lake Taymyr with lake levels up to 60 m above present existed prior to 16.8 ka., it seems to be likely that the deposits of T4 correlate with the above profile. This would imply that unit T4 is of Mid to Late Weichselian age. There is a further seismic indication that the deposits and the palaeo-environment of T4 sediments are indeed similar to that described by Møller et al. Both strong reflectors discontinuous to the stratification and back-scatter hyperbolae associated with pock marks indicate a relatively high sedimentary gas content in unit T4 (Figure 6). This suggests that T4 sediments are rich in organic matter. Thus, a glacial or proglacial origin for T4 seems unlikely because such sediments would be poor in organic matter. It is interesting to note that the organic-rich sediments found by Møller et al. (submitted) are terrestrial in origin and associated with a deltaic type of sedimentation. In our study, the seismic evidence of a high content of sedimentary gas and thus organic matter in T4 may also be terrestrial in origin because the location of profile 10 (fig. 6) is close to a larger stream delta.

It is difficult to assume that the drastic changes of the Lake Taymyr level from 60 m above the present level prior to 16.8 ka BP to about 40 m b.p.l.l. thereafter, followed by a Holocene transgression, were controlled by climatically induced changes in the hydrology. Since the present outlet of the lake is only a few metres above sea level, and more than 70% of the present lake area is less than 3 m deep (Overduin et al., 1996), a drop of water level by 40 m would reduce the lake size to only a very small fraction of the present area and volume. If the outlet was similar to that of today, the low-stand Lake Taymyr would have been a closed basin system. Considering seasonal runoff from the large catchment of the lake in combination with low evaporation rates under cold Arctic conditions, this scenario seems to be unlikely. Thus, tectonic control on the altitude of the outlet of lake Taymyr might be invoked. Also, if our interpretation of the chronology of the different units in Lake Taymyr and Levinson Lessing is more or less correct, the low-stand units L2 and T3 did not occur at the same time. L2 probably comprises the entire Mid to Late Weichselian whereas T3 represents only a relatively short period during the Latest Weichselian. Because both lakes are part of a larger drainage system, it is difficult to imagine that Lake Levinson-Lessing had a low stand during the Mid-Weichselian

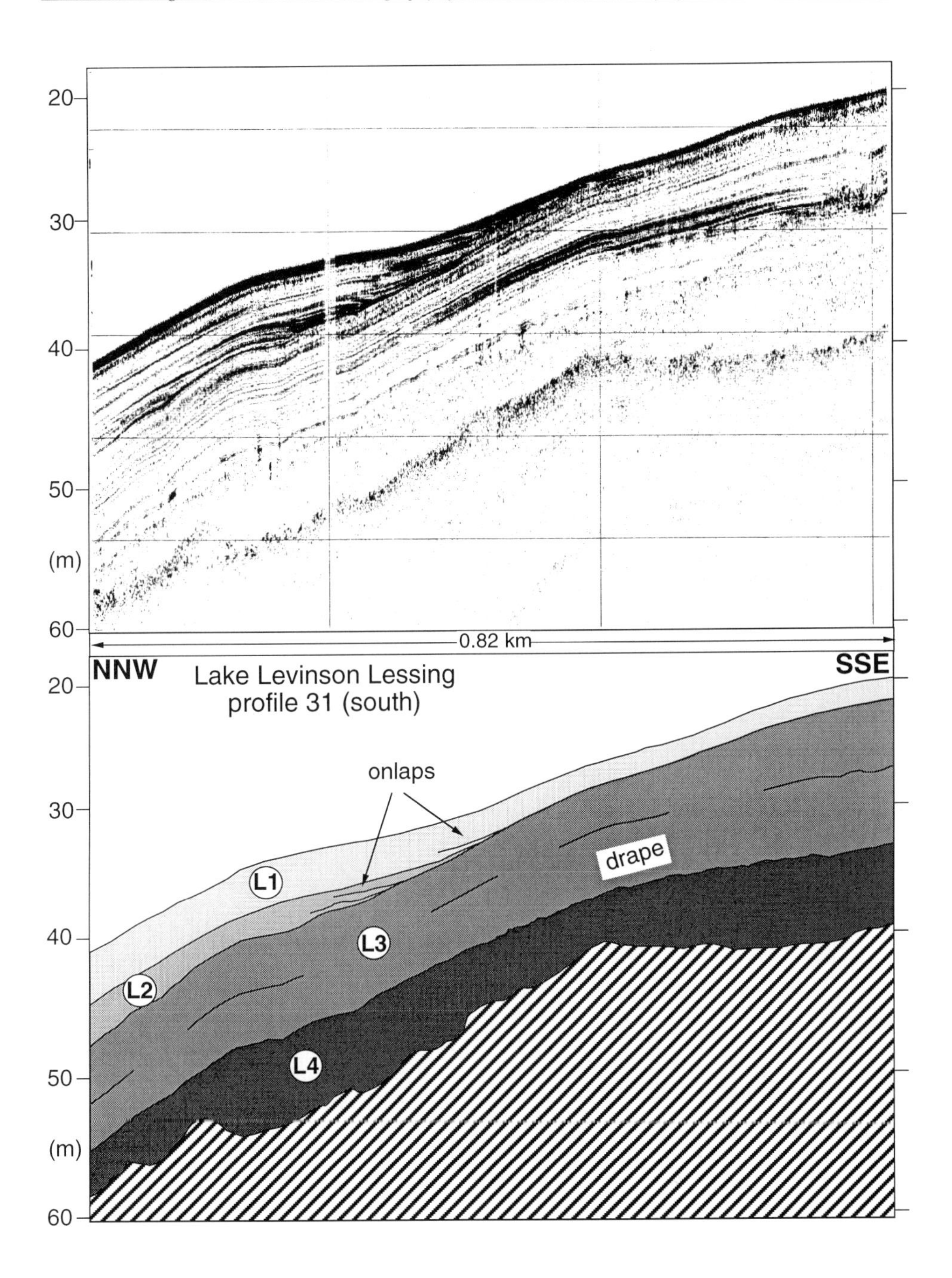

Figure 9: Southern end of profile 31 and seismic units from Lake Levinson Lessing.

while Taymyr had lake levels 60 m above present unless there was tectonic control on the size of Lake Taymyr.

Early Weichselian - the last glacial event in the area?

If the the assumption is correct that unit T4 is of Mid Weichselian age, the oldest units observed in this study are probably located in the lower part of the fill in Lake Levinson Lessing (L3, L4). The interpretation has to be considered as preliminary and speculative because the units L3 and L4 were not cored. Therefore, no direct lithological and chronological information is available yet. Because pollen evidence suggest that the overlying unit L2 is of Mid and Late Weichselian age, units L3 and L4 may date back to Early Weichselian. Here the question arises whether these oldest units could be associated with the last glaciation in the area. Since the valleys of the Lake Levinson Lessing catchment show typical U-shaped cross-sections indicative of glacier erosion, the early history of the lake was probably dominated by proglacial sedimentation which occurred during deglaciation. If so, deposition of relative thick units of lacustrine proglacial muds can be expected.

The seismic character of unit L3 is similar to that of proglacial muds in peri-alpine lakes. Giovanoli et al. (1984) describe parallel reflectors and weak backscatter in the subbottom of Lake Zurich which drape and conform to the rolling topography of an acoustically "solid" basement. A core penetrated through this unit and revealed laminated silt and clays (proglacial varves, Lister 1984). Similar deposits are described from numerous peri-alpine lakes from which seismic profiles and sediment cores exist (Finckh et al., 1984; Niessen and Kelts 1989; Niessen and Schroeder, 1990). Deposition of lacustrine varves is often part of a depositional sequence related to deglaciation. For example, relatively thick units of glacial varves were deposited during the relatively short time of a few centuries to millennias after the last glacial maximum (18 ka BP; Hsü et al. 1984; Niessen and Kelts, 1989). The termination of the varve deposition is often associated with a drop in lake level (e.g. Lister, 1984).

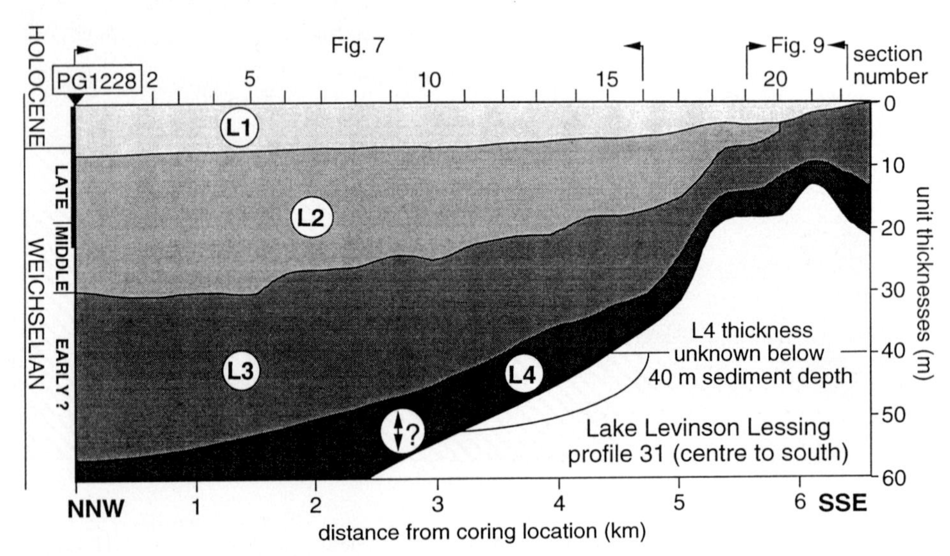

Figure 10: Thicknesses of seismic units in Lake Levinson Lessing based on the correlation of 23 vertical sections in profile 31. The location of section 1 correspondes to the coring site PG1228. Note that there is no lake-bottom topography in this graph.

Similar to T4, L3 represents a time of relatively high lake levels in Lake Levinson Lessing (Figure 11). In contrast to T4, back scatter of L3 is weak and any indication of sedimentary gas is missing (compare Figure 6 with 9). This suggests that the deposits consist of relatively fine-grained well stratified muds, possibly poor in organic matter. The seismic character of L3 implies deposition from suspension by which settling from "pelagic rain" is important. One is tempted to speculate that the deposition of unit L3 occurred in a proglacial environment during the last major deglaciation in the Byrranga Mountains. A higher lake level (Figure 11) can be explained by increased runoff rates from melting glacier ice and/or possible blocking of lake outlets by relic ice. The lowermost unit L4 in Levinson Lessing may then be interpreted as a transition from moraine or till deposits (strong basal reflector in Figure 9) to proglacial muds. Filling of pockets and diffuse transparent acoustic character suggest deposition of relatively fine-grained material with high sedimentation rates, which might have been dominant during the early phase of the deglaciation.

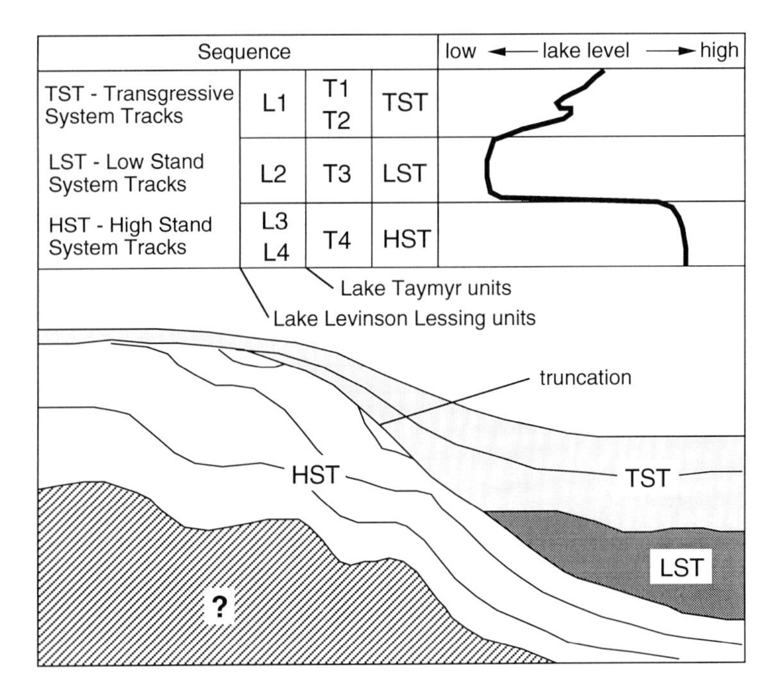

Figure 11: Idealized section from shallow to deep water in both Lakes Taymyr and Levinson Lessing as observed in seismic profiles (Figures 4 and 9). The interpretation of system tracks is based on the sequence stratigraphical interpretation of seismic profiles by Bally (1987). The interpretation of lake level changes is relative and valid for both lakes. Note that there is no chronological interpretation in this graph.

It is not evident from this study whether there are further depositional units of significant thickness below the basal reflectors in both Lake Levinson Lessing and Lake Taymyr. Formerly glaciated lake basins usually contain thick units of deformed muds and tills (e.g. Lister et al., 1984; Finckh et al., 1984). Such deposits are strongly consolidated by the weight of glacier ice and can therefore not be penetrated by high resolution seismic pulses such as transmitted by the Chirp system.

From this study there is no statement possible about the deposition which occurred in Lake Taymyr during Early Weichselian because the deeper fill of the basins was not penetrated by the GeoChirp system. However, because all sediments of Mid-Weichselian age or younger are non-glacial in origin, we suggest that the last glacial episode on the Taymyr Peninsula probably occurred during Early Weichselian (marine isotope stage 4, about 60 ka B.P., Figure 10). This is consistent with findings of fossil glacier ice in permafrost of possibly Early Weichselian age near Lake Labaz, about 200 km south of Lake Taymyr (Melles et al., 1996; Siegert et al., this volume).

Conclusions

There are 3 major conclusions which can be drawn from this study:

1 Thicknesses of at least 60 m of muddy unconsolidated sediments have accumulated in the central part of Lake Levinson Lessing and probably date as far back as Early Weichselian. Only the top 24 m have been cored (PG1228). In both lakes, however, the total thickness of the unconsolidated fill remain unknown because sound penetration is insufficient. In Lake Levinson Lessing two units (L3 and L4) older than the cored record of PG1228 are found. In Lake Taymyr, one unit (T4) older than the cored record of PG1227 (basal age: Pleistocene) is identified for which the original thickness and lateral distribution remain uncertain because major post-depositional erosion truncated large parts of this unit. Evidence of gas in unit T4 suggests accumulation of organic matter which might have occurred during a warmer climate of Mid Weichselian times.

2 Both lakes show similar trends in the variation of unit geometry and seismic character which can be explained by lake level changes. Similar to the approach of sequence stratigraphy in marine deposits, High System Tracks overlain by Low System Tracks overlain by Transgressive System Tracks can be defined. There is only one depositional cycle seen in both lakes Taymyr and Levinson Lessing because the units 1 to 4 show individual characteristics and none of the units are repeated in the sequence. The total variation of lake levels was probably more than 60 m in Lake Taymyr and is associated with tectonic control of the outlet. Lake level variations were less in Levinson Lessing and probably did not always occur synchronously to those in Lake Taymyr.

3 Glaciation of the Levinson Lessing and Taymyr lake basins during the last glacial maximum (Late Weichselian, marine isotope stage 2) can be excluded on the basis of this study. This implies that large parts of the Taymyr Peninsula were probably ice-free during the above period as suggested by Dunayev et al. (1988). However, an earlier glacial excavation of the basin of Levinson Lessing is suggested which probably occurred during Early Weichselian time (Marine Isotope Stage 4) and might have affected large areas of the Taymyr Peninsula, similarto the reconstruction suggested by Elverhøi et al. (1993) for Marine Isotope Stage 2.

Acknowledgements

We are grateful to all participants of the expedition Taymyr 1995 for their help in transporting the equipment and setting up the Levinson Lessing camp. M. Melles, D. Bolshyanov and A. Oufimzev are thanked for the logistic organisation of the field work. W. Korth is acknowledged for post-processing the GPS data. Useful comments of the two reviewers (A. Pugin and J. Merkt) helped to improve the manuscript. Financial support was provided by the German Ministry of Education, Science, Research and Technology (BMBF; Grant No. 03PL014A, B). This is contribution number 1483 of the Alfred Wegener Institute.

References

Bally, A.W. (1987) Atlas of Seismic Stratigraphy.- AAPG Studies in Geology, 27 (Vol.1). The American Association of Petroleum Geologists, Tulsa 125pp.

Bolshiyanov, D.Y. and H.-W. Hubberten (1995) Russian German Cooperation: The Expedition TAYMYR 1995 and the Expedition KOLYMA 1995 of the Pushchino Group. Berichte zur Polarforschung, 211, 208pp.

Bond, G. and R. Lotti (1995): Iceberg discharges into the North Atlantic on millennial time scales during the last glaciation. Science, 267: 1005-1010

Dunayev, N.N. & Pavlidis, J. A. (1988) A Model of the Late Pleistocene Glaciation Eurasiatic Arctic Shelf. In: Arctic Research - Advances and Prospects, Vol. 2, Proceeedings of the conference of arctic and nordic countries on cooperation of research in the Arctic, Kotlyakov, V.M. and V.E. Sokolov (eds.), Academy of Sciences of the USSR, Leningrad, 70-72p

Ebel, T., M. Melles and F. Niessen (this volume) A 22.4 m long core from Levinson-Lessing Lake, northern Taymyr Peninsula - description and first analytical results.

Elverhøi. A., W. Fjeldskaar, A. Solheim, M. Nyland-Berg and L. Russwurm (1993) The Barents Sea Ice Sheet - A model of its growth and decay during the last ice maximum. Quaternary Science Reviews, 12, 863-873

Finckh, P., K. Kelts and A. Lambert (1984) Seismic stratigraphy and bedforms in perialpine lakes. Geological Society of America Bulletin, Vol. 95, 1118-1128.

Giovanoli, F., K. Kelts, P. Finckh and K.J. Hsü (1984) Geological framework, site survey and seismic stratigraphy. In: Quaternary Geology of Lake Zurich: An Interdisciplinary Investigation by Deep-Lake Drilling, Hsü, K.J. And K. Kelts (eds.), Contributions to Sedimentology, 13, 5-20.

Grosswald, M. G. (1988) Late Pleistocene ice sheet in the Sovjet Arctic.- In: Arctic Research - Advances and Prospects, Kotlyakov, V.M. and V.E. Sokolov (eds.), Vol. 2, Proceeedings of the conference of arctic and nordic countries on cooperation of research in the arctic, Academy of Sciences of the USSR, Leningrad, 18-23.

Hahne, J. and M. Melles (1997) Late- and post-glacial vegetation and climate history of the south-western Taymyr Peninsula, central Siberia, as revealed by pollen analysis of sediment from Lake Lama. Vegetation History and Archaeobotany, 6(1), 1-8.

Hahne, J. and M. Melles (this volume) Climate and vegetation development on the Taymyr Peninsula since Middle Weichselian time - palynological investigations on lake sediment cores

Hubberten, H.-W., H. Grobe, W. Jokat, M. Melles, F. Niessen and R. Stein (1995) Glacial History of East Greenland Explored. EOS, Transactions, American Geophysical Union, Vol. 76, 36, 353-356.

Hsü, K.J., K.J. Kelts and F. Giovanoli (1984) Quaternary geology of the Lake Zurich region. In: Quaternary Geology of Lake Zurich: An Interdisciplinary Investigation by Deep-Lake Drilling, K.J. Hsü & K. Kelts (eds.), Contributions to Sedimentology, 13, 187-203.

Johnsen, S., H.B. Clausen, W. Daansgard, K. Fuhrer, N.S. Gundestrup, C.U. Hammer, P. Iversen, J. Jouzel, B. Stauffer and J.P. Steffensen (1992) Irregular glacial interstadials recorded in a new Greenland ice core. Nature, 359, 311-313.

Leemann, A. and F. Niessen (1994) Holocene glacial activity and climatic variations in the Swiss Alps: reconstructing a continuous record from proglacial lake sediments. The Holocene, 4/3, 259-268.

Leonard, E.M. (1985) Use of lacustrine sedimentary sequences as indicators of Holocene glacial history. Banff National Park, Alberta, Canada. Quaternary Research, 26, 218-231.

Lister, G.S. (1984) Deglaciation of the Lake Zurich area: a model based on the sedimentological record. In: Quaternary Geology of Lake Zurich: An Interdisciplinary Investigation by Deep-Lake Drilling, K.J. Hsü & K. Kelts (eds.), Contributions to Sedimentology, 13, 177-186

Melles, M., C. Siegert, J. Hahne and H.-W. Hubberten (1996) Klima- und Umweltgeschichte des nördlichen Mittelsibiriens im Spätquartär - erste Ergebnisse. Geowissenschaften, 14(9), 28-32.

Melles, M., D.Yu. Bolshiyanov, V. Samarkin, T. Müller-Lupp M. Wilmking (1997) Lake sediment coring on the Severnaya Zemlya Archipelago, Central Siberia, in 1996 - a preliminary report. In: S. Horie (ed.), IPPCCE Newsletter, No. 10, 106-112.

Møller, P., D.Yu. Bolshiyanov and H. Bergsten, H. (submitted) Weichselian geology and palaeoenvironmental history of the central Taymyr Peninsula, Siberia, indicating ice-free conditions during the Last Glacial Maximum. Boreas.

Niessen, F. and K. Kelts (1989) The deglaciation and Holocene sedimentary evolution of southern perialpine Lake Lugano - implications for Alpine paleoclimate. Eclogae geol. Helv., 82/1, 167-182.

Niessen, F. and H.G. Schröder (1990) Seismische Stratigraphie im Bodensee - Untersee. In: Kolloquium der Deutschen Forschungsgemeinschaft "Siedlungsarchäologische Untersuchungen im Alpenvorland". Ber. d. Röm. Germ. Kom., 71, 259-264.

Niessen, F., A. Lami and P. Guilizzoni (1993) Climatic and tectonic effects on sedimentation in Central Italian volcano Lakes (Latium) - implications from high resolution seismic profiles.- In: Paleolimnology of European Maar Lakes, Negendank, J.F.W. and B. Zolitschka (eds), Lecture Notes in Earth Sciences, 49, 129-148.

Niessen, F., C. Kopsch, T. Ebel and G.B. Fedorov (1997) Subbottom Profiling in the Lakes Levinson Lessing and Taymyr. In: Russian German Cooperation: The Expedition Taymyr and Zevernaya Zemlya 1996, Melles, M., B. Hagedorn And D.Y. Bolshiyanov (eds.), Berichte zur Polarforschung, 237, 70-78.
Overduin, P.P., D.Yu. Bolshiyanov and T. Ebel (1996) Lacustrine Geological Studies. In Russian German Cooperation: The Expedition Taymyr 1995 and the Expedition KOLYMA 1995 of the Pushchino Group, Bolshiyanov, D.Yu. and H.-W. Hubberten (eds), Berichte zur Polarforschung, 211, 111-121.
Siegert, C., A.Yu. Derevyagin, G.N. Shilova, W.-D. Hermichen and A. Hiller (this volume) Paleoclimatic indicators from permafrost sequences in the eastern Taymyr Lowland.
Svendsen, J.I., J. Mangerud and G.H. Miller (1989) Denudation rates in the Arctic estimated from lake sediments on Spitsbergen, Svalbard. Palaeogeography, Palaeoclimatology, Palaeoecology, 76, 153-168.
Velichko, A.A., L.L. Isayeva, V. Makeyev, G.G. Matishov and M.A. Faustova (1984) Late Pleistocene glaciation of the arctic shelf, and the reconstuctions of Eurasian ice sheets. In: Late Quaternary Environments of the Soviet Union, Velichko, A.A. (ed.), London.
Weber, M.E., F. Niessen, G. Kuhn and M. Wiedicke (1997) Calibration and Application of Marine Sedimentary Physical Properties using a Multi-Sensor Core Logger. Marine Geology, 136, 151-172.

Archaeological Survey in Central Taymyr

V.V. Pitul'ko

Institute for History of Material Culture, Russian Academy of Sciences, 18 Dvortsovaya nab., 191186 St. Petersburg, Russia

Received 14 February 1997 and accepted in revised form 10 October 1997

Abstract - There are very few sites covering all periods of human occupation in Taymyr, where surface finds predominate mainly, and the recent sites were largely unknown before the discovery of Oleny Brook. The results of the first excavations at Oleny Brook on the Upper Taymyra River in Central Taymyr were reported first as a contribution at the Laptev Sea System workshop in St. Petersburg (Pitul'ko, 1996). The cultural layer of the site is within the permafrost and contains organics materials including well-preserved wood pieces, bark, fish bones and even fish scales. Artifacts of the Oleny Brook site assemblage are made mainly of antler and bone. The collection contains artifacts of the most general kind which are typical of most northern aboriginal sites. They include bone/antler knifehandles, spear- and arrowheads, fish spear points, a part of a swivel block, and a decorated peice of bone. Ceramics and stone flakes were found, too. The site is carbon-dated to 1,880 +/- 75 BP, LE 5176 (uncalibrated age). Some features of the geology of the site are supposed to be rather remarkable, especially keeping in mind the results of carbon dating. It could be said, that the survey has raised more questions than answers if any – both archaeological and geological.

Introduction

When speaking on the Taymyr archaeology it is impossible to avoid Leonid P. Khlobystin, my teacher and the "Godfather" of archaeological studies in Taymyr. Although his fundamental monograph (Khlobystin, 1982) dedicated to Taymyr archaeology still remains unpublished, his main conclusions concerning the ancient habitation in Taymyr are well-known and are the only source of information on this subject. A series of excellent field projects undertaken by Khlobystin in 1967-81 years ended in 1981 and had never been resumed. His survey was focused first of all on the main river systems of the Taymyr region – the Pyasina and Kheta/Khatanga Rivers.

On his expeditions, numerous archaeological sites located in the Pyasina, Dudypta, Kheta and Khatanga River basins were found but a great number of those are relatively late. Concerning the earliest peopling of the territory, Khlobystin points out first of all assemblages from the Pyasina River sites (Pyasina I, III, IV, V; Lantoshka II site; Malaya Korennaya II, III; Kapkannaya II) and Tagenar VI site, the latter being "the only site dated exactly", to 6,020 +/- 100 BP (LE 884) and marking the upper chronological boundary of the Taymyr Mesolithic (Khlobystin, 1982). At the same time, there is a high probability that the peninsula was occupied much earlier. Such a suggestion is based on a find collected from the surface of the 2nd terrace in the Pyasina River valley near the confluence of the latter with the Polovinka River. Here was found a chopper-like tool made of greenstone flinty rock. Tools looking similar to the Polovinka artifact in style and made of the same (or almost the same) raw material, are very characteristic to the Afontovo and Kokorevo cultures of the Yenisey province of the Siberian Late Palaeolithic, as was noted by L.P. Khlobystin. Assuming that Tagenar VI and the other Mesolithic sites are not actually the most ancient on the Peninsula, Khlobystin considered the possibility of the peopling of that area (or at least the peopling of some territories) about 12,000 BP (Khlobystin, 1982), and this now is confirmed by the recent data

----
In: Kassens, H., H.A. Bauch, I. Dmitrenko, H. Eicken, H.-W. Hubberten, M. Melles, J. Thiede and L. Timokhov (eds.) Land-Ocean Systems in the Siberian Arctic: Dynamics and History. Springer-Verlag, Berlin, 1999, 457-467.

on the natural history of the Taymyr – North Land Islands Region in the Late Quaternary thatrefutes the idea on an extensive final glaciation (Badinova et al., 1976; Danilov and Parunin, 1982; Makeyev et al., 1979; Makeyev et al., 1989; Bolshiyanov and Makeyev, 1996).

The cultural development of the Taymyr Peninsula beginning from the Mesolithic is thought to be connected with the influence of cultural traditions spreading westward from modern Yakutia, although some groups of the West Siberian origin penetrated into the area in the Late Neolithic, and participated in the formation of the later aboriginal culture(s) also.

Only once Khlobystin did survey territory which could be considered as more or less in the central part of the peninsula. This took place in 1981, when he surveyed the Verkhnyaya (Upper) Taymyra River valley as far as its confluence with the Logata River (the latter is the greatest right tributary on the Upper Taymyra River). Very few artifacts were found during this survey, and the site location discovered near the Logata River mouth, where artifacts representing different cultures were collected, was the most important find. Unfortunately, the location was unsuitable for further excavation because it lacked stratigraphy. But the finds made there, especially microblades, indicated that the Upper Taymyra valley has been populated since at least 6,000 BP, as well as the territories bordering central Taymyr. Thus, the central and northern part of the Taymyr Peninsula still remained a "blank spot" on the archaeological map of the Arctic.

Archaeological research in Taymyr in 1993 and 1996

A chance to improve this situation occured when the long-term German-Russian interdisciplinary Taymyr project began in 1993. Although it was expected that rather early sites would be found there, keeping in mind present knowledge of the regional palaeogeography and palaeoenvironmental dynamics (Bolshiyanov and Makeyev, 1996), and, especially, the recent results from the investigations on Zhokhov Island (New Siberian Archipelago, Laptev Sea Region) where the most ancient site in the Arctic dated to 8,000 BP had been excavated (Pitul'ko, 1993), the real results were more modest. Thus, in the 1993 field season the most interesting results were obtained surveying the southerneast coast of the Engelgardt Lake where Pleistocene (?) fauna remains were exposed in the section of the 1st lake terrace (Pitul'ko, 1994). Numerous fragmented bones were collected in an extremely small area (3-4 m). Unfortunately it was impossible to identify clearly all of the species because of extreme fragmentation of bones; nevetheless it is obvious that the bone remains are of animals which were markedly different in size. Besides non-diagnostic bones some fragments of cervical vertebrae belonging undoubtedly to mammoth (according to examination by A. K. Kasparov, Institute for Material Culture History) were found. The composition of species represented by fragmented bones and a specific character of fragmentation make it possible to assume that, the assemblage resulted from human activity. Some fragments were collected in cutting the stratigraphic horizon containing the bone pieces. Carbon dates obtained for the latter – 10,020 +/- 80 (LU 3152) and 9,680 +/- 130 (LU 3153) – to my mind, could be considered as additional supporting data for the the artificial origin of the complex which occurred near the Pleistocene/Holocene boundary and was contemporaneous to the most favourable period of the Holocene climatic optimum. The latter is dated in the Asian Arctic from 10,000 to 9,000 BP and has been repeatedly confirmed by research in the High Arctic – on North Land and New Siberian archipelagos and on the continental polar territories as well (Bolshiyanov and Makeyev, 1996; Makeyev et al., 1979; Makeyev et al., 1989; Tomskaya, 1989; Boyarskaya et al., 1989; Kaplina and Lozhkin, 1982; and others). But, unfortunately, there is still no direct archaeological evidence confirming human habitation in Taymyr earlier than 6,000 BP. Nevertheless, I believe it will be found some day.

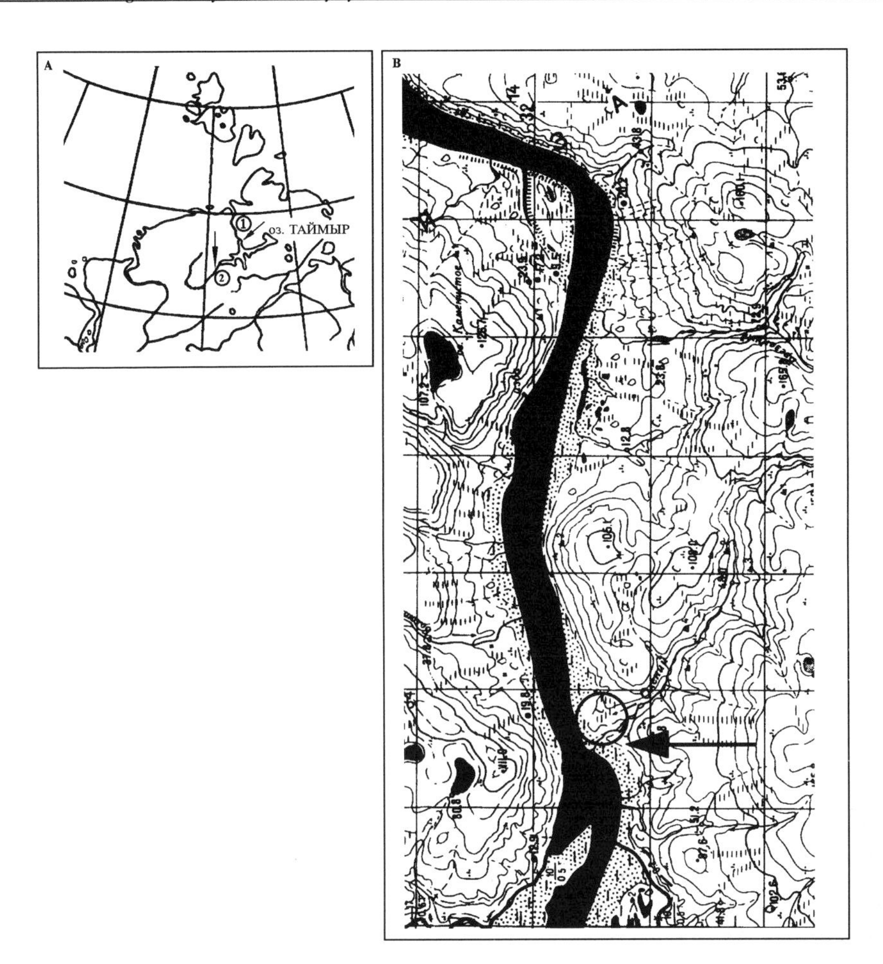

Figure 1: Map of the area surveyed: (A) – schematic map showing both the Engelgardt findings location (1) and the Oleny Brook site (2); (B) – Upper Taymyra River downstream area with the Oleny Brook site shown.

The Oleny Brook site

At the same time, a very interesting location similar to the one described above was discovered in 1994 by geologists from Khatanga Geological Survey in the Upper Taymyra River near the Oleny brook (a small right tributary of the river) mouth, at 74°08' North and 99°06 East (Figures 1 and 2). Although the cultural layer exposed in the right river bank (Figure 3) was supposed to be dated to 5-9,000 BP, it is much younger actually. I had a chance to survey the location in the 1996 field season, and although only small test excavations were undertaken, it is possible to introduce some preliminary information and conclusions concerning the site. Here, the cultural layer is exposed along a 60 m long abraded river bank, 6-7 m high, on what is supposed to be the terrace level, covered by 1-1,2 m thick stratified accumulation, composed by yellow or light yellow-brownish colored sandy soil (Figures 4 and 5). Sediments, both underlying and covering, are undoubtedly of fluvial origin. The cultural layer described formerly as "peat-bog sediments", is very solid, consisting of peat containing large number of fragmented bones. Judging by cultural remains discovered during the excavations (we excavated about 6 sq. m of cultural layer found both *in situ* and in sediment blocks of diverse

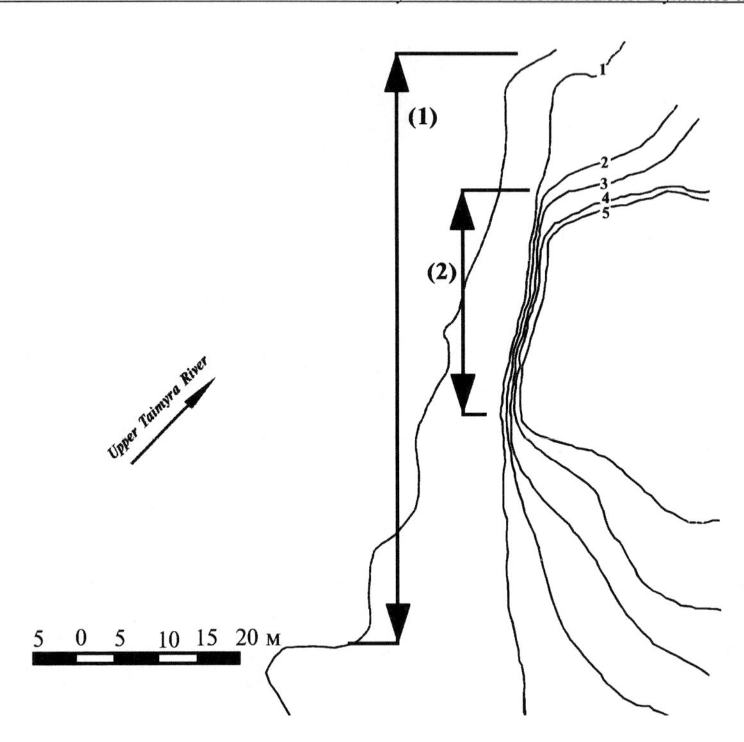

Figure 2: Oleny Brook site: (1) – area where re-deposited bone fragments and artifacts as well were collected; (2) – a section of the river bank where the cultural layer is exposed.

size (Figure 4) sloping down from the edge of the river bank), the site location had been chosen in the typical, most traditional aboriginal style. Diverse food sources such as fish, birds and reindeer were available here, but the latter was the main resource, and reindeer bones were found in huge quantities. It sould be said that the cultural layer "consists" of fragmented and crushed reindeer bones. Of course, the site location was especially favourable for reindeer hunting, and this place is still very reach in this food source. During the excavations we were lucky to see uncountable numbers of reindeer moving along the river all day long. And even the name of the brook itself – "Oleny" – means "reindeer brook". Obviously the site location chosen because of the abundance of the migrating reindeer, was occupied for some periods in spring or in late summer, or both. It was neither a "kill" nor a "living" site, most likely it was a "combination" of both, as all the animals were mainly taken somewhere in the surrounding area, and then butchered and consumed at the site. The composition of reindeer bone remains supports that view: almost all categories of reindeer skeleton remains are represented including hoof phalanges and even skulls, which are almost intact (the latter – both the presence of skulls and their good condition – is rather unusual). At the same time, the vertebrae and hoof phalanges are rather rare. As mentioned above, the bones are extremely fragmented, or crushed, including the smallest such as phalanges I and II.

Due to preservation qualities of permafrost (the cultural level is permanently frozen much deeper than the bottom level of the active layer), organics including wooden pieces, bark, fish bones and scales are well-preserved. Artifacts composing the Oleny Brook site assemblage, were both excavated and collected on the beach and are mainly made of bone and antler. The collection contains artifacts which are typical for northern aboriginal sites. They include bone and antler knife handles (Figures 6; 8: 5), spearpoints (Figure 7: 2, 3) and arrowheads (Figures

Figure 3: Oleny Brook site viewed from the West.

7: 5; 8: 3), fragmented unilaterally barbed fish spear (Figure 7: 4), worked bone and antler pieces (Figure 8: 1, 2, 7), part of a swivel block (Figure 8: 6) for a sledge harness (?), and a decorated piece of bone (Figure 8: 4) used as a pendant(?). Many ceramic fragments were found also. The latter represents two kinds of pottery (both thick and thin-walled) which were in use at the site. Although the inhabitants undoubtedly used metal tools as can be seen from the character of marks on the surfaces of worked bone and antler pieces (and it was iron most likely), stone tools such as axes or large scrapers were presumably in use too, since waste flakes (solid coarsed grey stone, about 20 pieces), and some pieces of raw material was found. In this way we have a composition which is rather typical of sites which are generally dated to 2,000 BP or a little younger, and this is confirmed by carbon dating. The date obtained from charcoal collected from the cultural layer is 1,880 +/- 75 BP, LE 5176 (uncalibrated age). And, it is very interesting that the location definitely was not in use for a long time even though the site area has diverse food sources available. In general, the assemblage (Figures 6–8) comprises the artifacts of the most general types which can be found in any Eurasian arctic site of this period, or even later, from West to East. Some of these pieces, such as knife handles, can be found even in today's living arctic cultures. Thus, exactly the same types of handles as well as the multi-pointed arrowhead similar to the unfinished one from Oleny (Figure 7: 5) were excavated from Mys Vkhodnoy site located on the mainland coast opposite Vaigach island, and from Karpova Guba site found on Vaigach (Pitul'ko, 1991). Projectile heads looking as a copy of that from the Oleny brook site (Figure 8: 3) are known, for example, from Eskimo sites on Cape Krusenshtern, Alaska (Giddings and Anderson, 1986: pl. 104, items m, n, and o). The others, such as a harpoon point, have a stable shape for a long time. Thus, the same types of harpoons are known in South Siberian Mesolithic sites (Medvedev, 1971). At the same time, there is one piece of special interest. This is a primitive flat swivel block (Figure 8: 6) which is,

Figure 4. A cultural layer exposed in the river bank. Andrew Ivanov excavates blocks of the cultural layer sloping down the bank.

of course, of the very general type, too. Findings of these blocks are extremely wide-spread – from the Ust-Polui site in West Siberia (Moshinskaya, 1953: pl. VI: 2, 3, 5, 7; 1965) to Eskimo sites, – and they were in use even at the beginning of our century (Birket-Smith, 1929: 181, Fig. 54a) . But these harness parts are of special interest because they give us evidence of sledge transportation. The only (and important) question is – of which type of sled? In the Ust-Polui site where a knife handle with a depiction of a harnessed dog was found, it probably signifies dog traction, and it was definitely dog traction in the Eskimo sites; but it remains unclear in the case of Oleny Brook site, or in the finds from Mys Vkhodnoy in the far northeastern portion of European Russia.

Figure 5: One meter thick bedded accumulation of sandy soil covering the cultural layer. The scale is standing on the upper stratigraphic boundary of the cultural layer.

Concluding remarks

Although the data from the test excavations at Oleny Brook site are of very preliminary nature, they are nevertheless very significant – thus, the Oleny Brook site may be the northernmost site ever found in Taymyr, and it was found in an area which was supposed to have remained unpopulated for centuries. The indigenous people inhabitting the Taymyr autonomous area today, live in the south, moving along the main rivers while the central and the northern portions of Taymyr are sparcely populated, or almost an unpopulated territory and has been for the last three hundred years. Leonid Khlobystin (1982) supposed this area to have

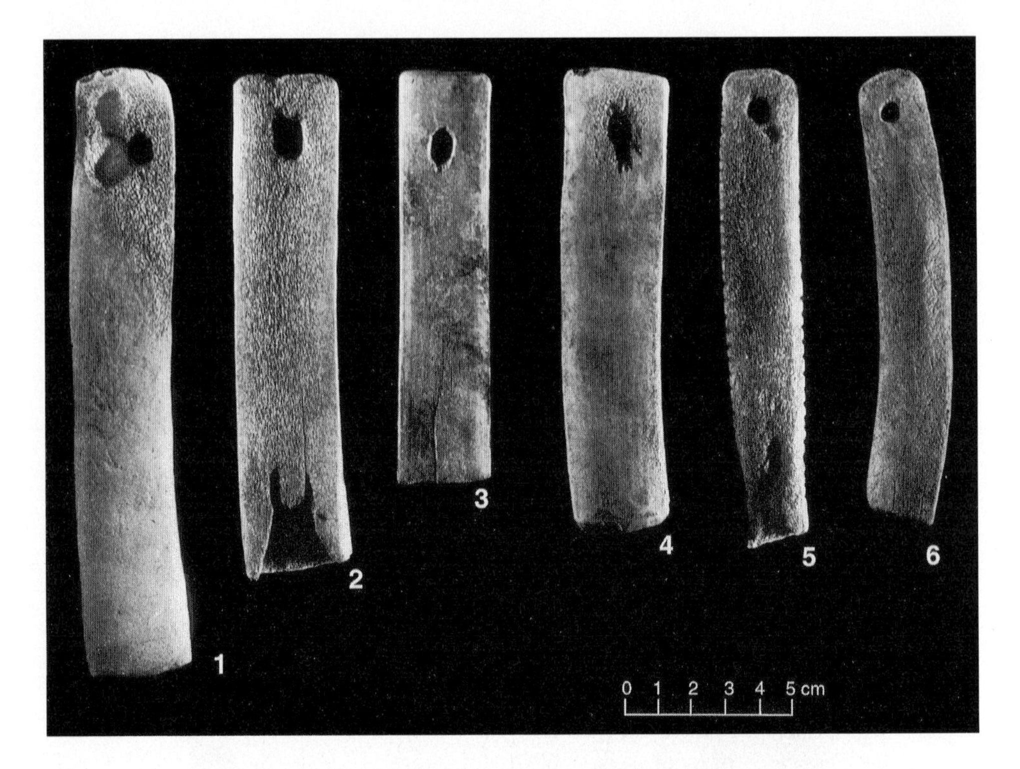

Figure 6: Artifacts recovered from the Oleny Brook site: 1 – 6. Knife handles made of antler.

archaeological cultures of another type, in comparison with the sites studied in the southern territories. Unfortunately, there is nothing to support or to argue that view now.

At the same time, and I would like to emphisize this, Oleny Brook is a very good archaeological site which can provide us with important information, if excavated. It is well-known that despite the abundance of archaeological sites in the Eurasian Polar area in the tundra zone very few are stratified, and the period covering the last 2000 yrs is almost lacking in such sites, apart from the coastal sites of maritime cultures in Kola and Chukchi Peninsula. Basically, there is a gap in archaeological data between sites dated from the past 3-2,500 yrs, and ethnographical data since XVII AD. Probably this is because of a very slow rate of sedimentation in the Arctic. But these archaeological data are of special importance because of the formation of the "ethnic map" of the contemporary known Arctic. The ethnic history of the Taymyr area was shown by Khlobystin (1982) to be of very complicated character as a result of the interaction here of both of West and East Siberian ethnic groups.

And, to conclude, I would like to note some features of the geology of the site which are rather remarkable and probably will provide us with the very important information on the recent environmental history of the area. As shown on Figures 4 and 5, the cultural layer is covered by approximately 1 meter thick bedded alluvium stratum of sandy soils deposited definitely after the formation of the cultural layer, i.e. in the last 2,000 yrs. However, the river bank is at least 6 meters high now, and this terrace exists in the Upper Taymyra River valley in many locations. What caused this stratum to accumulate?

It is interesting to note in this connection that while the terrace was described as the Early Holocene by geologists, Leonid Khlobystin surveyed that level in 1981 as far as the Logata

Figure 7: Artifacts recovered from the Oleny Brook site: 1 – a piece of worked antler (a spearhead preform?); 2, 3 – spearheads; 4 – unilaterally barbed fish spear point, and unfinished multi-pointed arrowhead of antler (5).

River mouth looking for the archaeological sites and nothing really ancient was found. The only place where redeposited surface finds were collected was near the Logata mouth. And this is nothing to be surprised at, because the terrace is much younger than it was expected. In this case, the observations from the Oleny brook site are going to change the strategy of further surveying in that area. At the same time, the accumulation could be (at least at the Oleny Brook) of the lake origin, marking some catastrophic fluctuation of the Taymyr Lake (personel communication from Dmitri Bolshiyanov during the IIIrd Laptev Sea System Workshop,

St.Petersburg) which means that some drastic environmental changes took place in the last 2,000 yrs. Did this cause the area to become unpopulated? These are questions beyond the "regular" archaeological evidence coming from the test research of the Oleny Brook site. Perhaps, some day we will get answers ...

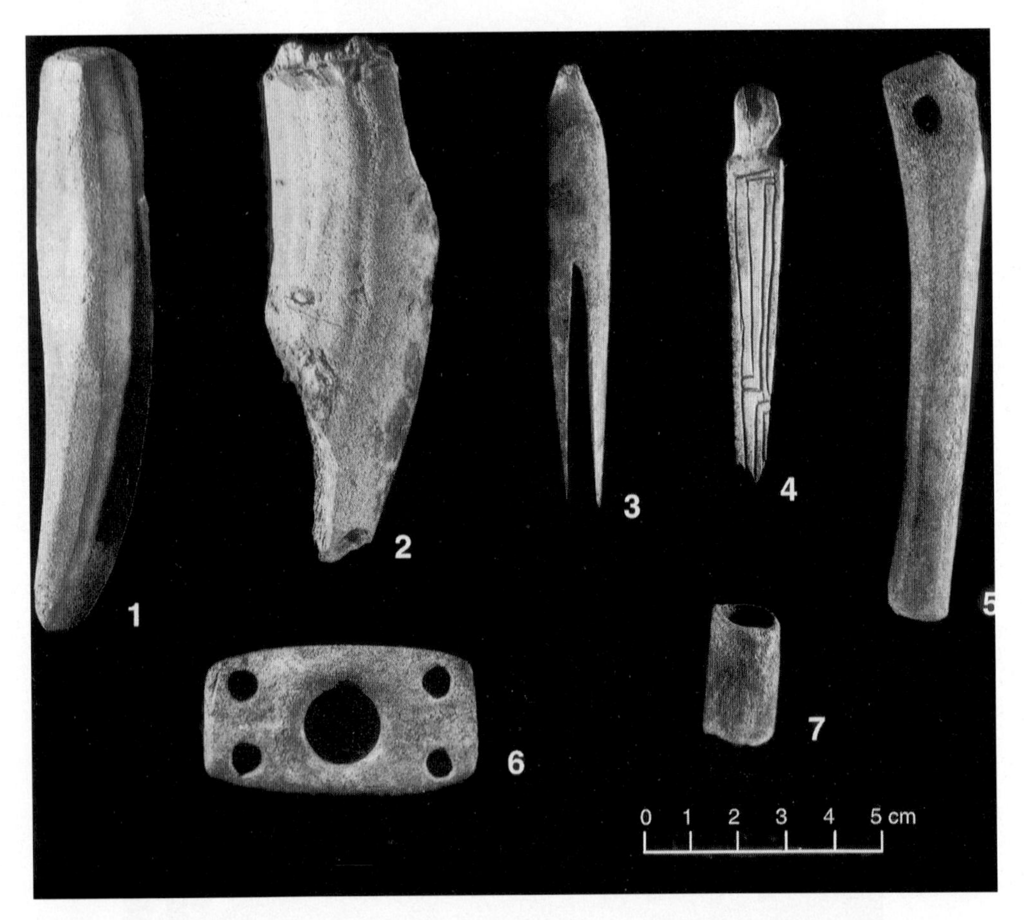

Figure 8: Artifacts recovered from the Oleny Brook site: 1,2, and 7 – worked antler pieces; 3 – arrowhead; 4 – a decorated bone pendant (?) with engravings; 5 –a knife handle of antler, and detail of a swivel block made of antler (6).

Acknowledgments

I would like to thank Dmitri Bolshiyanov (AARI, St.Petersburg, Russia) who involved me in the project and participated in discussion of the results; to William Fitzhugh (Arctic Studies Centre, Smithsonian Institution, US) and Ted Carpenter (Rock Foundation, US) for useful discussion of finds; Anatoly Kler (IMHC RAS, St.Petersburg, Russia) and Marcia Barky (Smithsonian Institution, Washington D.C., US) for technical assistance with the illustrations. My special thanks should be given to Andrew Ivanov (AARI, St. Petersburg, Russia), my co-traveller in the Upper Taymyra survey and assistant in the excavations of the site.

References

Badinova, V.P., V.A. Zubakov and Ye.M. Itsykson (1976) Carbon-14 Dates of the VSEGEI Radiocarbon Laboratory, LG index, List III (in Rusian). Bulletin of Committee for Quaternary Studies, 45, 154-157.

Birket-Smith, K. (1929) The Carobou Eskimos; Material and Social Life and Their Cultural Position. Report of the Fifth Thule Expedition 1921-24. National Museum, Copenhagen, Denmark, 226pp.

Bolshiyanov, D. Yu. and V.M. Makeyev (1996) Severnaya Zemlya archipelago: glaciation and the history of environmental changes (in Russian). Gidrometeoizdat, St.Petersburg, Russia. 213pp.

Boyarskaya, T.D. (1989) Correlation of the Late Pleistocene – Holocene Climatic Trends of the Different Territories of the USSR (in Rusian). In: Paleoklimaty pozdnelednikovya i golotsena, Khotinsky, N.A. (ed.), Nauka, Moscow, Russia, 85-90.

Danilov, I.D. and O.B. Parunin (1982) Comparative results of the carbon-14 dating of carbonate concretions and carbofossils from Upper Pleistocene deposits composing the Karga terrace of the Yenisei River downstream area (in Russian). DAN SSSR, 262 (2), 402-404.

Kaplina, T.V. and A.V. Lozhkin (1982) History of the development of Yakutian Coastal Lowlands in Holocene (in Russian). In: Razvitiye prirody territorii SSSR v pozdnem pleistotsene i golotsene, Velichko, A.A. (ed.), Nauka, Moscow, Russia, 207-220.

Khlobystin, L. P. (1982) Ancient History of the Taymyr Trans Polar Region and Questions of the Formation of Cultures of the Eurasia Extreme North (in Russian). Avtoreferat dokt. diss. Moscow, Russia. 1982. 36pp.

Makeyev, V.M., Kh.A. Arslanov and V.E. Garutt (1979) Age of mammoth skeleton remains from Severnaya Zemlya Islands and some questions of the Late Pleistocene palaeogeography (in Russian). DAN SSSR, 245(2), 421-424.

Makeyev, V.M., Kh.A. Arslanov, O.F. Baranovskaya, A.V. Kosmodamianskiy, D.P. Ponomaryova and T.V. Tertychnaya (1989) Stratigraphy, geochronology and palaeogeography of Late Pleistocene and Holocene of Ostrov Kotel'nyi (in Russian). Bulletin of Committee for Quaternary Studies, 58, 58-69.

Medvedev, G.I., ed. (1971) The Mesolithic of Southerneast Siberia (in Russian). Irkutsky Univ., Irkutsk, Russia. 242pp.

Moshinskaya, V.I. (1953) Material Culture and Economy of the Ust-Polui site (in Russian). In: Drevnyaya istoriya Nizhnego Priobya, Zbrueva, A.V. (ed.). Materialy i issledovaniya po arkheologii SSSR, 35, 72-106.

Moshinskaya, V.I. (1965) Archaeological sites of West Siberian North (in Russian). Nauka, Moscow, Russia. 88pp.

Pitul'ko, V. V. (1991) Archaeological data on the Maritime cultures of the West Arctic. Fennoscandia archaeologika, VIII, 23-34.

Pitul'ko, V. V. (1993) An Early Holocene Site in the Siberian High Arctic. Arctic Anthropology, 30(1), 13-21.

Pitul'ko, V. V. (1994) Natural Environment of the Arctic and the problem of Initial Human Occupation (in Rusian). In: Scientific Results of the LAPEX-93 project, Timokhov, L.A. (ed.), Gidrometeoizdat, St.Petersburg, Russia, 360-376.

Pitul'ko, V. V. (1996) Oleny Brook Site: A New Contribution to Taymyr Archaeology. – Terra Nostra. Schriften der Alfred-Wegener-Stittung, 96/9, 82-83.

Tomskaya, A.I. (1989) Climatic conditions of the Yakut district in the Late Glacial Age and in the Holocene (according to pollen core analysis) (in Rusian). In: Paleoklimaty pozdnelednikov'ya i golotcena, Khotinsky, N.A. (ed.), Nauka, Moscow, Russia, 109-116.

Marine Pleistocene Deposits of the Taymyr Peninsula and their Age from ESR Dating

D. Bolshiyanov[1] and A. Molodkov[2]

(1) State Research Center - Arctic and Antarctic Research Institute, 38 Bering St., 199226 St. Petersburg, Russia

(2) Institute of Geology of the Academy of Sciences of Estonia, 7 Boulevard Estonia, Tallinn, Estonia

Received 3 March 1997 and accepted in revised form 3 March 1998

Abstract - Electron spin resonance (ESR) dating and lithostratigraphic studies of marine deposits on the TaymyrTaimyr Peninsula suggest that a marine sedimentary environment dominated during isotope Stages 4, 5, 7, 8, 9 and 15. The marine basins were predominantly cold and influenced by freshwater.

Introduction

Marine Quaternary deposits are widespread on the Taymyr Peninsula but not well exposed due to active slope processes. It is thus important that studies of natural outcrops of Quaternary deposits are combined with information from bore-holes.

Based on the results of drilling carried out by the Central Arctic Geological Exploration Expedition (Norilsk) and the Polar Geological Exploration Expedition (Khatanga), a general understanding of the structure of the TaymyrTaimyr Pleistocene sediments has been gained. It has also been possible to retrieve samples for dating purposes from several of these bore-holes.

This paper summarizes preliminary results of studies of marine deposits on the Taymyr Peninsula. These studies were carried out during the four field seasons (1993-1996) of the joint Russian-German expedition „Taymyr-Severnaya Zemlya" and in previous expeditions to the western part of Taymyr (1986), to northern Taymyr (1988) and to the south-eastern part of the peninsula (1989-1990).

ESR dating methods

The electron spin resonance method (ESR) was used to date marine mollusc shells, found in investigated exposures and cores. This is an efficient tool for determining the age of exoskeleton remnants of malaco-fauna within a time interval of several hundred years to about 1 million years (Molodkov, 1989, 1992, 1993). The dating of subfossil mollusc shells with ESR is based on the ability of the mollusc shell material to accumulate the absorbed dose of natural radiation and preserve it over a long period of time (Ikeya and Ohmura, 1981). An estimate of the paleodose, accumulated during shell burial in embedding deposits, is performed by comparing the natural intensity of the shell signal in the ESR-spectrum (caused by paramagnetic carbonate centers) with the intensity of radioactive irradiation induced by a laboratory source. The radiation background consists of three components: (1) radiation from natural radionuclides scattered in the embedding deposits, with the main contribution to the exposure dose of shells made by ^{238}U, ^{232}Th, ^{40}K and the decay products of uranium and thorium (88%, on average, of the total dose); (2) radiation from the uranium and products of its decay incorporated in the crystalline matrix of the shell carbonate material ($\cong$ 7%); and (3) cosmic radiation ($\cong$ 5%). The radiation dose absorbed by the mollusc shell material is proportional to the concentration of

---

In: Kassens, H., H.A. Bauch, I. Dmitrenko, H. Eicken, H.-W. Hubberten, M. Melles, J. Thiede and L. Timokhov (eds.) Land-Ocean Systems in the Siberian Arctic: Dynamics and History. Springer-Verlag, Berlin, 1999, 469-475.

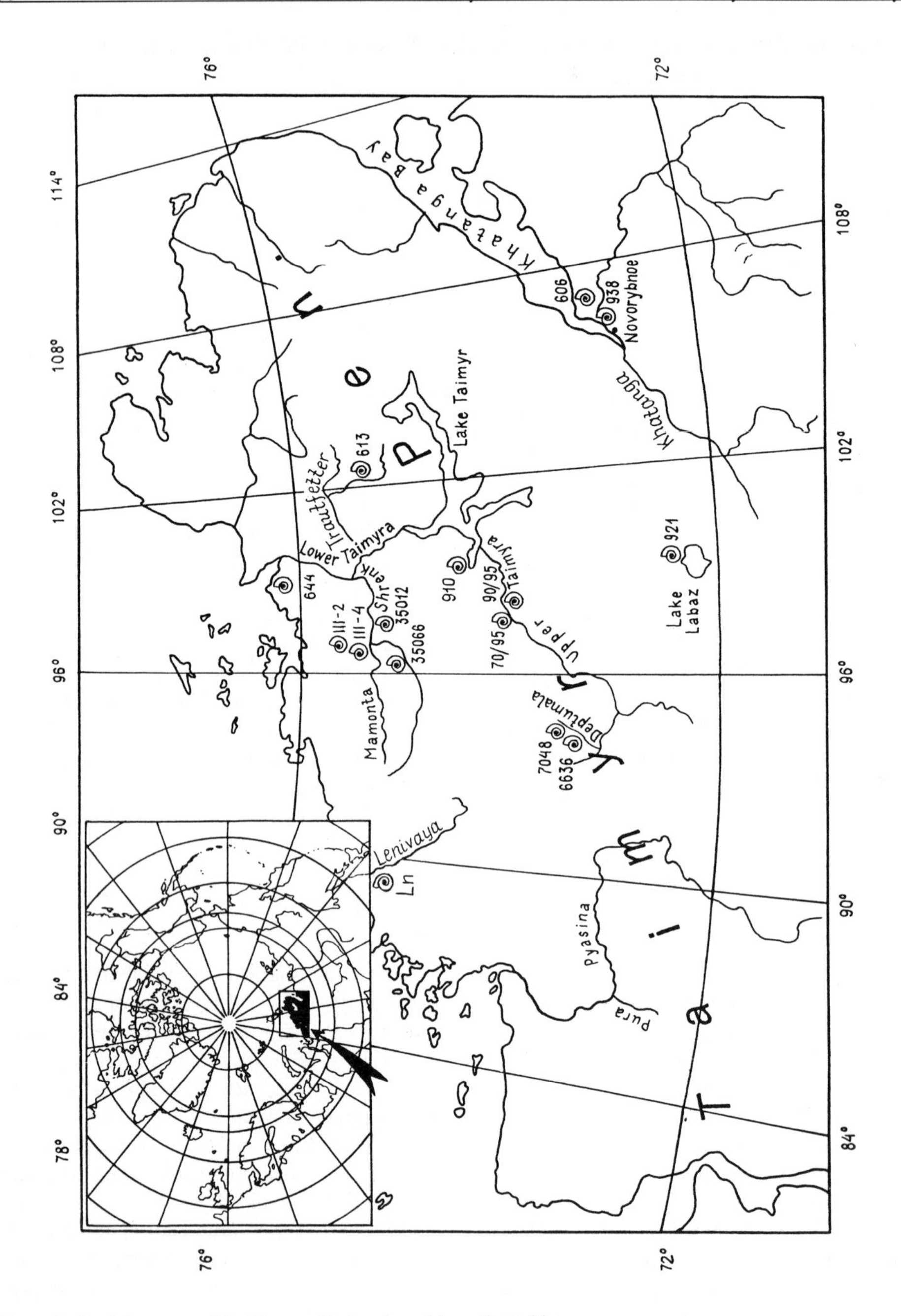

Figure 1: Overview map of the Taymyr Peninsula and investigated logs.

radioactive elements in the environment and in the shell itself, as well as to the time of exposure (shell age). Shell dating is performed by measuring the paleodose accumulated from the time of formation of the mollusc exoskeleton and the strength of natural radiation dose that affected the shell during its burial. To perform palaeodosimetric analysis of the shell material, the analytical line at 2.0012 (line-width $\Delta B_{pp} \cup 0.22$ mT, Molodkov, 1988, 1993) was separated. The dose-response curves with the use of this signal conformed most closely to the single exponential function.

The height of original absorption signal was used as an equivalent of the g=2.0012 centre concentration in the shell. Quantification of the 2.0012 centre concentration was obtained from the peak-to-peak amplitude of the relevant signal in derivative spectra of the shells by using an overmodulation (OM) detection method (Molodkov, 1988, 1993). The microwave power used for dosimetric reading was 2 mW with 100 kHz magnetic field modulation at 1 mT. The palaeodose for each sample was obtained by fitting with the reciprocal exponential function $-\ln(-I/I_{max})$, where I and I_{max} are the ESR signal intensity and the intensity of the level at saturation dose, respectively. The accumulated palaeodose, P_S, was estimated by extrapolation of the regression line to zero ESR intensity. The saturation value of ESR intensity, I_{max}, was found by iterative optimization.

The ages of the shell fossils from Taymyr were derived from the following equation (Molodkov, 1988, 1989):

$$T = \tau\left[-\ln\left(1-\frac{P_s}{\tau \dot{D}_\Sigma(t)}\right)\right],$$

where τ is the mean lifetime of the 2.0012 centre in shell carbonate, P_s is the accumulated palaeodose since the mollusc exoskeleton formation, $\dot{D}_\Sigma(t)$ is the total radiation dose rate as a function of time, and T is the shell age.

The dose rate, $\dot{D}_\Sigma(t)$, is a sum of the doses due to different radiations

$$\dot{D}_\Sigma(t) = \dot{D}_c + W_\gamma \dot{D}_{\text{ext}\,\gamma} + W_\beta k_\beta \dot{D}_{\text{ext}\,\beta} + \dot{D}_{\text{int}\,\alpha,\beta}(t),$$

where $\dot{D}_c$ is the cosmic dose rate proportional to the sample,s latitude, altitude, and burial depth; $\dot{D}_{\text{ext}\gamma,\,\beta}$ is the external dose rate depending on radioactive element concentration in the sediment surrounding the shells; W_α and W_β are the correction factors for water (ice); k_β is the beta-attenuation correlation factor; $\dot{D}_{\text{int}\,\alpha,\beta}(t)$ is the time-dependent component of the internal dose rate originating from the uranium incorporated in the shell substance.

In total, 22 samples were dated from 15 sections and bore-holes (Figure 1, Table 1). Some samples were dated two or three times as a control. The final ESR-age for samples with double or triple dating was determined as the mean arithmetic age.

Stratigraphy of Pleistocene deposits on Taymyr Peninsula

In our opinion, the Quaternary deposits of the Taymyr Peninsula can be roughly divided into two units: a lower unit up to 200 m thick, consisting predominantly of dark grey clayey-silty sediments with a large ice content, and an upper unit, 30-40 m thick and predominantly of a sand-pebbly composition. The lower unit forms part of the contemporary relief in the lowland valleys. This is especially clear in the western part of the peninsula where clayey silts are often eroded along the Pur, Pyasina and other rivers.

These exposures often show that the fine-grained sediments form rhythmic deposits of varved

Table 1: ESR results and radioactivity data for samples from Taymyr Peninsula

	Lab No.	Site/ Field No.	Locality	d (mm)	U_{in} (ppm)	U (ppm)	Th (ppm)	K (%)	D_c (mGy/a)	D_{int} (mGy/a)	D_{sed} (mGy/a)	D_S (mGy/a)	P_s (Gy)	ESR-age, T (ka)
1	126-109	644/8	Taimyr	0,50	0,60	1,18	2,40	1,25	50	325	1397	1772	140.0 ± 7.7	85.0 ± 15.0
2	127-109	606/1	Khatanga	0,50	0,60	1,06	5,95	1,74	20	380	2008	2408	270.0 ± 12.1	112.0 ± 20.0
3	128-109	613/5	Taimyr	1,60	0,60	0,80	1,34	0,76	140	354	597	1091	60.0 ± 1.8	70.0 - 37.0
4	138-051	6636/3	Taimyr	1,20	0,20	0,89	2,45	0,77	60	60	759	879	102.0 ± 4.6	116.0 ± 11.0
5	139-051	6636/3	Taimyr	1,60	0,85	0,89	2,45	0,77	60	227	685	972	105.0 ± 4.2	108.0 ± 8.0
6	140-051	6636/3	Taimyr	1,20	0,80	0,89	2,45	0,77	60	207	759	1026	115.0 ± 6.3	112.0 ± 18.0
7	141-051	6636/1	Taimyr	0,60	0,13	0,98	2,87	0,99	130	20	1124	1274	120.0 ± 6.0	94.0 ± 9.0
8	142-051	6636/1	Taimyr	0,70	0,68	0,98	2,87	0,99	130	162	1093	1385	126.0 ± 3.8	91.0 ± 8.0
9	143-051	606/13	Khatanga	0,30	0,23	1,46	6,90	1,91	60	64	2446	2570	360.0 ± 21.6	140.0 ± 11.0
10	145-051	7048	Taimyr	0,80	0,64	1,59	5,67	1,23	70	226	1507	1803	137.0 ± 5.5	76.0 ± 6.0
11	146-051	7048	Taimyr	0,40	0,67	1,59	5,67	1,23	70	158	1772	2000	160.0 ± 5.6	80.0 ± 6.0
12	211-065	910/1	Taimyr	1,45	0,60	1,61	6,17	1,22	110	177	1214	1500	165.7 ±9.9	111.0 ± 11.0
13	212-065	910/1	Taimyr	0,80	0,45	1,61	6,17	1,22	110	121	1433	1664	164.3 ± 9.8	99.5 ± 9.8
14	213-065	910/1	Taimyr	0,80	0,57	1,69	5,43	1,19	110	147	1381	1638	147.1 ± 8.1	90.3 ± 8.8
15	215-065	921/5	Taimyr	0,40	0,65	1,24	3,89	1,23	93	242	1459	1794	499.9 ± 11.0	283.0 ± 26.6
16	216-065	921/5	Taimyr	0,30	0,84	1,24	3,89	1,23	93	303	1536	1932	505.1 ± 12.6	265.0 ± 24.9
17	219-065	921/8	Taimyr	1,00	0,70	1,39	6,97	1,62	79	286	1579	1944	605.4 ± 12.1	316.5 ± 31.4
18	232-086	3501201	Taimyr	2,00	0,59	0,75	4,87	1,66	188	279	1086	1553	813.3 ± 14.8	535.5 ± 48.6
19	233-086	35066/03	Taimyr	0,80	0,44	1,32	4,86	1,34	132	126	1370	1628	187.8 ± 5.6	116.0 ± 11.1
20	234-086	35066/02	Taimyr	1,50	0,58	1,09	4,61	1,30	87	247	1061	1395	456.8 ± 9.0	332.0 ± 30.4
21	236-086	III-4/195	Taimyr	0,49	0,79	1,54	7,33	1,70	60	293	2025	2377	631.0 ± 7.2	268.4 ± 25.1
22	237-086	III-2/26	Taimyr	0,85	0,64	1,10	5,59	1,77	139	152	1605	1896	136.4 ± 10.4	72.2 ± 6.9
23	238-086	70/95	Taimyr	1,40	1,00	0,78	6,45	1,53	188	303	1254	1744	208.2 ± 2.4	120.0 ± 11.1
24	239-086	90/95	Taimyr	0,25	0,98	1,06	4,17	1,68	148	241	1936	2325	222.3 ± 4.1	96.0 ± 9.1
25	240-086	938/1	Khatanga	0,65	0,51	0,64	2,82	1,07	60	185	1012	1257	293.2 ± 16.7	235.5 ± 25.6
26	241-086	938/10	Khatanga	0,40	0,54	1,57	7,74	1,83	72	210	2238	2520	834.9 ± 46.0	336.0 ± 31.8
27	256-107	Lbz/ 1	Taimyr	0,51	0,55	0,81	4,78	1,45	148	214	1516	1878	592.1 ± 5.7	319.5 ± 30.3
28	257-107	Ln 1 /2	Taimyr	0,89	0,22	1,58	4,60	1,39	103	56	1386	1545	132.5 ± 6.0	86.1 ± 8.5

Notes:
d is the shell thickness; U_{in} is the uranium content in shells; P_s is the palaeodose; U, Th, K are the uranium, thorium and potassium content in sediments; D_c is the cosmic dose rate; D_{int} is the time-averaged internal dose rate; D_{sed} is the sediment dose rate; D_S is the total dose rate. Uncertainties: determination of thickness, ± 40mm; U determination, ± 2-3%; Th determination, ± 3-4%; K determination, ± 1-2%; U determination in the shells, ± 1-3%; gamma irradiation, ± 3%.

clay. The lower unit is less frequently exposed in the lower reaches of the Upper Taymyra and Lower Taymyra rivers and in the valleys of the Khatanga, Shrenk and Trautfetter rivers. The clayey-silty sediments of the lower unit often include boulders and pebbles. However, the frequency of these coarse clasts does not exceed 5-10% by volume of the embedding sediments. The occurrence of coarse clasts has been taken as an argument for interpreting these sediments as being of glacial origin (Urvancev, 1931; Troitskii,1966; Kind and Leonov,1982, Zolnikov and Shevko, 1989). However, in all studied outcrops it was found that the sediments are clearly laminated. The lower unit sediment often contained a considerable amount of ice, in particular ice wedges and ice bodies of unknown origin up to 60 m in thickness.

The sediments are usually poor in fossils, sometimes totally barren. However, some fossils, though rarely observed, can be found in the upper part of the lower unit sediments, e.g. the marine mollusc *Portlandia arctica*, foraminifers of the genus *Hynesina* sp., freshwater and estuary species of diatoms (*Åunotia diodon, Eunotia pracrupta, Tabellaria fenestrata, Pinnularia sp, Cyclotella striata, Stephanodiscus astrea*), and marine species of diatoms (*Paralia sulcata)*. Most diatoms were redeposited from Cenozoic sediments. The palaeoenvironmental indications point towards cold and freshwater basins. The transition from more massive clayey silts to varved clays also indicates that brackish basins were replaced by freshwater basins. The pelecypod fauna of the lowermost part of the lower sedimentary unit is boreal in composition, indicating warmer sea conditions. The lower sedimentary unit consists of not only clay and silt but of more coarse deposits too. The roof of the lower unit lies at an altitude of 160 to -30 m a.m.s.l. and its relief is very contrasting (Figure 2).

The upper sedimentary unit is present mostly as erosional remnants of a marine deposit that once covered significant parts of the TaymyrTaimyr Peninsula. These deposits have often been considered glacial in origin due to their sharp morphology and sandy to gravely composition (Kind and Leonov, 1982). However, the internal structure of these deposits, although rarely exposed due to a lack of recent erosion and active slope processes, indicates a coastal deltaic facies of these sediments. Molluscs (*Hiatella arctica, Astarte borealis, Mya truncata)*, foraminifera (*Haynesina orbicularis, Retroelphidium ex gr. clavatum, Cribroelphidium granatum, Astrononion gallowayi, Bucella frigida, Cassidulina neoteretis*) and diatom (*Navicula placentula, N. radiosa, Stephanodiscus astrea, Tabellaria fenestrata, Pinullaria borealis, Cocconeis disculus, Amphora ovalis var lybica, Cyclotella kuetzingiana var. schumannnii, Caloneis bacillum)* assemblages, found in some sections belonging to this upper sedimentary unit, indicate a delta-marine, coastal-marine, cold and brackish sedimentary environment. The altitude of bedding of upper unit is from -45 to 175 m a.m.s.l. The most widespread level of marine terraces consisting of sand and pebbles of the upper unit sediments is at 100 m a.m.s.l. in the central part of Taymyr.

Discussion and conclusions

The deposits can be divided into a number of groups on the basis of their composition and their age, as determined by ESR dating:

1) sandy-gravely erosional remnants with a **scattered** distribution. Dating falls within a time frame of 70 to 100 ka BP.

2) Clayey silts of a deep-water facies of marine sediments of the same age - 86 ka BP.

3) Clayey silts aged between 116 and 140 ka BP.

4) Clayey silts aged between 235 and 268 ka

5) Sand of shallow-water marine facies aged 289 ka BP.

6) Sandy silts aged between 316 and 336 ka BP.

7) Sand and gravels aged 535 ka BP.

This is a first attempt to date marine Quaternary deposits of the Taymyr Peninsula. We did not have an opportunity to date all geomorphological levels. However, the bedding of studied sediments at very different altitudes suggests that their modern position is under the influence of

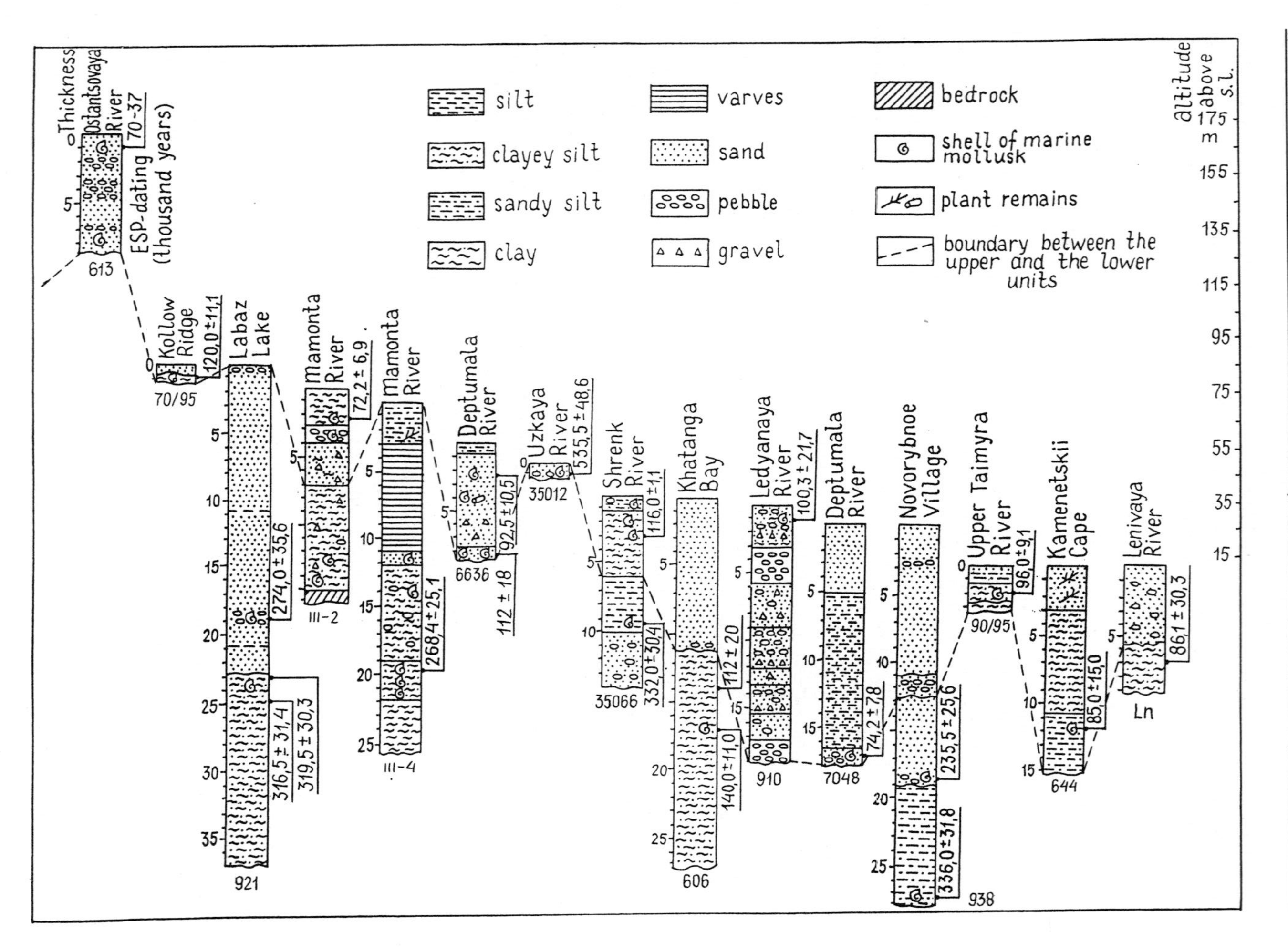

Figure 2: Sediment logs with ESR datings from the Taymyr Peninsula.

differential tectonic movements of the Taymyr Peninsula.

The ESR dates obtained thus suggest active marine sedimentation over the Taymyr Peninsula during isotope stages 4, 5, 7, 8, 9 and 15. These dates correlate well with the marine events of the Severnaya Zemlya archipelago during the same half million year time interval (Bolshiyanov and Makeev, 1995).

Acknowledgements

The authors are grateful to G. V. Schneider and V. V. Mezhubovsky for providing samples of marine deposits for dating, to Ye. A. Kirienko, who carried out foraminifera analysis, to V. S. Zarkhidze, who determined the malaco-fauna species, and to G. V. Stepanova, who carried out diatom analysis.

References

Bolshiyanov, D.Yu. and V.M. Makeev (1995) Severnaya Zemlya archipelago: the glaciation, the history of the environment. Gidrometeoizdat, St.Petersburg (In Russian), 217 pp.

Zolnikov, I.D. and A.Ya. Shevko (1989) On the stratigraphy of Quaternary deposit sequences near Novorybnoe village, Khatanga Rriver. In: Cenozoic group in Siberia and north-eastern part of the USSR, Nauka, Novosibirsk (In Russian), 172-180.

Ikeya, M. and K. Ohmura (1981) Dating of fossil shells with electron spin resonance. J.Geol., 89, 247-251.

Kind, N.V. and B.N. Leonov (1982) The anthropogene of the TaymyrTaimyr Ppeninsula. Nauka, Moscow (in Russian), 183 pp.

Molodkov, A. (1988) ESR dating of Quaternary shells: recent advances. Quat. Sci. Rev., 7, 477-484.

Molodkov, A. (1989) The problem of long fading of absorbed palaeodose on ESR-dating of Quaternary mollusc shells. Appl. Radiat. Isot., 40, 1087-1093.

Molodkov, A. (1992) ESR-analysis of Molluscan skeletal remains in Late Cenozoic Chronostratigraphic research. Tartu (In Russian), 33 pp.

Molodkov, A. (1993) ESR-dating of non-marine mollusc shells. Appl. Radiat. Isot. 44, 145-148.

Troitskii, S. L. (1966) Quaternary deposits and the relief of plains costs of the Yenisey Bbay and adjacent part of Byrranga Mmountains. Nauka, Moscow (In Russian), 208 pp.

Urvancev, N. N. (1931) Traces of Quaternary glaciation in the central part of North Siberia. Geological Publishing House, Moscow, Leningrad (In Russian), 55 pp.

Paleoclimatic Indicators from Permafrost Sequences in the Eastern Taymyr Lowland

C. Siegert[1], A.Y. Derevyagin[2], G.N. Shilova[2], W.-D. Hermichen[1] and A. Hiller[3]

(1) Alfred-Wegener-Institut für Polar- und Meeresforschung, Forschungsstelle Potsdam, Telegrafenberg A43, D 14473 Potsdam, Germany

(2) Faculty of Geology, Moscow State University, 119899 Moscow, Russia

(3) Universität Leipzig, Forschungsgruppe Paläoklimatologie, Permoserstrasse 15, D 04303 Leipzig, Germany

Revised 5 May 1997 and accepted in revised form 15 May 1998

Abstract - Permafrost sequences in the Labaz Lake area were investigated to reconstruct their paleoenvironmental history during the Late Quaternary using geocryological field studies, sedimentological, geochemical, stable isotopic and palynological analyses in connection with radiocarbon dating. The current results show that during post-Zyryan (Early Weichselian) times denudation and accumulation in this area was dominated by processes connected with the decay of an ice cover, formed at least during the Zyryan stadial. As a result of the specific peculiarities of glacier ice thawing in a territory with continuous, low-temperature permafrost, buried glacier ice bodies were preserved until the present time. The most intensive decay of glacier ice occurred during the first warming stage of the Kargin (Middle Weichselian) Interstadial more than 40000 yr BP. The formation of extensive lake depressions within the glacial deposits is connected with these times. About 40000 yr BP, a surface stabilization occurred, and syncryogenic deposits with thick polygonal ice wedges (Ice Complexes) and peat beds were formed. Subsequent warming phases, apparently connected with increases in precipitation, led to a new activation of glacier ice decay, and to the accumulation of additional lacustrine and fluvio-lacustrine sediments. Palynological and stable isotope data from ground ice suggest a continual trend to a cooler and more continental climate from the Kargin Interstadial to the Sartan stadial times. During the maximal cooling period, polygonal ice wedges in dried-up lacustrine deposits formed. At the termination of the Pleistocene, lacustrine sedimentation seems to have reactivated in reaction to short climate improvements. Thermokarst lakes were formed in this period. Extensive peat accumulation started at the end of the Boreal and spread to flat lacustrine depressions during the Atlantic. Since the Subboreal, a drying up of lakes and gradual increasing of permafrost aggradation can be inferred.

Introduction

Controversy exists concerning both the extent and the age of the last glaciation in Central Siberia. In particular, the environmental conditions during Post-Zyryan (Post-Early-Weichselian) times are only slowly becoming clear (Astakhov, 1992; Faustova and Velichko, 1992; Isaeva and Kind, 1989; Velichko, 1993). The solution to these problems is quite important as it would affect, firstly, the character of atmospheric circulation models during the Late Pleistocene, and secondly, specific features of sediment transport to the Arctic Ocean.

Paleogeographical studies of permafrost sequences have been carried out in the Labaz Lake region as part of the German-Russian research project "Taymyr,, during two expeditions (Siegert et al., 1995; Siegert et al., 1996). The study area represents the typical lowland surrounding the Laptev Sea to the southwest. According to the maximal-extent reconstruction of the Late Weichselian (Sartan) glaciation, this area was located to the east of the glaciated territory (Velichko, 1993). Based on remote sensing and the study of sediment sequences in surrounding river valleys, Isaeva (1982) assumed that the Late Pleistocene section at Labaz Lake included deposits of a vast ice-dammed lake ("Pre-Labaz") that existed during the Sartan glaciation. According to this interpretation, the level of the "Pre-Labaz" had reached a maximum

----

In: Kassens, H., H.A. Bauch, I. Dmitrenko, H. Eicken, H.-W. Hubberten, M. Melles, J. Thiede and L. Timokhov (eds.) Land-Ocean Systems in the Siberian Arctic: Dynamics and History. Springer-Verlag, Berlin, 1999, 477-499.

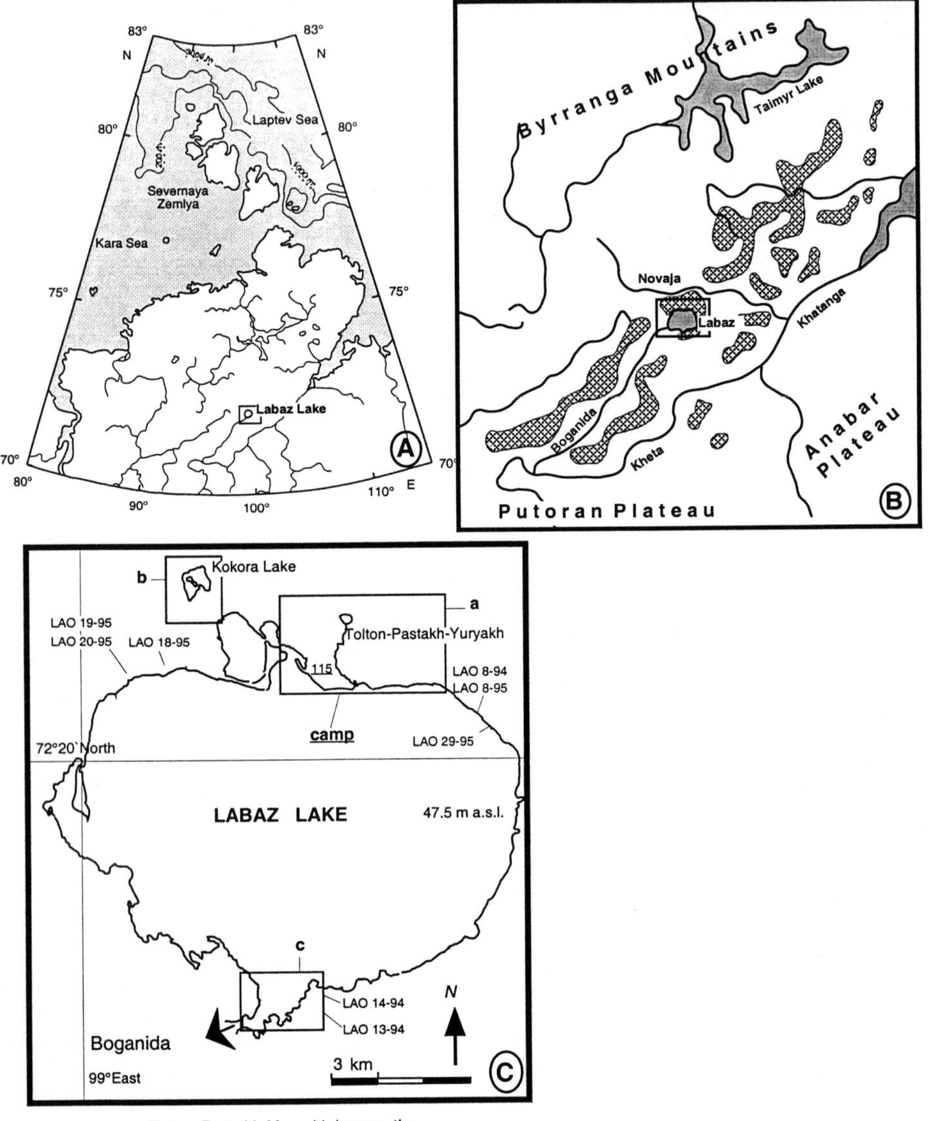

Figure 1: (A) Map of the Taymyr-Severnaya Zemlya region; (B) Map of the central part of the Taymyr Peninsula with distribution of marginal formations of the Zyryan (Early Weichselian) ice sheet (according to Kind and Leonov, 1982). (C) Overview map of the Labaz Lake area with location of the study sites.

of 70-80 m a.s.l. during the Nyapansk phase at the end of the Sartan glaciation. The modern Labaz Lake, covering 27 x 30 km^2, and the neighbouring Tonskoe and Kharga Lakes are regarded as relicts of the "Pre-Labaz".

Limited information exists about the climatic and environmental development during the Kargin (Middle Weichselian) Interstadial period, which is believed to have lasted from about 55000 to 24000 years BP (Alexeev et al., 1984). According to Andreeva and Kind (1982), this period was marked by unstable climatic conditions with alternating warmer and cooler phases in the Taymyr region.

For this reason, the objectives of our studies were to obtain records which permit the reconstruction of the environmental history in the study area since Post-Zyryan times, and to test the proposed concept of a "Pre-Labaz" during the Sartan period.

Table 1: Physical geographical characteristics of the Labaz Lake area. Data from: Atlas Arctici, 1985; Ershov, 1989; Vasilevskaya, 1980. Climatic data from the Khatanga Meteorological Station.

Botanical geographical zone	Boundary zone between southern tundra and typical (northern) tundra
Mean air temperature	-13.4°C
Mean air January temperature	-33.8°C
Mean air July temperature	+12.3°C
Frost free period	73 days
Precipitation/year	237 mm
Permafrost thickness	300-600 m
Rock temperature on the zero amplitude	the zero amplitude
Mean active layer thickness	0.20 to 0.50 m
Relief	low-to-moderate relief hummocky topography with altitudes of 47-130 m a.s.l.
Hydrology	Territory with a large number of lakes and swamplands strongly regulating the river runoff

Material and methods

Study area

Labaz Lake (72°N, 99°E) is one of the large lakes of the North Siberian Lowland. The morphology of the territory was primarily formed by glacial, lacustrine and cryogenic processes. Wide flat lacustrine depressions are situated within the glacial relief elements. Numerous small lakes of thermokarst genesis occur at different levels. Holocene peat beds are widespread. The soil cover is dominated by weakly drained cryosols/Pergelic Cryaquepts (Pfeiffer et al., 1996) underlain by ice-rich permafrost. This near-surface position of ice-rich permafrost, as well as the modern climatic conditions, are of major importance for modern exogenic processes. The main characteristics of the study area are given in Table 1.

Perennially frozen sediments were studied at numerous natural exposures. The majority of the exposures were situated in the northern shore zone of Labaz Lake, primarily at the key section ,a, (Figure 1C). The conditions on the steep northern shore were adequate for investigations. The shore bank rises up to 60 meters above lake level and was heavily influenced by thermoerosion, solifluction and thermokarst phenomena. In addition, three profiles were studied and sampled on shore terraces of the Kokora Lake depression (section ,b,). This territory differs from section ,a, as it has a pronounced glacial relief with numerous hillocks and ridges. According to Andreeva and Isaeva (1982), Kokora Lake is located on territory covered by sediments formed primarily during the decay of the Early Weichselian ice sheet. Finally, two exposures were investigated in section ,c, in the outflow area of the Boganida River in the southern surroundings of Labaz Lake. These profiles represent sediments of relatively young accumulation terraces. In addition, deposits of different geomorphological levels were investigated by core drilling within the key section ‚a‚. Altogether, 20 profiles up to 7 m depth were drilled.

Study methods

As a basic method for the investigation of permafrost profiles, the "frozen ground facial analysis„ (Katasonov, 1978) was used. This method is based on the relationship between the cryogenic structure of sediments, as well as on other cryogenic phenomena within sedimentary complexes, and the geocryological conditions at the time of sediment accumulation and freezing. Using this method, permafrost sequences can be stratified and information on the climatic conditions during syngenetic permafrost formation can be obtained (Romanovsky, 1978). Samples were taken for manifold analyses at all study sites. In this publication, results from radiocarbon dating, palynological investigations and selected results from stable isotope analyses of ground ice are presented and discussed.

The material which was dated consisted of various plant remains, preserved in permafrost: mainly peat, peaty soil and wood (the remains of trees and shrubs), and occasionally grass roots, twigs and/or seeds. The ^{14}C activities were measured in the AWI-Potsdam (lab. no. indicated as AWI) and in the Leipzig (LZ) laboratories by liquid scintillation counting (LSC), using Packard Tri-Carb 2560 TR/XL spectrometers in both cases. The ^{14}C ages are given as conventional ages (Stuiver and Polach, 1977; Geyh and Schleicher, 1990), i. e. the ages - given in ‚yr BP' (years before present, i.e. 1950) - were calculated on the basis of the measured activity using a ^{14}C half-life of 5568 years. The activity was corrected for isotopic fractionation by normalizing $\delta^{13}C$ value to -25‰.

In order to minimize contamination of the sample material, the evidently autochthonous plant material was removed manually at first. Afterwards, all samples were treated with diluted NaOH (0.5 mol., 80°C) to extract (allochthonous, migrating) humic acids (Shore et al., 1995) and then with diluted HCl (0.5 mol., 80 °C) to remove possible traces of carbonates. The washed and dried (neutralized) organic residue of each sample was converted into benzene (3 ml) in the standard manner (Polach et al., 1972). The radiocarbon ages and $\delta^{13}C$-values of samples from Labaz Lake area determined in the AWI-Potsdam and in the Leipzig laboratories are compiled in Table 2. In addition, some AMS dating of radiocarbon material from selected samples were carried out at the Leibniz Laboratory of the University Kiel. These results are shown in Table 3.

Samples for isotopic studies were taken in the field and stored and transported in 30 ml polyethylene bottles. Measurements of $\delta^{18}O$ values were carried out at the Alfred Wegener Institute Bremerhaven (A. Mackensen) and measurements of $\delta^{2}H$ at the Freie University Berlin (K. Friedrichsen).

Standard methods (maceration using HCl, H_2SO_4 with acetolysis and separation in gravity liquid by centrifuge) were used for preparing the pollen samples. The results are presented in percentage pollen diagrams including: (i) the ratio of four groups: 1 - arboreal pollen (AP) of trees, 2 - AP of shrubs, 3 -herbaceous pollen (NAP), 4 -spores; (ii) the dominant pollen and spores as a percentage of the sum of Quaternary pollen and spores; and (iii) the redeposited pre-Quaternary pollen and spores based on the total pollen and spores sum. A zonation of the diagrams was done by visual inspection.

Table 2: Conventional Radiocarbon ages and $\partial^{13}C$ values of fossil plant material from the Labaz Lake area, Taymyr Peninsula. Detection limit of Leipzig laboratory (LZ): about 48000 yr BP; Alfred Wegener Institute (AWI) laboratory: about 40000 yr BP.

No.	Sample No.	Material	Sampling site, depth [m]	Site character	14C-Age [yr BP]	Std.Dev. s (yr BP)	∂13C (‰ PDB)	Laboratory No.
1	S95-15	wood remains	LAO22; 2.5-2.7	lacustrine sediments, 60-65 m terrace	>48 000		-28,2	LZ-1271
2	S95-16	wood remains	LAO22; 5.7-6.1	lacustrine sediments, 60-65 m terrace	>46 000		-29,3	LZ-1272
3	S95-17	wood remains	LAO22; 6.4	fossile frost cracke in lacustrine sediments			-29,6	LZ-1273
4	S95-18	wood remains	LAO22; 6.8-7.2	lacustrine sediments, 60-65 m terrace	>48 000		-28,5	LZ-1274
5	S94-11	mixed plant remains (grass, grass roots,twigs, seeds, moss)	LAO15-94; 10 a. Labaz Lake l.	sediments of the "Ice complex", 60-65 m terrace	43 900	2900	-27,5	LZ-1164
6	S94-07	wood remains	LAO9; 1.2 a. Labaz Lake level	sediments at the basis of the "Ice complex", 60-65 m ter.	>42 000		-26,5	LZ-1163
7	S95-08	wood remains	LAO15-95; 0.8	covering horizon on the "Ice complex"	5 710	100	-26,5	AWI-089
8	S95-07	wood remains	LAO15-95; 0.8	covering horizon on the "Ice complex"	5 410	50	-26,5	AWI-094
9	S95-05	tree branch	LAO18; 1,0	covering horizon on the "Ice complex" (50 m level)	6 120	80	-27,3	AWI-091
10	S94-01	tree branches from floating peat	LAO1-94; 1,1-1,3	peat bed, top of a erosional remnant (65 m terrace)	>47 000		-29,3	LZ-P1
11	S95-10	peat, moderatly decomposited	LAO1-95; 0.45-0.85	peat bed, top of a erosional remnant (65 m terrace)	>40 000		-26,3	AWI-084
12	S95-11	peat, slighly decomposited	LAO1-95; 0.85-1.2	peat bed, top of a erosional remnant (65 m terrace)	>40 000		-27,4	AWI-082
13	S95-12	shrub branches	LAO1-95; 1.2-1.45	peat bed, top of a erosional remnant (65 m terrace)	>40 000		-28,7	AWI-083
14	S95-13	wood remains	LAO1-95; 1.2-1.45	peat bed, top of a erosional remnant (65 m terrace)	>48 000		-29,0	LZ-1275
15	S95-14	small mixed plant remains	LAO1-95/2; 1.5-1.6	fluviatil deposits adjacent to the erosional remnant	34 000	700	-27,0	AWI-099
16	S95-23	allochthonous woody peat	LAO25; 2.8	fluvio-lacustrine deposits at the 50 m level	28 500	400	-28,1	AWI-096
17	S95-24	allochthonous woody peat	LAO25; 3.5	fluvio-lacustrine deposits at the 50 m level	38 000	600	-27,2	AWI-097
18	S95-37	shrub branches	LAO25; 3.5	fluvio-lacustrine deposits at the 50 m level	>40 000		-29,4	AWI-122
19	S95-25	tree branches	LAO25; 4.3	fluvio-lacustrine deposits at the 50 m level	33 600	400	-28,1	AWI-103
20	S95-38	tree branches	LAO25; 4.3	fluvio-lacustrine deposits at the 50 m level	38 900	1300	-27,8	AWI-123
21	S95-36	shrub branches	LAO25; 5.1	fluvio-lacustrine deposits at the 50 m level	>40 000		-28,8	AWI-121
22	S95-33	tree branches	LAO33; 3.0 a.l.Guba bay	lacustrine deposits, 55 m level	40 000	1000	-27,2	AWI-100
23	S94-05	peat	LAO6-94; 0.6-0.65	peat bed, top of 70 m terrace-like remnant	7 860	90	-23,5	LZ-P5
24	S94-06	peat	LAO6-94; 1.1-1.2	peat bed, top of 70 m terrace-like remnant	8 760	90	-31,1	LZ-P6
25	S95-20	small plant remains	LAO6-95; 3.4-3.6	lacustrine deposits, 70 m terrace-like remnant	34 500	900	-30,5	LZ-1269
26	S95-21	small plant remains	LAO6-95; 6.6-6.8	lacustrine deposits, 70 m terrace-like remnant	38 000	1600	-33,2	LZ-1270
27	S94-09	peat	LAO13; 0.7-0.8	low terrace, outflow of the Boganida River	9 280	100	-26,8	LZ-P10
28	S94-08	peaty soil	LAO13; 0.1	low terrace, outflow of the Boganida River	930	60	-26,6	LZ-P9
29	S94-10	peaty soil	LAO14; 0.4-0.45	flood plain, outflow of the Boganida River	6 730	80	-28,9	LZ-P11
30	S95-01	small tree stump(Larix) in situ	meadow near LAB2-95	thermokarst depression on unit 1(90-100 m level)	7 790	60	-27,3	AWI-093
31	S95-2B	mixed peaty mater al	LAB2-95; 0.92-1.04	deposits of the thermokarst lake	9 150	130	-27,7	LZ-1279
32	S95-2A	branch remains	LAB2-95; 0.92-1.04	deposits of the thermokarst lake	8 850	115	-29,1	LZ-1276

Table 2 (continued):

No.	Sample No.	Material	Sampling site, depth [m]	Site character	14C-Age [yr BP]	Std.Dev. s (yr BP)	∂13C (‰ PDB)	Laboratory No.
33	S95-27	mixed peaty material	LAO26; 6.5 a.Kokora Lake level	steep slope of the lake terrace	9 130	90	-27,0	AWI-108
34	S95-26	tree trunk (Larix)	LAO26; 6.5 a.Kokora Lake level	steep slope of the lake terrace	7 170	100	-26,0	AWI-102
35	S95-28	tree remains	LAO27; 4.0 a.Kokora Lake level	steep slope of the lake terrace	8 390	70	-26,7	AWI-104
36	S94-04	shrub branchs from peat bed	LAO5; 1.3-1.4	peat bed, top of the 50-55 m terrace	7 890	90	-27,0	LZ-1162
37	S95-19	peat	LAO21; 2.5 a.Labaz Lake level	peat bed, top of the 50-55 m terrace	7 590	80	-28,8	AWI-098
38	S94-13	tree trunk (Larix)	LAO17; 10.5 a. Labaz lake level	basis of a peat bed, 50-55 m terrace	7 230	90	-26,7	LZ-P15
39	S94-12	shrub remains (branchs, twigs)	LAO17; 10.5 a. Labaz lake level	basis of a peat bed, 50-55 m terrace	6 900	90	-27,8	LZ-P14
40	S95-30	mixed peaty material	LAO30; 2.5	first terrace of the Tolton-Pastakh-Yuryakh River	6 640	50	-27,4	AWI-085
41	S95-29	tree branch	LAO30; 2.5	first terrace of the Tolton-Pastakh-Yuryakh River	5 780	60	-27,2	AWI-101
42	S95-31	tree wood (Larix)	LAO31; 1.5	first terrace of the Tolton-Pastakh-Yuryakh River	6 360	80	-28,0	AWI-087
43	S94-02	tree stump (Larix)in situ	recent surface	first terrace of the Tolton Pastakh-Yuryakh River	4 780	80	-26,0	LZ-P2
44	S94-03	tree trunk (Larix)	LAO3	bench of a thermokarst lake on deposits of unit 1	5 220	80	-25,9	LZ-P3
45	S95-04	shrub branchs	LAB12-95; 1.0	small flat thermokarst depression on 60 m terrace	4 700	70	-28,9	LZ-1280
46	S94-14	peat on the permafrost table	LAB1-94/3; 0.26-0.32	valley slope,Tolton-Pastakh-Yuryakh River	4 630	70	-29,1	LZ-P16
47	S94-17	peat	LAB1-94/2; 1.7	valley slope,Tolton-Pastakh-Yuryakh River	5 810	70	-28,8	LZ-P19
48	S94-15	shrub branchs	LAB1-94/2; 1.7	valley slope,Tolton-Pastakh-Yuryakh River	5 970	80	-26,3	LZ-P17
49	S94-19	peat	LAB1-94/2; 2.5	valley slope,Tolton-Pastakh-Yuryakh River	6 790	80	-28,8	LZ-P18
50	S94-20	tree trunk (Larix)	LAB1-94/2; 2.6	valley slope,Tolton-Pastakh-Yuryakh River	7 010	80	-26,4	LZ-P21
51	S95-34	peat on the permafrost table	LAB2-94/3; 0.36-0.40	peat bog at the 60 m terrace	4 200	65	-29,4	LZ-1278
52	S95-35	peat on the permafrost table	LAB2-94/2; 0.40	peat bog at the 60 m terrace	2 900	50	-28,4	AWI-105
53	S95-32	Tree wood	LAO29; 0.5	peat layer above dead ice body in a thermocirque	3 700	70	-27,5	AWI-088
54	S95-22	tree stump (Larix) in situ	meadow near LAO24	open dry depression (thermocirque)	3 680	70	-26,0	AWI-092
55	S95-09	tree stump (Larix) in situ	meadow tundra near LAB13-95	dry peripheral zone of a thermokarst depression	2 880	60	-27,6	AWI-090
56	S95-03	peat	LAB3-95; 0.3-0.6	peat bog, thermokarst depression on unit 1(100-110 m l.)	2 230	60	-27,9	LZ-1277

Table 3: AMS Radiocarbon ages and $\partial^{13}C$ values of fossil plant material from the Labaz Lake area, Taymyr Peninsula.

No.	Sample No.	Material	Sampling site depth [m]	Site description	14C-Age [yr BP]	Std.Dev. s [yr BP]	∂13C [‰ PDB]	Laboratory No.
1	LA-286-95	mixed plant remanants, (grass roots, seeds, moss)	LAO28; 1.5	lacustrine deposits with polygonal ice wedges, steep bank of the Kokora Lake	17 320	+220/-220	-27,74	KIA1413
2	LAB2-21-94	mixed plant remnants (moss, grass, small wood pieces	LAB2-94; 2.33	lacustrine deposits, flat lacustrine depression on the 60-65 m terrace	7360	+60/-60	-25,05	KIA1408
3	LAB2-26-94	mixed plant remnants	LAB2-94; 3.0	lacustrine deposits, flat lacustrine depression on the 60-65 m terrace	8960	+90/-90	-27,92	KIA1409
4	LAB2-33-94	sand with mixed plant remnants	LAB2-94; 3.7	lacustrine deposits, flat lacustrine depression on the 60-65 m terrace	20 400	+300/-290	-29,79	KIA1411
5	LAB2-34-94	sand with mixed plant remnants	LAB2-94; 3.73	lacustrine deposits, flat lacustrine depression on the 60-65 m terrace	24 990	+520/-480	-26,64	KIA1412
6	LAO6-17-94	sand with mixed plant remnants	LAO6-94; 2.00	lacustrine deposits, 70 m terrace-like remnant	26 240	+580/-540	-26,46	KIA1414
7	LAO6-19-94	sand with mixed plant remnants	LAO6-94; 2.40	lacustrine deposits, 70 m terrace-like remnant	14 390	+150/-140	-27,66	KIA1415
8	LAO13-11-94	mixed fine plant remains, leaves	LAO13-94; 1.35	lacustrine terrace at the southern	11 810	+140/-140	-27,97	KIA1416

Results and interpretations

On the basis of the obtained results, four stratigraphic units representing different historical periods of the Late Quaternary environmental development can be differentiated in the study area (Figure 2).

Unit 1: Glacial sediments

Unit 1 consists of different glaciogenic sediments most probably formed during the Zyryan (Early Weichselian) stadial. These deposits built up the highest relief levels in the study area. Their observed thickness reaches a maximum of about 60 m.

The bottom of unit 1 consists of dense dark grey or grey-brown boulder clay with a massive or fine lens-shaped cryostructure. In places, the horizon was exposed up to a height of 5-7 m above the modern lake level. Upwards, the boulder clay is overlain by glaciolacustrine sediments. This horizon consists of grey rhythmically interbedded and partially interlaminated clay, silt and sand, in parts with gravel and single boulders. The cryostructure of these ice-rich (more than 50 weight%) deposits is characterized by fine to coarse reticulate networks of ice lenses typical for perennially frozen lacustrine deposits transformed epigenetically into permafrost under the influence of water migration processes. Such sediments were studied and sampled during the summer of 1994 in a large thermocirque with a maximum depth of 31 m (LAO8, Figures 1 and 2). The analytic results confirmed a glacio-lacustrine origin: first, all sediments were characterized by a very low content of organic matter. The total amounts of C_{org} and N did not exceed 0.12% and 0.08%, respectively. Secondly, the pollen assemblages of the studied samples were strongly dominated by redeposited Mesozoic and Tertiary pollen and spores, while Quaternary pollen were found only as single grains. The observed pollen and spores from the Pleistocene belong to tundra vegetation.

Within unit 1, buried glacier ice bodies were found at two sites (LAO24, LAO29; Figures 1 and 2) during the field work in 1995. The first (thickness about 8 m) started to become exposed after an earth slide in a newly activated recent thermocirque located on the SW exposed slope of glaciogenic sediments (see Figure 2). In the ice body, layers of various ice structures including characteristic glacier ice crystals could be observed. The stable isotopic composition in this ice profile ranged from -22.35‰ to -26.90‰ and -172.1‰ to -201.0‰ for $\delta^{18}O$ and $\delta^{2}H$, respectively. Towards the end of the field work in 1995, a second buried glacier ice body became exposed on the shore bank near the stabilized thermocirque of LAO8. This glacier ice was observed from the beach up to an altitude of more than 10 m above lake level, and in a bowl-like thermokarst depression along a deep erosion valley, oriented almost vertically to the beach (Figure 3). The wall of the thermocirque consisted of dark-grey layered loamy-sandy deposits with gravel and small boulders, which belong to the upper horizon of unit 1. The stable isotope composition of ice sampled at the near-surface horizon varied between -24.55‰ and -29.51‰, and -188.3‰ and -225.6‰ for $\delta^{18}O$ and $\delta^{2}H$, respectively. All values obtained from the two sites plot on the Meteoric Water Line (Figure 4).

The upper horizon of unit 1 was built up of crudely stratified loamy sands with gravel. These deposits are interpreted as flow till, probably formed during deglaciation They contained a small amount of ground ice and were characterized by massive cryostructures. At the surface, these sediments were under the influence of aeolian processes - boulder pavements were observed.

Unit 2: Sediments of the Kargin Interstadial

Unit 2 represents a complex strata which comprises primarily lacustrine and fluvio-lacustrine sediments, overlying deposits of unit 1 with an erosional unconformity. These sediments fill extended depressions within unit 1 as well as erosional forms, developed later within sediments

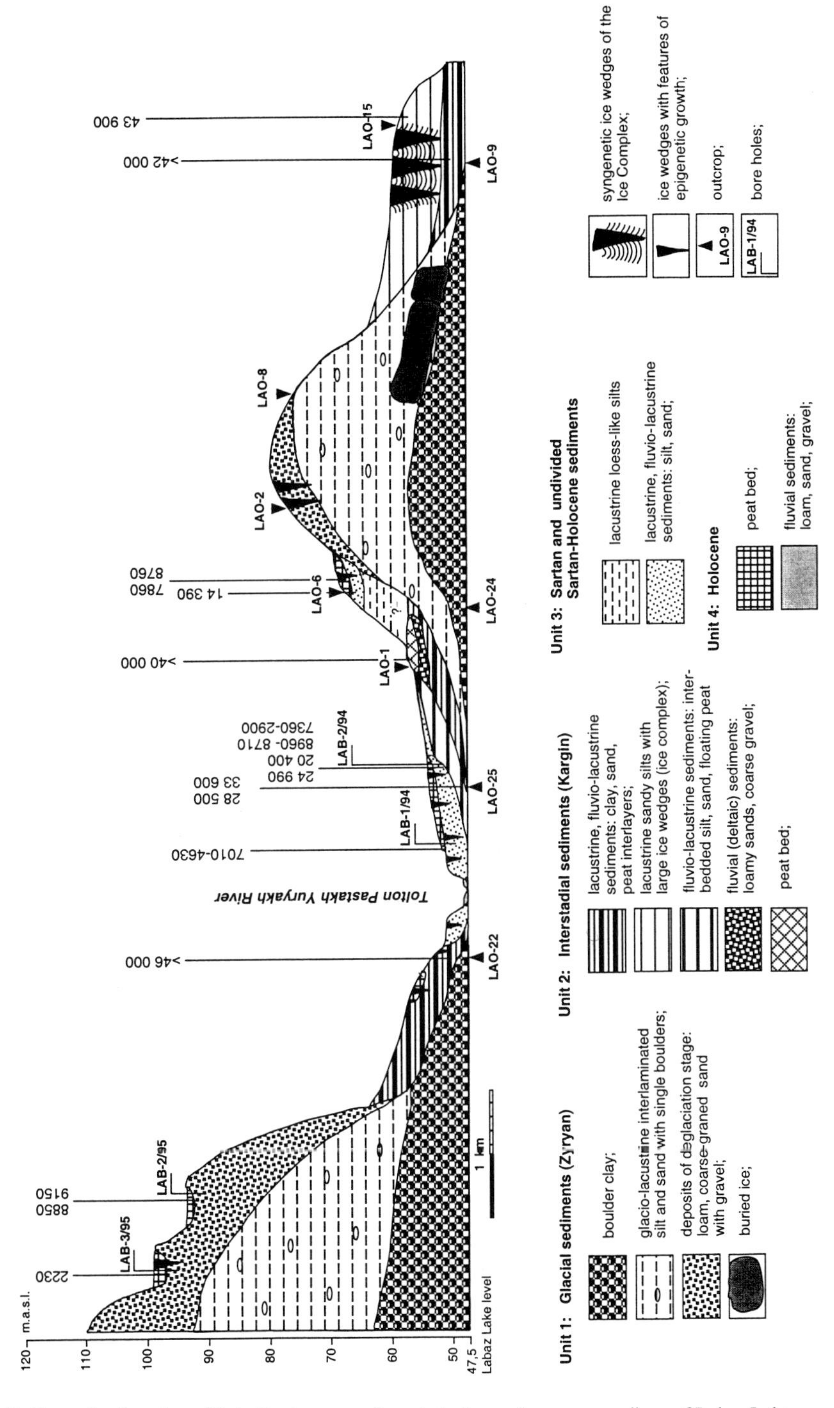

Figure 2: Generalized section of Late Quaternary sediments in the northern surroundings of Labaz Lake.

of post-glacial times (Figure 2). Fluvio-lacustrine deposits often contain floating peat layers. At one location, an autochthonous peat bed was found. In places, lacustrine sediments give way to overlying ice-rich permafrost deposits with polygonal ice wedges of up to 15 m thickness and 5 m width - the so-called ‚Ice Complex‚ (Figure 5). Radiocarbon data (Table 2: No 5, 6) from this exposed Ice Complex showed ages of more than 42000 yr BP for the lower part (LAO9) as well as for the top layer below a thermo-erosional unconformity (LAO15).

Deposits of the Ice Complex consist of weakly layered sandy silts with fine plant detritus and, in places, with subvertical grass roots. A banded lens-shaped cryostructure interlocked with the lateral ice wedge boundaries is characteristic. As a whole, these features are typical for a syngenetic transition of deposits to permafrost under subaerial conditions, perhaps in a seasonally drying lacustrine environment. Recently, analogous ice wedge growth occurred in the very flat shore zones of Labaz and other lakes. The stable isotopic composition in two investigated ice wedges varied between -24.3‰ and -30.7‰, and -184‰ and -234‰ for $\delta^{18}O$ and $\delta^{2}H$, respectively. As a rule, the ice wedges are covered with an unconformity by Holocene, very ice-rich loam (‚ice-ground‚). In places, peat beds with a younger Holocene ice wedge generation were formed above the Pleistocene Ice Complex. The stable oxygen and hydrogen isotope ratio of Holocene generation are clearly different from those of Late Pleistocene ice wedges as shown in Figure 4.

Figure 3. Dish-shaped thermokarst depression on a buried glacier ice body. The top of ice is exposed after an earth slide. LAO29, 29. 08.1995.

The geomorphological position as well as the complicated cryolithogenic structure of unit 2 suggest that it was formed during climatically different stages of the Kargin (Middle Weichselian) Interstadial period. In particular, formation was likely during the first phases of the Kargin Interstadial period, because the major part of the obtained radiocarbon data indicated ages greater than 40000 yr BP. Only some data pointed to sediment formation during later interstadial times, from around 34000 to 25000 yr BP (Table 2 and 3). To reconstruct the changing environmental conditions during this period, three dated sites were studied by pollen analysis, namely sites LAO22, LAO25, and LAO6. The results obtained from the first two sections characterize vegetation of the Kargin Interstadial, while the pollen diagram of the LAO6 section in part contradicts the radiocarbon datings and the stratigraphic classification thus remains partially problematic.

Site LAO22 is located on the 60-65 m lake terrace to the west of the Tolton-Pastakh-Yuryakh mouth (Figures 1 and 2). In its lower portion, the section consisted of ice-rich grey clayey silt alternating with brownish grey sand. Sandy layers had a massive or very fine lens-shaped cryostructure, whereas the silt layers were characterized by a coarse lattice-like structure. Rust-coloured zones enriched with iron hydroxides were observed at the contact between ice lenses and ground aggregates. A small ice wedge was found within the succession (Figure 6). These cryogenic features suggest a syngenetic sediment freezing, likely under conditions found in shallow lacustrine environments which dry up seasonally or during periodic lake level changes. The upper part of the section consists of light grey or grey-brown sandy deposits with interlayers of silt, clay and floated peat. The low ice content, the fine lens-shaped cryostructure of the clay and silt layers, and the massive cryostructure of the sand layers also argue for a formation under the above mentioned conditions. All sediments contained wood - primarily branches and twigs of shrubs. Radiocarbon measurements of these remains gave ages of greater than 46000 yr BP (Table 2, No 1-4). The top sediments seem to have been reworked by slope wasting, and samples were therefore not taken for radiocarbon dating.

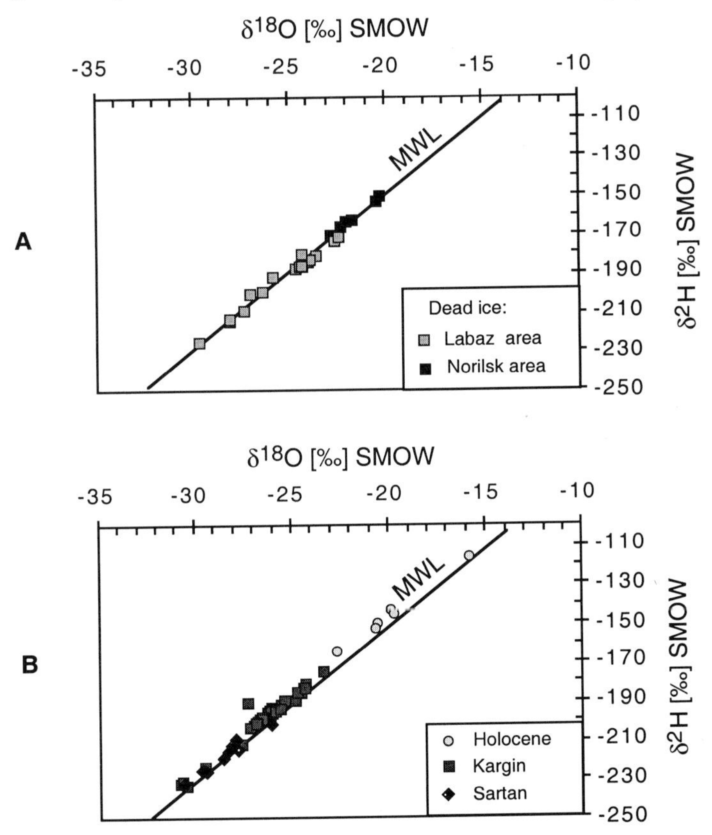

Figure 4. Stable isotope composition of ground ice bodies in the Labaz Lake area. (A) buried glacier ice; (B) Ice wedges.

Pollen assemblages obtained from profile LAO22 (Figure 7) indicated warmer climatic conditions for the accumulation period of this section than we have at present. A rich shrub-tundra vegetation with open larch and birch woods dominated. During the formation of zone 1 and 3, a more humid climatic character can be assumed. The pollen assemblages of zone 2 suggested drier conditions. There was a significant increase of shrub pollen percentage with dwarf birch dominance. Tree pollen, including both alder types as well as spores, decreased. A small increase of Poaceae and other herbs was observed in the NAP. The typical xerophyte pollen of *Ephedra* was also observed. The high amounts of redeposited Mesozoic but apparently Cretaceous pollen in all samples is significant. They indicate that glaciogenic deposits of unit 1 were strongly eroded and redistributed. Because all radiocarbon data indicated ages greater than 46000 yr BP, we suppose that these sediments were likely formed not later than the first warming phase of the Kargin Interstadial. Without a doubt, active glacier ice decay took place during this time, leading to a strong activation of erosional processes and to a rise of accumulation in adjacent lakes.

Site LAO25 represents fluvio-lacustrine (and likely deltaic) sediments which were exposed at the basis of the 55 - 60 m lake terrace to the east of the Tolton-Pastakh-Yuryakh River mouth (Figures 1 and 2). Exposed sediments showed a clear bedding with an inclination of about 10° towards the W-WNW. The lower part of the section consisted of light grey sand interlaminated by silt and clay with characteristic rust-coloured layer boundaries. Among various allochthonous plant remains, tree branches were present. From 3.6 to 2.8 m depth, layered sands containing coal detritus and small gravels were interbedded by numerous floating peat layers. The overlying horizon consisted of sand, silt and clay interbedding with several tree and shrub remains as well as crushed angular gravels and pebbles. Subvertical grass roots were found in the upper two meters. The latter sediments seemed to be influenced by recent or Holocene slope and soil forming processes. For wood remains from this section, radiocarbon ages between about 34000 yr and more than 40000 yr BP were recorded including some inverse data. A dating of 28500±400 yr BP was obtained from floating peat material at a depth of 2.8 m. These results indicate that the sediments contained potentially redeposited plant material, especially wood remains.

The pollen diagram of site LAO25 (Figure 8) showed similarities to that of site LAO22 (Figure 7). Small percentages of NAP and a dominance of *Betula* sect. *Nanea* and *Alnaster* are characteristic features. However, in contrast to site LAO22, the pollen percentage of trees, especially of *Betula Alba* and *Alnus*, was slightly higher at site LAO25. Similarities between both sites also existed in the composition of NAP and spores. Poaceae and Ericaceae were found in appreciable amounts, with less *Artemisia* and Cyperaceae. For other herbs, Asteraceae and Cichoriaceae were continuously present, and Carophyllaceae, *Pedicularis, Dryas* and others less frequently. Spores observed belonged to *Bryales,* Polypodiaceae and *Lycopodium* (especially *L. alpinum, less L. pungens,* and least of all *L. annotium* , *L. complanatum* and *L. appressum*). The spore pollen assemblages of zones 1 and 3 in the diagram correspond to a shrub tundra vegetation growing under humid and warmer climatic conditions than those of today. The pollen assemblages of zone 2 indicate the development of an almost pure birch-shrub tundra, sparsely distributed, with open larch forest under drier but still warmer climatic conditions than today. The considerable amounts of redeposited Cretaceous pollen in all sediments should be noted again. With that, the pollen assemblages of section LAO25 indicate a shrub tundra vegetation with tree occurrence, characteristic for the „Pre-Sartan„ warmer phase of the Kargin Interstadia in the Taymyr region. Analogous pollen diagrams are published by Belorusova and Ukraintseva (1980) and Andreeva and Kind (1982) from exposed interstadial fluvio-lacustrine sediments, dated between 32000 and 24000 yr BP, in river valleys to the north of the Labaz area. This means that a part of the available radiocarbon datings obtained from wood material in the section LAO25 cannot been used for determination of the age of

Figure 5: Lacustrine Ice Complex at the northern shore bank of Labaz Lake. LAO15.

sediment formation. This material was likely redeposited in younger sediments.

The study site LAO6 is located on a WSW-exposed steep slope of a terrace-like remnant with an altitude of around 70 m a.s.l.(Figure 2). In all other directions, the ‚lake-terrace' gradually passes into the surrounding flat lake plain with an altitude of 60 m a.s.l. The upper part of the section was studied and sampled in 1994, while the lower part was sampled in the summer of 1995 following some earth slides.

The lower part of section LAO6 (Figure 9) consisted of vaguely laminated loess-like silt. Small colour changes from dark grey to brownish-grey and varying amounts of organic detritus caused the lamination. In beds enriched with fine plant material, vivianite segregations occur. The grain size distribution in the sequence was very uniform and characterized by domination of silt particles (60-83%) while the clay content amount to 40%. Only in the upper 1.5 m was some sandy material present (< 250 µm; up to 2-8%). A high ground-ice content (60-80 weight%) and a specific lattice-like cryostructure with overlapping subvertical ice streaks as shown in Figure 10 characterized the lower 4 m of the silt sequence. Quite similar cryostructures have been obtained via experimental freezing of sapropelic sediments under relatively high temperature gradients (Ershov et al., 1995). From above 2.4 m depth, the loss-like silts showed a finer lattice-like cryostructure and a slightly lower ice content. Altogether, sedimentary and cryogenic characteristics suggested that accumulation of these loess-like silts occurred continually in a shallow lacustrine environment and that permafrost was formed by freezing of water-saturated sediments shortly after their accumulation. The dominant frosted sand-size quartz grains as well as the loess grain-size spectrum indicate the presence of material subject to aeolian transport in the lacustrine environment.

The pollen diagram of the section LAO6 (Figure 9) distinctly differs from the others discussed above. Pollen assemblages in these lacustrine sediments indicated a strong dominance of herbaceous plants while trees and shrubs as well as sporophytes were in the minority. Within the herbaceous plants, *Artemisia* predominated and reaches a maximum of 54 - 59% of the total pollen sum. In addition, high amounts of Poaceae, and less Caryophyllaceae, Compositeae and Cyperaceae, characterized the pollen assemblages. Shrubs were dominated by *Betula sect. Naneae*, while *Alnaster* and *Salix* were continuously present in small amounts. A generally low content of redeposited pre-Quaternary pollen in sediments of this site is important and indicates that the erosion of glacial deposits was significantly diminished at this time. The divided pollen zones had relatively small distinctions. Zones 1, 3 and 5 were characterized by a weak increase of both tree and shrub pollen percentages. For the herbs, the pollen percentage of *Artemisia* clearly decreased in favour of an increase of Poaceae and some other taxa. In all, the pollen assemblages suggest a wide-spread occurrence of grass-moss tundra and somewhat less shrub tundra. Single larch populations with alder underbrush seems to have grown in the area, but only sparsely distributed, probably similar to a gallery forest. The climate during these periods was slightly more humid than that during the formation of pollen zones 2 and 4, the pollen assemblages of which indicated distinctly cooler and drier climatic conditions. AP percentages including shrub pollen (*Betula nana, Alnaster, Salix*) reached a minimum here. Moss spores (Bryales) diminished. Other evidence for a significant change in the environmental conditions during the accumulation of these loess-like deposits has not been found. In general, the palynological data of section LAO6 corresponded more closely with tundra-steppe vegetation considered to be typical for non-glaciated permafrost areas of Northern Siberia during the Sartan (Late Weichselian) glacial period (Kaplina, 1979; Tomskaya, 1981). With a depth of 1.4 m upwards, the pollen diagram of the profile LAO6 (see Figure 9) represents typical assemblages for the beginning of Holocene climate warming similar to those presented by Hahne and Melles (this volume) from two lake sediment cores drilled on the Taymyr Peninsula.

According to the available radiocarbon datings, the lower part of the loess-like silts in section LAO6 accumulated at a time related to the middle warming phase of the Kargin Interstadial (called the Malaya Kheta period by Kind (1974) and occurring about 35000 yr BP). That would mean that they were formed during periods similar to those in which the sediments of profile LAO25 were formed. However, the significant differences between pollen assemblages of the two sections contradict this concept. The following interpretations are possible: 1. we accept the reliability of datings. This means that the lower part of the loess-like silts accumulated during the second half of the Kargin Interstadial period. The occurrence of relatively high pollen percentages for *Larix* (up to 10% of the total pollen and spore sum) as well as for *Alnus* trees in this part of the section could argue for such an interpretation; 2. we assume that the radiocarbon ages are not correct. This means that the dated fine plant remnants were mixed with redeposited material. It is generally accepted that fresh plant remnants, and especially wood material, can be very well preserved and repeatedly deposited under permafrost conditions, i.e. that plants preserved in situ and redeposited plant material become very similar in appearance.

Sulerzhitsky (1982) was also concerned about the influence of such dating problems on the study of permafrost sections from the lowland areas of Taymyr. Two datings in the upper part

Figure 6: Lattice-like cryostructure of clayey silts in interbedded lake sediments (early Kargin period); on the right a small ice wedge; LAO22.

of the silt sequence in LAO6 show an inversion as well. From this geochemically striking horizon radiocarbon ages of 26240 +580/-540 and 14390 +150/-140 yr BP were obtained for depths of 2.0 and 2.4 m, respectively (see Table 3, No 6, 7). Because fine plant remains selected by microscope for the AMS-dating showed no signs of decomposition, it can be assumed that older allochthonous plant detritus from the neighbouring area was mixed with plant remains from the younger sedimentary environment which existed at the termination of the Sartan. Geochemical characteristics of the frozen lacustrine deposits at this depth prove that during a certain period the sedimentation was interrupted. Relatively high contents of Na, Ca, Mg, Sr, Ba and SO_4^{2-} in segregated ground ice and the presence of small amounts of newly formed carbonates at a depth of about 2 m seem to indicate a drying-up of the lake. Salt accumulation could be connected with evaporation and freezing-out processes in the active layer within the dried lake depression during climate cooling in the Sartan period. More detailed studies of the geochemical and mineralogical characteristics of such syncryogenic shallow lake deposits for which formation times remains problematic must be carried out.

Unit 3: Sartan and undivided Sartan-Holocene sediments

Four radiocarbon ages indicating sediment accumulation during the Sartan period have been obtained by AMS dating (Table 3): 20400 +300/-290 yr BP, 17320±220 yr BP, 14390+150/-140 yr BP and 11810 ±140 yr BP. The first datum quoted is related to silty and sandy deposits recovered at some sites by core drilling at the broad, dried lake depression to the west of the Tolton-Pastakh-Yuryukh River (see Figure 2). The investigated sediments consist of laminated light grey, yellow-grey or brownish grey fine layered silts and silty sands with trace fine plant remains (mosses, seeds, fine branches of dwarf shrubs). The ground-ice contents range from 30 to 50%. Fine lens-shaped and massive cryostructures are common. In filled lake depressions, ice wedges were found. Such ice wedges were drilled in LAB4/94, LAB5/94 and LAB9/95. Their stable isotope ratio range from -26.04‰ to -28.54‰ and from -202.7‰ to 218.54‰ for $\delta^{18}O$ and $\delta^{2}H$, respectively.

The age of 17.3 ka BP (Table 3, No 1) was obtained from the upper part of vaguely laminated sandy deposits with ice-wedge polygons in the surroundings of Kokora Lake. The „ice complex„ here forms an esker-like ridge along the eastern lake shore. Primarily fine residues of seeds, grass roots and branches were found as plant remains. The pollen assemblages from these sediments represent a dry grass-moss-tundra vegetation dominated by NAP with significant amounts of Poaceae, *Artemisia*, Caryophyllaceae, Saxifragaceae, other herbs and also cryophytic sporophytes, while Aboreal pollen (nearly only shrub pollen) were present in quite small amounts (10 to 20 %).

Sediments formed at 11800 ka (No 8 in Table 3) were studied in the southern shore zone of Labaz lake (LAO 13-94 in Figure 1, C). Plant remains are very rare in these well-sorted fine-grained sands. However, the pollen assemblage from this horizon, although dominated by Poacea and Artemisia, indicated that shrubs (primarily Betula sect. Nanaea) and Larix began to spread in the surroundings at this time. In the underlying sand horizon, the pollen assemblage suggested a significantly cooler and drier environment which likely existed during the Sartan. Peat accumulation already took place at this site during the Preboreal, at 9300 yr BP (see No 27 in Table 2).

Unit 4: Holocene deposits

Holocene deposits occur at different geomorphological levels. They are presented primarily by lacustrine silts and peats formed in thermokarst depressions, lakes and by different slope deposits. The latter overlie different relief elements and, together with fluvial sediments, fill the recent river valleys in this area. The wide-spread occurrence of slope deposits blur the older relief elements and complicate the reconstruction of former lake level stages.

Numerous radiocarbon ages were obtained from Holocene plant material (see Table 2, 3). At first, these data suggested that the development of thermokarst processes (thawing of buried ice bodies, other ice-rich permafrost) already started during the warming events at the termination of the Pleistocene. For example, silty sediments in section LAB2.95, which were characterized by pollen assemblages relating to the termination of the Sartan, accumulated in a thaw-lake (thermokarst lake) and were overlain by a peaty horizon dated around 9000 yr BP (Table 2, No 31, 32). The wide-spread occurrence of thick tree trunks, especially of larch at the base of peat beds with ages of about 7000-8000 yr BP, showed that optimum conditions for their growth in this area were from the Boreal to the beginning of the Atlantikum. Younger larch trunks were characterized by a smaller diameter and signs of deformation. The youngest larch wood had a ^{14}C-age of 2880±60 yr BP and was found as a stump at the slope of a thermokarst depression near the drilling site LAB 13-95 (see Table 2, No 55). Several pollen diagrams of dated Holocene deposits (for example, Figure 9) also indicated that optimum climatic conditions for an expansion of trees towards the north in this region existed during the Boreal time. This corresponds with the results of other investigators (Hahne and Melles, this volume; Nikolskaya, 1982; Velichko et al., 1994). Apparently, the progressive swamping of the flat

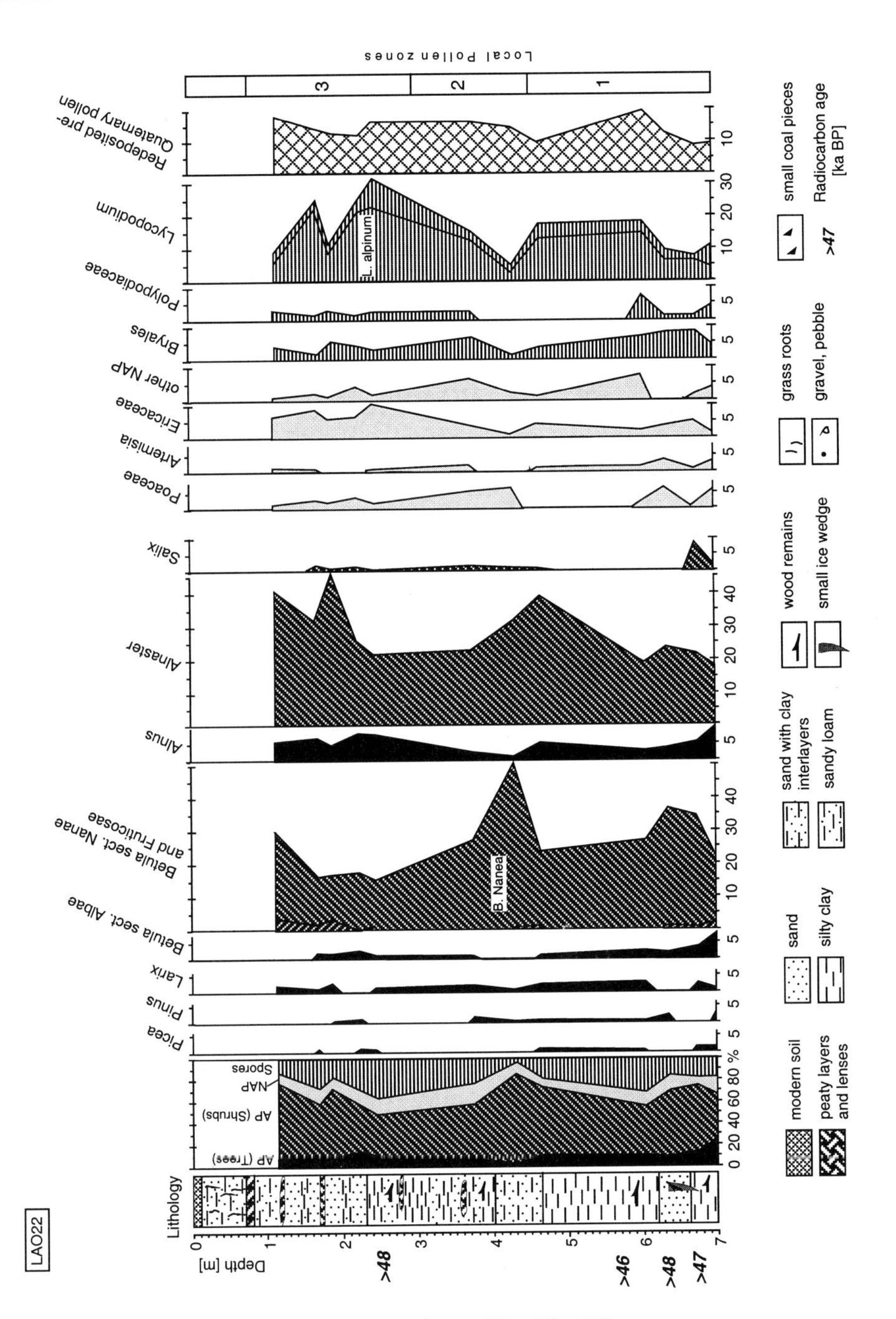

Figure 7: Pollen diagram of section LAO22. (For location see Figs. 1C and 2).

lowland connected with wide-spread peat accumulation in the middle Holocene led to a decrease of both the active layer thickness and permafrost temperature, that is, to a deterioration of the

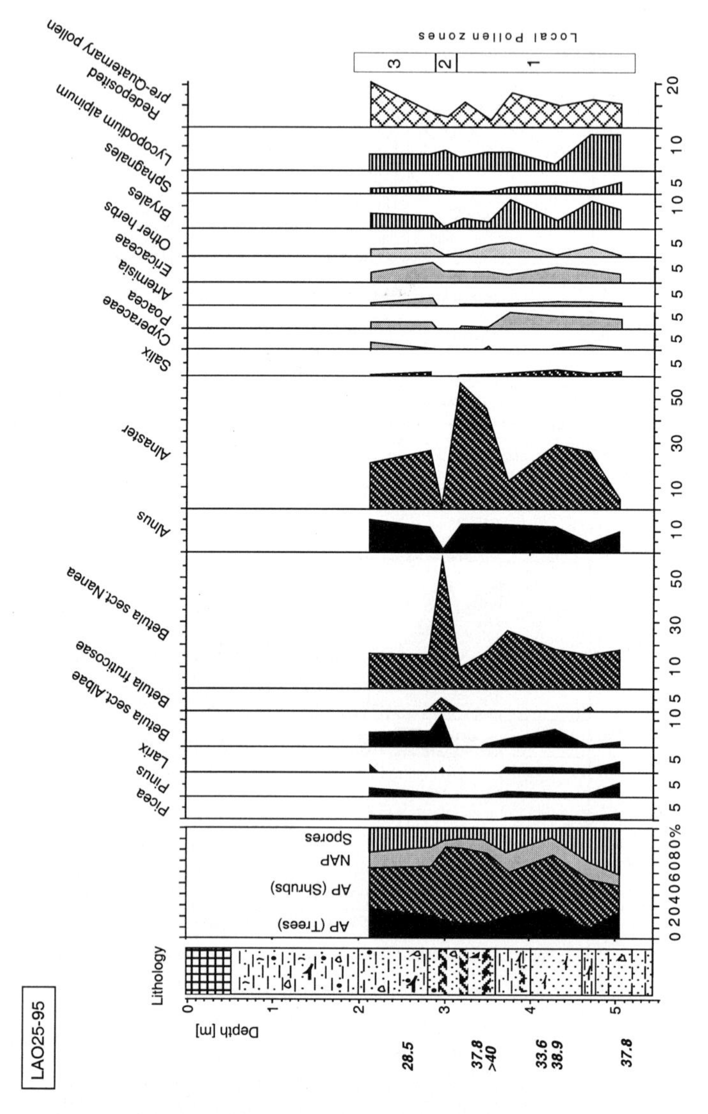

Figure 8: Pollen diagram of section LAO25. (For location see Figs. 1C and 2).

soil microclimate. A general climate cooling can be observed during the Subboreal. It seems to have been accompanied by a decrease in precipitation. Peat bogs dried up. Aquatic environments underwent a transformation from low-centred polygons to high-centred polygons with vegetation similar to high bog peats. In all palynologically investigated Holocene sections, a significant rise of Ericaceae pollen percentages and a simultaneous decrease of Hydrophythes was observed for this period. Suprapermafrost soil horizons were dated to 4630±70, 4200±65 and 2900±50 yr BP at some peat sections (see Table 2, No 46, 51 and 52, respectively).

Altogether, the Holocene climate warming did not change the geocryological conditions in the Labaz Lake area radically. The area was, as it is today, characterized by a domination of cold

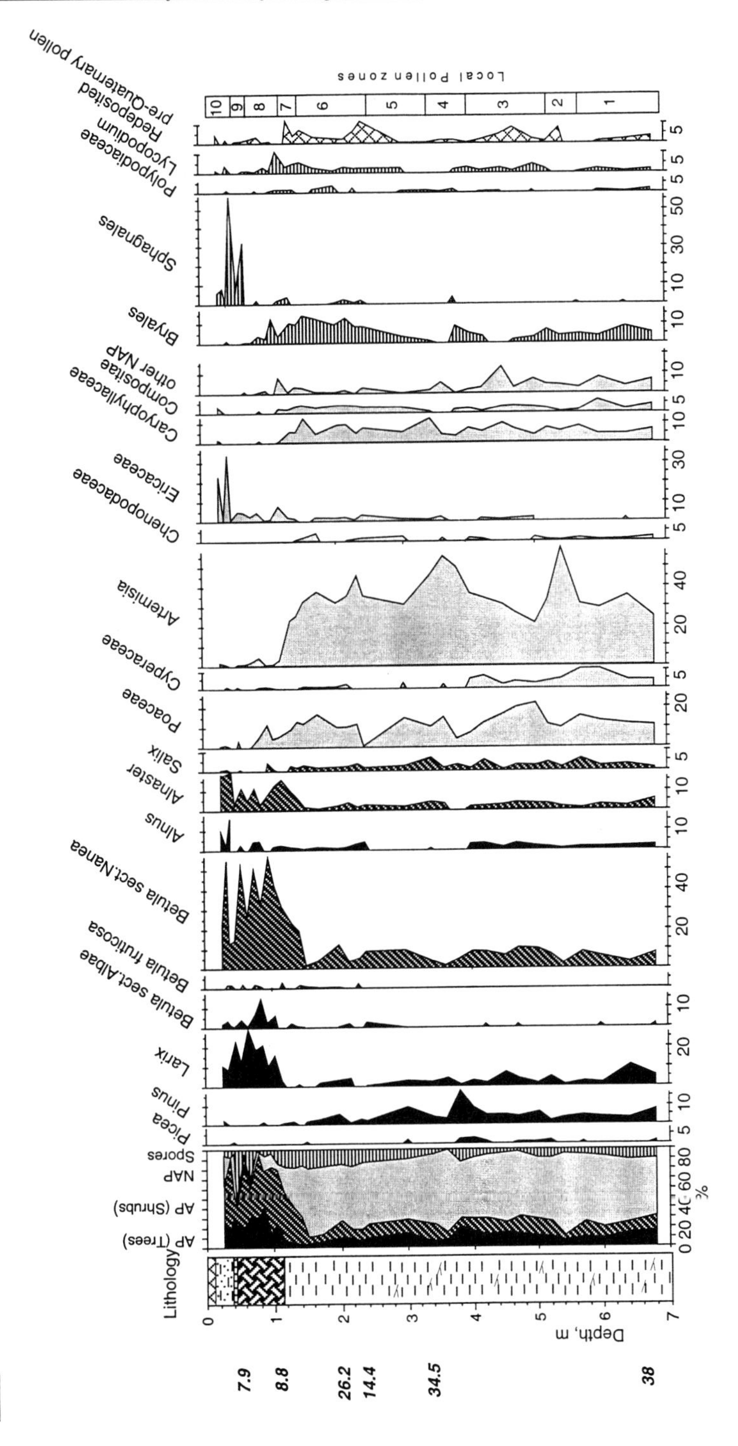

Figure 9: Pollen diagram of section LAO6. (For location see Figs. 1C and 2).

ice-rich permafrost, slight drainage and low relief energy. The calculated increase of the active layer depth during the Holocene climatic optimum was between 0.5 to 1.0 m on the watersheds and 0.2 to 0.4 m in depressions (Derevyagin et al, 1996). The cryostructures of permafrost lying below investigated Holocene lacustrine deposits suggest that the thickness of sub-lake taliks rose to maximum values of about 2 m. The near-surface location of Pleistocene polygonal ice wedges and the conservation of large buried glacier ice bodies confirm this result. Polygonal ice wedge systems also continue to grow in flat wet areas of the Labaz region at the present time. Nonetheless, their stable isotope ratio ranges from -15.8 ‰ to -23.3 ‰ and from -117 ‰ to -165 ‰ for $\delta^{18}O$ and $\delta^{2}H$, respectively.

Conclusion

The landscape evolution in the Labaz Lake area during early post-Zyryan times was primarily dominated by processes connected with the decay of the Zyryan/Early Weichselian ice cover. The character of these processes was strongly influenced by the specific peculiarities of glacier ice thawing in a territory with underlying continuous, low-temperature permafrost. Climate fluctuations in post-Zyryan times have led to a varying intensity of ice decay, denudation, and accumulation. Complicated thermokarst and thermoerosional processes have taken place. The most intensive glacial ice decay seems to have occurred during the first warming stage of the Kargin/Middle Weichselian Interstadial before more than 40000 yr BP. The formation of extensive lake depressions within the glacial deposits and the accumulation of sediments which built the widespread lacustrine 60 m terrace seems to be connected with these times.

A surface stabilization occurred about 40000 yr BP, probably caused by a cooler and more continental climate. Glacier ice disintegration was slowed and large flat areas of lacustrine sediments became subaerially exposed. As a consequence, deposits with thick polygonal ice wedges (Ice Complex) and peat beds could be formed. Subsequent warming phases, apparently connected with an increase of precipitation, led to a new activation of thermokarst and glacier ice decay, and to the accumulation of the second horizon of Kargin lacustrine and fluvio-lacustrine sediments. Glacier ice was buried progressively deeper by ablation deposits.

Although the paleoenvironmental data in general suggest strong continental climatic conditions, we have not obtained evidence permitting a decision on the existence of a vast ice-dammed lake ‚Pre-Labaz' during the Sartan period, as supposed by Isaeva (1982). Occurrence of lacustrine sediments dated around 17000 yr BP at an altitude of more than 80 m at the Kokora Lake argue for this concept. If we assume that the loess-like silts in the section LAO6 also relate to the Sartan time, then we have a second indicator for the existence of lakes during this time in the Labaz Lake area.

The pollen assemblages from the investigated post-Zyryan sediment profiles indicate a continuous tundra vegetation. From the presence of significant amounts of shrub pollen and especially larch throughout the documented time interval, sufficiently high summer temperatures are evidenced not only during the Kargin Interstadial period, but also during the Sartan Glacial. In this context, the ‚cold' stable isotope signature of the studied Late Pleistocene ice wedges suggests a strong continental climate, characterized by relative warm summer and very low winter temperatures.

Altogether, it is most probable that after 17-15000 yr BP lacustrine sedimentation was interrupted in wide areas. Wide spread polygonal ice-wedges, partially with signs of epigenetic growth in lacustrine sediments, support such an assumption. Lacustrine sediments formed during reactivation of lacustrine sedimentation after the drying of lake depressions were dated between 14400 and 11800 yr BP and likely belong to the late glacial warming events Bølling and Allerød. To achieve a clarification of this problem, the use of other dating methods in addition to the radiocarbon dating and additional multidisciplinary studies of sediments relating to the Sartan are necessary.

Figure 10: Lattice-like cryostructure with overlapping subvertical ice streaks in lacustrine silts; layers with fine cryostructure are enriched with organic matter. LAO6.

Acknowledgments

Our investigations were financially supported by the German Federal Ministry of Education, Science, Research and Technology (BMBF). We thank Bettina Beeskow (AWI Potsdam) who prepared most of the samples for radiocarbon dating. We also thank Dr. Mackensen and Prof. K. Friedrichsen (stable isotope analyses), as well as Prof. P. M. Grootes (AMS dating) and the staff of their laboratories for their efforts. A. Sher is thanked for the very constructive review. This is contribution No. 1356 of the Alfred Wegener Institute for Polar and Marine Research, Bremerhaven and Potsdam.

References

Alexeev M.N., E.V. Devyatkin., S.A. Arkhipov, S.A. Laukhin, O.V. Grinenko and V.A. Kameletdinov (1984) Problems of Quaternary Geology of Siberia (in Russian). In: Reports presented at the 27th International

Geological Congress, Moscow, USSR. August 4-14, 1984, Nikiforova K.V. et al. (eds.), Nauka, Moscow, vol 3, 3-12.

Andreeva S.M. and L.L. Isaeva (1982) Muruktin (Nizhne Zyryanka) deposits of the North-Siberian Lowland (in Russian). In: The Antropogen of the Taimyr Peninsula, Kind N.V. and B.N. Leonov (eds.), Nauka, Moscow, 51-78.

Andreeva S.M. and N.V. Kind (1982) Karginsk Deposits (in Russian). In: The Antropogen of the Taimyr Peninsula, Kind N.V. and B.N. Leonov (eds.), Nauka, Moscow, 78-114.

Astakhov V.I. (1992) The last glaciation in West Siberia. Sveriges Geologiska Undersökning, Ser. Ca 31, 21-30.

Astakhov V-I. and L.L Isayeva.(1988) The Ice Hill - an example of the retarded deglaciation in Siberia. Quaternary Sci. Rev., 6, 152-174.

Belorusova J.M. and V.V Ukraintseva. (1980) Paleogeography of the Late Pleistocene and Holocene of the Basin of the river Novaya (Taimyr) (in Russian). Botanichesky Zhurnal, 65, No 3, 368-379.

Chizhov A.B., A.Yu. Derevyagin, E.F. Simov, H.-W. Hubberten and Ch. Siegert (1997) Isotopic ground ice composition in the Labaz Lake region (Taymyr) (in Russian). Earth Cryosphere, vol 1, No 3.

Derevyagin A. Yu., Ch. Siegert, E.V. Troshin and G.N. Shilova (1996) New data on Quaternary geology and permafrost of Taymyr (Labaz Lake region) (in Russian). In: Proceedings of the First Conference of Russian Geocryologists, June 3-5, 1996, Moscow, vol. 1. Moscow Univ. Press, 204-212.

Ershov E.D. (ed.) (1995) Fundamentals of Geocryology. Part 1: Physico-chemical basics of Geocryology (in Russian). Moscow Univ. Press, 262-276.

Faustova M.A. and A.A. Velichko (1992) Dynamics of the last glaciation in northern Eurasia. Geological Survey of Sweden, Series Ca, 81, 113-118.

Geyh, M.A. and H. Schleicher (1990) Radiocarbon (^{14}C) Method. In: Absolute Age Determination. Springer-Verlag Berlin etc., 162-180.

Hahne J. and M. Melles (this volume) Climate and vegetation history of the Taymyr Peninsula since Middle Weichselian time - palynological evidence from lake sediments.

Isaeva L.L. (1982) Sartan deposits of the North-Siberian Lowland (in Russian). In: The Antropogen of the Taimyr Peninsula, Kind N.V. and B.N. Leonov (eds.), Nauka, Moscow, 114-136.

Isaeva L.L and N.V. Kind (1989) On the probleme of the dimensions of the Sartan glaciation in NW of Central Siberia (in Russian). In: The Late Quaternary Glaciations in Central Siberia. Nauka, Moskau, 52-59.

Kaplina T.N. (1979) Spore-Pollen assemblages of the ‚Ice complex‘ deposits of the Yakutian Coastal plans : An overview (in Russian). Izvestiya AN USSR, Ser. Geography, No 2, 85-93.

Katasonov E.M. (1978) Permafrost-Facies Analysis as the main method of Cryolithology. In: International Conference on Permfrost, 2nd, Yakutsk, 1973, Proceedings, Sanger F.J. and Hyde P.J. (eds.), Washington, DS, Nat. Acad. Sciences, Nat. Acad. Press, 171-176.

Kind N.V. (1974) Geochronology of the Late Quaternary according to isotopic data (in Russian). Nauka, Moscow, 255 pp.

Nikolskaya M.V. (1982) Paleobotanic and paleoclimatic reconstruction of the Holocene of the Taimyr Peninsula (in Russian). In: The Antropogen of the Taimyr Peninsula, Kind N.V. and B.N. Leonov (eds.), Nauka, Moscow, 148-157.

Pfeiffer, E.-M., A. Gundelwein, T. Nöthen, H. Becker and G. Guggenberg (1996) Characterization of the organic matter in Permafrost soils and sediments of the Taymyr Peninsula/Siberia and Severnaya Zemlya/Arctic region. In: Russian German Cooperation: The Expedition Taymyr 1995 and the Expedition Kolyma 1995 of the ISSP Pushchino Group, Bol‘shiyanov D. and H.W. Hubberten (eds.), Reports on Polar Research, 211, 46-63.

Polach, H.A., J. Gower and I. Frazer (1972) Synthesis of high purity benzene for radiocarbon dating by the liquid scintillation method. In: Proc. of the 8th International Radiocarbon Conference, Rafter, T.A. and T. Taylor (eds), Wellington, Royal Society of New Zealand, 147-157.

Romanovsky N.N. (1977) The formation of polygonal ice wedge structures (in Russian). Nauka Novosibirsk, 216 pp.

Siegert C., S.F. Khrutsky and A.Yu. Derevyagin (1995) Investigations in the Labaz Lake area: Paleogeographical permafrost studies. In: Russian German Cooperation: The Expedition TAYMYR 1994, Siegert C. and D. Bol‘shiyanov (eds.), Reports on Polar Research, 175, 27-36.

Siegert C., A.Yu. Derevyagin and G. Vannahme (1996) Geocryological and paleogeographical studies in the Labaz Lake area. In: Russian German Cooperation: The Expedition TAYMYR 1995 and the Expedition KOLYMa 1995 of the ISSP Pushchino Group, Bol‘shiyanov D .and H.W. Hubberten (eds.), Reports on Polar Research, 211, 28-45.

Shore, J.S., D.D. Bartley and D.D. Harkness (1995) Problems encountered with the ^{14}C dating of peat. Quaternary Science Reviews (Quaternary Geochronology), vol. 14, 373-383.

Stuiver, M. and H. Polach (1977) Discussion reporting ^{14}C data. Radiocarbon 19 (3), 355-363.

Tomskaya A.I. (1981) The Palynology of the Cenozoicum of Yakutia (in Russian). Nauka, Novosibirsk, 222 pp.

Vasilevskaya, V.D. (1980) Soil formations in the Central Siberian Tundra (in Russian). Nauka, Moscow, 235 pp.
Velichko A.A. (ed.) (1993) Evolution of landscapes and climates of the northern Eurasia. Late Pleistocene-Holocene; Elements of prognosis. 1. Regional Paleogeography (in Russian). Nauka, Moscow, 102 pp.
Velichko A.A., A.A. Andreev and V.A. Klimanov (1994) Vegetation and Climate Dynamics during Late Glacial and Holocene in Tundra and Forest Zones of Northern Eurasia (in Russian). In: Short-Term and Sharp Landscape and Climate Changes during the last 15000 yrs., Velichko, A.A. (ed.), Moscow, 4-60.

Section E

Marine Depositional Environment - Past and Present

Stable Oxygen Isotope Ratios in Benthic Carbonate Shells of Ostracoda, Foraminifera, and Bivalvia from Surface Sediments of the Laptev Sea, Summer 1993 and 1994

H. Erlenkeuser[1] and U. von Grafenstein[2]

(1) Leibniz-Labor für Altersbestimmung und Isotopenforschung, Universität Kiel, Max-Eyth-Strasse 11, D 24118 Kiel, Germany

(2) Laboratoire des Sciences du Climat et de l'Environnement (LSCE), Domaine du CNRS, Ave de la Terrasse, F - 91199 Gif-sur-Yvette Cedex, France.

Received 3 April 1997 and accepted in revised form 5 May 1998

Abstract - Shells of benthic carbonate producers - ostracoda, foraminifera and bivalvia, with emphasis lain on the ostracoda - have been picked for stable isotope analysis from surface sediments sampled on the Transdrift I and II expeditions in summer 1993 and 1994, respectively, all over the middle and southern Laptev Sea. This work presents mean $\delta^{18}O$ values of single shell analyses made on a total of 19 benthic species - bivalvia: 3; foraminifera (Elphidium): 3; ostracoda: 13 - from 49 stations. The $\delta^{18}O$ figures are related to water depth and reflect the competition of isotopically light riverine waters added from the south and normal marine waters from the north-west. Although a numerical link to actual salinities has not been evaluated, the results nevertheless attest benthic isotopes from sediment cores an interesting potential to reconstruct the lateral field of the riverine impact on the hydrography of the Laptev Sea for past climates. - The vital offsets in $\delta^{18}O$ were roughly quantified for most of the benthic species.

Introduction

The Laptev Sea is one of the large shallow seas in the Arctic domain of the Siberian shelf. Its meridional extension measures about 800 km, from 113° to 140 °E, and the latidudinal range covers a similar distance, roughly from 71° to 78°N, the northern shelf break. Water depth is low, at about 20 to 40 m, in most parts of the shelf sea, but the bottom relief shows some deeper submarine valleys, running northward, which have been formed by the Siberian rivers in Glacial and Postglacial time when sea level was low.

The wide dimension of the shallow water province of the Laptev Sea largely controls the hydrographic coupling with the open Arctic Ocean in the north. Salinity and nutrients, two major factors in the shelf water ecology, are strongly related to the fresh water discharged by the Siberian rivers from the south and to their pronounced seasonally modulated impact. In the south-eastern Laptev Sea, far distant and oceanographically shielded from the high marine environment in the north (see Hass et al., 1995, for a circulation pattern), the fresh water discharge of the rivers Lena and Yana make salinity of the surface waters drop to 10 psu or less in the surface waters (Karpiy, 1994). The freshwater contribution also imprints on the physical and chemical properties of the bottom waters, both directly by seasonal convection and shelf brine water formation and by the biogeochemical response of the sediment-water interface as to the planktonic productivity in the surface waters. This response is readily seen in the light isotopic composition of the dissolved inorganic carbon (DIC) of bottom waters (Erlenkeuser et al., 1995), a feature largely resulting from remineralisation of organic matter. In the northern Laptev Sea, where the advection of high marine waters is more effective, bottom water salinity may attain almost normal marine values. Sea ice formation may have additional effects on salinity and in detail could have contributed to the often found complex stratification of the water column (Karpiy et al., 1994; Churun and Timokhov, 1995).

In: Kassens, H., H.A. Bauch, I. Dmitrenko, H. Eicken, H.-W. Hubberten, M. Melles, J. Thiede and L. Timokhov (eds.) Land-Ocean Systems in the Siberian Arctic: Dynamics and History. Springer-Verlag, Berlin, 1999, 503-514.

Broadly speaking, the lateral distribution of salinity in the Laptev Sea results from the competition of high marine waters advected from the Arctic Ocean and fresh waters discharged by the rivers from the south. The spatial distribution pattern of salinity hence traces the balance of these hydrological momenta and may provide an interesting aspect from the paleoclimatic/ paleohydrographic point of view which also relates to the hydrological development in the catchment areas in inner Asia under the regime of central continental climate. The salinity signal of the bottom waters is preserved in the sediments through the stable oxygen isotope composition of carbonate in the shells of the benthic fauna.

Of particular interest are the ostracoda which inhabitate a broad variety of aquatic environments, such as - among others - fresh water to shelf and deep sea marine habitats (DeDeckker, Colin, and Peypouquet (Eds), 1988). These benthic crustaceans may provide through their carapaces which are composed of crystalline calcite embedded in layers of cuticular chitin (Benson, 1988), isotopic information on the physicochemical settings of the benthic boundary layer, such as soft sediments, turbid bottom waters, low salinity conditions, which are not usually tolerated by foraminifera and bivalvia.

This work studies the significance of the stable oxygen isotopes in the biogenic benthic carbonates and the lateral distribution of the isotope signal under modern conditions. A brief overview was presented in Erlenkeuser et al. (1997).

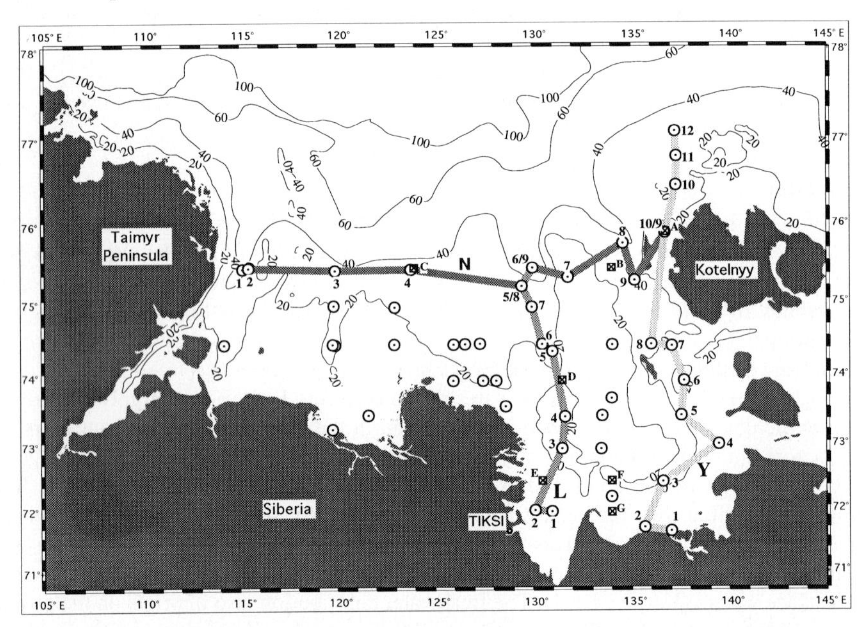

Figure 1: Sampling stations in the Laptev Sea.. Stations are labeled (numbered) for reference along each transect. T,S-stations presented in Figure 2, are labeled by capitals.

Methods

Benthic carbonate producers have been sampled from surface sediments collected from numerous stations all over the Laptev Sea in August/ September 1993 and 1994 on the Transdrift expeditions I and II with RV Ivan Kireyev (code: IK) and RV Professor Multanovskiy (PM), respectively (Kassens et al., 1994; Kassens and Dmitrenko, 1995). Sampling locations are listed in Tables 1a, 1b and are shown in Figure 1.

The bulk samples were collected as 0,5 cm thick surficial layers scraped off from the sediments retrieved by spade box coring and have been stored unfrozen in plastic bags at 4°C. Staining to mark living organisms was used for the IK-series only. Some of the species, like Cytheropteron, showed poor staining only or none.

The bulk sediments were oven-dried at 60°C, weighed, and soaked with a mixture of 10 % hydrogenperoyide, water and ammonia for 1 hr to remove organic compounds. Care was taken to keep foaming low. The soaked samples were gently wet-sieved on a 63µm-mesh, using tap water as washing liquid, the coarse fraction washed in ethanol, dried at room temperature and the fraction >125 µm separated. The fine fraction in the soakage was concentrated for further studies by reverse filtration and was repeatedly washed with ammonia water to remove residual tap water and peroxide. From the >125µm-fraction, the fragile shells of the ostracods were picked under the binocular and were gently though carefully cleaned, using a fine brush, ethanol, and a steel needle, from adhered sedimentary particles which easily settle behind the inner calcified lamella. Other taxa such as bivalvia and foraminifera (*Elphidium*) were separated as well.

From 49 stations, 316 ostracod carapaces of 13 species were separated for isotope analysis on the species level, 48 samples of bivalvia of 3 species yet undetermined, and 143 samples of foraminifera of the genus *Elphidium* (3 species). A summary of sample statistics is given in Table 2. Reference specimens were archivated separately and the bulk of the shells splitted into single shell samples for isotope analysis to learn and possibly quantify the isotopic variability in this environment of the Laptev Sea. The carapaces of the ostracods were well sufficient for reliable isotope measurements on single specimens. Similarly Elphidium and bivalvia species could be measured as single shells. The present work discusses the average isotope figures as calculated for each station for each taxon rsp. species if determined.

For isotope analysis, the carbonate sample was reacted with an individual aliquot of (4 drops of) 100% Orthophosphoric acid under vacuum at 73°C in the Kiel carbonate device which is on-line-coupled to a Finnigan MAT 251 gas isotope mass-spectrometer. Minimum sample size, normally at 12 µg of $CaCO_3$, was reduced for this study to 6 µg applying a specially designed valve/ cold finger gas provider assembly at the gas inlet capillary of the spectrometer (Cordt and Erlenkeuser, in prep.). The external error amounts to less than 0.08 ‰ on the $\delta^{18}O$-scale (1-sigma value).

The isotope results are given in the usual δ-notation quoting the relative difference in permille of the sample's isotope abundance ratio from that ratio of the PDB-standard. The PDB-standard is represented by the NBS 20 secondary isotope standard. (Recently NBS 19 was said to have the better isotopic reproducibility and was chosen to represent the international isotope scale, then termed VPDB (Vienna PDB; Coplen, 1995). There is no systematic offset between the PDB and VPDB measures).

The effect of river discharge to the Laptev Sea relates the $\delta^{18}O$ of the water and its salinity with a coefficient of 0.6 ‰/psu (Erlenkeuser, unpubl.), with the Lena water showing $\delta^{18}O$ about 20 ‰ lower than the open ocean source waters advecting onto the shelf (33.6 psu at 40 m water depth on the outer Laptev shelf and slope; Karpiy et al., 1994). The effect of water temperature on $\delta^{18}O$ of a calcite precipitate is about -0.25 ‰/°K (Shackleton, 1974).

Results and discussion

The sampling stations are listed in Tables 1a, 1b and are shown in Figure 1. They cover the southern and middle Laptev Sea and represent benthic environments with widely differing oceanographic disposition, ranging from the prominent seasonally scheduled fresh water regime in the southeast to almost full permanent marine conditions in the northwest. The

Table 1a: Station list with sampling dates and water depths. Station counting helps identify locations evaluated in the diagramms. Also shown are sample and analysis statistics: Frequencies of benthic carbonate shells, available and analysed for stable isotope composition, according to station and species. The shells were collected from samples of surface sediments, Laptev Sea, Aug./Sept. 1993 and 1994. Asterisks label samples used to determine vital offsets in $\delta^{18}O$ against expected equilibrium-$\delta^{18}O$, dots and underlined dots indicate species and reference species, respectively, estimating offsets from inter-species comparison. **Abbreviations: Coll D**: Collection Date, **Lat**: Latitude, **Long**: Longitude, **D**: Water depth, **EL1*, EL2*, EL3***: Elphidium spp., **C1***: Cytheropteron pralatissimum, **C2**: Cytheropteron elaeni, **C3**: Cytheropteron camplainum, **E1***: Rabimilis aramirabilis, **H***: Hetercyprideis sorbyana, **SP1***: Semicytherura complanata, **SP3**: Elofsonella concinna, **SP6**: Acanthocythereis dunelmensis, **SP7***: Krithe cf. glacialis, **X1***: Sarsicytheridea pseudopunctillata female, **X2***: Sarsicytheridea pseudopunctillata male, **X3**: Normanicythere leidoderma, **No**: Number of species.

					Station No. - Profile -			Bivalvia Species			Foraminifera		
Core	Coll D	Lat. N°	Long. E°	D	N	L	Y	M1	M2	M3	EL1*	EL2*	EL3
IK9301-6	29.8.93	71,753	135,660	16			2						
IK9306-6	27.8.93	72,012	130,987	18		1							
IK9309-4	30.8.93	72,490	136,587	24			3				3		
IK9313-6	22.8.93	73,070	139,427	16			4	3					
IK9315-1	23.8.93	73,000	133,472	18				5			1		
IK9316-6	23.8.93	73,002	131,502	28		3					1		
IK9318-3	21.8.93	73,498	137,523	24			5				2		3
IK9320-1	21.8.93	73,495	133,520	18						2			
IK9321-5	20.8.93	73,478	131,648	25		4		2					
IK9323-6	20.8.93	73,630	128,653	17						1	3		
IK9324-4	14.8.93	73,502	121,668	13				4					
IK9330-2	16.8.93	74,000	127,505	27				1	1	1	3		3
IK9334-7	19.8.93	74,000	137,663	22			6	2					
IK9338-5	14.8.93	74,492	119,955	34				2					1
IK9340-6	15.8.93	74,500	122,993	16						1			3
IK9342-6	1.1.04	74,510	127,348	34						1	1		3.
IK9344-8	17.8.93	74,405	131,013	30		5							2
IK9346-4	17.8.93	74,495	134,035	14							1		1
IK9348-6	18.8.93	74,480	137,045	22			7				4		4
IK9353-0	4.9.93	75,000	129,977	37		7		2	2		2		3
IK9356-2	5.9.93	74,997	123,015	33				2					4.
IK9358-5	5.9.93	75,017	119,888	34						1			3.
IK9365-6	6.9.93	75,468	119,962	43	3					2	2.		
IK9367-2	7.9.93	75,482	123,842	44	4			2					3.
IK9370-7	7.9.93	75,270	129,465	44	5	8				1			2
IK9371-1	8.9.93	75,383	131,803	20	7			1					
IK9373-8	9.9.93	75,342	135,167	43	9				1		2		
IK9373A-6	2.9.93	75,810	134,583	46	8			1.	1.		1		
IK9382-6	2.9.93	76,503	137,272	25			10		1		3		3
IK9384-1	13.9.93	77,112	137,225	33			12				2		
IK93K1-1	9.9.93	75,940	136,708	20	10		9		2		1		2
IK93K2-1	13.9.93	76,835	137,295	30			11						1
IK93Z3-3	13.8.93	73,295	119,832	11				1			3		1
IK93Z4-4	24.8.93	72,033	130,127	14		2		1			3		
IK93Z5-3	29.8.93	71,690	137,007	11			1	1					
PM9402-3	3.9.94	75,491	115,249	47	1						1		3
PM9417-4	6.9.94	75,503	130,014	51	6	9					2*	3*	
PM9441-4	10.9.94	74,000	125,988	14									
PM9442-3	10.9.94	74,499	126,003	40									6
PM9451-7	13.9.94	74,503	130,495	25		6					1		3
PM9462-1	12.9.94	74,502	136,004	27			8				3		3
PM9463-8	14.9.94	74,504	126,582	36							3		3
PM9475-3	18.9.94	72,250	133,995	21							2		3

Table 1a (continued):

					Station No. - Profile -			Bivalvia Species			Foraminifera		
Core	Coll D	Lat. N°	Long. E°	D	N	L	Y	M1	M2	M3	EL1*	EL2*	EL3
PM9481-2	19.9.94	73,750	134,004	17							12		2
PM9482-1	20.9.94	73,999	128,175	27							3		3
PM9492-3	22.9.94	74,501	119,835	34							3		3
PM9494-5	23.9.94	74,501	114,284	37									
PM9499-1	24.9.94	75,501	115,545	48	2								
PM94T3-2	1.9.94	77,069	99,220	110									1

Table 1b: for Abbreviations see Table 1a.

		Ostracoda												
Core	Coll D	C1*	C2	C3	E1*	H*	SP1*	SP3*	SP6	SP7*	X1*	X2*	X3	No.
IK9301-6	29.8.93					4								1
IK9306-6	27.8.93										2			1
IK9309-4	30.8.93					2								2
IK9313-6	22.8.93													2
IK9315-1	23.8.93					2					3			4
IK9316-6	23.8.93										3			2
IK9318-3	21.8.93					2								3
IK9320-1	21.8.93													1
IK9321-5	20.8.93					1					3			3
IK9323-6	20.8.93					2								3
IK9324-4	14.8.93													1
IK9330-2	16.8.93													5
IK9334-7	19.8.93					2					2			3
IK9338-5	14.8.93													2
IK9340-6	15.8.93					6					6			4
IK9342-6	16.8.93							2.			6			5
IK9344-8	17.8.93													1
IK9346-4	17.8.93					1					1			4
IK9348-6	18.8.93				2	3			3	3	3			7
IK9353-0	4.9.93													4
IK9356-2	5.9.93							3.			10.			4
IK9358-5	5.9.93							2.	2.				3	5
IK9365-6	6.9.93													2
IK9367-2	7.9.93		1											3
IK9370-7	7.9.93							2						3
IK9371-1	8.9.93					2					4			3
IK9373-8	9.9.93										7			3
IK9373A-6	2.9.93								1		4.			5
IK9382-6	2.9.93					2		1			6			6
IK9384-1	13.9.93													1
IK93K1-1	9.9.93					4					3	3		6
IK93K2-1	13.9.93	2							1					3
IK93Z3-3	13.8.93										3			4
IK93Z4-4	24.8.93					4					3			4
IK93Z5-3	29.8.93													1
PM9402-3	3.9.94					4*					4	2		5
PM9417-4	6.9.94	2*			2*		1*	2*						6
PM9441-4	10.9.94					25					8	5		3

Table 1b (continued):

Core	Coll D	Ostracoda												
		C1*	C2	C3	E1*	H*	SP1*	SP3*	SP6	SP7*	X1*	X2*	X3	No.
PM9442-3	10.9.94						3							2
PM9451-7	13.9.94					2								3
PM9462-1	12.9.94				4	5					4	3		6
PM9463-8	14.9.94													2
PM9475-3	18.9.94													3
PM9481-2	19.9.94					12					5	2		5
PM9482-1	20.9.94						2					2		4
PM9492-3	22.9.94			3										3
PM9494-5	23.9.94			2.		5.	1	5			15			5
PM9499-1	24.9.94					22*		2*		3*	7*	3*		5
PM94T3-2	1.9.94	3			4		2							4

species analysed, their spatial distribution over the Laptev Sea, and the absolute abundances sampled for this study are listed in Tables 1a, 1b.

The strong discharge of riverine waters into the Laptev Sea has a pronounced effect on salinity and temperature stratification of the water column (Karpiy et al. 1994; Churun et al., 1995). This fresh water impact is shown for a few stations in Figure 2. The impact is largest in the south, but is well recognized also at northern latitudes. It concentrates to the upper 15 or 20 m and extends down to almost 40 m.

Although the profiles shown in Figure 2 for the late summer, will vary with season, and the long term significance of these temporary measurements is not readily seen, water depth plays a key role in any case to control the temperature and salinity conditions of the benthic boundary layer in the Laptev Sea (Dmitrenko et al, 1995).

Accordingly the isotope results have been compared with water depth (Figure 3). The isotope figures represent mean values calculated from a variable number of single shell measurements (Table 1a, 1b) and thus have a variable statistical confidence. Figure 3 includes the mean isotope composition of bivalvia, foraminifera of the genus *Elphidium*, and of ostracoda on the species level. Basically, the role of water depth is evident, with the lowest (‚lightest') $\delta^{18}O$-values seen at the shallower depths and the heaviest oxygen isotope composition in the deep waters. Figure 3 reveals a significant scatter in the isotopic data from a given water depth. Part of this scatter is due to the vital offset in $\delta^{18}O$ which is different for the various taxa and ranges between 0 and -1 ‰ (see below). Apparently the major part of the variance cannot be accounted for by this effect and clearly reveals further influences on the isotopic composition. The variability is particularly high for the shallower depths, say 10 to 30 m, where salinity and temperature gradients in the water column are greatest and even smaller local and temporal variations of the vertical structure have a large imprint on the isotopic settings in the benthic boundary layer.For closer inspection, parts of the stations have been grouped along different transects allowing to study the fresh water imprint on the isotopic situation in provinces of different oceanographic disposition. On the northern, west-east orientated transect N, most stations show water depths of 40 to 50 m and comparatively uniform isotopic composition of the benthos, close to the expected $\delta^{18}O$ of a hypothetical calcite formed in isotopic equilibrium with a subsurface water at about this depth representing pelagic conditions over the deeper outer shelf and slope. An average of -1.5°C and 33.6 psu was taken for the 39 m sampling depth from the T,S-data of Polarstern cruise ARK XI-1 (1995), stations 16, 17, 19 on the outer Laptev Sea shelf. For mixing models, these waters and their isotope characteristics are

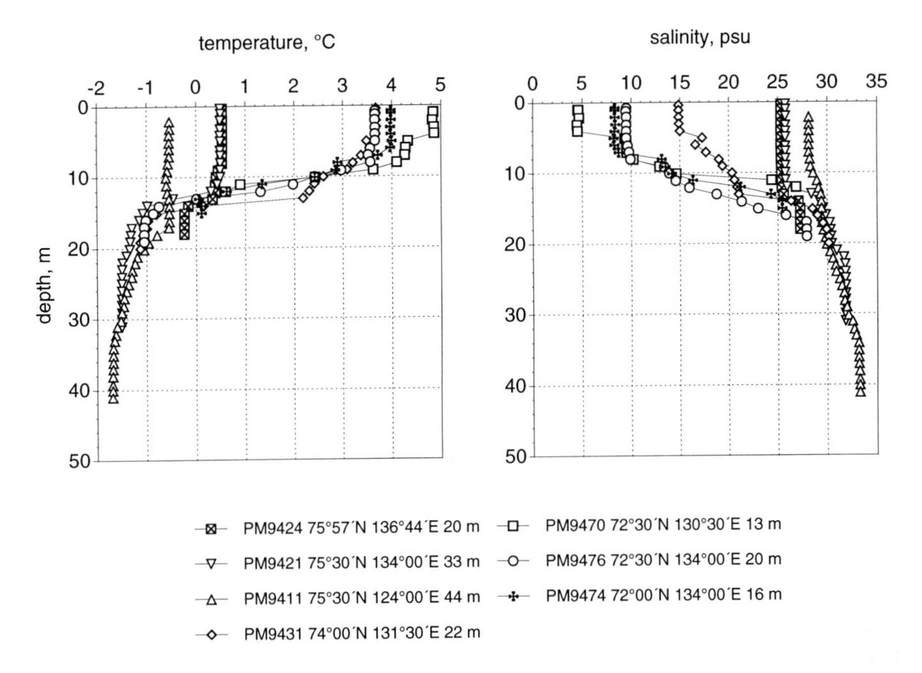

Figure 2: Examples of temperature and salinity distribution with water depth in the Laptev Sea. See Figure 1 for positions.

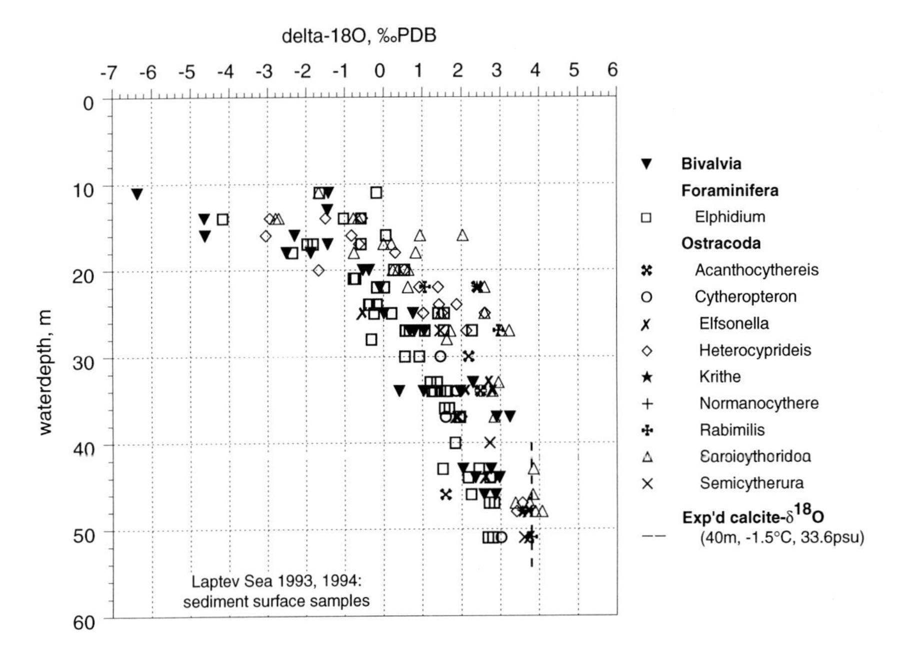

Figure 3: Stable oxygen isotope composition of benthic carbonate shells, collected Aug./Sept. 1993 and 1994 from surface sediment samples of the Laptev Sea, in relation to water depth. The data represent mean values of single shell measurements of variable number. Vital offsets are not corrected for. - The expected equilibrium $\delta^{18}O$ for calcite precipitated in the 40 m water layer is indicated.

considered the marine source waters advected onto the Laptev Sea shelf. This marine source applied here does not represent the possible ultimate high marine water, as its salinity is significantly lower at about 33.6 psu than in the core of Atlantic Water found at greater depths in the boundary current along the upper Laptev Sea slope (Schauer et al., 1995).

With regard to the prominent presence of Atlantic water, the 40 m-shelf-water' may be considered to recruit from this Atlantic realm and be mixed up with Siberian riverine waters. Given $\delta^{18}O = 0.26$ ‰ SMOW and S = 35.2 psu for Atlantic water far ,upstream' off Mid-Norway (Craig and Gordon, 1965) and accounting for the effect of Siberian riverine waters on $\delta^{18}O$ as 0,6 ‰/psu, $\delta^{18}O = -0,7$ ‰ SMOW is expected for the shelfwater salinity of S = 33.6 psu. This value is corroborated by measured data of -0.6 to -0.7 ‰ SMOW (Stein, 1996; profile 4, Laptev Sea shelf at ca. 40 m; S = 33.5 psu). The calcite equilibrium $\delta^{18}O$ (termed δ_c in Shackleton 1974) is calculated as $\delta_c = 3.8$ ‰ PDB for t = -1.5 °C from Shackleton's (1974) paleotemperature equation, and we refer to this value to quantify the isotopic vital offsets.

As the seasonal variation in temperature and salinity presumably plays a minor role at the deeper stations of transect N, the remaining isotope differences among the studied species ofthe carbonate benthos likely reflect vital offsets, i.e. species-dependent isotopic compositions other than in a calcite precipitated in thermodynamic isotopic equilibrium with the surrounding water.

The phenomenon of ,vital offsets' in the benthic isotopic composition of biogenic carbonates is known for long for foraminiferal carbonates (e.g. Duplessy et al, 1970) and is commonly corrected for, deducing environmental parameters from isotope results. For ostracods, vital effects are largely unknown.

In this work, the vital offsets were roughly quantified, for the present state of evaluation, in units of 0.5 ‰ (Table 2), for the deep samples from transect N by refering to the calculated equilibrium-$\delta^{18}O$ of the ,40m-shelf-water' and for a few more species from other, shallower stations by inter-species isotope comparison. Directly calculated offsets, i.e. differences between measured and expected isotopes values, and resulting average figures are indicated in Table 1a, 1b through specimen number and species key each labeled by an asterisk. Reference species and counterpart in inter-species comparison are marked by a dot and underlined dot, respectively. Occasionally, with too poor a data basis, a uniform offset on the species level was hypothesized, e.g. for species C2 (*Cytheropteron elaeni*). Of the ostracoda, 7 of the 11 species analysed reveal isotopic equilibrium, while the other 4 show $^{18}O/^{16}O$-ratios too light by 0.5 ‰. The bivalvia species appear to calcify out of equilibrium as well, by about -0.5 ‰, while the benthic foraminfer *Elphidium* is even more biased in $\delta^{18}O$, by about -1 ‰.

Correcting for vital offsets greatly reduces the inter-species isotope scatter for the deep stations of the northern transect N (Figure 4). This is no longer true for the shallower stations and the stations along the eastern transects L and Y, in particular its southern stations in the Yana Bight (Figure 4). This finding probably relates to the interference of the species' life rhythm, i.e. the seasonal schedule of shell growth, and the seasonal variation of salinity and temperature. Ostracods can calcify their carapaces in a short time, possibly within days, thereby flash-lighting the isotopic situation in the water. On the other hand, we expect bivalvia to build their shells in a more gradual way. In the Baltic, for instance, isotope studies on *Mytilus edulis* revealed calcification throughout the year, though at a rate varying with the season (Erlenkeuser et al., 1980). The bivalve shells from the Yana Bay stations, where the riverine regime is dominant, surprisingly reflect the seasonal low salinity phase (or higher summer temperature) much more pronounced than the ostracoda. Offset values gained by interspecies comparison in particular are sensitive to differences in life rhythms, and these values, such as for the bivalve species (Table 2), still have an unknown reliability.

The different oceanographic disposition of the benthic localities studied shows in the

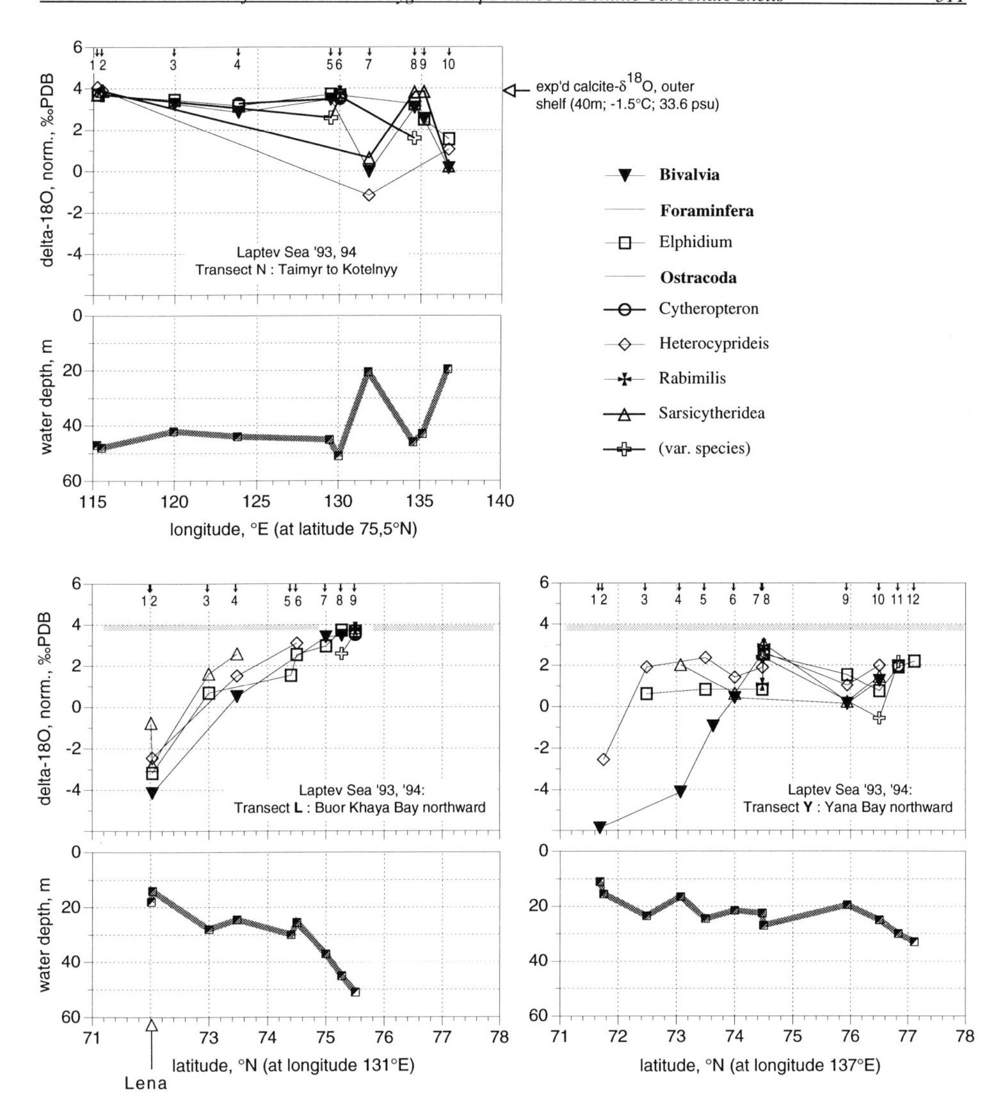

Figure 4: Stable oxygen isotope composition of benthic carbonate producers from surface sediment samples collected Aug./Sept. 1993 and 1994 in the Laptev Sea along transects N, L, and Y (Figure 1). Station labels (numbers) given at the upper margins serve for reference (cf. Figure 1). Water depths at the stations are shown for comparison. The data represent mean values of single shell measurements of variable number. Vital effects have been corrected for.

distribution of $\delta^{18}O$, corrected for vital offsets, with depth along the transects N, L, Y (Figure 5). Station counting along the transects, beginning in the west (transect N) or in the south (L, Y) is displayed in the digramms and indicates the geographical relation of the stations. Differences in the mean benthic $\delta^{18}O$ between stations of about the same water depth are evident in transects N and L. $\delta^{18}O$ is slightly lower for the 120 to 125°E stations (no. 3 and 4) and for stations 8 and 9 in the outer Yana Valley at about 135°E. As an aspect of broader interest, brine waters may be thought of to form on the shallower southern shelf, flowing off and forming a bottom water layer sufficiently persistent to contribute to the benthic isotope signature.

On transect L, station 6 in the upper outer Eastern Lena Valley is clearly affected by the pelagic influence in contrast to station 4 which is situated only 1° south beyond an extended submarine topographic high and from the bathymetrical point of view belongs to the channel system of the inner southeastern Laptev Sea. Also noteworthy, $\delta^{18}O$ is relatively uniform in the 19 to 25 m water depth range along the eastern transect Y from station 3 through station 10, a long distance ranging from 72,5 °N in the Yana Bight to 76,5°E north of Kotelnyy. This finding possibly relates to the oceanographic conditions in the eastern Laptev Sea, i.e. the effect of bottom topography and bathymetric boundary conditions of the eastern margin on the outflow and the density structure of the eastern Laptev Sea waters.

Table 2: Species of benthic carbonate producers, analysed for stable isotope composition, from surface sediments, coll. in Aug./Sept. 1993 and 1994, of the Laptev Sea, and approximative values of vital offsets in $\delta^{18}O$. Species of bivalvia and foraminifera have not yet been determined.

	Specie key (cf. Table 1)	Vital Offset $\Delta d^{18}O$ (species-equil)
Bivalvia		
Bivalvia sp1	M1	-0.5
Bivalvia sp2	M2	-0.5
Bivalvia sp3	M3	-0.5
Foraminifera		
Elphidium sp1	EL1	-1
Elphidium sp2	EL2	-1
Elphidium sp3	EL3	-1
Ostracoda		
Acanthocythereis dunelmensis	SP6	0
Cytheropteron champlainum	C3	-0.5
Cytheropteron elaeni	C2	-0.5
Cytheropteron paralatissimum	C1	-0.5
Cytheropteron pseudomontrosiense		n.d.
Elofsonella concinna	SP3	0
Heterocyprideis sorbyana	HS	-0.5
Krithe cf. glacialis	SP7	0
Normanicythere leidoderma	X3	n.d.
Rabimilis paramirabilis	E1	0
Sarsicytheridea pseudopunctillata female	X1	0
Sarsicytheridea pseudopunctillata male	X2	0
Semicytherura complanata	SP1	0

Conclusions

The present study presents some evidence that the isotopic composition of the benthic carbonate fauna from the Laptev Sea likely has the potential to help reconstruct the lateral pattern of the riverine impact for past climates from adequately dated sediment cores. Moreover, if the biological rhythm of shell growth is better known for the different species, the fresh water signal in the isotope compositions may be studied and evaluated as an interesting signal on seasonal aspects of the climate.

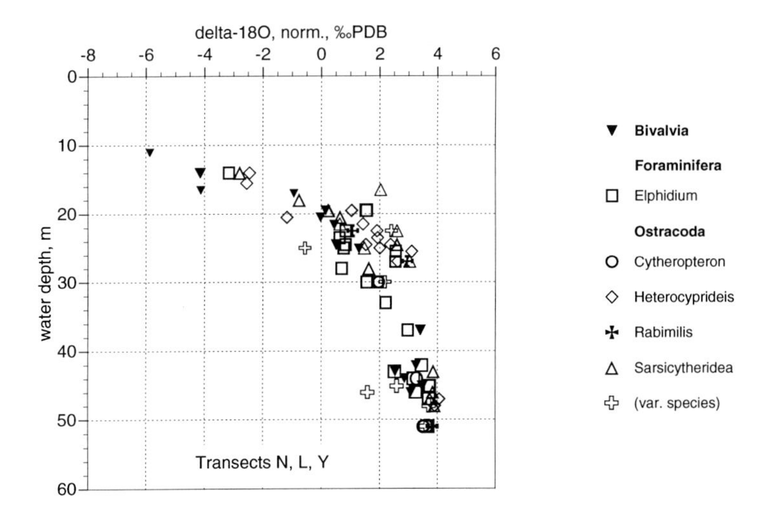

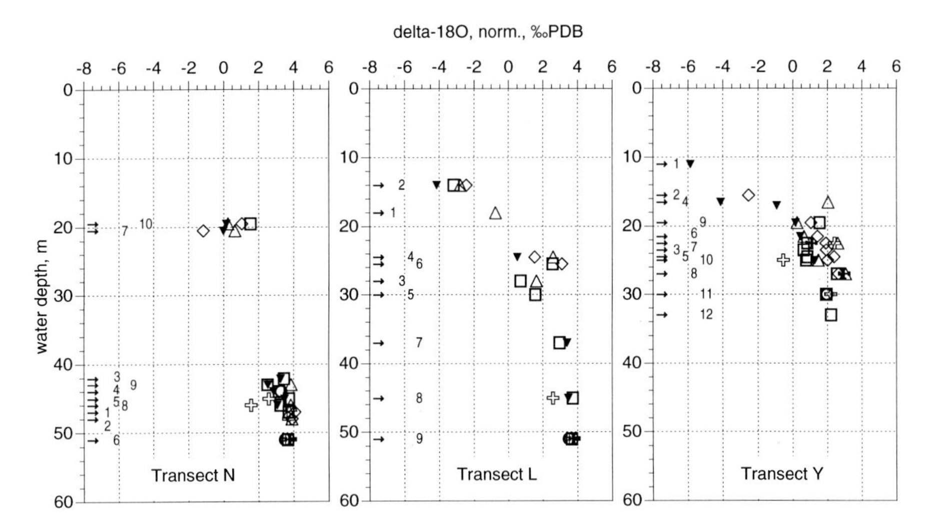

Figure 5: Stable oxygen isotope composition in benthic shell carbonates, collected Aug./Sept. 1993 and 1994 from surface sediment samples of the Laptev Sea, in relation to water depth. The data represent mean values of single shell measurements of variable number. Vital offsets have been corrected for. - Arrows and labels (numbers) identify sampling stations along the resp. transects (cf. Figure 1, Figure 4).

Acknowledgements

Our colleagues of the Transdrift expeditions did a hard days' work to gain the surface sediments from the Laptev Sea. The isotope analyses, in particular those on the the sub-minimum samples, would not have succeeded without the work of Hans H. Cordt and Heinz Heckt. We gratefully acknowledge all their help. We are obliged to the Russian partners who provided the temperature and salinity data shown in Figure 2. And we appreciate the experienced advice and help of T. Cronin with the taxonomy of the ostracoda. This study is part of the Joined Research Project System Laptev Sea at Geomar, Kiel, and is supported by the German Ministry of Education, Science, Research and Technology (BMBF).

References

Benson, R.H. (1988) Ostracods and palaeoceanography. In: Ostracoda in the Earth Sciences. De Deckker, P., J.P. Colin and J.P. Peypouquet (eds.), Elsevier, Amsterdam, 1-26.

Churun, V.N. and L. Timokhov (1995) Cold bottom water in the southern Laptev Sea. In: Russian-German Cooperation: Laptev Sea System. Kassens, H. et al. (eds.), Reports on Polar Research, 182, 107-113.

Coplen, T.B. (1995): Reporting of stable carbon, hydrogen, and oxygen isotopic abundances. In: Reference and intercomparison materials for stable isotopes of light elements. Proceedings of a consultants meeting held in Vienna, 1-3 December 1993. IAEA, Vienna (IAEA-TECDOC-825), 31-34.

Craig, H. and L.I. Gordon (1965) Deuterium and oxygen 18 variations in the ocean and the marine atmosphere. In: Stable Isotopes in Oceanographic Studies and Paleotemperatures, Spoleto 1965. Tongiorgi, E. (ed.), Consiglio Nazionale delle richerche, Laboratorio di Geologica Nucleare, Pisa, 9-129.

De Deckker, P., J.P. Colin and J.P. Peypouquet (1988) Ostracoda in the Earth Sciences. Elsevier, Amsterdam, 302pp.

Dmitrenko, I.A. and the Transdrift II Shipboard Scientific Party (1995) The distribution of river run-off in the Laptev Sea: the environmental effect. In: Russian-German Cooperation: Laptev Sea System. Kassens, H. et al. (eds.), Reports on Polar Research, 182, 114-120.

Erlenkeuser, H., G. Wefer, R.B. Temple (1980) Growth of bivalves, gastropods, and foraminifera recorded in their ^{18}O and ^{13}C profiles. 15th European Symposium on Marine Biology, Kiel, (poster, unpubl. manuscript).

Erlenkeuser, H. and the Transdrift II Shipboard Scientific Party (1995) Stable carbon isotope ratios in the waters of the Laptev Sea/ Sept.94. Reports on Polar Research, 176, 170-177.

Erlenkeuser, H, U. von Grafenstein and I. Dmitrenko (1997) Die Sauerstoff-Isotopenverhältnisse im Schalenkarbonat von Ostrakoden aus Oberflächensedimenten der Laptev See 1993, 1994. Kiel Univ., Mitteilungen zur Kieler Polarforschung, 13, (ISSN 0939-5083), 15-19.

Golovin, P.N., V.A. Gribanov and I.A. Dmitrenko (1995) Macro- and Mesoscale hydrophysical structure of the outflow zone of the Lena river water to the Laptev Sea. In: Russian-German Cooperation: Laptev Sea System. Kassens, H. et al. (eds.), Reports on Polar Research, 182, 99-106.

Hass, H.C., M. Antonov and Shipboard Scientific Party (1995) Movement of Laptev Sea shelf waters during the Transdrift II expedition. In H. Kassens et al. (Ed's.) Russian-German Cooperation: Laptev Sea System. Kassens, H. et al. (eds.), Reports on Polar Research, 182, 121-134.

Kassens, H., V. Karpiy and the Shipboard Scientific Party (1994) Russian-German Cooperation: The Transdrift I Expedition to the Laptev Sea. Reports on Polar Research 151, 168pp.

Karpiy, V, N. Lebedev and A. Ipatov (1994) Thermohaline and dynamic water structure in the Laptev Sea. In: Russian-German Co-operation: The Transdrift I Expedition to the Laptev Sea. Kassens, H. and V. Karpiy, (eds.), Reports on Polar Research, 151, 16-47.

Kassens, H. and I. Dmitrenko (1995) The Transdrift II Expedition to the Laptev Sea. In Laptev Sea System: Expeditions in 1994. Kassens, H. et al. (eds.), Reports on Polar Research, 182, 1-180.

Schauer, U., B. Rudels, R.D. Muench and L. Timokhov (1995) Circulation and water mass modifications along the Nansen Basin slope. In: Russian-German Cooperation: Laptev Sea System. Kassens, H. et al. (eds.), Reports on Polar Research, 182, 94-98.

Shackleton, N.J. (1974) Attainment of isotopic equilibrium between ocean water and the benthonic foraminifera Genus Uvigerina: isotopic changes in the ocean during the last glacial. In: Variation du climat au cours du Pleistocène, Labeyrie, L. (ed.), Colloques Internationaux du C.N.R.S. No. 219, Paris, 203-209.

Stein, W. (1996) ^{18}O- and ^{3}H-Untersuchungen im Nordpolarmeer. Univ. Heidelberg (Diplomarbeit), 101pp.

Determination of Depositional Beryllium-10 Fluxes in the Area of the Laptev Sea and Beryllium-10 Concentrations in Water Samples of High Northern Latitudes

C. Strobl[1], V. Schulz[1], S. Vogler[1], S. Baumann[1], H. Kassens[2], P.W. Kubik[3], M. Suter[4] and A.Mangini[1]

(1) Heidelberger Akademie der Wissenschaften, Im Neuenheimer Feld 366, D 69120 Heidelberg, Germany

(2) GEOMAR Forschungszentrum für marine Geowissenschaften, Wischhofstrasse 1-3, D 24148 Kiel, Germany

(3) Paul Scherrer Institut, c/o Institut für Partikelphysik, ETH Hönggerberg, 8093 Zürich, Switzerland

(4) Institut für Partikelphysik, ETH Hönggerberg, 8093 Zürich, Switzerland

Received 20 February 1997 and accepted in revised form 9 May 1998

Abstract - Present day accumulation rates of nine sediment cores recovered during the Russian-German Expedition Transdrift II (1994) from the shelf area of the Laptev Sea were determined by $^{210}Pb_{ex}$ dating and vary from 0.05 to 0.24 g cm^{-2} a^{-1}. In addition, the sedimentation rates during the isotopic stages 2, 3 and 5 of the sediment core PS 2471-4 from the continental slope of the Laptev Sea were determined via $^{230}Th_{ex}$ dating. The ^{10}Be concentrations together with the accumulation rates (or sedimentation rates) yield the depositional ^{10}Be fluxes in the shelf area of [(10 - 150) $\bullet 10^6$ at cm^{-2} a^{-1}] and of [(0.9 - 4.1) $\bullet 10^6$ at cm^{-2} a^{-1}] on the continental slope from the Laptev Sea. They are clearly higher than the recent atmospheric input determined in Greenland ice cores [(0.2 - 0.5) $\bullet 10^6$ at cm^{-2} a^{-1}]. We conclude that large amounts of continental ^{10}Be are delivered to the Laptev Sea through the rivers (e.g. Lena,Yana) and that the major fraction of ^{10}Be is deposited directly in the shelf area. The distinctly higher concentrations of ^{10}Be in water samples from the shelf area of the Laptev Sea [1000 to 6000 at/g] compared to the concentrations measured in the Norwegian and Greenland Sea [300 to 1000 at/g] and the Central Arctic Ocean [500 at/g] are further evidence that rivers are an important source for the input of ^{10}Be from the Siberian hinterland to the Arctic Ocean.

Introduction

The cosmogenic radionuclide ^{10}Be ($t_{1/2}$ = 1.5 Ma) is a sensitive stratigraphic tool for sediments from the Arctic Ocean with low or negligible content of biogenic carbonate. ^{10}Be records from sediment cores from the Norwegian and Greenland Sea exhibit high concentrations of ^{10}Be during the interglacials in contrast to lower values during glacial periods (Eisenhauer et al., 1994). These distinct changes enable a glacial/interglacial stratigraphy of Arctic sediments. Better knowledge of the pathways of ^{10}Be from the Laptev Shelf into the sediments of the Arctic Ocean could render the records from sediment cores of the Arctic Ocean more reliable. As part of the Russian-German cooperative research project "System Laptev Sea" we focussed on the evaluation of supply and export fluxes of ^{10}Be in the shelf area and on the continental slope of the Laptev Sea. The atmospheric input was measured in ice cores from the Greenland Ice Sheet (Stanzick, 1996). The supply of ^{10}Be with the river was determined from the measurement of water samples from the estuary of the river Lena. For comparison we also measured profiles of the ^{10}Be contentration at three localities in the central Arctic Ocean. The amount of riverine ^{10}Be deposited on the shelf was evaluated from the concentration of ^{10}Be in shelf sediments. The accumulation rates of the sediments were determined from profiles of $^{210}Pb_{ex}$.

In: Kassens, H., H.A. Bauch, I. Dmitrenko, H. Eicken, H.-W. Hubberten, M. Melles, J. Thiede and L. Timokhov (eds.) Land-Ocean Systems in the Siberian Arctic: Dynamics and History. Springer-Verlag, Berlin, 1999, 515-532.

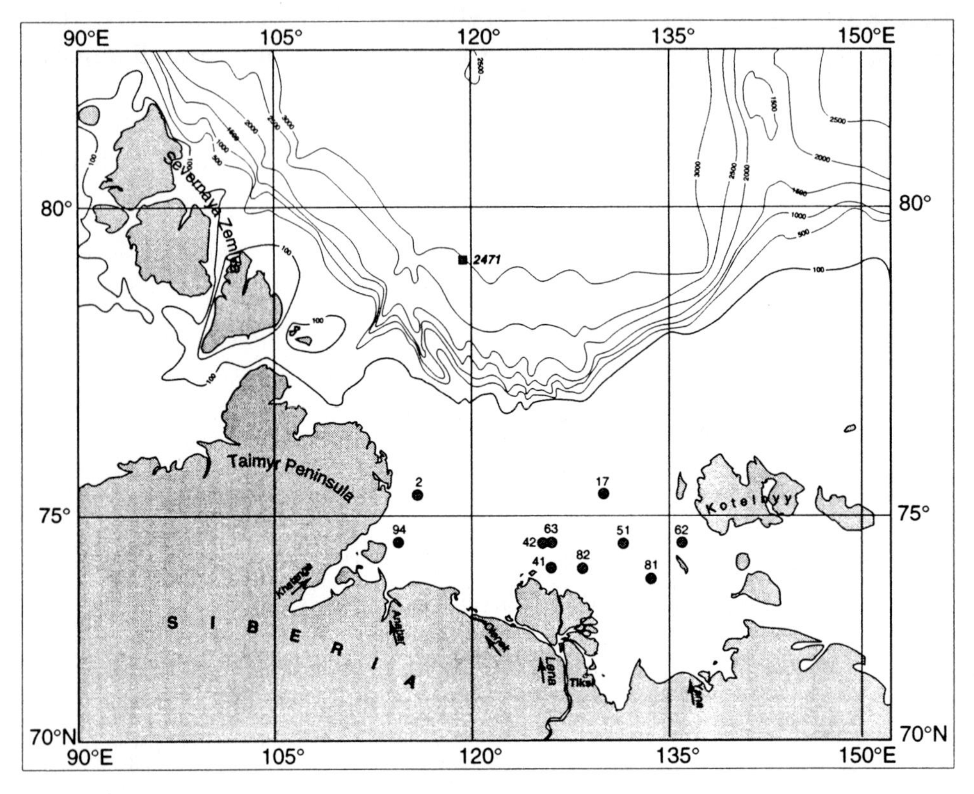

Figure 1: Map showing the locations of the investigated sediment cores and water profiles in the Laptev Sea.

Table 1: Locations of the sediment cores from the shelf area and from the continental slope

Station	Location	Water Depth [m]	Core Length [cm]
PM 9402-3 (GKG)	75°29.44`N, 115°14.94`E	47	39
PM 9417-4 (GKG)	75°30.17`N, 130°00.83`E	51	45
PM 9441-4 (GKG)	74°00.01`N, 125°59.35`E	14	19
PM 9442-3 (GKG)	74°30.05`N, 126°00.20`E	40	47
PM 9451-7 (GKG)	74°30.16`N, 130°29.70`E	25	18
PM 9462-1 (GKG)	74°30.13`N, 136°00.23`E	27	53
PM 9462-4 (VC)	74°30.18`N, 136°00.32`E	27	467
PM 9463-8 (GKG)	74°30.21`N, 126°34.91`E	36	44
PM 9481-2 (GKG)	73°45.00`N, 134°00.25`E	17	35
PM 9482-1 (GKG)	73°59.94`N, 128°10.47`E	27	56
PS 2471-4 (KAL)	79°09.07`N, 119°47.55`E	3047	417

Material and methods

The sediment cores (Table 1) investigated in our study were recovered during the Expeditions Transdrift II (Kassens,1995) and ARK IX/4 (Fütterer, 1994). Further we determined the ^{10}Be concentrations of 4 water profiles from the Norwegian and Greenland Sea (Thiede and Hempel, 1991), 3 profiles from the central Arctic Ocean (Fütterer, 1992) and water samples from the shelf area of the Laptev Sea (Kassens,1995). Their locations are listed in Table 2. All locations are plotted in Figures 1 and 2.

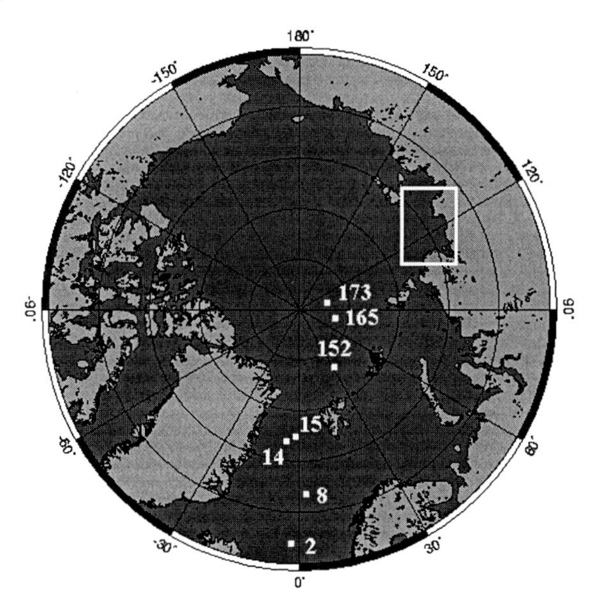

Figure 2: Map showing the locations of the investigated water profiles in the Norwegian- and Greenland Sea and the Central Arctic Ocean.

Preparation and measurement of sediment samples

The activities of the radionuclides ^{214}Bi, ^{210}Pb and of the anthropogenic ^{137}Cs were measured by γ-spectroscopy. For the γ-spectroscopy the dried samples (10- 30 g) were filled into 50 ml polyethylene vials and counted on two low-level HPGe detectors. The detector efficiency was calibrated applying standards of known activity but differing geometries and densities prepared from a multi-nuclide standard (QCY44, Amersham). The radioisotopes ^{230}Th, ^{232}Th, ^{234}U, ^{238}U were measured by α-spectroscopy. The sample material (0.5 g) was dried and homogenized. The chemical separation of these isotopes followed the procedure of Mangini (1984) and is described by Frank et al. (1994). The concentration of the radioisotope ^{10}Be was measured via accelerator mass spectrometry (AMS) at the tandem facility of the ETH Zürich and calibrated to an internal standard (S555) with a ^{10}Be/^{9}Be ratio of $95.5 \cdot 10^{-12}$. The chemical preparation of ^{10}Be followed the method described by Henken-Mellies et al. (1990) with minor modifications.

Preparation of water samples

Water samples from the Norwegian- and Greenland Sea, the central Arctic Ocean and the Laptev Sea of about 30 l were acidified with HCl to pH $\cong$ 2 and spiked with the stable ^{9}Be-Isotope (1 ml). The chemical preparation followed the method described by Segl et al. (1987).

Table 2: Locations of the water profiles where ^{10}Be measurements have been performed.

Sample	Location	Water Depth [m]	Sampled Depths [m]
Norwegian- and Greenland Sea			
2	69°47`N, 15°39`W	1189	6, 50, 250, 750
8	70° 45`N, 05°25`W	2387	6, 100, 200, 500, 1000, 1500, 1950
14	75°25`N, 07°20`W	3360	6, 50, 100, 200, 500, 1000, 1500, 2000
15	75°50`N, 08°10`W	1970	6, 50, 100, 200, 500, 1000, 1500
Central Arctic Ocean			
152	83°58.5`N, 30°24.8`E	3890	50, 500, 1500, 2250, 3000, 3500
165	87°34.4`N, 60°23.1`E	4300	50, 500, 1300, 2300, 3300, 4300
173	87°45.2`N, 108°59.1`E	4220	50, 320, 1220, 2220, 3220, 4220
Laptev Sea			
PM 9463-1	74°30.07`N, 126°35.06`E	36	2, 30
PM 9472-3	71°59.85`N, 130° 30.77`E	16	2
PM 9494-3	74°30.06`N, 114°17.05`E	36	2

Results

Determination of the Accumulation Rates in the Shelf Area of the Laptev Sea

^{210}Pb

The Constant-Flux $^{210}Pb_{ex}$ method has been often applied for dating sediment cores from lakes and estuaries (Dominik et al., 1981; Doerr et al., 1991; von Gunten et al., 1993; Bollhoefer et al., 1994). It relies on the assumption that the atmospheric ^{210}Pb deposition remained constant over the last century. This atmospheric component of ^{210}Pb ($^{210}Pb_{ex}$), decays in the sediment column with a half life of 22.3 years. It can be determined from the measured ^{210}Pb ($^{210}Pb_{tot}$) specific activity minus a component which is produced by the in situ-decay of ^{226}Ra in detritic material ($^{210}Pb_{supp}$) and another component produced by the decay of dissolved ^{226}Ra in the water column. Because of the low water depth of 50 m in the shelf area of the Laptev Sea this last component can be neglected. Thus the $^{210}Pb_{ex}$ activity of each sample was determined as:

$$^{210}Pb_{ex} = {}^{210}Pb_{tot} - {}^{210}Pb_{supp} \quad (1)$$

The activity of the supported ^{210}Pb corresponds to the specific activity of the ^{214}Bi ($t_{1/2}$ = 20 min) which is assumed to be in radioactive equilibrium with its mother ^{226}Ra. Therefore the $^{210}Pb_{ex}$ activity can be calculated as:

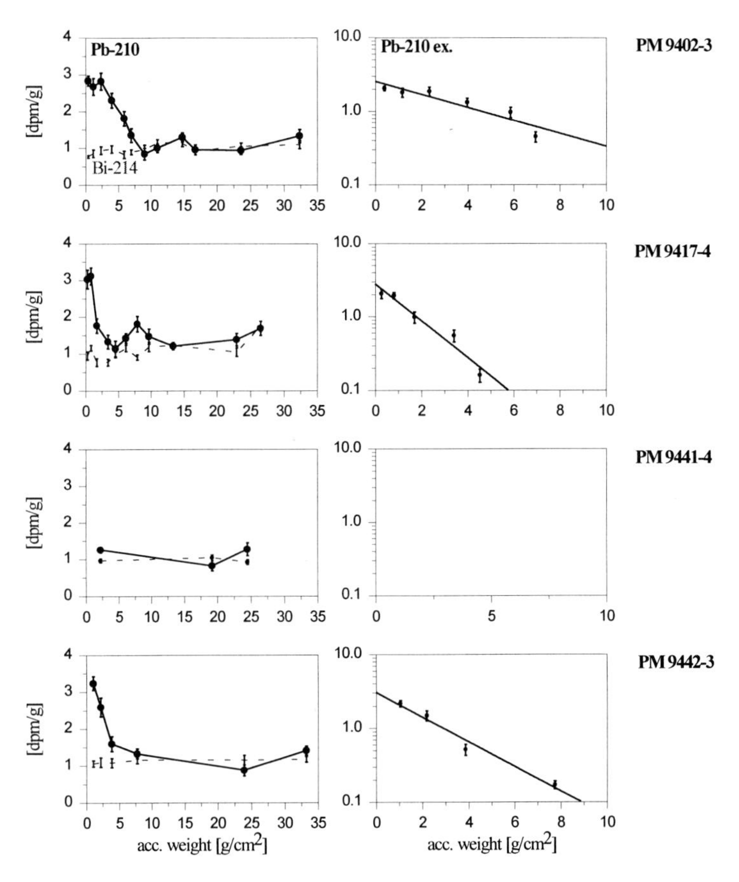

Figure 3: Activities of $^{210}Pb_{tot}$, $^{210}Pb_{ex}$ and ^{214}Bi plotted against compaction-corrected depth. On the plots on the right the $^{210}Pb_{ex}$ activities are presented on a logarithmical scale.

$$^{210}Pb_{ex} = {}^{210}Pb_{tot} - {}^{214}Bi \qquad (2)$$

The activities of $^{210}Pb_{tot}$, $^{210}P_{ex}$, and ^{214}Bi are plotted against compaction-corrected depth in Figures 3 and 4. The right diagrams show the profiles of the $^{210}Pb_{ex}$ activities on a logarithmical scale. The mean accumulation rate (R) and sedimentation rate (S) can be calculated from the slope of the best exponential fit to $^{210}P_{ex}$ as:

$$^{210}Pb_{ex}(G) = {}^{210}Pb_{ex}(0) \cdot e^{-\lambda \cdot G/R} \qquad (3)$$

$$R = \lambda \cdot G \cdot (\ln ({}^{210}Pb_{ex}(G)/{}^{210}Pb_{ex}(0)))^{-1} \qquad (4)$$

$$S = R / \rho \qquad (5)$$

where:
G : compaction corrected depth [g cm^{-2}]
$^{210}Pb_{ex}(0)$: $^{210}Pb_{ex}$ activity at the surface

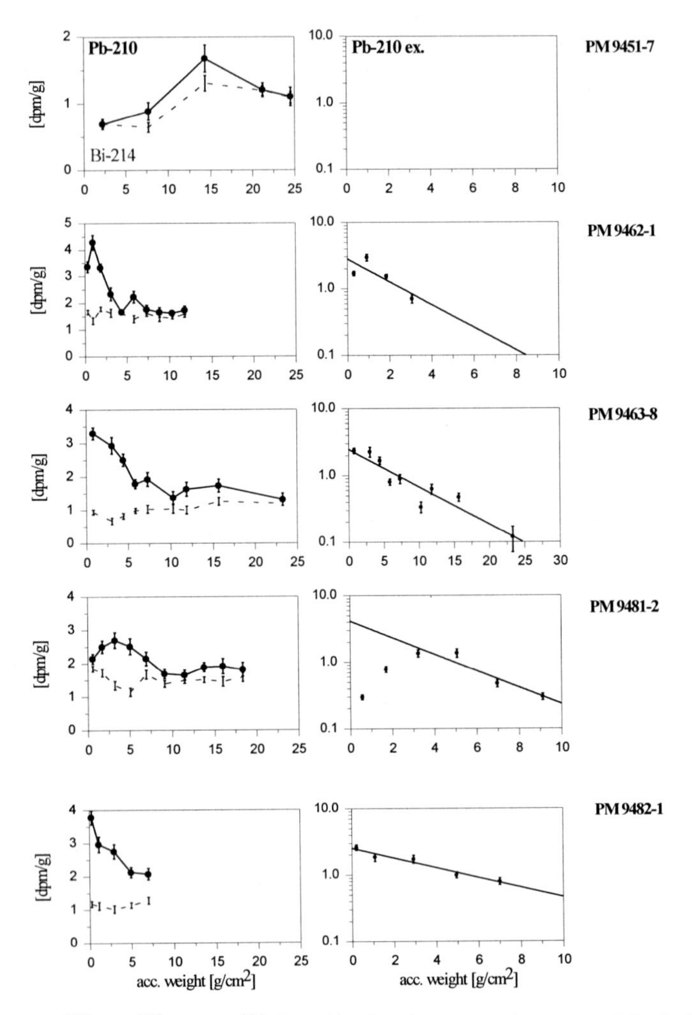

Figure 4: Activities of $^{210}Pb_{tot}$, $^{210}Pb_{ex}$ and ^{214}Bi are plotted against compaction-corrected depth. On the plots on the right the $^{210}Pb_{ex}$ activities are presented on a logarithmical scale.

λ :	decay constant of ^{210}Pb [0.031 a^{-1}]
ρ :	dry bulk density [g cm^{-3}]
R :	mean accumulation rate [g $cm^{-2}a^{-1}$]
S :	mean sedimentation rate [cm a^{-1}]

The mean accumulation rates [g cm^{-2} a^{-1}] , the average dry bulk densities (pers. comm. Kassens, 1995), and the sedimentation rates [cm a^{-1}] of the investigated cores are listed in Table 3. The sediment cores PM 9441-4 and PM 9451-7 could not be dated because we did not find any $^{210}Pb_{ex}$ at all. This lack of $^{210}Pb_{ex}$ could be ascribed either to dilution with sediments accumulating at a very high rate or to conditions preventing accumulation of recent sediment. The sandy composition of these sediments suggests possible export of the fine grain fraction. However, bioturbative mixing of ^{210}Pb at the sediment water interface may distort the profiles of the radionuclides and lead to apparent larger accumulation rates (Berner, 1980; Benninger et al., 1980; Mangini et al., 1986). The apparent contribution due to mixing can be evaluated as an additional accumulation rate (S_{diff}) (Mangini et al., 1988) given as:

$$S_{diff} = D \cdot \lambda / S \qquad (6)$$

where S is the accumulation rate derived from the depth profile of the radioisotope and D is the effective diffusion coefficient.

Table 3: Average bulk densities, accumulation and sedimentation rates of the investigated sediment cores from the shelf area of the Laptev Sea.

Station	Accumulation rate (R)	Bulk density	Sedimentation rate (S)
	[$g\ cm^{-2}\ a^{-1}$]	[$g\ cm^{-3}$]	[$cm\ a^{-1}$]
PM 9402-3	0.15 ± 0.03	0.95	0.16 ± 0.03
PM 9417-4	0.05 ± 0.02	0.73	0.08 ± 0.03
PM 9441-4	no dating		
PM 9442-3	0.08 ± 0.01	0.62	0.13 ± 0.02
PM 9451-7	no dating	1.62	
PM 9462-1	0.08 ± 0.02	0.69	0.12 ± 0.03
PM 9463-8	0.24 ± 0.04	0.69	0.35 ± 0.06
PM 9481-2	0.19 ± 0.08	1.15	0.17 ± 0.07
PM 9482-1	0.19 ± 0.01	0.49	0.39 ± 0.02

Mixing coefficients below the topmost layers of Long Island Sound sediments, are in the range of 0.6 $cm^2\ a^{-1}$ (Benninger et al., 1980), and are significantly lower in pelagic sediments from the Norwegian Sea (range 0.02 — 0.5 $cm^2\ a^{-1}$, Arnold, 1989).

Obviously, the smallest distortion will be in those cores where the profile of $^{210}Pb_{ex}$ can be detected to the deepest depths into the sediments and where the apparent accumulation rates derived from the depth distribution are highest. In the two cores, PM9463-8 and PM9482-1, where excess ^{210}Pb is detected up to about 30 to 40 cm depth, distortion due to bioturbative mixing should be smallest.

Applying a value of 0.6 $cm^2\ a^{-1}$ and an accumulation rate of 0.4 cm a^{-1}, the apparent contribution to accumulation rate can be evaluated at 0.05 cm a^{-1}. Thus the distortion due to mixing in these two cores lies within the uncertainty of the accumulation rate.

^{137}Cs

The artificial radionuclide ^{137}Cs delivers an independent time mark to test our datings with $^{210}Pb_{ex}$. The first appearance of this isotope in sediments corresponds to the beginning of the nuclear tests in 1954. The maximum fall-out corresponds to 1963. The function of atmospheric input of ^{137}Cs is plotted in Figure 5. The ^{137}Cs records of the investigated sediment cores are plotted against $^{210}Pb_{ex}$ ages (Figure 6) calculated from the mean sedimentation rates listed in Table 3. The top of the sediment cores was fixed to 1994. The two cores displaying the fastest accumulation rates, from localities PM 9463-8 and PM 9482-1, show very good agreement between the $^{210}Pb_{ex}$ timescale and the first appearance of ^{137}Cs. In the other 6 cores the agreement is rather good. However, in these cores we detected ^{137}Cs activities in layers older than 1940 (dated with $^{210}Pb_{ex}$), which are in evident discrepancy to the timescale of bomb ^{137}Cs (Figure 5). This discrepancy originates either from a wrong (to small) evaluation of the accumulation rate with $^{210}Pb_{ex}$, or from postdepositional migration of ^{137}Cs in the sediments. The latter was suggested to happen in anoxic sediments (Evans et al. 1983). As the sediments from the shelf area become suboxic within few cm depth below the surface (Langner et al., 1995), we cannot exclude some migration of ^{137}Cs to have occurred. Because of the larger uncertainty of the accumulation rates from $^{210}Pb_{ex}$ in these cores we cannot address this question properly.

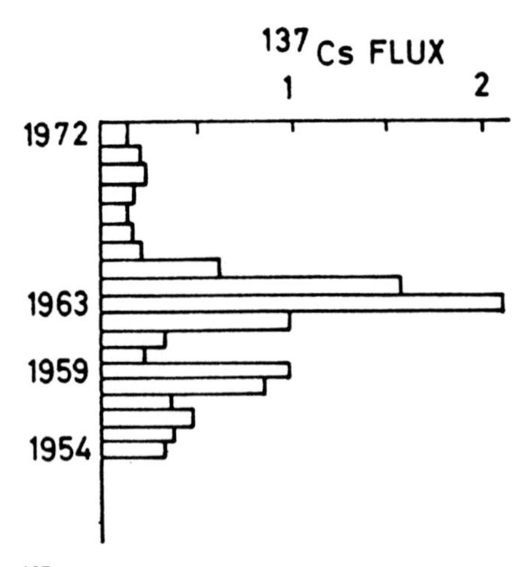

Figure 5: Atmospheric flux of ^{137}Cs in relative units (after Robbins and Edgington, 1976).

Standing Crop of $^{210}Pb_{ex}$

Comparison of the Standing crop of $^{210}Pb_{ex}$ in cores with uncertain accumulation rates with the one of core PM 9463-8 allows an additional test of their average accumulation rates. The Standing crop (SC) of $^{210}Pb_{ex}$ is defined as:

$$SC = \int_0^\infty \rho(x)A(x)dx \qquad (7)$$

where:

x : core depth [cm]

ρ : dry bulk density [g cm^{-3}]

A: $^{210}Pb_{ex}$ activity [dpm g^{-1}]

In the other cores, where ^{210}Pb does not penetrate as deep, we are aware that mixing may significantly distort the depth profiles of $^{210}Pb_{ex}$. In these other cores we evaluate the accumulation rate by comparison of their SC with the one of core PM9463-8 under the

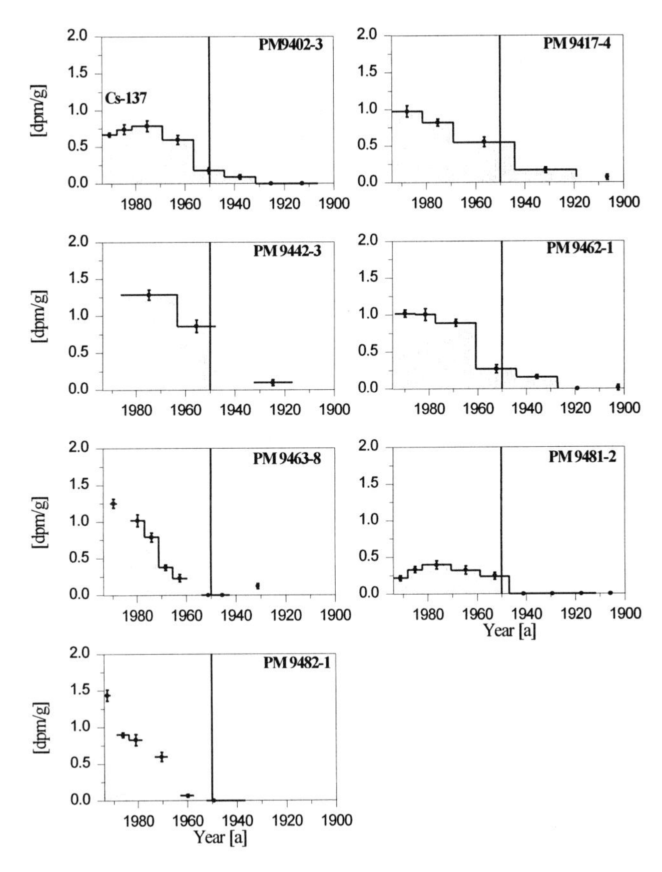

Figure 6: ^{137}Cs records of the investigated sediment cores against ^{210}Pb ages. The line represent the time mark 1950.

assumption that the flux density ^{210}Pb into the sediments and the concentration of sedimenting material has remained constant. A smaller SC then the SC of the sediment core PM9463-8 corresponds to a smaller accumulation due to sediment winnowing, a larger Sc to sediment focussing.

$$R_x = \frac{R_{63}}{Sc_{63}} \cdot Sc_x \qquad (8)$$

R_{63}: mean accumulation rate of core PM 9463-8 [g $cm^{-2}a^{-1}$]
R_x: mean accumulation rate of core x [g cm^{-2} a^{-1}]
SC_{63}: Standing Crop of core PM 9463-8 [dpm cm^{-2}]
SC_x: Standing Crop of core x [dpm cm^{-2}]

As listed in Table 4 the rates derived with this equation are similar to the ones derived from the depth profiles of $^{210}Pb_{ex}$. The deviations range within the uncertainties, suggesting that the rates derived from the $^{210}Pb_{ex}$ profiles are not significantly distorted by bioturbative mixing of the sediment even at those localities with lower accumulation rates. Further the comparison of the average accumulation rates with the $^{210}Pb_{ex}$ method of 0.12 ± 0.02 cm a^{-1} (PM9462-1) is in the same range as the accumulation rate derived by ^{14}C dating of 0.06 ± 0.02 cm a-1 (PM9462-4) (Bauch, pers. comm.) therefore the bioturbative processes can be only of small influence.

Table 4: Accumulation rates derived from the depth profiles of $^{210}Pb_{ex}$ and the Standing crop method

Station	Accumulation Rate $^{210}Pb_{ex}$ [g cm^{-2} a^{-1}]	Accumulation Rate Standing Crop [g cm^{-2} a^{-1}]
PM9402-3	0.15 ± 0.03	0.14 ± 0.04
PM9417-4	0.05 ± 0.02	0.09 ± 0.03
PM9442-3	0.08 ± 0.01	0.10 ± 0.04
PM9462-1	0.08 ± 0.02	0.13 ± 0.04
PM9481-2	0.19 ± 0.08	0.11 ± 0.04

Table 5: Depositional $^{210}Pb_{ex}$ fluxes of sediment cores from the shelf area of the Laptev Sea.

Station	Depositional $^{210}Pb_{ex}$ Fluxes [dpm cm^{-2} a^{-1}]
PM 9402-3	0.35 ± 0.02
PM 9417-4	0.22 ± 0.02
PM 9442-3	0.25 ± 0.04
PM 9462-1	0.31 ± 0.02
PM 9463-8	0.60 ± 0.05
PM 9481-2	0.26 ± 0.02

Depositional $^{210}Pb_{ex}$ fluxes in the shelf area of the Laptev Sea

At steady state, the Standing crop of excess ^{210}Pb must be balanced by the net flux of $^{210}Pb_{ex}$ into the sediment. The depositional $^{210}Pb_{ex}$ flux can be determined as:

$$F = \lambda \cdot SC \qquad (9)$$

F : depositional ^{210}Pb flux [dpm cm^{-2} a^{-1}]
SC : Standing Crop [dpm cm^{-2}]
λ : decay constant of ^{210}Pb [0.031 a^{-1}]

The depositional $^{210}Pb_{ex}$ fluxes of the investigated locations are listed in Table 5. The atmospheric supply of ^{210}Pb in different regions of the world ranges between 0.15 and 1.5 dpm cm^{-2} a^{-1} (Graustein and Turekian, 1986; Gopalakrishnan et al., 1973; Turekian et al.,

1977). Our data suggest an atmospheric flux of ^{210}Pb for the area of the Laptev Sea ≤ 0,6 dpm cm^{-2} a^{-1}, which is in agreement with model results of Rehfeld (1994).

Sedimentary ^{10}Be-Fluxes in the Shelf Area of the Laptev Sea

The sedimentation rates from Table 3 were used to calculate the depositional ^{10}Be flux into the sediments of the shelf area from the Laptev Sea. The depositional ^{10}Be flux is defined as:

$$F(x) = C(x) \bullet S(x) \bullet \rho(x) \tag{10}$$

where:

F : depositional ^{10}Be flux [at cm^{-2} a^{-1}]
C : concentration of ^{10}Be [atoms g^{-1}]
S : sedimentation rate [cm ka^{-1}]
ρ : dry bulk density [g cm^{-3}]

In Table 6 the ^{10}Be concentrations and the depositional ^{10}Be fluxes of the investigated sediment cores are listed. These ^{10}Be fluxes of [(10 - 150) $\bullet 10^6$at cm^{-2} a^{-1}] in the shelf area of the Laptev Sea, are by two orders of magnitude higher than the recent atmospheric input [(0.2 - 0.5) $\bullet 10^6$ at cm^{-2} a^{-1}] in Greenland (Stanzick, 1996) and other world regions (Table 7). This surplus of ^{10}Be indicates a significant supply of continental ^{10}Be with the rivers into the Laptev Sea during the last century.

Table 6: Measured ^{10}Be concentrations and the calculated ^{10}Be depositional fluxes.

Station	Depth [cm]	^{10}Be [10^8 atoms g^{-1}]	Depositional ^{10}Be Flux [10^6 atoms cm^{-2} a^{-1}]
PM 9402-3	0 - 1	2.95 ± 0.23	36.8 ± 8.0
PM 9402-3	14 - 16	3.28 ± 0.16	48.8 ± 10.0
PM 9402-3	32 - 36	3.53 ± 0.14	49.1 ± 10.0
PM 9417-4	0 - 1	4.62 ± 0.32	19.9 ± 3.0
PM 9451-7	1 - 2	0.56 ± 0.11	
PM 9462-1	0 - 1	3.53 ± 0.16	26.0 ± 3.0
PM 9462-1	14 - 16	4.49 ± 0.21	40.5 ± 5.0
PM 9463-8	1 - 2	3.93 ± 0.30	72,9 ± 15,0
PM 9463-8	14 - 16	4.02 ± 0.31	105.6 ± 21.0
PM 9463-8	32 - 36	4.78 ± 0.23	105.4 ± 21.0
PM 9481-2	1 - 2	1.87 ± 0.19	36.2 ± 8.0
PM 9481-2	14 - 16	1.99 ± 0.10	38.9 ± 7.0
PM 9482-1	0 - 1	4.10 ± 0.44	68.6 ± 12.0

Table 7: Atmospheric ^{10}Be fluxes as determined at other locations.

Atmospheric ^{10}Be fluxes	References	Investigated material
$1.21 \pm 0{,}26\ \Sigma 10^6$ at $cm^{-2}\ a^{-1}$	Monaghan, 1985/86	North America, Precipitation
$1.50 \pm 0.50\ \Sigma 10^6$ at $cm^{-2}\ a^{-1}$	Lao et al., 1992 a/b	Pacific, Sediments
~ $0.7\ \Sigma 10^6$ at $cm^{-2}\ a^{-1}$	Southon et al., 1987	North Atlantic, Sediments
~ $0.3\ \Sigma 10^6$ at $cm^{-2}\ a^{-1}$	Finkel et al., 1977	Arctic, Sediments

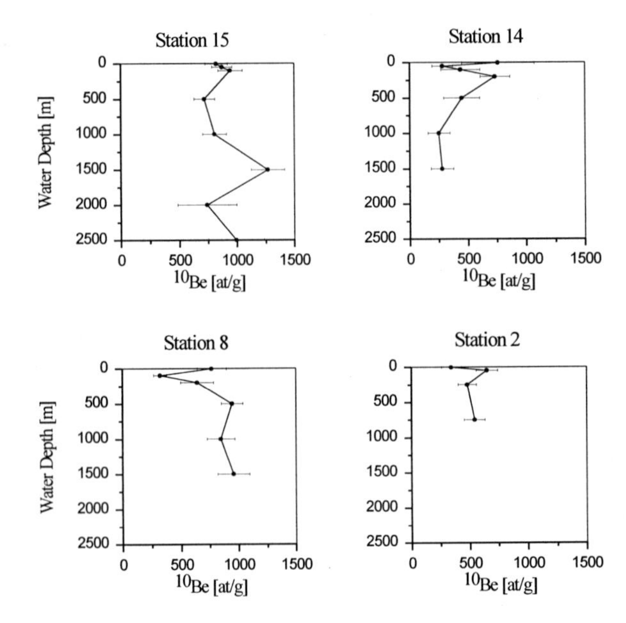

Figure 7: ^{10}Be concentrations versus water depth [m].

^{10}Be in the Arctic water column and in the Laptev Sea

We measured ^{10}Be profiles at 4 stations in the Norwegian- and in the Greenland Sea. The localities of the water profiles are listed in Table 2 and shown in Figure 7. The deep water concentrations of ^{10}Be at stations 8 and 15 (Norwegian current) come close to the value of Atlantic deep water (ranging between 1000 and 1500 at/g, Segl et al., 1987), whereas stations 2 and 14 (East Greenland Current) show deep water concentrations lower than the NADW. The deep water concentration at station 14 probably reflects Arctic deep water conditions. Three further profiles of ^{10}Be from the central Arctic Ocean sampled during the campaign ARK VIII/3 are plotted in Figure 8. They show nearly constant ^{10}Be concentration around 500 at/g. The values are lower than the concentrations in the Atlantic deep water and at least one order of magnitude lower than the concentrations near the mouth of the Lena of 5000 at/g. The high concentrations in the water masses of the Laptev Sea near the mouth of the river Lena (PM

9463-1 and PM 9472-3; Table 8) reflect the input of continental ^{10}Be with the rivers. Comparison with the profiles in the central Arctic Ocean clearly shows that the high concentration of ^{10}Be near the mouth of the Laptev Sea does not reach the central part of the Arctic Ocean. We therefore conclude that it is being scavenged close to its continental source. This conclusion corroborates the very high fluxes of ^{10}Be into shelf sediments derived above, suggesting that riverine ^{10}Be is adsorbed on aluminosilicates (Southon et al., 1987, Jansen et al., 1987) and rapidly sedimented.

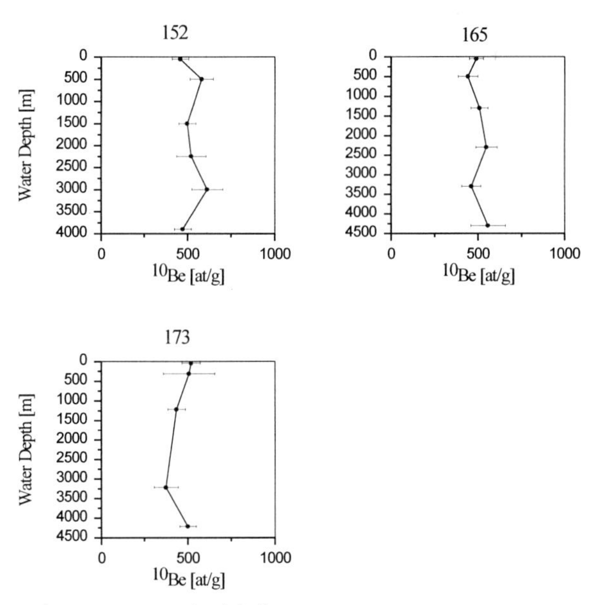

Figure 8: ^{10}Be concentrations versus water depth [m].

Table 8: ^{10}Be concentrations of water samples from the Laptev Sea.

Station	Water Depth [m]	^{10}Be [atoms/g]
PM 9463-1	2	2150 ± 300
PM 9463-1	30	3135 ± 520
PM 9472-3	2	5740 ± 600
PM 9494-3	2	895 ± 200

Sedimentary ^{10}Be-Fluxes in the continental slope of the Laptev Sea

To test if significant amounts of terrigenic ^{10}Be may still be transferred into the central Arctic Ocean, we measured the ^{10}Be concentrations and $^{230}Th_{ex}$ activities of the sediment core PS 2471-4 from the continental slope of the Laptev Sea. This core is located about 500 km from the mouth of the Lena. The average sedimentation rates for each oxygen isotope stage (Table 8) of this core were determined by biostratigraphy (Fahl, pers. comm.) and from the depth profile of the $^{230}Th_{ex}$ activity. As presented by Nürnberg et al. (1995) the upper 50 cm of this sediment core can be associated with the oxygen isotope stage 1. Further the abundance of Gephyrocapsa spp. in the core section between 200 and 300 cm was related to the isotope oxygen stage 5. However Baumann (1990) showed that in sediments from the Nansen Basin, the coccolithophoride Gephyrocapsa spp. was also recorded in the oxygen isotope stage 3. $^{230}Th_{ex}$, ^{10}Be, microfossils and amino acids deliver further stratigraphic information.

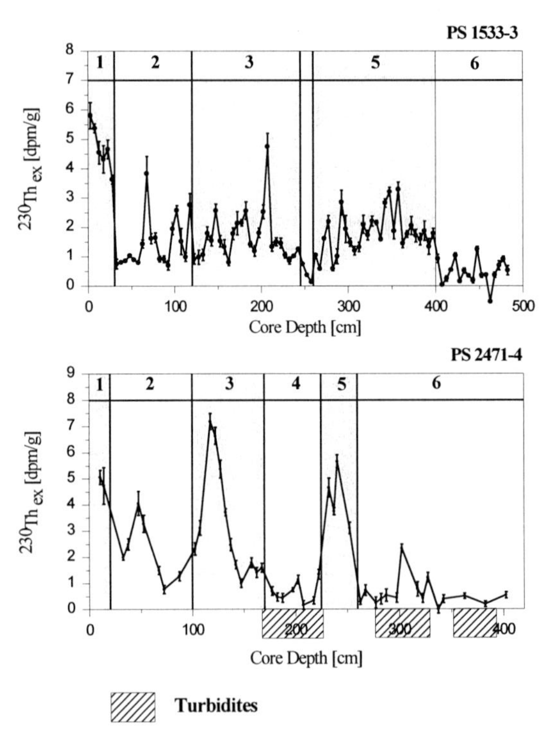

Figure 9: $^{230}Th_{ex}$ activities of the sediment cores PS 1533-3 and PS 2471-4 versus core depth [cm].

$^{230}Th_{ex}$ and ^{10}Be

In this study we compare the $^{230}Th_{ex}$- and ^{10}Be profiles of PS 2471-4 (Figure 9, Figure 10) with the radionuclide profiles of PS 1533-3 from the Yermak Plateau (Eisenhauer et al., 1994), where an age-depth model is available. Characteristic for the $^{230}Th_{ex}$ profile of core PS 1533-3 and of other sediment cores from high northern latitudes are two dominant features in the $^{230}Th_{ex}$ profiles. These are marked by a rather abrupt drop of the $^{230}Th_{ex}$ activity at the stage boundaries 6/5 and 2/1 (Scholten et al., 1994). By analogy we related the stage boundary 6/5 of PS 2471-4 at a core depth of 265 cm. Because of missing $^{230}Th_{ex}$ data for the upper 15 cm of the core, the transition at 2/1, fixed at a depth of 20 cm, is rather uncertain. According to this model the three turbitide layers were deposited during the glacial stages 4 and 6. The comparison of the ^{10}Be-profile to that of core PS 1533-3 confirms the age-depth model derived from $^{230}Th_{ex}$. From the ^{10}Be concentrations we calculated the depositional ^{10}Be fluxes during the isotopic stages 2, 3 and 5 (Table 9). The average ^{10}Be flux amounts to $[(2.32 \pm 0.50) \cdot 10^6$ at cm^{-2} $a^{-1}]$. It is by at least one order of magnitude lower than the flux in the shelf area, but higher than the fluxes of $[(0.1 - 0.6) \cdot 10^6$ at cm^{-2} $a^{-1}]$ in the central part of the Arctic Ocean (Strobl et al., in prep.). We did not evaluate the flux during stage 4 because the sedimentation

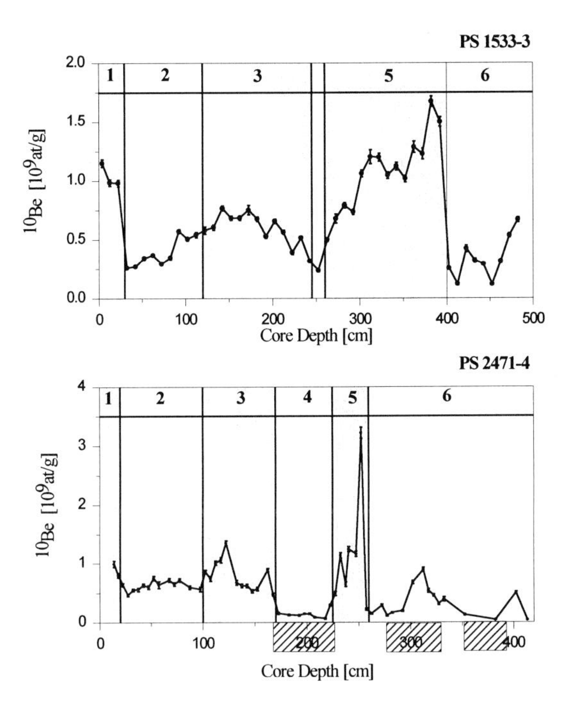

Figure 10: ^{10}Be concentrations of the sediment cores PS 1533-3 and PS 2471-4 versus core depth [cm].

Table 9: Sedimentation rates and the depositional ^{10}Be fluxes during the isotope stages 2, 3 and 5 of the sediment core PS 2471-4.

Oxygen Isotope Stage	Sedimentation rate [cm ka^{-1}]	Depositional ^{10}Be Flux [10^6 atoms cm^{-2} a^{-1}]
2	5.80 ± 0.80	4.10 ± 0.50
3	1.86 ± 0.30	1.96 ± 0.30
5	0.54 ± 0.20	0.90 ± 0.10

rate during this oxygen isotope stage is highly uncertain. As the fluxes of $^{230}Th_{ex}$ are close to the expected production flux, we can exclude that the ^{10}Be fluxes were distorted by processes of sediment redistribution or enhanced scavenging. The depositional ^{10}Be-fluxes of the cores PM 9482-1, PM 9463-8, PM 9462-1 and PS 2471-4 (oxygen isotope stage 3) are presented im Figure 11. The conclusion from the rather low ^{10}Be flux on the continental slope is that only a smaller part of the terrigenic ^{10}Be finds its way to the Arctic Ocean.

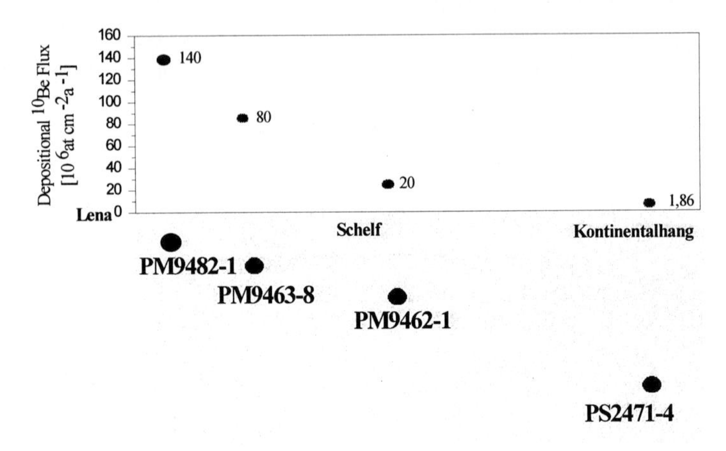

Figure 11: Depositional ^{10}Be fluxes in the shelf area of the Laptev Sea and on the continental slope.

Conclusions

We conclude that the rivers are the main source of ^{10}Be in the shelf areas of the Arctic Ocean. From the comparison of the fluxes of ^{10}Be in the central Arctic Ocean, on the continental slope and in the shelf area of the Laptev Sea we further conclude that most of the ^{10}Be delivered to the shelf area by the rivers Lena, Yana and Kathanga must be deposited directly in the shelf area.

Acknowledgments

The financial support by the German Ministry of Education, Science, Research and Technology (BMBF) is gratefully acknowledged. The authors want to thank M. v. de Loeff, J. Scholten, T. Eisenhauer and T. Billen for sampling and the availabiltity of water samples from the central Arctic Ocean, the Norwegian- and Greenland Sea. The chemical preparation and measurement of the water samples from the Norwegian- and Greenland Sea were done by H.J. Rutsch and M. Frank.

References

Arnold, T. (1989) Bestimmung von Mischungskoeffizienten in der Norwegischen See mit ^{210}Pb. Diplomarbeit Universität Heidelberg.

Baumann, M. (1990) Coccoliths in sediments of the eastern Arctic Basin. In: Geological History of the Polar Ocean: Arctic versus Antarctic, NATO Series C, Mathematical and Physical Sciences, Bleil, U. and J. Thiede (eds), 308, 437-445.

Benninger, L.K, R.C. Aller, J.K. Cochran, and K.K. Turekian (1979) Effects of biological sediment mixing on the ^{210}Pb chronology and trace metal distribution in a Long Island Sound sediment core. Earth Planet. Sci. Lett., 43, 241-259.

Berner, R.A. (1980) Early diagenesis - a theoretical approach. Princeton University Press, Princeton.

Bollhoefer, A., A. Mangini, A. Lenhard, M. Wessels, F. Giovanoli and B. Schwarz (1994) High-Resolution ^{210}Pb-Dating of Lake Constance Sediments; Stable Lead in Lake Constance. Environ. Geol., 24, 267-274.

Doerr, H. and K.O. Muennich (1991) Lead and Cesium Transport in European Forest Soils. Water Air and Soil Pollution, 57-58, 808-818.

Dominik, J., A. Mangini and G. Mueller (1981) Determination of recent deposition rates in Lake Constance with radioisotopic methods. Sedimentology, V28, 653-677.

Eisenhauer, A., R.F. Spielhagen, M. Frank, G. Hentzschel, A. Mangini, P.W. Kubik, B. Dittrich-Hannen and T. Billen, ^{10}Be records of sediment cores from high northern latitudes, Earth and Planet. Sci. Lett., 124, 171-184, 1994.

Evans, D.W., J.J. Alberts and A.C. Clark (1983) Reversible ion-exchange fixation of cesium-137 leading to mobilization from reservoir sediments. Geochimica et Cosmochimica Acta, Vol. 47, 1041-1049.

Finkel, R., S. Krishnaswami and D.L. Clark (1977) ^{10}Be in Arctic ocean sediments. Earth and Planet. Sci. Lett,. 35, 199-204.

Frank, M. (1996) Reconstruction of Late Quaternary environmental conditions applying the natural radionuclides ^{230}Th, ^{10}Be, ^{231}Pa and ^{238}U: A study of deep-sea sediments from the eastern sector of the Antarctic Current System. Ber. Polarforschung, 186, 136 pp.

Fütterer, D.K. (1992) Arctic'91: Die Expedition ARK-VIII/3 mit FS Polarstern. Reports on Polar Research, 107, 142pp.

Fütterer, D.K. (1994) Die Expedition Arctic'93. Der Fahrtabschnitt ARK-IX/4 mit FS Polarstern. Reports on Polar Research, 149, 244 pp.

Gopalakrishnan, S., C. Rangarajan, L.U. Joshi, D.K. Kapoor and C.D. Eapen (1973) Measurements on airborne and surface fallout radioactivity in India. Government of India, Atomic Energy Commission, Bhabha Atomic Research Centre, Bombay, India.

Graustein, W.C. and K.K. Turekian (1989) The effects of forests and topography on the deposition of sub-micrometer aerosols measured by lead-210 and cesium-137 in soils. Agricultural and Forest Meteorology, 47, 199-220.

Henken-Mellies, W.-U., J. Beer, F. Heller, K.-J. Hsü, C. Shen, G. Bonani, H.-J. Hofmann, M. Suter and W. Wölfli (1990) ^{10}Be and ^{9}Be in South Atlantic DSDP site 519: Relation to geomagnetic reversals and to sediment composition. Earth Planet. Sci. Lett., 98, 267-276.

Jansen, J.H.F., C. Alderliesten, A.J. van Bennekom, K. van der Borg and A.F.M. de Jong (1987) Terrigenous supply of ^{10}Be and dating with ^{14}C and ^{10}Be in sediments of the Angola Basin (SE Atlantic), Nucl. Instr. Meth. Vol., B29, 311-316.

Kassens, H. (1995) Laptev Sea System: Expeditions in 1994. Reports on Polar Research, 182, 195pp.

Langner, C. and Transdrift I Shipboard Party (1995) Distribution of Fe and Mn in pore waters and sediments of the Laptev Sea-Results of the expedition Transdrift I. In: Russian-German Cooperation: Laptev Sea System, Reports on Polar Research, 176, H. Kassens, D. Piepenburg, J. Thiede, L. Timokhov, H.-W. Hubberten, and S.M. Priamikov (eds), 387pp.

Lao, Y., R.F. Anderson, W.S. Broecker, S.E. Trumbore, H.J. Hofmann and W.Wölfli (1992a) Increased production of cosmogenic ^{10}Be during the last glacial maximum. Nature, 357, 576-578.

Lao, Y., R.F. Anderson, W.S. Broecker, S.E. Trumbore, H.-J. Hofmann and W. Wölfli (1992b) Transport and burial rates of ^{10}Be and ^{231}Pa in the Pacific Ocean during the Holocene period. Earth and Planet. Sci. Lett., 113, 173-189.

Mangini, A., M. Segl, G. Bonani, H.-J. Hofmann, E. Morenzoni, M. Nessi, M. Suter, W. Wölfli and K.K. Turekian (1984) Mass spectrometric Beryllium dating of deep-sea sediments applying the Zürich tandem accelerator. Nucl. Instrum. Methods Phys. Res., B5, 353-357.

Mangini, A., M. Segl, H. Kudrass, M. Wiedicke, G. Bonani, H.-J. Hofmann, E. Morenzoni, M. Nessi, M. Suter, and W. Wölfli (1986) Diffusion and supply rates of ^{10}Be and ^{230}Th radioisotopes in two manganese encrustation from the South China Sea. Geochim. et Cosmochim. Acta 50, 149-156.

Monaghan, M.C., S. Krishnaswami and K.K Turekian (1985/1986) The global-average production rate of ^{10}Be. Earth and Planet. Sci. Lett., 76, 279-287.

Nürnberg, D., D. Fütterer, F. Niessen, N. Nörgaard-Petersen, C. Schubert, R.F. Spielhagen and M. Wahsner (1995) The depositional environment of the Laptev Sea continental margin: Preliminary results from the RV „Polarstern„ ARK-IX/4 cruise. Polar Research, 14, 43-53.

Rehfeld, S. (1994) Deposition radioaktiver Tracer in einem Transportmodell der Atmosphäre. Thesis, Univ. Hamburg, 144pp.

Robbins, J.A. and D.N. Edgington (1976) Depositional processes and the determination of recent sedimentation rates in Lake Michigan. Proc. 2nd Federal Conf. Great Lakes, 378-390.

Scholten, J.C., R. Botz, H. Paetsch and P. Stoffers (1994) $^{230}Th_{ex}$ flux into Norwegian-Greenland sediments: Evidence for lateral sediment transport during the past 300,000 years. Earth and Planet. Sci. Lett., 121, 111-124.

Segl, M, A. Mangini, J. Beer, G. Bonani, M. Suter and W. Wölfli (1987) ^{10}Be in the Atlantic Ocean, a transect at 25°N. Nucl. Instr. Meth., B29, 332-334.

Southon , J.R., T.L. Ku, D.E. Nelson, J.L. Reyss, J.S. Vogel (1987) ^{10}Be in a deep sea core: implications regarding ^{10}Be production changes over the past 420 ka. Earth Planet. Sci. Lett., 85, 356-364.

Stanzick, A. (1996) Räumliche und zeitliche Depositionsvariationen der Radioisotope ^{10}Be und ^{210}Pb in Eisbohrkernen Zentralgrönlands. Diploma thesis, Institut für Umweltphysik Heidelberg, Germany, 96pp.

Thiede, J. and G. Hempel (1991) Die Expedition Arktis-VII/1 mit FS Polarstern. Reports on Polar Research, 80, 137pp.

Turekian, K.K., Y. Nozaki and L.K. Benninger (1977) Geochemistry of atmospheric radon and radon products. Ann. Rev. Earth Planet. Sci., 5, 227-255.

von Gunten, H.R. and R.N. Moser (1993) How reliable is the ^{210}Pb dating method? Old and new results from Switzerland. Journal of Paleolimnology, 0, 1-18.

Spatial Distribution of Diatom Surface Sediment Assemblages on the Laptev Sea Shelf (Russian Arctic)

H. Cremer

GEOMAR Forschungszentrum für marine Geowissenschaften, Wischhofstrasse 1-3, D 24148 Kiel, Germany

Received 14 February 1997 and accepted in revised form 18 December 1997

Abstract - The spatial distribution of diatom assemblages was investigated in surface sediments from the Laptev Sea shelf. Diatoms are an abundant and diverse microfossil component in these sediment samples. A total of 345 diatom taxa from 56 genera could be identified with freshwater diatoms contributing up to 62% of the total taxa number. However, diatom assemblages are dominated by a few planktic species. A factor analysis of 21 species and species groups from 75 surface sediment samples was carried out. Four factors (= diatom assemblages) could be extracted. They are the ice algae assemblage, the freshwater diatom assemblage, the *Chaetoceros* assemblage, and the *Thalassiosira nordenskioeldii* assemblage. The spatial distribution pattern of these assemblages in sediments of the Laptev Sea shelf shows a distinct relationship to oceanography. Parameters mainly affecting the composition and distribution of diatom sediment assemblages on the Laptev Sea shelf are the sea-ice conditions, the freshwater influx by Siberian rivers, and, controlled by the river water influx, the salinity of surface waters.

Introduction

The Laptev Sea (Figure 1) is one of the largest Arctic shelf seas and has great importance for the oceanographic situation of the Arctic Ocean. Particularly both the huge freshwater input by the large rivers (especially by the Lena River) and its significance for the freshwater balance of the Arctic Ocean, and the importance of the Laptev Sea as a source for new sea-ice which is supplied to the Transpolar Drift (Figure 1) have been recently pointed out (e.g. Aagaard and Carmack, 1989; Gordeev and Sidorov, 1993; Kassens et al., 1995; Nürnberg et al., 1994; Gordeev et al., 1996; Steele et al., 1996; Eicken et al., 1997). The freshwater supply and the ice cover which is present from November until May affect the biology and the evolution of organism communities (Clark, 1990). However, compared to other Eurasian shelf seas, scientific and especially paleoceanographic investigations in the Laptev Sea are sparse. The reason was the unaccessibility of the Laptev Sea for scientists during Soviet times.

Diatoms in the Arctic have been studied for a long time. The first descriptions of Arctic diatom species were given by Ehrenberg (1932) who was also the first refering to the presence of diatoms in sea-ice (Ehrenberg 1943). Comprehensive investigations of Arctic diatoms were first carried out by Cleve and Grunow (1880), Cleve (1883), Grunow (1884), Østrup (1895), Gran (1904a) and Meunier (1910). However, the knowledge about the diatom flora of the Laptev Sea is very scarce. First descriptions of plankton and sediment diatoms were published from the famous expeditions of the "Vega" 1878-79 (Cleve, 1883) and the "Fram" 1893-96 (Gran 1904b). Later, Kisselew (1932), Shirshov (1937), and Usachev (1946) contribute descriptions of the diatom plankton of the Laptev Sea. Above all, these publications concentrate on a listing of identified genera and species.

The present study investigates the spatial distribution of diatoms in surface sediments of the Laptev Sea shelf (Cremer, 1998). The purpose of this study is to describe distinct distribution patterns of diatom assemblages and its species composition in surface sediments, and moreover to recognize a relationship between the biogeographic distribution of diatom assemblages

---

In: Kassens, H., H.A. Bauch, I. Dmitrenko, H. Eicken, H.-W. Hubberten, M. Melles, J. Thiede and L. Timokhov (eds.) Land-Ocean Systems in the Siberian Arctic: Dynamics and History. Springer-Verlag, Berlin, 1999, 533-551.

and oceanographic conditions of the surface waters. In several studies the distribution of diatoms in surface sediments has been related to oceanographic parameters of the overlying water masses (e.g. Maynard, 1976; Williams, 1986; Koç Karpuz and Schrader, 1990; Zielinski, 1993).

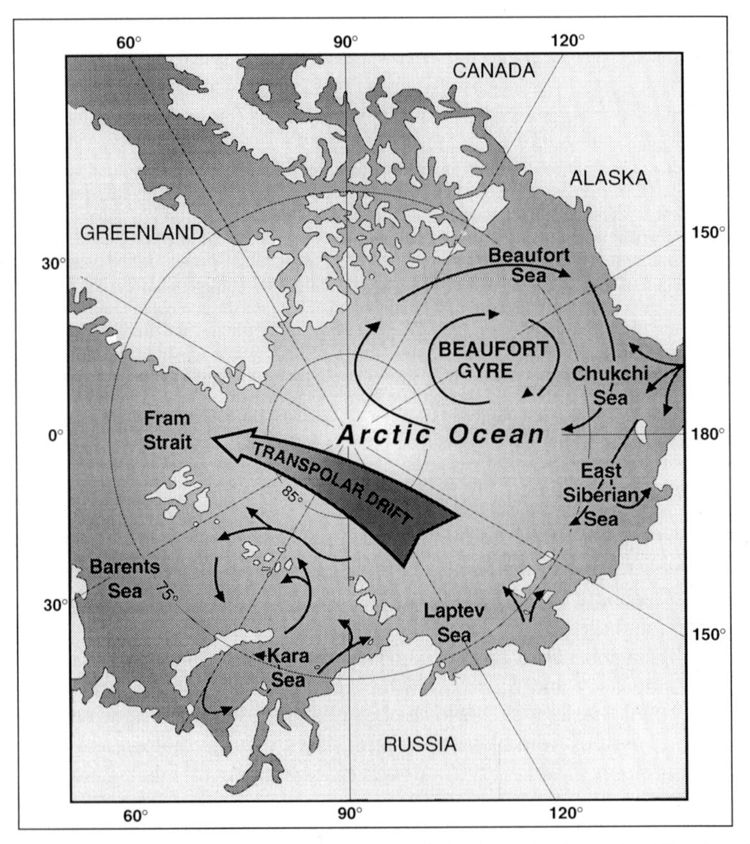

Figure 1: Geography of the Eurasian shelf seas and sea-ice drift in the Arctic Ocean (modified from Gordienko and Laktionov, 1969).

Physiographic and oceanographic conditions in the Laptev Sea

The Laptev Sea is part of an extensive shelf zone along the coasts of northeast Siberia (Figure 1) and covers an area of about 460,000 km^2. The shelf is extremely shallow (the average slope is 0-5 m/km; Holmes and Creager, 1974) and extents 300-500 km northwards to the shelf edge at the 60 m bathymetric contour line. Five submarine valleys transverse the Laptev Sea in a northerly and northeasterly direction. These valleys run along tectonic faults (Drachev et al., 1995) and were eroded during times with a low sea level stand. Physiographic conditions and the geological history of the Laptev Sea have already been summarized by Holmes and Creager (1974).

Modern sediment distribution have been studied by Holmes (1967), Lindemann (1994), and Rossak (1995). The western part of the shelf is characterized by deltaic sands from the Anabar, Khatanga, and Olenek Rivers whereas the sediment load of the Lena and Yana Rivers mainly accumulate in the eastern Laptev Sea shelf region. Aswell, the distribution of clay minerals lead to a subdivision of the Laptev Sea shelf into a western and an eastern province (Rossak, 1995). According to Ivanov and Piskun (1995), the annual sediment supply is 19 million tons (Lena River), 4 million tons (Yana River), and 1.5 million tons (Olenek River).

The hydrologic situation of the Laptev Sea shelf is marked by a strong freshwater input. The annual overall input to the Laptev Sea shelf by the Khatanga, Anabar, Olenek, Lena, and Yana Rivers is about 730 km^3/a (Treshnikov, 1985; Gordeev et al., 1996), but just the Lena River discharges about 520 km^3/a via its vast delta. The maximum input occurs in late June and early July. From the end of October until mid June the Lena River is frozen up. Large quantities of freshwater are released to the water column during ice melting in summer. This, together with the riverine input, results in low salinity surface waters and creates large brackish water areas (Figure 2), especially in the southeastern region of the Laptev Sea (Codispoti and Richards, 1968; Létolle et al., 1993; Martin et al., 1993). Due to this large freshwater flux the Laptev Sea shelf has a halocline in 5-15 m water depth with underlying cold and saline shelf water and overlying low-saline surface water (Karpiy et al., 1994; Churun and Timokhov, 1995). The Lena Delta is also the main source area for sediment and suspension load (e.g. Gordeev and Sidorov, 1993; Peulvé et al., 1996) and for nutrients (Pivovarov and Smagin, 1995; Cauwet and Sidorov, 1996).

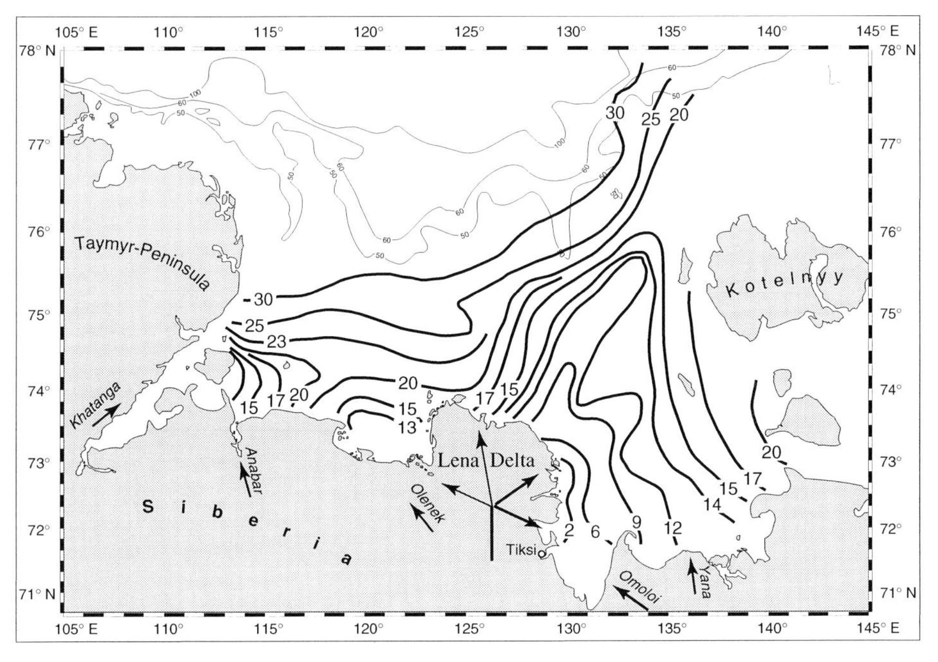

Figure 2: Schematic salinity pattern of surface waters of the Laptev Sea shelf (combined after Codispoti and Richards, 1968; Karpiy et al., 1994, and Heiskanen and Keck, 1996).

The Laptev Sea is covered by sea-ice for eight months of the year with a maximum thickness of two meters (Barnett, 1991). The winter situation is characterized by an up to 100 km wide polynya which separates the coastal fast ice from the drift ice (Reimnitz et al. 1994; Dethleff, 1995). This open water sector has a relatively stable position and represents a source for new sea-ice which is supplied to the Transpolar Drift (Dethleff, 1995; Nürnberg et al., 1994). The sea-ice in the Laptev Sea and in the Arctic bears remarkable amounts of sediment and suspension particles which are incorporated during sea-ice formation processes on the shelf and are exported into the Arctic Ocean by the Transpolar Drift (Nürnberg et al., 1994; Eicken et al., 1997).

Materials and methods

Sediment material

Samples were recovered during the Transdrift I, II, and III expeditions which took place during the summer months in 1993, 1994 and in autumn 1995. Surface sediment samples were collected aboard Russian research vessels (1993, 1994) and the Russian icebreaker "Kapitan Dranitsyn" (1995). For the present study a total of 75 sediment samples were investigated (Figure 3, Table 3). These had been collected mainly by a spade box corer. At all stations general hydrologic features like temperature, salinity, and water depth were documented (cruise reports of Kassens and Karpiy, 1994; Kassens, 1995; Kassens, 1997). From each box corer the uppermost centimeter was sampled. Based on ^{210}Pb- and ^{137}Cs-activity measurements the age of the surface sediments can assumed to be younger than 50 years (Strobl, pers. comm.).

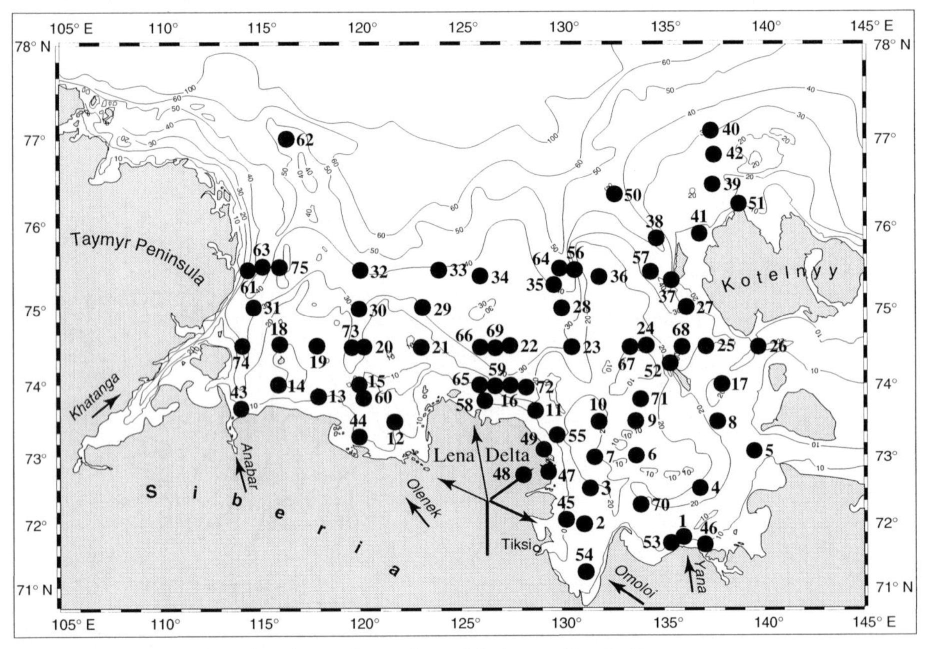

Figure 3: Bathymetry and numbered sampling stations of the Laptev Sea shelf.

Methods

For investigations of the diatom flora the uppermost centimeter of the sediment column was extruded and dry-freezed. Of the dried sediment 0.5 to 3 grams were heated for 20 minutes with 30% hydrogen peroxide and 30% hydrochloric acid. After oxidation of the organic material and the carbonate, the excess acid was removed by seven sedimentation procedures in demineralized water always after 24 h. with a water jet pump. During further sedimentation processes the clay fraction was removed from the acid insoluble residue. For this purpose the suspension was regularly mixed with a 0.3% solution of sodiumhexametaphosphate in order to get a better dispersion of the clay material.

Permanent slides were prepared according to the method of Battarbee (1973) with the high resolution diatom resin MOUNTEX®. Counting of slides was carried out after Schrader and Gersonde (1978) using a Zeiss-Axiophot microscope (Plan-Neofluar 100x/1.30 oil objective) at a magnification of x1000. Generally 300 valves were counted without resting spores of the genus *Chaetoceros*. The total abundance of diatoms was determined and the relative abundances of the most important taxa and taxa groups were expressed in percentages of the total diatom assemblage in each sediment sample.

For description and definition of diatom surface sediment assemblages the data set was treated with a factor analysis realized with the program StatView (Abacus Concepts, Inc., Berkeley, 1992). The steps, results and interpretation of a factor analysis can be checked in Imbrie and Kipp (1971) and Maynard (1976).

Table 1: Number of genera and taxa identified in surface sediment samples of the Laptev Sea shelf.

Genus	No. of identified taxa	No. of polar taxa	Genus	No. of identified taxa	No. of polar taxa
Achnanthes	16	0	*Gomphonema*	11	0
Actinocyclus	1	0	*Grammatophora*	2	0
Amphora	2	0	*Gyrosigma*	5	2
Asterionella	1	0	*Hantzschia*	1	0
Aulacoseira	3	0	*Haslea*	1	1
Bacteriosira	1	1	*Melosira*	4	1
Biremis	1	0	*Meridion*	1	0
Caloneis	7	0	*Minidiscus*	1	0
Chaetoceros	10	1	*Navicula*	72	25
Cocconeis	6	0	*Neidium*	5	0
Coscinodiscus	1	0	*Nitzschia*	21	9
Craspedopleura	1	1	*Opephora*	1	0
Cyclotella	4	0	*Paralia*	1	0
Cymatopleura	1	0	*Pinnularia*	27	10
Cymbella	20	2	*Placoneis*	2	0
Denticula	1	0	*Pleurosigma*	1	1
Diatoma	5	0	*Porosira*	1	0
Diatomella	1	1	*Pseudogomphonema*	4	4
Didymosphaenia	2	1	*Pseudo-Nitzschia*	2	2
Diploneis	12	4	*Rhopalodia*	1	0
Entomoneis	4	3	*Stauroneis*	7	1
Epithemia	2	0	*Stenoneis*	2	2
Eunotia	20	1	*Stephanodiscus*	3	0
Fallacia	2	0	*Surirella*	5	0
Fossula	1	1	*Tabellaria*	1	0
Fragilaria	20	0	*Tetracyclus*	2	0
Fragilariopsis	3	1	*Thalassiosira*	11	4
Frustulia	1	0	*Thalassiothrix*	1	0
			56 Genera	**345**	**78**

Observations and discussion

Taxonomy

A total of 345 taxa from 56 genera were identified from surface sediment samples of the Laptev Sea shelf with the genera *Navicula, Pinnularia, Nitzschia, Fragilaria, Cymbella,* and *Eunotia* being the most diverse ones (Table 1). A complete taxa list with remarks to synonymy, references, ecology and biogeographic distribution is given in Cremer (1998). Most of the taxa identified (214 taxa) are freshwater diatoms all of which usually inhabit Siberian lakes and rivers in the hinterland (Komarjenko and Vasiljeva, 1975). They are transported by the rivers into the Laptev Sea. The remaining taxa prefer marine or marine-brackish conditions. Most of the taxa normally inhabit epibenthic and/or epiphytic habitats (272 taxa) whereas only 47 taxa are typically planktic diatoms. 16 taxa are meroplanktic diatoms which settle in both planktic and epibenthic environments.

Distribution of ecologically important species

Diatom valves are common in all sediment samples. Total abundance values range from 0.001 x 10^6 to 6.7 x 10^6 valves per gram dry sediment. The distribution of total abundance values is relatively uneven on the Laptev Sea shelf and does not correlate with any hydrologic or sedimentologic parameters. The preservation of diatom valves in the surface sediments is generally moderate to excellent.

In order to document the composition of diatom assemblages in the sediment samples three west-east-transects have been choosen (Figure 4) along which the relative abundance of the most important species was determined (Figures 5, 6, and 7). In the following these species and species groups and their distribution are shortly described.

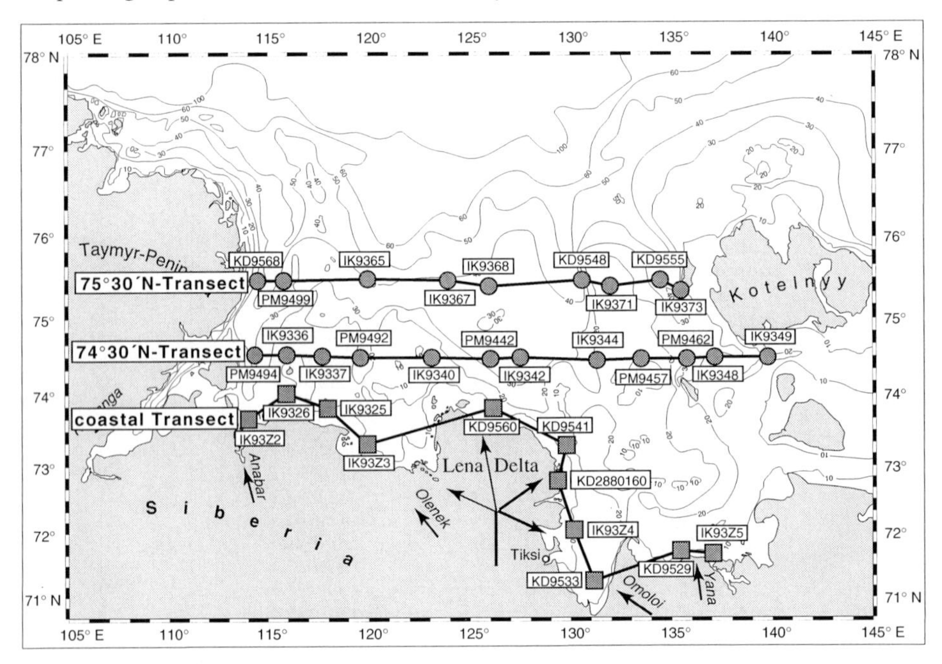

Figure 4: Map of the Laptev Sea shelf showing three west-east-transects along which the relative abundance of the most important diatom species and species groups was determined (see Figures 5, 6 and 7).

Aulacoseira spp.: species of this genus are presently known only from freshwater habitats. Because this genus has great importance in the surface sediments *Aulacoseira* was counted separately and not together with the freshwater diatoms group (see below). Three species could be clearly identified (*A. granulata, A. islandica,* and *A. subarctica*) but soleley *A. subarctica* is dominant in some coastal samples. Like the other freshwater diatoms *Aulacoseira* spp. is less abundant on the two northern transects but is significantly abundant in the region east of the Lena delta (Figures 5, 6, and 7).

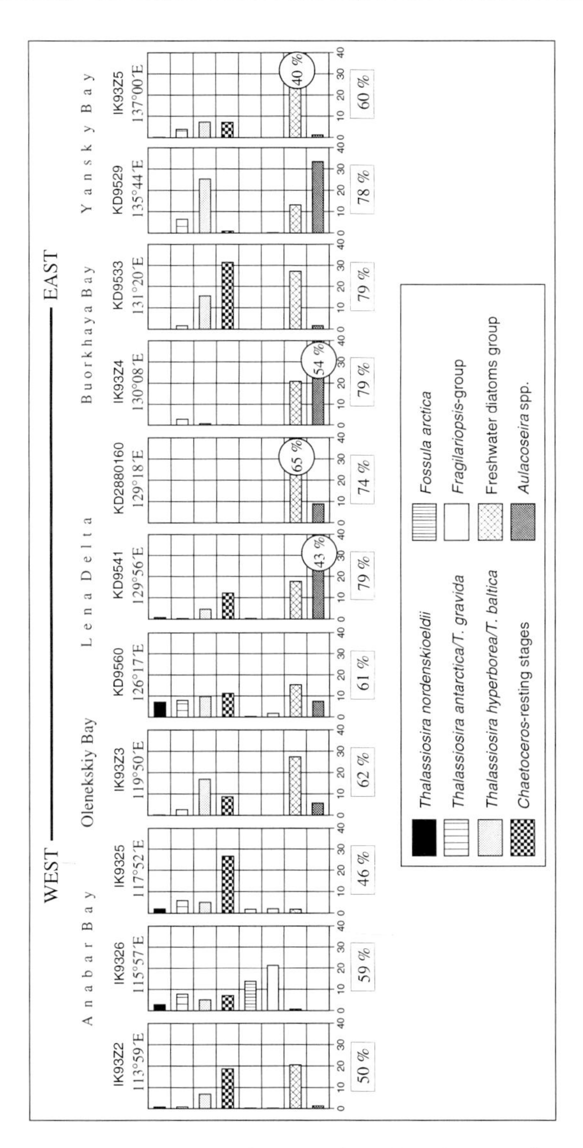

Figure 5: Relative abundance (in %) of important diatom species and species groups in coastal surface sediment samples (see Figure 4) of the Laptev Sea shelf. The percent value at the bottom of each diagram indicates the portion of these species and species groups of the total diatom content.

Chaetoceros-resting spores: nearly all valves of *Chaetoceros* spp. in the sediments are resting spores. As the classification of vegetative valves and resting spores to distinct species is extremely difficult (Hustedt, 1930) all valves of this genus have been treated together as *Chaetoceros* spp.. The majority of the taxa prefer marine/brackish-neritic conditions but their biogeographic distribution varies from cosmopolitan to polar-subpolar (see species list in Cremer, 1998). Resting spores of *Chaetoceros* are among the most common diatoms in shelf sediments and occur with a relative abundance from 8 to 65% (Figures 6 and 7), whereas in front of the eastern Lena delta they are totally absent (Figure 5).

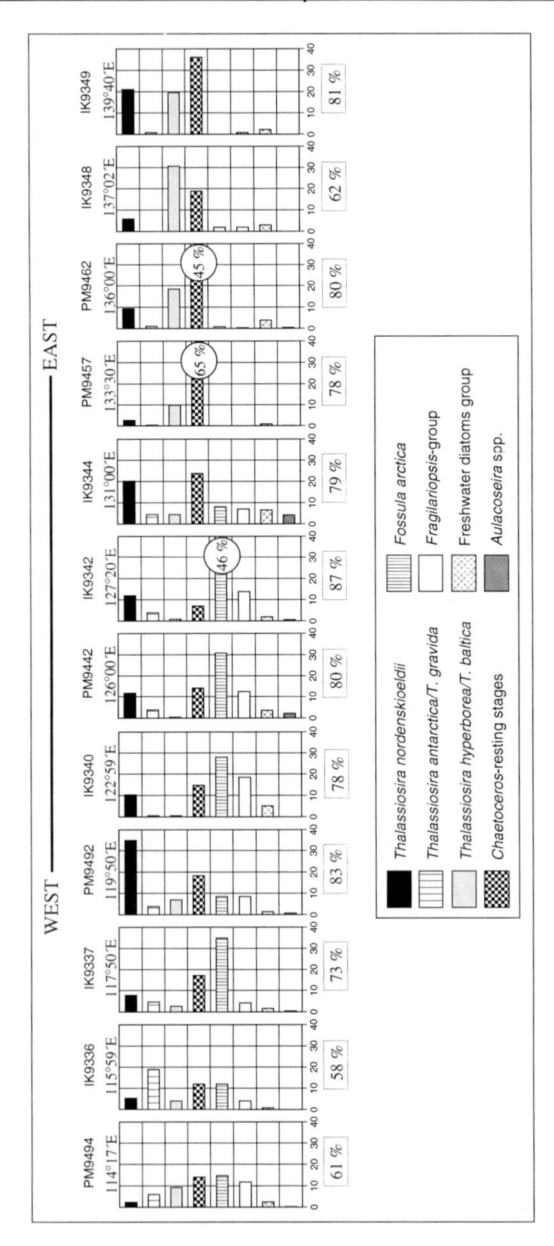

Figure 6: Relative abundance (in %) of important diatom species and species groups in surface sediment samples along the 74°30'N-transect (see Figure 4). The percent value at the bottom of each diagram indicates the portion of these species and species groups of the total diatom content.

Fossula arctica: this diatom species was recently described by Hasle et al. (1996). In the samples of the southern and middle transects *Fossula arctica* is present just in the western region of the shelf (Figures 5 and 6). On the most northern transect this taxon also occurs in the eastern region with a maximum abundance of 15% (Figure 7). According to Hasle et al. (1996)

Fossula arctica is a marine, planktic diatom but also is a typical sea-ice species.

Fragilariopsis group: this group consists of the species *Fragilariopsis cylindrus* (synonym *Nitzschia cylindrus*) and *Fragilariopsis oceanica* (synonym *Nitzschia grunowii*). Both species are cold water diatoms in circumarctic waters (Hasle and Syvertsen, 1996). *F. cylindrus* is bipolar and thrives in circumantarctic waters too and its biogeographic distribution extends to 50°N and 50°S respectively, whereas *F. oceanica* appears down to 45°N (Hasle, 1976). Both species occur in large quantities in sea-ice and are often the predominating diatoms in this environment (Hasle, 1976; Poulin, 1990). The relative abundance pattern of the *Fragilariopsis*-group is as the distribution of *Fossula arctica*, even though abundance values are lower with 2 to 28% (Figures 5, 6, and 7).

Freshwater diatoms group: in this group all species and genera other than *Aulacoseira* spp. are integrated which have their main distribution in freshwater habitats (Table 2). Solely these genera and species do not occur in an appreciable abundance but as a group the freshwater diatoms are an important element in the surface sediment samples of the Laptev Sea shelf. The relative abundance of freshwater diatoms is low on the two northern transects (Figures 6 and 7). In coastal sediments values increase to 65% (Figure 5).

Thalassiosira antarctica and *Thalassiosira gravida*: both species live in similar ecological habitats and show a similar morphology. They are therefore counted together. However, *T. gravida* has a lower abundance in the sediments. Both taxa are marine-neritic species and show a bipolar distribution in circumpolar waters. *T. antarctica* is known to form heavily silicified resting spores (Hasle and Syvertsen, 1996). Vegetative and resting spore valves were counted together, but resting spores predominate. The biogeographic range of *T. antarctica* is between 80-60°N and 76-58°S respectively (Hasle, 1976). This species prefers temperature conditions between -1.3 and 7°C as well as salinities between 24.9 and 34.3 (Hasle and Heimdal, 1968). *T. antarctica*/*T. gravida* occur in offshore sediment samples (Figures 6 and 7) with an abundance of 0-19%, in coastal sediments (Figure 5) with a maximum abundance of 8%. Their abundance is similar to that of sediments of the Bering and Chukchi Seas (Sancetta, 1982).

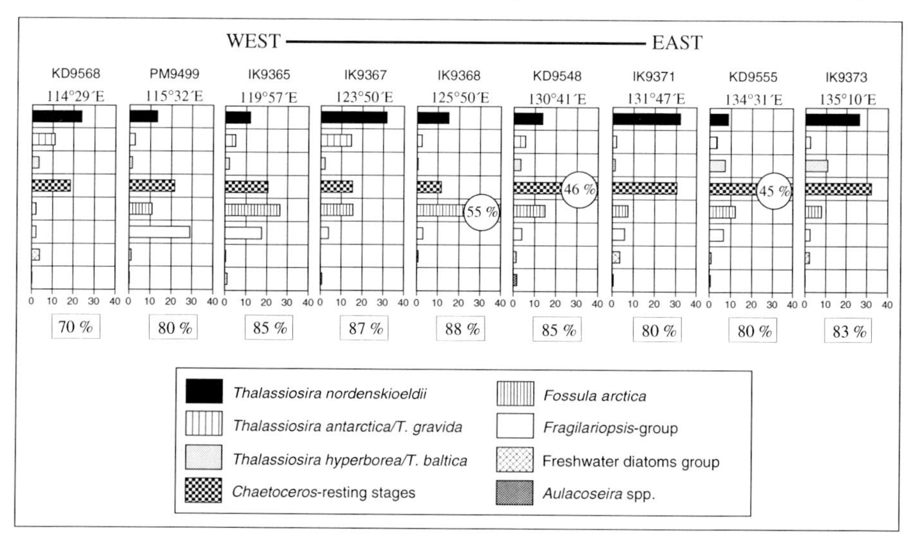

Figure 7: Relative abundance (in %) of important diatom species and species groups in surface sediment samples along the 75°30'N-transect (see Figure 4). The percent value at the bottom of each diagram indicates the portion of these species and species groups of the total diatom content.

Thalassiosira hyperborea and *Thalassiosira baltica*: these two species were also counted together because of their similarities in ecology. Both species inhabit brackish environments (mesohalobous-euryhaline habitats; Pankow, 1990) in circumpolar to temperate waters of the northern hemisphere. They are especially abundant in the vicinity of the large Siberian and Canadian rivers (Hasle and Lange, 1989). *T. hyperborea/T. baltica* have their highest abundance in the western region of the 74°30`N-transect, reaching between 20 and 30% (Figure 6). In coastal sediments both species occur with 5 to 25% (Figure 5) whereas their portion on the northernmost transect has a maximum abundance of 10% (Figure 7).

Thalassiosira nordenskioeldii: this species is a typical cold water diatom (marine-neritic) of polar to temperate waters of the northern hemisphere. According to Hasle (1976) this species occurs down to 35°N. *T. nordenskioeldii* is a resting spore forming diatom and grows best at temperatures between 0 and 5°C (Durbin, 1974, 1978). In samples of the southern transect *T. nordenskioeldii* is nearly missing (Figure 5). On the other hand this species shows a generally high abundance in sediments of the central Laptev Sea shelf (Figures 6 and 7).

Factor analysis

In order to describe well defined diatom assemblages and their occurrence in sediments of the Laptev Sea shelf a factor analysis was carried out. Factor analyses were used to investigate micropaleontological datasets (e.g. Imbrie and Kipp, 1971; Maynard, 1976; Williams, 1986; Koç Karpuz and Schrader, 1990; Pichon et al., 1992; Koç et al., 1993; Matthiessen, 1994). A factor analysis reduces a great number of variables (here: relative abundance of diatom species) to a small number of factors (here: diatom assemblages), which can replace the original variables in further investigations. Thus, the dataset becomes reduced and simplified, and interpretations are easier.

The factor analysis was carried out on 75 surface sediment samples and 21 species and species groups (Figure 3, Table 2). To reduce the data set species with similar environmental characteristics were grouped. After this clustering all species and groups with a relative abundance generally less than 2 % of the total assemblage were removed. Table 2 lists the 21 species and species groups used in this factor analysis. Results of the factor analysis are given in tables 3 and 4. Table 3 contains the varimax factor matrix in which the portion of each factor in each sediment sample is expressed as a factor loading. The communality indicates to what degree the factor model explains the original information in each sediment sample. The factor score matrix (Table 4) explains the composition of each factor (of each diatom assemblage) determined by the analysis. The factor score value is a measure for the significance of each species and species group in the defined assemblages. The distribution of the diatom sediment assemblages is shown in figure 8.

The factor analysis yielded four factors which explain 81,4% of the variance in the original dataset (Table 3). The communality is higher than 0.70 in 60 (= 80%) of the samples. However, the species composition of 15 sediment samples (= 20%) cannot be sufficiently explained by the model of four factors, indicated by relatively low communalities (Table 3). Reasons may be local differences in ecological conditions, selective opal dissolution processes, local species blooms, or changes of the species composition through resuspension and input of fossil taxa (Imbrie and Kipp, 1971; Matthiessen, 1994).

Species assemblages

Factor 1 (ice algae assemblage) explains 43.9% of the variance and consists mainly of the *Fragilariopsis*-group and *Fossula arctica* (Table 4). These species are the main components of sea-ice communities in the Arctic (Poulin, 1990; Hasle et al., 1996). This assemblage shows highest factor loadings in sediments from the central and northeastern regions of the Laptev Sea shelf (Figure 8). In the southeastern and coastal regions this factor has no importance and factor loadings are generally low (Table 3). As the species of this assemblage are typical sea-ice diatoms their distribution may be associated with sea-ice conditions on the Laptev Sea shelf.

Factor 2 (freshwater diatoms assemblage) consists of the freshwater diatoms group and *Aulacoseira* spp.. This assemblage dominates the coastal regions of the Laptev Sea and especially the region east of the Lena delta (Figure 8). In sediments from the central and northern shelf factor loadings are low. This factor explains 19.9% of the total variance.

Factor 3 (*Chaetoceros* assemblage) is dominated, in order, by resting spores of *Chaetoceros* spp., by *Nitzschia* spp. (Arctic-marine species, Table 2), and by *Thalassiosira hyperborea*. This assemblage is mainly present in the eastern, southeastern, and northeastern parts of the Laptev Sea shelf (Figure 8) and explains 11.8% of the variance. In some samples of the western shelf factor loadings of this assemblage are relatively high too.

Table 2: Twenty-one species and species groups (bold-faced) and composition of the species groups used in the factor analysis of 75 surface sediment samples.

Achnanthes taeniata	**Freshwater diatoms group**
Arctic-marine-epibenthic species	*Asterionella* spp.
Craspedopleura kryophila	*Cymbella* spp.
Entomoneis gigantea	*Denticula* spp.
Entomoneis kjellmannii	*Diatoma* spp.
Entomoneis paludosa var. *borealis*	*Didymosphenia* spp.
Navicula algida	*Epithemia* spp.
Navicula directa	*Eunotia* spp.
Navicula gelida	*Fragilaria* spp. (freshwater species)
Navicula glacialis	*Gomphonema* spp.
Navicula imperfecta	*Meridion circulare*
Navicula impexa	*Navicula* spp. (freshwater species)
Navicula kariana	*Neidium* spp.
Navicula kryokonites	*Pinnularia* spp. (freshwater species)
Navicula lineola var. *perlepida*	*Placoneis* spp.
Navicula novadecipiens	*Stephanodiscus* spp.
Navicula oestrupii	*Tabellaria* spp.
Navicula pagophila	***Melosira* spp.**
Navicula superba	***Navicula peregrina***
Navicula transitans	***Nitzschia linearis***
Navicula trigonocephala	***Nitzschia* (Arctic-marine species)**
Navicula valida	*Nitzschia arctica*
Navicula vanhoeffeni	*Nitzschia brebissonii*
Pinnularia polaris	*Nitzschia delicatisima*
Pinnularia quadratarea	*Nitzschia gelida*
Pinnularia semiinflata	*Nitzschia hudsonii*
Pseudogomphonema arcticum	*Nitzschia hybrida*
Pseudogomphonema kamtschaticum	*Nitzschia laevissima*
Pseudogomphonema septentrionale	*Nitzschia lanceolata*
Stenoneis inconspicua var. *baculus*	*Nitzschia polaris*
Stenoneis obtuserostrata	*Nitzschia scabra*
***Aulacoseira* spp.**	*Nitzschia seriata*
***Chaetoceros* spp. (resting stages)**	***Paralia sulcata***
Coscinodiscus oculus-iridis	***Porosira glacialis***
***Cyclotella* spp.**	***Thalassiosira antarctica/T. gravida***
Fossula arctica	***Thalassiosira baltica***
***Fragilariopsis* group**	***Thalassiosira bulbosa***
Fragilariopsis cylindrus	***Thalassiosira hyalina***
Fragilariopsis oceanica	***Thalassiosira hyperborea***
	Thalassiosira nordenskioeldii

Table 3: Sample number, geographical position, and varimax factor loadings for four factors and 75 samples. Factors characterizing the surface sediment samples are underlined.

No.	Sample	Latitude (°N)	Longitude (°E)	Communality	Factor 1	Factor 2	Factor 3	Factor 4
1	IK93 01	71°45.2	135°39.6	0.335	-0.081	0.443	0.251	-0.045
2	IK93 06	72°00.6	130°59.2	0.990	-0.002	0.984	-0.127	-0.044
3	IK93 07	72°33.0	131°17.8	0.962	-0.038	0.942	0.041	-0.032
4	IK93 09	72°29.4	136°35.2	0.963	-0.076	0.310	0.776	-0.099
5	IK93 13	73°04.2	139°25.6	0.748	-0.083	0.131	0.746	-0.094
6	IK93 15	73°00.0	133°28.3	0.899	0.028	0.222	0.743	-0.073
7	IK93 16	73°00.1	131°30.1	0.717	-0.009	0.820	-0.033	0.087
8	IK93 18	73°29.9	137°31.4	0.843	-0.050	-0.002	0.843	-0.163
9	IK93 20	73°29.7	133°31.2	0.902	-0.029	0.512	0.624	-0.123
10	IK93 21	73°28.7	131°38.9	0.799	-0.026	0.834	0.053	0.103
11	IK93 23	73°37.8	128°39.2	0.867	0.048	0.927	-0.161	0.067
12	IK93 24	73°30.1	121°40.1	0.414	-0.258	0.500	0.169	0.099
13	IK93 25	73°49.9	117°52.3	0.778	0.014	-0.134	0.783	-0.098
14	IK93 26	73°59.6	115°57.5	0.627	0.737	-0.180	0.012	-0.264
15	IK93 27	73°59.9	119°51.6	0.564	-0.136	0.009	0.654	0.000
16	IK93 30	74°00.0	127°30.2	0.916	0.473	0.375	-0.226	0.458
17	IK93 34	74°00.0	137°39.8	0.769	0.013	-0.096	0.527	0.277
18	IK93 36	74°29.4	115°59.0	0.422	0.374	-0.080	0.057	0.191
19	IK93 37	73°30.1	117°50.9	0.815	0.779	-0.040	0.056	-0.085
20	IK93 38	74°29.5	119°57.3	0.871	-0.094	0.025	0.053	0.741
21	IK93 40	74°30.0	122°59.6	0.945	0.871	0.008	-0.002	-0.106
22	IK93 42	74°30.6	127°20.9	0.913	0.886	-0.003	-0.274	-0.044
23	IK93 44	74°24.3	131°00.8	0.936	0.235	0.131	0.251	0.436
24	IK93 46	74°29.7	134°02.1	0.439	-0.298	-0.148	0.337	0.384
25	IK93 48	74°28.8	137°02.7	0.508	-0.192	-0.096	0.572	0.136
26	IK93 49	74°30.0	139°40.1	0.955	-0.161	-0.081	0.558	0.428
27	IK93 50	75°00.8	136°01.7	0.459	-0.204	0.005	0.575	0.055
28	IK93 53	74°57.6	129°45.9	0.864	0.148	0.100	-0.011	0.641
29	IK93 56	74°59.8	123°00.9	0.949	0.449	-0.047	-0.018	0.465
30	IK93 58	75°00.1	119°53.3	0.739	0.201	-0.065	0.197	0.434
31	IK93 61	74°59.8	114°32.0	0.881	0.824	-0.047	0.053	-0.116
32	IK93 65	75°28.1	119°57.7	0.989	0.780	-0.065	0.082	0.028
33	IK93 67	75°28.9	123°50.5	0.952	0.236	-0.068	-0.200	0.709
34	IK93 68	75°25.5	125°49.8	0.773	0.783	-0.012	-0.219	0.025
35	IK93 70	75°16.2	129°27.9	0.943	0.801	-0.060	-0.027	0.039
36	IK93 71	75°23.0	131°47.9	0.914	0.118	-0.019	0.184	0.575
37	IK93 73	75°20.5	135°10.0	0.960	0.053	-0.061	0.341	0.519
38	IK93 73A	75°48.6	134°35.0	0.515	0.477	-0.090	0.125	0.089
39	IK93 82	76°30.2	137°16.3	0.978	0.770	-0.096	0.050	0.057
40	IK93 84	77°06.7	137°13.5	0.890	0.156	-0.063	0.723	-0.020
41	IK93 K1	75°56.4	136°42.5	0.932	0.112	-0.083	0.304	0.501
42	IK93 K2	76°50.1	137°17.7	0.918	0.401	-0.065	0.532	0.028
43	IK93 Z2	73°39.9	113°59.8	0.506	-0.028	0.407	0.464	-0.141
44	IK93 Z3	73°17.7	119°49.9	0.773	-0.053	0.776	0.224	-0.108
45	IK93 Z4	72°02.0	130°07.6	0.739	-0.011	0.848	-0.276	0.082
46	IK93 Z5	71°41.4	137°00.4	0.599	0.031	0.713	0.129	-0.155
47	KD 2880160	72°48.4	129°18.6	0.661	0.045	0.802	-0.098	-0.090
48	KD 2880201	72°45.0	128°07.7	0.770	0.043	0.868	-0.106	-0.074
49	KD 2880307	73°07.3	129°09.9	0.592	0.049	0.756	-0.075	-0.103
50	KD95 02	76°11.5	133°06.9	0.904	0.805	-0.017	0.069	-0.067
51	KD95 17	76°14.2	138°50.1	0.691	0.039	-0.093	0.662	0.028
52	KD95 23	74°18.4	135°26.9	0.934	0.131	-0.052	0.778	-0.062
53	KD95 29	71°45.1	135°44.1	0.848	0.022	0.486	0.600	-0.119

Table 3 (continued):

No.	Sample	Latitude (°N)	Longitude (°E)	Communality	Factor 1	Factor 2	Factor 3	Factor 4
54	KD95 33	71°14.0	131°20.6	0.683	-0.110	0.804	-0.167	0.056
55	KD95 41	73°22.8	129°56.5	0.732	-0.007	0.835	-0.046	0.063
56	KD95 48	75°28.7	130°41.7	0.934	0.282	-0.028	0.534	0.150
57	KD95 55	75°36.0	134°31.3	0.959	0.261	-0.066	0.636	0.060
58	KD95 60	73°47.7	126°17.6	0.750	-0.126	0.629	0.191	0.305
59	KD95 61	73°54.1	126°54.8	0.958	0.271	0.431	0.307	0.253
60	KD95 65	73°50.7	120°19.0	0.947	0.106	0.019	0.728	0.035
61	KD95 68	75°29.1	114°28.9	0.894	-0.086	-0.024	0.090	0.730
62	KD95 72	77°01.4	116°03.0	0.940	0.294	-0.060	0.663	-0.026
63	PM94 02	75°29.4	115°14.9	0.856	0.748	-0.033	0.067	-0.002
64	PM94 17	75°30.1	130°00.8	0.955	0.873	-0.037	-0.219	0.011
65	PM94 41	74°00.0	125°59.2	0.872	0.405	-0.010	0.461	0.077
66	PM94 42	74°29.9	126°00.2	0.972	0.849	0.051	-0.112	0.033
67	PM94 57	74°30.1	133°29.9	0.933	-0.002	-0.060	0.801	0.004
68	PM94 62	74°30.1	136°00.2	0.967	-0.072	-0.035	0.750	0.147
69	PM94 63	74°30.2	126°34.9	0.924	0.621	0.025	-0.310	0.399
70	PM94 75	72°15.0	133°59.7	0.858	0.011	0.489	0.559	-0.010
71	PM94 81	73°45.0	134°00.2	0.965	-0.039	-0.054	0.868	-0.078
72	PM94 82	73°59.9	128°10.4	0.938	0.907	0.249	-0.192	-0.124
73	PM94 92	74°30.0	119°50.1	0.930	0.110	-0.065	-0.067	0.726
74	PM94 94	74°30.0	114°17.0	0.895	0.739	-0.063	0.348	-0.265
75	PM94 99	75°30.0	115°32.7	0.695	0.563	-0.101	0.205	0.039
				Variance (%):	43.900	19.900	11.800	5.800
				Cumulative Variance (%):	43.900	63.800	75.600	81.400

Table 4: Factor score matrix for 21 species and species groups explaining the composition of each diatom assemblage. Species and species groups which are characteristic for the factors are underlined.

Species/species group	Factor 1	Factor 2	Factor 3	Factor 4
Achnanthes taeniata	-0.257	-0.484	-0.063	-0.402
Arctic-marine-epibenthic species	-0.407	-0.351	0.577	-0.201
Aulacoseira spp.	-0.158	2.720	-1.365	0.654
Chaetoceros spp. (resting spores)	0.322	-0.194	4.130	-0.251
Coscinodiscus oculus-iridis	-0.440	-0.519	-0.200	-0.198
Cyclotella spp.	-0.365	-0.383	0.001	-0.437
Fossula arctica	2.354	-0.306	-0.433	-1.183
Fragilariopsis group	4.238	0.036	-1.443	-1.136
Freshwater diatoms group	0.282	3.273	-0.357	-0.598
Melosira spp.	-0.578	-0.217	0.189	0.273
Navicula peregrina	-0.438	-0.525	-0.175	-0.211
Nitzschia linearis	-0.430	-0.548	0.030	-0.274
Nitzschia spp. (Arctic-marine species)	0.058	-0.348	1.201	-1.262
Paralia sulcata	-0.457	-0.574	-0.026	-0.262
Porosira glacialis	-0.484	-0.444	-0.279	0.155
Thalassiosira antarctica/gravida	-0.100	0.038	-0.780	1.064
Thalassiosira baltica	-0.420	0.482	0.399	-0.511
Thalassiosira bulbosa	-0.386	-0.539	-0.140	-0.301
Thalassiosira hyalina	-0.052	-0.544	-0.125	-0.523
Thalassiosira hyperborea	-1.480	-0.364	1.134	0.612
Thalassiosira nordenskioeldii	-0.802	-0.209	-2.277	4.992

Factor 4 (*Thalassiosira nordenskioeldii* assemblage) explains only 5.8% of the total variance. The dominant species is *Thalassiosira nordenskioeldii* (marine-neritic species), and *T. antarctica* (marine-neritic) is of secondary importance. This factor shows highest factor loadings in some samples of the central and northeastern parts of the Laptev Sea shelf (Figure 8).

Diatom sediment assemblages and their relationship to oceanography

Based on the factor model of diatom sediment assemblages explained above the Laptev Sea shelf can be subdivided into three provinces (Figure 8): the eastern province with dominating *Chaetoceros* spp. and brackish water species, the central shelf province of which sediments are mainly characterized by ice algae and a patchy occurrence of *Thalassiosira nordenskioeldii*, and the coastal province where freshwater taxa are the dominant components. This distribution pattern approximately traces the position of the polynya during winter months (Figure 8; Reimnitz et al., 1994; Dethleff, 1995). North of the polynya the sediments are dominated by the ice algae and *T. nordenskioeldii* assemblages, whereas the sediments south of this polynya are characterized by the freshwater diatoms and *Chaetoceros* assemblages.

Statistically, the ice algae assemblage is the most significant factor in the Laptev Sea and explains 43.9% of the total variance. This assemblage particularly dominates the sediments of the central shelf to the north of the wintry polynya (Figure 8). Characteristic for this region is also the patchy dominance of the marine-brackish planktic diatom *Thalassiosira nordenskioeldii* which is a typical species of the spring phytoplankton bloom in polar waters (Paasche, 1975). A similar co-occurrence pattern of ice algae and *T. nordenskioeldii* in surface sediments is also reported from the Bering and Chukchi Seas (Sancetta, 1981; Schandelmeier and Alexander, 1981). A reason for this could be an ice algae and *T. nordenskioeldii* bloom that occurs along the ice edge which retreats to the north during ice melting in spring and early summer. In sediments from deeper regions (e.g. the continental slope) ice algae show generally a low abundance, due to strong dissolution of the thin and delicate valves of the ice algae (Cremer, 1998). Thus, the co-occurrence of ice algae and *T. nordenskioeldii* assemblages points to shallow shelf regions with a seasonal ice cover and ice edge blooms during ice melting in spring.

The freshwater diatoms assemblage is the second-significant factor and explains 19.9% of the total variance. This assemblage predominantly consists of a freshwater diatom flora rich in genera (see Table 2) and species. However, few species occur with higher abundances, e.g. *Aulacoseira subarctica*, *Asterionella formosa*, *Diatoma tenuis* (Cremer, 1998). The freshwater diatom flora is characteristic for coastal sediments and especially for the region to the east of the Lena delta (Figure 8). In front of the Yana and Anabar Rivers diatom sediment assemblages not only consist of freshwater diatoms but of other, marine and brackish water species (Figure 8). The distribution of freshwater diatoms in sediments of the Laptev Sea reflects the intensity of riverine freshwater supply by the rivers on the one hand and the mixing of river water with brackish-marine water masses from the shelf on the other hand. In all, the occurrence of freshwater diatoms in high quantities points to a strong riverine freshwater supply in the vicinity of river mouths and river deltas.

The eastern region of the Laptev Sea shelf is dominated by the *Chaetoceros* assemblage. This assemblage consists particularly of *Chaetoceros* resting spores. Arctic-marine-epibenthic species of *Nitzschia* and *Thalassiosira hyperborea* are of secondary importance. The regions showing a main significance of the *Chaetoceros* assemblage correspond with regions where salinity of surface waters is reduced to brackish conditions due to strong river water influx during summer (Figures 2 and 8). Resting spores of *Chaetoceros* are the most abundant diatoms in sediments of the Laptev Sea. The chain forming vegetative stages of this genus

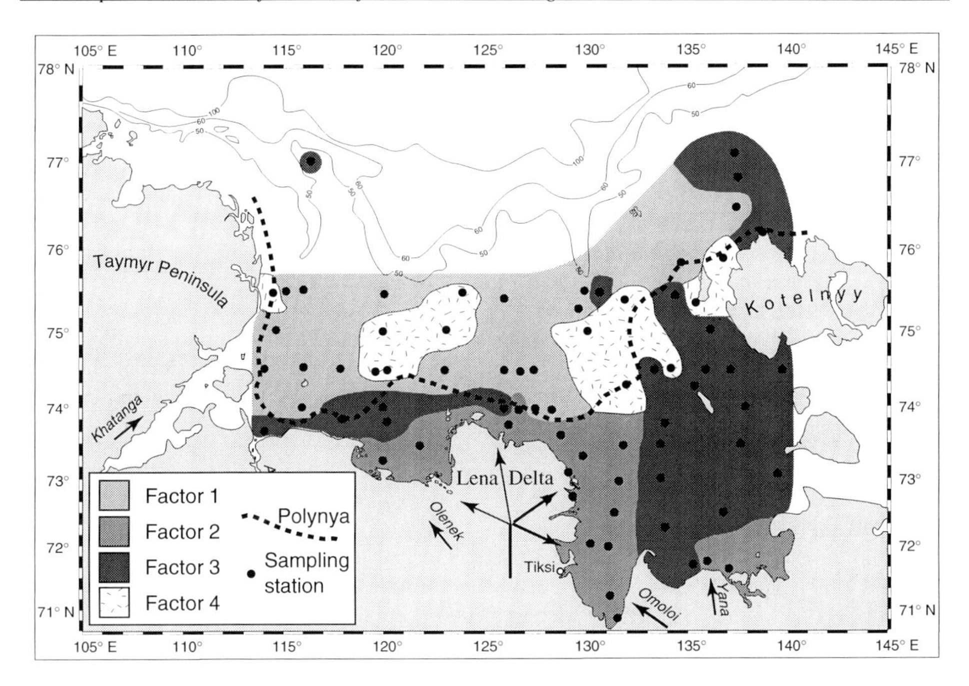

Figure 8: Geographical distribution of diatom sediment assemblages on the Laptev Sea shelf according to maximum varimax factor loadings from factor analysis. Factor 1 - ice algae assemblage, factor 2 - freshwater diatoms assemblage, factor 3 - *Chaetoceros* assemblage, factor 4 - *Thalassiosira nordenskioeldii* assemblage. The dashed line indicates to the course of the wintry polynya according to Reimnitz et al. (1994).

dominate the plankton of the neritic regions of the Laptev Sea shelf (Cremer, 1998). A similar high abundance of *Chaetoceros* resting spores has been reported from other polar regions, e.g. from the North Pacific (Sancetta, 1982) and the Greenland-Iceland-Norwegian-Sea (Schrader and Koç Karpuz, 1990). A high relative abundance of *Chaetoceros* resting spores is assumed to be an indication for high productivity (Schuette and Schrader, 1979; Sancetta, 1981, 1982; Williams, 1986). Species of *Chaetoceros* are obviously able to use the short, ice-free summer period for intense growth. If growth conditions deteriorate *Chaetoceros* forms resting spores which may retain their vitality and form the seed for the next growth period (Smetacek, 1985; McQuoid and Hobson, 1995). *T. hyperborea* is a brackish water species being typical for the vicinity of river mouths and river deltas of the Arctic Ocean (Hasle and Lange, 1989). According to Pankow (1990) *T. hyperborea* is a mesohalobous-euryhaline species and prefers salinities between 2 and 30. Highest abundance of this species occurs in the eastern part of the Laptev Sea shelf where brackish conditions exist (Figures 5 and 6). *T. hyperborea* is also reported to be a typical under-ice species which grows under the melting ice during spring and summer (Syvertsen, 1990). Thus, the *Chaetoceros* assemblage may indicate to occasionally ice covered shelf regions with reduced salinity and a relatively high productivity in surface waters during summer.

Concluding remarks

The results of the study of surface sediment diatom assemblages from the Laptev Sea shelf lead to the following conclusions:

a Diatoms are very abundant in surface sediments of the Laptev Sea shelf. The diatom flora consists to a high degree of freshwater species which are transported to the Laptev Sea shelf by several Siberian rivers.
b In all, 345 taxa from 56 genera could be identified from the sediment material.
c Only a few planktic and sea-ice species dominate the sediment assemblages.
d Diatom assemblages on the Laptev Sea shelf are mainly consisted of eight species and species groups.
e A factor analysis yielded four diatom assemblages. These are the ice algae assemblage, the freshwater diatoms assemblage, the *Chaetoceros* assemblage, and the *Thalassiosira nordenskioeldii* assemblage.
f The distribution pattern of these assemblages show a distinct relationship to oceanographic conditions. Important parameters influencing the biogeographic distribution of the four diatom assemblages are the sea-ice conditions, the freshwater influx, and the salinity conditions of surface waters of the Laptev Sea shelf.

Acknowledgements

This study was carried out within the Russian-German cooperation project "Laptev Sea System" which is financially supported by the German Ministry of Education, Science, Research and Technology (BMBF). I wish to thank my colleagues of the Laptev Sea Group for many fruitful discussions and especially Jörn Thiede and Heidemarie Kassens (all GEOMAR, Kiel) who enabled my participation in the Transdrift II and III Expeditions as well as in two Laptev Sea workshops in St. Petersburg. I thank Jens Hölemann, Bjorg Stabell, and Nalân Koç for reviewing this paper and providing valuable comments.

References

Aagaard, K. and E.C. Carmack, (1989) The role of sea ice and other fresh water in the Arctic circulation. Journal of Geophysical Research, 94, 14485-14498.

Barnett, D. (1991) Sea ice distribution in the Soviet Arctic. In: Barnett, D. (ed.), The Soviet Maritime Arctic, Belhaven Press, London, 47-62.

Battarbee, R.W. (1973) A new method for estimation of absolute microfossil numbers, with reference especially to diatoms. Limnology and Oceanography, 18, 647-653.

Cauwet, G. and I. Sidorov (1996) The biogeochemistry of Lena River: organic carbon and nutrients distribution. Marine Chemistry, 53, 211-227.

Churun, V.N. and L.A. Timokhov (1995) Cold bottom water in the southern Laptev Sea. In: Russian-German Cooperation: Laptev Sea System, Kassens, H., D. Piepenburg, J. Thiede, L. Timokhov, H.-W. Hubberten and S.M. Priamikov (eds.), AWI Bremerhaven, Bremerhaven (Reports on Polar Research 176), 107-113.

Clark, D.L. (1990) Arctic Ocean ice cover; geologic history and climatic significance. In: The Arctic Ocean Region, Grantz, A., L. Johnson and J.F. Sweeney (eds.), The Geological Society of America, Boulder (The Geology of North America, Vol. L), 53-62.

Cleve, P.T. (1883) Diatoms collected during the expedition of the Vega. Ur Vega-Expeditionens Vetenskapliga Jakttagelser, 3, 457-517.

Cleve, P.T. and A. Grunow (1880) Kenntnis der arktischen Diatomeen. Kongliga Svenska Vetenskaps-Akademiens Handlingar, 17, 1-121.

Codispoti, L.A. and F.A. Richards (1968) Micronutrient distributions in the East Siberian and Laptev Seas during summer 1963. Arctic, 21, 67-83.

Cremer, H. (1998) Diatoms in the Laptev Sea (Arctic Ocean): Taxonomy and biogeographic distribution. AWI Bremerhaven, Bremerhaven (Reports on Polar Research 260), 225 pp.

Dethleff, D. (1995) Die Laptevsee - eine Schlüsselregion für den Fremdstoffintrag in das arktische Meereis. Kiel, 111 pp (unpublished PhD thesis).

Drachev, S.S., L.A. Savostin and I.E. Bruni (1995) Structural pattern and tectonic history of the Laptev Sea region. In: Russian-German Cooperation: Laptev Sea System, Kassens, H., D. Piepenburg, J. Thiede, L. Timokhov, H.-W. Hubberten and S.M. Priamikov (eds.), AWI Bremerhaven, Bremerhaven (Reports on Polar Research 176), 348-366.

Durbin, E.D. (1974) Studies on autecology of the marine diatom *Thalassiosira nordenskioeldii* Cleve. 1. The influence of daylight, light intensity, and temperature on growth. Journal of Phycology, 10, 220-225.
Durbin , E.D. (1978) Aspects of the biology of resting spores of *Thalassiosira nordenskioeldii* and *Detonula confervacea*. Marine Biology 45: 31-37.
Ehrenberg, C.G. (1832) Beiträge zur Kenntnis der Organisation der Infusorien und ihrer geographischen Verbreitung besonders in Sibirien. Abhandlungen der Königlichen Akademie der Wissenschaften zu Berlin, 1830, 1-88.
Ehrenberg, C.G. (1943) Verbreitung und Einfluß des mikroskopischen Lebens in Süd- und Nordamerika. Abhandlungen der Königlichen Akademie der Wissenschaften zu Berlin, 1841, 291-446.
Eicken, H., E. Reimnitz, V. Alexandrov, T. Martin, H. Kassens and T. Viehoff (1997) Sea-ice processes in the Laptev Sea and their importance for sediment export. Continental Shelf Research, 17, 205-233.
Gordeev, V.V. and I.S. Sidorov (1993) Concentrations of major elements and their outflow into the Laptev Sea by the Lena River. Marine Chemistry, 43, 33-45.
Gordeev, V.V., J.M. Martin, I.S. Sidorov and M.V. Sidorova (1996) A reassessment of the Eurasian river input of water, sediment, major elements, and nutrients to the Arctic Ocean. American Journal of Science, 296, 664-691.
Gordienko, P.A. and A.F. Laktionov (1969) Circulation and physics of the Arctic Basin waters. Annals of the International Geophysical Year, Oceanography, 46, 94-112.
Gran, H.H. (1904a) Die Diatomeen der arktischen Meere. Die Diatomeen des Planktons. Fauna Arctica, 3, 510-554.
Gran, H.H. (1904b) Diatomaceae from the ice-floes and plankton of the Arctic Ocean. In: The Norwegian North Pole Expedition 1893-1896, Scientific results, Nansen, F. (ed.), F. Nansen Fund for the Advancement of Science, London, 1-85.
Grunow, A. (1884) Die Diatomeen von Franz Josefs-Land. Denkschriften der Kaiserlichen Akademie der Wissenschaften, Mathematisch-Naturwissenschaftliche Classe, 48, 53-112.
Hasle, G.R. (1976) The biogeography of some marine planktonic diatoms. Deep Sea Research, 23, 319-338.
Hasle, G.R. and B.R. Heimdal (1968) Morphology and distribution of the marine centric diatom *Thalassiosira antarctica* Comber. Journal of the Royal Microscopical Society London, 88, 357-369.
Hasle, G.R. and C.B. Lange (1989) Freshwater and brackish water *Thalassiosira* (Bacillariophyceae): taxa with tangentially undulated valves. Phycologia, 28, 120-135.
Hasle, G.R. and E.E. Syvertsen (1996) Marine diatoms. In: Identifying Marine Diatoms and Dinoflagellates, Tomas, C.R. (ed.), Academic Press, San Diego, 5-386.
Hasle, G.R., E.E. Syvertsen and C.A. von Quillfeldt (1996) *Fossula arctica* gen. nov., spec. nov., a marine Arctic araphid diatom. Diatom Research, 11, 261-272.
Heiskanen, A.-S. and A. Keck (1996) Distribution and sinking rates of phytoplankton, detritus, and particulate biogenic silica in the Laptev Sea and Lena River (Arctic Siberia). Marine Chemistry, 53, 229-245.
Holmes, M.L. (1967) Late Pleistocene and Holocene history of the Laptev Sea. University of Washington, Washington, D.C., 99 pp (unpublished PhD thesis).
Holmes, M.L. and J.S. Craiger (1974) Holocene History of the Laptev Sea. In: Marine Geology and Oceanography of the Arctic Ocean, Herman, Y. (ed.), Springer Verlag, New York, 211-229.
Hustedt, F. (1930) Die Kieselalgen Deutschlands, Österreichs und der Schweiz. Koeltz Scientific Books, Champaign, 920 pp (Reprint 1991).
Imbrie, J. and N.G. Kipp, N.G. (1971) A new micropaleontological method for quantitative paleoclimatology: Application to a late Pleistocene Caribbean core.- In: The late Cenozoic glacial ages, Turekian, K. (ed.), Yale University New Press, New Haven, 71-181.
Ivanov, V.V. and A.A. Piskun (1995) Distribution of river water and suspended sediments in the river deltas of the Laptev Sea. In: Russian-German Cooperation: Laptev Sea System, Kassens, H., D. Piepenburg, J. Thiede, L. Timokhov, H.-W. Hubberten and S.M. Priamikov (eds.), AWI Bremerhaven, Bremerhaven (Reports on Polar Research 176), 142-153.
Karpiy, V.Y., N. Lebedev and A. Ipatov (1994) Thermohaline and dynamic water structure in the Laptev Sea.- In: Russian-German Cooperation: The Transdrift I Expedition to the Laptev Sea, Kassens, H. and V.Y. Karpiy (eds.), AWI Bremerhaven, Bremerhaven (Reports on Polar Research 151), 16-47.
Kassens, H. (1995) Laptev Sea System: Expeditions in 1994. AWI Bremerhaven, Bremerhaven (Reports on Polar Research 182), 195 pp.
Kassens, H. (1997) Laptev Sea System: Expeditions in 1995. AWI Bremerhaven, Bremerhaven (Reports on Polar Research 248), 210 pp.
Kassens, H. and V.Y. Karpiy (1994) Russian-German Cooperation: The Transdrift I Expedition to the Laptev Sea. AWI Bremerhaven, Bremerhaven (Reports on Polar Research 151), 168 pp.
Kassens, H., D. Piepenburg, J. Thiede, L. Timokhov, H.-W. Hubberten and S.M. Priamikov (1995) Russian-German Cooperation: Laptev Sea System. AWI Bremerhaven, Bremerhaven (Reports on Polar Research 176), 387 pp.
Kisselew, J.A. (1932) Das Planktonmaterial aus dem südöstlichen Teile des Laptev-Meeres. Explorations des mers d´USSR, 15, 67-103 (in Russian with German abstract).

Koç, N., E. Jansen and H. Haflidason (1993) Paleoceanographic reconstructions of surface ocean conditions in the Greenland, Iceland and Norwegian Seas through the last 14 ka based on diatoms. Quaternary Science Reviews, 12, 115-140.
Koç Karpuz, N. and H. Schrader (1990) Surface sediment diatom distribution and Holocene paleotemperature variations in the Greenland, Iceland and Norwegian Sea. Paleoceanography, 5, 557-580.
Komarjenko, L.E. and I.I. Vasiljeva (1975) Freshwater diatoms and blue-green algae of Yakutia. Nauka, Moskau, 422 pp (in Russian).
Létolle, R., J.M. Martin, A.J. Thomas, V.V. Gordeev, S. Gusarova and I.S. Sidorov (1993) ^{18}O abundance and dissolved silicate in the Lena delta and Laptev Sea (Russia). Marine Chemistry, 43, 47-64.
Lindemann, F. (1994) Sonographische und sedimentologische Untersuchungen in der Laptevsee, sibirische Arktis. Universität Kiel, Kiel, 75 pp (unpublished diploma thesis).
Martin, J.M., D.M. Guan, F. Elbaz-Poulichet, A.J. Thomas and V.V. Gordeev (1993) Preliminary assessment of the distributions of some trace elements (As, Cd, Cu, Fe, Ni, Pb and Zn) in a pristine aquatic environment: the Lena river estuary (Russia). Marine Chemistry, 43, 185-199.
Matthiessen, J. (1994) Distribution of marine palynomorph assemblages in recent sediments from the Norwegian-Greenland Sea.- Neues Jahrbuch für Geologie und Paläontologie Abhandlungen, 194, 1-24.
Maynard, N.G. (1976) Relationship between diatoms in surface sediments of the Atlantic Ocean and the biological and physical oceanography of overlying waters. Paleobiology, 2, 99-121.
McQuoid, M.R. and L.A. Hobson (1995): Importance of resting stages in diatom seasonal succession. Journal of Phycology, 31, 44-50.
Meunier, A. (1910) Campagne Arctique de 1907. Microplankton des Mers de Barents et de Kara. Imprimerie Scientifique, Bruxelles, 355 pp.
Nürnberg, D., I. Wollenburg, D. Dethleff, H. Eicken, H. Kassens, T. Letzig, E. Reimnitz and J. Thiede (1994) Sediments in the Arctic sea ice: Implications for entrainment, transport and release. Marine Geology, 119, 185-214.
Østrup, E. (1895) Marine Diatomeer fra Ostgronland. Meddelelser om Gronland, 18, 396-476.
Paasche, E. (1975) Growth of the plankton diatom *Thalassiosira nordenskioeldii* Cleve at low silicate concentrations. Journal of Experimental Marine Biology and Ecology, 18, 173-183.
Pankow, H. (1990): Ostsee-Algenflora. Gustav Fischer Verlag, Jena, 648 pp.
Pichon, J.J., L.D. Labeyrie, J. Duprat and J. Jouzel (1992) Surface water temperature changes in the high latitudes of the southern hemisphere over the last glacial-interglacial cycle. Paleoceanography, 7, 289-318.
Peulvé, S., M.-A. Sicre, A. Saliot, J.W. de Leeuw and M. Baas (1996) Molecular characterization of suspended and sedimentary organic matter in an Arctic delta. Limnology and Oceanography, 41, 488-497.
Pivovarov, S.V. and V.M. Smagin (1995) The distribution of oxygen and nutrients in the Laptev Sea in summer. In: Russian-German Cooperation: Laptev Sea System, Kassens, H., D. Piepenburg, J. Thiede, L. Timokhov, H.-W. Hubberten and S.M. Priamikov (eds.), AWI Bremerhaven, Bremerhaven (Reports on Polar Research 176), 135-141.
Poulin, M. (1990) Ice diatoms: the Arctic. In: Polar marine diatoms, L.K. Medlin and J. Priddle (eds.), British Antarctic Survey, Cambridge, 15-18.
Reimnitz, E., D. Dethleff and D. Nürnberg (1994) Contrasts in Arctic shelf sea-ice regimes and some implications: Beaufort Sea versus Laptev Sea. Marine Geology, 119, 215-225.
Rossak, B. (1995) Zur Tonmineralverteilung und Sedimentzusammensetzung in Oberflächensedimenten der Laptevsee, sibirische Arktis. Universität Kiel, Kiel, 95 pp (unpublished diploma thesis).
Sancetta, C. (1981) Oceanographic and ecologic significance of diatoms in surface sediments of the Bering and Okhotsk seas. Deep-Sea Research, 28 A, 789-817.
Sancetta, C. (1982) Distribution of diatom species in surface sediments of the Bering and Okhotsk Seas. Micropaleontology, 28, 221-257.
Schandelmeier, L. and V. Alexander (1981) An analysis of the influence of ice on spring phytoplankton population structure in the southeast Bering Sea. Limnology and Oceanography, 26, 935-943.
Schrader, H.-J. and R. Gersonde (1978) Diatoms. Utrecht Micropaleontological Bulletin, 17, 129-176.
Schrader, H. and N. Koç Karpuz (1990) Norwegian - Icealnd Seas: Transfer functions between marine planktic diatoms and surface water temperature.- In: Geological History of the Polar Oceans: Arctic versus Antarctic, Bleil, U. and J. Thiede (eds.), Kluwer Academic Publishers, Dordrecht (NATO ASI Series C), 337-361.
Schuette, G. and H. Schrader (1979) Diatom taphocoenosis in the coastal upwelling area off western South America. Nova Hedwigia Beihefte, 64, 359-378.
Shirshov, P.P. (1937) Seasonal changes of the phytoplankton of the polar seas in connection with the sea ice regime. Transactions of the Arctic and Antarctic Science Institute Leningrad, 82, 47-111 (in Russian).
Smetacek, V.S. (1985) Pole of sinking in diatom life-history cacles: ecological, evolutionary and geological significance. Marine Biology, 84, 239-251.
Steele, M., D. Thomas and D. Rothrock (1996) A simple model study of Arctic Ocean freshwater balance, 1979-1985. Journal of Geophysical Research, 101, 20833-20848.
Syvertsen, E.E. (1990) Ice algae in the Barents Sea: types of assemblages, origin, fate and role in the ice-edge phytoplankton bloom. Polar Research, 10, 277-287.

Treshnikov, A.F. (1985) Atlas of the Arctic. Main Department of Geodesy and Cartography under the Council of Ministers of the USSR, Moscow, 204 pp (in Russian).
Usachev, P.I. (1946) Phytoplankton collected by the "Sedov"-Expedition 1937-1939. Proceedings of the Glavsevmorput Drift Expedition on the G. Sedov Icebreaker, 3, 371-397 (in Russian with English abstract).
Williams, K.M. (1986) Recent Arctic marine diatom assemblages from bottom sediments in Baffin Bay and Davis Strait. Marine Micropaleontology, 10, 327-341.
Zielinski, U. (1993) Quantitative estimation of palaeoenvironmental parameters of the Antarctic surface water in the late Quaternary using transfer functions with diatoms. AWI Bremerhaven, Bremerhaven (Reports on Polar Research 126), 148 pp.

Diatoms from Surface Sediments of the Saint Anna Trough (Kara Sea)

R.N. Djinoridze[1], G. I. Ivanov[2], E. N. Djinoridze[2], and R. F. Spielhagen[3]

(1) St. Petersburg State University, 199004 St. Petersburg, Russia
(2) VNII Okeangeologia, 190121 St. Petersburg, Russia
(3) GEOMAR Forschungszentrum für marine Geowissenschaften, Wischhofstrasse 1-3, D 24148 Kiel, Germany

Received 7 April 1997 and accepted in revised form 3 March 1998

Abstract - 15 samples from surface sediments of the St. Anna Trough served as the basis of this study. 91 recent species (29 genera) were identified, of which 30 belong to the class *Centrophyceae* and 61 belong to the class *Pennatophyceae*. Besides that 11 reworked Paleogene species were found. True marine planktic diatom species constitute 92-98% of all specimens in the surface sediments. The biogeographical composition of the thanatocoenoses corresponds to biocoenoses inhabiting the water masses of the Arctic seas. In the surface sediments, arctoboreal and bipolar planktic and kryopelagic species are prevailing, which reflect the low temperatures and the seasonally ice-covered environment: *Melosira arctica, Fragilariopsis oceanica, Poroira glacialis, Thalassiosira gravida, T. antarctica* (spores), *T. kryophila, T. hyalina*, and *T. nordenskiöldii*. Warm-water species are absent in the sediments, probably because Atlantic waters have cooled significantly before entering the St. Anna Trough.

Introduction

The St. Anna Trough is a subbathyal depression with depths down to 620 m, separating the northern Kara Sea and Barents Sea shelves (Figure 1). The hydrological regime (Rudels et al., 1994) of the northern Kara Sea is controlled by the inflow of several water masses into the St. Anna Trough area. Atlantic waters derived from the North Atlantic Drift flow across the Barents Sea, further along the St. Anna Trough, and finally enter the intermediate waters of the Arctic Ocean. Another branch of Atlantic water, turning east around northern Svalbard and sinking below the low-salinity surface layer while flowing along the Barents Sea continental margin, reaches the St. Anna Trough from the North at depths of 100-450 m. Another important water mass is constituted from freshwater transported mainly by the large Siberian rivers Ob and Yenisei. According to data obtained during cruise 9 of RV "Professor Logachev" in 1994, the upper water layer in the central and southern St. Anna Trough has temperature between 1-2.4 °C, while in the North and Northeast they vary from -0.1 to -1.2 ° C (Ivanov et al., 1995).

The field of salinity on ocean surface tests significant freshening due to summer melting of floating ice. Salinity of surface waters in the trough changed in summer 1994 in the limits from 30.00 up to 33.75‰.

Materials and methods

Diatoms were studied in 15 samples of surface sediments collected during cruise 9 (1994) of RV "Professor Logachev" in the St. Anna Trough from stations with water depths of 100-605 m (Figure 1). The treatment of diatom valves was carried out only slightly changed according to the method described by Schrader and Gersonde (1978). Dried and weighed sediment (15-20 g) was boiled first with H_2O_2 (30%), then with HCl (30%) and then was washed out by distilled water. The sediment was cleaned of clay by washing out 5-6 times with Na_3PO_4 *

In: Kassens, H., H.A. Bauch, I. Dmitrenko, H. Eicken, H.-W. Hubberten, M. Melles, J. Thiede and L. Timokhov (eds.) Land-Ocean Systems in the Siberian Arctic: Dynamics and History. Springer-Verlag, Berlin, 1999, 553-560.

$12H_2O$ (5%). To extract the diatom frustules, the sediment was treated by heavy liquid (2.6 g/cm^3) according to the method described by Glejzer et al. (1974). The counting of diatoms per 1 gram of dry sediment was done according to a formula proposed by Kvasov and Jakovshikova (1971):

$K= b * n * y / a * m * x$
where
b = dilution in ml
n = the number of rows in a slide
y = the number of counted frustules
a = weight of sediment in grams
m = the volume of a drop (0.02-0.03 ml)
x = the number of analyzed rows.

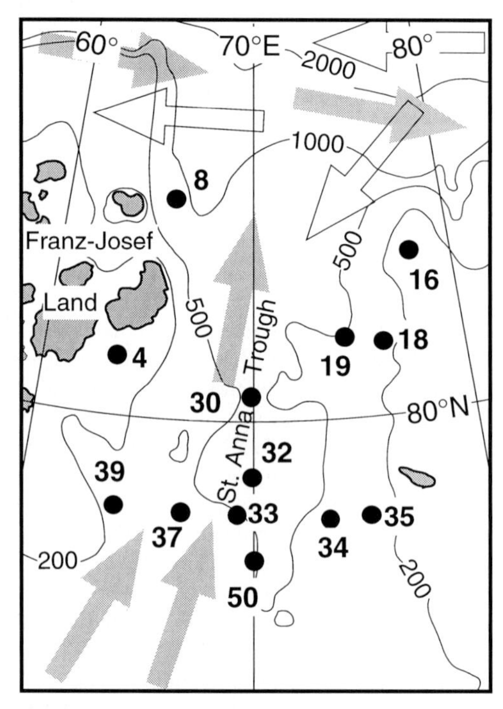

Figure 1: Core locations and currents in the St. Anna Trough area. Bathymetry is given in meters below sea level. White arrows mark the flow direction of low saline Arctic surface water; gray arrows mark the flow of Atlantic water (submerged north of 80°N).

Results and Discussion

Bottom sediments of the northern part of the Kara Sea contain relatively low amounts of diatoms. The number ranges from 10 to 750 frustules per gram of dry sediment. The silty sands from stations 4, 16, 19, and 30 contain rich assemblages of recent diatoms with minor admixtures of Paleogene species. Silty muds from stations 32, 33, 34, and 35 contain a poor composition of modern and Paleogene diatoms, and in cores 8, 18, 37, 39, and 50 they contain only reworked Paleogene species. 93 recent species and varieties (29 genera) and 11 reworked species (6 genera) were identified. Among the recent diatoms, 30 taxa belong to the class *Centrophyceae* and 63 taxa belong to the class *Pennatophyceae*.

The representatives of phytoplankton are mainly centric species such as: *Thalassiosira gravida* and *Th. antarctica* (16-25%), *T. nordenskiöldii* (10-16%),T. *kryophila* (16%), *Th. hyalina* (1-8%), *Porosira glacialis* (20%) and *Coscinodiscus oculus-iridis* (10-11%). The second group, present in lower amounts, is a group of kryopelagic and kryointersticial species, dwelling on lower and marginal sea ice surfaces: *Melosira arctica* (2-4%), *Fragilariopsis oceanica* (2-12%), *F. cylindrus* (>1%), *Nitzschia frigida* (2%), *N. polaris* (>1%), *Navicula cancelata* var. *gregori* (2%), *Nav. directa* (1%), *Nav. reinhardtii* var. *tschuktschorum* (4%). Species of the genera *Thalassiosira* (9 taxa), *Navicula* (11 taxa), *Diploneis* (10 taxa) and *Nitzschia* (10 taxa) are the most numerous.

All species found in sediments from the St. Anna Trough have been found previously already in the Kara Sea and are enclosed in the list of Kara Sea diatoms (148 species and varieties) compiled by Makarevich and Koltsova (1989) from own investigations and literature data (Zabelina, 1930, 1946; Kiselev, 1935; Usachev, 1938, 1949, 1968). Observations of ice diatoms the deep Arctic Basin by Horner (1982) showed that the bloom near in the ice margin mainly consists of pennatic diatoms, whereas the bloom of phytoplankton in ice-free areas is dominated by centric diatoms. *Fragilariopsis oceanica* and *F. cylindrus* are abundant in both habitats. In spring plankton, the most abundant species are *Thalassiosira gravida*, *T. nordensköldii*, *Fragilariopsis oceanica* and *Porosira glacialis* (Usachev,1968; Grant and Horner, 1976). All these species are dominant in the assemblage composition in surface sediments of the St. Anna Trough. They are also typical for diatom compositions found in sediments under ice packs (Sancetta, 1982; Williams, 1986).

The true marine diatoms in the described assemblages constitute 92-98% (Figure 2). Brackishwater euryhaline species, dwelling on sea ice are also present (up to 7%). Species of this group are: *Thalassiosira hyperborea* var. *septentrionalis*, *Achnanthes taeniata* , *A. brevipes*, *Diploneis interrupta*, *D. stroemii*, *D. smithii*, *Nitzschia hybrida* var. *kryokonites*, *Amphora laevis* var. *laevissima*, *Navicula kjellmanii*. Freshwater species such as: *Aulocoseira islandica* subsp. *helvetica*, *A. granulata*, and *Amphora ovalis* contribute up to 3% of the sediment assemblage.

The existence of brackishwater species may be explained by a freshening effect of melting sea ice. The salinity in a thin layer below the sea ice can be strongly variable from 15 to 29‰ (Melnikov, 1989). This fact may allow algae with different salinity tolerances to exist below sea ice. In 1994 during cruise 9 of RV "Professor Logachev" the surface water salinity at some stations had decreased to 13-16‰, probably from increased atmospheric precipitation. The presence of freshwater diatoms in the assemblages allows to suggest that they had lived in sea ice, which was formed in a proximal freshwater-dominated environment and then drifted to the open sea. A current transport from the southern Kara Sea to the St. Anna Trough seems rather unlikely.

In sediments from the majority of stations, the frustules of planktic, mostly neritic species are prevailing (92-98%; Figure 2). Panthalassic species constitute up to 12% (*Coscinodiscus oculus-iridis*, *C. curvatulus* var. *kariana*, *Actinocyclus curvatulus*). In samples from stations with a water depth of about 100 m, the littoral species constitute up to 30%. This may be explained by a mechanism of ice flora forming on account of planktic and benthic species (Melnikov, 1989). In shallow waters the species distribution and quantitative composition of ice algae is dominated by benthic species, whereas in deep sea regions it is dominated by planktic species.

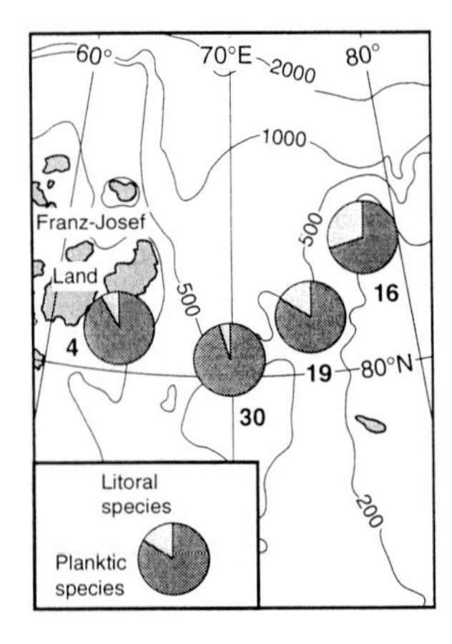

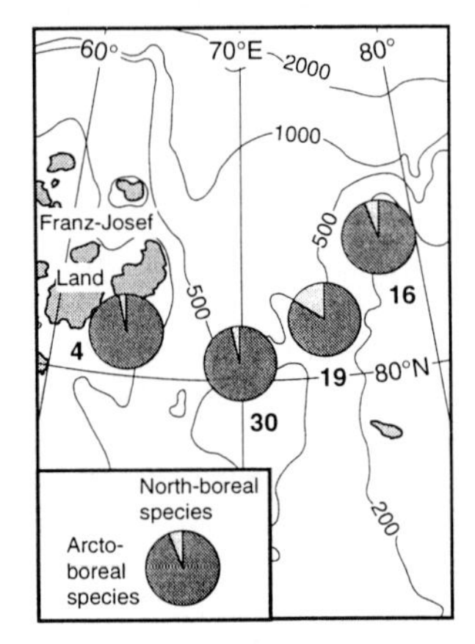

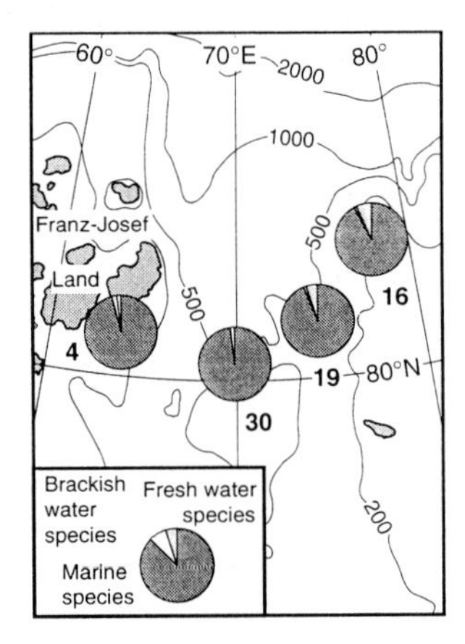

Figure 2: Diatom species composition of surface samples from stations 4, 16, 19, and 30 according to habitat (left), biogeography (center), and facies types (right).

In general, the composition of thanatocoenoses in the analyzed sediments corresponds to the biogeography of biocoenoses inhabiting the water masses of Arctic shelves. Bipolar and arctoboreal diatoms constitute 83-97% (Figure 2). In this region the temperatures of the surface waters reach 0-1°C in summer and sink to -1.8 ° C in winter.

The diatom assemblage of analogous species composition was established by Karpuz and Schrader (1990) in surface sediments of the Nordic Seas by factor analysis. They called it sea-ice assemblage (Factor 3) and it is confined to maximum ice expansion in the Norwegian and Greenland seas.

The structure of the diatom assemblage in surface sediments from the St. Anna Trough is similar to the structure of the thanatocoenosis found in the northern regions of the Barents Sea (Djinoridze,1986). We propose that this is a consequence of the strong water exchange between these basins and the very similar hydrological characteristics of the predominant water masses.

Some differences are traced in the structure of thanatocoenoses in the St. Anna Trough (Figure 2). In sediments of the western part arctoboreal, bipolar planktic and kryopelagic species are more abundant. In sediments of the eastern part, bordering shoal waters, the littoral species have relatively higher percentages.

The results of diatom analysis of surface layer of bottom sediments in the St. Anna Trough reveal their low satiation with diatoms. Only 30% of the studied samples contain representative diatom assemblages (38-60 taxa) in numbers of 100-750 frustules per gram of sediment. Only a small part (about 25%) of the known species variety in the planktic diatom flora of the modern Kara Sea is preserved in the sediments. As a result, spores of only 5 out of 34 species of the genus *Chaetoceros* found in modern plankton were registered in the sediments.

Spores of *Melosira arctica*, *Porosira glacialis*, *Thalassiosira gravida*, and *T. antarctica* are most abundant in sediments of the St. Anna Trough. The generally low amount of diatoms, which is also noted by Polyakova (1997) for the southern Kara Sea, is probably a result of the unfavorable accumulation conditions, which promote the dissolution of diatom valves during their precipitation: Low silica concentrations, a pronounced density stratification of the waters, and strong bottom currents.

The occurrence of reworked Paleogene and Cretaceous species in the sediments is a peculiarity of the diatom thanatocoenoses of surface sediments in the St. Anna Trough. It indicates that the modern bottom sediments to a strong extent are formed from erosion of reworked underlying rocks.

References

Djinoridze, R. N. (1986) Diatom complexes from Ice-Fiord sediments, Svalbard (in Russian).Vestnik Leningradskogo universiteta, 7, N1,108-111.

Glejzer, Z. I., A. P. Juze, I. V. Makarova, et al. (eds.) (1974) The diatoms of the USSR, Fossil and recent (in Russian). Nauka Publishing House, Leningrad, 402 p.

Grant, W. S., and R. A. Horner (1976) Growth responses to salinity variation in four arctic ice diatoms. Phycol., 12, 180-185.

Grebennikova, T.A. (1989). Paleogeographical analysis of Late Quaternary conditions of sedimentation in the Japan Sea (using data diatoms investigation). Unpubl. Ph.D. thesis, Novosibirsk.

Horner, R. A. (1982) Do ice algae produce the spring phytoplankton bloom in seasonally ice-covered waters?. In: Diatom Symposium, Mann, D. G. (ed.), Otto Koeltz Science Publ., 401-409.

Ivanov, G. I., V. P. Shevchenko, and A. V. Nesheretov (1995) Cruise in the Saint Anna Trough. Priroda Mag., 10, 56-62.

Karpuz, N. K., and H. Schrader (1990) Surface sediment diatom distribution and Holocene paleotemperature variations in the Greenland, Iceland and Norwegian Sea. Paleoceanography, 5 (4), 557-580.

Kiselev, I. A. (1935) Some data on the phytoplankton of the north-eastern part of the Kara Sea (in Russian). Trudy Tajmyr. Hydro. exped. 1932, 2, 191-202.

Kvasov, D. D., and T.K. Jakovshikova (1971) Subdivision and dating of Upper Quaternary Caspian deposits according to data of diatom analysis and paleohydrochemistry (in Russian). Problemi periodizacii pleistocena, L., 304-307.

Makarevich, P. R., and T. I. Koltsova (1989) Microplankton of the Kara Sea (in Russian). In: Ecologia i bioresursi Karskogo morya, G. G. Matishov (ed.), Apatiti, 38-45.

Melnikov, Y. A (1989) Ecosystems of Arctic Sea Ice (in Russian). Academy of Sciences of the USSR, Moscow, 191 p.

Polyakova, E. I. (1997) The Eurasian Arctic Seas during the Late Cenozoic (in Russian). Scientific World Publishing House, Moscow, 146 pp.

Rudels, B., E. P. Jones, L. G. Anderson and G. Kattner (1994) On the intermediate depth waters of the Arctic Ocean. In: The Polar Oceans and Their Role in Shaping the Global Environment, Johannessen, O. M., R. D. Muench, J. E. Overland (eds.), Washington, American Geophysical Union, 33-46.

Sanccetta, C. (1982) Distribution of diatom species in surface sediments of the Bering and Okhotsk seas. Micropaleontology, 28, 221-257.

Schrader, H., and R. Gersonde (1978) Diatoms and silicoflagellates in the eight meters sections of the lower Pleistocene at Capo Rossello. Micropal. Bull., 17, 129-176.

Usachev, P. I. (1938) Biological analysis of sea ice (in Russian). Doklady AN SSSR, 19 (8), 643-646.

Usachev, P. I. (1949) The microflora of polar sea ice (in Russian). Trudy Inst. okeanologii AN SSSR, 3, 216-259.

Usachev, P. I. (1968) Phytoplankton of the Kara Sea (in Russian). In: Plankton Tikhogo okeana, G. I. Semina (ed.), , Nauka, 6-28.

Williams, K. M. (1986) Recent arctic marine diatom assemblages from bottom sediments in Baffin Bay and Davis Strait. Mar. Micropal., 10, 327-341.

Zabelina, M. M. (1930) Some new data on phytoplankton of the Kara Sea (in Russian). Issledovanie morej SSSR, 13, 105-143.

Zabelina, M. M. (1946) Phytoplankton of the south-western part of the Kara Sea (in Russian). Trudy AANII, 193, 45-73.

Appendix 1: Modern Diatoms

Genus and species name	Facies	Habitat	Biogeo-graphy	Station numbers								
				4	16	19	30	32	33	34	34d	35
Achnanthes arctica	m	l	ar		1	1						
Achnanthes brevipes	b	l	cp		1							
Achnanthes septata	m	l	ab	1			1					
Achnanthes taeniata	e	n	ar				1					

Appendix 1 (continued):

Genus and species name	Facies	Habitat	Biogeo-graphy	Station numbers								
				4	16	19	30	32	33	34	34d	35
Actinocyclus curvatulus	m	p	nb		2	1						
Amphora eunotia	m	l	ab				1					
Amphora laevis var. laevissime	b	l	ab	1			1					
Amphora ovalis	fr	l	cp			1	4					
Amphora proteus	m	l	ab	5	1	1	1					
Amphora terroris	m	l	ab	1	1		3				1	
Aulacosira granulata	fr	l	cp		1	7	1			1		1
Aulacosira islandica subsp. helvetica	fr	l	cp				1				3	
Aulacosira italica	fr	l	cp		1	10	2					
Bacillaria socialis	m	l	cp	3	1	8				1		2
Bacterosira fragilis	m	n	ab	40			20					
Campylodiscus thuretii	m	l	nb	1	1	5	14					
Chaetoceros debilis	m	n	ab				1					
Chaetoceros furcellatus	m	n	ab	5	2							
Chaetoceros karianus	m	n	ab				3					
Chaetoceros mitra	m	n	ab			1	1					
Chaetoceros diadema	m	n	ab	2	2	3	10					
Cocconeis californica	m	l	ab	3								
Cocconeis costata	m	l	cp		2	1						
Cocconeis placentula	fr	l	cp		1							
Cocconeis scutellum	m	l	cp	1			2					
Coscinodiscus asteromphalus	m	p	sb		1							
Coscinodiscus curvatulus var. kariana	m	p	ab		7	8	4			2	5	
Coscinodiscus marginatus	m	p	nb		1							
Coscinodiscus oculus-iridis	m	p	ab		10		6	1	1	10	4	10
Diploneis coffaeformis	m	l	nb				1					
Diploneis interrupta	m	l	ab		2	1	1					
Diploneis litoralis	m	l	ab		1							
Diploneis litoralis var. clathrata	m	l	ar	1			2					
Diploneis smithii	b	l	cp	22	3	2	10					1
Diploneis smithii var. borealis	b	l	cp	2	1	1	20					
Diploneis smithii var. rhombica	b	l	cp	1	2	1	1					
Diploneis stroemii	m	l	nb	1								
Diploneis subcincta	m	l	ab	31	12	15	5					1
Diploneis suborbicularis	m	l	nb		4							
Entomoneis hyperborea	m	n	ar	2								
Entomoneis kjellmanii var. kariana	m	n	ab	1								
Fragilaria lapponica	fr	l	cp		1							
Fragilariopsis cylindrus	m	n	ar	10	1	1	3					
Fragilariopsis oceanica	m	n	ar	64	3	3	57				1	
Gomphonema exguum var. arctica	m	l	ab				2					
Grammatophora angulosa	m	l	ab				10					
Grammatophora arctica	m	l	ar	3	1							
Hantzschia weyprechtii	m	l	ab				1					1
Lycmophora jurgensii	m	l	ab				1					
Melosira arctica	m	n	ar	4	20	50	30			1	1	2
Navicula abrupta	m	l	ab				4					
Navicula cancellata	m	l	ab				1					
Navicula cancellata var. gregori	m	l	ab				14					
Navicula directa	m	l	ab	1			5					
Navicula distans	m	l	ab	2								
Navicula glacialis	m	l	ar	2			1					
Navicula kariana	m	l	ab		1							
Navicula kjelmannii	m	l	ab	2								

Appendix 1 (continued):

Genus and species name	Facies	Habitat	Biogeo-graphy	Station numbers								
				4	16	19	30	32	33	34	34d	35
Navicula palpebralis var. angulosa	m	l	nb		2							
Navicula reichardtii var. tschuktschorum	m	l	ab		2		20					
Navicula spectabilis	m	l	nb		2	1						
Nitzschia frigida	m	n	ar	2			14					
Nitzschia hybrida	b	l	ar				6					
Nitzschia hybrida var. kryokonites	b	l	ab				2					
Nitzschia insignis	m	l	nb				3					
Nitzschia insignis var. arctica	m	l	ar	2								
Nitzschia mitcheliana	m	l	ab				1					
Nitzschia polaris	m	l	ar	1	3	1	1					
Nitzschia seriata	m	n	ab				1					
Nitzschia triblionella var. levidensis	b	l	cp			1						
Nitzschia triblionella var. victoriae	b	l	ar			1	7					
Odontella aurita	m	n	ab		2							
Paralia sulcata	m	l	nb		5	10						
Pinnularia lata	fr	l	na			1						
Pinnularia quadratarea var. baltica	m	l	nb		1		1					
Pinnularia quadratarea var. stuxbergii	m	l	ab		1		1					
Porosira glacialis	m	n	bp	140	12	20	60				1	2
Rhizosolenia hebetata f. hiemalis	m	p	ab		1	3						
Rhizosolenia setigera var. arctica	m	n	ar	1								
Scoliotropis laterostrata	m	l	ab	1								
Stephanodiscus rotula	fr	p	cp	1	1	1						
Synedra kamtschatica var. finmarchica	m	l	ab	1			2					
Tetracyclus lacustris	fr	l	na			1	1					
Thalassiosira angulata	m	n	nb	5	2	1						
Thalassiosira anguste-lineata	m	p	cp	4	10	10	5					2
Thalassiosira antarctica	m	n	bp	50	10	30	40				2	
Thalassiosira hyperborea	m	n	ab	4	2	2	24					2
Thalassiosira gravida	m	n	bp	51	10	20	30			1	2	3
Thalassiosira hyalina	m	n	ab	24	2		3			2		
Thalassiosira kryophila	m	n	ab	14	10	2	10			2		
Thalassiosira latimarginata	m	n	ab	1	2	1						
Thalassiosira nordenskiodii	m	n	ab	32		1	74					
Thalassiothrix longissima	m	n	ab	4	1	4	2					
Trachyneis aspera	m	l	ab	5	1	1	1					

ABBREVIATIONS
Facies: m = marine; b = brackishwater; fr = freshwater; e = euryhaline.
Habitat: n = neritic; p = panthalassic; l = littoral.
Biogeography: ar = arctic; ab = arcto-boreal; bp = bipolar; nb = north-boreal; sb = south-boreal; cp = cosmopolitan; na = north-alpic.
Absolute abundance of diatoms is given per slide (18x18 mm)

Appendix 2: Paleogene Diatoms

Genus and species name	Facies	Habitat	Biogeo-graphy	Station Numbers													
				4	16	19	30	32	33	34	34d	35	37	39	50	8	18
Coscinodiscus payeri	m	p	(r)							1					1		1
Costopyxis broschii	m	n	(r)				1						1		1		
Costopyxis schulzii	m	n	(r)												1		
Grunowiella gemmata	m	n	(r)					1			1		1				
Hemiaulus polymorphus	m	l	(r)			1			1								1
Hyalodiscus radiatus	m	l	ab			2		1	1	10	2	3				10	2
Paralia ornata	m	l	(r)										1			1	1
Paralia sulcata var. biseriata	m	l	(r)	11	2	22	10	1	7	5	11	30	1	8	1	12	15
Pyxidicula polaris	m	n	(r)											1		3	
Pyxidicula turris	m	n	(r)		1	1		1		1		2	1	1			
Pyxilla gracilis	m	n	(r)					1								6	

ABBREVIATIONS
Facies: m = marine
Habitat: n = neritic; p = panthalassic; l = littoral.
Biogeography: ab = arctoboreal; (r) = resedimented.
Absolute abundances of diatoms are given per slide (18x18 mm)

Distribution of Aquatic Palynomorphs in Surface Sediments from the Laptev Sea, Eastern Arctic Ocean

M. Kunz-Pirrung

Alfred-Wegener-Institut für Polar- und Meeresforschung, Postfach 120161, D 27515 Bremerhaven, Germany.

Received 11 March 1997 and accepted in revised form 5 March 1998

Abstract - Aquatic palynomorphs were studied in surface sediments from the Laptev Sea shelf and the adjacent continental slope. The Laptev Sea is characterized by a strong salinity gradient owing to an extreme freshwater influx from the Siberian rivers in summer. The assemblages are composed of various organic-walled microfossils, in particular dinoflagellate cysts, chlorococcalean algae, acritarchs and several groups of zoomorphs. The species composition and the distribution pattern of dinoflagellate cysts and chlorococcalean algae is related to the salinity gradient of the surface water from the coast to the continental slope. A distinct change in the dinoflagellate cyst assemblages occurs across the continental slope corresponding roughly to a salinity interval of 28 to 30. Assemblage I from the shelf is dominated by *Brigantedinium* spp., *Algidasphaeridium? minutum* and related morphotypes. *Impagidinium? pallidum, Nematosphaeropsis labyrinthus* and *Operculodinium centrocarpum* characterize assemblage II from the continental slope. The chlorococcalean algae assemblages are composed of *Pediastrum* spp. and *Botryococcus* cf. *braunii* which usually live in freshwater. Therefore, their occurrence in the marine environment reflects the freshwater discharge into the Laptev Sea. The strong gradient in chlorococcalean algae concentrations suggests that the influence of freshwater is confined mainly to the shelf.

Introduction

The Laptev Sea is one of the most important sea-ice source areas for the Arctic Ocean. Large amounts of sea ice are produced here during the winter months (e.g., Kassens and Karpiy 1994b). This ice cover affects the surface circulation in the Arctic by restricting the heat exchange between the ocean and the atmosphere. The ice masses, which can be sediment-laden, are transported with the Transpolar Drift across the Arctic Ocean (Figure 1) and exit through the Fram Strait into the Greenland and Iceland seas (Dethleff et al., 1993). In these areas the sea ice melts out in spring and summer.

The Laptev Sea is a broad shelf area with an average water depth of less than 50 to 60 meters. Five south to north oriented submarine valleys cross the shelf (Holmes and Creager, 1974) and link the mouths of the large Siberian rivers Khatanga, Anabar, Olenek, Lena and Yana with the continental slope. These rivers are very important for the hydrographic conditions in the Laptev Sea and the Arctic Ocean. During the short Arctic summer, enormous amounts of freshwater are discharged into the Laptev Sea. Therefore, sea-surface salinities show a strong gradient (Figure 2) from the river mouths to the continental slope, increasing from approximately 6 to 30 in summer (Dmitrenko et al.,1994; Treshnikov,1985). These surface water masses are underlain by more saline water masses with an average salinity of approximately 34.

Within the Russian-German multidisciplinary research projekt "Laptev Sea System" planktonic microfossils, especially diatoms (Cremer, 1998) and palynomorphs are investigated in order to provide paleoenvironmental information about changes in surface water mass distribution and freshwater discharge during the Holocene. In particular, dinoflagellate cysts (e.g., Mudie, 1992; De Vernal et al., 1994, 1996; Matthiessen, 1996) and chlorococcalean algae are important tools for paleoceanographic reconstructions in high-latitude marine

----

In: Kassens, H., H.A. Bauch, I. Dmitrenko, H. Eicken, H.-W. Hubberten, M. Melles, J. Thiede and L. Timokhov (eds.) Land-Ocean Systems in the Siberian Arctic: Dynamics and History. Springer-Verlag, Berlin, 1999, 561-575.

environments as their cyst walls and coenobia consist of very stable sporopollenin-like organic material. Thus, they are more resistent to degradation and dissolution than skeletons of siliceous and calcareous microfossils. The goal of this study is to document the principal composition and distribution of aquatic palynomorph assemblages in surface sediments from the Laptev Sea. Detailed information about the distribution pattern of the individual taxa is given in Kunz-Pirrung (1998).

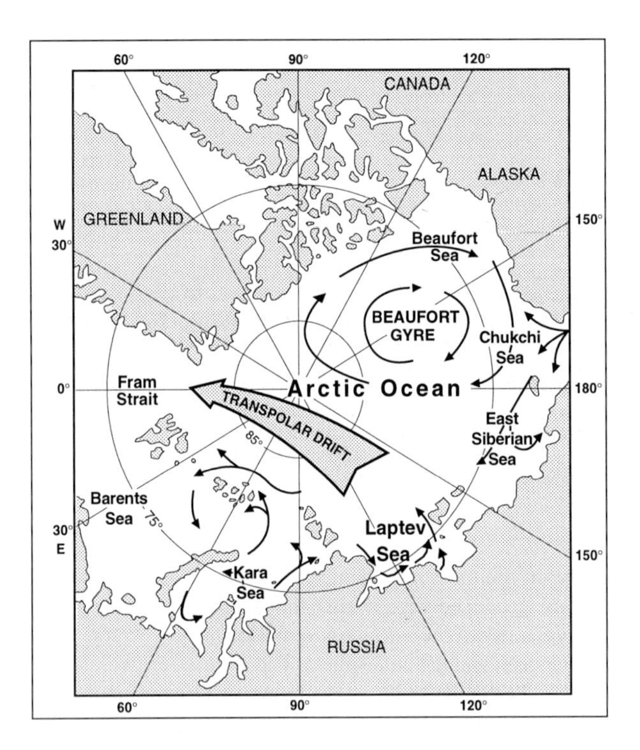

Figure 1: Location of the Laptev Sea and schematic surface currents in the Arctic Ocean (modified after Gordienko and Laktionov, 1969)

Material and methods

Ninetytwo surface sediment samples from the Laptev Sea shelf and the adjacent continental slope were investigated for their organic-walled microfossil contents. The samples were taken from box cores that were recovered during Transdrift Expeditions I/II/III in 1993/94/95 (Kassens and Karpiy, 1994a; Kassens, 1995; Kassens, 1997) and during the expedition ARK-IX/4 of RV "Polarstern" in 1993 (Fütterer, 1994). Samples from the uppermost centimeter of each of the cores were used for further analysis.

All samples were prepared using standard palynological preparation methods (Barss and Williams, 1973; Doher, 1980; Phipps and Playford, 1984). In order to avoid the loss of any peridinioid cyst no oxidation treatment was applied (Dale, 1976). During the preparation *Lycopodium clavatum* spore tablets were added to calculate cyst concentrations (cysts per gramme dry sediment) according to the method of Stockmarr (1971). A drop of the organic residue was mounted in glycerin jelly on microscropic slides.

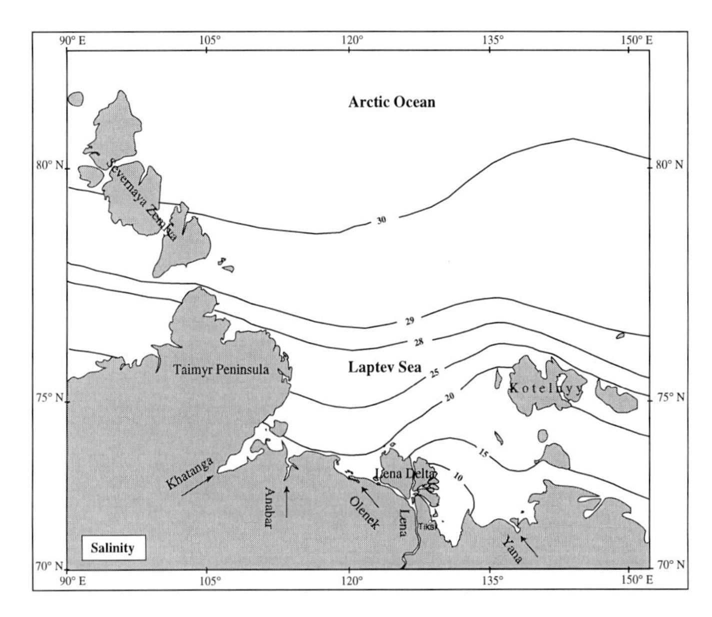

Figure 2: Principal distribution of sea surface salinity isobars during summer in the Laptev Sea (after Treshnikov, 1985)

Table 1: List of identified microfossils.

1. Dinoflagellate cysts
Algidasphaeridium? minutum s.l. (HARLAND and REID 1980) MATSUOKA and BUJAK 1988
Bitectatodinium tepikiense WILSON 1973
Brigantedinium spp. REID 1977
Brigantedinium cariacoense (WALL 1967) REID 1977
Brigantedinium simplex (WALL 1965) REID 1977
Impagidinium? pallidum BUJAK 1984
Nematosphaeropsis labyrinthus (OSTENFELD 1903) REID 1974
Operculodinium centrocarpum (DEFLANDRE and COOKSON 1955) WALL 1967
Cyst of *Pentapharsodinium dalei* INDELICATO and LOEBLICH 1986
Polykrikos? spp. BÜTSCHLI 1873
Cyst of *Protoperidinium denticulatum* (GRAN and BRAARUD 1935) BALECH 1974
Selenopemphix quanta (BRADFORD 1975) MATSUOKA 1985
Spiniferites elongatus/S. frigidus-group REID 1974, HARLAND and REID 1980

2. Chlorococcalean algae
Botryococcus cf. *braunii* KÜTZING 1849
Pediastrum boryanum (TURPIN 1828) MENEGHINI 1840
Pediastrum duplex MEYEN 1829
Pediastrum kawraiskyi SCHMIDLE 1897
Pediastrum simplex MEYEN 1829

3. Acritarchs
Acritarch Typ A
Halodinium spp. BUJAK 1984
Hexasterias problematica CLEVE 1900
Radiosperma corbiferum MEUNIER 1910

4. Several groups of zooplankton
foraminiferal test lining
ciliate cyst
tintinnid lorica

A minimum of 100 dinoflagellate cysts was analysed and counted under the light microscope using phase and differential interference contrasts. Taxa found in this study are listed in Table 1. Dinoflagellate cysts were classified according to the references cited in Lentin and Williams (1993). The taxonomy of the chlorococcalean algae followed Parra Barrientos (1979) and Matthiessen and Brenner (1996).

Distribution of assemblages

Palynomorph assemblages

The palynomorph assemblages consist of four main groups: dinoflagellate cysts, chlorococcalean algae, acritarchs, and several groups of zoomorphs (Figure 3). The chlorococcalean algae dominate the microfossil assemblages in the coastal area between the Anabar and the Yana river mouths. All groups are present in similar abundances on the shelf up to 75°30′N, whereas dinoflagellate cysts are more abundant on the continental slope.

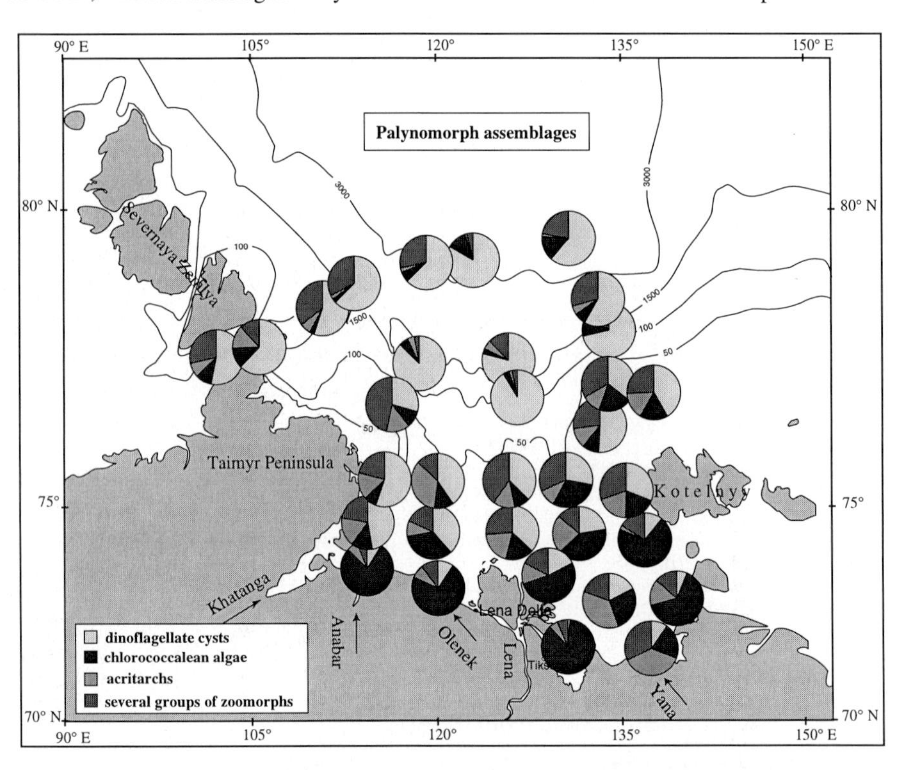

Figure 3: Composition of the palynomorph assemblages in surface sediments from the Laptev Sea. The relative abundances are calculated based on the sum of all organic-walled microfossils. Only selected stations are shown.

Together with the described four palynomorph groups there are, especially in front of the mouths of the large rivers a large number of terrestrial palynomorphs, e.g., pollen, spores, wood fragments and cuticules. Pollen and spores were transported fluviatile or eolian into the Laptev Sea. Therefore, together with the chlorococcalean algae they are useful indicators for the terrestrial input (Naidina and Bauch, in press).

This paper will focus on dinoflagellate cyst and chlorococcalean algae assemblages. Both groups show distinct distribution patterns which can be related to specific environmental conditions. In contrast, knowledge about the ecological requirements of the recorded acritarchs and zoomorph groups is relatively sparse and the distribution of most taxa in the Laptev Sea is heterogeneous.

Dinoflagellate cyst assemblages

Dinoflagellate cyst concentrations range from 3 to 4000 cysts per gramme dry sediment. The highest concentrations occur at the shelf break in the eastern part of the Laptev Sea (Figure 4). Dinoflagellate cysts are nearly absent off the Lena delta and the Yana river mouths. Relative abundances strongly decrease from the Kotelnyy Island to the Lena Delta and towards the southern coast (Figure 5).

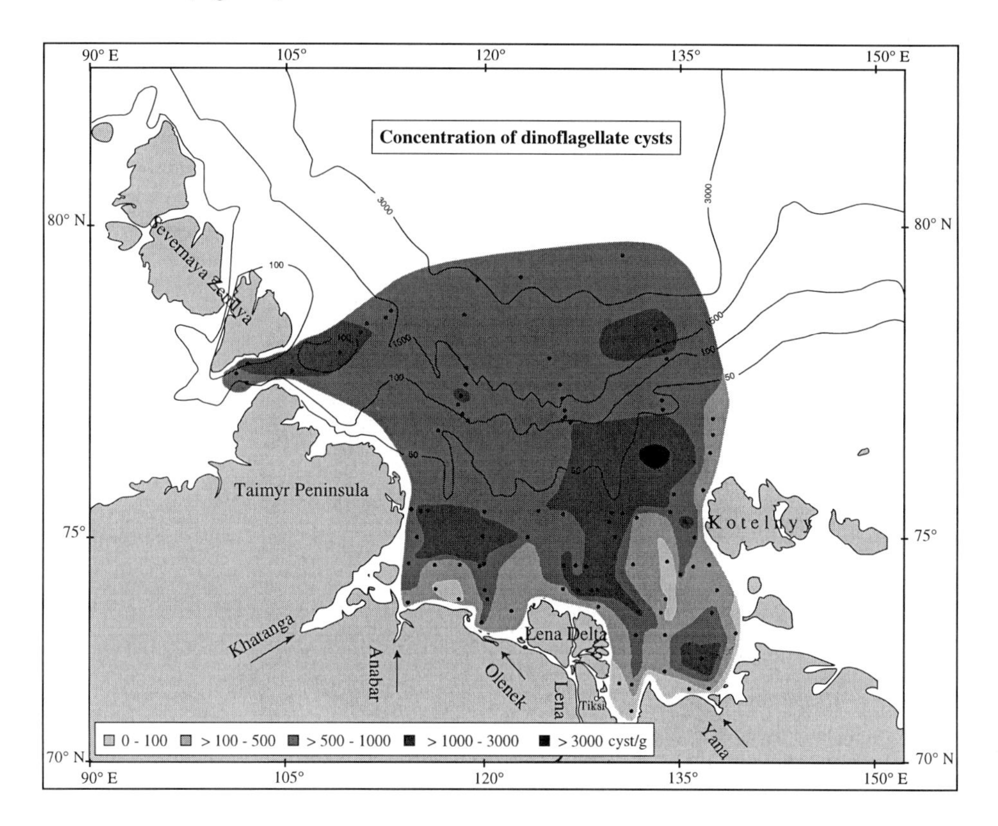

Figure 4: Concentrations of dinoflagellate cysts in surface sediments from the Laptev Sea.

Fourteen dinoflagellate cyst taxa were found in the surface sediments from the Laptev Sea shelf and the adjacent continental slope (Table 1). Relative abundances of individual species indicate the presence of two dinoflagellate cyst assemblages. The abundances of the more prominent taxa are shown in Figure 6.

Assemblage I from the shelf consists almost exclusively of *Brigantedinium* spp. (mainly *B. simplex*) and *Algidasphaeridium? minutum* and related morphotypes, with present to common occurrences of *Polykrikos?* spp. Other species are rare and occur mainly along the shelf break. Assemblage I is similar to assemblages that are characteristic for seasonally ice covered environments of the Arctic Ocean, the Labrador Sea, the Baffin Bay (Mudie, 1992), the Barents Sea (Harland, 1982), and the East Greenland continental shelf and slope (Matthiessen, 1995). However, assemblage I differs in having numerous morphotypes of *Algidasphaeridium?* and in containing morphotypes tentatively referred to *Polykrikos?*.

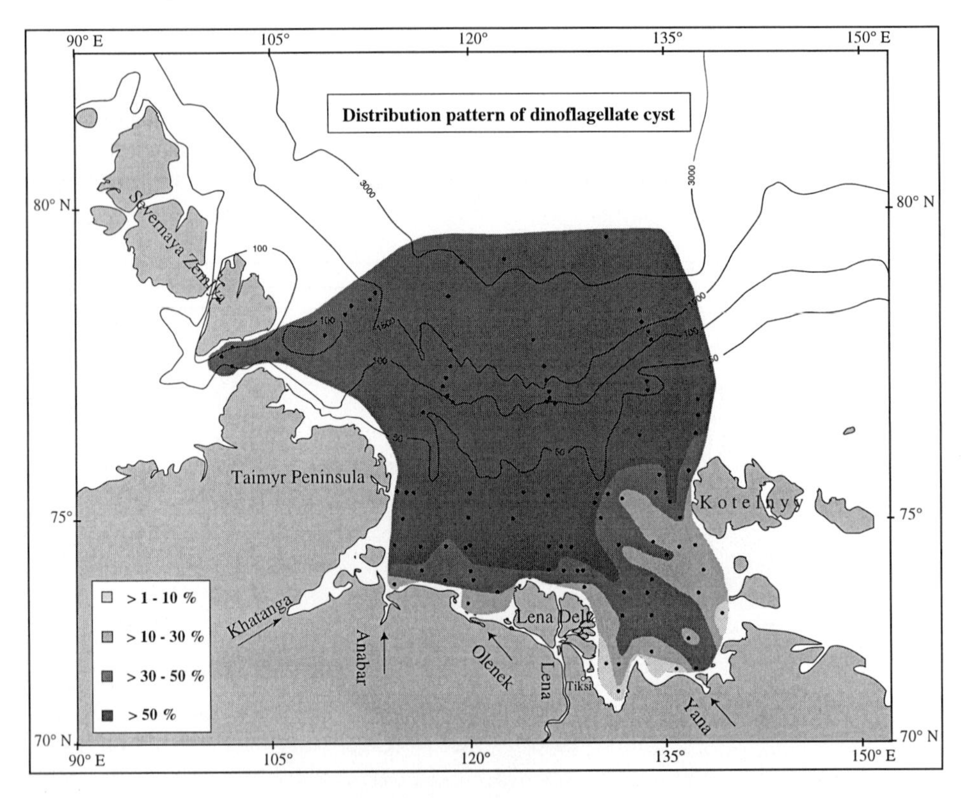

Figure 5: Distribution pattern of dinoflagellate cysts in surface sediments from the Laptev Sea. The relative abundances in figure 5 and 8 are calculated based on the sum of all dinoflagellate cysts and chlorococcalean algae. Following classes were distinguished: present (0-1%), rare (1-10%), common (10-30%), abundant (30-50%), dominant (>50%).

Assemblage II from the continental slope is characterized by *Impagidinium? pallidum*, *Nematosphaerosis labyrinthus* and *Operculodinium centrocarpum*. These species are inversely related to *Brigantedinium* spp. and *Algidasphaeridium*? spp. which are still abundant, but strongly decrease with distance from the shelf (Figure 6). Assemblage II resembles in composition assemblages from the Iceland and Greenland seas (Matthiessen, 1995).

Chlorococcalean algae assemblages

The chlorococcalean algae concentrations range from 7 to 3800 per gramme dry sediment. In contrast to the dinoflagellate cysts, highest concentrations are found in front of the mouths of the large rivers and within the submarine valleys. Concentrations steadily decrease towards the shelf break (Figure 7). A peculiar feature is the distinct west to east gradient across the Laptev Sea. High concentrations as well as high relative abundances (Figure 8) are associated with freshwater plumes that originate from the Anabar and Olenek rivers and in particular from the Lena and Yana rivers.

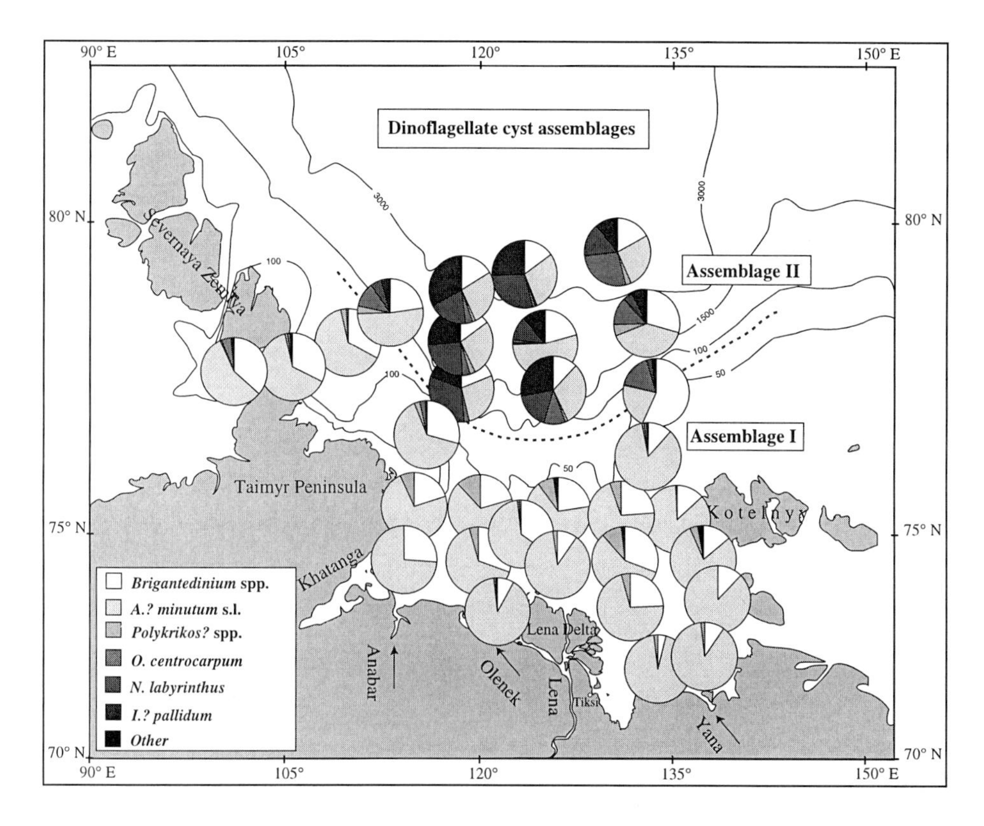

Figure 6: Composition of the dinoflagellate cyst assemblages in surface sediments from the Laptev Sea. Selected stations are only shown.

The chlorococcalean algae assemblages are dominated by *Pediastrum* spp. and *Botryococcus* cf. *braunii*. Within the genus *Pediastrum*, *P. boryanum* is dominant in most samples, *P. kawraiskyi* is common and *P. duplex* and *P. simplex* are rare. The genera *Pediastrum* and *Botryococcus* are freshwater forms of which *P. boryanum*, *P. kawraiskyi* and *B.* cf. *braunii* are the only species that tolerate slightly higher salinities (Matthiessen and Brenner, 1996). The high relative abundances and concentrations in the Laptev Sea are due to *Pediastrum* spp. In contrast, *B.* cf. *braunii* is the dominant species in samples from the continental slope.

Other aquatic palynomorphs

Four acritarch taxa have been recognized in the surface sediments from the Laptev Sea. Acritarchs are organic-walled microfossils with unknown biological affinity (Evitt, 1985). In this study, *Halodinium* spp., *Radiosperma corbiferum* and *Hexasterias problematica* have been recognized. The highest concentrations of *Halodinium* occur on the shelf, whereas on the continental slope *Halodinium* appears sporadically. Another important acritarch taxon is *Radiosperma corbiferum* in the surface sediments from the Laptev Sea. The highest values are observed north and east of the Lena delta and in front of the Yana river mouth. In the western part of the shelf and on the continental slope *Radiosperma corbiferum* is rare. The zoomorph group consists of foraminiferal test linings, ciliate cysts and tintinnid loricae. Foraminiferal test linings are the inner organic layer of benthic calcareous foraminifers (e.g., De Vernal et al., 1992). The highest concentrations of foraminiferal test linings occur by water depth < 100 m on the Laptev Sea shelf.

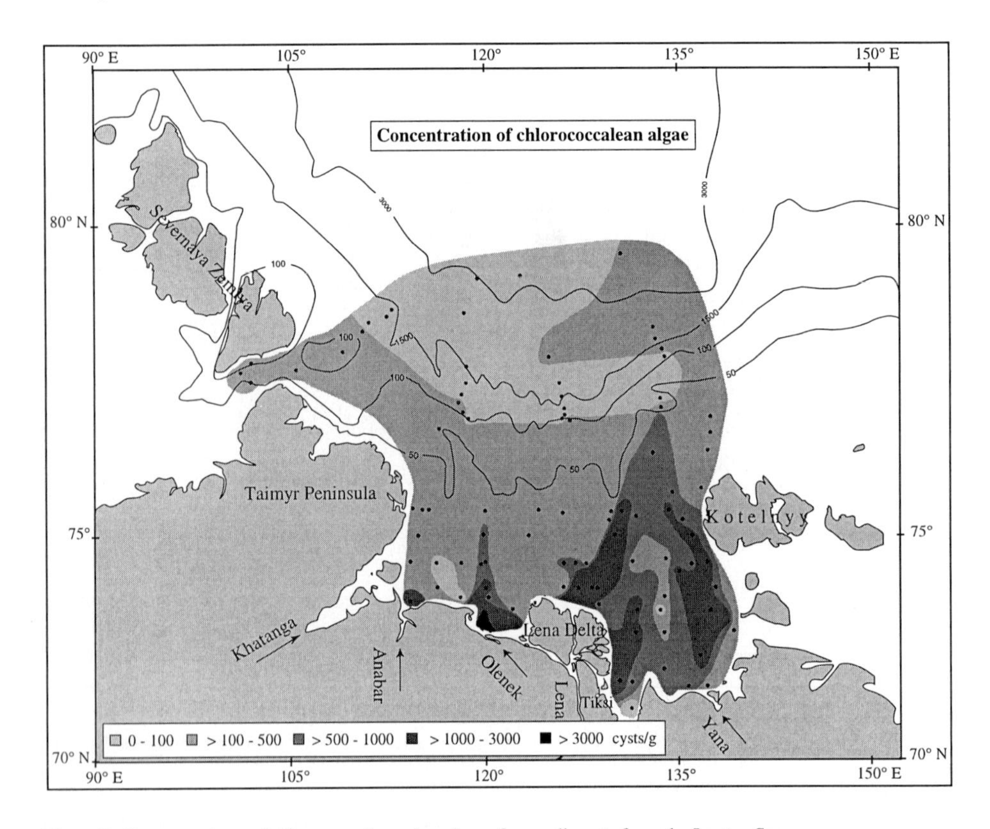

Figure 7: Concentrations of chlorococcalean algae in surface sediments from the Laptev Sea.

Discussion

The occurrence and distribution of dinoflagellate cysts and chlorococcalean algae in surface sediments of the Laptev See clearly reflect the strong salinity gradient in the surface water of the Laptev Sea and the adjacent Arctic Ocean (Figure 2). The change in the composition of the dinoflagellate cyst assemblages occurs mainly at the shelf break where salinities change from approximately 28 to 30 (compare Figures 2 and 6). Assemblage I is typical for the lower saline surface waters of the Laptev Sea, whereas assemblage II is characteristic for relatively higher saline surface waters. The occurrence of dinoflagellate cysts in surface sediments of the coastal area and in front of the large river mouths where salinities are well below 15 (Figure 2) does not necessarily imply an indigenous occurrence. Decreasing concentrations (Figure 4) towards the coast suggest transport from the relatively stable central Laptev Sea to the extremely variable environments off the river mouths. Additionally, there are no previous reports that the polar assemblage I which consists of *Algidasphaeridium? minutum*, *Brigantedinium* spp. and *Polykrikos?* spp. occur at these extremely low salinities.

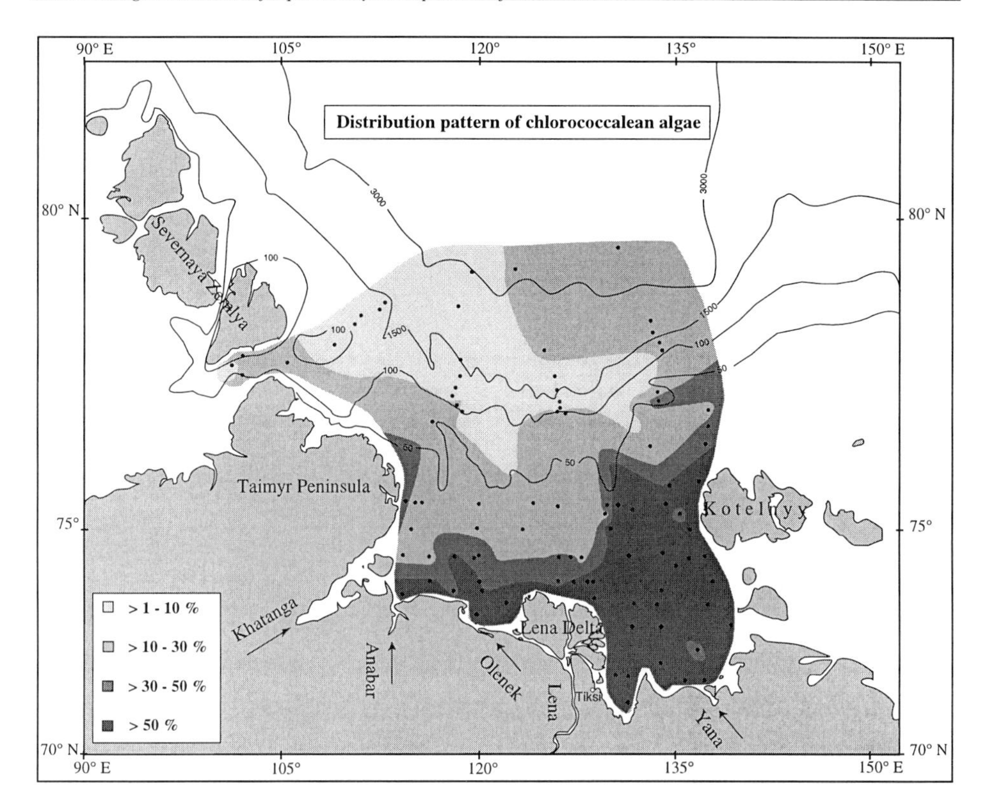

Figure 8: Distribution pattern of chlorococcalean algae in surface sediments from the Laptev Sea.

The concentrations and the distribution pattern of the chlorococcalean algae assemblages indicate that the Laptev Sea shelf is strongly influenced by the freshwater input of the rivers Olenek, Lena and Yana. In contrast, surface sediments from the mouth of the Khatanga River do not contain significant amounts of these algae. The Khatanga river may not transport any chlorococcalean algae into the Laptev Sea but the few samples taken in the western Laptev Sea do not allow to resolve this question unequivocally. The anti-clockwise surface circulation system then transports the freshwater algae out of the Laptev Sea (Figure 1). The strong decrease in the concentrations suggest that the influence of the Lena and Yana is confined to the eastern Laptev Sea. Therefore, the presence of chlorococcalean algae in deep-sea sediments may be an indicator of the freshwater influx to the open Arctic Ocean.

The occurrence of chlorococcalean algae in deep-sea sediments may be also explained by a different process. When sea-ice formation takes place in the Laptev Sea in autumn (Eicken et al., 1997), sediment and incorporated microfossils may be frozen into the sea ice. The sea ice is then transported with the Transpolar Drift from the Laptev Sea shelf across the Arctic Ocean to the Greenland and Iceland seas (e.g., Dethleff et al., 1993). Previous observations suggest that this is a likely process to explain the occurrence of *Pediastrum* and acritarchs in recent sediments from the Arctic Ocean and the Greenland and Iceland seas. Mudie (1992) observed *Pediastrum* sp. in deep-sea sediments from the Gakkel Ridge, underlying the Transpolar Drift and proposed that they may indicate melt-out of ice-rafted sediment which came from the Laptev Sea.

Abelmann (1992) reported brackish diatoms from sea-ice samples along a transect between the outer Barents Shelf and the northern flank of the Gakkel Ridge and suggested that they might be incorporated into sea ice in the Laptev Sea region. Matthiessen (1995) also assumed that the *Pediastrum* and *Halodinium* occurrence in surface sediments from the Iceland and Greenland seas is due to the transport of sea ice in the Transpolar Drift and the East Greenland current from the Laptev Sea across the eastern Arctic Ocean. Therefore, chlorococcalean algae can be useful indicators for both paleo-river discharge and sea-ice transport.

Acknowledgements

I thank J. Thiede and H. Kassens for the possibility to join the Russian-German project "System Laptev Sea", the opportunity to participate in the Transdrift Expeditions and their great interest in this investigation. I thank also P. Mudie, W. Brenner and J. Matthiessen for constructive reviews. The Alfred Wegener Institute is thanked for providing surface samples of "Polarstern" Expedition ARK-IX/4. This research is supported by the Ministry for Education, Science, Research and Technology (BMBF).

References

Abelmann, A. (1992) Diatom assemblages in Arctic sea ice - indicator for ice drift pathways. Deep-Sea Res., 39 (2), 525-538.

Barss, M.S. and G.L. Williams (1973) Palynology and nannofossil processing techniques. Geol. Surv. Can., paper 73-26, 1-25.

Cremer, H. (1998) Die Diatomeen der Laptevsee (Arktischer Ozean): Taxonomie, biogeographische Verbreitung und ozeanographische Bedeutung. Ber. Polarforsch., 260, 205 pp.

Dale, B. (1976) Cyst formation, sedimentation and preservation: factors affecting dinoflagellate assemblages in recent sediments from Trondheimsfjord, Norway. Rev. Palaeobot. Palynol., 22, 39-60.

Dethlef, D., D. Nürnberg, M. Reimnitz, M. Saarso and Y.P. Savchenko (1993) East Siberian Arctic Region Expedition `92: The Laptev Sea - Its significance for Arctic sea-ice formation and transpolar sediment flux. Ber. Polarforsch., 120, 1-44.

De Vernal, A., G. Bilodeau, C. Hillaire-Marcel and N. Kassou (1992) Quantitative assessment of carbonate dissolution in marine sediments from foraminifer linings vs. shell ratios: Davis Strait, northwest North Atlantic. Geology, 20, 527-530.

De Vernal, A., C. Hillaire-Marcel and G. Bilodeau (1996) Reduced meltwater outflow from the Laurentide ice margin during the Younger Dryas. Nature, 381, 774-777.

De Vernal, A., J.-L. Turon and J. Guiot (1994) Dinoflagellate cyst distribution in high-latitude marine environments and quantitative reconstruction of sea-surface salinity, temperature and seasonality. Can. J. Earth Sci., 31, 48-62.

Dmitrenko, I.A., V.I. Karpiy and N.V. Lebedev (1994) The modern Environment of the Laptev Sea: Oceanographic studies. In: Laptev Sea System: Expeditions in 1994, Kassens, H. (ed.), Ber. Polarforsch.,182, 22-33.

Doher, L.I. (1980) Palynomorph preparation procedures currently used in the paleontology and stratigraphy laboratories, U. S. Geological Survey. Geol. Surv. Circ., 830, 29.

Eicken, H., E. Reimnitz, V. Alexandrov, T. Martin, H. Kassens and T. Viehoff (1997) Sea-ice processes in the Laptev Sea and their importance for sediment export. Conti. Shelf Res., 17 (2), 205-233.

Evitt, W.R. (1985) Sporopollenin dinoflagellates cysts: their morphology and interpretation. Am. Assoc. Strat. Palynol., 333 pp.

Fütterer, D.K. (1994) The Expedition ARCTIC `93 Leg ARK-IX/4 of RV "Polarstern" 1993. Ber. Polarforsch., 149, 244 pp.

Gordienko, P.A. and A.F. Laktionov (1969) Circulation and physics of the Arctic Basin water. In: Annals f the international geophysical year, oceanography, Gordon, A.L. and F.W.G. Baker (eds.), Pergamon, New York, 46, 94-112.

Harland, H. (1982) Recent dinoflagellate cyst assemblages from the southern Barents Sea. Palynology, 6, 9-18.

Holmes, M.L. and J.S. Creager (1974) Holocene history of the Laptev Sea continental shelf. In: Marine Geology and Oceanography of the Arctic Sea, Herman, Y. (ed.), Springer, New York, 211-229.

Kassens, H. (1995) Laptev Sea System: Expeditions in 1994. Ber. Polarforsch., 182, 195 pp.

Kassens, H. (1997) Laptev Sea System: Expeditions in 1995. Ber. Polarforsch., 248, 210 pp.

Kassens, H. and V.Y. Karpiy (eds., 1994a) Russian-German Cooperation: The Transdrift I Expedition to the Laptev Sea. Ber. Polarforsch., 151, 168 pp.

Kassens, H. and V.Y. Karpiy (1994b) The Transdrift I Expedition: A multidisciplinary Russian-German approach to study the complex Laptev Sea system. In: Russian-German Cooperation: The Transdrift I Expedition to the Laptev Sea, Kassens, H. and V.Y. Karpiy (eds.), Ber. Polarforsch., 151, 1-8.

Kunz-Pirrung, M. (1998) Rekonstruktion der Oberflächenwassermassen der östlichen Laptevsee im Holozän anhand von aquatischen Palynomorphen. Ber. Polarforsch., 281, 117 pp.

Lentin, J.K. and G.L. Williams (1993) Fossil dinoflagellates: Index to genera and species; 1993 edition. Am. Assoc. Strat. Palynol. Contr. Ser., 28, 1-856.

Matthiessen, J. (1995) Distribution pattern of dinoflagellate cysts and other organic-walled microfossils in recent Norwegian-Greenland Sea sediments. Mar. Micropaleontol., 24, 307-334.

Matthiessen, J. (1996) Dinoflagellate cyst evidence of Holocene environmental conditions off East Greenland. Zbl. Geol. Paläont., part I 1995 (1/2), 271-286.

Matthiessen, J. and W. Brenner (1996) Chlorococcalalgen und Dinoflagellaten-Zysten in rezenten Sedimenten des Greifswalder Bodden. Senckenbergiana Marit., 27 (1/2), 33-48.

Mudie, P.J. (1992) Circum-arctic Quaternary and Neogene marine palynoflora: paleoecology and statistical analysis. In: Neogene and Quaternary dinoflagellate cysts and acritarchs, Head, M.J. and J.H. Wrenn (eds.), Am. Ass. Stratigr. Palynol. Found., 347-390.

Naidina, O.D. and H.A. Bauch (this volume) Distribution on pollen and spores in surface sediments of the Laptev Sea. In: Kassens, H., H.A. Bauch, I. Dmitrenko, H. Eicken, H.-W. Hubberten, M. Melles, J. Thiede and L. Timokhov (eds.) Land-Ocean Systems in the Siberian Arctic: Dynamics and History, Springer-Verlag, Berlin.

Parra Barrientos, O.O. (1979) Revision der Gattung *Pediastrum* Meyen (Chlorophyta). Bibl. Phycol., 48, 186 pp.

Phipps, D. and G. Playford (1984) Laboratory techniques for extraction of palynomorphs from sediment. Pap. Dept. Geol. Uni. Qd., 11 (1), 1-23.

Stockmarr, J. (1971) Tablets with spores used in absolute pollen analysis. Pollen Spores, 13 (4), 616-621.

Treshnikov, A.F. (1985) Atlas of the Arctic. Arctic and Antarctic Institute, Moscow, 204 pp.

Plate I

Number after sample is slide number. The alphanumeric code is the England finder coordinate. Scale bar is 20 µm. Interference contrast (1-2, 5-7, 9, 10-12); phase contrast (3-4, 8).

1, 2. *Algidasphaeridium? minutum*: (1) Sample KD9568-7, 0-1 cm, 2, D 55/2 (2) Sample PM9462-4, 110 cm, 1, L 40/3.
3. *Algidasphaeridium?* Typ A: Sample IK9373-10, 75 cm, 1, D 50.
4. *Algidasphaeridium?* Typ B: Sample PM9462-4, 450 cm, 3, O 43/2.
5. *Algidasphaeridium?* Typ C: Sample KD9572-1, 0-1 cm, 2, D 54/1.
6. *Brigantedinium simplex*: Sample PM9499-1, 0-1 cm, 2, N 43.
7, 8. *Nematosphaeropsis labyrinthus:* (7) Sample PS2473-2, 0-1 cm, 1, Z 49/2 (8) Sample PS2464-2, 0-1 cm, 2, V 46/1.
9. *Impagidinium? pallidum:* Sample IK9373A-6, 0-1 cm,1, Y 33/4.
10. *Operculodinium centrocarpum*: Sample IK9316-6, 0-1 cm, 2 ´, S 45/4.
11, 12. *Polykrikos?*, Typ 1 (11): Sample PM9441-4, 0-1 cm, 5; O 47/3 (12) Sample PM9462-4, 430 cm, 4, D 41/1.

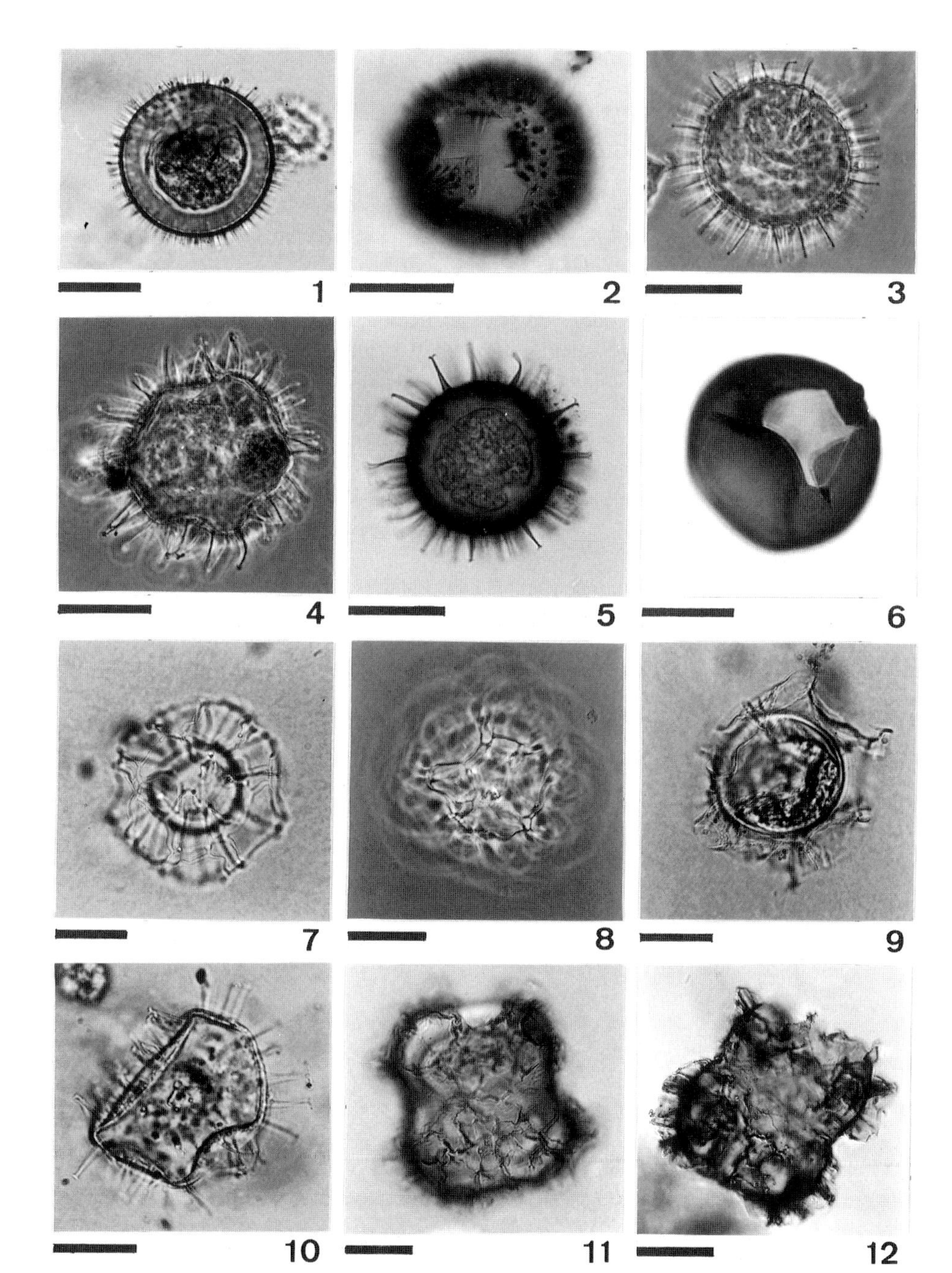
1
2
3
4
5
6
7
8
9
10
11
12

Plate II

Scale bar is 20 µm. Interference contrast (1-2, 4, 6-9, 11-12); phase contrast (3, 10).

1. *Polykrikos?*, Typ 2: Sample PM9463-8, 0-1 cm, 3, R 40.
2. *Polykrikos?*, Typ 3: Sample IK9348-6, 0-1 cm, 3, M 50 / 3.
3. *Pediastrum boryanum:* Sample IK9334-7, 0-1 cm, 3, O 42 / 3.
4. *Pediastrum duplex*: Sample IK9321-5, 0-1 cm, 1 ´, H 44 / 3.
5. *Pediastrum kawraiskyi:* Sample IK9371-1, 0-1 cm, 2, U 47/3.
6. *Botryocccus* cf. *braunii:* Samplc PM9462-4, 270 cm, 3, G 29.
7. *Radiosperma coriferum:* IK93Z5-3, 0-1 cm, 3 ´, W 58 / 3.
8. *Halcdinium* sp.: Sample IK9371-1, 0-1 cm, 2, G 57 / 4.
9, 10. Acritarch Typ A (9): Sample PM9475-3, 0-1 cm, 2 ´, N 53 / 4. (10) Proben-Nr.: IK9318-3, 0-1 cm, 1, X 39 / 4.
11. *Hexasterias problematica:* Sample KD9568-7, 0-1 cm, 2, D 55 / 2.
12. foraminiferal test lining: Sample PM9462-4, 210 cm, 5, V 50.

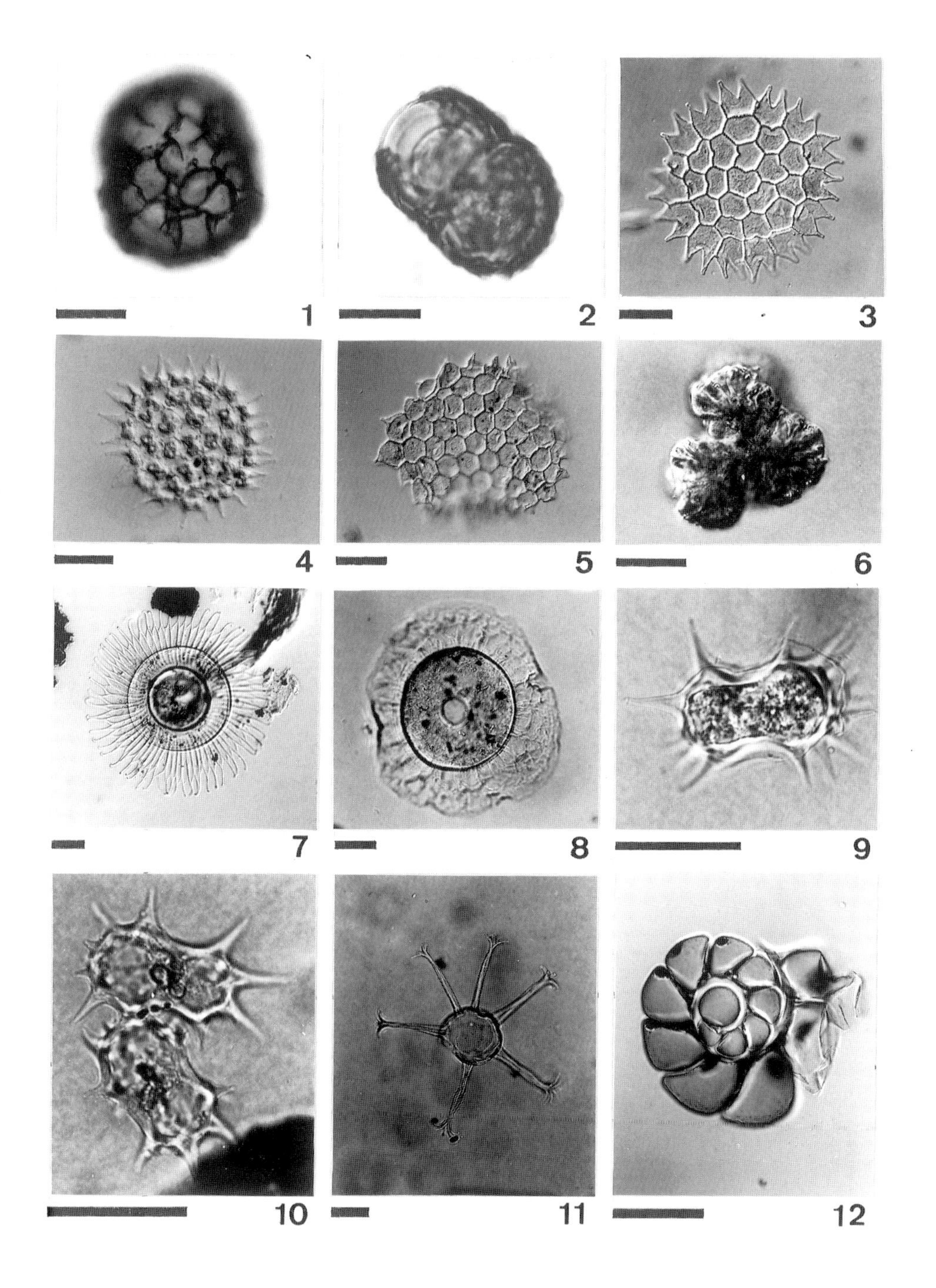
1
2
3
4
5
6
7
8
9
10
11
12

Distribution of Pollen and Spores in Surface Sediments of the Laptev Sea

O.D. Naidina[1] and H.A. Bauch[2]

(1) Institute of the Lithosphere, Russian Academy of Sciences, Staromonetny per. 22, 109180 Moscow, Russia
(2) GEOMAR Forschungszentrum für marine Geowissenschaften, Wischhofstrasse 1-3, D 24148 Kiel, Germany

Received 17 March 1997 and accepted in revised form 10 September 1998

Abstract - The palynological content of Laptev Sea surface sediments were studied. The pollen-spore spectra in these recent sediments are characterized by a predominance of coniferous pollen and moss spores. The pollen have been transported over long distances whereas the spores are of more local origin. The majority of these pollen and spore grains accumulate in the near-shore zone. The pollen-spore spectra in the eastern part of the Laptev Sea are characterized by a different taxonomical composition, which is linked to the huge perennial discharge of freshwater from the Lena and Yana rivers. Thus, the distribution of pollen and spores in the Laptev Sea, as well as their diversity, may be attributed to atmospheric and surface water circulation patterns but is mainly influenced by the intensity of freshwater runoff from the Lena river.

Introduction

The Laptev Sea is located at the northern Eurasian margin of Central Siberia. This shelf sea is bounded by the Taymyr Peninsula to the west and the New Siberian Islands to the east (Figure 1) and covers an area of about 660,000 km^2, most of which is relatively shallow water (less than 50 m deep). The recent Arctic vegetation on the adjacent land is characterized by rather treeless landscapes (Figure 2). However, tree pollen are very abundant in the pollen-spore spectra of continental as well as marine deposits. In general, it is established that pollen transportation into deposits of various origins is principally governed by aerial transportation (Semenov, 1973; Kabailene, 1976). A quantitative model of marine pollen transport, deposition processes (Mudie, 1984) as well as data from a coastal shelf box model (Mudie and McCarthy, 1994) shows that aerial transport is the main process by which pollen moves across the land adjacent to the western North Atlantic. These box model results indicate that wind is the most important marine pollen transport process off eastern Canada, where rivers are relatively small, have small runoff volumes and where strong westerly and southeasterly offshore winds prevail (Mudie and McCarthy, 1994). In regions where pollen grains are most abundant, i.e. in this case off the mouths of large rivers and particularly in near-shore marine sediments, however, fluvial transport is an important process for the accumulation of pollen (Cross et al., 1966; Heusser, 1985; Mudie, 1982; Muller, 1959; Traverse, 1988 and 1992).

The influence of fluvial pathway transport on the pollen distribution has been investigated in some large rivers: the Volga River (Fedorova, 1952), the Delaware River estuary (Groot, 1966) and the Mississippi River (Smirnov et al., 1996). Some investigators have concluded that the suspended pollen load in large rivers depends on the flow velocity, the concentration of pollen deposited aerially in surface waters, and the segregation of tributary water flow. Other studies (e.g. Smirnov et al., 1996) show no significant correlation between pollen load and velocity. Application of sediment mechanics to pollen grain transport demonstrates that such relationships should not be expected in a river and that pollen rain and resuspension of grains from the riverbed are more likely to control the distribution of pollen (Smirnov et al., 1996).

One of the main factors which shape the hydrography and, thus, the depositional environment

----

In: Kassens, H., H.A. Bauch, I. Dmitrenko, H. Eicken, H.-W. Hubberten, M. Melles, J. Thiede and L. Timokhov (eds.) Land-Ocean Systems in the Siberian Arctic: Dynamics and History. Springer-Verlag, Berlin, 1999, 577-585.

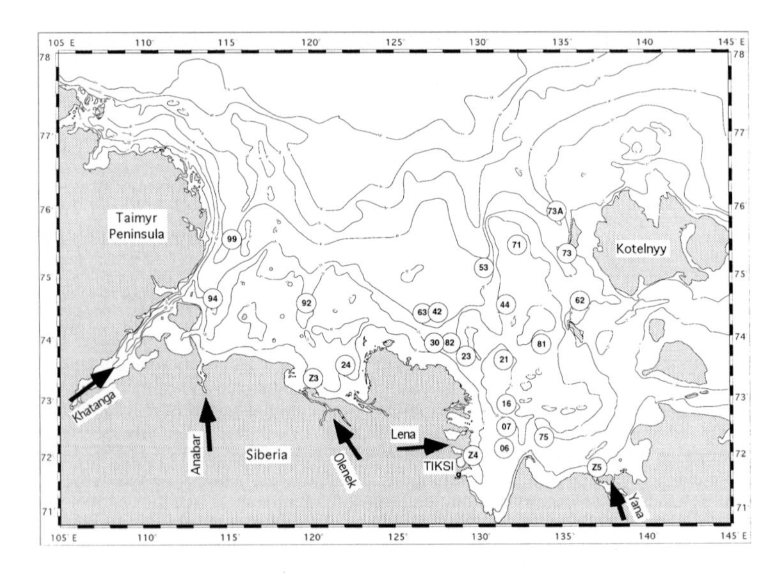

Figure 1: Location of studied surface sediments from the Laptev Sea containing sufficient numbers of pollen and spores grains.

of the Laptev Sea is the strong seasonal freshwater runoff by the main rivers (Figure 1) in the summer when the ice breaks (Timokhov, 1994). It is believed that sediment transport onto the Laptev Sea shelf during this time takes place either by ice or by suspension and that this material then accumulates mainly along the channels (Lindemann, 1994). The strong gradient in chlorophycean algae concentrations recorded by Kunz-Pirrung (this volume) suggests that the influence of freshwater is confined mainly to the shelf. The principal objective of this paper is to investigate the distribution of pollen and spores in surface sediments of the Laptev Sea, and to discuss their possible sources and transport pathways.

Material and methods

The palynological study was based mainly on surface samples taken during the Transdrift I and Transdrift II expeditions to the Laptev Sea on board the RV Ivan Kireyev (1993), RV Professor Multanovsky (1994) and RV Polarstern (1993).

The preparation of the samples for the study was carried out at the laboratory of the GEOMAR Research Center for Marine Geosciences in Kiel, Germany. Prior to the investigation, all samples were freeze-dried and then treated with cold acid (HCL, HF) to dissolve carbonates and silicates according to the method of Phipps and Playford (1984), Doher (1980) and Barss and Williams (1973).

For identification and quantification of the grains, a microscope was used at a constant magnification of 400x. A total of 86 slides from 45 surface samples were studied. Pollen and spore percentages were calculated based on the total number of grains.

Results

Pollen and spore diversity

In the sediments, a total of 20 palynomorph taxa were recognized. This number of taxa is quite typical for recent sediments of the high latitude regions of the Arctic (Kupriyanova, 1951). Among the arboreal plants, the pollen of spruce *Picea*, pine *Pinus sibirica,* cedar pine *Pinus pumila,* larch *Larix,* willow *Salix,* alder *Alnus,* and birch *Betula sect. Nanae* (*Betula exilis, B. middendorffii, B. nana)* were recorded. From the herbaceous plants, the pollen of Ericaceae, Gramineae, Cyperaceae, Asteraceae, Rosaceae, Saxifragaceae, Ranunculaceae, Caryophyllaceae were present. The spore plants were represented by green mosses *Bryales*, sphagnum *Sphagnales* and the fern Polypodiaceae.

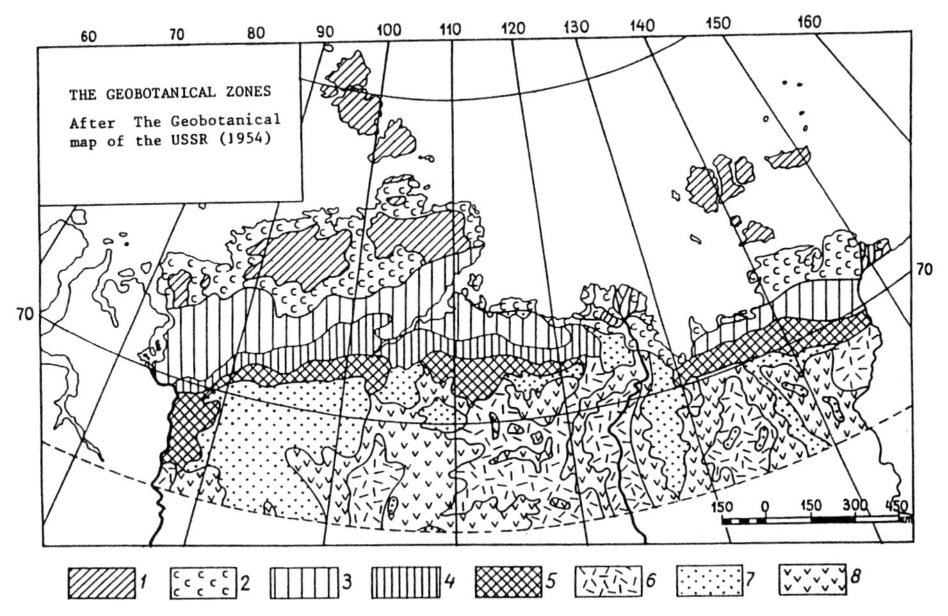

Figure 2: Map of geobotanical zones adjacent to the Laptev Sea (after Lavrenko and Sochava, 1954). 1) Arctic deserts and glaciers; 2) Arctic tundra; 3) typical tundra; 4) shrubs and tussocks tundra; 5) forest-tundra; 6) northern taigas; 7) mountain tundra and shrubs of the alpine taiga regions; 8) mountain forest-tundra of the northern taiga region.

Pollen and spores of Laptev Sea surface samples

From 45 investigated surface samples taken from the shelf, 24 samples (Figure 1) contained sufficient specimens to calculate percentage data. Pollen-spore spectra of these samples were dominated by arboreal pollen from coniferous trees (7-93%). Most of them belonged to *Pinus pumila*. The pollen of other conifers belonged to either *Pinus sibirica* or *Picea.* Occasionally, pollen grains of *Larix* were recognized. Deciduous trees comprised about 4% of the spectra and included *Salix, Alnus* and *Betula sect. Nanae* (*B. exilis, B. nana , B. middendorffii*). Among the herbaceous species, the pollen of Cyperaceae were dominant (up to 17%). The pollen of Ericaceae, Gramineae, Asteraceae, Rosaceae, Saxifragaceae, Ranunculaceae, Caryophyllaceae were observed, but remained low. Among the spore plants, *Sphagnales* (up to 40%), *Bryales* (up to 82%) and *Polypodiacea* (up to 4%) comprised the majority of most spectra.

The pollen-spore spectra of surface samples from sites nearer to the coast - sites 24, 23, 30 in the Lena Delta, sites Z4, 06, 07, 16 in Tiksi Bay and sites Z3 and Z5 in the estuaries of the Olenek and Yana rivers, respectively - are characterized by a decrease in the relative abundance of spores and an increase in arboreal coniferous pollen. In the sample from site Z5 at the estuary of the Yana river, the pollen-spore spectrum was characterized by the appearance of Ericaceae and birch dwarf shrubs. The composition of pollen-spore spectra in samples from site

82 just north of the Lena Delta showed the greatest quantity of coniferous tree pollen whereas the quantity of arboreal pollen was reduced in samples close to the coast of the Taymyr Peninsula (sites 94, 99). Pollen-spore spectra in these samples were characterized by a predominance of spores of the mosses *Bryales* and *Sphagnales*. The frequency of pollen grains decreased with increasing distance from the coast. In general, the special morphological texture of these pollen can be easily transported by wind and water. In the open sea, to the north of Kotelnyy Island (i.e. samples from sites 84, 73A, K1; Figure 1), pollen grains of pine are rare.

All 3 surface samples taken in the open sea (77°41´ - 78°22´ N; 125°00´-133°12´ E) contain only infrequent pollen grains of coniferous trees.

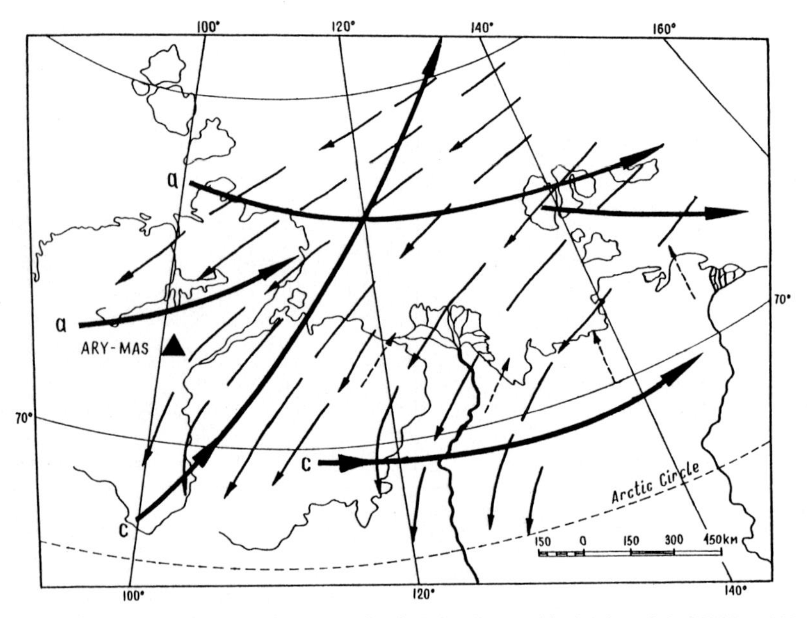

Figure 3: Atmospheric circulation over the Laptev Sea in July (Geographical Atlas of the USSR, 1983). Solid arrows represent prevailing surface winds, dotted small arrows indicate offshore winds; c - cyclones, a - anticyclones; Ary-Mas - northernmost forest in the world.

Summary of pollen-spore spectra

The overall distribution of the main components of the pollen-spore spectra is given in Figure 4. Based on this analysis, it follows that grains of pollen and spores concentrate in the off-shore as well as the near-shore zone. Pollen and spore distributions in the regions of the Lena Delta, Tiksi Bay and near the estuaries of the Olenek and Yana were characterized by a greater number of coniferous pollen grains (Figure 4a, b). In the area of Tiksi Bay, the summed percentage contents of *Pinus pumila* and *Pinus sibirica* reached values of up to 93%. Relatively high coniferous pollen contents were also found in the regions of the submarine valleys of the Olenek (up to 57%) and Lena (up to 74%) rivers.

Among the spore plants, the *Bryales* mosses were dominant and reached 82% in the submarine valley of the Lena (Figure 4c, d). *Sphagnales* percentages increased near the Taymyr Sea coast (up to 40%) and varied in the eastern part of the sea between 2-30% (Figure 4e).

Among the herbaceous plants, the amount of Cyperaceae varied between 1 and 17%. Frequency of Cyperaceae and the rest of the herbaceous species were higher in the western part of the sea and varied between 23-32% (Figure 4f, g).

Farther off the coast of the eastern Laptev Sea, there were still relatively high proportions of cedar pine pollen and sphagnum spores whereas the coastal zone yielded herbaceous pollen and spores of green mosses. The latter two are of local origin and therefore much more closely linked to the coastal vegetation. This implies that specific hydrodynamic processes which occur along the pathway of the main river runoff may be responsible for this distinction and, further, that along the coastal zones and away from the river mouths such processes have less influence on the distribution.

Discussion and conclusions

The composition of pollen and spore spectra in surface sediments of the Laptev Sea can be regarded as a result of transportation. This particularly applies to the occurrence of coniferous pollen grains which do not reflect the typical vegetation on land adjacent to the Laptev Sea. Therefore, these grains were brought into the shelf region either by offshore winds (Figure 3) or by the riverine freshwater runoff. Owing to their special morphology, pollen grains of pine are easily transported for longer distances by wind and rivers. It has been shown that the boundary of the forest-tundra zone in the Lower Yana region shifted northward in more recent times (Aleksandrova, 1953). As a result, forest tundra is now located in this area to within 50 km south of the coast (Figure 2). The flowering period for trees along this northern forest boundary is often accompanied by northerly directed winds. These winds may then account for the transfer of pollen to the Arctic from the forest-tundra and northern taiga. Interestingly in this context, the Ary-Mas forest as a major potential source is situated just along the pathway of the cyclones (Figure 3). Organic matter, which is made up of pollen, spores and other organic fibres, is one of the main components of aerosols over the Laptev Sea (Shevchenko et al. 1995). Previous studies of pollen-spore numbers and composition in Arctic Ocean surface water and seabed sediments (Mudie and Matthiessen, 1988) indicate a primary aeolian transportation, with the circum-Arctic and boreal-tundra vegetation being the main source.

Figure 2 shows that there are several other possible sources of pollen and spores from various tundra areas: 1) terraces along the Lena River valley, 2) plains and terraces of the Lower Khatanga River (the eastern part of the Northern Siberian Lowland), and 3) the mouth lowland of the Yana River (the western part of the Yana-Indigirka Lowland). In this area, the particular vegetation cover connected with the existence of the perennially frozen ground and specific thermokarst forms of tundra-relief, so-called alases, form favourable conditions for the accumulation and preservation of pollen (Naidina, 1995). But pollen preservation is not always perfect. In Northern Yakutia, one of the most widespread trees is *Larix*. Single undersized specimens of *Larix* grow in the tundra of the Yana lowland (Khotinsky et al., 1971) and together with *Betula exilis* and *Salix reptans* form the majority of the northernmost part of the Ary-Mas forest (Anonymous, 1978). But *Larix* pollen are practically absent in palynological spectra (Khotinsky et al., 1971; Kupriyanova, 1951; this study). This accounts for the complete decomposition of these pollen grains due to maceration.

Siberian rivers transport large amounts of dissolved and particulate material onto the shelf area, where it accumulates and/or from which area it is further transported towards the Arctic Ocean by, for example, sea-ice (e.g. Eicken et al., 1997). Due to the influence of the Lena River, the eastern Laptev Sea receives considerably higher amounts of freshwater than the western part (Timokhov, 1994). This distinction is also recognized in many other sedimentological, geochemical and biological proxy data (e.g. Dehn et al., 1995; Höleman et al., 1995; Létolle at al., 1993). Therefore, it is reasonable to conclude that the river runoff also accounts for the observed higher relative abundances of pollen and spore grains within the eastern Laptev Sea .

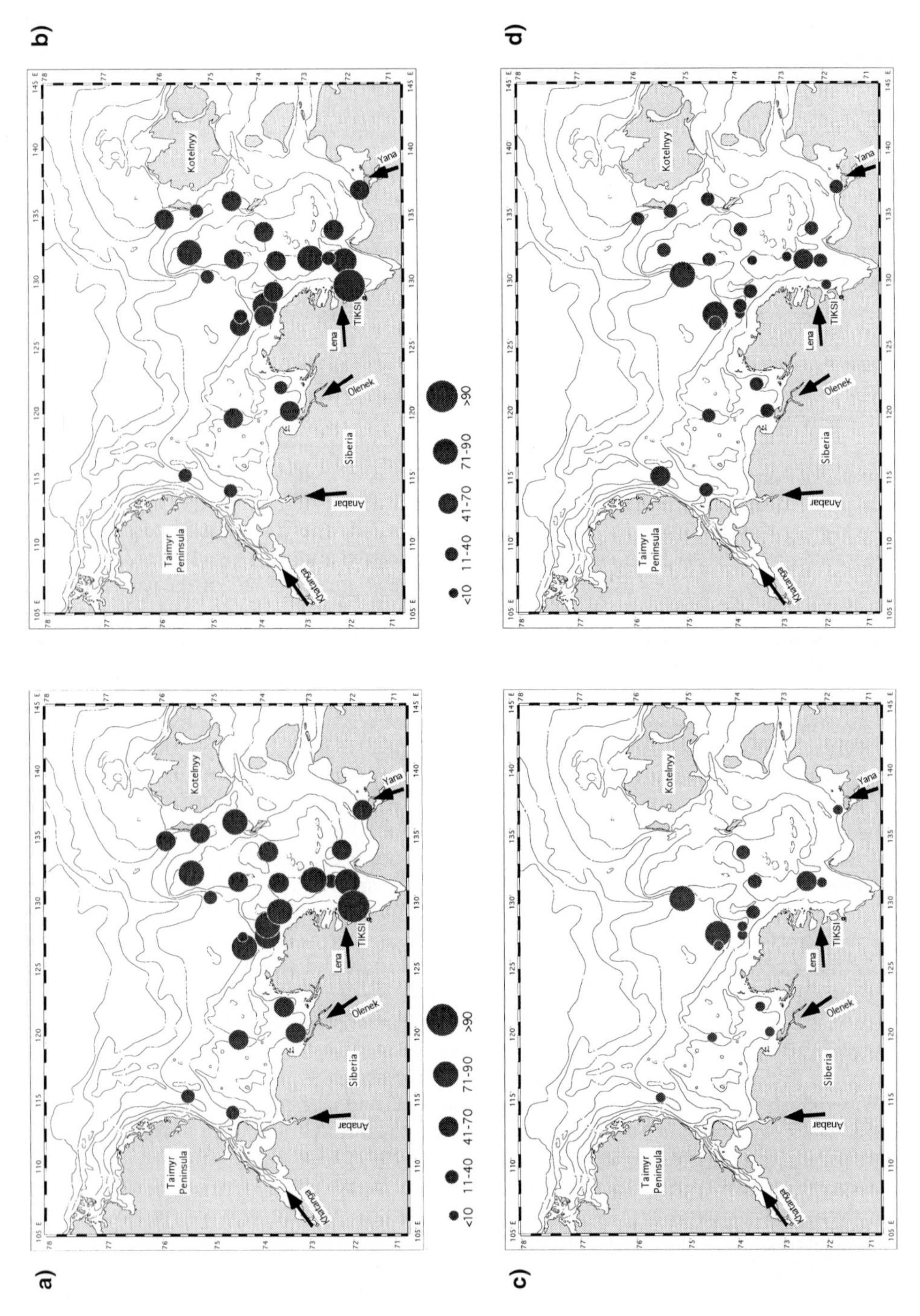

Figure 4: Relative distribution (%) of the main groups of pollen and spores in surface samples from the Laptev Sea. a) tree pollen (sum); b) tree pollen of *Pinus*; c) spores of *Bryales*; d) total spores (sum).

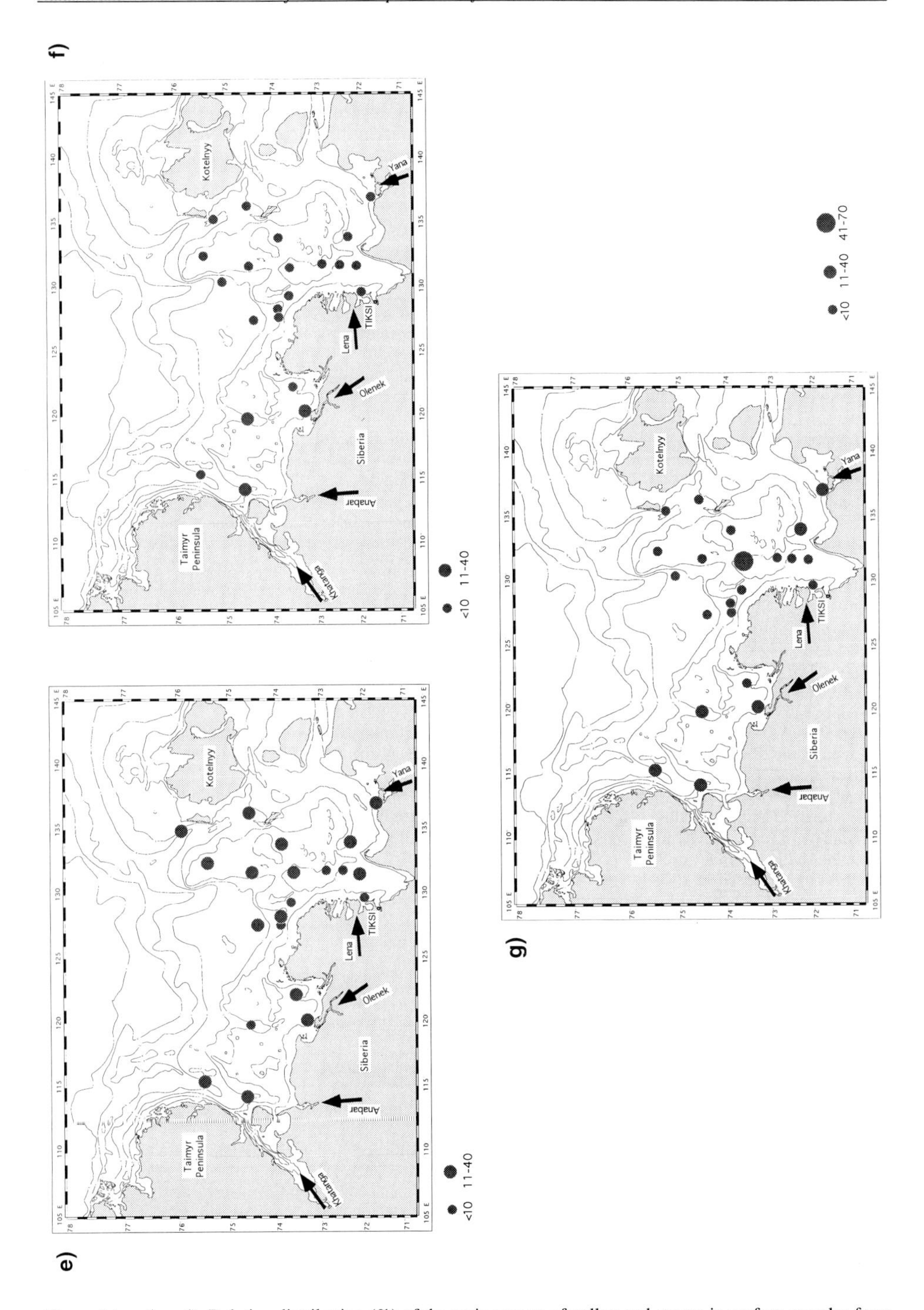

Figure 4 (continued): Relative distribution (%) of the main groups of pollen and spores in surface samples from the Laptev Sea. e) spores of *Sphagnales*; f) pollen of *Cyperaceae*; g) pollen of herbs (sum).

Acknowledgements

We express our special thanks to the shipboard parties for collecting the surface sediments during cruise Transdrift I and II. We are further grateful to M. Kunz-Pirrung for sharing her processed samples with us and to P. Overduin for revising the final text. M. Matthiessen and M. Saarnisto are thanked for their helpful review comments.

References

Aleksandrova, V.D. (1953) On the boundary of the vegetation zones in Lower Yana. Izvestija Vsesouznogo geographicheskogo obszestva, 85 (1): 98-101 (in Russian).

Anonymous (1978) Ary-Mas (1978), natural conditions, flora, vegetation. Nauka, Leningrad, 190 pp.

Barss, M.S. and G.L. Williams (1973) Palynology and nannofossil processing techniques. Geol.Surv.Can., paper 73-26: 25 S.; Ottawa (in Russian).

Cross, A.T., G.G. Thompson and J.B. Zaitzeff (1966) Source and distribution of palynomorphs in bottom sediments, southern part of Gulf California. Mar. Geol., 4: 461-466.

Dehn, J., H. Kassens and Transdrift 2 Shipboard Scientific Party (1995) The sedimentary environment of the Laptev Sea: Preliminary results of the Transdrift 2 Expedition. Rep. Polar Res., 176: 314- 323.

Doher, L.I. (1980) Palynomorph preparation procedures currently used in the paleontology and stratigraphy laboratories, U.S.G.S. Geol. Surv. Circ., 830: 29 S., Washington.

Eicken, H, E. Reimnitz, V. Alexandrov, T. Martin, H. Kassens and T. Viehoff (1997) Sea-ice processes in the Laptev Sea and their importance for sediment transport. Cont. Shelf Res., 17 (2): 205-233.

Fedorova, R. V. (1952) The spread of pollen and spores by water currents (in Russian). Trans. Inst. Geogr. Acad. Sci. USSR, 52: 46-73.

Geographical atlas of the USSR (1983) Glavnoe upravlenie geodezii i kartografii, Moscow, 259 pp., (in Russian).

Groot, J.J. (1966) Some observations on pollen grains in suspension in the estuary of the Delaware River. Mar. Geol., 4: 409-416 (in Russian).

Heusser, L.E. (1985) Quaternary palynology of marine sediments in the northeast Pacific, northwest Atlantic, and Gulf of Mexico. In: V.M. Bryant Jr. and R.G. Holloway (eds.), Pollen records of Late Quaternary North American Sediments. Am. Assoc. Stratigr. Palynol. Found., Tulsa, pp. 385-403.

Höleman, J.A., M. Schirmacher A. Prange and Transdrift 1 Shipboard Scientific Party (1995) Transport and distribution of trace elements in the Laptev Sea: first results of the Transdrift expeditions. Reports on Polar Research, 176, p. 297-302.

Kabailene , M.V., 1976. On pollen dispersion and methods of its determination. In: Palynology in continental and marine geologic sediments. Zinatne, Riga, p. 155-165 (in Russian).

Khotinsky, N.A., G.G. Kartashova, A.M. Velitsky (1971) On the Holocene vegetational history of the Lower Yana Region (by the results of pollen analysis of alas deposits). In: Holocene palynology, Nauka, Moscow: 159-169 (in Russian).

Kupriyanova , L.A. (1951) Investigations of the pollen and spores from surface of the soil in the high latitudes regions of Arctic. Botanical Journal, vol. 36 (3): 258-269 (in Russian).

Lavrenko , E.M. and V.B. Sochava (eds.) (1954) The Geobotanical map of the USSR. AN SSSR, Moscow-Leningrad, pt.1 (in Russian).

Létolle, R., J.M. Martin, A.J. Thomas, V.V. Gordeev, S. Gusarova and I.S. Sidorov (1993). 18-O abundance and dissolved silicate in the Lena delta and the Laptev Sea (Russia). Mar. Chem., 43: 47-64.

Lindemann, F. (1995) Sonographische und sedimentologische Untersuchungen in der Laptevsee, sibirische Arktis. Unpubl. MSc. thesis, University of Kiel, 75 pp.

Mudie, P.J., 1982. Pollen distribution in recent marine sediments, eastern Canada. Can. J. Earth Sci., 19: 729-747.

Mudie, P.J. (1984) Quantitative model of pollen transport in the marine environment. 6th Int. Palynol. Conf. (Calgary). Progr. Abstr., p. 113.

Mudie, P.J. and J. Matthiessen (1988) Dinoflagellates, pollen/spore assemblages and related studies. In: J. Thiede (Editor), Scientific Cruise Report of Arctic Expedition ARK IV/3. Rep. Polar Res., 43: 92-97.

Mudie, P.J. and F.M.G. McCarthy (1994) Late Quaternary pollen transport processes, western North Atlantic: Data from box models, cross-margin and N-S transects. Mar. Geol., 118: 79-105.

Muller, J. (1959) Palynology of Recent Orinoco delta and shelf sediments, reports of the Orinoco Shelf Expedition. Micropal., 5: 1-32.

Naidina, O.D. (1995) Holocene climatic, vegetation and pollen data of Siberia adjacent to the Laptev Sea. Rep. Polar Res., 176: 235-253.

Phipps, D. and G. Playford (1984) Laboratory techniques for extraction of palynomorphs from sediment. Pap. Dept. Geol. Uni. Qd., 11(1): 23 S., St.Lucia.

Smirnov, A., G.L. Chmura and M.F. Laponite (1996) Spatial distribution of suspended pollen in the Mississippi River as an example of pollen transport in alluvial channels. Rev. Paleobot. and Palynol., 92: 69-81.

Semenov, I.N. (1973) Stratigraphy of the antropogenic deposits of the Bolshezemelskoj tundram. Nauka, Moscow, 160 pp (in Russian).

Shevchenko ,V.P. A.P. Lisitzin, V.M. Kuptzov, G.I. Ivanov, V.N. Lukashin, J.M. Martin, V.Yu. Rusakov, S.A. Safarova, V.V. Serova, R. Van Grieken and H. Van Malderen (1995) The composition of aerosols over the Laptev, the Kara, the Barents, the Greeland and the Norvegian sea. Rep. Polar Res., 176: 7-16.

Timokhov ,L.A. (1994) Regional characteristics of the Laptev and the East- Siberian Seas. Rep. Polar Res., 144: 15-31.

Traverse, A. (1988) Paleopalynology, Allen and Unwin, Winchester, MA, 600 pp.

Traverse, A. (1992) Organic fluvial sediment: palynomorphs and "palynodebris" in the lower Trinity River, Texas. Ann. Mo. Bot. Gard., 79: 110-125.

Clay Mineral Distribution in Surface Sediments of the Laptev Sea: Indicator for Sediment Provinces, Dynamics and Sources

B.T. Rossak[1], H. Kassens[2], H. Lange[2] and J. Thiede[2]

(1) Geologisches Landesamt Nordrhein-Westfalen, De-Greiff-Str. 195, D 47803 Krefeld, Germany
(2) GEOMAR Forschungszentrum für marine Geowissenschaften, Wischhofstrasse 1-3, D 24148 Kiel, Germany

Received 11 March 1997 and accepted in revised form 1 December 1997

Abstract - Forty-eight surface sediment samples from the Laptev Sea taken during the Russian - German expedition Transdrift I in summer 1993 were analysed for their clay mineral composition (illite, smectite, chlorite, and kaolinite). Different clay mineral provinces, the role of fluvial sediment-supply, transport mechanisms, and possible source areas are discussed.
The distribution patterns of the clay minerals allow to distinguish between three different provinces: 1. In the western Laptev Sea sediments are particularly rich in smectite and are characterized by a slight enrichment in kaolinite. 2. Sediments in the eastern Laptev Sea are very poor in kaolinite. 3. The southeastern Laptev Sea is dominated by illite and chlorite.
The distribution of clay minerals in the Laptev Sea is controlled both by river run-off and summer surface currents. The Lena and Yana rivers mainly deliver illite and chlorite, smectite is supplied almost exclusively by the Anabar and Khatanga rivers, kaolinite by the Anabar, Khatanga, and Olenek rivers. From the river mouths, surface currents transport smectite and kaolinite hundreds of km eastward.
Illite and chlorite are most probably erosional products of Paleozoic slates cropping out in the drainage areas of the Lena and Yana rivers. Smectite originates from Mesozoic and Cenozoic weathering residues of the Permo- Triassic Putoran-Plateau flood basalts. Kaolinite is probably derived from the erosion of kaolinite-rich Mesozoic sediments of the Siberian Platform.

Introduction

Recent studies have shown that huge amounts of the sediment-laden sea ice in the Transpolar Drift (Figure 1) are formed over the shallow Siberian shelf areas (Colony and Thorndike, 1984; Pfirman et al., 1990). In particular, the smectite concentrations of sea ice sediment samples taken between 81° N and 83° N are in good correspondence with those in hitherto existing surface sediment samples of the Laptev Sea suggesting the Laptev Sea as one of the main sources for sea ice sediments transported by the Transpolar Drift (Silverberg, 1972; Wollenburg, 1993; Nürnberg et al., 1994).

This study presents detailed clay mineral distribution patterns of the Laptev Sea. It discusses different clay mineral provinces, the role of bathymetry, fluvial sediment supply, and transport mechanisms of the clay minerals. In addition, possible clay mineral source areas in the Siberian hinterland are shown.

Sedimentary environment and bathymetry of the Laptev Sea

Strongly different winter and summer ice conditions influence the sedimentary environment of the very broad and shallow Laptev Sea shelf, which has maximum water depths of 50 - 60 m at its edge. From October to mid-July the Laptev Sea is covered with ice. About 1.5 - 2 m thick fast ice builds up between shallow coastal areas and the 20 - 30 m isobath (Dethleff et al., 1993). An approximately 1800 km long and up to 100 km wide zone of open water (polynya) separates the fast ice from the northward drift ice zone (Zakharov, 1966; Barnett, 1991;

In: Kassens, H., H.A. Bauch, I. Dmitrenko, H. Eicken, H.-W. Hubberten, M. Melles, J. Thiede and L. Timokhov (eds.) Land-Ocean Systems in the Siberian Arctic: Dynamics and History. Springer-Verlag, Berlin, 1999, 587-599.

Dethleff et al., 1993). As a result of the steep increase in river run-off with beginning river ice breakup in May followed by the gradual melting of the Laptev Sea ice cover, a large brackish surface plume extending up to 350 km northward covers large parts of the Laptev Sea during the short summer period (Létolle et al., 1993).

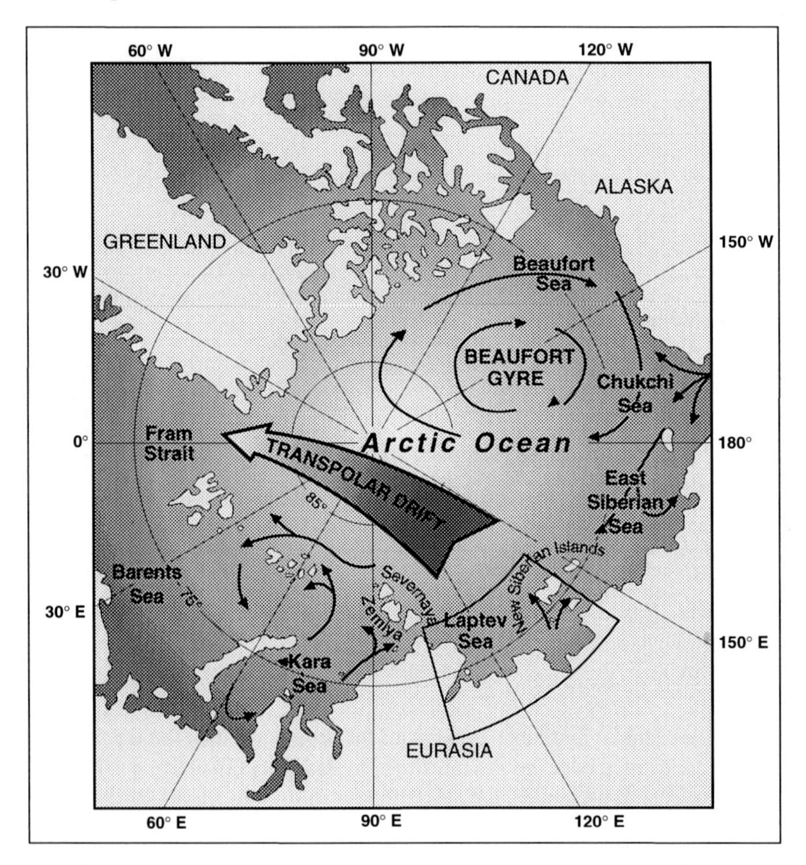

Figure 1: Generalized ice circulation in the Arctic Ocean (modified from Reimnitz et al., 1992).

As emphasised by Silverberg (1972), Holmes and Creager (1974), Kassens and Karpiy (1994), and Lindemann (1994), the Laptev Sea depositional environment is mainly controlled by terrigenous sediment supply due to large Siberian river systems. The Lena river for example has an average water discharge of 525 km^3/year and is discharging yearly about 17.6 x 10^6 t of suspended matter into the Laptev Sea (Gordeev and Sidorov, 1993). During the Pleistocene glacial periods five large northward extending valleys named after their eroding rivers incised into the shelf (Holmes and Creager, 1974) and characterize today's bathymetry of the Laptev Sea (Figure 2).

Main geological features of the Siberian hinterland

The eastern part of the Siberian hinterland is drained by the Yana, Lena, Olenek, and Anabar rivers and is predominated by Paleozoic to Mesozoic sedimentary rocks. Paleozoic rocks are cropping out on Taimyr Peninsula (Figure 2), on the Siberian platform, and in the Werchojansk-Fold-Zone, a giant Mesozoic anticline limiting the Siberian platform to the east.

Mesozoic, mostly clastic sediments are widespread in the western foreland basin of the Werchojansk-Fold-Zone and north of the Siberian Platform in the Jenissej-Khatanga-Basin. On the Siberian platform basement crops out in the Aldan-shield drained by the Lena river and in the Anabar-shield drained by the Anabar river (Dolginow and Kropatschjow, 1994; Nalivkin et al., 1965).

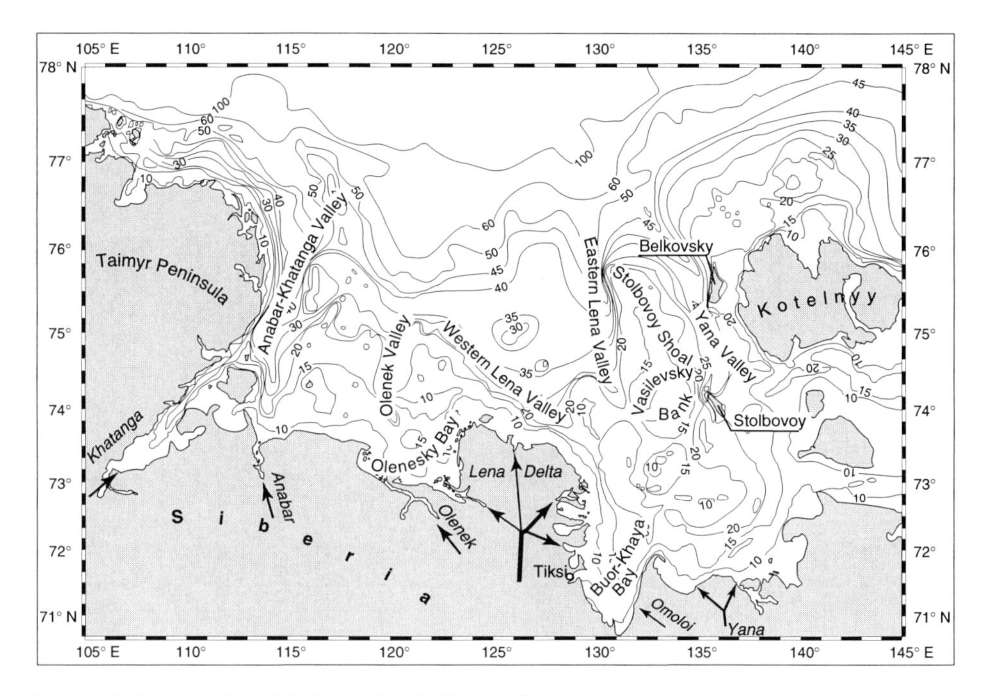

Figure 2: Bathymetric chart of the Laptev Sea Shelf (water depth in metres).

Volcanic rocks predominate the western part of the Siberian hinterland. Here, erosional remnants of formerly giant Permo-Triassic continental flood basalt sheets are forming the Putoran-Plateau drained by the Anabar and Khatanga rivers and by the Jenissej river which draines into the Kara sea (e. g., Nalivkin et al., 1965 and Coffin, 1992).

Materials and methods

During the Transdrift I-Expedition with the Russian RV Ivan Kireyev in summer 1993 (Kassens and Karpiy, 1994) 48 sediment samples were taken at 47 different locations (Figure 3 and Table 1) with a spade box corer, a gravity kasten corer or with a grab sampler. At each site the uppermost centimeter of the sediment was taken for sample and freeze-dried (Lindemann, 1994).

Each sample was shaked in a 10 % hydrogenic peroxide solution for 24 hours to disaggregate the sample and to remove organic carbon. After removing the sand-fraction (> 63 µm) by wet sieving, silt (2 - 63 µm) and clay (< 2 µm) were separated by the settling method based on Stoke's Law, using a 0.25 % NH_3 solution to avoid coagulation of the clay particles. The clay fraction was charged with $MgCl_2$ (Lange, 1982) and to remove free ions the clay fractions then were dispersed with deionized water in an ultrasonic cleaner and centrifuged for 20 minutes at

4000 r.p.m., decanting the clear water. This washing-procedure was repeated until no more clear water could be decanted after centrifugation. The cloudy water then was evaporated and the coagulating clay minerals were returned to the centrifuged sample. Because of the very small carbonate and opal concentrations, removal of these substances was not necessary.

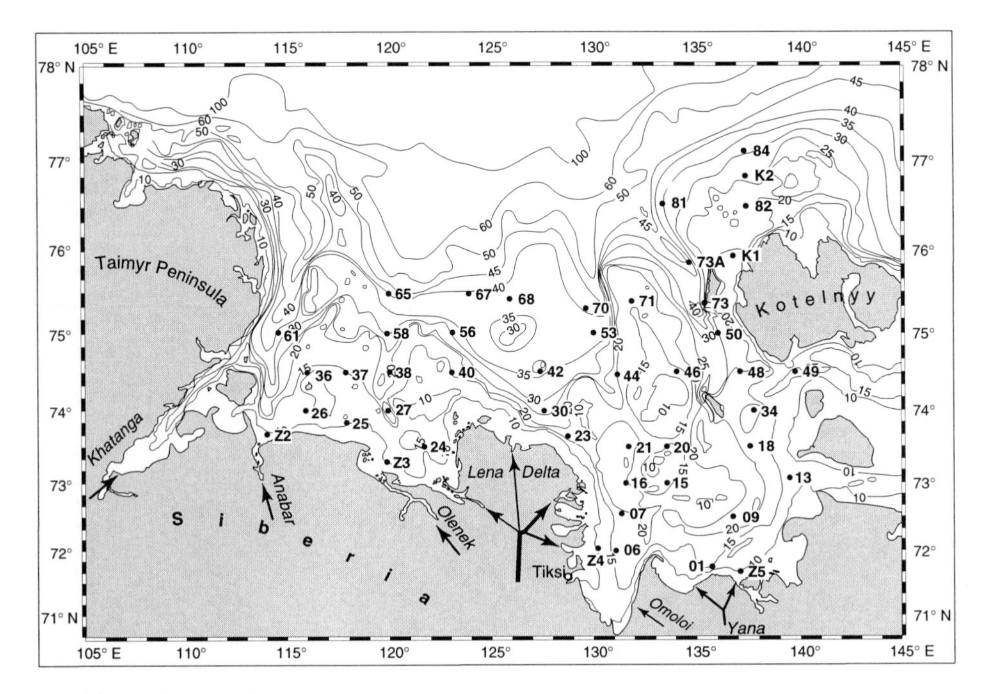

Figure 3: Stations of the expedition Transdrift I in 1993: surface sediment samples.

The dried clay fractions were carefully ground in an agate mortar and 200 mg of each sample were dispersed with 15 ml of deionized water. In order to prepare highly texturated slides without particle-size fractioning, 3 ml of the suspension (40 mg of clay) were sucked by vacuum filtration onto a membrane filter (cellulose nitrate, pore size 0.15 µm). The dried clay cakes were mounted on aluminium platelets using double-sided adhesive tape and carefully removing the filters. The thickness of the clay films was about 10 mg/cm^2.

All samples were analysed with a Philips PW 1830 X-ray difractometer (XRD) using copper radiation (Cu Kα, 45 KV, 40 mA), a theta-circle-integrated automatic divergence slit, a graphite monochromator, and an automatic sample changer. The samples were measured air dried and after solvating them with ethylene-glycol vapor at 60 °C for at least 12 hours. Scans were performed between 2° and 40° 2Θ with a step size of 0.01° 2Θ/s and between 24° and 26° 2Θ with a step size of 0.005° 2Θ/2s. For easier comparison with other studies (Wollenburg, 1993; Nürnberg et al., 1994; Nürnberg et al, 1995; Wahsner and Shelekhova, 1994) the scans were not converted into scans measured with fixed divergence slit as used by Biscaye (1964, 1965).

Table 1: Clay mineral concentrations and illite 'crystallinity' in Laptev Sea surface sediment samples. Latitude and longitude of sampling sites are indicated.

Sample	Latitude N	Longitude E	Water depth (m)	Illite (%)	Smectite (%)	Chlorite (%)	Kaolinite (%)	Illite 'crystallinity' (°Δ2Θ)
01	71°45.2´	135°39.6´	16	50	13	28	9	0.31
06	72°00.1´	131°00.0´	18	57	10	24	9	0.25
07	72°33.1´	131°17.6´	21	51	7	29	13	0.24
09	72°30.0´	136°40.0´	24	60	5	26	9	0.29
13	73°04.0´	139°22.2´	16	56	3	31	10	0.22
15	73°00.0´	133°29.9´	18	54	9	26	11	0.26
16	73°00.1´	131°30.0´	28	53	10	25	12	0.24
18	73°30.0´	137°30.0´	24	59	6	26	9	0.26
20	73°30.0´	133°30.0´	18	54	9	26	11	0.21
21	73°30.0´	131°40.4´	25	51	12	25	12	0.24
23	73°38.0´	128°39.8´	17	51	8	28	13	0.26
24	73°30.1´	121°40.1´	13	44	16	23	17	0.26
25	73°49.9´	117°52.3´	13	32	31	22	15	0.22
26	73°59.8´	115°54.0´	16	35	31	21	13	0.21
27	74°00.0´	119°51.7´	30	37	20	24	19	0.26
30	74°00.0´	127°30.0´	27	46	17	23	14	0.25
34	74°00.0´	137°39.8´	22	52	11	27	10	0.23
36	74°29.5´	115°59.6´	16	35	28	22	15	0.21
37	73°30.1´	117°50.9´	19	38	25	22	15	0.22
38	74°30.1´	119°58.5´	34	42	20	21	17	0.27
40	74°30.0´	122°59.6´	16	39	23	26	12	0.23
42	74°30.3´	127°19.8´	34	43	24	22	11	0.23
44	74°28.0´	131°05.9´	30	48	11	28	13	0.24
46	74°29.9´	134°00.7´	14	46	18	26	10	0.24
48-5	74°30.0´	137°01.0´	23	54	13	24	9	0.30
48-6	74°30.0´	137°01.0´	23	54	11	25	10	0.26
49	74°30.0´	139°40.0´	24	47	20	24	9	0.24
50	75°00.0´	136°00.0´	31	45	22	23	10	0.26
53	75°00.0´	129°57.3´	40	43	21	24	12	0.26
56	75°00.0´	123°00.0´	32	43	18	24	15	0.25
58	75°00.0´	119°50,0´	33	40	23	22	15	0.27
61	75°00.6´	114°32.2´	42	36	26	20	18	0.25
65	75°29.0´	119°54.0´	40	40	23	22	15	0.23
67	75°29.0´	123°50.5´	44	42	22	22	14	0.25
68	75°25.0´	125°51.0´	41	41	22	23	14	0.27
70	75°18.0´	129°34.0´	44	43	24	22	11	0.24
71	75°23.0´	131°48.2´	20	44	21	22	13	0.31
73	75°21.0´	135°22.0´	43	51	18	21	10	0.26
73A	75°51.1´	134°32.1´	47	48	18	23	11	0.26
81	76°31.6´	133°18.6´	37	48	19	22	11	0.25
82	76°29.9´	137°19.7´	25	51	12	26	11	0.25
84	77°06.7´	137°13.5´	33	54	11	24	11	0.26
K1	75°56.0´	136°42.0´	20	52	14	24	10	0.26
K2	76°50.1´	137°17.7´	30	54	9	27	10	0.25

Table 1 (continued):

Sample	Latitude N	Longitude E	Water depth (m)	Illite (%)	Smectite (%)	Chlorite (%)	Kaolinite (%)	Illit 'crystallinity' (°Δ2Θ)
Z2	73°39.9´	113°59.6´	9	31	30	20	19	0.27
Z3	73°17.5´	119°49.9´	12	42	15	24	19	0.29
Z4	72°02.0´	130°07.6´	14	51	14	26	9	0.27
Z5	71°41.4´	137°00.3´	11	56	1	35	8	0.24

For semi-quantitative calculations of the relative contents of smectite, illite, kaolinite, and chlorite the integrated peak areas of their basal reflections were multiplied by empirically estimated weighting factors (Biscaye, 1965). For smectite (here, scans of the glycolated samples were used) the 17 Å (001) peak area - after removing the 14 Å (001)-chlorite peak area - multiplied by one, for illite the 10 Å (001) peak area multiplied by four, and for kaolinite and chlorite the 7 Å (001/002) peak area multiplied by two were used for calculations. To determine the relative contents of kaolinite and chlorite the intensities of the 3.57 - 3.58 Å (001)-kaolinite peak and the 3.53 - 3.54 Å (002)-chlorite peak were considered (detailed scans), following Biscaye (1964, 1965).

For determination of the illite 'crystallinity' the Kübler index was applied. This index is defined as the half-height width of the (001)-illite basal reflection (Kübler, 1967, 1968, and 1984). Values for the low-grade and high-grade limit of the anchizone are 0.42° and 0.25° $\Delta 2\Theta$/Cu Kα (Kübler, 1984). Considering that sample preparation methods affect the measured peak width, parameters applied in this study (grain size, thickness of the slides, Mg-saturation, use of air dried samples for measurements, and the goniometer scanning rate) lead to an increase in peak width (Kisch, 1987; Kisch and Frey, 1987; Krumm and Buggisch, 1991, and Krumm, 1992). Only the use of ultrasonic disaggregation causes decrease in peak width, simulating higher illite 'crystallinity' (Kisch and Frey, 1987).

In the following paragraphs percentages (%) always represent relative percentages (rel.-%) of the clay minerals.

Results

Illite is the main clay mineral in the investigated sediment samples with concentrations of 30 - 60 % (Figure 4 and Table 1). Highest concentrations of up to 60 % occur in the southeastern part of the Laptev Sea and northwest of Kotelnyy Island. Illite concentrations ranging from 40 to < 50 % dominate in the central and eastern Laptev Sea. Only in the western Laptev Sea illite concentrations are < 40 % (Figure 4 and Table 1).

Smectite was detected in concentrations between 1 % and 31 % (Table 1). In contrast to the illite distribution, smectite is enriched in the western Laptev Sea, especially near the mouths of the Anabar and Khatanga rivers where highest concentrations were measured (Figure 5 and Table 1). Concentrations gradually decrease to 15 - < 25 % in the eastern and to < 15 and < 5 % in the southeastern Laptev Sea (Figure 5). In the Olenek- and the Western-Lena-Valleys smectite concentrations are reduced to values between 15 and 20 % (Table 1).

In almost all samples chlorite concentrations range between 20 and < 30 % (Figure 6 and Table 1). Concentrations > 25 % can be found mainly in the southeastern Laptev Sea and northwest of Kotelnyy Island with up to 31 % and 35 % near the eastern mouth of the Yana river (Figure 6 and Table 1).

In most regions kaolinite concentrations are < 15 % (Figure 7 and Table 1). Only in the western Laptev Sea concentrations are ≥ 15 %. Highest values of up to 19 % were measured in the Olenek- and the Anabar-Khatanga-Valleys (Table 1).

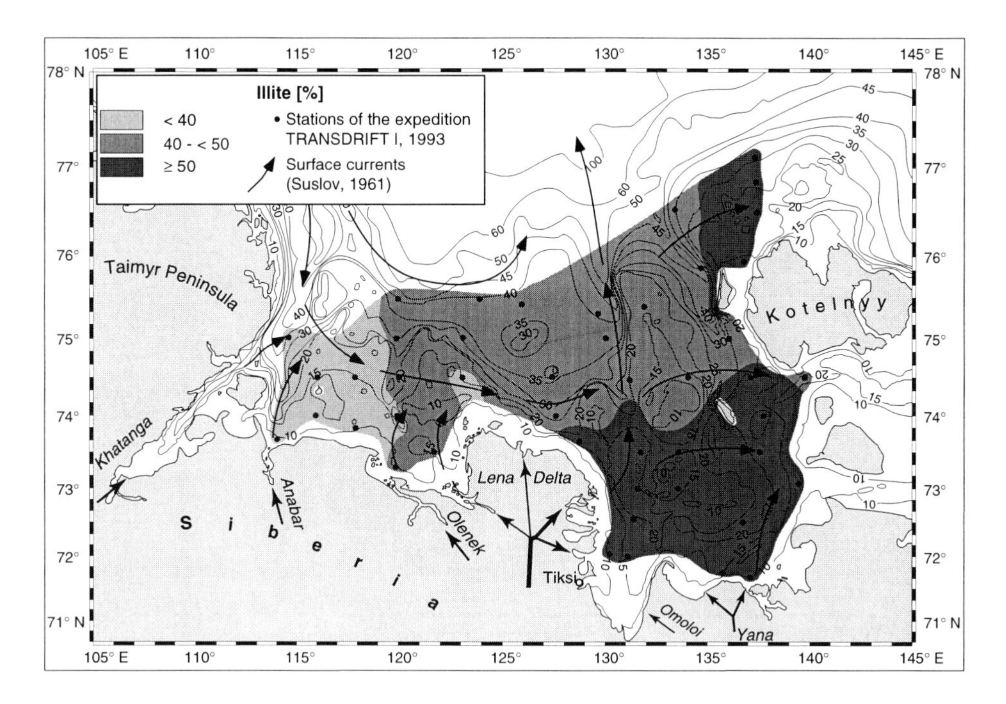

Figure 4: Distribution of illite in surface sediments of the Laptev Sea.

Discussion

Clay mineral provinces and sediment dynamics of the Laptev Sea Shelf

The clay mineral distribution of the Laptev Sea surface sediments (Figures 4 - 7) defines three provinces: 1. In the western Laptev Sea sediments are particularly rich in smectite and are characterized by a slight enrichment in kaolinite. 2. Sediments in the eastern Laptev Sea are very poor in kaolinite. 3. The southeastern Laptev Sea is dominated by illite and chlorite.

These clay mineral provinces correspond almost exactly with the benthic zonation of the Laptev Sea into a western and eastern marine region and a southern Estuarine-Arctic Region (Sirenko and Piepenburg, 1994). Furthermore, the distribution patterns of the surface sediment grain sizes (Lindemann, 1994) and of different sediment-echotypes (Benthien, 1994) allow to define two provinces: 1. The western Laptev Sea is dominated by sandy sediments which are characterized by a medium to low penetration of acoustic waves. 2. The eastern and southeastern Laptev Sea sediments are dominated by silt and clay, showing very high penetration of acoustic waves.

These zonations closely mirror the recent sediment dynamics of the Laptev Sea shelf which is strongly influenced by fluvial sediment supply (Silverberg, 1972; Holmes and Creager, 1974; Kassens and Karpiy, 1994; and Lindemann, 1994). The clay mineral distribution patterns (Figuress 4 - 7) indicate that all rivers, particularly Lena and Yana, deliver illite and chlorite - the most abundant clay minerals in cold regions. Smectite which today predominantly occurs in soils of warm-temperate and humid climate (e.g., Chamley, 1987) is mainly delivered by the

Anabar and Khatanga rivers. Kaolinite which cannot form under a polar climate is most probably supplied by the Anabar, Khatanga, and Olenek rivers.

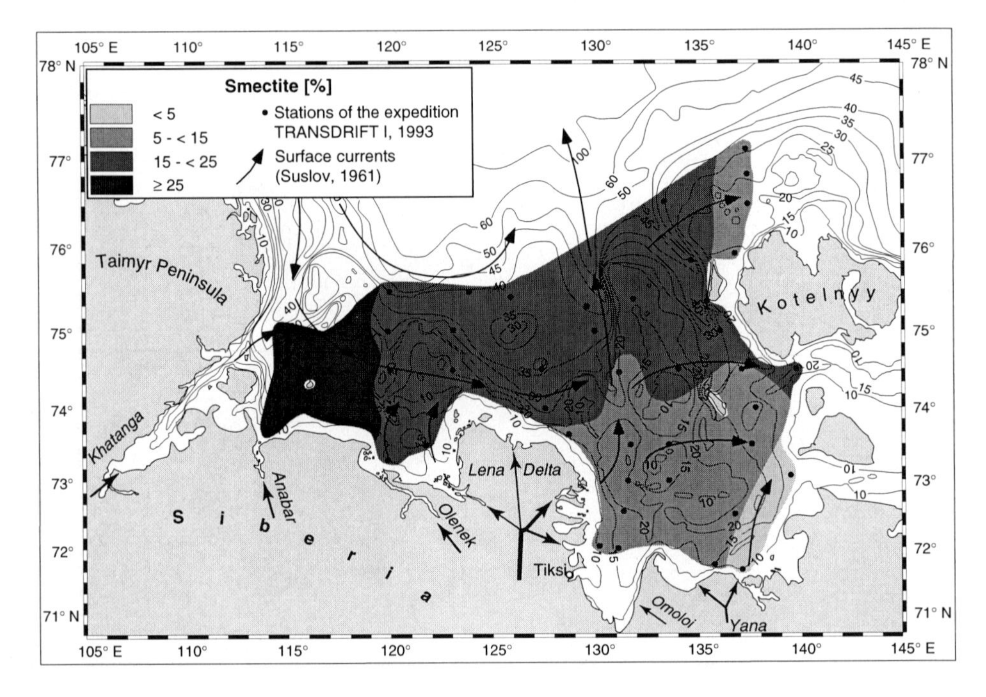

Figure 5: Distribution of smectite in surface sediments of the Laptev Sea.

Obviously, surface currents (Suslov, 1961) play an important role for the spatial distribution of the flaky clay minerals: eastward surface currents prevent westward transport of illite and chlorite, whereas kaolinite and in particular the small sized smectite cystals are transported in suspension hundreds of km to the east (Figure 8).

To explain the enrichment of illite and chlorite in the southeastern part of the Laptev Sea, bathymetry has to be considered: the southeastern part of the Laptev Sea is limited northward by the Vasilevsky Bank (Figure 2). The northern end of the brackish surface plume covering the southeastern Laptev Sea every summer is also located above this shoal (Létolle et al., 1993). Probably, the dominant clay minerals of the southeastern Laptev Sea, illite and chlorite, are settled out of a nepheloid layer described by Létolle et al. (1993) and Burenkov (1993) who observed this layer in close connection with the brackish surface plume in this area. The good correspondence of the clay mineral provinces and the benthic faunal zonation (Sirenko and Piepenburg, 1994) emphasises the combined supply of oxygen, nutrients, and sediment by the Yana and Lena rivers into the southeastern Laptev Sea.

Ice transport and aeolian input of clay minerals within the Laptev Sea shelf is not obvious from the clay mineral distribution patterns but cannot be excluded. After Pfirman et al. (1989, 1990) and Wollenburg (1993) aeolian transport of sediments into Arctic sea ice is generally negligible because of little fallout rates.

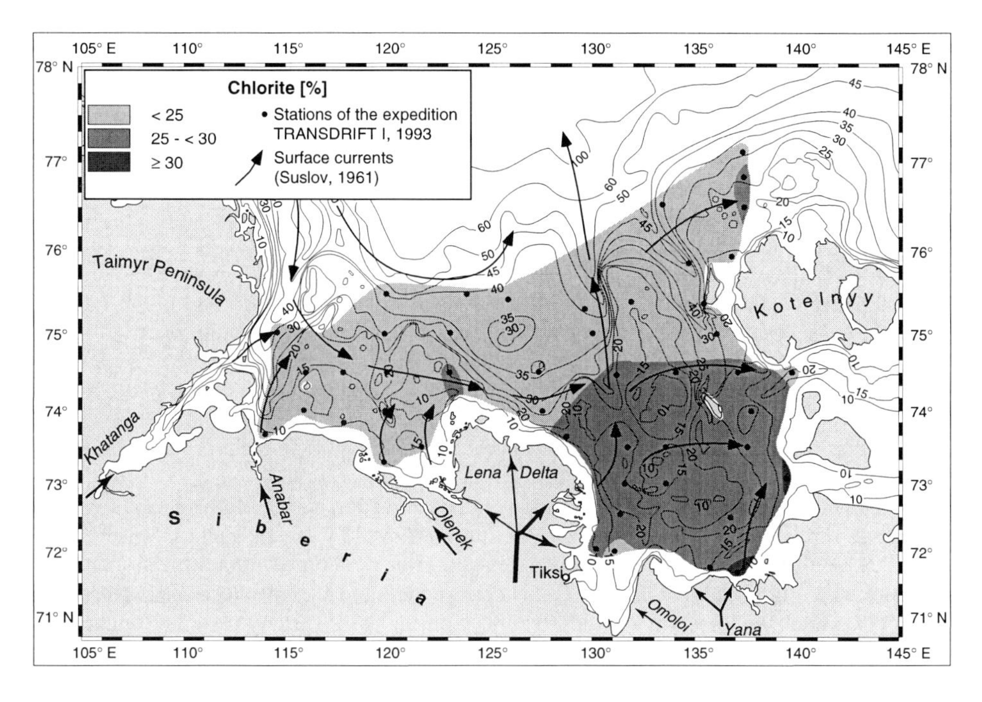

Figure 6: Distribution of chlorite in surface sediments of the Laptev Sea.

Potential clay mineral source areas in the Siberian hinterland

In arctic regions physical weathering is predominant. The formation of clay minerals by chemical weathering processes, for instance hydrolysis, is negligibly low so that the clay fraction in arctic soils only consists of - sometimes altered - fragments of the minerals of the parent rocks or soils (e. g., Chamley, 1987). Therefore, particularly in cold regions clay minerals are very good indicators for sediment origin.

Illite and chlorite

Due to alteration like mica foliation, feldspar sericitisation, and silica chloritization of widespread magmatic or metamorphic substrates illite and chlorite are the most abundant clay minerals in high latitudes (Chamley, 1987). This fact is clearly reflected by the high illite and chlorite concentrations in the Laptev Sea surface sediments (Figures 4 and 6, and Table 1). However, it is obvious that especially the Lena and Yana rivers deliver illite and chlorite (Figures 4, 6, and 8), two clay minerals which are main components of Paleozoic slates (e.g., Heim, 1990). Paleozoic slates occur on the Siberian platform and in the Werchojansk-Fold-Zone (Dolginow and Kropatschjow, 1994). Erosion of these rocks may therefore produce high amounts of illite and chlorite which are transported to the Laptev Sea by the Lena and Yana rivers. The illite 'crystallinity' measurements support this hypothesis: the 'crystallinity' of the Laptev Sea illites averages 0.25° $\Delta 2\Theta$ indicating that the illites originate from at least anchimetamorphic source rocks. Furthermore, the surface sediments near the Lena and Yana river mouths contain up to 5 wt.-% slate fragments in the coarse fraction (Rossak, 1995), confirming the assumed source area of chlorite and illite.

Smectite and kaolinite

Smectite and kaolinite can basically form by alteration of volcanic glass, feldspars, micas, and various FeMg (alumo-) silicates (Weaver, 1989). For the formation of smectite a warm-temperate and humid climate is necessary with lower drainage and rainfall and with lower temperatures than are necessary for kaolinite formation (Chamley, 1987 and Weaver, 1989). Today smectite is present in all soils of warm-temperate regions and is also abundant in Mesozoic and Tertiary sedimentary rocks, (Weaver, 1989).

Therefore, low concentrations of smectite in sediments of the Laptev Sea shelf could be derived from the erosion of Jurassic and Cretaceous sediments which are widespread in the foreland basin of the Werchojansk-Fold-Zone drained by the Lena river, and in the Jenissej-Khatanga basin drained by the Anabar and Khatanga rivers. The high smectite concentrations of the western Laptev Sea sediments, however, require a further source of smectite in the Siberian hinterland, namely volcanic rocks. Exposed to a temperate-humid climate the weathering of basaltic rocks may lead to the formation of smectite (e.g., Chamley, 1987 and Weaver, 1989). With increasing soil development under a tropical climate smectite formed on basaltic substrate transforms gradually into kaolin minerals (Caroll and Hathaway, 1963).

The existence of a warmer, non-polar climate in Arctic regions during Mesozoic and Cenozoic periods permitting intensive chemical weathering is prooved by several studies: Spicer (1987) for instance infered from late Cretaceous vegetation of northern Alaska a mean annual air temperature of 10 °C between 75° and 85° N corresponding to a temperate climate. Nilsen and Kerr (1978) document a warm and humid climate during early Tertiary period throughout the North Atlantic region and into the Arctic Ocean region: on a plateau basalt of the Iceland-Faeroe Ridge a laterite soil containing smectite at the bottom and kaolinite at the top of its profile (see Caroll and Hathaway, 1963) developed during early Tertiary period.

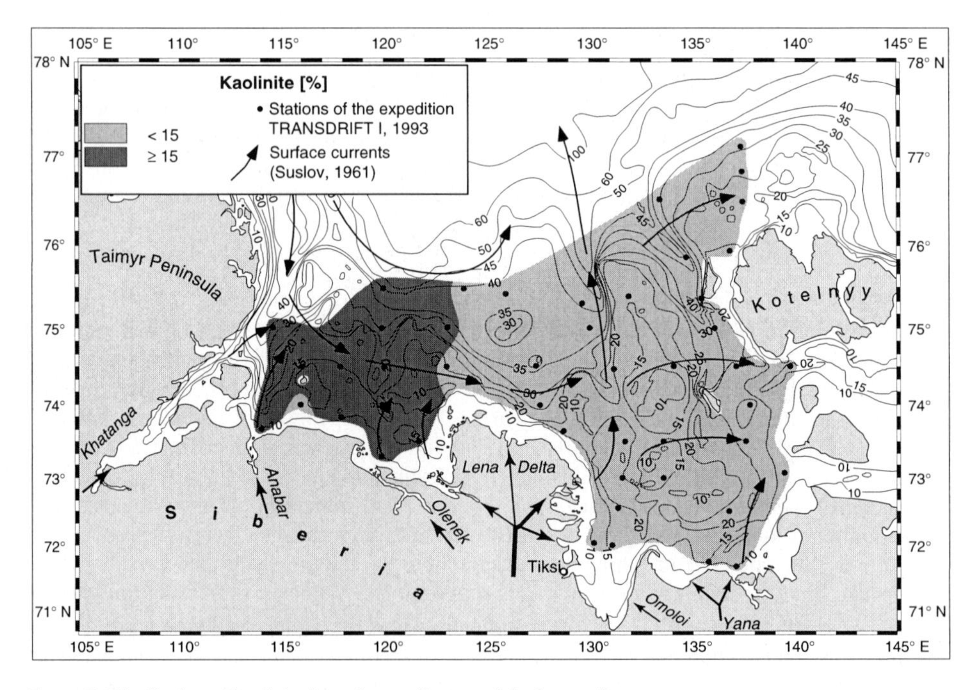

Figure 7: Distribution of kaolinite in surface sediments of the Laptev Sea.

For the Putoran Plateau flood basalts drained by the Anabar and Khatanga rivers the development of soils including the formation of smectite and possibly even of kaolinite therefore is probable during Mesozoic and Cenozoic periods. Erosion of these smectite-bearing remnants of paleosols explain the high concentrations of smectite in western Laptev Sea sediments. Since the Putoran Plateau is also drained by the Jenissej river, smectite concentrations of the Kara sea should also be high. In fact, concentrations of smectite even reach values of up to 70 % off the Jenissej river mouth (Wahsner and Shelekhova, 1994).

Since kaolinite cannot form under a polar climate, it has to be of detrital origin in Laptev Sea surface sediments, too. According to Darby (1975), kaolinite in Amerasian Arctic Ocean deep sea sediments is derived from reworked deposits and from relict soils of northern Alaska and Canada. Particularly in Mesozoic series kaolinitization is the most common weathering process indicating a warm and humid climate during this period (e.g., Chamley, 1987; Sigleo and Reinhardt, 1988, and Weaver, 1989). In addition to volcanic origin, kaolinite enrichment in western Laptev Sea surface sediments can therefore be explained by the erosion of kaolinite-bearing Jurassic and Cretaceous clastic series of the Siberian Platform and of the Jenissej-Khatanga basin acting as the drainage area of Olenek, Anabar, and Khatanga rivers.

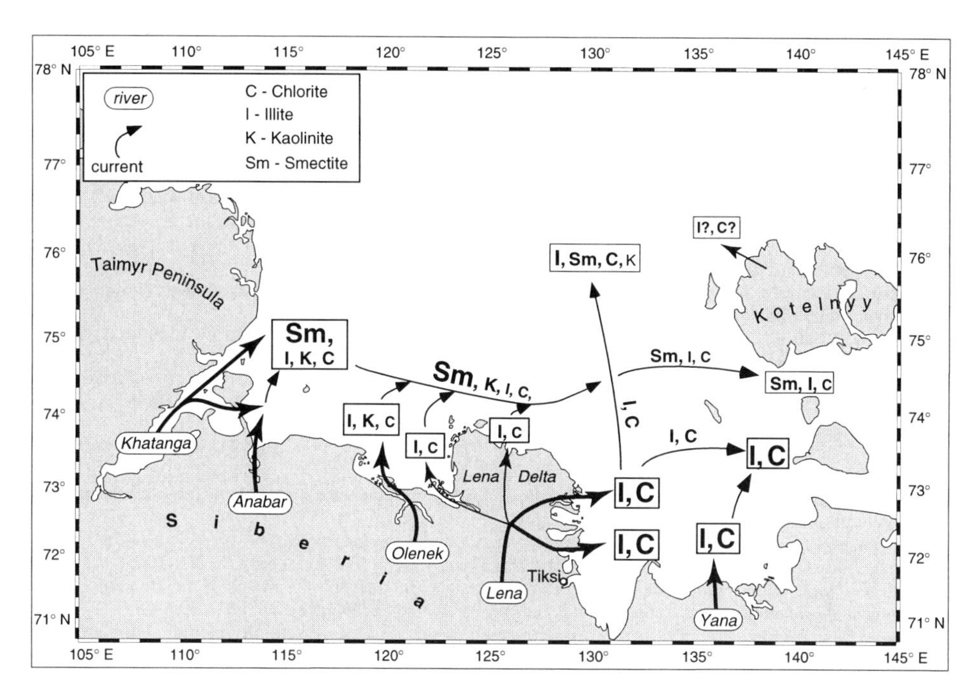

Figure 8: Relationship between fluvial input and surface currents (Suslov, 1961) in the distribution of clay minerals on the Laptev Sea shelf. The sizes of the symbols correspond to the regional significance of the clay minerals, not to their absolute concentrations.

Acknowledgements

This study is part of B.T. Rossak's diploma-thesis carried out at the GEOMAR Research Center for Marine Geosciences in Kiel, Germany, and was supported by the German Ministry of Education, Science, Research and Technology (BMBF), research project "Laptev Sea

System". We are grateful to the crew of RV Ivan Kireyev who carefully collected the sediment samples. H. Vossmerbäumer, University of Würzburg, is thanked for making this study possible. Thanks are due to F. Lindemann, GEOMAR, Kiel, for providing the samples and for carrying out the grain size separation. M. Wahsner, Alfred Wegener Institute for Polar and Marine Research, Bremerhaven, offered useful suggestions and allowed insight into her unpublished data.

References

Barnett, D. (1991) Sea ice distribution in the Soviet Arctic. In: The Soviet Maritime Arctic, Brigham, L.W. (ed.), Belhaven Press, London, 47-62.

Benthien, A. (1994) Echographiekartierung und physikalische Eigenschaften der oberflächennahen Sedimente in der Laptevsee. Unpubl. Diplomarbeit, Teil II, Geologisch-Paläontologisches Institut, Univ. Kiel, 80pp.

Biscaye, P.E. (1964) Distinction between kaolinite and chlorite in recent sediments by X-ray diffraction. Amer. Mineralogist, 49, 1281-1289.

Biscaye, P.E. (1965) Mineralogy and sedimentation of recent deep-sea clay in the Atlantic Ocean and adjacent seas and oceans. Geol. Soc. America Bull., 76, 803-832.

Burenkov, V.I. (1993) Distribution of suspended matter in the Laptev Sea waters (from optical data). Third International Symposium: The Arctic Estuaries and Adjacent Seas: Biogeochemical Processes and Interaction with Global Change, Abstracts, 9-10.

Caroll, D. and J.C. Hathaway (1963) Mineralogy of Selected Soils from Guam, with a section on description soil profiles by C.H. Stensland. Geol. Survey Prof. Paper, 403-F, 1-53.

Chamley, H. (1989) Clay Sedimentology. Springer, Berlin, 623pp.

Coffin, M.F. (1992) Large igneous provinces studied. EOS, 73, 66-67.

Colony, R. and A.S. Thorndike, (1984) An estimate of the mean field of Arctic sea ice motion. J. Geophys. Res., 89, 10623-10629.

Darby, D.A. (1975) Kaolinite and other clay minerals in Arctic Ocean sediments. Journ. Sed. Pet., 45, 272-279.

Dethleff, D., D. Nürnberg, E. Reimnitz, M. Saarso and Y.P.Savchenko (1993) East Siberian Arctic Region Expedition '92: The Laptev Sea - Its Significance for Arctic Sea-Ice Formation and Transpolar Sediment Flux. Reports on Polar Research, 120, 1-44.

Dolginow, J. and S. Kropatschjow (1994) Abriß der Geologie Rußlands und angrenzender Staaten. Schweizerbart, Stuttgart, 174pp.

Gordeev, B.B. and I.S. Sidorov (1993) Concentrations of major elements and their outflow into the Laptev Sea by the Lena River. Mar. Chem., 43, 33-45.

Heim, D. (1990) Tone und Tonminerale. Enke, Stuttgart, 157pp.

Holmes, M.L. and J.S. Creager (1974) Holocene history of the Laptev Sea continental shelf. In: Marine Geology and Oceanography of the Arctic Sea, Herman, Y. (ed.), Springer, New York, 211-229.

Kassens, H. and V.Y. Karpiy (1994) Russian-German Cooperation: The TransdriftI Expedition to the Laptev Sea. Reports on Polar Research, 151, 1-168.

Kisch, H.J. (1987) Correlation between indicators of very low-grade metamorphism. In: Low temperature metamorphism, Frey, M. (ed.), Blackie, Glasgow, 227-300.

Kisch, H.J. and M. Frey (1987) Effect of sample preparation on the measured 10 Å peak width of illite (illite 'crystallinity'). In: Low temperature metamorphism, Frey, M. (ed.), Blackie, Glasgow, 301-304.

Krumm, S. (1992) Illitkristallinität als Indikator schwacher Metamorphose - methodische Untersuchungen, regionale Anwendungen und Vergleiche mit anderen Parametern. Erlanger geol. Abh., 120, 1-75.

Krumm, S. and W. Buggisch (1991) Sample preparation effects on illite crystallinity measurement, grain size gradation and particle orientation. J. metamorphic Geol., 9, 671-677.

Kübler, B. (1967) La cristallinité de l'illite et les zones tout à fait supérieures du métamorphisme. In: Étages tectonipues, Colloque de Neuchâtel, 18-21 avril 1966, Institut de Géologie de l'Université de Neuchâtel (ed.), À la Baconnière, Neuchâtel, Suisse, 106-122.

Kübler, B. (1968) Evaluation quantitative du métamorphisme par la cristallinité de l'illite. Bull Centre Rech. Pau-SNPA, 2, 385-397.

Kübler, B. (1984) Les indicateurs des transformations physiques et chimiques dans la diagenèse, température et calorimétrie. In: Thérmométrie et barométrie géologiques, Lagache, M. (ed.), Soc. Franç. Minér. Crist., Paris, 489-596.

Lange, H. (1982) Distribution of chlorite and kaolinite in eastern Atlantic sediments off North Africa. Sedimentology, 29, 427-431.

Létolle, R., J.M. Martin, A.J. Thomas, V.V. Gordeev, S. Gusarova and I.S. Sidorov (1993) ^{18}O abundance and dissolved silicate in the Lena delta and Laptev Sea (Russia). Mar. Chem., 43, 47-64.

Lindemann, F. (1994) Sonographische und sedimentologische Untersuchungen in der Laptevsee, sibirische Arktis. Unpubl. Diplomarbeit, Teil II, Geologisch-Paläontologisches Institut, Univ. Kiel, 75pp.

Nalivkin, D.V., A.P. Markovskiy, S.A. Muzylev, E.T. Shatalov and L.P. Kolosova (eds.)(1965): Geological Map of the Union Of Soviet Socialist Republics, Scale 1 : 2 500 000. The Ministry of Geology of the USSR, Moscow.

Nilsen, T.H. and D.R. Kerr (1978) Paleoclimate and paleogeographic implications of a lower Tertiary laterite (latosol) on the Eceland-Faeroe Ridge, North Atlantic region. Geol. Magazine, 115, 153-184.

Nürnberg, D., M.A. Levitan, J.A. Pavlidis and E.S. Shelekhova (1995) Distribution of clay minerals in surface sediments from the eastern Barents and south-western Kara seas. Geol. Rundsch., 84, 665-682.

Nürnberg, D., I. Wollenburg, D. Dethleff, H. Eicken, H. Kassens, T. Letzig, E. Reimnitz, and J. Thiede (1994) Sediments in the Arctic sea ice: Implications for entrainment, transport and release. Mar. Geol., 119, 185-214.

Pfirman, S.L., M.A. Lange, I. Wollenburg and P. Schlosser (1990) Sea ice characteristics and the role of sediment inclusions in deep-sea deposition: Arctic - Antarctic comparisons. In: Geological History of the Polar Oceans: Arctic versus Antarctic, Bleil, U. and J. Thiede (eds.), Kluwer, Dordrecht (NATO ASI Series C), 187-211.

Pfirman, S.L., I. Wollenburg, J. Thiede and M.A. Lange (1989) Lithogenic sediment on Arctic pack ice: Potential aeolian flux and contribution to deep sea sediments. In: Paleoclimatology and Paleometeorology: Modern and Past Patterns of Global Atmospheric Transport, Leinen, M. and M. Sarnthein (eds.), Kluwer, Dordrecht (NATO ASI Series C), 463-493.

Reimnitz, E., L.J. Marincovich, M. McCormick and W.M. Briggs (1992) Suspension freezing of bottom sediment and biota in the Northwest Passage and implications for Arctic Ocean sedimentation. Can. J. Earth Sci., 29, 693-703.

Rossak, B. (1995) Zur Tonmineralverteilung und Sedimentzusammensetzung in Oberflächensedimenten der Laptevsee, sibirische Arktis. Unpubl. Diplomarbeit (Laborteil), Institut für Geologie/GEOMAR Forschungszentrum für marine Geowissenschaften, Univ. Würzburg/Univ. Kiel, 101pp.

Sigleo, W. and J. Reinhardt (1988) Paleosols from some Cretaceous environments in the southeastern United States. In: Paleosols and weathering through geologic time: principles and applications, Reinhardt, J. and W.R. Sigleo (eds.), Geol. Soc. Am., Boulder, Colorado (special paper 216), 123-142.

Silverberg, N. (1972) Sedimentology of the Surface Sediments of the East Siberian and Laptev Seas. Unpubl. PhD-thesis, Department of Oceanography, Washington D.C., 185pp.

Sirenko, B. and D. Piepenburg, (1994) Current knowledge on biodiversity and benthic zonation patterns of Eurasian Arctic shelf seas, with special reference to the Laptev Sea. Reports on Polar Research, 144, 69-77.

Spicer, R.A. (1987) The Significance of the Cretaceous Flora of Northern Alaska for the Reconstruction of the Climate of the Cretaceous. In: Das Klima der Kreide-Zeit, Kemper, E. (ed.), Schweizerbart, Hannover (Geol. Jb. A, 96), 265-291.

Suslov, S.P. (1961): Physical Geography of Asiatic Russia. Freeman & Co., San Francisco, 594pp.

Wahsner, M. and E.S. Shelekhova (1994) Clay-mineral distribution in Arctic deep sea and shelf surface sediments.- Greifswalder Geologische Beiträge, A (2), 234 (Abstract).

Weaver, C.E. (1989) Clays, Muds, and Shales. Elsevier, Amsterdam (Developments in Sedimentology 44), 819pp.

Wollenburg, I. (1993) Sedimenttransport durch das arktische Meereis: Die rezente lithogene und biogene Materialfracht. Reports on Polar Research, 127, 1-159.

Zakharov, V.F. (1966) The role of flaw leads off the edge of fast ice in the hydrological and ice regime of the Laptev Sea.- Acad. Sci. USSR Oceanology, 6, 815-821.

Planktic Foraminifera in Holocene Sediments from the Laptev Sea and the Central Arctic Ocean: Species Distribution and Paleobiogeographical Implication

H.A. Bauch

GEOMAR Forschungszentrum für marine Geowissenschaften, Wischhofstrasse 1-3, D 24148 Kiel, Germany

Received 10 March 1997 and accepted in revised form 24 March 1998

Abstract - Two sediment cores, one from the Siberian Laptev Sea shelf and another from the central Arctic Ocean were investigated with respect to the temporal distribution of planktic foraminifera as well as the species composition.
The assemblage in the Laptev Sea, which cover the time back to about 2.3 ka, is mainly comprised of species that can be related to a subtropical-subpolar biogeography and is dominated by the minute species *Turborotalita clarkei.* The origin of the non-polar species on the Siberian shelf remains unknown. It is suggested that these microfossils were either reworked from older sediments and/or entrained into sea-ice on other shallow Arctic shelfs, transporting them eventually to the Laptev Sea.
In contrast to the Laptev Sea the glacial and Holocene sediments from the perennially ice-covered central Arctic Ocean reveal a typical monospecific record, which is entirely made up of the polar species *Neogloboquadrina pachyderma* (right and left coiling forms). The relative variability of the two coiling varieties is of the order of 5%. Although this variability does not seem to be related to climate-induced environmental changes, total test concentration shows a clear signal of increasing deposition of foraminiferal tests after 7 ka. Since sea-ice and surface water conditions seem to be stable in this region, the observed increase in plankton productivity during the Holocene must be linked to changes in water mass circulation.

Introduction

The concept that past surface water properties can be deduced from assemblage studies of planktic foraminifera was established some time ago (Imbrie and Kipp, 1971). This finding is based on studies of sediments from the surface of the ocean floor which are thought to give a fairly good reflectance of the species composition in the overlying surface water masses. Previous faunal studies, in which a detailed mapping of the ocean's floor core-top sediments was conducted, revealed various foraminiferal assemblages (Kipp, 1976). The faunal composition could be linked to specific water mass properties such as sea surface water temperature (SST). In the modern ocean, the SST displays a latitudinal change towards the northern polar regions that is mirrored by a change in the foraminiferal assemblage from warm-water representing types to a typically subpolar or polar species dominated assemblage (Bradshaw, 1959; Kipp, 1976).

Studies of plankton tows and sediment samples of Holocene age from the Arctic Ocean indicate that the faunal assemblage is rather monospecific, consisting mainly of the polar species *Neogloboquadrina pachyderma* sinistral (Bé, 1960; Steuerwald and Clark, 1972; Vilks, 1975). In areas of the Arctic Ocean influenced today by inflowing non-Arctic waters, foraminfera of a subpolar provenance can comprise notable portions of the faunal assemblage (Carstens and Wefer, 1992). However, these species are not indigenous to the Arctic Ocean but are rather the result of water-mass advection from the south (Bauch et al., 1997).

Since planktic foraminifera are deep-ocean surface dwellers, they are commonly not present on shallow shelf areas unless they are being advected. In this paper the down-core species

----

In: Kassens, H., H.A. Bauch, I. Dmitrenko, H. Eicken, H.-W. Hubberten, M. Melles, J. Thiede and L. Timokhov (eds.) Land-Ocean Systems in the Siberian Arctic: Dynamics and History. Springer-Verlag, Berlin, 1999, 601-613.

composition is studied in a sediment core from a shallow Siberian shelf and from the Central Arctic Ocean. The purpose of this study is to document the occurrence of unusual species which were up to now never known to occur in Holocene sediments from this region, and to discuss various possibilities regarding the origin of these species.

Water masses and depositional environment

The modern oceanography of the Arctic Ocean is influenced by waters from three main sources. Through Fram Strait and across the Barents and Kara seas high-salinty water of Atlantic origin enters the Eurasian basin of the Arctic Ocean within the upper 600 m of the water column (Rudels, 1995; Hanzlick, 1983). To a much lesser degree Pacific surface water flows in via the Bering Strait (Coachman et al., 1975). Besides these marine sources, the Arctic Ocean is also fed by a vast fluvial freshwater runoff. This water mainly derives from large Siberian rivers such as the Ob, Jenisey, and Lena rivers. Although the contribution of the total mass of this river water to the Arctic Ocean is relatively low in comparison to the other two sources, it is of major significance for the formation of a distinct halocline. Furthermore, increased cooling processes on the shelf areas during fall season subsequently leading to sea ice formation favours the release of brines (Aagaard and Carmack, 1989). This process is responsible for much of the water mass transformation within the Arctic Ocean circulation system.

The bulk of Siberian shelves are rather shallow on average (<50 m) whereas the continental slope exhibits a steep break at ~100 m water depth. The topography of the Laptev Sea is marked by a gently northerly dipping plain cut by various submarine valleys (Holmes and Creager, 1974). These channels are linked to the mouths of the major rivers and run along preformed tectonic structures (Drachev et al, in press). The freshwater from these rivers not only reduces the salinity at the surface, it also influences the temperature of the surface water during summer, which can reach 8-10°C or more (Baskakov et al., 1987). Farther off the shelf, temperatures within the halocline are - 1.8° C on average. Below the halocline Atlantic core water with temperatures of ~1.5° C is noted (Timokhov, 1994). The Atlantic water flows along the continental slope of the Barents and Kara seas before reaching the Laptev Sea slope. Nevertheless, the Atlantic water does not seem to flow onto the Laptev Sea shelf itself (Schauer et al., 1997). The fate of the Pacific surface water in the Arctic Ocean is not very well known. This surface water has a distinct silicate signature (Walsh et al., 1989) and may be used as a suitable tracer (Bauch et al., 1994).

Previous investigations on grain-size analyses in surface sediments from the Laptev Sea showed that sediments are generally finer in the eastern part where silty clays dominate the surface pattern (Lindemann, 1995; Rossak 1995). To some extent this pattern is due to different amounts of suspended sediment material transported onto the shelf by the main rivers (Alabyan et al., 1995). Another source for sediment supply is coastal erosion (Aré, 1988). By incorporation into sea-ice these sediments can be transported far distances across the shelf (Eicken et al., 1997). Biogenic carbonate is rare in these sediments (Bauch et al., 1995) as compared to those Eurasian shelf areas which are under the influence of Atlantic surface water (Hald and Steinsund, 1996).

Sample processing and counting technique

The studied gravity core (IK 9373-10) from the Laptev Sea shelf was taken onboard RV *Ivan Kireyev* during the Russian-German Transdrift I expedition in 1993 (Figure 1). Core IK 9373-10 was retrieved from a water depth of 47 m and has an overall sediment recovery of 109 cm (Kassens and Karpiy, 1995). The site is located within the so-called Yana valley in the eastern

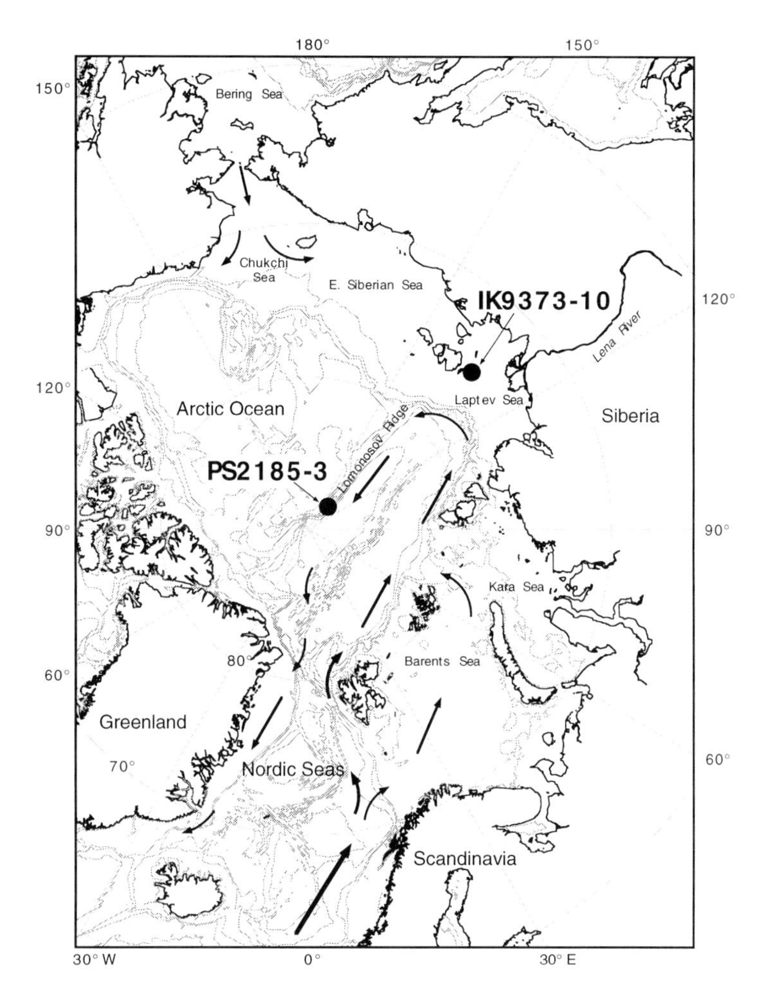

Figure 1: Map of the Arctic Ocean showing the general circulation pattern of the surface/sub-surface water mass flow into the Arctic basins (modified from Rudels, 1995). The position of the two studied cores are indicated.

Laptev Sea (75°20,8' N / 135°12,3' E). Samples were taken at 5 cm intervals as 2 cm thick slices. Prior to washing, all samples were frozen and dehydrated. In order to prevent the formation of aggregates during washing the sediments had to be treated with hydrogen peroxide. All samples were then wet-sieved over 63 µm meshsize, and eventually dried in an oven at 50° C. It was already realized during the early part of the sample processing that faunal constituents will remain rather low in this core due to small amounts of sample residues >125 µm. Therefore, the > 63µm size fraction was left unsplit in order to take into account every specimen for the qualitative and quantitative determination down to species level. Later quantifications are expressed as specimens per weight of the dried bulk sediment and as relative abundances.

The second core, PS2185-3, originates from the Lomonosov Ridge in the Central Arctic Ocean (87°32' N / 144°22.9' E) and was taken from 1051 m water depth during cruise ARCTIC '91 (Fütterer, 1992). Samples of this trigger box core were taken in 1 cm steps.

Faunal counts were carried out in two separate size-fractions (125-250 μm, and 250-500 μm). In order not to miss any small-sized species, which do occur in high-latitude glacial and interglacial sediments below the lower limit of the used mesh size (Kellogg, 1984; Bauch, 1994; Hebbeln et al., 1994), the size-fraction 63-125 μm was also carefully checked.

Table 1: List of radiocarbon measurements of core IK9373-10

Lab. No.	Depth in core (cm)	age (y BP-400y)
AAR-2253	32.5	1150±75
AAR-2254	53.5	1420±45
AAR-2255	62.5	1540±55
AAR-2256	71.5	1760±55
AAR-2257	80.5	1670±70
AAR-2258	89.5	1955±60
AAR-2259	100	2150±60

Age model

The age model of core IK 9373-10 is based on several radiocarbon datings. These datings were performed on bivalve shells (*Portlandia sp.*) at the AMS-laboratory of Aarhus University in Denmark (Table 1). All datings are corrected for a ^{14}C reservoir effect of 400 y (original ages are between 300-500 y). This reservoir age of 400 y, which seems slightly younger than reservoir ages recently published for the Barents and Kara seas (Forman and Polyak, 1997), is based on pre-bomb mollusc shells (Bauch and Heinemeier, unpubl.) and is additionally corroborated by the method of 'supported ^{210}Pb' performed on the same core (Erlenkeuser, unpubl. data).

As is revealed by the depth/dating relation in Table 1, there is an age reversal between 80-70 cm. This age reversal is marked by two datings, and because the errors of these two datings overlap an interpolated age of the mid-depth between the two datings was calculated and used for the final age model (Figure 2). Ages between all dated samples were calculated by assuming a constant sedimentation rates between these points. From the original ^{14}C-age measurements follows that the sedimentation rates above the uppermost dating steeply decreases towards the sediment surface. Because it was not possible to produce a dating for the upper section of core IK9373-10, it is assumed that either sedimentation ceased after 700 y BP or that this area became eroded after this time. Thus sediment ages in the upper- and lowermost section of the core were estimated by extrapolation. According to this modified age model, core IK9373-10 covers the time interval 2500-500y BP with sedimentation rates varying between 55-77 cm/ky.

The age model of the core from the Central Arctic Ocean is based on a series of 13 AMS ^{14}C datings performed on *N. pachyderma* sin. (Spielhagen et al., 1997; Nørgaard-Pedersen et al., in press). The shown record goes back to about 34 ka BP (at 20 cm core depth). Sedimentation rates for the Holocene section above the end of the last deglaciation are of the order of 1 cm/ky.

Faunal records

Species composition

A list with all identified planktic foraminiferal species together with the approximate time of

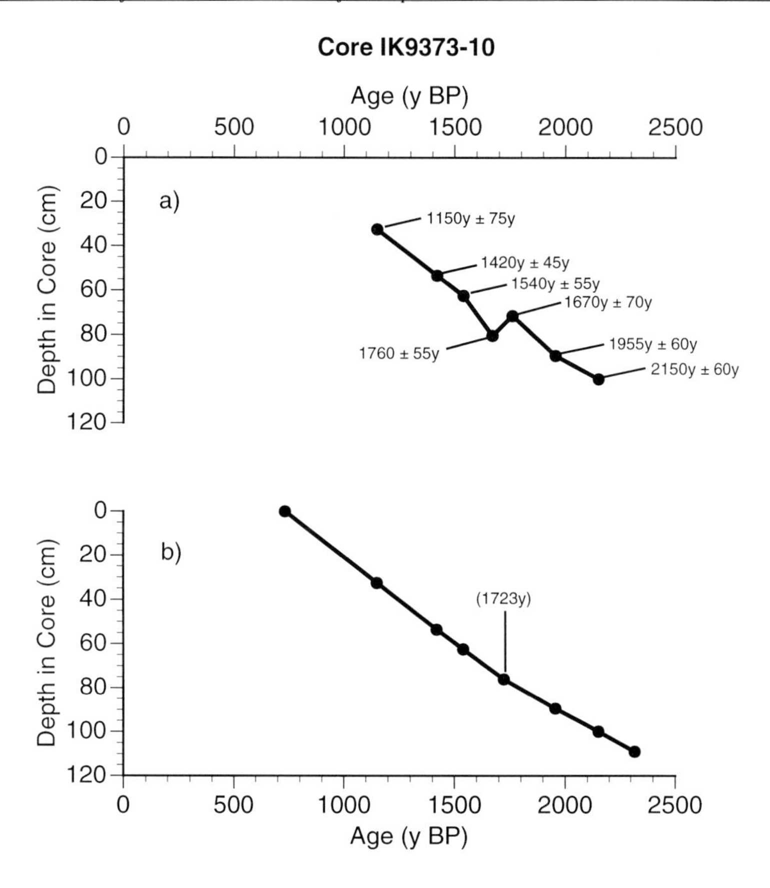

Figure 2: Age/depth relation of core IK9373-10 from the eastern Laptev Sea: a) original radiocarbon measurements (after subtracting 400 y to account for the reservoir effect) showing an age reversal; b) final age model used in this study. The grey dots at the lower and upper end of the curve are based on interpolation. The age in parentheses represents the calculated age between the two datings which marks the age reversal (see also Table 1).

their first stratigraphical occurrence as well as their presumed main latitudinal distribution is given in Table 2. From this list it follows that all species are extent. Although, individual stratigraphical records can go back to the Late Oligocene, most species first occurred some time during the Miocene or Pliocene. Apart from the polar species *N. pachyderma* and the subpolar species *G. bulloides* and *G. glutinata*, *Globoturborotalita tenella* and *Turborotalita clarkei* usually reflect much warmer water masses. All non-polar species were found on the Laptev Sea shelf. Samples from the Lomonosov Ridge only yielded the right and left coiling variety of *N. pachyderma*.

Species from core IK9373-10 are illustrated in Plate 1. Intriguing is the very small test sizes of *T. clarkei*, which range well below 100 µm in diameter. These generally small tests have been noted before and may be the major reason why this species is so often not recognized in most sediment studies (Boltovskoy, 1991).

Temporal variability on Laptev Sea Shelf

The record of total test concentration from core IK9373-10 reveals that planktic foraminifera occur during few specific time intervals only (Figure 3). Accordingly, highest abundances are recognized between 800-1000 a BP and between 1600-1900 a BP. Comparing the relative abundances of individual species with the record of total test concentration clearly indicates that *T. clarkei* is by far the dominant species during these intervals. All other species remain inferior to *T. clarkei* and, moreover, never reach significant numbers even during other intervals.

Table 2: List of identified species from the investigated cores. Species names according to Hemleben et al. (1989). Biogeographical distribution and stratigraphical first occurrence (FO) is based on Kipp (1976), Kennett and Srinivasan (1983), Bolli and Saunders (1984), and Boltovskoy (1991). * biogeographical distribution strongly depending on coiling direction.

Species	FO/Biogeography
Globigerina bulloides	Middle Miocene/subtropical-subpolar
Globigerinita glutinata	Late Oligocene/tropical-subpolar
Globoturborotalita tenella	Late Pliocene/tropical-temperate
Turborotalita clarkei	Late Miocene/tropical-temperate
Neogloboquadrina pachyderma *	Late Miocene/subtropical-polar

Temporal variability on Lomonosov Ridge

The record from the Central Arctic Ocean consists essentially of a two-species assemblage with *N. pachyderma* sin. being the dominant species (Figure 4). This species makes up 94-99 % of all tests. The remainder proportion is due to *N. pachyderma* dex. A comparison of the test concentrations of the two species indicates fairly good corresponding records throughout the past 35 ka. In general, the two records show significantly reduced deposition of tests during the last glaciation. The concentration of tests only begins to increase after the end of the deglaciation, and reaches highest numbers in the later Holocene. Despite the good conformity of the two records, the relative abundances reveal a less proportional relation between both species. A comparatively strong dominance of *N. pachyderma* sin. is found between 15-20 ka, i.e., during the last glacial maximum (LGM). During the deglacial phase (15-9 ka) this dominance becomes less until relatively low values are noted between 6-8 ka. The mid-Holocene is marked by a significant shift at 3-4 ka although test concentrations steadily increase already after about 7 ka.

Discussion

Origin of planktic foraminifera in Arctic sediments of Holocene age

The occurrence of unusually warm-water indicating planktic foraminifera in upper Holocene sediments from the Laptev Sea shelf is intriguing in itself, but cannot be easily answered considering the modern pathway of surface water flow in the Arctic Ocean. At present it is difficult to reconcile the fact that these species from a tropical-temperate provenance can be advected to the very shallow Laptev Sea shelf or that they have lived in this region during any time of the Holocene. So far, species such as *G. tenella* or *T. clarkei* have not been reported from areas such as the Nordic seas (e.g., Kellogg, 1984; Bauch, 1994) from where the main mass of non-polar water masses is advected to the Arctic Ocean. On the other hand species such as *G. bulloides* and *G. glutinata* may more likely invade the Laptev Sea via Atlantic derived waters across the Barents Sea or along the Eurasian continental margin. A recent downcore

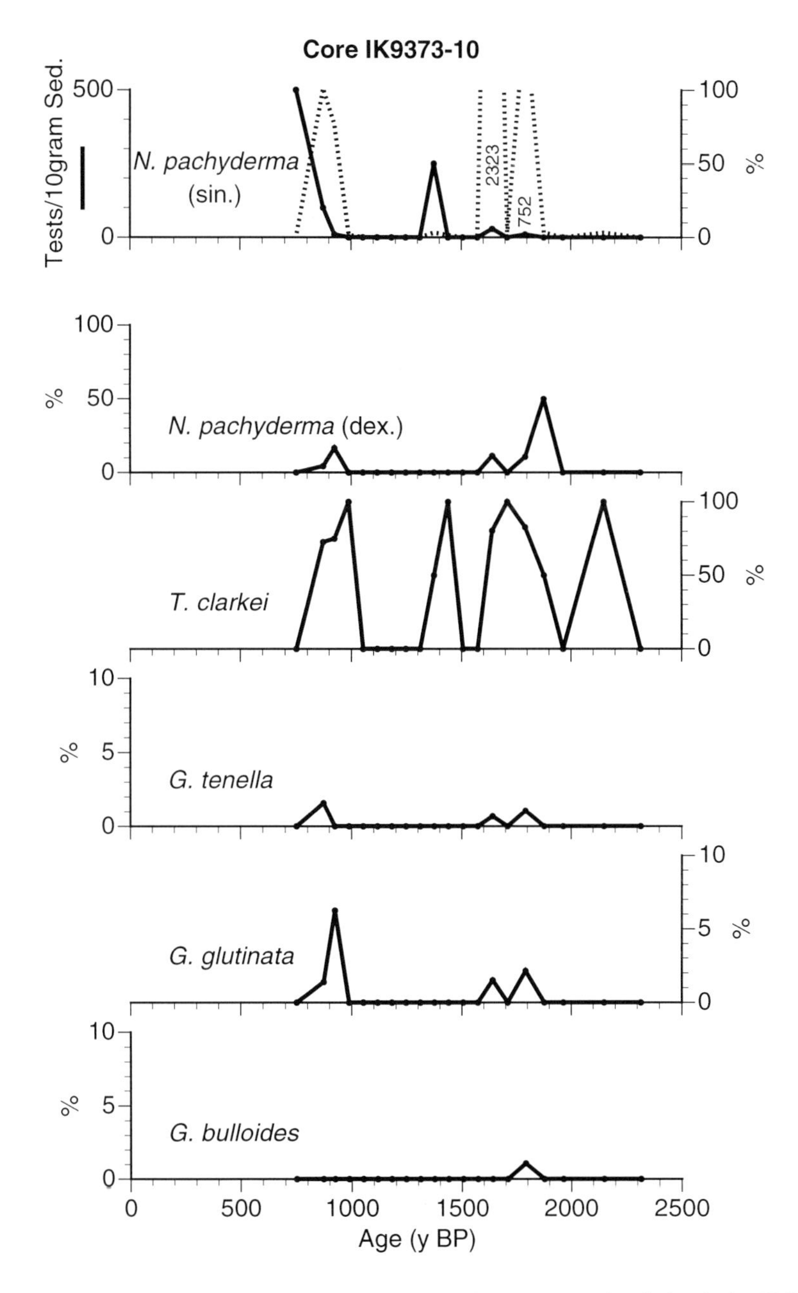

Figure 3: Variability of planktic foraminifera in core IK9373-10 from the Laptev Sea during the late Holocene.

study at the continental slope of the Laptev Sea has yielded some specimens of *G. ruber* in deglacial sediments (R.F. Spielhagen pers. com. 1997). Furthermore, ongoing investigations in this region has confirmed the presence of *G. ruber* even in surface sediments (R. Volkmann pers. com. 1997) - it should be noted at this point that species such as *G. ruber*, *G. tenella*, and

G. rubescens are evolutionary closely linked (Kennett and Srinivasan, 1983) and are morphologically also quite similar. In Laptev Sea surface sediments only very few specimens of *G. bulloides* and *N. pachyderma* have been reported so far (Tamanova, 1971). This generally low number of specimens appears to be typical in this area and is corroborated by more recent studies (Rossak, 1995; Bude, 1997).

Because a direct water mass influence from outside the Arctic Ocean seems unlikely to be responsible for the occurrence of warm species in Holocene sediments, other processes such as coastal erosion and reworking of older sediments may be considered. Although not mapped in detail, cliffs composed of Miocene and Pliocene sediments are known to occur along the eastern shores of the Laptev Sea (Okulitch et al., 1989 Arctic Map) and are confirmed by recent field observations (A. Basilyan, G. Ivanenko, P. Nikolskiy pers. com. 1997). Previous drilling in the SE Laptev Sea yielded sediment successions down to the lower Miocene (Sudakov et al., 1991). But these sediments are shallow water sediments - like those from the terrestrial outcrops - containing typically shallow-marine and freshwater microfossils. Furthermore, detailed studies of the diatom and palynomorph assemblages in Laptev Sea surface sediments as well as in core IK9373-10 could not confirm the possibility of reworking of older sediments (Cremer, 1997; Kunz-Pirrung, this volume). However, previous studies have shown that sediment transport by sea-ice is of major importance for sediment dynamics on Arctic shelfs (Kempema et al., 1989; Eicken et al., 1997). Thus, sediment entrainment from shallow Arctic shelfs and the later release of these sediments (Reimnitz et al., 1987) may be a possible mechanism to explain far-distance transport of recent or even much older fossil remains (Abelmann, 1992; Reimnitz et al., 1992). The mean field of sea-ice motion along the east Siberian shelfs is from Bering Strait towards the Laptev Sea (Rigor, 1992). So, if the findings of warm-water foraminifera in the Laptev Sea is due to sea-ice transport then the actual source may be sought further east. But 'exotic' (mostly subtropical) planktic foraminifera have also been reported from various Holocene and glacial deposits in the Barents and Kara seas area (Gudina, 1966; Blaschishin et al., 1985; Khusid et al., 1995).

Conclusions

With the data presently available it is not possible to decipher the origin of the warm water foraminiferal species in the Laptev Sea. It remains speculative if they had derived from sediments deposited during times when subtropical water masses are known to have reached the Arctic Ocean (Spiegler, 1996) or if they are of Holocene age. To eventually verify that the findings of warm foraminiferal species in the Laptev Sea are actually caused by water inflow during the Holocene further investigations are needed with regard to their true age. This certainly should involve large quantities of sediment sampling in the hope that these large samples may yield sufficient numbers of specimens for radiocarbon analyses.

Based on the foraminiferal record from the Lomonosov Ridge it is evident that the Arctic Ocean fauna is composed of a single genus. Although this area is marked today by a perennial coverage of sea-ice close to 100% with water temperatures in the upper 200 m well below 0° C (Rudels, 1995), foraminiferal bioproductivity has increased steadily since about 7 ka. It is therefore suggested that an inflow of non-polar water masses into the Arctic has a notable influence on planktic productivity even at remote sites such as investigated in the central Arctic Ocean. The relative changes of up to 5 % noted in the two coiling varieties of *N. pachyderma* cannot be related to variations in surface water temperatures and, thus, seems independent of major glacial-interglacial climate changes. The variability of the two coiling forms is more likely caused by unknown population dynamics. To what extent the relatively small amount of Pacific surface water inflow through Bering Strait (Coachman and Aagaard, 1988) is also influencing the central Arctic foraminiferal productivity is not known. There are no reports regarding planktic foraminiferal species in sediments from the Chukchi and East Siberian seas which could give information on this issue. But as can be deduced from the modern sub-surface circulation in the eastern Arctic Ocean (see Figure 1), the site of PS2185-3 is is under influence of Atlantic water. It is therefore suggested that the increase of planktic foraminifera after 7 ka appears to be related to circulation changes in the Nordic seas, i.e., the onset of Atlantic-derived surface water inflow to the eastern Arctic Ocean.

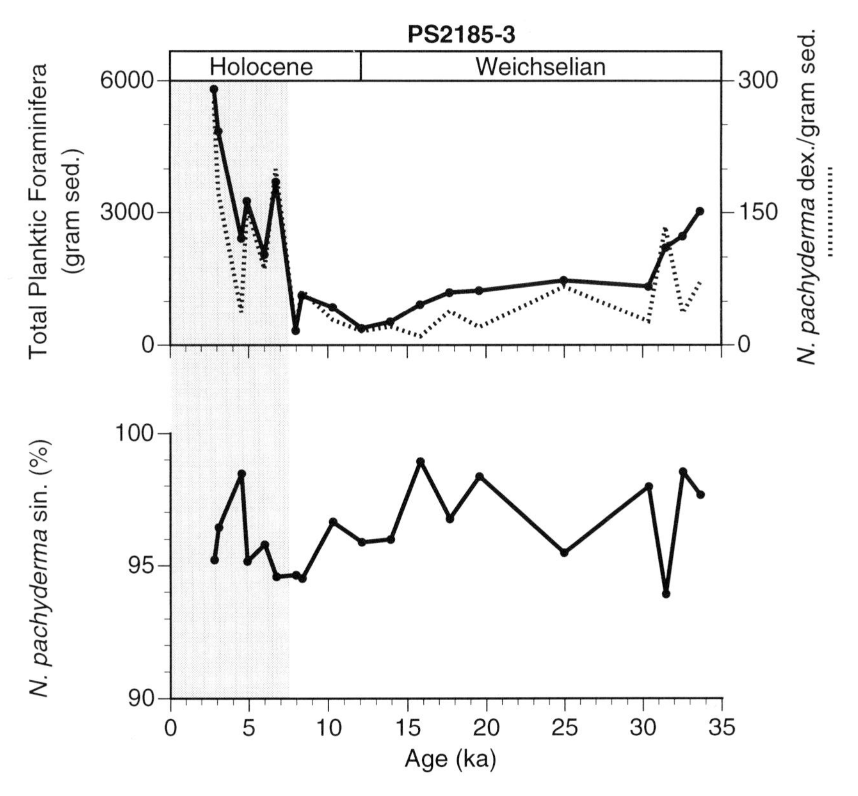

Figure 4: Variability of planktic foraminifera in the left and right coiling variety of *N. pachyderma* in core PS2185-3 from the Lomonosov Ridge during the past 35 ka. Shaded area marks the time interval after the last deglaciation when total number of test concentrations indicate enhanced foraminiferal productivity.

Acknowledgements

I am grateful to all my colleagues and the crews onboard RV „*Ivan Kireyev*" and RV „*Polarstern*" who helped collecting the two studied sediment cores. I am further indebted to R.F. Spielhagen and N. Nørgaard-Pedersen for supplying the age model of core PS2185-3 and to J. Heinemeier at the AMS-laboratory of Aarhus University (DK) for performing the age datings of core IK9373-10. Special thanks are expressed to L. Polyak and M.A. Kaminski for their careful reviews and useful hints. This study was financially supported by the German Ministry for Education, Science, and Technology (BMBF).

References

Aagaard, K. and E. Carmack (1989) The role of sea ice and other freshwater in the arctic circulation. J. Geophys. Res., 94, 14,485-14,498.

Abelmann, A. (1992) Diatom assemblage in Arctic sea-ice - indicator for ice-drift pathways. Deep-Sea Research, 39, S525-S538.

Alabyan, A.M., R.S. Chalov, V.N. Korotaev, A.Y. Sidorchuk and A.A. Zaitsev (1995) Natural and technogenic water and sediment supply to the Laptev Sea. Reports on Polar Research, 176, 265-271.

Aré, F.E. (1988) Thermal abrasions of sea coasts (parts I and II). Polar Geography and Geology, 12, 1-157.

Baskakov, G.A., Borodachev, V.Y., Dvorkin, Y.N., Mustafin, N.V. and Yanes, A.V., 1987. Hydrological and ice conditions of the shelf zone of the Arctic seas. In: Biological resources of the Arctic and the Antarctic. Nauka, 15-48.

Bauch, H.A. (1994) Significance of variability in *Turborotalita quinqueloba* (Natland) test size and abundance for paleoceanographic interpretations in the Norwegian-Greenland Sea. Marine Geology, 121, 129-141.

Bauch, D., Schlosser, P. and Fairbanks, R.G., 1995. Freshwater balance and the sources of deep and bottom waters in the Arctic Ocean inferred from the distribution of $H_2{}^{18}O$. Progress in Oceanography, 35, 53-80.

Bauch, D., J. Carstens and G. Wefer (1997) Oxygen isotope composition of living *Neogloboquadrina pachyderma* (sin.) in the Arctic Ocean. Earth and Planetary Science Letters, 146, 47-58.

Bé, A.W.H. (1960) Ecology of recent planktonic foraminifera: Part 1.- Areal distribution in the western North Atlantic. Micropaleontology, 6, 373-392.

Blaschishin, A.I., L.A. Digas and L.V. Polyak (1985) Exotic planktic microfauna in the diamict deposits of the Barents Sea and its paleogeographic significance. In: Complex studies of the Atlantic Ocean environments. 3rd Regional Conference Abstracts, Kaliningrad, 78-80 (in Russian).

Bolli, H.M. and J.B. Saunders (1985) Oligocene to Holocene low latitude planktic foraminifera. In: H.M. Bolli et al. (eds.), Plankton stratigraphy. Cambrigde University Press, 1, 155-262.

Boltovskoy, E., D. Boltovskoy, N. Correa and F. Brandini (1996) Planktic foraminifera from the southwestern Atlantic (30°-60°S): species-specific patterns in the upper 50 m. Marine Micrpaleontology, 28, 53-72.

Bradshaw, J.S. (1959) Ecology of living planktonic foraminifera in the North and Equatorial Pacific Ocean. Contributions from the Cushman Foundation for Foraminiferal Research, 10, 25-64.

Bude, S. (1997) Artengemeinschaften benthischer Foraminiferen in der Laptev-See, sibirische Arktis: Rezente Verteilungsmuster und Ökologie. Unpubl. M.Sc. thesis Kiel University, 1-46.

Coachman, L.K. and K. Aagaard (1988) Transport through Bering Strait: Annual and interannual variability. Journal of Geophysical Research, 93(C12), 15,535-15,539.

Coachman, L.K., K. Aagaard and R.B. Tripp (1975) Bering Strait: The regional physical oceanography. University of Washington Press.

Carstens, J. and G. Wefer (1992) Recent distribution of planktonic foraminifera in the Nansen Basin, Arctic Ocean. Deep-Sea Research, 39, 507-524.

Cremer, H. (1997) Die Diatomeen der Laptevsee (Arktischer Ozean): Taxonomie, biogeographische Verbreitung und ozeanographische Bedeutung. unpubl. M.Sc. thesis, Kiel University, 1-98.

Drachev, S.S., V.A. Savostin, V.G. Groshev and I.E. Bruni (in press) Structure and Geology of the continental shelf of the Laptev Sea, eastern Russian Arctic. Tectonophysics.

Eicken, H., E. Reimnitz, V. Alexandrov, T. Martin, H. Kassens and T. Viehoff (1997) Sea-ice processes in the Laptev Sea and their importance for sediment export. Continental Shelf Research, 17, 205-233.

Forman, S.L. and L. Polyak (1997) Radiocarbon content of pre-bomb marine mollusks and variations in the 14C reservoir age for coastal areas of the Barents and Kara seas, Russia. Geophysical Research Letters, 24, 885-888.

Fütterer, D.K. (1992) ARCTIC '91: The expedition ARK-VIII/3 of RV "Polarstern" in 1991. Reports on Polar Research, 107, 1-267.

Gudina, V.I. (1966) Foraminifers and stratigraphy of Quaternary deposits of the north-western Siberia. Moscow, Nauka, 132 pp.

Hald, M. and P.I. Steinsund (1996) Benthic foraminifera and carbonate dissolution in the surface sediments of the Barents and Kara seas. In: Stein, R. et al. (eds.), Reports on Polar Research. 212, 285-307.

Hebbeln, D., T. Dokken, E.S. Andersen, M. Hald and A. Elverhoi (1994) Moisture supply for northern ice-sheet growth during the Last Glacial Maximum. Nature, 370, 357-360.

Hemleben, C., M. Spindler O.R. and Anderson (1989) Modern planktonic foraminifera. Springer, New York.

Holmes, M.L. and J.S. Creager (1974) Holocene history of the Laptev Sea continental shelf. In: Arctic Ocean sediments, microfauna, and climatic record in the Late Cenocoic Time, Herman, Y. (ed.), 211-229.

Imbrie, J. and N.G. Kipp (1971) A new micropaleontological method for quantitative paleoclimatology: application to a Late Pleistocene Carribean core. In: The Late Cenocoic glacial ages, Turekian, K.K. (ed.), New Haven, 71-181.

Hanzlick, D. (1983) The West Spitsbergen Current: Transport, forcing and variability. PhD thesis, Univ. of Washington, 127 pp.

Kassens, H. and V.Y. Karpiy (1994) Russian-German Cooperation: The Transdrift I expedition to the Laptev Sea. Reports on Polar Research, 151, 1-168.

Kellogg, T.B. (1984) Paleoclimatic significance of subpolar foraminifera in high-latitude marine sediments. Canadian Journal of Earth Science, 21, 189-193.

Kempema, E.W., E. Reimnitz and P.W. Barnes (1989) Sea-ice sediment entrainment and rafting in the Arctic. Journal of Sedimentary Petrology, 59, 308-317.

Kennett, J.P. and M.S. Srinivasan (1983) Neogene planktonic foraminifera. Hutchinson Ross Publishing Company, Stroudsburg.

Khusid, T.A., N.V. Belyaeva and E.M. Potekhina (1995) Foraminiferal associations from the fjord glaciers of the Northern Island (Novaya Zemlya, Barents Sea). In: Modern and ancient microplankton of the World's Ocean, Barash, M.S. (ed.), Moscow, Russian Academy of Sciences, 116-118.

Kipp, N.G. (1976) New transfer function for estimating past sea-surface conditions from sea-bed distribution of planktonic foraminiferal assemblages in the North Atlantic. Geological Society Memoir, 145, 3-41.

Nørgaard-Pedersen, N., R.F. Spielhagen J. Thiede and H. Kassens (in press) Central Arctic surface ocean environment during the past 80,000 years. Paleoceanography

Okulitch, A.V., B.G. Lopatin and H.R. Jackson (1989) Circumpolar map of the Arctic. Geological Survey of Canada, Map 1765 A.

Reimnitz, E., E.W. Kempema and P.W. Barnes (1987) Anchor ice, seabed freezing, and sediment dynamics in shallow Arctic seas. Journal of Geophysical Research, 92, 14671-14678.

Reimnitz, E., L. Marincovich Jr., M. McCormick and W.M. Briggs (1992) Suspension freezing of bottom sediment and biota in the Northwest Passage and implications for Arctic Ocean sedimentation. Canadian Journal of Earth Sciences, 29, 693-703.

Rigor, I., (1992) Arctic Ocean buoy program. ARCOS Newsletter, 44, 1-3.

Rossak, B. (1995) Zur Tonmineralverteilung und Sedimentzusammensetzung in Oberflächensedimenten der Laptevsee, sibirische Arktis. unpubl. MSc. thesis Würzburg University/GEOMAR, 1-101.

Rudels, B. (1995) The thermohaline circulation of the Arctic and the Greenland seas. Philosophical Transactions Royal Society London, 352: 1-13.

Schauer, U., R.D. Muench, B. Rudels and L. Timokhov (1997) Impact of eastern Arctic shelf waters on the Nansen Basin intermediate layers. Journal of Geophysical Research, 102 (C2), 3371-3382.

Spiegler, D. (1996) Planktonic foraminifer Cenozoic biostratigraphy of the Arctic Ocean, Fram Strait (Sites 908-909), Yermak Plateau (Sites 910-912), and East Greenland margin (Sites 913). In: Thiede, J. et al. (eds.), Proceedings of Ocean Drilling Program, Scientific Results, 151, 153-167.

Spielhagen, R.F., G. Bonani, A. Eisenhauer, M. Frank, T. Frederichs, H. Kassens, P.W. Kubik, A. Mangini, N. Nørgaard-Pedersen, N.R. Nowaczyk, S. Schäper, R. Stein, J. Thiede, R. Tiedemann and M. Wahsner (1997) Arctic Ocean evidence for late Quaternary initiation of northern Eurasian ice sheets. Geology, 25, 738-786.

Steuerwald, B.A. and D.L. Clark (1972) Globigerina pachyderma in Pleistocene and Recent Arctic Ocean sediments. Journal of Paleontology, 46, 573-580.

Sudakov, S.B. (1991) Report on the results of accomplishment of geological survey in the scale 1:200000 of the shelf in the Dmitriy Laptev Strait. Vniiokeangeologiya, Rosgeolfond (State Geological Archives of the Russian Federation).

Tamanova, S.W. (1971) Foraminifera of the Laptev Sea. In: NIIGA-Leningrad (NIIGA-Leningrad), Geology of the Ocean. Ministery of Geology of the USSR, Leningrad, 1, 54-63.

Timokhov, L. (1994) Regional characteristics of the Laptev and East Siberian seas: climate, topography, ice phases, thermohaline regime, circulation. In: Russian-German cooperation in the Siberian shelf seas: Geo-system Laptev Sea, Kassens, H. et al. (eds.), Reports on Polar Research, 144, 15-31.

Vilks, G. (1975) Comparison of Globorotalia pachyderma (Ehrenberg) in the water column and sediments of the Canadian Arctic. Journal of Foraminiferal Research, 5, 313-325.

Walsh, J.J., C.P. McRoy, L.K. Coachman, J. Goering, J. Nihoul, T.E. Whitledge, T.H. Blackburn, P.L. Parker, C.D. Wirick, P.G. Shuert, J.M. Gebmeier, A.M. Springer, R.D. Tripp, D.A. Hansell, S. Djenidi, E. Deleersnijder, K. Henriksen, B.A. Lund, P. Anderson, F.E. Müller-Karger and K. Dean (1989) Carbon and nitrogen cycling within the Bering/Chukchi seas: Source regions for organic matter effecting AOU demands of the Arctic Ocean. Progress in Oceanography, 22, 277-359.

Plate I

SEM photographs of non-polar planktic foraminifera of core IK9373-10. Numbers in parentheses indicate magnification.

1. *Globoturborotalita tenella* (x 650)
2. *Globoturborotalita tenella* (x 250)
3. *Globoturborotalita tenella* (x 150)
4. *Globigerinita glutinata* (x 200)
5. *Globigerinita glutinata* (x 280)
6. *Globigerina bulloides* (x 100)
7. *Turborotalita clarkei* (x 360)
8. *Turborotalita clarkei* (x 380)
9. *Turborotalita clarkei* (x 360)
10. *Turborotalita clarkei* (x 310)
11. *Turborotalita clarkei* (x 350)

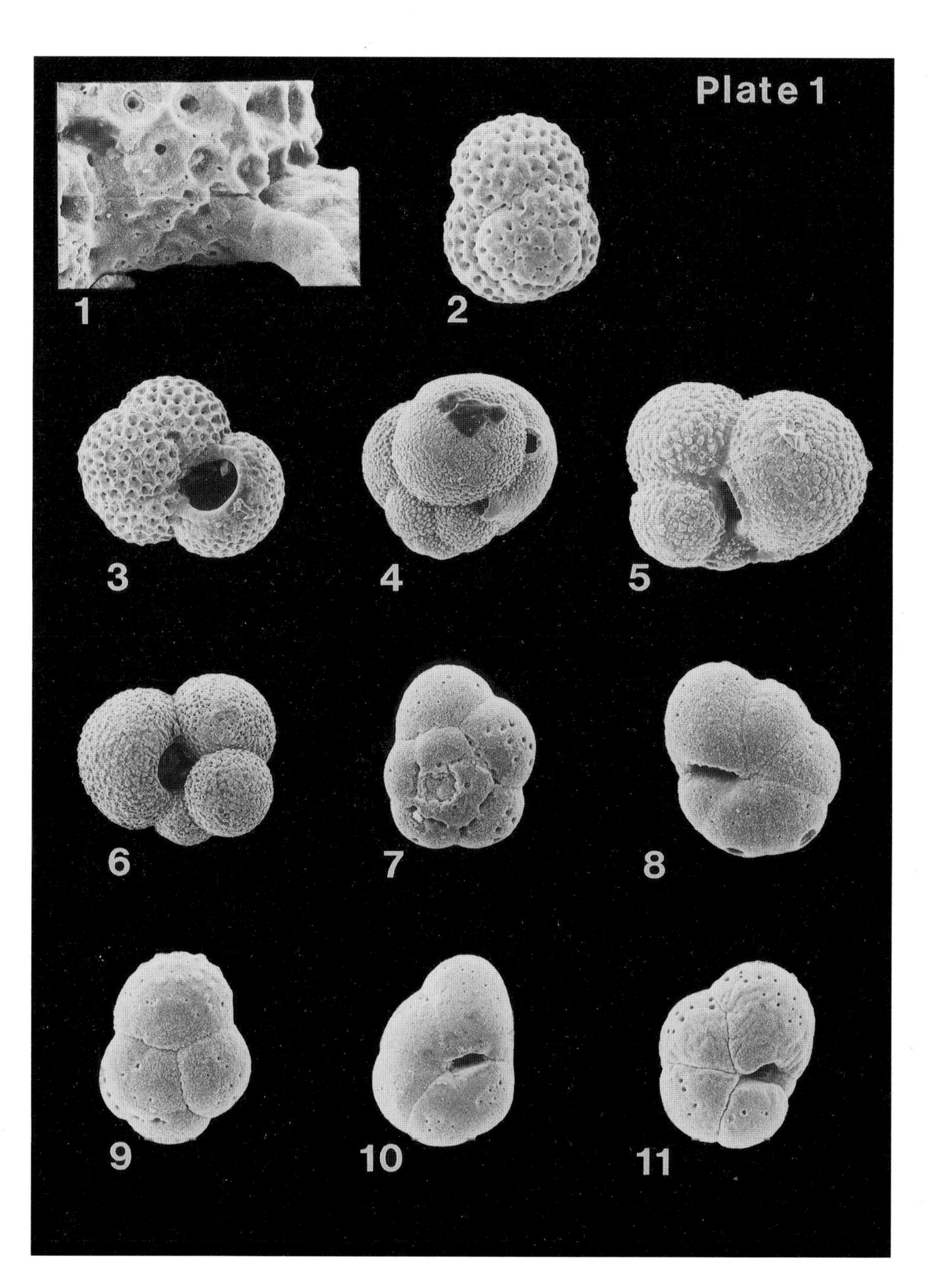
Plate 1
1
2
3
4
5
6
7
8
9
10
11

Holocene Diatom Stratigraphy and Paleoceanography of the Eurasian Arctic Seas

Y. Polyakova

Geographical Department, Moscow State University, 119899 Moscow, Russia

Received 11 March 1997 and accepted in revised form 3 May 1998

Abstract - Five ecological zones were established on the basis of changes in relative abundance of diatom valves, ecological and biogeographical composition of diatom assemblages from the Holocene sediments of the Eurasian Arctic seas. These diatom ecozones reflecting the main stages of paleoceanological evolution of the Arctic Seas during the Holocene may be used as stratigraphical markers for subdivision of the sediment sequences. It was revealed that in the eastern part of the Eurasian Arctic shelf the most significant paleoceanological changes took place about 12-11 Ka, caused by the opening of the Bering Strait and supply of warm highly productive Bering Sea waters into the Chukchi Sea. In the western part of the Arctic shelf this event was connected with intensifying of the the North Atlantic current about 9-8 Ka. During the Holocene in all Arctic Seas the most favourable hydrobiological conditions for the arctic phytocoenoses development ("hydrobiological or marine optimum") corresponding to the Ecozone III existed at the end of Atlantic to the beginning of Subboreal periods.

Introduction

The Holocene paleoceanography of the Arctic Ocean s still poorly known, yet it becomes increasingly clear that this very sensitive region played a key role in short- and long-term changes of global ocean circulation and climate formation. While much progress has been made toward reconstruction of the Late Pleistocene and Holocene paleoclimatic changes and their regional manifestations in the high latitudinal continental regions of the Northern hemisphere, paleoceanological fluctuations of the Arctic basin and marginal shelf seas are still not clearly understood. By now such important paleogeographical problems as correlation of paleoenvironmental changes in the Arctic Ocean (especially in shelf and adjacent land areas) and reconstructions of their inter-relations are still far from being solved. The Eurasian Arctic shelf comprising about 1/3 of the total area of the Arctic Ocean exhibits the most promise for solution of these problems. The Eurasian Arctic shelf stretching from the Atlantic to Pacific Oceans and fringed by the Arctic basin is a critical area for the flux of fresh water and sea ice formation that modulate the Earth's climate.

The data obtained to date give evidence for considerable paleoceanological fluctuations in the Arctic Ocean during the Late Pleistocene and Holocene, thus allowing us to draw the most general conclusions about the main paleoenvironmental changes during this period. The available core data from the deep Arctic basins relate faunal and sedimentological variables caused by changing paleoenvironmental conditions during theLatest Quaternary and Holocene. Some aspects have been emphasized in those studies: (i) two peak deglacial meltwater episodes in the Arctic Ocean at 14-12 Ka (Termination Ia) and 10-8 Ka (Termination Ib); (ii) high organic productivity during Termination I following the abundance minimum in last glacial stage associated with a strong inflowing of warm North Atlantic water (Cronin et al., 1994); (iii) the major abundance maxima of planktic and benthic foraminifers ascribed to mid- to late Holocene (Norgaard-Pedersen, 1996) and several mid-Holocene key faunal events indicating a period of warming and/or enhanced flow between the Canadian and Eurasian Basins (Cronin et al., 1994); (iiii) the seasonally open water in the central Arctic Ocean during the Holocene on

---

In: Kassens, H., H.A. Bauch, I. Dmitrenko, H. Eicken, H.-W. Hubberten, M. Melles, J. Thiede and L. Timokhov (eds.) Land-Ocean Systems in the Siberian Arctic: Dynamics and History. Springer-Verlag, Berlin, 1999, 615-634.

the basis of presence of photosynthetic algae.

Nevertheless, relatively low rates of sedimentation in the deep Arctic basins (Stein et al., 1994) do not allow for more detailed paleoceanological reconstructions for Holocene. At the same time, the restriction to the Eurasian Arctic shelf seas is very beneficial for studies of short-term paleoceanographic changes of the Arctic Ocean. Different from the central Arctic Ocean sedimentation rates proved to be very high during Holocene, for instance, sedimentation rates were up to 70 cm/ky in the eastern Laptev Sea shelf (Bauch et al., 1996) and about 50 cm/ky on the eastern Laptev Sea continental margin (Spielhagen et al., 1996).

The performed micropaleontological (mainly diatoms and foraminifers) investigations of the bottom sediments of the Eurasian Arctic Seas (Polyakova, 1990; 1997a; 1997b; Saidova, 1994; Dzhinoridze,1978; Samoilovich et al., 1988; Khusid, 1989) allow us to draw the following conclusions: (i) variations of the quantitative and qualitative composition of fossil assemblages may be the basis for both: ecostratigraphical subdivision of the Holocene sediments of the Arctic Seas and paleoceanological reconstructins; (ii) up to five diatom and foraminiferal ecozones corresponding to the main stages of paleogeographical evlution of the Eurasian Arctic Seas have been established in the Holocene sediments; (iii) the period of the most favorable hydrobiological conditions for planktonic and benthic microorganisms ("hydrobiological or marine optimum") has been established in all studied regions of the Eurasian Arctic shelf. It is manifested by increasing number of organisms and appearance of several warm-water species unknown in the modern flora and fauna of the Arctic Seas. However, age interpretation of this paleoenvironmental event is different: (i) marine optimum of the Arctic Seas either corresponds to the Holocene "climatic optimum", i.e. to the Atlantic period (according to Saidova, 1982; 1994; Khusid, 1989), or (ii) has different age estimations in various regions since changes in productivity and geographical distribution of species depend upon complicated interactions between hydrochemical and circulation processes in oceans and seas (Polyakova, 1997a, 1997b).

The purpose of this paper is to analyze the downcore changes in the Eurasian Arctic shelf sequences and to interpret variations of diatom assemblages and their abundance in terms of regional changes of paleoceanological conditions, e.g., sea-level fluctuations, the extent of various water masses and sea ice cover, changes in productivity through time. Most attention has been concentrated on the examining paleoceanological conditions within the context of the "hydrobiological optimum" (term suggested by Polyakova, 1997a, 1997b) or "marine optimum" (term suggested by Andrews, 1972; Williams, 1992) and its correlation with "climatic optimum" in the adjacent arctic regions of Eurasia. We used the terrestrial chronostratigraphical scale of Mangerud et al. (1974).

The performed reconstructions are based on the study of the modern diatom flora and distribution of diatom assemblages in the upper bottom sediment layer. It revealed strong dependence of the species composition and the number of diatom valves upon hydrological, hydrobiological and sedimentation conditions in the Arctic Seas (Polyakova, 1988, 1994, 1997b). This modern analogue approach has been used as the foundation for stratigraphical subdivision of the Holocene deposits of the Eurasian Arctic shelf and paleoceanological reconstructions.

Materials and methods

For the present study 45 sediment cores and 1 borehole from the Eurasian Arctic shelf were selected (Table 1, Figure 1). The cores were obtained during the last 15 years scientific expeditions of P.P. Shirshov Institute of Oceanology RAS and VNIIOceangeology through all over Eurasian Arctic seas.

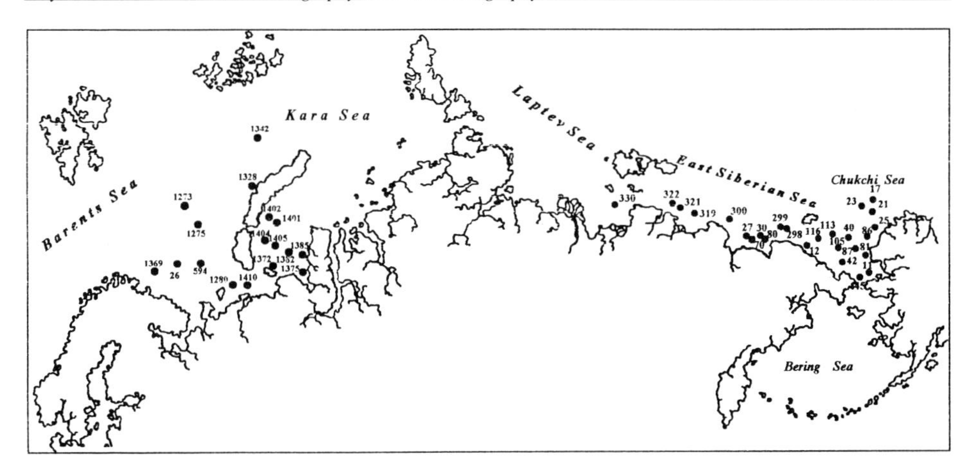

Figure 1: Location of the studied cores of the Eurasian Arctic seas.

All piston cores were sampled for diatoms at 5-10 cm intervales, whilst the long drill core (26) was sampled at 50-100 cm intervals. For the purpose of diatom analysis 1-2 to 20-30 gram sediment samples depending on the concentration of diatoms in the sediments were selected. The preparation methods for making diatom slides and determination of absolute abundances of diatoms per gram of dry sediment used in this study have been described by Glezer et al., 1974. According to this method diatoms were concentrated by treatment with 10% HCl in case sediments were carbonate (i), and with 20-30% H2O2(ii), with subsequent decantation with distilled water (iii). Then the procedure was followed by separation with heavy liquid (H2O:CdI2 :KI=1:2.5:2.25). Subsequently they were mounted on glass slides using a mounting medium having a high index of refraction (1.68). The valves were examined under a light microscope at x1000 magnification. The first 300-400 specimens encountered in each sample were identified, and their abundance was converted to percent. The following ecological groups were described: freshwater species, marine and brackish water species which were subdivided into sublittoral, planktonic neritic and panthalassic species (the latter one occupy both shelf and open-sea areas), and cryophilic or sea-ice species. Also, phytogeographical groups, e.g. arctic-boreal, bipolar, arctic-boreal-tropical, tropical-boreal and cosmopolitan, were distiguished according to Beklemishev and Semina (1986).

Observations and results

Downcore diatom assemblages

The cores covering the most complete Late Pleistocene and Holocene sequences reflecting regional peculiarities of paleoceanological variations within the Arctic shelf of Eurasia have been chosen to characterize diatom assemblages and diatom ecological zones. Under the term "ecozone" it is meant "assemblages of fossil organisms reflecting their ecological association in their lifetime. Changes of ecozones along the section are relate to changeing ecological-facial conditions" (Stratigraphical..., 1977). The latter manifest paleogeographical changes of different temporal and spatial extent.

Chukchi Sea

Bottom sediments of the Chukchi Sea are the most well studied among other Eurasian Arctic

Table 1: Key data and position of the core holes.

Core No.	Sea	Latitude	Longitude	Water depth (m)	Core length (m)
17	Chukchi	72°04.0 N	164°53.0 W	40.0	1.79
21	Chukchi	71°05.0 N	164°50.0 W	43.0	2.71
23	Chukchi	71°47.0 N	168°34.0 W	49.0	3.42
25	Chukchi	70°11.0 N	164°38.0 W	38.0	1.61
86	Chukchi	69°31.0 N	166°15.4 W	39.0	0.37
81	Chukchi	68°21.0 N	167°15.0 W	38.0	0.30
11	Chukchi	66°23.0 N	167°10.0 W	46.0	2.30
45	Chukchi	66°10.0 N	168°23.0 W	52.4	4.25
87	Chukchi	67°55.0 N	169°46.0 W	50.0	3.41
40	Chukchi	69°22.5 N	171°08.0 W	51.0	3.50
42	Chukchi	68°10.0 N	173°41.0 W	49.0	3.30
105	Chukchi	68°12.5 N	173°58.0 W	43.0	0.70
113	Chukchi	70°02.5 N	176°08.5 W	54.0	0.80
116	Chukchi	70°33.0 N	176°01.0 W	61.0	1.60
38	Chukchi	70°56.0 N	179°41.0 W	49.0	1.22
12	Chukchi	69°30.0 N	178°30.5 W	25.0	0.96
298	East Siberian	70°35.0 N	172°40.0 E	21.0	2.81
299	East Siberian	70°10.0 N	172°37.0 E	27.0	0.79
80	East Siberian	69°54.0 N	167°39.8 E	17.5	2.07
30	East Siberian	69°54.0 N	166°19.0 E	30.0	1.92
70	East Siberian	69°51.0 N	165°10.0 E	22.0	0.71
27	East Siberian	69°38.0 N	164°49.0 E	14.5	1.40
300	East Siberian	71°09.3 N	160°32.9 E	15.0	0.90
296	East Siberian	71°11.4 N	160°21.1 E	18.0	0.42
319	East Siberian	72°10.0 N	153°25.0 E	19.0	0.58
321	East Siberian	72°40.7 N	150°30.3 E	15.0	0.36
322	East Siberian	72°54.9 N	148°56.8 E	14.0	0.55
330	Laptev	72°27.5 N	135°24.8 E	28.0	1.28
1372	Kara	70°19.2 N	60°08.0 E	170.0	3.10
1375	Kara	69°44.5 N	65°49.5 E	30.0	2.70
1382	Kara	72°00.4 N	62°01.7 E	130.0	3.10
1385	Kara	71°27.3 N	64°52.1 E	120.0	2.90
1401	Kara	73°19.1 N	61°28.6 E	138.0	1.65
1403	Kara	73°48.4 N	58°43.4 E	290.0	1.56
1405	Kara	72°42.3 N	58°16.7 E	380.0	3.00
1407	Kara	72°15.0 N	60°44.1 E	125.0	3.30
1273	Barents	74°24.5 N	45°44.5 E	312.0	4.25
1275	Barents	73°42.0 N	46°35.4 E	330.0	1.35
1280	Barents	69°47.9 N	50°51.7 E	80.0	3.24
1328	Barents	75°29.2 N	57°17.3 E	91.0	2.55
1341	Barents	78°48.9 N	58°01.0 E	150.0	0.75
1369	Barents	70°28.2 N	33°21.3 E	250.0	2.85
1410	Barents	69°45.9 N	53°52.9 E	75.0	2.00
594	Barents	70°19.4 N	41°40.0 E	86.0	3.60
582	Barents	70°03.0 N	37°12.0 E	160.0	4.60
26	Barents	70°15.4 N	41°25.3 E	202.0	33.0

seas. Distribution of forams (Saidova, 1994) and diatoms (Polyakova, 1990, 1997a) in the Chukchi sea sediments are known sufficiently well. The author studied several cores (Table 1) recovered at various depths ranging from 9 m to 61 m. Maximum thickness of core sediments reached 4.2 m (Figure 1). Variations of species number and composition and correlation between ecological and biogeographical diatom groups allowed distinguishing six ecozones, which reflect paleogeographical changes in the Chukchi Sea during the end of Late Pleistocene and Holocene The latter are well manifested by Core 17 and Core 23 recovered from the north-eastern part of the sea.

Core 23. Late Pleistocene diatom assemblages (Ecozone VI) from the mainly sandy sediments of the lowermost parts of the cores are characterized by poor species composition (their taxonomic diversity is 2-3 times lower than that of the surface sediments) and relatively low (for this region) number of diatom valves (up to 0,4 million/g; Figure 2). High abundance of sublittoral diatoms, such as Paralia sulcata, Delphineis kippae and others (up to 40%) give evidence for shallow marine sedimentation conditions. Predominance of sea-ice species (Nitzschia grunowii, N.cylindrus and others, up to 52%) indicates the severe ice conditions of the sea basin. The relatively warm Bering Sea planktonic diatom species brought into the Chukchi Sea by the Bering Sea current (Ryzhov et al., 1984; Polyakova, 1997b) have not been found. This proves the Arctic Ocean to be isolated from the Pacific during the accumulation of the diatom assemblages of the Ecozone VI.

The opening of the Bering Strait and penetration of relatively warm highly productive Bering Sea waters into the Chukchi Sea was the only reason for the sharpest changes in diatom abundance and ecological and biogeographical structure of diatom assemblages established at the boundary of the Ecozones VI and V manifesting significant hydrobiological changes in the Chukchi Sea. These changes are characterized by (i) sharp increase in the total number of diatoms in sediments (by 10-20 times) probably related to increasing productivity of waters; (ii) general increase of taxonomic diversity of diatom flora; (iii) introduction of relatively warm-water species which are not typical of the arctic water mass.

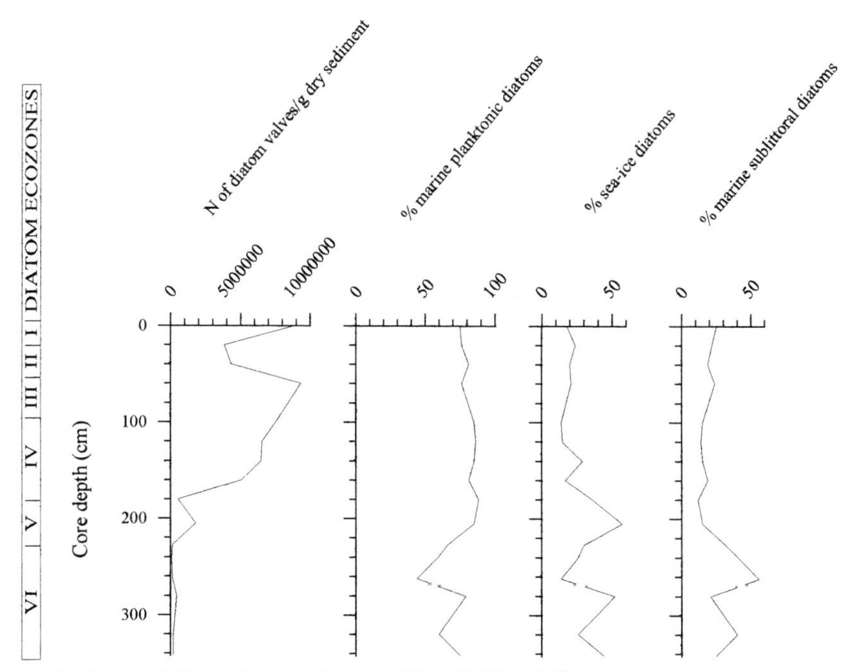

Figure 2: Distribution of diatoms in the sediments of Core 23, Chukchi Sea.

The Ecozone V diatom assemblages restricted to the silty and silty-clayey sediments of different density established in the lower part of the core are the "coldest" ones among Holocene assemblages. As compared with the Late Pleistocene assemblages of the Ecozone VI they are characterized by increase of both planktonic species number (up to 66%) and their taxonomic diversity (by 2 times), together with decrease of sublittoral species number (down to

32%), thus reflecting deepening of the sea basin. Abundance of the sea-ice species remains relatively high (up to 30%). There have been found several relatively warm-water species (*Proboscia alata, Pyxidicula nipponica*) uncommon for the arctic water masses.

Diatom abundance, as well as species composition and correlation between different ecological and biogeographical groups of diatom assemblages of the Ecozone IV are very close to the mordern ones of this region of the Chukchi Sea, thus indicating similar hydrological and hydrobiological conditions during accumulation of sediments of this ecozone.

Diatom assemblages of the Ecozone III are characterized by the maximum number of diatom valves (up to 9,3 million valves/g) and sharp increase of the total taxonomic diversity. The widely distributed species (*Thalassionema nitzschioides, Actinoptychus undulatus* and others) as well as the cold-water (*Thalassiosira nordenskioeldii, T.antarctica, T.gravida* and others) and even the sea-ice ones (*Nitzschia grunowii, Chaetoceros septentrionalis, Navicula vanhoeffenii* and others) increase in number. However, the percentage of sea-ice species remains the same. The characteristic feature of the Ecozone III diatom assemblages is appearance of both: diverse relatively warm-water species brought by the Bering Sea current (*Coscinodiscus asteromphalus, C.perforatus, C.radiatus, Proboscia alata, Neodenticula seminae* and others), and even boreal-tropical (*Chaetoceros dydimus*). This indicates considerable northward displacement of planktonic warm-water species caused by intensification of the Bering Sea current.

The subsequent short-time sedimentation period (Ecozone II, Figure 2) is characterized by low total diatom number (down to 3,8 million valves/g) and poor taxonomic composition of diatom assemblages in which relatively warm-water species are eliminated. Extremely cold-water diatom flora and very low number of relatively warm-water species give evidence for reduction of intensity of the Bering Sea current probably due to the sea level fall. Subsurface sediments include diatom assemblages similar to the mordern ones reflecting the present hydrological and sedimentological conditions.

Core 17. Though the total number of diatoms in sediments of this core is even 5 times as high as that of the above described Core 23, similar trends in variations of the total diatom abundance and correlations between different ecological and biogeographical diatom groups have been established along the core sequence (Figure 3), thus reflecting the same paleoenvironmental changes in the Chukchi Sea during the Holocene. Nevertheless, five Holocene diatom ecological zones distinguished in this core have several peculiarities comparable with those of the Core 23. Like in the case with the Core 23, diatom assemblages from the lowermost core layers (Ecozone V) are characterized by maximum abundance of sublittoral diatoms (up to 34%). However, sea-ice species are extremely abundant in assemblages of the Ecozone IV (up to 51%), thus reflecting more severe sea-ice conditions in this region. The Ecozone III is the marking one for this core as well, since it is characterized by the maximum number of diatoms (up to 48,1 million valves/g) and high taxonomical diversity. Occurrence of warm-water species with maximum total abundance up to 0,5 million valves/g gives evidence for their northward migration (up to 72(N and even more) during this time. The most characteristic feature of diatom assemblages of the Ecozone II is sharp decrease of both: the total diatom number (down to 2,9 million valves/g) and their taxonomic diversity (by 3 times). Against the background of extremely cold-water diatom flora, increase of the sea-ice species number (up to 74%) indicates severe sea-ice conditions and lowering of surface temperatures. The Ecozone I in this core is subdivided into two subzones. The lower subzone (Ib) is characterized by higher total diatom abundance due to the presence of ice-neritic and sea-ice species (up to 51%). Diatom assemblages of the upper subzone are similar to morden ones of this region thus reflecting similar hydrobiological conditions.

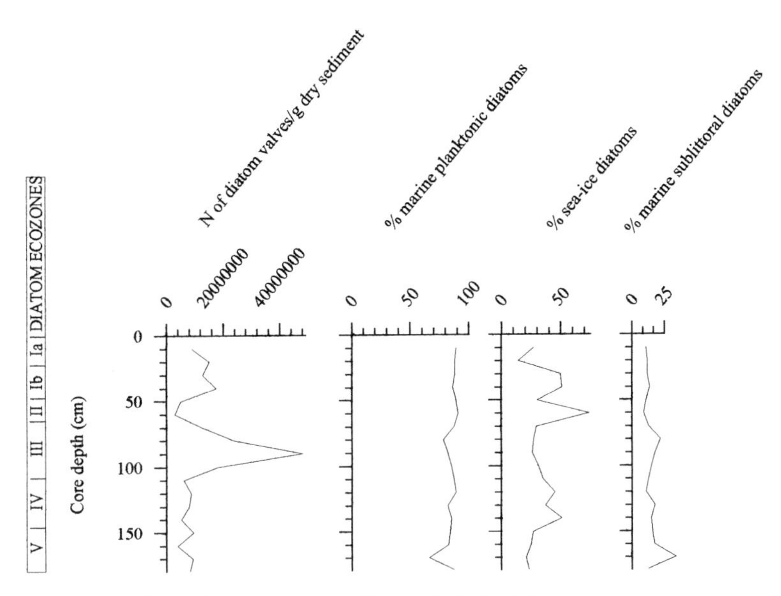

Figure 3: Distribution of diatoms in the sediments of Core 17, Chukchi Sea.

East Siberian Sea

As compared to the Chukchi Sea, bottom sediments of the East Siberian Sea still remain poorly studied. Diatom assemblages of the bottom sediments of the East Siberian Sea (with maximum thickness of 2.7 m) together with corresponding ecozones reflect regional peculiarities.

Core 299 was obtained in the south-eastern part of the sea between the Chauna Bay and the Long Strait. Five diatom ecological zones were established in it. Diatom assemblages from the dense silty clays at the core base (Ecozone V, Figure 4) characterize of shallow marine environments (with the depths of not more than 5 m). This supposition is also supported by high abundance of fresh-water (up to 47%) and marine sublittoral (up to 43%) species. The percentage of sea-ice species is low (1-29%) due to the warming effect of the river runoff.

Hydrological conditions similar to the modern ones existed in this area during accumulation of diatom assemblages of the Ecozone IV established in to the dense clayey silts. Ice-neritic and sea-ice species (*Thalassiosira nordenskioeldii, Nitzschia grunowii, Chaetoceros septentrionalis* and others) predominate in it, and the total number of fresh-water species does not exceed 1%. As in the case with the Chukchi Sea, the period of the most favorable hydrobiological conditions (Ecozone III) was established in the East Siberian Sea. Its diatom assemblages are characterized by the most diverse taxonomic composition and maximum diatom number (up to 3,3 million valves/g). The latter is an order of magnitude higher than the present one. Diatom number increases due to both: neritic species with predominance of different representatives of *Chaetoceros* (*C.compressus, C.ingolfianus, C.debilis, C.mitra, C.diadema* and others) and *Thalassiosira* (*T.antarctica, T.gravida, T.nordenskioeldii, T.hyalina, T.constricta* and others) genera, and sea-ice species (*Nitzschia grunowii, N.cylindrus, Detonula confervaceae, Chaetoceros septentrionalis, N.vanhoeffenii* and others). The number of sublittoral species (*Thalassiosira bramaputrae v.septentrionalis, Paralia sulcata, Delphineis kippae, Diploneis smithii* and others) decreases down to 9%. Planktonic group includes species that are now rare or unknown in the East Siberian Sea (*Neodenticula seminae, Coscinodiscus asteromphalus* and others). It is assumed that during accumulation of this ecozone the Long branch of the Bering Sea current reached this region. In the overlying sediments (Ecozone II) diatom assemblages are characterized by slight decrease of the total diatom number (down 0,4 million valves/g) together with increasing abundance of the ice-neritic and sea-ice species. Subsurface sediments include diatom assemblages accumulated under modern sedimentation and hydrological conditions (Ecozone I).

To the west, in the region between the Chauna Bay and the Kolyma River mouth (Cores 80, 30, 70, 27, Figure 1) taxonomic composition of diatom assemblages and their quantitative distribution in the sediments resemble those established in the south-eastern part of the sea. region.

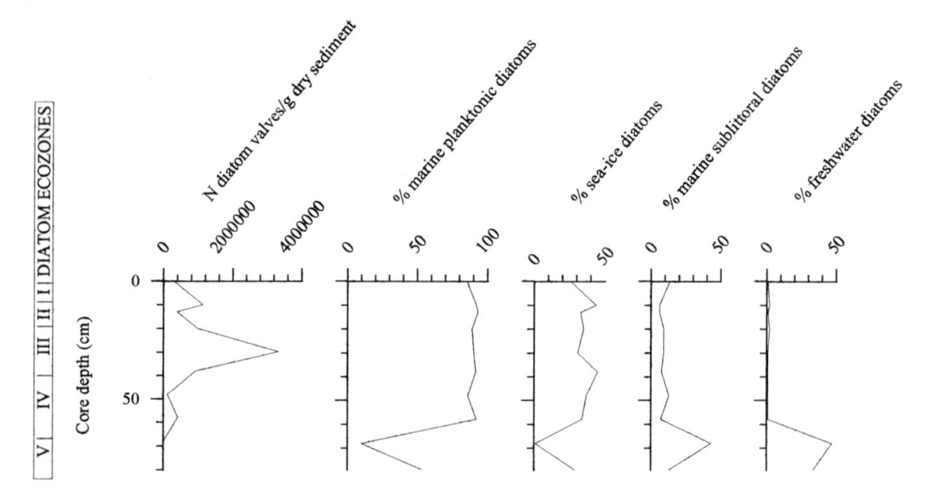

Figure 4: Distribution of diatoms in the sediments of Core 299, East Siberian Sea.

In the most shallow (water depth 14-19 m) regions freshened by the river runoff which are situated to the north-west of the Kolyma River mouth (Cores 296, 300, Figure 1, Table 1) and to the north and north-west of the Indigirka River mouth (Cores 319, 321, 322) the thickness of sediments sampled by a corer does not exceed 1 m (Figure 1). The core sediments are represented by dark gray silts of different density with hydrotroilite spots. The number of diatom valves is relatively low (up to 56 thousand valves/g) due to considerable supply of terrigenous sediments. The sublittoral eurihaline species Thalassiosira bramaputrae v.septentrionalis predominates. The number of fresh-water species increases westward and northward from the Indigirka mouth, where they are dominant in diatom assemblages comprising up to 100% in the lower sediment layers.

Core 296. Four diatom ecozones reflecting successive sea level rise and changes of hydrobiological conditions in course of the Holocene transgression have been established in this core. Diatom assemblages of the basal core layer (Ecozone IV; Figure 5) indicate the shallow marine sedimentation environment. They are characterized by low number of diatoms in sediments (35-85 valves/g) and predominance of sublittoral species (up to 100%) - Thalassiosira bramaputrae v.septentrionalis, Paralia sulcata. In the overlying sediments quantitative maximum of diatoms (up to 18,5 thousand valves/g) mainly represented by sublittoral species (*Thalassiosira bramaputrae v.septentrionalis*), and taxonomical divers marine planktonic species (*Chaetoceros diadema, C.compressus, Rhizosolenia hebetata f.hebetata, Thalassiosira antarctica+T.gravida, T.hyalina, T.trifulta, T.nordenskioeldii* and others) is observed. Abundance of fresh-water diatoms is low in this assemblages (1%).

Diatom assemblages of the Ecozone II display sharp decrease of both: the total diatom number (down to 1,1 thousand valves/g) and taxonomic diversity (by 2-3 times) accompanied by increase of the fresh-water species number (up to 10%). Diatom assemblages of the Ecozone I are similar to modern thanatocoenoses of this region.

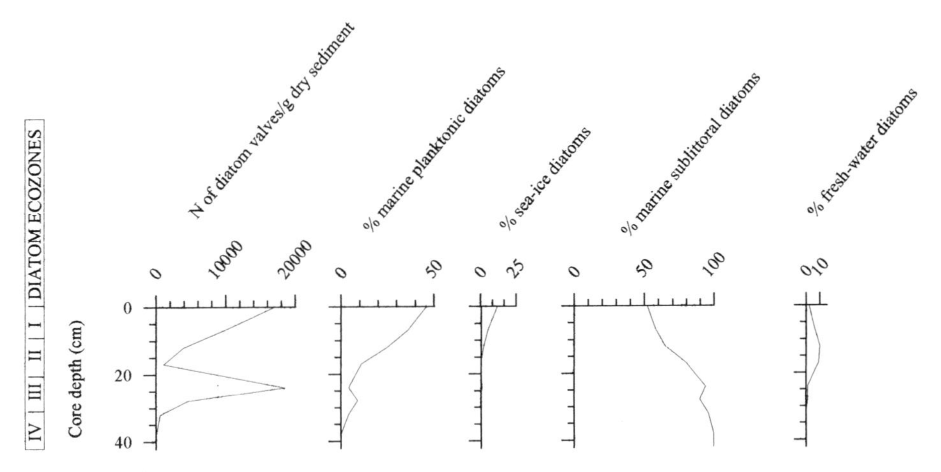

Figure 5: Distribution of diatoms in the sediments of Core 296, East Siberian Sea.

Laptev Sea.

Core 330. The studied core is situated at a distance of 200 km offshore from the Lena River delta. It is represented by unconsolidated fine-grained clayey mud devoid of any unconformities. Due to unstable hydrological and sedimentological conditions in this region the diatom number and correlations between different ecological groups display sharp and numerous changes along the core sequence. Variations of diatom species composition allowed distinguishing three diatom Ecozones (Figure 6). Diatom assemblages from the basal core layers (Ecozone III) are characterized by the high total diatom number (up to 334 thousand valves/g) and predominance of marine species (up to 97%). Among the latter euryhaline *Thalassiosira bramaputrae* v. *septentrionalis* is dominant, and rare *Porosira glacialis, Thalassiosira antarctica* as well as ice-neritic and sea-ice species *Thalassiosira nordenskioeldii, T.hyalina, Nitzschia grunowii* have been also marked (however, their total percentage does not exceed 6%). In the diatom assemblages of the Ecozone II freshwater species are the most abundant (up to 82%). They are mainly represented by planktonic species typical of the Lena and Yana Rivers. Diatom assemblages of the overlying sediments (Ecozone I) indicate the modern hydrological and sedimentological conditions.

Kara Sea

The Kara Sea sediments which are extremely poor in Quaternary microfossils include numerous redeposited Cretaceous and Paleogene forms. Rare Quternary diatoms were found only in the upper core layers. Their number sharply decreases downcore where only redeposited marine Paleogene and Cretaceous forms were observed.

Barents Sea

The Barents Sea bottom sediments are known to be poor in microfossils. Hence, its Holocene paleoceanography needs further investigations. The study of diatoms from the sediments of

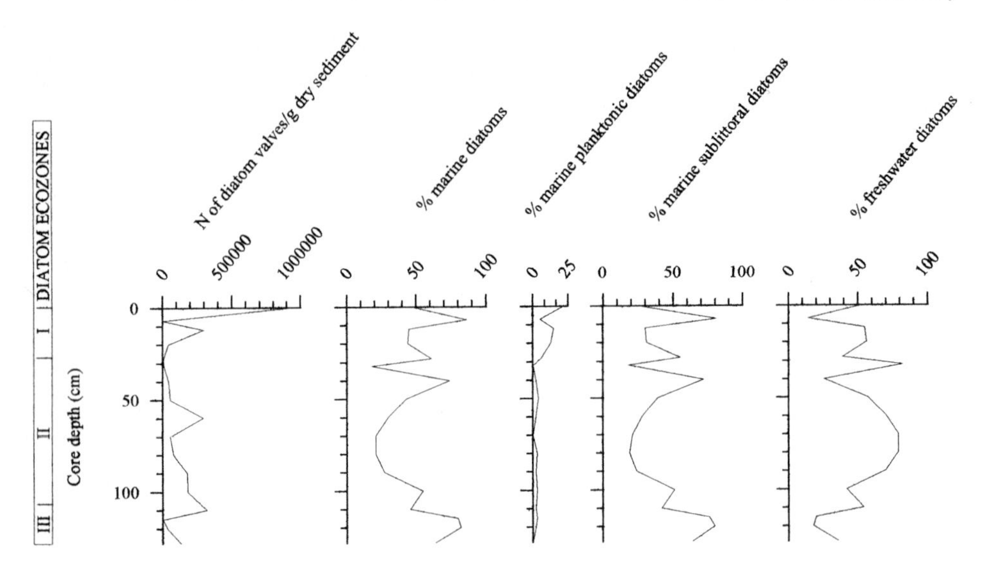

Figure 6: Distribution of diatoms in the sediments of Core 330, Laptev Sea.

piston cores (up to 4.3 m long) and borehole 26 (30 m long) recovered within the Kanin uplift in the southern part of the sea (Figure 1) revealed sedimentation processes in the central and north-eastern sea parts during Holocene to be unfavorable for accumulation of diatoms. Representative diatom assemblages were reported from the sediments of the southern part of the sea, where modern thanatocoenoses are also rich in diatoms (Polyakova et al., 1992).

Borehole 26. According to the data obtained, uninterrupted marine sedimentation was typical of the Kanin uplift during the Late Pleistocene and Holocene (Pavlidis et al., 1992). The boundary between the Upper Pleistocene and Holocene sediments is marked by sharp increase of diatom number (by 5-10 times) and definite change in the structure of diatom assemblages at the core depth of 7.5 m. Four ecozones corresponding to the main stages of the Holocene transgression were established.

The most ancient Holocene Ecozone IV (Figure 7) corresponds to the first influx of the Atlantic water into the Barents Sea. It resulted in appearance of various warm-water diatom species in its southern regions (*Coscinodiscos radiatus, C.asteromphalus, Actinocyclus divisus* and others) and increase of total diatom number in sediments (by 2-5 times). Cold-water ice-neritic species (*Thalassiosira atlantica, T.gravida, T.nordensioeldii* and others), sea-ice ones (*Nitzschia grunowii, Pleurosigma stuxbergii*), and species typical of divergence zones (genus *Chaetoceros*, *Thalassionema nitzschioides* and others) were the most abundant.

Diatom assemblages of the Ecozone III characterize the most favourable hydrobiological conditions in the Barents Sea during the Holocene. The total diatom number is 5-10 times as great as that of the surface bottom sediments and reached 10,4 thousand valves/g. Warm-water planktonic (*Coscinodiscus astermphalus, C.asteromphalus v.subbuliens, C.radiatus, C.perforatus* and others) and sublittoral (*Diploneis bombus, D.lyra* and others) elements are the most divers and abundant in this core interval. Diatom assemblages of the Ecozones II and I characterize formation of the present hydrobiological and hydrological conditions in the Barents Sea.

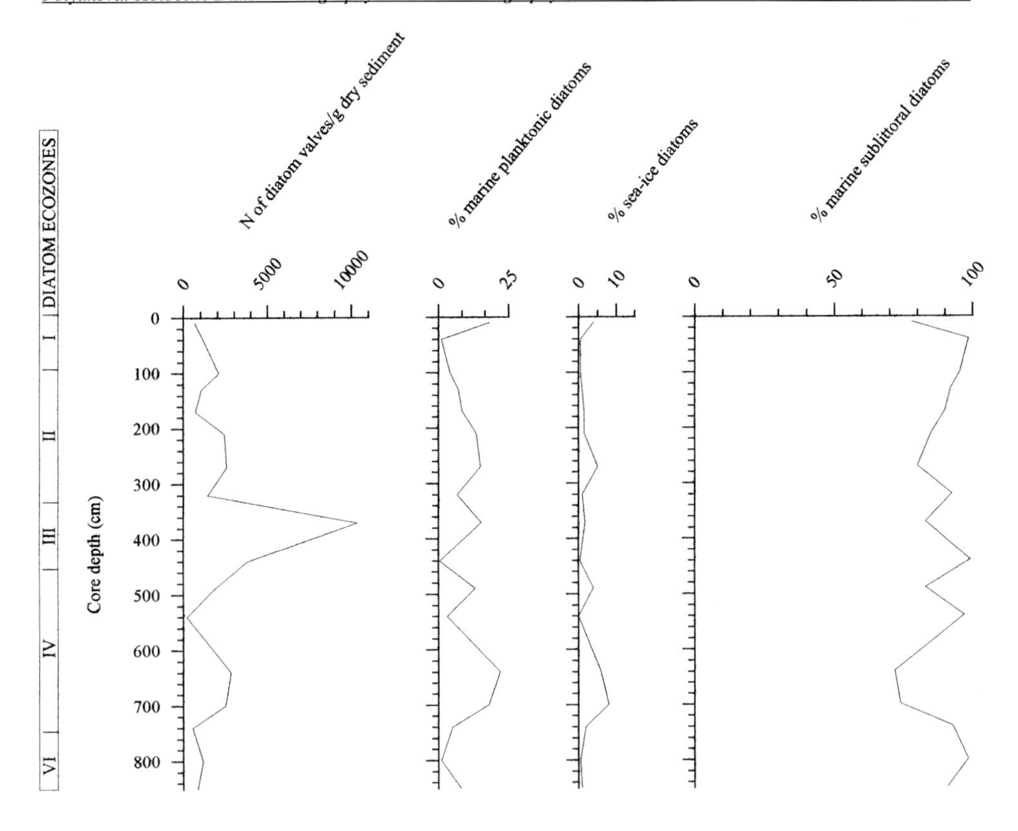

Figure 7: Distribution of diatoms in the sediments of Borehole 26, Barents Sea.

Estimated age of diatom ecozones

In order to determine the age of the above described diatom ecozones we used the published C^{14} datings and calculated sedimentation rates. By now numerous radiocarbon datings have been obtained for the Upper Pleistocene - Holocene deposits of the Beringia shelf area including the northern part of the Bering Sea adjacent to the Bering Strait and the Chukchi Sea shelf (Knebel et al., 1974; McManus and Creager, 1984; McManus et al., 1983). These data allow for reconstructing successive sea level changes in this region during the Holocene transgression and estimating the age of the established diatom ecozones (Polyakova, 1990). Analysis of radiocarbon and micropaleontological data revealed the connection between the Chukchi and Bering Seas to be absent until 12 Ka, because the youngest deposits yielding age datings in the Kotzebue Sound (Chukchi Sea) and marking sea level position at -40 m isobath are 12.600 years old (McManus et al., 1983). The Bering Strait is about 50 m deep, the depths of the Chirikov basin occupying the northern part of the Bering Sea are about 40 m. The Anadyr and Shpanberg Straits rounding St.Lawrence Island from the west and east correspondingly are slightly deeper than 40 m. The transformed Bering Sea waters flow through these straits to the Chirikov basin and then to the Chukchi Sea. The data of Hopkins (1973) suggest the Shpanberg Strait to be closed until 11.8 Ka. This allows supposing the Bering Land Bridge to exist until 12 Ka. It occupied the northern part of the Bering Sea including the modern Chirikov basin (reaching St.Lawrence Island) and the southern part of the Chukchi Sea to the modern -40 m isobath. In this connection we have assigned the basal sediment layers of the Cores 21 and 23 from the eastern part of the Chukchi Sea (Figure 1) together with diatom assemblages of the Ecozone VI to the time interval ranging from 14 to 12 Ka. These assemblages are characterized by absence of the Bering Sea elements of diatom flora, relatively high abundance of freshwater and marine sublittoral diatoms corresponding to coastal shallow marine environments, and predominance of cryophillic and ice-neritic species indicating more severe than present ice conditions in the Chukchi Sea.

Subsequent sea level rise resulted in opening of the Bering Strait 12 Ka and penetration of the Bering Sea waters into the Chukchi Sea. This caused sharp changes of the quantitative and qualitative composition of diatom assemblages (Ecozones V-I). If assuming sedimentation rates in the Chukchi Sea to be uniform during Holocene, then it is possible to estimate the age of sediments with diatom assemblages of different ecological zones. The Ecozone III is the marking one for all studied core sequences. All Holocene ecological zones were established in the 180 cm long Core 17 obtained at the depth of 40 m. If assuming accumulation of diatom assemblages to start since the onset of intensive water exchange between the Chukchi and Bering Seas, i.e. since 12-11 Ka, it is possible to estimate sedimentation rates in this region of the sea to be 16.8-18.3 cm per 1000 years. The Ecozone III was established in the core interval 110-70 cm, hence the age of its accumulation corresponds to the time interval 6.5-4.2 (or 6.0-3.8) Ka. The Ecozone II was established in the core interval of 55-70 cm, and the age of its accumulation corresponds to 4.2-3.3 (or 3.8-3.0) Ka. Similar results were obtained for the Core 23, where six ecozones corresponding to the end of the Late Pleistocene and Holocene were established. The age range of the Ecozone II accumulation was estimated at 1.1-2.7 Ka, and that of the Ecozone III - at 2.7-5.4 Ka. At the same time insignificant thickness of the subsurface layer (including the Ecozone I) in this core suggests it to be probably eroded. In the Chukchi Sea the thickness of sediments including diatom assemblages of the Ecozone I usually exceeds 50 cm, i.e. the calculated age estimations of the ecozones in this core might be slightly rejuvenated.

Shelf deposits of the East Siberian Sea yield scarce absolute age datings. However, if assuming sea level changes in the Eurasian Eastern Arctic Seas to be synchronous, and taking into account the data on eustatic sea level rise during the Holocene transgression (Morner, 1987; Selivanov, 1996) and scarce radiocarbon datings of the East Siberian Sea deposits, it becomes possible to estimate both: the age of the onset of marine sedimentation in separate regions of the East Siberian Sea shelf and the age of the established diatom ecozones. In the Chukchi Sea the pra-Hope deltaic deposits at the depths of 48-50 m yield age datings of 14.6 Ka. The Chukchi Sea level seems to stand below -20 m until 10 Ka (McManus and Creager, 1984; McManus et al., 1983). The data available for the East Siberian Sea suggest stabilization of sea level at -12 m to take place about 8.3 Ka (Badyukov and Kaplin, 1979).

The studied bottom sediment cores in the East Siberian Sea are restricted to shallow areas with the depths ranging from 14 to 27 m. The basal layers of the Cores 319, 321, and 322 situated northward and north-westward of the Indigirka mouth (sea depths of 14-19 m) are represented by deltaic sediments (according to the freshwater diatom assemblages). However, at the same depths (15-19 m) in the region between the Kolyma and Indigirka mouths only marine sediments have been found (Cores 296, 300). The onset of marine sedimentation in the East Siberian Sea at these depths probably started 8-9 Ka, i.e. the age of marine sequence should not exceed 8 thousand years. Successive changes in the structure of diatom associations allow suggesting that there have been no considerable breaks in sedimentation in this region of the sea. Calculation of sedimentation rates revealed them to be 5.3 cm/Kyr in the Core 296, and 11.3 cm/Kyr in the Core 300. On this basis the time span corresponding to accumulation of the sediments with diatom assemblages of the marking Ecozone III is 5.7-3.8 Ka in the Core 296 and 4.9-3.6 Ka in the Core 300 (in the last case the thickness of the third horizon has been probably reduced). In the Core 296 the sediments including diatom assemblages corresponding to the Ecozone II were accumulated 3.8-2.9 Ka. This is in a good correlation with the data obtained in the Chukchi Sea.

To the east in the region between the Chauna Bay and the Kolyma mouth (Cores 27, 70, 80) the sea depths in the region of drilling vary from 14.5 to 22 m. This suggests the onset of marine sedimentation in this region to start after 9 Ka. However, according to the results of diatom analysis, the marine Holocene sequence has not been recovered completely even in the longest cores (about 2 m). To the east of the Chauna Bay the sea depths increase up to 21-27 m (Cores 298 and 299). So, marine sedimentation in this region started about 10-9 Ka. The section of the Core 299 displays changes in the structure of diatom assemblages best of all. Diatom assemblages suggest the basal core layers to be of shallow marine origin (sea depths of not more than 5 m). On this basis, the calculated sedimentation rates in the Core 299 were about 8 cm/Kyr. Hence, the sediments of the Ecozone III were accumulated 5.7-2.5 Ka, and those of the Ecozone II - 2.5-1.3 Ka.

The radiocarbon datings of the bottom sediments of the Barents Sea and adjacent regions of the North Atlantic and Norwegian-Greenland basin allow for estimating the age of the established ecozones. In the North Atlantic transition from the cold Late Pleistocene environments to the warm Holocene ones was rapid, and the boundary between these two paleogeographical events was diachronous. In the North Atlantic the epoch of warming started 10.5 Ka, while in the Norwegian-Greenland basin - 9.8-9.3 Ka, i.e. during the Preboreal-Boreal periods of Holocene (Kellog et al., 1978; Koc et al., 1993). The age of this boundary in the Barents Sea is 8.7-8.0 Ka (Bjorlykke et al., 1978; Elverhoi and Bomstad, 1980). Warm waters reached Svalbard between 8.7 and 7.7 Ka (Salvigsen et al., 1992). Influx of warm North Atlantic waters into the Barents Sea caused north-eastward shift of the Polar front, thus sharply increasing the water productivity. We consider the sediments including diatom assemblages of the Ecozone IV (Borehole 26) to be accumulated since this epoch. The lower boundary of the Ecozone IV (7.5 m below the sea floor) corresponds to the general increase of diatom abundance in sediments and considerable changes in the structure of diatom assemblages. If assuming sedimentation rate to be uniform (88.2 cm/Kyr in the Borehole 26), then the age of the sediments including diatom assemblages of the Ecozone III (depth range 3.2-4.3 m) will be 5.7-3.7 Ka, i.e. the end of the Atlantic - beginning of the Subboreal periods of Holocene. Similar results were obtained for the core drilled to east of Kolguev Island (Okuneva and Stelle, 1986), where the age of the sediments with the most warm-water foraminiferal assemblages was determined according to radiocarbon datings of 4630(120 (Ri-309) and 5440(130 (Ri-309a) years.

Discussion

Results of the diatom analysis of the bottom sediments of the Eurasian Arctic seas allowed us to for interprete the paleoceanological variations in this region during Holocene and the end of Late Pleistocene. In the Eastern Arctic seas the most significant changes of paleoceanological conditions resulted from the opening of the Bering Strait (12-11 Ka) and influx of relatively warm high productive Pacific waters into the Chukchi and, partly, the East Siberian Seas. They caused sharp increase of diatom productivity and introduction of relatively warm-water species into diatom assemblages. The latter are uncommon for the arctic water masses. Gradual sea level rise up to its present position during the first half of Holocene (11-6 (5.5) Ka, ecozones V and IV) resulted in flooding of shallow Eastern Arctic seas and establishment of modern hydrological (temperature, circulation patterns, distribution of water masses) and hydrobiological conditions.

The period lasting from the end of Atlantic to the beginning of Subboreal was the most favorable for the phytocoenoses development ("hydobiological or marine optimum"; ecozone III). In all Eastern Arctic seas it was marked out by sharp increase (in 2-3 times) of the number of diatom valves in sediments. Since lithological composition of sediments is unchanged, this may be only caused by the burst of phytoplankton productivity. The area of the most warm-water diatoms considerably expanded, especially in the Chukchi Sea within the region influenced by the Alaskan branch of the Bering Sea current. The surface water temperatures in the Central Chukchi depression probably exceeded the modern ones by 2-4¯. Notwithstanding relatively stable paleotemperature conditions typical of the high latitudes, paleoproductivity increase have been marked out in the East Siberian and Laptev Seas, together with invasion of relatively warm-water Bering Sea species to the south-eastern regions of the East Siberian Sea. The latter indicates strengthening of the Long branch of the Bering Sea current.

All these changes were apparently caused by increasing advection of the relatively warm and highly productive Bering Sea waters. It is interesting to note that in the Holocene sequence of the North-Western Pacific Jouse (1961) distinguished an interval (diatom assemblage B) which had been characterized by significant changes in the structure of diatom assemblages suggesting considerable northward migration of warm-water diatoms (not less than 1000 km).

The next short period (ecozone II) in the Eastern Arctic seas is characterized by phytocoenoses production fall, elimination of the most warm-water species, and increase of the cryophillic species abundance in several regions related to decreasing afflux of the Bering Sea waters and severe ice conditions in the arctic seas. This period corresponds to the cooling established in some Arctic and Subarctic regions at 2.5 Ka (so called neoglacial). In general, environmental conditions during the last 1.5-2 thousand years were similar to the modern ones with slight increase of paleoproductivity about 2 Ka due to cryophilic and ice-neritic species. Similar paleohydrological changes were established in the Baffin Bay during the second half of Holocene (Williams, 1990). While the first burst of productivity ("marine optimum") had diachronous boundaries due to different termination of deglaciation of this area, the second one was practically simultaneous (about 2 Ka) being related, like in the case with the Chukchi Sea, to increasing abundance of cryophillic and ice-neritic diatoms.

Holocene paleoenvironmental changes in the Western Arctic seas had their own peculiar features. In the Norwegian,Greenland, and Iceland Seas abundant diatoms appear at the end of Pleistocene - beginning of Holocene. This boundary is diachronous and becomes younger in the north-western direction during the time period ranging from 14 to 7 Ka in accordance with the Polar front migration (Koc et al., 1993; Koc Karpuz and Jansen, 1992). Further regional variations of paleoproductivity and paleotemperatures were determined by the influence of different water masses. About 9 Ka surface water temperatures in the Norwegian-Greenland basin were close to the mordern ones, and during the interval between 8 and 5 Ka they exeeded the mordern ones. Hydrobiological or marine optimum was the most pronounced in the western and north-western coastal regions of Norway (Koc Karpuze and Jansen, 1992; Koc et al., 1993; Dzhinoridze et al., 1981; Samoilovich et al., 1988).

Intensive afflux of the North Atlantic waters to the southern regions of the Barents Sea which began about 9-8 Ka caused northward and north-westward migration of the Polar front and increasing productivity of waters. The main Holocene paleoceanographic event of the Barents Sea, the so called hydrobiologic or marine optimum, corresponded to the last significant intensification of the North Atlantic current. It was caused by the sharp increase of water productivity and surface water temperature at the end of Atlantic - beginning of Subboreal periods of Holocene.

Thus, the main Holocene paleoenvironmental event ("hydrobiological or marine optimum") through all over the Eurasian Arctic seas was established at the end of Atlantic - beginning of Subboreal periods. This rather short time period (about 2 - 2.5 Ka) with the most favorable hydrobiological conditions for marine phytocoenoses was marked out by sharp increase of productivity and spreading of warm-water diatom species. This was probably related to global changes in oceanic circulation patterns which expressed themselves in both North Atlantic and North Pacific.

Surface temperature rise was marked in many regions of the World Ocean, but the age of this event is different. In the northern regions of the Atlantic Ocean it occurred 7-6 Ka, in the Red Sea - about 8 Ka (Barash et al., 1989), in Subantarctic region - about 9 Ka (Hays, 1978). Production of the benthic foraminiferal assemblages of the Norwegian and south-western Barents Sea increased by an order of magnitude during the Atlantic and Subboreal periods of Holocene (Matishov and Pavlova, 1990).

Sometimes it is rather difficult to determine the influence of transgressions upon the paleogeographical conditions on the arctic coasts. History of the Holocene vegetation of the Chukchi Sea coast is one of the most striking examples of such influence. Now nearly all lowlands and highlands of Northern Chukotka are covered with northern hypoarctic tundra and only a narrow coastal region between Shelagin cape and the Amguema River mouth (the belt of dense fogs) is occupied by arctic tundra itself (Yurtsev, 1974).

"Holocene climatic optimum" at the arctic coast of Chukotka corresponds to the most favorable climatic conditions for terrestrial vegetation. This was a period of intensive accumulation of peat with shrub and tree remnants (1.5-2 and even 3-m thick). This remnants have several radiocarbon datings (Ivanov et al., 1984). Abundant paleobotanic and radiocarbon data show that favorable conditions for intensive peat accumulation existed at the arctic coast of Chukotka 11.5 Ka. Northward expansion of shrub and tree vegetation occurred 10-8 Ka. During the subsequent period, which is restricted to the time interval of 7-5 Ka, shrub and tree vegetation degraded, and modern vegetation of hypoarctic tundra occupied coastal lowlands.

The established history of the Holocene transgression of the Chukchi Sea shows that during the period of peat accumulation and expansion of tree and shrub vegetation the coastline stayed tens kilometers northward from its present position. Severe ice conditions of the sea in the early Holocene epoch allowed for existence of continental climatic conditions on the northern Chukotka coast. Growth of summer temperature at the beginning of Holocene caused northward migration of shrub-tree vegetation, as it has been previously shown for the North Western Siberia (Danilov and Polyakova, 1986).

At the same time "marine optimum" in the Chukchi Sea lasted since 5.5-6 to 3-2.5 Ka. By this time the Chukchi sea level reached its present position or was even several meters higher. Intensity of the Bering Sea current was greater than now. Surface water temperature exceeded the modern one by 2-4¯ C. This was a period of degradation of shrub and tree vegetation at the northern coastal lowlands resulting from southward shift of coastline and subsequent increase of marine features of climate. It was caused by frequent fogs that "ate coastal shrub and tree vegetation" on the arctic coasts (Yurtsev; 1974). Warmer sea and reduced ice cover favored appearance of dense fogs.

So, "climatic optimum" of Holocene at the arctic coast of Chukotka was represented by climatic conditions favoring vegetation growth. In the Chukchi Sea it coincided with surface temperature rise and activated biological processes ("hydrobiological or marine optimum"). These two phenomena were not simultaneous.

At the same time in the Western Arctic these paleogeographical events correlate with each other in time. At the northwestern edge of the Russian Plain in vicinity of the White Sea, maximum Holocene atmospheric humidity and average annual temperature rise were related to predominance of spruce in the forests during the first half of the Subboreal epoch (Devyatova, 1986; and others). In Karelia these events occurred at the end of the Atlantic - about 5.5 Ka (Klimanov, 1989). Study of the Barents Sea bottom sediments revealed that optimal hydrobiological conditions existed since 5.7-3.7 Ka (the end of the Atlantic - the first half of Subboreal time). So, optimum conditions on land and in the Barents Sea were simultaneous.

Similar results were obtained from the study of the sections in the southern part of the White Sea (Boyarskaya et al., 1986). Marine environments appeared in the inner part of the Onega Bay 9-8 Ka (Boreal time). Since the end of the Boreal time and during the whole Atlantic period (8-5 Ka) sea level position either coincided with the present one or was slightly higher. Sea level reached its maximum during the Subboreal. Transgression took place against the background of atmospheric humidity growth and average annual temperature rise manifested by succession of floral assemblages.

So, it was found out that Holocene paleogeographical events in the Eastern Arctic seas ("marine optimum") and on the adjacent land ("climatic optimum") were diachronous, while those in the Western Arctic were synchronous.

Conclusions

1 Five ecological diatom zones were established on the basis of the changes in relative abundance of diatom valves, ecologicaland biogeographical composition of diatom assemblages.

2 These diatom ecological zones reflecting the main stages of paleoceanological evolution of the Arctic Seas during the Holocene may be used as stratigraphical markers for subdivision of the sequences.

3 The Ecozone III serves as the reference one for subdivision of the Holocene sediments of the Arctic shelf. Its diatom assemblages reflect the most favourable hydrobiological conditions for the arctic phytocoenoses evolution ("hydrobiological or marine optimum"). Age estimations for the Chukchi Sea are 6.5-2.7 Ka, for the East Siberian Sea - 5.0-2.5 Ka, for the Barents Sea - 5.7-3.7 Ka.

4 The main Holocene paleogeographical events in the Eastern Arctic seas ("marine optimum") and on the ajacent land ("climatic optimum") were diachronous, while those in the Western Arctic were synchronous.

Acknowledgments

I am greatfull to Prof.Yu.A.Pavlidis, P.P.Shirshov`s Institute of Oceanology RAS, who provided me with the core samples for my study. I also thank Dr.E.E.Taldenkova for translation of this manuscript into English. This work was completed under financial support of GEOMAR Research Center for Marine Geosciences in Kiel (Project "The Laptev Sea System").

References

Andrews, J.T. (1972) Recent and fossil growth rates of marine bivalves, Canadian Arctic, and Late-Quaternary Arctic marine environments. Paleogeography, Paleoclimatology, Paleoecology, 11, 157-176.

Badyukov, D.D. and P.A. Kaplin (1979) Sea-level changes on the Far Eastern and Arctic coasts during the last 15000 years. Okeanologiya, 19 (4), 674-690 (in Russian).

Barash, M.S., N.S. Blyum, I.I. Burmistrova, O.G. Dmitrenko, E.V. Ivanva, G.H. Kazarina, S.B. Kruglikova, N.P. Lukashina, V.V. Mukhina, N.S. Oskina, L.V. Polyak, S.A. Safarova, E.A. Sokolova, and T.A. Khusid (1989) Neogene-Quaternary paleoceanlogy according to micropalentolgical data. Moscow: Nauka, 285 pp. (in Russian).

Bauch, H.A., J. Heinemeier, and P.M. Grootes (1996) Radiocarbon (AMS, 14C) ages of sediments from the Laptev Sea. Third Workshop on Russian-German Cooperation: Laptev-Sea System. Program and Abstracts, 90.

Beklemishev, K.V. and G.I. Semina (1986) Geography of plaktonic diatoms of the high and middle latitudes of the World Ocean. In: Skarlato, O.A. and V.V. Krylov (Eds.), Bilogical basis of the distribution of marine organisms. Moscow: Nauka, 7-23 (in Russian).

Boyarskaya, T.D., Ye.I. Polyakova, and A.A. Svitoch (1986) New data on Holocene transgression of the White

Sea. Doklady Akademii Nauk SSSR, 240 (4), 964-968 (in Russian).

Bjorlykke, K., B. Bue and A.Elverhoi (1978) Quaternary sediments in the north-western part of the Barents Sea and their relation to the underlying Mezozoic bedrock. Sedimentology, 25, 227-246.

Cronin, T.M., T.R. Holtz and R.C.Whatley (1994) Quaternary paleoceanography of the deep Arctic Ocean based on quantitative analysis of Ostracoda. Marine Geology, 119, 305-332.

Danilov, I.D. and Ye.I. Polyakova (1986) Climate and ground ice in the North of Western Siberia in Late Pleistocene and Holocene. Polar Geography and Geology, 10 (4), 303-308.

Devyatova, E.I. (1986) Environments and their changes (coasts of the northern and central Onego Lake). Petrozavodsk: Karelia Publ. House, 108 pp. (in Russian).

Dzhinoridze, R.N. (1978) Diatoms in the bottom sediments of the Barents Sea. In: Marine micropaleontology: diatoms, radiolarian, foraminiferes, carbonate nannoplankton, Jouse A.P. (ed.), Moscow: Nauka, 41-44 (in Russian).

Dzhinoridze, R.N., G.S. Golikova and G.A Nagaeva (1981) Materials and researches of diatoms in the Late Pleistocene and Holocene sediments of the Norwegian Sea. In: Pleistocene and Holocene Palynology, Seliverstov, Yu.P. (ed.), Leningrad: Leningrad State Univ.Publ.House, 159-165 (in Russian).

Elverhoi, A. and K.Bomstad (1980) Late Weichselian glacial and marine sedimentation in the western, central Barents Sea. Norsk Polarints. Papp., 3, 1-29.

Glezer, Z.I., A.P. Jouse, I.V. Makarova, A.I. Proshkina-Lavrenko and V.S. Sheshukova-Poretskaya (1974) Diatoms of the USSR. Leningrad: Nauka, I, 403 pp. (in Russian).

Hays, L.D. (1978) A review of the Late Quaternary climatic history at the Atlantic Seas. In: Antarctic glacial hystory and world paleoenvironments. Rotterdam: Balkema, 57-71.

Hopkins, D.M. (1973) Sea-level history in Beringia during the past 250000 years. Quaternary Research, 3, 520-540.

Ivanov, V.F., A.V. Lozhkin and S.S. Kalnichenko (1984) Late Pleistocene and Holocene of the Chukchi Peninsula and Northern Kamchatka. In: Geology and minerals of the North-East of Asia, Gorchakov, V.I. et al. (eds.), Vladivostok: Far Eastern Scientific Center of the Academy of Sciences of USSR, 33-42 (in Russian).

Jouse A.P. (1961) Stratigraphy of the bottom sediments on the North-Western of the Pacific Ocean. Proceedings of the Institute of Geology, Acad. of Sci. Estonia, 8, 183-195 (in Russian).

Kellog, T.B., J.C. Duplessy and N.S. Shackleton (1978) Planktonic foraminiferal and oxygen isitopic stratigraphy and paleoclimatology of Norwegian Sea deep-sea cores. Ibid., 7(1), 61-73.

Khusid T.A. (1989) Paleoecology of the Barents Sea during the Quaternary time according to foraminifers. Bulletin of the Quaternary Commission of the Acad. of Sci. USSR, 58b 105-116.

Klimanov, V.A. (1989) Cyclic recurrence and quasiperiodicity of climatic oscillations in Holocene. In: Paleoclimates of the late glacial and Holocene times, Khotinskii, N.A. (ed.), Moscow: Nauka, 29-33 (in Russian).

Knebel, H.J., J.S. Creager and R.J. Echols (1974) Holocene sedimentary framework of the east-central Bering Sea continental shelf. In: Marine geology and oceanography of the Arctic Seas, Herman, Y. (ed.), N.-Y.: Springer-Verland, 157-172.

Koc Karputz, N. and E.Jansen (1992) A high-resolution diatom records of the last glaciation from the SE Norwegian Sea: documentation of rapid climatic changes. Paleoceanography, 7(4), 499-520.

Koc, N., E. Jansen and H. Haflidason (1993) Paleogeographic reconstructions of surface ocean conditions in the Greenland, Iceland and Norwegian Seas through the last 14 Ka based on diatoms. Quaternary Science Reviews, 12, 115-140.

Mangerud, J., S.T. Andersen, B.E. Berglund and J.J. Donner (1974) Quaternary stratigraphy of Norden, a proposal for terminology and classification. Boreas, 3, 109-128.

Matishov, G.G. and L.G. Pavlova (1990) General ecology and paleogeography of the Polar seas. Leningrad: Nauka, 223 pp. (in Russian).

McManus, D.A. and J.S. Creager (1984) Sea-level data for parts of Bering-Chukchi shelves of Beringia from 19.000 to 10.000 yr. B.P. Quaternary Research, 21(3), 317-325.

McManus, D.A., J.S.Creager, R.J.Echols and M.L.Holmes (1983) The Holocene transgressin on the Arctic flank of Beringia: Chukchi valley to Chukchi estuary to Chukchi Sea. In: Quaternary coastlines and marine archaeology, Masters, P.M. and N.C. Fleming (eds.), 365-388.

Mode, W.H. and J.D.Jacobs (1987) Surficial geology and palynology, inner Frobisher Bay. In: Cumberland Sound and Frobisher Bay, southeastern Baffin Island, N.W.T, French, H.M. and P. Richard (eds.), 12th International Union for Quaternary Research. Field Excursion C-2, 53-62.

Morner, N.A. (1987) Late Quaternary sea-level changes. Sea-Surface Studies: A Global View. L.e.a.: Croom Helm, 242-264.

Norgaard-Pedersen, N. (1996) Late Quaternary Arctic Ocean sediment records: Dissertation zur Erlangung des Doktorgrades der Mathematisch-Naturwissenschaftlichen Fakultat der Christian-Albrechts-Universitat zu Kiel. 115 pp.

Okuneva, O.G. and V.Ya. Stelle (1986) New biostratigraphical data from the engineering-geological boreholes of the eastern coast of Kolguev Island. In: Engineering-geological conditions of the shelf and methods of their

investigations, 8-12 (in Russian).
Pavlidis, Yu.A., F.A. Scherbakov, T.D. Boyarskaya, N.N. Dunaev, Ye.I. Polyakova and T.A. Khusid (1992) New data on Quaternary stratigraphy and palegeography of the southern part of the Barents Sea. Okeanologiya, 32 (5), 917-923 (in Russian).
Polyakova, Ye.I. (1990) Stratigraphy of the Late Pleistocene-Holocene sediments of the Beringia Shelf on the basis of diatom complexes. Polar Geography and Geology, 14, 271-278.
Polyakova, Ye.I. (1994) Features of diatom thanatocoenoses formation in the bottom sediments of the Eurasian Arctic Seas. Okeanologiya, 44 (5), 346-352 (in Russian).
Polyakova, Ye.I. (1997a) Holocene of the Eurasian Arctic Seas (diatom sratigraphy and paleoceanology). Okeanologiya, 37 (2), 269-278 (in Russian).
Polyakova, Ye.I. (1997b) The Eurasian Arctic seas during the Late Cenozoic. Moscow: Scientific World, 145 pp. (in Russian).
Polyakova, Ye.I., Yu.A. Pavlidis and A.I. Levin (1992) The features of diatom thanatocoenoses formation in the bottom sediments of the Barents Sea . Okeanologiya, 32 (1), 166-175 (in Russian).
Rhyzhov, V.M., V.P. Rusanov and V.S. Latyshev (1984) Chemical-biological indications of water masses of the Chukchi Sea. Proceedings of the Arctic and Antarctic Scientific Research Institute, 368, 26-40 (in Russian).
Saidova, Kh.M. (1994) Ecology of the shelf foraminiferal assemblages and paleoenvironmental of the Holocene of the Bering and Chukchi Seas. Moscow: Nauka, 94 pp. (in Russian).
Salvigsen, O., S.L. Forman and G.H. Miller (1992) Thermophilous molluscs on Svalbard during the Holocene and their paleoclimatic implications. Polar Research, 11, 1-10.
Samoilovich, Yu.G., R.M. Lebedeva, L.Ya. Kagan, L.V. Ivanova and O.S.Chapina 1988) Experience and perspectives of the complex stratigraphical methods use for the Barents Sea Quaternary deposits study. In: Quaternary paleoecology and paleogeography of the North Seas, Matishov, G.G. and G.A.Tarasov (eds.), Moscow: Nauka, 150-162 (in Russian).
Spielhagen, R.F., H. Erlenkeuser and J. Heinemeier (1996) Variability of freshwater export from the Laptev Sea to the Arctic Ocean during the last 14.000 years. Third Workshop on Russian-German Cooperation: Laptev Sea System. Program and Abstracts, 107.
Stein, R., C. Schubert, C. Vogt and D. Futterer (1994) Stable isotope stratigraphy, sedimentation rates, and salinity changes inthe Latest Pleistocene to Holocene eastern central Arctic Ocean. Marine Geology, 119, 333-355.
Stratigraphical code of the USSR (1977). Leningrad:MSK, 79 pp.
Williams, K.M. (1992) Late Quaternary paleogeography of the western Baffin Bay region: evidence from fossil diatoms. Can.J.Earth Sci., 27, 1487-1494.
Yurtsev, B.A. (1974) Problems of the botanical geography of the North-East of Eurasia. Leningrad: Nauka, 159 pp. (in Russian).

Appendix: Marine and brackish water diatoms in the Holocene sediments of the Eurasian Arctic Seas. Key: Ecology: p – planctonic neritic and panthalassic species; s – sublittoral benthic and semi-benthic species; s-i – sea-ice species. Phytogeographical distribution: a-b – arctic-boreal; b – bipolar; a-b-t – arctic-boreal-tropical; t-b – tropical-boreal; c – cosmopolitan.

Current name	Ecology	Phytogeographical groups	Synonyms
Achnanthes taeniata Grun.	p	a-b	
Actinocyclus divisus (Grun.) Hust.	p	a-b	
A. ehrenbergii Ralfs	p	c	
A. ochotensis Jouse	p	a-b	
Actinoptychus undulatus (Beil.) Ralfs	p	c	Actinoptychus senarius (Ehr.) Ehr.
Bacterosira concava-convexa Makar	p	a-b	
Bacterosira fragilis Gran	p	a-b	Bacterosira bathyomphala (Cl.) Syvert. and Hasle
Caloneis aemula Schm.	s	a-b	
Chaetoceros compressus Lauder	p	c	
C.diadema (Ehr.) Gran	p	a-b	C.subsecundus (Grun.) Hust.
C.debilis Cl.	p	c	
C.didymus Ehr.	p	b-t	
C.compressus Lauder	p	c	

Appendix (continued):

Current name	Ecology	Phytogeographical groups	Synonyms
C.furcellatus Bail.	p	a-b	
C.ingolfianus Ostenf.	p	a-b	
C.mitra (Bail.) Cl.	p	a-b	
C.septentrionalis Oestr.	p, s-i	a-b	
Cocconeis scutellum Ehr.	s	c	
C.pediculus Ehr.	s	c	
C.placentula Ehr.	s	c	
Coscinodiscus asteromphalus Ehr.	p	c	
C. centralis Ehr.	p	c	
C.curvatulus Grun.	p	a-b	
C.marginatus Ehr.	p	a-b-t	
C.oculus-iridis Ehr.	p	a-b	
C.perforatus Ehr.	p	c	
C.radiatus Ehr.	p	c	
Cyclotella striata (Kutz.) Grun.	p	c	
Delphineis kippae Sancet.	p	a-b	
D. surirella (Ehr.) Andrews	s	a-b	Rhaphoneis surirella (Ehr.) Grun.
Detonula confervaceae (Cl.) Grun.	p	a-b	
Diploneis bombus Ehr.	s	c	
D.elliptica (Kutz.) Cl.	s	c	
D. interrupta (Kutz.) Cl.	s	c	
D.smithii (Breb.) Cl.	s	c	
D.smithii var.pumila (Grun.) Hust.	s	c	
D.subcincta (Schm.) Cl.	s	a-b	
D.suborbicularis (Greg.) Cl.	s	c	
Fallacia forcipata (Grev.) Stickle et Mann	s	c	Navicula forcipata Grev.
F. pygmaea (Kutz.) Stickle and Mann	s	c	Navicula pygmaea Kutz.
Grammatophora arctica Cl.	p	b	
Haslea kjelmanii (Cl.) Sim.	s	a-b	
Hyalodiscus scoticus (Kutz.) Grun.	p	c	
Melosira arctica (Ehr.) Dickie	p,s-i	a-b	
M.juergensii Agardh	s	c	
M. nummuloides (Dillw.) Agardh.	s	c	
M.sol (Ehr.) Kutz.	s	b	
Navicula cancellata var.gregorii Ralfs	s	a-b	
N. digitoradiata (Greg.) Ralfs	s	c	
N. directa (Smith.) Ralfs	s,s-i	b	
N.distans Smith	s	a-b	
N.gelida Grun.	s	a-b	
N.gregaria Donk.	s	c	
N.kariana var. detersa Grun.	s	a-b	
N.peregrina (Ehr.) Kutz.	s	c	
N.superba Cl.	s	a-b	
N.transitans Cl.	s,s-i	a-b	
N.transitans var.asymmetrica (Cl.) Cl.	s	a-b	
N.transitans var.derasa (Grun.) Cl.	s	a-b	
N.vanhoeffenii Gran	p,s-i	a-b	
Neodenticula seminae (Simons.et Kan.) Ak. et Yan.	p	b-t	Denticulopsis seminae
Nitzschia arctica Cl.	s	a-b	
N. cylindrus (Grun.) Hasle	p,s-i	b	Fragilariopsis cylindrus (Grun.) Krieger
N.grunowii Hasle	p, s-i	a-b	Fragilariopsis oceanica (Cl.) Hasle
N. frigida Grun.	s	a-b	
N. laevissima Grun.	s	a-b	
N.polaris Grun.	p,s-i	a-b	

Appendix (continued):

Current name	Ecology	Phytogeographical groups	Synonyms
N.scabra Cl.	s	a-b	
N.sigma (Kutz.) Smith	s	c	
Odontella aurita (Lyngb.) Agardh	s	a-b-t	Biddulfia aurita (Lyngb.) Breb.
Opephora marina (Greg.) Petit.	s	c	
Paralia sulcata (Ehr.) Cl.	s	c	
Pinnularia quadrataraea A.S.	p	a-b	
P.quadrataraea v.baltica Grun.	p	a-b	
P.quadrataraea v.constricta Heiden	p	a-b	
Porosira glacia;is (Grun.) Jorg.	p	b	Podosira glacialis (Grun.) Cl.
Pleurosigma stuxbergii Cl. Et Grun.	p	a-b	
Pseudogomphonema arcticum (Grun.) Medlin	s,s-i	b	
Pseudogomphonema groenlandicum (Oestr.) Medlin	s	a-b	
Pyxidicula nipponica (Gran et Yendo) Streln. et Nikol.	p	a-b	Stephanopyxis nipponica Gran and Yendo
Proboscia alata (Bright.) Sundsstr.	p	c	Rhizosolenia alata Bright., Simonseniella alata (Bright.) Fenner
Rhizosolenia hebetata f. Hebetata Bailey	p	c	
R.hebetata f. Semispina (Hensen) Gran	p	c	
R.setigera Bright.	p	c	
R.setigera var.arctica I.Kiss.	p	a-b	
R.styliformis Bright.	p	c	
Stenoneis inconspicua var.baculus (Cl.) Cl.	s	a-b	
Synedra tabulata (Agardh) Kutz.	s	c	
Thalassionema nitzschioides Grun.	p	c	
Thalassiosira angulata (Greg.) Hasle	p	a-b	
T.anguste-lineata (A.Schm.) Fryx. and Hasle	p	a-b-t	Thalassiosira polychorda (Gran) Jorg.
T.antarctica Comber	p	b	
T.baltica (Grun.) Ostenf.	p	a-b	Coscinodiscus polyacanthus var.baltica Grun.
T.bioculata (Grun.) Ostenf.	p, s-i	a-b	
T.bramaputrae (Ehr.) Hakans.	s	a-b	Thalassiosira hyperborea (Grun.) Hasle and Lange
T.bramaputrae v.septentrionalis (Grun.) Makar.	s	a-b	Thalassiosira hyperborea v.septentrionalis (Grun.) Hasle
T.bulbosa Syvert.	p	a-b	
T.constricta Gaard.	p	a-b	
T.eccentrica (Ehr.) Cl.	p	c	
T.gravida Cl.	p	b	
T.hyalina (Grun.) Gran	p	a-b	
T.latimarginata Makar.	p	b	Thalassiosira trifulta Fryx.
T.nordenskioeldii Cl.	p	a-b	
T.oestrupii (Ostenf.) Prosh.-Lavren.	p	c	
T.punctigera (Castr.) Hasle	p	c	
Thalassiothrix longissima (Cl.) Cl. and Grun.	p	a-b-t	
T.frauenfeldii Grun.	p	a-b-t	
Trachyneis asperas (Ehr.) Cl.	s	a-b	

Late Quaternary Organic Carbon and Biomarker Records from the Laptev Sea Continental Margin (Arctic Ocean): Implications for Organic Carbon Flux and Composition

R. Stein, K. Fahl, F. Niessen and M. Siebold

Alfred-Wegener-Institut für Polar- und Meeresforschung, Postfach 120161, D 27515 Bremerhaven, Germany

Received 10 April 1997 and accepted in revised form 1 October 1997

Abstract - In order to understand the processes controlling organic carbon deposition (i.e., primary productivity vs. terrigenous supply) and their paleoceanographic significance, three sediment cores (PS2471, PS2474, and PS2476) from the Laptev Sea continental margin were investigated for their content and composition of organic carbon. The characterization of organic matter includes the determination of bulk parameters (hydrogen index values and C/N ratios) and the analysis of specific biomarkers (*n*-alkanes, fatty acids, alkenones, and pigments).

Total organic carbon (TOC) values vary between 0.3 and 2%. In general, the organic matter from the Laptev Sea continental margin is dominated by terrigenous matter throughout. However, significant amounts of marine organic carbon occur. The turbidites, according to a still preliminary stratigraphy probably deposited during glacial Oxygen Isotope Stages 2 and 4, are characterized by maximum amounts of organic carbon of terrigenous origin. Marine organic carbon appears to show enhanced relative abundances in the Termination I (?) and early Holocene time intervals, as indicated by maximum amounts of short chain *n*-alkanes, short-chain fatty acids, and alkenones. The increased amounts of fatty acids, however, may also have a freshwater origin due to increased river discharge at that time. The occurrence of alkenones is suggested to indicate an intensification of Atlantic water inflow along the Eurasian continental margin starting at that time.

Oxygen Isotope Stage 1 accumulation rates of total organic carbon are 0.3, 0.17, and 0.02 g $C/cm^2/ky$ in cores PS2476, PS2474, and PS2471, respectively.

Introduction

The Arctic Ocean and the surrounding continental margins (Figure 1) are key areas for understanding the global climate system and its change through time (for overviews see ARCSS Workshop Steering Committee, 1990; NAD Science Committee, 1992; and further references therein). The present state of the Arctic Ocean itself and its influence on the global climate system strongly depend on the large river discharge. The fluvial freshwater supply contributes significantly to the strong stratification of the near-surface water masses of the Arctic Ocean and, thus, favors sea-ice formation. The melting and freezing of sea ice result in distinct changes in the surface albedo, the energy balance, the temperature and salinity structure of the upper water masses, and the biological processes. Furthermore, large amounts of dissolved and particulate (organic and inorganic) material are transported by the major Arctic rivers onto the shelves. The annual discharge of total suspended sediments by the Lena River alone, for example, is $17.6*10^6$ tons, and the amount of dissolved organic carbon reaches maximum values of 11 mg/l during summer floods (Martin et al., 1993). Particulate organic carbon supply by the Lena River is estimated to reach about $1.3 * 10^6$ tons per year (Rachold et al., 1996). Although most of this material is deposited in the inner Laptev Sea (Kuptsov and Lisitzin, 1996), significant amounts are further transported by different mechanisms onto the outer continental margin and the open ocean (Figure 2), which may contribute significantly to the entire Arctic Ocean sedimentary and chemical budgets. Riverine sediments are incorporated into the sea ice in the shelf areas and then transported as ice-rafted debris (IRD) through the

In: Kassens, H., H.A. Bauch, I. Dmitrenko, H. Eicken, H.-W. Hubberten, M. Melles, J. Thiede and L. Timokhov (eds.) Land-Ocean Systems in the Siberian Arctic: Dynamics and History. Springer-Verlag, Berlin, 1999, 635-655.

central Arctic Ocean via the Transpolar Drift (Figure 1; Pfirman et al., 1989; Nürnberg et al., 1994). In areas of extensive melting, sediment particles are released and deposited on the sea floor. In these areas, this process may dominate the supply and accumulation of terrigenous material in the polar environment. Downslope-transport by turbidity currents may control the sedimentation in the lower continental slope and deep sea (Figure 2).

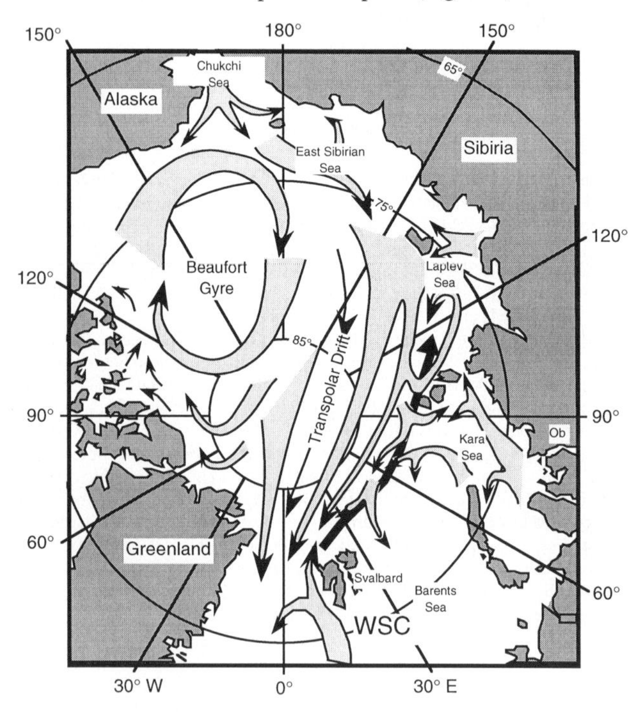

Figure 1: Overview map of the Arctic Ocean with major physiogeographic features and main surface-water-current systems (BG = Beaufort Gyre; TD = Transpolar Drift; WSC = Westspitsbergen Current). The Westspitsbergen Current, an extension of the North Atlantic-Norwegian Current, carries warm, relatively saline water into the Arctic Ocean where it cools down and extends in intermediate water depths along the Eurasian continental margin as countour current into the eastern Arctic Basins (black arrow) (Gordienko and Laktionov, 1969; Aagard et al., 1985; Rudels et al., 1994).

In relation to the world´s ocean, the Arctic Ocean is rather low-productive due to the permanent ice-cover (Subba Rao and Platt, 1984). In marginal seas (such as the Laptev Sea) characterized by an increased fluvial nutrient supply as well as the proximity to ice edges, however, significantly raised primary production rates and, thus, increased concentrations of marine organic carbon in surface sediments were observed (c.f., Sakshaug and Holm-Hansen, 1984; Nelson et al., 1989; Boetius et al., 1996; Fahl and Stein, 1997a).

This study focuses on the identification of geochemical parameters indicative for marine organic-carbon (i.e., surface-water productivity) and terrigenous organic-carbon supply in the Laptev Sea continental margin area. To obtain a preliminary estimate of the composition of the organic matter, hydrogen index (HI) values from Rock-Eval pyrolysis and total organic carbon/total nitrogen (C/N) ratios can be used. In immature organic-carbon-rich (TOC> 0.5%) sediments, HI values of < 100 mg HC / g C are typical of terrigenous organic matter (kerogen type III), whereas HI values of 300 to 800 mg HC / g C are typical of marine organic matter (kerogen types I and II) (Tissot and Welte, 1984). Carbon/nitrogen (C/N) ratios of marine organic matter (mainly phytoplankton and zooplankton) are around 6, whereas terrigenous organic matter (mainly from higher plants) has C/N ratios of >15 (e.g., Bordowskiy, 1965; Scheffer and Schachtschabel, 1984; Hedges et al., 1986).

For more precise informations about the marine and terrigenous proportions of the organic-carbon fraction in marine sediments and for paleoclimatic and paleoceanographic reconstructions, specific biomarkers can be used. The distribution of *n*-alkanes, for example, may allow an identification of contributions of land-derived vascular plant material (characterized by long-chain C_{27}, C_{29} and C_{31} *n*-alkanes) and of marine phytoplankton (dominated by C_{17} and C_{19} *n*-alkanes) (e.g., Blumer et al., 1971; Kollatukudy, 1976; Simoneit et al., 1977; Prahl and Muehlhausen, 1989). Other biomarkers used as a marine source indicator are long-chain unsaturated alkenones ($C_{37:2}$, $C_{37:3}$, and $C_{37:4}$) (e.g., Marlowe et al., 1984; Brassell et al., 1986; Prahl et al., 1988, 1993) and short-chain fatty acids (16:0, 16:1, 18:0, 18:1) (e.g., Ackman et al., 1968; Orcutt and Patterson, 1975; Mayzand et al., 1989; Kattner and Brockmann, 1990; Fahl and Stein, 1997a). Furthermore, the alkenones which are synthezised by prymnesiophytes (Volkman et al., 1980), provide an attractive geochemical tool for reconstructions of surface-water paleotemperature in lower latitude areas characterized by temperatures >10°C (e.g., Marlowe et al., 1984; Brassell et al., 1986; Prahl and Wakeham, 1987; Prahl et al., 1988, 1993; Sikes et al., 1991; Rostek et al., 1993; Rosell-Melé et al., 1995; Hinrichs et al., 1997). Concerning the alkenones as paleotemperature indicator in high latitudes with low temperatures << 10°C we are working, however, the interpretation of the data has much more uncertainties (cf., Sikes and Volkman, 1993; Rosell-Melé et al., 1995).

In the study area of the Laptev Sea it also has to be considered that the strong fluvial supply may also provide freshwater organic matter which may cause difficulties in interpreting some of the biomarker signals (Fahl and Stein, 1997b). Short-chain fatty acids, for example, can also be synthezised by freshwater (lacustrine) phytoplankton (Ahlgren et al., 1990, 1992; Léveillé et al., 1997).

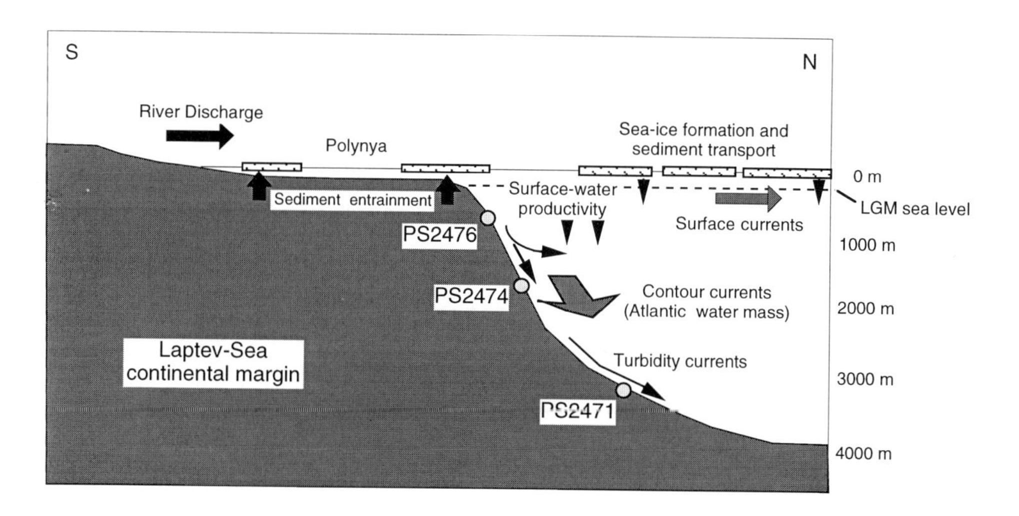

Figure 2: Scheme indicating major processes controlling sedimentation and organic-carbon flux along the Laptev-Sea continental margin and position of studied sediment cores PS2471, PS2474, and PS2476. Broken line marks sea level during the Last Glacial Maximum (LGM).

Material and stratigraphy

During RV *Polarstern* Cruise ARK-IX/4 sediment cores were recovered on four transects from the Laptev Sea continental margin (Fütterer, 1994; Nürnberg et al., 1995). The three cores PS2471, PS2474, and PS2476 discussed in this paper, were obtained from the western continental slope at water depths of 3047 m, 1494 m, and 524 m, respectively (Figures 2 and 3; cf. Table 2). In all cores, the sediments mainly consist of brown (upper part of the sedimentary sequences), olive, gray to dark gray sandy, silty clays with abundant black spots. Occasionally, sandy turbidites are intercalated in cores PS2471 and PS2474. Based on the visual core description (Fütterer, 1994), the sediments contain relatively high amounts of terrigenous organic matter (plant fibers). Despite the turbidites, the sediments do not show any major lithological change, thus preventing visual core correlation. Magnetic susceptibility records, however, provide an excellent tool for core correlation (Figure 4; Fütterer, 1994; Nürnberg et al., 1995), which is the basis for the correlation shown in Figures 5 and 6.

Based on the abundance maximum of the coccolithophoride *Gephyrocapsa* spp. which is an indicator for Arctic interglacials (Gard, 1988), the upper 50 cm of Core PS2471 (Figure 5, Unit A) is assumed to be of Oxygen Isotope Stage 1 (Holocene and late Termination 1) age (Nürnberg et al., 1995). This is also suported by correlation of magnetic susceptibility records of cores PS 2471, PS2474, and PS2476 with the AMS ^{14}C-dated magnetic susceptibility record of the nearby Core PS2778 (Figures 3 and 4). This would result in Oxygen Isotope Stage 1 (0-12 ka) mean linear sedimentation rates (LSR) of approximately 4 cm/ky, 22 cm/ky, and 50 cm/ky in cores PS2471, PS2474, and PS2476, respectively. The occurrence of *Gephyrocapsa* spp. between 200 and 300 cm in the sedimentary sequence of Core PS2471 is related by Nürnberg et al. (1995) to the Last Interglacial (Oxygen Isotope Stage 5). In sedimentary records from the Nansen Basin, however, *Gephyrocapsa* spp. was also recorded in Oxygen Isotope Stage 3 (Baumann, 1990).

Based on a detailed sedimentological study performed in the western Laptev Sea continental margin area, including PARASOUND profiles, magnetic susceptibility and clay mineral records as well as AMS ^{14}C datings, intervals with major occurrences of turbidites and debris flows are correlated with glacial stages of lowered sea level (Weiel, 1997). This may suggest that incores PS2471 and PS2474 the (upper) turbidite may fall either into Oxygen Isotope Stage 4 or Oxygen Isotope Stage 2. Especially during Oxygen Isotope Stage 4, a deposition of thick sequences of debris flows in the western Laptev Sea, triggered by a (in comparison to the Last Glacial Maximum much more extended) major ice sheet on Taymyr Peninsula and thus drastically increased supply of terrigenous matter, has been postulated by Weiel (1997) and Niessen et al. (1997). If the depositional area of cores PS2471 and PS2474 is connected to this sedimentary regime, the turbidites in Unit C are probably of Oxygen Isotope Stage 4 age or older. If, on the other hand, the two cores are more influenced by processes controlling the eastern Laptev Sea (i.e., influence by river Lena), a younger age of the upper turbidites is possible. Taking the mean Holocene sedimentation rates and assuming that drastically reduced pre-Holocene sedimentation rates are rather unrealistic (which is supported by the high pre-Holocene sedimentation rates recorded in several other cores from the central and eastern Laptev-Sea area; Bauch et al., 1996; Spielhagen et al., 1996; Fahl and Stein, 1997b), an Oxygen Isotope Stage 2 (LGM) age seems to be more probable for the upper turbidites. That means, Unit B (Figure 5) may represent post-LGM/Termination I, overlying the turbidites of Unit C. The two lower turbidites in Unit C of Core PS2471, occurring below the interval characterized by significant abundances of (Oxygen Isotope Stage 3 ?) *Gephyrocapsa* spp., may have been deposited during older (Oxygen Isotope Stage 4?) glacial periods.

Based on the present data base, the stratigraphic framework of the three cores is still preliminary; it is not possible yet to prefer one age model and reject the other. The stratigraphy

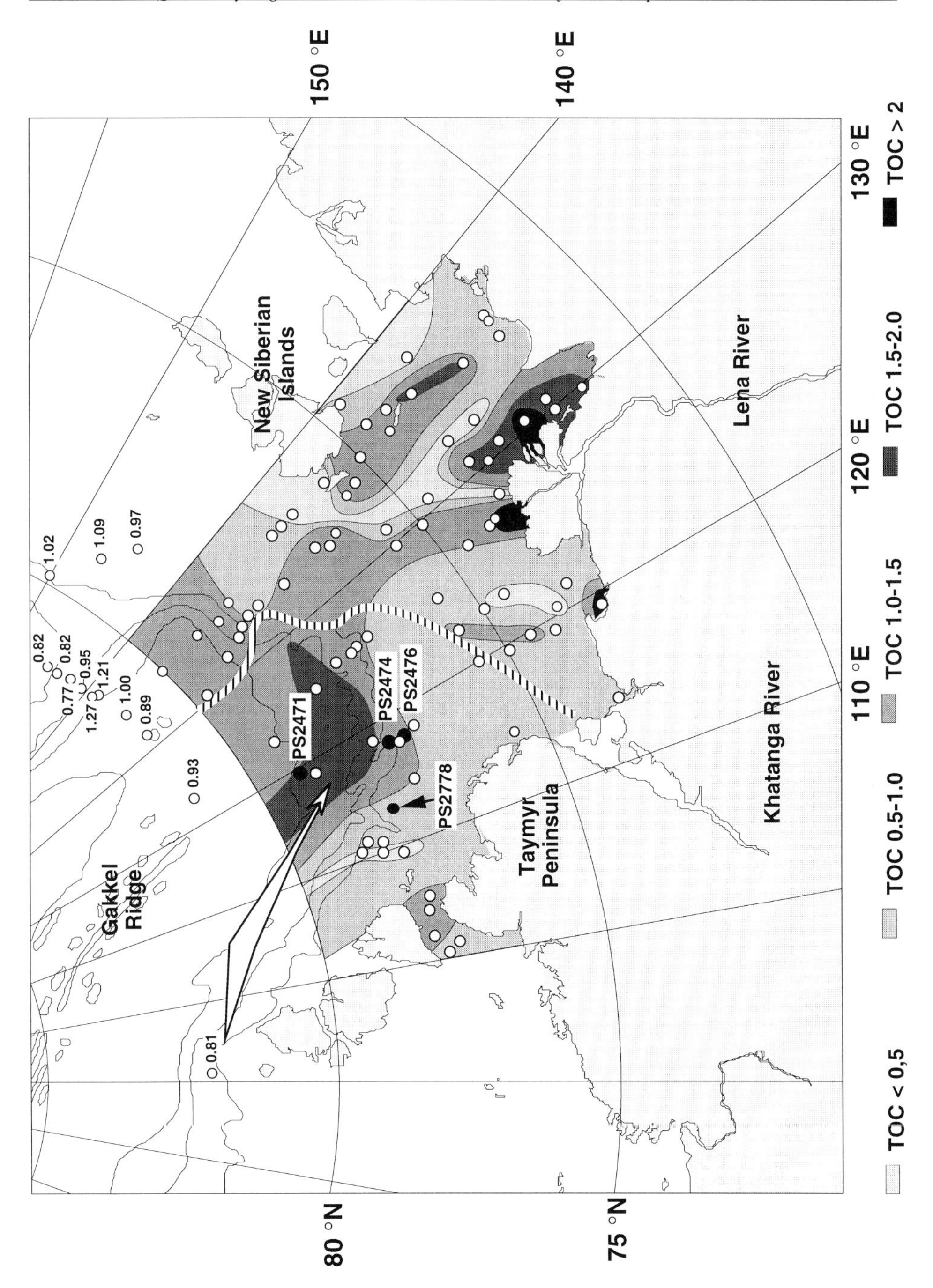

Figure 3: Distribution map of total organic carbon content in Laptev Sea surface sediments and location of sediment cores PS2471, PS2474, and PS2476. In addition, the location of Core PS2778 is shown. The hatched line symbolizes the ice margin during September 1993 (Eicken et al., 1995). The open arrow indicates Atlantic water inflow.

will be proven, however, by future determinations of other, more precise stratigraphic parameters such as ^{10}Be and ^{234}Th, microfossils, and amino acids. In this paper, both age models will be discussed and suggestions about the stratigraphy will be made according to the organic geochemical parameters.

Methods

Determination of bulk parameters

Total carbon, total nitrogen, and total organic carbon (TOC) contents were determined on both ground bulk samples and HCl-treated carbonate-free samples, using a HERAEUS CHN analyser (for details see Stein, 1991). C/N ratios were calculated as total organic carbon/total nitrogen ratios based on weight percentages. The Rock-Eval parameters hydrogen index (HI in mg HC / g TOC) and oxygen index (mg CO_2 / g TOC) were determined as described by Espitalié et al. (1977). Although HI values and C/N ratios determined on organic-carbon-poor sediments (TOC <0.5%) have to be interpreted with caution (e.g., Stein, 1991 and further references therein), the records are used as indicators for organic-carbon composition because they are strongly supported by the biomarker records (see results and discussion).

Determination of biomarkers

For the determinations of specific biomarkers gas chromatography (GC) and gas chromatrography/mass spectrometry (GC/MS) techniques were used.

For the lipid analyses the sediment samples were stored at -80°C or in dichloromethane: methanol (2:1, by vol.) at -23 °C until further treatment. The sediment (2 g) was homogenised, extracted and purified as recommended by Folch et al. (1957) and Bligh and Dyer (1959). An aliquot of the total extract was used for analysing *n*-alkanes and alkenones. Both lipid classes were separated from the other fractions by column chromatography (*n*-alkanes with hexane, alkenones with hexane:ethyl acetate, 95:5 and 90:10, by vol.). To provide a separation of the alkenone compounds ($C_{37:2}$, $C_{37:3}$, and $C_{37:4}$) from co-occurring alkenoates by gas chromatography, a saponification step with 1 M potassium hydroxide in 95 % methanol for 2 hours at 90°C was applied. The alkenone composition was analysed with a Hewlett Packard gas chromatograph (HP 5890, column 30 m x 0.25 mm; film thickness 0.25 µm; stationary phase: HP 1) using a temperature program for analysing *n*-alkanes as follows: 60°C (1 min), 150°C (rate: 10°C/min), 300°C (rate: 4°C/min), 300°C (45 min isothermal); for alkenones: 60°C (1 min), 270°C (rate: 20°C/min), 320°C (rate: 1°C/min), 320°C (20 min isothermal). A volume of 1 µl was injected (cold injection system: 60 °C (5 s), 300 °C (60 s, rate: 10°C/s) and helium was used as carrier gas. As internal standards squalane and octacosanoic acid methyl ester were added to the samples before any analytical step. Because of the low alkenone concentrations results obtained by GC were checked for a selected set of samples by GC/MS technique. The GC/MS consists of a gaschromatograph (HP 5890, column 30 m x 0.25 mm; film thickness 0.25 µm; stationary phase: HP 5) and a mass spectrometer (MSD, HP 5972, 70 eV electron-impact-ionisation, Scan 50-650 m/z, 1 scan/s, ion source temperature 175°C). The injection volume is 1 µl (splitless). Helium was used as carrier gas (1 ml/min at 60°C).

For preparing fatty acid methyl esters and free alcohols by transesterification an aliquot of the total extract was treated with 3 % concentrated sulfuric acid in methanol for 4 hours at 80°C. After extraction with *n*-hexane the product was analysed by GC (as above) but using DB-FFAP as stationary phase. The temperature program was as follows: 160°C, 240°C (rate: 4°C/min), 240°C (15 min isothermal) (modified after Kattner and Fricke, 1986). The injection volume was 1 µl. The fatty acids and alcohols were identified by standard mixtures and quantified using

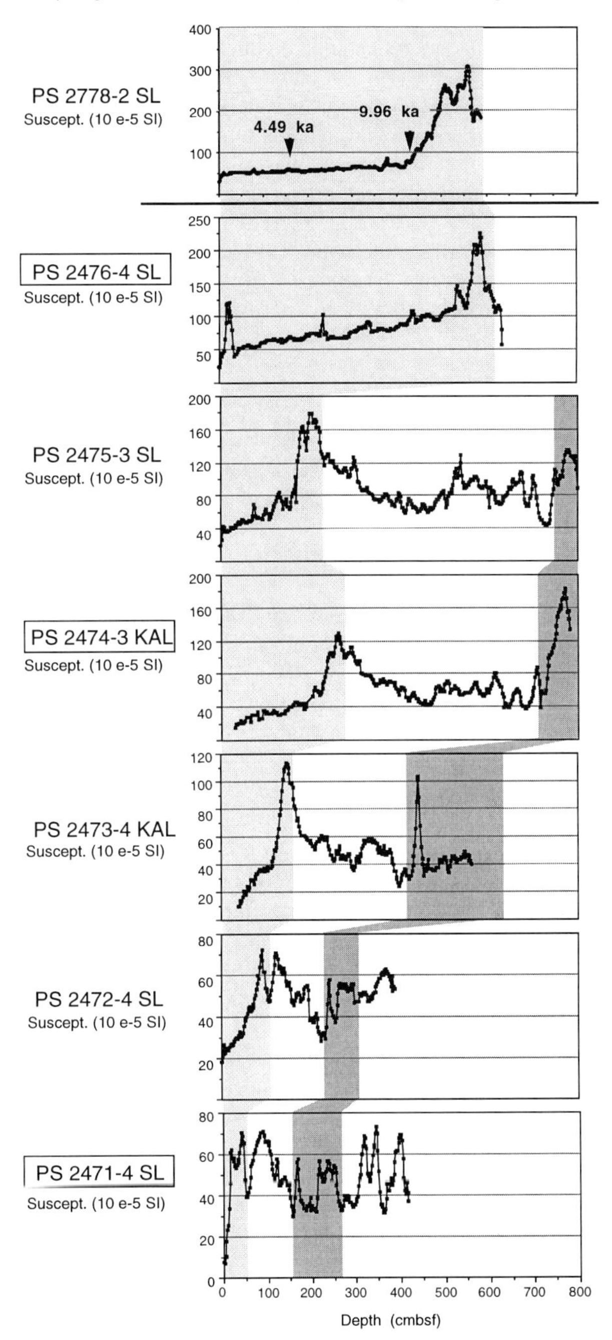

Figure 4: Magnetic susceptibility records of a north-south profile from the Laptev-Sea continental margin including the records from cores PS2471, PS2474, and PS2476 (Fütterer, 1994; Nürnberg et al., 1995). In addition the AMS ^{14}C-dated magnetic susceptibility record of nearby Core PS2778 (Rachor, 1997; Weiel, 1997) is shown (for location of the core see Figure 3). The AMS ^{14}C dates determined on marine bivalves, are corrected by a reservoir effect of 550 years.

19:0 fatty acid methyl ester as internal standard which was added to the samples before any analytical step.

The abundance of the tetrapyrrolic pigments was determined by measuring the absorbance of the solvent extract (80% acetone) at a wavelength of 410 nm (Rosell-Melé, 1994; Rosell-Melé and Koc, 1997). In addition, the measurement was carried out at 645 and 663 nm to determine chlorophyll abundances. The turbidity factor (absorbance at 750 nm) has been substracted.

The carbon-preference-index (CPI), i.e., the relation of odd-to-even chain lengths of *n*-alkanes between C_{21} and C_{32} (Bray and Evans, 1961), is used as a measure for the maturity of terrigenous organic matter. Fresh terrigenous organic material has a CPI-index of 3 to 10 (Brassell et al., 1978; Hollerbach, 1985), whereas fossil material varies around 1 depending on the state of diagenesis.

Calculation of accumulation rates

Bulk (BAR) and total organic carbon (TOCAR) accumulation rates were calculated for the Holocene using a mean wet bulk density (WBD) value of 1.43 g/cm^3 determined in Core PS2474 (Fütterer, 1994), and a grain density (GD) of 2.65 g/cm^3 (cf., Table 2):

(1) $PO\ (\%) = ((GD - WBD) / (GD - 1.026)) * 100$ (Weber et al., 1997)

(2) $BAR\ (g / cm^2 / ky) = LSR * (WBD - 1.026\ PO/100)$ (van Andel et al., 1975)

(3) $TOCAR\ (g\ C / cm^2 / ky) = BAR * TOC/100$

For calculation of mean linear sedimentation rates (LSR) see above.

Results

The TOC record of Core PS2476 displays very constant values ranging between 0.9 and 1.2%. Only in the lowermost part of the record (600 - 625 cmbsf), lower values of about 0.5% occur (Figure 5).Based on the TOC values, the sedimentary record of Core PS2474 can be divided into three intervals (Figure 5). In the lower part (500 - 780 cmbsf) TOC values vary between 0.6 and 1.1% with one single spike of 1.6% at 660 cmbsf. In the middle part most of the TOC values are around 0.5%. The uppermost 200 cm of the record are dominated by high TOC values of 1-1.5%.

In Core PS2471 from the lower slope, maximum TOC values of 0.6 to 2% occur in the three turbidites at 350 - 390 cmbsf, 280 - 320 cmbsf, and 180 - 230 cmbsf (Unit C, Figure 5). The intervals inbetween the turbidites as well as the interval between 50 and 180 cmbsf (Unit B) are characterized by low TOC values around 0.3%. The uppermost 50 cm (Unit A) display higher TOC values of 0.5 - 0.9%.

The correlation of the three cores based on the magnetic susceptibility records (Figure 4), is supported by the TOC data, suggesting a Holocene interval of elevated TOC values in all three cores (Unit A, Figure 5).

The total organic carbon/total nitrogen (C/N) ratios which may give informations about the organic carbon source (i.e., marine vs. terrigenous; see above), do not show any major variations in the upper slope Core PS2476 sediments; most of the values are around 8 (Figure 6). In Core PS2474, C/N ratios fluctuate around 10 (total range between 5 and 13), with one very high ratio of > 20 at 660 cmbsf. In Core PS2471, C/N ratios are low in the (pelagic) silty clay intervals (4 to 8), whereas the three turbidites are characterized by significantly higher ratios of 10 to 17, with one single value of up to 28 (Figure 6). Based on the C/N ratios, the

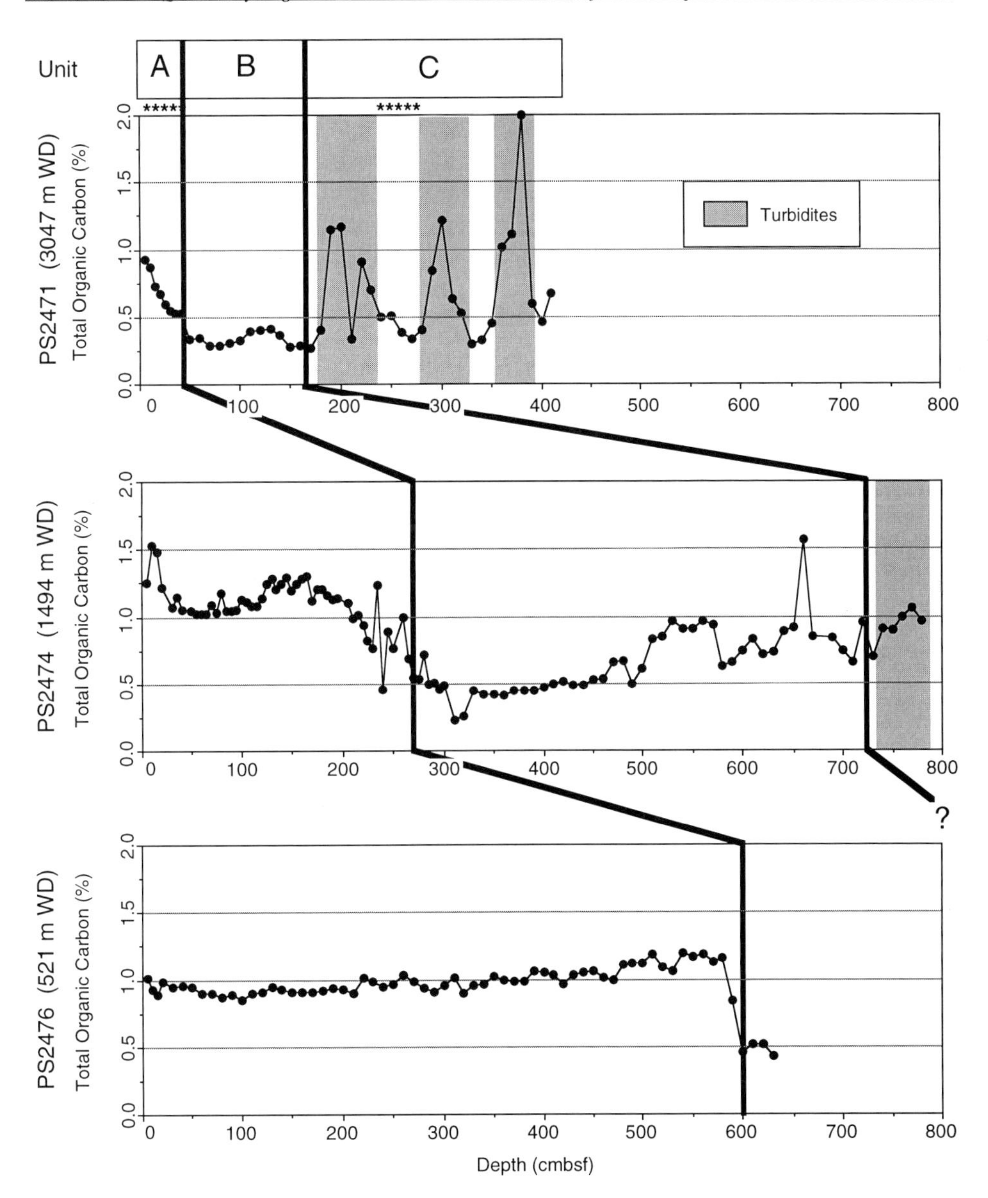

Figure 5: Total organic carbon records from sediment cores PS2471, PS2474, and PS2476. Correlation between cores (black lines) is based on magnetic susceptibility records (Fütterer, 1994; Nürnberg et al., 1995). Gray intervals indicate turbidites. A, B, and C mark stratigraphic/lithologic units discussed in the text. Asterisks indicate occurrences of coccolithophoride *Gephyrocapsa* sp. (according to Fütterer, 1994; Nürnberg et al., 1995).

organic matter seems to be a mixture of terrigenous and marine material with a stronger terrigenous dominance in the turbidite intervals.

When interpreting the C/N ratios in terms of organic carbon source we assume that the total nitrogen predominantly consists of organic nitrogen. In organic-carbon-poor sediments, however, the amount of inorganic nitrogen (fixed as ammonium ions in the interlayers of clay minerals, especially illite) may become a significant proportion of the total nitrogen (Stevenson and Cheng, 1972; Müller, 1977), causing (too) low C/N ratios. At Core PS2471, illite contents vary between 40 and 65% (C. Müller, unpubl. data 1997). A total organic carbon versus total nitrogen may help to estimate the amount of inorganic nitrogen (Figure 7; Ruttenberg and Goni, 1997). The positive total nitrogen intercept at TOC=0 recorded in the turbidite facies of Core PS2471 may indicate a background of inorganic nitrogen (i.e., 0.02% absolute). In the "pelagic" facies of all three cores, on the other hand, inorganic nitrogen seems to be a minor proportion of the total nitrogen pool, as based on the total organic carbon/total nitrogen diagrams. This is surprising because the illite contents of the "pelagic" intervals in Core PS2471 are higher than those determined in the turbidites (45-65 % and 40-45 %, respectively; C. Müller, unpubl. data 1997). Thus, the terrigenous proportion may be underestimated in the C/N ratios, and the C/N data should only be interpreted in combination with other organic-source indicators (see below).

In order to get some more detailed qualitative informations about the marine and terrigenous proportions of the total organic carbon fraction in more detail, organic geochemical bulk parameters (i.e., hydrogen indices and C/N ratios; see above) and specific biomarkers were presented for Core PS2471 sediments (Figure 8). The intervals of maximum TOC contents of the turbidites (Unit C) correlate well with minimum hydrogen index values (< 100 mg HC / g TOC), high C/N ratios (> 10), and low amounts of short-chain *n*-alkanes and alkenones, indicating the dominance of terrigenous organic matter. On the other hand, elevated hydrogen index values (100-250 mg HC / g TOC), low C/N ratios (< 10) as well as high amounts of short-chain *n*-alkanes, alkenones, and fatty acids suggest relatively increased marine organic matter proportions in the intervals between the turbidites and in the upper silty clay sequence of Unit B. In the uppermost, near-surface sediments (Unit A), the proportion of terrigenous material increases as suggested by low hydrogen index values and low amounts of short-chain *n*-alkanes, short-chain fatty acids, and alkenones.

The relative abundances of photosynthetic pigments (tetrapyrroles) determined as phaeopigment absorbance at a wave length of about 410 nm, are low in the non-turbidite intervals, but display maximum values in the turbidites (Figure 8). Because of the coincidence of the pigment maxima with low hydrogen index values, high C/N ratios, minima of short-chain *n*-alkanes and fatty acids, and minima in alkenones (Figure 8), as well as maxima in lignin phenols (J. Lobbes, unpubl. data, 1997), the pigment data are not interpreted as a signal for diagenetic products of marine-derived chlorophyll-a (e.g., Rosell-Melé, 1994). Possibly the maxima in the turbidites are caused by chlorophyll-b, which has been synthezised by terrestrial plants like Spermato-, Pterido-, or Bryophyta (Stryer, 1987).

The carbon preference index (CPI) as a measure for the maturity of the terrigenous organic matter, varies from 1.5 to 3 in the sedimentary record of Core PS2471, showing neither a general trend with depth nor any difference between turbidite and non-turbidite intervals.

Discussion

Modern processes controlling organic carbon deposition in the Laptev Sea continental margin area

As shown in studies of surface sediments (Stein and Nürnberg, 1995; Stein, 1996; Fahl and Stein, 1997a), the organic-carbon enrichments in the Laptev Sea and the adjacent continental margin and deep sea (Figure 3) are mainly controlled by increased supply of terrigenous organic matter. Especially the TOC maxima in the coastal zones off the rivers Lena and Olenek are directly related to riverine material. This is supported by low hydrogen index values and high C/N ratios (Stein, 1996) as well as high concentrations of long-chain *n*-alkanes (Fahl and

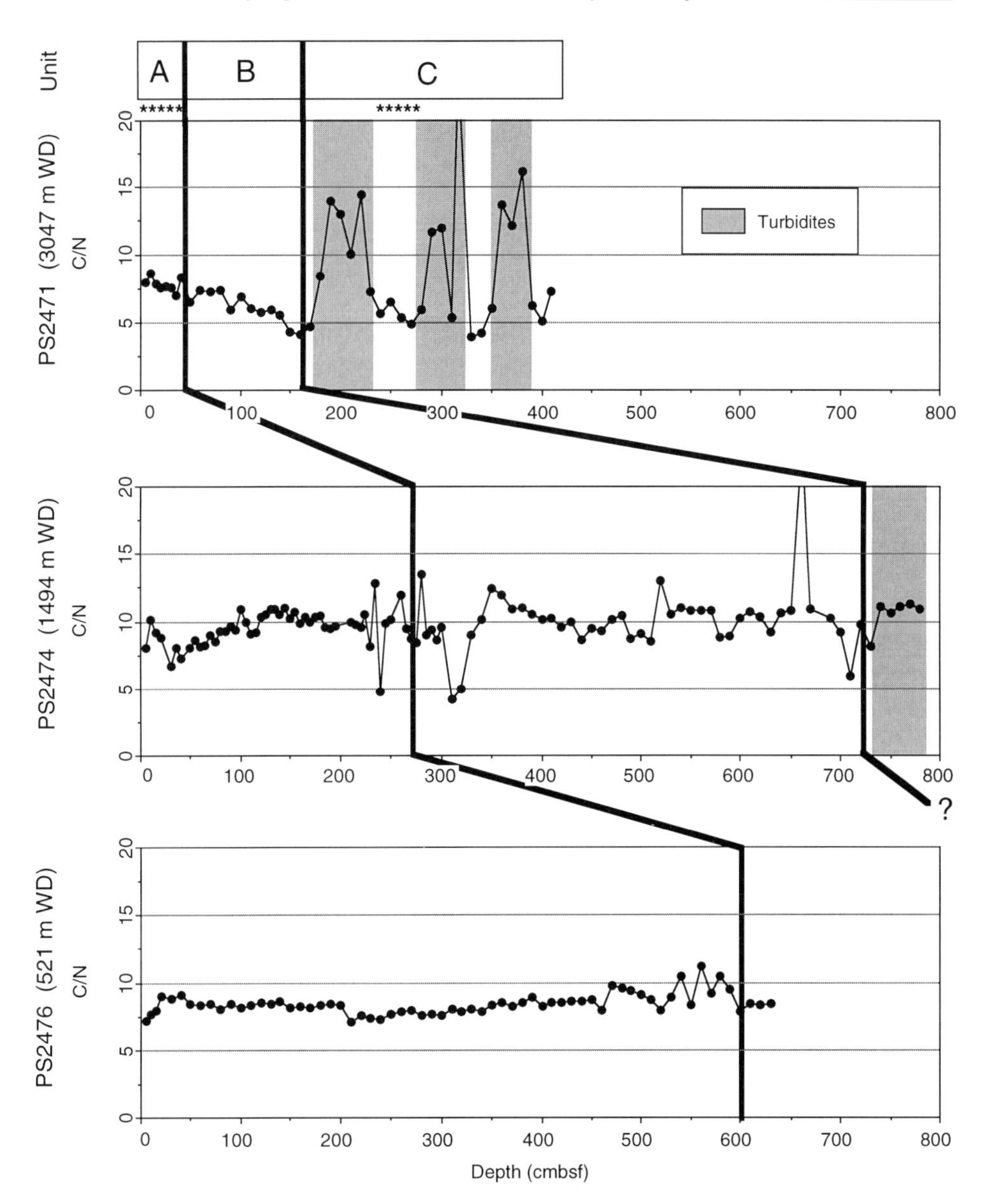

Figure 6: Total organic carbon/total nitrogen (C/N) ratios from sediment cores PS2471, PS2474, and PS2476. Correlation between cores (black lines) is based on magnetic susceptibility records (Fütterer, 1994; Nürnberg et al., 1995). Gray intervals indicate turbidites. A, B, and C mark stratigraphic/lithologic units discussed in the text. Asterisks indicate occurrences of coccolithophoride *Gephyrocapsa* sp. (according to Fütterer, 1994; Nürnberg et al., 1995).

Stein, 1997a). In addition, the elevated TOC values of the continental slope (Figure 3) may be related to the inflow of Atlantic water masses laterally transporting organic-carbon-enriched suspended matter (Fahl and Stein, 1997). Based on CPI indices varying between 2 and 5 in the surface sediments, the terrigenous organic matter appears to be dominantly fresh material. In the seasonally ice-free shelf areas of the Laptev Sea, significant amounts of marine organic matter were determined in surface sediments, with a maximum near the ice edge as indicated by increased hydrogen index and biogenic opal values (Stein and Nürnberg, 1995; Stein, 1996) and high concentrations of unsaturated fatty acids (Fahl and Stein, 1997a).

The surface sediment data of cores PS2471, PS2474, and PS2476 (Table 1; Fahl and Stein, 1997a) fall into the general trends described in the previous paragraph. Fresh terrigenous organic matter seems to be dominant as indicated by hydrogen index values < 100 mg HC / g TOC, the dominance of long-chain *n*-alkanes in the total *n*-alkane fraction, the occurrence of long-chain wax esters, and CPI values between 2.6 and 4.1. Marine (and freshwater ?) organic matter is present as shown by the occurrence of short-chain *n*-alkanes and fatty acids, but only occurs in minor amounts. The significant occurrences of alkenones (as an indicator for prymnesiophytes; Volkman et al., 1980; Marlowe et al., 1984; Prahl et al., 1988) at the location of Core PS2471 (0.57 µg / g TOC) is surprising because in all other surface sediments from the Laptev Sea - probably caused by the very low abundance of prymnesiophytes in the Laptev Sea - only very small amounts of alkenones were determined (< 0.07 µg/g TOC; Fahl and Stein, 1997a). Fahl and Stein (1997a) interpreted the occurrence of alkenones in the surface sediment of Core PS2471 as beeing derived from coccolithophorides or other prymnesiophytes transported by Atlantic water masses along the continental slope (cf., Figures 1 and 2).

Late Quaternary changes in organic carbon composition at the Laptev Sea continental margin and their paleoenvironmental significance

The records of cores PS2474 and PS2471 also comprising pre-Holocene time intervals display significant variations in total organic carbon content, whereas the Holocene time interval represented in all three cores, is characterized by less variable TOC values (Figure 5). Based on the C/N ratios (Figure 6), the organic matter in the non-turbidite sections is generally a mixture between marine and terrigenous material. The TOC contents of the two cores from the upper slope are generally higher than those in the non-turbidite sequences of the deepest, distal Core PS2471, which may reflect a higher supply of terrigenous organic matter in nearshore areas (decreasing towards the open ocean) and/or better preservation of organic matter because of distinctly higher sedimentation rates in the shallower cores (Table 2). In order to understand the processes controlling the organic carbon deposition and their change through late Quaternary times, i.e., to distinguish between (autochthonous) marine organic carbon produced in the water column and allochthonous terrigenous organic matter supplied by rivers, currents, and/or down-slope transport (Figure 2), the organic carbon and biomarker records of Core PS2471 are discussed in more detail.

The oldest time interval (Unit C) is dominated by turbidites characterized by relatively high organic carbon contents. The exact age of these turbidites is still under discussion as already mentioned. It is, however, likely that the turbidites were deposited during glacial intervals (Oxygen Isotope Stage 2 or 4) when the sea level was low and high amounts of terrigenous material were transported close to the shelf edge resulting in an instability of the upper slope and gravitational down-slope transport of material. With this process huge amounts of organic matter may have been transported down-slope and deposited in the deep sea environment where normally much less organic matter is preserved. Such a process has also been described for several other deep-sea regions in the world´s ocean, e.g., in the Canadian Basin (Grantz et al., 1996) and in the Northeast Atlantic (Stein et al., 1989; Stein, 1991). The composition of the organic matter buried in the turbidites depends on the origin of the turbidite. While the turbidites in the Northeast Atlantic originated from a shelf edge influenced by coastal upwelling and increased primary productivity where the sediments contain large amounts of marine organic matter (Stein et al., 1989; Stein, 1991), the turbidites in Core PS2471 originate from the Laptev

Table 1: Water depths, total organic carbon (TOC) contents, organic-geochemical bulk parameters (hydrogen index and C/N ratio), and biomarker composition in surface sediments from cores PS2471, PS2474, and PS2476 (Fahl and Stein, 1997a).

	Surface sediments		
	PS2476	**PS2474**	**PS2471**
Parameter	**Laptev Sea Continental Margin**		
TOC (%)	1.10	1.36	1.52
HI (mgHC/gTOC)	97	68	52
C/N	7.2	8.0	8.2
C17+C19 (μg/gTOC)	28.0	20.8	27.9
C27+C29+C31 (μg/gTOC)	179.4	100.8	117.9
(C27+C29+C31)/(C17+C19)	6.4	4.9	4.2
CPI	4.1	3.4	2.6
long-chain wax ester (C42) (μg/gTOC)	3.5	4.2	2.5
fatty acids (μg/gTOC)	264	249	47
alkenones (C37:2+C37:3) (μg/gTOC)	n.d.	n.d.	0.57

Table 2: Core position and water depths and accumulation rates of total organic carbon. Data from cores PS2163 and PS2170 from Stein et al. (1994). For further explanation see text (methods).

	PS2476	**PS2474**	**PS2471**	**PS2163**	**PS2170**
	Laptev Sea Continental Margin			**Central Arctic Ocean**	
Core Position					
Latitude	77°23.51′ N	77°40.15′ N	79°09.7′ N	86°14.5′ N	87°35.8′ N
Longitude	118°11.45′ E	118°34.5′ E	119°47.55′ E	59°12.9′ E	60°53.7′ E
Water depth (m)	524	1494	3047	3040	4083
Oxygen Isotope Stage 1 (0 - 12 ka)	0 - 600 cm	0 - 270 cm	0 - 50 cm		
Mean LSR (cm/ky)	50	22	4	0.7	0.85
Mean TOC (%)	0.99	1.14	0.68	0.55	0.7
Mean Grain Density (g/cm3)	2.65	2.65	2.65		
Mean Wet Bulk Density (g/cm3)	1.43	1.43	1.43		
Mean Dry Density (g/cm3)	0.66	0.66	0.66	0.75	0.75
Mean Porosity (%)	75	75	75		
Mean Bulk Accumulation Rate (g/cm2/ky)	33	14.5	2.6	0.53	0.64
Mean TOC Accumulation Rate (gC/cm2/ky)	0.33	0.17	0.02	0.003	0.004

Sea shelf which is characterized by the deposition of dominantly terrigenous organic matter supplied by the major Siberian rivers (Stein, 1996; Fahl and Stein, 1997a). The predominance of freshwater (riverine) diatom valves occurring in the turbidites of Core PS2471, also supports this explanation (Y. Polyakova, unpubl. data 1995). Thus, the organic matter in the turbidites displays a strong terrigenous signal indicated by low hydrogen index values (<100 mg HC / g TOC) and high C/N ratios (>10) (Figure 8). The terrigenous origin is also indicated by a maximum occurrence of lignin phenols in the three turbidite intervals of Core PS2471 (J. Lobbes, unpubl. data, 1997). Surprisingly, the long-chain *n*-alkanes (C_{27}, C_{29} and C_{31}) indicative for higher land plants and commonly used in sediments as a biomarker for terrigenous organic matter supply (e.g., Simoneit et al., 1977; Prahl and Muehlhausen, 1989), do not correlate with the hydrogen index and C/N records, i.e., they do not show any clear maximum in these turbidites (Figure 8). The same phenomenon has also been described by Schubert and Stein (1997) in sediments from the central Arctic Ocean. That means, the long-chain *n*-alkane concentrations cannot be used to estimate the terrigenous organic carbon supply into Arctic Ocean sediments in general. Other terrigenous components, such as lignin or carbohydrates, may contribute significantly to the terrigenous organic carbon proportion (cf., Eggers, 1994).

The intervals inbetween the turbidites of Unit C are rather organic-carbon-poor (0.3%) and display higher hydrogen index values, lower C/N ratios, and higher abundances of marine biomarkers (short-chain *n*-alkanes and, although less pronounced, alkenones) (Figure 8). This relative increase in the marine signal is probably caused by a reduced supply of terrigenous organic matter.

The most prominent change in the organic carbon and biomarker composition is observed in Unit B and Unit A (Figure 8). Maximum abundances of long-chain *n*-alkanes, the clear dominance of long-chain over short-chain *n*-alkanes, and hydrogen index values of 100-200 mg HC / g TOC still suggest a dominance of organic matter of terrigenous origin. Maximum abundances of short-chain *n*-alkanes, short-chain fatty acids, and alkenones, however, were also observed in the same interval suggesting significant amounts of marine organic matter were preserved in the sediments, especially in Unit B. In general, elevated amounts of marine organic matter in deep-sea sediments can either be explained by high primary productivity or increased preservation rates (e.g., Berger et al., 1989; Stein, 1991 and further references therein). At least from a comparison with measured low modern primary productivity values of about 25 mg $C/m^2/d$ (Gleitz, pers. comm. 1996) of the presently seasonally ice-covered study area, distinctly enhanced primary productivity seems to be less probable. A low productivity is also suggested from the almost complete absence of marine planktonic diatoms in Core PS2471 (Y. Polyakova, unpubl. data 1995). Instead, a reduced supply of terrigenous organic matter and better preservation of marine organic matter due to higher sedimentation rates is a more reasonable explanation.

It cannot be excluded, however, that part of the short-chain fatty acids are of freshwater origin due to a high supply of lacustrine/fluviatile organic matter by the rivers draining into the Laptev Sea. This possibility is supported by the distribution pattern of freshwater palynomorphs in Laptev Sea surface sediments (Kunz-Pirrung, 1997) and the frequent occurrence of the freshwater algae *Botryococcus* spp. in the sedimentary record of Core PS2471 (B. Boucsein, pers. comm., 1997). Further detailed kerogen and palynomorph microscopy work will allow a more definite interpretation of the biomarker signals.

The maximum abundances of alkenones may be related to the influence of Atlantic water masses (cf. Figure 1; Fahl and Stein, 1997a). Since coccolithophorides as the main producers of alkenones, are absent in the fossil record of this interval of Core PS2471 (Nürnberg et al., 1995), carbonate dissolution is assumed to have affected the sediments in this time interval. Carbonate dissolution may have been caused by increased CO_2 production during the decomposition of marine organic matter (de Vernal et al., 1992; Steinsund and Hald, 1994).

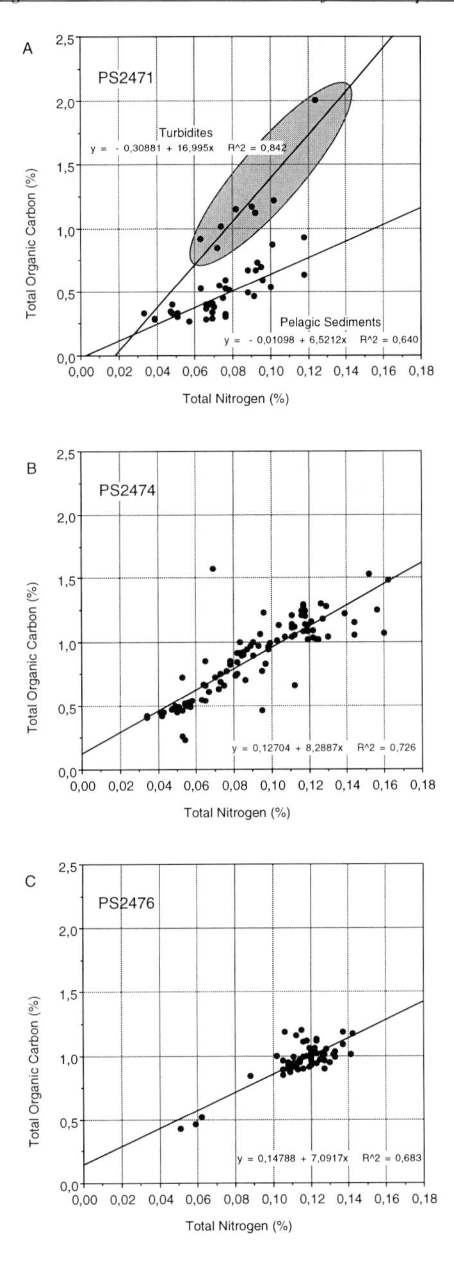

Figure 7: Total organic carbon versus total nitrogen concentrations at Core PS2471 (A), Core PS2474 (B), and Core PS2476 (C). In all three diagrams (at Core PS2471 separately for the turbidite and "pelagic" facies), linear least-squares regression lines are shown. The correlation coefficients of these linear fits are given as R^2 values.

If higher sedimentation rates are postulated for Unit B, this interval was probably deposited during post-LGM/Termination I rather than during Oxygen Isotope Stage 2 and 3 (which would imply reduced sedimentation rates in comparison to the Holocene time interval) (see discussion above). The correlation of the magnetic susceptibility records of the sediment cores discussed in this paper with the AMS^{14}C-dated magnetic susceptibility record of Core PS2778 (Figure 4) also supports a post-glacial age for Unit B (i.e., younger than Oxygen Isotope Stage 2). That means, the alkenone signal at Core PS2471 may be used as tracer for Atlantic water inflow along the Eurasian continental margin (cf., Figs 1 and 2) intensified during Termination I and the early Holocene. Increased Atlantic water inflow connected with strong carbonate dissolution and high marine organic carbon abundances during the last 12 ka has also been derived from an organic-geochemical study of a well-dated gravity core from the northern Barents continental margin (Knies and Stein, 1997).

Organic-carbon flux on the Laptev Sea continental margin during Oxygen Isotope Stage 1

In order to estimate the amount of total organic carbon stored in continental slope sediments of the Laptev Sea during Oxygen Isotope Stage 1, accumulation rates of organic carbon were calculated. The TOC accumulation rates are 0.3, 0.17, and 0.02 g C / cm^2 / ky in cores PS2476, PS2474, and PS2471, respectively (Table 2). Assuming Unit B of Core PS2471 to represent post-LGM/Termination I, TOC accumulation rates of Unit B would be higher by a factor of about two than the Oxygen Isotope Stage 1 rates. These accumulation rates mainly represent the flux of terrigenous organic carbon; marine organic carbon is only of secondary importance. This is also reflected in the decrease in accumulation rates in offshore direction. These rates are significantly higher than those calculated for the central Arctic Ocean (0.003 g C / cm^2 / ky; Table 2; Stein et al., 1994). That means, on the upper Laptev Sea continental slope where maximum accumulation rates were recorded, the burial of organic matter is higher by a factor of 100 than in the central Arctic. On the other hand, the upper slope TOC accumulation rates are in the same order of magnitude as those calculated for coastal upwelling areas where, however, most of the organic matter is of marine origin (cf. Stein, 1991 and further references therein). Neverthless, the amount of organic carbon stored in the wide circum Arctic shelf and slope areas are certainly of importance for calculation of organic carbon budgets on a global scale.

Conclusions

1 Organic geochemical bulk parameters (hydrogen index values and C/N ratios) and specific biomarkers (*n*-alkanes, fatty acids, and alkenones) can be used to distinguish between marine and terrigenous organic carbon sources. In general, terrigenous organic matter is dominant on the Laptev Sea continental margin throughout. Marine organic carbon, however, may occur occasionally in significant amounts.
2 The turbidites deposited during glacial Oxygen Isotope Stages 2 and/or 4 are clearly dominated by terrigenous organic matter.
3 During Termination I (?) and the early Holocene, relatively high amounts of marine organic matter were preserved in the sediments, as reflected be elevated hydrogen index values and increased abundances of short-chain n-alkanes, short-chain fatty acids, and alkenones. It is possible, however, that significant amounts of freshwater organic matter are also preserved in this interval.
4 The alkenone signal may be used as a tracer for the inflow of Atlantic water masses along the Eurasian continental margin, which probably started to intensify during Termination I. This date, however, has to be proven by a more detailed stratigraphic framework.
5 Oxygen Isotope Stage 1 accumulation rates of total organic carbon vary between 0.33 and 0.02 gC / cm^2 / ky, decreasing with increasing distance from the shelf and increasing water depth.

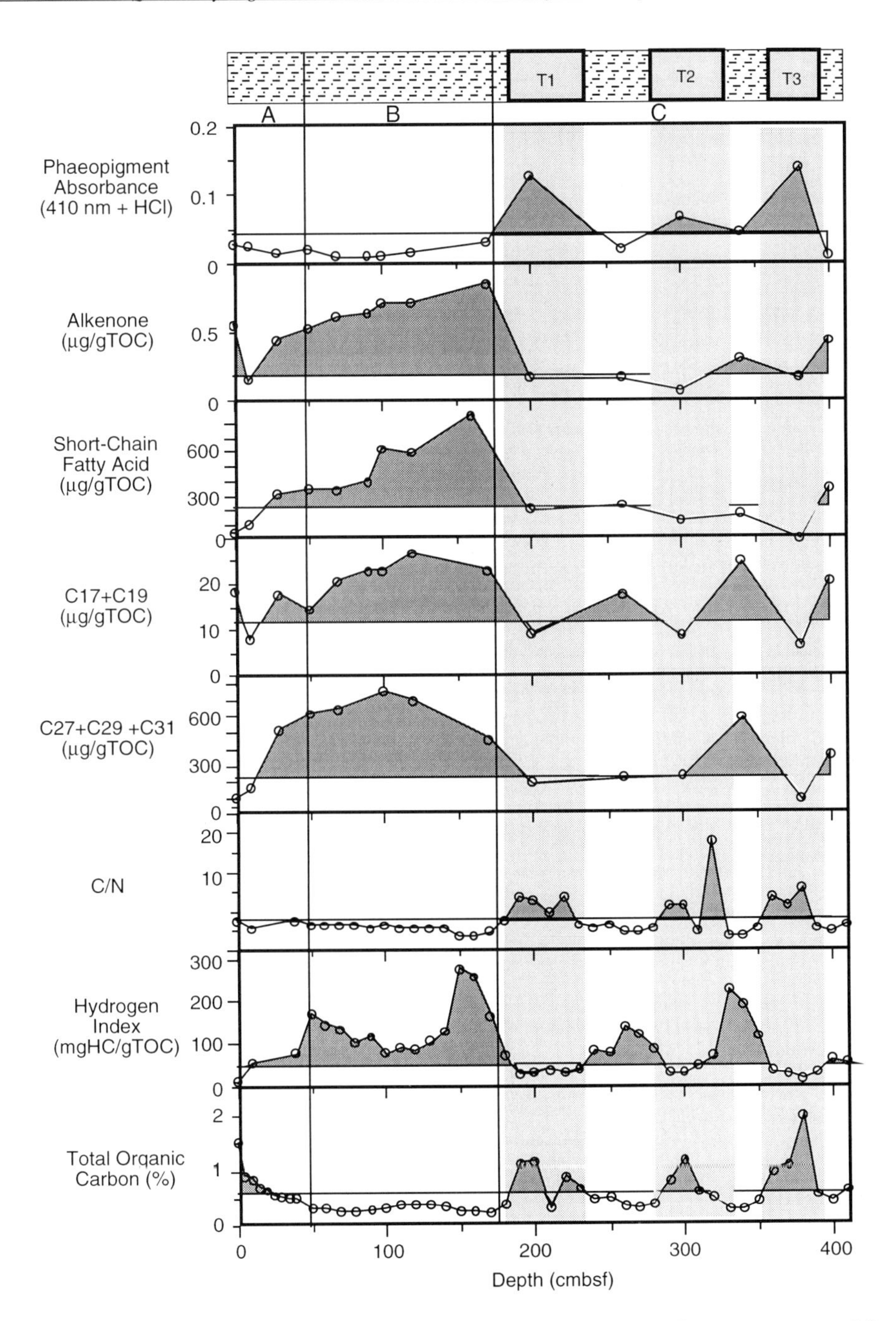

Figure 8: Organic-geochemical bulk parameters and specific biomarkers in the sedimentary sequence of Core 2471: Total organic carbon contents (%), hydrogen index values (mgHC/gTOC), total organic carbon/total nitrogen (C/N) ratios, short-chain *n*-alkanes (C_{17}+C_{19}) (µg/gTOC), short-chain fatty acids (µg/gTOC), alkenones (µg/gTOC), and phaeopigment absorbance values. Gray intervals indicate turbidites (T1, T2, and T3). A, B, and C mark stratigraphic/lithologic units discussed in the text.

Acknowledgements

We gratefully thank Peter Müller (Bremen University) and Jürgen Rullkötter (ICBM, Oldenburg) for the numerous constructive suggestions for improvement of the manuscript. This study was performed within the bilateral Russian-German multidisciplinary research project "System Laptev Sea". The financial support by the German Ministry of Education, Science, Research and Technology (BMBF) is gratefully acknowledged. This is contribution No. 1340 of the Alfred-Wegener-Institute for Polar and Marine Research.

References

Aagaard, K., J. Swift, and K. Carmack (1985) Thermohaline circulation in the Arctic Mediterranean Seas. Journ. Geophys. Res., 90, 4833-4846.

Ackman, R.G., C.S. Tochter, and J. Mc Lachlan (1968) Marine phytoplankter fatty acids. J Fish Res Board Can 25, 1603-1620.

Ahlgren, G., I.-B. Gustafsson and M. Boberg (1992) Fatty acid content and chemical composition of freshwater microalgae. Journ. Phycol., 28, 37-50.

Ahlgren, G., L. Lunstedt, M. Brett and C. Forsberg (1990) Lipid composition and food quality of some freshwater phytoplankton for cladoceran zooplankters. Journ. Plankt. Res., 12, 809-818.

ARCSS Workshop Steering Committee (1990) Arctic System Science: Ocean-Atmosphere-Ice Interactions. Lake Arrowhead Workshop Report, JOI Inc., Washington DC, 132pp.

Bauch, H.A., H. Cremer, H. Erlenkeuser, H. Kassens and M. Kunz-Pirrung (1996) Holocene paleoenvironmental evolution of the northern central Siberian shelf sea. Quaternary Environment of the Eurasian North (QUEEN), First Annual Workshop, Strasbourg, Abstract Volume.

Baumann, M. (1990) Coccoliths in sediments of the eastern Arctic Basin. In: Bleil, U. and J. Thiede (Eds.), Geological History of the Polar Oceans: Arctic versus Antarctic, NATO Series C, Mathematical and Physical Sciences, 308, 437-445.

Berger, W.H., V. Smetacek and G. Wefer (1989) Productivity of the Ocean: Past and Present. Life Sciences Research Report, Vol. 44, Wiley & Sons, New York, 471pp.

Bligh, E.G. and W.J. Dyer (1959) A rapid method of total lipid extraction and purification. Can. Journ. Biochem. Physiol., 37, 911-917.

Blumer, M., R.R.L. Guillard and T. Chase (1971) Hydrocarbons of marine phytoplankton. Mar. Biol., 8, 183-189.

Boetius, A., C. Grahl, I. Kröncke, G. Liebezeit E. and Nöthig (1996) Concentrations of chloroplastic pigments in Eastern Arctic sediments: Indication of recent marine organic matter input. In: R. Stein, G. Ivanov, M. Levitan and K. Fahl (Eds.), Reports on Polar Research, 212, 213-218.

Bordowskiy, O.K. (1965) Sources of organic matter in marine basins. Mar. Geol., 3, 5-31.

Brassell, S.C., G. Eglinton, I.T. Marlowe, U. Pflaumann and M. Sarnthein (1986) Molecular stratigraphy: a new tool for climatic assessment. Nature, 320, 129-133.

Brassell, S.C., G. Eglinton, J.R. Maxwell and R.P. Philip (1978) Natural background of alkanes in the aquatic environment. In: Hutzinger, O., I.H. Lelyveld and B.C.J. Zoetman (Eds.), Aquatic Pollutants, Transformation and Biological Effects. (Pergamon Press), 69-86.

Bray, E.E. and E.D. Evans (1961) Distribution of n-paraffins as a clue to recognition of source beds. Geochim. Cosmochim. Acta, 22, 2-15.

de Vernal, A., G. Bilodeau, C. Hillaire-Marcel and N. Kassou (1992) Quantitative assessment of carbonate dissolution in marine sediments from foraminifer living vs. shell ratios, Davis Strait, northwest North Atlantic. Geology, 20, 527-530.

Eicken, H., T. Viehoff, T. Martin, J. Kolatschek, V. Alexandrov and E. Reimnitz (1995) Studies of clean and sediment-laden ice in the Laptev Sea. In: Kassens, H., et al. (Eds.), Second Workshop on "Russian-German Cooperation in and Around the Laptev Sea", St. Petersburg, November 1994, Reports on Polar Research, 176, 62-70.

Eggers, J. (1994) Charakterisierung des Eintrages von organischen Substanzen anhand von Markerverbindungen in marinen Ökosystemen. Unpubl. Diploma Thesis, Bremen University, 68pp.

Espitalié, J., J.L. Laporte, M. Madec, F. Marquis, P. Leplat, J. Paulet and A. Boutefeu (1977) Méthode rapide de caractérisation des roches mères, de leur potentiel pétrolier et de leur degre d'évolution. Rev. Inst. Franc. Pétrol., 32, 23-42.

Fahl, K. and R. Stein (1997a) Modern organic-carbon-deposition in the Laptev Sea and the adjacent continental slope: Surface-water productivity vs. terrigenous supply. Org. Geochem., 26, 379-390.

Fahl, K. and R. Stein (1997b) Biomarkers in the Late Quaternary Arctic Ocean: Problems and perspectives, Mar. Chem., subm.

Folch, J., M. Lees and G.H. Sloane-Stanley (1957) A simple method for the isolation and purification of total lipids from animal tissues. Journ. Biol. Chem., 226, 497-509.

Fütterer, D.K. (ed.) (1994) The Expedition ARCTIC'93 Leg ARK IX/4 of RV "Polarstern" 1993. Reports on Polar Research, 149/94, 244pp.

Gard, G. (1988) Late Quaternary calcareous nannofossil biochronology and paleoceanography of Arctic and Szbarctic seas. Medd. Stockholms Univ. Geol. Inst., 275, 8-45.

Gordienko, P.A., and A.F. Laktionov (1969) Circulation and physics of the Arctic Basin waters. Ann. Int. Geophys. Yearb., 46, 94-112.

Grantz, A., R.L. Phillips, M.W. Mullen, S.W. Starratt, G.A. Jones, A. Sathy Naidu and B.P. Finney (1996) Character, paleoenvironment, rate of accumulation, and evidence for seismic triggering of Holocene turbidites, Canada Abyssal Plain, Arctic Ocean. Mar. Geol., 133, 51-73.

Hedges, J.I., W.A. Clark, P.D. Quay, J.E. Richey, A.H. Devol and U.D.M. Santos (1986) Composition and fluxes of particulate organic material in the Amazon River. Limnol. Oceanogr., 31, 717-738.

Hinrichs, K.-U., J. Rinna, J., Rullkötter and R. Stein (1997) A 160 kyr record of alkenone-derived sea-surface temperatures from Santa Barbara Basin sediments. Naturwissenschaften, 84, 126-128.

Hollerbach, A. (1985) Grundlagen der organischen Geochemie. Berlin (Springer Verlag). 190pp.

Kassens, H. and V.Y. Karpiy (1994) Russian-German Cooperation: The Transdrift I Expedition to the Laptev Sea. Reports on Polar Research, 151, 168pp.

Kattner, G. and U.H. Brockmann (1990) Particulate and dissolved fatty acids in an enclosure containing a unialgal *Skleletonema costatum* (Greve.) Cleve culture. J. Exp. Mar. Bio. Ecol., 114, 1-13.

Kattner, G. and H.S.G. Fricke (1986) Simple gas-liquid chromatographic method for simultaneous determination of fatty acids and alcohols in wax esters of marine organisms. J. Chromatogr. 361, 313-318.

Knies, J. and R. Stein (1997) Organic carbon composition and paleoenvironmental reconstructions along the northern Barents Sea margin during the last 30,000 years. Geology, subm.

Kollatukudy, P.E. (1976) Chemistry and Biochemistry of Natural Waxes, Elsevier (Amsterdam), 459pp.

Kunz-Pirrung, M. (1997) Distribution of aquatic palynomorphs in surface sediments from the Laptev Sea, Eastern Arctic Ocean. Lecture Notes in Earth Sciences, this vol.

Kuptsov, V. M. and A. Lisitzin (1996) Radiocarbon of Quaternary along shore and bottom deposits of the Lena and the Laptev Sea sediments. Mar. Chem., 53, 301-311.

Léveillé, J.-C., C. Amblard and G. Bourdier (1997) Fatty acids as specific algal markers in a natural lacustrian phytoplankton. J. Plankt. Res., 19, 469-490.

Marlowe, I.T., J.C. Green, A.C. Neal, S.C. Brassell, G. Eglinton and P.A. Course (1984) Long chain alkenones in the Prymnesiophyceae. Distribution of alkenones and other lipids and their taxonomic significance. Br. J. Phycol., 19, 203-216.

Martin, J.M., D.M. Guan, F. Elbaz-Poulichet, A.J. Thomas and V.V. Gordeev (1993) Preliminary assessment of the distributions of some trace elements (As, Cd, Cu, Fe, Ni, Pb and Zn) in a pristine aquatic environment: the Lena River estuary (Russia). Mar. Chem., 43, 185-199.

Mayzand, P., J.P. Chanut and R.G. Ackman (1989) Seasonal changes of the biochemical composition of marine particulate matter with special reference to fatty acids and sterols. Mar. Ecol. Prog. Ser., 56, 189-204.

Müller, P. (1977) C/N ratios in Pacific deep-sea sediments: Effect of inorganic ammonium and organic nitrogen compounds sorbed by clays. Geochim. Cosmochim. Acta, 41, 765-776.

NAD Science Committee (1992) The Arctic ocean record: Key to global change (Initial Science Plan of the Nansen Arctic Drilling Program). Polarforschung, 61, 1-102.

Nelson, D.M., W.O. Smith, R.D. Muench, L.I. Gordon, C.W. Sullivan and D.M. Husby (1989) Particulate matter and nutrient distribution in the ice-edge zone of the Weddell Sea: relationship to hydrography during late summer. Deep-Sea Res., 36, 191-209.

Niessen, F., D. Weiel, T. Ebel, J. Hahne, C. Kopsch, M. Melles and R. Stein (1997) Weichselian Glaciations in Central Siberia - Implications from Marine and Lacustrine High Resolution Seismic Profiles and Sediment Cores. EUG-9 Conference, European Union of Geosciences, Strasbourg, March 23-27, 1997, Abstract Volume.

Nürnberg, D., D. Fütterer, F. Niessen, N. Nörgaard-Petersen, C. Schubert, R. Spielhagen and M. Wahsner (1995) The depositional environment of the Laptev Sea continental margin: Preliminary results from the RV "Polarstern" ARK-IX/4 cruise. Polar Research, 14, 43-53.

Nürnberg, D., I. Wollenburg, D. Dethleff, H. Eicken, H. Kassens, T. Letzig, E. Reimnitz and J. Thiede (1994) Sediments in Arctic sea ice - Implications for entrainment, transport, and release. Mar. Geol., 119, 185-214.

Orcutt, D.M. and G.W. Patterson (1975) Sterol, fatty acids and elemantal composition of diatoms grow in chemical defined media. Comp. Biochem. Physiol., 50B, 579-583.

Pfirman, S., M.A. Lange, I. Wollenburg and P. Schlosser (1989) Sea ice characteristics and the role of sediment inclusions in deep-sea deposition: Arctic-Antarctic comparison. In: Bleil, U. and J. Thiede (Eds.), Geological History of the Polar Oceans: Arctic versus Antarctic, NATO ASI Ser., C308, 187-211.

Prahl, F.G. and L.A. Muelhlhausen (1989) Lipid biomarkers as geochemical tools for paleoceanographic study. In: W.H. Berger, V.S. Smetacek and G. Wefer (Eds.), Productivity of the Ocean: Present and Past. Wiley and Sons, Chichester, U.K, 271-289.

Prahl, F.G. and S.G. Wakeham (1987) Calibration of unsaturation patterns in long-chain ketone compositions for paleotemperature assessment. Nature, 320, 367-369.

Prahl, F.G., R.B. Collier, J. Dymond, M. Lyle and M.A. Sparrow (1993) A biomarker perspective on prymnesiophyte productivity in the northeast Pacific Ocean. Deep-Sea Res., 40, 2061-2076.

Prahl, F.G., L.A. Muehlhausen and D.L. Zahnle (1988) Further evaluation of long-chain alkenones as indicators of paleoceanographic conditions. Geochim. Cosmochim. Acta, 52, 2303-2310.

Rachold, V., R. Lara and H.-W. Hubberten (1996) Concentration and composition of dissolved and particulate organic material in the Lena River - Organic carbon transport to the Laptev Sea. Third Workshop on Russian-German Cooperation: Laptev Sea System, October 16-19, 1996, St. Petersburg, Abstract Volume, 24-25.

Rachor, E. (1992). Scientific Cruise Report of the 1991 Arctic Expedition ARK VIII/2 of RV "Polarstern", Reports on Polar Research, 115, 150pp.

Rosell-Melé, A. and N. Koc (1997) Paleoclimatic significance of stratigraphic occurrence of photosynthetic biomarker pigments in the Nordic seas. Geology, 25, 49-52.

Rosell-Melé, A. (1994) Long-chain alkenones, alkyl alkenoates and total pigment abundances as climatic proxy-indicators in the Northeastern Atlantic. Bristol University, PhD thesis, 164pp.

Rosell-Melé, A., G. Eglinton, U. Pflaumann and M. Sarnthein (1995) Atlantic core-top calibration of the U^{K}_{37} index as a sea-surface paleotemperature indicator. Geochim. Cosmochim. Acta, 59, 3099-3107.

Rostek, F., G. Ruhland, F. Bassinot, P.J. Müller, L. Labeyrie, Y. Lancelot and E. Bard (1993) Reconstructing sea-surface temperature and salinity using $\partial^{18}O$ and alkenone records. Nature, 364, 319-321.

Rudels, B., E.P. Jones, L.G. Anderson and G. Kattner (1994) On the Intermediate Depth Waters of the Arctic Ocean. In: Johannessen, O.M., Muench, R.D., and Overland, J.E. (Eds.), The Polar Oceans and Their Role in Shaping the Global Environment. American Geophysical Union, 33-46.

Ruttenberg, K.C. and M.A. Goni (1997) Phosphorus distribution, C:N:P ratios, and $\partial^{13}C_{OC}$ in arctic, temperate, and tropical coastal sediments: tools for characterizing bulk sedimentary organic matter. Mar. Geol., 139, 123-145.

Sakshaug, E. and O. Holm-Hansen (1984) Factors Governing Pelagic Production in Polar Oceans. In: Holm-Hansen, O., et al. (Eds.), Marine Phytoplankton and Productivity, Lect. Notes Coast. Est. Studies, 8, 1-17.

Scheffer, F. and P. Schachtschabel (1984) Lehrbuch der Bodenkunde. Enke Verlag Stuttgart, 442pp.

Schubert, C. and R. Stein (1996) Deposition of organic carbon in Arctic Ocean sediments: Terrigenous supply vs marine productivity. Org. Geochem., 24, 421-436.

Schubert, C.J. and R. Stein (1997) Lipid distribution in surface sediments from the eastern central Arctic Ocean. Mar. Geol., 138, 11-25.

Sikes, E.L. and J.K. Volkman (1993) Clibration of alkenone unsaturation ratios (U^{k}_{37}) for paleotemperature estimation in cold polar waters. Geochim. Cosmochim. Acta, 57, 1883-1889.

Sikes, E.L., J.W. Farrington and L.D. Keigwin (1991) Use of the alkenone unsaturation ratio U^{k}_{37} to determine past sea surface temperatures: core-top SST calibrations and methodology considerations. Earth Planet. Sci. Lett., 104, 36-47.

Simoneit, B. R. T., R. Chester and G. Eglinton (1977) Biogenic lipids in particulates from the lower atmosphere over the eastern Atlantic. Nature, 267, 682-685.

Spielhagen, R.F., H. Erlenkeuser and J. Heinemeier (1996) Deglacial changes of freshwater export from the Laptev Sea to the Arctic Ocean. Quaternary Environment of the Eurasian North (QUEEN), First Annual Workshop, Strasbourg, Abstract Volume.

Stein, R. and D. Nürnberg (1995) Productivity proxies. Organic carbon and biogenic opal in surface sediments from the Laptev Sea and the adjacent continental slope. In: Kassens, H., et al. (Eds.), Second Workshop on "Russian-German Cooperation in and Around the Laptev Sea", St. Petersburg, November 1994, Reports on Polar Research, 176, 286-298.

Stein, R. (1991) Accumulation of Organic Carbon in Marine Sediments: Lecture Notes in Earth Sciences 34, Springer, Heidelberg, 217pp.

Stein, R. (1996) Organic carbon and carbonate distribution in Eurasian continental margin and Arctic Ocean deep-sea surface sediments: Sources and pathways. In: Surface-sediment Composition and Sedimentary Processes in the Central Arctic Ocean and Along the Eurasian Continental Margin. In: R. Stein, G. Ivanov, M. Levitan, and K. Fahl (Eds.), Reports on Polar Research, 212, 243-267.

Stein, R., S.-I. Nam, C. Schubert, C. Vogt, D. Fütterer and J. Heinemeier (1994) The last deglaciation event in the eastern central Arctic Ocean. Science, 264, 692-696.

Stein, R., H.L. ten Haven, R. Littke, J. Rullkötter and D.H. Welte (1989) Accumulation of marine and terrigenous organic carbon at upwelling Site 658 and nonupwelling Sites 657 and 659: Implications for the reconstruction of paleoenvironments in the eastern subtropical Atlantic through late Cenozoic times. In: Ruddiman, W.F., M. Sarnthein, et al. (Eds.), Proc. ODP, Sci. Results, 108, College Station, Tx (Ocean Drilling Program), 361-386.

Steinsund, P.I. and M. Hald (1994) Recent calcium carbonate dissolution in the Barents Sea: Paleoceanographic applications. Mar. Geol., 117, 303-316.

Stevenson, F.J. and C.N. Cheng (1972) Organic geochemistry of the Argentine Basin sediments: Carbon-nitrogen relationships and Quaternary correlations. Geochim. Cosmochim. Acta, 36, 653-671.
Stryer, L. (1987) Biochemie. Vieweg Verlag 4. Auflage, 329-330.
Subba Rao, D.V. and T. Platt (1984) Primary production of Arctic waters. Polar Biol., 3, 191-201.
Tissot, B.P. and D.H. Welte (1984) Petroleum Formation and Occurrence, 2nd ed., Springer Verlag Heidelberg, 699pp.
van Andel, T.H., G.H. Heath and T.C. Moore (1975) Cenozoic history and paleoceanography of the Central Equatorial Pacific, Mem. Geol. Soc. Amer., 143, 134pp.
Volkman, J.K., G. Eglinton, E.D.S. Corner and T.E.V. Forsberg (1980) Long-chain alkenes and alkenones in the marine coccolithophorid *Emiliania huxleyi*. Phytochemistry, 19, 2619-2622.
Weber, M.E., F. Niessen, G. Kuhn and M. Wiedicke (1997) Calibration and application of marine sedimentary physical properties using a multi-sensor-core-logger. Mar. Geol., in press.
Weiel, D. (1997) Paläozeanographische Untersuchungen in der Vilkitsky-Straße und östlich von Severnaya Zemlya mit sedimentologischen und geophysikalischen Methoden. Unpubl. Diploma Thesis, Köln University, 138pp.

Late Pleistocene Paleoriver Channels on the Laptev Sea Shelf - Implications from Sub-Bottom Profiling

H.P. Kleiber and F. Niessen

Alfred-Wegener-Institut für Polar- und Meeresforschung, Postfach 120161, D-27515 Bremerhaven, Germany

Received 19 March 1997 and accepted in revised form 28 November 1997

Abstract - Bottom and sub-bottom reflection patterns received by the PARASOUND system (4 kHz) document 24 filled paleoriver channels in the uppermost sediments (1-13m) of the Laptev Sea shelf. The surfaces of the paleochannel fillings range from 32 to 97 m below the present sea level. The rivers are supposed to have been active during Weichselian time when the sea level was up to 120 m below that of today. They were probably filled during termination 1 until the transgression reached the present-day 30 m-isobath. The observed paleoriver channels are most likely related to the Olenek, Lena and Jana rivers and to local drainage systems on the Taymyr Peninsula and the New Siberian Islands. In the depth range of 2 to 20 m sub-bottom, a strong post-sedimentary reflector is commonly found, interpreted as the surface of submarine permafrost. The formation of paleoriver channels and permafrost is associated with subaerial exposure and suggest that most of the Laptev Sea shelf area was not covered by a large ice sheet during the last glacial maximum.

Introduction

Presently, central Siberian rivers (e.g. Lena) drain large amounts of freshwater loaded with sediments into the Laptev Sea which is one of three epicontinental seas located along the northern coast of central Asiatic Russia (Holmes and Creager, 1974) (Figure 1). Therefore, the Laptev Sea is a key location for understanding sediment transport processes and rates into the Arctic Ocean and their variability under different climatic conditions. Knowledge about existence and location of paleochannels on the shelf is fundamental because channels indicate major trajectories of sediment transport in Pre-Holocene time. Then channels control past sediment input and distribution on the upper continental slope of the Laptev Sea and thus local sediment budgets of the Arctic Ocean. Also, evidence for river drainage across the shelf would provide a valuable argument against the presence of a large ice sheet on the Laptev Sea shelf (Grosswald, 1990) and support the formation of permafrost on the exposed shelf during the period of late Pleistocene glacial-eustatic regression described by Romanovskii this (volume)

High resolution sub-bottom profiling surveys were carried out during two expeditions (ARK-IX/4 and ARK-XI/1) by the RV "Polarstern" (Fütterer, 1994; Rachor, 1997) (Figure 1). The profiles show evidences for the existence of numerous filled paleochannels on the Laptev Sea shelf. Examples of preliminary results are published in Fütterer (1994), Nürnberg et al. (1995) and Rachor (1997). In this study a complete investigation of occurrence of paleochannels in the Laptev Sea profiles of the ARK-IX/4 and ARK-XI/1 expeditions is presented. Different to earlier investigations (Holmes and Creager, 1974; Treshnikov, 1985), which suggested paleoriver channels based on bathymetry, we used sub-bottom structures for channel identification. The advantage is that the erosive incision and the different sedimentary structures of the channel fillings and the surrounding strata can be observed. Also, limited information about the depths and widths of the paleochannels can be obtained.

In: Kassens, H., H.A. Bauch, I. Dmitrenko, H. Eicken, H.-W. Hubberten, M. Melles, J. Thiede and L. Timokhov (eds.) Land-Ocean Systems in the Siberian Arctic: Dynamics and History. Springer-Verlag, Berlin, 1999, 657-665.

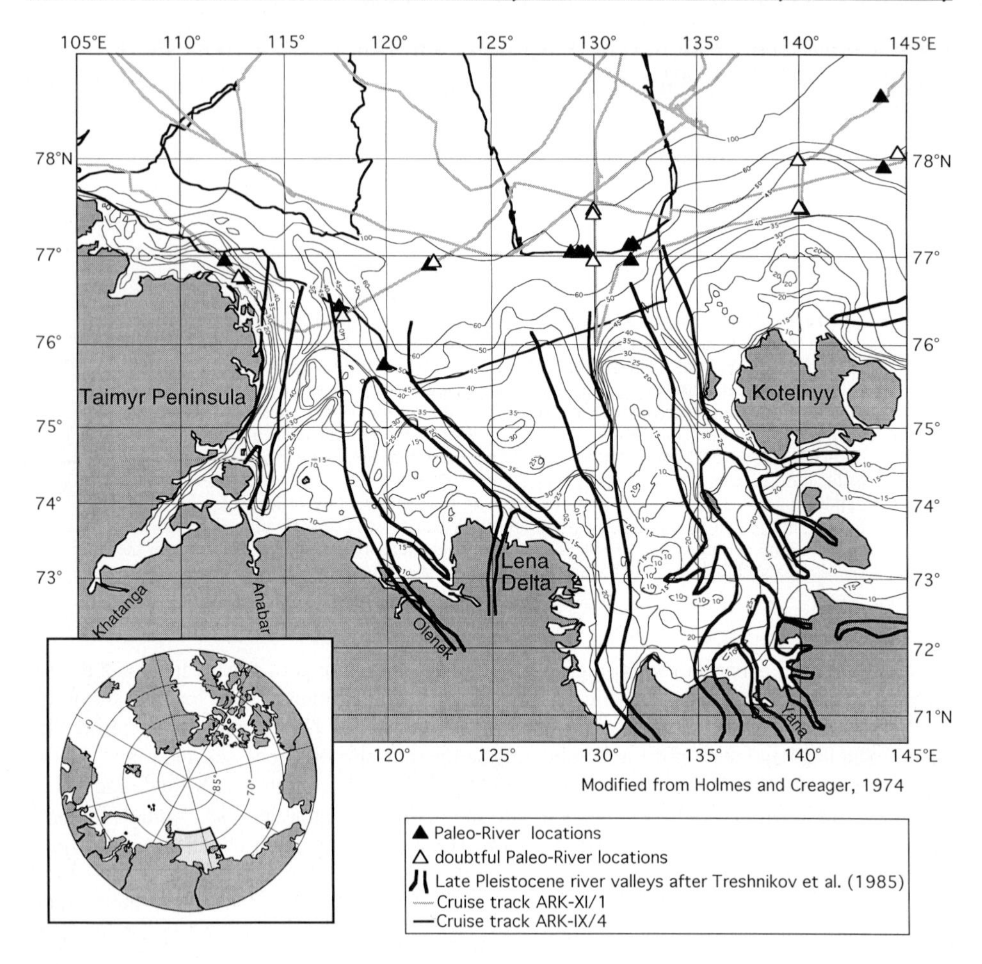

Figure 1: Bathymetry of the Laptev Sea modified from Holmes and Creager (1974), cruise tracks of the two RV "Polarstern" expeditions ARK-IX/4 and ARK-XI/1, and distribution of paleoriver channels identified in PARASOUND profiles.

Methods

High resolution sub-bottom profiling

The hull-mounted PARASOUND echosounder designed by STN Atlas Electronik GmbH (Bremen, Germany) radiates simultaneously two primary sonic frequencies. A constant frequency of 18 kHz and a second frequency, which can be selected by the operator between 20.5 and 23.5 kHz. As a result of the superimposition of the two primary frequencies in the water column, known as the parametric principle, a secondary frequency is created. The latter is equal to the difference between the two primary frequencies ranging between 2.5 and 5.5 kHz, respectively (Grant and Schreiber, 1990). The secondary frequency is suitable for continuos sub-bottom profiling of the uppermost unconsolidated sediment layers (Spiess, 1993). Because the secondary frequency is only generated in the central part of the beam, where the highest energy levels occur, the angle of the sounding cone is about 4° (Grant and Schreiber, 1990). The narrow beam angle results in a small acoustic footprint diameter on the sea floor, which allows high vertical as well as a lateral spatial resolution. The sub-bottom penetration is up to 100 m with a vertical resolution of 5 to 30 cm (Grant and Schreiber, 1990; Rostek, 1991). In this study the secondary frequency was set to 4 kHz.

Incising Pleistocene channels were identified using the following diagnostic criteria: (i), an erosive, u-shaped sub-bottom morphology and (ii), an undeformed fill geometry. Channel locations were mapped as doubtful when one of the criteria above was not clearly fulfilled.

The width of the paleochannels were calculated from the GPS-coordinates. Morphometric measurements, such as paleowidth and paleodepth, were carried out as defined in Figure 2.

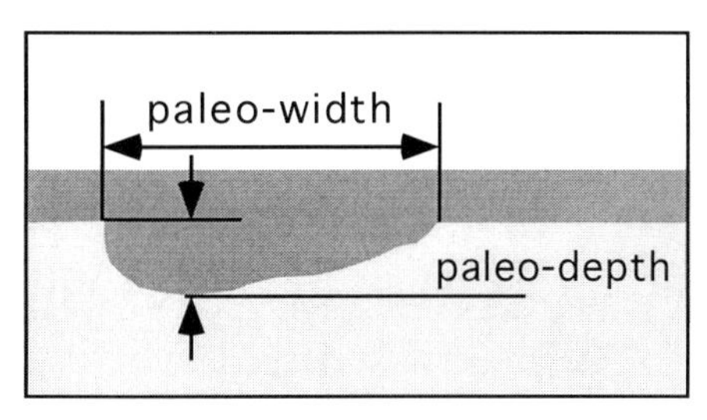

Figure 2: Definitions of morphometric paleochannel features.

Results

In the outer shelf region of the Laptev Sea a total of 24 filled paleochannel cross-sections are recorded, 9 of which are classified as doubtful (Figure 1). Geographically, four groups can be distinguished: (i), three paleochannels northeast of the Taymyr Peninsula, (ii), five paleochannels in the central Laptev Sea, (iii), ten paleochannels northwest of the New Siberian Islands and (iv), six paleochannels north of the New Siberian Islands (Figure 1). The positions as well as the internal reflection patterns on the individual cross-section suggest, that several images could originate from the same paleorivers. Sedimentological or morphological features of individual paleochannels cannot be assigned as typical to one of the four locations.

The surfaces of the paleochannel fillings are located between 32 to 97 m below the present sea level. The recorded channel widths range from 86 to 5331 m, the recorded channel-depths from 3.5 to 13 m. The channel fillings consist of stratified sediments (23 paleochannels). If the channels are cut into inclined stratified sediments, the filling of the channels do not show similar inclination but are parallel to the present sea floor (Figure 3). In addition, six sites are characterized by lateral accretional filling pattern (Figure 3, 4). Only one channel shows an acoustically transparent filling (Figure 5).

The filled paleochannels as well as the slightly deformed or "folded", stratified shelf sediments are overlain by a horizontal bed (Figure 3, 4, 5). This bed is mostly characterized by a continuous basal reflector which cuts unconformably the reflectors below. This top layer measures 0.5 to 2 m in thickness.

In the depth range of 2 to 20 m sub-bottom, a strong reflector is commonly observed, which causes an acoustic impedance high enough, to prevent further sound penetration. This reflector appears conformably as well as unconformably to the bedding with a flat or hummocky top (Figure 6). Below the paleochannels this reflector either diverges downward or cannot be observed because high acoustic impedance of the fillings prevent recording of reflectors from the underlying stratified sediments. In areas, where this reflector cannot be identified, the penetration reaches up to 65 m sub-bottom.

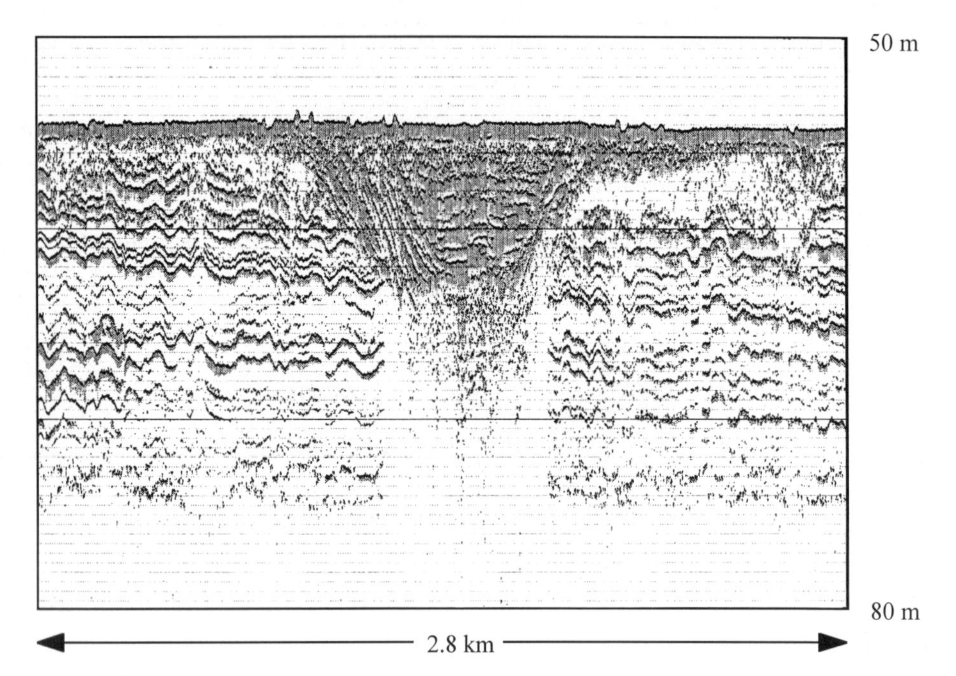

Figure 3: PARASOUND profile north of the New Siberian Islands between 77° 57.0´N / 143° 57.5´E and 77° 57.3´N / 144° 05.3´E, showing a filled paleochannel cutting into well stratified, inclined sediments. Note the accretional reflection pattern on the left hand side of the channel fill.

Discussion

Most likely the observed paleoriver channels are related to the Olenek, Lena and Jana rivers. It is interesting to note that for these rivers our channel locations match the reconstruction of major paleovalleys under the Laptev Sea (Treshnikov et al., 1985) (Figure 1). The paleovalley of the Anabar-Khatanga rivers assumed by Treshnikov et al. (1985) can neither be confirmed nor an other paleodrainage can be suggested based on our profiles. The existence of paleoriver channels implies that the Central Siberian freshwater drainage into the Arctic Ocean continued during the Weichselian glacial period. The mouths of these channels along the low-stand coast formed several point-sources of sediment input to the Arctic Ocean. Ongoing studies will show whether submarine fans are developed on the continental slope and rise which are related to the observed channels.

Channel systems directly north of the Taymyr Peninsula and the New Siberian Islands can hardly be linked to the drainage pattern of the major Siberian rivers. Nevertheless, the geometry of these channels imply drainage comparable to that of the larger river systems. Strong local drainage and channel formation could be the result of rivers formed at the termination of Weichselian glaciations caused by melting of local ice sheets and/or glaciers. Based on seismic and coring results from lakes on the Taymyr Peninsula, Niessen et al. (this volume) suggest that the last major glaciation in the area occurred during marine isotope stage 4. Thus, the channels north of the Taymyr Peninsula maybe related to drainage from such an ice sheet. On the New Siberian Islands, Flint (1971) reconstructed an ice sheet of about 200 m in thickness and 250 km in diameter of which the southern margin reached the mainland coast. The age of this glaciation is unknown. However, if the ice sheet existed and melted during some period of this last sea level lowstand, drainage channels north of the New Siberian Islands like those

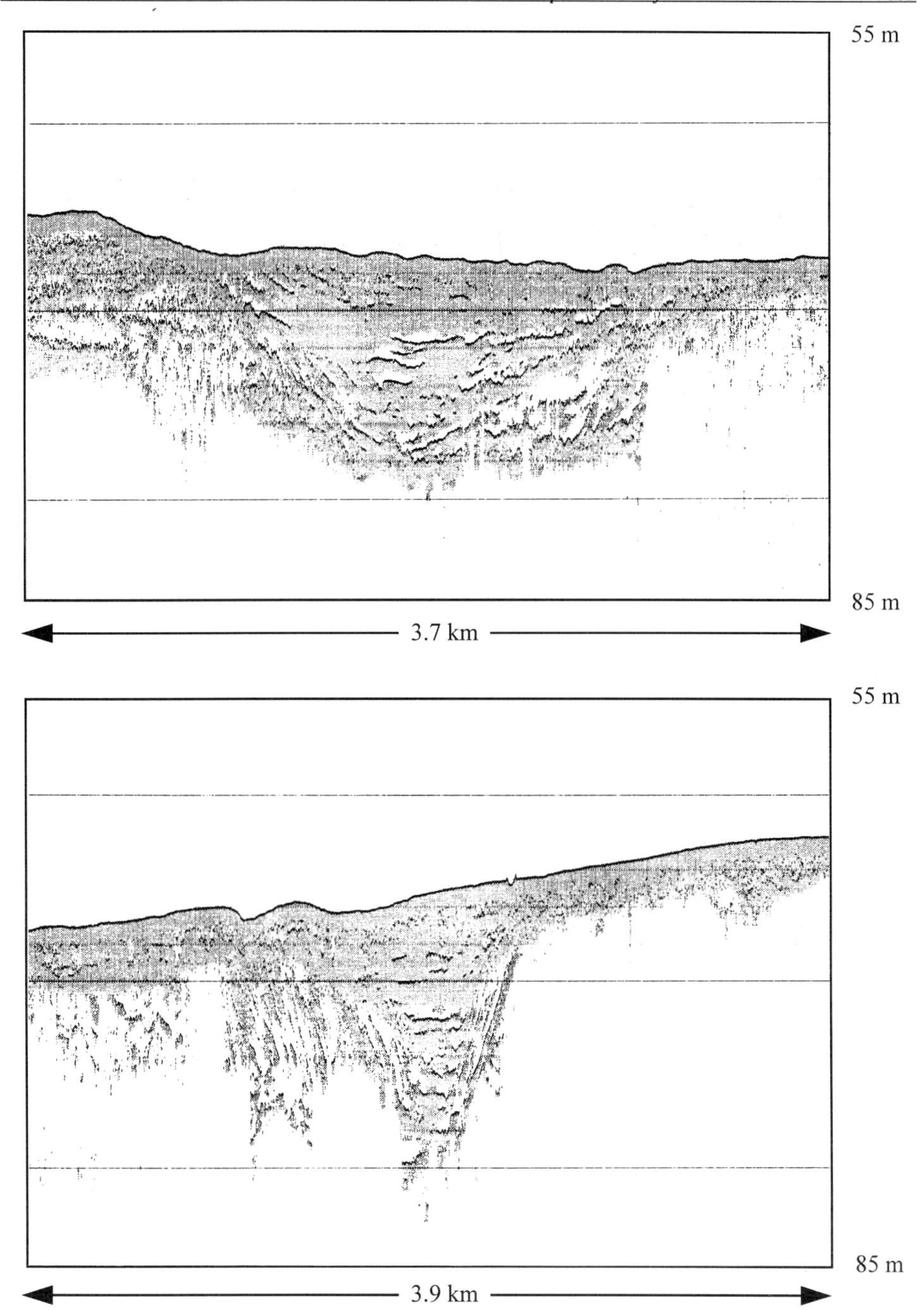

Figure 4: Two PARASOUND profiles northwest of the New Siberian Islands between 77° 03.0'N / 131° 05.3'E and 77° 02.0'N / 130° 57.5'E (a) resp. 77° 02.0'N / 130° 57.5'E and 77° 02.0'N / 130° 48.1'E (b). Due to the angular unconformity the fillings of the paleochannels can clearly be distinguished from the truncated stratified sediments below and from the thin homogeneous cover. The distance between the two paleochannel sites is only 3.7 km, thus they most likely represent two cross-sections of the same meandering river.

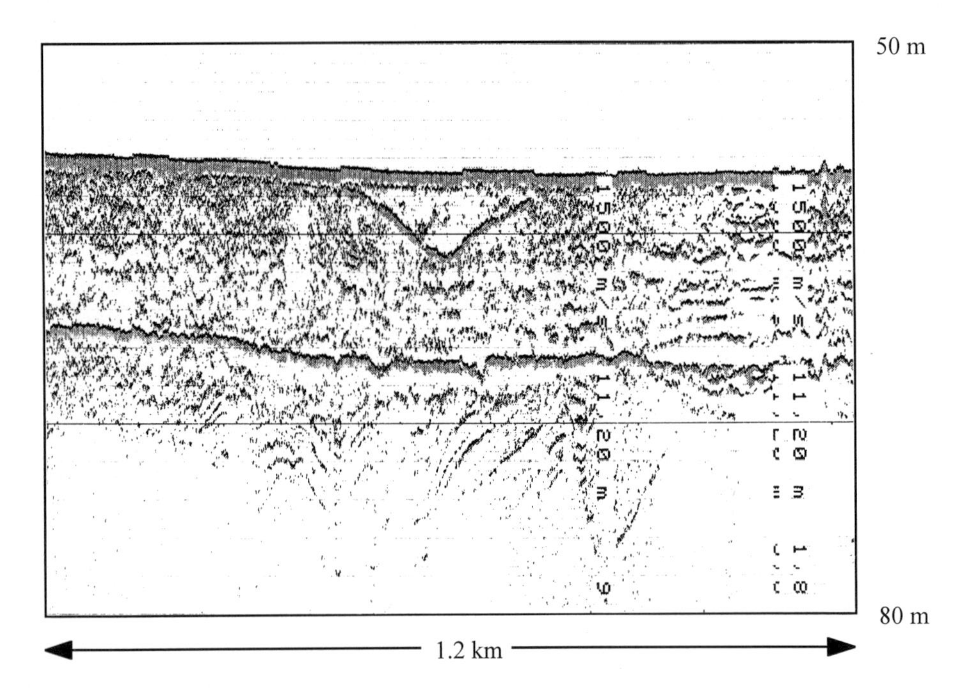

Figure 5: PARASOUND profile northwest of the New Siberian Islands between 76° 57.2´N / 131° 33.7´E and 76° 56.7´N / 131° 31.6´E, showing a filled paleochannel

observed in this study would be eroded into the exposed shelf. Since the glaciation history of the Laptev Sea area is still discussed controversially (e.g. Dunayev and Pavlidis, 1990; Grosswald, 1990; Romanovskii, 1996) further investigations are necessary to test these hypotheses.

Taken into account that the intersection angles of the cruise track and the paleochannels are unknown, the width/depth ratios and the relief on basal scour surface can hardly be used for separating the meandering from the braided river system (Füchtbauer, 1988; Schumm, 1977). The paleoriver cross-sections show erosive, single channels deeper than 3m, all characteristic features of a meandering river (Miall, 1977). The accretional reflection pattern of the stratified fillings, is interpreted as epsilon cross-bedded units caused by lateral accretion on an inclined surface. These sediment bodies are typical for point bars of a meandering channel, but can also be observed locally in a more braided river system (e.g. Allen, 1983). Crossbedded structures indicating longitudinal bars, common in a braided river system (Miall, 1977) have not been observed. Although most of the observations point to meandering rivers, a clear statement concerning the river type is hardly possible from this study because further important river variables like drainage pattern across the entire shelf, sinuosity or mode of sediment transport remain undetermined.

Sub-horizontal filling of channels cut into inclined sub-bottom layers of probably older Pleistocene age (Figure 3) clearly indicate that the channels were filled during morphological condition similar to that of today. This shows that the channels are relatively young and can be related to erosion during the last sea-level low-stand. The recorded depth of the paleochannel filling surfaces below the present sea level indicates that the rivers must have been active during Weichselian time when the sea level was at least 100 m lower than today (e.g. Chappell and Shackleton, 1986; Fairbanks, 1989). They were probably filled during termination 1 (onset ca.

18 ka B.P.) until the transgression reached the present-day 30 m-isobath approximately 10,000 years B.P. (Blanchon and Shaw, 1995). The uppermost bed which covers the filled paleorivers and the stratified sediments of probably Pleistocene age equally, is interpreted as Holocene deposits.

The unconformable character of the strong second sub-bottom reflector is indicative for a post sedimentary origin which is interpreted as the surface of submarine permafrost (e.g. Rogers and Morak, 1983). The formation of thick permafrost is only possible if subaerial exposure of the Laptev Sea shelf area occurred during times of the Weichselian sea-level low stand (Romanovskii et al., this volume). The permafrost situation under the paleochannels remains unclear because we cannot distinguish whether the reflector is not present or not seen because of strong backscatter from the channel base and fill.

The coexistence of paleorivers and permafrost is a strong argument against a glaciation of the Laptev Sea shelf during the Weichselian as suggested by Grosswald (1990). This is consistent with observed PARASOUND penetration into pre-Holocene Laptev Sea shelf deposits down to tens of meters which indicates lack of ice load during the Last Glacial. Generally, the surface of ice-consolidated sediments are excellent sound reflectors and little or no sound penetrates through such sediment interfaces (Damuth, 1978).

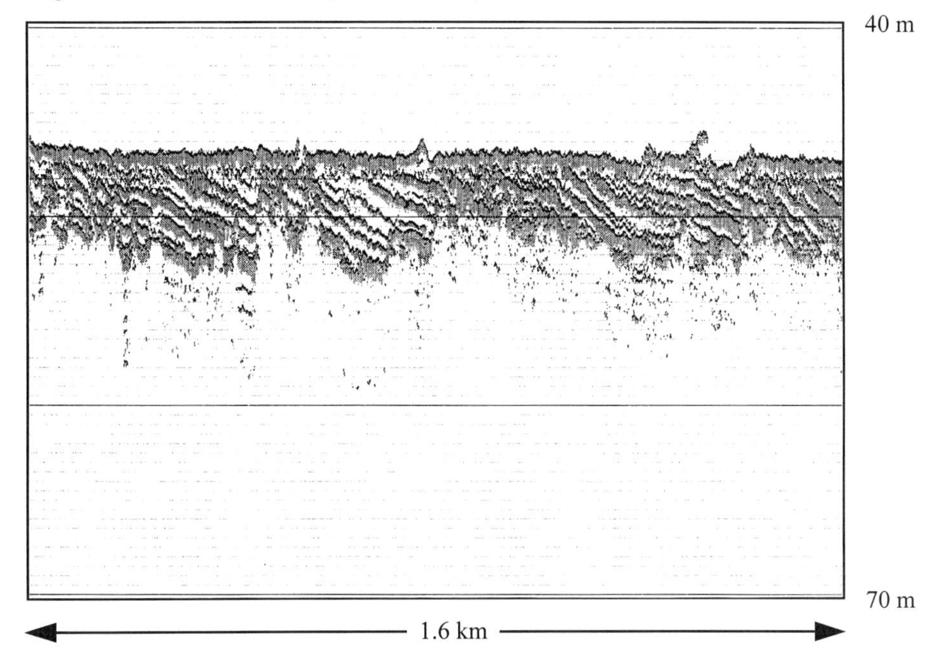

Figure 6: PARASOUND profile northwest of the Taymyr Peninsula between 76° 18.1´N / 117° 30.2´E and 76° 18.5´N / 117° 33.2´E, showing strong reflectors cutting the stratified sediments between 4 to 7 m sediment depth, thus indicating their post-sedimentary origin.

Conclusions

Laptev Sea paleochannels identified in this study match the location of paleovalleys reconstructed from other investigations. The rivers are supposed to have been active during the Weichselian sea-level low stand. This implies that they drained freshwater and sediment into the Arctic Ocean during marine isotope stage 2, 3 and 4. Epsilon cross-bedded units favor meandering Pleistocene river systems on the Laptev Sea shelf. Our study suggests that deltas were formed along the low-stand paleoshore line which functioned as point-sources of sediment input onto the upper continental slope of the Arctic Ocean. On the shelf, the river channels were probably filled between termination 1a and the early Holocene, when the transgression reached the present-day 30m-isobath. The filled channels were then covered by sediments of which the reflection geometry suggests that they are of Holocene age.

Some paleochannels north of the Taymyr Peninsula and north of the New Siberian Islands can geographically hardly be related to the large Siberian rivers. This suggests further investigations in order to test whether local glaciation and subsequent drainage from deglaciation can explain the observed pattern.

Nearly throughout the entire investigated area, the existence of a post-sedimentary reflector about 2 to 20m sub-bottom suggests that submarine permafrost is common on the Laptev Sea shelf. It is not evident from this study whether the permafrost continuous under the paleochannels or not.

There are three observations which indicate that most of the Laptev Sea shelf area was not covered by an extended ice sheet during the last glacial maximum: (i), the acoustic penetration into the Pleistocene sediments indicates no overconsolidation by an ice sheet, (ii), the existence of paleochannels are indicative for continuation of Eurasian river-drainage during the Weichselian and (iii), a strong post-sedimentary reflector indicates the existence of relict submarine permafrost.

Acknowledgements

We are indebted to the crew of RV "Polarstern" and our fellow scientists for their help in collecting the data. We thank K. Hinz, M. Block and N. Romanovskii for the constructive discussions. This is contribution no. 1342 of the Alfred Wegener Institute.

References

Allen, J.R.L. (1983) Studies in fluviatile sedimentation: Bars, bar-complexes and sandstone sheets (Low-sinuosity braided streams) in the Brownstones (L. Devonian), Welsh Borders. Sedim. Geol., 33, 237-293.

Blanchon, P. and J. Shaw (1995) Reef drowning during the last deglaciation: Evidence for catastrophic sea-level rise and ice-sheet collapse. Geology, 23 (1), 4-8.

Chappell, J. and N.J. Shackleton (1986) Oxygen isotopes and sea level. Nature, 324, 137-140.

Damuth, J.E. (1978) Echo character of the Norwegian-Greenland Sea: relationship to Quaternary sedimentation. Mar. Geol., 28, 1-36.

Dunayev, N.N. and J.A. Pavlidis (1990) A model of the late Pleistocene glaciation of Eurasiatic arctic shelf. In: Arctic research - Advances and prospects (Vol. 2), Proceedings of the conference of Arctic and Nordic Countries on coordination of research in the Arctic, Kotlyakov, V.M. and V.E. Sokolov (eds.), Academy of Sciences of the USSR, Leningrad, December 1988, 70-72.

Fairbanks, R.G. (1989) A 17,000-year glacio-eustatic sea level record: influence of glacial melting rates on the Younger Dryas event and deep-ocean circulation. Nature, 342, 637-642.

Flint, R.F. (1971) Glacial and Quaternary Geology. John Wiley and Sons, New York, 892pp.

Füchtbauer, H. (1988) Sedimente und Sedimentgesteine, Teil II: Sediment-Petrologie. Schweizerbart´sche Verlagsbuchhandlung, Stuttgart, 1141pp.

Fütterer, D.K. (ed.; 1994) The expedition ARCTIC´93. Leg ARK-IX/4 of RV "Polarstern" 1993. Ber. Polarforsch., 149, 244pp.

Grosswald, M.G. (1990) Late Pleistocene ice sheet in the Soviet Arctic. In: Arctic research - Advances and prospects (Vol. 2), Proceedings of the conference of Arctic and Nordic Countries on coordination of research in the Arctic, Kotlyakov, V.M. and V.E. Sokolov (eds.), Academy of Sciences of the USSR, Leningrad, December 1988, 18-23.

Holmes, M.L. and J.S. Creager (1974) Holocene history of the Laptev Sea continental shelf. In: Marine Geology and Oceanography of the Arctic Seas, Herman, Y. (ed.), Springer-Verlag, New York, 211-230.

Miall, D.A. (1977) A review of the braided-river depositional environment. Earth-Sci. Rev., 13, 1-62.

Niessen, F., D. Weiel, T. Ebel, J. Hahne, C. Kopsch, M. Melles, E. Musatov and R. Stein (1997) Weichselian glaciation in Central Siberia - Implications from marine and lacustrine high resolution seismic profiles and sediment cores. In: Abstract Supplement No 1, Terra Nova 9, Oxburgh, E.R. (ed.), European Union of Geosciences, Strasbourg, 208.

Nürnberg, D., D.K. Fütterer, N. Nørgaard-Pedersen, C.J. Schubert, R.F. Spielhagen and M. Wahsner (1995) The depositional environment of the Laptev Sea continental margin: Preliminary results from the R/V Polarstern ARK IX-4 cruise. Polar Res., 14 (1), 43-53.

Rachor, E. (1997) Scientific Cruise Report of the Arctic Expedition ARK-XI/1 of RV "Polarstern" in 1995. Ber. Polarforsch., 226, 157.

Rogers, J.C. and J.L. Morack (1983) Geophysical detection of subsea permafrost. In: Handbook of geophysical exploration at sea, Geyer, R.A. (ed.), CRC Press, Inc., Boca Raton, Florida, 187-210.

Romanovskii, N. (1996) Permafrost distribution on the Laptev Sea shelf. In: Third workshop on Russian-German cooperation: Laptev Sea system, Kassens, H., F. Lindemann and B. Rohr (eds.), Terra Nostra, Schriften der Alfred-Wegener-Stiftung, 96/6, 111.

Schumm, S.A. (1977) The fluvial system. John Wiley & Sons, New York, 337pp.

Spiess, V. (1993) Digitale Sedimentechographie - Neue Wege zu einer hochauflösenden Akustostratigraphie. Ber. Fachb. Geowiss. Univ. Bremen, 35, 199pp.

Treshnikov, A.F. (1985) Palaeogeographic atlas of the shelf regions of Eurasia for the Mesozoic and Cenozoic, Volume 2, Maps (in Russian). Main administration of geodesy and cartography of the USSR, Moscow, 204pp.

Main Structural Elements of Eastern Russian Arctic Continental Margin Derived from Satellite Gravity and Multichannel Seismic Reflection Data

S.S. Drachev[1,2], G.L. Johnson[3], S.W. Laxon[4], D.C. McAdoo[5] and H. Kassens[2]

(1) P.P. Shirshov Institute of Oceanology, 36 Nakhimovsky Prospekt, 117851 Moscow, Russia

(2) GEOMAR Forschungszentrum für marine Geowissenschaften, Wischhofstrasse 1-3, D 24148 Kiel, Germany

(3) University of Alaska, Fairbanks, Alaska 99775-7220, USA

(4) University College London, Space and Climate Physics Department, MSSL, Holmbury St. Mary, Dorking Surrey, RH5 6NT, UK

(5) NOAA, NODC, LSA, Silver Spring, MD 20910, US

Received 1 May 1997 and accepted in revised form 16 September 1997

Abstract - This paper presents an analysis of the altimeter derived marine gravity data and available multichannel seismic reflection profiles of the northeastern Siberia continental margin to provide a preliminary delineation of its geologic structure. Both data sets allow making a suggestion that the continental margin was strongly affected by the rifting process since mid of Mesozoic. The main rift zones extend south- and southeastward from the shelf edge toward the mainland. These are the Laptev Rift System, New Siberian and Vil'kitskii rifts in the East Siberian Sea, and North Chukchi rift basin in the Chukchi Sea. These structures may represent both active and aborted rifts related to extension episodes in the Arctic Ocean, namely: initial rifting in the Canada Basin (Middle Jurassic to Hauterivian), opening of the Makarov Basin (53-80 Ma ?) and spreading in the Eurasia Basin (0-56 Ma). Thus it is suggested that the spreading migrated westward during the Cretaceous-Cenozoic time and led to separation and fragmentation of the continental blocks (Lomonosov and Mendeleev ridges, Chukchi Plateau) and subsidence of the rift basins on the eastern Russian Arctic continental margin.

Introduction

The Arctic Ocean is unique among the oceans of the world in that half of its area is underlain by continental shelf, primarily the wide European and Siberian continental shelves. The Laptev, East Siberian and Chukchi seas overlie the extensive shelf of northeaster Russia which is up to 800 km wide and totals about 2,000,000 km^2 in area (Figure 1). To the north, this continental margin is bordered by the Eurasia and Amerasia oceanic basins which contain large submarine ridges and plateaus including: the Gakkel, Lomonosov, Alpha-Mendeleev and Chukchi Borderland.

The structure and tectonic history of the shelf are very poorly known and are primarily a postulated extension of the terrestrial geology of the New Siberian, De Long and Wrangel islands and the mainland (Vinogradov et al., 1977; Vinogradov, 1984; Fujita and Cook, 1990; Kos'ko, 1984; Kos'ko et al., 1990). Of the three seas, the Laptev Sea is better known as the result of the Russian seismic reflection studies which have defined the rift basins (Ivanova et al., 1990; Drachev et al., 1995a). The structural pattern and geological history of the remainder of the continental margin are still largely unknown and there are as many points of view as there are investigators. Since beginning of the 1970s, the Russian airborne magnetic data were the primary available means of deciphering the geology of this region (for more detail see Fujita and Cook (1990) and references contained therein). Also the northwestern part of the Chukchi Sea is crossed by eight U.S. Geological Survey multichannel seismic reflection (MCS) profiles (Grantz et al., 1986, 1990). In 1989, 1990, and 1993-1994 several seismic lines were acquired

----

In: Kassens, H., H.A. Bauch, I. Dmitrenko, H. Eicken, H.-W. Hubberten, M. Melles, J. Thiede and L. Timokhov (eds.) Land-Ocean Systems in the Siberian Arctic: Dynamics and History. Springer-Verlag, Berlin, 1999, 667-682.

in the western part of the East Siberian Sea by Russian and German research institutes (Sekretov, 1993a; Drachev et al., 1995b; Roeser et al., 1995). In this paper we present our hypothesis on the structure of the eastern Russian Arctic continental margin (ERAM) and its relationship to the evolution of the Arctic oceanic basins, based on combining of the recently obtained satellite altimeter gravity data with the results of the offshore MCS surveys.

Tectonic setting

The ERAM occupies an area where the Siberian Craton, Early Mesozoic Taimyr Fold Belt, and Late Mesozoic New Siberian-Chukchi and Verkhoyansk-Kolyma fold belts come together (Figure 1). The latter two are divided by the narrow and highly deformed Lyakhov-South Anyui ophiolitic suture which has been suggested by Savostin et al. (1984b), Parfenov and Natal'in (1986), Drachev and Savostin (1993) to be a zone of collision between northeastern edge of the Laurasia supercontinent and New Siberian-Chukchi microplate in the mid Cretaceous time. The deformed Paleozoic and Mesozoic complexes of these large tectonic elements form a poorly known heterogeneous basement overlain everywhere with a sharp angular unconformity by almost undeformed clastic sequences of Late Cretaceous to Cenozoic in age. These sequences fill the extensive offshore sedimentary basins which are believed to be initially originated by rifting and post-rift thermal subsidence during the opening of the Arctic oceanic basins. The Laptev Sea rifts reveal a strong link to sea-floor spreading axis (Gakkel Ridge) in the Eurasia Basin (Grachev, 1982; Drachev et al., 1995a; Roeser et al., 1995). The kinematics of the opening of this basin is well understood because it contains a readily decipherable sea-floor magnetic pattern suggesting that sea-floor spreading may have begun as early as 54-64 Ma (Karasik, 1968; Karasik et al., 1983; Kovacs et al., 1983; Savostin et al., 1984a). The poorly studied rifts of the East Siberian Sea and northwestern part of the Chukchi Sea probably originated in the similar tectonic conditions in response to earlier spreading episodes in the Amerasia Basin.

Satellite derived altimeter data

Satellite radar altimetry is the means to derive marine gravity field through accurate measurement of the average sea surface topography. The ERS-1 satellite, launched in 1991, provided the first altimetric observations of the Arctic Ocean to latitudes of 82° N.

However the ERAM suffers coverage by both seasonal and, in some areas, permanent sea ice cover. Satellite altimeter data over sea ice areas are normally excluded from global marine gravity fields as the onboard estimates of surface height suffer greatly increased noise due to changes in the return echo waveform (Laxon, 1994). To overcome this problem, techniques have been developed to reprocess the individual return echoes in order to significantly reduce noise and allow marine gravity anomalies to be extracted in the usual manner (McAdoo and Marks, 1992). This process was used to derive the first satellite marine gravity field of the Arctic Ocean from a single 35-day repeat cycle of ERS-1 giving an unprecedented view of its tectonic fabric (Laxon and McAdoo, 1994).

The gravity field presented in this paper supplants that first Arctic field. This new, higher resolution field is derived from much more data including data from the ERS-1 geodetic mission whose a 168-day repeat cycle yielded a dense grid of observations with a spacing of less than 4 km at altitudes above 60° N. The resolution of this new gravity field, estimated by comparisons with airborne gravity data, is approximately 30-35 km (Laxon and McAdoo, 1998).

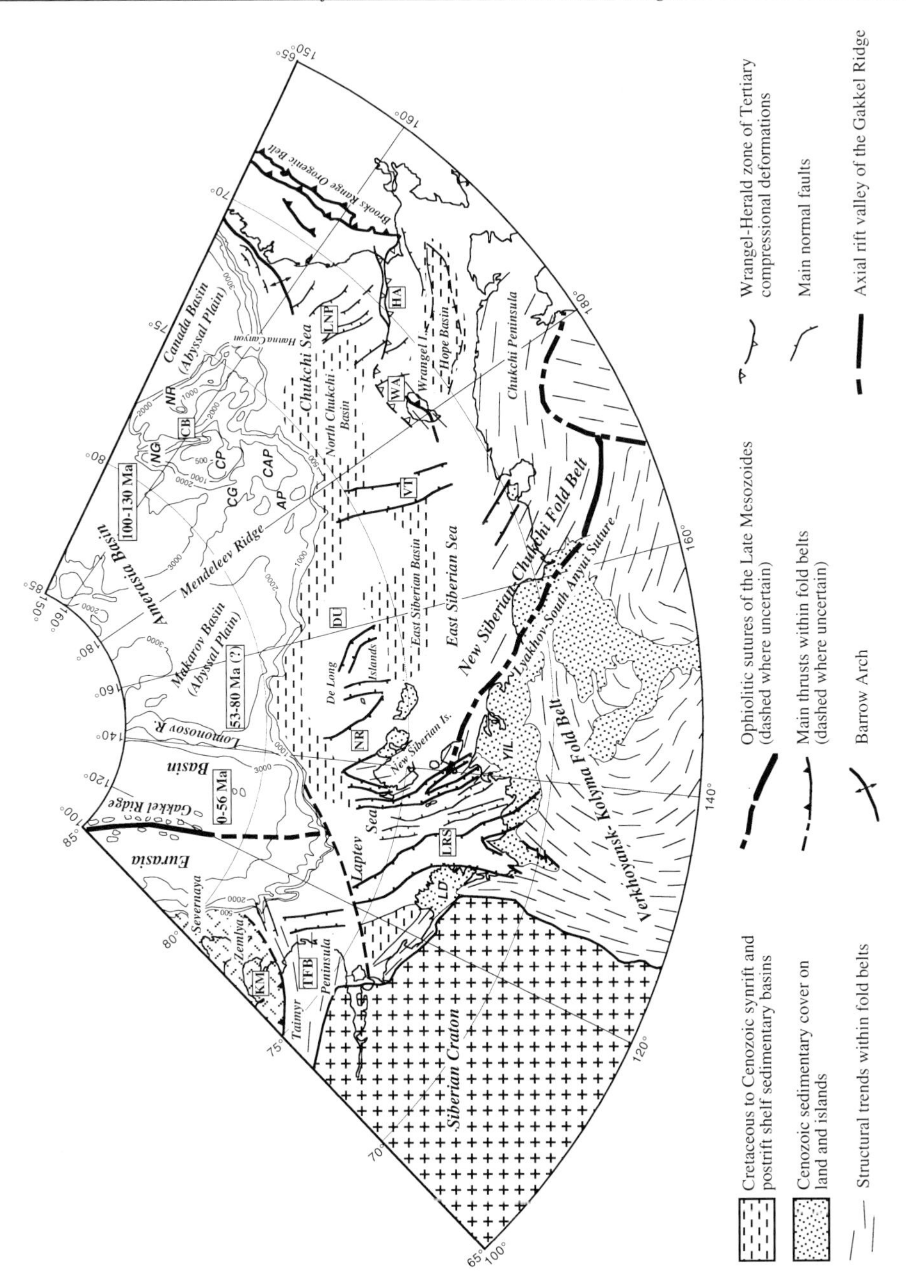

Figure 1: Geographical features and the main structural elements of the eastern Russian Arctic continental margin and adjacent regions. Polar stereographic projection. The italic capital letters show location of the physiographical features: LD - Lena Delta, YIL - Yana-Indigirka Lowland, AP - Arlis Plateau, CAP - Chukchi Abyssal Plain, CG - Charlie Gap, CP - Chukchi Plateau, NG - Nautilus Gap, NR - Northwind Ridge. Outlined bold capital letters denote the structural elements: KM - Kara Massif, TFB - Taimyr Fold Belt, LRS - Laptev Rift System, NR - New Siberian Rift, DU - De Long Uplift, VT - Vil'kitskii Trough, WA - Wrangel Arch, HA - Herald Arch, LNP - Listric-normal faults province, CB - Chukchi Borderland. Outlined bold numbers denote the time of the spreading events in the Arctic oceanic basins in millions of years.

Laptev Sector (Laptev Sea and southern Eurasia Basin)

This sector of ERAM is bounded by the Taimyr Peninsula and Severnaya Zemlya Archipelago on the west and New Siberian Islands on the east (Figure 1). It contains the southern termination of the Gakkel spreading ridge and thus its geologic structure is the result of a continental margin/spreading ridge interaction. The new gravity data (Figure 2) confirm the map-view geometry of the main elements of the rift system, which were previously mapped by MCS surveys (Drachev et al., 1995a), and provide more details where the seismic profiles are lacking.

The Laptev Rift System consists of several deep subsided rift basins and uplifted or high-standing blocks of the basement (Figures 2 and 3). From the west to the east there are: the West Laptev and South Laptev rift basins, Ust' Lena Rift, Stolbovoi and East Laptev horsts, Bel'kov-Svyatoi Nos and Anisin rifts. The structural pattern of the central and eastern parts of the shelf are clearly expressed in the gravity field with the low-to-high variations from -35 mGal over the rifts to 50 mGal over high-standing blocks. The Ust' Lena, Bel'kov-Svyatoi Nos and Anisin rifts are particularly prominent. The total preliminary estimated thickness of the rift-related sediments varies between 4 and 8-9 km while on the horsts the sedimentary cover is significantly reduced and its thickness does normally not exceed 2-2.5 km. The rift sedimentary fill contains several seismic stratigraphic units predicted to be of Late Cretaceous to Early Pliocene in age and its internal structure reflects the different stages of the spreading ridge/continental margin interaction (Figure 3). The whole shelf is covered by uppermost seismic unit. Onshore stratigraphy supports the Late Pliocene to Holocene age for them. This succession probably reflects a deceleration of the rifting during the last reorganization of the North American/Eurasian plates interaction about 2 Ma as shown by Cook et al. (1986) who calculated the present-day Euler Pole of the Eurasian/North American plates to be near the coast of the Laptev Sea.

The rifts strike SE-NW from the coast toward the shelf edge. The Bel'kov-Svyatoi Nos Rift contains a long and narrow axial horst in the southern part that divides it into two parallel grabens continued on the land. To the north this rift merges the Anisin Rift reaching the shelf edge at about 77° N 130° E. The northern part of the rift is seismically active (Fujita and Cook, 1990; Avetisov, 1993) and can be considered present-day manifestation of the extension axis of the Gakkel Ridge on the shelf. The Ust' Lena Rift and South Laptev Rift Basin, however, are not continuous across entire shelf but are terminated by a NE-SW structural line which is well-expressed in the gravity field. This line is coincident with the continental slope of Amundsen Basin, then follows previously outlined Severnyi (Northern) Graben (Vinogradov, 1984) and extends further toward the Khatanga Bay between Taimyr Peninsula and the mainland (Figure 2). The existence of such lineament was inferred by Fujita et al. (1990) who described it as Severnyi Transfer. In our opinion this lineament which we call the Northern Fracture could have acted as a transform fault during some stage of opening of the Eurasia Basin and displacement of the Lomonosov Ridge relative to the Laptev Shelf.

The decrease of the rift sedimentary fill as one moves eastward from 10-12 km in South Laptev Basin to 4-5 km in Bel'kov-Svyatoi Nos Rift and simplification of the structure as a whole suggest migration of the rifting in an eastward direction. The West Laptev and South Laptev rift basins may represent failed rifts associated with beginning of the rifting prior the continental breakup and initiation of sea-floor spreading at about 58 Ma. Such migration of the rifting could take place since the spreading axis in the Eurasia Basin had to migrate eastward with respect to the Barents-Kara continental margin.

The structural pattern of the Eurasia oceanic basin and Gakkel Ridge southward of 80° N is not morphologically expressed as they are buried under a thick sedimentary cover. The first MCS data were acquired there in 1990 by the Russian Marine Arctic Geologic Expedition

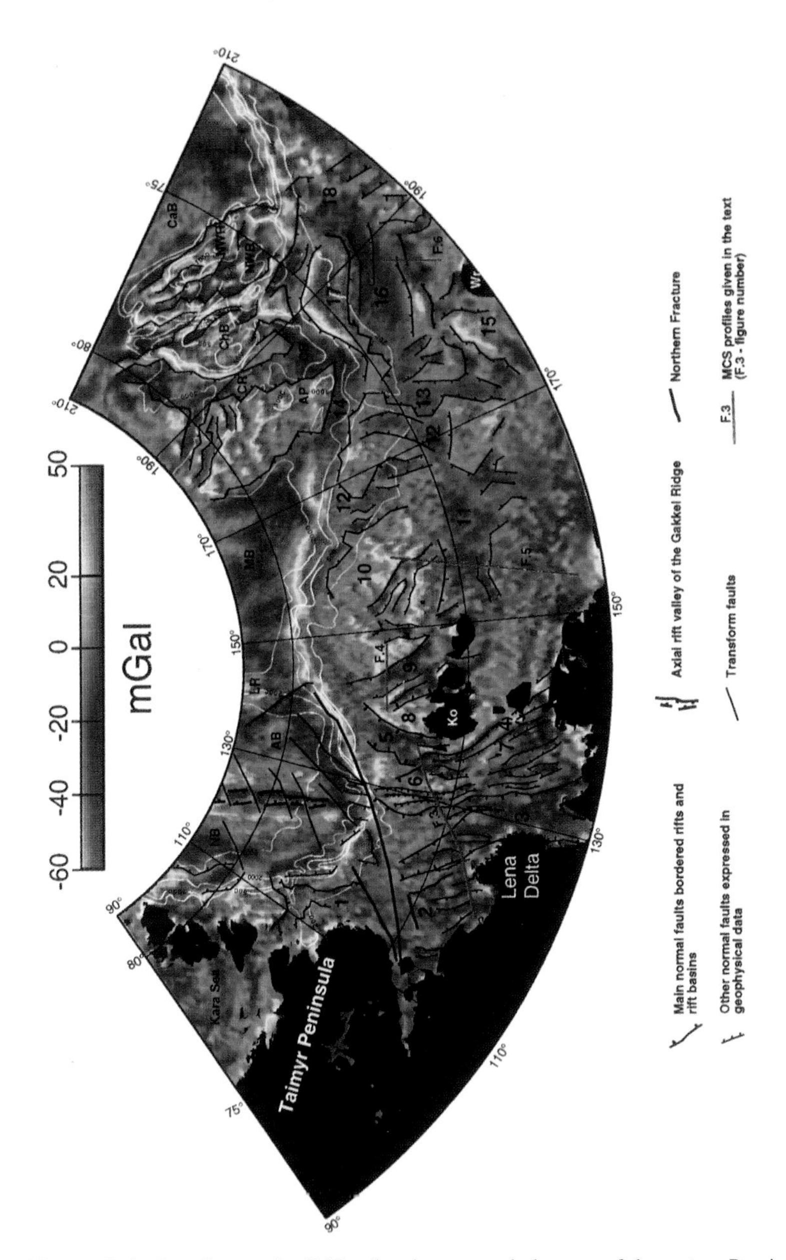

Figure 2: Altimeter derived marine gravity field and main structural elements of the eastern Russian Arctic continental margin. Polar stereographic projection. The bold numbers denote the geological structures on the continental shelf: 1 - West Laptev Rift Basin, 2 - South Laptev Rift Basin, 3 - Ust' Lena Rift, 4 - Bel'kov-Svyatoi Nos Rift, 5 - Anisin Rift, 6 - East Laptev Horst, 7 - Stolbovoi Horst, 8 - Kotel'nyi Uplift, 9 - New Siberian Rift, 10 - De Long Uplift, 11 - East Siberian Basin, 12 - Vil'kitskii Rift System, 13 - Vil'kitskii Trough, 14 - Toll' Rift, 15 - Wrangel Block (Arch), 16 - North Chukchi Rift Basin, 17 - North Chukchi Uplift, 18 - Listric-normal faults province. The capital bold letters show the structural features of the oceanic areas: NB - Nansen Basin, AB - Amundsen Basin, LR - Lomonosov Ridge, MB - Makarov Basin, AP - Arlis Plateau, CB - Chukchi Basin, CR - Charlie Rift, ChB - Chukchi Borderland, NWB - Northwind Basin; NWR - Northwind Ridge, CaB - Canada Basin. The white letters are: Ko - Kotel'nyi Island, Wr - Wrangel Island.

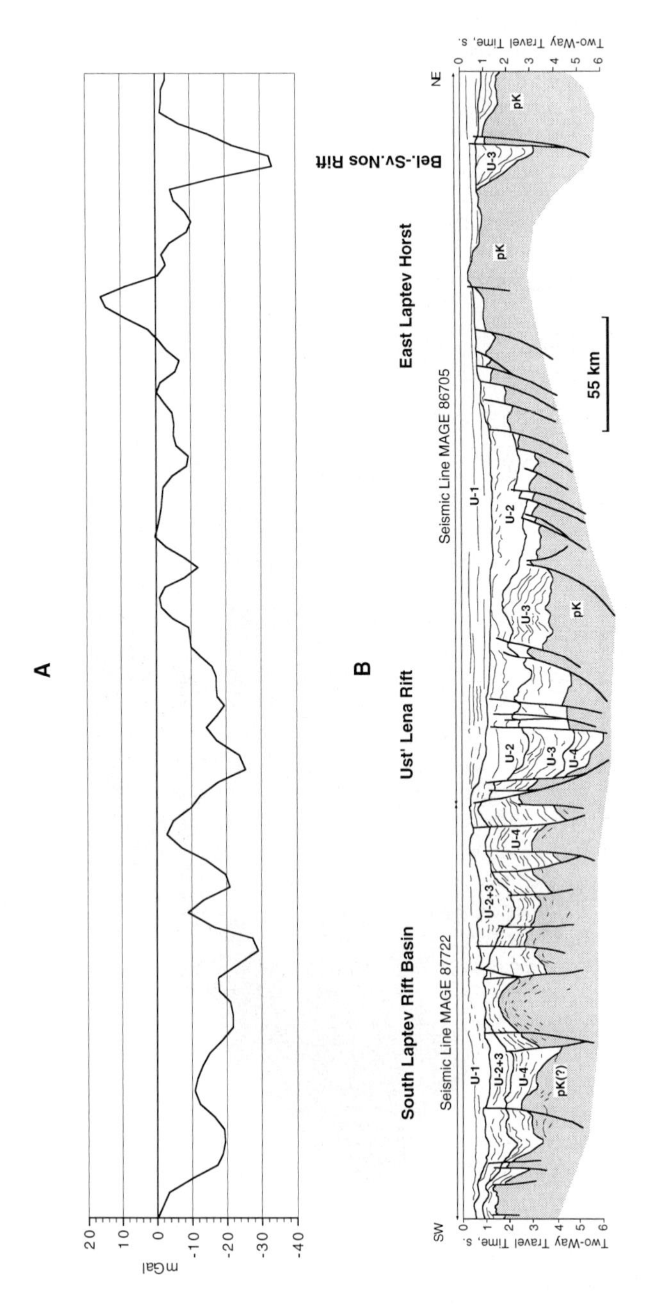

Figure 3: Free air gravity chart (A) and line drawing of the MCS profiles MAGE 87722 and 86705 by Marine Arctic Geologic Expedition (B) showing inferred geologic structure and stratigraphy across the Laptev Rift System (see Figure 2 for location). U-1 to U-4 denote the seismic stratigraphic units of following predicted age: U-1 - Late Pliocene to Holocene, U-2 - Oligocene to Pliocene, U-3 - Late Paleocene to Eocene, U-4 - Late Cretaceous to Paleocene (within the South Laptev Rift Basin this unit might be older (Cretaceous ?) than in the Ust‘ Lena Rift), U-2+3 - Eocene to Late Miocene (within the South Laptev Basin only).

(Sekretov, 1993b) but they are not published yet. The new gravity data give an unique insight into structure of this very interesting spreading ridge/continental margin conjugation.

The Gakkel Spreading Ridge is sharply displayed in the gravity field by a narrow low about -20 to -50 mGal and surrounding gravity highs of 10 to 30 mGal, which reflect an axial rift valley and both uplifted flanks of the buried ridge correspondingly. The rift valley is traced onto the continental slope up to 78° 30' N 126° 15' E, where it is cut by a set of oblique lineaments which appear to be transform-like fractures. Several transform faults can also be recognized within the ridge. They segment the ridge but do not reveal a visible offset and can be considered zero-transform faults.

The southern Eurasia Basin has been crossed by several MCS profiles in 1990 (Sekretov, 1993b). These data show a thick (about 5-8 km) sedimentary sequence suggested to be a Late Cretaceous to Cenozoic in age. The Gakkel Ridge is buried under its upper (Miocene to Holocene) part.

East Siberian Sector (East Siberian Sea and Southern Makarov Basin)

This sector is located between 140° and 180° E and thus is the central part of the ERAM. There is a limited amount of available MCS and seismic refraction data obtained by the Russian and German scientists in 1989-1994 in the western part of the East Siberian shelf, mainly northward of New Siberian Islands (Gramberg et al., 1993; Sekretov, 1993a; Drachev et al., 1995b; Roeser et al., 1995). Satellite altimeter data give us the opportunity to extend the structural contours, localized previously by seismic surveys, and to provide a preliminary delineation of the structure of the seismically unstudied areas (Figure 2).

The 500 km long and 100-120 km wide New Siberian Rift is a most prominent negative gravity feature of the East Siberian shelf. It occurs between the De Long and Kotel'nyi uplifts extending NW-SE toward central part of the shelf where it merges with the East Siberian sedimentary basin. The triangle-shaped Kotel'nyi Uplift separates the New Siberian Rift from the Laptev Rift System and consists of folded Paleozoic to Mesozoic clastic and carbonate complexes covered with thin, less than 1 km of thick, Upper Cenozoic sediments. The northern part of the rift was studied by joint German-Russian geophysical survey in 1993 (Roeser et al., 1995). It includes two deep subsided parallel grabens divided by a narrow horst, as shown on Figure 4. These features are well distinguished in the gravity field (Figure 4A). The preliminary estimated thickness of the Cenozoic sediments within the rift exceeds 5 km.

De Long Uplift, the large ellipsoid-shaped positive feature of the region, occurs between 145° and 175° E extending northward of 75° N up to the shelf edge. It comprises high-standing block of the folded basement (Bennett and Henrietta terranes; for more details see Fujita and Cook, 1990) covered by subaerial Cretaceous (Albian) and Late Cenozoic basaltic flows and Upper Cenozoic sediments. The boundaries and internal structural pattern of the uplift are clearly expressed both in gravity (Figure 2) and magnetic (Macnab, 1993; Verhoef et al., 1996) fields. The latter reveals a pattern typical for a volcanic plateau consisting of high amplitude short wavelength anomalies. Gravity and MCS data show small grabens and horsts which complicate the internal parts of the plateau (Drachev et al., 1995b; Roeser et al., 1995). In the northeast the uplift is bordered by the Vil'kitskii Rift System, whose western branch enters the uplift and merges with the East Siberian Basin. The northeastern slope of the uplift has been crossed by Marine Arctic Geologic Expedition MCS profile (Sekretov, 1993a). According to these data, the thickness of the sediments in the area of the continental slope northward of the uplift reaches 6-8 km and their age is predicted to be Aptian to Cenozoic.

In the area between 150° and 160° E the New Siberian Rift joins the East Siberian sedimentary basin. The latter occupies the central part of the East Siberian Shelf, bordering the De Long

Uplift on the south. The MCS profile LARGE 8901 (by Laboratory of Regional Geodynamics, Moscow) crosses this area and provides more details about internal structure and relationship between structural elements displayed in the gravity field (Figure 5). The southwestern part of the shelf is occupied by a structural terrace that represents a gently northward dipping of the acoustic basement and seismic reflectors toward the East Siberian Basin. Previously this area was suggested to be the deep subsided Blagoveshchensk sedimentary basin underlain by thinned continental or even oceanic crust and separated by a linear basement uplift (Anzhu Ridge) from the New Siberian Basin (Fujita and Newberry, 1982; Kos'ko, 1984; Fujita and Cook, 1990). Our data show neither Blagoveshchensk Basin nor Anzhu Ridge in this area.

The East Siberian Basin is bounded by the normal faults both from the shelf terrace and De Long Uplift (Figure 5B). The marginal parts of it are the deeper with the top of the acoustic basement at a depth of 3.5 s TWT, whereas in the central part of the basin the basement is occurred at 2.0-2.3 s TWT and affected by the numerous low-amplitude near-vertical normal faults (Figure 5B). A boundary between the De Long composite terrane and the rest of the basin basement is well expressed in the MCS data. It is a narrow (about 5 km) zone consisting of several anticline-shaped folds accompanied by thrusts and reverse faulting within the lowermost part of sedimentary cover. It can be considered either a compression or transpression fracture. The sedimentary cover of the basin is composed of 5 to 7 seismic stratigraphic units which were suggested to be Cretaceous to Holocene in age (Drachev et al., 1995b). The two uppermost units form an uninterrupted cover while others were deposited by both northward (from the mainland) and southward (from the De Long Uplift) sediment transport.

The eastern part of the East Siberian Shelf is occupied by the Vil'kitskii Rift System comprised of several NE-SW trending rifts. It has not been studied by offshore MCS profiling but can be delineated by satellite gravity and airborne magnetic data (Figure 2; for magnetic field see Macnab, 1993; Verhoef et al., 1996). The previously delineated Vil'kitskii Trough (Vinogradov, 1984; Fujita and Cook, 1990) is one of the most prominent members of the rift system and represents a natural boundary between East Siberian and Chukchi sectors of ERAM. There are no data to infer the age of the sedimentary fill of this rift system and the nature of the basement. However, by an analogy with the East Siberian Basin, we suggest the former to be of a Cretaceous (Middle to Late Cretaceous ?) to Quaternary age and the latter to be composed of deformed complexes of the Late Mesozoic New Siberian-Chukchi Fold Belt.

The southern Makarov Basin appears gravitationally as a smooth, low area, outlined by the 2000 m isobath (Figure 2). It does not reveal any clear structural zonality and is probably underlain by a thinned continental crust covered with a thick (8-9 km) sedimentary cover as it shown by seismic refraction and MCS data (Gramberg et al., 1993; Sekretov, 1993a). The southeastern wedge-shaped end of the basin separates the continental margin and Arlis Plateau (southern part of the Mendeleev Ridge). Further to the southeast, there is a prominent linear gravity low in the area of the continental slope. We consider this feature to be an expression of an extensional structural feature that we call the Toll' Rift (Figure 2). This rift is structurally bounded by the Arlis Plateau on the NE and De Long Uplift on the SW and merges the Vil'kitskii Rift System on the shelf.

Chukchi Sector (Northwestern Chukchi Sea and Chukchi Borderland)

The Chukchi Sector spans the eastern part of the ERAM and adjacent territories of the Alaskan continental margin, surrounded on the north by high-standing submarine blocks of the Amerasia Basin: Arlis Plateau and Chukchi Borderland (Figures 1 and 2). The principal source of the present-day geological understanding of this area is the ice- and aircraft based geophysical measurements carried out during many years by U.S.S.R., US and Canadian

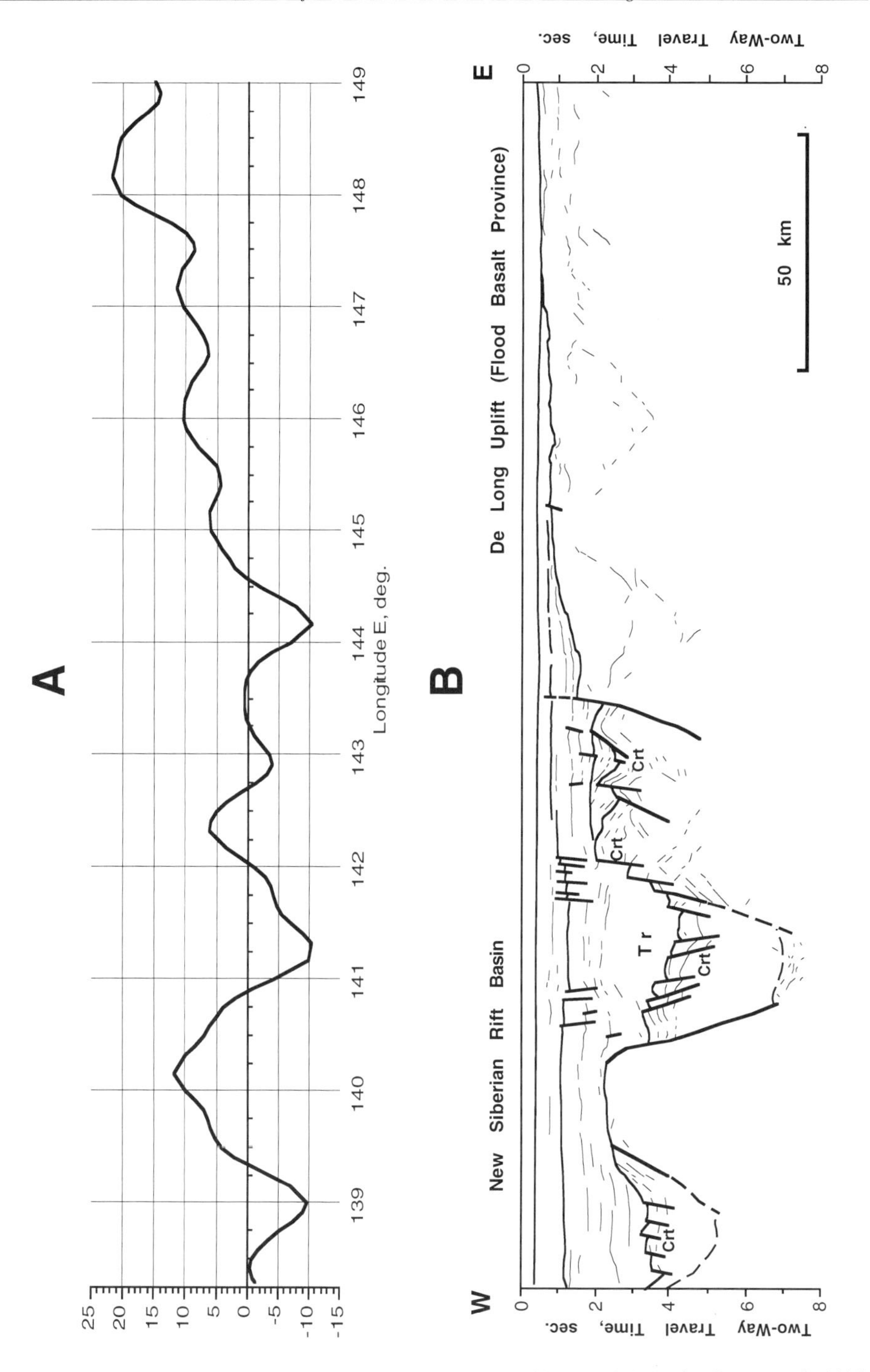

Figure 4: Free air gravity chart (A) and line drawing of the MCS profile BGR93-012 (after Roeser et al., 1995) (B) across the New Siberian Rift (see Figure 2 for location). Tr - Tertiary seismic unit, Crt - Cretaceous seismic unit.

institutes and published MCS data from the northwestern part of the Chukchi Shelf, gained by US Geological Survey (Grantz et al., 1986 and 1990). The surface observations of the gravity field were compiled by Sobczak et al. (1990), and the latest tectonic synthesizes were published by Hall (1990) and Grantz et al. (1990). The satellite gravity data shed more light on structure of the region.

The sector has the clear structural boundaries in the northwestern Chukchi Sea. These are the Vil'kitskii Rift System on the west and eastern limit of the Listric-normal faults province on the east. The latter was delineated by Grantz et al. (1990) and is located along the apparent shelf continuation of the Northwind Escarpment. The Wrangel Block, North Chukchi rift basin and North Chukchi Uplift represent the main structural elements of the shelf area which are well-expressed in the gravity field (Figure 2).

The Wrangel Block represents an uplift (Wrangel Arch by Grantz et al., 1990) of the folded basement. Recently it was studied on the Wrangel Island by Russian and Canadian geologists (Kos'ko et al., 1993) who reported the island to be composed of pentratively deformed complexes of Upper Proterozoic metamorphic rocks, Paleozoic (Late Silurian to Permian) and Triassic clastic, carbonate and flysch strata, overlain by undeformed of tens of meters thick Tertiary and Quaternary sediments. The main thrust-and-fold deformations are suggested to have occurred during a Middle Jurassic to Early Cretaceous orogeny. The gravity data suggest the Wrangel Arch extends northerly and northwesterly of the Wrangel Island up to 73° N.

The 500 km long and 200 km wide North Chukchi rift basin lies north of Wrangel Arch and represents one of the main negative structural features of the ERAM. It extends nearly W-E from 180° E toward Barrow Arch and Chukchi Platform of the Alaskan continental margin (Grantz et al., 1990). The MCS data show this basin is filled with a very thick, as much as 12-14 km, sequence of marine and shelf marine strata, interpreted to be of probable Cretaceous to Quaternary age (Figure 6B). However, the U.S.G.S. profiles are located on the southern slope of the basin just southwest of its most subsided part, marked by a gravity low of -40 to -60 mGal around 73° N, 170° W (Figure 6A). This suggests that the sediment thickness is even higher in the basin depocenter and supports the hypothesis of Grantz et al. (1990) that an oceanic or very highly thinned continental crust is present in the base of the North Chukchi Basin. The northern flank of the basin is marked by a steep gravity gradient. This feature was noted by Sobczak et al. (1990) and probably reflects some asymmetry in basin's structure with a gently sloping southern side and a steep, fault related (?), northern side.

The eastern part of the basin is affected by an array of northerly-striking listric normal faults, predicted to be of latest Cretaceous or earliest Tertiary age (Grantz et al., 1990). This Listric-normal fault province is well-expressed in the gravity data and obviously represents a shelf continuation of the extension-related structures of the Chukchi Borderland.

The North Chukchi Uplift occurs in the outer northwestern Chukchi Shelf (Figure 2). It is marked by one of the highest gravity peaks, which is as much as 50 mGal in intensity, and separates structurally the North Chukchi Basin and western part of the Chukchi Borderland (Chukchi Cap). This structure has not previously been described and there are no data to infer its geology. We only can suggest some similarity between this feature and Chukchi Cap (Plateau) since both structures display a similar gravity pattern.

The Chukchi Borderland has the most spectacular expression in the gravity field (Figure 2). Morphologically it represents an ensemble of the N-S trending narrow ridges and basins, divided by the steep gravity gradients and marked by gravity highs (40-50 mGal) and lows (-40 to -60 mGal), correspondingly. Such a characteristic pattern has previously been established and has been said to reflect rift-related structure of the borderland (Hall, 1990).

There are two main crustal blocks, which differ significantly in their gravity field. The more solid western block is represented by the Chukchi Plateau, whose top part consists of a 25 km wide N-S trending graben, delineated previously (Hall and Hunkins, 1968; Hall, 1990). The

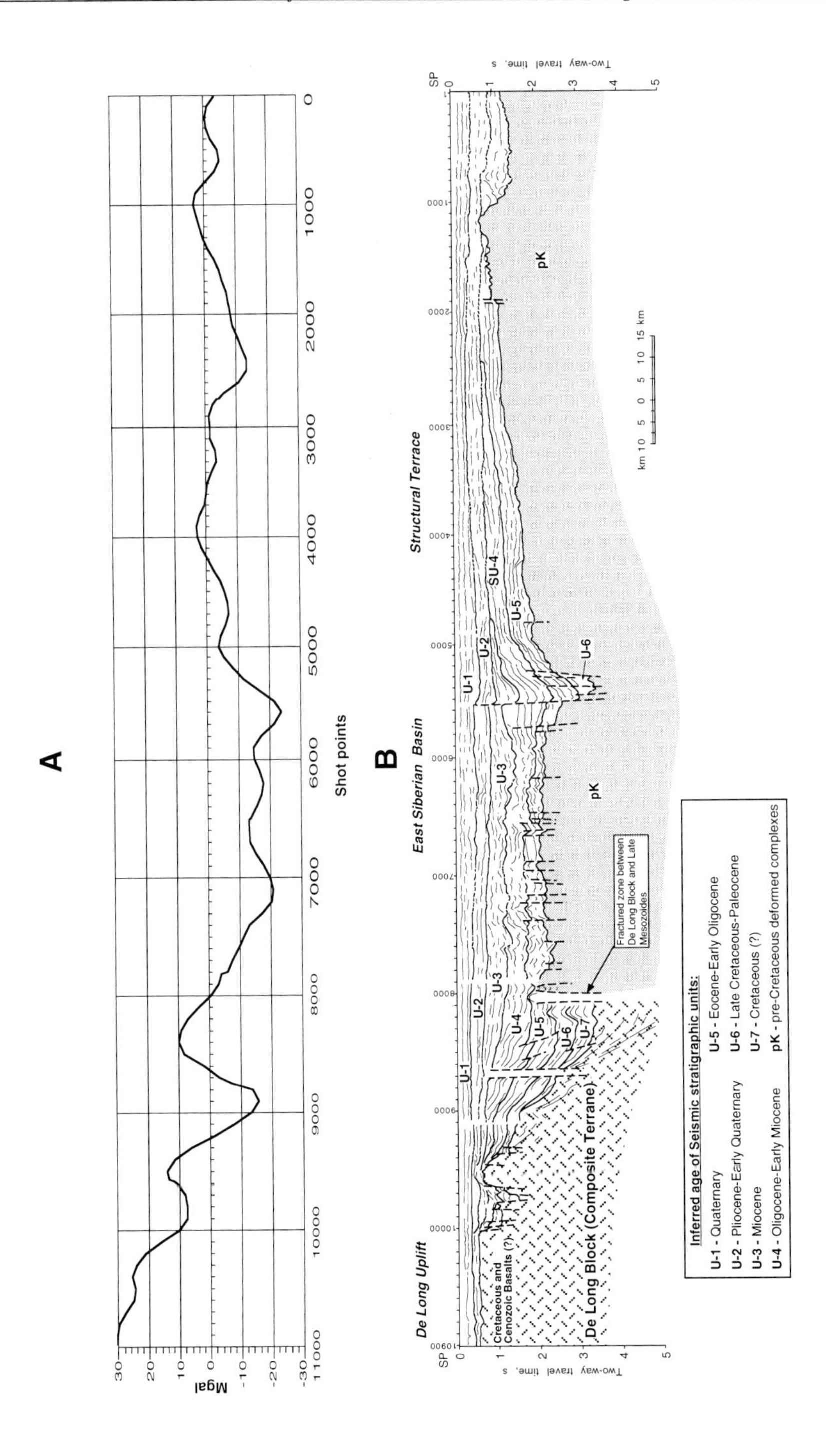

Figure 5: Free air gravity chart (A) and line drawing of the MCS profile LARGE 89001 (B) showing inferred geologic structure and stratigraphy across East Siberian Shelf (see Figure 2 for location).

eastern block reveals more complex ridge-to-basin structure and is composed of the Northwind Basin and Northwind Ridge. A boundary between the blocks is marked by a chain of narrow NE-SW and N-S trending echelon-like troughs, coinciding with the Nautilus Gap on the north. The Northwind Basin contains a central horst-like block, surrounded by several narrow steep-sided grabens. The eastward facing Northwind escarpment, marked by the sharpest gravity gradient, represents the boundary between the Chukchi Borderland and Canada abyssal basin.

It is now broadly accepted that the Chukchi Borderland is underlain by thinned continental crust, covered with the sediments of an uncertain age of less than 2 km thick on the uplifts and more than 4 km thick within the troughs (Hall, 1990). As it was shown by Grantz et al. (1993) the Northwind Ridge is underlain by continental crust from a Cretaceous continental margin or shelf basin.

The Mendeleev Ridge and Arlis Plateau appear in the gravity field as an unified block within the 2000 m isobath. The gravity pattern suggests some similarity with De Long Uplift, while both blocks are separated by the Toll‘ Rift.

The Chukchi Borderland and the Mendeleev Ridge are morphologically separated by Charlie Gap, that links the Mendeleev and Chukchi abyssal plains. This feature is shown in the gravity field by a linear minimum, that disintegrates into two orthogonal branches on the north and merges on the south with sharp gravity low about -50 mGal over the Chukchi Abyssal Plain. To compare this gravity pattern over the Charlie Gap with the similar ones over the other MCS studied structures of the ERAM we suggest it is a manifestation of an extension structure called Charlie Rift. A merger of this rift and the Toll‘ Rift in area of the Chukchi abyssal plain appears to have led to significant crustal thinning and formation over this junction a deep subsided Chukchi sedimentary basin. The latter, based on gravity data, might be filled with a sedimentary succession not less than 8 to 9 km in thickness.

Summary

New geophysical, particularly MCS and satellite marine gravity data collected during the last decade over the broad area of the ERAM provide a good basis for revision of the previous hypotheses on the structure and tectonic genesis of the region. The data presented in this paper allow us to suggest that the continental margin was strongly affected by the rifting process which produced and is still creating rift systems and rift-related deeply subsided sedimentary basins which are: the presently active Laptev Rift System, the inactive New Siberian Rift and Vil‘kitskii Rift System in the East Siberian Sea, North Chukchi rift basin in the Chukchi Sea and several rifts and rift basins in the area of the continental slope and adjacent marginal plateaus (Toll‘ and Charlie rifts, Nautilus Trough, Chukchi and Northwind rift basins). The older rifts and rift basins which are now abandoned may have commenced activity as early as end of Jurassic to beginning of Cretaceous. The first episode of the oceanic related rifting probably took place in the Chukchi Sector in response to initial breakup between North American plate and North Alaska-Chukchi Block (Grantz et al., 1990; Embry and Dixon, 1994). This event resulted in formation of the North Chukchi Basin and movement of the Chukchi Borderland out of the Chukchi continental margin, as was suggested by Grantz et al. (1990). The East Siberia Basin may be a basin downwarped by the sedimentary load or an analogue of the Sverdrup Basin in northern Canada (Sweeney et al., 1990).

Next, in the Late Cretaceous to Early Tertiary time, the large extensional basins were formed in East Siberian Sector. We speculate this event to be connected with a crustal dilatation/sea-floor spreading in the Makarov Basin. It was also suggested by Grantz et al. (1990) that this "Laramide" (latest Cretaceous-earliest Tertiary) event created a system of the N-S trending ridges and basins of the Chukchi Borderland and province of the listric-normal faults in the

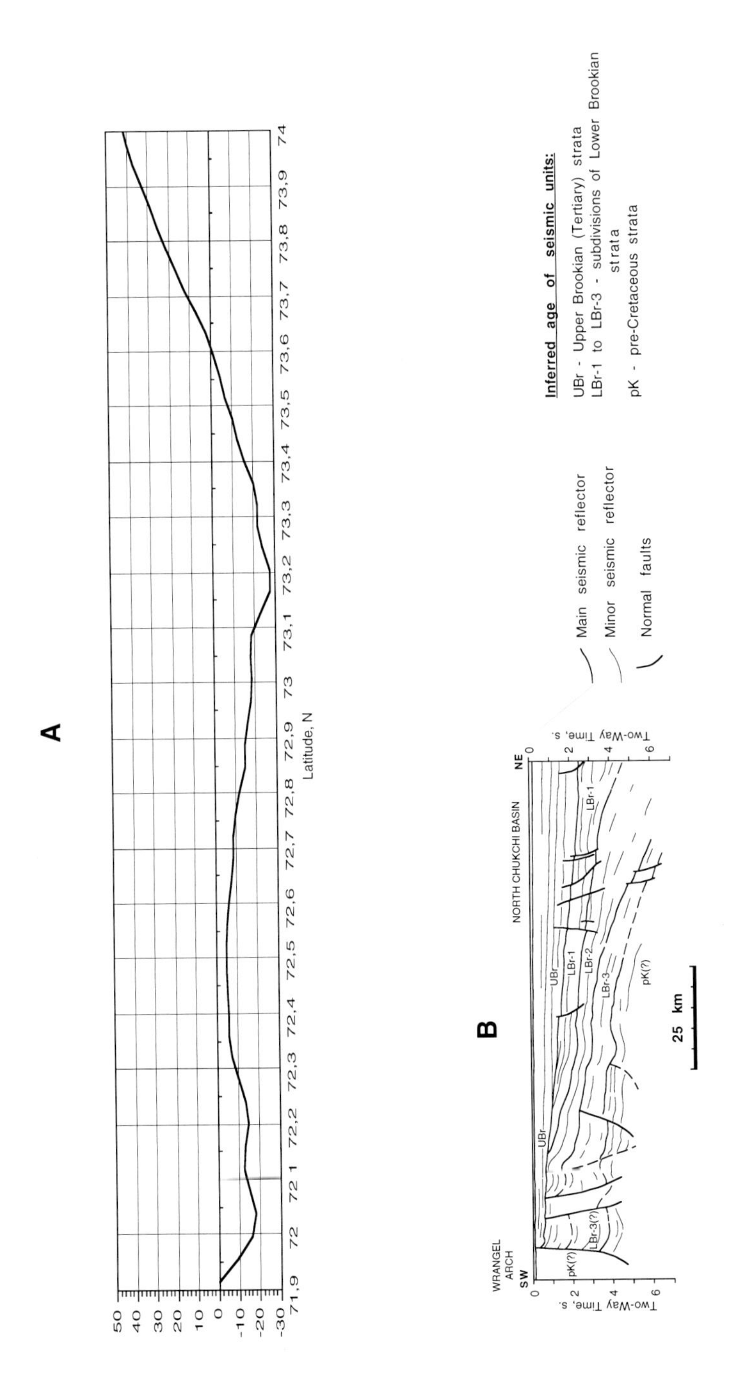

Figure 6: Free air gravity chart (A) and line drawing of the MCS profile U.S.G.S. 816 (from Grantz et al., 1990) (B) showing inferred geologic structure and stratigraphy across North Chukchi Basin (see Figure 2 for location).

eastern North Chukchi Basin. The main extension axis might be a continuation from the Mid-Atlantic Ridge via Baffin Bay and Makarov Basin toward the Mendeleev Ridge and Chukchi Borderland. We also speculate that in that time the southern part of the Mendeleev Ridge was rifted away from the De Long Block and the Toll' Rift, Chukchi Rift Basin and Charlie Rift were formed.

Finally, in the Tertiary to Quaternary time the Laptev Rift System was formed in response to the opening of the Eurasia Basin. This rifting started as early as 60-56 Ma or even earlier, and the main rifts were formed by the end of the Eocene since that period is characterized by the highest spreading rates (Savostin et al., 1984a). Present-day extension still occurs along Bel'kov Svyatoi Nos Rift but, probably, at a very low rate.

Thus we suggest that the spreading in the Arctic oceanic basins migrated westward during Cretaceous-Cenozoic time and led to rifting and subsidence of the sedimentary rift basins on the continental side as well as to separating and fragmentation of the marginal continental blocks: the Chukchi Borderland, the Mendeleev and Lomonosov ridges.

Acknowledgments

This paper was prepared during the visits of S. Drachev at GEOMAR Research Center for Marine Geosciences in the Kiel, Germany, in 1996 and 1997. We are grateful to Prof. J. Thiede and Dr. H. Kassens who generously provided this opportunity. This work was partly supported by the International Science Foundation (grant number M5U000), the German Ministry of Education, Science, Research and Technology, (grant number 525 4003 03 G 0517 A) and the Executive Committee of the Nansen Arctic Drilling Program. G. L. Johnson was supported by the Office of Naval Research.

References

Avetisov, G.P. (1993) Some questions of dynamic of the lithosphere of the Laptev Sea. Fisika Zemli, 5, 28-38 (in Russian).

Cook, D.B., K. Fujita and C.A. McMullen (1986) Present day plate interaction in northeast Asia; North American, Eurasian and Okhotsk plates. Journal of the Geodynamics, 6, 33-51.

Drachev, S.S. and L.A. Savostin (1993) Ophiolites of the Bol'shoi Lyakhov Island (New Siberian Islands). Geotektonika, 6, 33-51 (in Russian).

Drachev, S.S., L.A. Savostin and I.E. Bruni (1995a) Structural pattern and tectonic history of the Laptev Sea region. In: Kassens et al. (Editors), Reports on Polar Research, 175. Alfred Wegener Institute for Polar and Marine Research, Bremerhaven, Germany, 348-366.

Drachev, S.S., L.A. Savostin, A.V. Elistratov and I.E. Bruni (1995b) New multichannel seismic results from East Siberian Sea: Indigirka Bay to Jeanette Island profile. In: 5th Zonenshain Conference on Plate Tectonics, Moscow, November 1995, Programme and Abstracts. Institute of Oceanology, Moscow, GEOMAR Research Center, 42-43.

Embry, A.F. and J. Dixon (1994) The age of the Amerasia Basin. In: D.K. Thurston and K. Fujita (eds.), Proceedings of the 1992 International Conference on Arctic Margins. U.S. Dept. of the Interior Minerals Management Service, OCS Study, 289-294.

Fujita, K. and J.T. Newberry (1982) Tectonic evolution of northeastern Siberia and adjacent regions: Tectonophysics, 89, 337-357.

Fujita, K., F.W. Cambray and M.A. Velbel (1990) Tectonics of the Laptev Sea and the Moma rift systems, northeastern USSR. In: A.F. Weber, D.A. Forsyth, A.F. Embry and S.M. Blanco (eds.), Arctic Geoscience. Mar. Geol., 93, 95-118.

Fujita, K. and D. Cook (1990) The Arctic continental margin of eastern Siberia. In: A. Grantz, L. Johnson and J.F. Sweeney (eds.), The Geology of North America, vol. L, The Arctic Ocean Region. Geol. Soc. of Amer., Boulder, Co., 257-288.

Grachev, A.F. (1982) Geodynamics of the transitional zone from the Moma rift to the Gakkel ridge. In: J.S. Watkins and C.L. Drake (eds.), Studies in Continental Margin Geology. Am. Assoc. Pet. Geol. Mem., 34, 103-113.

Gramberg, I.S., V.V. Verba, G.A. Kudryavtzev, M.Yu. Sorokin and L.J. Kharitonova (1993) Crustal structure of the Arctic Ocean along the De Long Islands-Makarov Basin Transect. Reports of the Russian National Academy of Sciences, 328/4, 484-486 (in Russian).

Grantz, A., D.M. Mann and S.D. May (1986) Multichannel seismic-reflection data collected in 1978 in the eastern Chukchi Sea. U.S. Geological Survey Open-File Report 86-206, 3 p.

Grantz, A., S.D. May and P.E. Hart (1990) Geology of the Arctic Continental Margin of Alaska. In: A. Grantz, G.L. Johnson and J.F. Sweeney (eds.), The Geology of North America, vol. L, The Arctic Ocean Region. Geol. Soc. of Amer., Boulder, Co., 257-288.

Grantz, A. and shipboard party (1993) Cruise to the Chukchi Borderland, Arctic Ocean. Eos, Trans., AGU, 74, 250.

Hall, J.K. (1990) Chukchi Borderland. In: A. Grantz, G.L. Johnson and J.F. Sweeney (eds.), The Geology of North America, vol. L, The Arctic Ocean Region. Geol. Soc. of Amer., Boulder, Co., 337-350.

Hall, J.K. and K.L. Hunkins (1968) A geophysical profile across the southern half of the Chukchi Rise, Arctic Ocean. EOS Transactions of the American Geophysical Union, 49/1, 207.

Ivanova, N.M., S.B. Sekretov and S.N. Shkarubo (1990) Geologic structure of the Laptev Sea Shelf according to seismic studies. Oceanology, 29, 600-604.

Karasik, A.M. (1968) Magnetic anomalies of the Gakkel Ridge and the origin of the Eurasia Subbasin of the Arctic Ocean. Geofizicheskie metody razvedki v Arktike, 5, 8-19 (in Russian).

Karasik, A.M., L.A. Savostin and L.P. Zonenshain (1983) Parameters of the lithospheric plate movements within Eurasia Basin of the North Polar Ocean. Doklady Akademii Nauk SSSR, Earth Sci. Sect. 273, 1191-1196 (in Russian).

Kos'ko, M.K. (1984) East Siberian Sea. In: I. S. Gramberg and Y. E. Pogrebitsky (Editors), Geologic Structure of the USSR and its Relationship to the Distribution of Mineral resources, Vol. 9, Seas of the Soviet Arctic. Nedra, Leningrad, 51-60 (in Russian).

Kos'ko, M.K., B.G. Lopatin and V.G. Ganelin (1990) Major geological features of the islands of the East Siberian and Chukchi seas and the northern coast of Chukotka. Mar. Geol., 93, 349-367.

Kos'ko, M.K., M.P. Cecile, J.C. Harrison, V.G. Ganelin, N.V. Khandoshko and B.G. Lopatin (1993) Geology of Wrangel Island between Chukchi and East Siberian Seas, Northeastern Russia. Canad. Geol. Surv. Bull., 461, 101 p.p.

Kovacs, L.C. and five others (1983) Residual magnetic anomaly chart of the Arctic Ocean region. U. S. Naval Research Laboratory and Naval Ocean Research and Development Activity, scale 1:6,000,000.

Laxon, S.W. (1994) Sea ice altimeter processing scheme at the EODC. Int. Journal of Remote Sensing, 15/4, 915-924.

Laxon, S.W. and D.C. McAdoo (1994) Arctic Ocean Gravity Field Derived from ERS-1 Satellite Altimetry. Science, 265, 621-624.

Laxon, S.W. and D.C. McAdoo (1998) Satellites provide new insights into Polar geophysics. Eos, Trans., AGU, 79, 69, 72-73.

Macnab, R. (1993) Russia and the Arctic Ocean, Magnetic Field and Tectonic Structures, Provisional diagram based on preliminary data and interpretations. Atlantic Geoscience Centre chart.

McAdoo, D.C. and K.M. Marks (1992) Gravity Fields of the Southern Ocean From Geosat Data. J. Geophys. Res., 97, (B3), 3247-3260.

Parfenov, L.M. and B.A. Natal'in (1986) Mesozoic tectonic evolution of northeastern Asia. Tectonophysics, 127, 291-304.

Roeser, H.A., M. Block, K. Hinz and C. Reichert (1995) Marine Geophysical Investigations in the Laptev Sea and the Western part of the East Siberian Sea. In: Kassens et al. (eds.), Reports on Polar Research, 176. Alfred Wegener Institute for Polar and Marine Research, Bremerhaven, Germany, 367-377.

Savostin, L.A., A.M. Karasik and L.P. Zonenshain (1984a) The history of the opening of the Eurasia basin in the Arctic. Trans. USSR Acad. Sci. Earth Sci. Sect., 275, 79-83.

Savostin, L.A., L.M. Natapov and A.P. Stavsky (1984b) Mesozoic paleogeodynamics and paleogeography of the Arctic region. In: Arctic geology. Reports. Vol. 4. 27th International Geological Congress, Moscow. Nauka, Moskva, 217-237.

Sekretov, S.B (1993a) The continental margin north of East Siberian Sea: some geological results and conclusions on the base of CDP seismic-reflection data. In: 4th Zonenshain Conference on Plate Tectonics, Moscow, November 1993. Programme and Abstracts. Institute of Oceanology, Moscow, GEOMAR Research Center, Kiel, Germany, 124.

Sekretov, S.B (1993b) Junction of Gakkel Oceanic Ridge and Laptev Sea continental margin: General features of tectonics according to the data of multichannel seismic profiling. In: 4th Zonenshain Conference on Plate Tectonics, Moscow, November 1993, Programme and Abstracts. Institute of Oceanology, Moscow, GEOMAR Research Center, Kiel, Germany, 124-125.

Sobczak, L.W., D.B. Hearty, R. Forsberg, Y. Kristoffersen, O. Eldholm and S.D. May (1990) Gravity from 64° N to the North Pole. In: A. Grantz, G.L. Johnson and J.F. Sweeney (eds.), The Geology of North America, vol. L, The Arctic Ocean Region. Geol. Soc. of Amer., Boulder, Co., 101-118.

Sweeney, J.F., L.W. Sobczak and D.A. Forsyth (1990) The continental margin northwest of the Queen Elizabeth Islands. In: A. Grantz, G.L. Johnson and J.F. Sweeney (eds.), The Geology of North America, vol.

L, The Arctic Ocean Region. Geol. Soc. of Amer., Boulder, Co., 227-238.
Verhoef, J., R.R. Walter, R. Macnab, J. Arkani-Hamed and Members of the Project Team (1996) Magnetic anomalies of the Arctic and North Atlantic oceans and adjacent land areas. Geological Survey of Canada. Open File 3125a.
Vinogradov, V.A. (1984) Laptev Sea. In: I. S. Gramberg and Y. E. Pogrebitsky (Editors), Geologic Structure of the USSR and its Relationship to the Distribution of Mineral resources, Vol. 9, Seas of the Soviet Arctic. Nedra, Leningrad, 51-60 (in Russian).
Vinogradov, V.A., G.I. Gaponenko I.S. Gramberg and V.N. Shimaraev (1977) Structural-associational complexes of the Arctic shelf of eastern Siberia. International Geology Review, 19, 1331-1343.

High Resolution Seismic Studies in the Laptev Sea Shelf: First Results and Future Needs

B. Kim, G. Grikurov and V. Soloviev

All-Russia Research Institute for Geology and Mineral Resources of the World Ocean, 1 Angliisky Ave., 190121 St. Petersburg, Russia

Received 8 July 1997 and accepted in revised form 20 May 1998

Abstract - High resolution seismic (HRS) data were obtained in the central and eastern Laptev Sea Shelf (LSS) along several widely spaced regional profiles. Analysis of seismic records reveals four regional subbottom reflectors which generally correspond to major erosional unconformities and related stratigraphic breaks in the uppermost Cenozoic (Pliocene to Holocene) sequences averaging 100-130 m, locally up to 250 m in total thickness. The stratification proposed for this depth interval is based on correlation of HRS data with onshore geological evidence and includes four units of Pliocene, EoPleistocene to Middle Pleistocene, Upper Pleistocene, and Holocene age. HRS evidence were further interpreted in terms of permafrost distribution and cryogeothermal history of the LSS. Additional HRS surveys are critical for LSS paleoenvironmental reconstructions and selection of Nansen Arctic Drilling Program (NADP) sites.

Introduction

HRS data were first obtained in the LSS in 1986 during offshore geophysical survey conducted by the Murmansk Marine Geological Expedition (MAGE). The survey included regional multichannel seismic reflection (MCS) profiling combined with gravity and magnetic measurements and HRS profiling along some of MCS lines (Figure 1). These data, as well as MCS evidence obtained in the same area by the Laboratory of Regional Geodynamics (LARGE), formed the basis for several published interpretations of the Cenozoic stratigraphy of the LSS (Ivanova et al., 1989; Kim, 1994; Drachev, 1994; Drachev et al., 1995). More recently R/V „Polarstern" acquired PARASOUND records north of 76° but as yet only geomorphological interpretation of this new data has been published (Niessen and Musatov, 1997).

Interpretation by Drachev et al. (1995) was mainly based on MCS data and implied predominantly Pliocene-Quaternary age of the upper seismic sequences recognized in LARGE's MCS profiles. Consequently, Drachev attributed a very significant thickness (400 m and more) to the Pleistocene-Holocene deposits. Other authors derived their evidence predominantly from correlation of MAGE´s HRS data with geological observations on the New Siberian Islands and coastal mainland (including drilling results), and their assessment of thickness of the youngest Cenozoic strata was considerably lower.

Agreed interpretation of available HRS records becomes increasingly important in view of development of NADP planning which requires improved understanding of the uppermost Cenozoic stratigraphy of the LSS (Implementation plan..., 1997). In this paper we propose a seismic stratigraphic concept that may, in our opinion, serve as a starting point for future correlation and help to stimulate additional HRS seismic surveys needed for selection of NADP drill sites.

In: Kassens, H., H.A. Bauch, I. Dmitrenko, H. Eicken, H.-W. Hubberten, M. Melles, J. Thiede and L. Timokhov (eds.) Land-Ocean Systems in the Siberian Arctic: Dynamics and History. Springer-Verlag, Berlin, 1999, 683-692.

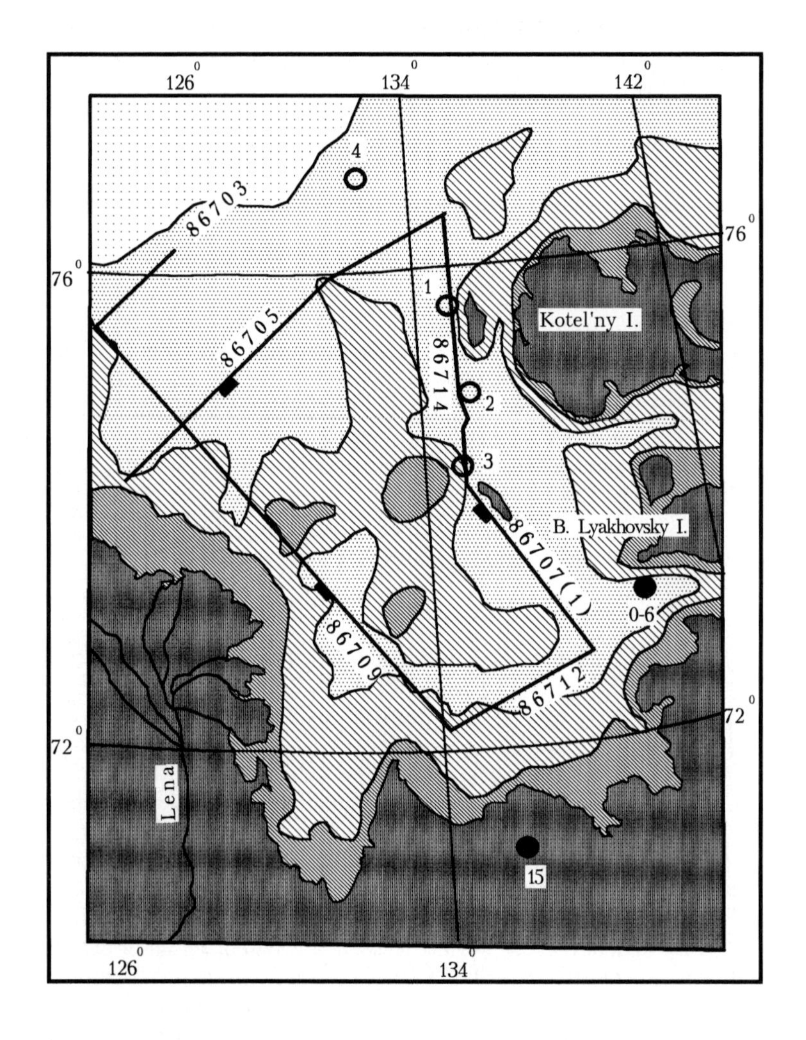

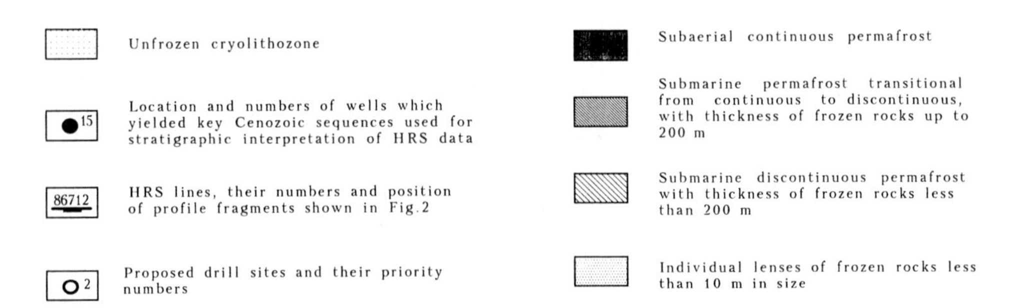

Figure 1: Zonation of permafrost in the Laptev sea Shelf inferred from cryogeothermal modeling (modified from Soloviev et al., 1987).

Data base for the proposed seismic stratigraphy

Single-channel HRS data were acquired simultaneously with CDP seismic regional profiling along some of MCS lines in the south-eastern part of the LSS. The acquisition was performed in analog form aboard R/V „Geolog Dmitry Nalivkin" cruising at a speed 4.5-4.8 knots. Observation interval was 17 m. A „Sparker" wave source with 10.5 K-joules power capability was towed at 0.5 m depth, and a standard streamer 48NS301 was applied as receiving device. Depth resolution of the survey was within 3 m, with total depth penetration not exceeding 200-250 ms TWT.

For interpretation of obtained HRS data, a seismic stratigraphic concept was implied based on the assumption that seismic reflectors recognized in the HRS sections represented offshore continuations of regional disconformities which had either been known from bedrock outcrops on coastal mainland and in the New Siberian Islands, or penetrated by shallow wells drilled both on land and from the sea ice (Kim, 1992). The key sections for stratigraphic correlation were chosen from two sites nearest to the southernmost extent of HRS lines: the Cenozoic section documented in well # 0-6 immediately south of the western promontory of the Bol'shoi Lyakhovsky Island, and another section drilled in well # 15 in the south-eastern coastal rim of the Laptev Sea (Figure 1).

Interpretation of the uppermost sub-bottom deposits was additionally verified by abundant gravity coring data. The latter throughout much of the LSS allowed to assess only a minimal thickness of the Holocene veneer, but several cores reached the top of the Pleistocene, thus providing the information about total thickness of the Holocene sediments.

The results of detailed paleogeographic analysis performed in VNIIOkeangeologia for the whole of the Russian East Arctic shelf were also considered. With few exceptions (e.g. Alekseev et al., 1992; Kim and Slobodin, 1991; Kim, 1992), these results remain largely unpublished and are available for analysis only as 1:5,000,000 maps (authors' drafts) showing paleoenvironmental reconstructions for several crucial age levels in the Cenozoic history.

Seismic stratigraphy

Four regional reflectors were recognized in HRS profiles (Table 1):

The lowest *Reflector 1* is clearly imaged in HRS records only on profiles 86707(1) (Figure 2a) and 86712. The horizon has a subdued relief with amplitudes not exceeding 35-40 ms TWT.

Reflector 2 has been recognized on HRS sections of lines 86707/1/ and 86705 (Figure 2a, 2b), as well as on lines 86703 and 86712. It is marked by a rugged relief most likely indicating an erosional nature of this seismic boundary.

Reflector 3 on lines 86707/1/ and 86705 (Figure 2a, 2b) appears to form the top of a sedimentary sequence which was accumulated in depressions of the erosional surface marked by Reflector 2. Reflector 3 is characterized by a relatively smooth relief and almost permanent depth at about 80 ms TWT which on lines 86707/1/ and 86705 is equivalent to the level of highest elevations of Reflector 2. On line 86709 (Figure 2c) identification of Reflector 3 is less confident due to proximity of the first sea-floor multiple and lack of evidence of underlying boundaries, and its recognition here is based mostly on analogy with other lines suggesting persistent presence at this depth of a real seismic horizon.

Reflector 4 is the uppermost regional seismic marker horizon recognized along all profiles at constant depths close to 60 ms TWT. It is represented by a flat, almost strictly horizontal and very clearly defined base of an acoustically homogenous sequence bounded at the top by the sea bottom.

Four seismic sequences separated by reflection horizons discussed above are labeled (in ascending order) units A, B, C, and D.

Unit A was identified along the southernmost seismic profiles showing notable variations in its thickness in the range between 50 and 140 m (see Table 1). It is characterized by an acoustically nearly transparent pattern lacking distinct coherent reflections.

In *Unit B* variations in thickness seem to be generally less conspicuous as compared to Unit A, except on profile 86712, but this may be an exception (see Table 1). The seismic pattern is consistent with interpretation of Unit B as weakly layered sequence without prominent reflectors.

Unit C exhibits, as a rule, the most persistent thickness whose maximum values are comparable with average thickness of underlying Unit B (see Table 1). It is well layered and shows several continuous internal seismic reflectors.

Unit D constitutes an easily discernible thin veneer on all HRS profiles. Average thickness of this unit is in the order of 4 to 7 m which is consistent with evidence obtained by coring for the Holocene sediments. Larger values are rare, while the lowest recorded thickness or even total absence of the Holocene sequence are usually associated with bathymetric highs. Direct sampling data indicate that Unit D is largely composed of marine sands, silts and muds.

Table 1: Generalized seismic stratigraphy from HRS data

Seismic reflectors and units	Numbers of seismic acoustic profiles and observed thickness of seismic units (m)					
	86703	86705	86709	86707(1)	86714	86712
D_4	4- 8	4-10	6-7	7-15	5-12	2-4
C_3	10-25	15-20	25	18-30	14-15	8-25
B_2	17-34	20-45	50	25-42	>35	50-55 (up to 80)
A_1				70-80 (up to 140)		50-70

Discussion

Interpretation of HRS data

Position of Reflector 1 on HRS profiles is in good agreement with boundary "L" recognized on MCS time sections as an erosional unconformity at the top of the Upper Miocene (Ivanova et al., 1989). This unconformity was observed in key sections documented in wells ## 15 and 0-6 and chosen as the basis for correlation of HRS data (Figure 3). In well # 15 a stratigraphic discontinuity in a lithologically uniform sequence was established by microfossil data, whereas in well # 0-6 it was also accompanied by a distinct lithological boundary. In both wells the Pliocene deposits, which are believed to constitute Unit A, are predominantly composed of lacustrine and near-coast marine sands and silts with intercalations of gravels and grits. Paleogeographic reconstructions suggest that farther to the north composition of Unit A may become more influenced by marine environment with domination of silty facies over sands and clays.

Configuration of Reflector 1 seems to depend to some extent on the position of graben structures established by MCS data in the underlying sedimentary sequences and the basement. This may suggest that during the time of accumulation of Unit A the depositional environment was still partly affected by tectonic movements. This, in turn, gives reason to believe that

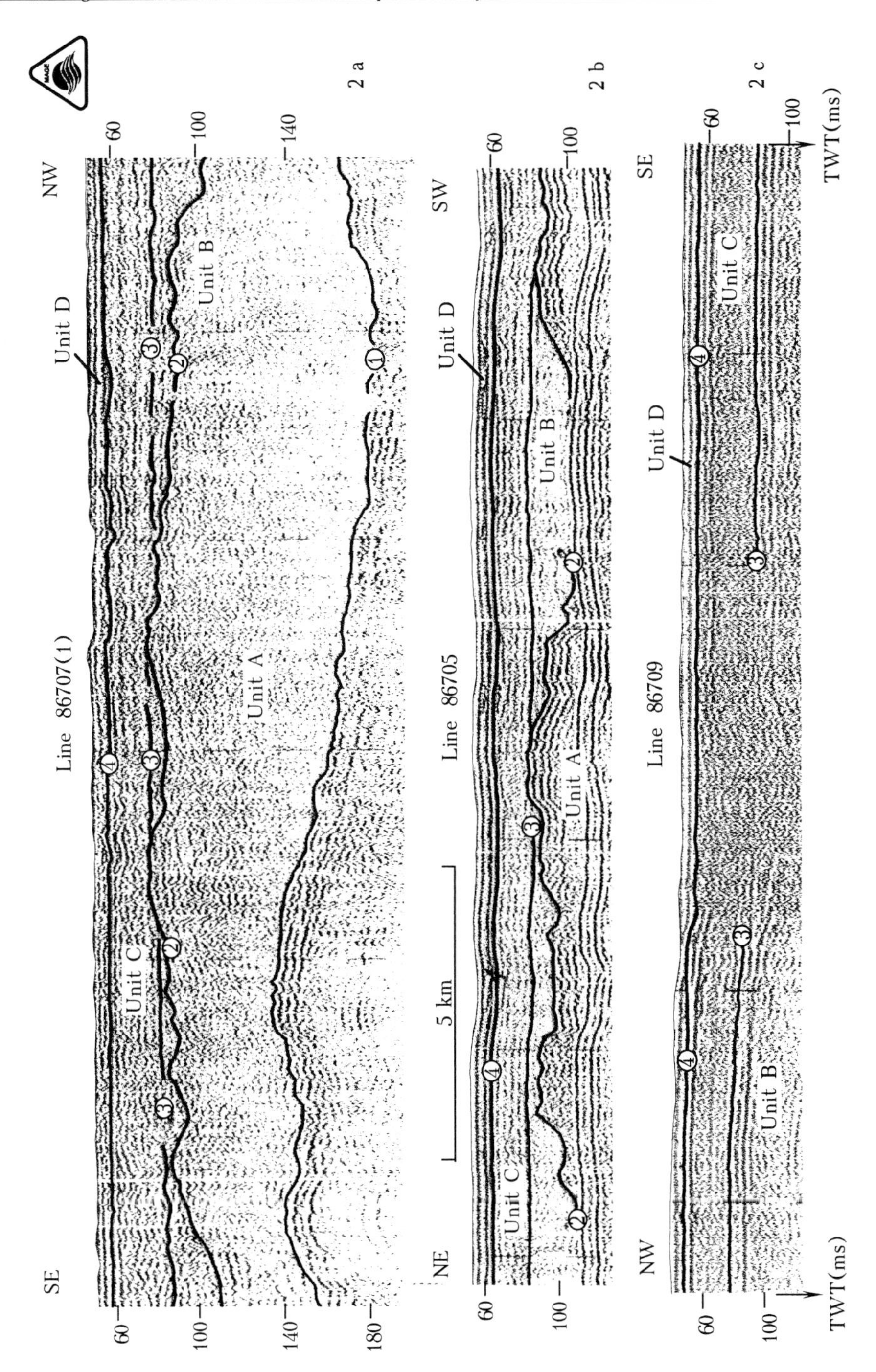

Figure 2: Fragments of interpreted HRS time sections (by courtesy of MAGE). See Figure 1 for location of HRS lines, and text for explanations. Numbers of reflectors correspond to those in Table and column III in Figure 3.

further to the north, where the role of rifting in Cenozoic history of the LSS has probably been more pronounced, one can expect even stronger structural control of distribution of Unit A and its much greater thickness in structural lows, despite anticipated increase in the same direction of influence of marine sedimentation. The existence of tectonically controlled sharp gradients in Cenozoic sections is illustrated in Figure 3 by much greater thickness of sediments in well # 15 which is located within a major graben.

Reflector 2 is correlated with the unconformity at the top of the Upper Pliocene section which is well exposed in coastal and island outcrops and also documented in well sections presented in Figure 3 where a transgressive nature of this unconformity is confirmed by transition of predominantly continental Pliocene facies to essentially marine environment during Eopleistocene-Middle Pleistocene interval. A rugged relief of Reflector 2 suggests that transgression was accompanied by differentiated erosion of Unit A. In coastal outcrops and boreholes the change of facies across the Pliocene/Pleistocene time boundary is mainly recorded in microfossil data, whereas offshore it is likely to be also reflected in a higher proportion in Unit B of marine silts and clays as compared to more coarse material in underlying Unit A.

Reflector 3 is correlated with the base of the Upper Pleistocene sequence. A flat relief of this seismic interface may indicate the absence of significant erosion during its formation. Such conclusion is confirmed by stratigraphic record in well # 15 demonstrating the absence of a disconformity associated with this boundary. Therefore reflector 3 may to a large extent mark a change in depositional environment which has, in fact, been observed at the Middle/Upper Pleistocene boundary in onshore sections and in both wells shown in Figure 3. Lithology of the Upper Pleistocene Unit C is presumably dominated by silts, sands and silty sands with subordinate coarser sediments. It is believed that formation of Unit C was controlled by multiple sea level oscillations which occurred in a generally regressive environment. The latter led to gradual retreat of the sea and replacement of marine and shallow marine facies by lacustrine and alluvial deposits.

Despite distinct morphology of Reflector 4, interpretation of its geological nature is not simple, and direct correlation of this horizon with erosional unconformity observed in onshore outcrops and wells at the top of the Upper Pleistocene deposits is, perhaps, not quite as apparent as it may seem at the first glance.

It is commonly believed that during pre-Holocene regression the sea retreated almost to the edge of modern continental slope, and the whole shelf became exposed to a relatively short-lived subaerial freezing. During subsequent Holocene transgression, this subaerial frozen surface remained well preserved on non-flooded land, but over greater part of the shelf it was considerably modified by advance of the sea and accompanying thermal-abrasive reworking. As the result, Reflector 4 observed in HRS profiles may correspond not to the original post-Pleistocene subaerial surface but to a somewhat younger interface of a more complex nature. It may, perhaps, mark the base of a layer at the top of the uppermost Pleistocene deposits in which the latter were thawed, reworked and mixed with clastic material supplied to the LSS during the Holocene.

Evidence of permafrost distribution

Direct evidence of the existence of submarine permafrost was obtained only from numerous shallow wells (not exceeding 200 m depth) drilled close to islands coastline. Not only the presence of subbottom cryolithic zone was confirmed, but it also appeared possible to establish its subcontinuous to discontinuous distribution, as well as to determine the depth to the surface of frozen rocks and their thermal condition in the drilled interval (Soloviev et al., 1987).

Outside the area studied by drilling, the presence of submarine permafrost may, perhaps, be exemplified by an apparent break in correlation of Reflector 3 on Line 86709 (see Figure 2c).

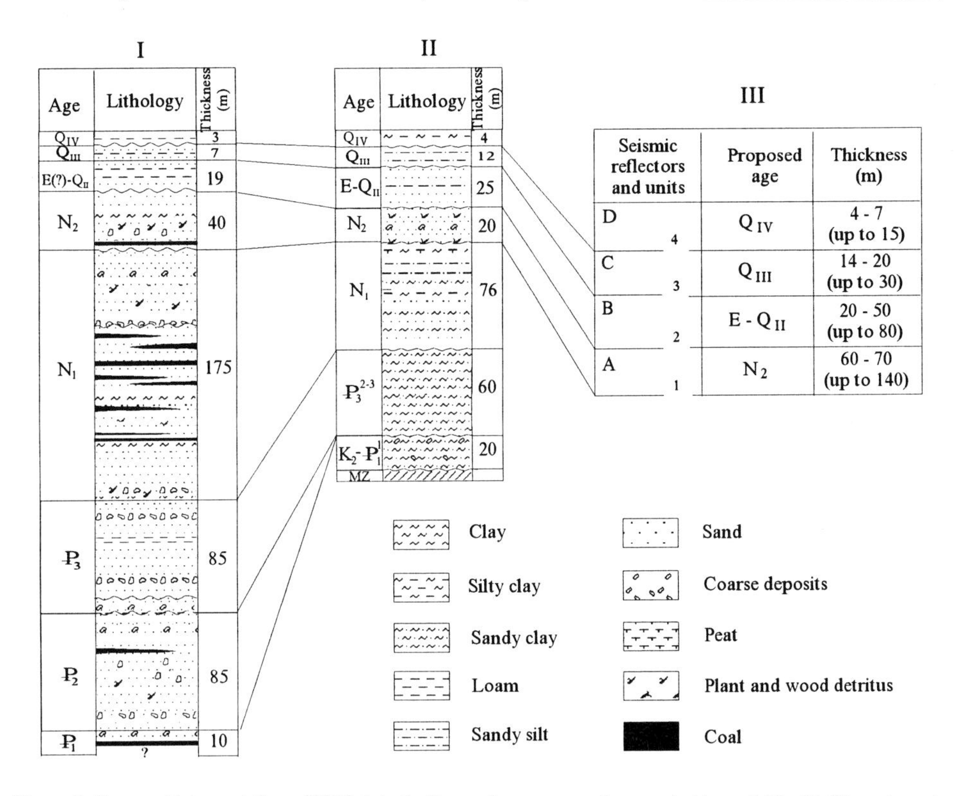

Figure 3: Proposed interpretation of HRS data. I - Cenozoic sequences documented in well No. 15 (Strepetova et al., 1981; Gusev and Grinenko, 1985); II - Full Cenozoic section penetrated in well No. 0-6 (Sudakov et al., 1991); III - Generalized seismic stratigraphy integrated from HRS data and its correlation with geological record in drill holes. For location of wells see Fig. 1. Note: Holocene sediments in columns I-II and all units in column III are not to scale.

Interpretation of this feature in terms of the presence here of discontinuous frozen rocks implies that their surface corresponds to Reflector 4, and that this surface has been disintegrated because it does not cause distinct dynamic features which would be expected in case of continuous solid permafrost. The results of cryogeothermal modeling performed for the Pleistocene epoch for the whole LSS on the basis of paleogeographic data and theoretical calculations (Soloviev, 1981; Soloviev et al., 1987) are shown in Figure 1 and consistent with the inferred presence of discontinuous permafrost at the location characterized by Figure 2c.

Alternatively, the lack of correlation observed in Figure 2c may be caused by gas emanations of either biogenic or crustal nature. Such possibility is confirmed by numerous methane shows established in bottom sediments throughout the LSS. Similar seismic records associated with gas-saturated sediments were also observed in many proven oil and gas provinces.

The model developed by Soloviev for the LSS (Soloviev, 1981; Soloviev et al., 1987) proposes the existence of the areas with different cryogeothermal history. The area subjected to the longest freezing is recognized in a relatively narrow coastal zone adjacent to the mainland and New Siberian Islands, as well as in small isolated localities east of the Lena River delta; this area is believed to have remained in subaerial conditions since the earliest Pleistocene. The second area with shorter freezing history (i.e. the land which did not emerge until the beginning of Late Pleistocene) borders the first cryolithozone and also forms a longitudinal contour in

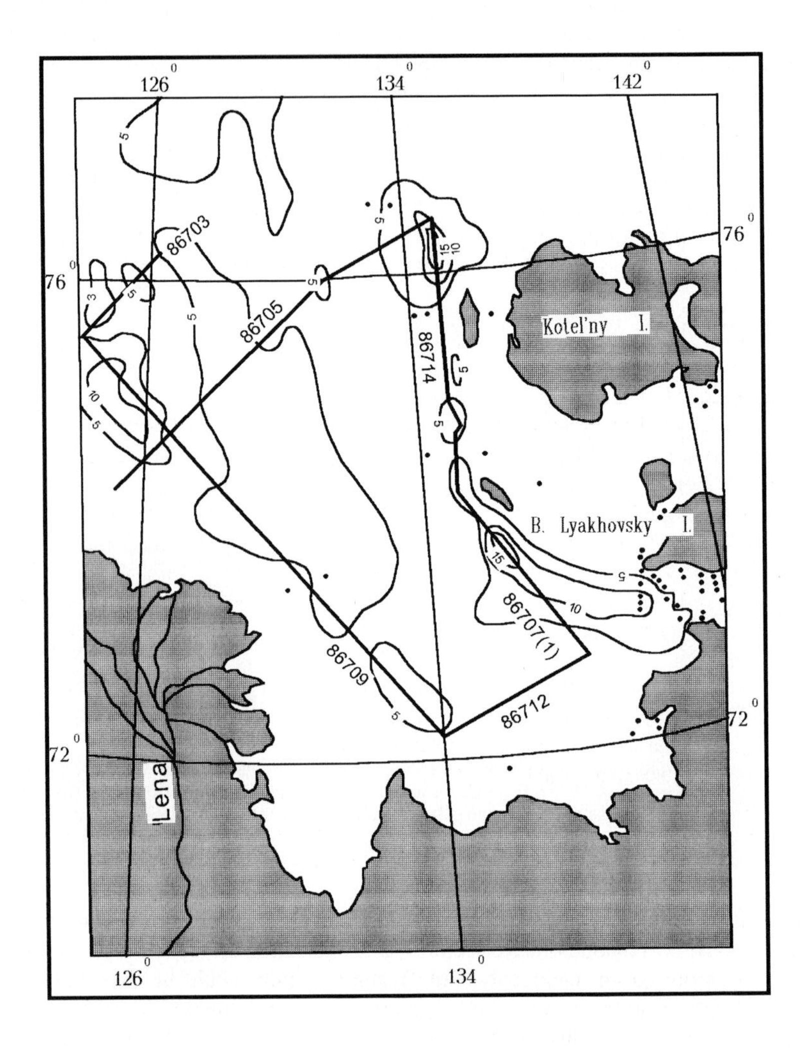

Figure 4. Thickness of the Holocene deposits in the south-eastern LSS (isopachs in m). Black dots indicate the sites where the base of the Holocene was reached by gravity cores and/or shallow wells. HRS lines and their numbers are as in Figure 1.

central eastern LSS. The remaining part of the LSS is characterized by a very short exposure to subaerial freezing which lasted, perhaps, only for a few thousand years close to the Pleistocene/Holocene boundary.

Thickness of the Holocene sediments

A generalized scheme of distribution of the Holocene sediments on the LSS was for the first time compiled on the basis of available HRS data and the results of gravity coring (Figure 4). The comparison of this very preliminary image with permafrost zonation shown in Figure 1 demonstrates that the greatest thickness of the Holocene sediments (in excess of 10 m) is associated with the areas characterized by a minimal distribution of submarine permafrost.

According to paleogeographic reconstruction for the Pleistocene/Holocene boundary, at this time the whole LSS was occupied by a lacustrine-fluvial plain dissected by numerous paleoriver valleys controlling accumulation of the Holocene sediments. HRS lines 86707(1) and 86714 where the thickness of the Holocene deposits was locally observed to exceed 15 m follow approximately the Yana River paleovalley, whereas another depression in the base of the Holocene recorded at about intersection of HRS lines 86709 and 86705 seems to correspond to a major right tributary of the paleo-Lena River. Position of paleovalleys and related thickest Holocene accumulations is broadly consistent with most prominent bathymetric lows.

Conclusions

1. The stratigraphy of the uppermost part of the LSS sedimentary cover was studied within penetration depth of HRS system used in the survey; this depth averaged between 100 and 130 m and locally exceeded 250 m. Four units corresponding to Pliocene, EoPleistocene-Middle Pleistocene, Upper Pleistocene, and Holocene intervals were identified.

2. Reflectors bounding the lowest Unit A are correlated with erosional unconformities of Late Miocene and Late Pliocene age. Reflector at the top of unit B is more likely to represent predominantly lithofacial change accompanied by moderate erosion at the Middle/Late Pleistocene boundary. The uppermost subbottom reflector at the base of the Holocene deposits (unit D) has a composite nature related to thermal and abrasive activity of advancing Holocene transgression.

3. Permafrost zonation in the central and eastern LSS is defined by the existence of three major categories of frozen rocks differing in duration of Pleistocene subaerial exposure and related freezing history. Available HRS evidence is too limited to allow unambiguous recognition of permafrost in seismic data. The areas least subjected to latest Pleistocene subaerial freezing seem to contain the thickest Holocene deposits. It may further be speculated that in the same areas no significant permafrost was formed in earlier epochs as well. If proved correct, this assumption may appear very important for selection of NADP drill sites.

4. The thickest Holocene sediments (and paleoriver valleys controlling their distribution) seem to follow the major structural depressions filled by older (pre-Quaternary) sedimentary sequences, although they do not necessarily coincide with the deepest parts of earlier grabens. A preliminary proposal for selection of NADP drill sites shown in Figure 1 is essentially based on this structural consideration and related assumption that the proposed sites may provide most complete late Cenozoic sections.

5. Additional high resolution seismic surveys in the LSS are absolutely critical for better understanding of regional distribution and paleoenvironmental significance of Late Cenozoic sequences. In particular, the areas of short-term permafrost formation and its subsequent degradation during the Holocene must be delineated with greater confidence for final selection of optimal sites for paleoenvironmental drilling and recovery of continuous sedimentary record of climate change.

Acknowledgements

Thanks are due to Prof. Karl Hinz and Dr. Sergei Drachev for their constructive review of the manuscript and well-meant critical comments which helped to improve the paper. We appreciate the courtesy of Dr. Rinat Murzin, Director General of MAGE, whose permit enabled us to reproduce in this paper some of MAGE‘s unpublished data. We are also grateful for scientific advice and technical assistance which we received from our colleagues in VNIIOkeangeologia during preparation of this contribution.

References

Alekseev, M.N., A.A. Arkhangelov, N.M. Ivanova, B.I. Kim et al. (1992). Laptev Sea. In: Paleogeographic atlas of the shelf regions of the Eurasia for the Mesozoic and Cenozoic. Vol. 1. By the Robertson Group plc., UK, and Geological Institute, Academy of Sciences, USSR, 1-14 - 1-33.

Drachev, S.S. (1994). Laptev Sea: The NAD drill sites. The Nansen Ice Breaker, 6, 1-11.

Drachev, S.S., L.A. Savostin and I.E. Bruni (1995). Structural pattern and tectonic history of the Laptev Sea region. Rep. Polar Res., 176, 348-366.

Gusev, G.S. and O.V. Grinenko (1985). The Cenozoic megacomplex. In: Structure and evolution of the earth's crust of Yakutia (in Russian). Moskva, „Nauka", 182-203.

Implementation plan for the Nansen Arctic Drilling Program (1997). JOI, 42pp.

Ivanova, N.M., S.B. Sekretov and S.I. Shkarubo (1989). Data on geological structure of the Laptev Sea shelf from the results of seismic investigations. Okeanologia, 29 (5), 789-795.

Kim, B.I. and V.Ya. Slobodin (1991). Main stages in the development of the East Arctic shelf of Russia and Arctic Canada during the Paleogene and Neogene (in Russian). In: Geology of folded rim of the Amerasian Subbasin. Sankt-Peterburg, VNIIOkeangeologia, 104-116.

Kim, B.I. (1992). The Cenozoic sedimentogenesis and paleogeography of the East Arctic shelf (in Russian). In: Geological history of the Arctic in Mesozoic and Cenozoic (2). Sankt-Peterburg, VNIIOkeangeologia, 47-55.

Kim, B.I. (1994). New stratigraphic correlation of platform cover reflectors on the Laptev Sea shelf (in Russian). Abstracts of VI Conference on new achievements in marine geology. Sankt-Peterburg, VNIIOkeangeologia, 60-61.

Niessen, F. and E.E. Musatov (1997). Marine sediment echosounding using PARASOUND. In: E.Rachor (ed.), Scientific cruise report of the Arctic expedition ARK-XI/1 of RV „Polarstern" in 1995, Rep. Polar Res., 226, 118-128.

Soloviev, V.A. (1981). Predicted distribution of relict submarine frozen zone (East Arctic seas example) (in Russian). In: Cryolithozone of the Arctic Shelf. Inst. Merzlotoved., SO Acad. Nauk SSSR, Yakutsk, 28-38.

Soloviev, V.A., G.D. Ginsburg, Eu.V. Telepnev and Yu.N. Mikhalyuk (1987). Cryogeothermal parameters and hydrates of near-bottom gases in the Arctic Ocean (in Russian). PGO "Sevmorgeologia", Leningrad, 150 pp.

Strepetova, Z.V., S.A. Laukhin, B.V. Ryzhov, A.I. Dubinchik (1981). Cenozoic key section on Yana-Omoloi watershed area (in Russian). Izvestiya Acad. Nauk SSSR, Ser. Geol., 7, 48-63.

Sudakov L.A., E.K. Serov, O.M. Il'in, P.S. Davydov et al. (1991). Results of geological survey at 1:200,000 scale in Dmitry Laptev Strait (in Russian). Unpublished report, PMGRE archive, Lomonosov.

Section F

Summary

Dynamics and History of the Laptev Sea and its Continental Hinterland: A Summary

J. Thiede[1/4], L. Timokhov[2], H.A. Bauch[1], D.Y. Bolshiyanov[2], I. Dmitrenko[2], H. Eicken[3], K. Fahl[4], A. Gukov[5], J. Hölemann[1], H.-W. Hubberten[6], K. v. Juterzenka[7], H. Kassens[1], M. Melles[6], V. Petryashov[8], S. Pivovarov[2], S. Priamikov[2], V. Rachold[6], M. Schmid[7], C. Siegert[6], M. Spindler[7], R. Stein[4] and Scientific Party

(1) GEOMAR Forschungszentrum für marine Geowissenschaften, Wischhofstrasse 1-3, D 24148 Kiel, Germany

(2) State Research Center - Arctic and Antarctic Research Institute, 38 Bering St., 199226 St.Petersburg, Russia

(3) Geophysical Institute, University of Alaska Fairbanks, 903 Koyukuk Dr., P.O. Box 757320, Fairbanks, AK 99775-7320, USA

(4) Alfred-Wegener-Institut für Polar- und Meeresforschung, Postfach 120161, D 27568 Bremerhaven, Germany

(5) Lena-Delta State Reserve, 28 Fiodorova St., 678400 Tiksi, Yakutia, Russia

(6) Alfred-Wegener-Institut für Polar- Meeresforschung, Forschungsstelle Potsdam, Telegrafenberg A43, D 14473 Potsdam, Germany

(7) Institut für Polarökologie, Universität Kiel, Wischhofstrasse 1-3, D 24148 Kiel, Germany

(8) Zoological Institute, Russian Academy of Sciences, 1 Universitetskaya, 199034 St. Petersburg, Russia

Abstract - Russian and German scientists have investigated the extreme environmental system in and around the Laptev Sea in the Siberian Arctic. For the first time a major comprehensive research program combining the efforts of several projects addressed both oceanic and terrestrial processes, and their consequences for marine and terrestrial biota, landscape evolution as well as land-ocean interactions. The primary scientific goal of the multidisciplinary program was to decipher past climate variations and their impact on contemporary environmental changes. Extensive studies of the atmosphere, sea ice, water column, and sea-floor on the Laptev Sea Shelf, as well as of the vegetation, soil development, carbon cycle, permafrost behaviour and lake hydrology, and sedimentation on Taymyr Peninsula and Severnaya Zemlya Archipelago were performed during the past years under a framework of joint research activities. They included land and marine expeditions during spring (melting), summer (ice free), and autumn (freezing) seasons. The close bilateral cooperation between many institutions in Russia and Germany succeeded in drawing a picture of important processes shaping the marine and terrestrial environment in northern Central Siberia in Late Quaternary time. The success of the projects, which ended in late 1997, resulted in the definition and establishment of a new major research effort which will concentrate on establishing a better understanding of the paleoclimatic and paleoenvironmental record of the area. This is important because it allows to be able to judge rates and extremes of potential future environmental changes.

Introduction

Mankind has an eminent interest in establishing forecasts within whose ranges its own living conditions may change in the foreseeable future. Geological synoptic reconstructions and climate modelling have shown independently that the polar latitudes of the northern hemisphere, in particular its marginal seas and adjacent continents, are apt to drastic and fast changes. There is growing concern about the reaction of the Arctic system to global environmental change and the impact of this change on future climate development. In this sense the shallow Laptev Sea Shelf, ice free only during few summer months, and its Siberian hinterland are of particular interest, because here riverine outflow acts as important freshwater source for the halocline and sea-ice cover in the Arctic Ocean proper. Climate dependent environmental changes in the Laptev Sea itself are strongly controlled by changes in the terrestrial environment of its

In: Kassens, H., H.A. Bauch, I. Dmitrenko, H. Eicken, H.-W. Hubberten, M. Melles, J. Thiede and L. Timokhov (eds.) Land-Ocean Systems in the Siberian Arctic: Dynamics and History. Springer-Verlag, Berlin, 1999, 695-711.

hinterland, particularly via the quantity and quality of the riverine water and sediment transport. Reversely, sea-level and ice-cover changes in the marine environment have direct influence on the precipitation and temperature regime on land, both of which have a strong impact on the vegetation and permafrost behaviour and thus on the water and carbon balances.

In order to address the complexity and interdependencies of processes both in the marine and in the terrestrial environment, three bilateral research projects were established in Russian-German collaboration in northern Central Siberia. Whilst part of the cooperation specifically focuses on the reconstruction of the Late Quaternary climatic and environmental history of the Laptev Sea Shelf and Taymyr Peninsula/Severnaya Zemlya Archipelago, other investigations were devoted to the study of modern processes, describing ecosystems and different environmental settings of the study area.

Strategy and history

Marine studies

The drastic and fast environmental reaction on climate change in the Arctic (e.g. CLIMAP, 1976; Broecker, 1994; Imbrie and Imbrie, 1980) is potentially of great importance because the modes and rates of present deep- and intermediate-water renewal of the Norwegian Greenland Sea and the Arctic Ovean have an important control on the global environment, while the heat exchange between the Arctic Ocean and North Atlantic Ocean generates a regional anomaly of the northern hemisphere climatic zonation resulting in habitable areas in northwestern Europe, including its highest North.

The sensitivity of the Arctic sea-ice cover to climate change is well established and provides a strong motivation for research in this area. Furthermore, however, the recent record minima in summertime ice extent observed over the Siberian shelves in the early and mid-1990's (Maslanik et al., 1996) underscore the importance of the Laptev Sea as an indicator as well as an agent of change. While the Laptev Sea is most remote from Atlantic and Pacific oceanographic influences, its importance as the prime ice production area in the Arctic Basin and as source area of the Transpolar Drift (Rigor and Colony, 1997) provides a linkage that reaches as far as the freshwater budget of the Greenland Sea (Figure 1). Global climate change potentially amplified by anthropogenic emission of greenhouse gases is assumed to considerably affect the Arctic Ocean's oceanographic circulation and the waxing and waning of its sea-ice cover (Wadhams, 1995). Unfortunately, the knowledge about present and past processes driving these changes is limited, and it is not entirely clear to what an extent such regional processes are tied into a global picture. We believe that the interpretation of paleoclimatic records in conjunction with the study of modern processes will further the understanding and prediction of the Arctic System.

Scientific investigations within the scope of the bilateral projects were planned as part of a system approach. They comprise marine and terrestrial investigations, and a suite of modelling experiments and theoretical considerations. While studying a complicated system such as the Laptev Sea region, it is expedient to divide it into subsystems. It will be most natural to delineate the modern (e.g., hydrophysical, chemical, biological) and past subsystems which are constrained by both marine and terrestrial boundary conditions. The boundary conditions of these subsystems are quite variable. They can be treated as constant (in as far as modern processes are concerned), faintly varying (interannual and climatic variations) and strongly varying (paleoclimatic variations) depending on temporal scales. While describing marine system evolution, its boundaries can be considered as a subsystem. Parameters of this subsystem play different roles depending on the temporal scale of process integration.

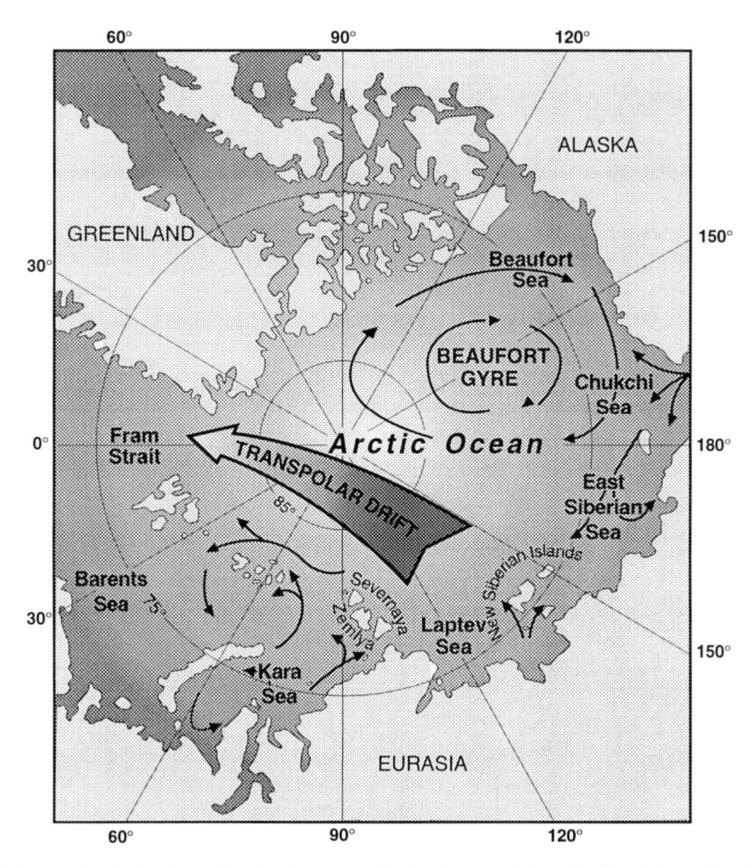

Figure 1: Major sea-ice drift paths in the Arctic Ocean. The Laptev Sea, at the tail of the Transpolar Drift, is considered a major ice factory for the Arctic Ocean.

Many properties of the water mass boundaries of the Laptev Sea are dependant on the interaction of the Laptev Sea with the continental hinterland. The hydrography of the Laptev Sea water masses and the extent of its sea-ice cover are controlled by the interaction of the open Arctic Ocean water masses and the influx of freshwater from a number of major river systems. The sea floor sediments in the Laptev Sea consist chiefly of biogenic and terrigenous components, and their distributional pattern directly depends on the water mass boundaries. The biogenic components consist in part of materials derived from freshwater organisms living in the rivers draining into the Laptev Sea and detrital plant remains, as well as of marine/brackish organisms specialized to the highly variable living conditions in the Laptev Sea. The terrigenous components consist of fine-grained sediments which can be linked to several major sources: Detrital suspended material transported to the Laptev Sea by the rivers draining its continental hinterland, and products of coastal and sea floor erosion.

All of these variables allow to define biological and sedimentological provinces in the Laptev Sea, and to determine the source regions. Detailed investigations of the riverine depositional environments as well as of the lacustrine sedimentary sequences and experiments on land adjacent to the Laptev Sea have additionally contributed to a comprehensive understanding of the entire Laptev Sea system. The compilation of results described in this book as well as the scientific approach of the investigations are henceforth of considerable general interest, even though they are based mainly on the bilateral Russian-German efforts under the projects mentioned above.

Due to both the political change and the support from modern technology, the Laptev Sea as well as its hinterland became only recently accessible to the international scientific community. In close cooperation with the State Research Center - Arctic and Antarctic Research Institute (AARI) in St. Petersburg (Russia), the GEOMAR Research Center for Marine Geosciences, Kiel (Germany), originally carried out two land-based expeditions to the Laptev Sea. Both expeditions, the Ameis '91 and the Esare '92 (Dethleff et al., 1993), were conducted during the early spring of the respective years and focussed on the role of the shallow Laptev Sea shelf in sediment entrainment into sea ice. These studies and furthermore the expeditions Transdrift I aboard RV "Ivan Kireyev" to the inner Laptev Sea in 1993 (Kassens and Karpiy, 1994) and Arctic 93 aboard RV "Polarstern" to the ice-covered continental slope of the Laptev Sea (Fütterer, 1994) revealed that the complex environmental system Laptev Sea required a multidisciplinary approach as was realized within the scope of the research program devoted to the "Laptev Sea System". These efforts were supplemented by the projects "Ecology of the Marginal Seas of the Eurasian Arctic" and "Late Quaternary Environmental Evolution of Central Taymyr". All these projects have been funded by the Russian and German Ministries of Science and Technology.

In 1994, activities continued in the line of the Transdrift II expedition aboard RV "Professor Multanovskiy" (Kassens et al., 1995a) and the first river expedition Lena 94 (Rachold et al., 1995). The recovery of few long sediment cores from the inner Laptev Sea allowed reconstructions of the paleoenvironment, i.e., changes in Lena River runoff. Contemporaneous investigations on the Lena River itself contributed to the identification of the river's fingerprint within the shelf deposits.

The expedition Transdrift III (1995) aboard the icebreaker "Kapitan Dranitsyn" met the most difficult logistical demands, since research topics focused on processes which occur during the extreme change from the ice-free conditions in late summer to the onset of freeze-up during autumn (Kassens et al., 1997). This expedition was accompanied by both the RV "Polarstern" ARK-XI/1 expedition (Rachor, 1997) and the land-based Lena-Yana expedition (Rachold et al., 1997a). RV "Polarstern" completed investigations during July to September 1995 within the pack ice of the northernmost Laptev Sea, the adjacent deep-sea basin and on the Lomonosov Ridge, thereby addressing relationships between shelf areas and adjacent deep-sea basins. The river studies attempted to quantify the river's contribution to the sedimentary, chemical, and hydrodynamic balance of the Laptev Sea (Rachold et al., 1996).

The primary objective of the Transdrift IV expediton was to study the Lena River break-up and its influence on the environmental system of the Laptev Sea (Kassens et al., 1996). The international biological Station "Lena-Nordenskiöld", located in the eastern Lena Delta, was the base for the scientists of the expedition. From here the field program in the Lena Delta and the Laptev Sea was carried out. The river investigations were completed during the Khatangar 96 expedition in 1996 (Rachold et al., in press).

All joint research activities were performed in a close formal cooperation between many Russian and German scientific institutions (but with liaisons to many international programs and institutions) and succeeded to draw a detailed picture of most important processes shaping the Laptev Sea system (Figure 2). Beside national efforts, the multinational expedition "Spaciba" (Martin et al., 1996) and the US-Russian expedition aboard the RV "Smirnitsky" in 1995 (Johnson, 1996) are the most important ones.

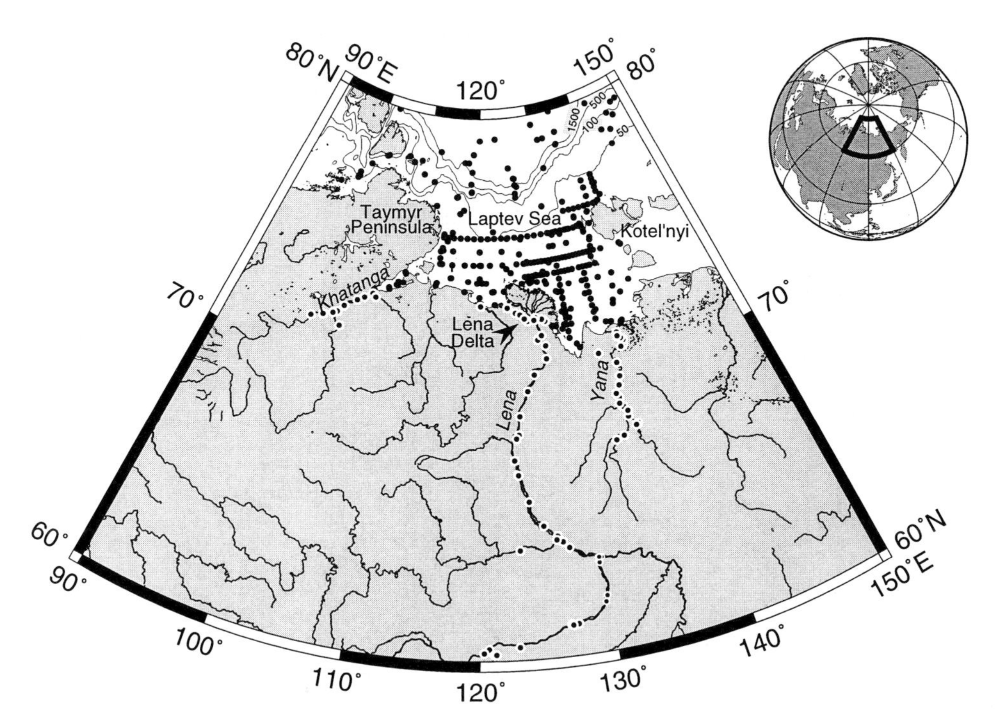

Figure 2: Station map of the Laptev Sea System Project. For the past six years, a multidisciplinary working program was carried out at almost 500 stations in the Laptev Sea and along the Siberian rivers draining into the Laptev Sea.

Terrestrial studies

The terrestrial investigations were carried out within the scope of the Russian-German research project "Late Quaternary Environmental Evolution of Central Taymyr" (Taymyr), which was established in 1993. Following a pilot phase, funding by the German Ministry of Education, Science, Research, and Technology was provided from 1994 to 1997. During these years, five expeditions with more than fifty participants from six Russian and five German institutions were carried out in the project study area: Norilsk/Taymyr 1993 (Melles et al., 1994), Taymyr 1994 (Siegert and Bolshiyanov, 1995), Taymyr 1995 (Bolshiyanov and Hubberten, 1996), Taymyr/Severnaya Zemlya 1996 (Melles et al., 1997), and Norilsk 1997 (Melles et al., in prep.).

The overall objective of the Taymyr project was the reconstruction of the Late Quaternary climatic and environmental history of the Taymyr Peninsula and the Severnaya Zemlya Archipelago (Figure 3). For a number of reasons, this region is a key area for understanding the modern and past environmental dynamics of northern Siberia.

For example, the Taymyr region is located in the transition zone between the West Siberian marine and East Siberian continental climates. The higher precipitation in West Siberia and Europe during Late Quaternary glacial times led to the formation of the Eurasian ice sheet, whilst large areas in East Siberia, despite lower temperatures, remained unglaciated. The eastern extension of the Eurasian ice sheet is still under discussion - for the Last Glacial Maximum, a maximalistic hypothesis with almost complete ice coverage of the Taymyr region (Grosswald, 1998) contradicts with a minimalistic hypothesis with glaciations being restricted to the

Severnaya Zemlya Archipelago and high-altitude areas of the Putoran Plateau and the Byrranga Mountains (Velichko et al., 1997, 1997a). In the Taymyr region, therefore, both the Late Quaternary changes in relative influences of the West and East Siberian climates and their influences on the kind and extension of the Eurasian glaciation can be investigated.

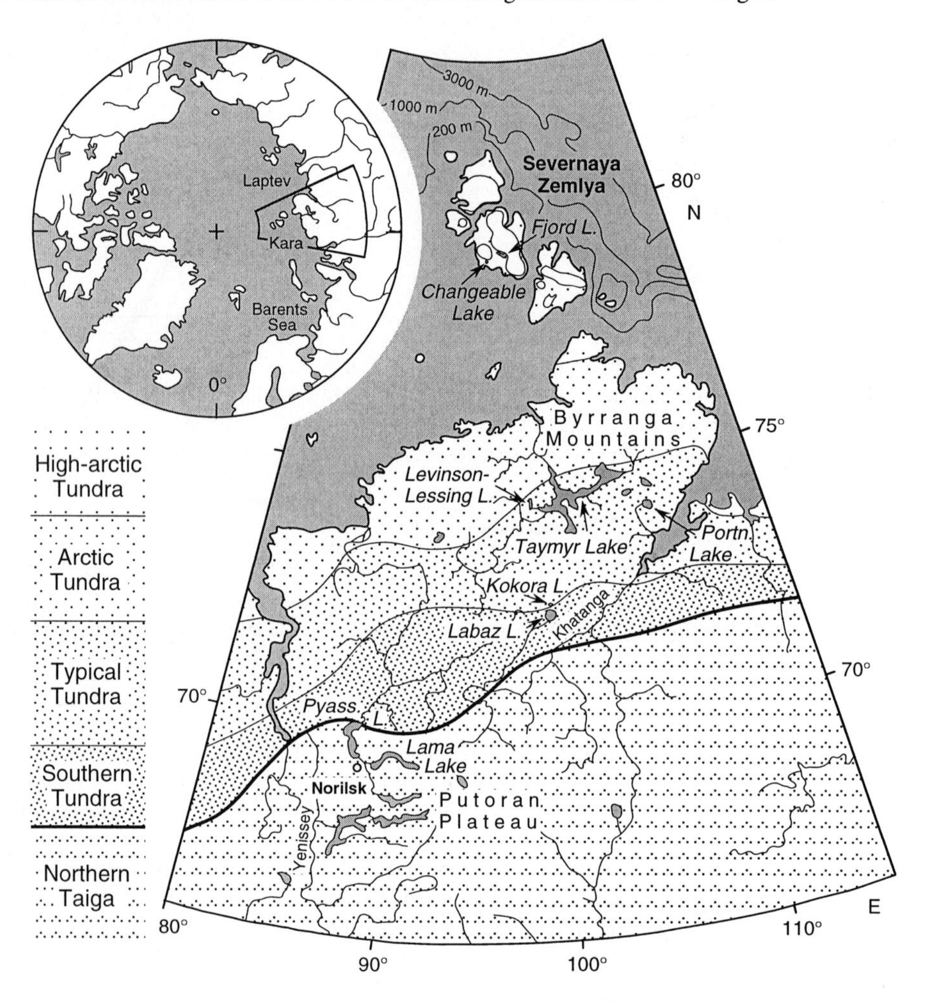

Figure 3: Location and map of the Taymyr Project study area in Central Siberia showing the modern distribution of the vegetation zones and geographic terms mentioned in the text.

In addition, the Taymyr Peninsula forms the northernmost region of the Eurasian continent, being widely unpolluted by human activities. Together with the Severnaya Zemlya Archipelago, it covers the entire spectrum of Arctic landscapes and vegetation zones, leading from the Northern Taiga in the south via different tundra zones on the central and northern peninsula to the high-Arctic tundra in the north. This ca. 1400 km long transect reflects a summer temperature gradient of more than 10°C (Matveyeva, 1998). Hence, the Taymyr region is predestined for reconstructing the development and changes in the location and extension of the vegetation zones in dependence on the Late Quaternary climatic variations.

Due to the proximity to the Laptev Sea, the results from the Taymyr region can directly be linked to the marine geological data. From this, a better understanding of the often problematic correlation of land and ocean records is expected. In addition, the comparison may supply comprehensive information concerning the interaction of the marine and terrestrial histories. The effects of sea-level and marine ice-cover changes on the precipitation and temperature development on land, both of which have strong impact on the vegetation and permafrost behaviour are of particular interest are.

In order to adress the complexity of climate dependent processes in the permafrost landscape of the Taymyr region and to obtain a most comprehensive understanding of the climatic and environmental history, a multidisciplinary approach was applied to the Taymyr Project (Figure 4). Investigations of the seasonal processes operating in the lithosphere, hydrosphere, and biosphere in the different climatic and environmental settings of the study area indicate the interaction between climate and permafrost. They form the basis for interpreting the development of the landscape and the composition of ancient deposits with respect to the climatic and environmental conditions in the past. Special emphasis is put on syngenetic permafrost deposits on land with included ground ice bodies and on unfrozen sediment sequences in presently existing lakes. Due to their individual formation processes, these natural archives gather individual information concerning the kind of paleoenvironmental evidence, and the length, completeness, and resolution of the documented time interval. The information derived from the terrestrial archives shall be complemented in the near future by data from an ice core which is planned to be drilled on the Severnaya Zemlya Archipelago.

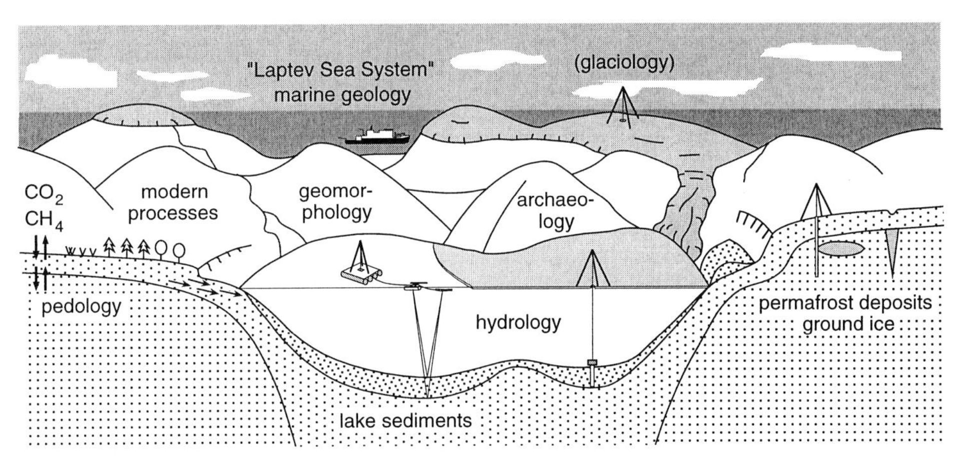

Figure 4: Schematic section through the permafrost landscape of northern Siberia illustrating the multidisciplinary character of the Taymyr project in the neighbourhood of the System Laptev Sea project. Complex investigations of the modern seasonal dynamic processes, taking place in the different landscapes and climates of the study area, form the basis for interpreting the composition of ancient deposits with respect to past environmental and climatic conditions. Major natural data archives, investigated for the Late Quaternary history, are the sediment fill in presently existing lakes and syngenetic permafrost sequences on land with incised ground ice bodies. Additional information comes from geomorphological and archaeological work. An ice core drilling on the Severnaya Zemlya Archipelago is planned to complement the existing data sets in the coming years.

Major results of marine and terrestrial investigations in the Siberian Arctic

On the following pages some of the major scientific results of the projects are summarized (see also Kassens et al., 1995; 1998; Melles et al., 1996). Many of these are described in the subsequent papers of this volume. Some of the discoveries are providing first insights into the different processes controlling Arctic climate and linking the land and shelves to the deeper basins.

Modern sea-ice and ocean processes

The Laptev Sea is one of the regions with the highest net-ice production rates in the Arctic Ocean (Rigor and Colony, 1997). One of the aims of the project was to assess the spatial patterns and variability of ice growth in the Laptev Sea including its linkage to atmospheric and oceanic processes as well as its importance for sediment transport through the entrainment of particulate matter.

The time period 1992 - 1996 is of particular interest since summer ice extent attained a record minimum in 1995. As shown by meteorological data and a large-scale sea ice model, this minimum is mostly a result of atmospheric circulation anomalies and their impact on ice dynamics, along with enhanced ice melting due to advection of warm air masses from central Siberia (Kolatschek et al., 1996). As a result of advective heat transport and storage of solar short-wave energy in the water column, the fall freeze-up 1995 was delayed by two weeks compared to the long-term mean of ice-formation onset. This is corroborated by hydrographic measurements which detected a warm water layer in the eastern Laptev Sea in the fall of 1995 (Kassens et al., 1997).

Given the importance of sea ice circulation and export (Dmitrenko et al., this volume), a more detailed study of the ice circulation regime in the Laptev Sea was carried out in the period of 1979 to 1995. With a lack of buoy data over the shelf regions, where ice motion may be quite complex, the ice velocity field was determined for selected years from remote-sensing data (Russian Okean Side-looking Radar, Special Sensor Microwave/Imager (SSM/I) 85 GHz passive microwave data, and Advanced Very High Resolution Radiometer (AVHRR) data). This was achieved by both manual tracking of conspicuous features in the ice pack (leads, boundaries between multi-year and first-year sea ice as visible in Okean and AVHRR data) as well as automated tracking using maximum-likelihood correlation techniques (for SSM/I scenes). Comparison with the available buoy data showed consistent results and these were utilized to validate and improve a large-scale dynamic-thermodynamic sea-ice model. From the model, the areal ice flux to the Arctic Ocean and the adjacent shelf seas was derived for the period of 1979 to 1995. Except for the summer months of 1979 and 1981, total net export during summer and winter is positive for the entire period, ranging between 3 and 7 x 10^5 km^2 in winter (October-May). Along the northern boundary, ice was imported into the Laptev Sea during the summer period of 12 out of 18 years. Export into the East Siberian Sea is considerable, with mean winter values of approximately 1 x 10^5 km^2.

Freshwater input through river discharge is an important component of the freshwater balance of the Laptev Sea and the Arctic Ocean (Bareiss, 1996; Dmitrenko et al., this volume; Golovin et al., this volume). Hydrographic surveys carried out between 1993 and 1996 have been able to trace the influence of the plume from the Lena River throughout the eastern Laptev Sea. Historical data indicate that salinity anomalies in the Transpolar Drift may also be directly affected by continental runoff. While a study based on remote-sensing and hydrological data has shown that the impact of river discharge (through its effect on short-wave radiation balance and heat advection) on summer ice retreat is of importance only in the near-delta regions (Bareiss, 1996), river discharge considerably affects the ice freeze-up.

We were able to show that the establishment of a fast-ice cover and its areal extent are mainly controlled by the dispersal of river water in the eastern Laptev Sea (Dmitrenko et al., this volume). High resolution studies also indicate that frazil-ice production in the boundary layer between the river plume and water of higher salinity may constitute an important component of the ice-mass balance and play a role in sediment entrainment and transport (Dmitrenko et al., 1998; Golovin et al., this volume). Given the variability on decadal and in particular on longer timescales, the dispersal and the fate of river discharge and its impact on the ice regime are a central issue in understanding the long-term changes in the Laptev Sea and the Arctic Ocean.

Ice export from the Laptev Sea is of great importance for sedimentation processes on the shelf, in the central Arctic Ocean and the Nordic Seas (e.g. Wollenburg, 1993; Nürnberg et al., 1994; Pfirman et al., 1995; Eicken et al., 1997). Field studies have shown that substantial quantities of sediments are entrained into sea-ice growing over the broad, shallow Siberian shelves. Part of this ice grows throughout the winter season in a system of coastal polynyas and flaw leads (Dethleff et al., 1993; 1995; Rigor and Colony, 1997). To date, the relative contribution of polynyas, both to total ice export and transport of sediments, has not yet been fully understood, however. These sediments are transported with the ice and released during ice melting, most of which occurs in the Greenland Sea area. A major entrainment event was identified in the eastern Laptev and western East Siberian Sea in 1994. Through a combination of remote-sensing and field measurements, the evolution and drift of this ice field were studied. Estimates of the total sediment load from this single event, surviving ice melt in 1995 and hence exported to the Arctic Basin, amount to between 7 and 10 million tons. In conjunction with entrainment of sediments over the remaining shelf areas, sea ice transport may represent an important component of the Laptev Sea sediment budget, in particular with respect to short-term, long-range transport.

The marine ecosystem

Biological studies focused on the abundance and community structure of phytoplankton (Juterzenka and Knickmeier, this volume; Cremer, this volume), on zooplankton and benthos (Abramova, this volume; Petryashov et al., this volume; Bauch, this volume) as well as on benthic life in relation to environmental changes (Grahl et al., this volume). Three distinct faunal and floral provinces have been identified by multivariate analysis in the central, the northern, and the southeastern part of the Laptev Sea. The overall phytoplankton biomass during the summer period (August/September 1993) given by carbon content of the different taxa was relatively low compared to other Arctic shelf regions, despite a maximum of chlorophyll *a*, revealing the influence of the Lena river runoff. The late freezing period in autumn 1995 was characterized by a locally high chlorophyll *a* biomass, which can be used by zooplankton as a last food source before the winter period.

In newly formed sea ice, pigment concentrations varied considerably. However, total algal biomass within the new and young ice was relatively high and within the range of the water column standing stock in October 1995. Therefore, this algal biomass may serve as food for the developing micro- and meiofaunal ice community. Endoscopic observations showed significant differences in the three-dimensional small-scale ice morphology as well as enclosures of biotic and abiotic origin, which are characterizing the habitat.

Zooplankton on the Laptev Sea Shelf is dominated by *Calanus glacialis* and *C. finmarchicus*, but in areas of freshwater influence in the south and east, brackish water taxa contributed as much as 27% of the total biomass. The export of *Calanus* species from the Nansen Basin onto the Laptev Shelf appears to be of great importance for the shelf communities. On the other hand, the eastern outer shelf and slope area of the Laptev Sea are thought to have a pronounced effect on the deep basin, modifying the populations entering the central Arctic Ocean (Kosobokova et al., 1998).

Benthic faunas can be abundant but show little diversity. Except for some shallow areas (< 20m) with low bottom water salinity (< 30), the brittle star *Ophiocten sericeum* dominates the megabenthic shelf assemblage and reaches a maximum density of more than 500 individuals per

m^2 in river valleys (> 30m water depth). Gross estimates of brittle star respiration, productivity, and organic carbon demand suggest that a substantial portion of the energy flow is channeled through these dense brittle star assemblages (Piepenburg and Schmid, 1997). In large isopods *(Saduria entomon* var. *sibirica)*, respiration rates as indicators of metabolic activity were significantly higher at low salinities, indicating metabolic costs of osmotic stress in a seasonally changing brackish environment. The findings show that energy flow models of this highly variable environment must be based on the occurrence of different biological communities, which also represent regional differences in biomass and, therefore, most likely, productivity and carbon demand.

Pathways of nutrients, trace elements and chlorinated biphenyls

Nutrients, trace elements, and chlorinated biphenyls were studied in sea ice, water column and surface sediments of the Laptev Sea and in river water in order (i) to trace shelf-ocean pathways of river discharge and sea ice, (ii) to study processes of ice formation, and (iii) to estimate the natural and anthropogenic input into the Laptev Sea (Hölemann et al., this volume; Lara et al., 1998; Utschakowski, 1998; Pivovarov et al., this volume). Input and distribution of nutrients, trace elements, and chlorinated biphenyls are controlled by river runoff. Remarkable is a strong increase in the content of chlorinated biphenyls and metals and a corresponding decrease in dissolved silicon in Lena River water during the onset of spring break-up, as shown by daily concentration measurements in 1996. As a result, approximately half of the annual discharge is drained into the Laptev Sea during this short period of less than one month. The most important factor controlling the transport of dissolved and suspended matter during spring break-up is the pronounced density stratification of the water column in the Laptev Sea due to freshwater input via river runoff. Working much like a conveyor belt driven by the intensity of river runoff, dissolved and particulate matter are transported offshore to the north (Hölemann et al., this volume; Ivanov and Piskin, this volume; Pivovarov et al., this volume). Therefore, possible depositional centers, although not clearly identified, are far away from the Lena Delta.

Another important process affecting the transport pathways of particles was studied in the autumn of 1995. During freeze-up, particulate matter was effectively incorporated into newly forming ice off the Lena Delta. Unexpectedly high concentrations of dissolved Fe, Mn, Zn, Cd, and Pb were measured in the sediment laden new ice (Hölemann et al., this volume). For instance, the concentration of Fe was up to 25 times higher as compared to the average concentration of Lena River water. We suggest that this is caused by a redox controlled remobilization of metals from ice-bound sediments. Further geochemical investigations of sea ice and surface sediments from different regions in the Laptev Sea have also shown that the chemical signature can be directly related to different fluvial sources: e.g., high Mg/Al ratios found in sea ice sediments in the southwestern Laptev Sea reflect the geochemical signature of flood basalts in the catchment area of the Khatanga River. Dissolved and particulate heavy metals and chlorinated biphenyls in sediments, water column, and sea ice are at very low levels. In particular, the content of heavy metals in sea ice and surface sediments are comparable to those found in unpolluted marine sediment and thus give no indication of anthropogenic pollution. High contents of As in surface sediments along the river valleys are caused by strong suboxic diagenesis.

Land-ocean transfer of sediments

To identify modern and past sediment transport from the Siberian hinterland across the Laptev Sea to the Arctic Ocean, geochemical, i.e., major, trace, and rare earth element (Hölemann et al., this volume; Rachold, this volume) and Sr isotope (Rachold et al., 1997b), and mineralogical, i.e., clay and heavy mineral (Behrends et al., this volume; Rossak et al., this volume), investigations were performed. These analyses concentrated on river suspended particulate matter as well as on surface sediments from all major river systems draining into the Laptev Sea and the adjacent Arctic Ocean. In addition, ice samples from the Laptev Sea and the Arctic Ocean were studied. Accordingly, the Laptev Sea can be divided into two different sedimentary provinces. While the eastern part is controlled by Lena River discharge, the western part is controlled by sediments supplied by the Khatanga River. Sea ice and sea-floor sediments from the western Laptev Sea can be distinguished from those from the eastern Laptev Sea by (i) the dominance of pyroxene in contrast to the amphibole-rich sediments of the eastern Laptev Sea, (ii) high amounts of the clay mineral smectite, which are lower in the eastern Laptev Sea, and (iii) significantly lower $^{87}Sr/^{86}Sr$ isotope ratios. In addition, sea-floor sediments of the Arctic Ocean indicate that Sr isotope ratios and heavy mineral assemblages can be used to identify sediment transportation pathways from Central Siberia to the Arctic Ocean.

The Lena River is considered to be the main sediment source for the Laptev Sea. About 150 km upstream from the Lena Delta measurements of suspended sediment transport were made during the past decades. But the quantitative evaluations of suspended sediment discharge made by different investigators using the same data sets range from 11.8 to 21 million tons per year. Hence, the exact amounts of sediment actually accumulating in the Lena Delta and finally reaching the sea is difficult to estimate. However, data on suspended sediments along a transect from the southeastern Lena Delta to the Laptev Sea have shown no concentration gradient during the onset of the river break-up in 1996.

Another important sediment source for the Laptev Sea is coastal erosion. Field investigations as well as air and satellite images indicate that coastal retreat varies between 2.5 and 6 m per year. Rough estimations of the amount of sediments released to the sea due to erosion are of the same order compared to data on sediment discharge of the rivers feeding the Laptev Sea (Are, 1996; Are, this volume). However, reliable estimates for land-ocean sediment transport will be the focus of further investigations in the Lena Delta.

The modern organic-carbon cycle

To understand the processes controlling organic-carbon flux and composition in the Laptev Sea, biological, micropaleontological, organochemical, and sedimentological investigations were performed on water samples and surface sediments of the Laptev Sea (Fahl and Stein, 1997; Stein et al., this volume). Data from the water column indicate a higher biological activity in the eastern Laptev Sea. This is also reflected in surface sediments, with increased contents of organic matter in this area. Micropaleontological parameters and specific biomarkers can be used to distinguish between marine and terrigenous sources of organic matter of surface sediments. Organic carbon maxima off the river mouths correlate with increased abundances of freshwater algae, plant debris, and terrigenous biomarkers (e.g., long-chain n-alkanes and wax esters, light $\partial^{13}C_{org}$) as well as high clay content. This points to the terrigenous source of this organic matter. Concentrations of terrigenous markers decrease with increasing distance from the source. The distribution of the marine biomarkers (e.g., short-chain fatty acids) correlates with sea ice distribution. The lowest concentrations of short-chain fatty acids are in ice-covered areas, whereas the highest amounts are located near the ice edge. Fatty-acid distribution correlates with the distribution of chlorophyll *a* and biogenic opal content, indicating increased surface-water productivity near the ice edge (Fahl and Stein, 1997).

To understand the modern organic carbon cycle, variations in modern organic carbon accumulation (flux) rates must also be considered. Accumulation rates of total organic carbon may reach 0.2 to 2 $gCcm^{-2}\ ky^{-1}$, decreasing to 0.02 $gCcm^{-2}\ ky^{-1}$ at the lower slope. A decrease in accumulation rates from the Laptev Sea towards the lower slope reflects a decrease in terrigenous organic matter supply in offshore direction.

The Laptev Sea since the Last Glacial

Paleoceanographic investigations based on radiocarbon datings (AMS ^{14}C) of bivalve shells in sediment cores (up to 5 m) have provided new insights into the history of the Laptev Sea since the last glacial. The history of the Laptev Sea can be subdivided into various phases inherently linked to sea level rise following the last glaciation.

During the last glaciation, large parts of the Laptev Sea Shelf were dry due to low global sea level. Sedimentation was probably governed by the deposition of syngenetic sediments (ice complexes), although light planktic oxygen isotope ratios from the central Arctic Ocean may indicate continuing river runoff during this time (Nørgaard-Pedersen et al., 1998). During deglaciation, records from the Laptev Sea continental slope reveal a major depletion in $\partial^{18}O$ at 11,000 years before present (yBP), which seems to correlate in age with the onset of the Younger Dryas cold spell. A major change towards marine conditions on the shelf due to rising sea level is noted at 9,500 yBP with increasing $\partial^{13}C$ values for organic matter and marine biomarker concentrations. This is coeval with an onset in the lateral distribution of heavy minerals from east to west on the shelf and along the continental slope. The outer parts of the Laptev Sea Shelf ($\geq$70 m water depth) are marked by an abrupt decrease in sedimentation rates at 9,000 yBP. In the inner-shelf areas, the sea level continued to rise until flooding of the shelf terminated near 6,000 yBP, accompanied by a major drop in sedimentation rates between paleodepth 30 to 50 m (Bauch et al., in press).

The later Holocene (since 6,000 yBP) situation appears to be rather stable in that sedimentation rates varied between 1.2 mm per year in the Lena Valley and 0.1 mm per year in the Khatanga Valley (Bauch et al., in press). This is about 5 times lower in comparison to the continental slope. Surface sediments, freshwater diatoms, and chlorophycean algae dominated the eastern shelf, whereas marine diatoms and dinoflagelate cysts are prominent on the western parts of the shelf (Cremer, 1998; this volume; Kunz-Pirrung, 1998; this volume). Ice algae species become more abundant particularly towards the north, along the average position of the summer ice edge. Grounding ice contributed considerably to sediment transport on the Laptev Sea Shelf. Side-scan sonar records and 30 kHz echograms indicate that at some locations (e.g., north of the Lena Delta) sediments are disturbed as a result of grounding ice (Benthien, 1995; Lindemann, 1995). These plough marks can be as deep as ten meters. Such grounded icebergs were actually occasionally observed between 15 and 25 m water depth in summer.

Another interesting feature of the Laptev Sea is the existence of offshore permafrost, which was verified by ice-bonded Holocene sediments found in areas as shallow as 12 cm below surface in the central Laptev Sea (Kassens and Karpiy, 1994; Romanovskii et al., 1997). The impact of offshore permafrost on the environmental system is unknown and will be a primary goal during further investigations.

Modern processes in permafrost landscapes

In Arctic regions, hydrological, geochemical, and biological processes are directly controlled by the presence of permafrost and the seasonal thawing of the active layer. Most of the processes occur during the short summer season from May to August and cease during the winter. As the active layer thaws with the progress of summer, the infiltration and storage capacity of the ground are increased. Consequently, geochemical, biological, and hydrological activities in the active layer are enhanced and influence the hydrology and geochemistry of surface and ground waters.

Modern seasonal dynamic processes were studied on various scales in the different climatic settings of the Levinson-Lessing and Labaz Lake catchments, and of northern Bolshevik Island, Severnaya Zemlya (Figure 3). For example, pedological and biological studies were conducted and replenished by seasonal measurements of CO_2 and CH_4 fluxes. In addition, the temporal and spatial variations in active layer hydrological and thermal processes and solute fluxes were quantified and compared with climatological data. A complete spring to autumn record of water and sediment discharge by streams entering Levinson-Lessing Lake and of the sedimentation in the lake, was measured in order to get an impression of the seasonal dynamic processes of sediment transport and accumulation under the present environmental and climatic conditions. The representative field data were put into a wider context by landscape mapping and subsequent connection with remote sensing data from Landsat TM images in order to extrapolate the field data to larger areas.

The results discussed in this book were obtained in the catchment of the Levinson-Lessing Lake. Mapping of vegetation, soils, and geomorphology in the field, supported by interpretations of air photos, enabled Anisimov and Pospelov (this volume) to classify 34 landscape units. Surface hydrological investigations were carried out by Zimichev et al. (this volume). They quantified seasonal fluxes of water and sediment for a typical high Arctic-nival streamflow regime entering the Levinson-Lessing Lake. The temporal dynamic of water, heat, and solute fluxes in the active layer was studied for a complete freeze-thaw cycle by Boike and Overduin (this volume) using Time Domain Reflectometry. The results indicate that the dominant heat sinks during spring and summer are the sensible and latent heat fluxes into the atmosphere. The soil heterogeneity strongly impacted hydrologic and thermal processes in the active layer.

The role of permafrost affected soils as sinks or sources of carbon is determined by labile balance between the production and the decay of organic matter. The relationship between the carbon-cycle and soil substrate, hydrological and thermal regime of the active layer, vegetation, and relief, was studied at different polygonal tundra sites. Seasonal trace gas emission of CH_4 and CO_2 was quantified (Sommerkorn et al., this volume) and, using carbon isotopes, the recent decomposition processes of carbon in permafrost-affected soils were studied (Samarkin et al., this volume). Such data are important for the understanding of soil generation and carbon accumulation processes in the present and past and thus for knowledge about the development of the Siberian Tundra.

Environmental history of the permafrost landscape

The development of permafrost is a characteristic process in non-glaciated regions with cold, continental climate. The deposits formed in permafrost landscapes contain complex information concerning the regional and local environmental conditions of the time of sediment accumulation and freezing. Besides chronological, sedimentological, mineralogical, geochemical, and paleontological data, specific information can be obtained from the cryostructure of syngenetic permafrost sequences as well as the kind and composition of ground ice bodies.

Widely continuous permafrost development takes place in lowland areas where accumulation distinctly exceeds denudation and where sufficient moisture leads to well expressed ground ice formation. Investigations of permafrost sequences in the Taymyr region, therefore, have focused on the Taymyr lowland being bordered by the Putoran Plateau to the south and the Byrranga Mountains to the north. Early work in the western and southwestern lowland had supplied first important information concerning the climatic history and paleogeography of the region (e.g., Karpov, 1986; Tumel, 1985). On the central and northern lowland, in contrast, permafrost data were restricted to some geothermic measurements in boreholes. The latter areas, therefore, were the major focus for permafrost investigations within the scope of the Taymyr Project. Special attention was drawn to the surroundings of the Labaz Lake, to Cape Sabler at the western shore of the Taymyr Lake, and to the western rim of the Putoran Plateau, where running work was continued. Additional information was obtained from paleogeographic, geomorphologic, and archaeologic work carried out in the foreland of the western Putoran Plateau, in the western Byrranga Mountains, and on the Severnaya Zemlya Archipelago.

First results from permafrost deposits and ground ice bodies at the Labaz Lake have shown that the last glaciation of the area predates the Middle Weichselian and that the East Siberian anticyclon had gained in significance during Middle and Late Weichselian times, leading to a distinctly higher continentality of the climate (Melles et al., 1996; Derevyagin et al., 1996; Chizhov et al., 1997). Evidence for a restricted ice advance during the Last Glacial Maximum was also found on the Severnaya Zemlya Archipelago, based on geochronological and paleogeographic investigations (Bolshiyanov and Makeev, 1995).

A more detailed presentation and discussion of the permafrost development in the Labaz Lake area, and its dependence on the Late Quaternary climatic history, is presented in this book by Siegert et al. (this volume). The results support the suggestion that the last ice advance to the central Taymyr lowland took place prior to the Middle Weichselian. The preservation of fossil glacier ice in near-surface permafrost deposits evidence a strongly retarded deglaciation of this territory with continuous, low-temperature permafrost. Climate warming during Middle Weichselian interstadials and at the Pleistocene/Holocene transition led to distinct vegetation changes and enhanced thermokarst processes, and subsequent peat accumulation. Following an early Holocene climatic optimum, climatic deterioration since about 3 ka led to the drying of the lakes and a gradual increase of permafrost aggregation. Additional information concerning the latest Holocene climate and environment comes from the archaeological findings on the northern Taymyr Peninsula (Pitul'ko, this volume). Limited information about the pre-Weichselian history, in contrast, is available from the sampled permafrost deposits. Some data about the fluctuations in relative sea level were obtained by ESR dating and lithostratigraphic studies of marine deposits (Bolshiyanov and Molodkov, this volume).

Paleoenvironmental reconstructions by lake sediments

The sediment fill in lakes functions as one of the best natural data archives for paleoenvironmental reconstructions because lakes act as sediment traps. Limnic sediments generally represent more complete depositional sequences than other terrestrial deposits. With multi-proxy data from lake sediments, complex information concerning the environmental setting both in the catchment area and in the water column of lakes can be obtained. In addition, detailed age determinations can often be achieved by radiocarbon dating of organic matter. Hence, lake sediments may supply high-resolution reconstructions of the environmental history, with good stratigraphic control.

Within the scope of the "Taymyr" Project, long sediment cores were recovered from the lakes Lama (18.9 m), Pyassino (6.0 m), Kokora (5.2 m), Portnyagino (4.0 m), Taymyr (14.3 m), Levinson-Lessing (22.4 m), Changeable (12.7 m), and Fjord (3.2 m). These lakes cover the entire project study area (Figures 3 and 4). In the lakes Lama, Taymyr, and Levinson-Lessing, in addition, the large-scale sediment architecture was investigated by sub-bottom profiling. With the existing sample and data set, therefore, both spatial and temporal variations of the climatic and environmental histories can be reconstructed.

The analytical work on the lake sediment cores is manifold, being used irregularly in dependence on individual objectives and core qualities. Special emphasis is put on the sediment chronology, sedimentology, palynology, micropaleontology, geochemistry, and mineralogy. First results, obtained on sediment cores from the Lama Lake and the Levinson-Lessing Lake, had shown that palynological analyses supply comprehensive information on the climatic evolution and related vegetation history (Hahne and Melles, 1997) and that investigations of lake sediment cores in the Taymyr region ideally complement investigations of permafrost profiles and ground ice bodies on land (Melles et al., 1996).

Results from sub-bottom profiling in the lakes Taymyr and Levinson Lessing, presented in this book, indicate that glaciers were present in the western Byrranga Mountains for the last time during the Early Weichselian glacial (Niessen et al., this volume). In addition, the seismic data are discussed with respect to the postglacial depositional histories of these lakes, with particular attention to their dependence on lake-level fluctuations. Sedimentological data from the Levinson-Lessing Lake core support the seismic interpretation and supply more detailed information concerning the kind and temporal variations of sedimentary processes since late Middle Weichselian time (Ebel et al., this volume).

Palynological analyses on sediment cores from four lakes, forming a transect from the southern Taymyr Peninsula to Severnaya Zemlya, were employed to reconstruct the vegetation history in the catchment areas of the lakes since Middle to Late Weichselian times (Hahne and Melles, this volume). The palynological results, compared with published ice core isotope data from Severnaya Zemlya, give a first impression of the development and of spatial and temporal variations of vegetation zones in dependence on climatic changes since the latest Pleistocene. The Holocene climatic evolution at the Lama Lake is also mirrored by the planktonic to benthonic diatom ratio in the lake sediments (Kienel, this volume). In addition, the diatom assemblages enable to reconstruct variations in the trophic conditions in the lake water. Hagedorn et al. (this volume) used ^{210}Pb dating and measurements of heavy metal concentrations in near-surface sediments of the Lama Lake to determine geogenic and anthropogenic sources and calculate recent accumulation rates.

References

Are, F.E. (1996) Dynamics of the littoral zone of the Arctic seas (state of the art and goals). Polarforsch., 64 (3), 123 - 131.

Bauch H.A., H. Kassens, H. Erlenkeuser, P.M. Grootes, J. Dehn, B. Peregovich and J. Thiede (in press) Holocene depositional history of the western Laptev Sea (Arctic Siberia), Boreas.

Bareiss J. (1996) Betrachtungen zur frühsommerlichen Meereisverteilung in der Laptewsee in Abhängigkeit vom kontinentalen Süßwassereintrag und der Witterung unter Nutzung passiver Mikrowellendaten. Unpublished diploma thesis, University of Trier, 164 pp.

Benthien, A. (1995) Echographiekartierung und physikalische Eigenschaften der oberflächennahen Sedimente in der Laptevsee. Unpublished diploma thesis, University of Kiel, 80 pp.

Bolshiyanov, D.Yu. and H.-W. Hubberten (1996) Russian-German Cooperation: The Expedition Taymyr 1995. In: The Expedition Taymyr 1995 and the Expedition Kolymy 1995 of the ISSP Pushchino Group, Bolshiyanov, D.Yu. and H.-W. Hubberten (eds.), Rep. on Polar Res., 211, 5-198.

Bolshiyanov, D.Yu. and V.M. Makeev (1995) The Archepelago Severnaya Zemlya: glaciation, history, environment. Gidrometizdat, St. Petersburg, 216 pp.

Broecker, W.S. (1994) Massive iceberg discharges as triggers for global climate change. Nature, 372, 421- 424.

Chizhov, A.B., A.Yu. Derevyagin, E.F. Simonov, H.-W. Hubberten and C. Siegert (1997) Isotopic composition of ground ice in the Labaz Lake region (Taymyr) (in Russian). Earth Cryosphere, 1(3), 79-84.

CLIMAP Project Members (1976) The surface of the ice-age earth. Science, 191, 1131- 1144.

Cremer H. (1998) Die Diatomeen der Laptevsee (Arktischer Ozean): Taxonomie und biogeographische Verbreitung. Rep. on Polar Res., 260, 225 pp.

Derevyagin A.Yu., C. Siegert, E.V. Troshin and N.G. Shilova (1996) New data on Quaternary Geology and cryogenic construction of the permafrost deposits in the Taymyr Peninsula (Labaz area) (in Russian). Proceedings of the I Conference of Russian Geocryologists, Moscow, Vol. 1, 204-212.

Dethleff, D., D. Nürnberg, E. Reimnitz, M. Saarso, Y.P. Savchenko (1993) East Siberian Arctic Region Expedition '92: The Laptev Sea - its significance for Arctic sea-ice formation and transpolar sediment flux. Rep. on Polar Res., 120, 3- 37.

Dethleff, D. (1995) Die Laptevsee - eine Schlüsselregion für den Fremdstoffeintrag in das arktische Meereis. Unpublished PhD thesis, University of Kiel, 111 pp.

Dmitrenko I., J. Dehn, P. Golovin, H. Kassens and A. Zatsepin (1998) Influence of sea ice on under-ice mixing under stratified conditions: potential impacts on particle distribution. Estuarine, Coastal and Shelf Science, 46, 523-529.

Eicken, H., E. Reimnitz, V. Alexandrov, T. Martin, H. Kassens and T. Viehoff (1997) Sea-ice processes in the Laptev Sea and their importance for sediment export. Continental Shelf Res., 17, 205- 233.

Fahl, K. and R. Stein (1997) Modern organic-carbon-deposition in the Laptev Sea and the adjacent continental slope: Surface-Water productivity vs. terrigenous input. Organic Geochem., 26, 379- 390.
Fütterer, D.K. (1994) Die Expedition ARCTIC '93. Der Fahrtabschnitt ARK-IX/4 mit FS "Polarstern" 1993. Rep. on Polar Res., 149, 244 pp.
Grosswald, M.G. (1998) Late-Weichselian ice sheet in arctic and pacific Siberia. Quaternary Internat., 45/46, 3-18.
Hahne, J. and M. Melles (1997) Late and postglacial vegetation and climate history of the south-western Taymyr Peninsula (Central Siberia), as revealed by pollen analysis of sediments from Lake Lama. Vegetation History and Archaeobotany, 6, 1-8.
Imbrie, J. and J.Z. Imbrie (1980) Modelling the Climate Response to Orbital Variations. Science, 207, 943-953.
Johnson L.G. (1996) Joint U.S. and Russian survey of Laptev and East Siberian seas conducted. The Nansen Icebreaker, 8, 11.
Karpov, E.G. (1986) Ground ice in the northern Yenissey region (in Russian). Nauka, Novosibirsk, 134 pp.
Kassens, H. and V.Y. Karpiy (1994) Russian-German Cooperation: The Transdrift I Expedition to the Laptev Sea. Rep. on Polar Res., 151, 168 pp.
Kassens, H. (1995a) Laptev Sea System: Expeditions in 1994. Rep. on Polar Res., 182, 195 pp.
Kassens, H., D. Piepenburg, J. Thiede, L. Timokhov, H.-W. Hubberten and S.M. Priamikov (1995) Russian-German Cooperation: Laptev Sea System. Rep. on Polar Res., 176, 387 pp.
Kassens, H., I. Dmitrenko and the Transdrift IV Shipboard Scientific Party (1996) Transdrift IV explores the Lena Delta . The Nansen Icebreaker, 9, 1 a. 6.
Kassens, H., I. Dmitrenko, L. Timokhov and J. Thiede (1997) The Transdrift III Expedition: freeze-up studies in the Laptev Sea. Rep. on Polar Res., 248, 1-192.
Kassens, H., I. Dmitrenko, V. Rachold, J. Thiede and L. Timokhov (1998) Russian and German scientists explore the Arctic's Laptev Sea and its climate system. EOS, 79 (27), 317, 322-323.
Kolatschek J., H. Eicken, V. Yu Alexandrov, and M. Kreyscher (1996) The sea-ice cover of the Arctic Ocean and the Eurasian marginal seas: a brief overview of present-day patterns and variability. Rep. on Polar Res., 212, 2-18.
Kosobokova, K.N., H. Hanssen, H.J. Hirche and K. Knickmeier (1998) Composition and distribution of zooplankton in the Laptev Sea and adjacent Nansen Basin during summer 1993. Polar Biol., 19, 63- 76.
Kunz-Pirrung M. (1998) Rekonstruktion der Oberflächenwassermassen der östlichen Laptevsee im Holozän anhand von aquatischen Palynomorphen. Rep. on Polar Res., 281, 117 pp.
Lara, R., V. Rachold, G. Kattner, H.-W. Hubberten, G. Guggenberger, A. Skoog and D.N. Thomas (1998) Dissolved organic matter and nutrients in the Lena River, Siberian Arctic: characteristics and distribution. Mar. Chem., 59, 301-309.
Lindemann, F. (1995) Sonographische und sedimentologische Untersuchungen in der Laptevsee, sibirische Arktis. Unpublished diploma thesis, University of Kiel, 74 pp.
Lindemann F. (1998) Sedimente im arktischen Meereis - Charakterisierung, Quantifizierung und Inkorporation. Unpublished PhD thesis, University of Kiel, 123 pp.
Martin, J.M., V.V. Gordeev and E. Emelyanov (1996) Introduction. Marine Chemistry, 53, 209.
Maslanik J. A., M. C. Serreze, and R. G. Barry (1996) Recent decreases in Arctic summer ice cover and linkages to atmospheric circulation anomalies. Geophys. Res. Lett., 23, 1677-1680.
Matveyeva, N.V. (1998) Zonation in plant cover of the Arctic (in Russian). Proc. Komarov Botan. Inst., 21, 220 pp.
Melles, M., D.Yu. Bolshiyanov and O.M. Lisitzina (in prep.) Russian-German Cooperation: The Expedition Norilsk 1997. Rep. on Polar Res.
Melles, M., B. Hagedorn and D.Yu. Bolshiyanov (1997) Russian-German Cooperation: The Expedition Taymyr/Severnaya Zemlya 1996. Rep. on Polar Res., 237, 170 pp.
Melles, M., C. Siegert, J. Hahne and H.-W. Hubberten (1996) Klima- und Umweltgeschichte des nördlichen Mittelsibiriens im Spätquartär - erste Ergebnisse. Geowissenschaften, 14(9), 376-380.
Melles, M., U. Wand, W.-D. Hermichen, B. Bergemann, D.Yu. Bolshiyanov and S.F. Khrutsky (1994): The Expedition Norilsk/Taymyr 1993 of the AWI Research Unit Potsdam. In: The Expeditions Norilsk/ Taymyr 1993 and Bunger Oasis 1993/94 of the AWI Research Unit Potsdam, M. Melles (ed.), Rep. on Polar Res., 148, 1-25.
Nørgaard-Petersen, N., R.F. Spielhagen, J. Thiede and H. Kassens (1998) Central Arctic surface ocean environment during the past 80,000 years. Paleoceanography, 13(2), 193- 204.
Nürnberg, D., I. Wollenburg, D. Dethleff, H. Eicken, H. Kassens, T. Letzig, E. Reimnitz and J. Thiede (1994) Sediments in Arctic sea-ice: Implications for entrainment, transport and release. Marine Geology, 119 (3-4), 185-214.
Pfirman S., H. Eicken, D. Bauch, and W. F. Weeks (1995) Potential transport of radionuclides and other pollutants by Arctic sea ice. Sci. Tot. Environm., 159, 129-146.

Piepenburg, D. and M.K. Schmidt (1997) A photographic survey of the epibenthic megafauna of the Arctic Laptev Sea shelf: distribution, abundance, and estimates of biomass and organic carbon demand. Marine Ecol. Progr. Ser., 147, 63- 75.

Rachold, V., J. Hermel and V.N. Korotaev (1995) Expedition to the Lena river in July/August 1994. In: Laptev Sea System: Expeditions in 1994, Kassens, H. (ed.), Rep. on Polar Res., 182, 181-195.

Rachold, V., A. Alabyan, H.-W. Hubberten, V.N. Korotaev and A.A. Zaitsev (1996) Sediment transport to the Laptev Sea - hydrology and geochemistry of the Lena river. Polar Res., 15, 183-196.

Rachold, V., E. Hoops, A.M. Alabyan, V.N. Korotaev and A.A. Zaitsev (1997a) Expedition to the Lena and Yana Rivers. In: Laptev Sea System: Expeditions in 1995, Kassens, H. (ed.), Rep. on Polar Res., 248, 197-210.

Rachold, V., A. Eisenhauer, H.-W. Hubberten and H. Meyer (1997b) Sr isotopic composition of suspended particulate material (SPM) of East Siberian rivers - sediment transport to the Arctic Ocean. Arctic and Alpine Res., 29, 422-429.

Rachold, V., E. Hoops and A.V. Ufimzev (in press) Expedition to the Khatanga river July-August 1996. In: Laptev Sea System: Expeditions in 1996, Kassens, H. (ed.), Rep. on Polar Res.

Rachor, E. (1997) Scientific Cruise Report of the Arctic Expedition ARK-XI/1 of RV Polarstern in 1995. Rep. on Polar Res., 226, 157 pp.

Rigor, I. and R. Colony (1997) Sea-ice production and transport of pollutants in the Laptev Sea, 1979-1993. Sci. Tot. Environm., 202, 89-110.

Romanovskii, N., A.V. Gavrilov, A.L. Kholodov, H. Kassens, H.-W. Hubberten and F. Niessen (1997) Offshore permafrost distribution on the Laptev Sea Shelf. Earth Cryosph., 3, 9- 18.

Siegert, C. and D. Bolshiyanov (1995) Russian-German Cooperation: the expedition "Taymyr" 1994. Rep. on Polar Res., 175, 82 pp.

Tumel, N.V. (1985) Development of permafrost during Late Pleistocene and Holocene times in the northen Yenissey (in Russian). In: The evolution of the permafrost zone in Eurasia during Late Cenozoic times, Popov, A.I. (ed.), Nauka, Moscow, 43-51.

Utschakowski S. (1998) Anthropogene organische Spurenstoffe im Arktischen Ozean. Untersuchungen chlorierter Biphenyle und Pestizide in der Laptevsee, technische und methodische Entwicklungen zur Probenahme in der Arktis und zur Spurenstoffanalyse. Unpublished PhD thesis, University of Kiel, 126 pp.

Velichko, A.A., Yu.N. Kononov and M.A. Faustova (1997) The last glaciation of earth: size and volume of ice sheets. In: Quaternary of northern Eurasia: Late Pleistocene and Holocene landscapes, stratigraphy and environments, Velichko, A.A. et al. (eds.) Quaternary Internat., 41/42, 43-51.

Velichko, A.A., A.A. Andreev and V.A. Klimanov (1997 a) Climate and vegetation dynamics in the tundra and forest zone during the Late Glacial and Holocene. In: Quaternary of northern Eurasia: Late Pleistocene and Holocene landscapes, stratigraphy and environments, Velichko, A.A. et al. (eds.) Quaternary International, 41/42, 71-96.

Wadhams P. (1995) Arctic sea ice extent and thickness. Philosoph. Trans. Royal Soc., Ser. A, 352, 301-319.

Wollenburg, I. (1993) Sediment transport by Arctic sea ice: The recent load of lithogenic and biogenic material. Rep. on Polar Res., 127, 159 pp.

Springer and the environment

At Springer we firmly believe that an international science publisher has a special obligation to the environment, and our corporate policies consistently reflect this conviction.

We also expect our business partners – paper mills, printers, packaging manufacturers, etc. – to commit themselves to using materials and production processes that do not harm the environment. The paper in this book is made from low- or no-chlorine pulp and is acid free, in conformance with international standards for paper permanency.

Printing: Saladruck, Berlin
Binding: Buchbinderei Lüderitz & Bauer, Berlin